Butter Tests of Registered Jersey Cows up to August 1st of 1898

by American Jersey Cattle Club

with an introduction by Jackson Chambers

Self Reliance Books

Get more historic titles on animal and stock breeding, gardening and old fashioned skills by visiting us at:

http://selfreliancebooks.blogspot.com/

Introduction

I am pleased to present another title in the "Cattle" series.

The work is in the Public Domain and is re-printed here in accordance with Federal Laws.

As with all reprinted books of this age that are intended to perfectly reproduce the original edition, considerable pains and effort had to be undertaken to correct fading and sometimes outright damage to existing proofs of this title. At times, this task is quite monumental, requiring an almost total "rebuilding" of some pages from digital proofs of multiple copies. Despite this, imperfections still sometimes exist in the final proof and may detract from the visual appearance of the text.

I hope you enjoy reading this book as much as I enjoyed making it available to readers again.

Jackson Chambers

PREFACE.

---✤---

THE butter records in this volume (except those designated "official" or "confirmed," and those made at the World's Columbian Exposition) have been received for publication on the affidavits of the managers of the tests or the certificates of the owners of the cows tested.

The tests designated "official" were made under the supervision of committees appointed by the President of the Club.

The tests designated "confirmed" were made in accordance with the rules of the Club requiring the application of the Babcock method of ascertaining the butter-fat to each milking by an appointee of the Club, or by the officials of any State, Provincial or national experiment station.

The tests are arranged alphabetically. When a number of tests of the same animal occur, they are arranged in order of date, if for seven days, and according to period, if for longer, the shorter periods coming first.

All tests are for the period of seven days, unless otherwise stated.

In the list of Tested Cows in Order of Merit as butter-producers, the best record only of each animal is given.

In the list of Sires with their Tested Daughters, arranged alphabetically, the names of the tested cows are arranged under their sires' names according to the amounts of their tests, the largest coming first. When an animal has more than one test, the largest only is given. The same arrangement has been followed in the list of Dams with their Tested Daughters.

CONTENTS.

BUTTER TESTS OF JERSEYS.

Abbia H. 24221.—Sire, Alaric of Lakeside 6621; dam, Ardell H. 12330.

Butter, 15 lbs. 6 oz. Milk, 326 lbs. 1 oz.

Test made from June 13 to 19, 1885; age, 2 yrs. 3 mos.; property of William H. Hayden, Albany, Vt.

Abbie K. 13227.—Sire, Duke F. 6134; dam, Sylvia 687.

Butter, 14 lbs. 4 oz. Milk, 224 lbs.

Test made from Sept. 3 to 10, 1888; age, 10 yrs. 5 mos.; actual weight, 1150 lbs.; fed, daily, 15 qts. bran, fine feed, oil meal and corn meal, mixed; property of C. E. Sunderlin, Rochester, N. Y.

Abbie Z. 14002.—Sire, Comet 130; dam, Lupar 14001.

Butter, 14 lbs. 11 oz. Milk, 329 lbs.

Test made from June 6 to 12, 1876; age, 6 yrs. 1 mo.; property of Harvey Newton, Southville, Mass.

Abbie Z. 14002.—Sire, Comet 130; dam, Lupar 14001.

Butter, 61 lbs. 2 oz. in 30 days. Milk, 1406 lbs.

Test made June, 1876; age, 6 yrs. 1 mo., property of Harvey Newton, Southville, Mass.

Abbie Z. 3d 14742.—Sire, Deerfoot Boy 1926; dam, Abbie Z. 14002.

Butter, 17 lbs. Milk, 288 lbs.

Test made from May 5 to 12, 1883; age, 4 yrs. 10 mos.; property of J. R. Gaston & Son, Normal, Ill.

Acacia of Home Farm 3d 78203.—Sire, Victory 23051; dam, Acacia of Home Farm 12094.

Butter, 17 lbs. 14½ oz. Milk, 271 lbs. 8 oz.

Test made from Dec. 20 to 27, 1897; age, 7 yrs. 6 mos.; actual weight, 840 lbs.; fed 4 qts. corn meal, 8 qts. ground oats and 8 qts. bran, daily; property of J. Gerow Dutcher, Pawling, N. Y.

Acme's Bonnie Bess 31509.—Sire, Johnson's Acme 3656; dam, Palestine's Last Daughter 12602.

Butter, 14 lbs. 5 oz. Milk, 221 lbs. 10 oz.

Test made from June 5 to 12, 1892; age, 9 yrs. 3 mos.; estimated weight, 1000 lbs., fed 3 gals. corn, oats and barley chop and 1 qt. oil meal, daily—good pasture, but wet; property of Belmont Jersey Cattle Co., Columbus, Ohio.

Actress 2311.—Sire, Emperor 287; dam, Annie 2d 776.

> Butter, 14 lbs., unsalted.

Test made from Dec. 24 to 31, 1882; age, 10 yrs. 9 mos.; property of W. B. Dinsmore, Staatsburg, N. Y.

Ada Fy 45479.—Sire, Ely Fy 18244; dam, Imperial's Omega 2d 19004.

> Butter, 17 lbs. 4 oz. Milk, 208 lbs. 4 oz.

Test made from May 23 to 30, 1891; age, 3 yrs. 11 mos.; estimated weight, 850 lbs.; fed 5 lbs. bran and 5 lbs. cotton-seed cake meal, daily—timothy, clover and blue grass pasture; property of Bordwell & Cochran, Batavia, Ohio.

Ada of Elmwood 98728.—Sire, Prince of Elmwood 20994; dam, Bertie Don 2d 74088.

> Butter, 14 lbs. 2 oz. Milk, 166 lbs. 2 oz.

Test made from Aug. 25 to Sept. 1, 1897; age, 3 yrs. 10 mos.; estimated weight, 550 lbs.; fed 21 lbs. cotton-seed meal, 21 lbs. ground oats, 28 lbs. corn meal, 210 lbs. corn ensilage and 35 lbs. clover hay; property of Charles Lautz, Buffalo, N. Y.

Ada of Orange 2d 65246.—Sire, Albert Fair 15920; dam, Ada of Orange 16749.

> Butter, 17 lbs. 3 oz. Milk, 292 lbs.

Test made from April 21 to 28, 1895; age, 6 yrs. 8 mos.; estimated weight, 900 lbs.; fed 126 lbs. corn-hearts and 42 lbs. cotton-seed meal—blue-grass pasture; property of Lucius P. Brown, Spring Hill, Tenn.

Adaphine 101425.—Sire, Mint 23600; dam, Adonia 64699.

> Butter, 15 lbs. 3 oz. Milk, 227 lbs. 12 oz.

Test made from Mar. 29 to Apr. 5, 1897; age, 3 yrs. 2 mos.; estimated weight, 900 lbs.; fed 42 lbs. ground oats, 7 lbs. oil meal, 42 lbs. bran, 21 lbs. ground rye, 14 lbs. corn meal and 40 lbs. corn ensilage; property of H. C. Taylor, Orfordville, Wis.

Ada S. 18366.—Sire, Mardi Gras 2927; dam, Lorette 7393.

> Butter, 16 lbs. 9 oz. Milk, 214 lbs. 1 oz.

Test made from Aug. 17 to 23, 1884; age, 3 yrs. 6 mos.; property of J. L. Shallcross, Anchorage, Ky.

Addais 92923.—Sire, Baron Hugo 15208; dam, Callie Rex 21672.

> Butter, 14 lbs. 3 oz. Milk, 210 lbs. 4 oz.

Test made from Mar. 7 to 14, 1894; age, 1 yr. 8 mos.; estimated weight, 650 lbs.; fed 3 lbs. coarse wheat bran, 3 lbs. corn meal, 2 lbs. gluten meal, 2 lbs. rye and oats provender, 30 lbs. corn ensilage and 8 lbs. cut clover hay, daily; property of E. Stevens Henry, Rockville, Conn.

Adelaide Saufley 96284.—Sire, Fancy's Jubilee 25205; dam, Miss Belinda 16638.

> Butter, 16 lbs. 5 oz. Milk, 243 lbs. 15 oz.

Test made from Feb. 26 to Mar. 4, 1896; age, 3 yrs. 6 mos.; actual weight, 1080 lbs., fed 56 lbs. cotton-seed hulls, 63 lbs. wheat bran and 32 lbs. shelled oats—mesquite hay *ad lib.*, half-hour daily on barley patch; property of R. B. Cousins, Mexia, Tex.

Adelpha Marigold 133768.—Sire, Stoke Pogis of Prospect 29121; dam, Ida of Oakland 2d 72061.

> Butter, 15 lbs. 11½ oz. Milk, 236 lbs. 10 oz.

Test made from July 6 to 12, 1898; age, 2 yrs. 10 mos.; actual weight, 800 lbs.; fed 9 qts. bran, 9 qts. ground oats, 3 qts. corn meal and 3 pts. oil meal, mixed, wet, daily —good clover pasture; property of Charles A. Sweet, East Aurora, N. Y.

Adina 1942.—Sire, Czar 251; dam, Irma 1298.

Butter, 14 lbs. 4 oz. Milk, 234 lbs.

Test made from May 2 to 9, 1881; age, 9 yrs.; property of James Cloud & Son, Kennett Square, Pa.

Adonia 64699.—Sire, Alphadron 16460; dam, Anollalla 19208.

Butter, 19 lbs. 1 oz. Milk, 264 lbs. 4 oz.

Test made from Mar. 3 to 10, 1898; age, 7 yrs. 11 mos.; estimated weight, 1000 lbs.; fed 35 lbs. bran, 42 lbs. oats, 35 lbs. corn meal and 14 lbs. oil meal—corn ensilage; property of H. C. Taylor, Orfordville, Wis.

Adonia 2d 103302.—Sire, Mint 23600; dam, Adonia 64699.

Butter, 17 lbs. 6 oz. Milk, 296 lbs. 4 oz.

Test made from Jan. 11 to 18, 1897; age, 3 yrs. 9 mos.; actual weight, 925 lbs.; fed 12 qts. wheat bran, 9 qts. ground oats, 3 qts. corn meal and 3 qts. oil meal, daily—plenty hay and ensilage; property of C. A. Sweet, East Aurora, N. Y.

Adora 18569.—Sire, The Hub 1009; dam, Sunny South 6830.

Butter, 14 lbs. 3 oz. Milk, 294½ lbs.

Test made from May 21 to 27, 1883; age, 3 yrs. 8 mos.; property of William E. Oates, Vicksburg, Miss.

Aggie of St. Lambert 37085.—Sire, Rubano 8806; dam, Polly of St. Lambert 28665.

Butter, 14 lbs. 2½ oz. Milk, 134 lbs. 4 oz.

Test made from May 28 to June 4, 1888; age, 3 yrs. 1 mo.; estimated weight, 800 lbs.; fed, daily, 20 qts. of a mixture of corn, oats, middlings and oil meal; property of P. J. Cogswell, Rochester, N. Y.

Agnes Clover 86378.—Sire, Plymouth Champion 27227; dam, Kate Clover 16952.

Butter, 14 lbs. 12 oz. Milk, 200 lbs. 11 oz.

Test made from Mar. 1 to 8, 1896; age, 3 yrs. 10 mos.; actual weight, 750 lbs.; fed 112 lbs. bran, 140 lbs. corn chop and 98 lbs. ensilage of poor quality—prairie hay *ad lib.*; property of J. D. Gray, Terrell, Tex.

Agnes of Clovernook 72553.—Sire, Sherbet's Koffee 18512; dam, Hildah of Clovernook 51597.

Butter, 16 lbs. 9½ oz. Milk, 192 lbs. 3½ oz.

Test made from June 19 to 26, 1892; age, 2 yrs. 7 mos.; estimated weight, 650 lbs.; fed 20 lbs. per day of ground oats and corn, in equal parts—very good blue and orchard grass pasture; property of Matt. M. Gardner, Nashville, Tenn.

Agnes of Glen Duart 82475.—Sire, Chief of Glen Duart 24340; dam, Kathleen of Glen Duart 50982.

Butter, 15 lbs. 11 oz. Milk, 283 lbs. 10 oz.

Test made from May 31 to June 7, 1896; age, 4 yrs. 4 mos.; estimated weight, 900 lbs.; fed 56 qts. corn meal and 56 qts. oat chop—blue grass and timothy pasture; property of A. T. Macartney, Dunnville, Ont., Can.

Ajax Riotress 113709.—Sire, St. Lambert Boy 17408; dam, Lady Ajax 2d 89637.

Butter, 16 lbs. Milk, 266 lbs. 7 oz.

Test made from June 10 to 17, 1898; age, 3 yrs. 6 mos.; estimated weight, 900 lbs.; fed 16 qts. wheat bran, 8 qts. corn meal and 3 pts. oil meal, daily—pasture; property of Brooks & Pidgeon, Salem, Ohio.

Alabama 7690.—Sire, Dainty Boy 2955; dam, Ruby 787.

        Butter, 15 lbs. 11 oz.   Milk, 209 lbs. 1 oz.

    Test made from Aug. 5 to 12, 1888; age, 10 yrs. 4 mos.; estimated weight, 775 lbs.; fed 14 qts. corn meal, 70 qts. bran and 14 qts. cotton-seed meal—hay and good mixed pasture; property of James Crook, Jacksonville, Ala.

Alaric's Millicent 35363.—Sire, Alaric of Lakeside 6621; dam, Millicent H. 14445.

        Butter, 18 lbs. 5½ oz.   Milk, 163 lbs. 8 oz.

    Test made from Aug. 30 to Sept. 6, 1890; age, 5 yrs. 4 mos.; estimated weight, 800 lbs.; fed, daily, 4 gals. ground corn and 2 gals. bran; property of Mayes & Lipscomb, Columbia, Tenn.

Alarm of Highland 41864.—Sire, Rioter's Combination 10363; dam, Miss Sharpless 24352.

        Butter, 16 lbs.   Milk, 252 lbs. 4 oz.

    Test made from June 1 to 8, 1895; age, 9 yrs.; estimated weight, 850 lbs.; fed 1 qt. oil meal, 1 pt. cotton-seed meal and 6½ qts. of a mixture of corn meal and middlings in equal parts, daily—timothy and clover pasture; property of A. B. Darling, Ramsey's, N. J.

Alberta d'Or 76452.—Sire, Eddystone 23413; dam, Countess Hebe 8102.

        Butter, 14 lbs. 14 oz.   Milk, 262 lbs. 2 oz.

    Test made from May 11 to 18, 1897; age, 7 yrs. 11 mos.; estimated weight, 900 lbs.; fed 28 qts. corn meal, 21 qts. linseed meal and 28 qts. bran—clover pasture; property of E. A. Tyrrell, Marysville, Ohio.

Alberta d'Or 2d 124269.—Sire, Alberta's Fancy Harry 30472; dam, Alberta d'Or 76452.

        Butter, 14 lbs. 13 oz.   Milk, 206 lbs. 4 oz.

    Test made from June 20 to 27, 1898; age, 3 yrs.; estimated weight, 750 lbs.; fed 35 lbs. wheat bran—green oats *ad lib.*, white clover and timothy pasture; property of E. A. Tyrrell, Marysville, Ohio.

Alberta Hugo 35424.—Sire, Prince of Melrose 2d 11015; dam, Cream of Springvale 16621.

        Butter, 17 lbs. 9 oz.   Milk, 228 lbs.

    Test made from Apr. 21 to 28, 1891; age, 5 yrs. 8 mos.; estimated weight, 900 lbs.; fed 70 qts. corn meal, 14 qts. bran, 14 qts. cotton-seed meal and 30 lbs. rich corn ensilage—hay, poor pasture; property of James Crook, Jacksonville, Ala.

Alberta Signal 18611.—Sire, Black Diamond V. 4970; dam, Aureola 8617.

        Butter, 20 lbs. 11 oz.   Milk, 224 lbs. 15 oz.

    Test made from Aug. 8 to 14, 1885; age, 3 yrs. 4 mos.; fed 18 qts. bran, shelled oats, pea meal and corn meal, mixed, daily—poor blue grass pasture; property of R. Mc-Michael, Lexington, Ky.

Albert's Alba 111721.—Sire, Albert of Amherst Villa 33165; dam, Helen of Bay View 66030.

        Butter, 14 lbs. 2 oz.   Milk, 117 lbs. 4 oz.

    Test made from Jan. 5 to 12, 1898; age, 1 yr. 11 mos.; actual weight, 495 lbs.; fed 21 lbs. ground oats, 21 lbs. corn meal, 14 lbs. oil meal, 21 lbs. wheat bran, 200 lbs. corn ensilage and 49 lbs. clover hay; property of Charles Lautz, Buffalo, N. Y.

Albert's Comtesse 111718.—Sire, Albert of Amherst Villa 33165; dam, Princess Rosalind 87288.

Butter, 21 lbs. 4 oz.   Milk, 121 lbs. 12 oz.

Test made from Jan. 11 to 18, 1898; age, 2 yrs.; actual weight, 500 lbs.; fed 21 lbs. ground oats, 21 lbs. corn meal, 14 lbs. oil meal, 28 lbs. wheat bran, 190 lbs. corn ensilage and 49 lbs. hay; property of Charles Lautz, Buffalo, N. Y.

Albert's Cretesia 22881.—Sire, Albert Rex 7724; dam, Cretesia 13657.

Butter, 16 lbs. 2 oz.

Test made from June 12 to 18, 1886; age, 3 yrs. 2 mos.; property of Cyrus W. Sprague, West Stockbridge, Mass.

Albert's Gem 34006.—Sire, Albert Rex 7724; dam, Sultana 2d 11798.

Butter, 165.777 lbs. in 90 days.   Milk, 2666.4 lbs.

Test made from May 31 to Aug. 28, 1893, at World's Columbian Exposition; age, 9 yrs. 5 mos. at beginning of test; actual weight, 970 lbs.; fed 310.8 lbs. old hay, 123 lbs. ensilage, 523.1 lbs. new hay, 191 lbs. oil meal, 477 lbs. bran, 85 lbs. oats, 563.5 lbs. corn-hearts, 124.25 lbs. cotton-seed, 385.5 lbs. middlings, 34 lbs. granogluten, 1116.4 lbs. clover and 10 lbs. swale-grass; property of F. A. Schermerhorn, Lenox, Mass.

Albert's Ideal 111719.—Sire, Albert of Amherst Villa 33165; dam, Lady Aurora 98734.

Butter, 15 lbs. 10 oz.   Milk, 125 lbs. 15 oz.

Test made from Feb. 1 to 8, 1898; age, 2 yrs.; actual weight, 460 lbs.; fed 14 lbs. ground oats, 21 lbs. corn meal, 21 lbs. oil meal, 28 lbs. wheat bran, 210 lbs. corn ensilage and 35 lbs. hay; property of Charles Lautz, Buffalo, N. Y.

Albert's Lilley 19489.—Sire, Albert Rex 7724; dam, Lilley Rex 9852.

Butter, 15 lbs. 2 oz.   Milk, 21½ qts. per day.

Test made from Mar. 23 to 29, 1887; age, 4 yrs. 5 mos.; estimated weight, 700 lbs.; fed 2 qts. corn meal, 2 qts. pea meal, 8 qts. ground oats, 1 qt. oil meal and 8 qts. middlings, daily; property of D. F. Appleton, Ipswich, Mass.

Albert's Lobelia 111902.—Sire, Albert of Amherst Villa 33165; dam, Clover of Bay View 65424.

Butter, 18 lbs. 1 oz.   Milk, 106 lbs. 9 oz.

Test made from Dec. 15 to 22, 1897; age, 2 yrs. 1 mo.; actual weight, 510 lbs.; fed 21 lbs. ground oats, 21 lbs. corn meal, 21 lbs. oil meal, 28 lbs. wheat bran, 175 lbs. corn ensilage and 49 lbs. hay; property of Charles Lautz, Buffalo, N. Y.

Alchesleda 103710.—Sire, Hugo Lofty 29176; dam, Countess of Lorne 20822.

Butter, 15 lbs. 6 oz.   Milk, 235 lbs. 4 oz.

Test made from Feb. 28 to Mar. 7, 1898; age, 3 yrs. 7 mos.; estimated weight, 700 lbs.; fed 4 lbs. spring wheat bran, 4 lbs. Chicago gluten meal, 2 lbs. mixed rye and oats provender, about 5 lbs. clover hay and 30 lbs. to 35 lbs. ensilage, daily; property of E. Stevens Henry, Rockville, Conn.

Aldarine 5301.—Sire, Signal 1170; dam, Alda 3873.

Butter, 15 lbs. 1½ oz., unsalted.   Milk, 232 lbs. 15 oz.

Test made from May 28 to June 3, 1882; age, 6 yrs. 3 mos.; no grain fed—good blue grass pasture only; property of W. J. Chinn, Frankfort, Ky.

Aldarine 2d 14196.—Sire, Fearnaught 1854; dam, Aldarine 5301.

> Butter, 14 lbs. Milk, 168 lbs. 8 oz.

Test made from Apr. 20 to 26, 1886; age, 7 yrs. 10 mos.; property of Campbell Brown, Spring Hill, Tenn.

Aldarine 3d 27482.—Sire, Wanderer 3014; dam, Aldarine 5301.

> Butter, 15 lbs. 3¼ oz. Milk, 121 lbs. 8 oz.

Test made from Nov. 20 to 27, 1890; age, 6 yrs. 9 mos.; estimated weight, 750 lbs.; fed, daily, 11 lbs. corn meal, 5 lbs. wheat bran, 4 lbs. ground oats and 2 lbs. linseed meal; property of William Hart Dexter, Springfield, Mass.

Aldarine 3d 27482.—Sire, Wanderer 3014; dam, Aldarine 5301.

> Butter, 28 lbs. 1 oz. Milk, 248 lbs. 12 oz.

Test made from May 3 to 10, 1891; age, 7 yrs. 3 mos.; estimated weight, 750 lbs.; fed 9 lbs. corn meal, 3 lbs. ground oats, 6 lbs. bran and 6 lbs. linseed meal, daily, on 18 lbs. cut hay; property of Wm. Hart Dexter, Springfield, Mass.

Aleph Judea 11389.—Sire, John Knox 3289; dam, Juliette of St. Lambert 5483.

> Butter, 15 lbs. 1¾ oz. Milk, 241 lbs. 8 oz.

Test made from Nov. 13 to 19, 1883; age, 3 yrs. 10 mos.; property of S. M. Neel, Shelbyville, Ky.

Aleph Judea 11389.—Sire, John Knox 3289; dam, Juliette of St. Lambert 5483.

> Butter, 18 lbs. 1½ oz., unsalted. Milk, 314 lbs. 13 oz.

Test made from Sept. 3 to 9, 1885; age, 5 yrs. 8 mos.; property of W. H. Corning, Cleveland, Ohio.

Alexa 64924.—Sire, Diploma 16219; dam, Mint Julep 64571.

> Butter, 14 lbs. 15 oz. Milk, 307 lbs. 1 oz.

Test made from June 15 to 22, 1893; age, 4 yrs. 1 mo.; estimated weight, 750 lbs.; fed 15 qts. daily of a mixture of ground oats, bran and a little corn meal—a little hay, good pasture; property of Richardson Bros., Davenport, Iowa.

Alfaretta C. 34928.—Sire, Alfonso 3013; dam, Annorette 20178.

> Butter, 14 lbs. 11½ oz. Milk, 288 lbs. 3 oz.

Test made from Apr. 5 to 12, 1895; age, 9 yrs. 6 mos.; estimated weight, 950 lbs.; fed 14 gals. ground corn, 7 gals. bran, 3½ gals. cotton and linseed meal, mixed—rye pasture; property of J. M. Hund, Hund's Station, Kan.

Alfieda 6744.—Sire, Harpado 1859; dam, Amaranth 6200.

> Butter, 16 lbs. 4 oz. Milk, 250 lbs. 4 oz.

Test made from Aug. 19 to 26, 1883; age, 5 yrs. 7 mos.; fed 3 qts. corn meal and 9 qts. wheat bran, daily—pasture, short aftermath of clover and timothy, with green cornstalks; property of Thomas Beer, Bucyrus, Ohio.

Alfritha 13673.—Sire, Niobe Grand Duke 4510; dam, Alfieda 6744.

> Butter, 15 lbs. 3 oz. Milk, 194 lbs. 4 oz.

Test made from June 1 to 7, 1884; age, 2 yrs. 10 mos.; property of Thomas Beer, Bucyrus, Ohio.

Alhena 15995.—Sire, Sir Samuel Cunard 2231; dam, Daffodil of Maplewood Farm 4853.

Butter, 16 lbs. 3 oz.

Test made from Jan. 1 to 7, 1883; age, 4 yrs. 8 mos.; property of Lyman Sperry, Watertown, Conn.

Alicealda 43604.—Sire, Witch's Token 9966; dam, Clover Bud 5th 27743.

Butter, 14 lbs. 1 oz. Milk, 248 lbs. 8 oz.

Test made from June 1 to 8, 1890; age, 4 yrs. 3 mos.; estimated weight, 850 lbs.; fed, daily, between 16 lbs. and 24 lbs. ground corn and oats; property of S. H. Godman, Muncie, Ind.

Alice Donnal 12726.—Sire, Jo Jones 3441; dam, Flora Lee of Tennessee 7694.

Butter, 14 lbs. Milk, 283 lbs. 11 oz.

Test made from June 13 to 20, 1883; age, 4 yrs. 4 mos.; property of S. A. Rodgers, Loudon, Tenn.

Alice McClellan 25237.—Sire, Snedens 4882; dam, Vesper of Woodstock 6916.

Butter, 15 lbs. 2 oz. Milk, 233 lbs. 12 oz.

Test made from Mar. 23 to 30, 1888; age, 5 yrs.; estimated weight, 850 lbs.; fed, daily, 8 lbs. corn meal, 4 lbs. oil meal, 2 lbs. middlings and 4 lbs. bran; property of James Stillman, Sing Sing, N. Y.

Alice of Clovernook 77532.—Sire, Dixie Prince 18379; dam, Toltec's Alice 31885.

Butter, 15 lbs. 3 oz. Milk, 138 lbs.

Test made from Nov. 29 to Dec. 6, 1893; age, 3 yrs. 11 mos.; estimated weight, 750 lbs.; fed 4 qts. ground oats, 4 qts. bran, 2 qts. cotton-seed meal and 2 qts. corn-hearts, daily—oat straw, poor pasture; property of S. N. Warren, Spring Hill, Tenn.

Alice of Salem 5053.—Sire, Clement 115; dam, Gunilda 662.

Butter, 14 lbs. 8 oz.

Test made Summer, 1874; age, about 3½ yrs.; property of William Cooper, Easton, Md.

Alice of the Meadows 20748.—Sire, Duke of Daffodil 1662; dam, Alice 540.

Butter, 14 lbs. 12 oz. Milk, 243 lbs. 13 oz.

Test made from May 23 to 29, 1883; age, 3 yrs. 9 mos.; property of Richard Rowett, Carlinville, Ill.

Alice Rex 3d 63907.—Sire, Matilda's Perfection 17100; dam, Alice Rex 33535.

Butter, 15 lbs. 6½ oz. Milk, 291 lbs. 2 oz.

Test made from June 20 to 27, 1896; age, 7 yrs. 4 mos.; estimated weight, 900 lbs.; fed 42 lbs. ground oats, 38½ lbs. pea meal, 42 lbs. gluten meal and 17½ lbs. wheat bran—rather poor hill pasture; property of Henry S. Redfield, Elmira, N. Y.

Allie Minka 2982.—Sire, Milo 590; dam, Minka 951.

Butter, 14 lbs. 6½ oz. Milk, 208 lbs. 8 oz.

Test made from Sept. 13 to 19, 1882; age, 8 yrs. 10 mos.; fed 3 lbs. wheat bran, 6 lbs. corn meal and 2 lbs. cotton-seed meal, daily; property of Campbell Brown, Spring Hill, Tenn.

Allie of Glen Rouge 129123.—Sire, Two Hundred Per Cent. 33592; dam, Madge of Glen Rouge 112455.

  Butter, 15 lbs. 14 oz. Milk, 206 lbs. 8 oz.

Test made from Apr. 11 to 18, 1898; age, 2 yrs.; estimated weight, 650 lbs.; fed 15 lbs. oats and peas mixed, chopped, 4 lbs. bran, 2 lbs. oil cake and 30 lbs. corn ensilage, daily—hay *ad lib.*; property of William Rolph, Markham, Ont., Can.

Allie of St. Lambert 24991.—Sire, Stoke Pogis 3d 2238; dam, Kathleen of St. Lambert 5122.

  Butter, 26 lbs. 12 oz. Milk, 329 lbs.

Test made from Feb. 5 to 12, 1886; age, 5 yrs. 11 mos.; fed 34 to 38 lbs. daily of mixed feed, as follows: 10 parts ground oats, 4 parts bran and 6 parts oil meal—no pasture; property of George Smith, Grimsby, Ont., Can.

Allie St. Helier 45794.—Sire, Corporal St. Helier 13397; dam, Cad Lee 4th 35721.

  Butter, 16 lbs. 13½ oz. Milk, 220 lbs. 8 oz.

Test made from July 13 to 20, 1889; age, 3 yrs. 10 mos.; actual weight, 850 lbs.; fed, daily, about 22 lbs. corn meal, fine middlings and wheat bran, in equal parts; property of P. J. Cogswell, Rochester, N. Y.

Allie's Emblem 45620.—Sire, King Allie 14290; dam, Signal Emblem 39473.

  Butter, 17 lbs. 7 oz. Milk, 209 lbs.

Test made from May 3 to 10, 1891; age, 4 yrs.; estimated weight, 700 lbs.; fed 12 lbs. of a mixture of corn meal and ship stuff, daily—short pasture; property of George M. Jewett, Glenville, Md.

Allie's Emblem 45620.—Sire, King Allie 14290; dam, Signal Emblem 39473.

  Butter, 20 lbs. Milk, 221 lbs.

Test made from May 17 to 24, 1891; age, 4 yrs. 1 mo.; estimated weight, 900 lbs.; fed 16 lbs. daily of a mixture of corn meal and ship stuff—clover, blue and orchard grass pasture; property of George M. Jewett, Glenville, Md.

Allie's Leaf 45617.—Sire, King Allie 14290; dam, Geranium Leaf 30875.

  Butter, 16 lbs. 3 oz. Milk, 228 lbs. 12 oz.

Test made from Nov. 8 to 15, 1891; age, 5 yrs. 5 mos.; estimated weight, 750 lbs.; fed 22 lbs. ship stuff, daily—hay, fair pasture; property of George M. Jewett, Glenville, Md.

Allie's Queen 45618.—Sire, King Allie 14290; dam, Fairy Queen of St. Brelade's 7464.

  Butter, 15 lbs. 2 oz. Milk, 216 lbs. 11 oz.

Test made from May 17 to 24, 1891; age, 4 yrs. 10 mos.; estimated weight, 900 lbs.; fed 16 lbs. corn meal and ship stuff, mixed, in two feeds, daily—clover, blue and orchard grass pasture; property of George M. Jewett, Glenville, Md.

Allie Victor Pogis 126907.—Sire, Melia Ann's Victor Pogis 20697; dam, May Pedro 50876.

  Butter, 22 lbs. 8 oz. Milk, 300 lbs. 8 oz.

Test made from Oct. 19 to 25, 1897; age, 5 yrs. 1 mo.; estimated weight, 900 lbs.; fed 12 lbs. corn and wheat bran, mixed in equal parts by weight, and 4 lbs. oil meal, daily—clover pasture; property of J. H. Martin, Granville, N. Y.

Alluring 5541.—Sire, Columblad 2d 1515; dam, Purity 1408.

Butter, 19 lbs. 5 oz.

Test made from July 20 to 27, 1882; age, 5 yrs. 6 mos.; test made on grass alone; property of William N. McConnell, Dartford, Wis.

Almah of Oakland 11102.—Sire, Thorndale 2582; dam, Pandora of Staatsburgh 3d 6497.

Butter, 16 lbs. 14 oz. Milk, 253 lbs. 9 oz.

Test made from May 1 to 7, 1883; age, 8 yrs. 2 mos.; fed 2 qts. corn meal, 2 gals. wheat bran, 1 pt. oil meal. daily, with hay—rather poor blue-grass pasture; property of R. McMichael, Lexington, Ky.

Alma of Springvale 26951.—Sire, Alpheus of Springvale 5645; dam, Alabama 7690.

Butter, 16 lbs. 14 oz. Milk, 229 lbs. 5 oz.

Test made from Apr. 2 to 9, 1888; age, 4 yrs. 2 mos.; estimated weight, 750 lbs.; fed 28 qts. corn meal, 28 qts. ground oats, 28 qts. cotton-seed meal and 175 lbs. ensilage —hay, pasture of winter rye; property of James Crook, Jacksonville, Ala.

Almeda Rex 10416.—Sire, Rex 1330; dam, Almeda 3842.

Butter, 15 lbs. 12 oz. Milk, 290 lbs. 4 oz.

Test made from May 26 to June 1, 1885; age, 6 yrs.; property of Moulton Bros., West Randolph, Vt.

Almond Blossom 59709.—Sire, Earl of Eastmore 18595; dam, Almond 11747.

Butter, 17 lbs. 3¾ oz. Milk, 163 lbs. 8 oz.

Test made from Sept. 8 to 15, 1895; age, 7 yrs. 11 mos.; estimated weight, 1000 lbs.; fed corn meal, bran and cut sorghum, not weighed—fall pasture; property of John C. Hooper, Murfreesborough, Tenn.

Alnora 112132.—Sire, Evangelist 25700; dam, Granda 92755.

Butter, 16 lbs. 6 oz. Milk, 242 lbs. 1 oz.

Test made from Feb. 20 to 27, 1896; age, 4 yrs. 7 mos.; estimated weight, 800 lbs.; fed 20 lbs., daily, of a mixture of corn meal, ground oats and bran—4 qts. roots daily; property of Richardson Bros., Davenport, Iowa.

Aloha's Tormentor 90239.—Sire, Tormentor's Jersey King 26928; dam, Aloha of St. Lambert 63445.

Butter, 14 lbs. 6 oz. Milk, 224 lbs.

Test made from May 8 to 15, 1896; age, 4 yrs. 7 mos.; estimated weight, 600 lbs.; fed 72 qts. corn and oats—feed of rye in morning and baited on June grass about two hours in afternoon; property of E. G. Silvus, Athens, Ohio.

Alphea 171.—Sire, Saturn 94; dam, Rhea 166.

Butter, 15 lbs. 6 oz.

Property of R. M. Hoe, New York, N. Y.

Alphea Jewel 22331.—Sire, Mercury 432; dam, Hartwick Belle 7722.

Butter, 14 lbs. Milk, 187 lbs. 6 oz.

Test made from May 10 to 16, 1885; age, 2 yrs. 2 mos.; property of William Simpson, New York, N. Y.

Alphea Marigold 117632.—Sire, Stoke Pogis of Prospect 29121; dam, Niobe's Alpheanette 3d 39459.

Butter, 14 lbs. 5 oz.   Milk, 223 lbs. 3 oz.

Test made from May 27 to June 3, 1898; age, 2 yrs. 5 mos.; actual weight, 895 lbs.; fed 6 qts. bran, 6 qts. ground oats, 3 qts. corn meal, 3 qts. middlings and 1½ qts. oil meal, mixed, fed wet, daily—grass in barn; property of Charles A. Sweet, Buffalo, N. Y.

Alphea Rajah 20605.—Sire, Topris 4135; dam, Martha M. 10867.

Butter, 23 lbs. 2 oz.   Milk, 247 lbs. 7 oz.

Test made from May 19 to 26, 1893; age, 10 yrs. 11 mos.; estimated weight, 850 lbs.; fed 42 qts. of a mixture of two-fifths corn, two-fifths oats and one-fifth peas, 28 qts. bran and 14 qts. cotton-seed meal—good June grass pasture, hay night and morning; property of L. H. Turner, Oakfield, N. Y.

Alphea Star 16532.—Sire, Mercury 432; dam, Starbeam 11846.

Butter, 14 lbs. 4½ oz.   Milk, 180 lbs. 4 oz.

Test made from June 30 to July 6, 1884; age, 2 yrs. 3 mos.; property of William Simpson, New York, N. Y.

Alphea Vaudrada 17468.—Sire, Black Diamond V. 4970; dam, Black Diamond's Queen 11865.

Butter, 19 lbs. 7 oz.   Milk, 212 lbs. 6 oz.

Test made from Aug. 31 to Sept. 7, 1890; age, 8 yrs. 5 mos.; estimated weight, 950 lbs.; fed, daily, 16 qts. ground oats, corn and peas and 2 qts. shorts; property of Charles E. Hill, Denver, Col.

Alpheon's Belle 27194.—Sire, Alpheon 6082; dam, Beatrice Cenci 16629.

Butter, 19 lbs. 11 oz. (official).   Milk, 217 lbs. 12 oz.

Test made from Mar. 1 to 8, 1888; age, 4 yrs. 2 mos.; estimated weight, 800 lbs.; fed, daily, 5 lbs. corn-hearts, 4 lls. rice meal, 2 lbs. oil meal and 6 lbs. bran. Chemical analyses: of butter, fats 86.20, casein 3.10, salts 2.90, water 7.80; of buttermilk, fats 0.45, casein 2.98, sugar 2.04, salts 0.43, water 94.10; property of John Boyd, Elmhurst, Ill.

Alphetta 16531.—Sire, Mercury 432; dam, Ideal 11842.

Butter, 14 lbs. 2½ oz.   Milk, 186 lbs. 5 oz.

Test made from July 7 to 13, 1884; age, 2 yrs. 7 mos.; property of William Simpson, New York, N. Y.

Alpheus Mostar 20037.—Sire, Clifton Hero 6759; dam, Queen Mostar 15256.

Butter, 14 lbs. 9 oz.   Milk, 240 lbs. 5 oz.

Test made from Apr. 20 to 27, 1885; age, 3 yrs. 6 weeks; property of James Cloud & Son, Kennett Square, Pa.

Alta B. 73003.—Sire, Bem 18867; dam, Hild 46148.

Butter, 14 lbs. 13 oz.   Milk, 215 lbs. 6 oz.

Test made from May 29 to June 5, 1898; age, 7 yrs. 2 mos.; estimated weight, 950 lbs.; fed 6 lbs. bran, 3 lbs. Buffalo gluten and 2 lbs. new process linseed meal, daily—medium pasture; property of Charles H. Ellsworth, Worcester, Mass.

Alteration 56436.—Sire, Little Harry 8808; dam, Analysis 29162.

Butter, 24 lbs. ½ oz. Milk, 264 lbs. 3 oz.

Test made from July 15 to 22, 1892; age, 4 yrs. 10 mos.; estimated weight, 800 lbs.; fed 4 gals. corn meal, 2 gals. ground oats and 1 gal. wheat bran, daily, on cut green corn—short pasture; property of Matthews & Moore, Huntsville, Ala.

Alteration 56436.—Sire, Little Harry 8808; dam, Analysis, 29162.

Butter, 179.879 lbs.* in 90 days. Milk, 3115.7 lbs.*

Test made from May 31 to Aug. 28, 1893, at World's Columbian Exposition; age, 5 yrs. 8 mos. at beginning of test; actual weight, 800 lbs.; fed 302.7 lbs. old hay, 210 lbs. ensilage, 256.6 lbs. new hay, 133.5 lbs. oil meal, 335 lbs. bran, 52 lbs. oats, 374.5 lbs. corn-hearts, 93.25 lbs. cotton-seed, 311 lbs. middlings, 33 lbs. grano-gluten, 1184.4 lbs. clover and 10 lbs. swale-grass*; property of W. E. Matthews, Huntsville, Ala.

Alvoretta 93817.—Sire, Oonan's Tormentor 22280; dam, Witching 42252.

Butter, 15 lbs. 14 oz. Milk, 178 lbs. 8 oz.

Test made from Dec. 21 to 28, 1896; age, 3 yrs. 5 mos.; estimated weight, 775 lbs.; fed 16 lbs. daily of a mixture of corn meal, ground oats, bran, pea meal, cotton-seed meal and oil meal, 40 lbs. ensilage and about 10 lbs. hay; property of R. F. Shannon, Pittsburg, Pa.

Amalthea 11319.—Sire, Miramon 1451; dam, Rioter 2d's Venus 3658.

Butter, 15 lbs. Milk, 241 lbs. 14 oz.

Test made from June 30 to July 6, 1885; age, 5 yrs. 1 mo.; property of Frederick Billings, Woodstock, Vt.

Amanda May 83011.—Sire, Jersey King 9458; dam, Amy Valentine 68378.

Butter, 14 lbs. 15½ oz. Milk, 261 lbs.

Test made from July 19 to 26, 1897; age, 5 yrs. 8 mos.; actual weight, 980 lbs.; fed 7 lbs. cotton-seed meal, 7 lbs. flaxseed meal, 49 lbs. wheat bran and 42 lbs. corn meal—millet hay, hay stubble pasture; property of Biltmore Farms, Biltmore, N. C.

Amanda of St. Lambert 64955.—Sire, Prince of Melrose 2d 11015; dam, Maid of Tupelo 41753.

Butter, 18 lbs. 14 oz. Milk, 197 lbs.

Test made from Nov. 5 to 12, 1893; age, 5 yrs.; estimated weight, 850 lbs.; fed 12 qts. bran and 8 qts. ground oats and corn meal, daily—prairie hay ad lib.; property of Terrell & Harris, Terrell, Tex.

Amanda Pogis 79412.—Sire, Count Rioter Pogis 21388; dam, Ampelis 7th 29851.

Butter, 18 lbs. 7¼ oz. Milk, 294 lbs. 4½ oz.

Test made from Nov. 21 to 28, 1896; age, 6 yrs. 6 mos.; actual weight, 910 lbs.; fed 2 gals. ground meal, 2 gals. ground oats and corn, 2 gals. wheat bran and 1 pt. pea or linseed meal, daily—dry pasture; property of James P. Craver, Harleton, Tex.

Amelia Davenport 81044.—Sire, Wasp 15358; dam, June 3d 11644.

Butter, 14 lbs. ⅝ oz. Milk, 250 lbs.

Test made from June 24 to July 1, 1898; age, 9 yrs. 8 mos.; actual weight, 1013 lbs.; fed 56 lbs. corn meal and 70 lbs. bran—blue-grass pasture; property of Kentucky Agricultural Experiment Station, Lexington, Ky.

* Estimated.

Amelia Pogis 53796.—Sire, St. Lambert's John Bull 16618; dam, Pogis' Dewdrop 40373.

Butter, 16 lbs. 14 oz. Milk, 238 lbs. 11 oz.

Test made from Nov. 1 to 8, 1892; age, 4 yrs. 6 mos.; estimated weight, 750 lbs.; fed 56 qts. corn meal, 28 qts. bran, 28 qts. cotton-seed meal and 30 lbs. ensilage—hay, pasture in very good rye lot; property of James Crook, Jacksonville, Ala.

Amelia S. King 112862.—Sire, Naiad's St. Lambert King 30645; dam, Exile's Majel 112861.

Butter, 16 lbs. 2 oz. Milk, 243 lbs. 14 oz.

Test made from Jan. 3 to 10, 1898; age, 3 yrs. 10 mos.; estimated weight, 800 lbs.; fed 12 qts. bran, 16 qts. crushed corn and cob and oats (one-third oats), 2 qts. oil meal, 5 lbs. clover hay and 30 lbs. ensilage, daily; property of Brooks & Pidgeon, Salem, Ohio.

Amna 35588.—Sire, Nancy's Rioter 8390; dam, Allie Minka 3d 27137.

Butter, 14 lbs. 12½ oz. Milk, 229 lbs. 8 oz.

Test made from Jan. 20 to 27, 1894; age, 8 yrs. 3 mos.; estimated weight, 800 lbs.; fed 2 qts. corn-hearts, 3 qts. ground oats, 1 qt. cotton-seed meal and 2 qts. bran, daily—clover hay and corn stover, poor blue-grass pasture; property of Samuel N. Warren, Spring Hill, Tenn.

Amna 2d 98269.—Sire, Wrangler 20881; dam, Amna 35588.

Butter, 18 lbs. 9 oz. Milk, 283 lbs. 8 oz.

Test made from Dec. 18 to 25, 1897; age, 4 yrs.; estimated weight, 750 lbs.; fed 25 lbs. per day of ground oats and corn, in equal parts, mixed with 4 bucketfuls of cut sheaf oats, wetted—millet hay *ad lib.* at night; property of Matt. M. Gardner, Nashville, Tenn.

Ampelis 5th 17548.—Sire, De Gruchy 3652; dam, Ampelis 5622.

Butter, 15 lbs. Milk, 194 lbs. 12 oz.

Test made from Mar. 15 to 21, 1885; age, 2 yrs. 11 mos.; property of I. D. Risher, Hope Church, Pa.

Amrah of Ingleside 56832.—Sire, Duke of Ingleside 14274; dam, Amrah J. 23284.

Butter, 14 lbs. 2 oz. Milk, 156 lbs. 2 oz.

Test made from Aug. 3 to 10, 1895; age, 7 yrs. 3 mos.; estimated weight, 1020 lbs.; fed 28 lbs. coarse wheat bran, 28 lbs. ground oats, 28 lbs. corn meal and 14 lbs. cotton-seed meal—poor pasture; property of John C. McClintock, Meadville, Pa.

Amy Landseer 74826.—Sire, Leonette's Landseer 20399; dam, Amy of Glenoir 47047.

Butter, 16 lbs. 1 oz. Milk, 219 lbs. 12 oz.

Test made from Jan. 26 to Feb. 2, 1896; age, 6 yrs.; estimated weight, 850 lbs.; fed 126 lbs. ground corn and oats, 98 lbs. bran and 76 qts. ensilage; property of J. D. Gray, Terrell, Tex.

Analysis 29162.—Sire, Surpry 2d (P. S. 459 J. H. B.); dam, Ava (F. S. 6217 J. H. B.)

Butter, 16 lbs. 8 oz. Milk, 234 lbs. 7 oz.

Test made from Jan. 15 to 22, 1892; age, 8 yrs.; estimated weight, 750 lbs.; fed 2 gals. corn meal, 3 gals. boiled cotton-seed, 2 gals. bran—ensilage and clover hay; property of Matthews & Moore, Huntsville, Ala.

Analysis 2d 55834.—Sire, Hilarious 7651; dam, Analysis 29162.

Butter, 20 lbs. 7½ oz. Milk, 242 lbs. 11 oz.

Test made from Mar. 25 to Apr. 1, 1891; age, 4 yrs. 6 mos.; estimated weight, 850 lbs.; fed 2 gals. corn-hearts, 2 gals. wheat bran and 2 gals. ground oats, daily, also good corn ensilage—clover hay *ad lib.*; property of Matthews & Moore, Huntsville, Ala.

Analysis 3d 82870.—Sire, Signalda's Torment 14673; dam, Analysis 29162.

Butter, 14 lbs. 1 oz. Milk, 155 lbs. 6 oz.

Test made from Mar. 17 to 24, 1893; age, 2 yrs. 3 mos.; estimated weight, 600 lbs., fed 48 qts. corn meal, 24 qts. corn and oats (one-third oats), 56 qts. wheat bran, 6 qts. linseed meal, 6 qts. cotton-seed meal and 2 qts. pea meal—hay; property of Matthews & Moore, Huntsville, Ala.

Angela 1682.—Sire, Roxbury 247; dam, Europa 121.

Butter, 14 lbs. 2 oz.

Property of Silas Betts, Camden, N. J.

Angela of Coal Hill 79380.—Sire, Ida's Rioter of Coal Hill 19661; dam, Rosalep Green 66753.

Butter, 15 lbs. 7 oz. Milk, 243 lbs. 9 oz.

Test made from June 13 to 20, 1898; age, 6 yrs. 5 mos.; estimated weight, 900 lbs.; fed 84 qts. wheat bran and 56 qts. corn meal—clover and timothy pasture; property of T. B. Kibler, Jenera, Ohio.

Angela's Pet 69776.—Sire, Angela's Rioter 20120; dam, Drosera's Pet 42427.

Butter, 15 lbs. 12 oz. Milk, 228 lbs.

Test made from Dec. 14 to 21, 1895; age, 6 yrs. 5 mos.; estimated weight, 1000 lbs.; fed 350 lbs. ensilage, 113 lbs. bran and 14 lbs. oil meal—bright oat straw *ad lib.*; property of Harmon Austin, Warren, Ohio.

Angelina D. 65514.—Sire, Catono's Harry 20096; dam, Miss Semple P. 56948.

Butter, 15 lbs. 3½ oz. Milk, 232 lbs. 13 oz.

Test made from Apr. 29 to May 6, 1898; age, 9 yrs. 1 mo.; estimated weight, 750 lbs.; fed 5 lbs. corn meal, 3 lbs. wheat bran, 2 lbs. ground oats, 2 lbs. linseed meal, 2 lbs. cotton-seed meal and 1 lb. malt sprouts, daily—blue-grass pasture; property of E. W. Sewell, Anchorage, Ky.

Angelina Pogis 40367.—Sire, Prince of Melrose 2d 11015; dam, Zenobia of Springvale 20094.

Butter, 18 lbs. 13 oz. Milk, 218 lbs.

Test made from Jan. 12 to 19, 1890; age, 3 yrs. 8 mos.; estimated weight, 800 lbs.; fed 70 qts. ground oats, 42 qts. corn meal, 14 qts. cotton-seed meal and 245 lbs. ensilage—hay, pasture of rye lot; property of James Crook, Jacksonville, Ala.

Angelo's Torment 39390.—Sire, Michael Angelo 10116; dam, Little Torment 15581.

Butter, 22 lbs. 8 oz. Milk, 237 lbs. 1 oz.

Test made from Nov. 14 to 21, 1890; age, 5 yrs. 8 mos.; estimated weight, 800 lbs.; fed, daily, 28 lbs. corn meal and middlings; property of George M. Jewett, Glenville, Md.

Angetta 19404.—Sire, Mahkeenac 3290; dam, Lady Anerly 10595.

Butter, 15 lbs. 4 oz.   Milk, 275 lbs. 8 oz.

Test made from June 9 to 15, 1887; age, 5 yrs. 2 mos.; estimated weight, 600 lbs.; fed, daily, 4 lbs. corn meal, 2 lbs. oil meal and 16 lbs. bran; property of A. H. Cooley, Little Britain, N. Y.

Anita H. 12334.—Sire, Homer H. 3683; dam, Gilda 2779.

Butter, 17 lbs. 4 oz.   Milk, 331 lbs. 4 oz.

Test made from Nov. 9 to 15, 1885; age, 5 yrs. 5 mos.; property of William H. Hayden, Albany, Vt.

Anita Lawrence 63909.—Sire, Matilda's Perfection 17100; dam, Sparks' Maid 45575.

Butter, 15 lbs. 6½ oz.   Milk, 232 lbs. 5 oz.

Test made from June 29 to July 6, 1895; age, 6 yrs. 2 mos.; estimated weight, 900 lbs.; fed 28 lbs. corn meal and 21 lbs. wheat bran—timothy and clover pasture; property of Henry S. Redfield, Elmira, N. Y.

Anklet 93500.—Sire, Major Appel Pogis 17861; dam, Beauty of Shady Side 49193.

Butter, 14 lbs. 6 oz.   Milk, 219 lbs. 8 oz.

Test made from Dec. 8 to 15, 1894; age, 3 yrs. 9 mos.; actual weight, 860 lbs.; fed 56 qts. ground oats, 28 qts. corn meal, 14 qts. wheat bran and 14 qts. ground oil cake; property of Miller & Sibley, Franklin, Pa.

Annabel Lee 55833.—Sire, Signalda's Torment 14673; dam, Mette D. 47433.

Butter, 22 lbs. 8 oz.   Milk, 279 lbs. 4 oz.

Test made from June 27 to July 4, 1893; age, 5 yrs. 4 mos.; estimated weight, 800 lbs.; fed 84 qts. oats and corn ground in equal parts, 28 qts. wheat bran, 14 qts. cotton-seed meal, fed on about 2 lbs. cut mixed hay, moistened—42 lbs. hay, short pasture; property of Wm. E. Matthews, Huntsville, Ala.

Anna Paris 67989.—Sire, Pride's Pogis 23459; dam, Bemona 52994.

Butter, 14 lbs. ¾ oz.   Milk, 237 lbs. 14 oz.

Test made from June 6 to 13, 1898; age, 7 yrs. 9 mos.; estimated weight, 850 lbs., fed 30 lbs. corn chop, 32 lbs. bran, 27 lbs. oil meal, 15 lbs. cotton-seed meal, 27 lbs. gluten meal and 10 lbs. H.O.—blue-grass and white clover pasture; property of Filston Farm, Glencoe, Md.

Anna Smith 10324.—Sire, Smith of Darlington 2458; dam, Violet of Darlington 5573.

Butter, 15 lbs. 6 oz., unsalted.   Milk, 278 lbs. 4 oz.

Test made from Jan. 24 to 31, 1882; fed 3 qts. corn and 3 qts. oats, daily; property of A. B. Darling, Ramsey's, N. J.

Annice Magnet 60256.—Sire, Champion Magnet 6480; dam, Annie Bropho 39052.

Butter, 119.284 lbs.* in 90 days.   Milk, 2064 lbs.*

Test made from May 31 to Aug. 28, 1893, at World's Columbian Exposition; age, 4 yrs. 2 mos. at beginning of test; actual weight, 846 lbs.; fed 274.5 lbs. old hay, 210 lbs. ensilage, 514.7 lbs. new hay, 174 lbs. oil meal, 334.5 lbs. bran, 150 lbs. oats, 321 lbs. corn-hearts, 92.75 lbs. cotton-seed, 233 lbs. middlings, 31 lbs. grano-gluten and 447.1 lbs. clover*; property of John Boyd, Elmhurst, Ill.

---

* Estimated.

Annie Dale's Princess 12664.—Sire, Prince of Scituate 3888; dam, Annie Dale 12662.

Butter, 19 lbs. 2½ oz. Milk, 335 lbs. 1 oz.

Test made from May 31 to June 6, 1886; age, 7 yrs. 10 mos.; property of E. Parmly, Oceanic, N. J.

Annie Hines 84038.—Sire, Maquilla's Harry 22701; dam, Violet Denison 49997.

Butter, 18 lbs. 6¾ oz. Milk, 284 lbs. 3 oz.

Test made from May 24 to 31, 1898; age, 6 yrs. 1 mo.; estimated weight, 750 lbs.; fed 24 lbs. per day of ground oats and corn, in equal parts—extra good orchard and blue-grass pasture; property of Matt. M. Gardner, Nashville, Tenn.

Annie L. 12934.—Sire, Champion of America 1567; dam, Annie Landers 2d 7670.

Butter, 17 lbs. 15¼ oz. Milk, 228 lbs. 4 oz.

Test made from Sept. 16 to 23, 1886; age, 5 yrs. 8 mos.; property of W. B. Montgomery, Starkville, Miss.

Annie Lane 45345.—Sire, Catono 3761; dam, Elsie Lane 13302.

Butter, 15 lbs. 6 oz. Milk, 259 lbs.

Test made from Apr. 17 to 24, 1890; age, 3 yrs. 10 mos.; actual weight, 750 lbs.; fed, daily, between 16 lbs. and 24 lbs., consisting principally of corn meal; property of William Morrow & Son, Nashville, Tenn.

Annie Linn 82892.—Sire, Ona's Koffee 2d 17609; dam, Dell's Daughter 48990.

Butter, 15 lbs. 2 oz. Milk, 240 lbs. 10 oz.

Test made from May 19 to 26, 1895; age, 2 yrs. 9 mos.; estimated weight, 700 lbs.; fed 28 qts. ship stuff, 14 qts. shelled oats, 14 qts. oil meal and 21 qts. cracked corn—blue-grass pasture; property of John S. Shannon, Shelbyville, Ky.

Annie's Prize 86548.—Sire, One Hundred Per Cent. 16590; dam, Ernestine's Annie 85645.

Butter, 19 lbs. 11¾ oz. Milk, 261 lbs.

Test made from Apr. 22 to 29, 1898; age, 5 yrs. 11 mos.; estimated weight, 875 lbs., fed 84 qts. ground oats, 49 qts. corn meal and 14 qts. ground oil cake—good old meadow pasture; property of Miller & Sibley, Franklin, Pa.

Ann Victor Pogis 77813.—Sire, Melia Ann's Victor Pogis 20697; dam, Carrie Pedroller 48966.

Butter, 14 lbs. 7 oz. Milk, 245 lbs. 12 oz.

Test made from Feb. 2 to 9, 1895; age, 5 yrs.; estimated weight, 900 lbs.; fed 10 lbs. oats, peas and corn, ground together, 2 lbs. oil meal and 6 lbs. bran, daily—hay; property of Mrs. J. H. Martin, Granville, N. Y.

Antigua 109426.—Sire, Tennessee's Tormentor 23498; dam, Peggy Scituate 76986.

Butter, 14 lbs. 2¾ oz. Milk, 203 lbs. 12 oz.

Test made from Jan. 26 to Feb. 2, 1896; age, 2 yrs. 1 mo.; estimated weight, 725 lbs.; fed 126 lbs. of a mixture of cotton-seed meal, crushed corn ears, crushed oats and wheat bran, and 56 lbs. roots—hay; property of D. Wolfe Bishop, Jr., Lenox, Mass.

Arawana Buttercup 6052.—Sire, Norajah 812; dam, Coe's Stella 3930.

Butter, 15 lbs. 5 oz. Milk, 343 lbs.

Test made from May 15 to 21, 1881; age, 4 yrs.; property of T. Alexander Seth, Baltimore, Md.

Arawana Chèvre Feuille 15097.—Sire, Atamasco 2781; dam, Coe's Stella 3930.

Butter, 14 lbs. 9 oz.   Milk, 214 lbs. 4 oz.

Test made from Dec. 27, 1885, to Jan. 2, 1886; age, 4 yrs. 1 mo.; property of Watts & Seth, Baltimore, Md.

Arawana Poppy 6053.—Sire, Norajah 812; dam, Poppy 4842.

Butter, 15 lbs. 2 oz., unsalted.   Milk, 236 lbs. 8 oz.

Test made from May 30 to June 5, 1883; age, 5 yrs. 10 mos.; fed 2 qts. corn meal and 4 qts. bran, daily, with green rye and pasture; property of Watts & Seth, Baltimore, Md.

Arawana Queen 5368.—Sire, Rex 1330; dam, Arawana Rose 3810.

Butter, 16 lbs. 9 oz.   Milk, 367 lbs.

Test made from June 15 to 21, 1883; age, 6 yrs. 7 mos.; property of John E. Phillips, Baltimore, Md.

Archie 1112, imp.

Butter, 15 lbs.

Test made from June 17 to 23, 1878; age, 10 yrs.; property of James A. Hayt, Patterson, N. Y.

Arietta 5115.—Sire, Young Baron 702; dam, Cowslip 1778.

Butter, 15 lbs., unsalted.

Age, 8 yrs. 8 mos.; property of James R. Crane, Washington, Ill.

Arletta 3d 14274.—Sire, New Year's 4352; dam, Arletta 14264.

Butter, 14 lbs. 13½ oz.   Milk, 295 lbs. 12 oz.

Test made from June 10 to 16, 1885; age, 4 yrs. 6 mos.; property of Marion Nash, Martinsburg, N. Y.

Armon 10862.—Sire, Butter Maker 3098; dam, Mary Azuline 6314.

Butter, 16 lbs. 13½ oz.   Milk, 227 lbs. 5 oz.

Test made from Apr. 26 to May 8, 1883; age, 3 yrs. 7 mos.; fed 12 lbs. boiled cottonseed, 7½ lbs. wheat bran, 4½ lbs. corn meal—blue-grass and red clover pasture; property of A. H. French, Aberdeen, Miss.

Arnold's Lulu 7328.—Sire, Lord Bronx 2d 1730; dam, Coreopsis 4186.

Butter, 15 lbs. 3 oz.   Milk, 273 lbs.

Test made from April 28 to May 4, 1884; age, 8 yrs. 2 mos.; property of Beech Grove Farm, Beech Grove, Ind.

Arthur's Frolic 4438, imp.

Butter, 15 lbs.   Milk, 241 lbs.

Test made from June 5 to 12, 1882; age, 13 yrs.; property of A. H. Cooley, Little Britain, N. Y.

Ashantee's Brunette 65946.—Sire, King of Ashantee 6677; dam, Brunette Star 27270.

Butter, 16 lbs. 7 oz.   Milk, 185 lbs. 8 oz.

Test made from May 27 to June 3, 1893; age, 3 yrs. 8 mos.; estimated weight, 800 lbs.; fed 70 lbs. bran, 30 lbs. oil meal and 60 lbs. corn meal; property of Cornelius Easthope, Niles, Ohio.

Ashantee's Lady 35951.—Sire, King of Ashantee 6677; dam, Lady Bountiful 17946.

Butter, 16 lbs. Milk, 129 lbs. 12 oz.

Test made from Sept. 24 to 30, 1887; age, 2 yrs. 5 mos.; estimated weight, 700 lbs.; fed, daily, 4 qts. corn meal and 6 qts. bran; property of M. Erskine Miller, Staunton, Va.

Ashantee's Rosy 59258.—Sire, King of Ashantee 6677; dam, Belle Rosy 27291.

Butter, 14 lbs. 11 oz. Milk, 160 lbs.

Test made from May 28 to June 4, 1893; age, 4 yrs. 8 mos.; estimated weight, 700 lbs.; fed 65 lbs. bran, 30 lbs. oil meal and 60 lbs. corn meal; property of Cornelius Easthope, Niles, Ohio.

Ashwood Pet 2d 84408.—Sire, Toltec's Prince 21961; dam, Ashwood Pet 59355.

Butter, 16 lbs. 11½ oz. Milk, 244 lbs.

Test made from June 3 to 10, 1895; age, 4 yrs.; estimated weight, 850 lbs.; fed 14 qts., daily, corn-hearts and bran, in equal parts—blue-grass pasture; property of Miss E. E. Polk, Spring Hill, Tenn.

Aspirante 9272.—Sire, Columbiad 2d 1515; dam, Perfection 1897.

Butter, 14 lbs. 7 oz. Milk, 238 lbs.

Test made from June 20 to 26, 1881; age, 3 yrs.; property of Seth L. Hoover, Columbus, Ohio.

Athlete 28137.—Sire, Scituate's Prize 7206; dam, Alphalia 13820.

Butter, 14 lbs. 6 oz. Milk, 213 lbs. 10 oz.

Test made from Sept. 8 to 10, 1886; age, 2 yrs. 3 mos.; estimated weight, 775 lbs.; fed, daily, 12 qts. ground peas and oats and oil meal; property of Richardson Bros., Davenport, Iowa.

Atlanta's Beauty 12949.—Sire, Smoke Smith 4844; dam, Laura Hill 9084.

Butter, 21 lbs. 3 oz. Milk, 146 lbs.

Test made from Feb. 8 to 10, 1885; age, 3 yrs. 6 mos.; fed, in the morning, 1 gal. corn meal, 1 gal. wheat middlings and about 15 lbs. corn fodder—ensilage; at noon, 1 peck carrots cut up and sprinkled with wheat middlings; at night, 1 gal. cotton-seed meal, 1 gal. wheat middlings, and plenty of hay—no pasture; property of L. J. & A. W. Hill, Atlanta, Ga.

Atricia 6029.—Sire, Blondin 3d 1935; dam, Patricia 4th 4579.

Butter, 15 lbs. 3 oz.

Test made from Apr. 30 to May 6, 1882; age, 5 yrs.; fed 2 qts. corn meal, 4 qts. wheat bran and 8 qts. oats, daily—hay; property of H. G. Westlake, Hillsdale, N. Y.

Attractive Maid 16925.—Sire, Don Pedro of Binghamton 2974; dam, Florence 1043.

Butter, 16 lbs. 13 oz. Milk, 240 lbs.

Test made from June 20 to 27, 1883; age, 3 yrs. 4 mos.; fed 2 qts. cotton-seed meal, daily—fair pasture; property of William H. Burr, Redding Ridge, Conn.

Attractive Maid 16925.—Sire, Don Pedro of Binghamton 2974; dam, Florence 1043.

Butter, 22 lbs. 5 oz. Milk, 254 lbs.

Test made from Oct. 9 to 16, 1885; age, 5 yrs. 7 mos.; fed 10 lbs. ground oats and corn and 6 lbs. cotton-seed meal, mixed, daily—good pasture; property of William H. Burr, Redding Ridge, Conn.

Augeres Girl 17015, imp.

Butter, 14 lbs. 6 oz., unsalted.

Test made Jan., 1883; age, 3 yrs. 8 mos.; property of S. W. Robbins, Wethersfield, Conn.

Augustella 63018.—Sire, Allegany Chief 2918; dam, Glenida 2d 5770.

Butter, 14 lbs. 2 oz.   Milk, 126 lbs.

Test made from Feb. 12 to 19, 1893; age, 5 yrs.; estimated weight, 850 lbs.; fed moderate quantities of bran, corn meal, cotton-seed and hay; property of J. D. Gray, Terrell, Tex.

Aunt Lora 77855.—Sire, King Albert's Rioter 14533; dam, Oaklands Fairy 34273.

Butter, 14 lbs. 14⅞ oz.   Milk, 273 lbs. 4 oz.

Test made from May 5 to 12, 1895; age, 6 yrs. 5 mos.; actual weight, 940 lbs.; fed 13 lbs. bran and 10 lbs. ground corn and oats, daily—hay, blue-grass, timothy and clover pasture; property of Mrs. Isa Bayler, Washington, Ill.

Auntybel 12582.—Sire, Guy Fawkes (F. S. 251 J. H. B.); dam, Aunty 16648.

Butter, 14 lbs. 9 oz., unsalted.   Milk, 120 lbs.

Test made from June 9 to 15, 1882; age, 3 yrs. 2 mos.; property of S. M. Burnham, Saugatuck, Conn.

Auraria 10688.—Sire, Manchester's Prospect 2817; dam, Jersey Cream 2d 8519.

Butter, 15 lbs. 10 oz.   Milk, 184 lbs. 8 oz.

Test made from May 20 to 27, 1890; age, 10 yrs. 2 mos.; estimated weight, 900 lbs.; fed, daily, 2 qts. cotton-seed meal, 2 qts. oil meal, 4 qts. corn meal, 6 qts. ground oats and 4 qts. wheat middlings—hay *ad lib.*; property of E. L. Van Deusen, Ashley Falls, Mass.

Aurated Pogis 2d 65666.—Sire, King of St. Lambert 2d 17823; dam, Aurated Pogis 48220.

Butter, 16 lbs. 6 oz.   Milk, 280 lbs. 5 oz.

Test made from Oct. 8 to 15, 1893; age, 4 yrs. 8 mos.; estimated weight, 850 lbs.; fed 28 qts. ship feed, 28 qts. corn meal, 14 qts. ground oats and about 7 bushels cut corn-stalks—upland timothy and clover pasture; property of H. G. Westlake, Hillsdale, N. Y.

Aurora Pogis 53798.—Sire, St. Lambert's John Bull 16618; dam, Alabama 7690.

Butter, 15 lbs. 9 oz.   Milk, 166 lbs. 14 oz.

Test made from May 7 to 14, 1893; age, 4 yrs. 10 mos.; estimated weight, 800 lbs.; fed 56 qts. ground oats, 14 qts. corn meal, 14 qts. cotton-seed meal—hay, pasture of good winter rye; property of James Crook, Jacksonville, Ala.

Auselma 37798.—Sire, Happy Boy 9846; dam, Otella 11918.

Butter, 14 lbs. 1 oz.   Milk, 212 lbs. 5 oz.

Test made from Apr. 17 to 24, 1895; age, 9 yrs. 4 mos.; estimated weight, 850 lbs.; fed 56 lbs. corn meal, 80 lbs. bran and 13 lbs. cotton-seed meal; property of Austin & Probert, Florence, Mich.

Austie May 7185.—Sire, Kit Carson 1772; dam, May 1390.

Butter, 14 lbs. 3 oz.

Test made from Dec. 10 to 17, 1880; age, 3 yrs. 6 mos.; property of Luther C. Powers, La Porte, Ind.

Azelda 2d 7022.—Sire, Monsieur 1723; dam, Azelda 3872.

Butter, 15 lbs. 2 oz. Milk, 216 lbs. 5 oz.

Test made from Apr. 7 to 13, 1883; age, 5 yrs. 2 mos.; property of William Craik, Frankfort, Ky.

Azelda's Valhalla 45234.—Sire, Azelda's Wanderer 12997; dam, Edna Browning 12292.

Butter, 14 lbs. 5 oz. Milk, 187 lbs.

Test made from Mar. 7 to 14, 1893; age, 6 yrs. 1 mo.; estimated weight, 800 lbs.; fed 60 lbs. bran, 56 lbs. corn meal and 40 lbs. oil meal; property of Cornelius Easthope, Niles, Ohio.

Azuline 2d 3888.—Sire, The Hub 1009; dam, Azuline 3360.

Butter, 14 lbs. 15¾ oz. Milk, 238 lbs.

Test made from Oct. 8 to 14, 1886; age, 11 yrs. 10 mos.; fed boiled cotton-seed, corn meal and bran; property of W. B. Montgomery, Starkville, Miss.

Baby Mar 115089.—Sire, Tormentor Catono 17004; dam, Lord Mar's Lady 48307.

Butter, 16 lbs. 6 oz. Milk, 287.3 lbs.

Test made from May 3 to 10, 1896; age, 6 yrs.; estimated weight, 850 lbs.; fed 28 gals. wheat bran, 28 qts. corn meal and 14 qts. cotton-seed meal—good blue-grass pasture; property of W. L. Reynolds, Cynthiana, Ky.

Baby Princess 57493.—Sire, Prince of Avon 17544; dam, Wilson's Beauty 27671.

Butter, 14 lbs. Milk, 229 lbs. 4 oz.

Test made from Dec. 16 to 23, 1890; age, 3 yrs. 6 mos.; estimated weight, 800 lbs.; fed 9 lbs. corn meal, 4 lbs. ground oats, 3 lbs. wheat bran and 2½ lbs. linseed meal, daily; property of C. H. Stevens, Canton Centre, Conn.

Baby Stanton 72397.—Sire, Coomassie Dick 9815; dam, Nellie Stanton 21069.

Butter, 15 lbs. 4 oz. Milk, 226 lbs.

Test made from Dec. 5 to 12, 1894; age, 8 yrs. 4 mos.; estimated weight, 900 lbs.; fed 16 lbs. corn and cob meal, 4 lbs. bran and 2 lbs. oil meal, daily—hay, 8 lbs. potatoes; property of M. S. King, Newark, Ohio.

Badger Girl 21463.—Sire, Caen Jr. 6522; dam, Fannie Bugler 19962.

Butter, 16 lbs. 1 oz. Milk, 260 lbs. 4½ oz.

Test made from June 24 to July 1, 1888; age, 5 yrs. 1 mo.; estimated weight, 800 lbs.; fed, daily, 7 lbs. wheat bran and 3 lbs. corn and oat meal; property of N. N. Palmer, Brodhead, Wis.

Badger Girl 4th 87488.—Sire, Gold of St. Lambert 16744; dam, Badger Girl 21463.

Butter, 15 lbs. 13 oz. Milk, 208 lbs. 4 oz.

Test made from Nov. 6 to 13, 1897; age, 4 yrs. 7 mos.; estimated weight, 900 lbs.; fed 10½ lbs. ground oats, 2 lbs. oil meal and 21 lbs. ensilage, daily—corn stover *ad lib.*; property of N. N. Palmer & Son, Brodhead, Wis.

Ballet Girl 18750.—Sire, Mercury 432; dam, Butterfly 12020.

Butter, 14 lbs. 1 oz. Milk, 186 lbs. 13 oz.

Test made from Sept. 20 to 26, 1884; age, 2 yrs. 1 mo.; property of William Simpson, New York, N. Y.

Balmoral 2d 13574.—Sire, Napier 2d (P. S. 244 J. H. B.); dam, Balmoral 13464.

Butter, 14 lbs. 10 oz. Milk, 101 lbs. 10 oz.

Test made from Mar. 1 to 7, 1886; age, 4 yrs. 6 mos.; property of F. C. Sayles, Paw-tucket, R. I.

Barbara Lambert 105208.—Sire, St. Lambert Boy 17408; dam, Barberry 2d's Prin-cess 69073.

Butter, 15 lbs. 11 oz. Milk, 264 lbs.

Test made from July 5 to 12, 1898; age, 4 yrs. 4 mos.; actual weight, 750 lbs.; fed 6 qts. wheat bran, 2½ qts. ground oats and 1½ pts. corn meal, daily—good pasture; property of Mrs. Jennie E. Hall, Columbus, Ohio.

Barbara of Melrose 90486.—Sire, Newton Prince 15221; dam, Liz 79114.

Butter, 14 lbs. 7 oz. Milk, 183 lbs. 15 oz.

Test made from Aug. 13 to 20, 1897; age, 5 yrs. 4 mos.; estimated weight, 900 lbs.; fed 27 lbs. 11 oz. cotton-seed meal, 27 lbs. 11 oz. wheat shorts, 25 lbs. 11 oz. wheat bran, 9 lbs. 3 oz. ground oats and 11 lbs. 11 oz. corn meal—Bermuda, lespedeza and striata pasture; property of J. H. Wright, Meridian, Miss.

Baroness Argyle 40498.—Sire, Baron Hugo 15208; dam, Dollie Argyle 26987.

Butter, 18 lbs. 12 oz. Milk, 310 lbs.

Test made from Apr. 18 to 25, 1893; age, 4 yrs. 7 mos.; estimated weight, 950 lbs.; fed 3 lbs. coarse wheat bran, 3 lbs. corn and rye meal, about 35 lbs. ensilage, with 10 lbs. clover rowen, daily; property of E. S. Henry, Rockville, Conn.

Baroness Argyle 40498.—Sire, Baron Hugo 15208; dam, Dollie Argyle 26987.

Butter, 56.215 lbs. in 30 days. Milk, 925.5 lbs.

Test made from Aug. 29 to Sept. 27, 1893, at World's Columbian Exposition; age, 7 yrs.; actual weight, 1005 lbs.; fed 366.2 lbs. hay, 90 lbs. oil meal, 112 lbs. ensilage, 40 lbs. corn meal, 180 lbs. bran, 120 lbs. oats, 199.5 lbs. corn-hearts, 44.5 lbs. cotton-seed, 60 lbs. middlings, 18 lbs. carrots and 5 lbs. old hay; property of E. S. Henry, Rockville, Conn.

Baroness Argyle 40498.—Sire, Baron Hugo 15208; dam, Dollie Argyle 26987.

Butter, 194.400 lbs. in 90 days. Milk, 3266.2 lbs.

Test made from May 31 to Aug. 28, 1893, at World's Columbian Exposition; age, 6 yrs. 8 mos. at beginning of test; actual weight, 958 lbs.; fed 361 lbs. old hay, 191.5 lbs. ensilage, 569.3 lbs. new hay, 202 lbs. oil meal, 605 lbs. bran, 134 lbs. oats, 578 lbs. corn-hearts, 130.25 lbs. cotton-seed, 448 lbs. middlings, 42 lbs. grano-gluten, 1193.4 lbs. clover and 10 lbs. swale-grass; property of E. S. Henry, Rockville, Conn.

Baroness Hugo 47563.—Sire, Baron Hugo 15208; dam, Clematis 3d 6653.

Butter, 15 lbs. 10¼ oz. Milk, 237 lbs. 1½ oz.

Test made from Nov. 21 to 28, 1896; age, 9 yrs. 11 mos.; estimated weight, 900 lbs.; fed 87½ lbs. corn, oats and wheat bran, in equal parts by weight (corn and oats ground together), 14 qts. oil meal and about 18 bush. beets—little fodder and straw and frosted pasture; property of J. P. Murdoch, Cool Spring, Pa.

Baronet's Victory 2d 39176.—Sire, Prince of Maple Lane 14361; dam, Baronet's Victory 20740.

Butter, 17 lbs. 5 oz. Milk, 231 lbs.

Test made from Oct. 20 to 27, 1890; age, 4 yrs. 11 mos.; estimated weight, 750 lbs.; fed 12 qts. peas and oats, daily—hay *ad lib.*, clover pasture during day-time; prop-erty of William Rolph, Markham, Ont., Can.

Baronetti 8425.—Sire, Baronet 2240; dam, Marquise 2d 2868.

Butter, 16 lbs. 14½ oz. Milk, 258 lbs.

Test made from May 11 to 17, 1885; age, 6 yrs. 7 mos.; property of W. Gettys, Athens, Tenn.

Baron's Lorna 101623.—Sire, Baron's Hugo Pogis 14008; dam, Lorna 2d 33634.

Butter, 21 lbs. 3 oz. Milk, 260.5 lbs.

Test made from May 27 to June 3, 1898; age, 4 yrs. 9 mos.; actual weight, 960 lbs., fed 4 lbs. oil meal, 4 lbs. ground oats, 5 lbs. wheat bran and 8 lbs. corn meal, mixed, daily—clover pasture; property of F. W. Hart, Cleveland, Ohio.

Baron's Rosette 25988.—Sire, Baron (P. S. 289 J. H. B.); dam, Rosette, on I. of J.

Butter, 15 lbs. 4 oz. Milk, 181 lbs. 13 oz.

Test made from Aug. 24 to 31, 1884; age, 2 yrs. 6 mos.; property of J. M. McMillan, Paris, Ky

Baron's Sophie 17615.—Sire, Baron (P. S. 289 J. H. B.); dam, Sophie (F. S. 434 J. H. B.)

Butter, 19 lbs. 15⅝ oz. Milk, 271 lbs. 8 oz.

Test made from May 2 to 8, 1887; age, 5 yrs. 1 mo.; estimated weight, 875 lbs.; fed 14 lbs. of mixed feed twice daily; property of M. C. Campbell, Spring Hill, Tenn.

Bay View Queen 60429.—Sire, Chief 10663; dam, Early Rosette 32070.

Butter, 14 lbs. 2 oz. Milk, 212 lbs. 8 oz.

Test made from Feb. 17 to 24, 1892; age, 4 yrs. 3 mos.; estimated weight, 850 lbs.; fed 10½ lbs. oil meal, 7 lbs. ground oats, 7 lbs. ground ear corn, 105 lbs. corn refuse, 70 lbs. dry corn-stalks and 17½ lbs. hay; property of Charles Lautz, Buffalo, N. Y.

B. D. Exile 105286.—Sire, Wells' Exile 20170; dam, Bessie Dunedin 73665.

Butter, 18 lbs. 10 oz. Milk, 240 lbs.

Test made from July 10 to 17, 1898; age, 6 yrs. 3 mos.; estimated weight, 1000 lbs.; fed 3 qts. corn meal, 4 qts. bran and 1 pt. oil cake meal, daily—short hill pasture; property of E. G. Silvus, Athens, Ohio.

Beatrice of Elmarch 11367.—Sire, Lucifer 2696; dam, Cupid of Lee Farm 5997.

Butter, 15 lbs. 2 oz. Milk, 176 lbs. 4 oz.

Test made from June 5 to 11, 1885; age, 4 yrs. 10 mos.; property of George V. Green, Hopkinsville, Ky.

Beautiful Blondine 82051.—Sire, Crown Prince Melrose 21790; dam, Southern Belle 18570

Butter, 16 lbs. 5 oz. Milk, 241 lbs. 4 oz.

Test made from May 2 to 9, 1897; age, 6 yrs. 3 mos.; estimated weight, 850 lbs.; fed 14½ lbs. corn meal, 2 lbs. oil meal, 3 lbs. molasses feed and 6 lbs. ground oats, daily—rye and Bermuda pasture; property of William E. Oates & Son, Vicksburg, Miss.

Beauty 2076.—Sire, Peter 835; dam, Diana 2d 1718.

Butter, 15 lbs. 7 oz. Milk, 205 lbs. 6 oz.

Test made from Aug. 1 to 7, 1885; age, 16 yrs. 5 mos.; property of Cornelius Wellington, East Lexington, Mass.

Beauty 17414.—Sire, Browny (P. S. 158 J. H. B.); dam, Rosy (F. S. 1006 J. H. B.)

  Butter, 15 lbs. Milk, 194 lbs.

Test made from July 24 to 30, 1883; age, 5 yrs.; property of Nathan Brownell, Hubbardsville, N. Y.

Beauty Bismarck 4967.—Sire, Litchfield 674; dam, Daisy Bismarck 2697.

  Butter, 14 lbs. 1 oz.

Test made in 1880; age, 6 yrs.; property of F. F. Fuessenich, Torrington, Conn.

Beauty Blücher 308.—Sire, Blücher 49; dam, Beauty 309.

  Butter, 14 lbs. 8 oz. Milk, 232 lbs. 5 oz.

Test made from May 8 to 15, 1881; age, 18 yrs.; property of Edward E. Barney, Homewood, Va.

Beauty Dee 18065.—Sire, Eupidee 4097; dam, Beauty of Lakeside 2d 11322.

  Butter, 23 lbs. 8 oz. Milk, 221 lbs. 8 oz.

Test made from Aug. 27 to Sept. 3, 1886; age, 4 yrs. 3 mos.; estimated weight, 700 lbs.; fed 28 qts. corn meal and 28 qts. shorts—short pasture; property of Joseph H. Walker, Worcester, Mass.

Beauty Messenger 17450.—Sire, Messenger (P. S. 223 J. H. B.); dam, Beauty (F. S. 1159 J. H. B.)

  Butter, 15 lbs. 9 oz. Milk, 245 lbs.

Test made from Dec. 7 to 13, 1885; age, 4 yrs. 8 mos.; property of William Crozier, Northport, N. Y.

Beauty Montague 57923.—Sire, Lord Montague 12385; dam, Julia Belle 13233.

  Butter, 14 lbs. 4 oz. Milk, 258 lbs.

Test made from May 14 to 21, 1894; age, 5 yrs. 8 mos.; estimated weight, 800 lbs.; no grain fed—good timothy and clover pasture; property of J. W. L. Motherspaw, Newark, Ohio.

Beauty Montague 57923.—Sire, Lord Montague 12385; dam, Julia Belle 13233.

  Butter, 17 lbs. 8 oz. Milk, 257 lbs.

Test made from June 7 to 14, 1895; age, 6 yrs. 9 mos.; actual weight, 820 lbs.; fed 18 lbs. daily of a mixture of equal parts of ground corn, oats and bran—extra good clover and timothy pasture; property of J. W. L. Motherspaw, Newark, Ohio.

Beauty of Hamilton 55827.—Sire, Young Pedro 9033; dam, Bellis 38284.

  Butter, 15 lbs. 15 oz. Milk, 227 lbs. 8 oz.

Test made from Aug. 16 to 23, 1892; age, 3 yrs. 8 mos.; estimated weight, 650 lbs.; fed 6 qts. corn meal, 5 qts. ground oats and 5 qts. shorts, daily—hay *ad lib.*, very poor pasture; property of D. F. Appleton, Ipswich, Mass.

Beauty of Jersey 7850.—Sire, Dick (F. S. 223 J. H. B.); dam, Mignonne (F. S. 955 J. H. B.)

  Butter, 19 lbs. 2 oz., unsalted. Milk, 280 lbs. 15 oz.

Test made from July 19 to 25, 1882; age, 6 yrs. 4 mos.; fed 2 qts. corn meal, 3 qts. wheat bran and 1 qt. oil meal, daily—good blue-grass pasture; property of W. J. Chinn, Frankfort, Ky.

Beauty of Lee Farm 15694.—Sire, Prince Milan 2318; dam, Beauty 1319.

Butter, 20 lbs. 12¾ oz. Milk, 233 lbs. 12 oz.

Test made from June 17 to 24, 1887; age, 5 yrs. 8 mos.; estimated weight, 750 lbs.; fed, daily, 8 qts. ground oats and wheat, and 4 qts. oil cake; property of William Rolph, Markham, Ont., Can.

Beauty of Meridale 114744.—Sire, Ida of St. Lambert's Bull 19169; dam, Beauty Dee 18065.

Butter, 16 lbs. ¾ oz. Milk, 215 lbs. 15 oz.

Test made from Dec. 18 to 25, 1896; age, 2 yrs. 6 mos.; estimated weight, 875 lbs., fed 40 lbs. wheat bran, 32 lbs. corn meal, 14 lbs. cotton-seed meal, 14 lbs. linseed meal and about 80 lbs. hay; property of Ayer & McKinney, Philadelphia, Pa.

Beauty of Meridale 114744.—Sire, Ida of St. Lambert's Bull 19169; dam, Beauty Dee 18065.

Butter, 18 lbs. 12½ oz. Milk, 305 lbs. 2 oz.

Test made from Jan. 21 to 28, 1898; age, 3 yrs. 7 mos.; estimated weight, 950 lbs.; fed 56 lbs. wheat bran, 42 lbs. corn meal, 21 lbs. cotton-seed meal and 14 lbs. linseed meal—about 42 lbs. hay, 2 bush. carrots and beets—25 lbs. ensilage daily; property of Ayer & McKinney, Philadelphia, Pa.

Beauty of Ninon 2d 18444.—Sire, Pearl Rex 4438; dam, Beauty of Ninon 9691.

Butter, 15 lbs. 3½ oz. Milk, 308 lbs. 8 oz.

Test made from June 9 to 16, 1885; age, 4 yrs.; property of William S. Shannon, Pittsburg, Pa.

Beauty of Oakland 45009.—Sire, King Signal 12491; dam, Nal Day 26553.

Butter, 15 lbs. 7 oz. Milk, 197 lbs. 2 oz.

Test made from May 21 to 28, 1892; age, 4 yrs. 11 mos.; actual weight, 805 lbs.; fed 8 lbs. 12 oz. corn meal and 4 lbs. wheat bran, daily—pasture of mixed grasses; property of Columbus Dixon, Gillespieville, Ohio.

Beauty of Oakland 45009.—Sire, King Signal 12491; dam, Nal Day 26553.

Butter, 17 lbs. 1 oz. Milk, 233 lbs. 1 oz.

Test made from Apr. 30 to May 7, 1893; age, 5 yrs. 11 mos.; actual weight, 875 lbs.; fed 84 lbs. corn meal and 42 lbs. wheat bran, mixed with cut timothy hay—hay *ad lib.*, pasture of mixed grasses; property of Columbus Dixon, Gillespieville, Ohio.

Beauty of Seekonk 14651.—Sire, Admiral Farragut 2666; dam, Lady Theresa 2d 4253.

Butter, 21 lbs. 3 oz. Milk, 251 lbs.

Test made from Aug. 21 to 28, 1885; age, 4 yrs. 5 mos.; fed 16 lbs. Indian meal, daily —second growth meadow pasture; property of William H. Hopkins, Providence, R. I.

Beauty Romeril 26090.—Sire, Progress (F. S. 286 J. H. B.); dam, Lady (P. S. 214 J. H. B.)

Butter, 18 lbs. 9 oz. Milk, 219 lbs. 8 oz.

Test made from Nov. 29 to Dec. 5, 1884; age, 4 yrs. 7 mos.; property of Newton Frazier, Simpsonville, Ky.

Beauty's Crescent 26733.—Sire, Edward Earle 10462; dam, Welcome Beauty 1268.

Butter, 15 lbs. 15⅝ oz. Milk, 126 lbs.

Test made from Dec. 5 to 12, 1889; age, 5 yrs. 10 mos.; estimated weight, 800 lbs., fed, daily, 4 gallons of a mixture two-thirds oats and one-third corn, ground; property of Big Spring Jersey Herd, Columbia, Tenn.

Beauty Wynn 45482.—Sire, Dexter Montague 17294; dam, Mary Wynn 40782.

Butter, 14 lbs. 14½ oz. Milk, 204 lbs.

Test made from June 6 to 13, 1891; age, 3 yrs. 10 mos.; estimated weight, 800 lbs., fed 5 lbs. bran and 5 lbs. cotton-seed cake meal, daily—timothy, clover and blue-grass pasture; property of Bordwell & Cochran, Batavia, Ohio.

Beckie Y. Pogis 63069.—Sire, Major Appel Pogis 17861; dam, Beckie of Shady Side 39464.

Butter, 17 lbs. 15¼ oz. Milk, 205 lbs. 8 oz.

Test made from Mar. 14 to 21, 1896; age, 7 yrs. 11 mos.; actual weight, 860 lbs., fed 63 qts. ground oats, 63 qts. corn meal, 14 qts. ground oil cake, 10 lbs. carrots and 10 lbs. cabbage—good clover hay; property of Miller & Sibley, Franklin, Pa.

Becky Z. 71311.—Sire, Elmwood Carrie's Pogis 19494; dam, Petunia's Pet 28814.

Butter, 14 lbs. 2 oz. Milk, 223 lbs. 12 oz.

Test made from May 31 to June 7, 1895; age, 4 yrs. 9 mos.; actual weight, 906 lbs., no grain fed—pasture, with plenty of blue-grass and clover; property of John Alexander, Middle Grove, Mo.

Beech Grove Elsie 67758.—Sire, Cobonon 15744; dam, Emily Le Brocq 36610.

Butter, 14 lbs. 5½ oz. Milk, 188 lbs. 3 oz.

Test made from Oct. 17 to 24, 1892; age, 4 yrs. 9 mos.; actual weight, 700 lbs.; fed 105 lbs. ground corn and cob meal and 60 lbs. wheat bran—pasture of mixed grasses, property of Columbus Dixon, Gillespieville, Ohio.

Beech Leaf 99381.—Sire, Stoke Pogis of Prospect 29121; dam, Fair Leaf 81293.

Butter, 16 lbs. 3½ oz. Milk, 221 lbs.

Test made from Aug. 3 to 10, 1896; age, 2 yrs. 1 mo.; actual weight, 720 lbs.; fed 9 qts. bran, 6 qts. ground oats, 3 qts. corn meal and 3 qts. oil meal, daily—good pasture at night, green corn in stable during day; property of C. A. Sweet, East Aurora, N. Y.

Bee Princess 40345.—Sire, Prince of Melrose 4819; dam, Busy Bee 2d 25166.

Butter, 14 lbs. 2⅝ oz. Milk, 201 lbs. 7 oz.

Test made from May 20 to 27, 1888; age, 3 yrs.; estimated weight, 800 lbs.; fed, daily, 12 qts. of corn, bran, oats and cotton-seed, ground together; property of M. Lothrop, Marshall, Texas.

Beeswax 9807.—Sire, Top-Sawyer 1404; dam, Bisma 3d 1870.

Butter, 17 lbs. 5 oz. Milk, 187 lbs. 13 oz.

Test made from Mar. 17 to 23, 1884; age, 5 yrs. 8 mos.; property of Columbia Jersey Cattle Co., Columbia, Tenn.

Belbonnie 29630.—Sire, Combination 4389; dam, Hulda R. 16497.

Butter, 14 lbs. 9 oz., unsalted. Milk, 208 lbs.

Test made from June 23 to 30, 1885; age, 8 yrs.; property of Richardson Bros., Davenport, Iowa.

Bella Delaine 10356.—Sire, Music (P. S. 118 J. H. B.); dam, Daisy (F. S. 742 J. H. B.)

Butter, 14 lbs. 2 oz. Milk, 222 lbs. 1 oz.

Test made from Apr. 15 to 21, 1883; age, 6 yrs.; property of S. L. Hoover, Columbus, Ohio.

Bella Marigold 94124.—Sire, Stoke Pogis of Prospect 29121; dam, Bristol Bella 15697.

Butter, 17 lbs. 2 oz. Milk, 223 lbs. 8 oz.

Test made from Oct. 26 to Nov. 2, 1895; age, 2 yrs. 1 mo.; actual weight, 695 lbs.; fed 9 qts. ground oats, 3 qts. corn meal, 9 qts. wheat bran and 3 qts. oil meal, daily—hay; property of C. A. Sweet, East Aurora, N. Y.

Belle Bonair 33960.—Sire, Romeo de Bonair 4091; dam, Obella 5080.

Butter, 16 lbs. 3½ oz. Milk, 257 lbs. 7 oz.

Test made from Sept. 9 to 16, 1889; age, 4 yrs. 5 mos.; estimated weight, 950 lbs.; fed, daily, 11 lbs. ground oats, 11 lbs. middlings and 10 lbs. corn meal; property of D. F. Appleton, Ipswich, Mass.

Belle Cotton 96205.—Sire, King Cotton 25391; dam, Bess of Edgewood 71178.

Butter, 14 lbs. 8 oz. Milk, 226 lbs. 2 oz.

Test made from Sept. 26 to Oct. 3, 1897; age, 5 yrs. 4 mos.; estimated weight, 800 lbs.; fed 56 qts. corn meal and 56 qts. wheat bran—good blue-grass pasture; property of Columbus Dixon, Gillespieville, Ohio.

Belle Dame 2d 22043.—Sire, Carlo 5559; dam, Belle Dame 11951.

Butter, 18 lbs. 12 oz. Milk, 143 lbs.

Test made from Dec. 7 to 14, 1890; age, 10 yrs. 9 mos.; actual weight, 920 lbs.; fed, daily, 12 lbs. wheat middlings and 17 lbs. beets; property of Thomas Allen, Pittsfield, Mass.

Belle Dawson 8270.—Sire, Commodore Roxbury 1586; dam, Mount Lebanon 4457.

Butter, 18 lbs. 3 oz. Milk, 278 lbs.

Test made from Oct. 11 to 18, 1885; age, 7 yrs.; property of Wm. H. Burr, Redding Ridge, Conn.

Belle France 22697 (F. S. 4924 J. H. B.)

Butter, 15 lbs. 10½ oz.

Test made in June, 1884; age, 3 yrs.; property of Alfred M. Herkness & Co., Philadelphia, Pa.

Belle Garner 23682.—Sire, Litchfield 15th 5802; dam, Daisy Workman 19990.

Butter, 15 lbs. 3 oz. Milk, 172 lbs. 4 oz.

Test made from Feb. 1 to 7, 1886; age, 2 yrs. 9 mos.; property of W. H. Haley, North Wilmington, Mass.

Belle Grinnell 4073.—Sire, Monitor 878; dam, Grinnella 3d 2209.

Butter, 18 lbs. 8 oz., unsalted.

Test made from June 11 to 18, 1882; age, 7 yrs. 11 mos.; fed 8 qts. corn meal, daily—rough, bushy pasture; property of S. W. Robbins, Wethersfield, Conn.

Belle Grinnell 3d 16503.—Sire, Lord Anglesea 4537; dam, Belle Grinnell 4073.

Butter, 14 lbs. 2 oz.

Property of E. J. Robbins, Wethersfield, Conn.

Belle Lyman 28522.—Sire, General Lyman 6631; dam, Belle of Wolcott 8938.

Butter, 16 lbs. 3½ oz.   Milk, 220 lbs. 8 oz.

Test made from Feb. 21 to 28, 1890; age, 5 yrs. 7 mos.; estimated weight, 800 lbs.; fed, daily, 4 gals. ground oats and corn mixed in equal parts; property of Morgan & Brown, Columbia, Tenn.

Belle Lyman 3d 86346.—Sire, Landseer's Pogis 15847; dam, Belle Lyman 28522.

Butter, 18 lbs. 2½ oz.   Milk, 282 lbs. 15 oz.

Test made from Feb. 10 to 17, 1897; age, 5 yrs. 4 mos.; estimated weight, 850 lbs.; fed 112 qts. ground oats and corn, with cob, mixed in equal parts by weight, and 7 qts. oil meal—hay; property of C. M. Oliver, Bunker Hill, Ill.

Belle Mardi 18362.—Sire, Mardi Gras 2927; dam, Lady O'Neal 7391.

Butter, 18 lbs. ½ oz.   Milk, 215 lbs. 12 oz.

Test made from May 24 to 30, 1885; age, 6 yrs. 1 mo.; property of J. L. Shallcross, Anchorage, Ky.

Belle Melrose 61920.—Sire, Prince of Melrose 4819; dam, Southern Belle 18570.

Butter, 17 lbs. 1½ oz.   Milk, 260 lbs. 4 oz.

Test made from July 3 to 10, 1892; age, 4 yrs. 5 mos.; estimated weight, 700 lbs.; fed 35 lbs. bran, 14 lbs. cotton-seed meal, 28 lbs. shelled oats and 42 lbs. corn, oats and peas ground together—good Bermuda and clover pasture; property of M. Lothrop, Marshall, Tex.

Belle Miller 60488.—Sire, Valley King 19752; dam, Rose Newton 50148.

Butter, 15 lbs. 2¾ oz.   Milk, 228 lbs.

Test made from Oct. 26 to Nov. 2, 1895; age, 6 yrs. 1 mo.; actual weight, 920 lbs.; fed 35 lbs. hominy meal, 45½ lbs. light bran and about 490 lbs. ensilage; property of J. T. Polk, Greenwood, Ind.

Belle Noble's Bessie 105710.—Sire, Recorder 29239; dam, Belle Noble 2d 77233.

Butter, 19 lbs. 5 oz.   Milk, 285 lbs.

Test made from June 12 to 19, 1897; age, 3 yrs. 4 mos.; estimated weight, 1000 lbs.; fed 42 lbs. corn meal, 35 lbs. oats, 21 lbs. rye and 7 lbs. oil meal—mixed pasture; property of H. C. Taylor, Orfordville, Wis.

Bell Ensign 56702.—Sire, Champion's Ensign 9642; dam, Bell Earle 26732.

Butter, 15 lbs. 13½ oz.   Milk, 233 lbs. 7 oz.

Test made from June 20 to 27, 1893; age, 6 yrs. 4 mos.; actual weight, 800 lbs.; fed 17 lbs. per day of ground oats and corn, in equal parts—good orchard and bluegrass pasture; property of Matt M. Gardner, Nashville, Tenn.

Belle of Corfu 83731.—Sire, St. Lambert Perrot 23680; dam, Tulla St. Lambert 59992.

Butter, 15 lbs. 8 oz.   Milk, 209 lbs. 4 oz.

Test made from Aug. 3 to 10, 1894; age, 3 yrs. 2 mos.; estimated weight, 800 lbs.; fed 12 qts. ground corn and oats, 12 qts. bran and middlings and 3 qts. oil meal, daily—green oats and peas in stable during day, clover pasture at night; property of C. A. Sweet, East Aurora, N. Y.

Belle of Dorset 85905.—Sire, Melia Ann's Victor Pogis 20697; dam, Belle of Rydal Grange 41787.

Butter, 14 lbs. 9 oz. Milk, 232 lbs.

Test made from Apr. 9 to 16, 1896; age, 4 yrs. 9 mos.; estimated weight, 800 lbs.; fed 4 lbs. corn meal, 8 lbs. oats and peas, 4 lbs. wheat middlings and 2 lbs. oil meal, daily—hay and ensilage *ad lib.*; property of Mrs. J. H. Martin, Granville, N. Y.

Belle of Echo 12432.—Sire, John Rex 2761; dam, Belle Atwood 5907.

Butter, 14 lbs. 14½ oz. Milk, 222 lbs.

Test made from May 12 to 18, 1884; age, 3 yrs. 2 mos.; property of J. R. Gaston, Normal, Ill.

Belle of Lynwood 18364.—Sire, Mardi Gras 2927; dam, Cocotte 7392.

Butter, 17 lbs. 14 oz. Milk, 220 lbs. 12 oz.

Test made from Aug. 15 to 21, 1884; age, 3 yrs. 8 mos.; property of J. L. Shallcross, Anchorage, Ky.

Belle of Milford 7445.—Sire, Nelusko 479; dam, Juliet 3d 5163.

Butter, 15 lbs. 6 oz. Milk, 228 lbs. 8 oz.

Test made from Mar. 30 to Apr. 5, 1885; age, 6 yrs. 10 mos.; property of T. M. Mosely, West Point, Miss.

Belle of Ogden Farm 1570.—Sire, Don 611; dam, Belle 53.

Butter, 14 lbs.

Property of John H. Freeman, Jackson, Tenn.

Belle of Passaic 85846.—Sire, Eurotas' Victor Hugo 15689; dam, Helfre 15445.

Butter, 17 lbs. 11 oz. Milk, 305 lbs. 4 oz.

Test made from Mar. 20 to 27, 1897; age, 6 yrs. 6 mos.; estimated weight, 900 lbs.; fed 63 qts. bran, 42 qts. corn meal, 5½ qts. cotton-seed meal, 10½ qts. oil meal, 168 lbs. ensilage, and 84 lbs. hay; property of R. A. Sibley, Rochester, N. Y.

Belle of Patterson 5664.—Sire, Signal 1170; dam, Azelda 3872.

Butter, 16 lbs. 6 oz., unsalted. Milk, 241 lbs. 10 oz.

Test made from June 5 to 11, 1882; age, 5 yrs. 3 mos.; no grain fed—good blue-grass pasture; property of W. J. Chinn, Frankfort, Ky.

Belle of Prospect 2d 14326.—Sire, Shirley 1613; dam, Belle of Prospect 6627.

Butter, 19 lbs. Milk, 204 lbs.

Test made from July 11 to 18, 1885; age, 5 yrs. 1 mo.; property of W. M. Bell, Miami, Mo.

Belle of River's Bank 48023.—Sire, Ponterrin's Duke 14999; dam, Pride of the Farm 2d 28161.

Butter, 15 lbs. 10 oz. Milk, 231 lbs. 3 oz.

Test made from May 24 to 31, 1894; age, 6 yrs. 5 mos.; estimated weight, 750 lbs.; fed 10 lbs. corn and oats, ground together, 6 lbs. bran and 2 lbs. oil meal, daily—hay, good mixed pasture; property of Mrs. J. H. Martin, Granville, N. Y.

Belle of Riverview 21464.—Sire, Bravy 1923; dam, Buttercup of Riverview 19952.

Butter, 17 lbs. 12½ oz. Milk, 283 lbs. 4 oz.

Test made from May 21 to 28, 1886; age, 4 yrs. 9 mos.; property of E. J. Larrabee, Albany, N. Y.

Belle of Rydal Grange 41787.—Sire, Romp's Tormentor 7126; dam, Miss Peggotty 14593.

Butter, 16 lbs. 6 oz. Milk, 232 lbs.

Test made from June 24 to July 1, 1892; age, 7 yrs. 5 mos.; estimated weight, 900 lbs.; no grain fed—clover pasture only; property of Mrs. J. R. Sherman Walsh, Pawlet, Vt.

Belle of St. John's 2d 29829.—Sire, Duke of Normandy 3446; dam, Belle of St. John's 8078.

Butter, 15 lbs. ½ oz. Milk, 175 lbs.

Test made from Apr. 28 to May 4, 1886; age, 3 yrs. 3 mos.; property of Campbell Brown, Spring Hill, Tenn.

Belle of Scituate 7977.—Sire, Pharos 3552; dam, Jersey Belle of Scituate 7828.

Butter, 16 lbs.

Property of C. O. Ellms, Scituate, Mass.

Belle of Vermillion 8798.—Sire, Diamond Earl 3116; dam, Marigold 3d 3220.

Butter, 15 lbs. 14 oz. Milk, 252 lbs.

Test made from Oct. 11 to 17, 1882; age, 4 yrs. 4 mos.; property of I. D. Risher, Hope Church, Pa.

Belle of Wayne 16457.—Sire, Sir Farmington 3316; dam, Belle of Canandaigua 10139.

Butter, 17 lbs. 9½ oz. Milk, 230 lbs. 8 oz.

Test made from May 20 to 27, 1886; age, 5 yrs. 9 mos.; property of George P. Lapham, Macedon, N. Y.

Belle's Charity 82217.—Sire, Charity Boy 11765; dam, Belle of Kennett 17085.

Butter, 14 lbs. 2 oz. Milk, 232 lbs. 8 oz.

Test made from July 1 to 8, 1896; age, 5 yrs. 7 mos.; estimated weight, 650 lbs.; fed 4 lbs. ground peas and oats, 5 lbs. corn meal and 6 lbs. bran, daily—good mixed pasture; property of J. H. Martin, Granville, N. Y.

Belle's Charity 82217.—Sire, Charity Boy 11765; dam, Belle of Kennett 17085.

Butter, 17 lbs. 1 oz. Milk, 314 lbs. 8 oz.

Test made from June 27 to July 4, 1898; age, 7 yrs. 7 mos.; estimated weight, 900 lbs.; no grain fed—pasture, white and red clover and mixed grasses; property of J. H. Martin, Granville, N. Y.

Belle's Esperanza 12053.—Sire, Uproar 4609; dam, Belle Warren 7978.

Butter, 15 lbs. 9½ oz. Milk, 154 lbs. 12 oz.

Test made from Nov. 22 to 29, 1887; age, 6 yrs. 8 mos.; estimated weight, 775 lbs.; fed, daily, 10 lbs. corn meal, 3 lbs. oil meal and 2 lbs. middlings; property of James Stillman, Sing Sing, N. Y.

Belle Smith 65807.—Sire, Dalton 20117; dam, Duke's Naomi 51629.

Butter, 22 lbs. 8¼ oz. Milk, 300 lbs.

Test made from May 2 to 9, 1895; age, 5 yrs.: actual weight, 900 lbs.; fed 42 qts. ground oats, 42 qts. corn meal, 28 qts. ground oil-cake and 28 qts. wheat bran—good old meadow pasture; property of Miller & Sibley, Franklin, Pa.

Belle Steuben 20115.—Sire, Steuben 4751; dam, Elsie Dinsmore 5834.

Butter, 16 lbs. 10 oz. Milk, 230 lbs.

Test made from Aug. 16 to 23, 1886; age, 5 yrs. 1 mo.; estimated weight, 850 lbs.; fed, daily, ground oats, corn and peas mixed with bran, all she would eat; property of Mrs. E. M. Jones, Brockville, Ont., Can.

Belle Thorne 13369.—Sire, Niobe Duke 2364; dam, Thorndale Belle 2d 6421.

Butter, 14 lbs. 11 oz. Milk, 215 lbs.

Test made from May 30 to June 5, 1884; age, 4 yrs. 1 mo.; property of Edwin Thorne, Millbrook, N. Y.

Belle Williamson 8386.—Sire, Pertinax 1965; dam, Josey Brown 7533.

Butter, 19 lbs. 10½ oz. Milk, 157 lbs. 6½ oz.

Test made from June 3 to 10, 1887; age, 8 yrs. 11 mos.; estimated weight, 800 lbs.; fed, daily, 11 lbs. ground corn and oats; property of Mrs. E. T. Allen, Columbia, Tenn.

Belle Williamson 3d 104290.—Sire, Ethol 19104; dam, Belle Williamson 8386.

Butter, 15 lbs. 3½ oz. Milk, 205 lbs. 10 oz.

Test made from May 19 to 26, 1895; age, 3 yrs. 9 mos.; estimated weight, 700 lbs.; fed 112 qts. mixed corn chop and oats—blue-grass pasture; property of J. H. Carpenter, Columbia, Tenn.

Belle Yakout 38020.—Sire, Yakout 6842; dam, Belle Dawson 8270.

Butter, 16 lbs. 13 oz. Milk, 303 lbs. 8 oz.

Test made from Apr. 15 to 22, 1888; age, 3 yrs. 10 mos.; estimated weight, 800 lbs.; fed, daily, 5 lbs. corn and oats, ground, and 2 lbs. cotton-seed meal; property of Wm. H. Burr, Redding Ridge, Conn.

Bellita 4553.—Sire, Hamilton 1074; dam, Little Bella 3693.

Butter, 17 lbs. 2 oz. Milk, 16 qts. per day.

Test made from June 11 to 18, 1885; age, 9 yrs. 5 mos.; property of John H. Taylor, Thomaston, Conn.

Bell Rex 11700.—Sire, Rex 1330; dam, Jessie Leavenworth 8248.

Butter, 14 lbs. 10 oz. Milk, 201 lbs. 8 oz.

Test made from Feb. 7 to 13, 1885; age, 4 yrs. 10 mos.; property of Moulton Bros., West Randolph, Vt.

Belmeda 6229.—Sire, Superb 1956; dam, Orphean 4636.

Butter, 18 lbs. 12 oz. Milk, 202 lbs. 12 oz.

Test made from Apr. 3 to 9, 1883; age, 6 yrs.; property of G. R. Dykeman, Shippensburg, Pa.

Belmina 19189.—Sire, Tormentor 3533; dam, Belle of Tennessee 9573.

Butter, 17 lbs. 3¼ oz. Milk, 214 lbs.

Test made from Aug. 15 to 22, 1888; age, 6 yrs.; weight, 947 lbs.; fed, daily, 5 lbs. bran, 5 lbs. corn meal and 6 lbs. oat meal; property of Jacob L. Thomas, Knoxville, Tenn.

Benefit's Daughter 63932.—Sire, Pedro 3d 12583; dam, Benefit 15218.

Butter, 20 lbs., unsalted. Milk, 370 lbs.

Test made from Feb. 20 to 27, 1895; age, 8 yrs. 8 mos.; estimated weight, 1200 lbs.; fed 18 lbs. per day corn meal, oil meal, middlings and cotton-seed meal, and 1 qt. per day wheat middlings and 1 qt. shelled corn; 84 lbs. hay and 1 bush. carrots during test; property of Mrs. H. E. Tremain, Lake George, N. Y.

Benetia 21511.— Sire, Duke of Milford 3820; dam, Laurena of Meadow Brook 4123.

Butter, 16 lbs 5⅓ oz. Milk, 162 lbs. 12 oz.

Test made from June 18 to 24, 1886; age, 3 yrs. 2 mos.; property of J. T. & W. S. Shields, Bean's Station, Tenn.

Ben's Bijou 86599.—Sire, Signal Ben Signal 27186; dam, Signal's Bijou 60800.

Butter, 18 lbs. 6 oz. Milk, 180 lbs. 7 oz.

Test made from Nov. 25 to Dec. 2, 1895; age, 3 yrs.; estimated weight, 700 lbs.; fed 18 lbs. corn meal, 18 lbs. ground oats, 18 lbs. wheat bran, and 16 lbs. oil meal; property of Thomas Friant, Marcellus, Mich.

Ben's Nettie 118137.—Sire, Signal Ben Signal 27186; dam, Nettie Signal 3d 67493.

Butter, 15 lbs. 3 oz. Milk, 239 lbs. 15 oz.

Test made from Jan. 1 to 8, 1898; age, 2 yrs. 10 mos.; estimated weight, 800 lbs.; fed 28 qts. bran, 28 qts. crushed corn and cob, 14 qts. ground oats and 14 qts. oil meal—clover hay *ad lib.*; property of Bishop Clay, Lexington, Ky.

Bermuda R. 67534.—Sire, Diploma 16219; dam, Upper Ten 37603.

Butter, 14 lbs. 1 oz. Milk, 197 lbs. 14 oz.

Test made from Aug. 3 to 10, 1890; age, 2 yrs.; estimated weight, 775 lbs.; fed, daily, 15 qts. ground corn, oats, peas and oil meal; property of Richardson Bros., Davenport, Iowa.

Bernice of Clovernook 72805.—Sire, Odelio 20223; dam, Signalda Silver Rose 33864.

Butter, 17 lbs. 1 oz. Milk, 189 lbs. 12 oz.

Test made from July 7 to 14, 1893; age, 3 yrs. 5 mos.; estimated weight, 900 lbs.; fed 56 lbs. bran, 14 lbs. corn meal, 14 lbs. cotton-seed meal and 7 lbs. pea meal—hay *ad lib.*; property of J. C. Munden, Marshall, Tex.

Bertha 18912 (F. S. 2315 J. H. B.)

Butter, 15 lbs. 10 oz. Milk, 261 lbs. 8 oz.

Test made from Feb. 17 to 24, 1886; age, 6 yrs. 8 mos.; property of William H. Burr, Redding Ridge, Conn.

Bertha Black 26275.—Sire, Czar of New York 4049; dam, Delle 3789.

Butter, 16 lbs. 14 oz. Milk, 256 lbs.

Test made from June 7 to 13, 1885; age, 4 yrs. 10 mos.; property of Mrs. E. M. Jones, Brockville, Ont., Can.

Bertha Morgan 4770.—Sire, Lopez 313; dam, Patterson's Beauty 4760.

Butter, 19 lbs. 6 oz. Milk, 294 lbs.

Test made from Jan. 10 to 17, 1882; age, 9 yrs.; property of Edward Worth, Wawa, Pa.

Bertha's Brunette 82336.—Sire, Mattituck's Bachelor 20920; dam, Madam F. 82335.

Butter, 18 lbs. 1 oz. Milk, 279 lbs. 12 oz.

Test made from Dec. 15 to 22, 1896; age, 6 yrs. 1 mo.; estimated weight, 700 lbs.; fed 49 lbs. gluten meal, 49 lbs. corn meal, 24½ lbs. middlings, 24½ lbs. bran and 140 lbs. roots—little hay; property of D. Wolfe Bishop, Jr., Lenox, Mass.

Bertha Scituate 75977.—Sire, Exile of Scituate 19235; dam, Nora Scituate 67571.

Butter, 14 lbs. 7 oz. Milk, 202 lbs. 4 oz.

Test made from May 12 to 19, 1895; age, 3 yrs. 8 mos.; estimated weight, 580 lbs.; fed 77 lbs. corn and cob meal—four hours daily on short blue-grass pasture; property of H. A. Simpson, Pana, Ill.

Bertha's Royal Daisy 104901.—Sire, Bertha's Pogis 2d 27065; dam, Royal Bomba's Daisy 35996.

Butter, 16 lbs. 2½ oz. Milk, 239 lbs.

Test made from July 1 to 8, 1897; age, 4 yrs. 5 mos.; actual weight, 690 lbs.; fed 9 qts. bran, 9 qts. ground oats, 3 qts. corn meal and 3 qts. oil meal, daily—plenty of good grass in barn; property of C. A. Sweet, Buffalo, N. Y.

Bertha Stewart 99882.—Sire, Chieftain Pogis 25035; dam, Panftie 18442.

Butter, 17 lbs. Milk, 262 lbs. 13 oz.

Test made from Dec. 21 to 28, 1897; age, 5 yrs. 1 mo.; estimated weight, 800 lbs., fed 42 lbs. ensilage, 5 lbs. bran, 2½ lbs. corn meal, 2½ lbs. oats, 1¼ lbs. cotton-seed meal, 1½ lbs. oil meal and 10 lbs. hay, daily; property of Pennsylvania Reform School, Morganza, Pa.

Bertha Stoke Pogis 35206.—Sire, Exile of St. Lambert 13657; dam, Lady Bullard 24648.

Butter, 14 lbs. 15½ oz. Milk, 142 lbs.

Test made from July 1 to 8, 1891; age, 6 yrs.; estimated weight, 775 lbs.; fed 8 lbs. corn meal, 8 lbs. fine feed and bran, 4 lbs. cotton-seed meal and oil meal, daily—dry hay, mixed with green oats and peas, three times daily; property of P. J. Cogswell, Rochester, N. Y.

Bertha Thorne 74824.—Sire, King Coomassie 2d 19545; dam, Madame Thorne 64197.

Butter, 16 lbs. 5 oz. Milk, 210 lbs. 4 oz.

Test made from May 30 to June 6, 1894; age, 4 yrs. 1 mo.; actual weight, 840 lbs., fed 42 qts. corn meal, 28 qts. oats and 56 qts. wheat bran—clover and blue-grass pasture; property of C. Dixon, Gillespieville, Ohio.

Bertie Briggs 5213.—Sire, Young Baron 702; dam, Hester D. 2283.

Butter, 14 lbs. 4½ oz. Milk, 234 lbs.

Test made from Aug. 20 to 27, 1885; age, 10 yrs. 5 mos.; property of S. W. Tallaferro, Guthrie, Ky.

Bertie Don 2d 74088.—Sire, Barry St. Helier 22065; dam, Bertie Don 32071.

Butter, 16 lbs. 14 oz.   Milk, 146 lbs. 8 oz.

Test made from Dec. 5 to 12, 1894; age, 3 yrs. 2 mos.; actual weight, 817 lbs.; fed 14 lbs. ground wheat, 10½ lbs. oil meal, 10½ lbs. cotton-seed meal, 21 lbs. oats and 224 lbs. ensilage; property of Charles Lautz, Buffalo, N. Y.

Berylla 2d 15117.—Sire, Golden Lion 5239; dam, Berylla 8957.

Butter, 15 lbs. 13 oz.   Milk, 237 lbs.

Test made from May 14 to 20, 1886; age, 4 yrs. 8 mos.; property of A. G. Goodlett, Clarksville, Tenn.

Bessie Bradford 7269.—Sire, King Pin 1878; dam, Edith 4th 817.

Butter, 14 lbs. 2 oz.

Test made from Nov. 4 to 11, 1883; age, 7 yrs. 8 mos.; property of L. S. Sprague, Austerlitz, N. Y.

Bessie Bradford 2d 7271.—Sire, Bluetooth 1821; dam, Bessie Bradford 7269.

Butter, 15 lbs. 2 oz.

Test made last week in June, 1883; age, 5 yrs. 6 mos.; property of Alice M. Bradford, West Chester, N. Y.

Bessie Landseer 58103.— Sire, Landseer's Pogis 15847; dam, Bessie Russ 2d 14649.

Butter, 15 lbs. 1 oz.   Milk, 147 lbs. 8 oz.

Test made from May 7 to 14, 1896; age, 8 yrs. 4 mos.; estimated weight, 850 lbs., fed 56 qts. ground oats and 56 qts. corn meal—woodland pasture; property of Morgan & Brown, Columbia, Tenn.

Bessie of Clover Leaf 13033.—Sire, Diamond Earl 3116; dam, Ellen of Collingwood 8285.

Butter, 14 lbs. 2 oz., unsalted.   Milk, 207 lbs. 7½ oz.

Test made from Nov. 23 to 30, 1885; age, 4 yrs. 10 mos.; property of J. W. Ford, Huntington, Ind.

Bessie of Montverde 18496.—Sire, Lord Ducie 2500; dam, Elainette 18314.

Butter, 15 lbs. 5 oz.   Milk, 233 lbs. 8 oz.

Test made from June 1 to 7, 1886; age, 4 yrs. 1 mo.; property of G. R. Hill, Oxford, Miss.

Bessie Pope St. Lambert 80352.—Sire, Pope St. Lambert 19850; dam, Bessie Morris 26730.

Butter, 16 lbs. 1¼ oz.   Milk, 234 lbs.

Test made from Feb. 10 to 17, 1898; age, 5 yrs. 10 mos.; estimated weight, 850 lbs., fed 56 lbs. cotton-seed, 49 lbs. bran, 49 lbs. unbolted corn meal, 10 lbs. old process oil meal and 49 lbs. hay—short wheat pasture; property of John D. Page, McKinney, Tex.

Bessie Ridgely 8293.—Sire, John Ridgely 3045; dam, Grace Davy 8292.

Butter, 14 lbs. 11½ oz.

Property of C. S. S. Baron, Bethesda, Ohio.

Bessie Russ 2d 14649.—Sire, Bullion 2d 5246; dam, Bessie Russ 14648.

Butter, 15 lbs. 1½ oz. Milk, 196 lbs. 15 oz.

Test made from May 1 to 7, 1887; age, 6 yrs.; estimated weight, 900 lbs.; fed, daily, 16 qts. ground oats and corn; property of Morgan & Brown, Columbia, Tenn.

Bessie Signalda 2d 86341.—Sire, Landseer's Pogis 15847; dam, Bessie Signalda 38178.

Butter, 15 lbs. 2½ oz. Milk, 182 lbs. 12 oz.

Test made from Oct. 5 to 12, 1896; age, 7 yrs. 2 mos.; estimated weight, 800 lbs., fed 56 qts. corn meal and 56 qts. ground oats—woodland pasture; property of Morgan & Brown, Columbia, Tenn.

Bessie Signalda 3d 86349.—Sire, Landseer's Pogis 15847; dam, Bessie Signalda 38178.

Butter, 15 lbs. ½ oz. Milk, 236 lbs. 12 oz.

Test made from June 18 to 25, 1895; age, 3 yrs.; estimated weight, 750 lbs.; fed 56 qts. ground oats and 56 qts. corn meal—poor pasture; property of T. C. Lipscomb, Shelbyville, Tenn.

Bessie's Mabel 40939.—Sire, Duke of Hamburg 5527; dam, Wal-Oak Bessie 8180.

Butter, 18 lbs. Milk, 281 lbs.

Test made from May 13 to 20, 1898; age, 13 yrs. 5 mos.; estimated weight, 875 lbs., fed 8 qts. daily of corn meal and ground oats—mixed pasture; property of L. C. Read & Son, East Pembroke, N. Y.

Bessie Wolcott 91417.—Sire, Duke of La Grange 27682; dam, Mollie Mack 30750.

Butter, 15 lbs. 1 oz. Milk, 241 lbs. 7 oz.

Test made from Apr. 29 to May 6, 1898; age, 5 yrs. 1 mo.; estimated weight, 900 lbs.; fed 30 qts. wheat bran and 30 qts. corn meal—timothy and clover pasture, property of T. B. Kibler, Jenera, Ohio.

Bess of Edgewood 71178.—Sire, King Coomassie 2d 19545; dam, Beauty of Oakland 45009.

Butter, 20 lbs. 6 oz. Milk, 276 lbs. 2 oz.

Test made from May 31 to June 7, 1893; age, 4 yrs.; actual weight, 775 lbs.; fed 50 lbs. corn meal and 28 lbs. wheat bran—mixed pasture; property of Columbus Dixon, Gillespieville, Ohio.

Bess of Ingleside 2d 59648.—Sire, Duke of Ingleside 14274; dam, Bess of Ingleside 42172.

Butter, 19 lbs. 10½ oz. Milk, 240 lbs. 8 oz.

Test made from Dec. 18 to 25, 1896; age, 8 yrs. 5 mos.; actual weight, 910 lbs.; fed 63 qts. ground oats, 21 qts. corn meal, 14 qts. wheat bran and 14 qts. ground oil-cake—half-bushel of carrots; property of Miller & Sibley, Franklin, Pa.

Bess Pogis of Prospect 87045.—Sire, Gipsy's Lorne Pogis 20346; dam, Bess of Ingleside 42172.

Butter, 16 lbs. 13½ oz. Milk, 245 lbs. 8 oz.

Test made from June 27 to July 4, 1893; age, 4 yrs. 2 mos.; actual weight, 890 lbs., fed 5 qts. corn meal, 8 qts. ground oats and 3 qts. ground oil-cake, daily—good old meadow pasture; property of Miller & Sibley, Franklin, Pa.

Bess Pogis of Prospect 87045.—Sire, Gipsy's Lorne Pogis 20346; dam, Bess of Ingleside 42172.

Butter, 29 lbs. 1¾ oz. Milk, 408 lbs.

Test made from July 8 to 15, 1895; age, 6 yrs. 2 mos.; actual weight, 880 lbs.; fed 56 qts. ground oats, 42 qts. corn meal, 42 qts. ground oil-cake and 28 qts. wheat bran —sweet corn *ad lib.*, old meadow pasture; property of Miller & Sibley, Franklin, Pa.

Bet Arlington 8970.—Sire, Fitz 1988; dam, Hattie Parks 3776.

Butter, 18 lbs. 11 oz. Milk, 236 lbs.

Test made from June 15 to 21, 1883; age, 5 yrs. 3 mos.; fed 10 lbs. corn meal, daily— early cut hay, rather short pasturage; property of N. C. Stoughton, Riverside, Mass.

Betsona 16776.—Sire, Catono 3761; dam, Betsy B. 9413.

Butter, 14 lbs. 3 oz. Milk, 105½ qts.

Test made from June 1 to 7, 1885; age, 3 yrs. 1 mo.; property of Lyman A. Mills, Middlefield, Conn.

Betsy Marjoram 65145.—Sire, Marjoram's Signal 11183; dam, Betsy Rain 37454.

Butter, 15 lbs. 7 oz. Milk, 301 lbs.

Test made from Sept. 26 to Oct. 3, 1896; age, 7 yrs. 8 mos.; estimated weight, 975 lbs.; fed 56 qts. bran and 28 qts. cracked corn, with warm water—alfalfa hay *ad lib.*, pumpkins and dry pasture in corral; property of G. B. Renfro, Savannah, Cal.

Betsy of Rochester 23562.—Sire, Catoctin 2692; dam, Cassia Catoctin 23561.

Butter, 14 lbs. 10½ oz. Milk, 220 lbs. 9 oz.

Test made from June 17 to 23, 1885; age, 4 yrs. 3 mos.; property of Edward Austen, Glencoe, Md.

Betsy's Glory 45604.—Sire, Arawana's Glory 8001; dam, Betsy 3d 8162.

Butter, 14 lbs. 1 oz. Milk, 219 lbs. 13 oz.

Test made from June 11 to 18, 1893; age, 7 yrs. 9 mos.; estimated weight, 950 lbs., fed 21 qts. bran, 7 qts. cotton-seed meal, 7 qts. old process oil meal—blue-grass, timothy and white clover pasture; property of A. Plummer & Son, Tacoma, Ohio.

Bettie Dixon 4527.—Sire, Aldine 1136; dam, Cenie Wallace 2663.

Butter, 15 lbs. Milk, 281 lbs. 8 oz.

Test made from June 28 to July 4, 1882; age, 6 yrs. 11 mos.; property of W. B. Montgomery, Starkville, Miss.

Bettie of Edgewood 31162—Sire, Antrascine 4747; dam, Bettie of Hillside 3864.

Butter, 14 lbs. 2 oz. Milk, 217 lbs. 8 oz.

Test made from Aug. 13 to 20, 1889; age, 6 yrs. 4 mos.; actual weight, 700 lbs.; fed ground corn and oats, no record kept; property of Wm. Morrow & Son, Nashville, Tenn.

Bettie Pogis 44542.—Sire, Prince Victor Pogis 14257; dam, Holyoke's Belle 30453.

Butter, 16 lbs. 7½ oz. Milk, 263 lbs. 4 oz.

Test made from July 12 to 19, 1896; age, 9 yrs. 4 mos.; actual weight, 1060 lbs.; fed 8 qts. ground oats, 5 qts. corn meal, 4 qts. wheat bran and 3 qts. oil meal, daily— poor pasture, with green oats and hay in stable; property of C. A. Sweet, East Aurora, N. Y.

Bettie's Pet 22550.—Sire, Prince of the Realm 7842; dam, Mazie Bate 3364.

Butter, 16 lbs. 1 oz. Milk, 180 lbs. 8 oz.

Test made from Oct. 20 to 26, 1885; age, 2 yrs. 5 mos.; property of T. W. McNeely, Petersburg, Ill.

Between 86664.—Sire, Billy Russell 26384; dam, Cocoa Butter 64983.

Butter, 18 lbs. 9¼ oz. Milk, 317 lbs. 8 oz.

Test made from July 24 to 31, 1897; age, 5 yrs. 5 mos.; actual weight, 810 lbs.; fed 42 qts. corn meal, 42 qts. ground oats, 14 qts. wheat bran and 14 qts. ground oilcake—good old meadow pasture; property of Miller & Sibley, Franklin, Pa.

Beulah De Gruchy 13460.—Sire, Farmer's Glory 5196; dam, Miss Beulah 24358

Butter, 22 lbs. 2 oz. Milk, 308 lbs.

Test made from June 20 to 27, 1885; age, 5 yrs. 2 mos.; fed 3 qts. corn meal, 1 qt. oil meal and 1 qt. pea meal, morning and evening, and at noon 1 qt. pea meal and 2 qts. oat meal—good pasture; property of Cornelius Wellington, East Lexington, Mass.

Beulah of Baltimore 3270.—Sire, Sir Davy 84; dam, Nellie 1507.

Butter, 14 lbs. 6½ oz.

Test made from May 25 to 31, 1880; age, 7 yrs. 4 mos.; property of Clarke & Jones, Baltimore, Md.

Bianca Lass 14997.—Sire, Etzel 3593; dam, Aldabella 3928.

Butter, 14 lbs. 3½ oz. Milk, 223 lbs. 8 oz.

Test made from June 13 to 19, 1886; age, 4 yrs. 4 mos.—wild grass pasture only; property of V. Lowe, Palmyra, Wis.

Bicora 2d 91188.—Sire, Biveno of Ogston 13253; dam, Bicora 88461.

Butter, 16 lbs. 3 oz. Milk, 218 lbs. 8 oz.

Test made from Apr. 7 to 14, 1896; age, 4 yrs. 2 mos.; actual weight, 900 lbs.; fed 10½ lbs. bran, 10½ lbs. oat meal, 19 lbs. oil meal and 21 lbs. corn meal; property of L. T. Becker, Galesburg, Mich.

Biederkeit 52257.—Sire, Southern Prince 10760; dam, Lady Julia G. 16199.

Butter, 14 lbs. 10 oz. Milk, 134 lbs.

Test made from Dec. 10 to 17, 1890; age, 2 yrs. 9 mos.; estimated weight, 825 lbs., fed, daily, 16 lbs. of a mixture of corn and cob meal, cotton-seed meal and oil meal, property of Campbell Brown, Spring Hill, Tenn.

Biederkeit 52257.—Sire, Southern Prince 10760; dam, Lady Julia G. 16199.

Butter, 18 lbs. 2 oz. Milk, 236 lbs.

Test made from Sept. 18 to 25, 1893; age, 5 yrs. 6 mos.; estimated weight, 925 lbs., fed 60 lbs. corn-hearts, 20 lbs. bran and 108 lbs. ground oats—blue-grass pasture, property of Est. of Campbell Brown, Spring Hill, Tenn.

Biederkeit 52257.—Sire, Southern Prince 10760; dam, Lady Julia G. 16199.

Butter, 22 lbs. 1¼ oz. Milk, 178 lbs.

Test made from Jan. 3 to 10, 1895; age, 6 yrs. 10 mos.; actual weight, 810 lbs.; fed 18 lbs. per day of ground oats and corn in equal parts, mixed with 6 bucketfuls of cut fodder corn—clover hay *ad lib.*; property of Matt. M. Gardner, Nashville, Tenn.

Big Wyuka 58345.—Sire, Regina's Sir George 13569; dam, Weston's Paragon 33297.

Butter, 17 lbs. 11 oz. Milk, 245 lbs.

Test made from June 1 to 8, 1895; age, 7 yrs. 2 mos.; estimated weight, 875 lbs.; fed 28 lbs. coarse bran, 28 lbs. ground oats, 28 lbs. corn meal, and 14 lbs. cotton-seed meal—common pasture; property of John C. McClintock, Meadville, Pa.

Bijou of St. Lambert 5112.—Sire, Lord Lisgar 1066; dam, Cupid of St. Lambert 5104.

Butter, 15 lbs. 4 oz. Milk, 150 lbs. 10 oz.

Test made from Apr. 19 to 26, 1891; age, 16 yrs.; estimated weight, 850 lbs.; fed 14 lbs. oil meal, 14 lbs. pea meal, 35 lbs. corn meal, 30 lbs. crushed oats, 5 lbs. oatmeal, 14 lbs. bran and 60 lbs. hay; property of Ayer & McKinney, Philadelphia, Pa.

Bijou Ogston 8210.—Sire, Duke (P. S. 76 J. H. B.); dam, Countess (F. S. 1302 J. H. B.)

Butter, 18 lbs. 15 oz. Milk, 217 lbs. 4 oz.

Test made from June 14 to 21, 1887; age, 11 yrs.; estimated weight, 1000 lbs.; fed, daily, 4 qts. corn and oats and 1 qt. pea meal; property of M. Erskine Miller, Staunton, Va.

Bintana 9837.—Sire, Oxoli 1922; dam, Kalmia 4561.

Butter, 14 lbs. 3½ oz. Milk, 183 lbs. 15 oz.

Test made from Apr. 15 to 22, 1880; age, 3 yrs.; property of Joseph Gavin, Chester, N. Y.

Birdie 2611, imp.

Butter, 20 lbs. in 10 days.

Property of G. Dawson Coleman, Brickerville, Pa.

Birdie Le Brocq 17263.—Sire, Le Brocq's Prize 3350; dam, Queen Bess of St. Brelade's 7462.

Butter, 14 lbs. Milk, 167 lbs. 8 oz.

Test made from Oct. 15 to 21, 1884; age, 5 yrs. 1 mo.; property of Beech Grove Farm, Beech Grove, Ind.

Birdie Nicholson 31676.—Sire, Signalda 4027; dam, Lady Mary Hampton 2d 26184.

Butter, 18 lbs. 13 oz. Milk, 214 lbs. 10 oz.

Test made from June 20 to 27, 1889; age, 4 yrs. 11 mos.; estimated weight, 800 lbs.; fed, daily, 4 gals., consisting of two-thirds meal and one-third oats; property of Maury Jersey Farm, Columbia, Tenn.

Birdsey's Surprise 48326.—Sire, Pedro of the Valley 8750; dam, Flashy Jessie 2d 23332.

Butter, 14 lbs. 2 oz. Milk, 255 lbs. 7 oz.

Test made from June 16 to 23, 1891; age, 5 yrs. 4 mos.; actual weight, 870 lbs.; fed 118 lbs. corn and oats, ground together—pasture; property of Lyman A. Mills, Middlefield, Conn.

Bissonise 110476.—Sire, Bisson's Landseer 27520; dam, Panise 56676.

Butter, 14 lbs. 11 oz. Milk, 212 lbs. 12 oz.

Test made from May 26 to June 2, 1896; age, 2 yrs. 5 mos.; estimated weight, 650 lbs.; fed 56 qts. corn meal and 56 qts. ground oats—woodland pasture; property of C. A. Morgan, Columbia, Tenn.

Bisson's Beauty 95023.—Sire, Bisson's Landseer 27520; dam, Belle Lyman 2d 82424.

Butter, 16 lbs. 4 oz. Milk, 166 lbs.

Test made from Apr. 17 to 24, 1897; age, 4 yrs. 2 mos.; actual weight, 718 lbs.; fed 7 qts. cooked cotton-seed, 12 qts. corn and oat meal, in equal parts, and 6 qts. wheat bran, daily—sheaf oats, short Bermuda pasture; property of J. M. Trosper, Bethany, La.

Bisson's Belle 31144.—Sire, Carlos, on I. of J.; dam, Purity, on I. of J.

Butter, 21 lbs. 15½ oz. Milk, 156 lbs. 7 oz.

Test made from Oct. 30 to Nov. 6, 1888; age, 5 yrs. 8 mos.; estimated weight, 1000 lbs.; fed, daily, 5 gals. ground corn and oats, in equal parts; property of Maury Jersey Farm, Columbia, Tenn.

Bisson's Belle 31144.—Sire, Carlos, on I. of J.; dam, Purity, on I. of J.

Butter, 28 lbs. 10 oz. Milk, 257 lbs. 12 oz.

Test made from July 21 to 28, 1890; age, 7 yrs. 4 mos.; estimated weight, 900 lbs.; fed, daily, 5 gals. corn meal and 2 gals. bran; property of Maury Jersey Farm, Columbia, Tenn.

Bisson's Belle 31144.—Sire, Carlos, on I. of J.; dam, Purity, on I. of J.

Butter, 1028 lbs. 15⅝ oz., in one year. Milk, 8412 lbs. 7 oz.

Test made from July 15, 1890, to July 15, 1891; age, 7 yrs. 4 mos. at beginning of test; estimated weight, 950 lbs.; fed, daily, 24 qts. grain; property of Maury Jersey Farm, Columbia, Tenn., Wm. J. Webster, President.

Bisson's Emma 100408.—Sire, Bisson's Landseer 27520; dam, Emma Landseer 58104.

Butter, 14 lbs. 3 oz. Milk, 160 lbs.

Test made from Dec. 18 to 25, 1895; age, 2 yrs. 3 mos.; estimated weight, 650 lbs.; fed 70 qts. ground oats and 70 qts. corn meal—millet hay; property of Morgan & Brown, Columbia, Tenn.

Bisson's Ethleel 80921.—Sire, Bisson's Landseer 27520; dam, Ethleel's Ida 66512.

Butter, 14 lbs. 11¾ oz. Milk, 211 lbs. 8 oz.

Test made from Jan. 28 to Feb. 4, 1896; age, 3 yrs. 11 mos.; estimated weight, 650 lbs.; fed 51½ lbs. bran and 84 lbs. ground corn and oats—hay *ad lib.*; property of G. W. Bayler, Washington, Ill.

Bisson's Fancy Pogis 106415.—Sire, Bisson's Landseer 27520; dam, Donney Pogis 2d 82426.

Butter, 14 lbs. 3½ oz. Milk, 232 lbs. 13 oz.

Test made from Apr. 8 to 15, 1897; age, 3 yrs. 9 mos.; estimated weight, 700 lbs.; fed 42 lbs. bran, 28 lbs. corn meal, 21 lbs. ground oats, 10½ lbs. oil meal, 105 lbs. roots and 210 lbs. ensilage—hay *ad lib.*; property of C. I. Hood, Lowell, Mass.

Bisson's Fancy Pogis 106415.—Sire, Bisson's Landseer 27520; dam, Donney Pogis 2d 82426.

Butter, 17 lbs. 1½ oz. Milk, 240 lbs. 12 oz.

Test made from Mar. 2 to 9, 1898; age, 4 yrs. 8 mos.; estimated weight, 850 lbs.; fed 42 lbs. bran, 21 lbs. corn meal, 21 lbs. ground oats, 14 lbs. oil meal, 210 lbs. ensilage and 70 lbs. roots—hay *ad lib.*; property of C. I. Hood, Lowell, Mass.

Bisson's Idex 133952.—Sire, Bisson's Pogis 30781; dam, Toridex 76873.

 Butter, 17 lbs. 11 oz. Milk, 270 lbs.

 Test made from May 26 to June 2, 1898; age, 3 yrs. 11 mos.; estimated weight, 900 lbs.; fed 2 lbs. cotton-seed meal, 2 lbs. linseed meal, 4 lbs. wheat bran and 8 lbs. corn meal, daily—blue-grass pasture; property of George Campbell Brown. Spring Hill, Tenn.

Bisson's Oonan 108518.—Sire, Bisson's Landseer 27520; dam, Pearl's Oonan 2d 86105.

 Butter, 14 lbs. 9 oz. Milk, 163 lbs. 8 oz.

 Test made from Mar. 23 to 30, 1897; age, 5 yrs. 3 mos.; estimated weight, 650 lbs.; fed 70 qts. ground oats and 70 qts. corn meal—clover hay; property of P. W. Middleton, Greenville, Ky.

Bisson's Perfection 103628.—Sire, Bisson's Landseer 27520; dam, Harry's Signaline 82419.

 Butter, 14 lbs. 9½ oz. Milk, 201 lbs. 11 oz.

 Test made from Nov. 11 to 18, 1896; age, 2 yrs. 1 mo.; actual weight, 650 lbs.; fed 14 lbs. per day of ground oats and corn, in equal parts, mixed with 4 bucketfuls of cut sheaf oats, wetted—millet hay *ad lib.* at night, little pasture; property of Matt. M. Gardner, Nashville, Tenn.

Bisson's Pride 108517.—Sire, Bisson's Landseer 27520; dam, Landseer's Signaline 2d 86343.

 Butter, 14 lbs. ½ oz. Milk, 157 lbs. 8 oz.

 Test made from Dec. 11 to 18, 1896; age, 1 yr. 11 mos.; estimated weight, 700 lbs.; fed 60 qts. ground oats and 60 qts. corn meal—poor pasture; property of G. M. Sherrill, Marshall, Tex.

Bisson's Rosa 97830.—Sire, Bisson's Pogis 30781; dam, Rosa Oxford 66678.

 Butter, 15 lbs. 1¾ oz. Milk, 176 lbs. 7½ oz.

 Test made from Feb. 17 to 24, 1897; age, 3 yrs. 2 mos.; estimated weight, 725 lbs.; fed 20 lbs. per day of ground corn and oats, in equal parts, mixed with 4 bucketfuls of cut sheaf oats, wetted—millet hay *ad lib.* at night; property of Matt. M. Gardner, Nashville, Tenn.

·Bisson's Signaline 93844.—Sire, Bisson's Landseer 27520; dam, Landseer's Signaline 58101.

 Butter, 14 lbs. 3 oz. Milk, 153 lbs. 8 oz.

 Test made from Mar. 29 to Apr. 5, 1895; age, 2 yrs. 2 mos.; estimated weight, 625 lbs.; fed 70 qts. ground oats and 70 qts. corn meal—clover and millet hay; property of Morgan & Brown, Columbia, Tenn.

Bisson's Southern Daisy 112992.—Sire, Bisson's Landseer 27520; dam, Kate Oxford 58102.

 Butter, 14 lbs. 2½ oz. Milk, 177 lbs. 10 oz.

 Test made from Dec. 29, 1897, to Jan. 5, 1898; age, 2 yrs. 4 mos.; estimated weight, 650 lbs.; fed 20 lbs. per day of ground oats and corn, in equal parts, mixed with 4 bucketfuls of cut sheaf oats, wetted—millet hay *ad lib.* at night; property of M. M. Gardner, Nashville, Tenn.

Bisson's Wandrina 100409.—Sire, Bisson's Landseer 27520; dam, Wandrina 2d 82420.

 Butter, 18 lbs. Milk, 227 lbs. 15 oz.

 Test made from May 2 to 9, 1897; age, 2 yrs. 6 mos.; estimated weight, 800 lbs.; fed 4 qts. cotton-seed, 4 qts. ground corn, 4 qts. ground oats and 10 qts. wheat bran, daily—oats pasture; property of Mrs. E. M. Mirick, Cleburne, Tex.

Black Teat 32932.—Sire, Royalist 6th 4977; dam, Maid of Fancy Creek 11431

Butter, 16 lbs. 11 oz. Milk, 239 lbs. 9 oz.

Test made from Apr. 15 to 22, 1895; age, 11 yrs. 3 mos.; estimated weight, 800 lbs., fed 28 gals. corn and cob meal and 14 qts. linseed oil meal—blue-grass pasture, property of Gus Aaron, Kickapoo, Kan.

Blanche Duncan 2d 73333.—Sire, Tormentor Stoke Pogis 20485; dam, Blanche Duncan 73077.

Butter, 14 lbs. 11¾ oz. Milk, 221 lbs. 9½ oz.

Test made from Oct. 20 to 27, 1893; age, 2 yrs. 9 mos.; actual weight, 865 lbs.; fed 45 lbs. corn meal, 43 lbs. wheat middlings, 12 lbs. new process oil meal and 10 lbs. cotton-seed meal—good blue-grass pasture; property of J. P. Bradbury, Pomeroy, Ohio.

Blanche Ferry 108501.—Sire, Kindergarten's Duke 30065; dam, Matina of Riverside 51783.

Butter, 16 lbs. 5 oz. Milk, 244 lbs. 4 oz.

Test made from Aug. 5 to 12, 1897; age, 2 yrs. 8 mos.; actual weight, 965 lbs.; fed 9 qts. ground oats, 12 qts. wheat bran, 3 qts. corn meal and 3 qts. oil meal, mixed, daily—hay, good clover and timothy pasture; property of C. A. Sweet, East Aurora, N. Y.

Blanche of Castile 43793.—Sire, Pascarrelle 16603; dam, Floriole 13069.

Butter, 22 lbs. 1½ oz. Milk, 299 lbs.

Test made from June 24 to July 1, 1892; age, 5 yrs. 7 mos.; actual weight, 1090 lbs.; fed 30 qts. mixed feed daily—good second season clover pasture; property of Miller & Sibley, Franklin, Pa.

Blanche Pomeroy 105013.—Sire, St. Lambert Boy 17408; dam, Blanche Duncan 2d 73333.

Butter, 16 lbs. 9½ oz. Milk, 245 lbs. 8 oz.

Test made from Dec. 16 to 23, 1897; age, 4 yrs. 2 mos.; actual weight, 912 lbs.; fed 24 lbs. corn meal, 40 lbs. wheat bran, 24 lbs. middlings, 16 lbs. old process oil meal and 8 lbs. cotton-seed meal—mixed hay and corn stover *ad lib.*; property of J. P. Bradbury, Pomeroy, Ohio.

Bliss Pogis 82873.—Sire, Kathy's Stoke Pogis 17566; dam, Blizz 55835.

Butter, 14 lbs. 9½ oz. Milk, 276 lbs. 12 oz.

Test made from May 21 to 28, 1897; age, 5 yrs. 9 mos.; estimated weight, 750 lbs., fed 42 lbs. bran, 35 lbs. corn, 31½ lbs. oats, 10½ lbs. oil meal, 210 lbs. ensilage and 70 lbs. roots—hay *ad lib.*, some green rye in barn; property of C. I. Hood, Lowell, Mass.

Blonde 2d 9268.—Sire, Khedive (P. S. 103 J. H. B.); dam, Blonde (F. S. 214 J. H. B.)

Butter, 14 lbs. 4 oz. Milk, 204 lbs.

Test made from Sept. 7 to 13, 1883; age, 6 yrs. 8 mos.; property of William Rolph, Markham, Ont., Can.

Bloomer 2d 2321.—Sire, Hector of Plymouth Rock 886; dam, Bloomer 2262.

Butter, 18 lbs. 6 oz. Milk, 312 lbs. 3 oz.

Test made from May 10 to 17, 1885; age, 13 yrs. 1 mo.; property of Henry Pierce, San Francisco, Cal.

Bloomfield Lady 6912.—Sire, Albert 2d 1835; dam, Lady Brown 4th 6911.

Butter, 14 lbs. 12 oz., unsalted.

Test made from May 2 to 9, 1882; age, 6 yrs. 6 mos.; property of J. H. Walker, Worcester, Mass.

Bloom of Amherst Villa 70507.—Sire, G. F. Train 18868; dam, Allie L. 32073.

Butter, 16 lbs. 10 oz. Milk, 184 lbs. 10 oz.

Test made from Jan. 8 to 15, 1895; age, 3 yrs. 10 mos.; actual weight, 930 lbs.; fed 14 lbs. wheat, 10½ lbs. oil meal, 10½ lbs. cotton-seed meal, 21 lbs. oats, 14 lbs. corn meal, 180 lbs. corn ensilage and 35 lbs. hay; property of Charles Lautz, Buffalo, N. Y.

Blossie Reynolds 6082.—Sire, Duke of Lebanon 1880; dam, Miss Vesper 4460.

Butter, 16 lbs. 3½ oz.

Test made from June 1 to 8, 1883; age, 6 yrs.; property of G. H. Reynolds, Canton, Pa.

Blossom's Belle 97433.—Sire, Prince of Tennessee 20772; dam, Forest Blossom 63501.

Butter, 15 lbs. 6 oz. Milk, 219 lbs. 10 oz.

Test made from May 26 to June 2, 1896; age, 3 yrs. 2 mos.; estimated weight, 800 lbs.; fed 140 lbs. wheat bran, 28 lbs. corn meal and 14 lbs. cotton-seed meal—wild grass pasture; property of Platter & Foster, Denison, Tex.

Blossom's Niobe 13855.—Sire, Duffee 2083; dam, Niobe Hampton 5343.

Butter, 16 lbs. 7½ oz. Milk, 243 lbs. 13 oz.

Test made from Oct. 17 to 24, 1893; age, 14 yrs. 4 mos.; estimated weight, 1000 lbs., fed 14 gals. corn bran, 28 qts. cotton-seed meal and 28 qts. corn chop, mixed, with cotton-seed hulls; property of A. J. Hawes & Son, Montgomery, Ala.

Bluster's Pip 101165.—Sire, Bluster 22798; dam, Pip 86038.

Butter, 14.41 lbs., confirmed; estimated butter on basis of 85 per cent. fat, 14.05 lbs. Milk, 210.2 lbs.

Test made from Jan. 14 to 20, 1897; age, 2 yrs. 3 mos.; weight at end of test, 720 lbs., fed 7 lbs. cotton-seed meal, 7 lbs. oil meal, 14 lbs. bran and 84 lbs. corn-hearts—clover hay *ad lib.*; property of Mrs. M. L. Sayre, Lexington, Ky.

Analyses of Butter.

First churning—Fat 83.56 per cent., water 13.13 per cent., ash 2.61 per cent., casein .70 per cent.

Second churning—Fat 82.07 per cent., water 12.95 per cent., ash 3.91 per cent., casein 1.07 per cent.

Third churning—Fat 80.40 per cent., water 14.99 per cent., ash 3.97 per cent., casein .64 per cent

Bobby's Lily 22055.—Sire, Bobby (P. S. 208 J. H. B.); dam, Duchess of Cherwell (F. S. 2955 J. H. B.)

Butter, 18 lbs. Milk, 202 lbs.

Test made from Mar. 17 to 24, 1890; age, 9 yrs. 3 mos.; actual weight, 950 lbs.; fed, daily, 12 lbs. wheat middlings; property of Mrs. Thomas Allen, Pittsfield, Mass.

Bobby's Magnolia 24401.—Sire, Bobby (P. S. 208 J. H. B.); dam on I. of J.

Butter, 15 lbs. 12 oz. Milk, 244 lbs.

Test made from Sept. 21 to 28, 1887; age, 6 yrs. 6 mos.; actual weight, 1000 lbs.; property of Mrs. Thomas Allen, Pittsfield, Mass.

Bomba 10330.—Sire, Duke of Darlington 2460; dam, Beauty of Darlington 5736.

Butter, 21 lbs. 11½ oz., unsalted (official). Milk, 205 lbs. 6 oz.

Test made from Oct. 6 to 12, 1882; age, 4 yrs.; fed 26 qts. wheat middlings, 12 qts. corn meal and 5 qts. linseed; property of A. B. Darling, Ramsey's, N. J.

Bomba 10330.—Sire, Duke of Darlington 2460; dam, Beauty of Darlington 5736.

Butter, 89 lbs. 14 oz. in 31 days.

Test made in 1882; age, 4 yrs.; property of A. B. Darling, Ramsey's, N. J.

Bonnie 2d 5742.—Sire, Baronet 2240; dam, Bonnie 491.

Butter, 14 lbs. 11½ oz., unsalted. Milk, 200 lbs. 2 oz.

Test made from Dec. 11 to 17, 1883; age, 6 yrs. 8 mos.; property of S. E. Gillett, Ravenna, Ohio.

Bonnie Grisette 2d 19526.—Sire, Duke of Darlington 2460; dam, Bonnie Grisette 6979.

Butter, 16 lbs. 12½ oz. Milk, 191 lbs. 12 oz.

Test made from June 20 to 26, 1886; age, 4 yrs. 1 mo.; property of T. Alexander Seth, Baltimore, Md.

Bonnie Moll 95616.—Sire, Bonnie Blue Pogis 26285; dam, Maud of Jersey Lawn 78936.

Butter, 16 lbs. 3 oz. Milk, 245 lbs. 13 oz.

Test made from May 30 to June 6, 1897; age, 4 yrs. 2 mos.; estimated weight, 875 lbs.; fed 15 lbs., daily, of oats, peas and corn, ground—upland pasture; property of Schuyler Bros., Cobleskill, N. Y.

Bonnie Signaldina 108681.—Sire, Fancy's Jubilee 25205; dam, Bonnie Katrina 80509.

Butter, 18 lbs. 4 oz. Milk, 244 lbs. 1½ oz.

Test made from Apr. 1 to 8, 1898; age, 2 yrs. 11 mos.; estimated weight, 800 lbs.; fed 20 lbs. corn meal, 18 lbs. cotton-seed meal, 42 lbs. wheat bran and 24 lbs. ground oats—green oats pasture; property of J. M. Long, Mexia, Tex.

Bonnie Wherry 115088.—Sire, Tormentor Catono 17004; dam, Bonnie Gladesia 47792.

Butter, 20 lbs. 4 oz. Milk; 245 lbs.

Test made from May 23 to 30, 1896; age, 7 yrs. 1 mo.; estimated weight, 800 lbs.; fed 28 gals. bran, 28 qts. corn meal and 14 qts. cotton-seed meal—good blue-grass pasture; property of W. L. Reynolds, Cynthiana, Ky.

Bonnie Yost 7943.—Sire, Rector 1458; dam, Bo Peep 2850.

Butter, 18 lbs. 2 oz., unsalted. Milk, 226 lbs. 6 oz.

Test made from May 9 to 16, 1883; age, 4 yrs. 11 mos.; fed 6 qts. corn meal, daily—very good blue-grass pasture; property of M. M. Gardner, Nashville, Tenn.

Bonny of Woodford 58718.—Sire, Dolphin St. Lambert 16431; dam, Fir Shade Bonnibel 27235.

Butter, 17 lbs. 4 oz. Milk, 296 lbs. 11 oz.

Test made from Apr. 1 to 8, 1894; age, 5 yrs.; estimated weight, 900 lbs.; fed 140 lbs. corn and cob meal, 28 lbs. wheat bran, 84 lbs. ground oats, 14 lbs. linseed meal—timothy hay and corn fodder *ad lib.*; property of Gus Aaron, Kickapoo, Kan.

Bo-Peep's Maid 100538.—Sire, Smith's St. Lambert 30332; dam, Gipsy's Bo-Peep 90205.

> Butter, 14 lbs. 2 oz. Milk, 275 lbs. 12 oz.

Test made from Jan. 25 to Feb. 1, 1898; age, 3 yrs. 4 mos.; estimated weight, 950 lbs.; fed 8 qts. wheat bran, 4 qts. ground oats and 4 qts. corn meal, daily—hay and corn fodder; property of J. R. Smith, Quincy, Mich.

Besnia 87508.—Sire, Diploma 16219; dam, Comanca 19389.

> Butter, 20 lbs. 2 oz. Milk, 271 lbs. 8 oz.

Test made from Mar. 9 to 16, 1898; age, 6 yrs. 7 mos.; estimated weight, 950 lbs.; fed 35 lbs. bran, 42 lbs. oats, 35 lbs. corn meal and 14 lbs. oil meal—corn ensilage; property of H. C. Taylor, Orfordville, Wis.

Beauty 1606.—Sire, Duke of Wellington 608; dam, Bee 448.

> Butter, 14 lbs., unsalted.

Test made in June, 1877; age, 9 yrs. 6 mos.; no feed—good pasture; property of Beech Grove Farm, Beech Grove, Ind.

Box 2d 4297.—Sire, Hector 791; dam, Box 1982.

> Butter, 14 lbs. 3 oz. Milk, 199 lbs. 4 oz.

Test made from Sept. 13 to 19, 1882; age, 7 yrs. 9 mos.; property of O. S. Chaffee, Mansfield Centre, Conn.

Bramballetta 10451.—Sire, Revenue 1744; dam, Camilla 4th 2023.

> Butter, 16 lbs. 4 oz.

Test made in September, 1885; age, 6 yrs. 8 mos.; property of Charles H. Peckham, North Scituate, R. I.

Braw Lassie 36221.—Sire, Pedro of the Valley 8750; dam, Ephrella 20651.

> Butter, 17 lbs. 6 oz. Milk, 174 lbs.

Test made from Feb. 5 to 12, 1895; age, 9 yrs. 4 mos.; estimated weight, 950 lbs.; fed 14 lbs. wheat bran, 4 lbs. corn chop, 4 lbs. cotton-seed meal, 2 lbs. old process oil meal and 27 lbs. corn ensilage per day—28 lbs. hay; property of B. F. Fackenthal, Jr., Riegelsville, Pa.

Brenda of Elmhurst 10762.—Sire, Stoke Pogis 3d 2238; dam, Rosette of St. Lambert 5108.

> Butter, 20 lbs. 8 oz. Milk, 201 lbs.

Test made from Aug. 21 to 27, 1885; age, 7 yrs. 6 mos.; fed pea meal and chopped oats, with a little bran—pasture dry and parched; property of Valancey E. Fuller, Hamilton, Ont., Can.

Brenhilda 7649.—Sire, Granger 1591; dam, Gunilda 662.

> Butter, 16 lbs. 8 oz., unsalted. Milk, 145 qts.

Test made from May 10 to 17, 1883; age, 5 yrs. 5 mos.; property of Isaac Hunt, Hamilton Square, N. J.

Bridal Rose 74883.—Sire, Tenella's Alexis 13170; dam, Lakeview Gem 37042.

> Butter, 23 lbs. 8 oz. Milk, 316 lbs. 5½ oz.

Test made from May 11 to 18, 1895; age, 7 yrs. 2 mos.; estimated weight, 900 lbs.; fed 12 qts. ground oats and corn, 4 qts. bran and middlings, 3 qts. cotton-seed meal, 1 qt. linseed meal and 1 qt. whole oats, daily—woodland blue-grass pasture; property of E. A. Berauer, Waldron, Ind.

Bride of Evergreen 72022.—Sire, Gold of St. Lambert 16744; dam, Bride of Lakeside 44139.

Butter, 20 lbs.   Milk, 288 lbs. 7 oz.

Test made from Oct. 16 to 23, 1895; age, 5 yrs. 2 mos.; estimated weight, 900 lbs.; fed 112 lbs. ground oats and 14 lbs. old process oil meal—good blue-grass meadow pasture; property of N. N. Palmer & Son, Brodhead, Wis.

Bride of Evergreen 3d 90473.—Sire, Winder 29955; dam, Bride of Evergreen 72022.

Butter, 17 lbs. 1 oz.   Milk, 249 lbs. 8 oz.

Test made from Mar. 6 to 13, 1898; age, 4 yrs. 4 mos.; estimated weight, 900 lbs.; fed 7 lbs. ground oats, 5 lbs. wheat bran, 2 lbs. old process oil meal and about 20 lbs. ensilage—clover hay *ad lib.*; property of N. N. Palmer & Son, Brodhead, Wis.

Brier of St. Lambert 61750.—Sire, Canada's John Bull 3d 17750; dam, Rioter's Sweet Brier 30582.

Butter, 18 lbs. 4 oz.   Milk, 267 lbs. 1 oz.

Test made from July 2 to 9, 1893; age, 4 yrs. 5 mos.; estimated weight, 900 lbs., fed 14 qts. peas and oats, daily—clover pasture; property of William Rolph, Markham, Ont., Can.

Bright Eyes 2d 2290.—Sire, Napoleon 291; dam, Bright Eyes 1537.

Butter, 19 lbs. 6 oz.

Test made from July 1 to 7, 1883; age, 12 yrs. 3 mos.; property of E. M. Phelan, Cherry Valley, N. Y.

Bright Eyes St. Lambert 2d 72699.—Sire, Rioter of St. Lambert 16501; dam, Bright Eyes St. Lambert 53720.

Butter, 19 lbs. 5 oz.   Milk, 317 lbs.

Test made from Apr. 30 to May 7, 1896; age, 4 yrs. 11 mos.; actual weight, 860 lbs., fed 126 lbs. of a mixture of oats, corn and wheat bran, two parts oats to one of corn and one of bran—pasture in wood lot; property of J. R. Smith, Quincy, Mich.

Bright Lady 5938.—Sire, Jersey King 879; dam, Morning Glory 2d 1299.

Butter, 14 lbs. 12 oz.   Milk, 14 qts. per day.

Test made from June 11 to 18, 1885; age, 8 yrs. 2 mos.; property of J. H. Taylor, Thomaston, Conn.

Bright Tass 46173.—Sire, Bright Lad 11035; dam, Ramapo's Tass 41846.

Butter, 21 lbs. 1 oz.   Milk, 249 lbs. 9 oz.

Test made from Mar. 6 to 13, 1896; age, 8 yrs. 10 mos.; estimated weight, 1000 lbs., fed 35 lbs. corn meal, 21 lbs. bran, 21 lbs. cracked oats, 21 lbs. linseed meal and 14 lbs. Buffalo gluten meal—hay and corn fodder *ad lib.*; property of C. Delano, Mount Vernon, Ohio.

Bristol Bella 15697.—Sire, Pompus 2881; dam, Dolly P. 10129.

Butter, 14 lbs. 7 oz.   Milk, 162 lbs. 7 oz.

Test made from Oct. 11 to 18, 1893; age, 12 yrs. 8 mos.; estimated weight, 950 lbs.; fed 84 qts. ground corn and oats, 84 qts. wheat bran and 21 qts. oil meal—clover hay; property of George H. Sweet, East Aurora, N. Y.

Bristol Bella 15697.—Sire, Pompus 2881; dam, Dolly P. 10129.

   Butter, 15 lbs. 3½ oz. Milk, 257 lbs. 8 oz.

Test made from Mar. 28 to Apr. 4, 1896; age, 15 yrs. 1 mo.; actual weight, 1065 lbs., fed 15 qts. wheat bran, 3 qts. corn meal, 6 qts. ground oats and 3 qts. oil meal, daily —hay; property of George D. Briggs, East Aurora, N. Y.

Broadway 50883.—Sire, Lisgar's Pride 14534; dam, Tenella's Regina 37092.

   Butter, 16 lbs. 8 oz. Milk, 237 lbs. 3 oz.

Test made from Oct. 1 to 8, 1890; age, 3 yrs.; estimated weight, 800 lbs.; fed, daily, 18 qts. ground oats, corn and peas; property of Richardson Bros., Davenport, Iowa.

Bronx Clover Blossom 38676.—Sire, Meg's Lord Bronx 11537; dam, Burckel's Clover Blossom 15609.

   Butter, 15 lbs. 2½ oz. Milk, 214 lbs. 8 oz.

Test made from Feb. 19 to 26, 1892; age, 7 yrs. 2 mos.; estimated weight, 800 lbs., fed 9 lbs. corn meal, 6 lbs. bran, 2 lbs. cotton-seed meal, 40 lbs. ensilage and 6 lbs. cut hay, daily—dry hay *ad lib.*; property of H. A. Huntington, Higganum, Conn.

Bronze Leaf 14902.—Sire, Okubo 1876; dam, Gold Leaf 2d 5860.

   Butter, 15 lbs. 1 oz. Milk, 18 qts. per day.

Test made in July, 1884; age, 4 yrs. 9 mos.; property of G. W. Farlee, Cresskill, N. J.

Bronzie 9368.—Sire, Samson 1079; dam, Imogene 4552.

   Butter, 21 lbs. 6 oz. Milk, 138 lbs. 8 oz.

Test made from Aug. 27 to Sept. 3, 1889; age, 11 yrs.; estimated weight, 900 lbs., fed, daily, 8 qts. bran, 6 qts. corn meal, 3 qts. fine middlings, 7 qts. ground oats and 3 qts. cotton-seed meal; property of A. D. McBride, Rochester, N. Y.

Bronzie of St. Lambert 49001.—Sire, Exile of St. Lambert 13657; dam, Bronzie 9368.

   Butter, 14 lbs. 2 oz. (official). Milk, 234 lbs. 8 oz.

Test made from May 23 to 30, 1892; age, 5 yrs. 4 mos.; estimated weight, 1000 lbs.; fed 91 lbs. grain and 42 lbs. hay; property of P. J. Cogswell, Rochester, N. Y.

Brown Bessie 74997.—Sire, Volco 7890; dam, Brown Flora 74996.

   Butter, 20 lbs. 8 oz. Milk, 283 lbs. 2 oz.

Test made from Sept. 8 to 15, 1890; age, 5 yrs. 5 mos.; estimated weight, 950 lbs., fed 94 lbs. of a mixture of equal parts in weight of ground corn and oats, and 16 lbs. bran—good clover, timothy and June grass pasture; property of Richardson Bros., Orfordville, Wis.

Brown Bessie 74997.—Sire, Volco 7890; dam, Brown Flora 74996.

   Butter, 72.235 lbs. in 30 days. Milk, 1134.6 lbs.

Test made from Aug. 29 to Sept. 27, 1893, at World's Columbian Exposition; age, 8 yrs. 5 mos.; actual weight, 1026 lbs.; fed 411.4 lbs. hay, 87 lbs. oil meal, 112 lbs. ensilage, 41 lbs. corn meal, 202 lbs. bran, 116 lbs. oats, 183 lbs. corn-hearts, 54.5 lbs. cotton-seed, 58 lbs. middlings, 18 lbs. carrots and 6 lbs. old hay; property of C. I. Hood, Lowell, Mass.

Brown Bessie 74997.—Sire, Volco 7890; dam, Brown Flora 74996.

Butter, 216.640 lbs. in 90 days. Milk, 3634 lbs.

Test made from May 31 to Aug. 28, 1893, at World's Columbian Exposition; age, 8 yrs. 2 mos. at beginning of test; actual weight, 1048 lbs.; fed 326 lbs. old hay, 210 lbs. ensilage, 643.8 lbs. new hay, 187.5 lbs. oil meal, 562 lbs. bran, 76 lbs. oats, 597 lbs. corn-hearts, 142.50 lbs. cotton-seed, 405 lbs. middlings, 32 lbs. grano-gluten, 1209.4 lbs. clover and 10 lbs. swale-grass; property of C. I. Hood, Lowell, Mass.

Brown Bessie 7th of H. F. 119841.—Sire, Brown Bessie's Son 34550; dam, Portrait 32592.

Butter, 14 lbs. 5 oz. Milk, 226 lbs. 13 oz.

Test made from Sept. 25 to Oct. 2, 1897; age, 2 yrs. 7 mos.; estimated weight, 700 lbs.; fed 35 lbs. bran, 21 lbs. corn meal, 49 lbs. ground oats and 7 lbs. oil meal—timothy and clover pasture; property of C. I. Hood, Lowell, Mass.

Brown Bessie's Bonita 109265.—Sire, Brown Bessie's Son 34550; dam, Smack 64575.

Butter, 20 lbs. 3 oz. Milk, 241 lbs. 4 oz.

Test made from Apr. 21 to 28, 1897; age, 4 yrs. 9 mos.; estimated weight, 925 lbs., fed 21 lbs. rye, 28 lbs. bran, 7 lbs. oil meal, 28 lbs. corn meal and 42 lbs. oats—corn ensilage, corn fodder and clover hay; property of H. C. Taylor, Orfordville, Wis.

Brown Bessie's Brownie 114433.—Sire, Brown Bessie's Son 34550; dam, Alnora 112182

Butter, 17 lbs. Milk, 234 lbs. 4 oz.

Test made from Feb. 20 to 27. 1897; age, 3 yrs.; estimated weight, 825 lbs.; fed 28 lbs. corn meal, 28 lbs. bran, 28 lbs. oats and 21 lbs. oil meal—corn ensilage and clover hay; property of H. C. Taylor. Orfordville, Wis.

Brown Bessie's Modita 128259.—Sire, Brown Bessie's Duke 42166; dam, Chromo's Modita 121489.

Butter, 15 lbs. 3 oz. Milk, 252 lbs.

Test made from May 23 to 30, 1898; age, 3 yrs.; estimated weight, 800 lbs.; fed 28 lbs. corn meal, 28 lbs. bran, 28 lbs. oats and 14 lbs. oil meal—mixed pasture; property of H. C. Taylor, Orfordville, Wis.

Brown Bessie's Princess 120166.—Sire, Brown Bessie's Son 34550; dam, Diploma's Princess 104002.

Butter, 14 lbs. 4 oz. Milk, 224 lbs. 4 oz.

Test made from Apr. 27 to May 4, 1897; age, 4 yrs.; estimated weight, 900 lbs.; fed 28 lbs. bran, 7 lbs. oil meal, 21 lbs. rye, 14 lbs. corn meal and 42 lbs. oats—mixed pasture; property of H. C. Taylor, Orfordville, Wis.

Brown Bonnie 102406.—Sire, Brown Bessie's Son 34550; dam, Loreda 19801.

Butter, 18 lbs. 8 oz. Milk, 215 lbs. 8 oz.

Test made from Apr. 13 to 20, 1895; age, 4 yrs. 2 mos.; estimated weight, 1025 lbs.; fed 35 lbs. oats, 21 lbs. bran, 21 lbs. corn meal, 21 lbs. oil meal and 14 lbs. corn-hearts —hay and corn fodder; property of H. C. Taylor, Orfordville, Wis.

Brown Coomassie 20322.—Sire, King (P. S. 238 J. H. B.); dam, Sassagua (F. S. 4194 J. H. B.)

Butter, 14 lbs. 8 oz. Milk, 163 lbs. 8 oz.

Test made from May 2 to 8, 1887; age, 6 yrs. 2 mos.; fed 18 lbs., daily, of mixed corn, oats and middlings; property of George E. Jones, Litchfield, Conn.

Brown Elsie 96595.—Sire, Combination 3d 17576; dam, Brown Flora 2d 96594.

Butter, 21 lbs. 12 oz.   Milk, 272 lbs. 8 oz.

Test made from June 30 to July 7, 1895; age, 7 yrs. 4 mos.; estimated weight, 900 lbs.; fed 28 lbs. oil meal, 56 lbs. corn meal, 21 lbs. bran and 21 lbs. oats—good mixed pasture; property of H. C. Taylor, Orfordville, Wis.

Brown Flora 2d 96594.—Sire, Volco 7890; dam, Brown Flora 74996.

Butter, 20 lbs. 2 oz.   Milk, 269 lbs. 10 oz.

Test made from Feb. 1 to 7, 1894; age, 8 yrs.; estimated weight, 800 lbs.; fed 10 lbs. ground oats, 7 lbs. corn meal, 10 lbs. bran, 4 lbs. oil meal, in three feeds, daily, and 4 qts. roots, daily; property of Richardson Bros., Davenport, Iowa.

Brown Lassie 92950.—Sire, Brown Bessie's Son 34550; dam, Commilla 79614.

Butter, 15 lbs. 7 oz.   Milk, 223 lbs. 5 oz.

Test made from Jan. 29 to Feb. 5, 1894; age, 2 yrs. 2 mos.; estimated weight, 775 lbs.; fed 4 qts. corn meal, 4 qts. ground oats and 2 qts. oil meal, daily, on cut hay—4 qts. roots daily; property of Richardson Bros., Davenport, Iowa.

Brown Princess 30941.—Sire, Cœur de Lion (P. S. 140 J. H. B.); dam, Playmate (F. S. 1786 J. H. B.)

Butter, 14 lbs. 9 oz., unsalted.   Milk, 12 to 14 qts. per day.

Test made from Sept. 1 to 7, 1885; age, 7 yrs. 11 mos.; test made on grass alone; property of M. H. Messchert, Douglassville, Pa.

Brunette Hammond 7284, imp.

Butter, 18 lbs. 8 oz.   Milk, 254 lbs.

Test made from Apr. 7 to 15, 1885; age, 9 yrs.; property of F. C. Havemeyer, West Chester, N. Y.

Brunette Landseer 109904.—Sire, Toltec 6th 14507; dam, Alice Landseer 2d 76815.

Butter, 16 lbs. 13¼ oz.   Milk, 202 lbs.

Test made from July 18 to 25, 1897; age, 3 yrs. 5 mos.; estimated weight, 750 lbs.; fed 28 lbs. bran, 28 lbs. ground oats, 28 lbs. corn meal, 14 lbs. linseed meal, 7 lbs. cotton-seed meal, 35 lbs. cotton-seed hulls and 168 lbs. green cut corn; property of S. H. Butler, Como, Miss.

Brunette Lass 1780, imp.

Butter, 15 lbs. 10 oz.

Age, 10 yrs.; property of W. J. Webster, Columbia, Tenn.

Brunette Le Gros 9755.—Sire, Tom (P. S. 77 J. H. B.); dam, Daisy (P. S. 92 J. H. B.)

Butter, 15 lbs. 15 oz.

Test made in Jan., 1883; age, between 7 and 8 yrs.; property of S. W. Robbins, Wethersfield, Conn.

Brunette Le Gros 3d 23326.—Sire, John Le Brocq 5424; dam, Brunette Le Gros 9755.

Butter, 14 lbs. 9 oz.

Test made in Dec., 1885; age, 3 yrs. 1 mo.; property of S. W. Robbins, Wethersfield, Conn.

Brunette of Scarsdale 13276.—Sire, Astronomer 4150; dam, Beautiful 5981.

Butter, 17 lbs. Milk, 240 lbs.

Test made from June 22 to 28, 1885; age, 4 yrs. 1 mo.; test made on grass alone; property of F. L. Gaston, Normal, Ill.

Brunette Star 27270.—Sire, Cicero 7657; dam, Brunette (P. S. 101 J. H. B.)

Butter, 17 lbs. 10 oz. Milk, 279 lbs. 10 oz.

Test made from June 27 to July 4, 1886; age, 4 yrs. 3 mos.; property of Cornelius Easthope, Niles, Ohio.

Brunhilde Regina 41742.—Sire, Holyoke 8147; dam, Ida Regina 5204.

Butter, 16 lbs. 13 oz. Milk, 282 lbs. 4 oz.

Test made from May 22 to 29, 1894; age, 8 yrs. 9 mos.; estimated weight, 900 lbs.; no grain fed—upland pasture; property of E. L. Van Deusen, Ashley Falls, Mass.

Bryant 4193.—Sire, Omaha 482; dam, Metah 1295.

Butter, 14 lbs. 8 oz., unsalted. Milk, 235 lbs.

Test made from July 10 to 16, 1881; age, 6 yrs. 4 mos.; property of George E. Bryant, Madison, Wis.

Buckeye Lass 10355.—Sire, Dick (F. S. 171 J. H. B.); dam, Fairy (F. S. 964 J. H. B.)

Butter, 14 lbs. 4 oz. Milk, 208 lbs. 15 oz.

Test made from Apr. 15 to 21, 1882; age, 5 yrs.; property of S. L. Hoover, Columbus, Ohio.

Buckmentor 62846.—Sire, Little Tormentor 17995; dam, Princess Potoka 53572.

Butter, 14 lbs. 10 oz. Milk, 195 lbs. 4 oz.

Test made from Nov. 30 to Dec. 7, 1896; age, 7 yrs. 3 mos.; estimated weight, 900 lbs.; fed 21 lbs. cotton-seed meal, 70 lbs. corn and cob meal and 75 lbs. bran—hay *ad lib.*; property of James L. Cooper, Nashville, Tenn.

Bunker's Kitty 47103.—Sire, Bunker 9025; dam, Kitty of Jefferson 2d 12313.

Butter, 14 lbs. 11½ oz. Milk, 187 lbs.

Test made from June 13 to 20, 1890; age, 3 yrs. 11 mos.; estimated weight, 800 lbs.; fed, daily, 10 qts. ground oats, 4 qts. linseed meal and 1 qt. corn meal; property of T. R. Proctor, Utica, N. Y.

Bunker's Pride 38571.—Sire, Bunker 9025; dam, Stella's Pride 16855.

Butter, 18 lbs. 3½ oz. Milk, 264 lbs.

Test made from June 5 to 12, 1890; age, 4 yrs. 5 mos.; actual weight, 750 lbs.; fed, daily, 6 qts. corn meal, 8 qts. ground oats and 4 qts. oil meal; property of T. R. Proctor, Utica, N. Y.

Bunker's Utica Belle 38567.—Sire, Bunker 9025; dam, Belle of Utica 11356.

Butter, 15 lbs. 15 oz. Milk, 236 lbs.

Test made from June 8 to 15, 1892; age, 7 yrs. 2 mos.; estimated weight, 800 lbs., fed 2 qts. corn meal, 2 qts. ground oats and 4 qts. shorts, daily—good pasture; property of James B. Johnson, Richfield Springs, N. Y.

Busy Bee 6336.—Sire, Top-Sawyer 1404; dam, Bisma 3d 1870.

Butter, 16 lbs. 4 oz. Milk, 277 lbs.

Test made from May 20 to 26, 1882; age, 4 yrs. 10 mos.; fed 6 qts. boiled cotton-seed, 4 qts. meal, 2 qts. shorts, daily—Bermuda grass and white clover pasture; property of William E. Oates, Vicksburg, Miss.

Busy Bee 2d 25166.—Sire, Carnival 5110; dam, Busy Bee 6336.

Butter, 15 lbs. 8 oz. Milk, 211 lbs. 8 oz.

Test made from Apr. 12 to 18, 1885; age, 3 yrs.; property of William E. Oates, Vicksburg, Miss.

Busy Melrose 54456.—Sire, Prince of Melrose 4819; dam, Busy Bee 6336.

Butter, 21 lbs. 1¼ oz. Milk, 289 lbs.

Test made from July 29 to Aug. 5, 1892; age, 5 yrs. 3 mos.; estimated weight, 950 lbs.; fed 28 lbs. oil meal, 28 lbs. bran, 49 lbs. corn meal and 56 lbs. cotton-seed hulls, estimated—poor pasture; property of Jo. C. Elstner, Shreveport, La.

Busy Melrose 54456.—Sire, Prince of Melrose 4819; dam, Busy Bee 6336.

Butter, 85 lbs. 2¾ oz. in 31 days. Milk, 1311 lbs.

Test made from July 10 to Aug. 10, 1892; age, 5 yrs. 3 mos.; estimated weight, 950 lbs.; fed 124 lbs. oil meal, 124 lbs. bran, 217 lbs. corn meal and 248 lbs. cotton-seed hulls, estimated; property of Jo. C. Elstner, Shreveport, La.

Busy Princess 48202.—Sire, Prince of Melrose 4819; dam, Busy Bee 2d 25166.

Butter, 17 lbs. 1¾ oz. Milk, 257 lbs. 2 oz.

Test made from June 10 to 17, 1890; age, 4 yrs. 2 mos.; estimated weight, 750 lbs., fed, daily, 5 lbs. bran, 6 lbs. oats, 6 lbs. corn meal and 2 lbs. cotton-seed meal; property of M. Lothrop, Marshall, Tex.

Buttercup 17285.—Sire, Sans-Peur (F. S. 201 J. H. B.); dam, Rose (F. S. 1450 J. H. B.)

Butter, 16 lbs. 5 oz. Milk, 295 lbs.

Test made from Jan. 5 to 11, 1885; age, 9 yrs. 9 mos.; property of William Crozier, Northport, N. Y.

Butterfly of Riverview 3d 85499.—Sire, King of St. Lambert 15175; dam, Butterfly of Riverview 21466.

Butter, 17 lbs. Milk, 265 lbs.

Test made from June 3 to 10, 1898; age, 7 yrs. 1 mo.; estimated weight, 950 lbs., fed 12 qts. wheat bran, 4 qts. ground oats and 2 qts. corn meal, daily—good pasture, property of Mrs. Jennie E. Hall, Columbus, Ohio.

Butterstamp Lass 19517.—Sire, Ramapo 4679; dam, Schonemunk Lass 9126.

Butter, 16 lbs. 11 oz. Milk, 19 qts. per day.

Test made from Aug. 25 to 31, 1886; age, 3 yrs. 11 mos.; estimated weight, 750 lbs., fed 3 qts. corn meal, 8 qts. oat meal and 8 qts. middlings, daily; property of D. F. Appleton, Ipswich, Mass.

Butterstamp Pedro 56383.—Sire, Young Pedro 2d 15011; dam, Rioter Butterstamp 46088.

Butter, 14 lbs. 8 oz. Milk, 232 lbs. 6 oz.

Test made from May 31 to June 7, 1892; age, 3 yrs. 3 mos.; estimated weight, 950 lbs.; fed 8 qts. daily of a mixture of equal parts corn meal, ground oats and wheat bran—good pasture; property of Walter W. Law, Whitson, N. Y.

Butterstar 7799.—Sire, Butterstamp (P. S. 101 J. H. B.); dam, Starlight (P. S. 136 J. H. B.)

Butter, 18 lbs. 4½ oz. Milk, 316 lbs. 8 oz.

Test made from July 13 to 19, 1882; age, 4 yrs. 5 mos.; fed 6 qts. coarse corn chop, 4 qts. bran and 1 qt. cotton-seed meal, daily; property of Campbell Brown, Spring Hill, Tenn.

Buttery 3502.—Sire, Mack 722; dam, Marietta 1813.

Butter, 14 lbs. 1 oz. Milk, 203 lbs.

Test made from Jan. 22 to 28, 1882; age, 7 yrs. 9 mos.; property of Campbell Brown, Spring Hill, Tenn.

Buttery 2d 27138.—Sire, Lord Harry 3445; dam, Buttery 3502.

Butter, 15 lbs. ⅝ oz. Milk, 146 lbs.

Test made from Apr. 10 to 17, 1887; age, 2 yrs. 10 mos.; estimated weight, 1000 lbs.; fed, daily, 4 lbs. cotton-seed meal, 4 lbs. wheat bran and 8 lbs. corn meal; property of C. J. Foster, Shreveport, La.

Cabbie Oak 90061.—Sire, Live Oak of Brushy 16155; dam, Birdie Berchette 66773.

Butter, 14 lbs. 12 oz. Milk, 197 lbs.

Test made from Jan. 8 to 15, 1895; age, 3 yrs. 3 mos.; estimated weight, 750 lbs.; fed 32 qts. corn meal, 24 qts. wheat chop, 32 qts. cotton-seed, 8 qts. bran and 24 qts. mixed cotton-seed and bran—cured sorghum *ad lib.*; property of S. T. Howard, Quanah, Tex.

Cabinet 22662.—Sire, Pilot (P. S. 183 J. H. B.); dam on I. of J.

Butter, 15 lbs. 10 oz. Milk, 149 lbs. 13 oz.

Test made from Feb. 14 to 20, 1887; age, 6 yrs. 4 mos.; fed 8 qts. corn meal, 6 qts. oats, 1 qt. oil meal, 2 qts. bran, daily; property of Archer N. Martin, Summit, N. J.

Calcina 80702.—Sire, Herotas 26500; dam, Calcium 79733.

Butter, 15 lbs. 3 oz. Milk, 249 lbs. 1 oz.

Test made from Dec. 25, 1894, to Jan. 2, 1895; age, 2 yrs. 8 mos.; estimated weight, 800 lbs.; fed 14 qts. corn meal, ground oats, oil meal and bran, and 2 qts. roots, daily —hay; property of Richardson Bros., Davenport, Iowa.

Calcina 80702.—Sire, Herotas 26500; dam, Calcium 79733.

Butter, 18 lbs. 8 oz. Milk, 273 lbs. 4¾ oz.

Test made from Mar. 7 to 14, 1897; age, 4 yrs. 11 mos.; estimated weight, 1000 lbs.; fed 49 lbs. ground oats, 28 lbs. bran, 14 lbs. ground rye, 7 lbs. oil meal, 21 lbs. corn meal and 42 lbs. corn ensilage—corn fodder *ad lib.*; property of H. C. Taylor, Orfordville, Wis.

Calcium 79733.—Sire, Upright 6147; dam, Commilla 79614.

Butter, 16 lbs. 15 oz. Milk, 212 lbs. 5 oz.

Test made from May 1 to 8, 1891; age, 4 yrs. 4 mos.; estimated weight, 850 lbs.; fed 12 qts. corn meal, oatmeal and pea meal, daily, in three feeds—hay *ad lib.*; property of Richardson Bros., Davenport, Iowa.

Calculus 75185.—Sire, Ida's Rioter of St. L. 13656; dam, Sweet Blossom Pogis 36995.

Butter, 14 lbs. 2 oz. Milk, 198 lbs. 4 oz.

Test made from Sept. 20 to 27, 1894; age, 4 yrs.; estimated weight, 800 lbs.; fed 6 qts. corn meal, 5 qts. ground oats, 5 qts. wheat middlings and 3 qts. oil meal, daily— clover pasture at night, green oats and peas in stable during day; property of C. A. Sweet, East Aurora, N. Y.

Calculus 75185.—Sire, Ida's Rioter of St. L. 13656; dam, Sweet Blossom Pogis 36995.

Butter, 23 lbs. 11½ oz.   Milk, 347 lbs. 12 oz.

Test made from Sept. 7 to 14, 1895; age, 4 yrs. 11 mos.; actual weight, 960 lbs.; fed 12 qts. ground corn and oats, 12 qts. wheat bran and 3 qts. oil meal, daily—poor pasture at night, green corn in stable during day; property of C. A. Sweet, East Aurora, N. Y.

Calisida 76889.—Sire, Ida's Stoke Pogis 13658; dam, Callie Nan 2d 64207.

Butter, 14 lbs. 4 oz.   Milk, 176 lbs. 13 oz.

Test made from May 19 to 26, 1896; age, 4 yrs. 5 mos.; estimated weight, 750 lbs.; fed 14 qts., daily, of ground oats and corn, in equal parts—blue-grass pasture; property of M. M. Gardner, Nashville, Tenn.

Calista of Newark 13296.—Sire, Samson Jr. 2723; dam, Biddy of Bovina 6655.

Butter, 15 lbs. 9 oz.   Milk, 195 lbs. 12 oz.

Test made from June 24 to 30, 1885; age, 5 yrs. 2 mos.; property of Andrew Baker, West Dryden, N. Y.

Calla Europa 77810.—Sire, Othick Pedro 20511; dam, Cream Calla 40233.

Butter, 15 lbs. 4 oz.   Milk, 229 lbs.

Test made from Aug. 26 to Sept. 2, 1892; age, 3 yrs.; estimated weight, 700 lbs.; fed green corn fodder—ordinary pasture; property of Mrs. J. H. Martin, Granville, N. Y.

Calla Europa 77810.—Sire, Othick Pedro 20511; dam, Cream Calla 40233.

Butter, 20 lbs. 6 oz.   Milk, 346 lbs. 14 oz.

Test made from Oct. 15 to 22, 1894; age, 5 yrs. 2 mos.; estimated weight, 750 lbs.; fed 5 lbs. corn meal, 5 lbs. bran and 2 lbs. oil meal, daily—clover after feed; property of Mrs. J. H. Martin, Granville, N. Y.

Callie Nan 7959.—Sire, Callis 1696; dam, Oonan 1485.

Butter, 16 lbs. 2 oz.   Milk, 252 lbs. 8 oz.

Test made from July 27 to Aug. 2, 1882; age, 4 yrs. 4 mos.; fed 6 lbs. corn meal, 2 lbs. wheat bran, 3 lbs. cotton-seed meal, daily; property of Campbell Brown, Spring Hill, Tenn.

Callie Nan 2d 64207.—Sire, Sitanda 20221; dam, Callie Nan 7959.

Butter, 23 lbs. 1 oz.   Milk, 302 lbs.

Test made from Sept. 25 to Oct. 2, 1893; age, 4 yrs. 6 mos.; estimated weight, 950 lbs.; fed 196 lbs. ground oats—excellent blue-grass pasture; property of Est. of Campbell Brown, Spring Hill, Tenn.

Callinette Pogis 99700.—Sire, Noonday 12961; dam, Callinette 2d 38908.

Butter, 17 lbs. 8 4-5 oz.   Milk, 209 lbs. 15 1-5 oz.

Test made from July 12 to 19, 1895; age, 1 yr. 11 mos.; estimated weight, 650 lbs.; fed 35 lbs. corn and oats, 49 lbs. bran and 28 lbs. cotton-seed meal—clover pasture; property of Hecla Jersey Cattle Co., Earlington, Ky.

Calma of Briarcliff 52599.—Sire, Rioter's Combination 10363; dam, Cetewayo's Lily 18950.

Butter, 14 lbs. 7½ oz.   Milk, 200 lbs. 1 oz.

Test made from Nov. 13 to 20, 1892; age, 4 yrs. 2 mos.; estimated weight, 1050 lbs.; fed 20 qts. daily of a mixture of equal parts of corn meal, ground oats and wheat bran and 1 pt. oil meal—hay *ad lib.*; property of Walter W. Law, Whitson, N. Y.

Calypris 5943.—Sire, Marius 760; dam, René Calypso 3205.

Butter, 15 lbs. 4½ oz. Milk, 234 lbs.

Test made from June 2 to 9, 1884; age, 7 yrs. 1 mo.; property of Beech Grove Farm, Beech Grove, Ind.

Camelia 2d 11188.—Sire, Grey King (P. S. 169 J. H. B.); dam, Camelia (F. S. 1687 J. H. B.)

Butter, 20 lbs. 3 oz. Milk, 296 lbs.

Test made from May 27 to June 3, 1885; age, 6 yrs. 3 mos.; fed 6 qts. ground oats, corn and pea meal, in equal parts, mixed with cut hay and a handful of oil-cake, daily—fair pasture of timothy and clover; property of Adams Earl, La Fayette, Ind.

Campania 88475.—Sire, Diploma 16219; dam, Caressa 88474.

Butter, 28.127 lbs. in 21 days. Milk, 556.7 lbs.

Test made from Sept. 30 to Oct. 20, 1893, at World's Columbian Exposition; age, 3 yrs.; actual weight, 764 lbs.; fed 237.5 lbs. hay, 6 lbs. ensilage, 55 lbs. oil meal, 93 lbs. corn meal, 134 lbs. bran, 61 lbs. oats, 41.5 lbs. cotton-seed, 42 lbs. middlings and 17 lbs. corn-hearts; property of Richardson Bros., Davenport, Iowa.

Canata 10523.—Sire, Prince of Warren 1512; dam, Tamy 2d 7125.

Butter, 27 lbs. Milk, 248 lbs. 1 oz.

Test made from Apr. 22 to 28, 1886; age, 8 yrs.; fed 6 qts. corn meal, 4 qts. ground oats, 3 qts. shorts, daily—no pasture; property of Joseph H. Walker, Worcester, Mass.

Candelabrum 75187.—Sire, Stoke Pogis 5th 5987; dam, Ida Twinkle 36994.

Butter, 14 lbs. 14½ oz. Milk, 193 lbs.

Test made from June 5 to 12, 1893; age, 2 yrs. 7 mos.; actual weight, 975 lbs.; fed 6 qts. corn meal, 6 qts. ground oats and 2 qts. ground oil-cake, daily—good meadow pasture; property of Miller & Sibley, Franklin, Pa.

Candelabrum 75187.—Sire, Stoke Pogis 5th 5987; dam, Ida Twinkle 36994.

Butter, 17 lbs. 13 oz. Milk, 275 lbs.

Test made from Mar. 2 to 9, 1894; age, 3 yrs. 4 mos.; actual weight, 995 lbs.; fed 112 qts. ground oats, 56 qts. corn meal, 28 qts. wheat bran and 28 qts. ground oil-cake—hay; property of Miller & Sibley, Franklin, Pa.

Cannie M. 10698.—Sire, Cardinal of Rosland 3335; dam, Peggy Stewart 7407.

Butter, 15 lbs. 1 oz. Milk, 226 lbs. 8 oz.

Test made from May 3 to 10, 1883; age, 4 yrs.; property of Henry B. Kelley, Vinton, Iowa.

Canessa 65243.—Sire, Lord St. Helier 15071; dam, Lizzie T. 65134.

Butter, 15 lbs. Milk, 183 lbs. 8 oz.

Test made from Feb. 23 to Mar. 2, 1890; age, 2 yrs. 10 mos.; estimated weight, 900 lbs.; well fed, but no record kept; property of Campbell Brown, Spring Hill, Tenn.

Cara Mia 64224.—Sire, Sitanda 20221; dam, Carida 37322.

Butter, 15 lbs. 9¼ oz. Milk, 216 lbs.

Test made from Apr. 9 to 16, 1893; age, 3 yrs. 4 mos.; estimated weight, 1000 lbs.; fed 14 lbs. corn-hearts and wheat bran, mixed, twice daily—blue-grass pasture; property of Campbell Brown, Spring Hill, Tenn.

Cara Mia 2d 93827.—Sire, Tormentor 5th 21962; dam, Cara Mia 64224.

  Butter, 15 lbs. 5½ oz. Milk, 178 lbs.

  Test made from Oct. 12 to 19, 1896; age, 2 yrs. 8 mos.; estimated weight, 775 lbs., fed 84 lbs. coarse corn meal and 56 lbs. ground oats—pastured on pea-vines during daytime and blue-grass at night; property of George Campbell Brown, Spring Hill, Tenn.

Caressa 88474.—Sire, Upright 6147; dam, Mercurina 64920.

  Butter, 17 lbs. 6 oz. Milk, 229 lbs. 1 oz.

  Test made from Dec. 1 to 8, 1892; age, 4 yrs. 1 mo.; estimated weight, 800 lbs.; fed 15 lbs. corn meal, ground oats and bran, daily—clover hay *ad lib.*; property of Richardson Bros., Davenport, Iowa.

Careta Pogis 42784.—Sire, Rioter's Combination 10363; dam, Careta 19092.

  Butter, 15 lbs. 1¼ oz. Milk, 295 lbs. 2 oz.

  Test made from July 8 to 15, 1892; age, 5 yrs. 4 mos.; actual weight, 764 lbs.; fed 42 qts. oats, 70 qts. corn and 84 qts. bran—some green cut oats and peas; property of Mrs. E. F. Hawley, Pittsford, N. Y.

Carida 37322.—Sire, Ida's Stoke Pogis 13658; dam, Carrie Lena 3d 20077.

  Butter, 15 lbs. 6 oz. Milk, 190 lbs. 8 oz.

  Test made from Jan. 8 to 15, 1890; age, 3 yrs. 11 mos.; estimated weight, 925 lbs., well fed, but no record kept; property of Campbell Brown, Spring Hill, Tenn.

Carlo's Daisy 16702.—Sire, Carlo 5559; dam, Fair Daisy (F. S. 2591 J. H. B.)

  Butter, 14 lbs. 9½ oz. Milk, 153 lbs. 14 oz.

  Test made from Apr. 6 to 13, 1888; age, 7 yrs.; estimated weight, 700 lbs.; fed, daily, 8 lbs. corn meal, 4 lbs. bran, 2 lbs. middlings and 1 lb. oil meal; property of James Stillman, Sing Sing, N. Y.

Carlo's Lass 94405.—Sire, Mistletoe's Carlo 12166; dam, Hulabaloo's Lass 34680.

  Butter, 16 lbs. 15 oz. Milk, 265 lbs.

  Test made from July 15 to 22, 1898; age, 9 yrs. 6 mos.; estimated weight, 900 lbs.; fed about 20 qts. daily of a mixture of ground oats, middlings, wheat bran, corn meal and some pea meal—clover hay and ensilage, good meadow and blue-grass pasture; property of T. S. Cooper, Coopersburg, Pa.

Carlo's Last Polly 94628.—Sire, Mistletoe's Carlo 12166; dam, Polly of Deerfoot 15328.

  Butter, 17 lbs. 6 oz. Milk, 273 lbs.

  Test made from May 15 to 22, 1898; age, 6 yrs. 10 mos.; estimated weight, 900 lbs., fed about 22 qts. daily of a mixture of ground oats, middlings, wheat bran, corn meal and some pea meal—clover hay and ensilage, clover and meadow pasture; property of T. S. Cooper, Coopersburg, Pa.

Carlo's Polly Rex 94627.—Sire, Mistletoe's Carlo 12166; dam, Polly J. Rex 33278.

  Butter, 16 lbs. 12 oz. Milk, 265 lbs.

  Test made from Apr. 22 to 29, 1897; age, 6 yrs. 3 mos.; estimated weight, 950 lbs., fed about 22 qts. daily of a mixture of ground oats, middlings, wheat bran, corn meal and some pea meal—clover hay and ensilage; property of T. S. Cooper, Coopersburg. Pa

Carlo's Rosebud 18223.—Sire, Carlo 5559; dam, Khedive's Rosebud 18173.

Butter, 15 lbs. 8 oz. Milk, 184 lbs. 8 oz.

Test made from Sept. 30 to Oct. 7, 1887; age, 7 yrs.; estimated weight, 850 lbs.; fed, daily, 8 lbs. corn meal and 4 lbs. bran; property of James Stillman, Sing Sing, N. Y.

Carlotta Tom 76228.—Sire, Champion of Silver Bluff 7378; dam, Vain Princess 46569.

Butter, 14 lbs. 5 oz. Milk, 229 lbs.

Test made from Jan. 17 to 24, 1897; age, 5 yrs. 2 mos.; estimated weight, 925 lbs., fed 4 lbs. oats, 3½ lbs. bran, 7 lbs. corn meal and 1¼ lbs. oil meal, daily—corn fodder *ad lib.*, short rye pasture; property of W. E. Johnson, Millican, Tex.

Carola's Crown Princess 90321.—Sire, Crown Prince of St. L. 20070; dam, Carola of Bois d'Arc 40198.

Butter, 14 lbs. 5 oz. Milk, 228 lbs. 9 oz.

Test made from Mar. 9 to 16, 1896; age, 3 yrs. 7 mos.; estimated weight, 900 lbs., fed 56 lbs. corn meal, 31 lbs. bran, 7 lbs. cotton-seed meal and 315 lbs. ensilage, property of Austin & Probert, Sandstone, Mich.

Carola's Crown Princess 90321.—Sire, Crown Prince of St. L. 20070; dam, Carola of Bois d'Arc 40198.

Butter, 14 lbs. 14 oz. Milk, 246 lbs. 14 oz.

Test made from Jan. 5 to 12, 1898; age, 5 yrs. 5 mos.; estimated weight, 900 lbs.; fed 56 lbs. bran, 42 lbs. corn meal, 18 lbs. gluten meal, 14 lbs. oil meal and 40 lbs. ensilage and hay; property of Austin & Probert, Sandstone, Mich.

Caroline 12019.—Sire, Remarkable (F. S. 229 J. H. B.); dam, Marie Spelterini (F. S. 1412 J. H. B.)

Butter, 14 lbs. 8 oz. Milk, 218 lbs. 12 oz.

Test made from Feb. 14 to 21, 1882; age, 6 yrs. 11 mos.; property of J. M. Richmond, Buffalo, N. Y.

Caroline Pauncefote 62919.—Sire, Sir Julian Pauncefote 22149; dam, Star Caroline 55610.

Butter, 16 lbs. 8 oz. Milk, 149 lbs. 14 oz.

Test made from Oct. 20 to 27, 1894; age, 5 yrs. 3 mos.; estimated weight, 1150 lbs.; fed 70 lbs. mixed grain (wheat middlings, shelled corn, whole oats, oil meal and ground rye), 84 lbs. hay and 1 bush. carrots; property of Mrs. H. E. Tremain, Lake George, N. Y.

Caro of St. Lambert 90875.—Sire, Meridale Pauline Pogis 25133; dam, Maria of St. Lambert 64908.

Butter, 17 lbs. 6 oz. Milk, 174 lbs. 8 oz.

Test made from June 1 to 8, 1895; age, 3 yrs. 2 mos.; estimated weight, 800 lbs., fed 49 lbs. wheat bran, 14 lbs. corn chop, 14 lbs. cotton-seed meal and 7 lbs. old process linseed meal—fair clover pasture; property of B. F. Fackenthal, Jr., Riegelsville, Pa.

Carria H. 14454.—Sire, Discard H. 5763; dam, Jenny Gray 3511.

Butter, 16 lbs. 10 oz. Milk, 358 lbs. 10 oz.

Test made from July 7 to 13, 1885; age, 4 yrs. 3 mos.; property of William H. Hayden, Albany, Vt.

Carrie 3894.—Sire, Sir Charles 131; dam, Mary Lowndes 273.

>   Butter, 16 lbs. 8 oz.

Property of John V. N. Willis, Marlborough, N. J.

Carrie Franklin 25635.—Sire, Dr. Ben Franklin 2d 7265; dam, Dioneince 19607.

>   Butter, 14 lbs. 1 oz.   Milk, 218 lbs.

Test made from June 10 to 17, 1895; age, 11 yrs. 9 mos.; estimated weight, 950 lbs.; fed 8 qts. gluten and bran in equal parts and 1 qt. oil meal, daily—hay *ad lib.*, short blue-grass pasture; property of W. J. Hussey, Mount Pleasant, Ohio.

Carrie Gold 56348.—Sire, Gold of St. Lambert 16744; dam, Carrie Ryan 47413.

>   Butter, 16 lbs. 8 oz.   Milk, 237 lbs. 8 oz.

Test made from Oct. 28 to Nov. 4, 1894; age, 5 yrs. 6 mos.; estimated weight, 1050 lbs.; fed 35 lbs. oats, 24 lbs. bran, 28 lbs. corn meal and 28 lbs. wheat—mixed pasture; property of Henry A. Egerton, Footville, Wis.

Carrie Lena 2d 19074.—Sire, Colonsay 3591; dam, Carrie Lena 3348.

>   Butter, 14 lbs. 5⅛ oz.   Milk, 219 lbs.

Test made from Aug. 10 to 17, 1887; age, 5 yrs. 5 mos.; estimated weight, 950 lbs.; fed, daily, 24 lbs. of a mixture consisting of corn-hearts, bran and ground oats; property of Campbell Brown, Spring Hill, Tenn.

Carrie Lena 3d 20077.—Sire, Lenox Cash Boy 6804; dam, Carrie Lena 3348.

>   Butter, 16 lbs. 5 oz.   Milk, 195 lbs.

Test made from Apr. 26 to May 2, 1885; age, 2 yrs. 1 mo.; property of Campbell Brown, Spring Hill, Tenn.

Carrie Lena 3d 20077.—Sire, Lenox Cash Boy 6804; dam, Carrie Lena 3348.

>   Butter, 16 lbs. 4 oz. (official).   Milk, 197 lbs.

Test made from May 21 to 27, 1885; age, 2 yrs. 2 mos.; actual weight, 660 lbs.; fed, daily, 2 lbs. ground oats, 2 lbs. wheat bran and 8 lbs. corn-hearts; property of Campbell Brown, Spring Hill, Tenn.

Carrie Lena 3d 20077.—Sire, Lenox Cash Boy 6804; dam, Carrie Lena 3348.

>   Butter, 18 lbs. 11¾ oz.   Milk, 245 lbs.

Test made from May 11 to 18, 1887; age, 4 yrs. 2 mos.; estimated weight, 900 lbs.; well fed, but no record kept; property of Campbell Brown, Spring Hill, Tenn.

Carrie Pogis 22568.—Sire, Orloff 3143; dam, Cowslip of St. Lambert 8349.

>   Butter, 15 lbs. 9 oz.   Milk, 219 lbs.

Test made from Jan. 10 to 17, 1885; age, 4 yrs. 5 mos.; property of V. E. Fuller, Hamilton, Ont., Can.

Carrie Rex 10271.—Sire, Rex 1330; dam, Carrie Clark 6584.

>   Butter, 16 lbs. 7½ oz.   Milk, 245 lbs. 12 oz.

Test made from Feb. 21 to 27, 1886; age, 6 yrs. 3 mos.; property of Moulton Bros., West Randolph, Vt.

Carrie's Beauty 14601.—Sire, Young Duke 138; dam, Carrie 3894.

Butter, 14 lbs. 7½ oz.

Test made from Feb. 23 to Mar. 1, 1885; age, 8 yrs.; property of Thomas S. Yocom, Richmond, N. Y.

Carrie's Daffodil 15232.—Sire, Monmouth Laddie 6996; dam, Carrie's Beauty 14601.

Butter, 14 lbs. 7¾ oz.

Test made from Feb. 28 to Mar. 6, 1886; age, 6 yrs.; property of Thomas S. Yocom, Richmond, N. Y.

Carrie's Wonder 15233.—Sire, Duke of Thornebrook 3832; dam, Carrie's Beauty 14601.

Butter, 14 lbs. 2½ oz.

Test made from June 4 to 10, 1886; age, 5 yrs.; property of Thomas S. Yocom, Richmond, N. Y.

Cassia 2d 21370.—Sire, Yankee 1003; dam, Cassia 3615.

Butter, 20 lbs. 10¼ oz.   Milk, 283 lbs. 8 oz.

Test made from June 22 to 29, 1885; age, 8 yrs. 11 mos.; fed about 23 lbs. ground oats, pea meal, corn meal and cotton-seed meal, daily—clover and timothy pasture, property of P. J. Cogswell, Rochester, N. Y.

Cassie Marjoram 120666.—Sire, One Hundred Per Cent. 16590; dam, Leclair's Marjoram 86355.

Butter, 16 lbs. 5 oz.   Milk, 250 lbs. 12 oz.

Test made from Nov. 9 to 16, 1895; age, 3 yrs.; estimated weight, 900 lbs.; fed 105 lbs. ground oats and peas, 21 lbs. oil-cake and 28 lbs. bran—hay *ad lib.*; property of Wm. Rolph, Markham, Ont., Can.

Cassie of Glen Rouge 112456.—Sire, One Hundred Per Cent. 16590; dam, Nellie of St. Lambert 69458.

Butter, 15 lbs. 9 oz.   Milk, 244 lbs. 12 oz.

Test made from Oct. 1 to 8, 1896; age, 2 yrs. 2 mos.; estimated weight, 650 lbs.; fed 105 lbs. ground peas and oats and 14 lbs. oil-cake—hay *ad lib.*, clover and timothy pasture; property of Wm. Rolph, Markham, Ont., Can.

Casta's Belle 56148.—Sire, Casta's Champion 11794; dam, Casta Diva 8154.

Butter, 14 lbs. 15 oz.   Milk, 244 lbs.

Test made from May 26 to June 2, 1891; age, 5 yrs. 3 mos.; estimated weight, 850 lbs.; fed 8 lbs. ground corn and oats, in equal parts, and 4 lbs. oil meal, daily—bluegrass pasture; property of A. S. Bell, London, Ohio.

Catchfly 25405.—Sire, Rocco 4517; dam, Phœbe 4th 2271.

Butter, 18 lbs. 4 oz.   Milk, 207 lbs.

Test made from Aug. 1 to 8, 1889; age, 8 yrs. 2 mos.; estimated weight, 750 lbs.; fed, daily, 6 qts. corn meal, 2 qts. oil meal and 6 qts. bran; property of W. C. Norton, Agent, Aldenville, Pa.

Catchfly 3d 35499.—Sire, Master Richmond 7429; dam, Catchfly 25405.

Butter, 15 lbs. 7 oz.   Milk, 280 lbs. 8 oz.

Test made from Aug. 19 to 26, 1890; age, 5 yrs. 3 mos.; estimated weight, 800 lbs.; fed, daily, 20 lbs. of a mixture consisting of two-sevenths bran, one-seventh old process oil meal and four-sevenths corn meal; property of W. C. Norton, Agent, Aldenville, Pa.

Catchfly 4th 47682.—Sire, Ramapo Chieftain 15324; dam, Catchfly 25405.

Butter, 17 lbs. 4 oz.   Milk, 247 lbs.

Test made from July 18 to 25, 1889; age, 3 yrs. 3 mos.; estimated weight, 750 lbs.; fed, daily, 4 qts. corn meal, 2 qts. oil meal and 4 qts. wheat bran; property of W. C. Norton, Agent, Aldenville, Pa.

Catchfly 5th 55483.—Sire, Ramapo Chieftain 15324; dam, Catchfly 25405.

Butter, 16 lbs. 2 oz.   Milk, 274 lbs. 8 oz.

Test made from June 17 to 24, 1890; age, 3 yrs. 2 mos.; estimated weight, 800 lbs.; fed, daily, 16 lbs. of a mixture consisting of two-sevenths bran, one-seventh oil meal and four-sevenths corn meal; property of W. C. Norton, Agent, Aldenville, Pa.

Catono's Kate 59976.—Sire, Gold Basis 5th 15725; dam, Gladdis Roy 18566.

Butter, 16 lbs. ½ oz.   Milk, 266 lbs.

Test made from July 22 to 29, 1893; age, 4 yrs. 2 mos.; estimated weight, 800 lbs.; fed 84 qts. meal, 7 qts. oil meal and 7 qts. cotton-seed meal—poor pasture; property of H. A. Huntington, Higganum, Conn.

Catono's Rosebud 23674.—Sire, Catono 3761; dam, Princess Rose 6249.

Butter, 14 lbs. 5 oz.   Milk, 132 lbs. 3 oz.

Test made from Mar. 17 to 23, 1886; age, 2 yrs. 3 mos.; property of the Columbia Jersey Cattle Co., Columbia, Tenn.

Caucus Maid 69525.—Sire, Bluster 22798; dam, Caucus 33462.

Butter, 15 lbs. 8 oz.   Milk, 229 lbs. 8 oz.

Test made from Sept. 27 to Oct. 4, 1894; age, 4 yrs. 3 mos.; estimated weight, 1050 lbs.; fed 28 lbs. oats, 28 lbs. bran, 35 lbs. corn meal and 21 lbs. oil meal—mixed pasture; property of H. C. Taylor, Orfordville, Wis.

Celebrity 52011.—Sire, Combination 3d 17576; dam, Goodbye 27366.

Butter, 14 lbs. 3 oz.   Milk, 212 lbs. 2 oz.

Test made from Jan. 15 to 22, 1889; age, 2 yrs. 1 mo.; estimated weight, 800 lbs., fed, daily, 15 qts. ground corn, oats, peas and oil meal, and 1 pk. roots; property of Richardson Bros., Davenport, Iowa.

Celestia 2d 29482.—Sire, Florinde's Duke 4368; dam, Celestia 1898.

Butter, 16 lbs. 13 oz.   Milk, 177 lbs. 8 oz.

Test made from Aug. 10 to 17, 1887; age, 3 yrs. 5 mos.; estimated weight, 850 lbs.; fed, daily, 7 qts. corn meal and 6 qts. bran; property of M. Erskine Miller, Staunton, Va.

Celia Belle 5865.—Sire, Duke of Portage 1270; dam, Belle of Chester 4442.

Butter, 14 lbs. 3 oz.   Milk, 246 lbs. 4 oz.

Test made from June 6 to 12, 1883; age, 6 yrs. 8 mos.; property of Campbell Brown, Spring Hill, Tenn.

Celia of Whitehall 68004.—Sire, King Peto 10541; dam, Lorena of Briarcliff 42817.

Butter, 17 lbs. 15 oz.   Milk, 196 lbs. 15 oz.

Test made from Mar. 31 to Apr. 7, 1893; age, 4 yrs.; estimated weight, 900 lbs.; fed 15 qts., daily, of a mixture of equal parts of corn meal, ground oats, wheat bran and wheat middlings, with a pint of oil meal, also 15 lbs. ensilage and hay; property of Walter W. Law, Whitson, N. Y.

Cenie Wallace 2d 6557.—Sire, Ralph 957; dam, Cenie Wallace 2663.

Butter, 15 lbs. 4½ oz. Milk, 239 lbs.

Test made from May 16 to 22, 1882; age, 4 yrs. 7 mos.; fed on bran, corn meal and cotton-seed—good pasture; property of W. B. Montgomery, Starkville, Miss.

Cerita of Meadow Brook 5056.—Sire, Troubadour 481; dam, Cyrene 4th 480.

Butter, 17 lbs. 8 oz. Milk, 119 qts.

Test made from Apr. 22 to 29, 1881; age, 5 yrs. 2 mos.; property of Newton T. Beal, Rogersville, Tenn.

Cetewayo's Daisy 18230.—Sire, Cetewayo (P. S. 224 J. H. B.); dam, Daisy of St. John's 18170.

Butter, 16 lbs. 4 oz. Milk, 188 lbs. 8 oz.

Test made from July 7 to 14, 1889; age, 8 yrs. 5 mos.; actual weight, 720 lbs.; fed, daily, 12 lbs. wheat middlings; property of Mrs. Thomas Allen, Pittsfield, Mass.

Cetewayo's Dorcas 20287.—Sire, Cetewayo (P. S. 224 J. H. B.); dam, Dorcas (F. S. 1851 J. H. B.)

Butter, 16 lbs. 2¼ oz. Milk, 223 lbs. 4 oz.

Test made from Aug. 3 to 9, 1885; age, 4 yrs. 3 mos.; property of Frederick Loeser, Somerville, N. J.

Cetewayo's Lily 18950.—Sire, Cetewayo (P. S. 224 J. H. B.); dam, Lily (P. S. 166 J. H. B.)

Butter, 17 lbs. (official). Milk, 261 lbs. 4 oz.

Test made from June 30 to July 7, 1886; age, 5 yrs. 5 mos.; weight, 950 lbs.; fed 45 lbs. corn meal, 21 lbs. crushed oats and 19½ lbs. pea meal; property of James Stillman, Briarcliff Farm, Sing Sing, N. Y.

Cetewayo's Silver Bell 18952.—Sire, Cetewayo (P. S. 224 J. H. B.); dam, Silver Bell (F. S. 1807 J. H. B.)

Butter, 17 lbs. 2½ oz. Milk, 214 lbs. 15 oz.

Test made from June 6 to 12, 1885; age, 4 yrs. 4½ mos.; property of D. A. Givens, Cynthiana, Ky.

Chamomilla 7552.—Sire, Baronet 2240; dam, Linda 2d 1927.

Butter, 16 lbs. 10 oz., unsalted. Milk, 223 lbs. 8 oz.

Test made from May 20 to 26, 1883; age, 4 yrs. 11 mos.; fed 12 lbs. corn meal and 8 lbs. cotton-seed meal, daily—orchard grass and white clover pasture; property of J. T. & W. S. Shields, Bean's Station, Tenn.

Championa Pogis M. 75926.—Sire, Mhoon Stoke Pogis 19193; dam, Champion Regina 21227.

Butter, 15 lbs. 15 oz. Milk, 255 lbs. 5½ oz.

Test made from May 20 to 27, 1895; age, 5 yrs. 7 mos.; actual weight, 760 lbs.; fed 21 qts. cotton-seed meal, 84 qts. bran, 112 qts. corn and oats ground together, two parts corn to one part oats—short Bermuda pasture; property of W. S. May, Ruston, La.

Champion Azuline 20542.—Sire, Champion of America 1567; dam, Azuline 2d 3888.

Butter, 18 lbs. ½ oz. Milk, 168 lbs. 5½ oz.

Test made from June 1 to 7, 1885; age, 3 yrs. 5 mos.; property of W. B. Montgomery, Starkville, Miss.

Champion Flower 20887.—Sire, Champion of America 1567; dam, April Flower 4421.

Butter, 14 lbs. 6 oz. Milk, 226 lbs. 8 oz.

Test made from July 5 to 11, 1886; age, 4 yrs. 8 mos.; fed a mixture of boiled cotton-seed, corn meal and bran; property of W. B. Montgomery, Starkville, Miss.

Champion's Chloe 12255.—Sire, Champion of America 1567; dam, Chloe 129.

Butter, 15 lbs. 5½ oz. Milk, 247 lbs.

Test made from Mar. 24 to 30, 1884; age, 6 yrs. 4 mos.; property of A. M. Turner, Northfield, Conn.

Chancery 37987.—Sire, Niburn 7658; dam, Chaffinch 16524.

Butter, 18 lbs. 10 oz. Milk, 140 lbs. 8 oz.

Test made from Nov. 17 to 24, 1889; age, 5 yrs. 2 mos.; actual weight, 930 lbs.; fed, daily, 12 lbs. wheat middlings; property of Mrs. Thomas Allen, Pittsfield, Mass.

Chansonnette 5695.—Sire, Westchester 1266; dam, Clochette d'Or 5696.

Butter, 16 lbs. 4 oz. Milk, 16 qts. per day.

Test made from Mar. 4 to 10, 1887; age, 10 yrs. 5 mos.; estimated weight, 1000 lbs.; fed 6 qts. corn meal, 2 qts. pea meal, 6 qts. oat meal, 6 qts. bran, daily; property of D. F. Appleton, Ipswich, Mass.

Chansonnette 2d 29672.—Sire, Young Pedro 9033; dam, Chansonnette 5695.

Butter, 16 lbs. 9 oz. Milk, 237 lbs. 14 oz.

Test made from Sept. 16 to 23, 1888; age, 4 yrs. 2 mos.; estimated weight, 900 lbs., fed, daily, 10 lbs. corn meal, 14 lbs. ground oats, 16 lbs. shorts and 2 lbs. oil meal; property of D. F. Appleton, Ipswich, Mass.

Charity of Argyle 35036.—Sire, Duke Thorne 7020; dam, Duchess of Argyle 4th 7571.

Butter, 14 lbs. 10 oz. Milk, 189 lbs. 10 oz.

Test made from Apr. 10 to 17, 1893; age, 7 yrs. 9 mos.; estimated weight, 850 lbs., fed 2 lbs. cotton-seed meal, 3 lbs. Chicago cream gluten, 3 lbs. spring wheat bran, 3 lbs. corn meal, mixed with 35 lbs. corn ensilage, daily, also 10 lbs. clover hay, property of E. S. Henry, Rockville, Conn.

Charity Victor Pogis 134217.—Sire, Melia Ann's Victor Pogis 20697; dam, Belle's Charity 82217.

Butter, 14 lbs. 6 oz. Milk, 215 lbs. 8 oz.

Test made from July 11 to 18, 1898; age, 2 yrs. 3 mos.; estimated weight, 500 lbs., fed 3 lbs. corn meal and 3 lbs. wheat bran, daily—clover pasture; property of J. H. Martin, Granville, N. Y.

Charm 3d's St. Lambert 96872.—Sire, Melia Ann's Stoke Pogis 22042; dam, Charm 3d 79876.

Butter, 14 lbs. 8 oz. Milk, 240 lbs. 2 oz.

Test made from May 15 to 22, 1898; age, 4 yrs.; estimated weight, 900 lbs.; fed 10 lbs. per day of a mixture of gluten feed, bran and Victor corn and oat feed—good blue-grass pasture; property of W. J. Hussey, Mt. Pleasant, Ohio.

Charmer 4771.—Sire, Scion 1033; dam, Clio 2d 1248.

Butter, 14 lbs. 12 oz. Milk, 257 lbs. 11 oz.

Test made from Aug. 1 to 7, 1883; age, 7 yrs. 5 mos.; fed 7 lbs., in equal parts, of corn meal, cotton-seed meal and wheat bran, daily—pasture; property of Henry C. Kelsey, Newton, N. J.

Charming of St. Lambert 69077.—Sire, Rioter's Pride 11694; dam, Girl of St. Lambert 20423.

Butter, 15 lbs. 4 oz. Milk, 137 lbs.

Test made from Dec. 8 to 10, 1890; age, 2 yrs. 3 mos.; estimated weight, 800 lbs.; fed, daily, 3 qts. pea meal, 3 qts. corn meal, 6 qts. bran and 6 qts. oats; property of Mrs. E. M. Jones, Brockville, Ont., Can.

Chautauqua Queen 26403.—Sire, Noble Rex 5174; dam, Dolly Daisy 2d 12005.

Butter, 14 lbs. 11 oz. Milk, 225 lbs. 13 oz.

Test made from July 5 to 11, 1885; age, 2 yrs. 5 mos.; property of J. M. Beebe, Cassadaga, N. Y.

Cheerful of St. Lambert 8348.—Sire, Stoke Pogis 3d 2238; dam, Jessamine of St. Lambert 5125.

Butter, 20 lbs. 8 oz. Milk, 239 lbs. 8 oz.

Test made from June 14 to 20, 1886; age, 8 yrs. 3 mos.; fed 6 qts. ground oats, 6 qts. bran and 3 qts. pea meal, mixed, daily—good pasture; property of Valancey E. Fuller, Hamilton, Ont., Can.

Cheerful of St. Lambert 2d 43745.—Sire, Canada's John Bull 8388; dam, Cheerful of St. Lambert 8348.

Butter, 22 lbs. 2 oz. Milk, 228 lbs. 8 oz.

Test made from June 9 to 16, 1890; age, 4 yrs. 2 mos.; estimated weight, 900 lbs.; fed 20 lbs. ground oats and peas, daily; property of William Rolph, Markham, Ont., Can.

Cheerful Pogis 61748.—Sire, Brier's Pogis 14163; dam, Cheerful of St. Lambert 2d 43745.

Butter, 21 lbs. 4½ oz. Milk, 286 lbs. 6 oz.

Test made from Feb. 12 to 19, 1897; age, 8 yrs. 10 mos.; estimated weight, 900 lbs., fed 136 lbs. ground oats, 21 lbs. ground oil-cake, 14 lbs. bran and 420 lbs. ensilage—hay *ad lib.*; property of William Rolph, Markham, Ont., Can.

Chelka 79033.—Sire, Tormentor 5th 21962; dam, Idalette Pogis 64220.

Butter, 14 lbs. 13¼ oz. Milk, 170 lbs.

Test made from Apr. 24 to May 1, 1897; age, 6 yrs.; estimated weight, 900 lbs.; fed 14 lbs. bran, 17¼ lbs. linseed meal, 14 lbs. cotton-seed meal, 14 lbs. ground oats, 31½ lbs. corn meal, 140 lbs. cotton-seed hulls and 70 lbs. cut oats; property of S. H. Butler, Como, Miss.

Chemung Maid 34704.—Sire. Tug Wilson 9680; dam, Cosette of Scituate 14726.

Butter, 14 lbs. 15 oz. Milk, 286 lbs. 10 oz.

Test made from June 5 to 12, 1895; age, 10 yrs. 2 mos.; estimated weight, 875 lbs.; fed 56 lbs. corn meal and 42 lbs. wheat bran—timothy and clover pasture; property of Henry S. Redfield, Elmira, N. Y.

Chenda 4599.—Sire, Marius 760; dam. Chatelaine 1916.

Butter, 15 lbs. 9½ oz. Milk, 284 lbs.

Test made from June 13 to 19, 1882; age, 6 yrs. 6 mos.; property of Campbell Brown, Spring Hill, Tenn.

Chenda 4599.—Sire, Marius 760; dam, Chatelaine 1916.

Butter, 17 lbs. 1⅛ oz. Milk, 251 lbs.

Test made from Apr. 30 to May 7, 1887; age, 11 yrs. 4 mos.; estimated weight, 1025 lbs.; no record of feed kept; property of Campbell Brown, Spring Hill, Tenn.

Chenda 3d 43246.—Sire, Tormentor 3533; dam, Chenda 4599.

Butter, 15 lbs. 7 oz. Milk, 197 lbs. 8 oz.

Test made from May 25 to June 1, 1891; age, 4 yrs. 8 mos.; estimated weight, 750 lbs.; fed 20 lbs. corn-hearts, daily—blue-grass pasture; property of Campbell Brown, Spring Hill, Tenn.

Chendado's Maid 45991.—Sire, Chendado 12955; dam, René of Hillside 12952.

Butter, 16 lbs. 7½ oz. Milk, 189 lbs.

Test made from Dec. 20 to 27, 1895; age, 8 yrs. 10 mos.; estimated weight, 1050 lbs.; fed 8 qts. ground oats, 4 qts. corn meal, 4 qts. bran and 2 qts. cotton-seed meal, daily —hay *ad lib.* during night, blue-grass pasture; property of D. P. Carter, Lynnville, Tenn.

Cherokee Rose 20921.—Sire, Beaconsfield 3416; dam, Rosebud of Belle Vue 7702.

Butter, 23 lbs. 10 oz. Milk, 227 lbs.

Test made from July 6 to 13, 1885; age, 3 yrs.; fed 18 qts. ground oats and corn, daily—good blue and orchard grass pasture; property of M. M. Gardner, Nashville, Tenn.

Cherry of Glynllyn 81145.—Sire, No License 16030; dam, Buff Nepte 35641.

Butter, 14 lbs. 4 oz. Milk, 247 lbs.

Test made from Mar. 1 to 8, 1895; age, 4 yrs. 2 mos.; estimated weight, 925 lbs.; fed 35 lbs. wheat bran, 49 lbs. corn and oats and 28 lbs. cotton-seed meal—hay; property of J. A. Griffith, Ashley Falls, Mass.

Cherry of Glynllyn 81145.—Sire, No License 16030; dam, Buff Nepte 35641.

Butter, 19 lbs. 4 oz. Milk, 214 lbs. 4 oz.

Test made from Jan. 6 to 13, 1897; age, 6 yrs. 1 mo.; estimated weight, 750 lbs.; fed 30 lbs. H. O. feed, 25 lbs. corn meal, 25 lbs. gluten, 30 lbs. bran, 16 lbs. oil meal and 155 lbs. carrots—hay *ad lib.*; property of D. Wolfe Bishop, Jr., Lenox, Mass.

Chestnuts 63449.—Sire, Upright 6147; dam, Modita 16626.

Butter, 15 lbs. 7¼ oz. Milk, 260 lbs.

Test made from May 26 to June 2, 1896; age, 9 yrs. 4 mos.; estimated weight, 825 lbs.; fed 42 lbs. bran, 45½ lbs. oats, 14 lbs. linseed oil meal, 28 lbs. corn meal, 4½ lbs. cotton-seed meal and 210 lbs. ensilage—hay *ad lib.*, timothy and clover pasture; property of C. I. Hood, Lowell, Mass.

Chestnut's Beauty 21576.—Sire, Napier (F. S. 275 J. H. B.); dam, Chestnut (F. S. 3125 J. H. B.)

Butter, 16 lbs. 10 oz. Milk, 258 lbs.

Test made from May 22 to 28, 1886; age, 6 yrs. 1 mo.; property of Campbell Brown, Spring Hill, Tenn.

Chestnut's Beauty 21576.—Sire, Napier (F. S. 275 J. H. B.); dam, Chestnut (F. S. 3125 J. H. B.)

Butter, 18 lbs. 4½ oz. Milk, 273 lbs.

Test made from June 5 to 12, 1887; age, 7 yrs. 1 mo.; estimated weight, 825 lbs.; fed, daily, 24 lbs. corn chop, bran and cotton-seed meal, mixed; property of Campbell Brown, Spring Hill, Tenn.

Chief's Beauty 39391.—Sire, Chief of the Miamis 2958; dam, Fan's Grouville Beauty 10079.

Butter, 20 lbs. 5½ oz. Milk, 305 lbs. 8 oz.

Test made from May 14 to 21, 1895; age, 9 yrs. 11 mos.; estimated weight, 800 lbs.; fed 140 lbs. ground corn and oats—blue-grass and clover pasture; property of George M. Jewett, Glenville, Md.

Chief's Charella 61999.—Sire, Chief 10663; dam, Charella of Bay View 48923.

Butter, 23 lbs. 10½ oz. Milk, 213 lbs.

Test made from Feb. 1 to 8, 1893; age, 4 yrs. 7 mos.; estimated weight, 625 lbs.; fed 14 lbs. oil meal, 245 lbs. corn refuse, 224 lbs. ensilage, 28 lbs. malt sprouts, 28 lbs. middlings and 28 lbs. clover hay; property of Charles Lautz, Buffalo, N. Y.

Chief's Fosie 61998.—Sire, Chief 10663; dam, Fosie 20255.

Butter, 14 lbs. 11 oz. Milk, 216 lbs. 6 oz.

Test made from Apr. 9 to 16, 1893; age, 5 yrs. 1 mo.; estimated weight, 700 lbs.; fed 224 lbs. ensilage, 14 lbs. oil meal, 28 lbs. middlings, 21 lbs. malt sprouts and 28 lbs. hay; property of Charles Lautz, Buffalo, N. Y.

Chief's Iona 111704.—Sire, Ellenetta's Chief 34896; dam, King's Fosie 80768.

Butter, 15 lbs. 12 oz. Milk, 142 lbs. 2 oz.

Test made from Jan. 25 to Feb. 1, 1898; age, 3 yrs. 2 mos.; actual weight, 660 lbs.; fed 21 lbs. ground oats, 21 lbs. corn meal, 21 lbs. oil meal, 28 lbs. wheat bran, 210 lbs. corn ensilage and 49 lbs. hay; property of Charles Lautz, Buffalo, N. Y.

Chief's Malita 25186.—Sire, Sultan Chief 3216; dam, Maffie 2d 4939.

Butter, 16 lbs. 8½ oz. Milk, 190 lbs. 4 oz.

Test made from June 20 to 27, 1892; age, 8 yrs. 3 mos.; actual weight, 900 lbs.; fed 115 lbs. grain (estimated)—short, rich blue-grass pasture; property of S. H. Godman, Muncie, Ind.

Chief's Queen 69967.—Sire, Geranium's Chief 18607; dam, Allie's Queen 45618.

Butter, 20 lbs. 3 oz. Milk, 246 lbs. 15 oz.

Test made from Dec. 7 to 14, 1892; age, 2 yrs. 11 mos.; estimated weight, 800 lbs.; fed 70 lbs. corn meal, 42 lbs. ground oats, 42 lbs. wheat bran, 42 lbs. cotton-seed meal, 280 lbs. ensilage and 70 lbs. hay; property of William Hart Dexter, Springfield, Mass.

Chief's Stevia 122187.—Sire, Ellenetta's Chief 34896; dam, King's Stevia 98737.

Butter, 16 lbs. 10½ oz. Milk, 149 lbs. 12 oz.

Test made from Apr. 5 to 12, 1898; age, 2 yrs. 1 mo.; actual weight, 520 lbs.; fed 14 lbs. ground oats, 14 lbs. corn meal, 21 lbs. oil meal, 21 lbs. wheat bran, 245 lbs. corn ensilage and 42 lbs. hay; property of Charles Lautz, Buffalo, N. Y.

Chilioness Queen 82973.—Sire, Real Queen's Stoke Pogis 20535; dam, Chilioness 44696.

Butter, 20 lbs. 8 oz. Milk, 323 lbs. 4 oz.

Test made from Aug. 18 to 25, 1897; age, 5 yrs. 3 mos.; estimated weight, 900 lbs.; fed 63 qts. bran, 78 qts. oats and 21 qts. corn meal—clover and green corn pasture; property of D. Wolfe Bishop, Jr., Lenox, Mass.

Chinqua 27384.—Sire, Telegraph 9457; dam, Chinquapin 4501.

Butter, 22 lbs. 9½ oz. Milk, 270 lbs.

Test made from May 28 to June 4, 1888; age, 4 yrs. 1 mo.; estimated weight, 950 lbs.; fed 14 lbs. corn meal, daily; property of J. R. Anderson, Jr., Lee, Va.

Chirp 65551.—Sire, Riot Act 20520; dam, Costa Rica 64570.

Butter, 19 lbs. 12 oz. Milk, 258 lbs. 8 oz.

Test made from Feb. 7 to 14, 1896; age, 6 yrs. 4 mos.; estimated weight, 900 lbs.; fed 35 lbs. bran, 48 lbs. corn meal, 21 lbs. oil meal and 21 lbs. oats; property of A. O. Auten, Jerseyville, Ill.

Chloe 4th 4612.—Sire, Yankee 1003; dam, Chloe 1540.

Butter, 17 lbs. 4 oz. Milk, 207 lbs.

Test made from Oct. 28 to Nov. 3, 1885; age, 9 yrs. 11 mos.; property of D. A. Givens, Cynthiana, Ky.

Chloe Beach 3931.—Sire, Colt Jr. 825; dam, Brighteyes 1517.

Butter, 14 lbs. 8 oz.

Property of Lyman A. Mills, Middlefield, Conn.

Chloe of Bolton 81340.—Sire, Viking of Bolton 24881; dam, Miriam of Bolton 50967.

Butter, 16 lbs. Milk, 259 lbs. 15 oz.

Test made from May 5 to 12, 1898; age, 7 yrs. 3 mos.; estimated weight, 900 lbs.; fed 25.2 lbs. shorts, 13.3 lbs. ground oats, 13.3 lbs. meal, 18.2 lbs. linseed, 175 lbs. roots and 110 lbs. hay; property of J. A. Cunningham, Bolton, Mass.

Chloe Tormentor of Lawn 71642.—Sire, Duke of Lincoln 15475; dam, Little Chloe Green 48767.

Butter, 14 lbs. 13½ oz. Milk, 199 lbs. 14 oz.

Test made from Sept. 15 to 22, 1897; age, 7 yrs. 1 mo.; estimated weight, 1000 lbs., fed 112 lbs. wheat bran, 112 lbs. corn and oat chop and 14 lbs. linseed meal—good crab-grass pasture; property of Platter & Foster, Denison, Tex.

Chlora 8325.—Sire, Planet 1130; dam, Rosette of Staatsburgh 8008.

Butter, 14 lbs. 6 oz., unsalted. Milk, 172 lbs.

Test made from June 28 to July 4, 1884; age, 8 yrs. 4 mos.; property of J. A. Fisher, Nashville, Tenn.

Chrissy 1448.—Sire on I. of J.; dam, Kitty Clover 1113.

Butter, 16 lbs. 8 oz.

Test made from June 21 to 27, 1878; age, 8 yrs. 5 mos.; property of James A. Hayt, Patterson, N. Y.

Chrissy 2d 7720.—Sire, Grand Duke Alexis 1040; dam, Chrissy 1448.

Butter, 16 lbs. 14 oz. Average of milk, 18 qts. per day.

Test made in May, 1884; age, 6 yrs. 1 mo.; property of George W. Farlee, Cresskill, N. J.

Chrissy 3d 15615.—Sire, Grand Duke Alexis 1040; dam, Chrissy 1448.

Butter, 14 lbs. 11 oz. Milk, 212 lbs.

Test made from Dec. 26, 1892, to Jan. 2, 1893; age, 11 yrs. 11 mos.; estimated weight, 900 lbs.; fed 5 qts. each of ground oats, wheat middlings and corn meal, with 20 lbs. ensilage, daily—hay *ad lib.*; property of George W. Sisson, Jr., Potsdam, N. Y.

Christel 6565.—Sire, King Philip of Mt. Hope 2399; dam, Idex 2d 5429.

Butter, 19 lbs. 5 oz. Milk, 205 lbs.

Test made from June 14 to 21, 1887; age, 10 yrs.; estimated weight, 850 lbs.; fed, daily, 4 qts. bran, 4 qts. ground oats and corn and 1 qt. pea meal; property of M. Erskine Miller, Staunton, Va.

Christina Pogis 35426.—Sire, Prince of Melrose 2d 11015; dam, Alabama 7690.

Butter, 17 lbs. 2 oz. Milk, 211 lbs. 1 oz.

Test made from Oct. 3 to 10, 1889; age, 3 yrs. 11 mos.; estimated weight, 800 lbs.; fed 56 qts. corn meal, 14 qts. cotton-seed meal, 28 qts. ground oats, 14 qts. pea meal and 280 lbs. ensilage—hay, poor pasture; property of James Crook, Jacksonville, Ala.

Christine of Side View 2d 94253.—Sire, Noble of Side View 28280; dam, Christine of Side View 47278.

Butter, 14 lbs. 15 oz. Milk, 211 lbs. 8 oz.

Test made from Dec. 26, 1896, to Jan. 2, 1897; age, 4 yrs. 9 mos.; estimated weight, 750 lbs.; fed 14 qts., daily, of a mixture in equal parts by weight of bran, corn and cob meal and Buffalo feed—ensilage and hay, not weighed; property of George E. Nichols, Afton, N. Y.

Christmas Nannie 4075.—Sire, Broker 873; dam, Princess 2205.

Butter, 19 lbs. 7 oz.

Test made from June 4 to 11, 1883; age, 9 yrs. 6 mos.; fed 1 qt. corn meal and 3 qts. wheat middlings, daily—good timothy and clover pasture; property of A. W. Sawyer, Sycamore, Ill.

Chroma 4572.—Sire, St. Heller 45; dam, Ianthe 4562.

Butter, 20 lbs. 6 oz.

Age, 7 yrs.; property of O. S. Hubbell, Stratford, Conn.

Chromatess 51578.—Sire, St. Heller of Idleside 12200; dam, Florence Pond 30980.

Butter, 25 lbs. ½ oz. Milk, 314 lbs.

Test made from May 24 to 31, 1895; age, 8 yrs. 3 mos.; actual weight, 950 lbs.; fed 6 qts. corn meal, 6 qts. ground oats and 4 qts. ground oil-cake, daily—good old meadow pasture; property of Miller & Sibley, Franklin, Pa.

Chrome Skin 7881.—Sire, Gilderoy 2107; dam, Regina 2d 2475.

Butter, 20 lbs. 10 oz. Milk, 226 lbs. 8 oz.

Test made from June 22 to 29, 1883; age, 5 yrs. 2 mos.; fed corn meal, ship stuff and bran—fair pasture; property of H. M. Howe, Bristol, R. I.

Chromia E. 97511.—Sire, Chromo 26113; dam, Nice 68027.

Butter, 17 lbs. 12 oz. Milk, 278 lbs. 4 oz.

Test made from May 30 to June 6, 1897; age, 4 yrs. 8 mos.; estimated weight, 950 lbs.; fed 35 lbs. corn meal, 7 lbs. oil meal, 28 lbs. rye and 63 lbs. oats; property of A. W. Sawyer, Chicago, Ill.

Chromoline of Argyle 103709.—Sire, Chromo 26113; dam, Baroness Argyle 40498.

Butter, 16 lbs. 10 oz. Milk, 227 lbs.

Test made from Feb. 22 to Mar. 1, 1898; age, 3 yrs. 8 mos.; estimated weight, 725 lbs.; fed 4 lbs. spring wheat bran, 4 lbs. Chicago gluten meal, 2 lbs. mixed rye and oats provender, about 5 lbs. clover hay and 30 lbs. to 35 lbs. ensilage, daily; property of E. Stevens Henry, Rockville, Conn.

Chromo's Modita 121439.—Sire, Chromo 26113; dam, Modita 16626.

Butter, 16 lbs. 7 oz. Milk, 250 lbs. 8 oz.

Test made from May 23 to 30, 1898; age, 5 yrs.; estimated weight, 900 lbs.; fed 28 lbs. corn, 28 lbs. oats, 28 lbs. bran and 14 lbs. oil meal—mixed pasture; property of H. C. Taylor, Orfordville, Wis.

Chromo's Pansy 114416.—Sire, Chromo 26113; dam, Diploma's Pansy 112871.

Butter, 17 lbs. 2½ oz. Milk, 274 lbs. 8 oz.

Test made from Feb. 25 to Mar. 4, 1898; age, 3 yrs. 11 mos.; estimated weight, 850 lbs.; fed 80 lbs. ground corn and oats, in equal parts, 34 lbs. wheat bran and 6 lbs. oil meal—timothy and prairie hay *ad lib.*; property of Eugene Stupfell, Quick, Iowa.

Chromo's Princess 121906.—Sire, Chromo 26113; dam, Upright's Princess 121905.

Butter, 14 lbs. 11 oz. Milk, 233 lbs. 4 oz.

Test made from Jan. 9 to 16, 1898; age, 3 yrs. 10 mos.; estimated weight, 775 lbs., fed 4 lbs. corn meal, 6 lbs. oat meal, 5 lbs. bran, 1½ lbs. oil meal and 2 qts. roots, daily; property of Richardson Bros., Davenport, Iowa.

Chrysippe 53373.—Sire, Signaldo Alexis 15051; dam, Signal Maiden 42793.

Butter, 14 lbs. 14 oz. Milk, 316 lbs.

Test made from Apr. 1 to 8, 1897; age, 9 yrs. 1 mo.; estimated weight, 900 lbs.; fed 275 lbs. ensilage and 120 lbs. corn meal—hay *ad lib.*; property of Baum & Hill, Frankfort, Ind.

Chuffy's Ethel 102195.—Sire, Ethleel's Landseer 22341; dam, Chuffy Mianella 70915.

Butter, 16 lbs. 1 oz. Milk, 250 lbs. 5 oz.

Test made from May 11 to 18, 1897; age, 4 yrs.; estimated weight, 800 lbs.; fed 56 qts. corn meal and 56 qts. wheat bran—blue-grass pasture; property of S. E. Du Bois, Vigo, Ohio.

Chump 81289.—Sire, Signalda's Torment 14673; dam, Genie Walker 38443.

Butter, 14 lbs. 13 oz. Milk, 149 lbs. 4 oz.

Test made from June 12 to 19, 1893; age, 2 yrs. 8 mos.; estimated weight, 550 lbs.; fed 12 qts. equal parts of ground oats and corn, 4 qts. bran and 1 qt. cotton-seed meal, daily, fed on cut hay—short pasture; property of William E. Matthews. Huntsville, Ala.

Cicero's Jolie 18246.—Sire, Cicero 7657; dam, Jolie (F. S. 1453 J. H. B.)

Butter, 18 lbs. 3 oz. Milk, 233 lbs. 7 oz.

Test made from Sept. 2 to 9, 1890; age, 8 yrs.; estimated weight, 900 lbs.; fed, daily. 8 lbs. corn meal, 8 lbs. shorts, 8 lbs. ground oats and 1 lb. oil meal; property of D. F. Appleton, Ipswich, Mass.

Cicero's Juno 16726.—Sire, Cicero 7657; dam, Juno Grey 16722.

Butter, 17 lbs. 2 oz. Milk, 20 qts. per day.

Test made from July 24 to 30, 1886; age, 4 yrs.; estimated weight, 850 lbs.; fed 8 qts. corn meal, 7 qts. ground oats and 7 qts. middlings per day; property of D. F. Appleton, Ipswich, Mass.

Cicero's Mabel 18238.—Sire, Cicero 7657; dam, Mabel 4th 18167.

Butter, 15 lbs. 2 oz. Milk, 247 lbs. 8 oz.

Test made from Dec. 25, 1885, to Jan. 1, 1886; age, 3 yrs. 9 mos.; property of William H. Burr, Redding Ridge, Conn.

Cicero's Ruby 29040.—Sire, Count Cicero (F. S. 398 J. H. B.); dam, Wonderful Lady 25108.

Butter, 14 lbs. 1½ oz. Milk, 174 lbs. 7 oz.

Test made from May 21 to 28, 1894; age, 9 yrs. 11 mos.; estimated weight, 950 lbs.; fed about 5 lbs. Buffalo gluten, 4 lbs. ground oats, 2 lbs. linseed meal and 4 lbs. bran, daily—fair blue-grass pasture; property of C. Delano, Mount Vernon, Ohio.

Cigarette 2849.—Sire, Milo 590; dam, Eloise 735.

Butter, 14 lbs. 4 oz. Milk, 182 lbs. 7 oz.

Test made from Apr. 1 to 7, 1884; age, 10 yrs. 8 mos.; property of W. B. Matthews & Son, Franklin, Tenn.

Cill of Glen Rouge 13818.—Sire, Jack Frost of St. Lambert 2419; dam, Pearl of St. Lambert 5527.

Butter, 16 lbs. 6 oz. Milk, 259 lbs.

Test made from July 21 to 27, 1885; age, 4 yrs. 4 mos.; property of Miller & Sibley, Franklin, Pa.

Cinderella's Oonan 108764.—Sire, Oonan's Tormentor 22280; dam, Tormentor's Cinderella 19564.

Butter, 14 lbs. 9¾ oz. Milk, 145 lbs. 3¼ oz.

Test made from May 30 to June 6, 1898; age, 3 yrs. 3 mos.; estimated weight, 650 lbs.; fed 20 lbs. per day of ground oats and corn, in equal parts—extra good orchard and blue-grass pasture; property of M. C. Campbell, Spring Hill, Tenn.

Cinxia St. Helier 46730.—Sire, O'Malley 6441; dam, Cinderella St. Helier 27241.

Butter, 17 lbs. 9 oz., unsalted. Milk, 238 lbs.

Test made from June 15 to 22, 1889; age, 5 yrs. 3 mos.; estimated weight, 800 lbs.; fed cut grass, with 1 qt. oil meal, daily—lowland pasture; property of Winslow S. Lincoln, Worcester, Mass.

Cinxia's Pavonia 64441.—Sire, L. D.'s Victor St. Helier 15286; dam, Cinxia St. Helier 46730.

> Butter, 14 lbs. 7½ oz.   Milk, 248 lbs.

Test made from May 23 to 30, 1894; age, 6 yrs. 1 mo.; estimated weight, 950 lbs.; fed 84 lbs. of a mixture consisting of two-thirds wheat middlings and one-third corn meal and oil meal, on cut hay—pasture of ordinary mixed grasses; property of H. C. Vanderhoef, Belmont, N. Y.

Cinxia's Pavonia 64441.—Sire, L. D.'s Victor St. Helier 15286; dam, Cinxia St. Helier 46730.

> Butter, 15 lbs. 9 oz.   Milk, 276 lbs. 3 oz.

Test made from Jan. 21 to 28, 1895; age, 6 yrs. 9 mos.; estimated weight, 1000 lbs.; fed about 56 lbs. wheat bran, middlings and buckwheat middlings and 14 lbs. oil meal and cotton-seed meal, on cut millet hay—hay *ad lib.*; property of H. C. Vanderhoef, Belmont, N. Y.

Cinxia's Pavonia 64441.—Sire, L. D.'s Victor St. Helier 15286; dam, Cinxia St. Helier 46730.

> Butter, 18 lbs. 3 oz.   Milk, 321 lbs. 8 oz.

Test made from Jan. 1 to 7, 1896; age, 7 yrs. 8 mos.; estimated weight, 1050 lbs.; fed 10½ lbs. oil meal, 10½ lbs. corn meal, 85 lbs. wheat middlings and 28 lbs. wheat bran—hay and carrots; property of H. C. Vanderhoef, Belmont, N. Y.

City Belle 16539.—Sire, Brother Jack 4042; dam, Zoedone 12034.

> Butter, 15 lbs. 5½ oz.   Milk, 245 lbs.

Test made from May 10 to 17, 1886; age, 3 yrs. 11 mos.; property of Edward Mayes, Oxford, Miss.

Clara C. Magnet 31563.—Sire, Champion Magnet 6480; dam, Caleri 7016.

> Butter, 14 lbs. 11 oz.   Milk, 190 lbs. 4 oz.

Test made from Feb. 3 to 9, 1886; age, 2 yrs.; property of John Boyd, Elmhurst, Ill.

Clara Oonan 78454.—Sire, Oonan's Tormentor 22280; dam, Clara Tormentor 41427.

> Butter, 21 lbs. 2 oz.   Milk, 145 lbs.

Test made from Jan. 22 to 29, 1896; age, 5 yrs. 1 mo.; estimated weight, 850 lbs.; fed 63 qts. corn meal, ground oats and wheat bran, in equal parts—barley pasture; property of M. C. Campbell, Spring Hill, Tenn.

Clara's Belle 16582.—Sire, Dexter of Jefferson 5681; dam, Clara of Jefferson 12311.

> Butter, 17 lbs. 9½ oz.   Milk, 219 lbs. 2 oz.

Test made from Oct. 1 to 8, 1889; age, 9 yrs. 6 mos.; estimated weight, 850 lbs.; fed 12 qts. corn and oat meal, daily—good pasture, hay; property of Richardson Bros., Davenport, Iowa.

Clara Signal of Lawn 85586.—Sire, Signal of Springwood 28545; dam, Denison's Pet 72901.

> Butter, 15 lbs. 15 oz.   Milk, 183 lbs. 14 oz.

Test made from Dec. 2 to 9, 1894; age, 3 yrs. 1 mo.; estimated weight, 850 lbs.; fed 70 lbs. wheat bran, 70 lbs. ground oats, 56 lbs. corn meal and 28 lbs. cotton-seed meal—sorghum hay once daily; property of Platter & Foster, Denison, Tex.

Clara Tormentor 2d 97279.—Sire, Oonan's Tormentor 22280; dam, Clara Tormentor 41427.

Butter, 17 lbs. 6 oz. Milk, 215 lbs. 12½ oz.

Test made from Dec. 8 to 15, 1897; age, 5 yrs. 1 mo.; estimated weight, 775 lbs.; fed 22 lbs. per day ground oats and corn, in equal parts, mixed with 4 bucketfuls cut sheaf oats, wetted—millet hay *ad lib.*; property of Matt. M. Gardner, Nashville, Tenn.

Clarina B. 98430.—Sire, Concord B. 29699; dam, Bertina B. 81715.

Butter, 14 lbs. 2 oz. Milk, 238 lbs.

Test made from Apr. 16 to 23, 1898; age, 5 yrs. 7 mos.; estimated weight, 750 lbs., fed 28 lbs. bran, 28 lbs. corn meal, 14 lbs. oat meal, 14 lbs. old process oil meal, 21 lbs. clover hay and 70 lbs. corn stover—corn fodder *ad lib.*; property of Henry O. Burchard, Fort Atkinson, Wis.

Clarintha 96919.—Sire, Fancy's Harry 9777; dam, Tormintha 37327.

Butter, 17 lbs. ¼ oz. Milk, 203 lbs. 13 oz.

Test made from Nov. 12 to 19, 1895; age, 2 yrs. 7 mos.; actual weight, 718 lbs.; fed 50 lbs. wheat bran, 35 lbs. wheat middlings, 20 lbs. corn meal and 15 lbs. old process oil meal—cut sweet corn fodder and millet hay *ad lib.*, short blue-grass pasture, property of J. P. Bradbury, Pomeroy, Ohio.

Clark's Gift 91708.—Sire, Pussie's Pogis 25935; dam, Bessie Niobe 55763.

Butter, 15 lbs. 4 oz. Milk, 244 lbs.

Test made from Mar. 26 to Apr. 2, 1898; age, 4 yrs. 6 mos.; estimated weight, 850 lbs.; fed 12 lbs. of a mixture of bran, middlings and corn meal and 2 lbs. oil meal, daily—clover hay *ad lib.*; property of H. C. Vanderhoef, Belmont, N. Y.

Classic 21402.—Sire, Combination 4389; dam, Income 19472.

Butter, 14 lbs. 13½ oz. Milk, 216 lbs.

Test made from Apr. 20 to 27, 1888; age, 5 yrs.; estimated weight, 850 lbs.; fed, daily, 4½ lbs. pea meal, 3 lbs. oil meal, 4½ lbs. oat meal and 4 lbs. bran; property of H. M. Baum, Frankfort, Ind.

Clematis 3d 6653.—Sire, Ready Money Jack 2986; dam, Clematis 3174.

Butter, 16 lbs. 1 oz. Milk, 209 lbs. 2 oz.

Test made from Sept. 29 to Oct. 6, 1884; age, 7 yrs.; property of E. S. Henry, Rockville, Conn.

Clematis of St. Lambert 5478.—Sire, Lord Lisgar 1066; dam, Snowdrop of St. Lambert 5119.

Butter, 14 lbs. 3 oz. Milk, 285 lbs.

Test made from May 2 to 8, 1882; age, 6 yrs.; property of William Rolph, Markham, Ont., Can.

Clem Pedro 73961.—Sire, Young Pedro 2d 15011; dam, Clem Pogis 49932.

Butter, 20 lbs. 8 oz. Milk, 259 lbs. 9 oz.

Test made from Apr. 10 to 17, 1897; age, 5 yrs. 9 mos.; estimated weight, 1050 lbs., fed 4 qts. ground oats, 3 qts. corn meal, 1½ qts. cotton-seed meal and 1½ qts. old process oil meal, daily—clover hay; property of John S. Clark, New Brunswick, N. J.

Clem Pogis 49932.—Sire, Victor Pogis 15592; dam, Clem's Lucy 26694.

Butter, 23 lbs. 4½ oz. Milk, 290 lbs. 5 oz.

Test made from Apr. 2 to 9, 1897; age, 9 yrs.; estimated weight, 1200 lbs.; fed 4 qts. corn meal, 5 qts. ground oats, 1½ qts. cotton-seed meal and 1½ qts. old process oil meal, daily—clover hay; property of John S. Clark, New Brunswick, N. J.

Cleodora 59311.—Sire, Clotaire 9884; dam, Clotaire's Annie 58319.

Butter, 22 lbs. 5 oz. Milk, 235 lbs. 2 oz.

Test made from Mar. 26 to Apr. 2, 1891; age, 4 yrs.; estimated weight, 900 lbs.; fed 16 qts. daily of a mixture of one part bran, two parts ground corn and three parts ground oats—bran mash daily, hay *ad lib.*; property of L. L. Tozier, Batavia, N. Y.

Cleora of Cornwall 63408.—Sire, Rioter's Combination 10363; dam, Rioter Carlotta 29667.

Butter, 18 lbs. 8 oz. Milk, 261 lbs. 5 oz.

Test made from June 4 to 11, 1893; age, 3 yrs. 5 mos.; estimated weight, 650 lbs.; fed 4 qts. bran, 2 qts. ground oats, 2 qts. corn meal and 2 qts. cotton-seed meal—hay *ad lib.*; property of James Stillman, Cornwall-on-Hudson, N. Y.

Clifty Beauty 30249.—Sire, Dickinson's Sea Gull 2d 7373; dam, May Indian 16075.

Butter, 15 lbs. 6½ oz. Milk, 173 lbs.

Test made from Aug. 2 to 9, 1890; age, 7 yrs. 4 mos.; estimated weight, 750 lbs.; fed, daily, 4 gals. ground corn; property of H. P. Webster, Columbia, Tenn.

Clotaire's Annie 58319.—Sire, Clotaire 9884; dam, Floribundus 2d 14949.

Butter, 19 lbs. 11½ oz. Milk, 212 lbs. 11 oz.

Test made from Jan. 30 to Feb. 6, 1890; age, 4 yrs. 10 mos.; estimated weight, 900 lbs.; fed 16 qts. daily of a mixture composed of one part bran, two parts ground corn and three parts ground oats, also bran slop daily—hay *ad lib.*; property of L. L. Tozier, Batavia, N. Y.

Clotaire's Beauty 28171.—Sire, Clotaire 9884; dam, Lady Bingo 24160.

Butter, 20 lbs. 4 oz. Milk, 262 lbs. 1½ oz.

Test made from Mar. 19 to 26, 1891; age, 7 yrs. 3 mos.; estimated weight, 950 lbs.; fed 16 qts. daily of a mixture of one part bran, two parts ground corn and three parts ground oats—bran slop daily; property of L. L. Tozier, Batavia, N. Y.

Clotaire's Daisy 45392.—Sire, Clotaire 9884; dam, Alphea Rajah 20605.

Butter, 18 lbs. Milk, 231 lbs. 15 oz.

Test made from May 28 to June 4, 1891; age, 5 yrs. 2 mos.; estimated weight, 950 lbs.; fed 16 qts. daily of a mixture of one part bran, two parts ground corn and three parts ground oats—bran slop daily; property of L. L. Tozier, Batavia, N. Y.

Clothilde of St. Lambert 72271.—Sire, Prince of Melrose 2d 11015; dam, Maggie May of Tupelo 71280.

Butter, 22 lbs. 7½ oz. Milk, 201 lbs. 1 oz.

Test made from Feb. 21 to 28, 1897; age, 7 yrs. 9 mos.; actual weight, 890 lbs.; fed 84 lbs. corn meal, 56 lbs. wheat bran, 35 lbs. ground oats and 21 lbs. cotton-seed meal—hay *ad lib.*, Bermuda grass pasture; property of S. B. Hopkins, Dallas, Tex.

Clover Bloom 2d 12736.—Sire, Lord Clive 3813; dam, Clover Bloom 9788.

Butter, 14 lbs., unsalted. Milk, 236 lbs.

Test made from Nov. 30 to Dec. 6, 1884; age, 2 yrs. 10 mos.; property of H. M. Howe, Bristol, R. I.

Clover Bud 4th 18992.—Sire, Lord Harry 3445; dam, Clover Bud 4074.

Butter, 16 lbs. 14 oz. Milk, 124 lbs. 2 oz.

Test made from Sept. 8 to 15, 1888; age, 5 yrs. 9 mos.; estimated weight, 800 lbs.; fed, daily, 16 qts. corn and oats, ground; property of W. J. Webster and C. Brown, Columbia, Tenn

Clover Bud 5th 27743.—Sire, Signalda 4027; dam, Clover Bud 4074.

Butter, 14 lbs. Milk, 134 lbs. 12½ oz.

Test made from May 8 to 14, 1886; age, 2 yrs. 8 mos.; property of S. W. Taliaferro, Guthrie, Ky.

Clover Mel 16159.—Sire, Gilderoy 2107; dam, Clover Bloom 9788.

Butter, 14 lbs. 9 oz. Milk, 190 lbs. 8 oz.

Test made from July 25 to Aug. 1, 1885; age, 3 yrs. 4 mos.; property of H. M. Howe, Bristol, R. I.

Clover of Argyle 76745.—Sire, Baron Hugo 15208; dam, Faith of Argyle 29765.

Butter, 15 lbs. 10 oz. Milk, 213 lbs. 4 oz.

Test made from Dec. 4 to 11, 1894; age, 3 yrs. 3 mos.; estimated weight, 750 lbs., fed 4 lbs. cream gluten meal, 1 lb. cotton-seed meal, 6 lbs. coarse wheat bran, 2 lbs. rye and oats provender, about 30 lbs. corn ensilage and 5 lbs. to 8 lbs. cut clover hay, daily; property of E. S. Henry, Rockville, Conn.

Clover of Bay View 65424.—Sire, Verte's Chief 16276; dam, Chief's Nattie 38469.

Butter, 18 lbs. 14 oz. Milk, 188 lbs.

Test made from Mar. 18 to 24, 1894; age, 4 yrs. 3 mos.; actual weight, 872 lbs.; fed 14 lbs. oil meal, 14 lbs. ground wheat, 7 lbs. rye bran, 21 lbs. ground oats, 14 lbs. corn meal, 175 lbs. ensilage and 50 lbs. hay; property of Charles Lautz, Buffalo, N. Y.

Clover of St. Lambert 60945.—Sire, Ruby's Star of St. Lambert 20719; dam, Daisy Deane of St. L. 35962.

Butter, 16 lbs. 14¼ oz. Milk, 203 lbs. 8 oz.

Test made from Feb. 6 to 13, 1893; age, 4 yrs. 1 mo.; actual weight, 836 lbs.; fed 124 qts. chop and 42 qts. wheat bran—clover hay and corn fodder; property of Mrs. A. M. Hallock, Columbus, Ohio.

Clover of St. Lambert 60945.—Sire, Ruby's Star of St. Lambert 20719; dam, Daisy Deane of St. L. 35962.

Butter, 18 lbs. 4¼ oz. Milk, 227 lbs. 8 oz.

Test made from Jan. 1 to 8, 1894; age, 5 yrs.; estimated weight, 900 lbs.; fed 19 qts., daily, of gluten. chop and wheat bran, mixed—hay and corn fodder; property of Mrs. A. M. Hallock, Columbus, Ohio.

Clover of St. Lambert 60945.—Sire, Ruby's Star of St. Lambert 20719; dam, Daisy Deane of St. L. 35962.

Butter, 37 lbs. 13¼ oz. in 15 days. Milk, 465 lbs. 12 oz.

Test made from Dec. 26, 1893, to Jan. 10, 1894; age, 5 yrs.; estimated weight, 900 lbs.; fed first 6 days of test 60 gals. cut hay, corn, bran and gluten, mixed in equal parts, wet; fed last 9 days 171 qts. gluten, chop and wheat bran, mixed; property of Mrs. A. M. Hallock, Columbus, Ohio.

Clover of St. Lambert 60945.—Sire, Ruby's Star of St. Lambert 20719; dam, Daisy Deane of St. L. 35962.

Butter, 70 lbs. 3 oz. in 31 days. Milk, 1028 lbs. 4 oz.

Test made from Dec. 26, 1893, to Jan. 26, 1894; age, 5 yrs. 1 mo.; estimated weight, 900 lbs.; fed first 6 days of test 60 gals. cut hay, corn, bran and gluten, mixed in equal parts, wet; next 23 days was fed 437 qts. gluten, chop and wheat bran; during last 2 days 40 qts. of same; property of Mrs. A. M. Hallock, Columbus, Ohio.

Clover of St. Lambert 60945.—Sire, Ruby's Star of St. Lambert 20719; dam, Daisy Deane of St. L. 35962.

Butter, 94 lbs. 6¼ oz. in 42 days. Milk, 1169 lbs. 6½ oz.

Test made from Jan. 2 to Feb. 13, 1893; age, 4 yrs. 1 mo.; actual weight, 836 lbs., fed same as herd for first 20 days; Jan. 21 to 24, fed 14 qts. chop and 6 qts. wheat bran per day; Jan. 25 to Feb. 7, 16 qts. chop and 6 qts. wheat bran; Feb. 8 to 13, 18 qts. chop and 6 qts. wheat bran—hay and corn fodder during test; property of Mrs. A. M. Hallock, Columbus, Ohio.

Clubs 2d 16725.—Sire, Traveller (P. S. 280 J. H. B.); dam, Clubs 16721.

Butter, 14 lbs. 7 oz. Milk, 208 lbs. 8 oz.

Test made from Feb. 16 to 23, 1890; age, 8 yrs.; estimated weight, 900 lbs.; fed, daily, 7 qts. corn meal, crushed oats and bran, mixed; property of Frederic Bronson, Greenfield Hill, Conn..

Clyde Landseer 74834.—Sire, Leonette's Landseer 20399; dam, Beauty of Four Pines 81788.

Butter, 14 lbs. Milk, 198 lbs.

Test made from Dec. 15 to 22, 1896; age, 6 yrs.; estimated weight, 850 lbs.; fed 70 lbs. bran, 56 lbs. corn meal and 28 lbs. raw cotton-seed—pastured on stalk field; property of D. M. Weatherford, Terrell, Tex.

Clytemnestra 2455.—Sire, Mercury 482; dam, Leda 799.

Butter, 15 lbs. 3½ oz. Milk, 216 lbs. 4 oz.

Test made from June 30 to July 6, 1884; age. 12 yrs. 3 mos.; property of William Simpson, New York, N. Y.

Cobweb 3d 21325.—Sire, Admiral Farragut 2666; dam, Cobweb 5006.

Butter, 18 lbs. 5 oz. Milk, 214 lbs.

Test made from Aug. 21 to 28, 1885; age, 3 yrs.; property of William H. Hopkins, Providence, R. I.

Cocotte 11958.—Sire, Hero (P. S. 90 J. H. B.); dam, Belle (F. S. 302 J. H. B.)

Butter, 14 lbs. 6 oz. (official). Milk, 239 lbs. 14 oz.

Test made from Jan. 5 to 12, 1886; age, 9 yrs. 10 mos.; fed, daily, 9 lbs. corn meal, 7 lbs. wheat bran, 6 lbs. crushed oats and 2 lbs. linseed oil-cake; property of James Stillman, Sing Sing, N. Y.

Cocotte 11958.—Sire, Hero (P. S. 90 J. H. B.); dam, Belle (F. S. 302 J. H. B.)

Butter, 16 lbs. 8½ oz. Milk, 172 lbs. 15 oz.

Test made from May 21 to 28, 1887; age, 11 yrs. 3 mos.; estimated weight, 825 lbs.; fed, daily, 12 lbs. corn meal, 4 lbs. bran, 4 lbs. oil meal; property of James Stillman, Sing Sing, N. Y.

Coix 85258.—Sire, Petit Victor 24788; dam, Catchfly 4th 47682.

Butter, 15 lbs. 4 oz. Milk, 223 lbs.

Test made from July 15 to 22, 1895; age, 4 yrs. 2 mos.; estimated weight, 1000 lbs., fed 3 qts. cotton-seed meal, 8 qts. corn meal and 4 qts. wheat bran, daily—rough, hilly pasture; property of W. C. Norton, Agt., Aldenville, Pa.

Colie 8309.—Sire, St. Valentine 2251; dam, Torfrida 3596.

Butter, 18 lbs. 4 oz. Milk, 241 lbs. 14 oz.

Test made from July 16 to 22, 1884; age, 5 yrs. 8 mos.; property of William Simpson, New York, N. Y

Colt's La Biche 6399.—Sire, Knave 1856; dam, La Biche 2d 4023.

Butter, 17 lbs. 2½ oz., unsalted. Milk, 282 lbs. 6 oz.

Test made from Feb. 11 to 17, 1882; age, 4 yrs. 8 mos.; fed bran, ship stuff, corn meal, oil meal, mixed—hay, blue-grass pasture; property of D. A. Givens, Cynthiana, Ky.

Columbia Beauty 30263.—Sire, Toltec 6831; dam, Heliotrope of Linwood 6952.

Butter, 15 lbs. 1 oz. Milk, 179 lbs. 5 oz.

Test made from Oct. 12 to 19, 1889; age, 5 yrs. 7 mos.; actual weight, 968 lbs.; fed, daily, 16 lbs. ground corn, oats and wheat bran, one-third of each; property of F. H. Bates, Hamburg, Ala.

Columbiana 74998.—Sire, Diploma 16219; dam, Lorita 33750.

Butter, 18 lbs. 6 oz. Milk, 216 lbs. 4 oz.

Test made from Oct. 5 to 12, 1895; age, 4 yrs. 5 mos.; estimated weight, 800 lbs.; fed 56 lbs. oats, 35 lbs. corn, 21 lbs. oil meal and 7 lbs. bran—pumpkins *ad lib.*, timothy pasture; property of H. C. Taylor, Orfordville, Wis.

Columbine of St. Lambert 8350.—Sire, Stoke Pogis 3d 2238; dam, Bijou of St. Lambert 5112.

Butter, 19 lbs. 1 oz. Milk, 263 lbs. 8 oz.

Test made from June 4 to 10, 1886; age, 8 yrs. 1 mo.; property of Valancey E. Fuller, Hamilton, Ont., Can.

Colza Melrose 93859.—Sire, Ida's Rioter of St. L.'s Son 28869; dam, Saucy Princess 40350.

Butter, 14 lbs. 13½ oz. Milk, 218 lbs. 12 oz.

Test made from Feb. 20 to 27, 1897; age, 3 yrs. 4 mos.; estimated weight, 900 lbs.; fed 35 lbs. bran, 56 lbs. meal, 14 lbs. linseed meal and 28 lbs. boiled whole cotton-seed—Bermuda grass and Japanese clover pasture; property of M. Lothrop, Marshall, Tex.

Coma 29330.—Sire, Combination 4389; dam, Metella 3905.

Butter, 15 lbs. 2½ oz. Milk, 223 lbs. 9 oz.

Test made from May 17 to 24, 1886; age, 4 yrs. 9 mos.; property of Richardson Bros., Davenport, Iowa.

Comanea 19389.—Sire, Combination 4389; dam, Miss Bianca 12517.

Butter, 16 lbs. 3 oz. Milk, 218 lbs.

Test made from June 13 to 20, 1887; age, 4 yrs. 4 mos.; property of Richardson Bros., Davenport, Iowa.

Combination of Lawn 76635.—Sire, Pansy Pogis Rioter 14064; dam, Castara 2d 46853.

Butter, 14 lbs. 2½ oz. Milk, 218 lbs. 3 oz.

Test made from July 7 to 14, 1895; age, 4 yrs. 4 mos.; estimated weight, 800 lbs., fed 42 lbs. cotton-seed meal, 14 lbs. linseed meal, 70 lbs. bran and 70 lbs. ground corn and oats—prairie pasture; property of Platter & Foster, Denison, Tex.

Combonnie 40260.—Sire, Combination 4389; dam, Belbonnie 29630.

Butter, 14 lbs. 2 oz. Milk, 227 lbs. 1 oz.

Test made from Dec. 20 to 27, 1887; age, 3 yrs. 1 mo.; estimated weight, 800 lbs.; fed, daily, 12 qts. corn meal, oat meal and pea meal; property of Richardson Bros., Davenport, Iowa.

Comedia 38885.—Sire, Combination 4389; dam, Loreda 19801.

Butter, 14 lbs. 2 oz. Milk, 223 lbs. 8 oz.

Test made from June 1 to 8, 1886; age, 2 yrs. 5 mos.; estimated weight, 800 lbs.; fed, daily, 12 qts. corn and oat meal; property of Richardson Bros., Davenport, Iowa.

Comely of St. Lambert 2d 41177.—Sire, Brier's Pogis 14163; dam, Comely of St. Lambert 6639.

Butter, 20 lbs. 10 oz. Milk, 266 lbs. 8 oz.

Test made from June 2 to 9, 1889; age, 3 yrs. 4 mos.; estimated weight, 700 lbs.; fed, daily, 12 qts. barley and oats and 2 qts. oil-cake; property of Wm. Rolph, Markham, Ont., Can.

Coming Girl 3d 71794.—Sire, St. Lambert's Hugo 17457; dam, Coming Girl 12772.

Butter, 17 lbs. 14 oz. Milk, 227 lbs. 10 oz.

Test made from June 16 to 23, 1893; age, 3 yrs. 8 mos.; estimated weight, 950 lbs.; fed 6 lbs. bran, 5 lbs. corn meal and 6 lbs. ground oats twice daily, with ½ bush. cut sorghum—prairie grass pasture; property of Platter & Foster, Denison, Tex.

Comlassa 40671.—Sire, Combination 4389; dam, Frankie's Lass 24900.

Butter, 14 lbs. Milk, 237 lbs. 10 oz.

Test made from Apr. 15 to 22, 1887; age, 3 yrs.; estimated weight, 850 lbs.; fed, daily, 18 qts. corn meal, ground oats and peas; property of Richardson Bros., Davenport, Iowa.

Commotion 52960.—Sire, Combination 4389; dam, Coma 29330.

Butter, 17 lbs. 6 oz. Milk, 227 lbs. 11 oz.

Test made from May 15 to 22, 1890; age, 7 yrs.; estimated weight, 850 lbs.; fed 18 qts. ground oats, corn and peas, daily—plenty of hay; property of Richardson Bros., Davenport, Iowa.

Como of Briarcliff 35849.—Sire, Domino of Darlington 2459; dam, Cocotte 11958.

Butter, 14 lbs. 6 oz. Milk, 155 lbs.

Test made from May 31 to June 7, 1888; age, 2 yrs. 6 mos.; estimated weight, 650 lbs.; fed, daily, 8 lbs. corn meal and 4 lbs. middlings; property of James Stillman, Sing Sing, N. Y.

Compact 52705.—Sire, Combination 3d 17576; dam, Mylra 18840.

Butter, 16 lbs. 3 oz. Milk, 226 lbs. 8 oz.

Test made from June 19 to 26, 1897; age, 9 yrs. 1 mo.; estimated weight, 900 lbs.; fed 35 lbs. bran, 28 lbs. oats, 35 lbs. corn meal and 14 lbs. oil meal—mixed pasture; property of H. C. Taylor, Orfordville, Wis.

Compass' Pansy 105706.—Sire, Compass 16958; dam, Upstart 87903.

Butter, 16 lbs. 10 oz. Milk, 267 lbs. 10 oz.

Test made from Sept. 28 to Oct. 5, 1897; age, 4 yrs. 3 mos.; estimated weight, 800 lbs.; fed 4 lbs. corn meal, 5 lbs. oat meal, 5 lbs. bran and 1½ lbs. oil meal, daily—some hay, ordinary pasture; property of Richardson Bros., Davenport, Iowa.

Complexia 56774.—Sire, Combination 4389; dam, Commotion 52960.

Butter, 15 lbs. 3 oz. Milk, 249 lbs. 1 oz.

Test made from Nov. 10 to 17, 1888; age, 3 yrs. 10 mos.; estimated weight, 775 lbs.; fed 18 qts. meal, daily; property of Richardson Bros., Davenport, Iowa.

Composite 58774.—Sire, Diploma 16219; dam, Damask Lawrence 14850.

Butter, 14 lbs. 15 oz. Milk, 217 lbs.

Test made from Nov. 9 to 16, 1890; age, 3 yrs. 2 mos.; estimated weight, 800 lbs.; fed, daily, 15 qts. ground oats, corn and peas; property of Richardson Bros., Davenport, Iowa.

Composite 58774.—Sire, Diploma 16219; dam, Damask Lawrence 14850.

Butter, 18 lbs. 10 oz. Milk, 288 lbs. 10 oz.

Test made from Oct. 14 to 21, 1894; age, 7 yrs. 1 mo.; estimated weight, 850 lbs.; fed 5 qts. corn meal, 7 qts. oat meal, 7 lbs. bran and 6 qts. roots, daily—clover pasture; property of Richardson Bros., Davenport, Iowa.

Compressa 55756.—Sire, Upright 6147; dam, Commotion 52960.

Butter, 14 lbs. 13½ oz. Milk, 248 lbs. 3 oz.

Test made from Apr. 21 to 28, 1889; age, 3 yrs. 3 mos.; estimated weight, 900 lbs.; fed 18 qts. of meal, daily; property of Richardson Bros., Davenport, Iowa.

Comtesse de St. Ouen 22600.—Sire, Sir Harry (F. S. 314 J. H. B.); dam, Rachel 24522.

Butter, 15 lbs. 8½ oz. Milk, 228 lbs. 3 oz.

Test made from Jan. 6 to 13, 1886; age, 4 yrs.; property of L. S. Sprague, West Stockbridge, Mass.

Comtesse D'Espagne 10308.—Sire, Ben Weston 3111; dam, Beletrice 7687.

Butter, 14 lbs. ½ oz. Milk, 204 lbs.

Test made from Aug. 15 to 21, 1884; age, 5 yrs. 11 mos.; property of Miller & Sibley, Franklin, Pa.

Conger's M. 32751.—Sire, Alphea Star 7487; dam, Mattie E. 16274.

Butter, 16 lbs. 2½ oz.   Milk, 183 lbs. 8 oz.

Test made from Sept. 24 to Oct. 1, 1890; age, 5 yrs. 4 mos.; estimated weight, 800 lbs.; fed, daily, 24 lbs. corn fodder, 8 lbs. shorts and ½ lb. oil-cake meal; property of Mrs. M. B. Pack, Edinburg, Mo.

Connoisseur 65644.—Sire, Ballad 20239; dam, Miralba 65002.

Butter, 20 lbs. 13 oz.   Milk, 304 lbs.

Test made from July 4 to 11, 1895; age, 6 yrs. 2 mos.; estimated weight, 1000 lbs.; fed 35 lbs. oats, 56 lbs. corn meal, 28 lbs. oil meal and 21 lbs. bran—mixed pasture; property of H. C. Taylor, Orfordville, Wis.

Conover's Beauty 2d 25315.—Sire, Prince of Warren 1512; dam, Conover's Beauty 12650.

Butter, 18 lbs. ¾ oz.   Milk, 252 lbs.

Test made from June 20 to 27, 1883; age, 5 yrs. 3 mos.; property of William A. Conover, Hackettstown, N. J.

Content of Linwood 6950, imp.

Butter, 14 lbs. 12 oz., unsalted.   Milk, 158 lbs.

Test made from Mar. 5 to 12, 1883; age, 7 yrs.; property of M. M. Gardner, Nashville, Tenn.

Coomassie Darlton 29519.—Sire, Duke of Darlington 2d 6948; dam, Augerez Georgiana 24421.

Butter, 16 lbs. 11 oz.   Milk, 221 lbs. 1 oz.

Test made from Mar. 25 to Apr. 1, 1891; age, 6 yrs. 6 mos.; estimated weight, 750 lbs.; fed 3 gals. corn-hearts, 2 gals. wheat bran and 2 gals. ground oats, on corn ensilage—clover hay *ad lib.*; property of Matthews & Moore, Huntsville, Ala.

Coomassie Fawkes 122445.—Sire, Duke of Hill View 22179; dam, Benefit's Daughter 63932.

Butter, 17 lbs. 8 oz.   Milk, 377 lbs.

Test made from Mar. 25 to Apr. 1, 1897; age, 2 yrs. 2 mos.; estimated weight, 1000 lbs.; fed 1 bush. night and morning of potatoes, apples and cut straw, with 8 lbs. middlings, corn meal, oil meal and wheat bran mixed in; ¼ bush. carrots daily—84 lbs. hay; property of Mrs. H. E. Tremain, Lake George, N. Y.

Coomassie L. 73273.—Sire, Sherbet's Coomassie 24115; dam, Lady Mabel 8133.

Butter, 15 lbs. 1 oz.   Milk, 296 lbs. 10 oz.

Test made from Feb. 1 to 8, 1896; age, 6 yrs. 4 mos.; estimated weight, 1000 lbs., fed 30 qts. daily of a mixture consisting of 4 parts corn, 4 parts oats, ground, and mixed with 1 part flaxseed meal—little hay, 14 qts. potatoes, sugar-cane fodder, property of Linscott Bros., Holton, Kan.

Coomassie of Ingleside 74045.—Sire, Duke of Ingleside 14274; dam, Coomassiella 2d 9860.

Butter, 17 lbs. 3½ oz.   Milk, 266 lbs. 8 oz.

Test made from June 16 to 23, 1894; age, 4 yrs. 4 mos.; actual weight, 790 lbs.; no record of feed kept—good old meadow pasture; property of Miller & Sibley, Franklin, Pa.

Coomassie Veda 55838.—Sire, Kathy's Stoke Pogis 17566; dam, Coomassie Darlton 29519.

Butter, 17 lbs. 3 oz. Milk, 237 lbs. 11 oz.

Test made from Jan. 15 to 22, 1892; age, 4 yrs. 1 mo.; estimated weight, 750 lbs.; fed 8 qts. corn meal, 6 qts. boiled cotton-seed and 6 qts. wheat bran, daily—ensilage, clover hay *ad lib.*; property of Matthews & Moore, Huntsville, Ala.

Copper 1979.—Sire, Hockanum 792; dam, Rose 3d 913.

Butter, 15 lbs. 7 oz.

Test made from June 6 to 13, 1878; age, 8 yrs.; property of Miller Ketchum, Westport, Conn.

Copper Queen 53659.—Sire, Golden Ray 10669; dam, Copper 2d 2960.

Butter, 14 lbs. 12 oz. Milk, 266.7 lbs.

Test made from Apr. 5 to 12, 1897; age, 8 yrs. 7 mos.; actual weight, 950 lbs.; fed 4 lbs. bran, 2 lbs. shorts, 3 lbs. gluten, 8 lbs. old process oil meal and 3 lbs. ground oats and 45 lbs. ensilage, daily; property of Storrs Agricultural College, Storrs, Conn.

Coquette of Glen Rouge 17559.—Sire, Jack Frost of St. Lambert 2419; dam, Sweet Brier of St. Lambert 5481.

Butter, 15 lbs. 1½ oz. Milk, 209 lbs.

Test made from June 17 to 23, 1884; age, 3 yrs. 2 mos.; property of Mrs. Emily Reesor, Toronto, Ont., Can.

Coquette of Glen Rouge 17559.—Sire, Jack Frost of St. Lambert 2419; dam, Sweet Brier of St. Lambert 5481.

Butter, 22 lbs. 14 oz. Milk, 232 lbs. 10 oz.

Test made from Feb. 10 to 17, 1891; age, 9 yrs. 11 mos.; actual weight, 1080 lbs.; fed, daily, 6 lbs. oil meal, 6 lbs. pea meal, 8 lbs. corn meal and 10 lbs. crushed oats; property of Ayer & McKinney, Philadelphia, Pa.

Coquette of Glen Rouge 17559.—Sire, Jack Frost of St. Lambert 2419; dam, Sweet Brier of St. Lambert 5481.

Butter, 93 lbs. 4 oz. in 31 days. Milk, 1014 lbs. 2 oz. Cream, 339 lbs.

Test made from Jan. 8 to Feb. 7, 1891; age, 9 yrs. 11 mos.; fed 930 lbs. grain, 800 lbs. hay and 800 lbs. ensilage; property of Ayer & McKinney, Philadelphia, Pa.

Coquette's Molly 23841.—Sire, Lofty 2856; dam, Coquette 2d 1933.

Butter, 14 lbs. 2 oz. Milk, 192 lbs. 13 oz.

Test made from Mar. 7 to 14, 1886; age, 6 yrs. 6 mos.; property of J. L. Morrow, Nashville, Tenn.

Cora Belmont 2d 48868.—Sire, Bunker 9025; dam, Cora Belmont 18869.

Butter, 19 lbs. 1 oz. Milk, 308 lbs.

Test made from June 22 to 29, 1890; age, 4 yrs. 3 mos.; actual weight, 820 lbs.; fed, daily, 8 qts. corn meal, 5 qts. oil meal and 10 qts. ground oats; property of T. R. Proctor, Utica, N. Y.

Cora of Arcadia 16151.—Sire, Rocco 4517; dam, Clio of Staatsburgh 2d 12540.

Butter, 15 lbs. Milk, 200 lbs. 4 oz.

Test made from July 14 to 21, 1886; age, 5 yrs. 4 mos.; estimated weight, 900 lbs.; pasture only, no grain; property of Jacob Lusk, East Palmyra, N. Y.

Cora of Hillside 25253.—Sire, Lucullus 2695; dam, Cora of Lebanon 11637.

Butter, 14 lbs. 7 oz. Milk, 156 lbs. 1 oz.

Test made from June 11 to 18, 1886; age, 4 yrs. 10 mos.; property of David Strong, Winsted, Conn.

Cora of Hillside 25253.—Sire, Lucullus 2695; dam, Cora of Lebanon 11637.

Butter, 15 lbs. 7 oz. Milk, 300 lbs. 7 oz.

Test made from May 13 to 19, 1887; age, 5 yrs. 9 mos.; estimated weight, 800 lbs.; fed 8 qts. per day, one-third corn and two-thirds oats, ground; property of David Strong, Winsted, Conn.

Cora of St. Lambert 8347.—Sire, Stoke Pogis 3d 2238; dam, Lucy of St. Lambert 5116.

Butter, 21 lbs. 6¾ oz. Milk, 239 lbs. 8 oz.

Test made from June 4 to 10, 1886; age, 8 yrs. 3 mos.; fed 6 qts. ground oats, 6 qts. bran and 3 qts. pea meal, mixed, daily—good pasture; property of Valancey E. Fuller, Hamilton, Ont., Can.

Cora of Sandusky 42038.—Sire, Primrose's Carlo 12174; dam, Rose of Spring Creek 25875.

Butter, 15 lbs. 4½ oz. Milk, 188 lbs.

Test made from Oct. 4 to 11, 1895; age, 9 yrs. 1 mo.; estimated weight, 900 lbs.; fed 4 qts. ground oats, 4 qts. wheat bran, 4 qts. corn meal, mixed with 6 qts. roots, in three feeds, daily—hay *ad lib.*; property of Flora A. Howe, Rochester, N. Y.

Cora's Beauty 92179.—Sire, Victor F. W. 16548; dam, Cora of Ashland 57500.

Butter, 15 lbs. 9 oz. Milk, 237 lbs. 5 oz.

Test made from Mar. 21 to 28, 1897; age, 4 yrs. 10 mos.; estimated weight, 850 lbs.; fed 49 lbs. bran, 49 lbs. corn, oats and buckwheat, ground together in equal parts by weight, 17½ lbs. corn meal and about 175 lbs. hay; property of J. C. Averill, Angelica, N. Y.

Cora Scituate 2d 94979.—Sire, Melia Ann's Stoke Pogis 22042; dam, Cora Scituate 47524.

Butter, 14 lbs. Milk, 245 lbs. 12 oz.

Test made from June 28 to July 5, 1897; age, 3 yrs. 4 mos.; estimated weight, 800 lbs.; fed about 10 lbs. gluten and bran, daily—best of blue-grass and white clover pasture; property of W. J. Hussey, Mount Pleasant, Ohio.

Cordelia Baker 8814.—Sire, Beeswax 1931; dam, Pussy Baker 6994.

Butter, 17 lbs. 9 oz.

Test made from May 1 to 7, 1883; age, 5 yrs. 2 mos.; property of James B. Wilder, Louisville, Ky.

Cordelia Signal 33452.—Sire, Wanderer 3014; dam, Second Cousin 12689.

Butter, 18 lbs. 15½ oz. Milk, 199 lbs. 8 oz.

Test made from July 15 to 22, 1889; age, 4 yrs. 8 mos.; estimated weight, 1000 lbs.; fed, daily, 4 qts. corn meal, 2 qts. ground oats and 2 qts. flaxseed meal; property of Carson & Bro., Crab Orchard, Ky.

Cordelia Signal 2d 44489.—Sire, Signal Lad 12199; dam, Cordelia Signal 33452.

Butter, 17 lbs. 6 oz. Milk, 237 lbs.

Test made from June 8 to 15, 1889; age, 2 yrs. 10 mos.; estimated weight, 750 lbs., fed, daily, about 9 lbs. corn meal, 6 lbs. wheat bran and 1½ lbs. flaxseed meal; property of Carson & Bro., Crab Orchard, Ky.

Cordelia Signal 3d 57808.—Sire, Signal Lad 12199; dam, Cordelia Signal 33452.

Butter, 17 lbs. 8 oz. Milk, 194 lbs. 8 oz.

Test made from Oct. 23 to 30, 1893; age, 6 yrs. 4 mos.; estimated weight, 900 lbs.; fed, daily, 3½ lbs. wheat, 8½ lbs. wheat bran, 6½ lbs. oats and 6½ lbs. corn—excellent blue-grass pasture; property of F. M. & O. P. Jones, Anderson, Ind.

Cordelia Signal 5th 83097.—Sire, Alfonso 3013; dam, Cordelia Signal 33452.

Butter, 14 lbs. 5 oz. Milk, 181 lbs. 8 oz.

Test made from May 24 to 31, 1894; age, 3 yrs. 5 mos.; estimated weight, 850 lbs., fed 4 qts. corn meal, 4 qts. ground oats and 2 qts. new process oil meal, daily—good blue-grass, timothy and clover pasture; property of J. E. Carson, Crab Orchard, Ky.

Corinna 2d 6594.—Sire, Graycoat 1105; dam, Corinna 3907.

Butter, 16 lbs. 7 oz.

Test made from July 1 to 7, 1883; age, 6 yrs. 2 mos.; property of E. M. Phelan, Cherry Valley, N. Y.

Corinne Melrose 54455.—Sire, Prince of Melrose 4819; dam, Corinne Moore 35748.

Butter, 16 lbs. 14 oz. Milk, 190 lbs. 8 oz.

Test made from Jan. 6 to 13, 1892; age, 4 yrs. 9 mos.; estimated weight, 800 lbs.; fed 14 lbs. ensilage, 8 lbs. bran, 4 lbs. ground corn, oats and peas, 2 lbs. cotton-seed meal and 4 lbs. hay, daily—pastured on green oats one hour per day; property of M. Lothrop, Marshall, Tex.

Corinne Moore 35748.—Sire, Sir Signal 3018; dam, Fall Leaf 8587.

Butter, 19 lbs. 8 oz. Milk, 270 lbs. 9 oz.

Test made from Dec. 6 to 13, 1891; age, 7 yrs. 9 mos.; estimated weight, 900 lbs., fed 126 lbs. ensilage, 56 lbs. bran, 42 lbs. corn meal, 14 lbs. oats and 14 lbs. cotton-seed meal; property of M. Lothrop, Marshall, Tex.

Corinne of St. Lambert 76579.—Sire, Pogis' Boom 18295; dam, Beauty of Springvale 26677.

Butter, 15 lbs. 10 oz. Milk, 205 lbs.

Test made from Dec. 5 to 12, 1893; age, 2 yrs. 11 mos.; estimated weight, 800 lbs., fed, daily, 1 gal. corn meal, 2 gals. ground oats, ½ gal. ground peas, ½ gal. raw cotton-seed and 30 lbs. ensilage—hay *ad lib.*; property of James Crook, Jacksonville, Ala.

Corinne's Dollie Rex 110987.—Sire, Chieftain Rex 29707; dam, Corinne Bright 66074.

Butter, 15 lbs. 4 oz. Milk, 233 lbs. 4 oz.

Test made from July 19 to 26, 1897; age, 3 yrs. 5 mos.; estimated weight, 700 lbs.; fed 10 lbs. daily of a mixture of ground corn and oats, in equal parts by weight—wild grass and clover pasture; property of S. B. Pierce, Woodlawn, Neb.

Corn 10504.—Sire, Orawapum 2833; dam, Veronica 6684.

Butter, 16 lbs. 2 oz.

Test made from June 4 to 10, 1883; age, 3 yrs. 8 mos.; no grain fed—good blue-grass and white clover pasture; property of J. Fisher, Urbana, Ohio.

Cornwall Maid 19024.—Sire, Ramapo 4679; dam, Lady Cornwall 7179.

Butter, 14 lbs. 9 oz. Average of milk, 17 qts. per day.

Test made from June 21 to 27, 1886; age, 4 yrs. 2 mos.; property of D. F. Appleton, Ipswich, Mass.

Cornwall Maid 19024.—Sire, Ramapo 4679; dam, Lady Cornwall 7179.

Butter, 29 lbs. 12 oz. Milk, 19½ qts. per day.

Test made from July 21 to 27, 1887; age, 5 yrs. 3 mos.; estimated weight, 900 lbs.; fed 6 qts. corn meal, 7 qts. oat meal, 2 qts. pea meal, 1 qt. oil meal and 6 qts. middlings, daily; property of D. F. Appleton, Ipswich, Mass.

Coronilla 8367.—Sire, Champion of America 1567; dam, Mayhew Belle 3883.

Butter, 14 lbs. 7 oz. Milk, 208 lbs.

Test made from June 30 to July 6, 1884; age, 5 yrs. 3 mos.; property of W. B. Montgomery, Starkville, Miss.

Cosmetic 75357.—Sire, Cadet 23958; dam, Flora Brown 75356.

Butter, 19 lbs. 1 oz. Milk, 239 lbs. 8 oz.

Test made from Mar. 6 to 13, 1896; age, 4 yrs. 8 mos.; estimated weight, 975 lbs.; fed 49 lbs. corn meal, 42 lbs. oats, 35 lbs. bran and 14 lbs. oil meal—corn ensilage and clover hay; property of H. C. Taylor, Orfordville, Wis.

Costa Rica 64570.—Sire, Upright 6147; dam, Modita 16626.

Butter, 21 lbs. 6½ oz. Milk, 282 lbs. 5 oz.

Test made from Nov. 11 to 18, 1896; age, 11 yrs. 9 mos.; estimated weight, 900 lbs.; fed 42 lbs. bran, 15 lbs. corn, 42 lbs. oats, 14½ lbs. linseed meal and 280 lbs. ensilage—hay ad lib.; property of C. I. Hood, Lowell, Mass.

Costa Rica 64570.—Sire, Upright 6147; dam, Modita 16626.

Butter, 87 lbs. 14¼ oz. in 30 days. Milk, 1225 lbs. 9 oz.

Test made from Nov. 4 to Dec. 4, 1896; age, 11 yrs. 10 mos.; estimated weight, 900 lbs.; fed 181 lbs. bran, 174 lbs. ground oats, 94 lbs. corn meal, 67 lbs. oil meal, 300 lbs. roots and 1050 lbs. ensilage; property of C. I. Hood, Lowell, Mass.

Costa Rica 64570.—Sire, Upright 6147; dam, Modita 16626.

Butter, 90 lbs. 11¼ oz. in 31 days. Milk, 1264 lbs. 9 oz.

Test made from Nov. 4 to Dec. 5, 1896; age, 11 yrs. 10 mos.; estimated weight, 900 lbs.; fed 187½ lbs. bran, 180 lbs. ground oats, 99 lbs. corn meal, 70 lbs. oil meal, 310 lbs. roots and 1085 lbs. ensilage—hay ad lib.; property of C. I. Hood, Lowell, Mass.

Cottage Lass 5332.—Sire, Guy Mannering 698; dam, Clytie Lass 3395.

Butter, 14 lbs. 8 oz. (official). Milk, 217 lbs. 12 oz.

Test made from May 16 to 23, 1883; age, 6 yrs. 9 mos.; property of Columbia Jersey Cattle Co., Columbia, Tenn

Cottage Lass 5332.—Sire, Guy Mannering 698; dam, Clytie Lass 3395.

Butter, 17 lbs. 7 oz. Milk, 201 lbs.

Test made from June 5 to 11, 1884; age, 7 yrs. 10 mos.; property of Columbia Jersey Cattle Co., Columbia, Tenn.

Cotton Patchwork 85926.—Sire, King Cotton 25391; dam, Gretna Coomassie 71177.

Butter, 16 lbs. 1 oz. Milk, 211 lbs. 11 oz.

Test made from Nov. 22 to 29, 1895; age, 4 yrs. 5 mos.; actual weight, 900 lbs.; fed 56 qts. corn and cob meal and 56 qts. coarse wheat bran, mixed with cut corn fodder—hay *ad lib.*, meadow pasture; property of Columbus Dixon, Gillespieville, Ohio.

Couch's Lily 3237.—Sire, Albert 44; dam, Lily Dale 3236.

Butter, 16 lbs. 5½ oz.

Test made June, 1874; age, 5 yrs. 4 mos.; property of J. O. Couch, Middlefield, Conn.

Couch's Lily 3237.—Sire, Albert 44; dam, Lily Dale 3236.

Butter, 71 lbs. in 31 days.

Test made in 1874; age, 5 yrs.

Countess 114, imp.

Butter, 16 lbs.

Property of Winslow S. Lincoln, Worcester, Mass.

Countess Alberta 42922.—Sire, Champion Albert 6023; dam, Countess Coomassie 19339.

Butter, 14 lbs. 2 oz. Milk, 245 lbs. 3 oz.

Test made from July 14 to 21, 1889; age, 3 yrs. 7 mos.; estimated weight, 800 lbs., fed 6 qts. crushed oats, 6 qts. chopped corn and 8 qts. wheat bran, daily, fed dry, with a little salt—fairly good blue-grass pasture; property of George E. McGill & Son, Leavenworth, Kan.

Countess Bee 32630.—Sire, Polonius De Pansy 6625; dam, Alphea Alexia 15951.

Butter, 14 lbs. 4 oz. Milk, 160 lbs.

Test made from June 10 to 16, 1885; age, 2 yrs. 1 mo.; property of Joseph C. Kiplinger & Co., Springfield, Ohio.

Countess Buttercup 13505.—Sire, Young Baltimore Boy 2048; dam, Glenmore Belle 4801.

Butter, 14 lbs. 3 oz. Milk, 175 lbs.

Test made from June 12 to 18, 1885; age, 5 yrs. 1 mo.; property of Andrew J. Fish, Lima, Ohio.

Countess Coomassie 19339.—Sire, Silver Mine 1658; dam, Russie Hill of Maxwell 8661.

Butter, 16 lbs. 10 oz. Milk, 260 lbs. 12 oz.

Test made from Feb. 11 to 17, 1885; age, 4 yrs. 4 mos.; property of M. G. Jacobs, Independence, Mo.

Countess Europa 35820.—Sire, Chief Modoc 1718; dam, Queenie 4696.

Butter, 17 lbs. 12 oz. Milk, 264 lbs. 5 oz.

Test made from Nov. 16 to 23, 1889; age, 9 yrs. 2 mos.; estimated weight, 900 lbs.; fed, daily, 6 lbs. corn meal, 4 lbs. bran and 2 lbs. linseed meal, mixed; property of C. W. Talmadge, Council Grove, Kan.

Countess Gilderine 29027.—Sire, Gold Basis 4038; dam, Countess Gisela of Belle Vue 9571.

Butter, 16 lbs. 10½ oz. Milk, 233 lbs. 12 oz.

Test made from Feb. 8 to 15, 1891; age, 6 yrs. 6 mos.; estimated weight, 850 lbs., fed 18 lbs. per day of ground oats and corn, in equal parts, mixed with 4 bucketfuls of cut sheaf oats, wetted—millet *ad lib.*; property of May Overton, Nashville, Tenn.

Countess Gisela of Belle Vue 9571.—Sire, Lord Lawrence 1414; dam, Mary Jane of Belle Vue 6956.

Butter, 15 lbs. 11 oz., unsalted. Milk, 213 lbs. 10 oz.

Test made from Oct. 17 to 31, 1883; age, 4 yrs. 6 mos.; fed 4 qts. corn meal, 4 qts. ground oats, 4 gals. cut sheaf oats, daily—short orchard and blue-grass pasture; property of M. M. Gardner, Nashville, Tenn.

Countess Godiva 10820.—Sire, Osiris 3792; dam, Calista Appel 10760.

Butter, 15 lbs. 7 oz. Milk, 210 lbs.

Test made from June 21 to 26, 1885; age, 5 yrs. 4 mos.; the milk of one day was accidentally lost; property of H. M. Baum, Frankfort, Ind.

Countess Hebe 2d 45899.—Sire, Rex of Jersey 12878; dam, Countess Hebe 8102.

Butter, 14 lbs. 3 oz. Milk, 241 lbs.

Test made from July 2 to 9, 1893; age, 8 yrs. 3 mos.; estimated weight, 900 lbs., fed 14 lbs. bran, 37 lbs. corn meal, 14 lbs. middlings, 14 lbs. oil meal and 55 lbs. oats —old pasture, not good; property of J. E. Tyrrell, Marysville, Ohio.

Countess Lowndes 26874.—Sire, Babylon 4723; dam, Tamy 3d 7127.

Butter, 17 lbs. 8 oz. Milk, 210 lbs.

Test made from June 18 to 24, 1884; age, 2 yrs. 3 mos.; property of W. A. Conover, Hackettstown, N. J.

Countess Matilda 74928.—Sire, Matilda's Perfection 17100; dam, Countess Dee 18061.

Butter, 15 lbs. Milk, 230 lbs. 4 oz.

Test made from July 6 to 13, 1897; age, 5 yrs. 9 mos.; estimated weight, 850 lbs.; fed 56 lbs. corn meal, 70 lbs. wheat bran and 28 lbs. linseed meal—timothy and clover pasture; property of Henry S. Redfield, Elmira, N. Y.

Countess Matilda 74928.—Sire, Matilda's Perfection 17100; dam, Countess Dee 18061.

Butter, 19 lbs. 11 oz. (confirmed); estimated butter on basis of 85 per cent. fat, 19.96 lbs. Milk, 270 lbs. 3 oz.

Test made from May 31 to June 6, 1898; age, 6 yrs. 8 mos.; estimated weight, 850 lbs.; fed 84 lbs. ground oats and 7 lbs. cotton-seed meal—cut clover and timothy night and morning, mixed timothy and clover hill pasture; property of Henry S. Redfield, Elmira, N. Y.

Analyses of Butter.

First churning—Fat 86.18. Second churning—Fat 87.89. Third churning—Water 9.70, fat 87.67, curd 1.31, ash 0.04, salt 1.28.

Countess Matilda 74928.—Sire, Matilda's Perfection 17100; dam, Countess Dee 18061.

Butter, 82 lbs. 15 oz. in 30 days. Milk, 532 lbs. 14 oz.

Test made from June 1 to July 1, 1898; age, 6 yrs. 7 mos.; estimated weight, 850 lbs.; fed 276 lbs. ground oats, 52 lbs. cotton-seed meal and 40 lbs. wheat bran—hill pasture, timothy and clover; property of Henry S. Redfield, Elmira, N. Y.

Countess Matilda 74928.—Sire, Matilda's Perfection 17100; dam, Countess Dee 18061.

Butter, 220 lbs. 6 oz. in 90 days. Milk, 3171 lbs.

Test made from Apr. 20 to July 19, 1898; age, 6 yrs. 6 mos.; estimated weight, 850 lbs.; fed 258 lbs. gluten meal, 39 lbs. corn meal, 674 lbs. ground oats, 122 lbs. wheat bran, 20 lbs. oil meal and 100 lbs. cotton-seed meal—hay, hill pasture; property of Henry S. Redfield, Elmira, N. Y.

Countess Micawber 1759.—Sire, Mr. Micawber 556; dam, Countess 3d 990.

Butter, 17 lbs. 1 oz. Milk, 152 lbs. 14 oz.

Test made from Apr. 21 to 28, 1884; age, 12 yrs. 2 mos.; property of Winslow S. Lincoln, Worcester, Mass.

Countess of Croton 5307.—Sire, Grand Duke Alexis 1040; dam, Chrissy 1448.

Butter, 15 lbs. 12 oz. Milk, 277 lbs. 4 oz.

Test made from June 20 to 27, 1881; age, 5 yrs.; property of William D. Burdett, Cooperstown, N. Y.

Countess of Lakeside 12135.—Sire, Dick Swiveller Jr. 276; dam, Betty 683.

Butter, 19 lbs. 7 oz., unsalted. Milk, 251 lbs.

Test made from Apr. 6 to 15, 1882; age, 14 yrs. 1 mo.; fed 8 qts. of a mixture of equal parts of oats, middlings and new process linseed meal, daily—hay; property of J. H. Walker, Worcester, Mass.

Countess of Lorne 20822.—Sire, Spangle's Boy 6229; dam, Maggie of Lawnfield 14153.

Butter, 14 lbs. 14 oz. Milk, 207 lbs. 10 oz.

Test made from June 15 to 22, 1885; age, 4 yrs. 3 mos.; property of E. Stevens Henry, Rockville, Conn.

Countess of Scarsdale 18633.—Sire, John Street 6156; dam, Beautiful 5981.

Butter, 14 lbs. 6 oz. Milk, 230 lbs.

Test made from Jan. 31 to Feb. 6, 1885; age, 2 yrs. 2 mos.; property of F. L. Gaston, Normal, Ill.

Countess of Warren 3896.—Sire, Hector 129; dam, Julia 3893.

Butter, 14 lbs.

Test made Mar., 1876; age, 5 yrs.; property of Chester Bordwell, Batavia, Ohio.

Countess Potoka 7496.—Sire, Fanchon's King 2637; dam, Carrie Lena 3348.

Butter, 18 lbs. 15 oz., unsalted. Milk, 220 lbs. 9 oz.

Test made from May 14 to 20, 1882; age, 4 yrs. 1 mo.; fed 12 qts. corn meal, daily—very good blue-grass pasture; property of Thomas H. Malone, Nashville, Tenn.

Countess Queen 13519.—Sire, Gen. Rosser 4189; dam, Queen of Jersey 4948.

Butter, 18 lbs. 3 oz.

Test made from May 29 to June 4, 1884; age, 3 yrs.; property of Mills & Walker, Greenville, S. C.

Countess Stoke Pogis 62540.—Sire, Garfield Stoke Pogis 15963; dam, Countess Snap Pogis 36807.

Butter, 16 lbs. 3 oz.   Milk, 194 lbs. 8 oz.

Test made from June 5 to 12, 1892; age, 3 yrs. 1 mo.; estimated weight, 850 lbs., fed 56 lbs. cut corn fodder, 7 lbs. corn meal, 14 lbs. gluten meal, 14 lbs. bran, 7 lbs. linseed meal, 56 lbs. hay and 210 lbs. roots; property of Est. of Frederick Billings, Woodstock, Vt.

Country Girl 3515.—Sire, Quaker 887; dam, Clara 4th 1541.

Butter, 15 lbs. 12 oz.   Milk, 174 lbs. 15 oz.

Test made from Oct. 9 to 16, 1890; age, 16 yrs. 6 mo.; estimated weight, 950 lbs.; fed, daily, 14 qts. ground corn, oats and peas; property of Charles E. Hill, Denver, Col.

Country Girl 3d 12538.—Sire, Copake 1387; dam, Country Girl 3515.

Butter, 17 lbs. 11 oz.   Milk, 219 lbs. 15 oz.

Test made from Oct. 7 to 14, 1890; age, 11 yrs. 6 mos.; estimated weight, 950 lbs.; fed, daily, 18 qts. ground corn, oats, peas and shorts; property of Charles E. Hill, Denver, Col.

Country Girl 4th 51877.—Sire, Magog 1868; dam, Country Girl 3515.

Butter, 18 lbs. 2 oz.   Milk, 232 lbs. 13 oz.

Test made from Oct. 15 to 22, 1890; age, 8 yrs. 7 mos.; estimated weight, 950 lbs.; fed, daily, 18 qts. ground oats, corn and peas, with a little shorts; property of Charles E. Hill, Denver, Col.

Count's Fillpail 30975.—Sire, Count Cicero (F. S. 398 J. H. B.); dam, Fille de l'Air (F. S. 3548 J. H. B.)

Butter, 24 lbs. 5 oz.   Milk, 170 lbs. 12 oz.

Test made from Feb. 23 to 29, 1888; age, 3 yrs. 9 mos.; weight, 830 lbs.; fed, daily, 6 lbs. corn and oats, 4 lbs. bran, 2 lbs. flax meal and 1 pt. condimental food; property of M. Erskine Miller, Staunton, Va.

Coupon Diamond 103559.—Sire, Prince of Montverde 2d 19975; dam, Daisy Diamond 103558.

Butter, 14 lbs. 5 oz.   Milk, 226 lbs. 14 oz.

Test made from Dec. 16 to 23, 1897; age, 3 yrs. 2 mos.; estimated weight, 850 lbs., fed 80 lbs. ear corn and oats, ground together, 52 lbs. wheat bran, 17 lbs. cotton-seed meal, 20 lbs. cotton-seed hulls and about 13 lbs. corn fodder; property of Gilbert C. Butler, Minden, La.

Cowles' Nonesuch 6199.—Sire, Pulaski 1932; dam, Cintra Cowles 6198.

Butter, 14 lbs. 12 oz.   Milk, 255 lbs. 4 oz.

Test made from Oct. 5 to 11, 1883; age, 7 yrs. 4 mos.; property of Mrs. L. M. Fair, Wallingford, Conn.

Cowry 4432.—Sire, The Squire 1298; dam, Cicely 997.

Butter, 14 lbs. 1 oz.   Milk, 188 lbs. 15 oz.

Test made from Dec. 21 to 27, 1884; age, 9 yrs. 3 mos.; property of Agricultural Experiment Station, Madison, Wis.

Cowslip 5th 849.—Sire, Paterson 11; dam, Cowslip 893.

Butter, 15 lbs. 4 oz.

Property of L. Henry Twaddell, West Philadelphia, Pa.

Cowslip of Rockwell 115354.—Sire, Warren's Prince 27973; dam, Addie Tormentor 78464.

Butter, 14 lbs. 13½ oz. Milk, 233 lbs. 12 oz.

Test made from Nov. 27 to Dec. 4, 1897; age, 3 yrs. 2 mos.; actual weight, 925 lbs.; fed 58 lbs. bran, 36 lbs. cotton-seed, 96 lbs. corn and cob meal, 42 lbs. hay and 280 lbs. ensilage; property of Biltmore Farms, Biltmore, N. C.

Cowslip of St. Lambert 8349.—Sire, Stoke Pogis 3d 2238; dam, Witch of St. Lambert 5479.

Butter, 17 lbs. 12 oz. Milk, 229 lbs. 8 oz.

Test made from Mar. 18 to 24, 1884; age, 5 yrs. 10 mos.; property of Valancey E. Fuller, Hamilton, Ont., Can.

Cream Calla 40233.—Sire, Europa's Duke 4832; dam, Creamlie Westfield 17818.

Butter, 16 lbs. 7 oz. Milk, 262 lbs. 12 oz.

Test made from Sept. 30 to Oct. 7, 1890; age, 5 yrs. 7 mos.; actual weight, 790 lbs.; no grain fed, pasture only; property of Mrs. J. H. Martin, Granville, N. Y.

Cream Caroline 17357.—Sire, McHenry 5890; dam, Charlton Caroline 11724.

Butter, 16 lbs. 2 oz. Milk, 252 lbs. 8 oz.

Test made from Jan. 16 to 23, 1893; age, 10 yrs. 9 mos.; estimated weight, 1000 lbs.; fed 14 lbs. cotton-seed meal, 14 lbs. Chicago gluten meal, 21 lbs. coarse wheat bran, 21 lbs. corn and cob meal, 250 lbs. corn ensilage and 70 lbs. chopped clover hay, daily, in two feeds; property of E. S. Henry, Rockville, Conn.

Cream Custard 97429.—Sire, Prince of Tennessee 20772; dam, Selita's Cream 50647.

Butter, 19 lbs. 6¼ oz. Milk, 229 lbs. 6½ oz.

Test made from June 1 to 8, 1897; age, 4 yrs. 4 mos.; estimated weight, 700 lbs.; fed 17 lbs. per day of ground oats and corn, in equal parts—very good orchard and blue-grass pasture; property of Matt. M. Gardner, Nashville, Tenn.

Creamer 2467.—Sire, Tom Dasher 420; dam, Creampot 460.

Butter, 14 lbs. 1 oz. Milk, 167 lbs. 13 oz.

Test made from June 5 to 12,. 1882; age, 9 yrs. 10 mos.; property of D. B. Dewolf, Lee, Mass.

Cream Lily 24015.—Sire, Silver (P. S. 287 J. H. B.); dam, Lutitia (F. S. 2919 J. H. B.)

Butter, 16 lbs. Milk, 221 lbs.

Test made from Mar. 4 to 10, 1886; age, 4 yrs. 1 mo.; property of James Crook, Jacksonville, Ala.

Cream of Sidney 17028.—Sire, Noble 901; dam, Rosetta of Sidney 4520.

Butter, 17 lbs. 2½ oz. Milk, 245 lbs. 14 oz.

Test made from June 26 to July 2, 1881; age, 5 yrs. 1 mo.; property of Andrew Baker, West Dryden, N. Y.

Cream of Springvale 16621.—Sire, Othello of Elmarch 3780; dam, Lady Pinckard of Springvale 11227.

Butter, 15 lbs. 10 oz. Milk, 229 lbs.

Test made from Sept. 24 to 30, 1885; age, 4 yrs. 1 mo.; property of James Crook, Jacksonville, Ala.

Cream's Pride 47570.—Sire, Baron Hugo 15208; dam, Cream Caroline 17357.

Butter, 17 lbs. 2 oz. Milk, 310 lbs.

Test made from Dec. 9 to 16, 1892; age, 5 yrs. 2 mos.; estimated weight, 800 lbs., fed 14 lbs. cotton-seed meal, 14 lbs. new process linseed meal, 21 lbs. coarse wheat bran, 21 lbs. corn meal, 245 lbs. corn ensilage, 75 lbs. chopped clover hay, mixed, daily; property of E. S. Henry, Rockville, Conn.

Crensa 74203.—Sire, Record 22801; dam, Cauline 15298.

Butter, 16 lbs. 2 oz. Milk, 276 lbs. 8 oz.

Test made from May 15 to 22, 1898; age, 7 yrs.; estimated weight, 850 lbs.; fed 28 lbs. bran, 42 lbs. oats, 12 lbs. oil meal and 28 lbs. corn meal—mixed pasture; property of H. C. Taylor, Orfordville, Wis.

Creole Maid 11017.—Sire, Braxton 1715; dam, Fleta 3859.

Butter, 16 lbs. 15 oz. Milk, 229 lbs.

Test made from Apr. 21 to 27, 1884; age, 4 yrs. 3 mos.; property of Columbia Jersey Cattle Co., Columbia, Tenn.

Creole Maid 2d 31881.—Sire, Toltec 6831; dam, Creole Maid 11017.

Butter, 14 lbs. 1 oz. Milk, 77 lbs. 11 oz.

Test made from Dec. 1 to 7, 1885; age, 1 yr. 9 mos.; property of Columbia Jersey Cattle Co., Columbia, Tenn.

Crescent's Rachael 74738.—Sire, Guy Mannering Pogis 16365; dam, St. Lambert's Crescent 39711.

Butter, 15 lbs. Milk, 278 lbs. 11 oz.

Test made from Feb. 26 to Mar. 4, 1896; age, 6 yrs. 10 mos.; estimated weight, 1000 lbs.; fed 14 lbs. ground oats and corn meal, in equal parts by weight, and 2½ lbs. oil meal, daily—ensilage *ad lib.*, clover hay at noon; property of John O. McClintock, Meadville, Pa.

Cretesia 13657.—Sire, Lord Dartmouth 6302; dam, Daisy Darling 6386.

Butter, 20 lbs. 1 oz. Milk, 269 lbs.

Test made from June 19 to 26, 1886; age, 6 yrs. 1 mo.; fed 6 qts. corn meal, 6 qts. crushed oats, 6 qts. wheat bran, 2 pts. pea meal, 2 pts. oil meal, daily—pasture; property of Cyrus W. Sprague, Austerlitz, N. Y.

Cretesia's Charma 81685.—Sire, Cretesia's Albert Pogis 21730; dam, Rura 21946.

Butter, 15 lbs. 8 oz. Milk, 215 lbs. 15 oz.

Test made from Oct. 25 to Nov. 1, 1895; age, 4 yrs. 6 mos.; actual weight, 850 lbs.; fed 35 lbs. cob meal, 42 lbs. bran and 10½ lbs. old process oil meal—2 bush. cut corn fodder and 5 lbs. hay, daily; property of B. F. Carper, Rosemond, Ill.

Cretesia's Dawson 89995.—Sire, Cretesia's Albert Pogis 21730; dam, Coomassie Dawson 52710.

Butter, 15 lbs. 1 oz. Milk, 247 lbs. 4 oz.

Test made from Oct. 5 to 12, 1897; age, 5 yrs.; estimated weight, 900 lbs.; fed 42 qts. bran, 21 qts. ground oats, 42 qts. corn meal, 10½ qts. oil meal, 136 lbs. ensilage and 136 lbs. hay; property of R. A. Sibley, Rochester, N. Y.

Cretesia's Matilda 90000.—Sire, Cretesia's Albert Pogis 21730; dam, Matilda of Maplewood 49723.

Butter, 14 lbs. 11½ oz. Milk, 265 lbs. 8 oz.

Test made from Apr. 11 to 18, 1896; age, 3 yrs.; estimated weight, 800 lbs.; fed 2 qts. bran, 2 qts. ground oats, 1 qt. corn meal, 1 pt. cotton-seed meal, 1 pt. oil meal, 30 lbs. ensilage, 6 lbs. hay and 8 qts. beets, daily; property of R. A. Sibley, Rochester, N. Y.

Cretesia's Matilda 90000.—Sire, Cretesia's Albert Pogis 21730; dam, Matilda of Maplewood 49723.

Butter, 15 lbs. 15 oz. Milk, 284 lbs.

Test made from Apr. 17 to 24, 1897; age, 4 yrs.; estimated weight, 900 lbs.; fed 70 qts. bran, 7 qts. ground oats, 49 qts. corn meal, 5½ qts. cotton-seed meal, 10½ qts. oil meal, 168 lbs. ensilage, 70 lbs. hay, 56 qts. beets; property of R. A. Sibley, Rochester, N. Y

Cretesia's Rachel 89998.—Sire, Cretesia's Albert Pogis 21730; dam, Rachel Spencer 50974.

Butter, 16 lbs. 8 oz. Milk, 279 lbs. 8 oz.

Test made from Mar. 27 to Apr. 3, 1897; age, 4 yrs. 3 mos.; actual weight, 870 lbs.; fed 49 qts. bran, 42 qts. oat meal, 35 qts. corn meal, 14 qts. oil meal, 105 lbs. ensilage, 105 lbs. hay and 42 qts. roots; property of P. J. Cogswell, Rochester, N. Y.

Cretesia's Silverette 74807.—Sire, Cretesia's Albert Pogis 21730; dam, Nutley Silverette 22410.

Butter, 17 lbs. 3 oz. Milk, 251 lbs.

Test made from Oct. 14 to 21, 1896; age, 6 yrs. 2 mos.; actual weight, 880 lbs.; fed 42 qts. ground oats, 42 qts. bran, 56 qts. corn meal, 14 qts. linseed meal, 110 lbs. ensilage and 126 lbs. hay; property of P. J. Cogswell, Rochester, N. Y.

Cretesia's Tamy 89994.—Sire, Cretesia's Albert Pogis 21730; dam, Theda's Tamy 75085.

Butter, 14 lbs. 8 oz. Milk, 245 lbs. 8 oz.

Test made from Feb. 18 to 25, 1896; age, 3 yrs. 6 mos.; estimated weight, 800 lbs.; fed 3 qts. bran, 3 qts. ground oats, 3 qts. corn meal, 21 lbs. ensilage, 1 pt. oil meal, 10 lbs. hay and 6 qts. beets, daily; property of R. A. Sibley, Rochester, N. Y.

Cretesia's Temisia 89996.—Sire, Cretesia's Albert Pogis 21730; dam, Temisia of Winnikee 46604.

Butter, 18 lbs. 7 oz. Milk, 294 lbs. 8 oz.

Test made from Oct. 10 to 17, 1896; age, 4 yrs. 1 mo.; estimated weight, 900 lbs.; fed 6 qts. bran, 9½ qts. to 10 qts. corn meal, 1½ qts. oil meal and 3 qts. ground oats, daily—18 lbs. hay; property of R. A. Sibley, Rochester, N. Y.

Cretesia's Violet A. 90051.—Sire, Cretesia's Albert Pogis 21730; dam, Violet of Maplewood 49724.

Butter, 14 lbs. 9 oz.   Milk, 270 lbs. 8 oz.

Test made from Oct. 23 to 30, 1897; age, 5 yrs. 2 mos.; actual weight, 960 lbs.; fed 42 lbs. bran, 18 lbs. ground oats, 31½ lbs. corn meal, 4½ lbs. cotton-seed meal, 10½ lbs. oil meal, 126 lbs. hay and 126 lbs. cut corn, from shock; property of R. A. Sibley, Rochester, N. Y.

Cricket's Minnie 26270.—Sire, Prince Harry 5176; dam, Cricket of Belle Vue 9570.

Butter, 15 lbs. 3½ oz.   Milk, 106 lbs.

Test made from Jan. 2 to 9, 1888; age, 4 yrs. 3 mos.; estimated weight, 850 lbs., fed, daily, 20 qts. corn and oats, ground in equal parts; property of Wm. J. Webster, Columbia, Tenn.

Cricket's Minnie 26270.—Sire, Prince Harry 5176; dam, Cricket of Belle Vue 9570.

Butter, 15 lbs. 14 oz.   Milk, 157 lbs. 9 oz.

Test made from Mar. 24 to 31, 1889; age, 5 yrs. 6 mos.; estimated weight, 750 lbs., fed, daily, 4½ gals. ground corn and oats, mixed equally; property of Maury Jersey Farm, Columbia, Tenn.

Crocus of St. Lambert 8351.—Sire, Stoke Pogis 3d 2238; dam, Lolly of St. Lambert 5480.

Butter, 17 lbs. 12 oz.   Milk, 282 lbs.

Test made from May 24 to 30, 1884; age, 6 yrs. 7 mos.; property of Valancey E. Fuller, Hamilton, Ont., Can.

Crotonia Signal 89308.—Sire, Signal Jr. 7166; dam, Duchess of Croton 6717.

Butter, 18 lbs. 8 oz.   Milk, 212 lbs. 4 oz.

Test made from May 17 to 24, 1896; age, 7 yrs. 1 mo.; estimated weight, 800 lbs., fed 12 lbs. per day of a mixture consisting of one-half bran, one-quarter ground oats, one-eighth corn meal and one-eighth cotton-seed meal—natural pasture; property of Geo. W. Sisson, Jr., Potsdam, N. Y.

Croton Maid 5305.—Sire, Signal 1170; dam, Lucilla 2735.

Butter, 21 lbs. 11½ oz.   Milk, 254 lbs. 6 oz.

Test made from June 28 to July 5, 1881; age, 5 yrs. 2 mos.; fed, 13 qts. bran and 5 qts. corn meal, daily—hay and pasture; property of Clark & East, Nashville, Tenn.

Croton Maid 4th 26729.—Sire, Tormentor 3533; dam, Croton Maid 5305.

Butter, 19 lbs. 7 oz.   Milk, 281 lbs. ½ oz.

Test made from Feb. 10 to 17, 1894; age, 10 yrs. 8 mos.; actual weight, 966 lbs.; fed 40 lbs. bran, 40 lbs. corn meal, 30 lbs. old process oil meal and 9 lbs. cotton-seed meal, mixed with shredded sweet corn fodder—clover hay *ad lib.*; property of J. P. Bradbury, Pomeroy, Ohio.

Crown 75198.—Sire, Ida's Rioter of St. L. 13656; dam, La Petite Pogis 28757.

Butter, 14 lbs. 14½ oz.   Milk, 173 lbs.

Test made from Mar. 10 to 17, 1894; age, 2 yrs. 10 mos.; estimated weight, 800 lbs.; fed 6 qts. ground corn and oats, 6 qts. wheat middlings, 9 qts. wheat bran and 1½ qts. oil meal, daily—hay; property of C. A. Sweet, East Aurora, N. Y.

Crewa 75198.—Sire, Ida's Rioter of St. L. 13656; dam, La Petite Pogis 28757.

Butter, 20 lbs. 12 oz. Milk, 310 lbs. 4 oz.

Test made from Oct. 3 to 10, 1895; age, 4 yrs. 5 mos.; actual weight, 975 lbs.; fed 12 qts. ground corn and oats, 12 qts. wheat bran and 3 qts. oil meal, mixed, daily—hay; property of C. A. Sweet, East Aurora, N. Y.

Crust 4775.—Sire, Alpheus 1168; dam, Countess of Windsor 2024.

Butter, 15 lbs. 7 oz.

Test made from Feb. 8 to 14, 1883; age, 7 yrs.; property of James B. Wilder, Louisville, Ky.

Crust 4775.—Sire, Alpheus 1168; dam, Countess of Windsor 2024.

Butter, 16 lbs. 8 oz

Test made from Apr. 1 to 7, 1884; age, 8 yrs. 2 mos.; property of James B. Wilder, Louisville, Ky.

Crusta 29637.—Sire, Combination 4389; dam, Exhibit 23245.

Butter, 16 lbs. 10 oz. Milk, 236 lbs. 10 oz.

Test made from July 1 to 8, 1892; age, 8 yrs.; estimated weight, 850 lbs.; fed 12 qts. corn and oats and pea meal, daily—orchard pasture; property of Richardson Bros., Davenport, Iowa

Cultured Cream 29196.—Sire, Rachel's Duke 7022; dam, Jersey Cream 2d 8519.

Butter, 20 lbs. 12 oz. Milk, 241 lbs. 5 oz.

Test made from Mar. 19 to 26, 1887; age, 3 yrs. 11 mos.; estimated weight, 800 lbs., fed, daily, about 8 qts. meal, 9 qts. ground oats, 8 qts. wheat middlings and 1½ qts. oil-cake; property of John P. Pomeroy, Housatonic, Mass.

Cupid of Lee Farm 5997.—Sire, Stoke Pogis 3d 2238; dam, Cupid of St. Lambert 5104.

Butter, 14 lbs. 6 oz., unsalted. Milk, 199 lbs. 8 oz.

Test made from July 16 to 22, 1882; age, 5 yrs. 3 mos.; property of D. A. Givens, Cynthiana, Ky.

Cupid of St. Lambert 5104.—Sire, Laval 506; dam, Amelia 484

Butter, 14 lbs. ½ oz. Milk, 185 lbs. 8 oz.

Test made from Nov. 16 to 23, 1886; age, 14 yrs. 5 mos.; estimated weight, 1075 lbs., fed, daily, 16 lbs. corn chop, 6 lbs. bran and 6 lbs. cotton-seed meal, mixed; property of Campbell Brown, Spring Hill, Tenn.

Cupid's Cloud 74040.—Sire, Signal Duke 12661; dam, Lady Cloud 2d 26825.

Butter, 17 lbs. 2 oz. Milk, 264 lbs.

Test made from July 31 to Aug. 7, 1895; age, 4 yrs. 8 mos.; actual weight, 1000 lbs.; fed 175 lbs. of a mixture of corn, oats and bran (one-third corn to two-thirds oats, and three-fifths corn and oats to two-fifths bran)—little hay, blue-grass and clover pasture; property of John A. Middelton, Shelbyville, Ky.

Cupid's Jersey Maid 35040.—Sire, Rioter Vulcan 5380; dam, Cupid of Collingwood 6867.

Butter, 55.163 lbs. in 30 days. Milk, 1028.7 lbs.

Test made from Aug. 29 to Sept. 27, 1893, at World's Columbian Exposition; age, 8 yrs.; actual weight, 886 lbs.; fed 372.9 lbs. hay, 76.5 lbs. oil meal, 94 lbs. ensilage, 48 lbs. corn meal, 174 lbs. bran, 67 lbs. oats, 163 lbs. corn-hearts, 58 lbs. cotton-seed, 56.5 lbs. middlings, 18 lbs. carrots and 5 lbs. old hay; property of C. S. Dole, Crystal Lake, Ill.

Curfew 16498.—Sire, Combination 4389; dam, Belinda R. 13150.

Butter, 14 lbs. 7½ oz., unsalted. Milk, 210 lbs. 11 oz.

Test made from Oct. 1 to 8, 1885; age, 3 yrs. 8 mos.; property of Richardson Bros., Davenport, Iowa.

Cymbelina 67201.—Sire, Combination 3d 17576; dam, Comballina 52701.

Butter, 14 lbs. 13 oz. Milk, 217 lbs.

Test made from Apr. 22 to 29, 1894; age, 4 yrs. 6 mos.; estimated weight, 900 lbs.; fed 35 lbs. oats, 21 lbs. bran, 28 lbs. corn meal and 14 lbs. oil meal—corn ensilage; property of Henry A. Egerton, Footville, Wis.

Cynthia A. 43721.—Sire, Alfonso 3013; dam, Chrissy Signal 27790.

Butter, 20 lbs. 10 oz. Milk, 246 lbs.

Test made from May 21 to 28, 1893; age, 7 yrs. 6 mos.; estimated weight, 950 lbs., fed 1 qt. oil meal, 2 qts. corn meal and 2 qts. ground oats, daily—good blue-grass, timothy and clover pasture; property of J. E. Carson, Crab Orchard, Ky.

Cyrene 4th 480.—Sire, Monmouth 210; dam, Cyrene 137.

Butter, 17 lbs. 1 oz. Milk, 112 qts.

Test made from June 2 to 9, 1879; age, 8 yrs.; property of Newton T. Beal, Rogersville, Tenn.

Czaretta 17358.—Sire, Czar of New York 4049; dam, Delle 2d 17787.

Butter, 14 lbs. 7 oz. Milk, 199 lbs. 7 oz.

Test made from June 8 to 15, 1885; age, 3 yrs. 2 mos.; property of E. S. Henry, Rockville, Conn

Czarina of Spring Hill 47568.—Sire, Baron Hugo 15208; dam, Czaretta 17358.

Butter, 15 lbs. 10 oz. Milk, 278 lbs. 8 oz.

Test made from Dec. 31, 1893, to Jan. 7, 1894; age, 6 yrs. 7 mos.; estimated weight, 900 lbs.; fed 3 lbs. coarse wheat bran, 2 lbs. corn meal, 2 lbs. cream gluten, 1 lb. cotton-seed, 35 lbs. corn ensilage and 10 lbs. cut clover hay, daily; property of E. Stevens Henry, Rockville, Conn.

Dacie's Lena 100796.—Sire, Lenox Pogis 27365; dam, Dacie St. Helier 86972.

Butter, 22 lbs. 12 oz. Milk, 280 lbs. 6 oz.

Test made from Feb. 23 to Mar. 2, 1897; age, 2 yrs. 11 mos.; estimated weight, 800 lbs.; fed 74 qts. oats, 73 qts. bran, 14 qts. meal, 14 qts. gluten and 140 lbs. roots; property of D. Wolfe Bishop, Jr., Lenox, Mass.

Daffy of St. Lambert 69894.—Sire, Canada's John Bull 5th 20092; dam, Empress of St. Lambert 56502.

Butter, 14 lbs. 7 oz. Milk, 211 lbs.

Test made from May 31 to June 7, 1892; age, 2 yrs. 9 mos.; estimated weight, 700 lbs.; fed 25 lbs. chopped peas and oats, daily—clover and timothy pasture; property of William Rolph, Markham, Ont., Can.

Daffy Wilcox 2d 18317.—Sire, Secretary 4074; dam, Daffy Wilcox 4046.

Butter, 15 lbs. 5 oz. Milk, 256 lbs. 7 oz.

Test made from July 9 to 16, 1886; age, 4 yrs. 2 mos.; weight, 820 lbs.; fed, daily, 4 gals. ground corn and oats; property of W. Gettys, Athens, Tenn.

Dagmar of Florence 120863.—Sire, Jubilee of Bois d'Arc 29041; dam, Vida's Crown Princess 76010.

Butter, 16 lbs. 8 oz.   Milk, 185 lbs. 10 oz.

Test made from Jan. 31 to Feb. 7, 1898; age, 4 yrs. 7 mos.; estimated weight, 750 lbs.; fed 42 lbs. bran, 35 lbs. corn meal, 21 lbs. gluten meal, 14 lbs. oil meal and 30 lbs. ensilage and hay; property of Austin & Probert, Sandstone, Mich.

Dainty Doris 85497.—Sire, King of St. Lambert 15175; dam, St. L. Gazelle 78870.

Butter, 16 lbs. 8½ oz.   Milk, 195 lbs.

Test made from Feb. 20 to 27, 1896; age, 4 yrs. 10 mos.; estimated weight, 900 lbs.; fed 5 qts. corn and cob-meal, 5 qts. oat meal and 2 qts. oat middlings, daily—cut corn fodder and hay *ad lib.*; property of E. C. Newton, Hudson, Ohio.

Dainty Rioter of St. L. 81168.—Sire, Nestor of St. Lambert 22385; dam, Dainty Bessie 31475.

Butter, 15 lbs. 10 oz.   Milk, 165 lbs. 8 oz.

Test made from July 4 to 11, 1894; age, 2 yrs. 11 mos.; estimated weight, 700 lbs.; fed 8 qts. chop, daily, for first 3½ days, and 10 qts. chop, daily, for rest of test; property of Mrs. A. M. Hallock, Columbus, Ohio.

Dairy 2d 3891.—Sire, The Hub 1009; dam, Dairy 2861.

Butter, 15 lbs. 5½ oz.   Milk, 226 lbs.

Test made from May 27 to June 2, 1883; age, 8 yrs. 4 mos.; property of W. B. Montgomery, Starkville, Miss.

Dairy C. 12227.— Sire, Champion of America 1567; dam, Dairy 2d 3891.

Butter, 15 lbs. ½ oz., unsalted.   Milk, 217 lbs.

Test made from Apr. 25 to May 1, 1883; age, 2 yrs. 9 mos.; property of W. B. Montgomery, Starkville, Miss.

Dairy Dolly 100605.—Sire, Major Hamilton 30164; dam, Pearl Darling 78002.

Butter, 15 lbs.   Milk, 253 lbs. 14 oz.

Test made from June 23 to 30, 1898; age, 6 yrs. 3 mos.; estimated weight, 900 lbs.; fed 17 qts. wheat bran, 8 qts. corn meal and 2 qts. oil meal, daily—pasture; property of Brooks & Pidgeon, Salem, Ohio.

Dairy Pride 4th 21681.—Sire, Victor (P. S. 148 J. H. B.); dam, Dairy Pride 2d (P. S. 37 J. H. B.)

Butter, 16 lbs.   Milk, 302 lbs. 4 oz.

Test made from May 29 to June 5, 1883; age, 3 yrs. 5 mos.; property of William S. Taylor, Burlington, N. J.

Daisette 64004.—Sire, Denise's Tormentor 11823; dam, Tormentor's Daisy 52496.

Butter, 17 lbs. 1 oz.   Milk, 264 lbs. 15 oz.

Test made from Oct. 24 to 31, 1893; age, 5 yrs. 2 mos.; actual weight, 846 lbs.; fed 40 lbs. corn meal, 47 lbs. ground wheat, 14 lbs. old process oil meal, 12 lbs. cottonseed meal and 65 lbs. hay—good meadow aftermath; property of J. P. Bradbury, Pomeroy, Ohio.

Daisette's Tormentress 106532.—Sire, Tormentor Stoke Pogis 20485; dam, Daisette 64004.

Butter, 21 lbs. 1⅛ oz. Milk, 289 lbs. 12 oz.

Test made from Nov. 1 to 8, 1897; age, 4 yrs. 1 mo.; actual weight, 916 lbs.; fed 60 lbs. wheat bran, 36 lbs. fine corn meal, 16 lbs. old process oil meal and 16 lbs. cotton-seed meal—good blue-grass and clover *ad lib.*; property of J. P. Bradbury, Pomeroy, Ohio.

Daisey Dixie 9469.—Sire, Henry Ward Beecher 2297; dam, Dixie 5341.

Butter, 14 lbs. 4 oz. Milk, 205 lbs. 7 oz.

Test made from July 5 to 11, 1884; age, 6 yrs. 6 mos.; property of J. B. Allen. Delavan, Ill.

Daisy 2d 15761.—Sire, Bee's Wing (P. S. 59 J. H. B.); dam, Daisy (F. S. 1260 J. H. B.)

Butter, 15 lbs. 8 oz. Milk, 231½ lbs.

Test made from July 19 to 26, 1883; age, 7 yrs.; property of J. E. Gillingham, Villa-nova, Pa.

Daisy Brown 12213.—Sire, Tormentor 3533; dam, Heedless 9522.

Butter, 17 lbs. 6½ oz. Milk, 237 lbs. 8 oz.

Test made from Apr. 22 to 28, 1884; age, 3 yrs. 5 mos.; property of Campbell Brown, Spring Hill, Tenn.

Daisy Brown 2d 37325.—Sire, Ida Stoke Pogis 13658; dam, Daisy Brown 12213.

Butter, 15 lbs. 10½ oz. Milk, 255 lbs. 8 oz.

Test made from June 13 to 20, 1893; age, 7 yrs. 4 mos.; estimated weight, 875 lbs.; fed 12 lbs. corn-hearts, 12 lbs. ground feed and 4 lbs. cut sheaf oats—blue-grass pasture; property of Campbell Brown, Spring Hill, Tenn.

Daisy Clementaise 77341.—Sire, Georgia 10606; dam, Daisy's Last 46376.

Butter, 14 lbs. 5 oz. Milk, 183 lbs.

Test made from Dec. 20 to 27, 1893; age, 4 yrs. 9 mos.; estimated weight, 975 lbs.; fed 12 qts. daily of a mixture in about equal parts of oats, peas, barley and corn—clover hay, cabbage; property of E. G. Youmans, Groton, N. Y.

Daisy Delafield 59230.—Sire, Stockholder 2d 15616; dam, Lillian of Sunswick 36616.

Butter, 20 lbs. 11 oz. Milk, 266 lbs. 7 oz.

Test made from Mar. 11 to 18, 1893; age, 5 yrs. 4 mos.; estimated weight, 850 lbs., fed 35 lbs. ground corn and oats, 28 lbs. bran, 14 lbs. linseed meal, 30 lbs. turnips, 30 lbs. potatoes and 126 lbs. hay and rowen; property of Elmer E. Maynard, Savoy Centre, Mass.

Daisy Dell 14562.—Sire, Micawber 4796; dam, Dove 2d 8742.

Butter, 14 lbs. 9 oz. Milk, 179 lbs.

Test made from Mar. 8 to 15, 1890; age, 13 yrs. 5 mos.; estimated weight, 875 lbs., fed, daily, 4 qts. corn meal, 4 qts. shorts, 4 qts. ground oats and 1 qt. oil meal; property of Terence Carrigan, Hopkinton, Mass.

Daisy Glen 71976.—Sire, Callie's Lad 21437; dam, Mallie 2d 38132.

Butter, 20 lbs. 3 oz. Milk, 222 lbs.

Test made from Nov. 19 to 26, 1895; age, 6 yrs. 9 mos.; estimated weight, 900 lbs.; fed 4 qts. ground oats, 4 qts. corn meal, 4 qts. wheat bran mixed with 2 pts. cotton-seed meal and 2 pks. cut sorghum, daily—hay *ad lib.* at night, blue-grass and winter oats pasture; property of D. P. Carter, Lynnville, Tenn.

Daisy Grant 1445, imp.

Butter, 15 lbs.

Test made from June 3 to 9, 1878; age, 9 yrs.; property of James A. Hayt, Patterson, N. Y.

Daisy Harris 92471.—Sire, Rosalie's Pogis 27654; dam, Dodona's Bee 88101.

Butter, 17 lbs. 4¼ oz. Milk, 235 lbs.

Test made from Dec. 31, 1896, to Jan. 7, 1897; age, 4 yrs. 10 mos.; estimated weight, 900 lbs.; fed 56 qts. corn meal, 42 qts. ground oats, 21 qts. ground oil-cake, 10 lbs. clover hay and 12 lbs. carrots; property of Miller & Sibley, Franklin, Pa.

Daisy Harrison 69253.—Sire, Erie Chief 13438; dam, Pride of Bovina 8050.

Butter, 25 lbs. 1 oz. Milk, 303 lbs. 4 oz.

Test made from Jan. 3 to 10, 1897; age, 8 yrs. 9 mos.; actual weight, 900 lbs.; fed 12 qts. wheat bran, 9 qts. ground oats, 3 qts. corn meal and 3 qts. oil meal, daily—hay; property of Charles A. Sweet, East Aurora, N. Y.

Daisy Hinman 61537.—Sire, Ida's Rioter of St. L. 13656; dam, Mary Hinman 17619.

Butter, 24 lbs. 10 oz. Milk, 296 lbs. 4 oz.

Test made from Feb. 25 to Mar. 4, 1891; age, 5 yrs. 2 mos.; estimated weight, 850 lbs.; fed 42 lbs. each of oil meal, pea meal, corn meal and crushed oats, 50 lbs. hay, 1½ bush. roots and 168 lbs. ensilage; property of Ayer & McKinney, Philadelphia, Pa.

Daisy Hinman 61537.—Sire, Ida's Rioter of St. L. 13656; dam, Mary Hinman 17619.

Butter, 102 lbs. in 31 days. Milk, 1308 lbs. 14 oz. Cream, 350 lbs.

Test made from Feb. 23 to Mar. 25, 1891; age, 5 yrs. 3 mos.; fed 804 lbs. grain, 300 lbs. hay and 420 lbs. ensilage; property of Ayer & McKinney, Philadelphia, Pa.

Daisy Hinman 61537.—Sire, Ida's Rioter of St. L. 13656; dam, Mary Hinman 17619.

Butter, 155.131 lbs. in 90 days. Milk, 2677.8 lbs.

Test made from May 31 to Aug. 28, 1893, at World's Columbian Exposition; age, 7 yrs. 5 mos. at beginning of test; actual weight, 903 lbs.; fed 304 lbs. old hay, 210 lbs. ensilage, 531.7 lbs. new hay, 186 lbs. oil meal, 388 lbs. bran, 159 lbs. oats, 438.5 lbs. corn-hearts, 124.25 lbs. cotton-seed, 277 lbs. middlings, 34 lbs. grano-gluten, 1126.4 lbs. clover and 10 lbs. swale-grass; property of Ayer & McKinney, Philadelphia, Pa.

Daisy Hudson 24647.—Sire, Catono 3761; dam, Rosabel Hudson 5704.

Butter, 17 lbs. 11 oz. Milk, 304 lbs. 4 oz.

Test made from May 7 to 14, 1889; age, 5 yrs. 3 mos.; actual weight, 900 lbs.; fed, daily, between 16 and 24 lbs. ground corn and oats; property of Wm. Morrow & Son, Nashville, Tenn.

Daisy Ida Torment 121824.—Sire, Lord Tormentor Pogis 29753; dam, Daisy Ida 95190.

> Butter, 15 lbs. 12 oz. Milk, 224 lbs.

Test made from Apr. 27 to May 4, 1898; age, 2 yrs. 2 mos.; estimated weight, 800 lbs.; fed 4 lbs. wheat bran, daily—blue-grass pasture; property of A. P. Walker, Rushville, Ind.

Daisy Leto 72404.—Sire, St. Helier Brave 17343; dam, Draxy Leto 47889.

> Butter, 18 lbs. 1 oz. Milk, 221 lbs. 8 oz.

Test made from May 23 to 30, 1898; age, 8 yrs. 5 mos.; estimated weight, 900 lbs.; fed 42 lbs. bran and 14 lbs. gluten meal—fine mixed pasture of blue grass, orchard grass, timothy and clover; property of H. A. Neff, Glencoe, Ohio.

Daisy Mekeel 70400.—Sire, Altaica's Perrot 21441; dam, Eva's Belle 20491.

> Butter, 15 lbs. 5 oz. Milk, 213 lbs.

Test made from Jan. 26 to Feb. 2, 1895; age, 4 yrs. 2 mos.; estimated weight, 900 lbs.; fed about 8 lbs., daily, of a mixture of wheat bran, buckwheat middlings, oats and peas, about 30 lbs. ensilage and 5 lbs. hay; property of W. & L. Mekeel, Jacksonville, N. Y.

Daisy Morrison 14035.—Sire, John Morrison 4551; dam, Melia Ann 5444.

> Butter, 25 lbs. 12½ oz. Milk, 254 lbs.

Test made from Jan. 4 to 11, 1886; age, 4 yrs. 4 mos.; fed 6 qts. corn meal, 4 qts. shorts and 4 qts. oat meal, daily; property of J. H. Walker, Worcester, Mass.

Daisy Morrison 3d 40300.—Sire, Winner's Lisgar 11557; dam, Daisy Morrison 14035.

> Butter, 14 lbs. 8 oz. Milk, 187 lbs. 8 oz.

Test made from Feb. 13 to 20, 1894; age, 7 yrs. 6 mos.; estimated weight, 950 lbs.; fed 12 lbs., daily, of bran, middlings, corn meal and oil meal, and 30 lbs. ensilage—hay *ad lib.*; property of George W. Sisson, Jr., Potsdam, N. Y.

Daisy Morrison 4th 46830.—Sire, L. D.'s Victor St. Helier 15286; dam, Daisy Morrison 14035.

> Butter, 14 lbs. 11 oz. Milk, 216 lbs.

Test made from Oct. 30 to Nov. 6, 1893; age, 6 yrs. 1 mo.; estimated weight, 800 lbs.; fed 42 qts. Buffalo gluten, 28 qts. Canada shorts and 28 qts. new process linseed meal—hay, poor pasture; property of C. H. Ellsworth, Worcester, Mass.

Daisy of Belhurst 3114.—Sire, Graylock 740; dam, Eudora 1863.

> Butter, 16 lbs. 8 oz.

Test made March, 1882; property of J. W. Whitenack, Dunellen, N. J.

Daisy of Chenango 18582.—Sire, Prince of Sidney 2665; dam, Daisy 684.

> Butter, 14 lbs. 7 oz. Milk, 252 lbs. 12 oz.

Test made from June 15 to 21, 1883; age, 5 yrs. 4 mos.; property of Hiram A. Jewell, Mount Upton, N. Y.

Daisy of Clermont 3492.—Sire, Gerald 895; dam, Jennie 5th 2269.

> Butter, 14 lbs.

Test made from June 13 to 20, 1881; age, 7 yrs.; property of Edward L. Clarkson, Tivoli, N. Y.

Daisy of Denison 53137.—Sire, Clara's Rioter Pogis 18317; dam, Laura Carrie Alphea 22311.

> Butter, 15 lbs. 12 oz. Milk, 223 lbs. 12 oz.

Test made from May 13 to 20, 1893; age, 5 yrs. 6 mos.; estimated weight, 1000 lbs., fed 10 lbs. bran, 10 lbs. corn meal, 12 lbs. ground oats and 2 lbs. cotton-seed meal, daily—prairie grass pasture; property of Platter & Foster, Denison, Tex.

Daisy of Dryden 77778.—Sire, Georgia 10606; dam, Daisy of Chenango 18582.

> Butter, 16 lbs. 4 oz. Milk, 196 lbs. 8 oz.

Test made from Dec. 20 to 27, 1893; age, 4 yrs. 10 mos.; estimated weight, 975 lbs.; fed 12 qts. daily of a mixture of oats, peas, barley and corn in about equal parts— clover hay, cabbage; property of E. G. Youmans, Groton, N. Y.

Daisy of Guilford 18583.—Sire, Prince of Sidney 2665; dam, Daisy 684.

> Butter, 14 lbs. 6 oz. Milk, 243 lbs. 8 oz.

Test made from Apr. 27 to May 3, 1884; age, 5 yrs. 2 mos.; property of Andrew Baker, West Dryden, N. Y.

Daisy of Hillside Farm 6025.—Sire, Blondin 3d 1935; dam, Daisy of Jersey 4th 4585.

> Butter, 16 lbs. Milk, 274 lbs.

Test made from May 4 to 11, 1884; age, 8 yrs.; property of Thomas L. Dodd, Nashville, Tenn.

Daisy of Malone 36505.—Sire, Hugo Chief of St. Anne's 12070; dam, Lily of Malone 30797.

> Butter, 18 lbs. 5 oz. Milk, 259 lbs.

Test made from Feb. 21 to 28, 1897; age, 11 yrs. 2 mos.; estimated weight, 850 lbs.; fed 12 lbs. oats, peas and corn, ground together, 4 lbs. wheat bran, 2 lbs. oil meal and 25 lbs. corn ensilage, daily—hay *ad lib.*; property of J. H. Martin, Granville, N. Y

Daisy of St. Peter's 18175.—Sire, Hero (P. S. 90 J. H. B.); dam, Nelly 6456.

> Butter, 20 lbs. 5½ oz., unsalted. Milk, 346 lbs. 1 oz.

Test made from June 19 to 25, 1883; age, 6 yrs.; fed 4 qts. corn meal and 6 qts. wheat bran, daily—clover and timothy pasture; property of C. Easthope, Niles, Ohio.

Daisy of Sunnyside 16338.—Sire, Louis of Highland 4854; dam, Marée 6592.

> Butter, 15 lbs. 1 oz. Milk, 267 lbs. 4 oz.

Test made from June 20 to 27, 1893; age, 11 yrs. 11 mos.; actual weight, 940 lbs., fed 21 lbs. bran, 14 lbs. middlings, 28 lbs. corn meal, 12 lbs. oil meal and 46 lbs. oats—natural pasture of indifferent quality; property of J. E. Tyrrell, Marysville, Ohio.

Daisy Pilotta 68852.—Sire, Pilot of Bowker Farm 6085; dam, Daisy Dell 14562.

> Butter, 15 lbs. 2½ oz. Milk, 237 lbs. 14½ oz.

Test made from June 11 to 18, 1894; age, 4 yrs. 6 mos.; actual weight, 870 lbs.; fed 28 lbs. corn meal, 26 lbs. ground oats, 6 lbs. oil meal and 24 lbs. bran—good woodland pasture day and night; property of Mrs. Ellen Sweeney, Clinton, Ill.

Daisy Pilotta 68852.—Sire, Pilot of Bowker Farm 6085; dam, Daisy Dell 14562.

Butter, 17 lbs. 12 oz. Milk, 274 lbs.

Test made from Jan. 23 to 30, 1897; age, 7 yrs. 2 mos.; estimated weight, 900 lbs.; fed 63 lbs. wheat bran, 21 lbs. ground corn and oats in equal parts, 10½ lbs. oil meal, 140 lbs. oat and pea straw and 210 lbs. roots; property of Mrs. Adda F. Howie, Elm Grove, Wis.

Daisy Queen 9619.—Sire, Khedive (P. S. 103 J. H. B.); dam on I. of J.

Butter, 16 lbs. 4 oz.

Test made on grass alone; property of H. B. Tatham, Philadelphia, Pa

Daisy-Seeker 43099.—Sire, Exile of St. Lambert 13657; dam, Minto Fay 27705.

Butter, 14 lbs. 1 oz. Milk, 156 lbs.

Test made from July 10 to 17, 1891; age, 4 yrs. 6 mos.; estimated weight, 850 lbs., fed grass, with 2 qts. bran and 1 pt. corn meal night and morning—old pasture and some new meadow; property of Robert Johnson, East Kendall, N. Y.

Daisy Staunton 46592.—Sire, Duke of Baldwin 7137; dam, Sœur Seraphine 29073.

Butter, 16 lbs. 14 oz. Milk, 224 lbs. 4 oz.

Test made from Mar. 5 to 12, 1897; age, 9 yrs. 7 mos.; estimated weight, 900 lbs., fed 12 qts. corn and oat meal in equal parts, 8 qts. good wheat bran and 2 qts. cotton-seed meal, daily—hay, pasture; property of H. J. Mitchell, Winnsborough, Tex.

Daisy Stillson 28174.—Sire, Seneca Chief 4098; dam, Nickel 2d 23352.

Butter, 15 lbs. 3 oz. Milk, 225 lbs.

Test made from Oct. 8 to 15, 1886; age, 4 yrs. 6 mos.; estimated weight, 950 lbs., fed, daily, 4 qts. oats, 2 qts. corn meal, 4 qts. middlings and 2 qts. bran; property of Peter D. Hulst, East Penfield, N. Y.

Daltonia Pogis 95203.—Sire, Dalton 20117; dam, Gracie Washington 95045.

Butter, 14 lbs. 14 oz. Milk, 284 lbs.

Test made from May 27 to June 3, 1896; age, 5 yrs. 9 mos.; estimated weight, 1000 lbs.; fed 100 lbs. ground corn and oats, mixed in equal quantities by weight—natural timothy and white clover pasture; property of John O. McClintock, Meadville, Pa.

Damara 65001.—Sire, North Pacific Prize 11459; dam, Mercurina 64920.

Butter, 16 lbs. 7 oz. Milk, 240 lbs. 2 oz.

Test made from May 1 to 8, 1890; age, 3 yrs. 6 mos.; estimated weight, 800 lbs., fed 12 qts. of a mixture of corn, oats and oil meal, daily—hay *ad lib.*, 4 qts. roots daily; property of Richardson Bros., Davenport, Iowa.

Damask Rose 22065.—Sire, Bobby (P. S. 208 J. H. B.); dam, Lily 2d (P. S. 147 J. H. B.)

Butter, 16 lbs. 3¼ oz. Milk, 218 lbs. 2 oz.

Test made from July 12 to 18, 1886; age, 5 yrs. 5 mos.; estimated weight, 800 lbs., fed, daily, 2 qts. corn meal, 2 qts. oat meal, 2 qts. bran and 1 qt. oil meal; property of Frederick Loeser, Somerville, N. J.

Dame Ida 76878.—Sire, Ida's Stoke Pogis 13658; dam, Dame Miller 59260.

Butter, 16 lbs. 7½ oz. Milk, 224 lbs. 12 oz.

Test made from Nov. 29 to Dec. 6, 1897; age, 6 yrs. 2 mos.; estimated weight, 750 lbs.; fed 8 lbs. ground oats, 8 lbs. corn meal and 4 lbs. wheat bran, daily—alfalfa, barley and rye pasture; property of S. Q. Hollingsworth, Coushatta, La.

Dame Lofty 47567.—Sire, Baron Hugo 15208; dam, Miss Lofty 9718.

Butter, 15 lbs. 10 oz. Milk, 296 lbs.

Test made from May 16 to 23, 1891; age, 4 yrs. 2 mos.; estimated weight, 850 lbs.; fed 3 lbs. coarse wheat bran, 2 lbs. cotton-seed meal, 2 lbs. linseed meal, 3 lbs. corn meal, 35 lbs. ensilage and 10 lbs. clover rowen, daily; property of E. S. Henry, Rockville, Conn

Dame Miller 59260.—Sire, Count Potoka 9831; dam, Mallie Miller 29862.

Butter, 16 lbs. 2½ oz. Milk, 203 lbs. 8 oz.

Test made from Nov. 22 to 29, 1890; age, 3 yrs. 10 mos.; estimated weight, 900 lbs., well fed, but no record kept; property of Campbell Brown, Spring Hill, Tenn.

Dame of Eau Claire 127862.—Sire, Melia Ann's Son 22041; dam, Ranolta 54292.

Butter, 14 lbs. 4 oz. Milk, 202 lbs. 10 oz.

Test made from May 25 to June 1, 1898; age, 2 yrs. 6 mos.; estimated weight, 900 lbs.; fed 6 qts. ground oats and corn, 6 qts. bran and 3 qts. cotton-seed meal, daily—clover and red-top pasture; property of E. L. Van Deusen, Ashley Falls, Mass.

Dandelion 2521, imp.

Butter, 16 lbs. 9 oz. Milk, 345 lbs. 6½ oz.

Test made from Aug. 25 to 31, 1883; age, 15 yrs. 6 mos.; fed 8 qts. wheat bran, 2 qts. corn meal and 2 qts. crushed oats, daily; property of John I. Holly, Plainfield, N. J.

Daphne of Arcadia 21710.—Sire, Talmadge 3992; dam, Daphne of Staatsburgh 2d 3027.

Butter, 28 lbs. 12 oz. Milk, 274 lbs.

Test made from May 1 to 8, 1889; age, 7 yrs. 3 mos.; estimated weight, 825 lbs.; fed, daily, 12 qts. corn meal, ground oats, bran and oil meal, mixed; property of Mrs. Lillie A. Morse, Newark, N. Y.

Daretta 62579.—Sire, Upright 6147; dam, Modita 16626.

Butter, 16 lbs. 8 oz. Milk, 217 lbs. 5 oz.

Test made from Nov. 11 to 18, 1889; age, 3 yrs. 10 mos.; estimated weight, 800 lbs., fed 15 qts. chop, corn meal, oat meal and pea meal, daily—plenty of hay; property of Richardson Bros., Davenport, Iowa.

Dariole 64106.—Sire, Diploma 16219; dam, Pledge 59214.

Butter, 16 lbs. 9 oz. Milk, 248 lbs. 6 oz.

Test made from Sept. 1 to 8, 1894; age, 4 yrs. 7 mos.; estimated weight, 850 lbs., fed 5 qts. corn meal, 6 qts. bran, with little oil and pea meal, daily—good pasture, property of Richardson Bros., Davenport, Iowa.

Dark and Fair 24468.—Sire, Nero 7266; dam on I. of J.

Butter, 16 lbs. 9 oz. Milk, 167 lbs. 8 oz.

Test made from June 7 to 14, 1887; age, 6 yrs. 2 mos.; estimated weight, 900 lbs.; fed, daily, 4 qts. bran, 4 qts. ground corn and oats, 1 qt. pea meal; property of M. Erskine Miller, Staunton, Va.

Dark Cloud 9364.—Sire, Othello 1114; dam, Light Cloud 2865.

Butter, 15 lbs. 3½ oz. Milk, 256 lbs. 2 oz.

Test made from Aug. 14 to 21, 1883; age, 6 yrs. 4 mos.; property of J. S. Rogers, Paterson, N. J

Dark Rose 101892.—Sire, Noble Paddy 31072; dam, Matilda's Rose 81397.

Butter, 15 lbs. 14½ oz.   Milk, 255 lbs. 8 oz.

Test made from Oct. 27 to Nov. 3, 1896; age, 3 yrs. 7 mos.; actual weight, 765 lbs.; fed 12 qts. bran, 3 qts. meal, 3 qts. ground oats and 3 qts. oil meal, daily—hay; property of C. A. Sweet, East Aurora, N. Y.

Darling of Neatham 20086.—Sire, Delaware Darling 3461; dam, Nellie 2d 2369.

Butter, 15 lbs. 3 oz.   Milk, 194 lbs.

Test made from Apr. 1 to 7, 1885; age, 2 yrs. 4 mos.; property of R. McMichael, Lexington, Ky.

Darling Pansy 44830.—Sire, Crystal 9324; dam, Pansy McGrew 26993.

Butter, 15 lbs. 1 oz.   Milk, 223 lbs. 4 oz.

Test made from June 19 to 26, 1892; age, 4 yrs. 11 mos.; estimated weight, 650 lbs., fed 3 gals. corn, oats and barley chop and 1 qt. oil meal, daily—good but wet pasture; property of Belmont Jersey Cattle Co., Columbus, Ohio.

Darlymora 37771.—Sire, Duke of Darlington 2d 6948; dam, Lady Gilmore 2d 12140.

Butter, 14 lbs. 7 oz.   Milk, 211 lbs. 12 oz.

Test made from June 5 to 12, 1889; age, 5 yrs. 1 mo.; estimated weight, 850 lbs.; fed, daily, 4 qts. corn meal, 4 qts. shorts and 4 qts. ground oats; property of Terence Carrigan, Hopkinton, Mass.

Dassa Argyle 21180.—Sire, Campbell's Rex 6812; dam, Duchess of Argyle 3d 7569.

Butter, 15 lbs. 6 oz.   Milk, 224 lbs. 11 oz.

Test made from June 20 to 26, 1886; age, 3 yrs. 7 mos.; property of E. Stevens Henry, Rockville, Conn.

Dassa Argyle 21180.—Sire, Campbell's Rex 6812; dam, Duchess of Argyle 3d 7569.

Butter, 16 lbs. 14 oz.   Milk, 275 lbs.

Test made from May 3 to 10, 1891; age, 8 yrs. 5 mos.; estimated weight, 1100 lbs., fed 2 lbs. linseed meal, 2 lbs. cotton-seed meal, 3 lbs. coarse wheat bran, 3 lbs. corn and cob meal, mixed with 35 to 40 lbs. ensilage, daily; property of E. S. Henry, Rockville, Conn

Davy's Dot 41693.—Sire, Augustus Davy 15583; dam, Dot 2d 621.

Butter, 14 lbs. 8 oz.   Milk, 183 lbs. 8 oz.

Test made from Dec. 15 to 22, 1893; age, 11 yrs.; estimated weight, 950 lbs.; fed 56 qts. ground oats and 56 qts. corn meal—hay; property of Morgan & Brown, Columbia, Tenn.

Davy's Dot 2d 85744.—Sire, Pet's Landseer 29392; dam, Davy's Dot 41693.

Butter, 14 lbs. 4 oz.   Milk, 168 lbs. 4 oz.

Test made from May 13 to 20, 1894; age, 2 yrs. 11 mos.; estimated weight, 650 lbs., fed 56 qts. ground oats and 56 qts. ground corn—little clover hay, fair pasture; property of Morgan & Brown, Columbia, Tenn.

D. D.'s Coomassie 63673.—Sire, Happy Coomassie 12142; dam, Duke's Del 37702.

Butter, 14 lbs. 2 oz., unsalted.   Milk, 195 lbs. 11 oz.

Test made from June 12 to 19, 1895; age, 7 yrs. 2 mos.; estimated weight, 800 lbs.; fed 56 qts. ground feed, composed of oats and barley in equal parts, mixed with wheat bran—timothy and clover pasture; property of Schuyler Bros., Cobleskill, N. Y

Deborana 4718.—Sire, Coomassie 1484; dam, Deborah 2536.

Butter, 14 lbs. 8 oz.

Test made from June 11 to 17, 1879; age, 3 yrs. 3 mos.; property of M. K. Gregg, Imperial, Pa

Deerlick Petra Pogis 99987.—Sire, Stoke Pogis' Perfection 5984; dam, Petra's Signal 53376

Butter, 14 lbs. 9½ oz. Milk, 241 lbs. 12 oz.

Test made from Feb. 1 to 8, 1897; age, 5 yrs. 1 mo.; estimated weight, 800 lbs.; fed 448 lbs. ensilage, 98 lbs. corn meal and 21 lbs. oil meal—hay *ad lib.*; property of Baum & Hill, Frankfort, Ind.

Deerlick Pojoram 99988.—Sire, Stoke Pogis' Perfection 5984; dam, Cœur d'Alene 49802.

Butter, 15 lbs. 7 oz. Milk, 238 lbs. 12 oz.

Test made from Apr. 25 to May 2, 1897; age, 4 yrs. 2 mos.; estimated weight, 850 lbs., fed 280 lbs. ensilage and 126 lbs. corn meal—hay *ad lib.*; property of Baum & Hill, Frankfort, Ind.

Delesline 44474.—Sire, Prince of Melrose 4819; dam, Saucylip 18574.

Butter, 15 lbs. 9 oz. Milk, 254 lbs. 12 oz.

Test made from Mar. 8 to 15, 1896; age, 9 yrs. 10 mos.; estimated weight, 800 lbs.; fed 28 lbs. bran, 42 lbs. corn chop, 63 lbs. peas, ground with hull, 52½ lbs. ground oats and 21 lbs. cotton-seed meal—little clover hay; property of M. Lothrop, Marshall, Tex.

Deletta 21305.—Sire, Councillor 4468; dam, Moquette 6673.

Butter, 14 lbs. 15½ oz. Milk, 145 lbs. 8 oz.

Test made from Oct. 25 to 31, 1885; age, 3 yrs. 8 mos.; property of Mrs. A. N. Martin, Summit, N. J.

Delia Martin 92358.—Sire, Nora's Champion 20739; dam, Delia Lawrence 61064.

Butter, 18 lbs. 12 oz. Milk, 274 lbs. 12 oz.

Test made from May 7 to 14, 1898; age, 7 yrs. 11 mos.; estimated weight, 800 lbs.; fed 71 lbs. bran and 28 lbs. gluten feed—poor upland pasture; property of Mrs. Adda F. Howie, Elm Grove, Wis.

Delia of Grouville 16428.—Sire, Progress (F. S. 286 J. H. B.); dam, Dahlia 22005.

Butter, 14 lbs. 4 oz.

Test made from Sept. 15 to 21, 1884; age, 4 yrs. 3 mos.; property of James B. Wilder, Louisville, Ky.

Delight's Darling 45700.—Sire, Grace's Duke of Darlington 8289; dam, Friga's Delight 26649.

Butter, 15 lbs. 8 oz. Milk, 282 lbs.

Test made from June 10 to 17, 1891; age, 5 yrs. 2 mos.; estimated weight, 850 lbs., fed 8 lbs. corn meal, 4 lbs. oil meal and 4 lbs. bran, daily—blue-grass, timothy and clover pasture; property of Garretson Bros., Pendleton, Ind.

Delphina 105895.—Sire, Exile's Rebel 28928; dam, Fall Leaf 2d 25171.

Butter, 14 lbs. 2¾ oz. Milk, 164 lbs. 6 oz.

Test made from May 15 to 22, 1897; age, 3 yrs. 3 mos.; estimated weight, 600 lbs., fed 14½ lbs. corn meal, 2 lbs. oil meal, 3 lbs. molasses feed and 6 lbs. ground oats, daily—rye and Bermuda pasture; property of William E. Oates & Son, Vicksburg, Miss.

Dempsey's Beauty 62808.—Sire, Fancy Toltec 21167; dam, Tenie Rex 55085.

Butter, 14 lbs. 6 oz. Milk, 112 lbs. 10 oz.

Test made from Oct. 8 to 15, 1891; age, 2 yrs. 2 mos.; estimated weight, 550 lbs., fed 2 gals. corn, oats and barley chop and 1 qt. oil meal, daily—fair pasture; property of Belmont Jersey Cattle Co., Columbus, Ohio.

Denise 8281.—Sire, Top-Sawyer 1404; dam, Lucy 4577.

Butter, 14 lbs. 4½ oz. Milk, 170 lbs. 8 oz.

Test made from Nov. 23 to 29, 1883; age, 4 yrs. 8 mos.; property of Campbell Brown, Spring Hill, Tenn.

Denise 8281.—Sire, Top-Sawyer 1404; dam, Lucy 4577.

Butter, 17 lbs. 8 oz. Milk, 218 lbs.

Test made from Feb. 21 to 27, 1886; age, 6 yrs. 11 mos.; property of Campbell Brown, Spring Hill, Tenn.

Denise Landseer of Lawn 84411.—Sire, Daisy's Harry 21500; dam, Louisa Belle 64013.

Butter, 18 lbs. 5 oz. Milk, 226 lbs. 10 oz.

Test made from Apr. 12 to 19, 1895; age, 3 yrs. 1 mo.; estimated weight, 650 lbs.; fed 77 lbs. bran, 56 lbs. cotton-seed meal and 35 lbs. corn meal—hay *ad lib.*, common pasture; property of Dennis Foley, Galveston, Tex.

Denise's Fancy 92500.—Sire, Denise's Tormentor 11823; dam, John's Fancy 58256.

Butter, 15 lbs. 10 oz. Milk, 210 lbs. 5 oz.

Test made from Mar. 9 to 16, 1894; age, 2 yrs. 1 mo.; estimated weight, 750 lbs., fed 42 qts. corn meal, 42 qts. bran, mixed, and clover hay *ad lib.*—pastured in wheat field; property of W. H. Roe, Shelbyville, Ky.

Denise's Ida 54942.—Sire, Ida's Stoke Pogis 13658; dam, Denise 8281.

Butter, 15 lbs. 3½ oz. Milk, 244 lbs.

Test made from June 29 to July 6, 1890; age, 2 yrs. 3 mos.; estimated weight, 775 lbs.; well fed, but no record of feed kept; property of Campbell Brown, Spring Hill, Tenn.

Denise's Ida 54942.—Sire, Ida's Stoke Pogis 13658; dam, Denise 8281.

Butter, 17 lbs. 15½ oz. Milk, 309 lbs.

Test made from Apr. 29 to May 6, 1894; age, 6 yrs. 1 mo.; estimated weight, 875 lbs.; fed 168 lbs. corn meal, with a small sprinkling of ground oats and 28 lbs. cut sheaf oats—blue-grass pasture; property of Est. of Campbell Brown, Spring Hill, Tenn.

Denise's Ida 54942.—Sire, Ida's Stoke Pogis 13658; dam, Denise 8281.

Butter, 90 lbs. 6¼ oz. in 30 days. Milk, 1243 lbs. 6½ oz.

Test made from Apr. 30 to May, 30, 1895; age, 7 yrs. 1 mo.; actual weight, 850 lbs.; fed 28 lbs. per day of ground oats and corn in equal parts—extra good orchard and blue-grass pasture; property of M. M. Gardner, Nashville, Tenn.

Denise's Pearl 95757.—Sire, Denise's Tormentor 11823; dam, Pearl Landseer 50322.

Butter, 20 lbs. 11 oz. Milk, 272 lbs. 15 oz.

Test made from Dec. 12 to 19, 1897; age, 4 yrs. 11 mos.; estimated weight, 950 lbs.; fed 6 qts. corn meal, 8 qts. ground oats and 8 qts. wheat bran, daily; property of J. Gerow Dutcher, Pawling, N. Y.

Deeine 6343.—Sire, Columbiad 2d 1515; dam, Neatness 1894.

Butter, 14 lbs. 3 oz.

Test made from June 29 to July 5, 1882; age, 5 yrs. 1 mo.; test made on grass alone; property of William N. McConnell, Ripon, Wis.

Derjava 59830.—Sire, Lord Harry 3445; dam, Renalda 20924.

Butter, 18 lbs. 14 oz. Milk, 204 lbs. 4 oz.

Test made from Oct. 31 to Nov. 7, 1890; age, 2 yrs.; estimated weight, 550 lbs.; fed, daily, 4 lbs. corn, oats and barley chop, 4 lbs. malt, 2 lbs. bran and 2 lbs. oil meal; property of Belmont Jersey Cattle Co., Columbus, Ohio.

Desda Pogis 58344.—Sire, Stoke Pogis Prince 11005; dam, Desdemona Belle 19366.

Butter, 14 lbs. 7½ oz. Milk, 247 lbs. 2 oz.

Test made from June 18 to 25, 1893; age, 4 yrs. 9 mos.; estimated weight, 800 lbs.; fed 67 lbs. ground corn and oats—blue-grass and white clover pasture; property of O. P. Bowers, Muncie, Ind.

Design 18481.—Sire, Finis Lawrence 4107; dam, Damask 10259.

Butter, 17 lbs. 2½ oz. Milk, 250 lbs. 4 oz.

Test made from Dec. 1 to 8, 1889; age, 7 yrs.; estimated weight, 820 lbs.; fed, daily, 18 qts. ground corn, oats and peas; property of Richardson Bros., Davenport, Iowa.

Dew 62474.—Sire, Ida's Rioter of St. L. 13656; dam, Lady Appel 8612.

Butter, 16 lbs. 7 oz. Milk, 181 lbs.

Test made from May 29 to June 5, 1893; age, 4 yrs. 2 mos.; actual weight, 940 lbs.; fed 6 qts. corn meal, 6 qts. ground oats and 4 qts. ground oil-cake, daily—good old meadow pasture; property of Miller & Sibley, Franklin, Pa.

Dewdrop Pansy 19736.—Sire, Platon (F. S. 310 J. H. B.); dam, Pansy (F. S. 1072 J. H. B.)

Butter, 19 lbs. 8 oz. Milk, 289 lbs.

Test made from Dec. 15 to 21, 1884; age, 4 yrs. 2 mos.; property of James Crook, Jacksonville, Ala.

Dia 13658.—Sire, Lord Dartmouth 6302; dam, Blossom of Hanover 13655.

Butter, 15 lbs. 13½ oz. Milk, 166 lbs. 10 oz.

Test made from Oct. 15 to 21, 1885; age, 4 yrs. 7 mos.; property of F. C. Sayles, Pawtucket, R. I

Diamond Rose 25944, imp.

Butter, 14 lbs. 7 oz. Milk, 190 lbs.

Test made from May 19 to 26, 1891; age, 9 yrs.; estimated weight, 850 lbs.; fed 56 qts. corn and cob meal—good blue-grass pasture; property of J. K. Sheets, Troy, Ohio.

Diamond's Lass 81767.—Sire, Muggy 22004; dam, Rex's Diamond 73244.

Butter, 16 lbs. 4 oz.   Milk, 172 lbs. 13 oz.

Test made from Apr. 25 to May 2, 1898; age, 7 yrs.; estimated weight, 850 lbs.; fed 8 qts. corn meal, 5 qts. bran, 3 qts. oat meal and 1 qt. cotton-seed meal, daily—hay at noon, about 20 lbs. ensilage daily; property of E. E. Benjamin, Greenfield, Mass.

Diana Deon 61539.—Sire, Diana's Stoke Pogis 13686; dam, Dora Doon 12909.

Butter, 15 lbs. 9 oz.   Milk, 236 lbs.

Test made from Dec. 21 to 28, 1893; age, 4 yrs. 5 mos.; actual weight, 780 lbs.; fed 112 qts. of a mixture of equal parts of cotton-seed, steamed oats and wheat, corn and oat chop—pastured on sorghum stubble and winter oats; property of S. T. Howard, Quanah, Tex.

Diana of St. Lambert 6636.—Sire, Stoke Pogis 3d 2238; dam, Pet of St. Lambert 5123.

Butter, 16 lbs. 8 oz.   Milk, 298 lbs.

Test made from June 6 to 12, 1882; age, 5 yrs. 2 mos.; fed 8 lbs. meal and 4 lbs. bran, daily—good old pasture; property of W. D. Reesor, Markham, Ont., Can.

Diana Victor Pogis 100763.—Sire, Melia Ann's Victor Pogis 20697; dam, Pink Pedro 77809.

Butter, 14 lbs. 6 oz.   Milk, 229 lbs. 8 oz.

Test made from Oct. 15 to 22, 1894; age, 1 yr. 10 mos.; estimated weight, 650 lbs.; fed 5 lbs. bran and 5 lbs. corn meal, daily—clover after feed; property of J. H. Martin, Granville, N. Y.

Diana Victor Pogis 100763.—Sire, Melia Ann's Victor Pogis 20697; dam, Pink Pedro 77809.

Butter, 28 lbs. 1 oz.   Milk, 364 lbs.

Test made from Nov. 21 to 28, 1897; age, 4 yrs. 11 mos.; estimated weight, 900 lbs.; fed about 95 lbs. corn meal, 28 lbs. wheat middlings, 28 lbs. wheat bran, 25 lbs. oil meal and about 1½ bush. corn ensilage per day—mixed hay *ad lib.*; property of J. H. Martin, Granville, N. Y.

Dicta 55221.—Sire, Young Pedro 9033; dam, Brenda Yorke 34460.

Butter, 20 lbs. 9 oz.   Milk, 255 lbs. 2 oz.

Test made from Apr. 30 to May 7, 1893; age, 4 yrs. 7 mos.; estimated weight, 850 lbs.; fed 18 qts. daily of a mixture of equal parts of corn meal, ground oats, wheat bran, 1 pt. oil meal and 5 lbs. cut hay—long hay *ad lib.*; property of Walter W. Law, Whitson, N. Y.

Dido B. 112926.—Sire, Queen's Legacy 29652; dam, Eufielda 62546.

Butter, 15 lbs.   Milk, 262 lbs. 3 oz.

Test made from May 15 to 22, 1898; age, 3 yrs. 4 mos.; estimated weight, 950 lbs.; fed 6 lbs. wheat bran, 4 lbs. Buffalo gluten and 4 lbs. new process linseed meal, daily —medium pasture; property of Charles H. Ellsworth, Worcester, Mass.

Dido Pogis 67260.—Sire, Proven Pogis 19594; dam, Frankie's Dido 22424.

Butter, 15 lbs. 9 oz., unsalted.   Milk, 202 lbs. 2 oz.

Test made from June 15 to 22, 1895; age, 5 yrs. 7 mos.; estimated weight, 850 lbs.; fed 56 qts. oats and barley and wheat bran, ground together in equal parts—timothy and clover pasture; property of Schuyler Bros., Cobleskill, N. Y.

Die Königin 124952.—Sire, Prince of Tennessee 20772; dam, Miss Mouse 92512.

Butter, 14 lbs. 12½ oz. Milk, 194 lbs. 7¾ oz.

Test made from May 17 to 24, 1898; age, 2 yrs. 6 mos.; estimated weight,, 575 lbs.; fed 20 lbs. per day ground corn and oats, in equal parts—very good orchard and blue-grass pasture; property of Thomas H. Malone, Nashville, Tenn.

Dilwa 30515.—Sire, Jersey Express 5771; dam, Wilda 21205.

Butter, 15 lbs. 10 oz. Milk, 179 lbs. 8 oz.

Test made from May 16 to 23, 1891; age, 7 yrs. 1 mo.; estimated weight, 900 lbs.; fed 36 lbs. corn meal, 36 lbs. crushed oats, 36 lbs. bran and 70 lbs. hay; property of Ayer & McKinney, Philadelphia, Pa.

Dimple 3248.—Sire, Wallace Barnes 1264; dam, Matchless 723.

Butter, 16 lbs. 11 oz. Milk, 20 qts. per day.

Test made from May 7 to 14, 1875; age, 3 yrs. 10 mos.; fed 6 lbs. chop and 5 lbs. bran, daily—timothy and blue-grass pasture; property of G. W. Felter, Williamsburg, Ohio.

Dinah Alexis 69345.—Sire, Diana's Rioter 10481; dam, Queen Alexis 33363.

Butter, 14 lbs. 8 oz. Milk, 237 lbs. 4 oz.

Test made from June 12 to 19, 1893; age, 4 yrs. 2 mos.; actual weight, 983 lbs.; fed 55 lbs. corn meal, 21 lbs. ground oats, 21 lbs. wheat middlings and 14 lbs. new process oil meal—good blue-grass pasture; property of J. P. Bradbury, Pomeroy, Ohio.

Diploma's Clara 103867.—Sire, Diploma 16219; dam, Clara's Belle 16582.

Butter, 15 lbs. 12½ oz. Milk, 233 lbs. 8 oz.

Test made from June 1 to 8, 1895; age, 2 yrs.; estimated weight, 750 lbs.; fed 4 lbs. corn meal, 5 lbs. oat meal, 3 lbs. bran and 1½ lbs. oil meal, daily—some hay, good meadow pasture; property of Richardson Bros., Davenport, Iowa.

Diploma's Coma 54070.—Sire, Diploma 16219; dam, Coma 29330.

Butter, 14 lbs. 10 oz. Milk, 137 lbs. 6 oz.

Test made from Sept. 9 to 16, 1890; age, 1 yr. 11 mos.; estimated weight, 750 lbs.; fed, daily, 15 qts. ground oats, corn and peas and 3 qts. shorts; property of Charles E. Hill, Denver, Col.

Diploma's Eureka 103865.—Sire, Diploma 16219; dam, Marissa 65285.

Butter, 15 lbs. 10 oz. Milk, 227 lbs. 2 oz.

Test made from May 8 to 15, 1895; age, 2 yrs. 11 mos.; estimated weight, 800 lbs.; fed 5 lbs. corn meal, 6 lbs. oat meal, 6 lbs. bran and 2 lbs. oil meal, daily—some hay, pasture; property of Richardson Bros., Davenport, Iowa.

Diploma's Loreda 101528.—Sire, Diploma 16219; dam, Loreda 19801.

Butter, 18 lbs. 1½ oz. Milk, 240 lbs. 14 oz.

Test made from Oct. 31 to Nov. 7, 1896; age, 4 yrs. 9 mos.; estimated weight, 900 lbs.; fed 42 lbs. bran, 35 lbs. oats, 31½ lbs. corn and 21 lbs. oil meal—ensilage, hay and rape *ad lib.*; property of H. C. Taylor, Orfordville, Wis.

Diploma's Marissa 95590.—Sire, Diploma 16219; dam, Marissa 65285.

Butter, 16 lbs. 5 oz.   Milk, 218 lbs. 4 oz.

Test made from Apr. 4 to 11, 1897; age, 5 yrs. 11 mos.; estimated weight, 875 lbs., fed 28 lbs. rye, 42 lbs. oats, 10½ lbs. oil meal, 30 lbs. bran and 21 lbs. corn meal— 45 lbs. corn ensilage and corn fodder, daily; property of H. C. Taylor, Orfordville, Wis.

Diploma's Pansy 112871.—Sire, Diploma 16219; dam, Parthia 64925.

Butter, 16 lbs. 3 oz.   Milk, 239 lbs. 8 oz.

Test made from Feb. 21 to 28, 1898; age, 6 yrs. 1 mo.; estimated weight, 850 lbs.; fed 35 lbs. bran, 35 lbs. oats, 35 lbs. corn meal and 14 lbs. oil meal—corn ensilage, property of H. C. Taylor, Orfordville, Wis.

Diploma's Princess 104002.—Sire, Diploma 16219; dam, Commilla 79614.

Butter, 17 lbs. 4 oz.   Milk, 263 lbs. 7 oz.

Test made from July 11 to 18, 1894; age, 4 yrs. 8 mos.; estimated weight, 800 lbs., fed 15 lbs. corn meal, ground oats and oil meal in three feeds, daily—4 qts. roots daily; property of Richardson Bros., Davenport, Iowa.

Diploma's Queen 98151.—Sire, Diploma 16219; dam, Commilla 79614.

Butter, 17 lbs. 8 oz.   Milk, 263 lbs. 7 oz.

Test made from Sept. 11 to 18, 1894; age, 6 yrs. 8 mos.; estimated weight, 800 lbs.; fed 15 lbs. corn meal, ground oats and oil meal, daily—4 qts. roots daily; property of Richardson Bros., Davenport, Iowa.

Dixie Landseer 88681.—Sire, Tormentor 5th 21962; dam, Dolly Landseer 70201.

Butter, 15 lbs. 6½ oz.   Milk, 151 lbs.

Test made from May 19 to 26, 1896; age, 3 yrs. 9 mos.; estimated weight, 775 lbs., fed 42 lbs. corn-hearts, 28 lbs. ground oats and 14 lbs. bran—blue-grass pasture; property of Jeff D. Clinard, Springfield, Tenn.

Dixie Princess 55658.—Sire, Dixie Prince 18379; dam, Duchess Cicero 41426.

Butter, 16 lbs. 8 oz.   Milk, 164 lbs.

Test made from Feb. 15 to 22, 1894; age, 5 yrs. 3 mos.; estimated weight, 650 lbs.; fed 6 qts. corn-hearts, 4 qts. ground oats, 2 qts. cotton-seed meal and 4 qts. bran, daily—clover hay, corn stover and poor pasture; property of H. A. Huntington, Nashville, Tenn.

Dodona A. 88102.—Sire, Sir Lawrence Angelo 19628; dam, Queen Dodona 42375.

Butter, 16 lbs. 10¾ oz.   Milk, 249 lbs.

Test made from July 10 to 17, 1894; age, 3 yrs. 10 mos.; actual weight, 880 lbs.; fed 42 qts. corn meal, 28 qts. wheat bran and 28 qts. ground oil-cake—good old meadow pasture; property of Miller & Sibley, Franklin, Pa.

Dollie's Valentine 105049.—Sire, Oonan's Tormentor Pogis 30505; dam, Dollie Fay 105047.

Butter, 16 lbs. 2¾ oz. (confirmed); estimated butter on basis of 85 per cent. fat, 15 lbs. 5⅝ oz.   Milk, 227.6 lbs.

Test made from Feb. 25 to Mar 3, 1898; age, 4 yrs.; average weight, 905 lbs.; fed 48 lbs. corn meal, 3 lbs. cotton-seed meal, 3 lbs. oil meal and 48 lbs. bran—15 lbs. beets, clover hay *ad lib.*; property of Mrs. Shelby Kinkead, Lexington, Ky.

Analyses of Butter.

First churning—Fat 82.58, water 13.39, ash 3.15, casein 0.88.
Second churning—Fat 77.56, water 14.54, ash 6.97, casein 0.93.

Dollie's Valentine 105049.—Sire, Oonan's Tormentor Pogis 30505; dam, Dollie Fay 105047

Butter, 16 lbs. 13 oz. (confirmed); estimated butter on basis of 85 per cent. fat, 16 lbs. 14 oz. Milk, 250.7 lbs.

Test made from Mar. 4 to 10, 1898; age, 4 yrs.; estimated weight, 900 lbs.; fed 64 lbs. corn meal, 14 lbs. cotton-seed meal, 14 lbs. oil meal, 56 lbs. bran and 70 lbs. beets—clover hay *ad lib.*; property of Mrs. Shelby Kinkead, Lexington, Ky.

Analyses of Butter.

First churning—Fat 82.04, water 12.76, ash 4.05, casein 1.15.
Second churning—Fat 82.58, water 13.27, ash 3.12, casein 1.03.

Dollie's Valentine 105049.—Sire, Oonan's Tormentor Pogis 30505; dam, Dollie Fay 105047.

Butter, 18 lbs. ⅞ oz. (confirmed); estimated butter on basis of 85 per cent. fat, 17 lbs. 12½ oz. Milk, 271.7 lbs.

Test made from Mar. 11 to 17, 1898; age, 4 yrs. 1 mo.; estimated weight, 900 lbs.; fed 70 lbs. corn meal, 14 lbs. cotton-seed meal, 14 lbs. oil meal, 56 lbs. bran and 70 lbs. beets—clover hay *ad lib.*; property of Mrs. Shelby Kinkead, Lexington, Ky.

Analyses of Butter.

First churning—Fat 80.93, water 14.82, ash 3.47, casein 0.78.
Second churning—Fat 80.19, water 15.00, ash 3.67, casein 1.14.

Dollie's Valentine 105049.—Sire, Oonan's Tormentor Pogis 30505; dam, Dollie Fay 105047.

Butter, 16 lbs. 9⅝ oz. (confirmed); estimated butter on basis of 85 per cent. fat, 16 lbs. 1⅝ oz. Milk, 273.8 lbs.

Test made from Mar. 18 to 24, 1898; age, 4 yrs. 1 mo.; estimated weight, 900 lbs., fed 70 lbs. corn meal, 14 lbs. oil meal, 14 lbs. cotton-seed meal, 56 lbs. bran and 105 lbs. beets—clover hay *ad lib.*; property of Mrs. Shelby Kinkead, Lexington, Ky.

Analyses of Butter.

First churning—Fat 81.89, water 14.70, ash 2.70, casein 0.71.
Second churning—Fat 83.16, water 13.44, ash 2.57, casein 0.83.

Dollie's Valentine 105049.—Sire, Oonan's Tormentor Pogis 30505; dam, Dollie Fay 105047.

Butter, 16 lbs. 5⅝ oz. (confirmed); estimated butter on basis of 85 per cent. fat, 16 lbs. 7⅞ oz. Milk, 271.4 lbs.

Test made from Mar. 25 to 31, 1898; age, 4 yrs. 1 mo.; estimated weight, 900 lbs.; fed 78 lbs. corn meal, 22 lbs. cotton-seed meal, 14 lbs. oil meal, 48 lbs. bran and 105 lbs. beets—clover hay *ad lib.*; property of Mrs. Shelby Kinkead, Lexington, Ky.

Analyses of Butter.

First churning—Fat 83.46, water 13.32, ash 2.37, casein 0.85.
Second churning—Fat 84.48, water 11.85, ash 2.58, casein 1.09.

Dollie's Valentine 105049.—Sire, Oonan's Tormentor Pogis 30505; dam, Dollie Fay 105047.

Butter, 18 lbs. 1⅝ oz. (confirmed); estimated butter on basis of 85 per cent. fat, 17 lbs. 13⅝ oz. Milk, 273.6 lbs.

Test made from Apr. 1 to 7, 1898; age, 4 yrs. 1 mo.; average weight, 893 lbs.; fed 84 lbs. corn meal, 28 lbs. cotton-seed meal, 14 lbs. oil meal, 42 lbs. bran and 105 lbs. beets—clover hay *ad lib.*; property of Mrs. Shelby Kinkead, Lexington, Ky.

Analyses of Butter.

First churning—Fat 82.53, water 13.85, ash 2.89, casein 0.73.
Second churning—Fat 83.15, water 13.28, ash 2.72, casein 0.85.

Dollie's Valentine 105049.—Sire, Oonan's Tormentor Pogis 30505; dam, Dollie **Fay** 105047.

> Butter, 16 lbs. 5¼ oz. (confirmed); estimated butter on basis of 85 per cent. fat, 16 lbs. 2¾ oz. Milk, 255.5 lbs.

Test made from Apr. 8 to 14, 1898; age, 4 yrs. 2 mos.; estimated weight, 893 lbs.; fed 70 lbs. corn meal, 28 lbs. cotton-seed meal, 14 lbs. oil meal and 42 lbs. bran— clover hay *ad lib.*; property of Mrs. Shelby Kinkead, Lexington, Ky.

Analyses of Butter.

First churning—Fat 82.77, water 13.61, ash 2.77, casein 0.85.
Second churning—Fat 82.27, water 14.38, ash 2.58, casein 0.77.

Dollie's Valentine 105049.—Sire, Oonan's Tormentor Pogis 30505; dam, Dollie **Fay** 105047.

> Butter, 16 lbs. 4¾ oz. (confirmed); estimated butter on basis of 85 per cent. fat, 16 lbs. 2¼ oz. Milk, 266.1 lbs.

Test made from Apr. 15 to 21, 1898; age, 4 yrs. 2 mos.; estimated weight, 890 lbs., fed 70 lbs. corn meal, 28 lbs. cotton-seed meal, 14 lbs. oil meal and 42 lbs. bran— clover hay *ad lib.*; property of Mrs. Shelby Kinkead, Lexington, Ky.

Analyses of Butter.

First churning—Fat 82.50, water 14.78, ash 2.00, casein 0.72.
Second churning—Fat 83.05, water 14.50, ash 1.33, casein 1.12.

Dollie's Valentine 105049.—Sire, Oonan's Tormentor Pogis 30505; dam, Dollie **Fay** 105047.

> Butter, 16 lbs. 11⅞ oz. (confirmed); estimated butter on basis of 85 per cent. fat, 16 lbs. 10⅝ oz. Milk, 265.7 lbs.

Test made from Apr. 22 to 28, 1898; age, 4 yrs. 2 mos.; estimated weight, 890 lbs.; fed 70 lbs. corn meal, 28 lbs. cotton-seed meal, 14 lbs. oil meal and 42 lbs. bran— clover hay *ad lib.*; property of Mrs. Shelby Kinkead, Lexington, Ky.

Analyses of Butter.

First churning—Fat 82.73, water 14.37, ash 2.11, casein 0.79.
Second churning—Fat 81.75, water 14.61, ash 2.58, casein 1.06.

Dollie's Valentine 105049.—Sire, Oonan's Tormentor Pogis 30505; dam, Dollie **Fay** 105047.

> Butter, 15 lbs. 6¼ oz. (confirmed); estimated butter on basis of 85 per cent. fat, 15 lbs. 10⅝ oz. Milk, 251.4 lbs.

Test made from Apr. 29 to May 5, 1898; age, 4 yrs. 2 mos.; estimated weight, 895 lbs.; fed 70 lbs. corn meal, 28 lbs. cotton-seed meal, 14 lbs. oil meal and 42 lbs. bran— clover hay *ad lib.* to May 1; property of Mrs. Shelby Kinkead, Lexington, Ky.

Analyses of Butter.

First churning—Fat 81.25, water 14.34, ash 2.98, casein 1.43.
Second churning—Fat 82.50, water 13.86, ash 2.76, casein 0.88.

Dollie's Valentine 105049.—Sire, Oonan's Tormentor Pogis 30505; dam, Dollie **Fay** 105047.

> Butter, 17 lbs. 4 oz. (confirmed); estimated butter on basis of 85 per cent. fat, 16 lbs. ⅜ oz. Milk, 255.8 lbs.

Test made from May 6 to 12, 1898; age, 4 yrs. 2 mos.; estimated weight, 900 lbs.; fed 70 lbs. corn meal, 28 lbs. cotton-seed meal, 14 lbs. oil meal and 42 lbs. bran; property of Mrs. Shelby Kinkead, Lexington, Ky.

Analyses of Butter.

First churning—Fat 82.79, water 14.12, ash 2.15, casein 0.94.
Second churning—Fat 82.60, water 14.72, ash 1.87, casein 0.81.

Dollie's Valentine 105049.—Sire, Oonan's Tormentor Pogis 30505; dam, Dollie Fay 105047.

Butter, 17 lbs. ⅞ oz. (confirmed); estimated butter on basis of 85 per cent. fat, 16 lbs. 2⅝ oz. Milk, 259.4 lbs.

Test made from May 13 to 19, 1898; age, 4 yrs. 3 mos.; estimated weight, 910 lbs.; fed 70 lbs. corn meal, 28 lbs. cotton-seed meal, 14 lbs. oil meal and 42 lbs. bran; property of Mrs. Shelby Kinkead, Lexington, Ky.

Analyses of Butter.

First churning—Fat 82.16, water 14.35, ash 2.43, casein 1.06.
Second churning—Fat 80.12, water 14.79, ash 4.09, casein 1.00.

Dollie's Valentine 105049.—Sire, Oonan's Tormentor Pogis 30505; dam, Dollie Fay 105047.

Butter, 16 lbs. 2⅛ oz. (confirmed); estimated butter on basis of 85 per cent. fat, 15 lbs. 14½ oz. Milk, 262.2 lbs.

Test made from May 20 to 26, 1898; age, 4 yrs. 3 mos.; estimated weight, 915 lbs.; fed 70 lbs. corn meal, 28 lbs. cotton-seed meal, 14 lbs. oil meal and 42 lbs. bran; property of Mrs. Shelby Kinkead, Lexington, Ky.

Analyses of Butter.

First churning—Fat 81.82, water 14.65, ash 2.65, casein 0.88.
Second churning—Fat 81.90, water 14.72, ash 2.69, casein 0.69.

Dollie's Valentine 105049.—Sire, Oonan's Tormentor Pogis 30505; dam, Dollie Fay 105047.

Butter, 16 lbs. 3 oz. (confirmed); estimated butter on basis of 85 per cent. fat, 16 lbs. 4⅝ oz. Milk, 264.8 lbs.

Test made from May 27 to June 2, 1898; age, 4 yrs. 3 mos.; estimated weight, 925 lbs.; fed 70 lbs. corn meal, 28 lbs. cotton-seed meal, 14 lbs. oil meal and 42 lbs. bran; property of Mrs. Shelby Kinkead, Lexington, Ky.

Analyses of Butter.

First churning—Fat 80.55, water 14.81, ash 3.50, casein 1.14.
Second churning—Fat 80.37, water 14.83, ash 4.06, casein 0.74.

Dollie's Valentine 105049.—Sire, Oonan's Tormentor Pogis 30505; dam, Dollie Fay 105047.

Butter, 16 lbs. 9½ oz. (confirmed); estimated butter on basis of 85 per cent. fat, 16 lbs. 9¾ oz. Milk, 265.3 lbs.

Test made from June 3 to 9, 1898; age, 4 yrs. 3 mos.; estimated weight, 930 lbs.; fed 70 lbs. corn meal, 28 lbs. cotton-seed meal, 14 lbs. oil meal and 42 lbs. bran; property of Mrs. Shelby Kinkead, Lexington, Ky.

Analyses of Butter.

First churning—Fat 80.38, water 15.21, ash 3.65, casein 0.76.
Second churning—Fat 79.85, water 15.33, ash 3.99, casein 0.83.

Dollie's Valentine 105049.—Sire, Oonan's Tormentor Pogis 30505; dam, Dollie Fay 105047.

Butter, 16 lbs. 15⅝ oz. (confirmed); estimated butter on basis of 85 per cent. fat, 15 lbs. 13⅝ oz. Milk, 265.9 lbs.

Test made from June 10 to 16, 1898; age, 4 yrs. 4 mos.; estimated weight, 935 lbs.; fed 70 lbs. corn meal, 28 lbs. cotton-seed meal, 14 lbs. oil meal and 42 lbs. bran; property of Mrs. Shelby Kinkead, Lexington, Ky.

Analyses of Butter.

First churning—Fat 78.85, water 15.29, ash 5.08, casein 0.78.
Second churning—Fat 78.98, water 16.43, ash 3.71, casein 0.88.

Dollie's Valentine 105049.—Sire, Oonan's Tormentor Pogis 30505; dam, Dollie Fay 105047.

Butter, 16 lbs. 13½ oz. (confirmed); estimated butter on basis of 85 per cent. fat, 16 lbs. 1¾ oz. Milk, 260.3 lbs.

Test made from June 17 to 23, 1898; age, 4 yrs. 4 mos.; estimated weight, 940 lbs., fed 70 lbs. corn meal, 28 lbs. cotton-seed meal, 14 lbs. oil meal and 42 lbs. bran; property of Mrs. Shelby Kinkead, Lexington, Ky

Analyses of Butter.

First churning—Fat 83.61, water 13.60, ash 2.01, casein 0.78.
Second churning—Fat 82.00, water 14.56, ash 2.77, casein 0.67.

Dollie's Valentine 105049.—Sire, Oonan's Tormentor Pogis 30505; dam, Dollie Fay 105047

Butter, 15 lbs. 12⅝ oz. (confirmed); estimated butter on basis of 85 per cent. fat, 15 lbs. 7 oz. Milk, 259.4 lbs.

Test made from June 24 to 30, 1898; age, 4 yrs. 4 mos.; estimated weight, 945 lbs., fed 70 lbs. corn meal, 28 lbs. cotton-seed meal, 14 lbs. oil meal and 42 lbs. bran, property of Mrs. Shelby Kinkead, Lexington, Ky.

Analyses of Butter.

First churning—Fat 81.22, water 14.08, ash 4.07, casein 0.63.
Second churning—Fat 81.54, water 14.65, ash 3.00, casein 0.81.

Dollie's Valentine 105049.—Sire, Oonan's Tormentor Pogis 30505; dam, Dollie Fay 105047.

Butter, 15 lbs. 15¼ oz. (confirmed); estimated butter on basis of 85 per cent. fat, 15 lbs. 5 oz. Milk, 258.6 lbs.

Test made from July 1 to 7, 1898; age, 4 yrs. 4 mos.; estimated weight, 950 lbs., fed 70 lbs. corn meal, 28 lbs. cotton-seed meal, 14 lbs. oil meal and 42 lbs. bran, property of Mrs. Shelby Kinkead, Lexington, Ky

Analyses of Butter.

First churning—Fat 82.04, water 14.64, ash 2.55, casein 0.77.
Second churning—Fat 80.23, water 14.96, ash 4.12, casein 0.69.

Dollie's Valentine 105049.—Sire, Oonan's Tormentor Pogis 30505; dam, Dollie Fay 105047

Butter, 16 lbs. (confirmed); estimated butter on basis of 85 per cent. fat, 15 lbs. 1¾ oz. Milk, 249.3 lbs.

Test made from July 8 to 14, 1898; age, 4 yrs. 5 mos.; estimated weight, 955 lbs., fed 70 lbs. corn meal, 28 lbs. cotton-seed meal, 14 lbs. oil meal and 42 lbs. bran, property of Mrs. Shelby Kinkead, Lexington, Ky

Analyses of Butter.

First churning—Fat 82.26, water 14.00, ash 3.00, casein 0.74.
Second churning—Fat 81.64, water 14.03, ash 3.67, casein 0.66.

Dollie's Valentine 105049.—Sire, Oonan's Tormentor Pogis 30505; dam, Dollie Fay 105047.

Butter, 14 lbs. ½ oz. (confirmed); estimated butter on basis of 85 per cent. fat, 13 lbs. 6½ oz. Milk, 220.1 lbs.

Test made from July 15 to 21, 1898; age, 4 yrs. 5 mos.; actual weight, 959 lbs., fed 70 lbs. corn meal, 28 lbs. cotton-seed meal, 14 lbs. oil meal and 42 lbs. bran; property of Mrs. Shelby Kinkead, Lexington, Ky

Analyses of Butter.

First churning—Fat 82.02, water 14.00, ash 3.26, casein 0.72.
Second churning—Fat 81.68, water 14.03, ash 3.68, casein 0.61.

Dollie's Valentine 105049.—Sire, Oonan's Tormentor Pogis 30505; dam, Dollie Fay 105047.

Butter, 345 lbs. 11⅝ oz. (confirmed) in 21 weeks; estimated butter on basis of 85 per cent. fat, 337.24 lbs. Milk, 5428.6 lbs.

Test made from Feb. 25 to July 21, 1898; age, 4 yrs.; average weight at beginning of test, 905 lbs.; average weight at end of test, 959 lbs.; fed 1464 lbs. corn meal, 515 lbs. cotton-seed meal, 283 lbs. oil meal, 936 lbs. bran and 470 lbs. beets—clover hay *ad lib.* to May 1, pasture from Apr. 15; property of Mrs. Shelby Kinkead, Lexington, Ky.

(For analyses of butter see weekly tests above.) •

Dolly Bond 82662.—Sire, Sumach 5249; dam, Olive E. 16032.

Butter, 18 lbs. 3 oz. Milk, 244 lbs. 8 oz.

Test made from Oct. 26 to Nov. 2, 1892; age, 4 yrs. 2 mos.; actual weight, 825 lbs., fed 42 qts. corn meal, 42 qts. shorts and 28 qts. ground oats—short pasture; property of C. H. Ellsworth, Worcester, Mass.

Dolly Doyle 22788.—Sire, Tom Doyle 4695; dam, Carrie H. 12182.

Butter, 15 lbs. 4½ oz. Milk, 153 lbs. 15 oz.

Test made from July 31 to Aug. 7, 1887; age, 5 yrs. 7 mos.; estimated weight, 750 lbs.; fed, daily, about 12 lbs. ground corn and oats; property of John Moore, Jr., Columbia, Tenn.

Dolly Landseer 70201.—Sire, Ida's Landseer 17745; dam, Dolly Doyle 2d 58043.

Butter, 14 lbs. 10 oz. Milk, 173 lbs.

Test made from Oct. 14 to 21, 1895; age, 5 yrs. 11 mos.; estimated weight, 900 lbs., fed 70 lbs. corn meal, 42 lbs. ground oats and 28 lbs. cotton-seed meal—short blue-grass pasture; property of Jeff D. Clinard, Springfield, Tenn.

Dolly of Edgewood 83345.—Sire, King Coomassie 2d 19545; dam, Pridalia 17249.

Butter, 15 lbs. 4 oz. Milk, 174 lbs. 11 oz.

Test made from Nov. 10 to 17, 1893; age, 2 yrs. 7 mos.; estimated weight, 700 lbs.; fed 35 qts. corn meal and 56 qts. wheat bran—hay *ad lib.*, poor pasture; property of Columbus Dixon, Gillespieville, Ohio.

Dolly of Edgewood 83345.—Sire, King Coomassie 2d 19545; dam, Pridalia 17249.

Butter, 20 lbs. 9½ oz. Milk, 289 lbs. 14.4 oz.

Test made from May 7 to 14, 1897; age, 6 yrs. 1 mo.; estimated weight, 1050 lbs.; fed 10 lbs. bran, 8 lbs. corn and oats, 2½ lbs. cotton-seed meal, 2½ lbs. linseed oil meal and 4 lbs. hominy feed, daily—hay *ad lib.*; property of John E. Robbins, Greensburg, Ind.

Dolly of Lakeside 10824.—Sire, Micawber 4796; dam, Dove 2d 8742.

Butter, 14 lbs. 8 oz., unsalted. Milk, 204 lbs. 8 oz.

Test made from Feb. 15 to 22, 1883; age, 7 yrs. 10 mos.; property of J. H. Walker, Worcester, Mass.

Dolly's Daughter 31392.—Sire, Duke of Darlington 2460; dam, Dolly Dunn 18298.

Butter, 16 lbs. Milk, 197 lbs. 10 oz.

Test made from Sept. 7 to 14, 1891; age, 6 yrs. 11 mos.; estimated weight, 800 lbs.; fed 8 qts. corn meal, 7 qts. ground oats and 5 qts. wheat middlings, daily—very good pasture; property of D. F. Appleton, Ipswich, Mass.

Demett 88179.—Sire, Bluster 22798; dam, Reif 58398.

Butter, 16 lbs. 2 oz. Milk, 205 lbs. 12 oz.

Test made from Nov. 3 to 10, 1895; age, 2 yrs. 7 mos.; estimated weight, 800 lbs.; fed 42 lbs. oats, 28 lbs. corn, 14 lbs. oil meal and 35 lbs. potatoes and pumpkins—corn ensilage, frosted clover pasture; property of H. C. Taylor, Orfordville, Wis.

Dom Pedro's Julian 8631.—Sire, Dom Pedro 2092; dam, Brunette Balliet 6733.

Butter, 16 lbs.

Test made from Dec. 12 to 18, 1881; age, 3 yrs. 8 mos.; property of Paul Balliet, Ballietsville, Pa

Doña Marina 7049.—Sire, King Arthur 2170; dam, Melia Ann 5444.

Butter, 18 lbs. 8 oz. Milk, 171 lbs. 4 oz.

Test made from Aug. 3 to 10, 1886; age, 8 yrs. 5 mos.; estimated weight, 900 lbs.; fed cut grass only, no grain—poor pasture; property of Joseph H. Walker, Worcester, Mass.

Donna Argyle 61456.—Sire, Don Hugo 19402; dam, Baroness Argyle 40498.

Butter, 15 lbs. 10 oz. Milk, 231 lbs. 8 oz.

Test made from Jan. 14 to 21, 1894; age, 4 yrs. 2 mos.; estimated weight, 800 lbs.; fed 2 lbs. coarse wheat bran, 2 lbs. corn meal, 2 lbs. gluten, 2 lbs. rye and oats provender, 1 lb. cotton-seed meal, 30 lbs. corn ensilage and 10 lbs. cut clover hay; property of E. S. Henry, Rockville, Conn.

Donna Le Brocq 99305.—Sire, Spokane 23440; dam, Le Brocq's May 59724.

Butter, 14 lbs. 6 oz. Milk, 211 lbs.

Test made from Oct. 25 to Nov. 1, 1897; age, 5 yrs. 7 mos.; estimated weight, 800 lbs.; fed 40 lbs. ensilage, daily—blue-grass pasture; property of Baum & Hill, Frankfort, Ind.

Donna Signal 29407.—Sire, Dunraven 7950; dam, Donna Fay 6294.

Butter, 16 lbs. 1 oz. Milk, 243 lbs.

Test made from July 1 to 8, 1888; age, 3 yrs. 11 mos.; estimated weight, 725 lbs.; fed, daily, 9 lbs. bran, 5 lbs. corn meal, 2 lbs. cotton-seed meal and 2 lbs. pea meal; property of Jacob L. Thomas, Knoxville, Tenn.

Donney Pogis 38907.—Sire, Pogis Chief 3998; dam, Don's Fanchonella 22175.

Butter, 16 lbs. 2⅛ oz. Milk, 143 lbs. 4 oz.

Test made from July 21 to 28, 1889; age, 4 yrs.; estimated weight, 950 lbs.; fed, daily, 4 gals. ground oats and corn, mixed equally; property of Morgan & Brown, Columbia, Tenn.

Donney Pogis 2d 82426.—Sire, Tennessee's Landseer 23015; dam, Donney Pogis 38907

Butter, 17 lbs. 1¼ oz. Milk, 280 lbs. 8 oz.

Test made from Nov. 29 to Dec. 6, 1896; age, 5 yrs. 10 mos.; estimated weight, 800 lbs.; fed 44½ lbs. bran, 44 lbs. oats, 29½ lbs. corn, 15½ lbs. linseed meal, 140 lbs. roots and 210 lbs. ensilage—hay *ad lib.*; property of C. I. Hood, Lowell, Mass.

Donney Pogis 3d 100199.—Sire, Bisson's Landseer 27520; dam, Donney Pogis 38907.

Butter, 16 lbs. ¾ oz. Milk, 198 lbs. 8 oz.

Test made from Aug. 11 to 18, 1897; age, 4 yrs. 8 mos.; estimated weight, 1000 lbs.; fed 42 lbs. bran, 42 lbs. corn meal, 7 lbs. cotton-seed meal, 14 lbs. linseed meal, 21 lbs. ground oats, 168 lbs. cut green corn and 35 lbs. cotton-seed hulls; property of S. H. Butler, Como, Miss.

Donnona's Nectarine 40093.—Sire, Champion Albert 6023; dam, Donnona 10120.

Butter, 16 lbs. 6 oz. Milk, 277 lbs. 2 oz.

Test made from Mar. 17 to 24, 1891; age, 6 yrs. 10 mos.; actual weight, 850 lbs.; fed 70 lbs. corn chop, 70 lbs. wheat bran and shorts, 35 lbs. new process oil meal, mixed, with chopped oat sheaves, salted and slightly wetted with warm water—clover hay *ad lib.*; property of George E. McGill, Leavenworth, Kan.

Dora 4th 3936.—Sire, Vermont 893; dam, Dora 1550.

Butter, 25 lbs. 2 oz.

Test made from Aug. 3 to 10, 1883; age, 8 yrs. 4 mos.; fed 2 gals. bran and 1 gal. corn meal, daily—clover hay *ad lib.*, good pasture; property of Thomas L. Dodd, Nashville, Tenn.

Dora Doon 12909.—Sire, Top-Sawyer 1404; dam, Daisy Dean 6855.

Butter, 15 lbs. Milk, 198 lbs.

Test made from June 21 to 27, 1884; age, 3 yrs. 2 mos.; property of G. R. Gwynn, Lebanon, Tenn.

Dora Lowndes 64136.—Sire, Warren Lowndes 17591; dam, Lady Madeline 2d 41854.

Butter, 16 lbs. 15 oz. Milk, 243 lbs. 4 oz.

Test made from Mar. 19 to 26, 1895; age, 7 yrs. 9 mos.; actual weight, 842 lbs.; fed 34 qts. bran, 56 qts. oat dust, 20 qts. Buffalo gluten feed, 14 qts. corn meal and 7 qts. linseed meal—28 lbs. hay, cut corn-stalks *ad lib.*; property of Henry A. Slack, Hurstville, N. Y.

Dora Neptune 20318.—Sire, Neptune 5368; dam, Dora (P. S. 155 J. H. B.)

Butter, 20 lbs. ½ oz. Milk, 324 lbs. 4 oz.

Test made from June 11 to 17, 1884; age, 3 yrs. 3 mos.; fed 13½ lbs. ground corn, oats and shorts, daily—clover pasture; property of C. E. Rowley, Cortland, N. Y.

Dora of Glynllyn 95993.—Sire, Star of Glynllyn 31097; dam, Dora Signal 2d 73628.

Butter, 17 lbs. 9 oz. Milk, 166 lbs. 9 oz.

Test made from Jan. 28 to Feb. 4, 1897; age, 3 yrs. 7 mos.; estimated weight, 700 lbs.; fed 80 lbs. H. O. feed, 25 lbs. middlings, 25 lbs. bran, 17 lbs. oil meal and 170 lbs. roots—hay *ad lib.*; property of D. Wolfe Bishop, Jr., Lenox, Mass.

Dora's Fawnie Exile 108696.—Sire, Exile of Jersey Lawn 29421; dam, Exile's Dora Lowndes 80523.

Butter, 15 lbs. 12 oz. Milk, 229 lbs.

Test made from Dec. 23 to 30, 1896; age, 3 yrs. 2 mos.; estimated weight, 825 lbs.; fed 3 lbs. bran and 4 lbs. middlings, in a warm mash each morning and evening, 4 lbs. ground oats, corn and oil-cake, mixed, each noon—clover and timothy hay *ad lib.*; property of G. W. Bayler, Washington, Ill.

Dora Victor Pogis 100921.—Sire, Melia Ann's Victor Pogis 20697; dam, Mooly Pedro 77811.

Butter, 18 lbs. 5 oz. Milk, 281 lbs. 12 oz.

Test made from Feb. 5 to 12, 1896; age. 3 yrs. 1 mo.; estimated weight, 700 lbs.; fed 3 lbs. oil meal, 6 lbs. corn meal, 5 lbs. peas and oat meal and 4 lbs. wheat bran, daily—hay, corn ensilage; property of J. H. Martin, Granville, N. Y.

Dorchester Maid 101385.—Sire, Gold Coast Jr. 9217; dam, Dorchester Lass 18297.

Butter, 16 lbs. 5 oz. Milk, 225 lbs. 12 oz.

Test made from Aug. 19 to 26, 1895; age. 2 yrs. 6 mos.; actual weight, 700 lbs., fed 9 qts. wheat bran, 6 qts. ground oats, 3 qts. corn meal, 1½ qts. oil meal and 1½ qts. cotton-seed meal, daily—fair pasture, green corn, oats and peas; property of C. A. Sweet, East Aurora, N. Y.

Dorine's Brunette 29309.—Sire, Brunette's Prince 7115; dam, Dorine 7456.

Butter, 20 lbs. 3 oz. Milk, 289 lbs. 8 oz.

Test made from Mar. 1 to 7, 1888; age. 3 yrs. 4 mos.; estimated weight, 800 lbs.; fed, daily, 7 lbs. corn meal, 10 lbs. bran and 3 lbs. linseed meal; property of J. R. Anderson, Jr., Lee, Va.

Doris of Mt. Pleasant 73330.—Sire, Sir Florian 11578; dam, Doris C. 33491.

Butter, 14 lbs. 7 oz. Milk, 235 lbs. 8 oz.

Test made from Feb. 24 to Mar. 3, 1897; age. 6 yrs. 9 mos.; actual weight, 785 lbs.; fed 28 lbs. bran, 35 lbs. oats and corn ground together, 14 lbs. cotton-seed meal, 14 lbs. linseed meal, 10 lbs. mixed feed, 140 lbs. hay, 70 lbs. roots and 70 lbs. carrots; property of George M. Haynes, Monmouth, Me.

Dorizella 20525.—Sire, Lenape 2732; dam, Kerima 7242.

Butter, 15 lbs. 7 oz. Milk, 239 lbs. 3 oz.

Test made from May 14 to 21, 1892; age. 9 yrs.; estimated weight, 950 lbs.; fed 35 lbs. corn meal and 63 lbs. wheat bran—Kentucky blue-grass pasture; property of N. S. Dudley, Flemingsburg, Ky.

Dorothy of Bovina 9373.—Sire, Ben Butler of Bovina 2024; dam, Daphne of Staatsburgh 2d 3027.

Butter, 15 lbs. 4 oz. Milk, 205 lbs.

Test made from June 20 to 26, 1883; age. 5 yrs. 3 mos.; no grain fed—pasture rather poor; property of W. L. Rutherford, Franklin, N. Y

Dorothy Taylor 99485.—Sire, Elmwood Nona's Pogis 24986; dam, Kismet 2d 22893.

Butter, 15 lbs. 10 oz. Milk, 280 lbs. 3 oz.

Test made from Apr. 5 to 12, 1897; age. 4 yrs. 5 mos.; estimated weight, 800 lbs.; fed 112 qts. bran, 56 qts. corn and cob meal and 32 qts. shelled oats—sheaf oats and clover hay at night, corn stover during day; property of Q. I. Simpson, Palmer, Ill.

Dosia Pogis 89049.—Sire, Sig Pogis 22806; dam, Maggie Cyrene 53462.

Butter, 16 lbs. 10¼ oz., unsalted. Milk, 195 lbs. 8 oz.

Test made from Apr. 4 to 11, 1895; age. 2 yrs. 5 mos.; estimated weight, 600 lbs.; fed 42 qts. ground oats, 7 qts. cotton-seed meal, 28 qts. crushed corn and 28 qts. bran—good rye pasture; property of D. P. Carter, Lynnville, Tenn.

Dot Badger Girl 101417.—Sire, Winder 29955; dam, Gold Badger Girl 74902.

Butter, 14 lbs. 1 oz. Milk, 201 lbs. 10 oz.

Test made from Jan. 4 to 11, 1898; age, 3 yrs. 4 mos.; estimated weight, 800 lbs.; fed 70 lbs. ground oats and wheat bran (one-third bran), 14 lbs. oil meal and 140 lbs. ensilage—corn stover and clover hay *ad lib.*; property of N. N. Palmer & Son, Brodhead, Wis.

Dot of Bear Lake 6170.—Sire, Prince of Warren 1512; dam, Monmouth Duchess 2d 4619.

Butter, 19 lbs. 4 oz. Milk, 213 lbs.

Test made from Feb. 4 to 10, 1883; age, 6 yrs. 11 mos.; fed 5 lbs. corn meal, 4 lbs. oats and 5 lbs. wheat bran, daily—hay *ad lib.*; property of John C. Drake, Marietta, Ohio

Dot of Rocky Ford 40763.—Sire, Steletho 10505; dam, Imperial's Omega 13241.

Butter, 15 lbs. 1 oz. Milk, 246 lbs. 8 oz.

Test made from May 23 to 30, 1891; age, 6 yrs. 5 mos.; estimated weight, 950 lbs.; fed 5 lbs. bran and 4 lbs. oil meal, daily—timothy, clover and blue-grass pasture, property of Bordwell & Cochran, Batavia, Ohio.

Dott Buttercup 16358.—Sire, Major Dott 7101; dam, Buttercup 3d 10253.

Butter, 16 lbs. 2 oz. Milk, 167 lbs. 6 oz.

Test made from June 11 to 17, 1885; age, 3 yrs. 8 mos.; property of Samuel McKeen, Terre Haute, Ind.

Douskha 63439.—Sire, Bloomfield's Pogis 17209; dam, Shekel of Gold 25169.

Butter, 15 lbs. 5¾ oz. Milk, 227 lbs. 14 oz.

Test made from Mar. 18 to 25, 1893; age, 4 yrs. 1 mo.; estimated weight, 800 lbs.; fed 56 qts. bran, 14 qts. ground corn, oats and peas, 42 qts. corn meal and 7 qts. cotton-seed meal—pasture; property of M. Lothrop, Marshall, Tex.

Dove 5th 14561.—Sire, Micawber 4796; dam, Dove 7824.

Butter, 16 lbs. 13 oz. Milk, 253 lbs.

Test made from Oct. 29 to Nov. 5, 1889; age, 14 yrs. 6 mos.; estimated weight, 875 lbs.; fed, daily, 4 qts. corn meal, 4 qts. ground oats and 4 qts. shorts; property of Terence Carrigan, Hopkinton, Mass.

Dove Dee 18059.—Sire, Eupidee 4097; dam, Dove 4th 10945.

Butter, 27 lbs. 9 oz. Milk, 197 lbs. 11 oz.

Test made from June 13 to 19, 1886; age, 4 yrs. 3 mos.; fed 4 qts. corn meal and 4 qts. shorts, mixed, daily—green cut clover and red-top, short pasture; property of J. H. Walker, Worcester, Mass.

Down of Chatsworth 15251.—Sire, Lord Rex 4113; dam, Bonnie Eloise 15244.

Butter, 16 lbs. 12 oz. Milk, 222 lbs. 14 oz.

Test made from Mar. 26 to Apr. 1, 1885; age, 3 yrs. 11 mos.; property of Edward Austen, Glencoe, Md.

Dubenna 65024.—Sire, Rajah Stoke Pogis 25276; dam, Zinokia 47786.

Butter, 14 lbs. 7½ oz. Milk, 184 lbs.

Test made from Sept. 23 to 30, 1893; age, 4 yrs. 4 mos.; estimated weight, 800 lbs.; fed 18 qts. chop and 12 ears sweet corn, daily, for first five days; 20 qts. chop and 12 ears sweet corn, daily, for last two days—pasture; property of Mrs. A. M. Hallock, Columbus, Ohio.

Dubenna 2d 74920.—Sire, Day's Romp 24344; dam, Dubenna 65024.

Butter, 18 lbs. ½ oz. Milk, 319 lbs. 4 oz.

Test made from July 14 to 21, 1896; age, 5 yrs. 1 mo.; estimated weight, 950 lbs.; fed 42 qts. corn meal, 28 qts. old process oil meal and 28 qts. wheat bran—blue-grass pasture; property of A. S. Bell, London, Ohio.

Duchess Caroline 3d 6039.—Sire, Aldine 1136; dam, Duchess Caroline 2022.

Butter, 15 lbs. 8 oz., unsalted. Milk, 224 lbs.

Test made from Apr. 9 to 15, 1883; age, 6 yrs. 2 mos.; property of W. B. Montgomery, Starkville, Miss.

Duchess Noble 27090.—Sire, Noble 4th 6246; dam, Rosalia of Sidney 4521.

Butter, 25 lbs. 8 oz. Milk, 268 lbs. 8 oz.

Test made from May 8 to 15, 1888; age, 5 yrs. 1 mo.; estimated weight, 800 lbs.; fed, daily, 3 qts. ground oats, 3 qts. corn meal, 2 qts. bran and 1 pt. oil meal; property of L. H. Morse, Newark, N. Y.

Duchess of Argyle 3758.—Sire, Jack Dasher 982; dam, Berlin Daisy 3759.

Butter, 14 lbs. 13 oz. Milk, 198 lbs. 12 oz.

Test made from Jan. 11 to 18, 1888; age, 9 yrs. 9 mos.; property of E. S. Henry, Rockville, Conn.

Duchess of Argyle 4th 7571.—Sire, Duke of Argyle 1517; dam, Duchess of Argyle 3758.

Butter, 14 lbs. 12 oz. Milk, 154 lbs.

Test made from Oct. 6 to 13, 1884; age, 6 yrs. 7 mos.; property of E. S. Henry, Rockville, Conn.

Duchess of Argyle 7th 47569.—Sire, Baron Hugo 15208; dam, Duchess of Argyle 3758.

Butter, 14 lbs. 8 oz. Milk, 229 lbs. 4 oz.

Test made from Feb. 7 to 14, 1893; age, 5 yrs. 4 mos.; estimated weight, 1000 lbs.; fed 2 lbs. cotton-seed meal, 2 lbs. Chicago gluten meal, 3 lbs. Pillsbury wheat bran, 3 lbs. corn and cob meal, about 35 lbs. corn ensilage and 10 lbs. cut clover hay, daily; property of E. S. Henry, Rockville, Conn.

Duchess of Bloomfield 3653.—Sire, Rioter 670; dam, Angela 1682.

Butter, 20 lbs. ½ oz. Milk, 317 lbs. 8 oz.

Test made from Apr. 21 to 27, 1882; age, 8 yrs. 2 mos.; fed 12 lbs. corn meal, 4 lbs. wheat bran and 4 lbs. cotton-seed meal, daily—good blue-grass pasture; property of Campbell Brown, Spring Hill, Tenn.

Duchess of Bloomfield 2d 12250.—Sire, Tormentor 3533; dam, Duchess of Bloomfield 3653.

Butter, 17 lbs. 4 7-16 oz. Milk, 111 lbs. 10 oz.

Test made from Mar. 6 to 13, 1886; age, 5 yrs.; property of Isaac Ross, Opelika, Ala.

Duchess of Bloomfield 3d 15580.—Sire, Lord Harry 3445; dam, Duchess of Bloomfield 3653.

Butter, 15 lbs. 1 oz. Milk, 159 lbs.

Test made from Jan. 3 to 9, 1886; age, 3 yrs. 11 mos.; property of A. K. Johnston, Stapleton, N. Y.

Duchess of Bloomfield 4th 35584.—Sire, Nancy's Rioter 8390; dam, Duchess of Bloomfield 3653.

Butter, 14 lbs. 8½ oz. Milk, 168 lbs. 4 oz.

Test made from Sept. 19 to 26, 1888; age, 3 yrs. 2 mos.; estimated weight, 950 lbs.; fed, daily, 2 gals. oats and 2 gals. corn meal; property of Campbell Brown, Spring Hill, Tenn.

Duchess of Bolton 98150.—Sire, Viking of Bolton 24881; dam, Bertha Polonius 24914.

Butter, 24 lbs. 4 oz. Milk, 212 lbs. 8 oz.

Test made from Sept. 25 to Oct. 2, 1897; age, 4 yrs. 8 mos.; actual weight, 930 lbs.; fed 63 lbs. 7 oz. grain (corn meal, wheat bran, ground oats and cotton-seed meal) and 2 bush. green sweet corn ears—good pasture; property of John A. Cunningham, Bolton, Mass.

Duchess of Darlington 13830.—Sire, Duke of Darlington 2460; dam, Minnie Stevens 13059.

Butter, 14 lbs. 11 oz. Milk, 274 lbs.

Test made from Oct. 17 to 23, 1886; age, 8 yrs.; weight, 780 lbs.; fed 30 lbs. daily of mixed corn-hearts, oats, oil meal and middlings; property of Miller & Sibley, Franklin, Pa.

Duchess of Jefferson 17144.—Sire, Beauty Boy of Jefferson 5753; dam, Frankie 4th 2253.

Butter, 16 lbs. 9 oz. Milk, 241 lbs. 10 oz.

Test made from Apr. 4 to 11, 1896; age, 15 yrs. 1 mo.; estimated weight, 900 lbs.; fed 28 lbs. bran, 35 lbs. oats, 42 lbs. corn meal and 21 lbs. oil meal—corn ensilage and hay *ad lib.*; property of H. C. Taylor, Orfordville, Wis.

Duchess of Jefferson 2d 118168.—Sire, Parade 30875; dam, Duchess of Jefferson 17144.

Butter, 14 lbs. 7 oz. Milk, 227 lbs.

Test made from June 18 to 25, 1898; age, 2 yrs. 3 mos.; estimated weight, 800 lbs.; fed 28 lbs. corn meal, 28 lbs. oats and 17½ lbs. oil meal—good mixed pasture; property of H. C. Taylor, Orfordville, Wis.

Duchess of St. Lambert 5111.—Sire, Lord Lisgar 1066; dam, Pride of Windsor 483.

Butter, 15 lbs. 11 oz. Milk, 216 lbs.

Test made from Dec. 19 to 26, 1883; age, 8 yrs. 9 mos.; property of V. E. Fuller, Hamilton, Ont., Can.

Dudu of Linwood 8336.—Sire, Rector 1458; dam, Little Duck 3651.

Butter, 16 lbs. 15 oz. Milk, 233 lbs. 8 oz.

Test made from Feb. 5 to 11, 1884; age, 5 yrs. 1 mo.; property of Campbell Brown, Spring Hill, Tenn

Duenna's Duchess 5508.—Sire, Duke of Grayholdt 1035; dam, Duenna 2716.

Butter, 16 lbs. 10 oz.   Milk, 239 lbs.

Test made from June 16 to 22, 1882; age, 5 yrs. 6 mos.; fed 8 qts. oats and 12 qts. bran, daily; property of G. H. & H. A. Grimmell, Jefferson, Iowa.

Duke of St. L.'s Vail's Pet 61970.—Sire, Duke of St. Lambert 16160; dam, Vail's Pet 12092.

Butter, 14 lbs. 3 oz.   Milk, 236 lbs. 14 oz

Test made from June 17 to 24, 1893; age, 5 yrs. 1 mo.; estimated weight, 850 lbs.; fed 5 qts. oat meal, 4 qts. corn meal, 5 qts. bran and 1 qt. cotton-seed meal. daily— clover and timothy pasture; property of Mrs. E. F. Hawley, Pittsford, N. Y.

Duke's Medusa 43511.—Sire, Duke of Cloverdale 6994; dam, Koffee's Medusa 26306.

Butter, 18 lbs.   Milk, 235 lbs. 8 oz.

Test made from Aug. 21 to 28, 1897; age, 10 yrs. 8 mos.; estimated weight, 1000 lbs.; fed 12 lbs., daily, of a mixture of corn meal, oats and bran, in equal parts by weight —wild grass and clover pasture; property of S. B. Pierce, Woodlawn, Neb.

Duke's Minnie 42785.—Sire, Duke of St. Lambert 16160; dam, Minnie Rioter 21336.

Butter, 20 lbs. 4 oz.   Milk, 190 lbs.

Test made from Sept. 23 to 30, 1894; age, 7 yrs. 6 mos.; actual weight, 860 lbs.; fed 16⅓ lbs. cotton-seed meal, 8 1-6 lbs. old process linseed meal, 32⅔ lbs. wheat bran, 40 5-6 lbs. corn chop, 98 lbs. cut green corn-stalks and 7 lbs. hay; property of B. F. Fackenthal, Jr., Riegelsville, Pa.

Duke's Pierrot Lady 31023.—Sire, Sweepstakes Duke 1905; dam, Pierrot's Lady Jennings 11675.

Butter, 21 lbs. 12 oz.   Milk, 266 lbs. 12 oz.

Test made from Jan. 11 to 18, 1890; age, 5 yrs. 6 mos.; estimated weight, 900 lbs., fed, daily, about 14 lbs., one-third crushed corn and two-thirds bran; property of Charles A. Galwith, Fulton, Mo.

Duke's Rowena 37190.—Sire, Rachel's Duke 7022; dam, Philene 9012.

Butter, 14 lbs. 8 oz.   Milk, 190 lbs. 12 oz.

Test made from Oct. 7 to 14, 1896; age, 11 yrs. 2 mos.; estimated weight, 800 lbs.; fed 4 qts. wheat bran, 2 qts. ground oats and 2 qts. corn meal, daily—river meadow pasture; property of E. L. Van Deusen, Ashley Falls, Mass.

Duke's Rowena 2d's Queen 79894.—Sire, Real Queen's Stoke Pogis 20535; dam, Duke's Rowena 2d 79877

Butter, 18 lbs. 12 oz.   Milk, 275 lbs. 14 oz.

Test made from July 13 to 20, 1896; age, 5 yrs. 2 mos.; estimated weight, 900 lbs.; fed 2 qts. corn meal, 2 qts. ground oats and 2 qts. wheat bran, daily—good upland pasture; property of E. L. Van Deusen, Ashley Falls, Mass.

Duke's Signal Queen 2d 82418.—Sire, Landseer's Pogis 15847; dam, Duke's Signal Queen 32323

Butter, 17 lbs. 8 oz.   Milk, 313 lbs. 15 oz.

Test made from Mar. 6 to 13, 1897; age, 7 yrs. 9 mos.; estimated weight, 900 lbs.; fed 52½ lbs. bran, 38½ lbs. corn meal, 45½ lbs. oats, 10½ lbs. linseed meal, 13½ lbs. cotton-seed meal, 140 lbs. roots and 280 lbs. ensilage—hay *ad lib.*; property of C. I. Hood, Lowell, Mass.

Dara 26001.—Sire, Prince of M. 2811; dam, Belle of Cherry Valley 15837.

Butter, 21 lbs. 5 oz. Milk, 282 lbs. 2 oz.

Test made from Oct. 18 to 25, 1890; age, 6 yrs. 7 mos.; actual weight, 1002 lbs.; fed, daily, 7½ lbs. corn meal, 7½ lbs. ground oats, 2 lbs. linseed meal and 2 lbs. bran; property of M. A. Scovell, Lexington, Ky.

Durex 66320.—Sire, Bob Dawson 18069; dam, Jennie's Pet 26728.

Butter, 15 lbs. 10¼ oz. Milk, 231 lbs. 4 oz.

Test made from Mar. 12 to 19, 1894; age, 4 yrs. 2 mos.; actual weight, 630 lbs.; fed 20 qts. boiled corn, 8 qts. boiled oats, 11 qts. boiled cotton-seed, 5 qts. cotton-seed meal, 29 qts. wheat bran and 125 lbs. fodder—clover pasture; property of W. N. Murphy, La Grange, Tex.

Durex 66320.—Sire, Bob Dawson 18069; dam, Jennie's Pet 26728.

Butter, 19 lbs. Milk, 344 lbs. 8 oz.

Test made from May 28 to June 4, 1896; age, 6 yrs. 4 mos.; actual weight, 630 lbs.; fed 28 qts. corn meal, 84 qts. ground oats, 112 qts. wheat bran, 14 qts. cotton-seed meal and 28 qts. boiled cotton-seed—hay *ad lib.*, good pasture; property of W. N. Murphy, La Grange, Tex.

Durwood's Lass 19710.—Sire, Durwood 4527; dam, Belle Williamson 8386.

Butter, 16 lbs. Milk, 176 lbs. 14 oz.

Test made from May 9 to 16, 1887; age, 6 yrs. 1 mo.; estimated weight, 600 lbs.; fed, daily, about 12 lbs. ground corn and oats; property of Mrs. E. T. Allen, Columbia, Tenn.

Dusky 2525.—Sire, Robbins 953; dam, Tidy 2520.

Butter, 16 lbs. 10 oz.

Test made in June, 1876; property of J. B. Williams, Glastonbury, Conn.

Dutsy Gordon 2d 108503.—Sire, Bessie's Toltec 19859; dam, Dutsy Gordon 108502.

Butter, 14 lbs. 7½ oz. Milk, 160 lbs. 4 oz.

Test made from Oct. 8 to 15, 1895; age, 5 yrs. 9 mos.; estimated weight, 650 lbs., fed 56 qts. ground oats and 56 qts. corn meal—millet hay, fair woodland pasture; property of Morgan & Brown, Columbia, Tenn.

Early Morn 50661.—Sire, Ruby's Harry 15664; dam, Pinkey Spry 18316.

Butter, 14 lbs. 12 oz. Milk, 236 lbs.

Test made from May 15 to 22, 1803; age, 6 yrs.; estimated weight, 900 lbs.; fed 95 lbs. corn meal, 58 lbs. bran, 18 lbs. oil meal and 4 lbs. shorts—blue-grass pasture; property of Purdue University, La Fayette, Ind.

Early Morn 50661.—Sire, Ruby's Harry 15664; dam, Pinkey Spry 18316.

Butter, 17 lbs. 13¼ oz. Milk, 248.4 lbs.

Test made from Mar. 12 to 19, 1897; age, 9 yrs. 11 mos.; actual weight, 1156 lbs.; fed 84 lbs. hominy meal, 21 lbs. oil meal, 55 lbs. hay and 193 lbs. corn ensilage; property of Purdue University, La Fayette, Ind.

Eastern Sultane 20298.—Sire, Eastern Chief (P. S. 171 J. H. B.); dam, Marie Louisa (F. S. 2472 J. H. B.)

Butter, 15 lbs. 3⅓ oz. Milk, 222 lbs. 14 oz.

Test made from Nov. 2 to 8, 1885; age, 7 yrs. 9 mos.; property of Frederick Loeser, Somerville, N. J.

Easter Ona 2d 97158.—Sire, Comus Rex 26804; dam, Easter Ona 68721.

Butter, 17 lbs. 5 oz. Milk, 283 lbs. 4 oz.

Test made from June 18 to 20, 1896; age, 3 yrs. 4 mos.; estimated weight, 800 lbs.; fed 6 lbs. bran, 8 lbs. corn meal, 3 lbs. oil meal, daily—green alfalfa *ad lib.* at night, mixed prairie, rye and clover pasture; property of F. M. Hosford, Bethany, Neb.

Easter Pogis 52371.—Sire, Eastwood's Pogis 15279; dam, Easter Belle 20806.

Butter, 15 lbs. 3 oz. Milk, 217 lbs. 14 oz.

. Test made from Mar. 5 to 12, 1895; age, 7 yrs. 1 mo.; estimated weight, 800 lbs.; fed 77 lbs. corn and oats, 56 qts. wheat bran and 21 lbs. cotton-seed meal—hay; property of F. M. Hosford, Bethany, Neb.

Easter Prize Rex 102166.— Sire, Comus Rex 26804; dam, Easter Prize 40324.

Butter, 15 lbs. Milk, 246 lbs. 5 oz.

Test made from Apr. 30 to May 7, 1897; age, 3 yrs. 8 mos.; estimated weight, 900 lbs.; fed 77 lbs. corn meal and 55 lbs. bran—timothy and clover hay *ad lib.*, timothy and clover pasture; property of H. A. McCauley, Highland, Kan.

Eastwood Clearwater 30445.—Sire, Perrot (P. S. 342 J. H. B.); dam, Clearwater 24382.

Butter, 27 lbs. Milk, 263 lbs.

Test made from June 10 to 17, 1888; age, 3 yrs. 11 mos.; fed, daily, 4 qts. corn meal and 4 qts. bran; property of M. Erskine Miller, Staunton, Va.

Edessa 21844.—Sire, Footstep 5163; dam, Effie of Verna 8928.

Butter, 16 lbs. 2 oz. Milk, 218 lbs. 5 oz.

Test made from June 18 to `25, 1890; age, 7 yrs. 2 mos.; estimated weight, 875 lbs.; fed, daily, 4 qts. of a mixture consisting of corn meal, crushed oats, bran and old process oil meal; property of Frederic Bronson, Greenfield Hill, Conn.

Edessa 3d 52043.—Sire, Halo 10517; dam, Edessa 21844.

Butter, 14 lbs. 6 oz. Milk, 173 lbs. 1 oz.

Test made from July 1 to 8, 1890; age, 2 yrs. 3 mos.; estimated weight, 750 lbs.; fed, daily, 4 qts. corn meal, crushed oats, bran and old process oil meal; property of Frederic Bronson, Greenfield Hill, Conn.

Editha Westray 89755.—Sire, Great Combination 20326; dam, Bombazine 2d 49327.

Butter, 19 lbs. 5 oz. Milk, 385 lbs. 6 oz.

Test made from Apr. 19 to 26, 1896; age, 3 yrs. 8 mos.; estimated weight, 850 lbs.; fed 70 lbs. wheat bran, 56 lbs. corn meal and 14 lbs. cotton-seed meal—prairie pasture; property of C. M. Bivins, Terrell, Tex.

Edith Campbell 23011.—Sire, Norman B. 7001; dam, Beauty of Snipsic 22909.

Butter, 21 lbs. 4½ oz. Milk, 256 lbs. 8 oz.

Test made from Jan. 6 to 13, 1888; age, 5 yrs. 3 mos.; estimated weight, 900 lbs.; fed, daily, 8 lbs. corn meal, 8 lbs. bran, 2 lbs. oil meal and 2 lbs. middlings; property of James Stillman, Sing Sing, N. Y.

Edith Darby 6246.—Sire, Prospect 2047; dam, Bell 1506.

Butter, 14 lbs. 2 oz. Milk, 233 lbs. 8 oz.

Test made from June 13 to 19, 1884; age, 7 yrs. 3 mos.; property of Clarke & Jones, Baltimore, Md.

Edith Golding 63709.—Sire, Ida's Stoke Pogis 13658; dam, Lady Feronia 17607.

Butter, 19 lbs. 1¼ oz. Milk, 250 lbs.

Test made from Feb. 23 to Mar. 2, 1894; age, 4 yrs. 8 mos.; actual weight, 950 lbs., fed 84 qts. ground oats, 56 qts. corn meal, 28 qts. wheat bran, 28 qts. ground oil-cake, 10 lbs. carrots and 12 lbs. cabbage—hay; property of Miller & Sibley, Franklin, Pa.

Edith Haley 92643.—Sire, Smith's St. Lambert 30332; dam, Edith of St. Lambert 2d 67327.

Butter, 22 lbs. 1½ oz. Milk, 304 lbs. 14 oz.

Test made from Jan. 2 to 9, 1898; age, 4 yrs. 6 mos.; actual weight, 995 lbs.; fed 40 lbs. corn meal, 40 lbs. wheat bran, 18 lbs. old process oil meal and 15 lbs. wheat middlings—ensilage and clover hay *ad lib.*; property of J. P. Bradbury, Pomeroy, Ohio.

Edith Hugo Pogis 105045.—Sire, Smith's St. Lambert 30332; dam, Edith of St. Lambert 2d 67327.

Butter, 15 lbs. 8½ oz. Milk, 290 lbs. 10 oz.

Test made from Jan. 19 to 26, 1898; age, 3 yrs. 7 mos.; estimated weight, 1000 lbs.; fed 21 lbs. ground oats, 21 lbs. corn meal, 42 lbs. wheat shorts, 14 lbs. barley meal and 14 lbs. old process oil meal—clover hay *ad lib.*; property of J. M. Clawson, Springport, Ind.

Edith of St. Lambert 2d 67327.—Sire, Ajax of St. Lambert 19295; dam, Edith of St. Lambert 27054.

Butter, 21 lbs. 11 oz. Milk, 308 lbs. 7 oz.

Test made from June 14 to 21, 1895; age, 5 yrs. 10 mos.; actual weight, 1012 lbs.; fed 40 lbs. wheat bran, 35 lbs. corn meal, 20 lbs. wheat middlings and 14 lbs. old process oil meal—excellent blue-grass pasture; property of J. P. Bradbury, Pomeroy, Ohio.

Edna F. of Lawn 73545.—Sire, Owen St. Lambert of Lawn 24992; dam, Daisy of Denison 53137.

Butter, 15 lbs. 3 oz. Milk, 209 lbs. 10 oz.

Test made from July 19 to 26, 1896; age, 5 yrs. 4 mos.; estimated weight, 1000 lbs.; fed 112 lbs. wheat bran, 28 lbs. corn meal and 28 lbs. cotton-seed meal—poor wild grass pasture; property of Platter & Foster, Denison, Tex.

Edna Hayes 90901.—Sire, Tormentor's Rioter 21964; dam, Nellie E. 66426.

Butter, 15 lbs. 6 oz. Milk, 227 lbs. 8 oz.

Test made from July 26 to Aug. 2, 1896; age, 4 yrs. 1 mo.; estimated weight, 700 lbs.; fed 280 lbs. green sorghum, 56 qts. bran, 48 qts. oats, 28 qts. corn meal and 7 qts. cotton-seed meal—pasture in dry lot; property of T. G. Bush, Anniston, Ala.

Edna of Verna 34537.—Sire, Halo 10517; dam, Edessa 21844.

Butter, 20 lbs. 2½ oz. Milk, 180 lbs. 2 oz.

Test made from June 13 to 19, 1888; age, 3 yrs. 3 mos.; weight, 875 lbs.; fed, daily, 2 qts. corn meal, 2 qts. ground oats, 4 qts. bran and 1 pt. oil meal; property of Frederic Bronson, Verna Farm, Greenfield Hill, Conn.

Edwina 6713.—Sire, Signal 1170; dam, Victorine Lachaise 2740.

Butter, 15 lbs. 18 oz.

Test made Autumn, 1881; age, 4 yrs. 6 mos.; property of J. B. Wade, Atlanta, Ga.

Edy Signal 19430.—Sire, Signal Jr. 7166; dam, Edy Bashan 2d 16098.

Butter, 16 lbs. 8 oz.   Milk, 250 lbs. 8 oz.

Test made from Sept. 21 to 28, 1887; age, 5 yrs.; fed, daily, 6 lbs. bran and shorts, 8 lbs. corn and oats and 4 lbs. oil meal; property of H. M. Baum, Frankfort, Ind.

Effie Baal 134222.—Sire, Eminent (P. S. 1842 J. H. B.); dam, Diplomate 2d (P. S. 3243 J. H. B.)

Butter, 14 lbs. 10¼ oz.   Milk, 165 lbs. 10¼ oz.

Test made from June 17 to 24, 1898; age, 2 yrs. 3 mos.; estimated weight, 650 lbs., fed 17 lbs. ground oats and corn, in equal parts, daily—orchard and blue-grass pasture; property of M. M. Gardner, Nashville, Tenn.

Effie of Hillside 1521.—Sire, Nero 13; dam, Evelina 446.

Butter, 16 lbs. 15 oz., unsalted.   Milk, 290 lbs.

Test made from May 23 to 29, 1882; age, 11 yrs. 3 mos.; property of Frederic Bronson, Greenfield Hill, Conn.

Effie of Verna 8928.—Sire, John Gilpin 2199; dam, Effie of Hillside 1521.

Butter, 14 lbs. 6 oz., unsalted.   Milk, 216 lbs. 8¼ oz.

Test made from June 22 to 29, 1883; age, 6 yrs. 1 mo.; property of Frederic Bronson, Greenfield Hill, Conn.

Effie's Girl 103452.—Sire, King Koffee Jr. 12317; dam, Bachelor's Effie 75678.

Butter, 14 lbs. 3½ oz.   Milk, 135 lbs., estimated.

Test made from Aug. 10 to 17, 1895; age, 2 yrs.; estimated weight, 600 lbs.; fed a small quantity of bran and corn meal and a goodly quantity of cut green corn, not weighed—dry fall pasture; property of Hooper & Christy, Murfreesborough, Tenn.

Elainette's Odelio 102115.—Sire, Odelio 20223; dam, Elainette's Myra 28233.

Butter, 14 lbs. 3½ oz.   Milk, 206 lbs. 8 oz.

Test made from Apr. 12 to 19, 1897; age, 2 yrs. 5 mos.; estimated weight, 700 lbs.; fed 28 lbs. bran, 17½ lbs. linseed meal, 14 lbs. cotton-seed meal, 28 lbs. ground oats, 31½ lbs. corn-meal, 140 lbs. cotton-seed hulls and 70 lbs. green cut oats; property of S. H. Butler, Como, Miss.

Elegantie 11943.—Sire, Wallenstein 2261; dam, Lilla Merritt 4924.

Butter, 14 lbs. 3 oz.   Milk, 253 lbs. 8 oz.

Test made from June 7 to 13, 1885; age, 5 yrs. 4 mos.; property of F. E. Warner & Son, Thomaston, Conn.

Elena of Oakdale 84162.—Sire, St. Ion 18876; dam, Meines 3d 7741.

Butter, 19 lbs. 4 oz.   Milk, 302 lbs. 13 oz.

Test made from July 13 to 20, 1896; age, 5 yrs.; actual weight, 875 lbs.; fed 49 lbs. corn meal, 14 lbs. bran, 35 lbs. ground oats and 21 lbs. oil meal—clover and timothy pasture; property of J. H. Smith & Son, Highfield, Ont., Can.

Elinor Wells 12068.—Sire, Le Brocq's Prize 3350; dam, Annie Wells 1947.

Butter, 14 lbs.   Milk, 183 lbs.

Test made from June 17 to 23, 1884; age, 3 yrs. 7 mos.; test made on grass alone; property of Beech Grove Farm, Beech Grove, Ind.

Elinor Wells 2d 72071.—Sire, St. Lambert Boy 17408; dam, Elinor Wells 12068.

Butter, 14 lbs. 1-16 oz. Milk, 198 lbs. 8 oz.

Test made from Aug. 12 to 19, 1894; age, 5 yrs.; estimated weight, 800 lbs.; fed 10 qts. chop and 10 stalks corn fodder, daily—pasture; property of Mrs. A. M. Hallock, Columbus, Ohio.

Elite 4299.—Sire, Hamilton 1074; dam, Esther 889.

Butter, 14 lbs. 2 oz.

Test made from Mar. 24 to 31, 1883; age, 7 yrs. 6 mos.; property of W. C. Raymond, Bridgewater, Vt.

Ella Europa 61062.—Sire, Rioter's Victory 16446; dam, Ella Esmond 35225.

Butter, 14 lbs. 15½ oz. Milk, 208 lbs. 2 oz.

Test made from Jan. 28 to Feb. 4, 1893; age, 5 yrs.; estimated weight, 775 lbs.; fed 168 lbs. of a mixture of equal parts of corn meal, oats and bran, and 7 lbs. oil meal—clover and corn fodder; property of Mrs. L. C. Smith, White Water, Wis.

Ella L. 84123.—Sire, Yumroar 19656; dam, Dusky Le Brocq 22967.

Butter, 14 lbs. 12 oz. Milk, 261 lbs. 3 oz.

Test made from May 14 to 21, 1897; age, 7 yrs. 3 mos.; estimated weight, 850 lbs.; fed 2¾ lbs. wheat bran, 1½ lbs. linseed meal, 2½ lbs. gluten feed, 1¾ lbs. ground oats, 1¾ lbs. corn meal and 4 qts. apples, daily—hay, clover and orchard grass pasture; property of C. N. Gilbert, Great Barrington, Mass.

Ella of Briarcliff 68939.—Sire, American Rioter 13363; dam, Etta of Briarcliff 68320.

Butter, 14 lbs. Milk, 201 lbs. 5 oz.

Test made from May 13 to 20, 1893; age, 5 yrs. 1 mo.; estimated weight, 900 lbs.; fed 15 qts., daily, of a mixture of equal parts corn meal, ground oats and wheat bran—hay and green rye *ad lib.*; property of Walter W. Law, Whitson, N. Y.

Ella of Endegeest 87376.—Sire, Lord Hugo Pogis 23673; dam, Nuella 53747.

Butter, 18 lbs. 8½ oz. Milk, 294 lbs. 12 oz.

Test made from Mar. 26 to Apr. 2, 1896; age, 4 yrs. 10 mos.; actual weight, 765 lbs.; fed 12 qts. wheat bran, 3 qts. corn meal, 6 qts. ground oats and 3 qts. oil meal, daily —hay; property of Charles A. Sweet, East Aurora, N. Y.

Ella of Endegeest 87376.—Sire, Lord Hugo Pogis 23673; dam, Nuella 53747.

Butter, 19 lbs. 12½ oz. Milk, 309 lbs. 5 oz.

Test made from Apr. 14 to 21, 1898; age, 6 yrs. 10 mos.; actual weight, 960 lbs.; fed 6 qts. bran, 6 qts. middlings, 3 qts. corn meal, 6 qts. ground oats and 3 qts. oil meal, wet, mixed, daily—hay and ensilage; property of C. A. Sweet, Buffalo, N. Y.

Ella of Ingleside 76414.—Sire, Duke of Ingleside 14274; dam, Coomassiella 2d 9860.

Butter, 15 lbs. 3½ oz. Milk, 167 lbs. 3 oz.

Test made from June 9 to 16, 1895; age, 4 yrs. 2 mos.; estimated weight, 850 lbs.; fed 28 lbs. winter wheat bran, 28 lbs. ground oats and 28 lbs. corn meal—common pasture; property of John C. McClintock, Meadville, Pa.

Ella Rosewood's Daisy 32787.—Sire, Darling Scituate 7499; dam, Ella Rosewood 2d 23602.

Butter, 15 lbs. Milk, 280 lbs.

Test made from June 6 to 13, 1891; age, 7 yrs.; estimated weight, 850 lbs.; fed 8 lbs. corn meal, 4 lbs. bran and 2 lbs. cotton-seed meal, daily—fair pasture; property of Frank E. Shaw, Dunkirk, N. Y.

Ella's Tina 68408.—Sire, Hulda's Duke 23810; dam, Laura Y. 2d 62333.

Butter, 14 lbs. 9¾ oz. Milk, 242 lbs. 14 oz.

Test made from Apr. 17 to 24, 1897; age, 6 yrs. 3 mos.; actual weight, 840 lbs.; fed 42 lbs. ground corn, 42 lbs. bran, 14 lbs. oil meal and 140 lbs. hay; property of Charles A. Foster, Trenton, Mo.

Ella Victor Pogis 100762.—Sire, Melia Ann's Victor Pogis 20697; dam, Empress Pedro 50878.

Butter, 19 lbs. 6 oz. Milk, 325 lbs. 8 oz.

Test made from Nov. 2 to 9, 1897; age, 4 yrs. 11 mos.; estimated weight, 800 lbs.; fed 70 lbs. corn meal, 45 lbs. wheat bran and about 30 lbs. oil meal—clover after feed; property of J. H. Martin, Granville, N. Y.

Ellenetta of Bay View 48921.—Sire, Chief 10063; dam, Ellenetta 9446.

Butter, 21 lbs. 11 oz. Milk, 211 lbs. 4 oz.

Test made from Mar. 5 to 12, 1893; age, 7 yrs. 10 mos.; estimated weight, 600 lbs.; fed 17½ lbs. oil meal, 175 lbs. corn refuse, 252 lbs. ensilage, 21 lbs. malt sprouts, 21 lbs. middlings and 28 lbs. hay; property of Charles Lautz, Buffalo, N. Y.

Elma Marigold 100630.—Sire, Stoke Pogis of Prospect 29121; dam, Ida of Oakland 2d 72061.

Butter, 17 lbs. 13½ oz. Milk, 257 lbs.

Test made from June 10 to 17, 1896; age, 2 yrs.; actual weight, 750 lbs.; fed 8 qts. ground oats, 5 qts. corn meal, 4 qts. wheat bran and 3 qts. oil meal, daily—hay, fair pasture; property of George D. Briggs, East Aurora, N. Y.

Elma's Star 102275.—Sire, Manifold Rioter Pogis 26158; dam, Comille 86371.

Butter, 14 lbs. 12 oz. Milk, 202 lbs. 10 oz.

Test made from June 3 to 10, 1897; age, 4 yrs. 3 mos.; actual weight, 700 lbs.; fed first two days 10 lbs. daily, next two days 5 lbs. daily and last three days 7 lbs. daily of corn, oats and bran, in equal parts by weight, and one-eighth each of gluten and oil meal—pasture; property of George E. Scott, Mount Pleasant, Ohio.

El Mora Mostar 15955.—Sire, Clifton Monarch 3546; dam, Mostar 6971.

Butter, 14 lbs., unsalted. Milk, 213 lbs.

Test made from May 2 to 9, 1883; age, 2 yrs. 3 mos.; property of James Cloud & Son, Kennett Square, Pa.

El Mora Mostar 15955.—Sire, Clifton Monarch 3546; dam, Mostar 6971.

Butter, 15 lbs. 4 oz. Milk, 231 lbs. 3 oz.

Test made from May 9 to 16, 1884; age, 3 yrs. 3 mos.; property of James Cloud & Son, Kennett Square, Pa.

Elmwood Daisy 2d 34193.—Sire, Royalist 3d 4500; dam, Elmwood Daisy 18402.

Butter, 16 lbs. 3 oz. Milk, 227 lbs.

Test made from May 31 to June 7, 1893; age, 10 yrs. 7 mos.; estimated weight, 900 lbs.; fed 84 qts. bran and chop, mixed half and half—timothy and red-top pasture; property of U. F. Harrison, Auxvasse, Mo.

Elmwood Duchess 8716.—Sire, Duke of Daffodil 1662; dam, Matin 5th 3399.

Butter, 14 lbs. Milk, 206 lbs. 4 oz.

Test made from May 22 to 28, 1885; age, 6 yrs. 9 mos.; property of T. W. McNeely, Petersburg, Ill.

Elmwood Melissa 4th 82578.—Sire, Elmwood Stoke Pogis 26552; dam, Elmwood Melissa 71275.

Butter, 14 lbs. 5 oz. Milk, 241 lbs. 10 oz.

Test made from May 24 to 31, 1896; age, 4 yrs. 6 mos.; estimated weight, 850 lbs.; fed 75 lbs. bran, corn and oats and cotton-seed meal, and 14 lbs. cut straw—rye pasture; property of A. Hebenstreit, Blue Mound, Ill.

Elodie 30222.—Sire, Count Coomassie 7542; dam, Rose of Sylvanhurst 16134.

Butter, 15 lbs. 5½ oz. Milk, 243 lbs. 8 oz.

Test made from Mar. 27 to Apr. 3, 1889; age, 5 yrs. 2 mos.; actual weight, 1055 lbs.; fed, daily, between 16 and 24 lbs. corn meal; property of William Morrow & Son, Nashville, Tenn.

Elorita 101742.—Sire, Ethleel 2d's Jubilee 18249; dam, Lena's Florence 2d 78688.

Butter, 16 lbs. 4 oz. Milk, 234 lbs. 12 oz.

Test made from Jan. 23 to 30, 1897; age, 3 yrs. 7 mos.; estimated weight, 900 lbs.; fed 46 lbs. corn meal, 46 lbs. bran, 18 lbs. cotton-seed meal, 225 lbs. ensilage and 50 lbs. oat straw; property of W. A. Ponder, Denton, Tex.

Elphie May 69396.—Sire, Sophie's Tormentor 20883; dam, Elsie Lane 2d 58629.

Butter, 16 lbs. 4½ oz. Milk, 227 lbs. 8 oz.

Test made from Feb. 14 to 21, 1898; age, 7 yrs. 9 mos.; estimated weight, 800 lbs.; fed 42 lbs. bran, 27½ lbs. corn meal, 21 lbs. ground oats, 16½ lbs. oil meal, 210 lbs. ensilage and 70 lbs. roots—hay *ad lib.*; property of C. I. Hood, Lowell, Mass.

Elsie Bonner 78864.—Sire, King of St. Lambert 15175; dam, Elsie Dee 18057.

Butter, 21 lbs. Milk, 299 lbs. 12 oz.

Test made from June 24 to July 1, 1895; age, 6 yrs. 11 mos.; actual weight, 1100 lbs., fed 12 qts. daily of a mixture of wheat bran and ground corn and oats (two-thirds corn to one-third oats) and 1 pt. oil meal—poor pasture; property of Hall & Lanman, Columbus, Ohio.

Elsie Coomassie 98081.—Sire, King Coomassie 2d 19545; dam, Beech Grove Elsie 67758.

Butter, 17 lbs. 12 oz. Milk, 231 lbs. 13 oz.

Test made from July 10 to 17, 1897; age, 4 yrs. 9 mos.; estimated weight, 800 lbs.; fed 42 qts. corn and cob meal and 56 qts. wheat bran—good blue-grass pasture; property of Columbus Dixon, Gillespieville, Ohio.

Elsie Harribee 91834.—Sire, Success of Pa. 23374; dam, Euro Polono 31962.

Butter, 18 lbs. 8½ oz.   Milk, 309 lbs. 12 oz.

Test made from Apr. 10 to 17, 1896; age, 4 yrs. 1 mo.; estimated weight, 750 lbs.; fed 24 lbs. daily of a mixture of ground corn, oats, bran, oil meal and cotton-seed meal, and 33 lbs. ensilage, daily, with some clover hay; property of R. F. Shannon, Pittsburg, Pa.

Elsie Lane 13302.—Sire, Catono 3761; dam, Tirzah Ann 10298.

Butter, 15 lbs. 4 oz.   Milk, 113⅞ qts.

Test made from June 19 to 25, 1885; age, 4 yrs.; property of Lyman A. Mills, Middlefield, Conn.

Elsie Lofty 61451.—Sire, Baron Hugo 15208; dam, Miss Lofty 9718.

Butter, 18 lbs. 3 oz.   Milk, 303 lbs.

Test made from Mar. 9 to 16, 1895; age, 6 yrs. 1 mo.; estimated weight, 925 lbs.; fed 4 lbs. gluten meal, 2 lbs. cotton-seed, 2 lbs. rye and oats, 4 lbs. wheat bran, 2 lbs. malt sprouts, 10 lbs. hay and 30 lbs. corn ensilage; property of E. S. Henry, Rockville, Conn.

Elsie of Shadyside 85500.—Sire, King of St. Lambert 15175; dam, Elsie Dee 18057.

Butter, 16 lbs. 8 oz.   Milk, 200 lbs. 4 oz.

Test made from Jan. 26 to Feb. 2, 1896; age, 4 yrs. 6 mos.; estimated weight, 900 lbs.; fed 5 qts. corn and cob meal, 5 qts. oat meal and 2 qts. oat middlings, daily—few raw potatoes, corn fodder and hay *ad lib.*; property of E. C. Newton, Hudson, Ohio.

Elsie Rowls of Lawn 71633.—Sire, Bonair St. Lambert 17600; dam, Lady Elkhorn 44407.

Butter, 14 lbs. 5 oz.   Milk, 211 lbs. 11 oz.

Test made from Mar. 10 to 17, 1895; age, 5 yrs. 9 mos.; estimated weight, 850 lbs.; fed 84 lbs. wheat bran, 98 lbs. ground oats and 86 lbs. corn meal—good prairie hay once a day; property of Platter & Foster, Denison, Tex.

Elsie's Lady 81493.—Sire, Nightingale's Rioter 14935; dam, Lady Elsie 33142.

Butter, 14 lbs. 9 oz.   Milk, 204 lbs. 4 oz.

Test made from Nov. 21 to 28, 1897; age, 7 yrs.; estimated weight, 1050 lbs.; fed 10 qts. wheat bran, 3 qts. buckwheat middlings, 2 qts. 1 pt. wheat middlings, 1 qt. cotton-seed meal and 20 lbs. corn ensilage, daily—hay *ad lib.*; property of Everett G. Campbell, Keister's, Pa.

Eltekeh 28266.—Sire, Footstep 5163; dam, Effie of Hillside 1521.

Butter, 16 lbs. 4 oz. (official).   Milk, 213 lbs. 11 oz.

Test made from May 25 to 31, 1886; age, 2 yrs. 2 mos.; weight, 850 lbs.; fed 64 lbs. crushed oats, 31 lbs. corn meal, 23 lbs. bran, 19 lbs. linseed oil cake meal, 21 lbs. pea meal; property of Frederic Bronson, Greenfield Hill, Conn.

Elturia 80701.—Sire, Diploma 16219; dam, Paradox 65003.

Butter, 24.137 lbs. in 21 days.   Milk, 483.4 lbs.

Test made from Sept. 30 to Oct. 20, 1893, at World's Columbian Exposition; age, 2 yrs. 11 mos.; actual weight, 830 lbs.; fed 235.5 lbs. hay, 6 lbs. ensilage, 55 lbs. oil meal, 72 lbs. corn meal, 117 lbs. bran, 63 lbs. oats, 38 lbs. cotton-seed, 59 lbs. middlings and 26 lbs. corn-hearts; property of Richardson Bros., Davenport, Iowa.

Embla 4799.—Sire, Saladin 447; dam, Elveta 2121.

Butter, 17 lbs. 8 oz.

Test made Sept., 1881; age, 6 yrs. 6 mos.; property of C. R. C. Dye, Troy, Ohio.

Embla Brick 15690.—Sire, Roy Brick 5637; dam, Embla 4799.

Butter, 14 lbs. 3 oz.　Milk, 199 lbs.

Test made from May 26 to June 2, 1885; age, 3 yrs. 10 mos.; property of George M. Jewett, Glenville, Md.

Emblem of Del Valle 102384.—Sire, Diana's Top-Sawyer 21310; dam, Forest Q. Pogis 63507.

Butter, 15 lbs. 8 oz.　Milk, 317 lbs. 8 oz.

Test made from Apr. 14 to 21, 1897; age, 5 yrs. 5 mos.; actual weight, 844 lbs.; fed 18 lbs. wheat bran, 20 lbs. oats, 15 lbs. corn meal, 7 lbs. cotton-seed meal, 21 lbs. corn and oats, ground together in equal parts, and about 20 lbs. hay—common pasture; property of R. F. Young, Georgetown, Tex.

Emily of Hillside 8073.—Sire, John Gilpin 2199; dam, Elfrida of Hillside 4056.

Butter, 16 lbs. 4 oz.　Milk, 229 lbs. 10 oz.

Test made from Feb. 7 to 14, 1883; age, 4 yrs. 11 mos.; property of Samuel Colgate, Orange, N. J.

Emma Hudson 12469.—Sire, Ben Nora 2262; dam, Rosabel Hudson 5704.

Butter, 16 lbs. 2 oz.

Test made from June 8 to 14, 1884; age, 4 yrs. 1 mo.; property of Elbert Miller, Middlefield, Conn.

Emma Landseer 58104.—Sire, Landseer's Pogis 15847; dam, Belle Lyman 28522.

Butter, 14 lbs. 2½ oz.　Milk, 162 lbs. 6 oz.

Test made from Dec. 1 to 8, 1890; age, 2 yrs. 9 mos.; estimated weight, 650 lbs.; fed, daily, 3½ gals. ground oats and corn and ½ gal. bran; property of Morgan & Brown, Columbia, Tenn.

Emma Landseer 2d 69220.—Sire, Tennessee's Landseer 23015; dam, Emma Landseer 58104.

Butter, 14 lbs. 2¾ oz.　Milk, 186 lbs.

Test made from May 2 to 9, 1893; age, 2 yrs. 7 mos.; estimated weight, 800 lbs.; fed 56 qts. oats, 56 qts. meal and 28 qts. bran—very little hay, blue-grass and white clover pasture; property of Morgan & Brown, Columbia, Tenn.

Emma Landseer 3d 110513.—Sire, Ida's Landseer 17745; dam, Emma Landseer 58104.

Butter, 37 lbs. 2¼ oz. in 14 days.　Milk, 572 lbs. 14½ oz.

Test made from June 19 to July 3, 1895; age, 5 yrs. 8 mos.; estimated weight, 1100 lbs.; fed 22 lbs. per day of ground oats and corn, in equal parts—good orchard and blue-grass pasture; property of Matt. M. Gardner, Nashville, Tenn.

Emma Landseer 4th 109233.—Sire, Bisson's Landseer 27520; dam, Emma Landseer 58104.

Butter, 14 lbs. 9½ oz.　Milk, 150 lbs. 8 oz.

Test made from Sept. 25 to Oct. 2, 1896; age, 2 yrs. 1 mo.; estimated weight, 650 lbs.; fed 56 qts. corn meal and 56 qts. ground oats—woodland pasture; property of Miss Irene Walker, Columbia, Tenn.

Emma of Eau Claire 127826.—Sire, Melia Ann's Son 22041; dam, Bina of Eau Claire 77585.

Butter, 21 lbs. 4 oz.   Milk, 267 lbs. 4 oz.

Test made from Dec. 10 to 17, 1897; age, 4 yrs. 11 mos.; estimated weight, 850 lbs.; fed 25 lbs. bran, 26 lbs. oil meal and 30 lbs. corn meal; property of J. A. Griffith, Ashley Falls, Mass.

Empress 6th 3203.—Sire, Vermont 893; dam, Empress 1552.

Butter, 17 lbs. 9¾ oz., unsalted.

Test made from July 1 to 7, 1882; property of M. H. Barnard, Forestville, Conn.

Empress of Ely 2d 6771.—Sire, Baritone 1075; dam, Empress of Ely 4823.

Butter, 16 lbs. 8 oz., unsalted.

Test made July, 1882; age, 5 yrs. 4 mos.; property of C. H. Carpenter, Troy, N. Y.

Empress Pedro 50878.—Sire, King Peto 10541; dam, Empress 6th 3203.

Butter, 15 lbs. 1 oz.   Milk, 194 lbs. 12 oz.

Test made from Oct. 9 to 16, 1893; age, 6 yrs. 9 mos.; estimated weight, 700 lbs.; fed green corn fodder—good pasture; property of Mrs. J. H. Martin, Granville, N. Y.

English Elm 17600.—Sire, Elmwood 2322; dam, Coralle 4446.

Butter, 14 lbs. 7 oz.   Milk, 114 lbs. 3 oz.

Test made from May 19 to 25, 1886; age, 3 yrs. 9 mos.; property of Columbia Jersey Cattle Co., Columbia, Tenn.

English Elm 3d 62906.—Sire, Fancy's Harry 9777; dam, English Elm 17600.

Butter, 17 lbs. 2 oz.   Milk, 233 lbs.

Test made from Jan. 2 to 9, 1894; age, 4 yrs.; estimated weight, 800 lbs.; fed 100 qts. mixed feed, consisting of cotton-seed, steamed oats, wheat and corn chop, in equal parts, fed with dry sorghum; property of S. T. Howard, Quanah, Tex.

Enid 2d 10782.—Sire, Duke of Oakland 1984; dam, Enid 1482.

Butter, 14 lbs. 7½ oz.   Milk, 187 lbs. 8 oz.

Test made from Mar. 31 to Apr. 6, 1882; age, 2 yrs. 3 mos.; fed corn meal, ground oats and bran and 1 pk. sugar beets, daily; property of H. M. Howe, Bristol, R. I.

Enigma 5360.—Sire, St. Martin 1482; dam, Little Emily 5356.

Butter, 15 lbs. 6 oz., unsalted.   Milk, 234 lbs. 4 oz.

Test made from Mar. 19 to 25, 1880; age, 6 yrs.; property of Edwin Thorne, Millbrook, N. Y.

Eoline of St. Lambert 104819.—Sire, St. Lambert Boy 17408; dam, Letty Coles 2d 48128.

Butter, 19 lbs. 1 oz.   Milk, 296 lbs. 12 oz.

Test made from Jan. 3 to 10, 1897; age, 3 yrs. 8 mos.; actual weight, 801 lbs.; fed 40 lbs. wheat bran, 40 lbs. wheat middlings, 40 lbs. corn meal and 10 lbs. old process oil meal—ensilage and cut corn fodder *ad lib.*; property of J. P. Bradbury, Pomeroy, Ohio.

Eppie Lady Pogis 110638.—Sire, King of St. Lambert Jr. 18808; dam, Lady Epple 72375.

Butter, 14 lbs. 3 oz. Milk, 263 lbs. 4 oz.

Test made from Apr. 11 to 18, 1898; age, 4 yrs.; estimated weight, 750 lbs.; fed 70 lbs. bran and 7 lbs. oil meal—2 bush. hay and corn fodder, cut, daily; property of Peter Raab, Brightwood, Ind.

Eppie Pogis 44972.—Sire, Prince of Melrose 2d 11015; dam, Dewdrop Pansy 19736.

Butter, 15 lbs. 6 oz. Milk, 226 lbs. 4 oz.

Test made from Oct. 31 to Nov. 7, 1890; age, 3 yrs. 5 mos.; estimated weight, 750 lbs.; fed 28 qts. corn meal, 28 qts. pea meal, 56 qts. bran and 210 lbs. corn ensilage—hay, short clover pasture; property of James Crook, Jacksonville, Ala.

Eric's Ida 69134.—Sire, R. F. Eric 21330; dam, Ida of Rocky Ford 40762.

Butter, 18 lbs. Milk, 223 lbs.

Test made from May 7 to 14. 1898; age, 7 yrs. 7 mos.; estimated weight, 1000 lbs.; fed 6 qts. corn meal and 10 qts. bran, daily—June grass pasture; property of E. G. Silvus, Athens, Ohio.

Eric Maid of River View 94307.—Sire, Nell's John Bull 21921; dam, Fawn of River View 78299.

Butter, 14 lbs. 12 oz. Milk, 207 lbs. 12 oz.

Test made from June 7 to 14, 1896; age, 3 yrs. 9 mos.; estimated weight, 850 lbs.; fed 56 qts. oat chop and 56 qts. corn meal—blue-grass and timothy pasture; property of Alfred T. Macartney, Dunnville, Ont., Can.

Ernestina 2d 77682.—Sire, Kathy's Stoke Pogis 17566; dam, Ernestina 38444.

Butter, 14 lbs. 8 oz. Milk, 222 lbs. 9 oz.

Test made from Sept. 9 to 16, 1891; age, 2 yrs. 7 mos.; estimated weight, 650 lbs.; fed 35 qts. corn meal and 35 qts. wheat bran, on green corn chop—short clover and orchard grass pasture; property of Matthews & Humes, Huntsville, Ala.

Esclarmonde 59509.—Sire, Dalton 20117; dam, Vernon Beauty 50687.

Butter, 17 lbs. 4 oz. Milk, 251 lbs. 8 oz.

Test made from July 20 to 27, 1893; age, 4 yrs. 4 mos.; actual weight, 1000 lbs.; fed 56 qts. ground oats, 56 qts. corn meal and 21 qts. ground oil-cake—good old meadow pasture; property of Miller & Sibley, Franklin, Pa.

Esther B. 24834.—Sire, Clusius H. 5781; dam, Efficious 2d 15204.

Butter, 14 lbs. 1 oz. Milk, 207 lbs. 12 oz.

Test made from Jan. 30 to Feb. 6, 1889; age, 5 yrs. 7 mos.; estimated weight, 950 lbs.; fed, daily, about 22½ lbs., consisting of one-half corn meal, one-fourth ground oats, one-eighth bran and one-eighth cotton-seed; property of Rufus A. Sibley, Spencer, Mass.

Esther Lofty 68754.—Sire, Baron Hugo 15208; dam, Miss Lofty 9718.

Butter, 14 lbs. 6 oz. Milk, 226 lbs. 2 oz.

Test made from Apr. 3 to 10, 1894; age, 3 yrs. 11 mos.; estimated weight, 800 lbs.; fed 4 lbs. coarse wheat bran, 2 lbs. corn meal, 2 lbs. cream gluten, 2 lbs. rye and oats provender, 30 lbs. corn ensilage and 10 lbs. cut clover hay, daily; property of E. Stevens Henry, Rockville, Conn.

Esther Thorne 35545.—Sire, Duke of St. Albans 11234; dam, Gertie Cowles 10736.

Butter, 14 lbs. 7½ oz.   Milk, 250 lbs. 8 oz.

Test made from May 6 to 13, 1893; age, 8 yrs. 2 mos.; estimated weight, 900 lbs.; fed 140 qts. wheat bran—roadside pasture; property of Peter Raab, Indianapolis, Ind.

Estrella 2831.—Sire, Meteor 453; dam, Ariadne 2d 1135.

Butter, 14 lbs. 12 oz.

Test made from May 20 to 27, 1877; age, 5 yrs. 8 mos.; property of J. L. Wells, Southport, Conn.

Etelka Tolemy 93807.—Sire, Tormentor 5th 21962; dam, Signora Reid 27125.

Butter, 14 lbs. 9 oz.   Milk, 157 lbs. 8 oz.

Test made from Aug. 3 to 10, 1896; age, 3 yrs. 8 mos.; estimated weight, 850 lbs.; fed 70 lbs. corn meal and 42 lbs. ground oats—blue-grass pasture; property of Mrs. H. M. Polk, Spring Hill, Tenn.

Eterna 27134.—Sire, Signalda 4027; dam, Su Lu 4705.

Butter, 14 lbs. 4¼ oz.   Milk, 169 lbs. 8 oz.

Test made from Dec. 11 to 18, 1889; age, 5 yrs. 10 mos.; estimated weight, 925 lbs.; fed, daily, 24 lbs. chopped corn and bran, mixed; property of Campbell Brown, Spring Hill, Tenn.

Ethelka 2d 14128.—Sire, Thomaston 4600; dam, Ethelka 9515.

Butter, 15 lbs. 4 oz.   Milk, 244 lbs. 8 oz.

Test made June, 1885; age, 4 yrs. 3 mos.; property of Jewett M. Richmond, Buffalo, N. Y.

Ethel McM. 100267.—Sire, King Coomassie 2d 19545; dam, Gertie McM. 64200.

Butter, 15 lbs. 13 oz.   Milk, 270 lbs. 14 oz.

Test made from Apr. 16 to 23, 1897; age, 4 yrs. 4 mos.; estimated weight, 800 lbs.; fed 70 qts. ground corn and cob and 28 qts. coarse wheat bran, mixed with cut corn fodder—hay *ad lib.*, blue-grass pasture; property of Columbus Dixon, Gillespieville, Ohio.

Ethel Marjoram 87266.—Sire, One Hundred per Cent. 16590; dam, Leclair's Marjoram 36355.

Butter, 16 lbs. 11 oz.   Milk, 248 lbs. 11 oz.

Test made from Oct. 20 to 27, 1896; age, 5 yrs. 8 mos.; estimated weight, 1000 lbs.; fed 84 lbs. ground peas and oats—hay *ad lib.*, four hours daily on timothy pasture; property of William Rolph, Markham, Ont., Can.

Ethel of Glen Rouge 131019.—Sire, Exile of Glen Rouge 37213; dam, Ethel Marjoram 87266.

Butter, 18 lbs. 9 oz.   Milk, 273 lbs.

Test made from Dec. 14 to 21, 1897; age, 2 yrs. 9 mos.; estimated weight, 600 lbs.; fed 18 qts. chopped peas and oats, 3 qts. bran and 2 lbs. oil-cake, daily—hay *ad lib.*; property of William Rolph, Markham, Ont., Can.

Ethel of Shelburne 46285.—Sire, Midas of Oxford 5986; dam, Ethel Hugo 30164.

Butter, 14 lbs. 1½ oz. Milk, 255 lbs. 1 oz.

Test made from Dec. 14 to 21, 1890; age, 3 yrs. 4 mos.; estimated weight, 850 lbs.; fed, daily, 3 lbs. ground oats, 3 lbs. bran and shorts, 7 lbs. corn meal and 2¾ lbs. oil meal; property of Mrs. S. E. K. Hudson, Alexandria Bay, N. Y.

Ethel of Shelburne 46285.—Sire, Midas of Oxford 5986; dam, Ethel Hugo 30164.

Butter, 14 lbs. 5½ oz. Milk, 253 lbs. 8 oz.

Test made from Jan. 18 to 25, 1897; age, 9 yrs. 5 mos.; estimated weight, 1000 lbs.; fed 21 lbs. bran, 35 lbs. ground peas and oats, 56 lbs. corn meal, 14 lbs. oil meal and 250 lbs. roots, mixed—corn fodder morning and evening, hay at noon; property of C. I. Hudson, Alexandria Bay, N. Y.

Ethel of Shelburne 46285.—Sire, Midas of Oxford 5986; dam, Ethel Hugo 30164.

Butter, 14 lbs. 8 oz. Milk, 270 lbs. 6 oz.

Test made from Jan. 25 to Feb. 1, 1897; age, 9 yrs. 5 mos.; estimated weight, 1000 lbs.; fed 21 lbs. bran, 35 lbs. ground peas and oats, 56 lbs. corn meal, 14 lbs. oil meal and 250 lbs. roots, mixed—corn fodder morning and evening, hay at noon; property of C. I. Hudson, Alexandria Bay, N. Y.

Ethel of Shelburne 46285.—Sire, Midas of Oxford 5986; dam, Ethel Hugo 30164.

Butter, 28 lbs. 13½ oz. in 14 days. Milk, 523 lbs. 14 oz.

Test made from Jan. 18 to Feb. 1, 1897; age, 9 yrs. 5 mos.; estimated weight, 1000 lbs.; fed 42 lbs. bran, 70 lbs. ground peas and oats, 112 lbs. corn meal, 28 lbs. oil meal and 500 lbs. roots, mixed—corn fodder morning and evening, hay at noon; property of C. I. Hudson, Alexandria Bay, N. Y.

Ethleel 18724.—Sire, Tormentor 3533; dam, Beeswax 9807.

Butter, 19 lbs. 14 oz. Milk, 163 lbs. 9 oz.

Test made from July 30 to Aug. 5, 1885; age, 4 yrs. 7 mos.; property of the Columbia Jersey Cattle Co., Columbia, Tenn.

Ethleel 2d 32291.—Sire, Lord Harry 3445; dam, Ethleel 18724.

Butter, 30 lbs. 15 oz. (official). Milk, 155 lbs. 11 oz.

Test made from July 1 to 7, 1885; age, 2 yrs. 8 mos.; fed 20 lbs. daily of corn-hearts, oats and bran (five-eighths corn-hearts, one-eighth oats, two-eighths bran)—little hay, good, rich pasture; property of J. A. McEwen, Nashville, Tenn.

Ethleel 4th 54275.—Sire, Fancy's Harry 9777; dam, Ethleel 18724.

Butter, 15 lbs. Milk, 156 lbs. 11½ oz.

Test made from Apr. 29 to May 6, 1897; age, 8 yrs. 6 mos.; actual weight, 1020 lbs.; fed 79 qts. grain—wheat bran, corn meal, ground oats and linseed meal—70 lbs. hay, 85 lbs. corn fodder, 1 pk. carrots and 2 qts. turnips; property of John A. Cunningham, Bolton, Mass.

Ethleel 5th 70652.—Sire, Lord Harry 3445; dam, Ethleel 18724.

Butter, 15 lbs. 6½ oz. Milk, 178 lbs. 15 oz.

Test made from May 7 to 14, 1896; age, 5 yrs. 10 mos.; estimated weight, 900 lbs.; fed 4 gals. daily of corn and oats, ground together—hay, blue-grass pasture; property of John Woodard, Nashville, Tenn.

Ethleel 6th 89802.—Sire, Fancy's Harry 9777; dam, Ethleel 18724.

Butter, 14 lbs. 15 oz.   Milk, 241 lbs. 1 oz.

Test made from July 5 to 12, 1896; age, 4 yrs. 6 mos.; estimated weight, 800 lbs.; fed 63 lbs. corn meal, 63 lbs. oat meal, 14 lbs. oil meal, 19 lbs. bran and 21 lbs. ground rye—mixed blue-grass pasture; property of H. M. Baum, Frankfort, Ind.

Ethleel D. 83006.—Sire, Ethleel 2d's Jubilee 18249; dam, Callinette 2d 38908.

Butter, 14 lbs. 1 3-5 oz.   Milk, 222 lbs. 10 oz.

Test made from May 30 to June 6, 1896; age, 3 yrs. 9 mos.; actual weight, 925 lbs.; fed 35 lbs. ground oats, 47 lbs. bran, 14 lbs. corn meal and 14 lbs. linseed meal—fair clover pasture; property of Hecla Jersey Cattle Co., Earlington, Ky.

Ethleel Khedive 99698.—Sire, Ethleel 2d's Jubilee 18249; dam, Malice 18988.

Butter, 15 lbs. 8 oz.   Milk, 182 lbs. 11 1-5 oz.

Test made from May 22 to 29, 1895; age, 2 yrs. 3 mos.; estimated weight, 600 lbs.; fed 28 lbs. ground corn and oats, 28 lbs. bran and 14 lbs. cotton-seed meal—clover pasture; property of Hecla Jersey Cattle Co., Earlington, Ky.

Ethleel Koffee of Lawn 116757.—Sire, Antoinette's Koffee 26010; dam, Signalda's Ethleel 77599.

Butter, 17 lbs. 7 oz.   Milk, 212 lbs. 9 oz.

Test made from June 1 to 8, 1898; age, 4 yrs. 1 mo.; estimated weight, 850 lbs.; fed 70 qts. wheat bran and 56 qts. corn and oats chop—good wild grass pasture; property of E. H. Hanna, Denison, Tex.

Ethleel's Nerviona 101300.—Sire, Ethleel's Landseer 22341; dam, Nerviona 43678.

Butter, 17 lbs. 12 oz.   Milk, 211 lbs. 3 oz.

Test made from Feb. 24 to Mar. 3, 1897; age, 3 yrs. 3 mos.; estimated weight, 800 lbs.; fed 84 qts. wheat bran, 56 qts. corn meal and 14 qts. oil meal—clover hay and corn fodder *ad lib.*; property of Henry DuBois, Vigo, Ohio.

Ethleel's Pet 85428.—Sire, Ethleel 2d's Son 25885; dam, Beauty Hugo 63143.

Butter, 17 lbs. 6 oz.   Milk, 209 lbs.

Test made from May 12 to 19, 1896; age, 4 yrs.; estimated weight, 840 lbs.; fed 3 qts. wheat middlings, 3 qts. bran and 3 qts. corn meal, daily—timothy and clover pasture; property of Alfred B. Darling, Ramsey's, N. J.

Ethlo's Belle 69053.—Sire, Ethlo Tormentor 19189; dam, Desdemona Belle 19366.

Butter, 15 lbs. 2 oz.   Milk, 206 lbs. 12 oz.

Test made from Nov. 5 to 12, 1894; age, 4 yrs. 3 mos ; actual weight, 800 lbs.; fed 56 qts. corn and cob meal and 112 qts. wheat bran—hay *ad lib.*, good meadow pasture; property of Columbus Dixon, Gillespieville, Ohio.

Ethlo's Jean 71954.—Sire, Ethlo Tormentor 19189; dam, Margaret of Lagonda 35127.

Butter, 24 lbs. 6 oz.   Milk, 261 lbs. 13½ oz.

Test made from June 29 to July 6, 1896; age, 5 yrs. 3 mos.; estimated weight, 1000 lbs.; fed 140 qts. of a mixture of corn and cob meal, ground wheat and ground oats, in equal parts by measure, and 21 qts. of oil-cake meal—timothy and blue-grass pasture; property of H. C. Schlegel, Daleville, Ind.

Etiquette 4300.—Sire, Hamilton 1074; dam, Etta 1756.

Butter, 15 lbs. 8 oz.

Property of Orestes Pierce, Baldwin, Me.

Etta Allman 12185.—Sire, Gen. Marion 2298; dam, Togus Belle 3252.

Butter, 15 lbs. 5 oz. Milk, 278 lbs. 7 oz.

Test made from May 4 to 11, 1884; age, 4 yrs. 1 mo.; property of Thos. L. Dodd, Nashville, Tenn.

Etta Bartlett 80254.—Sire, Ettamarius 17762; dam, Dorothea Day 43238.

Butter, 15 lbs. 13½ oz. Milk, 239 lbs. 4 oz.

Test made from May 3 to 10, 1898; age, 8 yrs. 1 mo.; estimated weight, 1025 lbs.; fed 6 lbs. bran, 4 lbs. ground corn and oats and 2 lbs. oil meal, daily—blue-grass pasture; property of Est. of C. Delano, Mount Vernon, Ohio.

Ettagem 80258.—Sire, Ettamarius 17762; dam, Gem of Ellasleigh 38532.

Butter, 16 lbs. 12 oz. Milk, 215 lbs. 6 oz.

Test made from Mar. 14 to 21, 1895; age, 3 yrs. 10 mos.; actual weight, 809 lbs.; fed 28 lbs. corn meal, 21 lbs. crushed oats, 21 lbs. linseed meal and 28 lbs. bran—hay and corn fodder *ad lib.*; property of C. Delano, Mount Vernon, Ohio.

Etta Lyne 83298.—Sire, Fadette's Son 16301; dam, Alice Fairfax 83297.

Butter, 20 lbs. 5 oz. Milk, 301 lbs. 6 oz.

Test made from May 23 to 30, 1896; age, 5 yrs. 11 mos.; estimated weight, 900 lbs.; fed 84 lbs. of a mixture of corn meal and ground oats, 42 lbs. wheat bran, 24½ lbs. cotton-seed meal, 24½ lbs. linseed oil meal and 168 lbs. green clover—short blue-grass pasture; property of E. W. Sewell, Anchorage, Ky.

Etta M. 2d 30820.—Sire, Gold Mine 7272; dam, Etta M. 15901.

Butter, 14 lbs. 14 oz. Milk, 287 lbs. 2 oz.

Test made from Apr. 23 to 30, 1888; age, 4 yrs. 5 mos.; estimated weight, 800 lbs.; fed, daily, 8 qts. ground oats and corn, in equal parts; property of D. D. Perry, Peabody, Kan.

Etta of Bay View 66028.—Sire, G. F. Train 18868; dam, Nattie 26842.

Butter, 22 lbs. 13 oz. Milk, 245 lbs. 7 oz.

Test made from Dec. 14 to 21, 1892; age, 3 yrs. 1 mo.; actual weight, 565 lbs.; fed 210 lbs. corn refuse, 210 lbs. ensilage, 2 lbs. clover hay and 14 lbs. linseed meal; property of Charles Lautz, Buffalo, N. Y.

Ettaroek 81103.—Sire, Ettamarius 17762; dam, Etta Blanche 80253.

Butter, 16 lbs. Milk, 233 lbs. 10 oz.

Test made from Apr. 21 to 28, 1898; age, 5 yrs. 11 mos.; estimated weight, 1050 lbs.; fed 8 lbs. bran, 2 lbs. corn meal, 2 lbs. oil meal, 2 lbs. ground oats, 2 lbs. Buffalo gluten meal and 20 lbs. ensilage, daily—blue-grass pasture; property of Est. of C. Delano, Mount Vernon, Ohio.

Ette 10315.—Sire, Duke of Magnolia 2826; dam, Queenette 5274.

Butter, 15 lbs. 7 oz. Milk, nearly 15 qts. per day.

Test made from Mar. 15 to 21, 1884; age, 4 yrs.; property of S. W. White, Hernando, Miss.

Ettie Palmer 4th 47481.—Sire, Troy's Duke 13003; dam, Ettie Palmer 14281.

Butter, 16 lbs. 5 oz. Milk, 241 lbs. 13¾ oz.

Test made from Mar. 10 to 17, 1897; age, 11 yrs. 1 mo.; estimated weight, 750 lbs.; fed 28 qts. bran, 28 qts. corn meal, 28 qts. cotton-seed meal, 28 qts. crushed oats and 70 lbs. hay—young pasture; property of Gebhart & Kauffman, Dallas, Tex.

Eudora 1863, imp.

Butter, 16 lbs. 2 oz. Milk, 268 lbs. 8 oz.

Test made from June 28 to July 4, 1885; **age, 18 yrs. 1 mo.;** property of Frederick Billings, Woodstock, Vt.

Eudora of Barre 12419.—Sire, Tasso of Mount Waite 2334; dam, Silky 2488.

Butter, 16 lbs. 9 oz. Milk, 249 lbs. 15 oz.

Test made from April 1 to 8, 1884; age, 4 yrs. 6 mos.; estimated weight, 900 lbs.; fed, daily, 18 qts. ground corn and oats and pea meal; property of Richardson Bros., Davenport, Iowa.

Eufielda 62546.—Sire, Garfield Stoke Pogis 15963; dam, Euphonia 5th 39124.

Butter, 14 lbs. 3 oz. Milk, 151 lbs. 8 oz.

Test made from June 5 to 12, 1892; age, 2 yrs. 8 mos.; estimated weight, 825 lbs.; fed 56 lbs. cut corn fodder, 7 lbs. corn meal, 14 lbs. gluten meal, 14 lbs. bran, 7 lbs. linseed meal, 56 lbs. hay and 210 lbs. roots; property of Est. of Frederick Billings, Woodstock, Vt.

Eugenie 2d 1623.—Sire, Rob Roy 17; dam, Eugenie 792.

Butter, 14 lbs.

Property of S. C. Colt, Farmington, Conn.

Eugenie 2d 12733.—Sire, Gilderoy 2107; dam, Eugenie 498.

Butter, 14 lbs. 2 oz. Milk, 256 lbs.

Test made from July 8 to 14, 1885; age, 4 yrs. 6 mos.; property of H. M. Howe, Bristol, R. I.

Eugénie Chouteau 6186.—Sire, Master 723; dam, Lucy Pope 5328.

Butter, 24 lbs. 8 oz. Milk, 262 lbs.

Test made from June 26 to July 3, 1885; age, 8 yrs.; fed 16 qts. bran and 1 pt. oil meal, daily—few small ears of corn, pasture; property of William M. Bell, Miami, Mo.

Euphonia 6783.—Sire, Gilroy 1653; dam, Eudora 1863.

Butter, 16 lbs. ½ oz. (official). Milk, 261 lbs. 6 oz.

Test made from Mar. 22 to 29, 1885; age, 7 yrs.; property of F. Billings, Woodstock, Vt.

Euphorbia 11229.—Sire, Allegany Chief 2918; dam, Nettle 3241.

Butter, 14 lbs. 9½ oz. Milk, 256 lbs. 3 oz.

Test made from June 17 to 24, 1884; age, 4 yrs. 3 mos.; property of J. B. Wallace, Lexington, Ky.

Euphorbia 11229.—Sire, Allegany Chief 2918; dam, Nettle 3241.

Butter, 23 lbs. 6¼ oz. Milk, 131 lbs. 12 oz.

Test made from Apr. 21 to 28, 1886; age, 6 yrs. 1 mo.; fed 6 qts. corn meal, 6 qts. shelled oats, 3 qts. oil meal and 3 qts. wheat bran, daily; property of John B. Wallace, Lexington, Ky.

Euphorbia 2d 27094.—Sire, Allegany Signal 10178; dam, Euphorbia 11229.

Butter, 15 lbs. ½ oz. Milk, 258 lbs. 8 oz.

Test made from June 9 to 16, 1889; age, 5 yrs. 3 mos.; estimated weight, 800 lbs.; fed, daily, 3 qts. corn meal, 3 qts. bran and 3 qts. oats; property of N. F. Berry, Lexington, Ky.

Euphorbia 3d 45359.—Sire, Heir of Signal 8080; dam, Euphorbia 11229.

Butter, 15 lbs. 7½ oz. Milk, 217 lbs. 12½ oz.

Test made from Oct. 24 to 31, 1894; age, 7 yrs. 3 mos.; estimated weight, 900 lbs.; fed 448 lbs. pumpkins, 49 lbs. corn and 49 lbs. bran—very short blue-grass pasture; property of F. G. Craft, Liberty, Ind.

Eupidee's Perfection 20175.—Sire, Eupidee 4097; dam, Countess of Lakeside 12135.

Butter, 15 lbs. 4 oz. Milk, 174 lbs.

Test made from Sept. 5 to 11, 1885; age, 2 yrs. 6 mos.; property of J. H. Walker, Worcester, Mass.

Eureen of Lawn 61372.—Sire, Lord Hugo 15830; dam, Eureen 40312.

Butter, 15 lbs. 8 oz. Milk, 222 lbs. 7 oz.

Test made from Mar. 17 to 24, 1895; age, 6 yrs.; estimated weight, 900 lbs.; fed 84 lbs. wheat bran, 84 lbs. ground oats and 70 lbs. corn meal—good prairie hay once a day; property of Platter & Foster, Denison, Tex.

Eureka McHenry 8341.—Sire, Hector 3814; dam, Custard 321.

Butter, 14 lbs., unsalted.

Test made from July 5 to 12, 1869; age, 6 yrs. 7 mos.; property of A. E. Kapp, Northumberland, Pa.

Eure Polono 2d 101052.—Sire, Wine 29954; dam, Euro Polono 31962.

Butter, 14 lbs. 2 oz. Milk, 222 lbs.

Test made from Oct. 21 to 28, 1896; age, 2 yrs. 6 mos.; estimated weight, 700 lbs.; fed 21 lbs. daily of a mixture of ground corn and oats, cotton-seed and oil meal and bran—hay and corn fodder, poor pasture; property of R. F. Shannon, Pittsburg, Pa.

Eurotas 2454.—Sire, Rioter 2d 469; dam, Europa 176.

Butter, 22 lbs. 7 oz. Milk, 216 lbs. 8 oz.

Test made from Feb. 26 to Mar. 4, 1879; age, 7 yrs. 6 mos.; property of A. B. Darling, Ramsey's, N. J.

Eurotas 2454.—Sire, Rioter 2d 469; dam, Europa 176.

Butter, 778 lbs. 1 oz. in one year.

Test made 1879-80; age, 8 yrs.; property of A. B. Darling, Ramsey's, N. J.

Eurotas' Blossom 63134.—Sire, Eurotas' Victor Hugo 15689; dam, Duke's Blossom 86669.

Butter, 15 lbs. 5 oz. Milk, 290 lbs. 4 oz.

Test made from June 10 to 17, 1895; age, 6 yrs. 2 mos.; estimated weight, 850 lbs.; fed 1 qt. oil meal, 1 pt. cotton-seed meal and 6½ qts. of a mixture of corn meal and middlings, in equal parts, daily—timothy and clover pasture; property of A. B. Darling, Ramsey's, N. J.

Eurotisama 29668.—Sire, Young Pedro 9033; dam, Amité 18877.

Butter, 20 lbs. 13½ oz. Milk, 16½ qts. per day.

Test made from Mar. 1 to 7, 1887; age, 2 yrs. 10 mos.; estimated weight, 600 lbs.; fed 4 qts. corn meal, 6 qts. ground oats, 1½ qts. pea meal, 1 qt. linseed meal and 7 qts. middlings; property of D. F. Appleton, Ipswich, Mass.

Eurotisama 29668.—Sire, Young Pedro 9033; dam, Amité 18877.

Butter, 27 lbs. 1½ oz. Milk, 267 lbs. 10 oz.

Test made from June 8 to 15, 1889; age, 5 yrs. 1 mo.; actual weight, 820 lbs.; fed, daily, 8 lbs. corn meal, 8 lbs. middlings and 7 lbs. ground oats; property of D. F. Appleton, Ipswich, Mass.

Eurotisama 29668.—Sire, Young Pedro 9033; dam, Amité 18877.

Butter, 945 lbs. 9 oz. in one year. Milk, 11 to 17½ qts. per day.

Test made from Apr. 22, 1889, to Apr. 21, 1890; age, 4 yrs. 11 mos. at beginning of test; weight, 820 lbs.; fed, daily, about 21 lbs. grain; property of D. F. Appleton, Ipswich, Mass.

Eurus 60801.—Sire, Diploma 16219; dam, Complexia 56774.

Butter, 14 lbs. 7 oz. Milk, 244 lbs. 3 oz.

Test made from Nov. 15 to 22, 1889; age, 2 yrs. 11 mos.; estimated weight, 850 lbs.; fed, daily, 18 qts. corn and oat meal; property of Richardson Bros., Davenport, Iowa.

Eva Horner 3d 24663.—Sire, Spotless 1860; dam, Eva Horner 2644.

Butter, 15 lbs. 2 oz. Milk, 132 lbs. 8 oz.

Test made from Aug. 22 to 28, 1885; age, 4 yrs. 2 mos.; property of Wallace L. Hubbs, Jonesville, N. Y.

Eva of Snipsic 17650.—Sire, Duke of Mansfield's Pierrot 6261; dam, Eliza A. Perkins 14604.

Butter, 14 lbs. 1 oz. Milk, 232 lbs.

Test made from Oct. 1 to 7, 1885; age, 3 yrs. 5 mos.; property of Miller & Sibley, Franklin, Pa.

Eva of Verna 15228.—Sire, Wanderer 3014; dam, Effie of Hillside 1521.

Butter, 21 lbs. 13 oz. Milk, 192 lbs. 8 oz.

Test made from Aug. 9 to 15, 1887; age, 6 yrs. 4 mos.; fed 6 qts. corn meal, 3 qts. oil meal, 6 qts. oats and 3 qts. bran per day; property of J. S. & W. M. Wallace, Lexington, Ky.

Evelina of Verna 10971.—Sire, Wanderer 3014; dam, Effie of Hillside 1521.

Butter, 19 lbs. 10½ oz. (official). Milk, 264 lbs. 13 oz.

Test made from June 23 to 29, 1885; age, 6 yrs. 2 mos.; property of F. Bronson, Greenfield Hill, Conn.

Eveline of Jersey 6781.—Sire, Grey Prince (F. S. 168 J. H. B.); dam, Daisy (F. S. 1355 J. H. B.)

Butter, 18 lbs. 6 oz. Milk, average of 18 qts. per day.

Test made from July 25 to 31, 1881; age, 5 yrs. 4 mos.; property of Edward L. Clarkson, Tivoli, N. Y.

Eveline Rose 80512.—Sire, Fancy's Jubilee 25205; dam, White's Dolly 65555.

Butter, 14 lbs. 6 oz., unsalted. Milk, 156 lbs.

Test made from Mar. 5 to 12, 1897; age, 5 yrs. 3 mos.; estimated weight, 675 lbs.; fed 4 lbs. bran, 4 lbs. corn meal, 2 lbs. cotton-seed meal and from 32 lbs. to 34 lbs. good ensilage, daily; property of T. S. Webb, Knoxville, Tenn.

Eve of Bolton 63665.—Sire, Chevalier 14748; dam, Nannie Harper 7248.

Butter, 14 lbs. 4½ oz. Milk, 177 lbs. 4 oz.

Test made from Sept. 23 to 30, 1896; age, 7 yrs. 10 mos.; actual weight, 1040 lbs.; fed 4 qts. shorts, 2 qts. corn meal, 2 qts. ground oats, 1 qt. cotton-seed meal, 1 qt. linseed meal, 3 bush. cut sweet corn stalks and 8 lbs. hay, daily; property of John A. Cunningham, Bolton, Mass.

Eve of Christmas 9473.—Sire, Cash Boy 2248; dam, Andrews' Ethel 4108.

Butter, 14 lbs. 5¼ oz. Milk, 257 lbs. 6 oz.

Test made from June 16 to 22, 1885; age, 5 yrs. 6 mos.; property of Frederick von Kapff, Govanstown, Md.

Evergreen Maid 11536.—Sire, Cushnoc Jr. 2358; dam, Fuchsia 2789.

Butter, 14 lbs.

Test made June, 1883; age, 4 yrs. 6 mos.; property of Freeman Partridge, Prospect, Me.

Evri 5282.—Sire, Marius 760; dam, Eve 456.

Butter, 15 lbs. 4 oz. Milk, 203 lbs. 8 oz.

Test made from Dec. 30, 1884, to Jan. 5, 1885; age, 8 yrs. 1 mo.; property of Beech Grove Farm, Beech Grove, Ind.

Exile's Aggie 52764.—Sire, Exile of St. Lambert 13657; dam, Aggie of St. Lambert 37085.

Butter, 14 lbs. 10¼ oz. Milk, 141 lbs. 8 oz.

Test made from Mar. 8 to 15, 1891; age, 3 yrs.; estimated weight, 650 lbs.; fed 5 lbs. corn meal, 5 lbs. middlings, 5 lbs. bran, 2 lbs. oil meal and 1 lb. cotton-seed meal, daily—30 lbs. roots during test; property of P. J. Cogswell, Rochester, N. Y.

Exile's Agnes 79796.—Sire, Exile of St. Lambert 4th 17196; dam, Angelo's Annie 58888.

Butter, 14 lbs. 8 oz. Milk, 302 lbs.

Test made from May 25 to June 1, 1894; age, 4 yrs. 8 mos.; estimated weight, 875 lbs.; fed 56 lbs. corn-cob meal, 35 lbs. bran and 14 lbs. oil meal—mixed clover pasture; property of John F. Avery, Saline, Mich.

Exile's Alphea 45393.—Sire, Exile of St. Lambert 13657; dam, Clotaire's Belle 28841.

Butter, 15 lbs. 14½ oz. Milk, 200 lbs. 6½ oz.

Test made from Apr. 2 to 9, 1891; age, 5 yrs.; estimated weight, 900 lbs.; fed from 8 to 12 qts. daily of a mixture of one part bran, two parts ground corn and three parts ground oats—also, bran slop daily and good hay; property of L. L. Tozier, Batavia, N. Y.

Exile's Anna 46881.—Sire, Exile of St. Lambert 13657; dam, Anna Gold Ear 21053.

Butter, 16 lbs. 12 oz. Milk, 219 lbs. 8 oz.

Test made from Dec. 23 to 30, 1890; age, 3 yrs. 9 mos.; estimated weight, 750 lbs.; fed 4 lbs. corn meal, 4 lbs. fine middlings, 6 lbs. bran, 3 lbs. oil meal, 3 lbs. pea meal and 16 lbs. ensilage, daily—clover hay; property of P. J. Cogswell, Rochester, N. Y.

Exile's Antoinette 111279.—Sire, Exile of St. Lambert 13657; dam, Miss Maude St. Lambert 54989.

Butter, 14 lbs. 7½ oz. Milk, 243 lbs. 8 oz.

Test made from June 24 to July 1, 1897; age, 2 yrs. 6 mos.; estimated weight, 880 lbs.; fed 42 qts. bran, 42 qts. oat meal, 28 qts. corn meal, 21 qts. oil meal, 140 lbs. alfalfa and 84 lbs. hay; property of P. J. Cogswell, Rochester, N. Y.

Exile's Arcadia 66201.—Sire, Exile of St. Lambert 13657; dam, Cora of Arcadia 16151.

Butter, 14 lbs. 4¼ oz. Milk, 243 lbs. 8 oz.

Test made from Jan. 24 to 31, 1893; age, 3 yrs. 6 mos.; estimated weight, 800 lbs.; fed 9 lbs. bran, 12 lbs. corn meal, 3¾ lbs. oil meal and 6 lbs. oat meal, daily—63 lbs. hay and 91 lbs. roots; property of Pierce J. Cogswell, Rochester, N. Y.

Exile's Beauty 61220.—Sire, Exile's Duke 19795; dam, Exile's Lucy 46883.

Butter, 16 lbs. 3 oz. Milk, 199 lbs.

Test made from Feb. 24 to Mar. 3, 1894; age, 4 yrs. 9 mos.; actual weight, 1020 lbs.; fed 42 lbs. corn meal, 14 lbs. oil meal, 56 lbs. bran, 42 lbs. ground oats and 14 lbs. cotton-seed meal—98 lbs. hay, 175 lbs. roots; property of H. L. S. Hall, Scottsville, N. Y.

Exile's Belle 40524.—Sire, Exile of St. Lambert 13657; dam, Success of St. Lambert 28489.

Butter, 24 lbs. 6 oz. Milk, 219 lbs.

Test made from July 12 to 19, 1891; age, 4 yrs. 10 mos.; actual weight, 1030 lbs.; fed 224 qts. of a mixture composed of 150 lbs. corn meal, 100 lbs. cotton-seed meal, 45 lbs. fine middlings, 20 lbs. bran and 25 lbs. ground oats—140 lbs. hay; property of A. D. McBride, Rochester, N. Y.

Exile's Belle 40524.—Sire, Exile of St. Lambert 13657; dam, Success of St. Lambert 28489.

Butter, 26 lbs. 7 oz. Milk, 240 lbs.

Test made from July 19 to 26, 1891; age, 4 yrs. 10 mos.; actual weight, 1030 lbs.; fed 224 qts. of a mixture composed of 150 lbs. corn meal, 100 lbs. cotton-seed meal, 45 lbs. fine middlings, 20 lbs. bran and 25 lbs. ground oats—140 lbs. hay; property of A. D. McBride, Rochester, N. Y.

Exile's Belle 40524.—Sire, Exile of St. Lambert 13657; dam, Success of St. Lambert 28489.

Butter, 32 lbs. 6 oz. Milk, 230 lbs. 8 oz.

Test made from July 26 to Aug. 2, 1891; age, 4 yrs. 11 mos.; actual weight, 1030 lbs.; fed 224 qts. of a mixture composed of 150 lbs. corn meal, 100 lbs. cotton-seed meal, 45 lbs. fine middlings, 20 lbs. bran and 25 lbs. ground oats—140 lbs. hay; property of A. D. McBride, Rochester, N. Y.

Exile's Belle 40524.—Sire, Exile of St. Lambert 13657; dam, Success of St. Lambert 28489.

Butter, 29 lbs. 3 oz. Milk, 165 lbs.

Test made from Aug. 2 to 9, 1891; age, 4 yrs. 11 mos.; actual weight, 1030 lbs.; fed 224 qts. of a mixture composed of 175 lbs. corn meal, 115 lbs. cotton-seed meal, 45 lbs. fine feed, 20 lbs. bran and 35 lbs. ground oats—140 lbs. hay; property of A. D. McBride, Rochester, N. Y.

Exile's Belle 40524.—Sire, Exile of St. Lambert 13657; dam, Success of St. Lambert 28489.

Butter, 122 lbs. 6½ oz. in 30 days. Milk, 902 lbs. 8 oz.

Test made from July 12 to Aug. 11, 1891; age, 4 yrs. 11 mos. at end of test; actual weight, 1030 lbs.; fed 960 qts. of grain (mixture of corn meal, cotton-seed meal, fine feed, bran and ground oats) and 600 lbs. of hay; property of A. D. McBride, Rochester, N. Y.

Exile's Bessie 49985.—Sire, Exile of St. Lambert 13657; dam, St. John's Daisy 28388.

Butter, 18 lbs. 12½ oz. Milk, 198 lbs. 8 oz.

Test made from Feb. 22 to 29, 1892; age, 4 yrs. 7 mos.; estimated weight, 950 lbs.; fed 35 qts. corn meal, 28 qts. fine feed, 28 qts. bran, 21 qts. cotton-seed meal, 42 lbs. hay and 84 qts. roots; property of P. J. Cogswell, Rochester, N. Y.

Exile's Butterball 65152.—Sire, Exile of St. Lambert 13657; dam, Exile's Myrtle 51351.

Butter, 15 lbs. 4 oz. Milk, 154 lbs.

Test made from Nov. 19 to 26, 1892; age, 2 yrs. 7 mos.; actual weight, 728 lbs.; fed 42 qts. corn meal, 42 qts. oats and peas, 28 qts. oil meal, 28 qts. bran, 70 lbs. hay and 84 lbs. roots; property of H. L. S. Hall, Scottsville, N. Y.

Exile's Claribel 107507.—Sire, Exile of St. Lambert 13657; dam, Rioter's Grace 61993.

Butter, 14 lbs. 5 oz. Milk, 224 lbs. 8 oz.

Test made from May 15 to 22, 1897; age, 2 yrs. 11 mos.; actual weight, 815 lbs.; fed 35 qts. bran, 35 qts. oat meal, 28 qts. corn meal and 14 qts. oil meal—84 lbs. hay; property of P. J. Cogswell, Rochester, N. Y.

Exile's Daisy 46885.—Sire, Exile of St. Lambert 13657; dam, Daisy Stillson 28174.

Butter, 16 lbs. 4 oz. Milk, 244 lbs.

Test made from Aug. 14 to 21, 1890; age, 2 yrs. 9 mos.; estimated weight, 800 lbs.; fed, daily, about 6 lbs. corn meal, 5 lbs. bran and 2 lbs. cotton-seed meal; property of Joseph W. Shurter, Gansevoort, N. Y.

Exile's Dell 67533.—Sire, Exile of St. Lambert 13657; dam, Puss Lurlene 21461.

Butter, 14 lbs. 10¼ oz. Milk, 191 lbs. 8 oz.

Test made from Apr. 10 to 17, 1891; age, 4 yrs. 2 mos.; estimated weight, 875 lbs.; fed 8 lbs. corn meal, 5 lbs. bran, 3 lbs. oil meal, 3 lbs. cotton-seed meal, 8 lbs. hay and 15 lbs. ensilage, daily; property of Pierce J. Cogswell, Rochester, N. Y.

Exile's Dewdrop 106104.—Sire, Exile of St. Lambert 13657; dam, Miss Bentley 51492.

Butter, 18 lbs. 4 oz. Milk, 326 lbs.

Test made from May 5 to 12, 1898; age, 4 yrs. 2 mos.; estimated weight, 950 lbs.; fed 4 qts. daily of bran and corn and cob meal, mixed in equal parts by bulk—good pasture of new grass and white clover; property of Joseph T. Hoopes, Bynum, Md.

Exile's Dolly 67452.—Sire, Exile of St. Lambert 13657; dam, Jennie Stoke Pogis 32010.

Butter, 19 lbs. 7½ oz. Milk, 327 lbs. 4 oz.

Test made from May 1 to 8, 1893; age, 5 yrs. 8 mos.; estimated weight, 900 lbs.; fed 94½ lbs. corn meal, 31½ lbs. cotton-seed meal, 42 lbs. bran, 15¾ lbs. oat meal, 64¼ lbs. hay and 147 lbs. roots; property of Pierce J. Cogswell, Rochester, N. Y.

Exile's Dumbella 79452.—Sire, Exile of St. Lambert 13657; dam, Dumbella 3d 11819.

Butter, 14 lbs. 11 oz. Milk, 277 lbs. 4 oz.

Test made from May 5 to 12, 1894; age, 6 yrs. 11 mos.; estimated weight, 1000 lbs.; fed 84 lbs. corn meal, 31½ lbs. bran, 31½ lbs. cotton-seed meal, 42 lbs. ground oats, 105 lbs. hay and 280 lbs. roots; property of P. J. Cogswell, Rochester, N. Y.

Exile's Echo 111276.—Sire, Exile of St. Lambert 13657; dam, Mabel A. 100601.

Butter, 15 lbs. 9¾ oz. Milk, 273 lbs.

Test made from Feb. 18 to 25, 1898; age, 3 yrs. 6 mos.; actual weight, 997 lbs.; fed 56 qts. bran, 56 qts. oat meal, 42 qts. corn meal, 14 qts. cotton-seed meal, 14 qts. linseed meal, 54 lbs. hay and 56 qts. roots; property of P. J. Cogswell, Rochester, N. Y.

Exile's Effie 40526.—Sire, Exile of St. Lambert 13657; dam, Lady Woodbine 26803.

Butter, 15 lbs. 2 oz. Milk, 202 lbs. 8 oz.

Test made from Feb. 25 to Mar. 4, 1890; age, 3 yrs. 5 mos.; estimated weight, 975 lbs.; fed 30 lbs. daily of a mixture composed of 125 lbs. corn meal, 80 lbs. oil meal, 30 lbs. fine feed, 15 lbs. bran and 15 lbs. ground oats; property of A. D. McBride, Rochester, N. Y.

Exile's Elf 111811.—Sire, Exile of St. Lambert 13657; dam, Duke's Flo 79109.

Butter, 14 lbs. 7¼ oz. Milk, 206 lbs. 8 oz.

Test made from Dec. 2 to 9, 1897; age, 3 yrs. 2 mos.; actual weight, 850 lbs.; fed 56 qts. bran, 42 qts. corn meal, 21 qts. cotton-seed meal, 105 lbs. ensilage, 84 lbs. hay and 56 qts. roots; property of P. J. Cogswell, Rochester, N. Y.

Exile's Fawn 40979.—Sire, Exile of St. Lambert 13657; dam, Saint Maggie 30363.

Butter, 15 lbs. 10 oz. Milk, 185 lbs.

Test made from May 2 to 9, 1890; age, 4 yrs. 1 mo.; estimated weight, 750 lbs.; fed about 18 qts. corn meal, bran and fine middlings and 12 qts. beets, daily—dry hay; property of Pierce J. Cogswell, Rochester, N. Y.

Exile's Flossie 116608.—Sire, Exile of St. Lambert 13657; dam, Pride's Beth 87106.

Butter, 16 lbs. 12 oz. Milk, 224 lbs. 8 oz.

Test made from May 14 to 21, 1898; age, 3 yrs. 9 mos.; estimated weight, 700 lbs.; fed 6 qts. corn meal and 10 qts. bran, daily—June grass pasture; property of E. G. Silvus, Athens, Ohio.

Exile's Gretchen 79245.—Sire, Exile of St. Lambert 13657; dam, Exile's Lucy 46883.

Butter, 16 lbs. 15½ oz. Milk, 230 lbs. 8 oz.

Test made from Mar. 7 to 14, 1895; age, 2 yrs. 10 mos.; estimated weight, 800 lbs.; fed 28 qts. corn meal, 14 qts. ground oats, 35 qts. bran, 21 qts. oil meal, 14 qts. cotton-seed, 8 qts. roots and 105 lbs. hay; property of P. J. Cogswell, Rochester, N. Y.

Exile's Harriet 100716.—Sire, Exile of St. Lambert 13657; dam, Exile's Success 49986.

Butter, 16 lbs. 6¼ oz. Milk, 221 lbs. 8 oz.

Test made from Feb. 12 to 19, 1895; age, 2 yrs.; actual weight, 900 lbs.; fed 28 qts. corn meal, 17½ qts. oil meal, 42 qts. bran, 24½ qts. ground oats, 42 qts. roots and 126 lbs. hay; property of P. J. Cogswell, Rochester, N. Y.

Exile's Harriet 100716.—Sire, Exile of St. Lambert 13657; dam, Exile's Success 49986.

Butter, 15 lbs. 1¼ oz. Milk, 226 lbs. 12 oz.

Test made from Feb. 19 to 26, 1895; age, 2 yrs.; actual weight, 900 lbs.; fed 28 qts. corn meal, 17½ qts. oil meal, 42 qts. bran and 24½ qts. ground oats, 126 lbs. hay and 42 qts. roots; property of P. J. Cogswell, Rochester, N. Y.

Exile's Harriet 100716.—Sire, Exile of St. Lambert 13657; dam, Exile's Success 49986.

Butter, 31 lbs. 7½ oz. in 14 days. Milk, 448 lbs. 4 oz.

Test made from Feb. 12 to 26, 1895; age, 2 yrs.; actual weight, 900 lbs.; fed 57 qts. corn meal, 35 qts. oil meal, 84 qts. bran, 49 qts. ground oats, 84 qts. roots and 252 lbs. hay; property of P. J. Cogswell, Rochester, N. Y.

Exile's Hazel 85181.—Sire, Exile of St. Lambert 13657; dam, Hazel Stoke Pogis 51793.

Butter, 15 lbs. 13 oz. Milk, 197 lbs. 6 oz.

Test made from May 19 to 26, 1896; age, 4 yrs. 4 mos.; estimated weight, 820 lbs.; fed 4 gals. wheat bran and middlings, mixed into a mash, in two feeds, daily—good timothy pasture; property of E. A. Berauer, Waldron, Ind.

Exile's Helen 61251.—Sire, Exile of St. Lambert 13657; dam, Helen Stoke Pogis 31947.

Butter, 14 lbs. 4½ oz. Milk, 198 lbs.

Test made from Jan. 22 to 29, 1892; age, 3 yrs. 11 mos.; estimated weight, 875 lbs.; fed 42 qts. corn meal, 21 qts. cotton-seed meal, 42 qts. bran and 42 lbs. clover hay; property of T. W. Palmer, Detroit, Mich.

Exile's Jessamine 111104.—Sire, Exile of St. Lambert 13657; dam, Aggie of St. Lambert 2d 77184.

Butter, 14 lbs. 1¾ oz. Milk, 211 lbs.

Test made from Apr. 14 to 21, 1897; age, 3 yrs. 2 mos.; actual weight, 885 lbs.; fed 42 qts. bran, 42 qts. oat meal, 35 qts. corn meal, 21 qts. oil meal, 105 lbs. ensilage, 84 lbs. hay and 42 qts. roots; property of P. J. Cogswell, Rochester, N. Y.

Exile's Kosi 107505.—Sire, Exile of St. Lambert 13657; dam, Elmanetta 2d 37945.

Butter, 16 lbs. 2½ oz. Milk, 256 lbs. 8 oz.

Test made from Mar. 24 to 31, 1898; age, 3 yrs. 11 mos.; actual weight, 865 lbs.; fed 42 qts. bran, 42 qts. oat meal, 28 qts. corn meal, 14 qts. cotton-seed meal, 42 lbs. hay and 84 qts. roots; property of P. J. Cogswell, Rochester, N. Y.

Exile's Lady Alexis 79112.—Sire, Exile of St. Lambert 13657; dam, Lady Alexis 26916.

Butter, 14 lbs. 4¾ oz. Milk, 208 lbs. 8 oz.

Test made from Jan. 27 to Feb. 3, 1895; age, 3 yrs. 10 mos.; estimated weight, 850 lbs.; fed 28 qts. corn meal, 28 qts. old process oil meal, 42 qts. bran, 42 qts. ground oats, 7 qts. cotton-seed, 42 qts. roots and 105 lbs. hay; property of P. J. Cogswell, Rochester, N. Y.

Exile's Lady Angela 46882.—Sire, Exile of St. Lambert 13657; dam, Eva Locust 21050.

Butter, 15 lbs. 11 oz. Milk, 263 lbs. 14 oz.

Test made from June 12 to 19, 1893; age, 6 yrs. 3 mos.; estimated weight, 875 lbs.; fed 5 qts. oat meal, 4 qts. corn meal, 1 qt. cotton-seed meal and 6 qts. bran, daily—clover and timothy pasture; property of Mrs. E. F. Hawley, Pittsford, N. Y.

Exile's Lady Palestine 107626.—Sire, Exile of St. Lambert 13657; dam, Exile's Nina 40522.

Butter, 14 lbs. 14 oz. Milk, 247 lbs.

Test made from May 23 to 30, 1897; age, 3 yrs. 4 mos.; actual weight, 905 lbs.; fed 42 qts. bran, 42 qts. oat meal, 35 qts. corn meal and 14 qts. oil meal—70 lbs. hay; property of P. J. Cogswell, Rochester, N. Y.

Exile's Lady Sarah 107938.—Sire, Exile's St. John 20202; dam, Exile's Phyllis 79111.

Butter, 17 lbs. 10 oz. Milk, 252 lbs. 8 oz.

Test made from Jan. 3 to 10, 1898; age, 3 yrs. 7 mos.; actual weight, 925 lbs.; fed 42 qts. bran, 42 qts. oat meal, 42 qts. corn meal, 14 qts. cotton-seed meal, 140 lbs. ensilage, 105 lbs. hay and 42 qts. roots; property of P. J. Cogswell, Rochester, N. Y.

Exile's Lady Star 63943.—Sire, Exile of St. Lambert 13657; dam, Lady Star 2d 15724.

Butter, 16 lbs. ¼ oz. Milk, 217 lbs. 12 oz.

Test made from June 8 to 15, 1894; age, 5 yrs. 3 mos.; estimated weight, 750 lbs.; fed 11 qts. chop and 2 qts. bran, daily—pasture; property of Mrs. A. M. Hallock, Columbus, Ohio.

Exile's Lady Star 63943.—Sire, Exile of St. Lambert 13657; dam, Lady Star 2d 15724.

Butter, 35 lbs. 7½ oz. in 16 days. Milk, 517 lbs. 8 oz.

Test made from May 30 to June 15, 1894; age, 5 yrs. 3 mos.; estimated weight, 750 lbs.; fed 8 qts. chop and 2 qts. bran, daily, for first 8 days; 11 qts. chop and 2 qts. bran, daily, for remainder of test—pasture; property of Mrs. A. M. Hallock, Columbus, Ohio.

Exile's Lizzie 61477.—Sire, Exile of St. Lambert 13657; dam, Lizzie Reeves 9222.

Butter, 14 lbs. 8½ oz. Milk, 168 lbs. 8 oz.

Test made from May 17 to 24, 1891; age, 2 yrs. 7 mos.; estimated weight, 750 lbs.; fed 7 lbs. corn meal, 8 lbs. bran, 3 lbs. oil meal, 2 lbs. cotton-seed meal and about 16 lbs. beets, daily—hay; property of Pierce J. Cogswell, Rochester, N. Y.

Exile's Lucile 65151.—Sire, Exile of St. Lambert 13657; dam, Exile's Lucy 46883.

Butter, 14 lbs. 6 oz. Milk, 283 lbs. 8 oz.

Test made from Sept. 30 to Oct. 7, 1895; age, 5 yrs. 6 mos.; estimated weight, 950 lbs.; fed 42 qts. corn and oats, 14 qts. bran, 14 qts. old process oil meal and 14 qts. middlings—clover hay *ad lib.*, poor pasture; property of H. & J. S. Wadsworth, Avon, N. Y.

Exile's Lucy 46883.—Sire, Exile of St. Lambert 13657; dam, Catoctin's Daughter 29262.

Butter, 15 lbs. 7¾ oz. Milk, 231 lbs. 8 oz.

Test made from May 10 to 17, 1891; age, 3 yrs. 10 mos.; estimated weight, 850 lbs.; fed 7 lbs. corn meal, 8 lbs. bran, 3 lbs. oil meal, 3 lbs. cotton-seed meal and about 16 lbs. beets, daily—hay; property of Pierce J. Cogswell, Rochester, N. Y.

Exile's Lulu 49984.—Sire, Exile of St. Lambert 13657; dam, Yankee Girl 2d 29261.

Butter, 16 lbs. 2 oz. Milk, 328 lbs. 8 oz.

Test made from Feb. 16 to 23, 1891; age, 4 yrs. 3 mos.; estimated weight, 850 lbs.; fed about 20 lbs. daily of a mixture composed of one-sixth oil meal, one-third corn meal and one-half bran, also about 20 lbs. beets daily—hay and ensilage; property of Pierce J. Cogswell, Rochester, N. Y.

Exile's Lulu 49984.—Sire, Exile of St. Lambert 13657; dam, Yankee Girl 2d 29261.

Butter, 54.017 lbs. in 30 days. Milk, 988.4 lbs.

Test made from Aug. 29 to Sept. 27, 1893, at World's Columbian Exposition; age, 6 yrs. 9 mos.; actual weight, 963 lbs.; fed 318.5 lbs. hay, 75 lbs. oil meal, 112 lbs. ensilage, 20 lbs. corn meal, 180 lbs. bran, 90 lbs. oats, 189 lbs. corn-hearts, 45 lbs. cotton-seed, 90 lbs. middlings, 18 lbs. carrots and 4 lbs. old hay; property of C. I. Hudson, New York, N. Y.

Exile's Lulu 49984.—Sire, Exile of St. Lambert 13657; dam, Yankee Girl 2d 29261.

Butter, 168.538 lbs. in 90 days. Milk, 3224.5 lbs.

Test made from May 31 to Aug. 28, 1893, at World's Columbian Exposition; age, 6 yrs. 6 mos. at beginning of test; actual weight, 934 lbs.; fed 298.5 lbs. old hay, 123 lbs. ensilage, 504.4 lbs. new hay, 189.5 lbs. oil meal, 445 lbs. bran, 96 lbs. oats, 556.5 lbs. corn-hearts, 124.50 lbs. cotton-seed, 467 lbs. middlings, 46 lbs. grano-gluten, 1153.4 lbs. clover and 10 lbs. swale-grass; property of C. I. Hudson, New York, N. Y.

Exile's May 52765.—Sire, Exile of St. Lambert 13657; dam, St. John's Daisy 28388.

Butter, 15 lbs. 4¼ oz. Milk, 157 lbs. 8 oz.

Test made from Jan. 4 to 11, 1892; age, 3 yrs. 7 mos.; estimated weight, 900 lbs.; fed 63 qts. corn meal, 21 qts. bran, 21 qts. fine feed, 14 qts. cotton-seed meal, 42 lbs. hay and 56 qts. roots; property of P. W. Arnold, Pawtucket, R. I.

Exile's Moss Rose 101155.—Sire, Exile of St. Lambert 13657; dam, Dorcas of Wayne 2d 60557.

Butter, 15 lbs. 4 oz. Milk, 231 lbs.

Test made from Feb. 25 to Mar. 4, 1897; age, 3 yrs. 6 mos.; actual weight, 805 lbs.; fed 42 qts. bran, 42 qts. oat meal, 42 qts. corn meal, 21 qts. oil meal, 140 lbs. ensilage, 84 lbs. hay and 56 qts. roots; property of P. J. Cogswell, Rochester, N. Y.

Exile's Myrtle 2d 94446.—Sire, Exile of St. Lambert 13657; dam, Exile's Myrtle 51351.

Butter, 17 lbs. 13½ oz. Milk, 255 lbs. 8 oz.

Test made from June 6 to 13, 1897; age, 5 yrs. 2 mos.; estimated weight, 950 lbs.; fed 28 qts. corn, 56 qts. bran, 28 qts. oats and 14 qts. oil meal—green oats and peas night and morning, poor pasture; property of M. E. Griffith, Manor Station, Pa.

Exile's Nancy 92030.—Sire, Exile of St. Lambert 13657; dam, May of Ashland 73925.

Butter, 14 lbs. 7 oz. Milk, 288 lbs.

Test made from Sept. 3 to 10, 1896; age, 3 yrs. 6 mos.; actual weight, 725 lbs.; fed 35 qts. bran, 35 qts. oat meal, 42 qts. corn meal and 28 qts. oil meal—105 lbs. hay; property of Pierce J. Cogswell, Rochester, N. Y.

Exile's Nina 40522.—Sire, Exile of St. Lambert 13657; dam, Oléo 38475.

Butter, 15 lbs. 11 oz. Milk, 237 lbs. 8 oz.

Test made from Feb. 21 to 28, 1895; age, 8 yrs. 9 mos.; estimated weight, 950 lbs.; fed 21 qts. corn meal, 10½ qts. oil meal, 42 qts. bran, 24½ qts. ground oats, 14 qts. cotton-seed, 42 qts. roots and 105 lbs. hay; property of P. J. Cogswell, Rochester, N. Y.

Exile's Nina 40522.—Sire, Exile of St. Lambert 13657; dam, Oléo 38475.

Butter, 15 lbs. 11½ oz. Milk, 232 lbs.

Test made from Feb. 28 to Mar. 7, 1895; age, 8 yrs. 9 mos.; estimated weight, 950 lbs.; fed 21 qts. corn meal, 10½ qts. oil meal, 42 qts. bran, 24½ qts. ground oats, 14 qts. cotton-seed, 42 qts. roots and 105 lbs. hay; property of P. J. Cogswell, Rochester, N. Y.

Exile's Nina 40522.—Sire, Exile of St. Lambert 13657; dam, Oléo 38475.

Butter, 31 lbs. 6½ oz. in 14 days. Milk, 469 lbs. 8 oz.

Test made from Feb. 21 to Mar. 7, 1895; age, 8 yrs. 9 mos.; estimated weight, 950 lbs.; fed 42 qts. corn meal, 21 qts. oil meal, 84 qts. bran, 49 qts. ground oats, 28 qts. cotton-seed, 84 qts. roots and 210 lbs. hay; property of P. J. Cogswell, Rochester, N. Y.

Exile's Ona 106212.—Sire, Exile of St. Lambert 13657; dam, Koffee's Duchess 2d 57988.

Butter, 14 lbs. 3¾ oz. Milk, 224 lbs. 8 oz.

Test made from Mar. 11 to 18, 1897; age, 2 yrs.; actual weight, 725 lbs.; fed 42 qts. bran, 42 qts. oat meal, 35 qts. corn meal, 21 qts. oil meal, 105 lbs. ensilage, 84 lbs. hay and 56 qts. roots; property of P. J. Cogswell, Rochester, N. Y.

Exile's Onnalinda 65189.—Sire, Exile of St. Lambert 13657; dam, Lady Alexis 26916.

Butter, 15 lbs. 1½ oz. Milk, 158.6 lbs.

Test made from Dec. 14 to 21, 1892; age, 2 yrs. 8 mos.; actual weight, 810 lbs.; fed 42 qts. corn meal, 28 qts. bran, 42 qts. peas and oats, 42 qts. oil meal, 210 lbs. corn ensilage and 28 lbs. hay; property of H. L. S. Hall, Scottsville, N. Y.

Exile's Pansy 79800.—Sire, Exile of St. Lambert 4th 17196; dam, Angelo's Annie 58888.

Butter, 17 lbs. 13 oz. Milk, 278 lbs. 12 oz.

Test made from May 6 to 13, 1896; age, 4 yrs. 3 mos.; estimated weight, 800 lbs.; fed 51 lbs. corn and cob meal and 10 lbs. oil meal—timothy and clover pasture; property of J. F. Avery, Saline, Mich.

Exile's Pauline 37530.—Sire, Exile of St. Lambert 13657; dam, Mollie Madden 25798.

Butter, 16 lbs. 15 oz. Milk, 186 lbs.

Test made from Jan. 21 to 28, 1890; age, 4 yrs. 3 mos.; estimated weight, 750 lbs.; fed, daily, 7 lbs. corn meal, 10 lbs. bran, 8 lbs. fine middlings and ½ lb. pea meal, mixed; property of P. J. Cogswell, Rochester, N. Y.

Exile's Penelope 77182.—Sire, Exile of St. Lambert 13657; dam, Queen of Berlin 2d 44834.

Butter, 15 lbs. 11 oz. Milk, 242 lbs. 12 oz.

Test made from Mar. 5 to 12, 1894; age, 4 yrs. 11 mos.; estimated weight, 800 lbs.; fed 84 lbs. corn meal, 42 lbs. ground oats, 31½ lbs. cotton-seed meal and 31½ lbs. bran—105 lbs. hay, 105 lbs. roots; property of P. J. Cogswell, Rochester, N. Y.

Exile's Penelope 77182.—Sire, Exile of St. Lambert 13657; dam, Queen of Berlin 2d 44834.

Butter, 18 lbs. 3¾ oz. Milk, 268 lbs. 8 oz.

Test made from Mar. 4 to 11, 1895; age, 5 yrs. 11 mos.; estimated weight, 900 lbs.; fed 35 qts. bran, 14 qts. oats, 28 qts. corn meal, 21 qts. oil meal, 14 qts. cotton-seed, 8 qts. roots and 105 lbs. hay; property of P. J. Cogswell, Rochester, N. Y.

Exile's Rosa 66552.—Sire, Exile of St. Lambert 13657; dam, Rosa's May Bud 47931.

Butter, 14 lbs. ½ oz. Milk, 159 lbs. 8 oz.

Test made from Mar. 10 to 17, 1892; age, 3 yrs.; estimated weight, 750 lbs.; fed 6 lbs. bran and fine feed, 4½ lbs. cotton-seed meal, 8½ lbs. corn meal, 5 lbs. hay and 6 qts. roots, daily; property of Pierce J. Cogswell, Rochester, N. Y.

Exile's Signal Belle 2d 106097.—Sire, Oonan's Harry 25877; dam, Exile's Signal Belle 82363.

Butter, 19 lbs. 13 oz. Milk, 197 lbs. 2½ oz.

Test made from June 14 to 21, 1897; age, 3 yrs. 9 mos.; estimated weight, 900 lbs.; fed 14 qts. old process oil-cake meal and 86 qts. of a mixture of corn and cob meal, ground wheat and ground öats, in equal parts by measure—timothy and blue-grass pasture; property of H. C. Schlegel, Daleville, Ind.

Exile's Success 49986.—Sire, Exile of St. Lambert 13657; dam, Success of St. Lambert 28489.

Butter, 17 lbs. 6½ oz. Milk, 283 lbs. 4 oz.

Test made from Mar. 8 to 15, 1895; age, 7 yrs. 8 mos.; estimated weight, 800 lbs.; fed 35 qts. bran, 14 qts. oats, 28 qts. corn meal, 21 qts. oil meal, 14 qts. cotton-seed—126 lbs. hay, 8 qts. roots; property of P. J. Cogswell, Rochester, N. Y.

Experience 62458.—Sire, Ida's Rioter of St. L. 13656; dam, Rioter's Zoe 19769.

Butter, 22 lbs. 13 oz. Milk, 276 lbs. 4 oz.

Test made from May 11 to 18, 1892; age, 4 yrs. 1 mo.; actual weight, 1060 lbs.; fed 224 qts. grain—second year clover and timothy pasture; property of Miller & Sibley, Franklin, Pa.

Ex Presage 128529.—Sire, Coomassie's Triple King 37939; dam, Grey Dame 114000.

Butter, 15 lbs. 5 oz. Milk, 212 lbs. 8 oz.

Test made from May 29 to June 5, 1898; age, 2 yrs. 3 mos.; estimated weight, 700 lbs.; fed 4 qts. daily of corn and oats, ground together, in equal parts—white clover and June grass pasture; property of George E. Nichols, Afton, N. Y.

Express 69574.—Sire, Gold Ridge 18951; dam, Eutoga 58027.

Butter, 18 lbs. 5 oz. Milk, 270 lbs. 12 oz.

Test made from June 7 to 14, 1893; age, 4 yrs.; estimated weight, 900 lbs.; fed 8 qts. corn meal and 8 qts. ground oats, daily—mixed pasture; property of George Fox, Torresdale, Pa.

Extract 3d 85261.—Sire, Petit Victor 24788; dam, Extract 58032.

Butter, 14 lbs. 6 oz. Milk, 212 lbs.

Test made from July 15 to 22, 1895; age, 3 yrs. 11 mos.; estimated weight, 700 lbs.; fed 2 qts. cotton-seed meal, 6 qts. corn meal and 4 qts. wheat bran, daily—rough, hilly pasture; property of W. C. Norton, Agt., Aldenville, Pa.

Fable 62520.—Sire, Ida's Rioter of St. L. 13656; dam, Joey of Prospect 27022.

Butter, 26 lbs. 5¼ oz. Milk, 304 lbs.

Test made from Mar. 17 to 24, 1894; age, 6 yrs. 1 mo.; actual weight, 1000 lbs.; fed 84 qts. ground oats, 56 qts. corn meal, 28 qts. wheat bran, 14 qts. ground oil-cake, 15 lbs. cabbage and 15 lbs. mixed carrots and sugar beets—hay; property of Miller & Sibley, Franklin, Pa.

Fable 62520.—Sire, Ida's Rioter of St. L. 13656; dam, Joey of Prospect 27022.

Butter, 26 lbs. 1¾ oz. Milk, 307 lbs.

Test made from Mar. 24 to 31, 1894; age, 6 yrs. 1 mo.; actual weight, 1000 lbs.; fed 84 qts. ground oats, 56 qts. corn meal, 28 qts. wheat bran, 14 qts. ground oil-cake, 15 lbs. cabbage and 15 lbs. mixed carrots and sugar beets—hay; property of Miller & Sibley, Franklin, Pa.

Fable 62520.—Sire, Ida's Rioter of St. L. 13656; dam, Joey of Prospect 27022.

Butter, 52 lbs. 7 oz. in 14 days. Milk, 611 lbs.

Test made from Mar. 17 to 31, 1894; age, 6 yrs. 1 mo.; actual weight, 1000 lbs.; fed 168 qts. ground oats, 112 qts. corn meal, 56 qts. wheat bran, 28 qts. ground oil-cake, 30 lbs. cabbage and 30 lbs. carrots and sugar beets, mixed; property of Miller & Sibley, Franklin, Pa.

Fadette of Canaan 14807.—Sire, Wanderer 3014; dam, Fadette of Verna 2d 9024.

Butter, 16 lbs. 1½ oz. Milk, 119 lbs. 12 oz.

Test made from Apr. 7 to 13, 1885; age, 3 yrs. 3 mos.; property of George M. Jewett, Glenville, Md.

Fadette of Verna 3d 11122.—Sire, Wanderer 3014; dam, Fadette of Verna 6814.

Butter, 22 lbs. 8½ oz. Milk, 22 qts. per day.

Test made from May 16 to 22, 1884; age, 3 yrs. 9 mos.; fed 31 lbs. daily of mixed feed, as follows: two-thirds ground corn and oats (one-third corn to two thirds oats) and one-third oil meal—good pasture; property of G. W. Farlee, Cresskill, N. J.

Fadette Signal 39472.—Sire, Chief of the Miamis 2958; dam, Fadette of Canaan 14807.

Butter, 20 lbs. 14½ oz. Milk, 234 lbs.

Test made from Jan. 18 to 25, 1893; age, 7 yrs. 10 mos.; estimated weight, 800 lbs.; fed 84 lbs. corn, 42 lbs. oats, 42 lbs. bran, 28 lbs. linseed, 280 lbs. ensilage and 70 lbs. hay; property of William Hart Dexter, Springfield, Mass.

Fadette Signal's Queen 109281.—Sire, Croton's Signal 17614; dam, Fadette Signal 39472.

Butter, 16 lbs. 3 oz. Milk, 206 lbs. 14 oz.

Test made from Feb. 5 to 12, 1898; age, 3 yrs. 1 mo.; estimated weight, 850 lbs.; fed 8 lbs. wheat bran, 11 lbs. corn meal, ½ lb. linseed meal, 3 lbs. cotton-seed meal and 2½ lbs. cotton-seed, daily—sorghum *ad lib.*, woods pasture; property of W. E. Johnson, Millican, Tex.

Fair Dairy-maid 29839.—Sire, Kenilworth 8091; dam, Fair Maid of Perth 13705.

Butter, 14 lbs. 8 oz. Milk, 182 lbs. 4 oz.

Test made from Feb. 13 to 19, 1887; age, 2 yrs. 4 mos.; estimated weight, 750 lbs.; fed, daily, 4 lbs. crushed oats, 1½ lbs. corn meal, 1½ lbs. pea meal, 1 lb. oil meal and 2 lbs. shorts; property of C. W. H. Eicke, West Monterey, Pa.

Fair Lady 6723.—Sire, Guy Mannering 698; dam, Fanny Fair 4136.

Butter, 18 lbs., unsalted (official). Milk, 232 lbs. 13 oz.

Test made from May 23 to 30, 1883; age, 6 yrs. 7 mos.; fed, daily, 6 qts. ground corn and oats on half a bucket of cut hay; property of Columbia Jersey Cattle Co., Columbia, Tenn.

Fair Lady 6723.—Sire, Guy Mannering 698; dam, Fanny Fair 4136.

Butter, 19 lbs. Milk, 233 lbs.

Test made from June 17 to 23, 1883; age, 6 yrs. 6 mos.; fed 6 qts. daily of oats and corn ground together, with cut hay—blue-grass and white clover pasture; property of the Columbia Jersey Cattle Co., Columbia, Tenn.

Fair Marigold 89387.—Sire, Major Appel Pogis 17861; dam, Ida Marigold 32615.

Butter, 16 lbs. 15 oz.   Milk, 230 lbs. 6 oz.

Test made from July 28 to Aug. 4, 1895; age, 2 yrs. 3 mos.; actual weight, 715 lbs.; fed 9 qts. wheat bran, 6 qts. ground oats, 3 qts. corn meal, 1½ qts. oil meal, 1½ qts. cotton-seed meal, daily—poor pasture, green oats and peas in stable; property of C. A. Sweet, East Aurora, N. Y.

Fair Starlight 7745.—Sire, King of Fairview 778; dam, Pinkie 2d 2987.

Butter, 17 lbs. 7½ oz., unsalted.   Milk, 250 lbs. 4 oz.

Test made from Apr. 10 to 17, 1883; age, 6 yrs. 9 mos.; fed 8 qts. of a mixture composed of one-third corn and two-thirds oats, ground together—hay *ad lib.*, no pasture; property of David Strong, Winsted, Conn.

Fairy Le Brecq 15283.—Sire, Le Brocq's Prize 3350; dam, Queen of the Fairies 8128.

Butter, 14 lbs. 3 oz.   Milk, 34 to 36 lbs. daily.

Test made from June 2 to 9, 1885; age, 3 yrs. 4 mos.; property of S. A. Fletcher, Indianapolis, Ind.

Fairy of Eau Claire 128596.—Sire, Melia Ann's Son 22041; dam, Angeline Pomeroy 2d 94684.

Butter, 14 lbs. 14 oz.   Milk, 240 lbs. 8 oz.

Test made from Feb. 19 to 26, 1898; age, 2 yrs. 9 mos.; estimated weight, 800 lbs.; fed 4 qts. corn meal, 6 qts. wheat bran, 2 qts. oil meal, 2 qts. cotton-seed meal and 20 lbs. corn ensilage, daily—hay *ad lib.*; property of E. L. Van Deusen, Ashley Falls, Mass.

Fairy of Verna 2d 10973.—Sire, Wanderer 3014; dam, Fairy of Verna 6813.

Butter, 20 lbs. 3¾ oz.   Milk, 280 lbs.

Test made from Sept. 11 to 18, 1884; age, 4 yrs. 5 mos.; fed first six days 14 lbs. 13 oz. ground corn and oats (one part corn to two of oats) and 10 lbs. 6½ oz. oil meal, daily; on last day 12 lbs. 5 oz. corn and oats, 5¾ lbs. oil meal and 1½ lbs. wheat bran—good pasture; property of George W. Farlee, Cresskill, N. J.

Fairy Oxford 66679.—Sire, Oxford Pride 19160; dam, Miss Remarkable 33446.

Butter, 17 lbs. 7½ oz.   Milk, 236 lbs. 8 oz.

Test made from June 30 to July 7, 1894; age, 4 yrs. 4 mos.; estimated weight, 925 lbs.; fed 168 lbs. corn-hearts and 28 lbs. cut sheaf oats—blue-grass pasture; property of Est. of Campbell Brown, Spring Hill, Tenn.

Fairy Queen of St. Brelade's 7464.—Sire, Duke (P. S. 76 J. H. B.); dam on I. of J.

Butter, 19 lbs. 7¼ oz.   Milk, 222 lbs. 7 oz.

Test made from Aug. 16 to 23, 1885; age, 9 yrs.; property of George M. Jewett, Glenville, Md.

Faith of Cloverdale 29277.—Sire, Duke of Cloverdale 6994; dam, Flo 3d 14754.

Butter, 14 lbs. 2 oz.   Milk, 128 lbs. 5½ oz.

Test made from Mar. 15 to 21, 1887; age, 2 yrs. 5 mos.; fed 6 qts. corn meal, 6 qts. oat meal, 2 qts. oil meal, daily; property of Archer N. Martin, Summit, N. J.

Faith of Eau Claire 120456.—Sire, Melia Ann's Son 22041; dam, Ruby of Eau Claire 93305.

Butter, 22 lbs. 12 oz.   Milk, 307 lbs. 9 oz.

Test made from Dec. 20 to 27, 1897; age, 3 yrs. 5 mos.; estimated weight, 900 lbs.; fed 3 qts. ground oats, 4 qts. corn meal, 4 qts. wheat bran, 1 qt. oil meal and 2 qts. cotton-seed meal, daily—clover hay *ad lib.*; property of E. L. Van Deusen, Ashley Falls, Mass.

Faith of Oaklands 19696.—Sire, Prince of the Valley (P. S. 88 J. H. B.); dam on I. of J.

Butter, 17 lbs. 4 oz.   Milk, 267 lbs.

Test made from Aug. 17 to 23, 1883; age, 6 yrs. 6 mos.; fed 8 qts. of ground oats, daily—second crop clover pasture; property of V. E. Fuller, Hamilton, Ont., Can.

Fall Leaf 8587.—Sire, Lord Lawrence 1414; dam, Sunny South 6830.

Butter, 14 lbs. 8 oz.   Milk, 256 lbs.

Test made from Sept. 15 to 21, 1882; age, 3 yrs. 10 mos.; property of W. E. Oates, Vicksburg, Miss.

Fall Leaf's Heiress 54454.—Sire, Prince of Melrose 4819; dam, Fall Leaf 2d 25171.

Butter, 16 lbs. 2 oz.   Milk, 210 lbs. 10 oz.

Test made from May 16 to 23, 1893; age, 6 yrs. 2 mos.; estimated weight, 850 lbs.; fed 17½ gals. corn meal, 17½ gals. bran, 3½ gals. cotton-seed meal and 5¼ gals. cob meal—Bermuda grass; property of M. Lothrop, Marshall, Tex.

False Step 69647.—Sire, Compass 16958; dam, Fausetta 29334.

Butter, 15 lbs. 7 oz.   Milk, 199 lbs. 2 oz.

Test made from June 7 to 14, 1894; age, 5 yrs. 1 mo.; estimated weight, 775 lbs.; fed 15 qts. of a mixture of ground corn, oats and oil meal, daily—good pasture; property of Richardson Bros., Davenport, Iowa.

Fancy Alice 106402.—Sire, Old Tenella's Signal 34802; dam, Malice 18988.

Butter, 14 lbs. 11½ oz.   Milk, 183 lbs. 15½ oz.

Test made from Jan. 9 to 16, 1898; age, 3 yrs.; estimated weight, 800 lbs.; fed 10 lbs. bran, 8 lbs. corn meal and oats and 3 lbs. cotton-seed meal, daily—clover hay *ad lib.*; property of A. L. Moseley, Calhoun, Ky.

Fancy Bee 37496.—Sire, Fancy's Harry 9777; dam, Beeswax 9807.

Butter, 15 lbs. 8 oz.   Milk, 116 lbs. 8 oz.

Test made from Aug. 22 to 29, 1888; age, 3 yrs. 7 mos.; estimated weight, 900 lbs.; fed, daily, 16 qts. of ground corn and oats; property of Maury Jersey Farm, Columbia, Tenn.

Fancy Juno 6086.—Sire, Red Cloud 2d 2260; dam, Fancy Fair 2858.

Butter, 15 lbs. 10 oz.

Test made from Nov. 29 to Dec. 6, 1883; age, 6 yrs. 5 mos.; property of R. S. Strader, Lexington, Ky.

Fancy Même 79748.—Sire, Fancy's Harry 6th 24768; dam, Même 34921.

Butter, 16 lbs. 12 oz.   Milk, 204 lbs. 12 oz.

Test made from Dec. 15 to 22, 1894; age, 2 yrs. 10 mos.; actual weight, 800 lbs.; fed 56 qts. wheat bran, 28 qts. ground oats and 28 qts. corn meal—clover hay *ad lib.*; property of Henry DuBois, Vigo, Ohio.

Fancy Millicent 71059.—Sire, Fancy's Landseer 12996; dam, Alaric's Millicent 2d 63588.

Butter, 14 lbs. 3½ oz.   Milk, 178 lbs. 8 oz.

Test made from Jan. 7 to 14, 1896; age, 5 yrs. 1 mo.; estimated weight, 850 lbs.; fed 112 qts. ground corn and oats, in equal quantities—hay; property of Mayes & Lipscomb, Ashwood, Tenn.

Fancy Nenita 62702.—Sire, Fancy's Harry 9777; dam, Creole Maid 2d 31881.

Butter, 17 lbs. 6½ oz.   Milk, 231 lbs. 13.4 oz.

Test made from Apr. 4 to 11, 1897; age, 7 yrs. 6 mos.; estimated weight, 850 lbs.; fed 7 lbs. bran, 7 lbs. corn and oats, 2 lbs. cotton-seed meal, 2 lbs. linseed meal and 24 lbs. corn ensilage, daily—hay *ad lib.*; property of John E. Robbins, Greensburg, Ind.

Fancy Ona 34867.—Sire, Pedro of the Valley 8750; dam, Emma Hudson 12469.

Butter, 15 lbs. ½ oz.   Milk, 224 lbs. 12 oz.

Test made from Nov. 13 to 20, 1888; age, 3 yrs. 8 mos.; estimated weight, 800 lbs.; fed, daily, between 16 lbs. and 22 lbs. ground corn and oats; property of Wm. Morrow & Son, Nashville, Tenn.

Fancy Phlox 49652.—Sire, Fancy's Harry 9777; dam, Phlox 3d 31882.

Butter, 14 lbs. 3 oz.   Milk, 166 lbs. 9 oz.

Test made from Oct. 28 to Nov. 4, 1891; age, 3 yrs. 11 mos.; estimated weight, 1000 lbs.; fed 8 qts. corn-hearts, 6 qts. ground oats, 6 qts. bran, 2 qts. cotton-seed meal and 2 qts. linseed meal, daily—pea and clover hay *ad lib.*, frost-bitten pasture; property of J. L. Thomas, Knoxville, Tenn.

Fancy's Gem 41797.—Sire, Fancy's Landseer 12996; dam, Dovie of Linwood 6857.

Butter, 14 lbs. 15 oz.   Milk, 171 lbs. 4 oz.

Test made from Feb. 18 to 25, 1894; age, 8 yrs.; estimated weight, 900 lbs.; fed 56 qts. ground oats and 56 qts. corn meal—hay; property of Morgan & Brown, Columbia, Tenn.

Fancy's Gem 2d 86342.—Sire, Landseer's Pogis 15847; dam, Fancy's Gem 41797.

Butter, 16 lbs. 4 oz.   Milk, 215 lbs. 8 oz.

Test made from May 15 to 22, 1896; age, 6 yrs. 4 mos.; estimated weight, 800 lbs.; fed 56 qts. ground oats and 56 qts. corn meal—good woodland pasture; property of Morgan & Brown, Columbia, Tenn.

Fancy's Gem 4th 86348.—Sire, Landseer's Pogis 15847; dam, Fancy's Gem 41797.

Butter, 14 lbs.   Milk, 191 lbs. 4 oz.

Test made from Aug. 6 to 13, 1894; age, 2 yrs. 9 mos.; estimated weight, 650 lbs.; fed 56 qts. corn meal and 56 qts. ground oats—badly burned pasture; property of C. W. Howard, Vernon, Tex.

Fancy's Harry's Sadie 77202.—Sire, Fancy's Harry 9777; dam, Sadie Tansil 36696.

Butter, 14 lbs. 14 oz.   Milk, 135 lbs., estimated.

Test made from Aug. 12 to 19, 1895; age, 5 yrs. 9 mos.; estimated weight, 850 lbs.; fed bran, corn meal and cut green corn, not weighed—dry pasture; property of Hooper & Christy, Murfreesborough, Tenn.

Fancy's Maybee 104383.—Sire, Fancy Ethleel's Pogis 30025; dam, Maybee H. 60620.

Butter, 15 lbs. 14 oz. Milk, 215 lbs. 8 oz.

Test made from Feb. 22 to Mar. 1, 1898; age, 3 yrs. 7 mos.; estimated weight, 750 lbs.; fed 10 lbs. of a mixture of corn meal, ground oats, pea meal and bran and 25 lbs. ensilage, daily—hay *ad lib.*; property of George W. Sisson, Jr., Potsdam, N. Y.

Fancy's Moyane 122270.—Sire, Oonan's Pogis 17165; dam, Moyane's Fancy 78452.

Butter, 15 lbs. ½ oz. Milk, 195 lbs. 10¾ oz.

Test made from Jan. 22 to 29, 1898; age, 2 yrs.; estimated weight, 600 lbs.; fed 18 lbs. per day of ground oats and corn, mixed with 4 bucketfuls cut sheaf oats, wetted; property of Matt. M. Gardner, Nashville, Tenn.

Fancy's Nienchen 104377.—Sire, Fancy Ethleel's Pogis 30025; dam, Nienchen 48698.

Butter, 16 lbs. 2 oz. Milk, 205 lbs.

Test made from June 12 to 19, 1898; age, 4 yrs. 4 mos.; estimated weight, 750 lbs.; fed 12 lbs. daily of a mixture of corn, oats, bran and pea meal—natural pasture; property of George W. Sisson, Jr., Potsdam, N. Y.

Fancy's Pet 36013.—Sire, Fancy's Landseer 12996; dam, Dovie of Linwood 6857.

Butter, 16 lbs. 4 oz. Milk, 165 lbs.

Test made from Sept. 3 to 10, 1893; age, 9 yrs. 4 mos.; estimated weight, 900 lbs.; fed 56 qts. ground oats and 56 qts. corn meal—little hay, short pasture; property of Morgan & Brown, Columbia, Tenn.

Fancy's Pet 2d 85912.—Sire, Landseer's Pogis 15847; dam, Fancy's Pet 36013.

Butter, 14 lbs. 1 oz. Milk, 173 lbs. 12 oz.

Test made from Aug. 20 to 27, 1893; age, 2 yrs. 8 mos.; estimated weight, 625 lbs.; fed 56 qts. ground oats and 56 qts. corn meal—little hay and short pasture; property of Morgan & Brown, Columbia, Tenn.

Fancy Torment 56273.—Sire, Fancy's Landseer 12996; dam, Tormentor's Favorite 36873.

Butter, 14 lbs. 8½ oz. Milk, 139 lbs. 4 oz.

Test made from Apr. 28 to May 5, 1896; age, 6 yrs. 11 mos.; estimated weight, 1100 lbs.; fed 70 qts. ground oats and 70 qts. ground corn—millet hay, woodland pasture; property of E. H. Hatcher, Columbia, Tenn.

Fancy Wax 37159.—Sire, Fancy's Harry 9777; dam, Beeswax 9807.

Butter, 19 lbs. 3½ oz. Milk, 116 lbs. 7 oz.

Test made from Dec. 3 to 10, 1888; age, 2 yrs. 11 mos.; estimated weight, 700 lbs.; fed, daily, 4 gals. ground oats and corn, in equal parts; property of Morgan & Brown, Columbia, Tenn.

Fancy Wax 2d 83357.—Sire, Landseer's Pogis 15847; dam, Fancy Wax 37159.

Butter, 14 lbs. 5½ oz. Milk, 158 lbs. 12 oz.

Test made from Oct. 12 to 19, 1893; age, 3 yrs. 2 mos.; estimated weight, 800 lbs.; fed 70 qts. ground oats and 70 qts. corn meal—little hay, fair pasture; property of Morgan & Brown, Columbia, Tenn.

Fancy Wax 3d 86345.—Sire, Landseer's Pogis 15847; dam, Fancy Wax 37159.

Butter, 14 lbs. 1 oz. Milk, 169 lbs.

Test made from June 5 to 12, 1894; age, 2 yrs. 10 mos.; estimated weight, 650 lbs.; fed 56 qts. ground oats and 56 qts. ground corn—woodland pasture; property of Morgan & Brown, Columbia, Tenn.

Fancy Wax 3d 86345.—Sire, Landseer's Pogis 15847; dam, Fancy Wax 37159.

Butter, 20 lbs. 3¾ oz. Milk, 206 lbs. 11½ oz.

Test made from Apr. 28 to May 5, 1896; age, 4 yrs. 9 mos.; estimated weight, 775 lbs.; fed 19 lbs. per day of ground oats and corn, in equal parts, mixed with 4 bucketfuls cut sheaf oats—extra good orchard and blue-grass pasture; property of Matt. M. Gardner, Nashville, Tenn.

Fancy Work 116422.—Sire, Ida's Rioter of St. L. 13656; dam, High Tea 65577.

Butter, 18 lbs. 2 oz. Milk, 168 lbs.

Test made from Dec. 21 to 28, 1897; age, 4 yrs. 1 mo.; actual weight, 1050 lbs.; fed 21 qts. wheat bran, 63 qts. ground oats, 21 qts. oat meal, 42 qts. corn meal, 21 qts. oil-cake, 21 qts. malt and 140 lbs. sugar beets; property of Miller & Sibley, Franklin, Pa.

Fandango 12908.—Sire, Top-Sawyer 1404; dam, Fannette 3070.

Butter, 18 lbs. 3 oz. Milk, 260 lbs. 3 oz.

Test made from June 12 to 18, 1886; age, 5 yrs. 4 mos.; property of Jordan Stokes, Jr., Nashville, Tenn.

Fanella 2d 64578.—Sire, Young Pedro 9033; dam, Fanella 8006.

Butter, 15 lbs. 6 oz. Milk, 235 lbs. 11 oz.

Test made from July 31 to Aug. 7, 1891; age, 4 yrs. 4 mos.; estimated weight, 675 lbs.; fed 6 qts. corn meal, 6 qts. ground oats, 6 qts. wheat middlings and 1 pt. oil meal, daily—good pasture; property of D. F. Appleton, Ipswich, Mass.

Fanfare 49927.—Sire, Daisy's Signal 12954; dam, Fandango 12908.

Butter, 16 lbs. Milk, 225 lbs. 10½ oz.

Test made from June 7 to 14, 1891; age, 3 yrs. 4 mos.; actual weight, 775 lbs.; fed 24 lbs. per day of ground oats and corn in equal parts—extra good orchard and blue-grass pasture; property of Matt. M. Gardner, Nashville, Tenn.

Fanfare 2d 119780.—Sire, Fancy's Harry 2d 20199; dam, Fanfare 49927.

Butter, 15 lbs. 12½ oz. Milk, 229 lbs. 15¾ oz.

Test made from Nov. 28 to Dec. 5, 1897; age, 4 yrs. 8 mos.; estimated weight, 725 lbs.; fed 20 lbs. per day of ground oats and corn, mixed with 4 bucketfuls cut sheaf oats, wetted—millet hay *ad lib.*, little pasture; property of Matt. M. Gardner, Nashville, Tenn.

Fannie Bugler 19962.—Sire, Bugler 2127; dam, Metah's Queen 4886.

Butter, 15 lbs. 2 oz. Milk, 162 lbs.

Test made from June 13 to 19, 1885; age, 4 yrs. 1 mo.; property of N. N. Palmer, Brodhead, Wis.

Fannie Landseer 1969.—Sire, Landseer 331; dam, Sylph 615.

Butter, 15 lbs. 9 oz. Milk, 97 lbs. 3 oz.

Test made from Feb. 9 to 16, 1886; age, 14 yrs. 5 mos.; property of the Columbia Jersey Cattle Co., Columbia, Tenn.

Fannie's Fairy 67304.—Sire, Gold of St. Lambert 16744; dam, Fannie Bugler 19962.

Butter, 19 lbs. 8 oz.   Milk, 254 lbs. 8 oz.

Test made from Apr. 18 to 25, 1896; age, 6 yrs. 5 mos.; estimated weight, 1000 lbs.; fed 56 lbs. ground oats, 56 lbs. wheat bran, 14 lbs. old process oil meal and 230 lbs. ensilage—hay and corn stover *ad lib.*; property of N. N. Palmer & Son, Brodhead, Wis.

Fanny Mason 45992.—Sire, Chendado 12955; dam, Esther Lee 38131.

Butter, 18 lbs. 5¼ oz.   Milk, 209 lbs.

· Test made from Nov. 17 to 24, 1895; age, 8 yrs. 7 mos.; estimated weight, 1000 lbs.; fed 4 qts. corn meal, 4 qts. wheat bran, 4 qts. ground oats and 2 qts. cotton-seed meal, daily—hay and cut sorghum at night, blue-grass and winter oats pasture; property of D. P. Carter, Lynnville, Tenn.

Fanny of Cream Brook 13930.—Sire, Waxy 2116; dam, Fanny Micawber 1804.

Butter, 17 lbs. 8 oz.   Milk, 259 lbs. 8 oz.

Test made from Jan. 4 to 11, 1886; age, 5 yrs. 7 mos.; property of E. Kent & Son, New Market, N. H.

Fanny of Yerba Buena 2d 10689.—Sire, Victor of Yerba Buena 3809; dam, Fanny of Yerba Buena 8151.

Butter, 14 lbs. 14 oz.   Milk, 244 lbs. 8 oz.

Test made from Apr. 27 to May 4, 1889; age, 9 yrs. 2 mos.; estimated weight, 900 lbs.; fed, daily, 6 qts. bran, 2 qts. old process oil meal and 3 qts. corn meal; property of E. J. Packard, Santa Barbara, Cal.

Fanny Taylor 6714.—Sire, Signal 1170; dam, Bonanza 3965.

Butter, 15 lbs. 12 oz., unsalted.   Milk, 223 lbs.

Test made from Dec. 28 to Jan. 3, 1883; age, 5 yrs. 8 mos.; property of John A. Middelton, Shelbyville, Ky.

Fan of Grouville 7458.—Sire, Sans-Peur (F. S. 201 J. H. B.); dam, Brunette (F. S. 1256 J. H. B.)

Butter, 15 lbs., unsalted.   Milk, 263 lbs. 12 oz.

Test made from June 11 to 17, 1883; age, 8 yrs. 4 mos.; fed 9 qts. ground oats and bran, in equal parts, daily—good clover and timothy pasture; property of Beech Grove Farm, Beech Grove, Ind.

Fan's Grouville Beauty 10079.—Sire, Le Brocq's Prize 3350; dam, Fan of Grouville 7458.

Butter, 19 lbs. 3 oz.   Milk, 234 lbs. 6 oz.

Test made from June 24 to July 1, 1885; age, 5 yrs. 3 mos.; property of George M. Jewett, Glenville, Md.

Fantine 1271.—Sire, Rajah 340; dam, Fancy 9.

Butter, 16 lbs. 6 oz.   Milk, 252 lbs. 4 oz.

Age, 11 yrs. 5 mos.; fed corn meal, ground oats, bran and pumpkins—very little pasture; property of H. M. Howe, Bristol, R. I.

Fantine of Guilford 64480.—Sire, Morgan Victor Hugo 15330; dam, Lucy Alba 22929.

Butter, 19 lbs. 1 oz. Milk, 266 lbs. 3 oz.

Test made from Oct. 2 to 9, 1895; age, 6 yrs. 1 mo.; actual weight, 990 lbs.; fed 35 lbs. corn and cob meal, 35 lbs. bran, 35 lbs. crushed oats and 7 lbs. oil meal—70 lbs. oat straw, short pasture; property of B. F. Carper, Rosemond, Ill.

Farmer's Pride 12284.—Sire, Farmer's Glory 5196; dam, Startled Fawn 7837.

Butter, 15 lbs. 4 oz. Milk, 252 lbs.

Test made from June 11 to 18, 1886; age, 5 yrs. 2 mos.; estimated weight, 1050 lbs.; fed 6 qts. shorts, daily; property of T. R. Proctor, Utica, N. Y.

Fashion Plate 62457.—Sire, Michael Angelo 10116; dam, Floribel Pogis 28759.

Butter, 15 lbs. 10½ oz. Milk, 268 lbs. 12 oz.

Test made from May 11 to 18, 1892; age, 4 yrs. 1 mo.; actual weight, 1000 lbs.; fed 12 qts. ground oats, 6 qts. corn meal and 3 qts. oil-cake, daily—good clover and timothy pasture; property of Miller & Sibley, Franklin, Pa.

Fauchon 2d 18200.—Sire, Pilot (P. S. 183 J. H. B.); dam, Fauchon 22001.

Butter, 14 lbs. 9 oz. Milk, 234 lbs. 9 oz.

Test made from Mar. 5 to 12, 1880; age, 6 yrs. 10 mos.; property of J. H. Walker, Worcester, Mass.

Faultless 12018.—Sire, Pasha (P. S. 64 J. H. B.); dam, Cérise (F. S. 737 J. H. B.)

Butter, 17 lbs. 5½ oz. Milk, 234 lbs. 13 oz.

Test made from July 14 to 20, 1883; age, 8 yrs. 6 mos.; property of William Simpson, New York, N. Y.

Faustine 10354.—Sire, Mopsus 1165; dam, Antianira 2457.

Butter, 14 lbs. 14½ oz. Milk, 225 lbs. 9 oz.

Test made from Aug. 6 to 12, 1883; age, 5 yrs. 5 mos.; property of William Simpson, New York, N. Y.

Favorite of Avon 13438.—Sire, Grey King (P. S. 169 J. H. B.); dam, Camelia (F. S. 1687 J. H. B.)

Butter, 17 lbs. 7¾ oz. Milk, 287 lbs. 4 oz.

Test made from May 31 to June 6, 1885; age, 8 yrs.; property of C. E. Rowley, Cortland, N. Y.

Favorite of the Elms 1656, imp.

Butter, 16 lbs. 4 oz. Milk, 314 lbs.

Test made from June 5 to 12, 1876; age, 6 yrs.; property of William S. Taylor, Burlington, N. J.

Favorite Rajah Rex 16153.—Sire, Rex 1330; dam, Duchess Corona 2d 8598.

Butter, 15 lbs. 4 oz. Milk, 108 lbs. 4 oz.

Test made from May 1 to 7, 1885; age, 8 yrs.; property of Joseph C. Kiplinger & Co., Springfield, Ohio.

Fawn of Ingleside 59649.—Sire, Duke of Ingleside 14274; dam, Ribbon's Fawn 26907.

Butter, 17 lbs. 9 oz. Milk, 231 lbs.

Test made from June 16 to 23, 1895; age, 6 yrs. 9 mos.; estimated weight, 1050 lbs.; fed 28 lbs. wheat bran, 28 lbs. ground oats and 28 lbs. corn meal—pasture; property of John C. McClintock, Meadville, Pa.

Fawn of St. Lambert 27942.—Sire, Bachelor of St. Lambert 4558; dam, Allie of St. Lambert 24991.

Butter, 15 lbs. 5½ oz. Milk, 353 lbs. 8 oz.

Test made from Mar. 5 to 11, 1886; age, 3 yrs. 9 mos.; property of Miller & Sibley, Franklin, Pa.

Fawn of St. Lambert 27942.—Sire, Bachelor of St. Lambert 4558; dam, Allie of St. Lambert 24991.

Butter, 571 lbs. 6 oz. for one year ending Feb. 25, 1887. Milk, 14535 lbs. 12 oz. for one year ending Feb. 13, 1887.

Age, 3 yrs. 9 mos. at beginning of test; (this butter test was incidental to test for milk yield, the cow being fed with reference to flow of milk); property of Miller & Sibley, Franklin, Pa.

Fawn Tormentor of Lawn 71639.—Sire, Tormentor's John 14715; dam, Daisy Fawn 34854.

Butter, 20 lbs. 2 oz. Milk, 294 lbs. 10 oz.

Test made from Apr. 21 to 28, 1895; age, 5 yrs. 2 mos.; estimated weight, 900 lbs.; fed 84 lbs. wheat bran, 56 lbs. corn meal, 42 lbs. cotton-seed meal and 84 lbs. crushed oats—hay *ad lib.*; property of Platter & Foster, Denison, Tex.

Fay Signal 2d 134018.—Sire, Moyane's Tormentor 21690; dam, Fay Signal 82159.

Butter, 14 lbs. 4 oz. Milk, 201 lbs. 8 oz.

Test made from June 3 to 10, 1898; age, 4 yrs. 9 mos.; estimated weight, 800 lbs.; fed 4 lbs. Quaker oats feed, 8 lbs. hominy meal, 2 lbs. oil meal and 6 lbs. wheat bran, daily—blue-grass pasture; property of Newton Frazier, Clark, Ky.

Fear Not 6059.—Sire, Sans-Peur (F. S. 201 J. H. B.); dam, Lady of the Isles (F. S. 992 J. H. B.)

Butter, 17 lbs. 3 oz. Milk, 255 lbs.

Test made from Aug. 6 to 12, 1882; age, 7 yrs.; fed 4 lbs. corn meal and 6 lbs. wheat middlings, daily—pasture; property of Woodside Farms Herd, Troy, N. Y.

Fear Not 2d 6061.—Sire, Bobby, on I. of J.; dam, Fear Not 6059.

Butter, 16 lbs. 2 oz. Milk, 216 lbs.

Test made from June 3 to 9, 1882; age, 5 yrs.; fed 4 lbs. corn meal and 6 lbs. wheat middlings, daily—pasture; property of Woodside Farms Herd, Troy, N. Y.

Federa of Greenwood 46892.—Sire, Kinsman of Greenwood 12881; dam, Fidette of Woodruff 40536.

Butter, 20 lbs. 7 oz. Milk, 280 lbs.

Test made from July 12 to 19, 1890; age, 3 yrs. 11 mos.; estimated weight, 700 lbs.; fed, daily, 7 lbs. ground oats, 5 lbs. corn meal and 1 lb. oil meal; property of James T. Henderson, Auvergne, Ark.

Felsticea 17083.—Sire, Squire Deerfoot 7835; dam, Olie's Lady Teazle 12307.

Butter, 16 lbs. 3 oz. Milk, 245 lbs. 14 oz.

Test made from May 25 to June 1, 1886; age, 3 yrs. 10 mos.; fed 76½ lbs. oats and barley, in equal parts, ground together, 54 lbs. corn meal and 8 lbs. wheat bran—short prairie grass; property of J. J. Gardner, Sibley, Iowa.

Fena Eric 69137.—Sire, R. F. Eric 21330; dam, Champion G.'s Duchess 40768.

Butter, 22 lbs. 10 oz. Milk, 323 lbs. 8 oz.

Test made from Mar. 12 to 19, 1896; age, 5 yrs. 3 mos.; estimated weight, 950 lbs.; fed 56 lbs. wheat bran, 70 lbs. corn and cob meal, 70 lbs. ground oats and 28 lbs. cotton-seed meal—prairie hay *ad lib.*, also corn fodder and turnip tops; property of B. L. Gill, Terrell, Tex.

Fenella of Bolton 2d 116768.—Sire, Tennessee of Bolton 27508; dam, Fenella of Bolton 49572.

Butter, 14 lbs. Milk, 198 lbs. 12 oz.

Test made from June 10 to 17, 1898; age, 3 yrs. 8 mos.; estimated weight, 850 lbs.; fed 3 qts. bran, 1½ qts. ground oats and 1½ qts. meal, daily—green clover fodder, good pasture; property of J. A. Cunningham, Bolton, Mass.

Fennel of Glynllyn 109755.—Sire, Pogis of Shadow Brook 27926; dam, Puin of Eau Claire 71261.

Butter, 17 lbs. 4 oz. Milk, 226 lbs. 5 oz.

Test made from Sept. 19 to 26, 1897; age, 2 yrs. 6 mos.; estimated weight, 700 lbs.; fed 63 qts. oats, 63 qts. bran and 21 qts. corn meal—green fodder corn and clover pasture; property of D. Wolfe Bishop, Lenox, Mass.

Fermentor 2d's Pet 114503.—Sire, Tancreda's Signal King 35742; dam, Fermentor 2d 74267.

Butter, 17 lbs. 8 oz. Milk, 183 lbs. 1 oz.

Test made from Feb. 28 to Mar. 7, 1898; age, 3 yrs. 8 mos.; estimated weight, 850 lbs.; fed 112 qts. of a mixture of corn and cob meal, ground wheat and ground oats, in equal parts by measure, and 14 qts. oil-cake meal—clover hay *ad lib.*; property of H. C. Schlegel, Daleville, Ind.

Fernside 113278.—Sire, Asterisk's Stoke Pogis 29182; dam, Mahairve Girl 25652.

Butter, 15 lbs. 3 oz. Milk, 226 lbs. 5 oz.

Test made from Feb. 23 to Mar. 2, 1898; age, 3 yrs. 5 mos.; estimated weight, 800 lbs.; fed 3 2-7 qts. corn meal, 4 2-7 qts. wheat bran, 6 qts. ground oats, ½ qt. cotton-seed meal, ½ qt. oil meal, 6 qts. beets, 4½ qts. carrots, 5 to 9 qts. apples and ½ bush. ensilage, daily—hay *ad lib.*; property of C. N. Gilbert, Great Barrington, Mass.

Fidelia 5817, imp.

Butter, 14 lbs.

Test made in 1881; test made on grass alone; property of Charles L. Sharpless, Philadelphia, Pa.

Fides 2d 1576.—Sire, Hannibal 618; dam, Fides 51.

Butter, 14 lbs. 8 oz.

Test made May, 1879; age, 7 yrs. 11 mos.; property of John E. Swallow, Hagerstown, Md.

Fillpail 16530.—Sire, Intrepid 5511; dam, Faultless 12018.

Butter, 15 lbs. 11 oz. Milk, 183 lbs. 13 oz.

Test made from Apr. 20 to 26, 1884; age, 2 yrs. 5 mos.; property of William Simpson, New York, N. Y.

Fill Pail 2d 24388.—Sire, King (P. S. 238 J. H. B.); dam, Fill Pail 24341.

Butter, 26 lbs. 2 oz. Milk, 207 lbs. 14 oz.

Test made from Apr. 15 to 21, 1885; age, 4 yrs. 2 mos.; fed on pea meal, ground oats, linseed cake, wheat bran, clover hay, beets and carrots; property of Mrs. A. C. E. Shoemaker, Stevenson, Md.

Fill Pail's Countess 24462.—Sire, Count St. George 8403; dam, Fill Pail 2d 24388.

Butter, 22 lbs. 8 oz. Milk, 223 lbs. 8 oz.

Test made from May 30 to June 5, 1887; age, 4 yrs.; fed 4 qts. bran, 4 qts. corn and oats and 1 qt. pea meal, daily; property of M. Erskine Miller, Staunton, Va.

Fill Pail's Nancy 95435.—Sire, Fill Pail's Duke 24650; dam, Nancy's Star 2d 70240.

Butter, 18 lbs. 3 oz. Milk, 238 lbs.

Test made from May 16 to 23, 1897; age, 4 yrs. 3 mos.; estimated weight, 750 lbs.; fed 52½ lbs. wheat bran, 45½ lbs. corn meal, 7 lbs. linseed oil-cake meal, 10½ lbs. cotton-seed meal and 140 lbs. ensilage—blue-grass pasture; property of Samuel McKeen, Terre Haute, Ind.

Fireside 75194.—Sire, June Pogis 19872; dam, Friendship 62460.

Butter, 17 lbs. 2½ oz. Milk, 250 lbs. 4 oz.

Test made from July 26 to Aug. 2, 1894; age, 3 yrs. 4 mos.; estimated weight, 900 lbs.; fed 12 qts. ground corn and oats, 12 qts. bran and 3 qts. oil meal, daily—green oats and peas, fair pasture during night; property of C. A. Sweet, East Aurora, N. Y.

First Prize 95323.—Sire, Spokane 23440; dam, Prize Classic 74900.

Butter, 15 lbs. 4 oz. Milk, 249 lbs. 12 oz.

Test made from Mar. 17 to 24, 1897; age, 4 yrs. 9 mos.; estimated weight, 850 lbs.; fed 210 lbs. ensilage, 126 lbs. corn meal and 35 lbs. hay—blue-grass pasture; property of Baum & Hill, Frankfort, Ind.

Flashy Jessie 9722.—Sire, Flash 2532; dam, Jessie 2d 1301.

Butter, 17 lbs. 15 oz.

Test made Mar., 1885; age, 7 yrs. 2 mos.; property of S. W. Robbins, Wethersfield, Conn.

Fleece 18568.—Sire, The Hub 1009; dam, Busy Bee 6336.

Butter, 17 lbs. 13½ oz. Milk, 284 lbs. 8 oz.

Test made from May 19 to 25, 1885; age, 5 yrs. 8 mos.; property of William E. Oates, Vicksburg, Miss.

Fleecy Princess 40342.—Sire, Prince of Melrose 4819; dam, Fleece 18568.

Butter, 14 lbs. 3 oz. Milk, 210 lbs. 14 oz.

Test made from Sept. 13 to 20, 1891; age, 6 yrs. 5 mos.; estimated weight, 900 lbs.; fed 18 lbs. corn meal, 7½ lbs. wheat bran and 2 lbs. cotton-seed meal, daily—short Japan clover and Bermuda grass; property of M. Lothrop, Marshall, Tex.

Fleecy Princess 40342.—Sire, Prince of Melrose 4819; dam, Fleece 18568.

Butter, 16 lbs. 11½ oz. Milk, 243 lbs. 13 oz.

Test made from Feb. 22 to Mar. 1, 1893; age, 7 yrs. 10 mos.; estimated weight, 800 lbs.; fed 3 qts. corn meal, 1½ gals. bran, 3 qts. corn-cob meal, 1 qt. cotton-seed meal, 15 lbs. ensilage and 5 lbs. hay, daily; property of M. Lothrop, Marshall, Tex.

Fleeter 61921.—Sire, Prince of Melrose 4819; dam, Fleetling 18576.

Butter, 16 lbs. 13½ oz. Milk, 234 lbs. 15½ oz.

Test made from June 4 to 11, 1892; age, 4 yrs. 4 mos.; estimated weight, 750 lbs.; fed 42 lbs. bran, 14 lbs. cotton-seed meal, 14 lbs. cotton-seed, 49 lbs. ground corn, oats and peas—good pasture of Bermuda grass and clover; property of M. Lothrop, Marshall, Tex.

Fleurette of Linwood 12918.—Sire, Top-Sawyer 1404; dam, Fannie G. 3829.

Butter, 16 lbs. Milk, 151 lbs. 12 oz.

Test made from June 8 to 15, 1885; age, 4 yrs.; property of M. M. Gardner, Nashville, Tenn.

Flight of Willow Farm 24783.—Sire, O'Malley 6441; dam, Floribel 2460.

Butter, 20 lbs. 12 oz. Milk, 199 lbs.

Test made from May 27 to June 2, 1886; age, 5 yrs. 1 mo.; fed 12 qts. meal and shorts, mixed in equal proportions—good pasture; property of Winslow S. Lincoln, Worcester, Mass.

Flora 113, imp.

Butter, 511 lbs. 2 oz. in 50 weeks.

Test made May 16, 1853, to Apr. 30, 1854; age, 4 yrs.; property of Thomas Motley, Jamaica Plain, Mass.

Flora B. 12446.—Sire, Uncle Toby Brown 5814; dam, Diana Franklin 12445.

Butter, 14 lbs. 2 oz. Milk, 201 lbs. 13 oz.

Test made from May 17 to 23, 1885; age, 4 yrs. 10 mos.; property of Webster & Weaver, Columbia, Tenn.

Flora Lee 13294.—Sire, Samson Jr. 2723; dam, Fancy of Bovina 8055.

Butter, 14 lbs. 1 oz. Milk, 199 lbs. 12 oz.

Test made from Apr. 20 to 26, 1884; age, 4 yrs. 2 mos.; property of Andrew Baker, West Dryden, N. Y.

Flora Lee of Tennessee 7694.—Sire, Dande 1556; dam, Nellie 3d 1928.

Butter, 16 lbs. 5 oz. Milk, 334 lbs.

Test made from July 15 to 22, 1885; age, 9 yrs. 9 mos.; property of S. A. Rodgers, Loudon, Tenn.

Flora of St. Peter's 8622.—Sire, Prince, on I. of J.; dam, Daisy, on I. of J.

Butter, 16 lbs. 5 oz., unsalted.

Test made in June, 1879; age, 3 yrs. 5 mos.; fed all the lucern she would eat—pasture at night; property of William Crozier, Northport, N. Y.

Flora Pansy 2d 78678.—Sire, Tormentor Jr. 15005; dam, Flora Pansy 19935.

Butter, 15 lbs. 10.8 oz. Milk, 215 lbs. 6.4 oz.

Test made from Dec. 3 to 10, 1895; age, 4 yrs. 7 mos.; estimated weight, 850 lbs.; fed 98 lbs. bran, 21 lbs. cotton-seed meal and 140 lbs. beets—hay *ad lib.*; property of Hecla Jersey Cattle Co., Earlington, Ky.

Flora Rex Carlo 2d 102140.—Sire, Fancy Farmer 12181; dam, Flora Rex Carlo 94407.

Butter, 18 lbs. 2 oz. Milk, 236 lbs.

Test made from Mar. 18 to 25, 1898; age, 5 yrs. 4 mos.; estimated weight, 775 lbs.; fed about 18 qts. daily of a mixture of ground oats, middlings, wheat bran, corn meal and some pea meal—clover hay and ensilage; property of T. S. Cooper, Coopersburg, Pa.

Flora Temple 3d 40086.—Sire, Color Blind 16861; dam, Flora Temple 3768.

Butter, 55.058 lbs. in 30 days. Milk, 923.6 lbs.

Test made from Aug. 29 to Sept. 27, 1893, at World's Columbian Exposition; age, 7 yrs. 3 mos.; actual weight, 1079 lbs.; fed 372.2 lbs. hay, 75 lbs. oil meal, 116 lbs. ensilage, 20 lbs. corn meal, 180 lbs. bran, 90 lbs. oats, 189 lbs. corn-hearts, 45 lbs. cotton-seed, 90 lbs. middlings, 18 lbs. carrots and 5 lbs. old hay; property of Frederic Bronson, Southport, Conn.

Flora Temple 3d 40086.—Sire, Color Blind 16861; dam, Flora Temple 3768.

Butter, 176.751 lbs. in 90 days. Milk, 3038.2 lbs.

Test made from May 31 to Aug. 28, 1893, at World's Columbian Exposition; age, 7 yrs. at beginning of test; actual weight, 1058 lbs.; fed 319.7 lbs. old hay, 210 lbs. ensilage, 556.9 lbs. new hay, 187 lbs. oil meal, 538 lbs. bran, 74 lbs. oats, 567 lbs. corn-hearts, 128.50 lbs. cotton-seed, 456 lbs. middlings, 34 lbs. grano-gluten, 1105.4 lbs. clover and 10 lbs. swale-grass; property of Frederic Bronson, Southport, Conn.

Florence Billot 7849.—Sire, Duke (P. S. 76 J. H. B.); dam on I. of J.

Butter, 14 lbs. 13 oz. Milk, 205 lbs. 6 oz.

Test made from Jan. 27 to Feb. 2, 1884; age, 7 yrs. 11 mos.; property of John P. Allen, Jr., Shelbyville, Ky.

Florence of St. Lambert 95192.—Sire, Meridale Pauline Pogis 25133; dam, Polly Bonair 42849.

Butter, 16 lbs. 12 oz. Milk, 190 lbs. 8 oz.

Test made from Mar. 17 to 24, 1897; age, 3 yrs. 5 mos.; estimated weight, 950 lbs.; fed 169 lbs. ensilage (made from corn-stalks without ears), 84 lbs. corn chop, 84 lbs. wheat bran, 7 lbs. old process linseed meal and 17½ lbs. hay; property of B. F. Fackenthal, Jr., Riegelsville, Pa.

Floret 9959.—Sire, Superb 1956; dam, Floralia 6230.

Butter, 17 lbs. 6 oz. Milk, 255 lbs. 8 oz.

Test made from Oct. 8 to 14, 1885; age, 6 yrs.; property of William R. Goodspeed, Goodspeed's Landing, Conn.

Floribundus 2d 14949.—Sire, Victor (P. S. 148 J. H. B.); dam, Floribundus (F. S. 659 J. H. B.)

Butter, 18 lbs. 8 oz.

Test made from June 24 to July 1, 1883; age, 4 yrs. 4 mos.; property of L. L. Tozier, Batavia, N. Y.

Florine Value 2d 80910.—Sire, Jersey King 9458; dam, Florine Value 58278.

Butter, 15 lbs. 1½ oz. Milk, 248 lbs.

Test made from Dec. 9 to 16, 1895; age, 3 yrs. 7 mos.; estimated weight, 650 lbs.; fed 42 lbs. corn meal, 42 lbs. wheat bran, 14 lbs. ground oats and 245 lbs. corn ensilage; property of E. B. C. Hambley, Rockwell, N. C.

Florrie May Baker 10728.—Sire, Thorndale 2582; dam, Fannie Baker 4274.

Butter, 14 lbs. 8 oz. Milk, 207 lbs. 2 oz.

Test made from July 14 to 20, 1885; age, 5 yrs.; property of R. McMichael, Lexington, Ky.

Florry Keep 6556.—Sire, Ralph 957; dam, Orange Flower 2d 3884.

Butter, 14 lbs. 14 oz., unsalted. Milk, 231 lbs. 8 oz.

Test made from May 2 to 8, 1883; property of W. B. Montgomery, Starkville, Miss.

Flory of the Oaks 8141, imp.

Butter, 14 lbs. 7 oz., unsalted.

Test made from June 8 to 15, 1883; age, 6 yrs. 4 mos.; property of W. B. Dinsmore, Staatsburg, N. Y.

Flower of Glen Rouge 17560.—Sire, Lord McDuff 5147; dam, Clematis of St. Lambert 5478.

Butter, 23 lbs. 14¾ oz. Milk, 262 lbs.

Test made from May 4 to 10, 1885; age, 3 yrs. 1 mo.; fed 20 qts. oats and peas (two parts of oats to one of peas) and 1 pk. potatoes, daily—clover hay, no pasture; property of William Rolph, Markham, Ont., Can.

Flower of Glen Rouge 2d 55559.—Sire, Canada's John Bull 3d 17750; dam, Flower of Glen Rouge 17560.

Butter, 18 lbs. 13½ oz. Milk, 257 lbs. 4 oz.

Test made from Jan. 13 to 20, 1895; age, 6 yrs. 11 mos.; estimated weight, 900 lbs.; fed 35 qts. corn meal, 35 qts. bran, 21 qts. old process oil meal, 35 qts. ground oats, 14 qts. cotton-seed, 56 qts. roots and 105 lbs. hay; property of P. J. Cogswell, Rochester, N. Y.

Flower of Meridale 64537.—Sire, Ida of St. Lambert's Bull 19169; dam, Flower of Glen Rouge 17560.

Butter, 14 lbs. 1 oz. Milk, 224 lbs. 2 oz.

Test made from Dec. 30, 1892, to Jan. 6, 1893; age, 3 yrs. 1 mo.; estimated weight, 800 lbs.; fed 14 lbs. oil meal, 42 lbs. pea meal, 80 lbs. corn, oats and shorts—some hay; property of Ayer & McKinney, Philadelphia, Pa.

Flower of Meridale 64537.—Sire, Ida of St. Lambert's Bull 19169; dam, Flower of Glen Rouge 17560.

Butter, 26 lbs. 11½ oz. Milk, 356 lbs. 8 oz.

Test made from Nov. 11 to 18, 1896; age, 6 yrs. 11 mos.; actual weight, 1030 lbs.; fed 12 lbs. corn meal, 8 lbs. bran, 5 lbs. ground oats and 2 lbs. oil meal, daily—hay, best rowen pasture; property of J. Gerow Dutcher, Pawling, N. Y.

Foolish 64069.—Sire, Romp's Tormentor 7126; dam, Princess of Mansfield 8070.

Butter, 17 lbs. ½ oz.   Milk, 306 lbs. 8 oz.

Test made from Oct. 18 to 25, 1893; age, 8 yrs. 2 mos.; estimated weight, 800 lbs.; fed 42 lbs. grain—green corn fodder and white clover pasture; property of Mrs. J. R. Sherman, Granville, N. Y.

Forest Queen 12229.—Sire, Champion of America 1567; dam, Glenn Forest Queen 4809.

Butter, 16 lbs. 15 oz.   Milk, 210 lbs.

Test made from Aug. 15 to 21, 1886; age, 5 yrs. 10 mos.; fed a mixture of corn meal, bran and cooked cotton-seed meal; property of W. B. Montgomery, Starkville, Miss.

Forget-me-not 5809.—Sire, Duke (P. S. 76 J. H. B.); dam, Trinitaise (F. S. 1343 J. H. B.)

Butter, 15 lbs. 8 oz.

Test made from June 1 to 8, 1882; age, 6 yrs. 1 mo.; property of B. F. & H. A. Harrington, Worcester, Mass.

Forget Me Not O 10564.—Sire, Compeer 2367; dam, Olive Branch 5324.

Butter, 15 lbs. 4 oz.   Milk, 220 lbs. 8 oz.

Test made from June 21 to 27, 1883; age, 3 yrs. 8 mos.; property of G. H. & H. A. Grimmell, Jefferson, Iowa.

Forsaken 7520.—Sire, Cockade 1979; dam, Forget-me-not 5809.

Butter, 15 lbs. 1 oz.

Test made from Apr. 9 to 15, 1882; property of Colin Cameron, Brickerville, Pa.

Founder's Pet 94107.—Sire, Founder 20928; dam, Epicure 33345.

Butter, 17 lbs. 8 oz.   Milk, 292 lbs. 2 oz.

Test made from May 13 to 20, 1896; age, 4 yrs. 2 mos.; estimated weight, 850 lbs.; fed 35 lbs. bran, 21 lbs. oats, 35 lbs. corn meal and 21 lbs. oil meal—mixed pasture; property of H. C. Taylor, Orfordville, Wis.

Fragrance 4059.—Sire, Albert 44; dam, Morning Glory 2d 1299.

Butter, 15 lbs. 3 oz.

Test made from Feb. 5 to 11, 1883; age, 10 yrs.; property of William Simpson, New York, N. Y.

Fragrance 4th 16509.—Sire, Cornelius 4204; dam, Fragrance 4059.

Butter, 14 lbs. 7 oz.

Test made in June, 1885; age, 3 yrs. 10 mos.; property of S. W. Robbins, Wethersfield, Conn.

Frances C. Magnet 22904.—Sire, Champion Magnet 6480; dam, Floy Le Premier 12468.

Butter, 14 lbs. 13½ oz.   Milk, 199 lbs. 4 oz.

Test made from Dec. 15 to 21, 1885; age, 2 yrs. 5 mos.; property of John Boyd, Elmhurst, Ill.

Francesville 53367.—Sire, Signaldo Alexis 15051; dam, Milkgood 27828.

Butter, 16 lbs. 6 oz. Milk, 232 lbs. 8 oz.

Test made from Dec. 4 to 11, 1896; age, 9 yrs. 6 mos.; estimated weight, 900 lbs.; fed 440 lbs. ensilage, 98 lbs. corn meal and 20 lbs. oil meal—hay *ad lib.*; property of Baum & Hill, Frankfort, Ind.

Francesville 2d 99307.—Sire, Spokane 23440; dam, Francesville 53367.

Butter, 15 lbs. 10 oz. Milk, 278 lbs. 8 oz.

Test made from June 10 to 17, 1897; age, 4 yrs. 6 mos.; estimated weight, 900 lbs.; fed 15 lbs. corn meal, daily—blue-grass pasture; property of Baum & Hill, Frankfort, Ind.

Frankie's Lass 24900.—Sire, Cobden 1871; dam, Frankie 5th 3542.

Butter, 17 lbs. 3½ oz. Milk, 217 lbs. 8 oz.

Test made from June 13 to 20, 1885; age, 6 yrs. 9 mos.; property of Richardson Bros., Davenport, Iowa.

Frankness 62451.—Sire, Stoke Pogis 5th 5987; dam, Yellow Emmie 28756.

Butter, 18 lbs. 4 oz. Milk, 198 lbs.

Test made from May 17 to 24, 1890; age, 2 yrs. 11 mos.; estimated weight, 800 lbs.; fed ordinary rations; property of Miller & Sibley, Franklin, Pa.

Fresh 91786.—Sire, Dalton 20117; dam, Dolly Darlo 32200.

Butter, 17 lbs. 11 oz. Milk, 231 lbs.

Test made from Mar. 10 to 17, 1895; age, 5 yrs. 9 mos.; actual weight, 960 lbs.; fed 42 qts. ground oats, 42 qts. corn meal, 28 qts. ground oil-cake, 28 qts. wheat bran and 120 lbs. cabbage; property of Miller & Sibley, Franklin, Pa.

Friar's Lady 36119.—Sire, Grey Friar 7627; dam, Lady Tip 2d 17799.

Butter, 14 lbs. 2½ oz. Milk, 155 lbs. 8 oz.

Test made from Dec. 24 to 31, 1885; age, 1 yr. 10 mos.; property of W. H. Walrath, Clayton, N. Y.

Fricka 86450.—Sire, Croton's Signal 17614; dam, Archie Signal of Croton 73831.

Butter, 15 lbs. 2½ oz. Milk, 229 lbs.

Test made from Apr. 11 to 18, 1898; age, 5 yrs. 5 mos.; estimated weight, 725 lbs.; fed 52 lbs. corn meal, 52 lbs. crushed oats, 42 lbs. bran and 10 lbs. ground linseed—short blue-grass pasture, some hay morning and evening; property of W. L. Alexander, Canton, Ohio.

Friendship 62460.—Sire, Ida's Rioter of St. L. 13656; dam, Little Pogis 32612.

Butter, 17 lbs. 3 oz. Milk, 276 lbs. 12 oz.

Test made from May 7 to 14, 1891; age, 2 yrs. 9 mos.; actual weight, 760 lbs.; fed 12 lbs. ground oats and 8 lbs. corn meal, daily, also grass—good old meadow pasture; property of Miller & Sibley, Franklin, Pa.

Fringe 16875.—Sire, Forget-me-not 6291; dam on I. of J.

Butter, 15 lbs. Milk, 298 lbs. 14 oz.

Test made from June 10 to 17, 1886; age, 4 yrs. 4 mos.; test made on grass alone; property of Henry M. Wood, Jeffersontown, Ky.

Frisky Elsie 81590.—Sire, Tamaroa Rex 18604; dam, Elfleda of Maple Dale 17554.

Butter, 17 lbs. 1½ oz.   Milk, 238 lbs.

Test made from Feb. 20 to 27, 1897; age, 6 yrs. 4 mos.; estimated weight, 875 lbs.; fed 105 lbs. corn and cob meal, 35 lbs. oats, 56 lbs. wheat bran, 14 lbs. oil meal and 140 lbs. carrots—corn stover and mixed hay *ad lib.*; property of P. M. Klinefelter, Vanderville, Ill.

Frisky Miss 18572.—Sire, Lord Longford 3997; dam, Lady Louisa 8586.

Butter, 14 lbs.   Milk, 194 lbs.

Test made from May 9 to 15, 1885; age, 4 yrs. 10 mos.; property of William E. Oates, Vicksburg, Miss.

Friz Cam 14655.—Sire, Camerlengo 3012; dam, Pansy of Willow Dale 7754.

Butter, 15 lbs. 7 oz.   Milk, 217 lbs.

Test made from Oct. 29 to Nov. 4, 1885; age, 4 yrs. 7 mos.; property of R. McMichael, Lexington, Ky.

Frolic of Chestnutwood 19405.—Sire, Mahkeenac 3290; dam, Arthur's Frolic 4438.

Butter, 16 lbs.   Milk, 251 lbs. 4 oz.

Test made from June 15 to 21, 1887; age, 5 yrs.; estimated weight, 800 lbs.; fed, daily, 4 lbs. corn meal, 2 lbs. oil meal and 12 lbs. bran; property of A. H. Cooley, Little Britain, N. Y.

Frolic's Pride 31667.—Sire, Mahkeenac 3290; dam, Arthur's Frolic 4438.

Butter, 17 lbs.   Milk, 201 lbs. 12 oz.

Test made from June 24 to 30, 1887; age, 4 yrs. 2 mos.; estimated weight, 900 lbs.; fed, daily, 4 lbs. corn meal, 2 lbs. oil meal and 16 lbs. bran; property of A. H. Cooley, Little Britain, N. Y.

Frugal 14925.—Sire, Pasha (P. S. 64 J. H. B.); dam, Polly (P. S. 69 J. H. B.)

Butter, 17 lbs. 2¾ oz.   Milk, 229 lbs. 4 oz.

Test made from June 25 to July 1, 1885; age, 9 yrs. 2 mos.; property of Frederick Loeser, Somerville, N. J.

Fruitland May 93002.—Sire, Robert of Green Lawn 27007; dam, May of Green Lawn 81426.

Butter, 14 lbs. 5 oz.   Milk, 248 lbs. 10 oz.

Test made from May 26 to June 2, 1898; age, 4 yrs. 8 mos.; estimated weight, 800 lbs.; fed 8 qts. Cerealine feed daily—clover hay *ad lib.* at night, good clover and blue-grass meadow pasture during day; property of Albert J. Roberts, Hartford, N. J.

Frushia Wilson 47432.—Sire, Little Harry 8808; dam, Guenn 25284.

Butter, 14 lbs. 4½ oz.   Milk, 181 lbs. 12 oz.

Test made from Jan. 15 to 22, 1892; age, 5 yrs. 10 mos.; estimated weight, 850 lbs.; fed 2 gals. corn meal, 4 gals. boiled cotton-seed and 2 gals. wheat bran—ensilage and clover hay *ad lib.*; property of Matthews & Humes, Huntsville, Ala.

Fulda Stoke Pogis 44992.—Sire, Yankee Stoke Pogis 15904; dam, Fulda 25687.

Butter, 15 lbs. 8 oz.   Milk, 207 lbs. 8 oz.

Test made from Apr. 17 to 24, 1893; age, 5 yrs. 11 mos.; actual weight, 860 lbs.; fed 12 qts. chop and 6 qts. wheat bran, daily—clover hay and pasture; property of Mrs. A. M. Hallock, Columbus, Ohio.

Gabriel Champion 14102.—Sire, Champion of America 1567; dam, Gilt Edge 3d 6041.

Butter, 17 lbs. 8 oz. Milk, 178 lbs. 8 oz.

Test made from June 17 to 23, 1885; age, 4 yrs. 2 mos.; property of John Boyd, Elmhurst, Ill.

Gala 1375, imp.

Butter, 16 lbs. 7 oz.

Test made from July 23 to 29, 1880; age, 11 yrs. 6 mos.; no grain fed—pasture of mixed grasses; property of Charles S. Dole, Crystal Lake, Ill.

Gambetta's Topsy 42376.—Sire, Gambetta M. 10012; dam, Dodona 4800.

Butter, 16 lbs. 8½ oz. Milk, 263 lbs.

Test made from Jan. 21 to 28, 1894; age, 7 yrs. 9 mos.; actual weight, 750 lbs.; fed 42 qts. corn meal, 63 qts. ground oats, 21 qts. ground oil-cake, 42 qts. bran and 20 lbs. carrots; property of Miller & Sibley, Franklin, Pa.

Gamma 4147.—Sire, Milo 590; dam, Petunia 1777.

Butter, 14 lbs. 12 oz. Milk, 200 lbs. 13 oz.

Test made from Nov. 30 to Dec. 6, 1885; age, 11 yrs. 2 mos.; property of W. B. Matthews & Son, Franklin, Tenn.

Gardiner's Ripple 11693.—Sire, Allegany Chief 2918; dam, Gardiner's Thistle 11338.

Butter, 19 lbs. 12½ oz. Milk, 264 lbs. 12 oz.

Test made from June 13 to 20, 1884; age, 5 yrs.; property of J. B. Wallace, Lexington, Ky.

Garella 62541.—Sire, Garfield Stoke Pogis 15963; dam, Matinella 32041.

Butter, 23 lbs. Milk, 301 lbs. 8 oz.

Test made from May 22 to 29, 1892; age, 3 yrs. 1 mo.; estimated weight, 875 lbs.; fed 56 lbs. cut corn fodder, 7 lbs. corn meal, 14 lbs. gluten meal, 14 lbs. bran, 7 lbs. linseed meal, 56 lbs. hay and 210 lbs. roots; property of Est. of Frederick Billings, Woodstock, Vt.

Garfield's Princess 62544.— Sire, Garfield Stoke Pogis 15963; dam, Lady Prince 48320.

Butter, 22 lbs. Milk, 283 lbs.

Test made from June 17 to 24, 1893; age, 3 yrs. 9 mos.; estimated weight, 820 lbs.; fed 56 lbs. cut corn fodder, 7 lbs. corn meal, 14 lbs. gluten meal, 14 lbs. bran, 7 lbs. linseed meal, 56 lbs. hay and 210 lbs. roots; property of Est. of Frederick Billings, Woodstock, Vt.

Garibaldi's Kate 2d 56478.—Sire, Boss of the Cedars 17712; dam, Garibaldi's Kate 25114.

Butter, 18 lbs. 2½ oz. Milk, 311 lbs.

Test made from May 19 to 26, 1897; age, 8 yrs. 3 mos.; estimated weight, 975 lbs.; fed 6 qts. corn meal, 6 qts. ground oats and 6 qts. wheat bran, daily—good pasture for season of year; property of J. Gerow Dutcher, Pawling, N. Y.

Gavotte 64205.—Sire, Ida's Stoke Pogis 13658; dam, Gildercream 39480.

Butter, 14 lbs. 14 oz. Milk, 225 lbs.

Test made from Apr. 22 to 29, 1894; age, 5 yrs. 2 mos.; estimated weight, 875 lbs.; fed 126 lbs. corn chop, 56 lbs. ground oats and 42 lbs. cut sheaf oats—blue-grass pasture; property of Est. of Campbell Brown, Spring Hill, Tenn.

Gavotte 64205.—Sire, Ida's Stoke Pogis 13658; dam, Gildercream 39480.

Butter, 15 lbs. 4 oz. Milk, 235 lbs. 5½ oz.

Test made from Apr. 12 to 19, 1896; age, 7 yrs. 2 mos.; actual weight, 800 lbs.; fed 17 lbs. per day of ground oats and corn, mixed with 4 bucketfuls of cut sheaf oats, wetted—millet hay *ad lib.* during night; property of Matt. M. Gardner, Nashville, Tenn.

Gay Orphan 25985.—Sire, Orphan (P. S. 256 J. H. B.); dam on I. of J.

Butter, 18 lbs. 1 oz. Milk, 269 lbs. 2 oz.

Test made from June 6 to 13, 1890; age, 8 yrs. 3 mos.; actual weight, 905 lbs.; fed, daily, 8 lbs. corn-hearts, average of 5 lbs. crushed oats and 2 lbs. equal mixture bran and shorts; property of Kentucky Agricultural Experiment Station, Lexington, Ky.

Gay Orphan 25985.—Sire, Orphan (P. S. 256 J. H. B.); dam on I. of J.

Butter, 138.973 lbs. in 90 days. Milk, 2175.9 lbs.

Test made from May 31 to Aug. 28, 1893, at World's Columbian Exposition; age, 11 yrs. 3 mos. at beginning of test; actual weight, 880 lbs.; fed 282 lbs. old hay, 206.5 lbs. ensilage, 446.8 lbs. new hay, 163.5 lbs. oil meal, 473.5 lbs. bran, 181 lbs. oats, 330 lbs. corn-hearts, 9.50 lbs. cotton-seed, 293 lbs. middlings, 33 lbs. grano-gluten, 926.4 lbs. clover and 10 lbs. swale-grass; property of Kentucky Agricultural Experiment Station, Lexington, Ky.

Gay Riotress 105999.—Sire, Imperial Rex Rioter 24096; dam, Lily of Elm Spring 51474.

Butter, 15 lbs. 9 oz. Milk, 189 lbs. 1 oz.

Test made from Feb. 2 to 9, 1898; age, 3 yrs. 3 mos.; estimated weight, 775 lbs.; fed 8 qts. corn meal, 8 qts. chopped sweet potatoes, 4 qts. cotton-seed and 2 qts. cotton-seed meal, daily—sorghum hay *ad lib.*, pasture during day; property of H. J. Mitchell, Winnsborough, Tex.

Gazania 4513.—Sire, Duke of Grayholdt 1035; dam, Beauvine 1593.

Butter, 14 lbs. 2 oz., unsalted. Milk, 198 lbs.

Test made from July 9 to 15, 1884; age, 9 yrs. 11 mos.; property of J. A. Fisher, Nashville, Tenn.

Gazella 3d 9355.—Sire, Compo Boy 2830; dam, Gazella 1880.

Butter, 16 lbs. 3 oz. Milk, 258 lbs. 8 oz.

Test made from Sept. 11 to 18, 1884; age, 5 yrs. 1 mo.; property of George W. Farlee, Cresskill, N. J.

Gazella 3d 9355.—Sire, Compo Boy 2830; dam, Gazella 1880.

Butter, 751 lbs. 6 oz. in one year. Milk, 10074 lbs.

Age, 8 yrs.; estimated weight, 900 lbs.; fed daily about 10½ lbs. grain; carried a calf during the entire year, excepting six weeks; property of George W. Farlee, Cresskill, N. J.

Gazellathirdella 89407.—Sire, Gazella's Signal 15600; dam, Vestvall 58372.

Butter, 16 lbs. 6 oz. Milk, 350 lbs.

Test made from Apr. 29 to May 6, 1898; age, 5 yrs.; estimated weight, 900 lbs.; fed 70 lbs. corn and cob meal, 28 lbs. oil meal and 14 lbs. cotton-seed meal—poor pasture; property of Davis Bros., Fairfield, Ill.

Gazelle 15961.—Sire, Koffee (F. S. 233 J. H. B.); dam on I. of J.

Butter, 14 lbs. Milk, 109⅗ qts.

Test made from June 8 to 14, 1884; age, 8 yrs.; property of S. M. Burnham, Saugatuck, Conn.

Gazelle of Mobile 1735, imp.

Butter, 14 lbs.

Property of W. B. Montgomery, Starkville, Miss.

Gazelle's Fawn 93704.—Sire, King of St. Lambert 15175; dam, Gazelle Dee 18056.

Butter, 25 lbs. 4½ oz. Milk, 333.1 lbs.

Test made from May 20 to 27, 1898; age, 6 yrs. 1 mo.; actual weight, 902 lbs.; fed 7 lbs. corn meal, 7 lbs. bran, 4 lbs. oil meal and 3 lbs. ground oats, mixed, daily—clover pasture; property of F. W. Hart, Cleveland, Ohio.

Gem Nehushta 73258.—Sire, Rioter of Brookside 18702; dam, Nehushta 4th 38379.

Butter, 15 lbs. 6 oz. Milk, 204 lbs. 8 oz.

Test made from Nov. 17 to 24, 1895; age, 6 yrs. 1 mo.; estimated weight, 875 lbs.; fed 12 qts. H. O. dairy feed, 1 qt. corn meal, 3 qts. bran, 1 pt. oil meal and 25 lbs. ensilage, daily—hay *ad lib.*; property of George W. Sisson, Jr., Potsdam, N. Y.

Gem of Ellasleigh 38532.—Sire, Signaldine 8556; dam, Alphea Hortense 14904.

Butter, 16 lbs. 10 oz. Milk, 221 lbs.

Test made from Jan. 24 to 31, 1896; age, 11 yrs.; actual weight, 1285 lbs.; fed 28 lbs. corn meal, 42 lbs. crushed oats, 28 lbs. gluten, 14 lbs. old process oil meal, 75 lbs. dry cut fodder and 70 lbs. roots; property of C. Delano, Mount Vernon, Ohio.

Gem of Hope 17102.—Sire, Knight of St. Louis 3680; dam, Lily of Les Niemes 7465.

Butter, 21 lbs. Milk, 253 lbs. 8 oz.

Test made from July 3 to 10, 1885; age, 4 yrs. 10 mos.; fed 4 qts. ground corn and oats, 10 qts. bran and 3 pts. oil meal, daily—poor pasture; property of W. M. Bell, Miami, Mo.

Gem of St. Cloud 7342.—Sire, Tristram Shandy 1767; dam, Amelia 2d 1730.

Butter, 14 lbs. 8½ oz., unsalted. Milk, 200 lbs.

Test made from Apr. 20 to 26, 1885; age, 8 yrs. 3 mos.; property of O. F. Cheatham, Edgefield C. H., S. C.

Gem of Sassafras 8434.—Sire, Duke of Sassafras 2431; dam, Belle of Kent 4371.

Butter, 14 lbs. 3½ oz. Milk, 253 lbs. 12 oz.

Test made from May 25 to 31, 1882; age, 4 yrs. 3 mos.; property of T. Howard Wilson, Lebanon, Del.

Gem's Louisa 39564.—Sire, Louis of St. Lambert 11013; dam, Gem of Oakland 2d 29756.

Butter, 17 lbs. Milk, 194 lbs. 10 oz.

Test made from Sept. 18 to 25, 1892; age, 6 yrs. 6 mos.; estimated weight, 900 lbs.; fed 15 lbs. oats, corn and bran and 3 lbs. oil meal, daily—84 lbs. ensilage, fresh pasture; property of Ayer & McKinney, Philadelphia, Pa.

Geneva 13220.—Sire, Farmer's Glory 5196; dam, Ristori 12701.

Butter, 15 lbs. 11 oz. Milk, 193 lbs. 10 oz.

Test made from May 15 to 21, 1884; age, 3 yrs. 5 mos.; property of William Simpson, New York, N. Y.

Geneva M.'s Fancy 104917.—Sire, Pride's Landseer 18590; dam, Geneva M. 62904.

Butter, 16 lbs. ½ oz. Milk, 195 lbs. 15 oz.

Test made from Mar. 12 to 19, 1898; age, 4 yrs. 10 mos.; estimated weight, 850 lbs.; fed 98 qts. of a mixture of corn and cob meal, ground wheat and ground oats, in equal parts by measure, and 14 qts. oil-cake meal—clover hay *ad lib.*; property of H. C. Schlegel, Daleville, Ind.

Geneva S. 57170.—Sire, Sue's Signal 14467; dam, Opportunity 25915.

Butter, 15 lbs. 15½ oz. Milk, 202 lbs. 9 oz.

Test made from Nov. 21 to 28, 1895; age, 8 yrs. 3 mos.; actual weight, 1010 lbs.; fed 84 lbs. corn meal, 56 lbs. bran and 14 lbs. cotton-seed meal—hay *ad lib.*, rich blue-grass pasture; property of Kentucky Agricultural Experiment Station, Lexington, Ky.

Gentle of Glastonbury 4651.—Sire, Hughes 954; dam, Dandelion 2521.

Butter, 14 lbs.

Property of James B. Williams, Glastonbury, Conn.

Georgiadear 11508.—Sire, Cargo 5370; dam, Normanda 3914.

Butter, 18 lbs. 3 oz.   Milk, 242 lbs. 12 oz.

Test made from Nov. 30 to Dec. 6, 1885; age, 5 yrs. 3 mos.; property of U. F. Shalter, Terre Haute, Ind.

Geranium 2d 7838.—Sire, Signal 1170; dam, Geranium 3963.

Butter, 26 lbs. 4¾ oz. Milk, 244 lbs. 12½ oz.

Test made from Sept. 28 to Oct. 4, 1885; age, 7 yrs. 6 mos.; fed 30 lbs. ground corn and oats and seconds, mixed with bran, daily—good pasture; property of George M. Jewett, Glenville, Md.

Geranium Leaf 2d 56859.—Sire, Leonidas' Germ 14289; dam, Geranium Leaf 30875.

Butter, 22 lbs. 7 oz. Milk, 261 lbs.

Test made from Jan. 25 to Feb. 1, 1893; age, 4 yrs. 10 mos.; estimated weight, 800 lbs.; fed 98 lbs. corn meal, 42 lbs. ground oats, 42 lbs. wheat bran, 42 lbs. linseed meal, 280 lbs. ensilage and 70 lbs. hay; property of William Hart Dexter, Springfield, Mass.

Gertie of Eau Claire 120019.—Sire, Melia Ann's Son 22041; dam, Cora of Eau Claire 77584.

Butter, 15 lbs. 7 oz. Milk, 228 lbs. 10 oz.

Test made from June 1 to 8, 1897; age, 2 yrs. 3 mos.; estimated weight, 800 lbs.; fed 6 qts. ground corn and oats, 2 qts. wheat middlings, 1 qt. oil meal and 1 qt. cotton-seed meal, daily—orchard pasture; property of E. L. Van Deusen, Ashley Falls, Mass.

Gertie of Glyallyn 74474.—Sire, Ramapose's Bachelor 25900; dam, Girletta 61472.

Butter, 14 lbs. 8 oz. Milk, 224 lbs. 8 oz.

Test made from Sept. 16 to 23, 1894; age, 3 yrs. 5 mos.; estimated weight, 750 lbs.; fed 6 qts. corn meal, 5 qts. wheat middlings, 5 qts. ground oats and 3 qts. oil meal, daily—clover pasture at night, green oats and peas in stable during day; property of Charles A. Sweet, East Aurora, N. Y.

Gertie of Glyallyn 74474.—Sire, Ramapose's Bachelor 25900; dam, Girletta 61472.

Butter, 18 lbs. 1½ oz. Milk, 316 lbs. 12 oz.

Test made from Feb. 28 to Mar. 6, 1896; age, 4 yrs. 11 mos.; actual weight, 850 lbs.; fed 9 qts. ground corn and oats, 12 qts. wheat bran and 3 qts. oil meal, daily—plenty hay; property of C. A. Sweet, East Aurora, N. Y.

Gertie of Glyallyn 74474.—Sire, Ramapose's Bachelor 25900; dam, Girletta 61472.

Butter, 21 lbs. 2½ oz. Milk, 359 lbs.

Test made from June 3 to 10, 1896; age, 5 yrs. 1 mo.; estimated weight, 875 lbs.; fed 8 qts. ground oats, 5 qts. corn meal, 4 qts. wheat bran and 3 qts. oil meal, daily, in three rations—hay, fair pasture four hours per day, fresh cut timothy and clover in stable; property of Charles A. Sweet, East Aurora, N. Y.

Gertie Victor Pogis 100922.—Sire, Melia Ann's Victor Pogis 20697; dam, Peggy Pedro 78146.

Butter, 16 lbs. 3 oz. Milk, 232 lbs. 12 oz.

Test made from Oct. 6 to 13, 1895; age, 2 yrs. 9 mos.; estimated weight, 700 lbs.; fed 7 lbs. peas, oats and corn, ground together, 2 lbs. oil meal and 3 lbs. wheat bran, daily—clover after feed; property of J. H. Martin, Granville, N. Y.

Gerty of Peach Hill 77773.—Sire, Young Weston 18439; dam, Golia of Peach Hill Farm 52931.

Butter, 14 lbs. 2½ oz. Milk, 224 lbs. 8 oz.

Test made from Nov. 21 to 28, 1897; age, 7 yrs. 8 mos.; estimated weight, 1000 lbs.; fed 10 qts. wheat bran, 3 qts. buckwheat middlings, 2 qts. 1 pt. wheat middlings, 1 qt. cotton-seed meal and 20 lbs. corn ensilage, daily—hay *ad lib.*; property of Everett G. Campbell, Keister's, Pa.

Gift 39901.—Sire, Rioter Hugo Pogis 13457; dam, Glint 12896.

Butter, 21 lbs. 6 oz. Milk, 249 lbs. 7 oz.

Test made from Oct. 23 to 30, 1890; age, 4 yrs. 8 mos.; estimated weight, 1000 lbs.; fed, daily, 18 qts. ground oats, corn and peas and 2 qts. shorts; property of Charles E. Hill, Denver, Col.

Gilda 2779.—Sire, Critic 540; dam, Lady Godfrey 678.

Butter, 14 lbs. 6 oz.

Test made in 1875; age, 3 yrs.; property of William H. Hayden, Albany, Vt.

Gilda Mercedes H. 14451.—Sire, Discard H. 5763; dam, Mercedes H. 12326.

Butter, 14 lbs. 8 oz. Milk, 336 lbs. 3 oz.

Test made from Sept. 12 to 18, 1885; age, 4 yrs. 8 mos.; property of William H. Hayden, Albany, Vt.

Gilded Pansy 20552.—Sire, Champion of America 1567; dam, Gilt Edge 3d 6041.

Butter, 15 lbs. 5 oz. Milk, 163 lbs.

Test made from June 1 to 7, 1885; age, 3 yrs.; property of W. B. Montgomery, Starkville, Miss.

Gilded Queen 42029.—Sire, Gilderoy 2107; dam, Regina 4th 12732.

Butter, 14 lbs. 9½ oz. Milk, 178 lbs. 8 oz.

Test made from May 26 to June 2, 1893; age, 7 yrs. 10 mos.; estimated weight, 900 lbs.; fed 14 lbs. corn-hearts, 4 lbs. cotton-seed meal and 2 lbs. cut sheaf oats, daily—blue-grass pasture; property of Campbell Brown, Spring Hill, Tenn.

Gildercream 39480.—Sire, Gilderoy 2107; dam, Christel 6565.

Butter, 15 lbs. 15 oz. Milk, 201 lbs.

Test made from Apr. 5 to 12, 1890; age, 4 yrs. 2 mos.; estimated weight, 850 lbs.; no record of feed kept; property of Campbell Brown, Spring Hill, Tenn.

Gildercream 39480.—Sire, Gilderoy 2107; dam, Christel 6565.

Butter, 17 lbs. Milk, 216 lbs.

Test made from Nov. 17 to 24, 1893; age, 7 yrs. 9 mos.; estimated weight, 925 lbs.; fed 168 lbs. ground corn and ground oats, mixed—blue-grass pasture; property of Est. of Campbell Brown, Spring Hill, Tenn.

Gilderine of Linwood 24488.—Sire, Gold Basis 4038; dam, Countess Gisela of Belle Vue 9571.

Butter, 15 lbs. 7 oz. Milk, 163 lbs. 12 oz.

Test made from Aug. 9 to 15, 1886; age, 3 yrs. 9 mos.; property of Columbia Jersey Cattle Co., Columbia, Tenn.

Gilderoy P. 99308.—Sire, Stoke Pogis' Perfection 5984; dam, Beech Grove 42709.

Butter, 15 lbs. 3 oz. Milk, 297 lbs. 4 oz.

Test made from Mar. 14 to 21, 1897; age, 4 yrs. 3 mos.; estimated weight, 800 lbs.; fed 10 lbs. corn meal, 6 lbs. bran, 2 lbs. oil meal and 4 lbs. hay, daily; property of H. M. Baum, Frankfort, Ind.

Gilderoy's Enid 32924.—Sire, Gilderoy 2107; dam, Enid 3d 19582.

Butter, 14 lbs. 4 oz. Milk, 235 lbs. 8 oz.

Test made from May 18 to 25, 1888; age, 4 yrs.; estimated weight, 800 lbs.; fed 18 lbs. ground oats and corn, daily; property of Webster & Morrow & Son, Nashville, Tenn.

Gilderoy's Idex 42027.—Sire, Gilderoy 2107; dam, Idex 2d 5429.

Butter, 19 lbs. Milk, 265 lbs. 8 oz.

Test made from May 11 to 18, 1889; age, 4 yrs. 2 mos.; estimated weight, 825 lbs.; fed, daily, 20 lbs. bran and corn-hearts, mixed; property of Campbell Brown, Spring Hill, Tenn.

Gilfilia Pogis 38688.—Sire, Stoke Pogis 5th 5987; dam, Gilfilia 2d 15990.

Butter, 14 lbs. 2½ oz. Milk, 278 lbs. 4 oz.

Test made from May 14 to 21, 1890; age, 4 yrs.; estimated weight, 900 lbs.; fed, daily, 8 lbs. corn meal, 6 lbs. crushed oats and 5 lbs. oil meal; property of Ayer & McKinney, Philadelphia, Pa.

Gilt Edge 2d 4420.—Sire, Nelusko 479; dam, Gilt Edge 2662.

Butter, 14 lbs.

Property of L. Q. C. Lamar, Oxford, Miss.

Gilt Edge C. 12223.—Sire, Champion of America 1567; dam, Gilt Edge 3d 6041.

Butter, 14 lbs. 3½ oz. (official).

Test made from July 20 to 27, 1884; property of W. B. Montgomery, Starkville, Miss.

Gilt Lady 2d 33969.—Sire, Lord Darlington 7285; dam, Gilt Lady 14981.

Butter, 14 lbs. 7 oz. Milk, 199 lbs.

Test made from Jan. 11 to 18, 1891; age, 5 yrs. 8 mos.; estimated weight, 900 lbs.; fed 42 lbs. bran, 42 lbs. corn meal and 42 lbs. cotton-seed meal; property of L. T. Hazen, Hazen's Junction, N. H.

Gingerbread 5th 37681.—Sire, St. Jacob 6438; dam, Gingerbread 7922.

Butter, 18 lbs. 9¾ oz. Milk, 248 lbs. 8 oz.

Test made from June 26 to July 3, 1895; age, 9 yrs. 2 mos.; actual weight, 900 lbs.; fed 42 qts. ground oats, 42 qts. corn meal, and 28 qts. ground oil-cake—good old meadow pasture; property of Miller & Sibley, Franklin, Pa.

Gipsy 5th 2252.—Sire, Napoleon 291; dam, Gipsy 319.

Butter, 17 lbs. 2 oz., unsalted.

Test made from Aug. 19 to 26, 1883; age, 13 yrs. 5 mos.; property of W. B. Dinsmore, Staatsburg, N. Y.

Gipsy Bess of Prospect 96928.—Sire, Gipsy's Lorne Pogis 20346; dam, Bess of Ingleside 42172.

Butter, 22 lbs. 10½ oz. Milk, 322 lbs.

Test made from May 24 to 31, 1895; age, 3 yrs. 11 mos.; actual weight, 820 lbs.; fed 42 qts. corn meal, 49 qts. ground oats and 28 qts. ground oil-cake—good old meadow pasture; property of Miller & Sibley, Franklin, Pa.

Gipsy Pogis 35422.—Sire, Prince of Melrose 2d 11015; dam, Miss Nellie Parker 24988.

Butter, 18 lbs. 7 oz. Milk, 199 lbs. 8 oz.

Test made from June 16 to 23, 1888; age, 3 yrs.; estimated weight, 875 lbs.; fed 56 qts. bran and shorts, 42 qts. corn meal and 14 qts. cotton-seed meal—hay, good mixed pasture; property of James Crook, Jacksonville, Ala.

Gipsy's Berry Duchess 86124.—Sire, Berry 26432; dam, Gipsy's Duchess 56833.

Butter, 18 lbs. 11½ oz. Milk, 312 lbs. 8 oz.

Test made from June 30 to July 7, 1894; age, 3 yrs. 7 mos.; actual weight, 890 lbs.; fed 28 qts. corn meal, 14 qts. white middlings, 14 qts. wheat bran and 14 qts. ground oil-cake—old meadow pasture; property of Miller & Sibley, Franklin, Pa.

Gipsy's Berry Duchess 86124.—Sire, Berry 26432; dam, Gipsy's Duchess 56833.

Butter, 28 lbs. 6¼ oz. Milk, 328 lbs.

Test made from Jan. 24 to Jan. 31, 1897; age, 6 yrs. 2 mos.; actual weight, 920 lbs.; fed 84 qts. corn meal, 84 qts. ground oats, 28 qts. ground oil-cake and 28 qts. wheat bran—some carrots; property of Miller & Sibley, Franklin, Pa.

Gipsy's Pride 2d 29365.—Sire, Tuscarora of Clermont 7232; dam, Gipsy's Pride 25290.

Butter, 27 lbs. 12 oz.   Milk, 295 lbs. 8 oz.

Test made from May 21 to 28, 1893; age, 9 yrs.; actual weight, 970 lbs.; fed 12 qts. corn meal, 6 qts. ground oats and 4 qts. oil meal, daily—good old meadow pasture; property of Miller & Sibley, Franklin, Pa.

Gipsy's Riotess 62482.—Sire, Duke of Ingleside 14274; dam, Gipsy's Pride, 25290.

Butter, 17 lbs. 8 oz.   Milk, 245 lbs. 4 oz.

Test made from May 22 to 29, 1894; age, 5 yrs. 2 mos.; estimated weight, 800 lbs.; fed 28 lbs. coarse bran, 42 lbs. corn meal, 14 lbs. cotton-seed meal and 14 lbs. oil meal —good pasture; property of John C. McClintock, Meadville, Pa.

Gipsy W. 132879.—Sire, King of St. Lambert's King 30752; dam, Lucile W. 117868.

Butter, 21 lbs. 4 oz.   Milk, 290 lbs.

Test made from June 6 to 13, 1898; age, 1 yr. 8 mos.; actual weight, 870 lbs.; fed 8 lbs. corn meal, 7 lbs. wheat bran and 5 lbs. ground oats, mixed, daily—clover pasture; property of F. W. Hart, Cleveland, Ohio.

Girletta 61472.—Sire, Crocker 19563; dam, Usilda's Pride 13921.

Butter, 15 lbs. 14 oz.   Milk, 247 lbs. 12 oz.

Test made from Nov. 12 to 19, 1893; age, 4 yrs. 6 mos.; estimated weight, 800 lbs.; fed 42 qts. ground corn and oats, 21 qts. oil meal and cotton-seed meal, 49 qts. wheat bran and 245 lbs. corn ensilage—hay *ad lib.*; property of E. L. Van Deusen, Ashley Falls, Mass.

Girletta 61472.—Sire, Crocker 19563; dam, Usilda's Pride 13921.

Butter, 17 lbs. 2 oz.   Milk, 262 lbs. 12 oz.

Test made from Oct. 9 to 16, 1894; age, 5 yrs. 5 mos.; estimated weight, 800 lbs.; no grain fed—meadow pasture only; property of E. L. Van Deusen, Ashley Falls, Mass.

Girletta 4th 109264.—Sire, Glynllyn Boy 22396; dam, Girletta 61472.

Butter, 15 lbs. 8½ oz.   Milk, 249 lbs. 6 oz.

Test made from Mar. 20 to 27, 1897; age, 3 yrs. 5 mos.; estimated weight, 900 lbs.; fed 6 qts. ground corn and oats, 6 qts. wheat bran, 2 qts. gluten meal, 2 qts. oil meal and 30 lbs. corn ensilage, daily—hay *ad lib.*; property of E. L. Van Deusen, Ashley Falls, Mass.

Girlie's Glory 46305.—Sire, Coomassie Welcome 9279; dam, Gilderoy's Girlie 24181.

Butter, 15 lbs.   Milk, 181 lbs. 12 oz.

Test made from Mar. 21 to 28, 1890; age, 3 yrs. 5 mos.; estimated weight, 800 lbs.; fed, daily, 2½ lbs. cotton-seed meal, 6 lbs. shorts and 12 lbs. corn meal; property of R. A. Sibley, Spencer, Mass.

Giulietta Cooke 32193.—Sire, Seneca Chief 4098; dam, Gigia 4447.

Butter, 21 lbs. 13½ oz.   Milk, 257 lbs. 4 oz.

Test made from July 28 to Aug. 4, 1888; age, 6 yrs.; estimated weight, 900 lbs.; fed, daily, about 24 qts. bran, 4 qts. oat meal, 3 qts. corn meal and 1 qt. oil meal; property of P. J. Cogswell, Rochester, N. Y.

Gladdis Roy 18566.—Sire, Catono 3761; dam, Cowles' Eirene 3926.

Butter, 14 lbs. 7½ oz. Milk, 287 lbs. 8 oz.

Test made from July 24 to 31, 1891; age, 9 yrs. 1 mo.; estimated weight, 850 lbs.; fed 9 lbs. meal, 4 lbs. bran and 3 lbs. oil meal, daily—ordinary pasture; property of H. A. Huntington, Higganum, Conn.

Gladdis Roy 18566.—Sire, Catono 3761; dam, Cowles' Eirene 3926.

Butter, 17 lbs. 13½ oz. Milk, 265 lbs. 8 oz.

Test made from Feb. 29 to Mar. 7, 1892; age, 9 yrs. 8 mos.; estimated weight, 1000 lbs.; fed 12 lbs. corn meal, 8 lbs. bran, 2 lbs. cotton-seed meal, 40 lbs. ensilage and 6 lbs. cut hay, daily—dry hay *ad lib.*; property of H. A. Huntington, Higganum, Conn.

Gladdis Roy 2d 55652.—Sire, Gold Basis 5th 15725; dam, Gladdis Roy 18566.

Butter, 18 lbs. 12 oz. Milk, 256 lbs.

Test made from Feb. 24 to Mar. 2, 1892; age, 4 yrs. 10 mos.; estimated weight, 900 lbs.; fed 9 lbs. corn meal, 6 lbs. bran, 2 lbs. cotton-seed meal, 6 lbs. cut hay and 40 lbs. ensilage, daily—hay *ad lib.*; property of H. A. Huntington, Higganum, Conn.

Gladdys Gaily 20878.—Sire, Tormentor 3533; dam, Sweet Alice 6402.

Butter, 16 lbs. 1⅝ oz. Milk, 102 lbs. 1 oz.

Test made from Mar. 6 to 13, 1886; age, 5 yrs. 3 mos.; property of John Moore, Jr., Columbia, Tenn.

Gladys of Belle Vue 9569.—Sire, Lord Lawrence 1414; dam, Stella of St. Ouen's 6955.

Butter, 16 lbs. 7 oz.

Test made from Sept. 22 to 28, 1884; age, 5 yrs. 6 mos.; property of James B. Wilder, Louisville, Ky.

Glen's Festina 25851.—Sire, May's Pedro 8588; dam, Glen Belma 17243.

Butter, 17 lbs. 15 oz., unsalted. Milk, 254 lbs. 10½ oz.

Test made from May 21 to 27, 1886; age, 2 yrs. 2 mos.; property of A. McClintock, Millersburg, Ky.

Glory of Elmarch 21521.—Sire, Philidor (P. S. 276 J. H. B.); dam, Dottie 15407.

Butter, 15 lbs. 13½ oz. Milk, 204 lbs. 12 oz.

Test made from June 16 to 22, 1885; age, 3 yrs. 1 mo.; property of D. A. Givens, Cynthiana, Ky.

Goddess of Staatsburgh 5252.—Sire, Vermont 893; dam, Gipsy 4th 782.

Butter, 14 lbs. 8 oz., unsalted.

Test made from June 30 to July 6, 1883; age, 7 yrs. 6 mos.; property of W. B. Dinsmore, Staatsburg, N. Y.

Gold Baby 105745.—Sire, Gold Bugler 20036; dam, Baby Ryan 42139.

Butter, 17 lbs. 12 oz. Milk, 233 lbs. 8 oz.

Test made from Apr. 20 to 27, 1898; age, 4 yrs. 10 mos.; estimated weight, 900 lbs.; fed 28 lbs. bran, 28 lbs. corn meal, 28 lbs. oats and 14 lbs. oil meal—corn ensilage; property of H. C. Taylor, Orfordville, Wis.

Gold Badger 67301.—Sire, Gold of St. Lambert 16744; dam, Badger Girl 2d 49717.

Butter, 16 lbs. 7 oz.   Milk, 226 lbs. 12 oz.

Test made from Nov. 14 to 21, 1896; age, 7 yrs. 4 mos.; estimated weight, 850 lbs.; fed 84 lbs. ground oats and wheat bran, mixed in equal parts, and 14 lbs. oil meal—corn fodder, poor pasture; property of N. N. Palmer & Son, Brodhead, Wis.

Gold Ear 2d 3592.—Sire, Brookside 1104; dam, Gold Ear 2200.

Butter, 18 lbs. 2 oz.

Test made from June 20 to 26, 1881; age, 7 yrs. 11 mos.; property of E. M. Phelon, Cherry Valley, N. Y.

Golden Dewdrop 96551.—Sire, Gold Coast Jr. 9217; dam, Countess Gazelle 4th 47729.

Butter, 23 lbs. 13¾ oz.   Milk, 269 lbs. 8 oz.

Test made from Dec. 18 to 25, 1896; age, 5 yrs. 4 mos.; actual weight, 790 lbs.; fed 63 qts. ground oats, 42 qts. corn meal, 21 qts. wheat bran and 21 qts. ground oil-cake—half-bushel carrots; property of Miller & Sibley, Franklin, Pa.

Golden Gem of St. L. 81167.—Sire, Nestor of St. Lambert 22385; dam, Golden Trudie 2d 38082.

Butter, 15 lbs. 3 oz.   Milk, 187 lbs.

Test made from Dec. 22 to 29, 1894; age, 3 yrs. 9 mos.; estimated weight, 700 lbs.; fed 8 qts. chop and 4 qts. wheat bran, daily—corn fodder and hay *ad lib.*; property of Mrs. A. M. Hallock, Columbus, Ohio.

Golden Gem of St. L. 81167.—Sire Nestor of St. Lambert 22385; dam, Golden Trudie 2d 38082.

Butter, 49 lbs. 4¼ oz., in 24 days.   Milk, 629 lbs. 8 oz.

Test made from Dec. 5 to 29, 1894; age, 3 yrs. 9 mos.; estimated weight, 700 lbs.; fed about 5 gals. daily of cut hay, corn chop, bran and gluten—hay and corn fodder *ad lib.*; property of Mrs. A. M. Hallock, Columbus, Ohio.

Golden Plover 22388.—Sire, Easter Boy 3032; dam, Nelly Pierpont 12531.

Butter, 20 lbs. 1¾ oz.   Milk, 309 lbs.

Test made from June 23 to 30, 1889; age, 6 yrs. 1 mo.; estimated weight, 900 lbs.; fed, daily, 6 qts. corn and oat chop, in equal parts; property of S. H. Evans, Tidioute, Pa.

Golden Princess 4557.—Sire, Marcot 726; dam, Daffodil 307.

Butter, 18 lbs. 14 oz.   Milk, 108 lbs. 8 oz.

Test made from Oct. 13 to 19, 1885; age, 9 yrs. 10 mos.; property of T. W. McNeely, Petersburg, Ill.

Golden Sheen 25561.—Sire, Happy Cicero 10601; dam, Lily of the Grove 24555.

Butter, 14 lbs. 4½ oz.   Milk, 205 lbs.

Test made from Dec. 20 to 27, 1888; age, 5 yrs. 2 mos.; weight, 1175 lbs.; fed, daily, about 6 lbs. corn meal, 6 lbs. oat meal, 6 lbs. wheat bran and 1 lb. pea meal; property of Jacob L. Thomas, Knoxville, Tenn.

Golden Skin 10861.—Sire, Roanoke 1448; dam, Delpha 2d 10713.

Butter, 16 lbs. 8 oz.   Milk, 254 lbs.

Test made from July 15 to 22, 1882; age, 3 yrs.; property of Raymond H. Perry, Bristol, R. I.

Golden Susette 96552.—Sire, Gold Coast Jr. 9217; dam, Albert's Susette 47727.

Butter, 17 lbs. 6 oz. Milk, 282 lbs. 8 oz.

Test made from Jan. 26 to Feb. 2, 1896; age, 4 yrs. 6 mos.; actual weight, 820 lbs.; fed 12 qts. wheat bran, 3 qts. corn meal, 6 qts. ground oats and 3 qts. oil meal, mixed, daily—hay; property of C. A. Sweet, East Aurora, N. Y.

Golden Trudie 34535.—Sire, Gold Finder 2225; dam, Trudie 277.

Butter, 14 lbs. 9 oz. Milk, 157 lbs. 10 oz.

Test made from Apr. 11 to 18, 1888; age, 7 yrs. 1 mo.; estimated weight, 700 lbs.; fed, daily, 2 lbs. corn meal, 2 lbs. corn-hearts, 2 lbs. bran, 1 lb. oil meal and 1 lb. middlings; property of James Stillman, Sing Sing, N. Y.

Golden Violet 30136.—Sire, Golden Rule 4785; dam, Violet 3d 3240.

Butter, 15 lbs. Milk, 170 lbs. 12 oz.

Test made from Jan. 25 to Feb. 1, 1891; age, 6 yrs. 4 mos.; estimated weight, 950 lbs.; fed, daily, 4 lbs. oil meal, 4 lbs. pea meal, 4 lbs. corn meal and 8 lbs. crushed oats; property of Ayer & McKinney, Philadelphia, Pa.

Golden Zoe 3975.—Sire, Golden Ear 1025; dam, Zoe Mon 2704.

Butter, 16 lbs. 3 oz. Milk, 282 lbs. 8 oz.

Test made from Jan. 28 to Feb. 3, 1885; age, 9 yrs. 9 mos.; property of Miller & Sibley, Franklin, Pa.

Gold Gem 72024.—Sire, Gold of St. Lambert 16744; dam, Little Germ 10965.

Butter, 15 lbs. 12 oz. Milk, 202 lbs. 6 oz.

Test made from Oct. 26 to Nov. 2, 1896; age, 5 yrs. 10 mos.; estimated weight, 900 lbs.; fed 35 lbs. wheat bran and 45½ lbs. ground oats—timothy hay at night, timothy meadow pasture day time; property of N. N. Palmer & Son, Brodhead, Wis.

Goldie La V. 2d 26319.—Sire, Hurrah 2814; dam, Goldie La V. 6983.

Butter, 14 lbs. 1½ oz. Milk, 157 lbs.

Test made from Dec. 20 to 26, 1885; age, 7 yrs. 2 mos.; property of T. Alexander Seth, Baltimore, Md.

Goldie Pogis 45423.—Sire, Pauline's Stoke Pogis 11859; dam, Goldie C. 8104.

Butter, 16 lbs. 2 oz. Milk, 206 lbs. 9 oz.

Test made from May 23 to 30, 1890; age, 4 yrs. 4 mos.; actual weight, 850 lbs.; fed, daily, about 7 lbs. bran and 3 lbs. corn and cob meal; property of Mrs. Kate M. Busick, Wabash, Ind.

Gold Lace 10726.—Sire, Roanoke 1448; dam, Delpha 2d 10713.

Butter, 14 lbs. 13 oz.

Test made from June 11 to 18, 1880; age, 4 yrs.; property of R. H. Perry, Bristol, R. I.

Gold Lace 10726.—Sire, Roanoke 1448; dam, Delpha 2d 10713.

Butter, 21 lbs. 1 oz. Milk, 285 lbs. 6½ oz.

Test made from June 7 to 13, 1885; age, 9 yrs. 1 mo.; fed 4 qts. pea meal, 4 qts. ground oats and 2 qts. linseed meal, daily—cut grass at night—woods pasture; property of Spencer Borden, Fall River, Mass.

174 BUTTER TESTS OF JERSEYS.

Gold Maid 98787.—Sire, Gold Base Boy 27116; dam, Rothey 18019.

Butter, 16 lbs. 14¾ oz. Milk, 176 lbs., estimated.

Test made from Aug. 12 to 19, 1895; age, 3 yrs. 10 mos.; estimated weight, 900 lbs.; fed bran, corn meal and cut green corn, not weighed—dry fall pasture; property of Hooper & Christy, Murfreesborough, Tenn.

Gold Mark 10727.—Sire, Gilderoy 2107; dam, Gold Lace 10726.

Butter, 14 lbs. 14 oz. Milk, 197 lbs.

Test made from June 15 to 22, 1882; age, 3 yrs. 4 mos.; property of Spencer Borden, Fall River, Mass.

Gold Princess 8809.—Sire, Charlie Kittredge 1247; dam, Goldie C. 8104.

Butter, 14 lbs. 12 oz.

Test made from Mar. 22 to 28, 1882; age, 3 yrs.; property of R. McMichael, Lexington, Ky.

Goldstraw 3d 14724.—Sire, Monarch of New Jersey 5199; dam, Goldstraw 85.

Butter, 14 lbs. 12 oz. Milk, 277 lbs.

Test made from May 19 to 25, 1885; age, 3 yrs. 10 mos.; property of Miller & Sibley, Franklin, Pa.

Goldthread 4945.—Sire, Norwood 1077; dam, Milwaukee 2920.

Butter, 17 lbs. 9 oz.

Test made from Jan. 14 to 21, 1881; age, 4 yrs. 7 mos.; property of Edward M. Burns, Middleville, N. Y.

Gold Trinket 9518.—Sire, Orange Skin 1216; dam, Azelda 3872.

Butter, 16 lbs. 2 oz., unsalted. Milk, 240 lbs. 12 oz.

Test made from July 13 to 19, 1882; age, 3 yrs. 6 mos.; fed 2 qts. corn meal, 1 gal. wheat bran and 2 pts. oil meal, daily; property of W. J. Chinn, Frankfort, Ky.

Gold Violet 85078.—Sire, Violetta's Hugo Pogis 15740; dam, Cloth of Gold 18490.

Butter, 17 lbs. 1 oz. Milk, 321 lbs. 8 oz.

Test made from Jan. 20 to 27, 1897; age, 4 yrs. 3 mos.; estimated weight, 900 lbs.; fed 12 lbs. oats, peas and corn, mixed in equal parts, ground, 4 lbs. wheat bran and 3 lbs. oil meal, daily—hay *ad lib.*; property of J. H. Martin, Granville, N. Y.

Golightly 25597.—Sire, Pedro 3187; dam, Bother 25595.

Butter, 18 lbs. 2 oz. Milk, 19 qts. per day.

Test made from Mar. 23 to 29, 1887; age, 4 yrs. 10 mos.; estimated weight, 800 lbs.; fed 8 qts. corn meal, 5 qts. ground oats, 5 qts. bran and 1 qt. pea meal, daily; property of D. F. Appleton, Ipswich, Mass.

Golightly 25597.—Sire, Pedro 3187; dam, Bother 25595.

Butter, 23 lbs. 4 oz. Milk, 275 lbs. 11 oz.

Test made from June 3 to 10, 1889; age, 7 yrs. 1 mo.; estimated weight, 850 lbs.; fed, daily, 8 lbs. corn meal, 9 lbs. ground oats and 9 lbs. middlings; property of D. F. Appleton, Ipswich, Mass.

Good Advice 62476.—Sire, Ida's Rioter of St. L. 13656; dam, Lady Mary of Prospect 19768.

Butter, 19 lbs. 5¼ oz. Milk, 215 lbs.

Test made from Mar. 23 to 30, 1894; age, 4 yrs. 10 mos.; actual weight, 995 lbs.; fed 84 qts. ground oats, 56 qts. corn meal, 28 qts. wheat bran and 14 qts. ground oil-cake —cabbage and carrots; property of J. C. McKinney, Titusville, Pa.

Goodbye 27366.—Sire, Auroraboreëllis 2408; dam, Frankie's Lass 24900.

Butter, 16 lbs. 13 oz. Milk, 223 lbs.

Test made from June 22 to 29, 1885; age, 4 yrs. 8 mos.; property of Richardson Bros., Davenport, Iowa.

Good Friday 20081.—Sire, Alexis of M. 7963; dam, Nellie Sullivan 11159.

Butter, 14 lbs. 12 oz. Milk, 197 lbs. 4 oz.

Test made from Jan. 21 to 27, 1886; age, 3 yrs. 9 mos.; property of John A. Middelton, Shelbyville, Ky.

Gordonetta 19570.—Sire, Lord Harry 3445; dam, Kate Gordon 8387.

Butter, 17 lbs. 1 oz. Milk, 200 lbs. 8 oz.

Test made from June 5 to 12, 1889; age, 6 yrs. 4 mos.; estimated weight, 900 lbs.; fed, daily, 4 gals. oats and corn meal, ground and mixed, in equal parts; property of M. C. Campbell, Spring Hill, Tenn.

Gowan 12272.—Sire, Lenape 2732; dam, Gemmi 2d 7314.

Butter, 14 lbs. 4 oz.

Test made from Feb. 24 to Mar. 2, 1886; age, 4 yrs. 11 mos.; property of Hugh DeHaven, Philadelphia, Pa.

Grace Davy 8292.—Sire, Young Sir Davy 3034; dam, Grace Darling 2d 304.

Butter, 14 lbs. 2 oz. Milk, 191 lbs. 1 oz.

Test made from Nov. 6 to 13, 1882; age, 6 yrs. 2 mos.; property of Mrs. George M. Jewett, Glenville, Md.

Grace Davy 8292.—Sire, Young Sir Davy 3034; dam, Grace Darling 2d 304.

Butter, 22 lbs. 5½ oz. Milk, 242 lbs. 11 oz.

Test made from Sept. 15 to 22, 1885; age, 9 yrs.; fed 15 lbs. ground corn and oats, shorts, and wheat bran, mixed, daily—pasture; property of George M. Jewett, Glenville, Md.

Grace Felch 8291.—Sire, Ike Felch 1292; dam, Grace Darling 2d 304.

Butter, 15 lbs. Milk, 192 lbs. 5 oz.

Test made from Aug. 21 to 27, 1882; property of George M. Jewett, Glenville, Md.

Grace G. Parks 29263.—Sire, Deerfoot Boy of Somerset 6944; dam, Estella Parks 15435.

Butter, 19 lbs. 3 oz. Milk, 257 lbs. 8 oz.

Test made from June 24 to July 1, 1888; age, 6 yrs. 1 mo.; estimated weight, 800 lbs.; fed, daily, 12 lbs. corn meal, 13½ lbs. oil meal, 1½ lbs. oats, 3¼ lbs. bran and middlings; property of A. D. McBride, Rochester, N. Y.

Grace Harribee 91836.—Sire, Grisette's Koffee 26008; dam, Beauty of Edgeworth 51361.

    Butter, 20 lbs. 2 oz. Milk, 260 lbs.

    Test made from Feb. 19 to 26, 1896; age, 2 yrs. 9 mos.; actual weight, 800 lbs.; fed 20 lbs. daily of a mixture of ground oats, corn bran, oil and cotton-seed meal, with 30 lbs. ensilage—hay; property of Robert F. Shannon, Pittsburg, Pa.

Grace M. 10697.—Sire, Cardinal of Rosland 3335; dam, Alexina 3836.

    Butter, 15 lbs. 2 oz. Milk, 225 lbs. 4 oz.

    Test made from May 3 to 10, 1883; age, 4 yrs.; property of Henry B. Kelley, Vinton, Iowa.

Grace Marigold 99377.—Sire, Stoke Pogis of Prospect 29121; dam, Lady Grace of Upholme 39569.

    Butter, 20 lbs. 1 oz. Milk, 257 lbs. 10 oz.

    Test made from May 23 to 30, 1896; age, 2 yrs. 2 mos.; actual weight, 715 lbs.; fed 8 qts. ground oats, 5 qts. corn meal, 4 qts. wheat bran and 3 qts. oil meal, mixed, in three rations, daily—hay, fair pasture about four hours daily; property of Charles A. Sweet, East Aurora, N. Y.

Grace of Glynllyn 67759.—Sire, Duke Thor 17323; dam, Astel 2d 10792.

    Butter, 14 lbs. 10 oz. Milk, 186 lbs. 14 oz.

    Test made from Dec. 9 to 16, 1895; age, 5 yrs. 7 mos.; estimated weight, 750 lbs.; fed 8 lbs. corn ensilage, 5 qts. bran, 7 qts. oat meal, 3 qts. oil meal and 3 qts. corn meal. 5 lbs. hay and 1 pk. roots, daily; property of E. E. & M. C. Harrington, Watertown, N. Y.

Grace Pansy 2d 18764.—Sire, Smoke Smith 4844; dam, Grace Pansy 12156.

    Butter, 17 lbs. 15 oz. Milk, 236 lbs. 15 oz.

    Test made from Nov. 2 to 9, 1886; age, 3 yrs. 10 mos.; estimated weight, 800 lbs.; fed, daily, 20 to 25 qts. corn, oats and wheat bran, in equal parts; property of G. V. Green, Hopkinsville, Ky.

Grace Pansy 2d 18764.—Sire, Smoke Smith 4844; dam, Grace Pansy 12156.

    Butter, 147.009 lbs. in 90 days. Milk, 2344.4 lbs.

    Test made from May 31 to Aug. 28, 1893, at World's Columbian Exposition; age, 10 yrs. 4 mos. at beginning of test; actual weight, 1050 lbs.; fed 330 lbs. old hay. 195 lbs. ensilage, 596.5 lbs. new hay, 177 lbs. oil meal, 395 lbs. bran, 98 lbs. oats, 437 lbs. corn-hearts, 115.25 lbs. cotton-seed, 366 lbs. middlings, 34 lbs. grano-gluten. 1122.4 lbs. clover and 10 lbs. swale-grass; property of George V. Green, Hopkinsville, Ky.

Grace Sheen 2d 79788.—Sire, King of St. Lambert Jr. 18808; dam, Grace Sheen 62319.

    Butter, 14 lbs. 9½ oz. Milk, 229 lbs. 13 oz.

    Test made from June 23 to 30, 1894; age, 3 yrs. 6 mos.; estimated weight, 900 lbs.; fed 2 lbs. each of ground oats, bran and corn meal, daily—timothy pasture; property of S. F. Gray, Greenfield, Ind.

Grace's Nightingale 19855.—Sire, Old Hickory 2d 9224; dam, Nightingale K. 16056.

    Butter, 14 lbs. 2 oz. Milk, 186 lbs. 8 oz.

    Test made from June 22 to 28, 1885; age, 6 yrs. 3 mos.; property of C. Easthope, Niles, Ohio.

BUTTER TESTS OF JERSEYS.

177

Grace's Pride 61018.—Sire, Geranium's Rajah 18605; dam, Miami Grace 39392.

Butter, 15 lbs. 7 oz. Milk, 211 lbs. 14 oz.

Test made from Apr. 19 to 26, 1891; age, 3 yrs. 3 mos.; estimated weight, 900 lbs.; fed 84 lbs. corn meal—good early pasture; property of George M. Jewett, Glenville, Md.

Grace Torment 87957.—Sire, King Torment 18609; dam, Miami Grace 39392.

Butter, 14 lbs. 10½ oz. Milk, 232 lbs. 8 oz.

Test made from Mar. 25 to Apr. 1, 1897; age, 5 yrs. 5 mos.; estimated weight, 800 lbs.; fed 70 lbs. corn and oats, ground together in equal bulks, 14 lbs. cotton-seed meal and 140 lbs. corn ensilage—clover hay *ad lib.*; property of Joseph T. Hoopes, Bynum, Md.

Gracie A. 54524.—Sire, Duke of Ingleside 14274; dam, Queen Dodona 42375.

Butter, 15 lbs. 4 oz. Milk, 235 lbs.

Test made from June 27 to July 4, 1893; age, 6 yrs. 2 mos.; actual weight, 925 lbs.; fed 6 qts. corn meal, 6 qts. ground oats, 2 qts. ground oil-cake, daily—good meadow pasture; property of Miller & Sibley, Franklin, Pa.

Grädde 22564.—Sire, Catono 3761; dam, Usilda's Creamlet 8817.

Butter, 14 lbs. 2½ oz. Milk, 187 lbs. 8 oz.

Test made from Feb. 21 to 27, 1888; age, 5 yrs.; weight, 910 lbs.; fed 3 lbs. oats, 12 lbs. corn meal, 3¾ lbs. middlings and 2¼ lbs. oil meal, daily; property of Mrs. A. N. Martin, Summit, N. J.

Granny's Baby 5th 95132.—Sire, Prince of Ferncliff 24294; dam, Granny's Baby 30478.

Butter, 15 lbs. 8½ oz. Milk, 214 lbs. 8 oz.

Test made from Dec. 4 to 11, 1897; age, 4 yrs. 8 mos.; actual weight, 1015 lbs.; fed 56 qts. bran, 56 qts. corn meal, 14 qts. cotton-seed meal, 7 qts. oil meal, 140 lbs. ensilage, 105 lbs. hay and 42 qts. roots; property of P. J. Cogswell, Rochester, N. Y.

Granny's Gem 30406.—Sire, King (P. S. 238 J. H. B.); dam, Granny (P. S. 495 J. H. B.)

Butter, 16 lbs. 5¼ oz. Milk, 164 lbs.

Test made from Oct. 22 to 28, 1885; age, 3 yrs. 7 mos.; property of J. V. & C. Ramsden, Morton, Pa.

Granny's Gem 30406.—Sire, King (P. S. 238 J. H. B.); dam, Granny (P. S. 495 J. H. B.)

Butter, 21 lbs. 13-16 oz. Milk, 179 lbs. 4 oz.

Test made from Oct. 8 to 15, 1887; age, 5 yrs. 8 mos.; estimated weight, 900 lbs.; fed, daily, 4 qts. corn meal, 6 qts. bran and 1 qt. flax meal; property of M. Erskine Miller, Staunton, Va.

Gray Beauty 3d 19845.—Sire, Nerontes 9222; dam, Gray Beauty 16053.

Butter, 19 lbs. 4 oz. Milk, 376 lbs.

Test made from June 22 to 28, 1885; age, 9 yrs. 4 mos.; property of Silas Shook, Youngstown, Ohio.

Gray Bessie 96596.—Sire, Combination 3d 17576; dam, Brown Flora 2d 96594.

Butter, 14 lbs. 4 oz. Milk, 215 lbs. 4 oz.

Test made from Aug. 3 to 10, 1895; age, 5 yrs. 6 mos.; estimated weight, 1050 lbs.; fed 35 lbs. corn meal, 28 lbs. oats, 35 lbs. bran and 28 lbs. oil meal—mixed pasture; property of H. C. Taylor, Orfordville, Wis.

Greatæna 85823.—Sire, Jersey Monarch 14084; dam, May Colt 23491.

Butter, 16 lbs. 10½ oz.   Milk, 247 lbs.

Test made from Mar. 27 to Apr. 3, 1897; age, 4 yrs. 6 mos.; actual weight, 875 lbs.; fed 49 qts. bran, 49 qts. oat meal, 49 qts. corn meal, 21 qts. oil meal, 105 lbs. ensilage, 84 lbs. hay and 56 qts. roots; property of P. J. Cogswell, Rochester, N. Y.

Gretchen of Pevely 46325.—Sire, Producer 9367; dam, Roe Webster 12562.

Butter, 15 lbs. 5½ oz.   Milk, 204 lbs. 13 oz.

Test made from June 1 to 8, 1890; age, 3 yrs. 6 mos.; estimated weight, 750 lbs.; fed, daily, 6 lbs. ground oats, 4 lbs. corn meal and 1 lb. oil meal; property of James T. Henderson, Auvergne, Ark.

Grey Friar's Bess 25128.—Sire, Grey Friar 7627; dam, Nerissa (F. S. 2641 J. H. B.)

Butter, 22 lbs.   Milk, 280 lbs. 8 oz.

Test made from May 11 to 18, 1892; age, 9 yrs. 4 mos.; estimated weight, 900 lbs.; fed 14 gals. corn meal—hay *ad lib.*, medium clover pasture; property of W. Cannon, Philadelphia, Tenn.

Grey Jessamine 17444.—Sire, Grey of the West (F. S. 317 J. H. B.); dam, Jessamine (F. S. 2188 J. H. B.)

Butter, 14 lbs. 4 oz.   Milk, 265 lbs. 8 oz.

Test made from Apr. 6 to 12, 1885; age, 4 yrs. 5 mos.; property of William Crozier, Northport, N. Y.

Grinnell Lass 11859.—Sire, The Marquis 3805; dam, Belle Grinnell 2d 9009.

Butter, 16 lbs. 10 oz., unsalted.   Milk, 252 lbs. 8 oz.

Test made from Apr. 20 to 26, 1883; age, 2 yrs. 10 mos.; property of William Crozier, Northport, N. Y.

Grinnell Lass 11859.—Sire, The Marquis 3805; dam, Belle Grinnell 2d 9009.

Butter, 16 lbs. 5 oz.   Milk, 308 lbs.

Test made from Feb. 9 to 15, 1885; age, 4 yrs. 8 mos.; property of William Crozier, Northport, N. Y.

Griselda Marjoram 111166.—Sire, Pedro Royal Marjoram 28560; dam, Carlo's Polly Rex 94627.

Butter, 15 lbs. 6 oz.   Milk, 230 lbs.

Test made from Mar. 3 to 10, 1897; age, 1 yr. 9 mos.; estimated weight, 725 lbs.; fed about 18 qts. daily of a mixture of ground oats, middlings, wheat bran, corn meal and some pea meal—clover hay and ensilage; property of T. S. Cooper, Coopersburg, Pa.

Guarder 90825.—Sire, Ethlo's Prince 24119; dam, Gipsy F. Eric 75462.

Butter, 21 lbs. 4 oz.   Milk, 279 lbs. 10 oz.

Test made from June 7 to 14, 1897; age, 3 yrs. 11 mos.; actual weight, 775 lbs.; fed 16 qts. germ feed and coarse bran first two days, 40 qts. corn meal during other five days—blue-grass pasture; property of S. E. DuBois, Vigo, Ohio.

Guess Not 57142.—Sire, Ruby's Harry 15664; dam, Countess Gilderine 29027.

Butter, 20 lbs. 8 oz.   Milk, 266 lbs. 8½ oz.

Test made from Apr. 12 to 19, 1896; age, 7 yrs. 5 mos.; actual weight, 775 lbs.; fed 20 lbs. per day of ground oats and corn, in equal parts, and 4 bucketfuls cut sheaf oats, wetted—good orchard and blue-grass pasture; property of Matt. M. Gardner, Nashville, Tenn.

Guess Not 2d 133252.—Sire, Odelio 2d 30299; dam, Guess Not 57142.

Butter, 16 lbs. 11¼ oz. Milk, 224 lbs. 11¾ oz.

Test made from June 17 to 24, 1898; age, 3 yrs. 8 mos.; estimated weight, 775 lbs.; fed 22 lbs. per day of ground oats and corn, in equal parts—very good orchard and blue-grass pasture; property of Matt. M. Gardner, Nashville, Tenn.

Guinevere Sinclair 11167.—Sire, Berlin Prince 3360; dam, Belle of Patterson 5664.

Butter, 14 lbs. 9 oz. Milk, 234 lbs. 4 oz.

Test made from Nov. 11 to 17, 1884; age, 4 yrs. 6 mos.; property of M. G. Jacobs, Independence, Mo.

Haidee Signal 86448.—Sire, Signal Jr. 7166; dam, Princess Haidee 34874.

Butter, 20 lbs. 2 oz. Milk, 304 lbs.

Test made from July 7 to 14, 1894; age, 6 yrs. 11 mos.; estimated weight, 800 lbs.; fed 14 lbs. to 16 lbs. per day of a mixture of bran, middlings, corn meal and oil meal —natural pasture; property of George W. Sisson, Jr., Potsdam, N. Y.

Haidee Signal 86448.—Sire, Signal Jr. 7166; dam, Princess Haidee 34874.

Butter, 79 lbs. 11 oz. in 31 days. Milk, 1264 lbs. 4 oz.

Test made from June 30 to July 31, 1894; age, 6 yrs. 11 mos.; estimated weight, 800 lbs.; fed 14 lbs. to 16 lbs. per day of bran, middlings, corn meal and oil meal—natural pasture; property of George W. Sisson, Jr., Potsdam, N. Y.

Hallie Rice 75777.—Sire, Aurora Chief 20039; dam, Belle of Aurora 67665.

Butter, 24 lbs. 8½ oz. Milk, 290 lbs. 12 oz.

Test made from July 22 to 29, 1895; age, 5 yrs. 5 mos.; actual weight, 920 lbs.; fed 12 qts. ground oats and corn, 12 qts. bran, 3 qts. oil meal and 1½ qts. cotton-seed meal, daily—green corn, oats and peas; property of C. A. Sweet, East Aurora, N. Y.

Hallie Signal 61726.—Sire, Signal Duke 12661; dam, Hallie's Jewel 17113.

Butter, 18 lbs. 12 oz. Milk, 251 lbs.

Test made from Aug. 28 to Sept. 4, 1892; age, 3 yrs. 6 mos.; estimated weight, 700 lbs.; fed 10 lbs. corn meal, 5 lbs. oat meal, 3 lbs. bran and 7 lbs. oil meal, daily— blue-grass pasture; property of John A. Middelton, Shelbyville, Ky.

Hallie's Jewel 17113.—Sire, Knight of St. Louis 3680; dam, Hallie Carter 5329.

Butter, 20 lbs. 10 oz. Milk, 197 lbs. 8 oz.

Test made from Mar. 6 to 12, 1886; age, 3 yrs. 7 mos.; fed 2 buckets cut oats, 14 lbs. corn meal, 5 lbs. oat meal and 6 lbs. oil meal, daily—clover hay *ad lib.*; property of John A. Middelton, Shelbyville, Ky.

Halsie McCurdy 12379.—Sire, Clive Duke 1901; dam, Diana 3d 263.

Butter, 14 lbs. 3½ oz. Milk, 212 lbs. 8 oz.

Test made from Nov. 24 to 30, 1885; age, 7 yrs.; property of M. J. Owings, Indianapolis, Ind.

Handsome Myra 14244.—Sire, Grey (F. S. 244 J. H. B.); dam, Daisy (P. S. 218 J. H. B.)

Butter, 21 lbs. Milk, 276 lbs.

Test made from Oct. 15 to 22, 1884; age, 5 yrs. 6 mos.; fed 6 qts. corn meal, 2 qts. oil meal and 4 qts. shorts, daily—good pasture; property of Cornelius Wellington, East Lexington, Mass.

Handy's Daisy 74017.—Sire, Handy's Stoke Pogis 18810; dam, Daisy Gray 12893.

 Butter, 16 lbs. 5 oz. Milk, 251 lbs. 12¾ oz.

 Test made from Mar. 6 to 13, 1897; age, 7 yrs. 1 mo.; estimated weight, 950 lbs.; fed 53.844 lbs. corn and cob meal, 43.071 lbs. gluten meal, 21.532 lbs. bran, 21.532 lbs. oil meal and 385 lbs. ensilage; property of James M. Stratton, Salem, Ohio.

Happy Blossom 18218.—Sire, Happy (P. S. 211 J. H. B.); dam, Cocette, on I. of J.

 Butter, 19 lbs. 5 oz. Milk, 247 lbs. 8 oz.

 Test made from Mar. 14 to 21, 1886; age, 5 yrs. 6 mos.; property of William H. Burr, Redding Ridge, Conn.

Happy May Girl 75588.—Sire, Happy Marquis 26801; dam, Silkey of Mt. Pleasant 68727.

 Butter, 14 lbs. 15½ oz. Milk, 240 lbs. 12½ oz.

 Test made from Mar. 17 to 24, 1898; age, 7 yrs. 10 mos.; estimated weight, 850 lbs.; fed 11 lbs. per day of the H. O. Co.'s dairy feed and 163 lbs. hay; property of Jasper S. Houdlette, Dresden, Me.

Happy Winn 35358.—Sire, Winner 5572; dam, Happy H. 20566.

 Butter, 18 lbs. 13 oz. Milk, 282 lbs.

 Test made from Apr. 1 to 8, 1889; age, 4 yrs.; estimated weight, 1000 lbs.; fed 22 lbs. feed meal, daily; property of Campbell Brown, Spring Hill, Tenn.

Hara of St. Lambert 42128.—Sire, Prince of St. Lambert 5287; dam, Hara Rex 19005.

 Butter, 14 lbs. 6 oz. Milk, 237 lbs. 11 oz.

 Test made from Apr. 18 to 25, 1890; age, 3 yrs. 7 mos.; estimated weight, 850 lbs.; fed, daily, about 9½ lbs. corn meal, 6½ lbs. pea meal, 2 lbs. cotton-seed meal and 4 lbs. oats; property of W. B. von Richthofen, Montclair, Col.

Hara of St. Lambert 42128.—Sire, Prince of St. Lambert 5287; dam, Hara Rex 19005.

 Butter, 17 lbs. 7½ oz. Milk, 332 lbs. 4 oz.

 Test made from Apr. 11 to 18, 1897; age, 10 yrs. 6 mos.; estimated weight, 950 lbs.; fed 60 lbs. wheat bran, 30 lbs. oat meal, 30 lbs. corn meal, 30 lbs. cotton-seed meal. 1½ bush. carrots and apples and about 80 lbs. hay; property of Ayer & McKinney. Philadelphia, Pa.

Harmintha 96920.—Sire, Fancy's Harry 9777; dam, Tormintha 37327.

 Butter, 16 lbs. 7 oz. Milk, 209 lbs. 6 oz.

 Test made from Nov. 30 to Dec. 7, 1895; age, 2 yrs. 7 mos.; actual weight, 693 lbs.; fed 52 lbs. wheat bran, 33 lbs. wheat middlings, 20 lbs. corn meal and 10 lbs. old process oil meal—corn fodder and millet hay *ad lib.*; property of J. P. Bradbury, Pomeroy, Ohio.

Harmony 2d 17118.—Sire, Alfonso 3013; dam, Harmony 4148.

 Butter, 19 lbs. 3 oz. Milk, 273 lbs. 3 oz.

 Test made from May 15 to 21, 1885; age, 5 yrs. 1 mo.; property of Richard Peters, Atlanta, Ga.

Harriet Neptune 52344.—Sire, Baron Neptune 10090; dam, Silistria 20264.

 Butter, 18 lbs. 6½ oz. Milk, 315 lbs.

 Test made from June 5 to 12, 1892; age, 5 yrs. 7 mos.; estimated weight, 900 lbs.; fed 8 qts. oats, peas, barley, corn-hearts and bran, mixed in equal parts and fed in two feeds, daily, on 15 lbs. potatoes - pasture of mixed grasses; property of E. G. Youmans, Groton City, N. Y.

Harry's Duchess 60289.—Sire, Fancy's Harry 9777; dam, Duchess of Bloomfield 2d 12250.

Butter, 16 lbs. 4 oz.   Milk, 206 lbs. 8 oz.

Test made from Nov. 4 to 11, 1896; age, 7 yrs. 3 mos.; estimated weight, 925 lbs.; fed 112 lbs. corn meal, 70 lbs. ground oats and 28 lbs. bran—good blue-grass pasture; property of Est. of Campbell Brown, Spring Hill, Tenn.

Harry's Ernestine 66232.—Sire, Fancy's Harry 9777; dam, Ernestine Mc. 10686.

Butter, 15 lbs. 6½ oz.   Milk, 190 lbs.

Test made from May 5 to 12, 1889; age, 2 yrs. 5 mos.; estimated weight, 750 lbs.; fed, daily, 4 gals., two-thirds corn and one-third oats, ground; property of Maury Jersey Farm, Columbia, Tenn.

Harry's Ethel 63970.—Sire, Ethleel's Harry 15962; dam, Fanny White 9804.

Butter, 14 lbs.   Milk, 177 lbs. 4 oz.

Test made from Sept. 18 to 25, 1894; age, 5 yrs. 7 mos.; estimated weight, 775 lbs.; fed 102 lbs. corn meal, 105 lbs. shorts, 19 lbs. oil meal and about 150 lbs. hay; property of F. S. Case, Logan, Ohio.

Harry's Fancita 96721.—Sire, Fancy's Harry 9777; dam, Pitapat 11136.

Butter, 20 lbs. 2½ oz.   Milk, 266 lbs. 8 oz.

Test made from Jan. 31 to Feb. 7, 1897; age, 3 yrs. 9 mos.; actual weight, 860 lbs.; fed 50 lbs. corn and cob meal, 50 lbs. wheat bran, 25 lbs. middlings and 10 lbs. old process oil meal—ensilage and cut corn fodder *ad lib.*; property of J. P. Bradbury, Pomeroy, Ohio.

Harry's Fancy 39114.—Sire, Fancy's Harry 9777; dam, Norella King 2d 23933.

Butter, 14 lbs. 6½ oz.   Milk, 233 lbs. 12 oz.

Test made from Apr. 26 to May 3, 1890; age, 5 yrs. 2 mos.; actual weight, 810 lbs.; fed, daily, between 16 and 24 lbs. ground corn and oats; property of Wm. Morrow & Son, Nashville, Tenn.

Harry's Fancy Belle 44764.—Sire, Fancy's Harry 9777; dam, Belle Lyman 28522.

Butter, 14 lbs. 8 oz.   Milk, 197 lbs.

Test made from May 18 to 25, 1893; age, 6 yrs. 2 mos.; estimated weight, 800 lbs.; fed 56 qts. ground oats, 56 qts. corn meal and 28 qts. wheat bran and a little clover hay—pasture of mixed white clover and blue-grass; property of Morgan & Brown, Columbia, Tenn.

Harry's Fancy Belle 2d 82416.—Sire, Landseer's Pogis 15847; dam, Harry's Fancy Belle 44764.

Butter, 14 lbs. 12½ oz.   Milk, 157 lbs.

Test made from Nov. 15 to 22, 1893; age, 4 yrs. 7 mos.; estimated weight, 750 lbs.; fed 56 qts. ground oats and 56 qts. corn meal—clover and millet, woodland pasture; property of Morgan & Brown, Columbia, Tenn.

Harry's Florrie 86263.—Sire, Fancy's Harry 6th 24768; dam, King's Florrie 2d 64455.

Butter, 14 lbs. 12 oz.   Milk, 198 lbs. 3 oz.

Test made from Apr. 3 to 10, 1897; age, 4 yrs. 7 mos.; estimated weight, 800 lbs.; fed 28 qts. wheat bran, 28 qts. corn meal and 28 qts. ground oats—clover and timothy hay, some blue-grass pasture; property of Henry DuBois, Vigo, Ohio.

Harry's May Bud 111817.—Sire, Little Harry's Wonder 35289; dam, Massie's Lula 69727.

Butter, 14 lbs. 15½ oz. Milk, 250 lbs. 3 oz.

Test made from Feb. 23 to Mar. 2, 1898; age, 2 yrs. 10 mos.; estimated weight, 850 lbs.; fed 112 qts. sorghum ensilage, 98 qts. bran. 42 qts. oats and corn, ground together, and 12 lbs. cotton-seed meal—prairie hay *ad lib.*, 8 hours on barley pasture during test; property of J. D. Gray, Terrell, Tex.

Harry's May Fair 37499.—Sire, Fancy's Harry 9777; dam, May Fair, 5184.

Butter, 16 lbs. 1 oz. Milk, 131 lbs. 4 oz.

Test made from Mar. 23 to 30, 1888; age, 2 yrs. 5 mos.; estimated weight, 800 lbs.; fed, daily, 16 qts. corn and oats; property of W. R. Rison, Huntsville, Ala.

Harry's Nervilette 86262.—Sire, Fancy's Harry 6th 24768; dam, Romp's Nervilette 72282.

Butter, 18 lbs. 1 oz. Milk, 213 lbs. 8 oz.

Test made from Jan. 31 to Feb. 7, 1897; age, 4 yrs. 9 mos.; estimated weight, 950 lbs.; fed 84 qts. ground corn and oats and 84 qts. wheat bran—good clover hay *ad lib.*; property of Henry DuBois, Vigo, Ohio.

Harry's Pet 52838.—Sire, Fancy's Harry 9777; dam, Pet of Wildwood 33395.

Butter, 22 lbs. 11 oz. Milk, 157 lbs. 10 oz.

Test made from June 7 to 14, 1891; age, 2 yrs. 9 mos.; estimated weight, 800 lbs.; fed 70 qts. corn meal. 35 qts. wheat bran, 14 qts. oil meal, on chopped clover hay—good clover and orchard grass pasture; property of Matthews & Humes, Huntsville, Ala.

Harry's Rovalti 115843.—Sire, Harry Torment's Signal 39846; dam, Rovalti's Maid 90089.

Butter, 17 lbs. ¼ oz. Milk, 241 lbs.

Test made from July 23 to 30, 1897; age, 3 yrs.; actual weight, 839 lbs.; fed 35 lbs. each of corn meal and wheat middlings, 15 lbs. old process oil meal and 25 lbs. wheat bran—good woods blue-grass pasture; property of J. P. Bradbury, Pomeroy, Ohio.

Harry's Ruth Morgan 107773.—Sire, Noonday 12961; dam, Ruth Morgan 3d 57043.

Butter, 14 lbs. 1.6 oz. Milk, 213 lbs. 8 oz.

Test made from Jan. 13 to 20, 1897; age, 3 yrs. 4 mos.; estimated weight, 750 lbs.; fed 210 lbs. roots, 21 lbs. corn meal, 70 lbs. bran, 28 lbs. cotton-seed meal and 10 lbs. cut sorghum and hay—some long hay; property of Hecla Jersey Cattle Co., Earlington, Ky.

Harry's Signaline 82419.—Sire, Fancy's Harry 9777; dam, Landseer's Signaline 58101.

Butter, 14 lbs. 8½ oz. Milk, 163 lbs. 4 oz.

Test made from Dec. 7 to 14, 1893; age, 4 yrs. 6 mos.; estimated weight, 775 lbs.; fed 56 qts. ground oats and 56 qts. corn meal—clover and millet, woodland pasture; property of Morgan & Brown, Columbia, Tenn.

Harry's Signaline 2d 82428.—Sire, Tennessee's Tormentor 23498; dam, Harry's Signaline 82419.

Butter, 20 lbs. 11 oz. Milk, 234 lbs. 12 oz.

Test made from Jan. 30 to Feb. 6, 1896; age, 4 yrs. 11 mos.; estimated weight, 850 lbs.; fed 27 qts. corn meal, 84 qts. wheat bran, 45 qts. cob meal, 13 qts. pea meal and 1½ qts. cotton-seed meal—hay *ad lib.*, green wheat pasture; property of Lindley H. Henley, Marshall, Tex.

Hartwick Belle 7722.—Sire, Grand Duke Alexis 1040; dam, Clytemnestra 2455.

Butter, 14 lbs. 8 oz. Milk, 189 lbs. 13 oz.

Test made from May 3 to 9, 1883; age, 4 yrs. 6 mos.; property of William Simpson, New York, N. Y.

Hatita 34538.—Sire, Footstep 5163; dam, Hilda D. 6683.

Butter, 14 lbs. 15 oz. Milk, 197 lbs. 8 oz.

Test made from June 27 to July 4, 1888; age, 3 yrs. 2 mos.; estimated weight, 875 lbs.; fed, daily, 5 lbs. grain, consisting of corn meal, bran and oil meal; property of Frederic Bronson, Greenfield Hill, Conn.

Hattie Adams 80909.—Sire, Jersey King 9458; dam, Jessica Fries 2d 60836.

Butter, 14 lbs. Milk, 226 lbs. 15 oz.

Test made from May 14 to 21, 1895; age, 3 yrs.; estimated weight, 770 lbs.; fed 84 qts. wheat bran, 42 qts. corn meal, 21 qts. cotton-seed meal, 14 qts. linseed meal and 18 lbs. ensilage—some green clover, short pasture; property of E. B. C. Hambley, Rockwell, N. C.

Hattie Douglass 24960.—Sire, Dainty Boy 2955; dam, Hattie 795.

Butter, 16 lbs. 5 oz. Milk, 262 lbs.

Test made from June 7 to 13, 1884; age, 5 yrs. 2 mos.; property of H. W. Douglass, Pevely, Mo.

Hattie Landseer 72719.—Sire, Pride's Landseer 18590; dam, Hattie D. Rex 38772.

Butter, 14 lbs. 4 oz. Milk, 246 lbs. 14 oz.

Test made from Mar. 1 to 8, 1898; age, 6 yrs. 11 mos.; estimated weight, 850 lbs.; fed 40 lbs. ensilage, 5 lbs. bran, 2½ lbs. corn, 2½ lbs. oats, 1¼ lbs. cotton-seed meal and about 10 lbs. hay, daily; property of Pennsylvania Reform School, Morganza, Pa.

Hattie N. 45433.—Sire, Roxana's Gilderoy 11585; dam, Bessie of Greenwood 9752.

Butter, 15 lbs. 14½ oz. Milk, 150 lbs. 9½ oz.

Test made from June 12 to 19, 1892; age, 5 yrs. 8 mos.; actual weight, 750 lbs.; fed 18 lbs. per day of ground oats and corn, in equal parts—good orchard and blue-grass pasture; property of Matt. M. Gardner, Nashville, Tenn.

Hattie Swett 109709.—Sire, Exile's Rebel 28928; dam, Southern Belle 18570.

Butter, 15 lbs. 10¼ oz. Milk, 155 lbs. 12½ oz.

Test made from May 2 to 9, 1897; age, 4 yrs. 2 mos.; estimated weight, 700 lbs.; fed 14½ lbs. corn meal, 2 lbs. oil meal, 3 lbs. molasses feed and 6 lbs. ground oats, daily—rye and Bermuda pasture; property of William E. Oates & Son, Vicksburg, Miss.

Hawthorn Minnie 26687.—Sire, Minnie's Duke of Darlington 6934; dam, Wash Maid 15638.

Butter, 18 lbs. 12 oz. Milk, 278 lbs.

Test made from Jan. 4 to 11, 1898; age, 13 yrs. 10 mos.; estimated weight, 950 lbs.; fed about 20 qts. daily of a mixture of ground oats, middlings, wheat bran, corn meal and some pea meal—clover hay and ensilage; property of T. S. Cooper, Coopersburg, Pa.

Hazalena's Butterfly 10123.—Sire, Mr. Guppy 993; dam, Hazalena 3275.

   Butter, 14 lbs.

  Test made from May 22 to 29, 1882; age, 8 yrs.; property of George W. Hulick, Batavia, Ohio.

Hazel Lee 76864.—Sire, Hulee 18209; dam, Witch-Hazel 4th 6131.

   Butter, 16 lbs. 1½ oz. Milk, 237 lbs.

  Test made from June 21 to 28, 1893; age, 2 yrs. 8 mos.; estimated weight, 825 lbs.; fed 14 lbs. corn-hearts and 4 lbs. cut sheaf oats, daily—blue-grass pasture; property of Campbell Brown, Spring Hill, Tenn.

Hazel St. L. 100035.—Sire, Homestead Victor Pogis 26535; dam, Zero St. L. 77230.

   Butter, 18 lbs. 10 oz. Milk, 314 lbs. 8 oz.

  Test made from May 19 to 26, 1897; age, 4 yrs. 4 mos.; actual weight, 800 lbs.; fed 84 qts. corn and cob meal—blue-grass pasture; property of J. F. Moore & Son, Moorefield, Ohio.

Hazelwood 52239.—Sire, Prince of Broadmoor 8005; dam, Grace Carpenter 30934.

   Butter, 14 lbs. 1 oz. Milk, 257 lbs.

  Test made from Apr. 13 to 20, 1895; age, 8 yrs. 4 mos.; estimated weight, 800 lbs.; fed 8 qts. daily of ground corn and oats, in equal parts—timothy hay of rather poor quality *ad lib.*; property of R. Porter Craig & Sons, Uniontown, Pa.

Hazen's Bess 7329.—Sire, Lord Bronx 2d 1730; dam, Zina 3d 4134.

   Butter, 24 lbs. 11 oz. Milk, 344 lbs. 13½ oz.

  Test made from Nov. 5 to 12, 1883; age, 7 yrs. 5 mos.; fed 6½ lbs. corn meal and 4 lbs. bran, daily, 7 ears of corn daily for three days, 5 lbs. cooked potatoes daily for two days, 8 lbs. beets and 8 lbs. turnips on one day—pasturage, short blue-grass; property of C. C. Crockett, Richmond, Ind.

Hazen's Nora 4791.—Sire, Norajah 812; dam, Zina 3d 4134.

   Butter, 20 lbs. 5 oz. Milk, 299 lbs.

  Test made from June 14 to 21, 1884; age, 9 yrs. 1 mo.; fed 2 qts. corn meal, 2 qts. ground oats, 2 qts. wheat bran, 1 qt. pea meal and 1 pt. oil meal, daily, with cut grass—cut grass at noon, poor pasture; property of Moulton Bros., West Randolph, Vt.

Hearty Pogis 81297.—Sire, Sig Pogis 22806; dam, Helen Walker 53465.

   Butter, 18 lbs. 13¾ oz. Milk, 220 lbs.

  Test made from Nov. 24 to Dec. 1, 1895; age, 5 yrs.; estimated weight, 800 lbs.; fed 8 qts. ground oats and corn meal, 4 qts. bran and 1 qt. cotton-seed meal, daily—hay *ad lib.* at night, blue-grass and winter oats pasture; property of D. P. Carter, Lynnville, Tenn.

Hebe's Fancy 84314.—Sire, King Jim 24346; dam, Hebe's Pride 76453.

   Butter, 14 lbs. 9 oz. Milk, 230 lbs. 13 oz.

  Test made from May 19 to 26, 1897; age, 5 yrs. 7 mos.; estimated weight, 750 lbs.; fed 28 qts. corn meal, 28 qts. linseed meal and 21 qts. bran—clover pasture; property of E. A. Tyrrell, Marysville, Ohio.

Heclatette 119403.—Sire, Old Tenella's Signal 34802; dam, Tette 20802.

Butter, 16 lbs. 9½ oz. Milk, 221 lbs. 3.2 oz.

Test made from Mar. 8 to 15, 1898; age, 3 yrs. 2 mos.; estimated weight, 675 lbs.; fed, 42 qts. corn and oats, 56 qts. bran and cotton-seed meal, 7 qts. oil meal and 168 lbs. roots—red-top hay *ad lib.*; property of Hecla Jersey Cattle Co., Earlington, Ky.

Heiress of St. Lambert 56507.—Sire, One Hundred per Cent. 16590; dam, Peony of St. Lambert 56501.

Butter, 16 lbs. 13½ oz. Milk, 209 lbs.

Test made from Oct. 10 to 17, 1891; age, 3 yrs.; estimated weight, 750 lbs.; fed 20 lbs. chopped peas and oats and 2 lbs. oil-cake, daily—hay, clover and timothy pasture; property of William Rolph, Markham, Ont., Can.

Helen Barry 55840.—Sire, Little Harry 8808; dam, Charlotte Brooks 29580.

Butter, 18 lbs. 7 oz. Milk, 225 lbs. 15 oz.

Test made from June 13 to 20, 1891; age, 3 yrs. 2 mos.; estimated weight, 700 lbs.; fed 2 gals. corn meal, 2 gals. bran and 3 gals. ground oats, daily, on green corn, cut short—short pasture; property of Matthews & Moore, Huntsville, Ala.

Helene of St. Lambert 73601.—Sire, St. Lambert's John Bull 16618; dam, Southern Pet 26675.

Butter, 20 lbs. 1½ oz. Milk, 207 lbs. 7 oz.

Test made from Jan. 30 to Feb. 6, 1894; age, 3 yrs. 10 mos.; estimated weight, 900 lbs.; fed 28 qts. cotton-seed meal, 82 qts. wheat bran, 40 qts. ground corn and oats and 140 lbs. cotton-seed hulls; property of W. W. Lipscomb, Luling, Tex.

Helen Stoke Pogis 31947.—Sire, Exile of St. Lambert 13657; dam, Lady Delphine 28460.

Butter, 17 lbs. 8 oz. Milk, 236 lbs. 8 oz.

Test made from July 9 to 16, 1888; age, 3 yrs. 3 mos.; estimated weight, 900 lbs.; fed, daily, 3 lbs. bran, 9 lbs. corn meal, 10½ lbs. oil meal, 3½ lbs. fine middlings and 3 lbs. ground oats; property of A. D. McBride, Rochester, N. Y.

Helen Walker 53465.—Sire, Chendado 12955; dam, René of Hillside 12952.

Butter, 19 lbs. 10½ oz. Milk, 241 lbs. 4 oz.

Test made from Mar. 31 to Apr. 7, 1895; age, 7 yrs. 3 mos.; estimated weight, 800 lbs.; fed 42 qts. ground oats, 14 qts. cotton-seed meal, 28 qts. crushed corn and 28 qts. bran—good rye pasture; property of D. P. Carter, Lynnville, Tenn.

Helpmeet 62459.—Sire, Ida's Rioter of St. L. 13656; dam, Sweet Blossom Pogis 36995.

Butter, 17 lbs. 15½ oz. Milk, 202 lbs. 4 oz.

Test made from May 30 to June 6, 1892; age, 3 yrs. 9 mos.; actual weight, 900 lbs.; fed 24 qts. grain daily—second season clover and timothy pasture; property of Miller & Sibley, Franklin, Pa.

Hennette 11624.—Sire, Fast Boy 2606; dam, Hennie 3335.

Butter, 14 lbs. 3½ oz. Milk, 144 lbs. 8 oz.

Test made from Dec. 11 to 18, 1887; age, 8 yrs. 7 mos.; estimated weight, 750 lbs.; fed, daily, 8 lbs. corn meal, 2 lbs. oil meal and 8 lbs. bran; property of James Stillman, Sing Sing, N. Y.

Herberta 8811.—Sire, Arnold's Bronx 3309; dam, Hazen's Bess 7329.

Butter, 16 lbs. 15 oz. Milk, 231 lbs. 15 oz.

Test made from June 4 to 10, 1885; age, 6 yrs. 2 mos.; property of Samuel McKeen, Terre Haute, Ind.

Heritage 50077.—Sire, Ida's Stoke Pogis 13658; dam, Narda 38012.

Butter, 15 lbs. 15 oz. Milk, 240 lbs. 10 oz.

Test made from Jan. 15 to 22, 1892; age, 4 yrs.; estimated weight, 850 lbs.; fed 2 gals. corn meal, 4 gals. boiled cotton-seed and 2 gals. wheat bran, daily—ensilage and clover hay *ad lib.*; property of Matthews & Humes, Huntsville, Ala.

Hermione C. 90598.—Sire, Fritz Pogis of Glynllyn 28042; dam, Bab 3d 40258.

Butter, 14 lbs. 11 oz. Milk, 222 lbs.

Test made from Apr. 6 to 13, 1896; age, 3 yrs. 7 mos.; estimated weight, 1000 lbs.; fed 10 lbs. daily of a mixture consisting of one-ninth oil meal, three-ninths corn meal, and five-ninths bran, also about 14 lbs. poor clover and 8 qts. mixed carrots and mangels; property of M. H. Olin, Perry, N. Y.

Hermosa Pogis 40369.—Sire, Prince of Melrose 2d 11015; dam, Litta Oaks' Alphea 26676.

Butter, 16 lbs. 3½ oz. Milk, 166 lbs. 14 oz.

Test made from Feb. 13 to 20, 1890; age, 3 yrs. 8 mos.; estimated weight, 700 lbs.; fed 42 qts. corn meal, 42 qts. ground oats, 28 qts. raw cotton-seed meal and 30 lbs. ensilage—hay, good pasture; property of James Crook, Jacksonville, Ala.

Herpa S. 127829.—Sire, Melia Ann's Son 22041; dam, Herpa 54293.

Butter, 18 lbs. 6 oz. Milk, 279 lbs. 2 oz.

Test made from June 28 to July 5, 1898; age, 2 yrs. 6 mos.; estimated weight, 875 lbs.; fed 8 qts. ground corn and oats, 4 qts. wheat middlings, 2 qts. cotton-seed meal and 1 qt. oil meal, daily—good pasture; property of E. L. Van Deusen, Ashley Falls, Mass.

Hertha 26102, imp.

Butter, 14 lbs. 12 oz. Milk, 274 lbs. 10 oz.

Test made from Apr. 4 to 11, 1886; age, 5 yrs.; property of W. A. Sudduth, Flemingsburg, Ky.

Hettie of Briarcliff 26621.—Sire, Domino of Darlington 2459; dam, Hennette 11624.

Butter, 18 lbs. 1 oz. Milk, 185 lbs.

Test made from May 31 to June 7, 1888; age, 4 yrs.; estimated weight, 700 lbs.; fed, daily, 8 lbs. corn meal and 4 lbs. middlings; property of James Stillman, Sing Sing, N. Y.

Highland Ida 38427.—Sire, Ida's Rioter of St. L. 13656; dam, Oakland Girl 11103.

Butter, 18 lbs. 1½ oz. Milk, 279 lbs. 8 oz.

Test made from Aug. 9 to 16, 1890; age, 4 yrs. 9 mos.; actual weight, 960 lbs.; fed, daily, 4 lbs. 8 oz. wheat bran, average of 11 lbs. corn meal and 2 lbs. 11 oz. ground oats; property of H. A. Slack, Hurstville, N. Y.

High Spirits of Lawn 61371.—Sire, Lord Hugo 15830; dam, High Spirits 35390.

Butter, 16 lbs. 9 oz. Milk, 208 lbs. 6 oz.

Test made from May 23 to 30, 1892; age, 3 yrs. 2 mos.; estimated weight, 750 lbs.; fed 12 lbs. bran, 16 lbs. corn meal and 4 lbs. cotton-seed meal, daily—wheat and oats pasture; property of Platter & Foster, Denison, Tex.

High Spirits of Lawn 61371.—Sire, Lord Hugo 15830; dam, High Spirits 35890.

Butter, 18 lbs. 7 oz. Milk, 237 lbs. 12 oz.

Test made from May 5 to 12, 1893; age, 4 yrs. 2 mos.; estimated weight, 800 lbs.; fed 10 lbs. bran, 8 lbs. corn meal, 10 lbs. ground oats and 2 lbs. cotton-seed meal, daily—prairie grass; property of Platter & Foster, Denison, Tex.

High Tea 65577.—Sire, Major Appel Pogis 17861; dam, Belle Clarendon 36334.

Butter, 17 lbs. 6½ oz. Milk, 202 lbs.

Test made from Jan. 24 to 31, 1894; age, 5 yrs. 7 mos.; actual weight, 925 lbs.; fed 84 qts. ground oats, 52 qts. corn meal, 24 qts. wheat bran and 22 qts. ground oil-cake; property of Miller & Sibley, Franklin, Pa.

Hilda 18178, imp.

Butter, 15 lbs. 4 oz. Milk, 196 lbs. 9 oz.

Test made from Jan. 6 to 13, 1886; age, 10 yrs.; estimated weight, 1025 lbs.; fed 3 qts. oats, 2 qts. middlings, 2 qts. oil meal and 1 qt. pea meal, daily; property of T. R. Proctor, Utica, N. Y.

Hilda 2d 5447.—Sire, Chief Justice 2d 1643; dam, Hilda 942.

Butter, 23 lbs. 5 oz. Milk, 289 lbs.

Test made from Sept. 3 to 9, 1884; age, 9 yrs. 4 mos.; fed 12 lbs. corn meal, 8 lbs. ground oats, 4 lbs. middlings and 8 lbs. oil meal, daily—hay *ad lib.*, no pasture; property of Miller & Sibley, Franklin, Pa.

Hilda 2d 14967.—Sire, Farmer's Glory 5196; dam, Hilda 18178.

Butter, 14 lbs. 12½ oz. Milk, 115 lbs. 6 oz.

Test made from Apr. 8 to 15, 1886; age, 5 yrs. 6 mos.; property of Webster & Morrow, Columbia, Tenn.

Hilda A. 2d 11120.—Sire, Wanderer 3014; dam, Hilda A. 3951.

Butter, 20 lbs. Milk, 280 lbs.

Test made from July 12 to 19, 1884; age, 4 yrs. 2 mos.; fed 27 lbs. daily, two-thirds ground corn and oats (one part of corn to two of oats), the remainder oil meal; property of George W. Farlee, Cresskill, N. J.

Hilda A. 3d 16636.—Sire, Footstep 5163; dam, Hilda A. 3951.

Butter, 17 lbs. 1 oz. (official). Milk, 246 lbs. 15 oz.

Test made from May 25 to 31, 1886; age, 4 yrs. 10 mos.; weight, 1240 lbs.; fed 86 lbs. crushed oats, 31 lbs. corn meal, 22 lbs. bran, 17 lbs. oil meal and 21 lbs. pea meal; property of Frederic Bronson, Greenfield Hill, Conn.

Hilda B-B-B 17640.—Sire, Prince of Bridgewater 2d 4489; dam, Hilda B-B 8599.

Butter, 15 lbs. 8 oz. Milk, 246 lbs. 8 oz.

Test made from July 29 to Aug. 5, 1884; age, 3 yrs. 1 mo.; property of William Craik, Frankfort, Ky.

Hilda D. 6683.—Sire, Chief Justice 2d 1643; dam, Hilda C. 3869.

Butter, 18 lbs. 5 oz., unsalted. Milk, 268 lbs. 8 oz.

Test made from May 27 to June 2, 1883; age, 6 yrs. 3 mos.; property of Frederic Bronson, Southport, Conn.

Hilda D. 6683.—Sire, Chief Justice 2d 1643; dam, Hilda C. 3869.

Butter, 21 lbs. 2½ oz. (official). Milk, 249 lbs. 8 oz.

Test made from June 23 to 29, 1885; age, 9 yrs. 4 mos.; property of Frederic Bronson, Greenfield Hill, Conn.

Hilda Glenn 81195.—Sire, Catono's Harry 20096; dam, Jane Glenn 56947.

Butter, 27 lbs. Milk, 239 lbs. 14 oz.

Test made from June 8 to 15, 1896; age, 5 yrs. 2 mos.; estimated weight, 750 lbs.; fed 105 qts. corn and cob meal, ground wheat and ground oats, in equal parts by measure, and 14 qts. old process oil-cake meal—timothy and blue-grass pasture; property of H. C. Schlegel, Daleville, Ind.

Hilda Glenn 81195.—Sire, Catono's Harry 20096; dam, Jane Glenn 56947.

Butter, 121 lbs. 7½ oz. in 35 days. Milk, 1181 lbs. 13 oz.

Test made from May 25 to June 29, 1896; age, 5 yrs. 2 mos.; estimated weight, 750 lbs.; fed 525 qts. corn and cob meal, ground wheat and ground oats, in equal parts by measure, and 70 qts. old process oil-cake meal—timothy and blue-grass pasture; property of H. C. Schlegel, Daleville, Ind.

Hildah of Clovernook 51597.—Sire, Dixie Prince 18379; dam, Trixket 16292.

Butter, 19 lbs. 2½ oz. Milk, 221 lbs.

Test made from Jan. 2 to 9, 1896; age, 7 yrs. 11 mos.; estimated weight, 700 lbs.; fed 2 qts. ground oats, 2 qts. ground peas, 2 qts. ground corn, 1 qt. bran and 1 qt. cotton-seed meal, daily—cut sheaf oats, corn stover, hay and poor blue-grass pasture; property of Samuel N. Warren, Spring Hill, Tenn.

Hindoo Rose 14602.—Sire, Duke of Thornebrook 3832; dam, Carrie's Beauty 14601.

Butter, 14 lbs. 7 oz.

Test made from Dec. 10 to 16, 1885; age, 6 yrs. 8 mos.; property of Thomas S. Yocom, Richmond, N. Y.

Hinds' Tormentor of Lawn 71640.—Sire, Tormentor's John 14715; dam, Mollie Pennebaker 28031.

Butter, 14 lbs. 7 oz. Milk, 203 lbs. 14 oz.

Test made from Dec. 16 to 23, 1894; age, 4 yrs. 5 mos.; estimated weight, 800 lbs.; fed 70 lbs. wheat bran, 84 lbs. ground oats, 56 lbs. corn meal and 42 lbs. cotton-seed meal—sorghum hay once a day; property of Platter & Foster, Denison, Tex.

Hoey Rex 82364.—Sire, Hazel's Rex 7443; dam, Hoey 6th 5315.

Butter, 14 lbs. 9 oz. Milk, 220 lbs. 1 oz.

Test made from Nov. 30 to Dec. 7, 1894; age, 8 yrs. 4 mos.; estimated weight, 950 lbs.; fed 200 qts. of a mixture of corn meal, oats, bran and oil meal, on moistened cut hay; property of R. F. Shannon, Pittsburg, Pa.

Hollie Pogissa 110490.—Sire, Sig Pogis 22806; dam, Coral's Calico 91825.

Butter, 16 lbs. 3¼ oz. Milk, 213 lbs.

Test made from Apr. 30 to May 7, 1896; age, 2 yrs. 3 mos.; estimated weight, 600 lbs.; fed 28 qts. ground oats, 28 qts. wheat bran, 28 qts. crushed corn and 7 qts. cotton-seed meal—rye pasture; property of M. M. Johnson, Lynnville, Tenn.

Holyoke's Leda 52812.—Sire, Holyoke 8147; dam, Dusky Le Brocq 22967.

Butter, 18 lbs. 11½ oz. Milk, 265 lbs. 12 oz.

Test made from Oct. 14 to 21, 1896; age, 10 yrs.; actual weight, 945 lbs.; fed 12 qts. bran, 3 qts. corn meal, 6 qts. ground oats and 3 qts. oil meal, daily—hay; property of C. A. Sweet, East Aurora, N. Y.

Home Matron 6707.—Sire, Duke of Lebanon 1880; dam, Lebanon Wife 6102.

Butter, 14 lbs. Milk, 153 lbs.

Test made from June 7 to 13, 1882; age, 5 yrs. 2 mos.; property of J. M. Hoover, Bradford, Ohio.

Homestead Lucy 96325.—Sire, Lucy's Stoke Pogis 4th 26533; dam, Homestead St. Lambert 56199.

Butter, 14 lbs. 10 oz. Milk, 228 lbs. 6 oz.

Test made from Nov. 20 to 27, 1895; age, 3 yrs. 3 mos.; estimated weight, 750 lbs.; fed 7 lbs. gluten, 7 lbs. bran, 2 lbs. oil meal and about 30 lbs. ensilage, daily—short blue-grass pasture, a few beets; property of W. J. Hussey, Mount Pleasant, Ohio.

Homestead Lucy 96325.—Sire, Lucy's Stoke Pogis 4th 26533; dam, Homestead St. Lambert 56199.

Butter, 17 lbs. 12 oz. Milk, 294 lbs. 8 oz.

Test made from Dec. 21 to 28, 1896; age, 4 yrs. 4 mos.; estimated weight, 800 lbs.; fed about 8 lbs. bran and gluten feed, 4 lbs. corn and oats, 2 lbs. oil meal and 35 lbs. ensilage, daily—clover hay *ad lib.*; property of W. J. Hussey, Mount Pleasant, Ohio.

Homestead Matilda 96448.—Sire, Lucy's Stoke Pogis 4th 26533; dam, Oaklands Nora 2d's Fille 94972.

Butter, 17 lbs. 8 oz. Milk, 220 lbs. 8 oz.

Test made from June 5 to 12, 1898; age, 4 yrs. 2 mos.; estimated weight, 700 lbs.; fed 12 lbs. daily of a mixture of bran, gluten feed, and "Victor" corn and oat feed—good blue-grass pasture; property of W. J. Hussey, Mount Pleasant, Ohio.

Homestead Pogis 44231.—Sire, Cill's Stoke Pogis 14149; dam, Lily of Riverside 19599.

Butter, 14 lbs. 11 oz. Milk, 228 lbs. 9 oz.

Test made from Nov. 6 to 13, 1889; age, 3 yrs. 9 mos.; estimated weight, 825 lbs.; fed, daily, 5 lbs. corn meal, 4 lbs. ship stuff and 2 lbs. oil meal; property of Horace G. Westlake, Hillsdale, N. Y.

Homestead Prue 82203.—Sire, King of St. Lambert 2d 17823; dam, Homestead Pogis 2d 57451.

Butter, 14 lbs. 13½ oz. Milk, 215 lbs. 6 oz.

Test made from Nov. 6 to 13, 1894; age, 4 yrs.; estimated weight, 850 lbs.; fed 9 lbs. grano-gluten, 2 lbs. oil meal and 30 lbs. ensilage, daily—hay *ad lib.*; property of W. J. Hussey, Mount Pleasant, Ohio.

Homestead Rowena Pogis 89591.—Sire, Lucy's Stoke Pogis 4th 26533; dam, Homestead Pogis 2d 57451.

Butter, 16 lbs. 7½ oz. Milk, 225 lbs. 8 oz.

Test made from Feb. 6 to 13, 1897; age, 3 yrs. 7 mos.; estimated weight, 800 lbs.; fed 30 lbs. ensilage and about 14 lbs. bran, gluten, corn and oats chop and oil meal, daily—hay *ad lib.*; property of W. J. Hussey, Mount Pleasant, Ohio.

Honey Belle 25824.—Sire, Duke of Darlington 2460; dam, Honeydrop 10033.

Butter, 20 lbs. 7½ oz.   Milk, 257 lbs. 3 oz.

Test made from Nov. 1 to 8, 1888; age, 5 yrs. 7 mos.; estimated weight, 900 lbs.; fed, daily, 14 lbs. of ground oats, 13 lbs. corn meal, 12 lbs. of shorts and 2 lbs. oil meal; property of D. F. Appleton, Ipswich, Mass.

Honeydrop 10033.—Sire, Guy Warwick 1450; dam, Lady Pauline 2651.

Butter, 14 lbs. ½ oz., unsalted.   Milk, 236 lbs. 12 oz.

Test made from May 1 to 7, 1883; age, 6 yrs.; property of A. B. Darling, Ramsey's, N. J.

Honeymoon of St. Lambert 11221.—Sire, Stoke Pogis 3d 2238; dam, Bijou of St. Lambert 5112.

Butter, 20 lbs. 5¼ oz.   Milk, 250 lbs. 8 oz.

Test made from Mar. 28 to Apr. 4, 1884; age, 4 yrs. 10 mos.; property of Valancey E. Fuller, Hamilton, Ont., Can.

Honeysuckle of St. Anne's 18674.—Sire, Jack Frost of St. Lambert 2419; dam, Clematis of St. Lambert 5478.

Butter, 14 lbs. 14 oz.   Milk, 285 lbs.

Test made from Dec. 3 to 9, 1883; age, 3 yrs. 9 mos.; property of William Rolph, Markham, Ont., Can.

Honeysuckle of St. Anne's 18674.—Sire, Jack Frost of St. Lambert 2419; dam, Clematis of St. Lambert 5478.

Butter, 14 lbs. 14 oz.   Milk, 285 lbs.

Test made from Oct. 14 to 20, 1884; age, 4 yrs. 7 mos.; property of William Rolph, Markham, Ont., Can.

Howdy 55119.—Sire, Young Pedro 9033; dam, Princess Chrysantha 34191.

Butter, 15 lbs. 3½ oz.   Milk, 237 lbs. 11 oz.

Test made from Apr. 22 to 29, 1892; age, 3 yrs. 6 mos.; estimated weight, 650 lbs.; fed 6 qts. corn meal, 6 qts. ground oats, 7 qts. shorts and 1 pt. oil meal, daily—no pasture; property of D. F. Appleton, Ipswich, Mass.

Howland Nightingale 33291.—Sire, Weston 6285; dam, Grace's Nightingale 19855.

Butter, 14 lbs. 1 oz.   Milk, 159 lbs.

Test made from Mar. 22 to 29, 1893; age, 8 yrs.; estimated weight, 950 lbs.; fed 70 lbs. bran, 35 lbs. oil meal and 63 lbs. corn meal; property of Cornelius Easthope, Niles, Ohio.

Hugo's Countess 68394.—Sire, Prince Pogis' Hugo 15060; dam, Countess Coomassie 19339.

Butter, 48.172 lbs. in 30 days.   Milk, 684.2 lbs.

Test made from Aug. 29 to Sept. 27, 1893, at World's Columbian Exposition; age, 5 yrs. 9 mos.; actual weight, 911 lbs.; fed 318.2 lbs. hay, 65.5 lbs. oil meal, 104.5 lbs. ensilage, 216 lbs. bran, 71 lbs. oats, 49 lbs. corn-hearts, 10.5 lbs. cotton-seed, 14 lbs. middlings, 18 lbs. carrots and 4.5 lbs. old hay; property of D. L. Heinshelmer, Glenwood, Iowa.

Hugo's Countess 68394.—Sire, Prince Pogis' Hugo 15660; dam, Countess Coomassie 19339.

Butter, 191.894 lbs. in 90 days. Milk, 3542.9 lbs.

Test made from May 31 to Aug. 28, 1893, at World's Columbian Exposition; age, 5 yrs. 6 mos. at beginning of test; actual weight, 1031 lbs.; fed 342 lbs. old hay, 210 lbs. ensilage, 575.5 lbs. new hay, 191 lbs. oil meal, 506 lbs. bran, 94.5 lbs. oats, 585 lbs. corn-hearts, 144.75 lbs. cotton-seed, 473.5 lbs. middlings, 34 lbs. grano-gluten, 1188.4 lbs. clover and 10 lbs. swale-grass; property of D. L. Heinsheimer, Glenwood, Iowa.

Hugo's Golden Sheen 61804.—Sire, St. Lambert's John Bull 16618; dam, Susie Pogis 35425.

Butter, 17 lbs. 9 oz. Milk, 241 lbs. 7 oz.

Test made from May 31 to June 7, 1892; age, 3 yrs. 6 mos.; estimated weight, 800 lbs.; fed 42 qts. corn meal, 42 qts. bran and shorts, 14 qts. cotton-seed meal and 100 lbs. rich corn ensilage—hay *ad lib.*, good mixed pasture; property of James Crook, Jacksonville, Ala.

Hugo's Joie de Vie 40372.—Sire, Prince of Melrose 2d 11015; dam, Southern Pet 26675.

Butter, 24 lbs. 5 oz. Milk, 243 lbs. 8 oz.

Test made from Apr. 14 to 21, 1893; age, 6 yrs. 9 mos.; estimated weight, 875 lbs.; fed 56 qts. bran and shorts, 42 qts. corn meal, 28 qts. cotton-seed and 30 lbs. ensilage —hay, pasture of good winter rye; property of James Crook, Jacksonville, Ala.

Hugo's Lass 107656.—Sire, L. D.'s Hugo 17374; dam, Hugo's Charm 107655.

Butter, 14 lbs. 1¾ oz. Milk, 282 lbs. 4¼ oz.

Test made from Sept. 4 to 11, 1897; age, 5 yrs. 6 mos.; estimated weight, 950 lbs.; fed 8 lbs. ground oats, 6 lbs. corn meal, 4 lbs. oil meal and 6 lbs. bran, daily—cut clover in barn, poor pasture during daytime; property of Joseph Lawrence, Palisades, N. Y.

Hulabaloo's Flora 111164.—Sire, Hulabaloo 6788; dam, Fancy Farmer's Flora 111163.

Butter, 15 lbs. 13 oz. Milk, 138 qts. 1 pt.

Test made from June 17 to 24, 1897; age, 6 yrs. 11 mos.; actual weight, 870 lbs.; fed 41 qts. ground feed, composed of one-half wheat middlings, one-quarter corn and one-quarter oats—green oats night and morning, white clover and mixed pasture; property of John K. Creevey, Westfield, N. J.

Hulla 7898.—Sire, Hector 2d 2837; dam, Hilpah 5879.

Butter, 19 lbs. 12 oz. Milk, 226 lbs.

Test made from Sept. 13 to 20, 1884; age, 5 yrs. 11 mos.; property of Edwin Thorne, Millbrook, N. Y.

Hurrah Pansy 12153.—Sire, Hurrah 2814; dam, Johnson's Daisy 8391.

Butter, 14 lbs. 1½ oz. Milk, 166 lbs.

Test made from Jan. 13 to 19, 1885; age, 4 yrs. 9 mos.; property of Garretson Bros., Pendleton, Ind.

Hyacinth Pogis 56020.—Sire, Ida's Stoke Pogis 13658; dam, Kitty Better 32911.

Butter, 15 lbs. 6 oz. Milk, 187 lbs. 8 oz.

Test made from Mar. 14 to 21, 1893; age, 4 yrs. 3 mos.; estimated weight, 700 lbs.; fed 48 qts. meal, 27 qts. bran, 6 qts. oil meal and 280 lbs. ensilage; property of H. A. Huntington, Higganum, Conn.

Hypathia 2d 14774.—Sire, Duke of Woodlawn 4160; dam, Hypathia 13358.

Butter, 19 lbs. 13½ oz. Milk, 208 lbs. 8 oz.

Test made from Oct. 9 to 16, 1885; age, 3 yrs. 8 mos.; property of John B. Wallace, Lexington, Ky.

Ianthe 4562.—Sire, St. Helier 45; dam, Blanche 594.

Butter, 16 lbs. 10 oz.

Age, 8 yrs.; property of O. S. Hubbell, Stratford, Conn.

Ida Bashan 4725.—Sire, Apis 1206; dam, Edy Bashan 1032.

Butter, 18 lbs., unsalted. Milk, 126 qts.

Test made from June 12 to 19, 1883; age, 7 yrs. 3 mos.; fed 8 qts. meal and cut hay, daily; property of John F. Maxfield, Bloomfield, N. J.

Idabel Landseer 88399.—Sire, Ida's Landseer 17745; dam, Belle of New York 2d 81731.

Butter, 15 lbs. 15¼ oz. Milk, 299 lbs. 8 oz.

Test made from May 12 to 19, 1896; age, 4 yrs. 11 mos.; estimated weight, 750 lbs.; fed 35 lbs. bran, 42 lbs. ground oats, 14 lbs. oil meal, 31½ lbs. corn meal, 10½ lbs. cotton-seed meal and 210 lbs. ensilage—hay *ad lib.*, timothy and clover pasture; property of C. I. Hood, Lowell, Mass.

Ida Bellman 48273.—Sire, Bellman's Boy 14003; dam, Derwood's Christine 41044.

Butter, 17 lbs. 5½ oz. Milk, 211 lbs. 8 oz.

Test made from May 7 to 14, 1891; age, 4 yrs.; estimated weight, 1000 lbs.; fed 10 lbs. corn and oats chop, 4 lbs. wheat bran, 7 lbs. corn meal, 2 lbs. oil meal, ½ bush. beets, daily—ensilage and hay; property of Mrs. Lucy J. Beebe, Cassadaga, N. Y.

Ida Eva Pogis 39271.—Sire, Ida's Rioter of St. L. 13656; dam, Eva of Snipsic 17650.

Butter, 18 lbs. 6 oz. Milk, 245 lbs. 8 oz.

Test made from July 28 to Aug. 4, 1891; age, 4 yrs.; actual weight, 1025 lbs.; fed 6 qts. ground oats, 3 qts. corn meal and 2 qts. oil meal, daily—good clover pasture; property of Miller & Sibley, Franklin, Pa.

Idalco 117140.—Sire, Noble Paddy 31072; dam, Ida of Clinton 25419.

Butter, 14 lbs. 10 oz. Milk, 173 lbs. 4 oz.

Test made from May 28 to June 4, 1897; age, 2 yrs. 8 mos.; estimated weight, 600 lbs.; no grain fed—good pasture only; property of George E. Nichols, Afton, N. Y.

Idalene 11841.—Sire, Cecco 1673; dam, Tyca 4559.

Butter, 15 lbs. 8½ oz. Milk, 234 lbs. 4 oz.

Test made from Aug. 6 to 12, 1883; age, 4 yrs. 10 mos.; property of William Simpson, New York, N. Y.

Idalette Pogis 64220.—Sire, Ida's Stoke Pogis 13658; dam, Eterna 27134.

Butter, 15 lbs. 4¼ oz. Milk, 223 lbs. 13½ oz.

Test made from July 4 to 11, 1896; age, 6 yrs. 8 mos.; estimated weight, 900 lbs.; fed 22 lbs. per day of ground oats and corn, in equal parts—very good orchard and blue-grass pasture; property of S. H. Butler, Como, Miss.

Ida Longfield 64227.—Sire, Ida's Stoke Pogis 13658; dam, Lady Longfield 23524.

Butter, 14 lbs. 7½ oz. Milk, 191 lbs. 8 oz.

Test made from Apr. 14 to 21, 1894; age, 4 yrs. 3 mos.; estimated weight, 900 lbs.; fed 56 lbs. corn chop and 28 lbs. ground oats—blue-grass pasture; property of Est. of Campbell Brown, Spring Hill, Tenn.

Ida Longfield 64227.—Sire, Ida's Stoke Pogis 13658; dam, Lady Longfield 23524.

Butter, 18 lbs. ½ oz. Milk, 241 lbs. 1¼ oz.

Test made from June 6 to 13, 1895; age, 5 yrs. 6 mos.; estimated weight, 870 lbs.; fed 17 lbs. daily of ground oats and corn, in equal parts—good orchard and blue-grass pasture; property of Matt. M. Gardner, Nashville, Tenn.

Ida Marigold 32615.—Sire, Ida's Rioter of St. L. 13656; dam, Arawana Marigold 9380.

Butter, 25 lbs. 2½ oz. Milk, 244 lbs. 8 oz.

Test made from Sept. 29 to Oct. 6, 1890; age, 5 yrs. 4 mos.; actual weight, 1160 lbs.; fed, daily, 12 qts. ground oats, 8 qts. corn meal and 4 qts. oil meal; property of Miller & Sibley, Franklin, Pa.

Ida Marigold 32615.—Sire, Ida's Rioter of St. L. 13656; dam, Arawana Marigold 9380.

Butter, 59.367 lbs. in 30 days. Milk, 985.8 lbs.

Test made from Aug. 29 to Sept. 27, 1893, at World's Columbian Exposition; age, 8 yrs. 2 mos.; actual weight, 1169 lbs.; fed 424.3 lbs. hay, 84 lbs. oil meal, 112 lbs. ensilage, 18 lbs. corn meal, 223 lbs. bran, 84 lbs. oats, 206 lbs. corn-hearts, 55.5 lbs. cotton-seed, 84 lbs. middlings, 18 lbs. carrots and 6 lbs. old hay; property of C. A. Sweet, Buffalo, N. Y.

Ida Marigold 32615.—Sire, Ida's Rioter of St. L. 13656; dam, Arawana Marigold 9380.

Butter, 199.756 lbs. in 90 days. Milk, 3448.3 lbs.

Test made from May 31 to Aug. 28, 1893, at World's Columbian Exposition; age, 7 yrs. 11 mos. at beginning of test; actual weight, 1184 lbs.; fed 361 lbs. old hay, 213 lbs. ensilage, 639.4 lbs. new hay, 207 lbs. oil meal, 604 lbs. bran, 92 lbs. oats, 614.5 lbs. corn-hearts, 146.50 lbs. cotton-seed, 470 lbs. middlings, 55 lbs. grano-gluten, 1209.4 lbs. clover and 10 lbs. swale-grass; property of C. A. Sweet, Buffalo, N. Y.

Ida Mary Pogis 56126.—Sire, Ida of St. Lambert's Bull 19169; dam, Lady Mary Pogis 27113.

Butter, 14 lbs. 2 oz. Milk, 150 lbs. 14 oz.

Test made from Oct. 20 to 27, 1891; age, 2 yrs. 10 mos.; estimated weight, 750 lbs.; fed 45 lbs. oat meal, 45 lbs. corn meal, 28 lbs. bran, 14 lbs. oil meal, about 80 lbs. hay and 1 bush. roots—millet, pumpkins and cured corn fodder; property of Ayer & McKinney, Philadelphia, Pa.

Idana 27974.—Sire, St. Jacob 6438; dam, Ida 9th 6528.

Butter, 14 lbs. 10 oz. Milk, 201 lbs. 7 oz.

Test made from Sept. 9 to 16, 1889; age, 5 yrs. 8 mos.; estimated weight, 850 lbs.; fed, daily, about 8 lbs. bran, 6 lbs. ground oats, 4 lbs. corn meal and 1¾ lbs. linseed meal; property of Joseph R. Anderson, Jr., Lee, Va.

Ida of Bear Lake 6169.—Sire, Prince of Warren 1512; dam, Countess of Warren 3896.

Butter, 16 lbs. Milk, an average of 12 qts.

Test made last week in Feb., 1881; age, 5 yrs.; well fed on hay, straw and different kinds of ground feed; property of Chester Bordwell, Batavia, Ohio.

Ida of Clinton 25419.—Sire, Noble of Sidney 5369; dam, Queen of Clinton 11500.

Butter, 15 lbs. 14 oz.   Milk, 236 lbs.

Test made from Sept. 30 to Oct. 7, 1892; age, 9 yrs. 6 mos.; estimated weight, 900 lbs.; fed 56 qts. of a mixture of 100 lbs. corn meal, 25 lbs. wheat bran and 10 lbs. cotton-seed meal—meadow pasture; property of G. E. Nichols, Afton, N. Y.

Ida of Coal Hill 12542.—Sire, Duke of Portage 1270; dam, Celia Belle 5865.

Butter, 15 lbs. ½ oz.   Milk, 232 lbs.

Test made from June 29 to July 5, 1885; age, 5 yrs.; property of Campbell Brown, Spring Hill, Tenn.

Ida of Elma 92316.—Sire, Stoke Pogis of Prospect 29121; dam, Ida of Oakland 2d 72061.

Butter, 14 lbs. 10 oz.   Milk, 210 lbs.

Test made from June 5 to 12, 1895; age, 1 yr. 11 mos.; actual weight, 640 lbs.; fed 12 qts. bran, 9 qts. ground oats, 3 qts. corn meal, 1½ qts. oil meal and 1½ qts. cotton-seed meal, daily—hay, poor pasture; property of George D. Briggs, Elma, N. Y.

Ida of Maple Glens 97785.—Sire, Ida's Harry Tormentor 27836; dam, Helen of Maple Glens 56339.

Butter, 14 lbs. 14 oz.   Milk, 219 lbs. 4 oz.

Test made from Jan. 12 to 19, 1896; age, 3 yrs. 4 mos.; estimated weight, 800 lbs.; fed 56 qts. bran, 28 qts. corn and cob meal, 28 qts. ground oats, 7 qts. old process oil meal, 14 bush. cut corn stover, 7 bush. clover hay, 3½ pks. turnips and 3½ pks. potatoes; property of J. E. Gray & Son, Youngstown, Ohio.

Ida of Oakland 2d 72061.—Sire, Ida's Spot of St. L. 15442; dam, Ida of Oakland 35798.

Butter, 15 lbs. ½ oz.   Milk, 185 lbs. 6 oz.

Test made from Sept. 4 to 11, 1893; age, 4 yrs.; actual weight, 912 lbs.; fed 84 qts. ground corn and oats, 105 qts. wheat bran and 21 qts. oil meal—hay; property of G. D. Briggs, East Aurora, N. Y.

Ida of St. Lambert 24990.—Sire, Stoke Pogis 3d 2238; dam, Kathleen of St. Lambert 5122.

Butter, 30 lbs. 2½ oz. (official).   Milk, 290 lbs. 8 oz.

Test made from Sept. 12 to 18, 1884; age, 6 yrs. 7 mos.; property of V. E. Fuller, Hamilton, Ont., Can.

Idarella 41433.—Sire, Ida's Stoke Pogis 13658; dam, Harrella 32912.

Butter, 15 lbs. ⅛ oz.   Milk, 204 lbs.

Test made from June 21 to 28, 1890; age, 3 yr. 7 mos.; estimated weight, 750 lbs.; fed, daily, about 3 gals. corn meal and 1 pt. cotton-seed meal; property of M. C. Campbell, Spring Hill, Tenn.

Ida's Alpha 64536.—Sire, Ida of St. Lambert's Bull 19169; dam, My Alpha 16892.

Butter, 15 lbs. 10 oz.   Milk, 199 lbs. 12 oz.

Test made from Dec. 30, 1892, to Jan. 6, 1893; age, 3 yrs. 3 mos.; estimated weight, 800 lbs.; fed 14 lbs. oil meal, 42 lbs. pea meal and 80 lbs. corn, oats and shorts, in equal parts—hay; property of Ayer & McKinney, Philadelphia, Pa.

Ida's Alpha 64536.—Sire, Ida of St. Lambert's Bull 19169; dam, My Alpha 16892.

Butter, 25 lbs. 15¾ oz. Milk, 280 lbs. 12 oz.

Test made from Dec. 12 to 19, 1896; age, 7 yrs. 2 mos.; actual weight, 980 lbs.; fed 63 qts. ground oats, 63 qts. corn meal, 42 qts. wheat bran and 21 qts. ground oil-cake—7 pks. carrots; property of Miller & Sibley, Franklin, Pa.

Ida's Dream 50070.—Sire, Ida's Stoke Pogis 13658; dam, Rosy Dream 2d 27127.

Butter, 16 lbs. ½ oz. Milk, 136 lbs. 14 oz.

Test made from Feb. 13 to 20, 1890; age, 2 yrs. 7 mos.; estimated weight, 850 lbs.; fed, daily, 3½ gals. ground corn and oats, mixed in equal parts; property of Maury Jersey Farm, Columbia, Tenn.

Ida's Dream 50070.—Sire, Ida's Stoke Pogis 13658; dam, Rosy Dream 2d 27127.

Butter, 40 lbs. 8½ oz. in 14 days. Milk, 570 lbs. 6½ oz.

Test made from Mar. 31 to Apr. 14, 1895; age, 7 yrs. 9 mos.; actual weight, 880 lbs.; fed 26 lbs. per day of ground oats and corn in equal parts, mixed with 6 bucketfuls cut millet, wetted—orchard and blue-grass pasture; property of Matt. M. Gardner, Nashville, Tenn.

Ida's Dream 3d 131899.—Sire, Oonan's Tormentor 22280; dam, Ida's Dream 50070.

Butter, 14 lbs. 7¼ oz. Milk, 199 lbs. 13¼ oz.

Test made from May 29 to June 5, 1898; age, 2 yrs. 5 mos.; estimated weight, 575 lbs.; fed 20 lbs. per day of ground oats and corn, in equal parts—extra good orchard and blue-grass pasture; property of Matt. M. Gardner, Nashville, Tenn.

Ida's Fawn 56018.—Sire, Ida's Stoke Pogis 13658; dam, Miss Rollo 21579.

Butter, 16 lbs. 1½ oz. Milk, 244 lbs. 13 oz.

Test made from July 13 to 20, 1892; age, 3 yrs. 8 mos.; estimated weight, 775 lbs.; fed 2 gals. corn meal and 3 gals. wheat bran, daily—good pasture; property of Matthews & Humes, Huntsville, Ala.

Ida's Grace 56019.—Sire, Ida's Stoke Pogis 13658; dam, Nancy's Grace 41421.

Butter, 14 lbs. ½ oz. Milk, 186 lbs. 1 oz.

Test made from May 16 to 23, 1894; age, 5 yrs. 6 mos.; actual weight, 850 lbs.; fed 14 lbs. per day of ground oats and corn, in equal parts—good orchard and blue-grass pasture; property of Matt. M. Gardner, Nashville, Tenn.

Ida's Landseer's Gold 121641.—Sire, Ida's Landseer 17745; dam, Diana's Gold 56141.

Butter, 14 lbs. 4 oz. Milk, 167 lbs.

Test made from May 8 to 15, 1897; age, 2 yrs. 11 mos.; estimated weight, 650 lbs.; fed 56 qts. ground oats and 56 qts. ground corn—woodland pasture; property of Morgan & Brown, Columbia, Tenn.

Ida's Lassie 77315.—Sire, Ida's Pogis 18000; dam, Dorchester Lass 18297.

Butter, 14 lbs. 9¼ oz. Milk, 238 lbs.

Test made from Mar. 15 to 22, 1896; age, 6 yrs. 5 mos.; estimated weight, 950 lbs.; fed 35 lbs. hominy meal, 45½ lbs. light bran and about 490 lbs. ensilage; property of J. T. Polk, Greenwood, Ind.

Ida's Little Wonder 104961.—Sire, Scales' Stoke Pogis 30098; dam, Royal Pogessa 49530.

Butter, 14 lbs. 7 oz. Milk, 209 lbs. 4 oz.

Test made from Dec. 30, 1896, to Jan. 6, 1897; age, 2 yrs. 9 mos.; estimated weight, 800 lbs.; fed 4 lbs. oats, 3¾ lbs. bran, 7¾ lbs. corn meal and 1¼ lbs. oil meal, daily—corn fodder *ad lib.*, short rye pasture; property of W. E. Johnson, Millican, Tex.

Ida's Lucille 99400.—Sire, Lord Harry 5th 29752; dam, Ida's Own 49959.

Butter, 17 lbs. 4¼ oz. Milk, 197 lbs.

Test made from Apr. 29 to May 6, 1898; age, 5 yrs. 8 mos.; estimated weight, 950 lbs.; fed 56 qts. ground oats and corn and 28 qts. bran—rye and blue-grass pasture; property of S. A. Appleby, Lewisburg, Tenn.

Ida's Myrrha 97939.—Sire, Ida's Landseer 17745; dam, Myrrha Landseer 65456.

Butter, 14 lbs. 11 oz. Milk, 216 lbs. 8 oz.

Test made from May 11 to 18, 1895; age, 3 yrs. 11 mos.; estimated weight, 750 lbs.; fed 56 qts. ground oats and 56 qts. corn meal—woodland pasture; property of R. P. Dodson, Columbia, Tenn.

Ida's Mysinda 64868.—Sire, Ida of St. Lambert's Bull 19169; dam, Mysinda 2d 42250.

Butter, 16 lbs. 11 oz. Milk, 213 lbs. 8 oz.

Test made from Dec. 15 to 22, 1892; age, 2 yrs. 7 mos.; estimated weight, 800 lbs.; fed 44 lbs. pea meal, 20 lbs. oil meal, 30 lbs. cotton-seed meal, 48 lbs. mixed feed and 150 lbs. ensilage; property of Ayer & McKinney, Philadelphia, Pa.

Ida's Nan 49957.—Sire, Ida's Stoke Pogis 13658; dam, Nanalda 32917.

Butter, 15 lbs. 6 oz. Milk, 239 lbs.

Test made from June 6 to 13, 1890; age, 2 yrs. 8 mos.; estimated weight, 800 lbs.; fed, daily, 12 qts. corn meal and 1 pt. cotton-seed meal; property of M. C. Campbell, Spring Hill, Tenn.

Ida's Rioter's Elopement 98412.—Sire, Ida's Rioter of St. L. 13656; dam, Elopement 65578.

Butter, 19 lbs. 15¼ oz. Milk, 263 lbs.

Test made from June 7 to 14, 1898; age, 4 yrs. 4 mos.; actual weight, 800 lbs.; fed 42 qts. corn meal, 42 qts. ground oats, 21 qts. wheat bran and 7 qts. oil meal—good old meadow pasture; property of Miller & Sibley, Franklin, Pa.

Ida's Rose of St. Lambert 71555.—Sire, Ida of St. Lambert's Bull 19169; dam, Lisgar's Rose 2d 56976.

Butter, 14 lbs. 15 oz. Milk, 231 lbs. 10 oz.

Test made from June 24 to July 1, 1893; age, 2 yrs. 2 mos.; estimated weight, 750 lbs.; fed 6 lbs. corn meal, 3 lbs. ground oats, 2 lbs. bran, 2 lbs. shorts and 2½ lbs. oil meal, daily—dry pasture; property of Mrs. S. E. K. Hudson, Alexandria Bay, N. Y.

Ida's Rose of St. Lambert 71555.—Sire, Ida of St. Lambert's Bull 19169; dam, Lisgar's Rose 2d 56976.

Butter, 19 lbs. 1 oz. Milk, 293 lbs. 4 oz.

Test made from May 10 to 17, 1896; age, 5 yrs. 1 mo.; estimated weight, 1000 lbs.; fed 4 lbs. bran, 3 lbs. ground oats, 6 lbs. corn meal and 1½ lbs. oil meal, daily—poor pasture; property of Mrs. S. E. K. Hudson, Alexandria Bay, N. Y.

Ida's Rose of St. Lambert 71555.—Sire, Ida of St. Lambert's Bull 19169; dam, Lisgar's Rose 2d 56976.

Butter, 21 lbs. 1½ oz. Milk, 313 lbs. 12 oz.

Test made from May 17 to 24, 1896; age, 5 yrs. 1 mo.; estimated weight, 1000 lbs.; fed 5 lbs. bran, 4 lbs. ground oats, 9½ lbs. corn meal and 2½ lbs. oil meal, daily—poor pasture; property of Mrs. S. E. K. Hudson, Alexandria Bay, N. Y.

Ida's Rose of St. Lambert 71555.—Sire, Ida of St. Lambert's Bull 19169; dam, Lisgar's Rose 2d 56976.

Butter, 17 lbs. 1½ oz. (confirmed); estimated butter on basis of 85 per cent. fat, 16 lbs. 15.45 oz. Milk, 280 lbs. 2 oz.

Test made from June 14 to 20, 1898; age, 7 yrs. 2 mos.; estimated weight, 1000 lbs.; fed 24½ lbs. oats, 70 lbs. corn meal, 21 lbs. bran and 24½ lbs. oil meal—pasture; property of C. I. Hudson, Alexandria Bay, N. Y.

Analyses of Butter.

First churning—Water 9.56, fat 85.71, curd 0.81, salt 3.92. Second churning—Fat 86.47. Third churning—Fat 89.16.

Ida's Rose of St. Lambert 71555.—Sire, Ida of St. Lambert's Bull 19169; dam, Lisgar's Rose 2d 56976.

Butter, 40 lbs. 2½ oz. in 14 days. Milk, 607 lbs.

Test made from May 10 to 24, 1896; age, 5 yrs. 1 mo.; estimated weight, 1000 lbs.; fed 63 lbs. bran, 49 lbs. ground oats, 108½ lbs. corn meal and 28 lbs. oil meal—poor pasture; property of Mrs. S. E. K. Hudson, Alexandria Bay, N. Y.

Ida's St. Jeannaise 2d 106316.—Sire, Millionaire Pogis 29688; dam, Ida's St. Jeannaise 83646.

Butter, 17 lbs. 4¼ oz. Milk, 295 lbs. 11 oz.

Test made from May 23 to 30, 1898; age, 4 yrs. 10 mos.; estimated weight, 700 lbs.; fed 31½ lbs. bran, 35 lbs. corn meal, 21 lbs. oat meal, 7 lbs. oil meal and 13½ lbs. cotton-seed meal—clover and timothy pasture; property of C. I. Hood, Lowell, Mass.

Ida's Signaline 108819.—Sire, Ida's Landseer 17745; dam, Bessie Signalda 2d 86341.

Butter, 15 lbs. 10 oz. Milk, 153 lbs.

Test made from Dec. 28 to Jan. 4, 1897; age, 2 yrs. 2 mos.; estimated weight, 800 lbs.; fed 70 lbs. cotton-seed hulls, 14 lbs. cotton-seed meal, 12¼ lbs. linseed meal, 21 lbs. bran, 31½ lbs. ground oats and 35 lbs. hay—turf oats pasture; property of S. H. Butler, Como, Miss.

Ida Twinkle 36994.—Sire, Ida's Rioter of St. L. 13656; dam, Twinkle 5th 18486.

Butter, 23 lbs. 2½ oz. Milk, 266 lbs. 4 oz.

Test made from Nov. 14 to 21, 1890; age, 4 yrs. 11 mos.; actual weight, 1090 lbs.; fed, daily, 6 qts. wheat middlings, 12 qts. ground oats and 9 qts. corn meal, with 1 bucket oat meal gruel at midnight; property of Miller & Sibley, Franklin, Pa.

Ida Zoe Pogis 38685.—Sire, Ida's Rioter of St. L. 13656; dam, Rioter's Zoe 19769.

Butter, 16 lbs. 2½ oz. Milk, 172 lbs. 6 oz.

Test made from Feb. 16 to 23, 1890; age, 3 yrs. 9 mos.; estimated weight, 750 lbs.; fed, daily, about 8 lbs. 2 oz. corn meal, 6 lbs. 9 oz. crushed oats, 2 lbs. 10 oz. cotton-seed meal and 1 lb. 12 oz. middlings; property of Ayer & McKinney, Philadelphia, Pa.

Ideal 11842.—Sire, Cecco 1673; dam, Lerna 3634.

    Butter, 14 lbs. 12½ oz. Milk, 204 lbs.

Test made from Feb. 15 to 21, 1883; age, 3 yrs. 6 mos.; property of William Simpson, New York, N. Y.

Ideal Alphea 18755.—Sire, Mercury 432; dam, Ideal 11842.

    Butter, 14 lbs. 6 oz. Milk, 182 lbs. 10 oz.

Test made from May 10 to 16, 1885; age, 2 yrs. 5 mos.; property of William Simpson, New York, N. Y.

Idelle's Idea 92510.—Sire, Prince of Tennessee 20772; dam, Idelle Pogis 56953.

    Butter, 14 lbs. 10¾ oz. Milk, 185 lbs. 12 oz.

Test made from Dec. 8 to 15, 1895; age, 4 yrs. 7 mos.; estimated weight, 700 lbs.; fed 17 lbs. per day of ground oats and corn, in equal parts, mixed with 4 bucketfuls cut sheaf oats, wetted—millet hay *ad lib.*; property of Thomas H. Malone, Nashville, Tenn.

Idessimus 50508.—Sire, Meg's Lord Bronx 11537; dam, Ides 14097.

    Butter, 16 lbs. Milk, 213 lbs. 8 oz.

Test made from Apr. 21 to 28, 1895; age, 10 yrs.; estimated weight, 900 lbs.; fed 105 lbs. provender and 43 lbs. cotton-seed meal—hay *ad lib.*, poor pasture; property of Miner C. Hazen, Haddam, Conn.

Idlewild Pogis 80271.—Sire, Kathy's Stoke Pogis 17566; dam, Annabel Lee 55883.

    Butter, 14 lbs. 11 oz. Milk, 218 lbs. 10 oz.

Test made from Sept. 11 to 18, 1892; age, 2 yrs. 1 mo.; estimated weight, 600 lbs.; no account of feed kept; property of Matthews & Humes, Huntsville, Ala.

Iliana 44594.—Sire, Oonan's Signal 11586; dam, Golden Sheen 25561.

    Butter, 15 lbs. 5 oz. Milk, 232 lbs.

Test made from Apr. 16 to 23, 1889; age, 2 yrs. 2 mos.; actual weight, 756 lbs.; fed, daily, about 19 lbs. corn meal, oat meal and bran, in equal parts; property of Jacob L. Thomas, Knoxville, Tenn.

Ilinka 91418.—Sire, Duke of La Grange 27682; dam, Mollie Mack 2d 46966.

    Butter, 17 lbs. 8 oz. Milk, 267 lbs. 2 oz.

Test made from May 21 to 28, 1898; age, 5 yrs. 1 mo.; estimated weight, 950 lbs.; fed 70 qts. wheat bran and 64 qts. corn meal—timothy and blue-grass pasture; property of T. B. Kibler, Jenera, Ohio.

Imogene Pogis 44967.—Sire, Prince of Melrose 2d 11015; dam, Lass de Grantez 26673.

    Butter, 18 lbs. 13 oz. Milk, 223 lbs. 14 oz.

Test made from Dec. 19 to 26, 1890; age, 3 yrs. 8 mos.; estimated weight, 800 lbs.; fed 56 qts. pea meal, 28 qts. corn meal, 14 qts. bran, 14 qts. cotton-seed meal and 245 lbs. ensilage—hay, pasture of good winter rye; property of James Crook, Jacksonville, Ala.

Imperial Pansy Pogis 49750.—Sire, Hugo Pogis of Elmarch 16318; dam, Imperial Pansy 26830.

Butter, 14.42 lbs. (confirmed); estimated butter on basis of 85 per cent. fat, 14.68 lbs. Milk, 283.1 lbs.

Test made from May 25 to 31, 1897; age, 9 yrs. 9 mos.; actual weight, 805 lbs.; fed 14 lbs. oil meal, 70 lbs. bran and 70 lbs. corn-hearts—clover hay *ad lib.*, short pasture; property of E. A. Farra, Nealton, Ky.

Analyses of Butter.

First churning—Fat 80.92, water 15.06, ash 2.92, casein 1.10.
Second churning—Fat 81.81, water 14.04, ash 3.47, casein 0.68.

Imperial Riotress 30259.—Sire, Tug Wilson 9080; dam, Lady Imperial 16019.

Butter, 18 lbs. 4 oz. Milk, 292 lbs. 6 oz.

Test made from Nov. 19 to 26, 1889; age, 4 yrs. 11 mos.; estimated weight, 800 lbs.; fed, daily, about 9 qts. bran, 3 qts. corn meal and 1 3-7 pts. old process oil meal; property of Mrs. Kate M. Busick, Wabash, Ind.

Imperial's Omega 2d 19664.—Sire, Imperial 2d 5254; dam, Imperial's Omega 13241.

Butter, 14 lbs. 14 oz. Milk, 207 lbs. 12 oz.

Test made from June 13 to 20, 1891; age, 9 yrs. 8 mos.; estimated weight, 900 lbs.; fed 6 lbs. bran and 6 lbs. cotton-seed cake meal, daily—short pasture, timothy, clover and blue-grass; property of Bordwell & Cochran, Batavia, Ohio.

· Income 19472.—Sire, Auroraboreëllis 2408; dam, Niva 7523.

Butter, 16 lbs. 9 oz. Milk, 251 lbs. 9 oz.

Test made from Nov. 15 to 22, 1885; age, 5 yrs. 8 mos.; estimated weight, 825 lbs.; fed, daily, 18 qts. corn, oats and pea meal in about equal proportions; property of Richardson Bros., Davenport, Iowa.

Indulgence 50105.—Sire, Young Combination 14550; dam, Islip Lenox 31703.

Butter, 17 lbs. 1 oz. Milk, 271 lbs. 8 oz.

Test made from Feb. 11 to 18, 1897; age, 9 yrs. 7 mos.; estimated weight, 900 lbs.; fed 45½ lbs. bran, 36½ lbs. corn, 32½ lbs. oats, 14 lbs. linseed meal, 280 lbs. ensilage and 140 lbs. roots—hay *ad lib.*; property of C. I. Hood, Lowell, Mass.

Inez of Ingleside 28977.—Sire, King Coomassie, on I. of J.; dam, Quaker's Lady 25110.

Butter, 14 lbs. 12 oz. Milk, 155 lbs. 8 oz.

Test made from Mar. 3 to 9, 1886; age, 1 yr. 10 mos.; property of Mrs. Ida A. Nelson, Loudon, Tenn.

Inez of Riverside 51781.—Sire, Holyoke 8147; dam, Clara of Riverside 26301.

Butter, 20 lbs. 13½ oz. Milk, 408 lbs. 8 oz.

Test made from July 28 to Aug. 4, 1897; age, 11 yrs.; actual weight, 920 lbs.; fed 9 qts. ground oats, 9 qts. bran, 6 qts. corn meal and 3 qts. oil meal, daily—hay, good timothy and clover pasture; property of C. A. Sweet, East Aurora, N. Y.

Inez of Riverside 51781.—Sire, Holyoke 8147; dam, Clara of Riverside 26301.

Butter, 24 lbs. 1½ oz. Milk, 437 lbs.

Test made from Aug. 4 to 11, 1897; age, 11 yrs.; actual weight, 920 lbs.; fed 9 qts. ground oats, 9 qts. bran, 6 qts. corn meal and 3 qts. oil meal, daily—hay, good clover pasture, and cut sweet corn in stable; property of C. A. Sweet, East Aurora, N. Y.

Ines of Riverside 51781.—Sire, Holyoke 8147; dam, Clara of Riverside 26301.

Butter, 26 lbs. 1½ oz.   Milk, 463 lbs. 12½ oz.

Test made from Aug. 11 to 18, 1897; age, 11 yrs.; actual weight, 920 lbs.; fed 9 qts. ground oats, 9 qts. bran, 6 qts. corn meal, and 3 qts. oil meal, daily—hay, good clover pasture morning and night, cut green corn in stable during heat of day; property of C. A. Sweet, East Aurora, N. Y.

Ines of Riverside 51781.—Sire, Holyoke 8147; dam, Clara of Riverside 26301.

Butter, 71 lbs. ½ oz. in 21 days.   Milk, 1309 lbs. 4½ oz.

Test made from July 28 to Aug. 18, 1897; age, 11 yrs.; actual weight, 920 lbs.; fed 189 lbs. ground oats, 189 lbs. bran, 126 lbs. corn meal and 63 lbs. oil meal—hay, good timothy and clover pasture, some cut sweet corn; property of C. A. Sweet, East Aurora, N. Y.

Infanta Pedro Marjoram 88584.—Sire, Pedro Royal Marjoram 28560; dam, Rioter's Lorne Primrose 88583.

Butter, 17 lbs. 6 oz.   Milk, 285 lbs.

Test made from June 15 to 22, 1897; age, 4 yrs. 2 mos.; estimated weight, 850 lbs.; fed 4 qts. bran, 4 qts. middlings, 4 qts. ground oats and 2 qts. sugar feed, daily—clover hay *ad lib.*, meadow pasture; property of T. S. Cooper, Coopersburg, Pa.

Insie of Riverside 23825.—Sire, Duke of Mansfield's Pierrot 6261; dam, Pride of Willimantic 13557.

Butter, 30 lbs. 5¾ oz.   Milk, 289 lbs. 12 oz.

Test made from May 6 to 12, 1886; age, 4 yrs.; fed 4 qts. corn meal, 2 qts. shorts and 1 pk. carrots, daily, with hay and short pasture; property of Joseph H. Walker, Worcester, Mass.

Io 5th 280.—Sire, Roger 121; dam, Io 3d 245.

Butter, 17 lbs. 8 oz.

Test made from June 22 to 29, 1881; age, 12 yrs. 8 mos.; property of I. W. Vance, Cantrall, Ill.

Iola 4627.—Sire, Mercury 432; dam, Gussie Richards 1673.

Butter, 15 lbs. 2½ oz.   Milk, 232 lbs. 10 oz.

Test made from Oct. 20 to 26, 1882; age, 7 yrs. 2 mos.; property of William Simpson, New York, N. Y.

Iola F. 85529.—Sire, Gold of St. Lambert 16744; dam, Iversnaid 8978.

Butter, 25.251 lbs. in 21 days.   Milk, 465.3 lbs.

Test made from Sept. 30 to Oct. 20, 1893, at World's Columbian Exposition; age, 2 yrs. 2 mos.; actual weight, 711 lbs.; fed 202 lbs. hay, 6 lbs. ensilage, 42 lbs. oil meal, 77 lbs. corn meal, 126 lbs. bran, 84 lbs. oats, 35 lbs. cotton-seed, 21 lbs. middlings and 7 lbs. corn-hearts; property of E. W. Fairman, Brodhead, Wis.

Irene of St. Lambert 2d 75759.—Sire, Sir George of St. Lambert 6036; dam, Irene of St. Lambert 27056.

Butter, 15 lbs. 5 oz.   Milk, 259 lbs. 8 oz.

Test made from Apr. 12 to 19, 1898; age, 6 yrs. 9 mos.; actual weight, 1000 lbs.; fed 3⅝ lbs. corn meal, 3⅝ lbs. linseed meal, 4 lbs. wheat middlings, 13½ lbs. ground oats, 20 lbs. hay and 8 qts. roots, daily; property of Horatio J. Gilbert, Milton, Mass.

Irene of Short Hills 5137.—Sire, Okubo 1876; dam, Jarto 1902.

Butter, 14 lbs. 6½ oz. Milk, 237 lbs. 8 oz.

Test made from Mar. 14 to 20, 1882; age, 5 yrs. 9 mos.; property of Campbell Brown, Spring Hill, Tenn.

Irondequoit Belle 74696.—Sire, Chief of St. Lambert 18803; dam, Exile's Belle 40524.

Butter, 14 lbs. 3 oz. Milk, 214 lbs. 1 oz.

Test made from July 25 to Aug. 1, 1894; age, 3 yrs. 8 mos.; actual weight, 945 lbs.; fed 12 lbs. to 16 lbs. daily of a mixture of bran, ground oats, corn meal, cotton-seed meal and oil meal, and 485 lbs. cut hay during test; property of A. D. McBride, Rochester, N. Y.

Iryl Dane 3d 38375.—Sire, Ned Dean 5387; dam, Iryl Dane 22842.

Butter, 14 lbs. 15 oz. Milk, 223 lbs. 12 oz.

Test made from June 23 to 30, 1894; age, 10 yrs.; estimated weight, 1000 lbs.; fed 12 lbs. daily of a mixture of bran, middlings, corn meal and oil meal—natural pasture; property of George W. Sisson, Jr., Potsdam, N. Y.

Isadore Stoke Pogis 37297.—Sire, Exile of St. Lambert 13657; dam, Grace G. Parks 29263.

Butter, 16 lbs. 9½ oz. Milk, 200 lbs.

Test made from July 3 to 10, 1889; age, 3 yrs. 5 mos.; estimated weight, 650 lbs.; fed, daily, about 22 lbs., one-half fine middlings, one-fourth corn meal and one-fourth wheat bran; property of P. J. Cogswell, Rochester, N. Y.

Island Chrissie 12007.—Sire, Duke of Burlington 1639; dam, Island Lily 6969.

Butter, 14 lbs. 14 oz., unsalted. Milk, 247 lbs. 12 oz.

Test made from Mar. 28 to Apr. 3, 1884; age, 4 yrs. 10 mos.; property of Moulton Bros., West Randolph, Vt.

Island Dots 17003.—Sire, Pride of the Island 5416; dam, Lady Lillie 11813.

Butter, 14 lbs. 9 oz. Milk, 202 lbs. 8½ oz.

Test made from July 21 to 27, 1883; age, 1 yr. 9 mos.; property of Frederick B. Simpson, New York, N. Y.

Island Star 11876.—Sire, Guy Fawkes (F. S. 251 J. H. B.); dam, Gazelle 15961.

Butter, 18 lbs. 10 oz., unsalted.

Test made from May 30 to June 5, 1883; age, 4 yrs. 2 mos.; fed 4 qts. corn meal and 4 qts. bran, daily—June grass, timothy and red-top pasture; property of S. M. Burnham, Saugatuck, Conn.

Island Star 11876.—Sire, Guy Fawkes (F. S. 251 J. H. B.); dam, Gazelle 15961.

Butter, 21 lbs. 3 oz. Milk, 150⅞ qts.

Test made from June 7 to 13, 1884; age, 5 yrs. 2 mos.; fed 6 qts. coarse corn meal and 6 qts. ground oats, mixed, daily—mixed grass pasture; property of S. M. Burnham, Saugatuck, Conn.

Islip Lenox 31703.—Sire, Czar of Lenox 6875; dam, Islip 4th 1884.

Butter, 47.699 lbs. in 30 days. Milk, 714.6 lbs.

Test made from Aug. 29 to Sept. 27, 1893, at World's Columbian Exposition; age, 9 yrs. 9 mos.; actual weight, 1016 lbs.; fed 332.9 lbs. hay, 67 lbs. oil meal, 107 lbs. ensilage, 196 lbs. bran, 84 lbs. oats, 134 lbs. corn-hearts, 16 lbs. cotton-seed, 45 lbs. middlings, 9 lbs. carrots and 5 lbs. old hay; property of C. I. Hood, Lowell, Mass.

Islip Lenox 31703.—Sire, Czar of Lenox 6875; dam. Islip 4th 1884.

Butter, 178.066 lbs. in 90 days. Milk, 3070 lbs.

Test made from May 31 to Aug. 28, 1893, at World's Columbian Exposition; age, 9 yrs. 6 mos. at beginning of test; actual weight, 1040 lbs.; fed 323 lbs. old hay, 126 lbs. ensilage, 525.2 lbs. new hay, 222 lbs. oil meal, 508 lbs. bran, 149 lbs. oats, 538.5 lbs. corn-hearts, 111.75 lbs. cotton-seed, 349 lbs. middlings, 21 lbs. grano-gluten, 1108.4 lbs. clover and 10 lbs. swale-grass; property of C. I. Hood, Lowell, Mass.

Islithanks 90544.—Sire, Trump 22048; dam, Islianne 19545.

Butter, 17 lbs. 2 oz. Milk, 285 lbs. 8 oz.

Test made from May 29 to June 5, 1898; age. 5 yrs. 6 mos.; estimated weight, 950 lbs.; fed 42 lbs. corn meal, 56 lbs. oats and 14 lbs. oil meal—mixed pasture; property of H. C. Taylor, Orfordville, Wis.

Italian 4096.—Sire, Maba Rajah 794; dam, Natali 1433.

Butter, 17 lbs. 13 oz. Milk, 274 lbs. 12 oz.

Test made from May 20 to 26, 1883; age, 8 yrs. 6 mos.; property of W. P. Ijams, Terre Haute, Ind.

Italica 58775.—Sire, Compass 16958; dam, Income 19472.

Butter, 14 lbs. 4½ oz. Milk, 208 lbs. 13 oz.

Test made from Aug. 15 to 22, 1891; age, 2 yrs. 2 mos.; estimated weight, 800 lbs.; fed 4 qts. ground corn and oats three times daily—orchard pasture; property of Richardson Bros., Davenport, Iowa.

Itura 61090.—Sire, Diploma 16219; dam, Petura 28775.

Butter, 14 lbs. 14 oz. Milk, 193 lbs. 14 oz.

Test made from Mar. 15 to 22, 1891; age, 3 yrs.; estimated weight, 775 lbs.; fed 12 qts. daily of corn, oats and peas, ground, and 1 pk. roots—plenty of hay; property of Richardson Bros., Davenport, Iowa.

Iuka of Riverview 2d 61271.—Sire, Flora Lee's Victor Hugo 17675; dam, Iuka of Riverview 40648.

Butter, 17 lbs. 12 oz. Milk, 211 lbs.

Test made from May 17 to 24, 1895; age, 5 yrs. 7 mos.; actual weight, 735 lbs.; fed 6 qts. cotton-seed meal and 6 qts. corn meal, daily—pastured on green wheat; property of S. T. Howard, Quanah, Tex.

Ivy Tormente 121700.—Sire, Harry Torment's Signal 39846; dam, Southern Ivy Leaf 95670.

Butter, 16 lbs. 1¼ oz. Milk, 234 lbs. 13 oz.

Test made from Dec. 26, 1897, to Jan. 2, 1898; age, 2 yrs. 9 mos.; actual weight, 713 lbs.; fed 40 lbs. wheat bran, 20 lbs. corn meal, 20 lbs. wheat middlings and 15 lbs. old process oil meal—hay and corn ensilage *ad lib.*; property of J. P. Bradbury, Pomeroy, Ohio.

Jane Neptune 52345.—Sire, Young Neptune 17656; dam, Favorite of Avon 13438.

Butter, 19 lbs. 4½ oz. Milk, 308 lbs.

Test made from Dec. 19 to 26, 1891; age, 4 yrs. 9 mos.; estimated weight, 1000 lbs.; fed 12 qts. daily of a mixture of oats, peas, barley, corn-hearts, bran and cotton-seed meal, in two feeds, on 1 pk. sweet apples—clover hay and corn fodder; property of E. G. Youmans, Groton City, N. Y.

Janie Caruth 73311.—Sire, Bobby Welcome Pogis 12514; dam, Janie Moore 17514.

Butter, 15 lbs. 4 oz. Milk, 240 lbs. 1 oz.

Test made from Mar. 8 to 15, 1897; age, 6 yrs. 4 mos.; estimated weight, 800 lbs.; fed 84 qts. wheat bran, 28 qts. corn meal and 56 qts. cotton-seed—prairie hay *ad lib.*, prairie pasture; property of J. W. Hardin, Terrell, Tex.

Jap Pogis 69122.—Sire, Bradley Pogis 22891; dam, Nell Pogis 31991.

Butter, 14 lbs. 7 oz. Milk, 264 lbs. 15 oz.

Test made from May 28 to June 4, 1895; age, 4 yrs. 10 mos.; estimated weight, 800 lbs.; fed 44 lbs. ground corn and oats, in equal parts, and 54 lbs. bran—clover and timothy pasture; property of Bowers & Sharp, Cross Roads, Ind.

Jaquenetta 10958.—Sire, Chief Baron 2984; dam, Blanche Amory 7592.

Butter, 14 lbs. 6 oz. Milk, 141 lbs. 12 oz.

Test made from Mar. 22 to 29, 1884; age, 3 yrs. 9 mos.; property of M. M. Gardner, Nashville, Tenn.

Jazella G. 14191.—Sire, Gold Basis 4038; dam, Jazel's Maid 11011.

Butter, 20 lbs. 7 oz.

Test made from July 8 to 14, 1885; age, 4 yrs. 2 mos.; fed about 20 lbs. ground corn and oats (two parts of oats to one of corn by measure) and about 3 qts. oil meal, daily—pasture dry; property of Thomas H. Malone, Nashville, Tenn.

Jazel's Maid 11011.—Sire, Jazel 3501; dam, Maid of Avranches 6959.

Butter, 14 lbs. 6 oz., unsalted. Milk, 250 lbs. 11 oz.

Test made from May 22 to 28, 1883; age, 3 yrs. 9 mos.; fed 5 qts. corn meal and 2 qts. bran, daily—very good blue-grass pasture; property of Thomas H. Malone, Nashville, Tenn.

Jean Ingelow 42515.—Sire, Chief of the Miamis 2958; dam, Dice 18182.

Butter, 17 lbs. 7 oz. Milk, 221 lbs. 14 oz.

Test made from June 11 to 18, 1893; age, 6 yrs. 5 mos.; estimated weight, 950 lbs.; fed 10 lbs. dry and 16 lbs. wet brewer's grains (barley waste) and 8 lbs. corn chop, daily, in two feeds—good blue and meadow grass pasture; property of George E. McGill, Leavenworth, Kan.

Jeanne Le Bas 2476.—Sire, Noble (F. S. 71 J. H. B.); dam, Dairy Pride (F. S. 348 J. H. B.)

Butter, 15 lbs. 8 oz.

Property of Holder Borden-Bowen, Bristol, R. I.

Jeannette Geranium 61023.—Sire, Geranium's Chief 18607; dam, Oonan's Jeannette 45616.

Butter, 14 lbs. 4 oz. Milk, 121 lbs. 4 oz.

Test made from June 25 to July 2, 1891; age, 2 yrs. 6 mos.; estimated weight, 700 lbs.; fed 8 lbs. corn meal, 6 lbs. bran and 4 lbs. linseed meal, daily, on moistened cut hay—scant pasture; property of William Hart Dexter, Springfield, Mass.

Jeannette of Pittsford 73185.—Sire, Doge of St. Lambert 17889; dam, Duke's Dream 42783.

        Butter, 29.973 lbs. in 21 days. Milk, 501.8 lbs.

        Test made from Sept. 30 to Oct. 20, 1893, at World's Columbian Exposition; age. 3 yrs.; actual weight, 686 lbs.; fed 195 lbs. hay, 6 lbs. ensilage, 53.5 lbs. oil meal. 93 lbs. corn meal, 138 lbs. bran, 70 lbs. oats, 43 lbs. cotton-seed, 35 lbs. middlings and 14 lbs. corn-hearts; property of Aaron O. Auten, Jerseyville, Ill.

Jeannie Platt 6005.—Sire, Rex 1330; dam, Jule 3640.

        Butter, 14 lbs. 4 oz. Milk, 120⅜ qts.

        Test made from July 31 to Aug. 6, 1882; age, 5 yrs. 4 mos.; property of Lyman A. Mills, Middlefield, Conn.

Jefferson Albina 12196.—Sire, Jersey Pride 2355; dam, Venice Beauty 2d 11008.

        Butter, 14 lbs. 13 oz. Milk, 186 lbs.

        Test made from June 21 to 27, 1885; age, 4 yrs. 3 mos.; property of G. H. Grimmell, Jefferson, Iowa.

Jemele 2d 101437.—Sire, Hugo Pogis of St. Anne's 19316; dam, Jemele 36524.

        Butter, 15 lbs. 7 oz. Milk, 247 lbs.

        Test made from Jan. 22 to 29, 1898; age, 4 yrs. 8 mos.; estimated weight, 700 lbs.; fed 2 lbs. bran, 4 lbs. middlings, 3 lbs. corn meal, 20 lbs. rye straw and 36 lbs. mangels, daily; property of William C. Browning, New York, N. Y.

Jennette Darling 10702.—Sire, Man of Ipswich 1510; dam, Jenny B. 4190.

        Butter, 16 lbs. 2 oz. Milk, 215 lbs. 7 oz.

        Test made from June 26 to July 2, 1887; age, 9 yrs.; weight, 950 lbs.; fed 18 lbs. equal parts corn meal, oat meal and bran, with 4½ lbs. oil meal, daily; property of E. L. Briggs, Wilton Junction, Iowa.

Jennie 766, imp.

        Butter, 14 lbs. 9 oz. Milk, 105 qts.

        Test made from Sept. 10 to 16, 1872; age, 6 yrs.; property of W. B. Dinsmore, Staatsburg, N. Y.

Jennie Cream 30622.—Sire, Rachel's Duke 7022; dam, Jersey Cream 2d 8519.

        Butter, 17 lbs. 4 oz. Milk, 237 lbs. 10 oz.

        Test made from June 8 to 15, 1889; age, 4 yrs. 10 mos.; estimated weight, 800 lbs.; fed, daily, 4 qts. meal, 1 qt. cotton-seed and 1 qt. oil-cake; property of John P. Pomeroy, Housatonic, Mass.

Jennie Cream 2d 47466.—Sire, Lucy's Stoke Pogis 11544; dam, Jennie Cream 30622.

        Butter, 20 lbs. 5 oz. Milk, 309 lbs. 10 oz.

        Test made from June 19 to 26, 1897; age, 9 yrs. 11 mos.; estimated weight, 950 lbs.; fed 3 qts. corn meal, 3 qts. ground oats, 2 qts. cotton-seed meal and 6 qts. wheat bran, daily—timothy and clover pasture; property of E. L. Van Deusen, Ashley Falls, Mass.

Jennie Fordyce 57169.—Sire, Signalda's Torment 14673; dam, Signal's Lily Flagg 31035.

        Butter, 15 lbs. 5½ oz. Milk, 218 lbs. 1 oz.

        Test made from May 7 to 14, 1891; age, 2 yrs. 5 mos.; estimated weight, 600 lbs.; fed 2 gals. corn-hearts, 2 gals. oats and 1 gal. bran, daily, on ensilage—good pasture; property of Matthews & Moore, Huntsville, Ala.

Jennie Myrtle 22977.—Sire, Mulberry Lad 3954; dam, Dorcas 4th 3533.

Butter, 18 lbs. 1 oz.

Test made from May 19 to 26, 1884; age, 4 yrs. 1 mo.; property of Ingleside Herd, Meadville, Pa.

Jennie of the Vale 9553.—Sire, Highland Blade 2164; dam, Elsie Brown 4026.

Butter, 14 lbs. 6½ oz.   Milk, 272 lbs.

Test made from Apr. 21 to 28, 1883; age, 4 yrs. 4 mos.; property of H. W. Douglass, Pevely, Mo.

Jennie of the Vale 9553.—Sire, Highland Blade 2164; dam, Elsie Brown 4026.

Butter, 17 lbs. 7½ oz.   Milk, 250 lbs.

Test made from May 1 to 8, 1884; age, 5 yrs. 5 mos.; property of H. W. Douglass, Pevely, Mo.

Jennie Stoke Pogis 32010.—Sire, Exile of St. Lambert 13657; dam, Puss Lurlene 21461.

Butter, 15 lbs. 10¾ oz.   Milk, 206 lbs. 8 oz.

Test made from Aug. 3 to 10, 1891; age, 6 yrs. 4 mos.; estimated weight, 750 lbs.; fed 8 qts. corn meal, 3 qts. oil meal, 4 qts. bran, 4 qts. fine feed, 2 lbs. hay and 30 lbs. green corn-stalks, daily; property of P. J. Cogswell, Rochester, N. Y.

Jenny Dodo H. 14448.—Sire, Homer H. 3683; dam, Jenny Gray 3511.

Butter, 21 lbs. 8 oz.

Test made from June 20 to 27, 1883; age, 5 yrs. 2 mos.; hill pasture only; property of W. H. Blasdell, Barton's Landing, Vt.

Jenny Gray 3511.—Sire, Major Adams 1044; dam, Lady Lightfoot 2745.

Butter, 17 lbs. 15 oz.   Milk, 368 lbs. 5 oz.

Test made from June 13 to 19, 1885; age, 12 yrs. 5 mos.; property of William H. Hayden, Albany, Vt.

Jenny Le Brocq 9757.—Sire, Hero (P. S. 90 J. H. B.); dam, Jeune Jenny (F. S. 706 J. H. B.)

Butter, 14 lbs. 14 oz.

Test made Jan., 1881; age, 4 yrs.; property of S. W. Robbins, Wethersfield, Conn.

Jersey 3260.—Sire, Dick Swiveller Jr. 276; dam, High Life 3259.

Butter, 15 lbs. 6 oz.   Milk, 212 lbs.

Test made from Oct. 29 to Nov. 4, 1882; age, 18 yrs. 6 mos.; property of Woodside Farm Herd, Troy, N. Y.

Jersey Belle of Belmont 68337.—Sire, Captain Fox 23942; dam, Flora G. 38119.

Butter, 14 lbs. 6 oz.   Milk, 243 lbs. 10 oz.

Test made from May 31 to June 7, 1894; age, 3 yrs. 6 mos.; estimated weight, 775 lbs.; fed 7 lbs. oil meal, 7 lbs. corn meal, 24 lbs. wheat middlings and 18 lbs. wheat bran—common mixed pasture; property of H. C. Vanderhoef, Belmont, N. Y.

Jersey Belle of Scituate 7828.—Sire, Victor 3550; dam, Jenny 7827.

Butter, 25 lbs. 3 oz.   Milk, 44 to 45 lbs. daily.

Test made from June 15 to 21, 1880; age, 9 yrs.; property of C. O. Ellms, Scituate, Mass.

Jersey Belle of Scituate 7828.—Sire, Victor 3550; dam, Jenny 7827.

Butter, 705 lbs. in one year.

Test made 1877-78; age, 6 yrs.; property of C. O. Ellms, Scituate, Mass.

Jersey Cream 3151.—Sire, Tom Dasher 420; dam, Creampot 460.

Butter, 17 lbs.   Milk, 266 lbs. 8 oz.

Test made from June 15 to 22, 1881; age, 7 yrs. 11 mos.; property of D. B. Dewolf, Lee, Mass.

Jersey Cream 2d 8519.—Sire, King of Fairview 778; dam, Jersey Cream 3151.

Butter, 14 lbs. 12 oz.

Test made from June 1 to 7, 1882; age, 5 yrs. 1 mo.; property of H. G. Westlake, Hillsdale, N. Y.

Jersey Cream 3d 8521.—Sire, Manchester's Prospect 2817; dam, Jersey Cream 3151.

Butter, 16 lbs. 5 oz.   Milk, 239 lbs. 12 oz.

Test made from May 22 to 29, 1885; age, 6 yrs. 1 mo.; property of H. G. Westlake, Hillsdale, N. Y.

Jersey Jane 38308.—Sire, Jersey Express 5771; dam, Jane Riley 11455.

Butter, 16 lbs. 4½ oz.   Milk, 222 lbs. 8 oz.

Test made from May 22 to 29, 1888; age, 2 yrs. 1 mo.; estimated weight, 750 lbs.; fed, daily, 6 lbs. corn meal and 3 lbs. oil meal; property of J. R. Anderson, Jr., Lee, Va.

Jersey Lily 14044.—Sire, Grey King (P. S. 169 J. H. B.); dam on I. of J.

Butter, 15 lbs.   Milk, 246 lbs. 12 oz.

Test made from May 22 to 29, 1886; age, 7 yrs. 4 mos.; estimated weight, 800 lbs.; fed, daily, 8 lbs. corn meal, 2 lbs. bran and malt sprouts; property of James Stillman, Sing Sing, N. Y.

Jessaline 26099.—Sire, Gillo, on I. of J.; dam, Columbine (F. S. 131 J. H. B.)

Butter, 16 lbs. 6 oz.   Milk, 264 lbs. 8 oz.

Test made from Mar. 7 to 13, 1886; age, 5 yrs.; property of G. V. Green, Hopkinsville, Ky.

Jessaline 2d 38295.—Sire, Denise's Tormentor 11823; dam, Jessaline 26099.

Butter, 16 lbs. 4½ oz.   Milk, 270 lbs. 5½ oz.

Test made from Aug. 3 to 10, 1894; age, 8 yrs. 7 mos.; actual weight, 850 lbs.; fed 15 lbs. per day of ground oats and corn, in equal parts, mixed with 4 bucketfuls of cut sheaf oats—orchard and blue-grass pasture; property of Matt. M. Gardner, Nashville, Tenn.

Jessie Belle B. 100285.—Sire, Sorona's Prince Pogis 16319; dam, Belle of Jessamine 79456.

Butter, 14 lbs. 2 oz. Milk, 197 lbs. 2 oz.

Test made from Dec. 2 to 9, 1894; age, 3 yrs. 10 mos.; estimated weight, 750 lbs.; fed 112 lbs. grain feed—hay *ad lib.*, short blue-grass pasture, some sorghum; property of J. S. Burrier, South Elkhorn, Ky.

Jessie Leavenworth 8248.—Sire, Champion of America 1567; dam, Bull's Kitty Clover 3897.

Butter, 14 lbs. 2 oz.

Test made in June, 1881; age, 4 yrs. 1 mo.; **test made on grass alone**; property of Warren F. Daniell, Franklin, N. Y.

Jessie Lee of Labyrinth 5290.—Sire, Tycoon Jr. 1212; dam, Beulah of Baltimore 3270.

Butter, 14 lbs. 7 oz.

Age, 4 yrs.; property of James Crook, Jacksonville, Ala.

Jessie Lee Pogis 44969.—Sire, Prince of Melrose 2d 11015; dam, Maggie of Springvale 15931.

Butter, 17 lbs. 4 oz. Milk, 180 lbs. 11 oz.

Test made from Jan. 30 to Feb. 6, 1892; age, 4 yrs. 9 mos.; estimated weight, 750 lbs.; fed 14 lbs. ensilage, 8 lbs. bran, 4 lbs. ground corn, oats and peas, and 2 lbs. cotton-seed meal, daily—28 lbs. hay and 98 lbs. ensilage during test, and one hour on green oats daily; property of M. Lothrop, Marshall, Tex.

Jessie Lorne 25213.—Sire, Lorne 5248; dam, Nina of St. Lambert 12963.

Butter, 16 lbs. 15 oz. Milk, 270 lbs. 8 oz.

Test made from June 1 to 7, 1885; age, 2 yrs. 4 mos.; property of William Rolph, Markham, Ont., Can.

Jessie's Sayda 105697.—Sire, Sayda's Prince 28501; dam, Jessie S. 59822.

Butter, 15 lbs. 10 oz. Milk, 207 lbs. 4 oz.

Test made from Dec. 21 to 28, 1896; age, 3 yrs. 5 mos.; estimated weight, 875 lbs.; fed 21 lbs. daily of a mixture of corn meal, ground oats, bran, pea meal, cotton-seed meal and oil meal, 40 lbs. ensilage and about 10 lbs. hay, daily; property of R. F. Shannon, Pittsburg, Pa.

Jessie Stoke Pogis 37291.—Sire, Exile of St. Lambert 13657; dam, Ethel's Nelly 19097.

Butter, 17 lbs. 2½ oz. Milk, 187.8 lbs.

Test made from Dec. 6 to 13, 1892; age, 7 yrs. 4 mos.; actual weight, 925 lbs.; fed 42 qts. corn meal, 42 qts. bran, 42 qts. oat and pea meal, 42 qts. oil meal, 210 lbs. corn ensilage and 28 lbs. hay; property of H. L. S. Hall, Scottsville, N. Y.

Jessie Wilson 76121.—Sire, Signalda's Torment 14673; dam, Frushia Wilson 47432.

Butter, 14 lbs. 8 oz. Milk, 197 lbs. 7 oz.

Test made from Dec. 4 to 11, 1893; age, 5 yrs. 3 mos.; estimated weight, 800 lbs.; fed 84 qts. corn and oats, 28 qts. wheat bran, 14 qts. linseed meal and 14 qts. cotton-seed meal—clover hay *ad lib.*; property of William E. Matthews, Huntsville, Ala.

Jetsam's May 62530.—Sire, One Hundred per Cent. 16590; dam, Jetsam 32893.

Butter, 14 lbs. 5 oz. Milk, 150 lbs. 8 oz.

Test made from Sept. 7 to 14, 1890; age, 2 yrs. 4 mos.; estimated weight, 700 lbs.; fed, daily, 14 qts. ground peas, oats and barley, mixed; property of Mrs. Eliza M. Jones, Brockville, Ont., Can.

Jetty d'Or 39845.—Sire, Rayon d'Or 7516; dam, Jetty Lee 7988.

Butter, 15 lbs. 2 oz. Milk, 208 lbs. 15 oz.

Test made from Sept. 24 to Oct. 1, 1892; age, 7 yrs. 11 mos.; estimated weight, 925 lbs.; fed 28 qts. corn meal, 42 qts. germ bran and 42 qts. common bran—pasture of mixed grasses; property of Fulton Dixon, Gillespieville, Ohio.

Jetty d'Or 2d 62403.—Sire, Taureau d'Or 11743; dam, Jetty d'Or 39845.

Butter, 17 lbs. 1 oz. Milk, 235 lbs. 6 oz.

Test made from June 14 to 21, 1893; age, 5 yrs. 3 mos.; actual weight, 705 lbs.; fed 49 lbs. corn meal and 28 lbs. common wheat bran—pasture of mixed grasses; property of Fulton Dixon, Gillespieville, Ohio.

Jewel of Argyle 68752.—Sire, Baron Hugo 15208; dam, Dassa Argyle 21180.

Butter, 18 lbs. 2 oz. Milk, 295 lbs. 8 oz.

Test made from May 20 to 27, 1893; age, 3 yrs. 3 mos.; estimated weight, 800 lbs.; fed 3 lbs. Chicago cream gluten, 3 lbs. corn meal, 3 lbs. coarse wheat bran, mixed with 10 lbs. corn ensilage, daily—common pasture; property of E. S. Henry, Rockville, Conn.

Jewel of Ingleside 54544.—Sire, Duke of Ingleside 14274; dam, Jewel of Ridley 17231.

Butter, 17 lbs. 1 oz. Milk, 294 lbs.

Test made from June 2 to 9, 1894; age, 6 yrs. 8 mos.; estimated weight, 1050 lbs.; fed 28 lbs. wheat middlings, 14 lbs. corn meal and 14 lbs. ground oats—orchard grass pasture; property of John C. McClintock, Meadville, Pa.

Jim's Glory 60527.—Sire, John Rex 6th 4579; dam, Jim's Pride 31376.

Butter, 14 lbs. 4 oz. Milk, 181 lbs. 6 oz.

Test made from Oct. 19 to 26, 1893; age, 5 yrs. 2 mos.; estimated weight, 800 lbs.; fed 14 gals. corn bran, 14 qts. cotton-seed meal and 14 qts. corn meal—poor pasture; property of A. J. Hawes & Son, Montgomery, Ala.

Jim's Pride 31376.—Sire, Jim (P. S. 346 J. H. B.); dam, Little Belle, on I. of J.

Butter, 18 lbs. 3 oz. Milk, 178 lbs. 8 oz.

Test made from Apr. 9 to 16, 1888; age, 4 yrs. 11 mos.; estimated weight, 700 lbs.; fed 14 gals. corn bran, 28 qts. cotton-seed meal and 28 qts. corn meal—pastured on growing oats; property of A. J. Hawes & Son, Montgomery, Ala.

Joan Montague 40787.—Sire, Lord Montague 12385; dam, Woodland Maid 2d 19666.

Butter, 16 lbs. 1 oz. Milk, 246 lbs.

Test made from June 6 to 13, 1891; age, 5 yrs. 3 mos.; estimated weight, 900 lbs.; fed 5 lbs. bran and 5 lbs. cotton-seed cake meal, daily—mixed timothy, clover and blue-grass pasture; property of Bordwell & Cochran, Batavia, Ohio.

Jocal 95182.—Sire, Exile of St. Lambert 23d 20712; dam, Judy Fagan 3d 37653.

Butter, 20 lbs. 14 oz. Milk, 268 lbs. 8 oz.

Test made from Feb. 16 to 23, 1897; age, 3 yrs. 11 mos.; actual weight, 775 lbs.; fed 56 qts. bran, 56 qts. oat meal, 56 qts. corn meal, 28 qts. oil meal, 100 lbs. ensilage, 112 lbs. hay and 112 qts. roots; property of P. J. Cogswell, Rochester, N. Y.

Jodie Jenks 76739.—Sire, Mississippi Rioter 20401; dam, Madam Jenks 34694.

Butter, 18 lbs. 7 oz. Milk, 244 lbs. 10 oz.

Test made from Apr. 12 to 19, 1896; age, 6 yrs.; estimated weight, 1000 lbs.; fed cotton-seed meal, shorts, cotton-seed hulls, bran, ground oats and hay, as much as she would eat; property of Mrs. Mattie A. Rubush, Meridian, Miss.

John Bull's Princess 49670.—Sire, Canada's John Bull 8388; dam, Lucilla Kent 8892.

Butter, 19 lbs. 14 oz. Milk, 233 lbs.

Test made from Oct. 21 to 28, 1890; age, 2 yrs. 10 mos.; estimated weight, 950 lbs.; fed, daily, 18 qts. ground oats, corn and peas, and 2 qts. shorts; property of Mrs. L. E. Hill, Denver, Col.

Jolie of St. Lambert 5126.—Sire, Lord Lisgar 1066; dam, Psyche of St. Lambert 5121.

Butter, 15 lbs. 13½ oz. Milk, 298 lbs.

Test made from May 2 to 8, 1883; age, 8 yrs. 10 mos.; property of W. A. Reburn, Ste. Anne de Bellevue, P. Q., Can.

Jonquil of Bingo Farm 41786.—Sire, Pickwick 2d 8915; dam, Pansy of Bingo Farm 41784.

Butter, 19 lbs. 3 oz. Milk, 289 lbs. 12 oz.

Test made from June 9 to 16, 1892; age, 9 yrs. 2 mos.; estimated weight, 800 lbs.; no grain fed—timothy and June grass pasture only; property of Mrs. J. R. Sherman, Granville, N. Y.

Josephina 64921.—Sire, Diploma 16219; dam, Nettie C. Magnet 35411.

Butter, 14 lbs. 1 oz. Milk, 218 lbs. 8 oz.

Test made from Oct. 1 to 8, 1890; age, 2 yrs. 2 mos.; estimated weight, 750 lbs.; fed, daily, 15 qts. ground corn, oats, peas and oil meal; property of Richardson Bros., Davenport, Iowa.

Josephine Noble 66441.—Sire, Moidore 15911; dam, Princess Amy 30185.

Butter, 18 lbs. 14 oz. Milk, 261 lbs. 4 oz.

Test made from Feb. 10 to 17, 1898; age, 7 yrs. 7 mos.; actual weight, 848 lbs.; fed 56 qts. each of cooked cotton-seed and shelled oats, 14 qts. cotton-seed meal, 28 qts. bran and 70 qts. corn meal—oat hay *ad lib.*, pastured on green oats; property of J. M. Trosper, Bethany, La.

Josephine O. 103635.—Sire, Eric J. 33246; dam, Martha Ethlo 89029.

Butter, 19 lbs. 12 oz. Milk, 234 lbs.

Test made from May 16 to 23, 1898; age, 4 yrs. 2 mos.; estimated weight, 950 lbs.; fed about 6 qts. ground corn, oats and bran, mixed in equal parts, daily—old sod pasture; property of S. F. Boyer, Finleyville, Pa.

Joy of Argyle 40495.—Sire, Duke Thorne 7020; dam, Duchess of Argyle 4th 7571.

Butter, 15 lbs. 9 oz. Milk, 281 lbs.

Test made from Apr. 26 to May 3, 1891; age, 4 yrs. 11 mos.; estimated weight, 950 lbs.; fed 3 lbs. wheat bran, 2 lbs. linseed meal, 2 lbs. cotton-seed meal, 3 lbs. corn meal and about 35 lbs. ensilage, daily, mixed, also 10 lbs. clover rowen; property of E. S. Henry, Rockville, Conn.

J. S. Exile 105495.—Sire, Wells' Exile 20170; dam, Jenny Sutliff 2d 73131.

Butter, 15 lbs. 4 oz. Milk, 255 lbs.

Test made from Mar. 25 to Apr. 1, 1896; age, 4 yrs.; estimated weight, 800 lbs.; fed 84 lbs. corn and cob meal, 21 lbs. linseed meal, 42 lbs. bran and 105 lbs. hay; property of E. G. Silvus, Athens, Ohio.

Jubilethleel 99697.—Sire, Ethleel 2d's Jubilee 18249; dam, Ruth Morgan 3d 57043.

Butter, 17 lbs. 12 oz. Milk, 199 lbs. 6.4 oz.

Test made from Apr. 17 to 24, 1895; age, 2 yrs. 5 mos.; estimated weight, 600 lbs.; fed 35 lbs. corn and oats ground together (two-thirds oats and one-third corn), 28 lbs. bran and 14 lbs. cotton-seed meal—fair rye pasture; property of Hecla Jersey Cattle Co., Earlington, Ky.

Judith Coleman 11391.—Sire, John Knox 3289; dam, Juliette of St. Lambert 5483.

Butter, 17 lbs. 5 oz. Milk, 208 lbs. 8 oz.

Test made from Oct. 30 to Nov. 5, 1883; age, 2 yrs. 11 mos.; fed 12 qts. wheat bran and 2 qts. corn meal, daily—timothy and blue-grass pasture; property of Rutherford Douglas, Lexington, Ky.

Judith Coleman 11391.—Sire, John Knox 3289; dam, Juliette of St. Lambert 5483.

Butter, 18 lbs. 4 oz., unsalted. Milk, 256 lbs. 10 oz.

Test made from Sept. 22 to 28, 1885; age, 4 yrs. 9 mos.; property of W. H. Corning, Cleveland, Ohio.

Judy of Riverside 16495.—Sire, Wessex 3638; dam, Floss of Lawnfield 16085.

Butter, 21 lbs. 4½ oz. Milk, 284 lbs.

Test made from Oct. 18 to 24, 1886; age, 5 yrs. 2 mos.; estimated weight, 1000 lbs.; fed 4½ lbs. oats, 2½ lbs. oil meal, 3½ lbs. shorts, 3¾ lbs. bran, 3¾ lbs. corn meal, daily; property of C. W. H. Eicke, West Monterey, Pa.

Julia Evelyn 6007.—Sire, Aldine 1136; dam, Mink 2d 3890.

Butter, 15 lbs. 15½ oz., unsalted. Milk, 233 lbs. 8 oz.

Test made from Mar. 23 to 29, 1883; age, 6 yrs. 5 mos.; property of W. B. Montgomery, Starkville, Miss.

Julia Landseer 56268.—Sire, Landseer's Fogis 15847; dam, Julia Walker 10133.

Butter, 14 lbs. 2½ oz. Milk, 91 lbs.

Test made from Dec. 11 to 18, 1890; age, 1 yr. 11 mos.; estimated weight, 550 lbs.; fed, daily, 3½ gals. ground corn; property of Morgan & Brown, Columbia, Tenn.

Julia of Eau Claire 120018.—Sire, Melia Ann's Son 22041; dam, Jolie of Linden 16695.

Butter, 17 lbs. 8 oz. Milk, 231 lbs. 10 oz.

Test made from Feb. 18 to 25, 1898; age, 3 yrs. 1 mo.; estimated weight, 850 lbs.; fed 8 qts. ground corn and oats, 6 qts. wheat middlings, 2 qts. oil meal and 2 qts. gluten meal, daily—hay *ad lib.*; property of E. L. Van Deusen, Ashley Falls, Mass.

Julia Rex of Ingleside 74046.—Sire, Duke of Ingleside 14274; dam, Amrah J. 23284.

Butter, 16 lbs. 3 oz. Milk, 210 lbs.

Test made from June 22 to 29, 1895; age, 5 yrs. 2 mos.; estimated weight, 900 lbs.; fed 28 lbs. coarse winter wheat bran, 28 lbs. ground oats and 28 lbs. corn meal—common pasture; property of John C. McClintock, Meadville, Pa.

Julia Tyre 45187.—Sire, Bonnie's Polonius 10138; dam, Orphan Duchess 4th 29006.

Butter, 14 lbs. 6 oz. Milk, 228 lbs. 6 oz.

Test made from Feb. 10 to 17, 1891; age, 5 yrs. 7 mos.; estimated weight, 650 lbs.; fed 56 lbs. mill feed, 42 lbs. corn meal and 14 lbs. oil meal—also hay; property of D. W. Voyles, Crandall, Ind.

Julia Walker 10133.—Sire, Pierrot 2d 1669; dam, Fannie Landseer 1969.

Butter, 15 lbs. 12 oz.

Test made June, 1880; age, 5 yrs.; property of Thomas Fitch, New London, Conn.

Juliette Guion 13143.—Sire, Walnut Chief 3130; dam, Pride of Walnut Farm 11501.

Butter, 14 lbs. 4 oz. Milk, 187 lbs. 8 oz.

Test made from Dec. 1 to 8, 1887; age, 6 yrs. 7 mos.; estimated weight, 750 lbs.; fed, daily, 8 lbs. corn meal, 8 lbs. bran and 2 lbs. oil meal; property of James Stillman, Sing Sing, N. Y.

Julippa 69000.—Sire, Diploma 16219; dam, Neonice 65004.

Butter, 15 lbs. 9 oz. Milk, 248 lbs. 2 oz.

Test made from Apr. 30 to May 7, 1897; age, 6 yrs. 8 mos.; estimated weight, 700 lbs.; fed 42 lbs. bran, 24½ lbs. corn, 24½ lbs. oats, 14 lbs. oil meal, 105 lbs. roots and 210 lbs. ensilage—hay *ad lib.*; property of C. I. Hood, Lowell, Mass.

June Sweets 59978.—Sire, Gold Basis 5th 15725; dam, Sweets 7891.

Butter, 14 lbs. 8½ oz. Milk, 188 lbs. 8 oz.

Test made from May 6 to 13, 1892; age, 2 yrs. 11 mos.; estimated weight, 700 lbs.; fed 4½ lbs. meal, 3 lbs. cotton-seed meal, 6 lbs. wheat feed and 10 lbs. cut hay, daily; property of H. A. Huntington, Higganum, Conn.

June Sweets 2d 87967.—Sire, Riverside Stoke Pogis 29235; dam, June Sweets 59978.

Butter, 15 lbs. 11 oz. Milk, 283 lbs. 6 oz.

Test made from Aug. 29 to Sept. 5, 1897; age, 4 yrs. 6 mos.; estimated weight, 900 lbs.; fed 56 lbs. bran, 28 lbs. corn meal and 21 lbs. cotton-seed meal—shocked sorghum; property of W. A. Ponder, Denton, Tex.

Juno of Milton 71528.—Sire, Majestic Pedro's Son 9492; dam, Sue Sweetser 31332.

Butter, 15 lbs. 2 oz. Milk, 252 lbs.

Test made from May 4 to 11, 1895; age, 5 yrs. 7 mos.; estimated weight, 950 lbs.; fed 2 bush. cut hay, 6½ lbs. ground oats, 4 lbs. shorts, 5 lbs. corn meal, 2 lbs. linseed meal, 6 qts. carrots and mangels, mixed, and 4 lbs. loose hay, daily; property of Horatio J. Gilbert, Milton, Mass.

Justa Pogis 64863.—Sire, Jupiter Pogis 18192; dam, Jenny Justa 43731.

Butter, 14 lbs. 15¼ oz. Milk, 262 lbs. 10½ oz.

Test made from May 23 to 30, 1892; age, 3 yrs. 8 mos.; estimated weight, 850 lbs.; fed 82 lbs. corn meal, 16 lbs. ground oats and 30 lbs. bran—blue-grass pasture; property of Kentucky Agricultural Experiment Station, Lexington, Ky.

Justa Pogis 64863.—Sire, Jupiter Pogis 18192; dam, Jenny Justa 43731.

Butter, 157.697 lbs. in 90 days.   Milk, 2745.3 lbs.

Test made from May 31 to Aug. 28, 1893, at World's Columbian Exposition; age, 4 yrs. 8 mos. at beginning of test; actual weight, 932 lbs.; fed 329 lbs. old hay, 210 lbs. ensilage, 518.2 lbs. new hay, 168 lbs. oil meal, 465 lbs. bran, 83.5 lbs. oats, 559.5 lbs. corn-hearts, 113.50 lbs. cotton-seed, 310 lbs. middlings, 32 lbs. grano-gluten, 1165.4 lbs. clover and 10 lbs. swale-grass; property of Kentucky Agricultural Experiment Station, Lexington, Ky.

Justa Pogis 2d 115494.—Sire, Oonan's Tormentor Pogis 30505; dam, Justa Pogis 64863.

Butter, 14 lbs. 9½ oz.   Milk, 256.5 lbs.

Test made from June 1 to 8, 1898; age, 3 yrs. 11 mos.; actual weight, 975 lbs.; fed 56 lbs. corn meal, 7 lbs. cotton-seed meal, 7 lbs. oil meal and 70 lbs. bran—blue-grass pasture; property of Kentucky Agricultural Experiment Station, Lexington, Ky.

Kaembe's Maud Eurota 91066.—Sire, Euroto Pogis Le Brocq 19503; dam, Maud of Maple View 39511.

Butter, 14 lbs. 2½ oz.   Milk, 171 lbs. 5 oz.

Test made from Nov. 16 to 23, 1893; age, 1 yr. 9 mos.; estimated weight, 650 lbs.; fed 16 qts. bran, 4 qts. ground oats and corn and 1 qt. old process oil meal, daily, also 150 lbs. clover hay during test; property of Mrs. Kate M. Busick, Wabash, Ind.

Kaleta 2d 38810.—Sire, Lord Harry 3445; dam, Kaleta 19711.

Butter, 15 lbs. 14 oz.   Milk, 188 lbs. 4 oz.

Test made from Jan. 19 to 26, 1898; age, 11 yrs. 11 mos.; estimated weight, 850 lbs.; fed 42 lbs. bran, 42 lbs. corn meal, 18 lbs. gluten meal, 18 lbs. oil meal and 40 lbs. ensilage and hay; property of Austin & Probert, Sandstone, Mich.

Kalma of Milford 52742.—Sire, Prince of Milford 7951; dam, Marcia Rex 27224.

Butter, 15 lbs. 12½ oz.   Milk, 217 lbs. 12 oz.

Test made from July 12 to 19, 1897; age, 10 yrs. 5 mos.; estimated weight, 950 lbs.; fed 28 lbs. bran, 28 lbs. ground oats, 42 lbs. corn meal, 7 lbs. cotton-seed meal, 35 lbs. cotton-seed hulls and 168 lbs. cut green corn; property of S. H. Butler, Como, Miss.

Kamaretta 80479.—Sire, Ettamarius 17762; dam, Kamris 5941.

Butter, 16 lbs. 7 oz.   Milk, 266 lbs. 8 oz.

Test made from Dec. 22 to 29, 1895; age, 5 yrs.; estimated weight, 1050 lbs.; fed 42 lbs. cracked oats, 35 lbs. bran, 14 lbs. old process oil meal, 28 lbs. corn meal, 4 lbs. cotton-seed meal, 70 lbs. uncut corn stover, 70 lbs. roots and about 140 lbs. millet hay; property of C. Delano, Mount Vernon, Ohio.

Kammerfrau 125093.—Sire, Prince of Tennessee 20772; dam, Beauty's Baby 92511.

Butter, 15 lbs. 4 oz.   Milk, 172 lbs.

Test made from June 12 to 19, 1898; age, 2 yrs. 5 mos.; estimated weight, 550 lbs.; fed 19 lbs. per day of ground oats and corn, in equal parts—very good orchard and blue-grass pasture; property of Thomas H. Malone, Nashville, Tenn.

Kate Crecraft 40783.—Sire, Champion's Loyal 12673; dam, Imperial's Omega 13241.

Butter, 14 lbs. 6 oz.   Milk, 173 lbs. 4 oz.

Test made from June 13 to 20, 1891; age, 5 yrs. 7 mos.; estimated weight, 850 lbs.; fed 5 lbs. bran and 5 lbs. cotton-seed meal, daily—timothy, clover and blue-grass pasture; property of Bordwell & Cochran, Batavia, Ohio.

Kate Crecraft 2d 75461.—Sire, Lord Montague 12385; dam, Kate Crecraft 40783.

Butter, 18 lbs. 3 oz. Milk, 225 lbs. 10 oz.

Test made from Nov. 25 to Dec. 2, 1893; age, 3 yrs. 10 mos.; estimated weight, 750 lbs.; fed 48 qts. ground corn and oats, 112 qts. bran and about 50 lbs. hay—good blue-grass pasture; property of L. F. Hulick, Batavia, Ohio.

Kate Daisy 8264.—Sire, Barker's Dandy 3758; dam, White Daisy 6739.

Butter, 14 lbs. 4 oz. Milk, 276 lbs. 4 oz.

Test made from Oct. 20 to 26, 1883; property of Mrs. L. M. Fair, Wallingford, Conn.

Kate Durwood 2d 82417.—Sire, Landseer's Pogis 15847; dam, Kate Durwood 48971.

Butter, 14 lbs. 1 oz. Milk, 152 lbs. 4 oz.

Test made from Nov. 23 to 30, 1893; age, 4 yrs. 6 mos.; estimated weight, 750 lbs.; fed 56 qts. ground oats and 56 qts. corn meal—clover and millet, poor pasture; property of Morgan & Brown, Columbia, Tenn.

Kate Gordon 8387.—Sire, Pertinax 1965; dam, Normanda 3914.

Butter, 15 lbs. 15 oz., unsalted. Milk, 194 lbs. 7 oz.

Test made from May 14 to 20, 1883; age, 4 yrs. 6 mos.; property of M. C. Campbell, Spring Hill, Tenn.

Kate Gordon 2d 83662.—Sire, Oonan's Tormentor 22280; dam, Kate Gordon 8387.

Butter, 15 lbs. 5½ oz. Milk, 251 lbs. 5 oz.

Test made from Mar. 26 to Apr. 2, 1897; age, 5 yrs. 2 mos.; estimated weight, 1000 lbs.; fed 50 lbs. bran, 41 lbs. corn meal, 26½ lbs. ground oats, 14 lbs. linseed meal, 105 lbs. roots and 280 lbs. ensilage—hay *ad lib.*; property of C. I. Hood, Lowell, Mass.

Kate Kyle 64434.—Sire, L. D.'s Victor St. Helier 15286; dam, Ori's Rose 64429.

Butter, 16 lbs. 4 oz. Milk, 244 lbs. 8 oz.

Test made from May 11 to 18, 1895; age, 7 yrs. 5 mos.; estimated weight, 850 lbs.; fed 55 lbs. wheat bran and middlings and 14 lbs. cotton-seed meal—hay, mixed pasture; property of H. C. Vanderhoef, Belmont, N. Y.

Kate Landseer of Lawn 84415.—Sire, Daisy's Harry 21500; dam, Kate Linn 77176.

Butter, 16 lbs. 5 oz. Milk, 216 lbs. 11 oz.

Test made from Nov. 15 to 22, 1896; age, 4 yrs. 7 mos.; estimated weight, 890 lbs.; fed 224 lbs. wheat bran, 140 lbs. corn and oats chop and 28 lbs. cotton-seed meal—hay *ad lib.*, fair pasture; property of Platter & Foster, Denison, Tex.

Kate McCollum 107306.—Sire, Buckstone 26035; dam, Kate Weston 47286.

Butter, 15 lbs. 12½ oz. Milk, 206 lbs.

Test made from Feb. 23 to Mar. 1, 1896; age, 1 yr. 9 mos.; estimated weight, 650 lbs.; fed 56 qts. wheat bran, 28 qts. corn and cob meal, 14 qts. ground oats and 14 qts. oil and cotton-seed meal—clover hay *ad lib.*; property of Ed. K. Cover, New Bedford, Pa.

Kate Oxford 58102.—Sire, Landseer's Pogis 15847; dam, Southern Daisy 38292.

Butter, 14 lbs. 15 oz. Milk, 96 lbs.

Test made from Nov. 11 to 18, 1890; age, 2 yrs. 10 mos.; estimated weight, 700 lbs.; fed 4 gals. corn meal daily; property of Morgan & Brown, Columbia, Tenn.

Kate Oxford 58102.—Sire, Landseer's Pogis 15847; dam, Southern Daisy 38292.

Butter, 16 lbs. 3 oz.   Milk, 153 lbs. 8 oz.

Test made from Nov. 20 to 27, 1894; age, 6 yrs. 11 mos.; estimated weight, 850 lbs.; fed 56 qts. corn meal, 56 qts. oats (ground together) and 28 qts. wheat bran—German millet; property of Morgan & Brown, Columbia, Tenn.

Kate Oxford 2d 69270.—Sire, Tennessee's Landseer 23015; dam, Kate Oxford 58102.

Butter, 14 lbs. 4 oz.   Milk, 188 lbs. 12 oz.

Test made from June 12 to 19, 1894; age, 3 yrs. 9 mos.; estimated weight, 700 lbs.; fed 56 qts. ground oats and 56 qts. ground corn—fair woodland pasture; property of Morgan & Brown, Columbia, Tenn.

Kate Pansy 15177.—Sire, Royalist 3d 4500; dam, Pansy Dixie 9470.

Butter, 15 lbs. 1 oz.   Milk, 204 lbs. 6 oz.

Test made from June 17 to 23, 1885; age, 4 yrs. 8 mos.; property of J. B. Allen & Son, Delavan, Ill.

Kate Ritchey 54283.—Sire, Earl of Russellville 11935; dam, Emma Bowling 23969.

Butter, 14 lbs. 7¼ oz.   Milk, 244 lbs. 6 oz.

Test made from Feb. 28 to Mar. 7, 1897; age, 9 yrs. 11 mos.; estimated weight, 875 lbs.; fed 5 lbs. whole cotton-seed, 3 lbs. bran, 3½ lbs. ground oats and 1 lb. shelled oats, daily—125 lbs. hay; property of John F. Dabney, Cleburne, Tex.

Katherine of Pittsford 73169.—Sire, Duke of St. Lambert 16160; dam, Kate Winslow 28055.

Butter, 54.107 lbs. in 30 days.   Milk, 1062.3 lbs.

Test made from Aug. 29 to Sept. 27, 1893, at World's Columbian Exposition; age, 4 yrs. 3 mos.; actual weight, 825 lbs.; fed 361.6 lbs. hay, 72.5 lbs. oil meal, 110 lbs. ensilage, 121 lbs. corn meal, 180 lbs. bran, 90 lbs. oats, 85 lbs. corn-hearts, 44.5 lbs. cotton-seed, 36 lbs. middlings, 18 lbs. carrots and 5 lbs. old hay; property of E. F. Hawley, Pittsford, N. Y.

Kathletta 19567.—Sire, Lord Harry 3445; dam, Kate Gordon 8387.

Butter, 22 lbs. 12½ oz.   Milk, 174 lbs. 6 oz.

Test made from Nov. 8 to 15, 1888; age, 6 yrs. 10 mos.; estimated weight, 900 lbs.; fed, daily, 2 gals. ground corn and 2 gals. ground oats; property of M. C. Campbell, Spring Hill, Tenn.

Kathletta's Fancy 60738.—Sire, Ida's Landseer 17745; dam, Kathletta 19567.

Butter, 17 lbs. 6¾ oz.   Milk, 344 lbs. 9 oz.

Test made from Jan. 2 to 9, 1897; age, 7 yrs. 4 mos.; estimated weight, 800 lbs.; fed 56 lbs. bran, 34½ lbs. corn, 55 lbs. oats, 14 lbs. linseed meal, 210 lbs. ensilage and 140 lbs. roots—hay *ad lib.*; property of C. I. Hood, Lowell, Mass.

Kathy Torment 32910.—Sire, Tormentor 3533; dam, Kathletta 19567.

Butter, 16 lbs. 13½ oz.   Milk, 134 lbs.

Test made from Feb. 23 to Mar. 2, 1889; age, 5 yrs. 5 mos.; estimated weight, 750 lbs.; fed, daily, 3½ gals. corn meal and ground oats; property of Maury Jersey Farm, Columbia, Tenn.

Katie Bisson 100200.—Sire, Bisson's Landseer 27520; dam, Kate Durwood 2d 82417.

Butter, 14 lbs. 7 oz. Milk, 137 lbs.

Test made from Dec. 27, 1895, to Jan. 3, 1896; age, 2 yrs. 2 mos.; estimated weight, 625 lbs.; fed 70 qts. ground oats and 70 qts. corn meal—millet hay; property of Morgan & Brown, Columbia, Tenn.

Katie Herr 39970.—Sire, Silver Duke 7125; dam, Mambrino Kate 14494.

Butter, 15 lbs. 8½ oz. Milk, 222 lbs. 10 oz.

Test made from June 7 to 14, 1893; age, 7 yrs. 4 mos.; actual weight, 870 lbs.; fed 60 lbs. corn meal and 25 lbs. bran—pasture of mixed grasses; property of Columbus Dixon, Gillespieville, Ohio.

Katie of St. Lambert 54776.—Sire, Duke of St. Lambert 16160; dam, Kate Winslow 28055.

Butter, 19 lbs. 7 oz. Milk, 235 lbs. 9 oz.

Test made from Apr. 23 to 30, 1893; age, 5 yrs. 10 mos.; estimated weight, 875 lbs.; fed 5 qts. corn meal, 5 qts. oatmeal, 7 qts. wheat bran, 2 qts. oil meal, 22 lbs. ensilage and 8 lbs. roots, daily; property of Mrs. E. F. Hawley, Pittsford, N. Y.

Katie Putnam 40366.—Sire, Prince of Melrose 2d 11015; dam, Lass de Grantez 26673.

Butter, 16 lbs. 4 oz. Milk, 207 lbs. 6 oz.

Test made from Feb. 19 to 26, 1890; age, 3 yrs. 10 mos.; estimated weight, 700 lbs.; fed 56 qts. ground oats, 28 qts. cotton-seed meal, 28 qts. corn meal and 30 lbs. ensilage —hay, good winter rye pasture; property of James Crook, Jacksonville, Ala.

Katie's Kind 97434.—Sire, Prince of Tennessee 20772; dam, May Katydid 63503.

Butter, 15 lbs. 5¼ oz. Milk, 190 lbs. 10 oz.

Test made from May 3 to 10, 1896; age, 3 yrs.; estimated weight, 625 lbs.; fed 16 lbs. per day of ground oats and corn, in equal parts—orchard and blue-grass pasture; property of Matt. M. Gardner, Nashville, Tenn.

Kattie 41431.—Sire, Tormentor 3533; dam, Kathletta 19567.

Butter, 14 lbs. 8½ oz. Milk, 159 lbs. 8 oz.

Test made from Apr. 7 to 14, 1895; age, 8 yrs. 6 mos.; estimated weight, 800 lbs.; fed 4 qts. ground oats, 2 qts. corn meal, 2 qts. bran and 1 qt. cotton-seed meal, daily— clover hay, sheaf oats, corn stover, blue-grass pasture; property of Sam. N. Warren, Spring Hill, Tenn.

Katydidn't 2734, imp.

Butter, 14 lbs. 12 oz.

Test made from June 19 to 25, 1878; age, 7 yrs.; property of James A. Hayt, Patterson, N. Y.

Katy Scituate's Beauty 100259.—Sire, Albert Johnson 29637; dam, Katy Scituate 29300.

Butter, 16 lbs. Milk, 257 lbs. 10 oz.

Test made from Feb. 22 to Mar. 1, 1898; age, 5 yrs. 3 mos.; estimated weight, 925 lbs.; fed 50 lbs. wheat bran, 35 lbs. corn meal, 14 lbs. cotton-seed meal, about 175 lbs. ensilage, 42 lbs. hay and 1 bush. beets; property of Ayer & McKinney, Philadelphia, Pa.

Katy Signal 2d 44673.—Sire, Pauline's Stoke Pogis 11859; dam, Katy Signal 24690.

Butter, 16 lbs. 8 oz. Milk, 244 lbs. 8 oz.

Test made from Mar. 15 to 22, 1892; age, 5 yrs. 6 mos.; actual weight, 850 lbs.; fed 8 lbs. wheat bran, 2 lbs. corn meal, 2 lbs. oil meal, 1½ lbs. cotton-seed meal, daily, also 210 lbs. ensilage and 120 lbs. hay during test—poor clover pasture; property of Mrs. Kate M. Busick, Wabash, Ind.

Khedive's Fairy 66849.—Sire, Khedive's Landseer 21926; dam, Landseer's Content 56985.

Butter, 15 lbs. 7 oz. Milk, 186 lbs. 8 oz.

Test made from Sept. 25 to Oct. 2, 1896; age, 6 yrs. 4 mos.; estimated weight, 750 lbs.; fed 10 qts. wheat bran, 3 qts. corn meal and 2 qts. cotton-seed, daily—green corn *ad lib.*. very good pasture; property of George V. Green, Hopkinsville, Ky.

Khedive's Fancy 18180.—Sire, Khedive (P. S. 103 J. H. B.); dam, Rose (F. S. 1158 J. H. B.)

Butter, 15 lbs. 3 oz. Milk, 219 lbs. 2 oz.

Test made from June 28 to July 4, 1886; age, 9 yrs. 3 mos.; estimated weight, 1000 lbs.; fed, daily, 7 qts. corn meal, 2 qts. oil meal, 4 qts. crushed oats and 4 qts. shorts; property of F. C. Sayles, Pawtucket, R. I.

Khedive's Fancy 3d 70282.—Sire, Marjoram's Hugo Pogis 18311; dam, Khedive's Fancy 18180.

Butter, 14 lbs. 6 oz. Milk, 211 lbs.

Test made from May 15 to 22, 1894; age, 4 yrs. 2 mos.; estimated weight, 900 lbs.; fed 6 qts. bran, 3 qts. oil-cake and 6 qts. ground corn and oats, daily—blue-grass pasture; property of John Alexander, Paris, Mo.

Khedive's Pearl 71056.—Sire, Khedive's Landseer 21926; dam, Landseer's Pearl 56986.

Butter, 14 lbs. 12 oz. Milk, 199 lbs. 12 oz.

Test made from July 6 to 13, 1894; age, 4 yrs. 4 mos.; estimated weight, 800 lbs.; fed 112 qts. corn and oats, ground, in equal parts—fair pasture; property of Mayes & Lipscomb, Lipscomb, Tenn.

Khedive's Princess 38568.—Sire, Bunker 9025; dam, Princess of Trinity 23641.

Butter, 15 lbs. 4 oz. Milk, 172 lbs.

Test made from Feb. 15 to 22, 1891; age, 5 yrs. 8 mos.; estimated weight, 900 lbs.; fed, daily, 6 lbs. oil meal, average of 8 lbs. pea meal, 7 lbs. corn meal and 7 lbs. crushed oats; property of Ayer & McKinney, Philadelphia, Pa.

Khelula 17970.—Sire, King (P. S. 238 J. H. B.); dam, Sophie (F. S. 434 J. H. B.)

Butter, 19 lbs. 6 oz. Milk, 212 lbs. 8 oz.

Test made from Nov. 23 to 30, 1885; age, 4 yrs. 8 mos.; property of James Stillman, Sing Sing, N. Y.

Khelula 17970.—Sire, King (P. S. 238 J. H. B.); dam, Sophie (F. S. 434 J. H. B.)

Butter, 21 lbs. 8 oz. Milk, 184 lbs. 13 oz.

Test made from Nov. 30 to Dec. 7, 1885; age, 4 yrs. 9 mos.; estimated weight, 800 lbs.; fed, daily, 9 lbs. corn meal, 3 lbs. oil meal, 2 lbs. middlings and 2 lbs. bran; property of James Stillman, Sing Sing, N. Y.

Khelula 17970.—Sire, King (P. S. 238 J. H. B.); dam, Sophie (F. S. 434 J. H. B.)

Butter, 14 lbs. 6½ oz. (official). Milk, 184 lbs. 3 oz.

Test made from Jan. 5 to 12, 1886; age, 4 yrs. 10 mos.; fed, daily, 9 lbs. corn meal, 3 lbs. oil-cake meal, 6 lbs. crushed oats and 6 lbs. wheat bran on first five days; 14 lbs. crushed oats and 14 lbs. wheat bran on last two days; property of James Stillman, Sing Sing, N. Y.

Khelula of Briarcliff 46799.—Sire, Young Garenne's Duke 6863; dam, Khelula 17970.

Butter, 18 lbs. 2½ oz. Milk, 241 lbs. 6 oz.

Test made from Jan. 4 to 11, 1893; age, 5 yrs. 1 mo.; estimated weight, 900 lbs.; fed 84 qts. of a mixture of equal parts of ground oats, corn meal, bran and cotton-seed meal—hay; property of James Stillman, Cornwall-on-Hudson, N. Y.

King of St. L.'s Allie 86797.—Sire, King of St. Lambert 15175; dam, King's Bravy 78868.

Butter, 20 lbs. 10 oz. Milk, 291 lbs. 11 oz.

Test made from July 18 to 24, 1808; age, 5 yrs. 5 mos.; estimated weight, 960 lbs.; fed 8 qts. oat meal and 6 qts. corn meal, daily—clover pasture; property of T. B. Kibler, Jenera, Ohio.

King's Antoinette 40456.—Sire, King 2d 11570; dam, Lady Antoinette 24391.

Butter, 15 lbs. 1 oz. Milk, 102 lbs.

Test made from July 17 to 24, 1887; age, 2 yrs. 3 mos.; estimated weight, 700 lbs.; fed, daily, 4 qts. bran, 4 qts. corn meal, 1 qt. pea meal; property of M. Erskine Miller, Staunton, Va.

King's Beauty 24397.—Sire, King (P. S. 238 J. H. B.); dam, Bornou (F. S. 4338 J. H. B.)

Butter, 16 lbs. 13 oz. Milk, 196 lbs. 8 oz.

Test made from May 14 to 20, 1886; age, 5 yrs. 2 mos.; property of George T. Hodgson, Athens, Ga.

King's Charella 80765.—Sire, King of Amherst 25902; dam, Lady Charella 66187.

Butter, 15 lbs. 9½ oz. Milk, 160 lbs. 11 oz.

Test made from Feb. 1 to 8, 1896; age, 4 yrs. 4 mos.; actual weight, 832 lbs.; fed 35 lbs. ground oats, 21 lbs. ground corn, 14 lbs. ground wheat, 14 lbs. oil meal and 200 lbs. corn ensilage; property of Charles Lautz, Buffalo, N. Y.

King's Erica 22096.—Sire, King (P. S. 238 J. H. B.); dam, Erica 25005.

Butter, 16 lbs. 3¼ oz. Milk, 248 lbs. 3¾ oz.

Test made from Jan. 8 to 15, 1886; age, 4 yrs. 6 mos.; property of J. V. & C. Ramsden, Morton, Pa.

King's Fosie 80768.—Sire, King of Amherst 25902; dam, Chief's Fosie 61998.

Butter, 15 lbs. 10 oz. Milk, 145 lbs. 12 oz.

Test made from Jan. 15 to 22, 1895; age, 2 yrs. 9 mos.; actual weight, 780 lbs.; fed 14 lbs. wheat, 21 lbs. oats, 14 lbs. corn meal, 21 lbs. oil meal, 182 lbs. corn ensilage and 28 lbs. hay; property of Charles Lautz, Buffalo, N. Y.

King's Gesnera 98727.—Sire, King of Amherst 25902; dam, Pansy of Amherst Villa 70503.

> Butter, 15 lbs. 10 oz.   Milk, 158 lbs. 10 oz.

Test made from Feb. 7 to 14, 1896; age, 2 yrs. 3 mos.; actual weight, 653 lbs.; fed 24½ lbs. ground oats, 7 lbs. ground wheat, 14 lbs. oil meal, 42 lbs. corn refuse and 140 lbs. corn ensilage; property of Charles Lautz, Buffalo, N. Y.

King's Lady Natica 80769.—Sire, King of Amherst 25902; dam, Lady Natica 66031.

> Butter, 15 lbs. 8 oz.   Milk, 144 lbs. 10 oz.

Test made from Jan. 30 to Feb. 6, 1895; age, 2 yrs. 8 mos.; actual weight, 727 lbs.; fed 14 lbs. ground wheat, 21 lbs. ground oats, 14 lbs. corn meal, 21 lbs. oil meal, 178 lbs. corn ensilage and 35 lbs. hay; property of Charles Lautz, Buffalo, N. Y.

King's Lassie 79103.—Sire, King of St. Lambert 15175; dam, Gazelle Dee 18056.

> Butter, 18 lbs. 5 oz.   Milk, 282 lbs.

Test made from June 3 to 10, 1898; age, 8 yrs. 3 mos.; actual weight, 950 lbs.; fed 16 qts. wheat bran, 6 qts. ground oats and 4 qts. corn meal, daily—good pasture; property of Mrs. Jennie E. Hall, Columbus, Ohio.

King's Lucy Pogis 111380.—Sire, Lucy's Stoke Pogis 26533; dam, Homestead Prue 82203.

> Butter, 18 lbs. 8 oz.   Milk, 249 lbs. 8 oz.

Test made from May 22 to 29, 1898; age, 3 yrs. 7 mos.; estimated weight, 800 lbs.; fed 10 lbs. per day of a mixture of gluten feed, bran and "Victor" corn and oat feed, in equal parts—good blue-grass pasture; property of W. J. Hussey, Mount Pleasant, Ohio.

King's Princess 30948.—Sire, King (P. S. 238 J. H. B.); dam, Grise, on I. of J.

> Butter, 24 lbs. 5 oz.   Milk, 232 lbs. 4 oz.

Test made from June 13 to 19, 1887; age, 6 yrs.; estimated weight, 900 lbs.; fed, daily, 5 qts. corn meal, 4 qts. bran and 1 qt. pea meal; property of M. Erskine Miller, Staunton, Va.

King's Queen 98712.—Sire, King Torment 18609; dam, Princess of Canaan 61020.

> Butter, 14 lbs. 12 oz.   Milk, 223 lbs. 8 oz.

Test made from Dec. 28, 1897, to Jan. 4, 1898; age, 5 yrs. 1 mo.; estimated weight, 750 lbs.; fed 6 lbs. middlings and 2 lbs. cotton-seed meal on 25 lbs. good corn ensilage, and about 6 lbs. cut clover hay, mixed, daily; property of Joseph T. Hoopes, Bynum, Md.

King's Rosella 80767.—Sire, King of Amherst 25902; dam, Rosella of Bay View 66032.

> Butter, 15 lbs. 12½ oz.   Milk, 205 lbs. 14 oz.

Test made from Dec. 11 to 18, 1894; age, 2 yrs. 9 mos.; actual weight, 800 lbs.; fed 14 lbs. wheat, 21 lbs. oats, 10½ lbs. oil meal, 10½ lbs. cotton-seed meal, 210 lbs. corn ensilage and 35 lbs. hay; property of Charles Lautz, Buffalo, N. Y.

King's Stevia 98737.—Sire, King of Amherst 25902; dam, Chief's Charella 61999.

> Butter, 19 lbs. 3½ oz.   Milk, 191 lbs.

Test made from Feb. 16 to 23, 1897; age, 2 yrs. 11 mos.; actual weight, 615 lbs.; fed 35 lbs. corn meal, 21 lbs. oil meal, 35 lbs. ground oats, 280 lbs. corn ensilage and 28 lbs. corn-stalks; property of Charles Lautz, Buffalo, N. Y.

King's Sultana 76657.—Sire, King Koffee Jr. 12317; dam, Lalla Ingleside 46129.

> Butter, 14 lbs. 8½ oz. Milk, 192 lbs.

Test made from June 4 to 11, 1896; age, 6 yrs. 9 mos.; estimated weight, 750 lbs.; fed 56 qts. ground oats and 56 qts. corn meal—good woodland pasture; property of Morgan & Brown, Columbia, Tenn.

King's Trust 18946.—Sire, King (P. S. 238 J. H. B.); dam, Trust (F. S. 3993 J. H. B.)

> Butter, 18 lbs. Milk, 137⅞ qts.

Test made from Apr. 23 to 29, 1884; age, 3 yrs. 4 mos.; fed ground corn, oats, wheat middlings, oil meal and a pail of parsnips, daily; property of S. M. Burnham, Saugatuck, Conn.

King's Trust 18946.—Sire, King (P. S. 238 J. H. B.); dam, Trust (F. S. 3993 J. H. B.)

> Butter, 36 lbs. 6¼ oz. in 15 days.

Test made in 1884; age, 3 yrs.; property of S. M. Burnham, Saugatuck, Conn.

Kisberine Rex 31436.—Sire, Easter Boy 3032; dam, Kisberine 23480.

> Butter, 16 lbs. 8 oz. Milk, 257 lbs. 12 oz.

Test made from Dec. 10 to 17, 1897; age, 13 yrs. 3 mos.; estimated weight, 800 lbs.; fed 15 lbs. daily of a mixture of ground corn and oats, in equal parts by weight, and 12 lbs. hay; property of S. B. Pierce, Woodlawn, Neb.

Kittie Green 88379.—Sire, Denise's Tormentor 11823; dam, Kittie Dolan 66834.

> Butter, 15 lbs. Milk, 217 lbs.

Test made from May 23 to 30, 1897; age, 4 yrs. 5 mos.; estimated weight, 800 lbs.; fed 52½ lbs. wheat bran, 45½ lbs. corn meal, 7 lbs. linseed oil cake meal and 10½ lbs. cotton-seed meal—blue-grass pasture; property of Samuel McKeen, Terre Haute, Ind.

Kittura of St. Lambert 78344.—Sire, Ruby's Star of St. Lambert 20719; dam, Maggie Clarendon 2d 23546.

> Butter, 18 lbs. 8¾ oz. Milk, 217 lbs. 4 oz.

Test made from Feb. 16 to 23, 1896; age, 6 yrs. 11 mos.; estimated weight, 900 lbs.; fed 11½ gals. daily of cut corn fodder and mill feed (one part gluten to two parts bran), in equal parts, wet; property of Mrs. A. M. Hallock, Columbus, Ohio.

Kitty Better 32911.—Sire, Lord Harry 3445; dam, Kate Gordon 8387.

> Butter, 15 lbs. 14⅜ oz. Milk, 221 lbs. 4 oz.

Test made from Dec. 16 to 23, 1889; age, 6 yrs.; estimated weight, 850 lbs.; fed, daily, about 3 gals. corn meal and 1 pt. cotton-seed meal; property of M. C. Campbell, Spring Hill, Tenn.

Kitty Black Prince 104472.—Sire, Wine 29954; dam, Belle of Allentown 83636.

> Butter, 14 lbs. ½ oz. Milk, 104 lbs. 12 oz.

Test made from Oct. 27 to Nov. 3, 1896; age, 2 yrs. 1 mo.; estimated weight, 750 lbs.; fed 25 lbs. daily of a mixture of ground corn and oats, bran, cotton-seed and oil meal—hay and corn fodder, poor pasture; property of R. F. Shannon, Pittsburg, Pa.

Kitty Clover 1113, imp.

> Butter, 14 lbs.

Test made from June 7 to 14, 1878; age, 10 yrs. 6 mos.; property of James A. Hayt, Patterson, N. Y.

Kitty Clover 2d 16099.—Sire, Grand Duke Alexis 1040; dam, Kitty Clover 1113.

Butter, 14 lbs. 15 oz. Milk, 231 lbs. 12 oz.

Test made from July 3 to 10, 1893; age, 13 yrs. 1 mo.; estimated weight, 850 lbs.; fed 4 qts. gluten meal and 4 qts. corn meal, ground oats and oil meal, mixed, daily—natural pasture; property of Geo. W. Sisson, Jr., Potsdam, N. Y.

Kitty Colt 2213.—Sire, Albert 44; dam, Kittie Lightfoot 2d 1300.

Butter, 15 lbs. 9½ oz. Milk, 214 lbs. 8 oz.

Test made from May 16 to 22, 1883; age, 11 yrs. 5 mos.; property of W. B. Montgomery, Starkville, Miss.

Kitty Kringel 57120.—Sire, Sir Pedro Pogis 16010; dam, Kitty of Cedar Glen 10948.

Butter, 16 lbs. 8 oz. Milk, 229 lbs. 12 oz.

Test made from Jan. 20 to 27, 1893; age, 5 yrs. 5 mos.; estimated weight, 750 lbs.; fed 6 lbs. feed (corn ground on cob, with one-fifth oats) and 1½ lbs. oil meal, daily—hay; property of Mrs. Evirilla Sherman Walsh, Granville, N. Y.

Kitty Livingston 34303.—Sire, Gen. James A. Garfield 14874; dam, Lady Livingston 33374.

Butter, 15 lbs. 2 oz. Milk, 145 lbs. 8-oz.

Test made from Nov. 22 to 29, 1887; age, 3 yrs.; estimated weight, 700 lbs.; fed, daily, 10 lbs. corn meal, 3 lbs. oil meal, 2 lbs. middlings; property of James Stillman, Sing Sing, N. Y.

Kitty of Jefferson 2d 12313.—Sire, Dexter of Jefferson 5681; dam, Kitty of Jefferson 12193.

Butter, 14 lbs. 3 oz. Milk, 196 lbs. 6 oz.

Test made from Jan. 3 to 10, 1891; age, 11 yrs. 8 mos.; estimated weight, 900 lbs.; fed 56 lbs. crushed oats, 56 lbs. corn meal, 28 lbs. pea meal, 28 lbs. oil meal, 80 lbs. hay and 168 lbs. ensilage; property of Ayer & McKinney, Philadelphia, Pa.

Kitty of Prospect 111121.—Sire, Nell's John Bull 21921; dam, Kit 34376.

Butter, 19 lbs. 4¾ oz. Milk, 230 lbs. 8 oz.

Test made from Jan. 24 to 31, 1897; age, 4 yrs. 3 mos.; actual weight, 910 lbs.; fed 63 qts. corn meal, 63 qts. ground oats, 42 qts. ground oil-cake and 21 qts. wheat bran —some carrots; property of Miller & Sibley, Franklin, Pa.

Kitty of St. Lambert 6637.—Sire, Stoke Pogis 3d 2238; dam, Rosette of St. Lambert 5108.

Butter, 16 lbs. 11 oz. Milk, 267 lbs. 8 oz.

Test made from May 31 to June 7, 1892; age, 15 yrs. 1 mo.; estimated weight, 800 lbs.; fed 84 lbs. chopped oats—clover and timothy pasture; property of William Rolph, Markham, Ont., Can.

Kitty of St. Lambert 2d 40106.—Sire, Brier's Pogis 14163; dam, Kitty of St. Lambert 6637.

Butter, 16 lbs. 5 oz. Milk, 204 lbs.

Test made from June 6 to 13, 1889; age, 4 yrs. 4 mos.; estimated weight, 800 lbs.; fed, daily, 12 qts. oats and barley and 2 qts. oil-cake; property of Wm. Rolph, Markham, Ont., Can.

Kitty of Seguin 74347.—Sire, St. Lambert Gilderoy 20210; dam, Dinah of Camp Oaks 58291.

Butter, 15 lbs. 6¼ oz. Milk, 298 lbs. 5 oz.

Test made from Mar. 27 to Apr. 3, 1897; age, 6 yrs.; estimated weight, 800 lbs.; fed 9 lbs. wheat bran, 4½ lbs. corn meal, 6 lbs. crushed shuck corn and 3 lbs. cotton-seed, daily—hay *ad lib.*, pastured on green oats; property of H. C. Schulz, Marion, Tex.

Kitty Ona 24209.—Sire, Catono 3761; dam, Kitty Floos 11539.

Butter, 14 lbs. 11 oz. Milk, 233 lbs. 4 oz.

Test made from May 14 to 21, 1880; age, 6 yrs.; estimated weight, 1000 lbs.; fed, daily. about 5½ qts. corn meal, ground corn and oats, with a little bran; property of J. Burckel, Durham, Conn.

Kitty Pedrothorn 2d 84562.—Sire, Highland Pogis 22146; dam, Kitty Pedrothorn 49581.

Butter, 15 lbs. 4 oz. Milk, 198 lbs.

Test made from May 8 to 15, 1898; age, 6 yrs. 11 mos.; estimated weight, 800 lbs.; fed about 8 qts. daily of corn, oats and bran, mixed—good old meadow pasture; property of S. F. Boyer, Finleyville, Pa.

Kitty Potter 9893.—Sire, Almont 2789; dam, Honeysuckle 1313.

Butter, 18 lbs. 5 oz.

Test made in June, 1885; age, 6 yrs.; property of C. T. Grannis, Terryville, Conn.

Kitty Pride Pogis 78134.—Sire, Lorne Victor Pogis 22214; dam, Pride of Ingleside 54545.

Butter, 15 lbs. 9 oz. Milk, 240 lbs. 8 oz.

Test made from Oct. 3 to 10, 1896; age, 5 yrs. 6 mos.; estimated weight, 800 lbs.; fed 28 lbs. corn meal, 28 lbs. ground oats, 28 lbs. wheat middlings and 14 lbs. oil meal— pasture not very good; property of John C. McClintock, Meadville, Pa.

Kitty Sands 77432.—Sire, Fenwick 26451; dam, Pride of Aurora 67066.

Butter, 18 lbs. 1½ oz. Milk, 230 lbs.

Test made from July 12 to 19, 1895; age, 4 yrs. 3 mos.; actual weight, 720 lbs.; fed 9 qts. bran, 6 qts. ground oats, 3 qts. corn meal and 3 qts. oil meal, daily—green oats and peas; property of C. A. Sweet, East Aurora, N. Y.

Kitty's Ida 97736.—Sire, Ida of St. L.'s Last Son 23601; dam, Kitty of Jefferson 2d 12313.

Butter, 18 lbs. 8 oz. Milk, 319 lbs. 6 oz.

Test made from Feb. 13 to 20, 1897; age, 4 yrs. 5 mos.; estimated weight, 900 lbs.; fed 38 lbs. wheat bran, 42 lbs. corn meal, 36 lbs. oat meal, 19 lbs. cotton-seed meal, 70 lbs. hay, 1 bush. apples and ½ bush. carrots; property of Ayer & McKinney, Philadelphia, Pa.

Kitty's Matilda 75349.—Sire, Matilda 4th's Son 20214; dam, Kitty of Jefferson 2d 12313.

Butter, 14 lbs. 3 oz. Milk, 199 lbs. 3 oz.

Test made from Jan. 30 to Feb. 6, 1893; age, 2 yrs. 4 mos.; estimated weight, 700 lbs.; fed 14 lbs. oil meal, 42 lbs. pea meal, 80 lbs. corn, oats and shorts, in equal parts— hay; property of Ayer & McKinney, Philadelphia, Pa.

Kitty Victor Pogis 85904.—Sire, Melia Ann's Victor Pogis 20697; dam, Miss Palliser 64073.

Butter, 19 lbs. 12 oz. Milk, 311 lbs. 4 oz.

Test made from Nov. 25 to Dec. 2, 1897; age, 6 yrs. 5 mos.; estimated weight, 900 lbs.; fed 12 lbs. wheat feed, 7 lbs. corn meal and 3 lbs. oil meal, daily—clover rowen *ad lib.*; property of Mrs. Frances Clayton, Granville, N. Y..

Kobe's Marcella 4597.—Sire, Star Neal 1495; dam, Kobe 3055.

Butter, 14 lbs. 13 oz. Milk, 234 lbs. 10 oz.

Test made from June 4 to 11, 1892; age, 16 yrs. 7 mos.; estimated weight, 900 lbs.; fed 7 lbs. wheat bran, 7 lbs. old process linseed meal and 7 lbs. corn meal—roots *ad lib.*, good natural pasture; property of Giltedge Farm Co., Fayetteville, N. Y.

Koffee's Clarinda 65216.—Sire, Olymph's Koffee 20087; dam, Clarinda 2d 51037.

Butter, 19 lbs. 9½ oz. Milk, 259 lbs. 14 oz.

Test made from Apr. 11 to 18, 1897; age, 8 yrs. 7 mos.; estimated weight, 800 lbs.; fed 21 lbs. bran, 14 lbs. corn meal, 14 lbs. oats, 7 lbs. oil meal and 280 lbs. ensilage--hay *ad lib.*; property of William Weis, Palmyra, Wis.

Koffee's Duchess 2d 57988.—Sire, Ida's Pogis 18000; dam, Koffee's Duchess 35113.

Butter, 15 lbs. 8 oz. Milk, 283 lbs. 12 oz.

Test made from Apr. 7 to 14, 1894; age, 5 yrs. 5 mos.; estimated weight, 900 lbs.; fed 14 lbs. corn meal, 28 lbs. bran and 56 lbs. ground oats; property of R. A. Sibley, Spencer, Mass.

Koffee's Duchess 2d 57988.—Sire, Ida's Pogis 18000; dam, Koffee's Duchess 35113.

Butter, 17 lbs. 1 oz. Milk, 362 lbs.

Test made from Apr. 11 to 18, 1896; age, 7 yrs. 5 mos.; actual weight, 750 lbs.; fed 35 qts. bran, 35 qts. oat meal, 21 qts. corn meal, 14 qts. oil meal, 56 qts. roots and 100 lbs. hay; property of Pierce J. Cogswell, Rochester, N. Y.

Koffee's Grisette 30433.—Sire, King Koffee 5522; dam, Bonnie Grisette 2d 19526.

Butter, 15 lbs. 7 oz. Milk, 147 lbs. 8 oz.

Test made from June 6 to 12, 1887; age, 2 yrs. 8 mos.; estimated weight, 700 lbs.; fed, daily, 4 qts. bran, 4 qts. ground corn and oats, and 1 qt. pea meal; property of M. Erskine Miller, Staunton, Va.

Koffee's Lady Pogis 65219.—Sire, Olymph's Koffee 20087; dam, Pogis of Walnut Grove 51962.

Butter, 23 lbs. 4 oz. Milk, 253 lbs. 11 oz.

Test made from May 2 to 9, 1897; age, 7 yrs. 7 mos.; estimated weight, 850 lbs.; fed 21 lbs. bran, 21 lbs. corn meal, 14 lbs. oats and 14 lbs. oil meal—hay *ad lib.*; property of Wm. Weis, Palmyra, Wis.

Koffee's Lily 25515.—Sire, King Koffee 5522; dam, Le Gros' Lily of the Valley 2d 13386.

Butter, 15 lbs. 3½ oz. Milk, 134 lbs. 8 oz.

Test made from June 7 to 14, 1887; age, 3 yrs. 7 mos.; estimated weight, 850 lbs.; fed, daily, 4 qts. bran, 4 qts. corn and oats, 1 qt. pea meal; property of M. Erskine Miller, Staunton, Va.

Koffee's May Belle 65220.—Sire, Olymph's Koffee 20087; dam, Le Brocq's May Belle 27980.

Butter, 20 lbs. 5 oz. Milk, 242 lbs. 12 oz.

Test made from June 6 to 13, 1897; age, 7 yrs. 6 mos.; estimated weight, 950 lbs.; fed 28 lbs. bran and 28 lbs. oat meal—pasture ordinary grass, with some clover; property of William Weis, Palmyra, Wis.

Koffmadge 2d 77790.—Sire, Chieftain Pogis 25035; dam, Koffmadge 63735.

Butter, 16 lbs. 14 oz. Milk, 253 lbs. 4 oz.

Test made from Nov. 19 to 26, 1897; age, 6 yrs. 4 mos.; estimated weight, 800 lbs.; fed 50 lbs. ensilage, 5 lbs. bran, 3 lbs. oil meal, 2½ lbs. corn, 1¼ lbs. cotton-seed meal, 2½ lbs. oats and 10 lbs. hay, daily; property of Pennsylvania Reform School, Morganza, Pa.

Ko Ko's Allie 59317.—Sire, Ko Ko of St. Lambert 16527; dam, Yum Yum of St. Lambert 39152.

Butter, 15 lbs. 8¾ oz. Milk, 191 lbs. 8 oz.

Test made from May 3 to 10, 1891; age, 3 yrs.; estimated weight, 800 lbs.; fed 8 to 12 qts. daily of a mixture of one part bran, two parts ground corn and three parts ground oats—bran slop daily, good hay; property of L. L. Tozier, Batavia, N. Y.

Kosi 3431, imp.

Butter, 14 lbs. 7 oz. Milk, 198 lbs. 3 oz.

Test made from June 10 to 17, 1882; age, 12 yrs.; no feed, short mixed pasture; property of W. H. Walrath, Clayton, N. Y.

La Belle Petite 5472.—Sire, Stoke Pogis 3d 2238; dam, La Petite Mère 5470.

Butter, 15 lbs. 8 oz.

Test made in June, 1881; age, 5 yrs.; property of Cooper & Maddux, Reading, Ohio.

Lactine 10680.—Sire, Gilderoy 2107; dam, Rosy of St. Martin's 6564.

Butter, 17 lbs. 1½ oz., unsalted. Milk, 197 lbs.

Test made from June 22 to 28, 1884; age, 4 yrs. 3 mos.; property of H. M. Howe, Bristol, R. I.

Lady Adams 2d 6529.—Sire, Moscow 2303; dam, Lady Adams 4919.

Butter, 15 lbs. 3 oz., unsalted.

Test made from June 2 to 8, 1883; age, 5 yrs. 6 mos.; fed 4 qts. corn meal, oat meal and wheat bran, mixed, daily—good pasture; property of W. B. Dinsmore, Staatsburg, N. Y.

Lady Alexis 26916.—Sire, Duke of Albany 3899; dam, Belle of Maple Grove 11334.

Butter, 16 lbs. 8 oz. Milk, 244 lbs. 14 oz.

Test made from June 18 to 24, 1887; age, 3 yrs. 6 mos.; weight, 840 lbs.; fed 2 qts. corn and oats, ground, daily; property of W. H. Kennedy, Lincoln, N. Y.

Lady Alice of Hillcrest 7450.—Sire, Jeweler 1385; dam, Mary Ann 2038.

Butter, 16 lbs. 3 oz.

Test made from Dec. 25 to 31, 1882; property of James Crook, Jacksonville, Ala.

Lady Alice of Hillcrest 7450.—Sire, Jeweler 1385; dam, Mary Ann 2038.

Butter, 18 lbs. 6 oz.   Milk, 215 lbs.

Test made from Oct. 23 to 29, 1885; age, 7 yrs. 2 mos.; property of James Crook, Jacksonville, Ala.

Lady Alice of the Wilderness 12207.—Sire, Gilderoy 2107; dam, Gold Lace 10726.

Butter, 15 lbs. 14 oz.   Milk, 266 lbs.

Test made from June 15 to 22, 1880; age, 2 yrs. 1 mo.; fed grass and green rye only; property of Raymond H. Perry, Bristol, R. I.

Lady Alice Pogis 30273.—Sire, Prince of Melrose 2d 11015; dam, Lady Alice of Hillcrest 7450.

Butter, 22 lbs. 8 oz.   Milk, 251 lbs. 8 oz.

Test made from Feb. 21 to 28, 1889; age, 4 yrs. 4 mos.; estimated weight, 1100 lbs.; fed 70 qts. corn meal, 42 qts. cotton-seed meal, 42 qts. ground oats and 350 lbs. corn ensilage—hay, fair pasture of winter rye; property of James Crook, Jacksonville, Ala.

Lady Anna Pogis 52734.—Sire, Leo Pogis 10785; dam, Anna of Mountain Side 15544.

Butter, 20 lbs. 4 oz.   Milk, 286 lbs. 8 oz.

Test made from Jan. 12 to 19, 1895; age, 6 yrs. 7 mos.; actual weight, 1090 lbs.; fed 84 qts. ground oats, 62 qts. corn meal and 21 qts. ground oil-cake; property of Miller & Sibley, Franklin, Pa.

Lady Antoinette 24391.—Sire, Garibaldi (P. S. 242 J. H. B.); dam, Castaledes (F. S. 2876 J. H. B.)

Butter, 21 lbs. 6 oz.   Milk, 272 lbs.

Test made from May 31 to June 6, 1887; age, 6 yrs. 3 mos.; estimated weight, 900 lbs.; fed, daily, 4 qts. bran, 4 qts. ground corn and oats and 1 qt. pea meal; property of M. Erskine Miller, Staunton, Va.

Lady Appel 8612.—Sire, Young St. Martin 2219; dam, Isolda 5900.

Butter, 18 lbs. 3 oz.   Milk, 288 lbs.

Test made from Aug. 31 to Sept. 6, 1884; age, 7 yrs. 3 mos.; property of I. D. Risher, Hope Church, Pa.

Lady Armington 7610.—Sire, Von Bismarck 3d 1780; dam, Lady Theresa 2d 4253.

Butter, 17 lbs. 8 oz.   Milk, 200 lbs.

Test made from July 10 to 17, 1885; age, 8 yrs.; property of William H. Hopkins, Providence, R. I.

Lady Aspinwall 8374.—Sire, Count Logan 1599; dam, Houri 2842.

Butter, 14 lbs.   Milk, 190 lbs.

Test made from June 13 to 20, 1883; age, 5 yrs. 6 mos.; property of E. D. Griswold, Orwell, Vt.

Lady Aurora 98734.—Sire, Hans of Amherst Villa 27451; dam, King's Lady Natica 80769.

Butter, 14 lbs. 10 oz.   Milk, 156 lbs. 8 oz.

Test made from Mar. 3 to 10, 1897; age, 3 yrs. 1 mo.; actual weight, 700 lbs.; fed 35 lbs. ground oats, 28 lbs. corn meal, 21 lbs. cotton-seed meal, 280 lbs. corn ensilage and 28 lbs. dry corn-stalks; property of Charles Lautz, Buffalo, N. Y.

Lady Bidwell 10303.—Sire, Paddy Wilson 3084; dam, Queen of Coventry 8065.

Butter, 15 lbs. 12 oz.   Milk, 265 lbs. 4 oz.

Test made Mar., 1883; age, 4 yrs.; fed 6 qts. wheat bran and corn meal, mixed, and 10 qts. of potatoes, daily—hay; property of George S. Phelps, Warehouse Point, Conn.

Lady Bingo 24160.—Sire, Bingo 2d 6749; dam, Mrs. Knickerbocker 19367.

Butter, 15 lbs. 4 oz.   Milk, 172 lbs. 8 oz.

Test made from Mar. 11 to 18, 1887; age, 6 yrs.; estimated weight, 800 lbs.; fed, daily, 8 lbs. corn meal, 6 lbs. bran, 2 lbs. middlings and 4 lbs. oil meal; property of James Stillman, Sing Sing, N. Y.

Lady Bloomfield 4704.—Sire, Rioter 670; dam, Bloomfield Fairy 2d 3210.

Butter, 14 lbs. 12½ oz.

Property of John B. Mills, Griffin, Ga.

Lady Bountiful 17946 (F. S. 3348 J. H. B.)

Butter, 14 lbs. 15½ oz.   Milk, 229 lbs. 14 oz.

Test made from June 8 to 14, 1885; age, 7 yrs. 3 mos.; property of C. Easthope, Niles, Ohio.

Lady Bowen 354, imp.

Butter, 16 lbs. 8 oz., unsalted.   Milk, 223 lbs.

Test made from May 10 to 17, 1877; age, 15 yrs.; property of James Cloud & Son, Kennett Square, Pa.

Lady Brown 433.—Sire, McClellan 4th 85; dam, Pansy 7th 130.

Butter, 14 lbs.

Test made Dec., 1876; age, about 8 yrs.; property of S. W. Robbins, Wethersfield, Conn.

Lady Brown 2d 2348.—Sire, Albert 44; dam, Lady Brown 433.

Butter, 14 lbs. 3 oz.   Milk, 195 lbs. 15 oz.

Test made from Sept. 18 to 24, 1884; age, 14 yrs. 1 mo.; property of William Simpson, New York, N. Y.

Lady Brown 4th 6911.—Sire, Albert 44; dam, Lady Brown 433.

Butter, 14 lbs. 12 oz.

Test made from May 16 to 23, 1878; age, 5 yrs.; property of James Woodruff, Terryville, Conn.

Lady Caroline of St. Aubin's 11372, imp.

Butter, 14 lbs. unsalted.   Milk (average), 16 qts. daily.

Test made in Oct. 1879; age, 4 yrs. 4 mos.; property of M. H. Messchert, Douglassville, Pa.

Lady Cecilia 24821.—Sire, Whip 2638; dam, Eugenie 5th 6573.

Butter, 15 lbs. 1 oz.   Milk, 257 lbs. 8 oz.

Test made from Jan. 8 to 15, 1885; age, 5 yrs. 6 mos.; property of H. W. Douglass, Pevely, Mo.

Lady Charella 66187.—Sire, Chief 10663; dam, Charella 12017.

Butter, 14 lbs. 7 oz.   Milk, 220 lbs. 8 oz.

Test made from Oct. 18 to 24, 1893; age, 4 yrs. 8 mos.; actual weight, 838 lbs.; fed 14 lbs. corn meal, 21 lbs. cotton-seed meal, 14 lbs. middlings and 21 lbs. ground oats —clover pasture; property of Charles Lautz, Buffalo, N. Y.

Lady Clarendon 3d 17578.—Sire, Camerlengo 3012; dam, Lady Clarendon 7030.

Butter, 14 lbs. 5½ oz.

Test made from May 28 to June 4, 1883; age, 3 yrs. 9 mos.; property of J. Horatio Earll, Skaneateles, N. Y.

Lady Cleveland 30251.—Sire, Lord Alfred 5215; dam, Belle of New York 6963.

Butter, 17 lbs. 1 oz.   Milk, 178 lbs. 1 oz.

Test made from July 21 to 28, 1890; age, 5 yrs. 8 mos.; estimated weight, 800 lbs.; fed 4 gals. ground corn, daily; property of H. P. Webster, Columbia, Tenn.

Lady Cloud 19358.—Sire, Fayette Boy 4737; dam, Lady Tilford 14467.

Butter, 16 lbs. 10 oz.   Milk, 183 lbs. 8 oz.

Test made from Apr. 8 to 14, 1884; age, 1 yr. 10 mos.; property of John A. Middelton, Shelbyville, Ky.

Lady Cloud 19358.—Sire, Fayette Boy 4737; dam, Lady Tilford 14467.

Butter, 27 lbs. 6 oz.   Milk, 235 lbs. 8 oz.

Test made from June 20 to 27, 1886; age, 4 yrs.; estimated weight, 800 lbs.; fed 25 lbs. corn meal, oil meal and bran, mixed, daily—blue-grass pasture; property of John A. Middelton, Shelbyville, Ky.

Lady Cloud 2d 26825.—Sire, Croton Maid's Duke 6658; dam, Lady Cloud 19358.

Butter, 16 lbs. 12 oz.   Milk, 229 lbs. 8 oz.

Test made from May 22 to 29, 1887; age, 3 yrs. 2 mos.; estimated weight, 725 lbs.; fed 14 lbs. corn meal, 5 lbs. oatmeal and 6 lbs. oil meal, daily—blue-grass pasture; property of John A. Middelton, Shelbyville, Ky.

Lady Creamly 4th 19077.—Sire, Blossom's Tennessee 6060; dam, Lady Creamly 1975.

Butter, 17 lbs. 6½ oz.   Milk, 215 lbs. 12 oz.

Test made from May 24 to 31, 1887; age, 5 yrs.; estimated weight, 800 lbs.; fed, daily, about 12 lbs. ground corn and oats; property of E. C. McDowell, Columbia, Tenn.

Lady Crouse 69336.—Sire, Know Nothing 11638; dam, Lady Adams 4th 11464.

Butter, 18 lbs. 1 oz.   Milk, 305 lbs.

Test made from Apr. 2 to 9, 1896; age, 5 yrs. 9 mos.; actual weight, 910 lbs.; fed 28 qts. corn meal, 35 qts. bran, 35 qts. oat meal, 21 qts. oil meal, 105 lbs. hay and 56 qts. roots; property of P. J. Cogswell, Rochester, N. Y.

Lady Dartmouth 23159.—Sire, Lord Dartmouth 6302; dam, Snowball of Deerfoot 2d 14457.

Butter, 21 lbs. 13 oz.   Milk, 243 lbs. 5 oz.

Test made from Aug. 11 to 18, 1889; age, 7 yrs. 6 mos.; estimated weight, 875 lbs.; fed, daily, 10 lbs. corn meal, 10 lbs. ground oats and 10 lbs. middlings; property of D. F. Appleton, Ipswich, Mass.

Lady Delphine 28460.—Sire, Baldwin's Frolic 1384); dam, Lady Sarah 4931.

Butter, 14 lbs. 4 oz. Milk, 277 lbs.

Test made from June 25 to July 1, 1888; age, 8 yrs. 3 mos.; estimated weight, 1000 lbs.; fed, daily, 9 lbs. corn meal, 9 lbs. oil meal, 4½ lbs. bran, 3 lbs. middlings and 3 lbs. ground oats; property of A. D. McBride, Rochester, N. Y.

Lady Dorcas 74348.—Sire, Roi Rene 23730; dam, Lady Elfled 40757.

Butter, 14 lbs. 10 oz. Milk, 243 lbs. 4 oz.

Test made from Dec. 6 to 13, 1896; age, 5 yrs. 5 mos.; actual weight, 800 lbs.; fed 56 qts. ground corn and cob meal and 56 qts. wheat bran, mixed with cut corn stover—hay *ad lib.*, short pasture; property of Columbus Dixon, Gillespieville, Ohio.

Lady Dorlena 36105.—Sire, Lady Mary's Chelten Duke 6002; dam, Lena of Pleasant View 13453.

Butter, 15 lbs. 7 oz. Milk, 249 lbs.

Test made from June 25 to July 2, 1889; age, 4 yrs.; estimated weight, 900 lbs.; fed, daily, 4 lbs. bran, 4 lbs. oat meal and 4 lbs. corn meal; property of H. M. Baum, Frankfort, Ind.

Lady Elfled 40757.—Sire, Baron of the Isles 9146; dam, Alfritha 13673.

Butter, 18 lbs. Milk, 213 lbs.

Test made from June 16 to 23, 1890; age, 4 yrs. 3 mos.; actual weight, 780 lbs.; fed, daily, 4 lbs. corn and 4 lbs. oats, ground, and 4 lbs. bran; property of Thomas Beer, Bucyrus, Ohio.

Lady Emily Kilduff 101053.—Sire, Grisette's Koffee 26008; dam, Beauty of Edgeworth 51361.

Butter, 14 lbs. 4 oz. Milk, 234 lbs. 8 oz.

Test made from Oct. 14 to 21, 1896; age, 2 yrs. 3 mos.; estimated weight, 775 lbs.; fed 24 lbs. ground corn and oats, bran, cotton-seed, pea and oil meal, daily—hay and dry corn fodder, poor pasture; property of R. F. Shannon, Pittsburg, Pa.

Lady Emma C. 57041.—Sire, Denise's Tormentor 11823; dam, Lady Lily Gray 21364.

Butter, 16 lbs. Milk, 272 lbs. 8 oz.

Test made from Apr. 3 to 10, 1894; age, 6 yrs. 11 mos.; estimated weight, 800 lbs.; fed 9 qts. wheat bran, 7 qts. corn-hearts and 1 qt. oil meal, with cut clover hay, daily—short blue-grass pasture; property of Thomas Green, Hopkinsville, Ky.

Lady Fair 22103.—Sire, Cicero 7657; dam, Charmante (F. S. 1866 J. H. B.)

Butter, 19 lbs. 10 oz.

Test made from Dec. 1 to 7, 1885; age, 3 yrs. 11 mos.; property of W. L. Conyngham, Wilkes Barre, Pa.

Lady Fawn of St. Anne's 10920.—Sire, Victor Hugo 197; dam, Lisette 492.

Butter, 16 lbs. 12½ oz. Milk, 263 lbs. 4 oz.

Test made from Sept. 8 to 14, 1885; age, 14 yrs. 11 mos.; fed 26 lbs. daily of ground oats and peas and oil-cake, two-thirds oats, one-third peas, and 2½ lbs. oil-cake—pasture; property of W. A. Reburn, Ste. Anne de Bellevue, P. Q., Can.

Lady Gilmore 12136.—Sire, Pilot 3549; dam, Topsey of Lakeside 10240.

Butter, 14 lbs. 14 oz. Milk, 226 lbs. 8 oz.
Test made from June 15 to 22, 1881; age, 9 yrs. 5 mos.; property of Terence Carrigan,

Hopkinton, Mass.

Lady Golddust 2d 19861.—Sire, Duke of Darlington 2460; dam, Lady Golddust 7718.

Butter, 14 lbs. 4 oz. Milk, about 18 qts. per day.

Test made from July 15 to 21, 1886; age, 5 yrs.; property of D. F. Appleton, Ipswich, Mass.

Lady Golddust 2d 19861.—Sire, Duke of Darlington 2460; dam, Lady Golddust 7718.

Butter, 23 lbs. 4 oz. Milk, 21 qts. per day.

Test made from July 15 to 21, 1887; age, 6 yrs.; estimated weight of cow, 900 lbs.; fed 5 qts. corn meal, 8 qts. oat meal, 2 qts. pea meal, 1 qt. oil meal and 5 qts. middlings, daily; property of D. F. Appleton, Ipswich, Mass.

Lady Grace of Upholme 39569.—Sire, Brie 6591; dam, Weetamoe 5428.

Butter, 18 lbs. 4 oz. Milk, 163 lbs. 1½ oz.

Test made from June 20 to 27, 1893; age, 7 yrs.; estimated weight, 900 lbs.; fed 12 qts. corn and oats (ground and mixed in equal parts), 6 qts. wheat bran and 3 qts. oil meal, daily—little hay, good clover pasture; property of Charles A. Sweet, East Aurora, N. Y.

Lady Grace of Upholme 39569.—Sire, Brie 6591; dam, Weetamoe 5428.

Butter, 25 lbs. 5½ oz. Milk, 319 lbs. 12 oz.

Test made from May 9 to 16, 1894; age, 7 yrs. 10 mos.; estimated weight, 950 lbs.; fed 12 qts. ground oats, 8 qts. corn meal, 5 qts. bran and 3 qts. oil meal, daily—hay, poor pasture; property of C. A. Sweet, East Aurora, N. Y.

Lady Grace Pogis of S. 79462.—Sire, Bertha's Pogis 17993; dam, Lady Grace of Upholme 39569.

Butter, 20 lbs. 5 oz. Milk, 248 lbs.

Test made from Jan. 4 to 11, 1897; age, 5 yrs.; actual weight, 952 lbs.; fed 12 qts. bran, 9 qts. ground oats, 3 qts. corn meal and 3 qts. oil meal, daily—hay; property of Charles A. Sweet, East Aurora, N. Y.

Lady Gray of Hilltop 6850.—Sire, Wethersfield 966; dam, Bess Lena 3349.

Butter, 18 lbs. 12 oz. Milk, 42 lbs. per day.

Property of R. J. Fair, Wallingford, Conn.

Lady Gray of Hilltop 2d 14641.—Sire, Champion of Hilltop 1839; dam, Lady Gray of Hilltop 6850.

Butter, 14 lbs. 12 oz. Milk, 243 lbs. 15 oz.

Test made from Jan. 20 to 26, 1883; age, 4 yrs. 3 mos.; property of Mrs. L. M. Fair, Wallingford, Conn.

Lady Gray of Hilltop 3d 14642.—Sire, Champion of Hilltop 1839; dam, Lady Gray of Hilltop 6850.

Butter, 14 lbs. 2 oz. Milk, 253 lbs. 4 oz.

Test made from June 5 to 11, 1883; age, 3 yrs. 6 mos.; property of Mrs. L. M. Fair, Wallingford, Conn.

Lady Greville 12930.—Sire, Champion of America 1567; dam, Ruth Duffee 2481.

Butter, 14 lbs. 6 oz. Milk, 30 gals. 1 pt.

Test made from May 5 to 11, 1884; age, 3 yrs. 5 mos.; property of L. Q. C. Lamar, Sr., Oxford, Miss.

Lady Hugo 29430.—Sire, Combination 4389; dam, Calpurnia 13267.

Butter, 16 lbs. 7 oz. Milk, 239 lbs. 8 oz.

Test made from July 15 to 21, 1887; age, 3 yrs. 6 mos.; estimated weight, 750 lbs.; fed 2 qts. corn meal, 6 qts. oats, 6 qts. bran and 1 pt. oil meal, daily; property of L. E. Hill, Denver, Col.

Lady Ives 3d 6740.—Sire, Success 2097; dam, Lady Ives 1708.

Butter, 14 lbs. 8 oz., unsalted.

Test made from Apr. 8 to 15, 1883; age, 8 yrs. 8 mos.; property of J. H. Walker, Worcester, Mass.

Lady Jane of St. Peter's 7475.—Sire, Grey King (P. S. 169 J. H. B.); dam, Matin 7768 (F. S. 1629 J. H. B.)

Butter, 15 lbs. Milk, 207 lbs.

Test made from Dec. 22 to 28, 1884; age, 6 yrs. 2 mos.; property of Beech Grove Farm, Beech Grove, Ind.

Lady Jane of St. Peter's 2d 62536.—Sire, Garfield Stoke Pogis 15963; dam, Lady Jane of St. Peter's 7475.

Butter, 14 lbs. 5 oz. Milk, 148 lbs. 12 oz.

Test made from June 12 to 19, 1892; age, 3 yrs. 8 mos.; estimated weight, 790 lbs.; fed 56 lbs. cut corn fodder, 7 lbs. corn meal, 14 lbs. gluten meal, 14 lbs. bran, 7 lbs. linseed meal, 210 lbs. roots and 56 lbs. hay; property of Est. of Frederick Billings, Woodstock, Vt.

Lady Josephine 11560.—Sire, Bertram 1883; dam, Josephine Beacon 3306.

Butter, 10 lbs. 2 oz. in 8 days.

Test made from June 18 to 25, 1882; property of E. L. Briggs, Wilton Junction, Iowa.

Lady Julia G. 16199.—Sire, Gold Basis 4038; dam, Rosa of Belle Vue 6954.

Butter, 18 lbs. 3 oz. Milk, 210 lbs. 12 oz.

Test made from Apr. 28 to May 4, 1886; age, 3 yrs. 10 mos.; property of Thomas H. Malone, Nashville, Tenn.

Lady Kate Rex 91751.—Sire, Peggy's Royal Signal 22613; dam, Flora du Coin 39492.

Butter, 14 lbs. 8 oz. Milk, 239 lbs.

Test made from May 6 to 13, 1895; age, 2 yrs. 5 mos.; estimated weight, 750 lbs.; fed 56 lbs. corn meal—prairie pasture; property of F. M. Hosford, Bethany, Neb.

Lady Kingscote 26085.—Sire, Vertumnus (P. S. 161 J. H. B.); dam, Cocotte (P. S. 38 J. H. B.)

Butter, 15 lbs. 10 oz. Milk, 239 lbs.

Test made from Oct. 15 to 22, 1885; age, 7 yrs. 8 mos.; property of J. L. Shallcross, Anchorage, Ky.

Lady K. of St. Lambert 112294.—Sire, King of St. Lambert's King 30752; dam, Lady Perfection 32116.

Butter, 24 lbs. 6 oz. Milk, 333.4 lbs.

Test made from Mar. 14 to 21, 1898; age, 4 yrs.; estimated weight, 1000 lbs.; fed 15 lbs. clover hay, 25 lbs. ensilage, 8 lbs. oil meal, 12 lbs. corn meal, 10 lbs. wheat bran and 15 lbs. hay, daily; property of Frank W. Hart, Cleveland, Ohio.

Lady Larkspur 56051.—Sire, Prince Pogis 10682; dam, Honeymoon of St. Lambert 11221.

Butter, 21 lbs. 2¾ oz. Milk, 234 lbs. 8 oz.

Test made from Nov. 11 to 18, 1897; age, 9 yrs. 6 mos.; estimated weight, 850 lbs.; fed 126 qts. ground oats, 42 qts. corn meal, 14 qts. wheat bran and 14 qts. ground oil-cake—clover hay; property of Miller & Sibley, Franklin, Pa.

Lady Livingston 33374.—Sire, Garibaldi H. 7106; dam, Rainbow 2d 13962.

Butter, 15 lbs. 2 oz. Milk, 202 lbs.

Test made from Dec. 30, 1886, to Jan. 5, 1887; age, 3 yrs. 9 mos.; estimated weight, 900 lbs.; fed 4 qts. corn meal, 4 qts. oats, 1 qt. oil meal and 8 qts. bran. daily; property of George E. Peer, Rochester, N. Y.

Lady Longfield 23524.—Sire, Cicero 7657; dam, Dappled (F. S. 2773 J. H. B.)

Butter, 17 lbs. 3½ oz. Milk, 203 lbs. 8 oz.

Test made from Nov. 23 to 30, 1886; age, 4 yrs. 9 mos.; estimated weight, 900 lbs.; fed, daily, 16 lbs. corn chop, 6 lbs. bran and 6 lbs. cotton-seed meal, mixed; property of Campbell Brown, Spring Hill, Tenn.

Lady Longfield 23524.—Sire, Cicero 7657; dam, Dappled (F. S. 2773 J. H. B.)

Butter, 20 lbs. 13 oz. Milk, 227 lbs. 8 oz.

Test made from Jan. 12 to 19, 1891; age, 8 yrs. 11 mos.; estimated weight, 950 lbs.; fed 16 lbs. corn chop and 8 lbs. wheat bran, mixed, daily; property of Campbell Brown, Spring Hill, Tenn.

Lady Louise 4339.—Sire, Graycoat 1105; dam, Gold Ear 2200.

Butter, 15 lbs.

Age, 8 yrs.; property of R. G. Skiff, Green's Farms, Conn.

Lady Louise D. 58056.—Sire, Moidore 15911; dam, Mobile Beauty 47365.

Butter, 22 lbs. 12 oz. Milk, 278 lbs.

Test made from July 10 to 17, 1896; age, 7 yrs. 8 mos.; estimated weight, 800 lbs.; fed 175 lbs. cut corn, 175 lbs. cut sorghum, 56 qts. bran, 42 qts. oats, 7 qts. cotton-seed meal and 7 qts. ship stuff—Bermuda grass pasture; property of T. G. Bush, Mobile, Ala.

Lady Love 2d 2212.—Sire, Albert 44; dam, Lady Love 1315.

Butter, 16 lbs. 8 oz.

Test made in May, 1883; age, 11 yrs. 6 mos.; fed ground corn and oats—very good pasture; property of William Simpson, New York, N. Y.

Lady McDowell 37414.—Sire, Truxton 4654; dam, Normandy Girl 7737.

Butter, 15 lbs. 12 oz. Milk, 159 lbs. 2 oz.

Test made from May 19 to 26, 1886; age, 3 yrs. 2 mos.; property of John A. McEwen, Nashville, Tenn.

Lady Marjoram 123887.—Sire, Canada's John Bull 5th 20092; dam, Ethel Marjoram 87266.

Butter, 21 lbs. 11 oz. Milk, 355 lbs.

Test made from Oct. 2 to 9, 1897; age, 4 yrs. 7 mos.; estimated weight, 950 lbs.; fed 21 qts. oats and peas, 2 qts. bran, and 2 lbs. oil-cake, daily—clover pasture; property of William Rolph, Markham, Ont., Can.

Lady Mary Hampton 4861.—Sire, Orange Peel 864; dam, Georgetta 93.

Butter, 14 lbs. 5 oz. Milk, 219 lbs. 8 oz.

Test made from May 29 to June 4, 1884; age, 9 yrs. 8 mos.; property of Campbell Brown, Spring Hill, Tenn.

Lady Mary Lenox 105574.—Sire, Combine 20238; dam, Lady Mary Combination 98442.

Butter, 16 lbs. 7½ oz. Milk, 252 lbs. 8 oz.

Test made from Dec. 15 to 22, 1894; age, 4 yrs. 2 mos.; estimated weight, 800 lbs.; fed 16 lbs. daily of a mixture of ground oats, corn meal, oil meal and bran—4 qts. roots, daily; property of Richardson Bros., Davenport, Iowa.

Lady Mary of Prospect 19768.—Sire, Stoke Pogis 5th 5987; dam, Lady Mary Linden 12800.

Butter, 19 lbs. 15½ oz. Milk, 261 lbs. 8 oz.

Test made from Aug. 23 to 29, 1886; age, 3 yrs. 5 mos.; weight, 725 lbs.; fed 22 lbs. daily of mixed corn-hearts, oats, oil meal and middlings; property of Miller & Sibley, Franklin, Pa.

Lady Matilda Pogis 36270.—Sire, Louis of St. Lambert 11013; dam, La Petite Mère 3d 12814.

Butter, 21 lbs. 9 oz. Milk, 282 lbs. 5 oz.

Test made from Apr. 16 to 23, 1890; age, 4 yrs. 3 mos.; estimated weight, 850 lbs.; fed, daily, about 8 lbs. corn, 6 lbs. bran, 7½ lbs. pea meal, 2½ lbs. cotton-seed meal and 4 lbs. oats; property of W. B. von Richthofen, Montclair, Col.

Lady Maud of Cinque Park 2d 71548.—Sire, Signal Sawyer 16786; dam, Lady Maud of Cinque Park 57572.

Butter, 14 lbs. 1¼ oz. Milk, 265 lbs.

Test made from May 4 to 11, 1897; age, 7 yrs. 2 mos.; estimated weight, 950 lbs.; fed 32 lbs. wheat bran, 50 lbs. corn and oat hulls chop and 18 lbs. oil meal—poor June grass pasture; property of C. L. Boyer, Finleyville, Pa.

Lady May Jefferson 96591.—Sire, Money Pogis 16674; dam, Rival's Kitty 61760.

Butter, 14 lbs. 12 oz. Milk, 235 lbs. 7 oz.

Test made from Apr. 7 to 14, 1896; age, 6 yrs. 1 mo.; estimated weight, 900 lbs.; fed 4 lbs. oil-cake meal, 4 lbs. bran, 4 lbs. gluten and 5 lbs. hay, daily; property of E. T. Walker, Emerson, Ohio.

Lady Mel 2d 1795.—Sire, Albert 44; dam, Lady Mel 429.

Butter, 183 lbs. in 61 days. Milk, average of 18 qts. per day.

Test made from Apr. 15 to June 15, 1875; age, 5 yrs. 4 mos. at end of test; property of C. P. Chapman, Pittsfield, Ill.

Lady Mitchell 29214.—Sire, Joe Lozer 7538; dam, Cymelia 18706.

Butter, 14 lbs. 12¼ oz. Milk, 252.6 lbs.

Test made from Aug. 6 to 13, 1894; age, 10 yrs.; actual weight, 825 lbs.; fed 40 qts. corn meal, 26 qts. oil meal, 56 qts. wheat bran, 12 qts. cotton-seed meal, 35 qts. ground oats and 84 lbs. cut clover hay, mixed with grain, wetted; property of Charles S. Ellis, Scottsville, N. Y.

Lady Monmouth 15173.—Sire, Prince of Warren 1512; dam, Monmouth Duchess 4th 7129.

Butter, 15 lbs. 3 oz. Milk, 258 lbs. 4 oz.

Test made from June 10 to 16, 1887; age, 7 yrs.; estimated weight, 900 lbs.; fed, daily, 4 lbs. corn meal, 2 lbs. oil meal and 12 lbs. bran; property of A. H. Cooley, Little Britain, N. Y.

Lady Musa 98730.—Sire, Hans of Amherst Villa 27451; dam, Bloom of Amherst Villa 70507.

Butter, 16 lbs. 9½ oz. Milk, 175 lbs. 3 oz.

Test made from Jan. 26 to Feb. 2, 1896; age, 2 yrs. 1 mo.; actual weight, 610 lbs.; fed 35 lbs. ground oats, 42 lbs. corn refuse, 14 lbs. ground wheat, 14 lbs. oil meal and 252 lbs. corn ensilage; property of Charles Lautz, Buffalo, N. Y.

Lady Negelia 98735.—Sire, Hans of Amherst Villa 27451; dam, Daisy S. of Amherst Villa 70508.

Butter, 16 lbs. 8 oz. Milk, 204 lbs.

Test made from Dec. 8 to 15, 1897; age, 3 yrs. 10 mos.; actual weight, 700 lbs.; fed 21 lbs. ground oats, 21 lbs. ground corn, 21 lbs. oil meal, 7 lbs. wheat, 210 lbs. corn ensilage and 49 lbs. hay; property of Charles Lautz, Buffalo, N. Y.

Lady Nolia 87972.—Sire, Denise's Tormentor 11823; dam, Grace Pansy 2d 18764.

Butter, 14 lbs. 6 oz. Milk, 178 lbs.

Test made from Nov. 20 to 27, 1894; age, 3 yrs. 1 mo.; estimated weight, 750 lbs.; fed 12 qts. wheat bran, 6 qts. corn meal and 3 qts. cotton-seed meal, in three feeds, daily—cut sorghum, little hay; property of George V. Green, Hopkinsville, Ky.

Lady Oaks 2d 5246.—Sire, Baltimore Boy 837; dam, Lady Oaks 2081.

Butter, 15 lbs. 2 oz.

Test made from May 18 to 24, 1881; age, 6 yrs. 6 mos.; fed 2 qts. corn meal and 2 qts. wheat middlings—good clover and orchard grass pasture; property of John E. Phillips, Baltimore, Md.

Lady of Bay View 66029.—Sire, Bay View Chief 23853; dam, Minnie B. Rex 9256.

Butter, 16 lbs. 12 oz. Milk, 213 lbs. 4 oz.

Test made from Feb. 8 to 15, 1898; age, 2 yrs. 11 mos.; estimated weight, 450 lbs.; fed 17½ lbs. oil meal, 252 lbs. corn refuse, 210 lbs. corn ensilage, 14 lbs. malt sprouts and 28 lbs. hay; property of Charles Lautz, Buffalo, N. Y.

Lady of Belle Vue 7705.—Sire, Lord Lawrence 1414; dam, Lady Burlington 1713.

Butter, 15 lbs. 11 oz., unsalted. Milk, 169 lbs. 8 oz.

Test made from Nov. 19 to 26, 1883; age, 5 yrs. 4 mos.; fed 5 qts. corn meal, 6 qts. ground oats and 4 gals. cut-up sheaf oats, daily—very poor blue-grass pasture; property of Matt. M. Gardner, Nashville, Tenn.

Lady of Dryden 27642.—Sire, Sultan of New York 6186; dam, Lady of Venice 13342.

Butter, 16 lbs. 3 oz. Milk, 328 lbs.

Test made from June 10 to 17, 1887; age, 6 yrs. 1 mo.; weight, 870 lbs.; fed 2 qts. corn meal, 2 qts. middlings and 1 qt. oil meal, daily; property of Wm. E. Brown, West Dryden, N. Y.

Lady of Malone 25734.—Sire, Tarquin 3d 8203; dam, Malone 4986.

Butter, 16 lbs. Milk, 173 lbs. 6 oz.

Test made from Dec. 31, 1884, to Jan. 8, 1885; age, 4 yrs. 6 mos.; property of P. P. Paddock, Malone, N. Y.

Lady of Union Valley 101733.—Sire, Lex Linmore 18623; dam, Little Dorrit C. 73100.

Butter, 20 lbs. 7½ oz. Milk, 264 lbs.

Test made from June 6 to 13, 1898; age, 5 yrs. 3 mos.; estimated weight, 900 lbs.; fed 8 qts. daily of corn chop, oats chop and bran, mixed in equal parts—old pasture; property of S. F. Boyer, Finleyville, Pa.

Lady Ona 50642.—Sire, Catono 3761; dam, Virginie Star 15662.

Butter, 14 lbs. Milk, 216 lbs.

Test made from Nov. 2 to 9, 1895; age, 8 yrs. 10 mos.; estimated weight, 900 lbs.; fed 35 lbs. hominy meal, 45½ lbs. light bran and about 490 lbs. ensilage; property of J. T. Polk, Greenwood, Ind.

Lady Oonan Pogis 119154.—Sire, Oonan's Tormentor 22280; dam, Lady Ida Pogis 94166.

Butter, 19 lbs. 6¾ oz. Milk, 231 lbs. 3 oz.

Test made from Feb. 9 to 16, 1898; age, 5 yrs. 3 mos.; estimated weight, 800 lbs.; fed 22 lbs. per day of ground oats and corn, in equal parts, mixed with 4 bucketfuls cut sheaf oats, wetted—millet hay; property of M. C. Campbell, Spring Hill, Tenn.

Lady Oxford 4860.—Sire, Excelsior of Jersey 949; dam, Olivia 1397.

Butter, 22 lbs. 2 oz. in 10 days. Milk, 310 lbs.

Test made Jan., 1878; age, 3 yrs. 8 mos.; property of Thomas C. Murphy, Thayer, Kan.

Lady Oxford 2d 15101.—Sire, Rodney 1941; dam, Lady Oxford 4860.

Butter, 14 lbs. 7 oz. Milk, 202 lbs. 13 oz.

Test made from Nov. 10 to 17, 1884; age, 5 yrs. 10 mos.; property of T. C. Murphy, Thayer, Kan.

Lady Palestine 2769.—Sire, Jersey Prince 1062; dam, Palestine 2d 1455.

Butter, 14 lbs. 5 oz.

Test made from Oct. 10 to 16, 1873; age, 5 yrs. 3 mos.; test made on grass alone; property of Frank Ford, Ravenna, Ohio.

Lady Panalphrex 17400.—Sire, Easter Boy 2d 5310; dam, Castelara 2d 9859.

Butter, 23 lbs. 9 oz. Milk, 21 qts. per day.

Test made from Apr. 15 to 21, 1885; age, 3 yrs. 3 mos.; fed 2 lbs. oil meal, 8 lbs. oat meal, 6 lbs. corn meal and 8 lbs. of shorts, with a little salt, daily—good blue-grass pasture; property of Joseph C. Kiplinger & Co., Springfield, Ohio.

Lady Pedrona 63173.—Sire, Maximum Rioter 18530; dam, Young Pedrona 46096.

Butter, 14 lbs. 9 oz. Milk, 206 lbs. 5 oz.

Test made from Dec. 31, 1892, to Jan. 7, 1893; age, 2 yrs. 11 mos.; estimated weight, 850 lbs.; fed 14 qts. daily of a mixture of equal parts of corn meal, ground oats, wheat bran and middlings, also 1 pt. oil meal and 15 lbs. ensilage—hay *ad lib.*; property of Walter W. Law, Whitson, N. Y.

Lady Petite 100562.—Sire, Leo of Glynllyn 32015; dam, Painted Lady 45278.

Butter, 15 lbs. 11 oz.   Milk, 196 lbs. 9 oz.

Test made from Jan 15 to 22, 1897; age, 2 yrs. 10 mos.; estimated weight, 700 lbs.; fed 25 lbs. gluten, 25 lbs. corn meal, 35 lbs. bran, 15 lbs. oats, 9 lbs. oil meal and 140 lbs. roots—hay; property of D. Wolfe Bishop, Jr., Lenox, Mass.

Lady Phillis 18240.—Sire, Forget-me-not 6291; dam, Phillis 2d 18198.

Butter, 18 lbs. 15 oz.   Milk, 245 lbs. 8 oz.

Test made from Mar. 14 to 21, 1886; age, 3 yrs. 11 mos.; property of William H. Burr, Redding Ridge, Conn.

Lady Phillis 2d 35629.—Sire, Koffee of Ridgeside 11659; dam, Lady Phillis 18240.

Butter, 18 lbs. 8 oz.   Milk, 260 lbs. 12 oz.

Test made from Jan. 29 to Feb. 5, 1888; age, 2 yrs. 9 mos.; estimated weight, 800 lbs.; fed 8 lbs. corn and oats, 4 lbs. cotton-seed meal and 3 lbs. middlings, daily; property of Wm. H. Burr, Redding Ridge, Conn.

Lady Ramaposa 26232.—Sire, Ramapo 4679; dam, Gray Therese 5322.

Butter, 17 lbs. 5½ oz.   Milk, 245 lbs. 3 oz.

Test made from Dec. 11 to 18, 1888; age, 4 yrs. 8 mos.; estimated weight, 675 lbs.; fed, daily, 15 lbs. middlings, 14 lbs. ground oats, 12 lbs. corn meal and 2 lbs. oil meal; property of D. F. Appleton, Ipswich, Mass.

Lady Rareripe 23081.—Sire, Tormentor 2d 7124; dam, Lady Dove 4418.

Butter, 16 lbs. 1 oz.   Milk, 199 lbs.

Test made from July 18 to 24, 1886; age, 3 yrs. 4 mos.; fed, daily, 9 qts. cooked cotton-seed, 9 qts. corn meal, 9 qts. shorts and 3 qts. oil meal; property of Mat. Mahorner, Macon, Miss.

Lady Robin 1119, imp.

Butter, 17 lbs. 13¼ oz.   Milk, 344 lbs. 9 oz.

Test made from May 1 to 7, 1882; age, 15 yrs. 6 mos.; property of C. C. Crockett, Richmond, Ind.

Lady St. Helier of S. 51380.—Sire, Lord Allison 16402; dam, Lady Grace of Upholme 39569.

Butter, 20 lbs. 9½ oz.   Milk, 271 lbs. 12 oz.

Test made from Nov. 2 to 9, 1895; age, 7 yrs. 4 mos.; actual weight, 1075 lbs.; fed 9 qts. ground oats, 3 qts. corn meal, 12 qts. wheat bran and 3 qts. oil meal, daily—hay, 10 lbs. ensilage per day; property of C. A. Sweet, East Aurora, N. Y.

Lady's Blossom 18491.—Sire, Hard Trials 5050; dam, Lady Ellen 11660.

Butter, 20 lbs. 15¾ oz.   Milk, 148 lbs. 8 oz.

Test made from July 15 to 22, 1886; age, 4 yrs. 10 mos.; fed 5 qts. bran daily; property of M. Erskine Miller, Staunton, Va.

Lady Superior 22865.—Sire, Grey (F. S. 244 J. H. B.); dam on I. of J.

Butter, 16 lbs. 5 oz.   Milk, 264 lbs.

Test made from Apr. 10 to 16, 1884; age, 5 yrs. 11 mos.; property of Heulings Lippincott, Cinnaminson, N. J.

Lady Thalma 54701.—Sire, Jersey Boy Thalma 12511; dam, Lady Gray of Hilltop 6th 49485.

Butter, 14 lbs. 12 oz. Milk, 243 lbs. 8 oz.

Test made from June 11 to 18, 1893; age, 6 yrs. 6 mos.; estimated weight, 750 lbs.; fed 10½ lbs. oil meal, 35 lbs. corn meal, 28 lbs. ground oats and 49 lbs. wheat bran—blue-grass pasture; property of Samuel McKeen, Terre Haute, Ind.

Lady Thekla 98270.—Sire, Little Harry 8808; dam, Thekla of Clover Nook 33445.

Butter, 15 lbs. 4¼ oz. Milk, 204 lbs. 3½ oz.

Test made from Apr. 23 to 30, 1896; age, 2 yrs. 4 mos.; estimated weight, 650 lbs.; fed 18 lbs. per day of ground oats and corn, in equal parts—extra good orchard and blue-grass pasture; property of Matt. M. Gardner, Nashville, Tenn.

Lady Thurlow 12410.—Sire, Leander 1184; dam, Mollie Dickey 10349.

Butter, 17 lbs. 10 oz. Milk, 256 lbs. 2 oz.

Test made from May 8 to 15, 1885; age, 5 yrs. 4 mos.; property of Flemister & Bro., Tunnel Hill, Ga.

Lady Torenia 98736.—Sire, Hans of Amherst Villa 27451; dam, Etta of Bay View 66028.

Butter, 14 lbs. 9 oz. Milk, 108 lbs. 10 oz.

Test made from Mar. 16 to 23, 1897; age, 3 yrs. 1 mo.; actual weight, 645 lbs.; fed 35 lbs. ground oats, 21 lbs. cotton-seed meal, 28 lbs. corn meal, 280 lbs. corn ensilage and 30 lbs. dry corn-stalks; property of Charles Lautz, Buffalo, N. Y.

Lady Vain 18573.—Sire, Lord Longford 3997; dam, Variella 6337.

Butter, 17 lbs. 9 oz. Milk, 302 lbs. 8 oz.

Test made from Apr. 4 to 10, 1886; age, 5 yrs. 6 mos.; property of William E. Oates, Vicksburg, Miss.

Lady Velvetine 15771.—Sire, Vertumnus (P. S. 161 J. H. B.); dam, Lady of the Isles (F. S. 992 J. H. B.)

Butter, 17 lbs. 2 oz., unsalted. Milk, 132 lbs. 8 oz.

Test made from June 19 to 26, 1883; age, 5 yrs. 2 mos.; property of W. R. McCready, Saugatuck, Conn.

Lady Wellington of Milton 12234.—Sire, Autocrat 1065; dam, Kitty Clover of Burlington 11788.

Butter, 18 lbs. 11 oz. Milk, 253 lbs.

Test made from May 14 to 21, 1886; age, 11 yrs. 3 mos.; property of James N. Smith, Highland, Conn.

Lady White 29213.—Sire, Creamer's Sir George 8721; dam, Mollie Dickey 2d 18704.

Butter, 14 lbs. 14¾ oz. Milk, 197 lbs. 8 oz.

Test made from Dec. 8 to 15, 1891; age, 7 yrs. 6 mos.; estimated weight, 850 lbs.; fed 63 qts. corn meal, 14 qts. fine feed, 17 qts. bran and 42 lbs. hay; property of H. L. S. Hall, Scottsville, N. Y.

Lady Woodbine 26803.—Sire, Stand Point 7th 7929; dam, Lady Madge 25541.

    Butter, 14 lbs. 2 oz. Milk, 107 lbs. 8 oz.

Test made from Feb. 14 to 21, 1890; age, 6 yrs. 1 mo.; estimated weight, 800 lbs.; fed 210 lbs. of a mixture of 125 lbs. corn meal, 80 lbs. oil meal, 30 lbs. fine feed, 15 lbs. bran and 15 lbs. ground oats—hay and ensilage; property of A. D. McBride, Rochester, N. Y.

La Fantine 24489.—Sire, Top-Sawyer 1404; dam, La Falaise 6961.

    Butter, 15 lbs. 4 oz. Milk, 120 lbs. 11 oz.

Test made from Nov. 14 to 21, 1885; age, 2 yrs. 9 mos.; property of Matt. M. Gardner, Nashville, Tenn.

La Financière 11970.—Sire, Grey King (P. S. 169 J. H. B.); dam, La Picote, on I. of J.

    Butter, 15 lbs. 5¼ oz. Milk, 238 lbs.

Test made from Jan. 22 to 28, 1886; age, 7 yrs. 1 mo.; property of George & H. B. Cromwell, New Dorp, N. Y.

La France's Pansy Rex 49141.—Sire, Prize La France 8162; dam, Pansy Rex 11559.

    Butter, 14 lbs. 10½ oz. Milk, 214 lbs. 12 oz.

Test made from June 4 to 11, 1890; age, 5 yrs. 7 mos.; estimated weight, 950 lbs.; fed. daily, 16 lbs. ground corn and oats and 4 lbs. bran; property of Henry M. Baum, Frankfort, Ind.

La France's Pogis 99986.—Sire, Stoke Pogis' Perfection 5984; dam, La France's Pansy Rex 49141.

    Butter, 21 lbs. ½ oz. Milk, 320 lbs. 11.2 oz.

Test made from May 26 to June 2, 1896; age, 5 yrs.; actual weight, 855 lbs.; fed 56 lbs. bran, 42 lbs. corn and oats, ground together (two-thirds oats to one-third corn), 35 lbs. hominy meal and 21 lbs. cotton-seed meal—short blue-grass pasture; property of J. E. Robbins, Jr., Greensburg, Ind.

La Jolie Lizette 16426.—Sire, Progress (F. S. 286 J. H. B.); dam, Lizette (P. S. 79 J. H. B.)

    Butter, 16 lbs. 1 oz.

Test made from Nov. 3 to 9, 1884; age, 3 yrs. 11 mos.; property of James B. Wilder, Louisville, Ky.

Lalla Rookh of Sugar Grove 15882.—Sire, Marquis of Lossie 3512; dam, Lalla Rookh of Woodlawn Home 12146.

    Butter, 20 lbs. 1 oz. Milk, 296 lbs. 3 oz.

Test made from June 1 to 7, 1885; age, 3 yrs. 9 mos.; fed 20 qts. corn meal, 62 qts. wheat bran and 3 qts. oil meal—good timothy and clover pasture; property of C. Easthope, Niles, Ohio.

Lalla Rookh of Sugar Grove 15882.—Sire, Marquis of Lossie 3512; dam, Lalla Rookh of Woodlawn Home 12146.

    Butter, 18 lbs. 9 oz. Milk, 286 lbs. 4 oz.

Test made from June 3 to 10, 1893; age, 11 yrs. 9 mos.; estimated weight, 850 lbs.; fed 70 lbs. bran, 65 lbs. corn meal and 35 lbs. oil meal—meadow and woods pasture; property of Cornelius Easthope, Niles, Ohio.

Lalla Rookh's Queen 65848.—Sire, King of Ashantee 6677; dam, Lalla Rookh of Sugar Grove 15882.

Butter, 15 lbs. 10 oz. Milk, 226 lbs.

Test made from May 8 to 15, 1893; age, 3 yrs. 9 mos.; estimated weight, 750 lbs.; fed 70 lbs. bran, 30 lbs. oil meal and 55 lbs. corn meal; property of C. Easthope, Niles, Ohio.

Lamba 79039.—Sire. St. Lambert Boy 17408; dam, Dairymaid 2d 13651.

Butter, 20 lbs. 3 oz. Milk, 272 lbs. 10 oz.

Test made from Nov. 1 to 8, 1893; age, 4 yrs. 1 mo.; actual weight, 835 lbs.; fed 60 lbs. corn meal, 30 lbs. ground wheat, 24 lbs. ground oats, 14 lbs. cotton-seed meal and 72 lbs. hay—good meadow aftermath; property of J. P. Bradbury, Pomeroy, Ohio.

Lampedo 46204.—Sire, Gold Plate 10733; dam, Laella 38791.

Butter, 21 lbs. Milk, 156 lbs. 8 oz.

Test made from Oct. 2 to 9, 1889; age, 3 yrs. 4 mos.; actual weight, 780 lbs.; fed, daily, 12 lbs. wheat middlings; property of Mrs. Thomas Allen, Pittsfield, Mass.

Landseer's Content 56985.—Sire, Fancy's Landseer 12996; dam, Content of Linwood 6950.

Butter, 16 lbs. 8 oz. Milk, 232 lbs. 2 oz.

Test made from June 8 to 15, 1893; age, 4 yrs. 11 mos.; estimated weight, 900 lbs.; fed 8 qts. wheat bran, 4 qts. corn meal and 2 qts. cotton-seed meal, daily—good mixed pasture; property of George V. Green, Hopkinsville, Ky.

Landseer's Croton Maid 82535.—Sire, Khedive's Landseer 21926; dam, Croton Maid 3d 26726.

Butter, 14 lbs. 15½ oz. Milk, 176 lbs. 12 oz.

Test made from July 17 to 24, 1892; age, 2 yrs. 2 mos.; estimated weight, 625 lbs.; fed 20 lbs. per day of ground oats and corn, in equal parts—good orchard and blue-grass pasture; property of Matt. M. Gardner, Nashville, Tenn.

Landseer's Ethel 74315.—Sire, Landseer's Toltec 25426; dam, Ethel Wanda 56144.

Butter, 15 lbs. Milk, 206 lbs. 12 oz.

Test made from May 20 to 27, 1896; age, 4 yrs. 11 mos.; estimated weight, 750 lbs.; fed 9 lbs. shorts and 13 lbs. corn meal, daily—dry timothy and clover pasture; property of F. S. Case, Logan, Ohio.

Landseer's Fancy 2876.—Sire, Landseer 331; dam, Young Fancy 97.

Butter, 21 lbs. 15 oz. (official). Milk, 123 lbs. 10 oz.

Test made from Dec. 14 to 20, 1883; age, 10 yrs.; property of Columbia Jersey Cattle Co., Columbia, Tenn.

Landseer's Fancy 2876.—Sire, Landseer 331; dam, Young Fancy 97.

Butter, 29 lbs. ½ oz. Milk, 152 lbs. 8 oz.

Test made from July 16 to 22, 1885; age, 11 yrs. 7 mos.; fed 8 qts. corn-hearts, 8 qts. oats and 4 qts. wheat bran, daily—good blue-grass pasture; property of Columbia Jersey Cattle Co., Columbia, Tenn.

Landseer's Fancy 2876.—Sire, Landseer 331; dam, Young Fancy 97.

Butter, 936 lbs. 14¾ oz. in one year.

Test made from Jan. 26, 1885, to Jan. 25, 1886; age, 11 yrs. 2 mos. at beginning of test; carried calves for nine months of the year; property of Columbia Jersey Cattle Co., Columbia, Tenn.

Landseer's Fancy 2d 43184.—Sire, Toltec 6831; dam, Landseer's Fancy 2876.

Butter, 15 lbs. ½ oz. Milk, 103 lbs. 3 oz.

Test made from May 25 to June 1, 1888; age, 1 yr. 10 mos.; estimated weight, 600 lbs.; fed, daily, 12 qts. corn and oats; property of Messrs. Webster & Morrow & Son. Columbia, Tenn.

Landseer's Fancy 3d 70653.—Sire, Toltec 6831; dam, Landseer's Fancy 2876.

Butter, 15 lbs. 14½ oz. Milk, 115 lbs. 1½ oz.

Test made from Nov. 8 to 15, 1894; age, 4 yrs. 3 mos.; estimated weight, 750 lbs.; fed 70 qts. crushed corn and barley, mixed, in equal portions—blue-grass pasture; property of John Woodard, Nashville, Tenn.

Landseer's Gilderine 68048.—Sire, Ida's Landseer 17745; dam, Gilderine of Linwood 24488.

Butter, 14 lbs. 7 oz. Milk, 177 lbs. 8 oz.

Test made from May 20 to 27, 1894; age, 4 yrs. 7 mos.; estimated weight, 800 lbs.; fed 56 qts. ground oats and 56 qts. corn meal—good woodland pasture; property of L. P. Padgett, Columbia, Tenn.

Landseer's Hilda 49788.—Sire, Landseer's Pogis 15847; dam, Hilda 2d 14967.

Butter, 17 lbs. Milk, 229 lbs. 12 oz.

Test made from Mar. 22 to 29, 1891; age, 3 yrs. 7 mos.; estimated weight, 950 lbs.; fed 17 lbs. oil meal, 17 lbs. pea meal, 42 lbs. corn meal, 48 lbs. crushed oats and 70 lbs. hay; property of Ayer & McKinney, Philadelphia, Pa.

Landseer's Lily 87294.—Sire, Landseer's Major 25759; dam, Ethel Wanda 56144.

Butter, 14 lbs. 2 oz. Milk, 177 lbs.

Test made from May 9 to 16, 1897; age, 4 yrs. 9 mos.; estimated weight, 650 lbs.; fed 14 lbs. shorts and middlings, mixed, and 2 lbs. oil-cake meal, daily—mixed pasture; property of F. S. Case, Logan, Ohio.

Landseer's Pogene 85743.—Sire, Landseer's Pogis 15847; dam, Davy's Dot 41693.

Butter, 15 lbs. Milk, 227 lbs. 12 oz.

Test made from July 1 to 8, 1893; age, 4 yrs. 6 mos.; estimated weight, 800 lbs.; fed 56 qts. ground oats, 56 qts. corn meal and 28 qts. wheat bran—short pasture; property of Morgan & Brown, Columbia, Tenn.

Landseer's Regina 85988.—Sire, Lord Landseer, 24228; dam, Diana Doon 61539.

Butter, 17 lbs. Milk, 140 lbs.

Test made from May 17 to 24, 1895; age, 4 yrs. 1 mo.; actual weight, 910 lbs.; fed 35 qts. cotton-seed meal and 35 qts. corn meal—pastured on green wheat; property of S. T. Howard, Quanah, Tex.

Landseer's Rosamond 56142.—Sire, Fancy's Landseer 12996; dam, Fair Rosamond 11135.

Butter, 15 lbs. 5 oz. Milk, 189 lbs. 8 oz.

Test made from Apr. 1 to 8, 1896; age, 7 yrs. 11 mos.; estimated weight, 1100 lbs.; fed 56 lbs. corn meal, 30 lbs. bran and 7 lbs. cotton-seed meal—hay and ensilage; property of Austin & Probert, Florence, Mich.

Landseer's Signaline 58101.—Sire, Landseer's Pogis 15847; dam, Bessie Signalda 38178.

Butter, 14 lbs. 11⅞ oz. Milk, 126 lbs. 2 oz.

Test made from Oct. 20 to 27, 1889; age, 2 yrs. 3 mos.; estimated weight, 750 lbs.; fed, daily, 4 gals. of a ground mixture consisting of one part corn and two parts oats; property of Morgan & Brown, Columbia, Tenn.

Landseer's Signal Queen 66363.—Sire, Landseer's Pogis 15847; dam, Duke's Signal Queen 32323.

Butter, 14 lbs. 14 oz. Milk, 173 lbs. 15 oz.

Test made from Sept. 1 to 8, 1890; age, 3 yrs. 2 mos.; estimated weight, 850 lbs.; fed, daily, 4 gals. ground oats and corn, mixed equally; property of Morgan & Brown, Columbia, Tenn.

Landseer's Signal Queen 66363.—Sire, Landseer's Pogis 15847; dam, Duke's Signal Queen 32323.

Butter, 16 lbs. 8 oz. Milk, 224 lbs. 8 oz.

Test made from Mar. 3 to 10, 1894; age, 6 yrs. 9 mos.; estimated weight, 1000 lbs.; fed 70 qts. ground oats and 70 qts. ground corn—clover hay; property of Morgan & Brown, Columbia, Tenn.

Lanison's Belle 84986.—Sire, Lanison 15283; dam, Clotaire's Belle 28841.

Butter, 20 lbs. 5 oz. Milk, 285 lbs. 8 oz.

Test made from Oct. 14 to 21, 1897; age, 6 yrs. 6 mos.; estimated weight, 1000 lbs.; fed 10 qts. ground feed (two-thirds wheat bran, corn meal and ground oats, in equal parts, and one-third gluten meal) and 10 qts. whole oats, daily—clover hay, short clover pasture; property of L. C. Read & Son, East Pembroke, N. Y.

Lanison's Cora 59315.—Sire, Lanison 15283; dam, Exile's Alphea 45393.

Butter, 14 lbs. 14 oz. Milk, 171 lbs. 2 oz.

Test made from July 2 to 9, 1890; age, 2 yrs. 2 mos.; estimated weight, 800 lbs.; fed 8 qts. of a mixture of one part bran, two parts ground corn and three parts ground oats, daily, also bran slop daily—old pasture; property of L. L. Tozier, Batavia, N. Y.

Lanison's Nora 59316.—Sire, Lanison 15283; dam, Clotaire's Alphea 32502.

Butter, 15 lbs. 6½ oz. Milk, 226 lbs. 11 oz.

Test made from May 20 to 27, 1890; age, 2 yrs. 1 mo.; estimated weight, 700 lbs.; fed 8 qts. of a mixture of one part bran, two parts ground corn and three parts ground oats, daily, also bran slop daily—old pasture; property of L. L. Tozier, Batavia, N. Y.

Lanison's Queen 2d 128773.—Sire, Lanison 15283; dam, Lanison's Queen 80097.

Butter, 15 lbs. 14 oz. Milk, 222 lbs.

Test made from Dec. 3 to 10, 1897; age, 6 yrs. 6 mos.; actual weight, 900 lbs.; fed 21 qts. per day, one-third gluten feed, ⅔ qt. oil meal and the rest corn meal, ground oats and wheat bran, in equal parts, and 4 qts. roots—hay; property of L. C. Read & Son, East Pembroke, N. Y.

La Pera 2d 13404.—Sire, Kahela 2859; dam, La Pera 7091.

Butter, 14 lbs. 8 oz., unsalted. Milk, 210 lbs. 8 oz.

Test made from July 14 to 20, 1883; age, 3 yrs. 11 mos.; property of W. L. & W. Rutherford, Waddington, N. Y.

La Petite Mère 2d 12810.—Sire, Stoke Pogis 1259; dam, La Petite Mère 5470.

Butter, 16 lbs. 7 oz. Milk, 346 lbs.

Test made from Nov. 20 to 26, 1885; age, 6 yrs. 8 mos.; property of Miller & Sibley, Franklin, Pa.

La Petite Mère 2d 12810.—Sire, Stoke Pogis 1259; dam, La Petite Mère 5470.

Butter, 660 lbs. 4 oz. for one year ending Dec. 15, 1886.. Milk, 16699 lbs. 8 oz. for one year ending Nov. 2, 1886.

Age, 6 yrs. 9 mos. at beginning of test; property of Miller & Sibley, Franklin, Pa. (This butter-test was incidental to test for milk-yield.)

La Petite Mère 3d 12814.—Sire, Midas of Oxford 5986; dam, La Petite Mère 5470.

Butter, 16 lbs. 9 oz. Milk, 227 lbs. 6 oz.

Test made from Feb. 12 to 18, 1886; age, 5 yrs. 11 mos.; property of George G. Creamer, Hamilton, Mass.

La Petite Pogis 28757.—Sire, Titan 8548; dam, La Petite Mère 2d 12810.

Butter, 20 lbs. 10¾ oz. Milk, 233 lbs.

Test made from July 27 to Aug. 3, 1890; age, 6 yrs.; estimated weight, 950 lbs.; fed, daily, 4 qts. bran, 4 qts. oil meal, 6 qts. ground oats and 4 qts. corn meal; property of Miller & Sibley, Franklin, Pa.

La Pucelle 16829.—Sire, Mercutio 4591; dam, Nannette of Allerton 8515.

Butter, 15 lbs. 8 oz. Milk, 271 lbs. 2 oz.

Test made from Sept. 20 to 26, 1886; age, 7 yrs. 5 mos.; estimated weight, 1100 lbs.; fed 8 qts. oats and 4 qts. shorts, daily; property of T. R. Proctor, Utica, N. Y.

La Reine Centennielle 4999.—Sire, Aquila 1208; dam, Garland 851.

Butter, 15 lbs. 9 oz. Milk, 209 lbs. 8 oz.

Test made from Nov. 10 to 17, 1885; age, 9 yrs. 4 mos.; property of A. R. Hill, North Washington, Pa.

La Septième 13432, imp.

Butter, 17 lbs. 10 oz. Milk, 166 lbs. 2 oz.

Test made from Mar. 15 to 21, 1884; age, 4 yrs. 3 mos.; property of C. E. Rowley, Cortland, N. Y.

Lassie 1134.—Sire, Duke of Wellington 35; dam, Newton Belle 355.

Butter, 15 lbs. 1½ oz., unsalted. Milk, 122 lbs. 2 oz.

Test made from Feb. 27 to Mar. 4, 1884; age, 14 yrs. 8 mos.; property of Winslow S. Lincoln, Worcester, Mass.

Lass of Burrwood 37437.—Sire, Pansy's Double Rex 9997; dam, Carma 6361.

Butter, 14 lbs. 1½ oz. Milk, 233 lbs. 8 oz.

Test made from June 13 to 20, 1890; age, 6 yrs. 2 mos.; estimated weight, 900 lbs.; fed, daily, 8 qts., consisting of two parts corn meal and one of oats; property of George Burr, Lodi, Ohio.

Lass of Scituate 9555.—Sire, Pharos Jr. 3621; dam, Jersey Belle of Scituate 7828.

Butter, 15 lbs. 14 oz. Milk, 196 lbs. 12 oz.

Test made June, 1883; age, 5 yrs. 1 mo.; fed 2 qts. bran. 2 qts. corn meal and 1 qt. cotton-seed meal, daily; property of Orestes Pierce, Baldwin, Me.

Lass Pogis 40740.—Sire, Sir George of St. Lambert 6036; dam, Orphan Lass 9646.

Butter, 14 lbs. 14 oz. Milk, 258 lbs. 12 oz.

Test made from Jan. 17 to 24, 1893; age, 6 yrs. 2 mos.; estimated weight, 1150 lbs.; fed 2 bush. cut hay, 6½ lbs. ground oats, 4 lbs. shorts, 6½ lbs. corn meal, 2 lbs. linseed meal, 6 qts. mixed carrots and mangels and 4 lbs. loose hay, daily; property of Horatio J. Gilbert, Milton, Mass.

Lass Rex Alphea 16965.—Sire, Rex Alphea 4509; dam, Sunny Lass 6033.

Butter, 16 lbs. 10¾ oz. Milk, 218 lbs.

Test made from Nov. 4 to 10, 1885; age, 4 yrs. 1 mo.; property of D. A. Givens, Cynthiana, Ky.

Lass Signal 16308.—Sire, Chief of the Miamis 2958; dam, Goldie C. 8104.

Butter, 20 lbs. 2 oz. Milk, 212 lbs.

Test made from Apr. 7 to 13, 1886; age, 4 yrs. 3 mos.; fed twice daily on 1 gal. bran. 1 gal. corn meal, 1 qt. oil-cake and good clover hay, with blue-grass and rye pasturage; property of R. McMichael, Lexington, Ky.

Laundress 2d 24649.—Sire, Sir Joseph Peck 4978; dam, Laundress 13867.

Butter, 14 lbs. 9 oz. Milk, 138 lbs.

Test made from Dec. 1 to 8, 1887; age, 4 yrs. 7 mos.; estimated weight, 750 lbs.; fed, daily, 8 lbs. corn meal, 8 lbs. bran and 2 lbs. oil meal; property of James Stillman. Sing Sing, N. Y.

Laura Lee 2d 128023.—Sire, Chromo 26113; dam, Laura Lee 8550.

Butter, 16 lbs. 4 oz. Milk, 244 lbs. 10 oz.

Test made from Jan. 2 to 9, 1897; age, 3 yrs. 10 mos.; estimated weight, 700 lbs.; fed 4½ lbs. corn meal, 5½ lbs. oat meal, 6 lbs. bran, 2 lbs. oil meal and 2 qts. roots, daily— some hay; property of Richardson Bros., Davenport, Iowa.

Laura of Eau Claire 110739.—Sire, Melia Ann's Son 22041; dam, Brown Lily 32770.

Butter, 18 lbs. Milk, 268 lbs. 8 oz.

Test made from Jan. 11 to 18, 1898; age, 3 yrs. 11 mos.; estimated weight, 800 lbs.; fed 25 lbs. wheat bran, 25 lbs. corn meal, 20 lbs. oil meal, 15 lbs. cotton-seed meal and 2 bush. roots—hay; property of J. A. Griffith, Ashley Falls, Mass.

Laurie Pansy Champion 20396.—Sire, Champion's Son 3286; dam, Annie Laurie 3d 13236.

Butter, 14 lbs. 5½ oz. Milk, 160 lbs. 8 oz.

Test made from June 1 to 7, 1886; age, 3 yrs. 8 mos.; property of Bordwell & Cochran, Batavia, Ohio.

La Violette 2d 52604.—Sire, Garfield Stoke Pogis 15963; dam, La Violette 36801.

Butter, 14 lbs. 7 oz. Milk, 158 lbs. 8 oz.

Test made from June 5 to 12, 1892; age, 4 yrs. 2 mos.; estimated weight, 900 lbs.; fed 56 lbs. cut corn fodder, 7 lbs. corn meal, 14 lbs. gluten meal, 14 lbs. bran, 7 lbs. linseed meal, 210 lbs. roots and 56 lbs. hay; property of Est. of Frederick Billings, Woodstock, Vt.

La Vivienne 2d 1324.—Sire on I. of J.; dam, La Vivienne 1068.

Butter, 16 lbs. 2 oz. Milk, 95 qts.

Test made from June 1 to 7, 1882; age, 12 yrs.; property of W. C. Stoughton, Riverside, Mass.

L. D.'s Helita 29366.—Sire, Lord Darlington 7285; dam, Helita 22113.

Butter, 14 lbs. 4½ oz. Milk, 199 lbs. 8 oz.

Test made from Nov. 22 to 29, 1890; age, 6 yrs. 2 mos.; estimated weight, 925 lbs.; fed, daily, 1½ lbs. shorts, 2 lbs. ground oats, 3 lbs. cotton-seed and 6 lbs. meal; property of L. T. Hazen, Hazen's Mills, N. H.

L. D.'s Mabel 29368.—Sire, Lord Darlington 7285; dam, Mabel Girl 22040.

Butter, 14 lbs. 6 oz. Milk, 157 lbs. 6 oz.

Test made from Oct. 16 to 23, 1886; age, 2 yrs. 1 mo.; actual weight, 748 lbs.; fed 28 qts. shorts and 28 qts. corn meal—good fall pasture; property of J. H. Walker, Worcester, Mass.

L. D.'s Melia 29087.—Sire, Lord Darlington 7285; dam, Melia Ann 5444.

Butter, 20 lbs. 6 oz. Milk, 222 lbs.

Test made from Jan. 1 to 8, 1888; age, 3 yrs. 5 mos.; estimated weight, 875 lbs.; fed 28 qts. bran, 28 qts. oat meal and 42 qts. roots—hay *ad lib.*; property of J. H. Walker, Worcester, Mass.

Leaf Pogis 130184.—Sire, Stoke Pogis of Prospect 20121; dam, Fair Leaf 81293.

Butter, 14 lbs. 5½ oz. Milk, 207 lbs. 1 oz.

Test made from May 1 to 8, 1898; age, 2 yrs.; actual weight, 700 lbs.; fed 6 qts. bran, 6 qts. ground oats, 3 qts. corn meal, 3 qts. middlings and 1½ qts. oil meal, mixed, daily—plenty good hay and ensilage; property of W. C. Scheanck, Angelica, N. Y.

Leah Darlington 13836.—Sire, Duke of Darlington 2460; dam, Leda 799.

Butter, 15 lbs. 5½ oz. Milk, 187 lbs. 14 oz.

Test made from Sept. 20 to 26, 1884; age, 3 yrs. 1 mo.; property of William Simpson, New York, N. Y.

Lebanon Daughter 6106.—Sire, Iron Bank 1120; dam, Emmie's Pet 3568.

Butter, 14 lbs. 4 oz. Milk, 16 to 18 qts. daily.

Test made from Dec. 10 to 16, 1881; age, 5 yrs. 9 mos.; property of G. Dawson Coleman, Brickerville, Pa.

Lebanon Lass 6108.—Sire, Iron Bank 1120; dam, Tipsey 3572.

Butter, 14 lbs. 2 oz. Milk, average of 18 qts. daily.

Test made from Dec. 2 to 8, 1881; age, 5 yrs. 8 mos.; property of G. Dawson Coleman, Brickerville, Pa.

Le Brocq's May Belle 27980.—Sire, Major Marion 6922; dam, Dolly Bloomfield 7842.

Butter, 14 lbs. 15 oz. Milk, 220 lbs. 4 oz.

Test made from Jan. 27 to Feb. 3, 1894; age, 9 yrs. 8 mos.; estimated weight, 1000 lbs.; fed 28 qts. crushed corn and 56 qts. bran—hay; property of Abraham McNutt, Lewisburg, Ohio.

Le Brocq's Pansy Rex 23789.—Sire, Le Brocq's Prize 3350; dam, Pansy Rex 11559.

Butter, 18 lbs. 6 oz.　Milk, 280 lbs.

Test made from May 29 to June 6, 1888; age, 4 yrs. 8 mos.; estimated weight, 890 lbs.; fed, daily, 2 lbs. 6 oz. oil meal, 4 lbs. 10 oz. ground oats, 2 lbs. 8 oz. bran and 3 lbs. 6 oz. pea meal; property of H. M. Baum, Frankfort, Ind.

Le Broemer 10670.—Sire, Le Brocq's Prize 3350; dam, Carma 6361.

Butter, 15 lbs. 7 oz.　Milk, 277 lbs. 8 oz.

Test made from June 5 to 12, 1889; age, 9 yrs. 9 mos.; estimated weight, 900 lbs.; fed, daily, 9½ lbs. corn meal and fine bran, equal parts of each, and 2½ lbs. oil meal; property of George Burr, Lodi, Ohio.

Leclair's Marjoram 36355.—Sire, Stoke Pogis 1259; dam, Marjoram 3239.

Butter, 15 lbs. 3 oz.　Milk, 110 lbs. 8 oz.

Test made from June 4 to 11, 1889; age, 11 yrs.; estimated weight, 700 lbs.; fed, daily, 6 qts. barley and oats and 2 qts. oil-cake; property of Wm. Rolph, Markham, Ont., Can.

Le Gros' Lily of the Valley 2d 13386.—Sire, Tom McGreevy 1692; dam, Le Gros' Lily of the Valley 11537.

Butter, 19 lbs. 10¼ oz.　Milk, 158 lbs. 4 oz.

Test made from Mar. 2 to 9, 1888; age, 6 yrs. 10 mos.; estimated weight, 950 lbs.; fed, daily, 8 lbs. corn and oats, 4 lbs. bran and 2 lbs. oil meal; property of M. Erskine Miller, Staunton, Va.

Leila of Briarcliff 24184.—Sire, Domino of Darlington 2459; dam, Jersey Lily 14044.

Butter, 17 lbs. 6½ oz.　Milk, 171 lbs. 6 oz.

Test made from May 21 to 28, 1886; age, 2 yrs. 6 mos.; estimated weight, 750 lbs.; fed, daily, 4 qts. corn meal, 2 qts. bran and 2 qts. malt sprouts; property of James Stillman, Sing Sing, N. Y.

Leila of Briarcliff 24184.—Sire, Domino of Darlington 2459; dam, Jersey Lily 14044.

Butter, 14 lbs. 10½ oz. (official).　Milk, 199 lbs. 6 oz.

Test made from June 30 to July 7, 1886; age, 2 yrs. 6 mos.; estimated weight, 800 lbs.; fed, daily, about 7 lbs. corn meal, 3 lbs. crushed oats and 3 lbs. pea meal; property of James Stillman, Sing Sing, N. Y.

Leila of Clovernook 72554.—Sire, Sherbet's Koffee 18512; dam, Lora of Clovernook 51599.

Butter, 14 lbs. 2 oz.　Milk, 206 lbs. 7½ oz.

Test made from Sept. 30 to Oct. 7, 1894; age, 4 yrs. 9 mos.; actual weight, 650 lbs.; fed 20 lbs. per day of ground oats and corn, in equal parts, mixed with 4 bucketfuls cut sheaf oats, wetted—good orchard and blue-grass pasture; property of Matt. M. Gardner, Nashville, Tenn.

Leila Stratton 25198.—Sire, Lord Lawrence 1414; dam, Petite Mère 8516.

Butter, 17 lbs. 4 oz.　Milk, 222 lbs. 3 oz.

Test made from Feb. 8 to 15, 1889; age, 5 yrs.; estimated weight, 825 lbs.; fed 24 lbs. per day of ground oats and corn, in equal parts, mixed with 4 bucketfuls of cut clover hay, wetted—clover hay *ad lib.*; property of Matt. M. Gardner, Nashville, Tenn.

Leila' Victor 85433.—Sire, Eurotas' Victor Hugo 15689; dam, Leila Stratton 25198.

Butter, 14 lbs. 1 oz. Milk, 203 lbs. 4 oz.

Test made from May 24 to 31, 1896; age, 3 yrs. 5 mos.; estimated weight, 650 lbs.; fed 3 qts. wheat middlings, 3 qts. wheat bran, 3 qts. corn meal and 1 qt. oil meal—timothy and clover pasture; property of Alfred B. Darling, Ramsey's, N. J.

Lelia Koffee of Lawn 120982.—Sire, Antoinette's Koffee 26010; dam, Moorey of Lawn 68847.

Butter, 17 lbs. 3½ oz. ·Milk, 211 lbs. 8 oz.

Test made from Apr. 14 to 21, 1898; age, 3 yrs. 1 mo.; estimated weight, 850 lbs.; fed 84 qts. wheat bran and 84 qts. corn and oats chop—good wild grass pasture; property of Mrs. F. W. Platter, Denison, Tex.

Lelia Talbot 96922.—Sire, Prince of Beech Grove 16234; dam, Lelia Douland 44473.

Butter, 14 lbs. 8½ oz. Milk, 205 lbs.

Test made from Mar. 30 to Apr. 6, 1897; age, 2 yrs. 11 mos.; estimated weight, 700 lbs.; fed 35 lbs. bran, 56 lbs. corn meal, 14 lbs. linseed meal and 28 lbs. boiled whole cotton-seed—Bermuda grass and Japanese clover pasture; property of M. Lothrop, Marshall, Tex.

Lemon Fern 22140.—Sire, Lemon Peel (P. S. 320 J. H. B.); dam, La Parleuse (F. S. 3393 J. H. B.)

Butter, 15 lbs. 1 oz. Milk, 223 lbs.

Test made from June 21 to 28, 1889; age, 6 yrs. 4 mos.; estimated weight, 800 lbs.; fed, daily, 11 lbs. ground corn and oats; property of G. L. & A. C. Davis, Port Jefferson, N. Y.

Lemon's Flossie 71214.—Sire, Nihilist of St. Lambert 18256; dam, Pearl's Lemon 41646.

Butter, 14 lbs. 1¾ oz. Milk, 250 lbs. 8 oz.

Test made from Apr. 27 to May 4, 1892; age, 2 yrs. 11 mos.; estimated weight, 700 lbs.; fed 56 lbs. corn meal and 14 lbs. ground oil-cake—rye pasture; property of Albert Stracke, Warsaw, Ill.

Lemon Spray 44532.—Sire, Coomassie's Gilderoy 14672; dam, Lemon Twig 42028.

Butter, 15 lbs. 1½ oz. Milk, 234 lbs. 8 oz.

Test made from May 9 to 16, 1892; age, 5 yrs. 1 mo.; estimated weight, 700 lbs.; fed 14 lbs. corn-hearts, daily—blue-grass pasture; property of Campbell Brown, Spring Hill, Tenn.

Lemon Twig 42028.—Sire, Gilderoy 2107; dam, Orange Twig 11796.

Butter, 16 lbs. 4½ oz. Milk, 246 lbs. 8 oz.

Test made from Mar. 23 to 30, 1889; age, 3 yrs. 11 mos.; estimated weight, 850 lbs.; fed, daily, 22 lbs. mixed wheat bran and corn-hearts on the first three days, and 15 lbs. corn-hearts on the last four days; property of Campbell Brown, Spring Hill, Tenn.

Lemon Twig 42028.—Sire, Gilderoy 2107; dam, Orange Twig 11796.

Butter, 17 lbs. 4 oz. Milk, 253 lbs. 8 oz.

Test made from May 28 to June 4, 1892; age, 7 yrs. 1 mo.; estimated weight, 850 lbs.; fed 12 lbs. corn-hearts, daily—good blue-grass; property of Campbell Brown, Spring Hill, Tenn.

Lena Hogan 41506.—Sire, Champion Royal 7983; dam, Mhoon Romp 4528.

Butter, 16 lbs. 15 oz. Milk, 212 lbs. 4 oz.

Test made from Dec. 9 to 16, 1894; age, 8 yrs. 9 mos.; estimated weight, 900 lbs.; fed 56 qts. corn meal, 56 qts. wheat bran, 7 qts. cotton-seed meal and 7 qts. pea meal, mixed—hay *ad lib.*; property of B. P. Holliday, Prairie Station, Miss.

Lena Lowndes 23202.—Sire, Prince of Warren 1512; dam, Monmouth Duchess 4th 7129.

Butter, 14 lbs. 9 oz. Milk, 168 lbs.

Test made from Oct. 10 to 17, 1884; age, 7 yrs. 5 mos.; property of R. S. McCracken, Hackettstown, N. J.

Lena's Florence 42460.—Sire, Fancy's Harry 9777; dam, Lena Woods 6687.

Butter, 14 lbs. 8 oz. Milk, 134 lbs. 7 oz.

Test made from July 3 to 10, 1889; age, 2 yrs. 9 mos.; estimated weight, 650 lbs.; fed, daily, 4 gals. of a mixture consisting of two-thirds corn and one-third oats, ground; property of Maury Jersey Farm, Columbia, Tenn.

Lena's Florence 2d 78688.—Sire, Toltec 6831; dam, Lena's Florence 42460.

Butter, 14 lbs. 9½ oz. Milk, 210 lbs. 6¼ oz.

Test made from May 13 to 20, 1897; age, 5 yrs. 8 mos.; estimated weight, 800 lbs.; fed 12 qts. daily ground oats and barley, in equal parts—blue-grass pasture four hours daily; property of J. E. Hart, Nashville, Tenn.

Lena Stoke Pogis 39756.—Sire, Exile of St. Lambert 13657; dam, Nellie B. Parks 20264.

Butter, 18 lbs. 8 oz. Milk, 259 lbs.

Test made from July 17 to 24, 1891; age, 5 yrs.; estimated weight, 850 lbs.; fed 4 qts. corn meal, 1 pt. oil meal and 2 qts. bran, daily; property of E. L. Earl, Earl, N. Y.

Leneca 17178.—Sire, Steamboat 4422; dam, Belle of Beaumont 2d 9209.

Butter, 15 lbs. ¼ oz., unsalted.

Test made from May 23 to 30, 1885; age, 3 yrs. 9 mos.; property of William H. Parks, Newnan, Ga.

Lenore H. 2d 29767.—Sire, Duke Thorne 7020; dam, Lenore H. 15798.

Butter, 16 lbs. 3 oz. Milk, 294 lbs. 8 oz.

Test made from Apr. 22 to 29, 1893; age, 8 yrs. 5 mos.; estimated weight, 800 lbs.; fed 2 lbs. cotton-seed meal, 3 lbs. Chicago cream gluten, 3 lbs. corn meal, 3 lbs. spring wheat bran, mixed, with 35 lbs. corn ensilage, daily—also 10 lbs. clover hay; property of E. S. Henry, Rockville, Conn.

Leoline 2d 18315.—Sire, Pouch of Gold 4121; dam, Leoline 9285.

Butter, 14 lbs. 4 oz. Milk, 173 lbs. 8 oz.

Test made from Jan. 30 to Feb. 5, 1884; age, 3 yrs.; property of R. W. Miller, Lebanon, Tenn.

Leonetta H. 67514.—Sire, Tamy's Rioter 15288; dam, Lesbia H. 35634.

Butter, 15 lbs. 1½ oz. Milk, 196 lbs. 12 oz.

Test made from Dec. 19 to 26, 1893; age, 4 yrs. 9 mos.; estimated weight, 900 lbs.; fed, daily, 9 qts. corn and gluten meal, 9 qts. oat meal and 3 pts. old process oil meal —hay, 8 qts. potatoes; property of E. C. Newton, Hudson, Ohio.

Leonetta H. 67514.—Sire, Tamy's Rioter 15288; dam, Lesbia H. 35634.

Butter, 62 lbs. 3 oz. in 30 days. Milk, 814 lbs. 12 oz.

Test made from Nov. 28 to Dec. 28, 1893; age, 4 yrs. 9 mos.; estimated weight, 920 lbs.; fed 216 qts. corn meal, 108 qts. wheat and oats ground together and 36 qts. oil meal—4 qts. daily of cut raw potatoes, no pasture; property of E. C. Newton, Hudson, Ohio.

Leonette 29752.—Sire, Toltec 6831; dam, Leoni 11868.

Butter, 15 lbs. 5 oz. Milk, 130 lbs. 8 oz.

Test made from July 31 to Aug. 7, 1887; age, 3 yrs. 4 mos.; estimated weight, 700 lbs.; fed, daily, about 10 lbs. ground corn and oats; property of John Moore, Jr., Columbia, Tenn.

Leonette's Orange 108521.—Sire, Leonette's Landseer 20399; dam, Orange Pogis 49531.

Butter, 15 lbs. Milk, 212 lbs. 4 oz.

Test made from Dec. 30, 1896, to Jan. 6, 1897; age, 2 yrs. 11 mos.; estimated weight, 850 lbs.; fed 4 lbs. oats, 3¾ lbs. bran, 7¾ lbs. corn meal and 1¼ lbs. oil meal, daily—corn fodder *ad lib.*, short rye pasture; property of W. E. Johnson, Millican, Tex.

Leoni 11868.—Sire, Rector 1458; dam, Brunette Lass 1780.

Butter, 18 lbs. 7 oz. Milk, 224 lbs.

Test made from Apr. 30 to May 6, 1884; age, 4 yrs.; property of H. P. Figuers, Columbia, Tenn.

Leoni 2d 29750.—Sire, Toltec 6831; dam, Leoni 11868.

Butter, 15 lbs. 9 oz. Milk, 148 lbs. 1 oz.

Test made from May 21 to 28, 1887; age, 4 yrs. 1 mo.; estimated weight, 750 lbs.; fed, daily, about 11 lbs. ground corn and oats; property of John Moore, Jr., Columbia, Tenn.

Leonice 2d 8342.—Sire, Gil Blas 1103; dam, Leonice 4401.

Butter, 16 lbs. 8 oz., unsalted.

Test made from Nov. 5 to 12, 1880; age, 2 yrs. 7 mos.; fed 1 gal. of bran scalded with 2 gals. of hot water—pasture, fine blue-grass; property of P. W. Hardin, Frankfort, Ky.

Leonie of Linwood 25319.—Sire, Gold Basis 4038; dam, Lorelle 12913.

Butter, 14 lbs. 6 oz. Milk, 152 lbs.

Test made from June 26 to July 2, 1886; age, 2 yrs. 5 mos.; property of W. S. Shields, Bean's Station, Tenn.

Leoni Landseer 65317.—Sire, Fancy's Harry 9777; dam, Leoni 2d 29750.

Butter, 16 lbs. Milk, 192 lbs. 12 oz.

Test made from Jan. 1 to 8, 1894; age, 7 yrs. 8 mos.; estimated weight, 825 lbs.; fed 56 qts. ground oats and 56 qts. corn meal—hay, poor pasture; property of Morgan & Brown, Columbia, Tenn.

Leonora of Canada 94906.—Sire, Canada's John Bull 5th 20092; dam, Kitty of St. Lambert 6637.

Butter, 22 lbs. 11 oz.   Milk, 284 lbs. 8 oz.

Test made from Apr. 14 to 21, 1898; age, 4 yrs. 10 mos.; estimated weight, 700 lbs.; fed 18 lbs. oats and peas, mixed, chopped, 2 lbs. wheat bran, 2 lbs. oil-cake and 30 lbs. corn ensilage, daily—hay *ad lib.*; property of William Rolph, Markham, Ont., Can.

Lerna 3634.—Sire, Mercury 432; dam, Gussie Richards 1673.

Butter, 15 lbs. 12 oz.   Milk, 242 lbs. 9 oz.

Test made from Sept. 29 to Oct. 5, 1880; age, 6 yrs.; property of William Simpson, New York, N. Y.

Lernella 22322.—Sire, Mercury 432; dam, Lerna 3634.

Butter, 14 lbs. 1½ oz.   Milk, 191 lbs. 11 oz.

Test made from May 10 to 16, 1885; age, 2 yrs. 3 mos.; property of William Simpson, New York, N. Y.

Le Rosa 10078.—Sire, Le Brocq's Prize 3350; dam, Rosari 5518.

Butter, 14 lbs.   Milk, 222 lbs. 8 oz.

Test made from May 26 to June 1, 1884; age, 4 yrs. 3 mos.; property of John A. J. Shultz, St. Louis, Mo.

Le Roy's Pet 94106.—Sire, Founder 20928; dam, Mamie Le Roy 52459.

Butter, 20 lbs. 7 oz.   Milk, 233 lbs.

Test made from Apr. 28 to May 5, 1896; age, 4 yrs. 1 mo.; estimated weight, 800 lbs.; fed 35 lbs. middlings, 35 lbs. bran, 21 lbs. oil meal and 157 lbs. roots—woods pasture; property of Mrs. Adda F. Howie, Elm Grove, Wis.

Lesbie 9179.—Sire, Oxoli 1922; dam, Ariella 9178.

Butter, 16 lbs. 3 oz.   Milk, 187 lbs. 8 oz.

Test made from Mar. 11 to 17, 1885; age, 8 yrs.; property of J. V. N. Willis, Marlborough, N. J.

Les Marais' Dell 20314.—Sire, Silver (P. S. 287 J. H. B.); dam, Picton (F. S. 4236 J. H. B.)

Butter, 15 lbs. 8½ oz.   Milk, 189 lbs. 4 oz.

Test made from Nov. 19 to 25, 1885; age, 4 yrs. 9 mos.; property of Frederick Loeser, Somerville, N. J.

Leta May 82171.—Sire, Tenella's Tormentor 22387; dam, Ina Kyle 71806.

Butter, 15 lbs. 2 oz.   Milk, 197 lbs.

Test made from Sept. 15 to 22, 1895; age, 4 yrs. 7 mos.; estimated weight, 750 lbs.; fed 8 qts. H. O. dairy feed, 2 qts. corn meal, 2 qts. bran and 1 qt. oil meal, daily—natural pasture; property of George W. Sisson, Jr., Potsdam, N. Y.

Leta of Llorac 70220.—Sire, Butterbred 18270; dam, Yone of the Hermitage 54652.

Butter, 15 lbs. 4½ oz.   Milk, 242 lbs.

Test made from Nov. 25 to Dec. 2, 1894; age, 4 yrs. 1 mo.; actual weight, 860 lbs.; fed 42 qts. ground oats, 42 qts. corn meal, 28 qts. wheat bran and 28 qts. ground oil-cake—10 lbs. hay, cabbage, old meadow pasture; property of Miller & Sibley, Franklin, Pa.

Letitia 3977.—Sire, Rioter 670; dam, Angela 1682.

Butter, 15 lbs. 5 oz.   Milk, 209 lbs. 4 oz.

Test made from Apr. 30 to May 6, 1884; age, 8 yrs. 5 mos.; property of Matt. M. Gardner, Nashville, Tenn.

Letitia Hunter 35575.—Sire, Black Diamond V. 4970; dam, Belle of Cork 17403.

Butter, 17 lbs. 5 oz.   Milk, 211 lbs. 4 oz.

Test made from June 7 to 14, 1890; age, 4 yrs. 11 mos.; estimated weight, 950 lbs.; fed, daily, about 16 lbs. ground oats, corn and bran; property of S. H. Godman, Muncie, Ind.

Letitia of St. Lambert 91054.—Sire, Meridale Pauline Pogis 25133; dam, Luteefy 46313.

Butter, 16 lbs.   Milk, 174 lbs. 8 oz.

Test made from Feb. 17 to 24, 1895; age, 3 yrs. 3 mos.; estimated weight, 850 lbs.; fed 126 lbs. corn ensilage, 73½ lbs. wheat bran, 21 lbs. corn chop, 21 lbs. cotton-seed meal and 10½ lbs. linseed meal—28 lbs. hay; property of B. F. Fackenthal, Jr., Riegelsville, Pa.

Letta of Eau Claire 120021.—Sire, Melia Ann's Son 22041; dam, Lily of Eau Claire 2d 77580.

Butter, 16 lbs. 3 oz.   Milk, 260 lbs. 6 oz.

Test made from July 5 to 12, 1897; age, 2 yrs. 3 mos.; estimated weight, 800 lbs.; fed 4 qts. ground corn and oats, 2 qts. cotton-seed meal and 4 qts. wheat middlings, daily—clover and red-top pasture; property of E. L. Van Deusen, Ashley Falls, Mass.

Lette Signal 26823.—Sire, Croton Maid's Duke 6658; dam, Signalette 14877.

Butter, 17 lbs. 2½ oz.   Milk, 224 lbs. 8 oz.

Test made from May 15 to 22, 1887; age, 3 yrs. 4 mos.; estimated weight, 700 lbs.; fed 14 lbs. corn meal, 5 lbs. oat meal and 6 lbs. oil meal, daily—blue-grass pasture; property of John A. Middelton, Shelbyville, Ky.

Letty Coles 2d 48128.—Sire, Diana's Rioter 10481; dam, Letty Coles 23351.

Butter, 21 lbs. 8 oz.   Milk, 300 lbs. 15 oz.

Test made from Apr. 12 to 19, 1895; age, 7 yrs. 10 mos.; actual weight, 998 lbs.; fed 50 lbs. wheat bran, 50 lbs. corn meal, 15 lbs. oil-cake meal and 20 lbs. ground oats—clover hay and oat straw *ad lib.*; property of J. P. Bradbury, Pomeroy, Ohio.

Letty Rioter 73475.—Sire, Diana's Rioter 10481; dam, Letty Coles 2d 48128.

Butter, 20 lbs. 1 oz.   Milk, 251 lbs. 6 oz.

Test made from Oct. 31 to Nov. 7, 1893; age, 4 yrs. 4 mos.; actual weight, 820 lbs.; fed 40 lbs. corn meal, 30 lbs. ground wheat, 24 lbs. ground oats, 14 lbs. cotton-seed meal, 10 lbs. old process oil meal and 72 lbs. hay—good meadow aftermath; property of J. P. Bradbury, Pomeroy, Ohio.

Letty Rioter 73475.—Sire, Diana's Rioter 10481; dam, Letty Coles 2d 48128.

Butter, 24 lbs. 2 oz.   Milk, 318 lbs. 14 oz.

Test made from May 26 to June 2, 1895; age, 5 yrs. 11 mos.; actual weight, 810 lbs.; fed 40 lbs. corn meal, 35 lbs. ground oats and 35 lbs. ground wheat—excellent blue-grass pasture; property of J. P. Bradbury, Pomeroy, Ohio.

Leurona 43271.—Sire, Catono 3761; dam, Valeur's Winnie 20007.

Butter, 14 lbs. 11 oz. Milk, 202 lbs. 8 oz.

Test made from Oct. 28 to Nov. 4, 1891; age, 5 yrs. 6 mos.; actual weight, 890 lbs.; fed 101 lbs. grain and 150 lbs. to 200 lbs. pumpkins—poor pasture; property of James T. Inglis, Middlefield, Conn.

Liberty 2d 16717.—Sire, Sir George 7656; dam, Liberty 16665.

Butter, 14 lbs. 6½ oz. Milk, 186 lbs. 10 oz.

Test made from June 23 to 29, 1885; age, 2 yrs. 11 mos.; property of J. L. Shallcross, Anchorage, Ky.

Lida Mullin 9198.—Sire, Superb 1950; dam, Lily of Maple Grove 5079.

Butter, 16 lbs. 8 oz. Milk, 269 lbs. 4 oz.

Test made from Mar. 24 to 30, 1882; age, 2 yrs. 10 mos.; property of G. R. Dykeman, Shippensburg, Pa.

Light Foot Pogis 57531.—Sire, Tombigbee Cowboy 18372; dam, Royal Ensigna 35410.

Butter, 14 lbs. 11 oz. Milk, 233 lbs. 4 oz.

Test made from Oct. 7 to 14, 1897; age, 8 yrs. 7 mos.; estimated weight, 1000 lbs.; fed 26½ lbs. cotton-seed meal, 26½ lbs. wheat shorts, 24 lbs. wheat bran, 29 lbs. corn meal, 140 lbs. cotton-seed hulls and 11 qts. roots—Bermuda, lespedeza and striata pasture; property of J. H. Wright, Meridian, Miss.

Light Foot Pogis 57531.—Sire, Tombigbee Cowboy 18372; dam, Royal Ensigna 35410.

Butter, 28 lbs. 15 oz. in 14 days. Milk, 459 lbs. 15 oz.

Test made from Sept. 30 to Oct. 14, 1897; age, 8 yrs. 7 mos.; estimated weight, 1000 lbs.; fed 53 lbs. cotton-seed meal, 53 lbs. wheat shorts, 48 lbs. wheat bran, 58¾ lbs. corn meal, 280 lbs. cotton-seed hulls and 22 qts. roots—mixed pasture; property of J. H. Wright, Meridian, Miss.

Lightning of St. Lambert 91392.—Sire, Baron of Riverside 23793; dam, Belle Marquise 34859.

Butter, 14 lbs. 2 oz. Milk, 271 lbs. 8 oz.

Test made from May 10 to 17, 1897; age, 5 yrs. 11 mos.; estimated weight, 900 lbs.; fed 84 qts. bran and 35 qts. corn meal—blue-grass pasture; property of A. R. Glascock & Son, Maysville, Ky.

Lilium Excelsum 24945.—Sire, Velpeau 2146; dam, Minerva of Canterbury 12130.

Butter, 17 lbs. ½ oz. Milk, 231 lbs.

Test made from Mar. 25 to Apr. 1, 1891; age, 8 yrs.; estimated weight, 900 lbs.; fed 14 qts. pea meal, 14 qts. corn meal, 28 qts. ground oats, 28 qts. bran and 3½ bush. roots; property of Mrs. E. M. Jones, Brockville, Ont., Can.

Lilium Excelsum 2d 66040.—Sire, Rioter of Rocky Farm 18183; dam, Lilium Excelsum 24945.

Butter, 14 lbs. 5 oz. Milk, 210 lbs. 8 oz.

Test made from June 28 to July 5, 1892; age, 3 yrs. 4 mos.; estimated weight, 700 lbs.; fed, 28 qts. pea meal and 56 qts. wheat bran—fair pasture; property of Mrs. E. M. Jones, Brockville, Ont., Can.

Lille Bonne 8108.—Sire, Plough Boy (P. S. 102 J. H. B.); dam, Queen (F. S. 1239 J. H. B.)

Butter, 19 lbs. 12 oz.   Milk, 275 lbs.

Test made from June 12 to 18, 1885; age, 8 yrs. 10 mos.; property of F. C. Havemeyer, West Chester, N. Y.

Lilley Rex 9852.—Sire, Prince of M. 2811; dam, Lilley Russ 2d 9514.

Butter, 14 lbs. 7 oz.   Milk, 257 lbs. 4 oz.

Test made from Dec. 10 to 16, 1885; age, 6 yrs. 6 mos.; property of Moulton Bros., West Randolph, Vt.

Lilley Rex Pogis 62545.—Sire, Garfield Stoke Pogis 15963; dam, Lilley Rex Rioter 48464.

Butter, 14 lbs. 9 oz.   Milk, 183 lbs. 12 oz.

Test made from June 5 to 12, 1892; age, 2 yrs. 9 mos.; estimated weight, 800 lbs.; fed 56 lbs. cut corn fodder, 7 lbs. corn meal, 14 lbs. gluten meal, 14 lbs. bran, 7 lbs. linseed meal, 56 lbs. hay and 210 lbs. roots; property of Est. of Frederick Billings, Woodstock, Vt.

Lillian Mostar 10364.—Sire, Diamond Earl 3110; dam, Mostar 6971.

Butter, 14 lbs. 3 oz.   Milk, 219 lbs. 12 oz.

Test made from Apr. 1 to 8, 1883; age, 4 yrs.; property of James Cloud & Son, Kennett Square, Pa.

Lillie Pope 8589.—Sire, Master 723; dam, Lucy Pope 5328.

Butter, 14 lbs. 5 oz.   Milk, 213 lbs. 15 oz.

Test made from May 18 to 25, 1885; age, 6 yrs. 10 mos.; property of G. B. & C. S. Smith, Eagle, Mich.

Lillie Victor Pogis 85471.—Sire, Melia Ann's Victor Pogis 20697; dam, Nelope Pedro 50879.

Butter, 14 lbs. 4 oz.   Milk, 206 lbs. 8 oz.

Test made from Aug. 25 to Sept. 1, 1893; age, 2 yrs. 5 mos.; estimated weight, 650 lbs.; fed corn fodder—good pasture; property of Mrs. J. H. Martin, Granville, N. Y.

Lilly Cross 13796.—Sire, Hector 791; dam, Lilly Burnside 4384.

Butter, 14 lbs. 3 oz.   Milk, 287 lbs. 8 oz.

Test made from June 1 to 7, 1883; age, 6 yrs. 6 mos.; property of E. A. Flagg, West Hartford, Conn.

Lilly Mac 22976.—Sire, Mulberry Lad 3954; dam, Miss Bos 10842.

Butter, 14 lbs. 2 oz.

Test made from May 19 to 26, 1884; age, 4 yrs. 1 mo.; property of J. C. McClintock, Meadville, Pa.

Lilly of Leeds 17059.—Sire, Hazel Eye 2d 7822; dam, Silkey Morrill 17058.

Butter, 14 lbs. 9 oz.

Test made from Apr. 28 to May 5, 1884; age, 4 yrs. 11 mos.; property of C. F. Cobb, South Vassalborough, Me.

Lilly Signalda 23227.—Sire, Signalda 4027; dam, Belle of Tennessee 9573.

Butter, 17 lbs. 10 oz. Milk, 158 lbs. 14 oz.

Test made from Aug. 22 to 29, 1888; age, 5 yrs. 2 mos.; estimated weight, 850 lbs.; fed, daily, 16 qts., two-thirds oats, one-third corn; property of M. P. Webster, Columbia, Tenn.

Lily Darling 11713, imp.

Butter, 15 lbs. 9 oz. Milk, 197 lbs. 8 oz.

Test made from Aug. 30 to Sept. 5, 1883; age, 5 yrs. 7 mos.; fed 12 qts. ground corn and oats, one-third corn and two-thirds oats, daily—no pasture; property of A. C. Jennings, Urbana, Ohio.

Lily Garfield 79819.—Sire, Garfield Stoke Pogis 15963; dam, Lilley Rex Rioter 48464.

Butter, 37.488 lbs. in 21 days. Milk, 562.7 lbs.

Test made from Sept. 30 to Oct. 20, 1893, at World's Columbian Exposition; age, 2 yrs.; actual weight, 794 lbs.; fed, 224.5 lbs. hay, 6 lbs. ensilage, 63 lbs. oil meal, 93 lbs. corn meal, 138 lbs. bran, 84 lbs. oats, 50.5 lbs. cotton-seed, 42 lbs. middlings and 14 lbs. corn-hearts; property of Est. of Frederick Billings, Woodstock, Vt.

Lily Martin 49954.—Sire, Ida's Stoke Pogis 13658; dam, Sigletta 32915.

Butter, 16 lbs. 8 oz. Milk, 256 lbs.

Test made from June 12 to 19, 1890; age, 3 yrs. 1 mo.; estimated weight, 700 lbs.; fed, daily, about 3 gals. corn meal and 1 pt. cotton-seed meal; property of M. C. Campbell, Spring Hill, Tenn.

Lily Martin 49954.—Sire, Ida's Stoke Pogis 13658; dam, Sigletta 32915.

Butter, 164.227 lbs. in 90 days. Milk, 3520.2 lbs.

Test made from May 31 to Aug. 28, 1893, at World's Columbian Exposition; age, 6 yrs. at beginning of test; actual weight, 1003 lbs.; fed 339 lbs. old hay, 210 lbs. ensilage, 594 lbs. new hay, 191 lbs. oil meal, 479 lbs. bran, 93.5 lbs. oats, 578 lbs. corn-hearts, 145.50 lbs. cotton-seed, 434 lbs. middlings, 34 lbs. grano-gluten, 1204.4 lbs. clover and 10 lbs. swale-grass; property of M. C. Campbell, Spring Hill, Tenn.

Lily Martin 2d 89080.—Sire, Oonan's Tormentor 22280; dam, Lily Martin 49954.

Butter, 14 lbs. 6½ oz. Milk, 268 lbs.

Test made from July 31 to Aug. 7, 1896; age, 4 yrs. 4 mos.; estimated weight, 900 lbs.; fed 15 lbs. corn meal, daily—blue-grass pasture; property of Baum & Hill, Frankfort, Ind.

Lily Niobe 55765.—Sire, Niobe's Stoke Pogis 13448; dam, Lily of St. Lambert 5120.

Butter, 21 lbs. 6½ oz. Milk, 225 lbs.

Test made from Oct. 29 to Nov. 5, 1894; age, 6 yrs. 6 mos.; estimated weight, 1100 lbs.; fed 56 qts. corn meal, 70 qts. bran and 21 qts. oil meal—millet pasture; property of George E. Peer, Rochester, N. Y.

Lily of Burr Oaks 11001.—Sire, Ike Felch 1292; dam, Lily Godfrey 3792.

Butter, 15 lbs. 13 oz. Milk, 192 lbs. 15 oz.

Test made from Nov. 9 to 15, 1883; age, 3 yrs. 11 mos.; property of Mrs. John Hamilton, Farmer City, Ill.

Lily of Elm Spring 51474.—Sire, Rex Alphea's Rioter 9190; dam, Minneola's Beauty 33038.

Butter, 19 lbs. 8 oz.   Milk, 223 lbs. 10 oz.

Test made from Dec. 5 to 12, 1897; age, 10 yrs. 7 mos.; actual weight, 850 lbs.; fed 8 qts. bran, 4 qts. corn meal, 4 qts. cotton-seed and 8 qts. chopped sweet potatoes, daily—corn fodder and sorghum hay *ad lib.*; property of H. J. Mitchell, Winnsborough, Tex.

Lily of Glynllyn 31340.—Sire, Holyoke 8147; dam, Lady Keith 20064.

Butter, 16 lbs.   Milk, 227 lbs. 8 oz.

Test made from Feb. 18 to 25, 1895; age, 10 yrs. 10 mos.; estimated weight, 800 lbs.; fed 42 lbs. wheat bran, 49 lbs. corn meal, 28 lbs. cotton-seed meal and 28 lbs. oil meal —hay; property of J. A. Griffith, Ashley Falls, Mass.

Lily of Maple Grove 5079.—Sire, Isaac B. 1951; dam, Symphonia 4635.

Butter, 16 lbs. 3 oz.   Milk, 204 lbs. 4 oz.

Test made from Mar. 13 to 19, 1882; age, 5 yrs. 10 mos.; property of G. R. Dykeman, Shippensburg, Pa.

Lily of Riverside 19599.—Sire, Blossom's Tennessee 6060; dam, Lily of Oxford 12820.

Butter, 14 lbs. 9 oz.   Milk, 214 lbs.

Test made from Mar. 28 to Apr. 3, 1887; age, 4 yrs. 1 mo.; estimated weight, 900 lbs.; fed, daily, 12 qts. ground corn, oats and ship stuff and 2 qts. oil meal, mixed; property of H. G. Westlake, Hillsdale, N. Y.

Lily of St. Lambert 5120.—Sire, Laval 506; dam, Pride of Windsor 483.

Butter, 14 lbs.

Property of T. S. Cooper, Coopersburg, Pa.

Lily of Staatsburgh 5427, imp.

Butter, 14 lbs. 2¼ oz., unsalted.

Test made from June 15 to 22, 1883; age, 8 yrs. 2 mos.; property of W. B. Dinsmore, Staatsburg, N. Y.

Lily Scituate 12665.—Sire, Prince of Scituate 3888; dam, Annie Dale 12662.

Butter, 24 lbs. 9½ oz.   Milk, 366 lbs. 8 oz.

Test made from June 21 to 27, 1885; age, 4 yrs. 11 mos.; fed 13 lbs. ground oats, 9 lbs. corn meal, 7 lbs. bran and 13 lbs. middlings, daily—good pasture of mixed grasses with clover; property of Ehrick Parmly, Oceanic, N. J.

Lily's Crown Princess 76011.—Sire, Crown Prince of St. L. 20070; dam, Lily's Pogis of Bois d'Arc 67776.

Butter, 15 lbs. 7 oz.   Milk, 214 lbs. 7 oz.

Test made from Nov. 18 to 25, 1895; age, 4 yrs. 4 mos.; estimated weight, 900 lbs.; fed 91 lbs. corn-cob meal, 35 lbs. bran, 20 lbs. cotton-seed meal and 350 lbs. ensilage; property of Austin & Probert, Florence, Mich.

Lily's Crown Princess 76011.—Sire, Crown Prince of St. L. 20070; dam, Lily's Pogis of Bois d'Arc 67776.

Butter, 64 lbs. 12 oz. in 30 days.   Milk, 879 lbs. 5 oz.

Test made from Nov. 8 to Dec. 8, 1895; age, 4 yrs. 4 mos.; estimated weight, 900 lbs.; fed 390 lbs. corn-cob meal, 150 lbs. bran, 86 lbs. cotton-seed meal and 1500 lbs. ensilage; property of Austin & Probert, Florence, Mich.

Lily's Daffodil 89469.—Sire, Muggy 22004; dam, Dot's Lily 49220.

Butter, 14 lbs. 8 oz. Milk, 190 lbs. 8 oz.

Test made from Jan. 12 to 19, 1898; age, 6 yrs. 2 mos.; estimated weight, 775 lbs.; fed 35 lbs. oats, peas and corn meal, mixed, 28 lbs. bran and 175 lbs. ensilage—hay *ad lib.*; property of A. D. Lee, Moravia, N. Y.

Limited Pedro 56382.—Sire, Young Pedro 2d 15011; dam, Miss Limited 37620.

Butter, 14 lbs. Milk, 259 lbs. 15 oz.

Test made from May 11 to 18, 1892; age, 3 yrs. 3 mos.; estimated weight, 900 lbs.; fed 6 qts. ground oats and 6 qts. corn meal, daily—pasture, timothy, common grass and some white clover; property of George Fox, Torresdale, Pa.

Limited Pedro 56382.—Sire, Young Pedro 2d 15011; dam, Miss Limited 37620.

Butter, 21 lbs. 7 oz. Milk, 288 lbs.

Test made from June 18 to 25, 1893; age, 4 yrs. 4 mos.; estimated weight, 850 lbs.; fed 8 qts. corn meal and 8 qts. ground oats, daily—mixed pasture; property of George Fox, Torresdale, Pa.

Lionette 18038.—Sire, Cicero 7657; dam, Zingaratta 17016.

Butter, 17 lbs. 1 oz. Milk, 19½ qts. per day.

Test made from July 4 to 10, 1887; age, 5 yrs. 1 mo.; estimated weight, 800 lbs.; fed 4 qts. corn meal, 7 qts. oat meal, 1½ qts. pea meal, 1 qt. oil meal and 3 qts. middlings, daily; property of D. F. Appleton, Ipswich, Mass.

Lisetta Johnson 5321.—Sire, Hamilton 1074; dam, Lizzie Johnson 2002.

Butter, 15 lbs. 10 oz. Milk, 219 lbs. 12 oz.

Test made from June 21 to 27, 1885; age, 8 yrs. 7 mos.; property of Frederick Billings, Woodstock, Vt.

Lisgar's Ella 24992.—Sire, Orloff 3143; dam, Witch of St. Lambert 5479.

Butter, 17 lbs. Milk, 314 lbs. 8 oz.

Test made from Aug. 16 to 23, 1887; age, 6 yrs. 6 mos.; estimated weight, 800 lbs.; fed peas, oats, corn meal and bran, all she would eat; property of Mrs. E. M. Jones, Brockville, Ont., Can.

Lisgar's Rose 2d 56976.—Sire, Canada's John Bull 8388; dam, Lisgar's Rose 26607.

Butter, 16 lbs. 11¾ oz. Milk, 213 lbs. 11 oz.

Test made from June 7 to 14, 1891; age, 5 yrs.; estimated weight, 900 lbs.; fed 15 lbs. oil meal, 16 lbs. meal and 38 lbs. middlings—good pasture; property of E. D. Halstead, Bartlett, N. Y.

Litta Oaks 19734.—Sire, Favori (P. S. 185 J. H. B.); dam, Lida-Arich, on I. of J.

Butter, 15 lbs. 5½ oz. Milk, 220 lbs.

Test made from May 2 to 8, 1884; age, 3 yrs. 7 mos.; property of James Crook, Jacksonville, Ala.

Litta Pogis 30275.—Sire, Prince of Melrose 2d 11015; dam, Litta Oaks 2d 20020.

Butter, 18 lbs. 6 oz. Milk, 236 lbs. 8 oz.

Test made from June 30 to July 7, 1888; age, 3 yrs. 7 mos.; estimated weight, 900 lbs.; fed 84 qts. ground oats, 28 qts. cotton-seed meal and 14 qts. corn meal—hay, mixed pasture; property of James Crook, Jacksonville, Ala.

Little Accident 15578.—Sire, Toltec 6831; dam, Odelle Sales 15564.

Butter, 14 lbs. 3½ oz. Milk, 147 lbs.

Test made from Dec. 5 to 11, 1885; age, 3 yrs. 11 mos.; property of A. K. Johnston, Stapleton, N. Y.

Little Accident 15578.—Sire, Toltec 6831; dam, Odelle Sales 15564.

Butter, 16 lbs. 10½ oz. Milk, 176 lbs. 8 oz.

Test made from Nov. 6 to 13, 1886; age, 4 yrs. 9 mos.; estimated weight, 750 lbs.; fed, daily, 20 lbs. of a mixture consisting of chopped corn, bran and cotton-seed meal; property of A. K. Johnston, Stapleton, N. Y.

Little Butterstar 20926.—Sire, Tormentor 3533; dam, Butterstar 7799.

Butter, 14 lbs. 12½ oz. Milk, 127 lbs.

Test made from Apr. 20 to 26, 1886; age, 2 yrs. 11 mos.; property of Campbell Brown, Spring Hill, Tenn.

Little Goldie 38671.—Sire, Little Harry 8808; dam, Alice Brand 17836.

Butter, 34 lbs. 8½ oz. Milk, 251 lbs. 5 oz.

Test made from June 21 to 28, 1892; age, 7 yrs. 7 mos.; estimated weight, 1000 lbs.; fed 140 qts. corn meal, 56 qts. wheat bran and 21 qts. linseed meal—pasture, short clover and orchard grass; property of Matthews & Humes, Huntsville, Ala.

Little Goldie 38671.—Sire, Little Harry 8808; dam, Alice Brand 17836.

Butter, 176.394 lbs. in 90 days. Milk, 3284.1 lbs.

Test made from May 31 to Aug. 28, 1893, at World's Columbian Exposition; age, 8 yrs. 7 mos. at beginning of test; actual weight, 989 lbs.; fed 322.3 lbs. old hay, 149.5 lbs. ensilage, 529.7 lbs. new hay, 186.5 lbs. oil meal, 510 lbs. bran, 89 lbs. oats, 533 lbs. corn-hearts, 126.25 lbs. cotton-seed, 424 lbs. middlings, 34 lbs. grano-gluten, 1182.4 lbs. clover and 10 lbs. swale-grass; property of C. I. Hood, Lowell, Mass.

Little Ida Hilda 97739.—Sire, Ida of St. L.'s Last Son 23601; dam, Matilda Hilda 75346.

Butter, 15 lbs. 8 oz. Milk, 239 lbs.

Test made from Dec. 22 to 29, 1896; age, 4 yrs. 2 mos.; estimated weight, 875 lbs.; fed 45 lbs. wheat bran, 40 lbs. corn meal, 20 lbs. cotton-seed meal and 14 lbs. linseed meal—about 85 lbs. hay; property of Ayer & McKinney, Philadelphia, Pa.

Little Lass 28512.—Sire, Tormentor 3533; dam, Bisma 3d 1870.

Butter, 17 lbs. 5 oz. Milk, 163 lbs. 8 oz.

Test made from June 9 to 15, 1886; age, 3 yrs. 3 mos.; property of W. S. Shields, Bean's Station, Tenn.

Little Meysie 84230.—Sire, Little Tormentor 17995; dam, Tillie Meysenburg 23148.

Butter, 15 lbs. 5 oz. Milk, 191 lbs. 8 oz.

Test made from Jan. 22 to 29, 1897; age, 5 yrs. 5 mos.; estimated weight, 800 lbs.; fed 70 qts. corn and cob meal, 70 qts. bran and 14 lbs. cotton-seed meal—clover hay *ad lib.*; property of James L. Cooper, Nashville, Tenn.

Little Mistiss 62849.—Sire, Little Tormentor 17995; dam, Tillie Meysenburg 23148.

Butter, 14 lbs. 3 oz. Milk, 220 lbs. 12 oz.

Test made from Dec. 25, 1893, to Jan. 1, 1894; age, 4 yrs. 3 mos.; estimated weight, 800 lbs.; fed 35 lbs. bran, 35 lbs. cotton-seed meal, 63 lbs. coarse corn meal and 21 lbs. ground oats—hay; property of James L. Cooper, Nashville, Tenn.

Little Myers 104650.—Sire, Leo of Glynllyn 32015; dam, Dewolf's Picture 85402.

Butter, 16 lbs. 8 oz. Milk, 196 lbs. 9 oz.

Test made from Mar. 10 to 17, 1897; age, 2 yrs. 8 mos.; estimated weight, 650 lbs.; fed 63 qts. oats, 56 qts. bran, 28 qts. corn meal and 70 lbs. roots; property of D. Wolfe Bishop, Jr., Lenox, Mass.

Little Pogis 32612.—Sire, Stoke Pogis 5th 5987; dam, La Petite Mère 5470.

Butter, 16 lbs. 1 oz. Milk, 200 lbs. 12 oz.

Test made from Nov. 15 to 22, 1890; age, 5 yrs. 8 mos.; estimated weight, 800 lbs.; fed, daily, about 10¼ lbs. crushed oats and 6 lbs. corn meal; 2 lbs. oil meal on each of the first two days; property of Ayer & McKinney, Philadelphia, Pa.

Little Princess Potoka 79285.—Sire, Little Tormentor 17995; dam, Princess Potoka 53572.

Butter, 14 lbs. 2 oz. Milk, 221 lbs. 8 oz.

Test made from June 17 to 24, 1894; age, 3 yrs. 11 mos.; estimated weight, 750 lbs.; fed 56 lbs. coarse corn meal, 35 lbs. ground oats and 14 lbs. cotton-seed meal—fair blue-grass pasture; property of James L. Cooper, Nashville, Tenn.

Little Texas 84229.—Sire, Little Tormentor 17995; dam, Tillie Texas 62840.

Butter, 17 lbs. 9½ oz. Milk, 177 lbs.

Test made from Dec. 22 to 29, 1896; age, 5 yrs. 6 mos.; estimated weight, 850 lbs.; fed 70 lbs. cotton-seed hulls, 22¾ lbs. cotton-seed meal, 17½ lbs. linseed meal, 21 lbs. bran, 31½ lbs. ground oats, 50¾ lbs. corn and cob meal and 35 lbs. hay—turf oats pasture; property of S. H. Butler, Como, Miss.

Little Tipsey 79398.—Sire, Little Tormentor 17995; dam, Tipsey C. 23147.

Butter, 15 lbs. 5 oz. Milk, 264 lbs. 8 oz.

Test made from May 28 to June 4, 1894; age, 3 yrs. 7 mos.; actual weight, 685 lbs.; fed 56 lbs. corn meal, 35 lbs. ground oats and 14 lbs. cotton-seed meal—good blue-grass pasture; property of James L. Cooper, Nashville, Tenn.

Little Torment 15581.—Sire, Tormentor 3533; dam, Méprise 4898.

Butter, 23 lbs. 2½ oz. Milk, 224 lbs. 8 oz.

Test made from Mar. 17 to 23, 1885; age, 3 yrs. 1 mo.; fed 6 qts. ground corn and oats and 1 pt. oil meal, daily, with all the hay she would eat; property of George M. Jewett, Glenville, Md.

Little Torment 15581.—Sire, Tormentor 3533; dam, Méprise 4898.

Butter, 227 lbs. 1 1-32 oz. in 13 weeks. Milk, 2,752 lbs. 6 oz. Cream, 638 lbs. 9 oz.

Test made from Mar. 17 to June 15, 1885; age, 3 yrs. 4 mos. at end of test; estimated weight, 800 lbs.; fed, daily, 18 lbs. ground corn and oats and 3 pts. oil meal, for nine weeks, with plenty of hay; 12 qts. ground corn and oats, with pasture for the rest of the time; property of George M. Jewett, Glenville, Md.

Litty 8017.—Sire, Duke of Maplehurst 2290; dam, Kitarene 4267.

Butter, 14 lbs. Milk, 258 lbs.

Test made from June 1 to 8, 1885; age, 6 yrs. 7 mos.; property of Ambler Bros., Chatham, N. Y.

Litza 6338.—Sire, Top-Sawyer 1404; dam, Roxana 2d 2532.

Butter, 14 lbs. 3 oz.   Milk, 268 lbs. 4 oz.

Test made from July 24 to 30, 1883; age, 5 yrs. 10 mos.; property of Mary B. Adams, Aberdeen, Miss.

Lizentris 81821.—Sire, Stokentor 25422; dam, Tormentor's Lizzie 64010.

Butter, 14 lbs. 13 oz.   Milk, 207 lbs.

Test made from Apr. 16 to 23, 1897; age, 6 yrs. 5 mos.; estimated weight, 800 lbs.; fed 70 qts. wheat bran, 28 qts. corn-hearts, 14 qts. cotton-seed meal and 7 qts. oil meal, fed wet—good pasture; property of Geo. V. Green, Hopkinsville, Ky.

Lizzette's Mary 12723.—Sire, Euroclydon 4789; dam, Millicent 2d 7229.

Butter, 14 lbs. 11 oz.

Test made from Sept. 8 to 14, 1884; age, 3 yrs. 2 mos.; property of James B. Wilder, Louisville, Ky.

Lizzie D. 10408.—Sire, Superb 1956; dam, Lily of Maple Grove 5079.

Butter, 16 lbs. 15 oz.   Milk, 207 lbs. 9 oz.

Test made from May 27 to June 2, 1884; age, 4 yrs.; property of John B. Trice, Hopkinsville, Ky.

Lizzie of Glynllyn 74274.—Sire, Ramapose's Bachelor 25900; dam, Arawana Iris 2d 10420.

Butter, 21 lbs. 15½ oz.   Milk, 322 lbs. 4 oz.

Test made from Aug. 2 to 9, 1896; age, 5 yrs. 1 mo.; actual weight, 1010 lbs.; fed 9 qts. bran, 9 qts. ground oats, 6 qts. corn meal and 3 qts. oil meal, daily—good pasture at night, fed cut green corn in stable during day; property of C. A. Sweet, East Aurora, N. Y.

Lobelia 2d 6650.—Sire, Burnside 1234; dam, Lobelia 4379.

Butter, 14 lbs. 6 oz.

Test made from June 14 to 20, 1883; age, 7 yrs. 7 mos.; no grain fed—timothy and clover pasture; property of R. S. Kingman, Sparta, Wis.

Loessin Fancy 91599.—Sire, Lisgar's Victor 25887; dam, Ella Smith 69540.

Butter, 18 lbs. 8 oz.   Milk, 229 lbs. 12 oz.

Test made from Sept. 28 to Oct. 5, 1896; age, 4 yrs. 2 mos.; estimated weight, 900 lbs.; fed 28 lbs. bran, 28 lbs. cotton-seed, 28 lbs. oil-cake, 17½ lbs. oats and about 75 lbs. hay; property of Fritz Loessin, La Grange, Tex.

Lois A. 107594.—Sire, Ramapogis 3d 26443; dam, My Queen's Lois 3d 81474.

Butter, 15 lbs. 14 oz.   Milk, 223 lbs. 1 oz.

Test made from Feb. 22 to Mar. 1, 1897; age, 4 yrs. 5 mos.; estimated weight, 700 lbs.; fed 43½ lbs. bran, 24½ lbs. corn, 34 lbs. oats, 7 lbs. linseed meal, 210 lbs. ensilage and 70 lbs. roots; property of C. I. Hood, Lowell, Mass.

Lois B. 107595.—Sire, Ramapogis 3d 26443; dam, My Queen's Lois 3d 81474.

Butter, 14 lbs. 2 oz.   Milk, 228 lbs. 10 oz.

Test made from Jan. 7 to 14, 1898; age, 4 yrs. 4 mos.; estimated weight, 800 lbs.; fed 42 lbs. bran, 35 lbs. corn meal, 21 lbs. ground oats, 14 lbs. oil meal, 49 lbs. roots and 210 lbs. ensilage—hay *ad lib.*; property of C. I. Hood, Lowell, Mass.

Lois of Pittsford Farms 88848.—Sire, Exile of St. Lambert 16th 19155; dam, Hope of Pittsford 73186.

Butter, 14 lbs. 9 oz. Milk, 249 lbs. 10 oz.

Test made from June 14 to 21, 1898; age, 5 yrs. 3 mos.; estimated weight, 800 lbs.; fed 58 qts. wheat bran and 40 qts. corn meal—timothy and clover pasture; property of T. B. Kibler, Jenera, Ohio.

Lolly Darling 78005.—Sire, Glorene's King 22813; dam, Buttercup Darling 28020.

Butter, 16 lbs. 6 oz. Milk, 255 lbs. 5 oz.

Test made from Apr. 18 to 25, 1898; age, 6 yrs. 4 mos.; estimated weight, 900 lbs.; fed 17 qts. corn and cob meal, 9 qts. ground oats, 3 qts. oil meal and 30 lbs. ensilage, daily—clover hay *ad lib.*; property of Brooks & Pidgeon, Salem, Ohio.

Lora Martin 88381.—Sire, Tony Lumpkin 31154; dam, Elise Fosdick 65515.

Butter, 19 lbs. 6 oz. Milk, 254 lbs.

Test made from Mar. 3 to 10, 1898; age, 5 yrs.; estimated weight, 800 lbs.; fed 63 lbs. bran, 20 lbs. oil meal, 18 lbs. corn and cob meal and 198 lbs. roots—timothy and clover hay; property of D. W. Howie, Jr., Elm Grove, Wis.

Lora Nicholson 27135.—Sire, Stockholder 7214; dam, St. Ello 3d 19078.

Butter, 19 lbs. 12¾ oz. Milk, 236 lbs.

Test made from Dec. 19 to 26, 1888; age, 4 yrs. 7 mos.; estimated weight, 925 lbs.; fed, daily, 24 lbs. corn and cob meal and cotton-seed meal; property of Campbell Brown, Spring Hill, Tenn.

Lora of Clovernook 51599.—Sire, Dixie Prince 18379; dam, Lera Williams 27139.

Butter, 16 lbs. 2 oz. Milk, 179 lbs. 8 oz.

Test made from Apr. 23 to 30, 1894; age, 6 yrs.; estimated weight, 700 lbs.; fed 2 qts. ground oats, 4 qts. corn meal, 4 qts. bran and 1 qt. cotton-seed meal, daily—hay and corn stover, fair pasture; property of Samuel N. Warren, Spring Hill, Tenn.

Lord Fife's Joliette 30591.—Sire, Lord Fife 5148; dam, Daisy of St. Lambert 25840.

Butter, 16 lbs. 2 oz. Milk, 172 lbs. 10 oz.

Test made from Apr. 19 to 26, 1891; age, 9 yrs.; estimated weight, 800 lbs.; fed 21 lbs. oil meal, 21 lbs. pea meal, 42 lbs. corn meal, 42 lbs. crushed oats, 14 lbs. bran and 50 lbs. hay; property of Ayer & McKinney, Philadelphia, Pa.

Lord Hallock's Princess 96924.—Sire, Lord Hallock 29589; dam, Winema 2d 64239.

Butter, 14 lbs. 1 oz. Milk, 207 lbs. 2 oz.

Test made from Feb. 20 to 27, 1898; age, 5 yrs. 8 mos.; estimated weight, 900 lbs.; fed 56 qts. corn and cob meal and 73 qts. light wheat bran—clover hay and corn fodder; property of S. E. Hanna, New Holland, Ohio.

Lord Harry's Lucy 86091.—Sire, Lord Harry 3d 27904; dam, Exile's Lucy 46883.

Butter, 16 lbs. 5 oz. Milk, 220 lbs.

Test made from Apr. 11 to 18, 1897; age, 4 yrs. 1 mo.; actual weight, 910 lbs.; fed 42 qts. bran, 42 qts. oat meal, 56 qts. corn meal, 28 qts. oil meal, 175 lbs. ensilage, 84 lbs. hay and 56 qts. roots; property of P. J. Cogswell, Rochester, N. Y.

Loreda 19801.—Sire, Glue 3960; dam, Adopted Girl 12767.

Butter, 17 lbs. 13 oz. Milk, 286 lbs. 10 oz.

Test made from June 1 to 8, 1888; age, 6 yrs. 6 mos.; estimated weight, 850 lbs.; fed 16 lbs. ground oats, corn meal, oil meal and bran, daily—clover pasture; property of Richardson Bros., Davenport, Iowa.

Lorelle 12913.—Sire, Lord Lawrence 1414; dam, Letitia 3977.

Butter, 14 lbs. 7 oz. Milk, 148 lbs. 8 oz.

Test made from Jan. 8 to 15, 1885; age, 3 yrs. 9 mos.; property of Matt. M. Gardner, Nashville, Tenn.

Lorilla 87505.—Sire, Toltec's Signal 29501; dam, Lorita 33750.

Butter, 16 lbs. 2 oz. Milk, 227 lbs. 4 oz.

Test made from Aug. 18 to 25, 1897; age, 4 yrs. 6 mos.; estimated weight, 900 lbs.; fed 28 lbs. oats, 35 lbs. corn meal, 21 lbs. bran and 14 lbs. oil meal—mixed pasture; property of H. C. Taylor, Orfordville, Wis.

Lorita 33750.—Sire, Combination 4389; dam, Moragina 26344.

Butter, 15 lbs. 6½ oz. Milk, 231 lbs. 10 oz.

Test made from Sept. 15 to 22, 1886; age, 3 yrs.; property of Richardson Bros., Davenport, Iowa.

Lorita 33750.—Sire, Combination 4389; dam, Moragina 26344.

Butter, 146.619 lbs. in 90 days. Milk, 2320.3 lbs.

Test made from May 31 to Aug. 28, 1893, at World's Columbian Exposition; age, 9 yrs. 5 mos. at beginning of test; actual weight, 926 lbs.; fed 305.9 lbs. old hay, 202.5 lbs. ensilage, 445.7 lbs. new hay, 179 lbs. oil meal, 463.5 lbs. bran, 69.5 lbs. oats, 450.5 lbs. corn-hearts, 116.25 lbs. cotton-seed, 336.5 lbs. middlings, 33 lbs. grano-gluten, 1088.4 lbs. clover and 10 lbs. swale-grass; property of C. A. Sweet, Buffalo, N. Y.

Lorna 2d 33634.—Sire, Stiletto 8320; dam, Lorna 19016.

Butter, 17 lbs. 9 oz. Milk, 203 lbs. 4 oz.

Test made from Jan. 11 to 18, 1895; age, 10 yrs. 4 mos.; actual weight, 1040 lbs.; fed 12 qts. ground corn and oats, 12 qts. wheat bran and 3 qts. oil meal, with 18 lbs. roots, daily; property of Frank W. Hart, Cleveland, Ohio.

Lorna of Maple Glens 48094.—Sire, Column 9622; dam, Lorna 2d 33634.

Butter, 14 lbs. Milk, 242 lbs. 12 oz.

Test made from June 2 to 9, 1891; age, 4 yrs. 8 mos.; estimated weight, 900 lbs.; fed 15 lbs. ground oats and 31½ lbs. corn meal during first three days, after that wood-land pasture only; property of J. E. Gray, Youngstown, Ohio.

Lorne's Diamond 66087.—Sire, Lorne 5248; dam, Gray Diamond 30946.

Butter, 19 lbs. Milk, 318 lbs.

Test made from Jan. 17 to 24, 1895; age, 7 yrs. 9 mos.; estimated weight, 1200 lbs.; fed 12 lbs. daily of a mixture of corn meal, bran, ground oats, oil meal, cotton-seed meal, wheat middlings and buckwheat meal, fed on cut fodder—49 lbs. hay and 4 bush. carrots; property of Mrs. H. E. Tremain, Lake George, N. Y.

Lorne's Marjoram 43327.—Sire, Lorne 5248; dam, Pedro's Marjoram 29585.

Butter, 16 lbs. 4½ oz.

Test made from Dec. 3 to 10, 1888; age, 2 yrs. 7 mos.; estimated weight, 750 lbs.; fed a mixture of corn meal, middlings, wheat bran, linseed meal, pea meal and whole oats—clover hay and good ensilage; property of T. S. Cooper, Coopersburg, Pa.

Lorne's Pretty Marjoram 77942.—Sire, Pedro 8187; dam, Lorne's Marjoram 43327.

Butter, 14 lbs. 14 oz. Milk, 265 lbs.

Test made from Dec. 22 to 29, 1894; age, 6 yrs. 2 mos.; estimated weight, 750 lbs.; fed about 18 qts. daily of a mixture of ground oats, middlings, wheat bran and some pea meal—clover hay and ensilage; property of T. S. Cooper, Coopersburg, Pa.

Lorne's Pretty Pearl 46683.—Sire, Lorne 5248; dam, Bliss 15219.

Butter, 19 lbs. 6 oz. Milk, 278 lbs.

Test made from Oct. 23 to 30, 1894; age, 8 yrs. 9 mos.; estimated weight, 1000 lbs.; fed about 24 qts. daily of a mixture of ground oats, middlings, bran, corn meal and some pea meal—clover hay and green corn, meadow blue-grass pasture; property of T. S. Cooper, Coopersburg, Pa.

Lorne's Primrose 66082.—Sire, Lorne 5248; dam, Pedro's Primrose 29586.

Butter, 16 lbs. 12 oz. Milk, 254 lbs.

Test made from May 23 to 30, 1896; age, 9 yrs. 5 mos.; estimated weight, 950 lbs.; fed about 20 qts. daily of a mixture of ground oats, middlings, wheat bran, corn meal and some pea meal—clover hay and ensilage, meadow pasture; property of T. S. Cooper, Coopersburg, Pa.

Lorraine 1435.—Sire on I. of J.; dam, Lady Rhine 1381.

Butter, 14 lbs. 8 oz.

Test made from May 7 to 13, 1874; age, 3 yrs. 9 mos.; property of Amanda Estes, Rogers Park, Ill.

Lotchem 19823.—Sire, Nero 7266; dam, Rebecca (F. S. 2625 J. H. B.)

Butter, 16 lbs. 7 oz. Milk, 248 lbs. 10 oz.

Test made from June 28 to July 4, 1885; age, 4 yrs. 4 mos.; property of Cornelius Wellington, East Lexington, Mass.

Lotta Pogis 40368.—Sire, Prince of Melrose 2d 11015; dam, Alma of Springvale 26951.

Butter, 16 lbs. 11 oz. Milk, 212 lbs. 1 oz.

Test made from June 20 to 27, 1889; age, 3 yrs. 1 mo.; estimated weight, 700 lbs.; fed 28 qts. pea meal, 28 qts. corn meal and 56 qts. bran—hay, good mixed pasture; property of James Crook, Jacksonville, Ala.

Lettie Derey 21578.—Sire, Signal (F. S. 278 J. H. B.); dam, Nancy (F. S. 2235 J. H. B.)

Butter, 16 lbs. 12½ oz. Milk, 175 lbs. 8 oz.

Test made from Jan. 21 to 28, 1889; age, 8 yrs. 10 mos.; estimated weight, 850 lbs.; fed, daily, about 3 gals. corn meal and 1 pt. cotton-seed meal; property of M. C. Campbell, Spring Hill, Tenn.

Lettie Rex 18757.—Sire, Rex 1330; dam, Lizzie Chaffin 8958.

Butter, 14 lbs. 4 oz. Milk, 164 lbs. 10 oz.

Test made from Jan. 20 to 26, 1885; age, 2 yrs. 1 mo.; property of William Simpson, New York, N. Y.

Lottie Signal 90039.—Sire, Alfonso 3013; dam, Miss M'liss 77747.

Butter, 14 lbs. 3½ oz. Milk, 203 lbs. 4 oz.

Test made from June 1 to 8, 1895; age, 3 yrs. 8 mos.; estimated weight, 750 lbs.; fed 70 qts. crushed corn, wheat and oats, ground, in equal parts, and 7 qts. of linseed meal—fair pasture; property of J. E. Carson, Crab Orchard, Ky.

Louisa D. 90959.—Sire, Lucy's Stoke Pogis 4th 26533; dam, Gilderoy's Mary 53004.

Butter, 16 lbs. 14 oz. Milk, 227 lbs. 12 oz.

Test made from Feb. 16 to 23, 1897; age, 5 yrs. 3 mos.; estimated weight, 900 lbs.; fed 35 lbs. ensilage, 14 lbs. gluten, bran and chop feed, 1½ lbs. oil meal and 5 lbs. corn stover, daily—hay *ad lib.*; property of W. J. Hussey, Mount Pleasant, Ohio.

Louisa Deming 23469.—Sire, Tunxes Chief 3705; dam, Lydia Deming 4399.

Butter, 14 lbs. 15 oz. Milk, 230 lbs. 4 oz.

Test made from Mar. 23 to 29, 1887; age, 3 yrs. 5 mos.; fed 4 lbs. ground oil-cake, 3 lbs. ground oats, 7 lbs. wheat bran, 4 lbs. corn meal, daily; property of Mrs. Mary A. Thomas, Bristol, Conn.

Louisa Hinman 12802.—Sire, Chelten Duke 924; dam, Bennie Hinman 7166.

Butter, 14 lbs., unsalted.

Test made from June 1 to 8, 1884; age, 3 yrs. 9 mos.; test made on grass alone; property of William H. Couch, Saugatuck, Conn.

Louisa of Riverside 69769.—Sire, Lord Harry 3445; dam, Spiralee 37960.

Butter, 15 lbs. 3½ oz. Milk, 205 lbs.

Test made from Mar. 13 to 20, 1894; age, 4 yrs. 1 mo.; estimated weight, 800 lbs.; fed 56 qts. wheat bran, 28 qts. corn meal, 14 qts. oil meal, 14 qts. cotton-seed meal and 210 lbs. ensilage; property of H. A. Huntington, Higganum, Conn.

Louisa of Riverside 69769.—Sire, Lord Harry 3445; dam, Spiralee 37960.

Butter, 23 lbs. 5 oz. Milk, 220 lbs. 8 oz.

Test made from Mar. 24 to 31, 1895; age, 5 yrs. 1 mo.; estimated weight, 800 lbs.; fed 103 lbs. provender and 40½ lbs. cotton-seed meal—hay *ad lib.*; property of H. A. Huntington, Higganum, Conn.

Louise Obella 37060.—Sire, Louis of St. Lambert 11013; dam, Obella 5080.

Butter, 17 lbs. 12 oz. Milk, 265 lbs. 14 oz.

Test made from Mar. 1 to 8, 1891; age, 4 yrs. 11 mos.; estimated weight, 875 lbs.; fed 18 lbs. oil meal, 18 lbs. pea meal, 53 lbs. corn meal, 53 lbs. crushed oats, 40 lbs. hay, 112 lbs. ensilage and ¾ bush. roots; property of Ayer & McKinney, Philadelphia, Pa.

Louise of Lawnfield 14151.—Sire, Duke of Argyle 1517; dam, Duchess of Argyle 2d 7568.

Butter, 14 lbs. 11½ oz. Milk, 268 lbs.

Test made from Mar. 31 to Apr. 7, 1885; age, 6 yrs.; property of Wm. S. Loomis, Holyoke, Mass.

Lou Tormentor of Lawn 84413.—Sire, Denise's Tormentor 11823; dam, Mattie P. 57042.

Butter, 15 lbs. 3 oz. Milk, 210 lbs.

Test made from May 19 to 26, 1896; age, 4 yrs. 2 mos.; estimated weight, 850 lbs.; fed 140 lbs. wheat bran, 28 lbs. corn meal and 14 lbs. cotton-seed meal—wild grass pasture; property of Platter & Foster, Denison, Tex.

Louvie 3d 6159.—Sire, Rex 1330; dam, Louvie 3899.

Butter, 14 lbs. 13 oz.

Test made from Feb. 1 to 8, 1883; age, 5 yrs. 6 mos.; property of W. S. & H. E. Savage, East Berlin, Conn.

L. Toltec's First 130688.—Sire, Little Toltec 24445; dam, Racket's Spot 55746.

Butter, 14 lbs. 13¼ oz. Milk, 154 lbs.

Test made from Mar. 23 to 30, 1898; age, 5 yrs. 2 mos.; estimated weight, 800 lbs.; fed 38 qts. bran, 28 qts. cotton-seed, 28 qts. corn meal and 2½ qts. linseed meal—14 lbs. to 20 lbs. hay, good pasture; property of Gebhart & Kauffman, Dallas, Tex.

Luani 37989.—Sire, Niburn 7658; dam, Lua 3d 8019.

Butter, 21 lbs. Milk, 190 lbs.

Test made from Sept. 20 to 27, 1889; age, 4 yrs. 5 mos.; actual weight, 880 lbs.; fed 12 lbs. wheat middlings daily; property of Mrs. Thomas Allen, Pittsfield, Mass.

Lucetta 6856.—Sire, Rector 1458; dam, Letitia 3977.

Butter, 14 lbs. 3 oz. Milk, 160 lbs. 8 oz.

Test made from Apr. 16 to 23, 1884; age, 6 yrs. 11 mos.; property of Matt. M. Gardner, Nashville, Tenn.

Lucia Rex 78326.—Sire, Comus Rex 26804; dam, Easter Prize 40324.

Butter, 14 lbs. 10 oz. Milk, 190 lbs. 8 oz.

Test made from Sept. 15 to 22, 1894; age, 5 yrs. 7 mos.; estimated weight, 950 lbs.; fed 91 lbs. corn meal, oat meal and wheat bran, in equal parts by measure, mixed—hay, prairie and white clover pasture; property of F. M. Hosford, Bethany, Neb.

Lucie of Bay View 66027.—Sire, G. F. Train 18868; dam, Ellenetta of Bay View 48921.

Butter, 14 lbs. 8 oz. Milk, 243 lbs.

Test made from Feb. 24 to Mar. 2, 1892; age, 2 yrs. 10 mos.; actual weight, 500 lbs.; fed 10½ lbs. oil meal, 7 lbs. ground oats, 7 lbs. ground corn, 105 lbs. corn refuse, 70 lbs. corn-stalks and 17½ lbs. hay; property of Charles Lautz, Buffalo, N. Y.

Lucile of Pearl Hill 91151.—Sire, Gray Beauty's Rioter 27131; dam, Madam Torpedo 84287.

Butter, 14 lbs. 4 oz. Milk, 256 lbs. 4 oz.

Test made from June 6 to 13, 1897; age, 4 yrs.; estimated weight, 800 lbs.; fed 28 qts. wheat bran, 14 qts. corn meal and 14 qts. ground oats, mixed—timothy hay at night, poor timothy pasture; property of A. M. Cope, Dick, Pa.

Lucilla 2735, imp.

Butter, 14 lbs.

Property of James A. Hayt, Patterson, N. Y.

Lucilla 3d 9786.—Sire, Dana 3620; dam, Lucilla 2735.

Butter, 17 lbs. 1 oz. Milk, 222 lbs.

Test made from Dec. 19 to 25, 1884; age, 4 yrs. 10 mos.; property of H. W. Douglass, Pevely, Mo.

Lucilla Kent 8892.—Sire, Conqueror (P. S. 89 J. H. B.); dam, Bessie (P. S. 232 J. H. B.)

Butter, 16 lbs. 6 oz. Milk, 149 lbs. 8 oz.

Test made from Mar. 7 to 13, 1883; age, 7 yrs.; property of V. E. Fuller, Hamilton, Ont., Can.

Lucky Belle 2d 6037.—Sire, Aldine 1136; dam, Lucky Belle 2214.

Butter, 16 lbs. 14 oz. Milk, 276 lbs. 8 oz.

Test made from May 6 to 12, 1882; age, 5 yrs. 4 mos.; fed liberally on cooked cotton-seed, corn meal and wheat bran—good pasture; property of W. B. Montgomery, Starkville, Miss.

Lucky Olive Germ 80256.—Sire, Lucky King 17350; dam, Olive Germ 54739.

Butter, 14 lbs. 3½ oz. Milk, 236 lbs. 4 oz.

Test made from June 30 to July 7, 1898; age, 7 yrs. 3 mos.; estimated weight, 650 lbs.; fed 84 lbs. corn meal, 84 lbs. oats and 56 lbs. bran—good pasture; property of R. H. Bulley, Canton, Ohio.

Lucy Cobb 15133.—Sire, Smoke Smith 4844; dam, Sida 4th 9284.

Butter, 18 lbs. 6½ oz. Milk, 229 lbs.

Test made from May 19 to 25, 1885; age, 3 yrs. 9 mos.; property of L. J. & A. W. Hill, Atlanta, Ga.

Lucy Gaines' Buttercup 5058.—Sire, Macgregor 2178; dam, Lucy Gaines 1711.

Butter, 14 lbs. Milk, 229 lbs.

Test made from June 3 to 9, 1881; age, 8 yrs.; property of A. H. Cooley, Little Britain, N. Y.

Lucy Gray 2746.—Sire, Major Adams 1044; dam, Lady Lightfoot 2745.

Butter, 15 lbs. 13 oz., unsalted.

Test made from June 17 to 23, 1874; property of William H. Hayden, Albany, Vt.

Lucy Lee H. 58106.—Sire, Landseer's Pogis 15847; dam, Tolteca 44762.

Butter, 14 lbs. 15½ oz. Milk, 229 lbs. 8 oz.

Test made from June 2 to 9, 1893; age, 4 yrs. 10 mos.; estimated weight, 750 lbs.; fed 56 qts. ground oats, 50 qts. corn meal and 28 qts. wheat bran—little clover hay, mixed white clover and blue-grass pasture; property of Morgan & Brown, Columbia, Tenn.

Lucy McClung 20368.—Sire, Monmouth Cyrene 6835; dam, Baronetti 8425.

Butter, 14 lbs. 3 oz. Milk, 204 lbs. 13 oz.

Test made from May 13 to 20, 1887; age, 5 yrs. 7 mos.; estimated weight, 850 lbs.; fed 3½ gals. ground oats and corn, daily; property of W. Gettys, Athens, Tenn.

Lucy of Wabash 9741, imp.

Butter, 15 lbs. ¼ oz.

Test made from May 8 to 14, 1883; age, 5 yrs. 8 mos.; property of B. F. Southworth, Defiance, Ohio.

Lucy S. King 106584.—Sire, Naiad's St. Lambert King 30645; dam, Exile's Beauty 61220.

Butter, 15 lbs. 14¼ oz. Milk, 212 lbs. 8 oz.

Test made from Jan. 26 to Feb. 2, 1896; age, 2 yrs. 1 mo.; estimated weight, 800 lbs.; fed 42 qts. corn meal, 21 qts. old process oil meal, 42 qts. bran, 21 qts. ground oats and 126 lbs. hay; property of H. & J. S. Wadsworth, Avon, N. Y.

Lucy's Love 20568.—Sire, Clusius H. 5781; dam, Lucy Gray 2746.

Butter, 14 lbs. 14 oz. Milk, 341 lbs. 7 oz.

Test made from Sept. 7 to 13, 1885; age, 3 yrs. 1 mo.; property of William H. Hayden, Albany, Vt.

Lucy T. 48200.—Sire, Sir Roger's T. 3938; dam, Ruby D. 2d 34627.

Butter, 16 lbs. 12.7 oz. Milk, 194 lbs. 3 oz.

Test made from Jan. 30 to Feb. 6, 1894; age, 6 yrs. 6 mos.; estimated weight, 975 lbs.; fed 1½ qts. wheat bran, 2 qts. boiled oats, 2 qts. cotton-seed meal, 3 qts. unbolted corn meal and 9 lbs. roots, daily; property of J. W. Largent, McKinney, Tex.

Luella Berlin 16927.—Sire, Berlin 4507; dam, Daisy Orton 16221.

Butter, 14 lbs. 6 oz.

Test made from Mar. 3 to 9, 1885; age, 3 yrs. 10 mos.; property of Andrew J. Fish, Findlay, Ohio.

Lula Gordon 2d 90067.—Sire, Bessie's Toltec 19859; dam, Lula Gordon 90066.

Butter, 15 lbs. 4½ oz. Milk, 222 lbs. 8 oz.

Test made from Mar. 13 to 20, 1895; age, 5 yrs. 2 mos.; estimated weight, 800 lbs.; fed 56 qts. ground oats and 84 qts. corn meal—clover and millet hay; property of Morgan & Brown, Columbia, Tenn.

Lula M. 32643.—Sire, Zalma's Mercury 6983; dam, Bisma 1669.

Butter, 15 lbs. 6 oz. Milk, 198 lbs.

Test made from June 7 to 14, 1887; age, 3 yrs. 5 mos.; estimated weight, 800 lbs.; fed shelled oats, corn meal and wheat bran—hay ad lib., pasture; property of H. L. Muldrow, Starkville, Miss.

Lulu of Riverside 19175.—Sire, Coventry Boy 5847; dam, Pride of Willimantic 13557.

Butter, 17 lbs. Milk, 198 lbs. 8 oz.

Test made from Jan. 13 to 20, 1888; age, 6 yrs. 6 mos.; actual weight, 790 lbs.; fed 21 qts. corn and cob meal, 21 qts. middlings, 21 qts. shorts and 21 qts. roots; property of J. H. Walker, Worcester, Mass.

Lulu Pedro 71902.—Sire, Young Pedro 2d 15011; dam, Lulu's Queen 40055.

Butter, 17 lbs. 4 oz. Milk, 328 lbs.

Test made from Apr. 19 to 26, 1895; age, 4 yrs. 1 mo.; actual weight, 870 lbs.; fed 42 lbs. ground oats, 42 lbs. corn meal and 14 lbs. linseed meal—about 100 lbs. hay and 56 lbs. carrots; property of W. C. McCallister, Monongahela, Pa.

Lulu Pedro 71902.—Sire, Young Pedro 2d 15011; dam, Lulu's Queen 40055.

Butter, 31 lbs. 4 oz. in 13 days. Milk, 611 lbs. 8 oz.

Test made from Apr. 12 to 26, 1895; age, 4 yrs. 1 mo.; actual weight, 870 lbs.; fed 6 lbs. ground oats, 6 lbs. corn meal and 2 lbs. linseed meal, daily—184 lbs. clover hay and 104 lbs. beets; property of W. C. McCallister, Monongahela, Pa.

Luna Dean 3d 84309.—Sire. Eddystone 23413; dam, Luna Dean 36380.

Butter, 15 lbs. Milk, 235 lbs. 12 oz.

Test made from June 8 to 15, 1898; age, 7 yrs. 2 mos.; estimated weight, 950 lbs.; fed 4 lbs. wheat bran daily—white clover, timothy and red-top pasture; property of J. E. Tyrrell, Marysville, Ohio.

Lustre 2062.—Sire, Orangepeel 502; dam, Locket 560.

Butter, 15 lbs. 8½ oz. Milk, 286 lbs.

Test made from June 12 to 18, 1876; age, 4 yrs.; no grain fed—pasturage of mixed grasses; property of C. S. Dole, Crystal Lake, Ill.

Luta Pogis 69501.—Sire, Beauty's Stoke Pogis 14175; dam, Mary Lou 35747.

Butter, 14 lbs. 1½ oz. Milk, 229 lbs.

Test made from Mar. 4 to 11, 1894; age, 5 yrs. 3 mos.; estimated weight, 850 lbs.; fed 56 qts. corn meal, 56 qts. bran and 28 qts. cotton-seed meal—oats pasture; property of W. W. Lipscomb, Luling, Tex.

Luteefy 46313.—Sire, Titan 8548; dam, Hamra 36522.

Butter, 14 lbs. 6 oz. Milk, 179 lbs.

Test made from Sept. 1 to 8, 1895; age, 8 yrs. 5 mos.; estimated weight, 900 lbs.; fed 56 lbs. wheat bran—good clover and timothy pasture; property of B. F. Fackenthal, Jr., Riegelsville, Pa.

Lydia D. 2d 97860.—Sire, Lord Gordon 24901; dam, Lydia D. 43735.

Butter, 14 lbs. 9¾ oz. (confirmed); estimated butter on basis of 85 per cent. fat, 14 lbs. 10⅞ oz. Milk, 246 lbs.

Test made from Mar. 12 to 18, 1898; age, 5 yrs. 6 mos.; estimated weight, 850 lbs.; fed 56 lbs. corn meal, 14 lbs. cotton-seed meal, 14 lbs. oil meal and 56 lbs. bran—clover hay *ad lib.*; property of Mrs. J. W. Sayre, Lexington, Ky.

Analyses of Butter.

First churning—Fat 81.70, water 14.63, ash 2.83, casein 0.84.
Second churning—Fat 80.67, water 15.71, ash 3.10, casein 0.52.

Lydia D. 2d 97860.—Sire, Lord Gordon 24901; dam, Lydia D. 43735.

Butter, 14 lbs. 6¼ oz. (confirmed); estimated butter on basis of 85 per cent. fat, 14 lbs. 1½ oz. Milk, 259.9 lbs.

Test made from Mar. 19 to 25, 1898; age, 5 yrs. 6 mos.; weight at end of test, 876 lbs.; fed 56 lbs. corn meal, 14 lbs. cotton-seed meal, 14 lbs. oil meal and 56 lbs. bran—clover hay *ad lib.*; property of Mrs. J. W. Sayre, Lexington, Ky.

Analyses of Butter.

First churning—Fat 81.97, water 14.55, ash 2.96, casein 0.52.
Second churning—Fat 81.60, water 15.04, ash ?.62, casein 0.74.

Lydia D. 2d 97860.—Sire, Lord Gordon 24901; dam, Lydia D. 43735.

Butter, 29 lbs. in 14 days (confirmed); estimated butter on basis of 85 per cent. fat, 28.77 lbs. Milk, 505.9 lbs.

Test made from Mar. 12 to 25, 1898; age, 5 yrs. 6 mos.; weight at beginning of test, 845 lbs.; weight at end of test, 876 lbs.; fed 112 lbs. corn meal, 28 lbs. cotton-seed meal, 28 lbs. oil meal and 112 lbs. bran—clover hay *ad lib.*; property of Mrs. J. W. Sayre, Lexington, Ky.

(For analyses of butter, see weekly tests above.)

Lydia Darrach 4903.—Sire, Doctor H. 2132; dam, Bertha Morgan 4770.

Butter, 17 lbs. 14 oz. (official). Milk, 238½ lbs.

Test made from May 4 to 11, 1883; property of Edward Worth, Wawa, Pa.

Lydia Darrach 2d 8056.—Sire, Duke of Brandywine 2213; dam, Lydia Darrach 4903.

Butter, 16 lbs.

Test made from May 24 to 30, 1885; age, 7 yrs. 2 mos.; test made on grass alone; property of H. P. Lewis, Brady, Pa.

Lydia Darrach 3d 10662.—Sire, Duke of Brandywine 2213; dam, Lydia Darrach 4903.

Butter, 16 lbs. 4 oz.

Test made from May 31 to June 6, 1885; age, 6 yrs. 5 mos.; test made on grass alone —timothy, red-top and clover pasture during the day, and blue-grass pasture at night; property of H. P. Lewis, Brady, Pa.

Lydia Darrach 5th 16577.—Sire, Duke of Brandywine 2213; dam, Lydia Darrach 4903

Butter, 15 lbs.

Test made from June 7 to 14, 1885; age, 3 yrs. 6 mos.; test made on grass alone— red-top, timothy and clover pasture; property of H. P. Lewis, Brady, Pa.

Lydia Libby 11698.—Sire, Jack Libby 3307; dam, Fawnette of Woodstock 3710.

Butter, 15 lbs. 3 oz. Milk, 266 lbs. 4 oz.

Test made from June 16 to 22, 1883; age, 4 yrs. 1 mo.; property of John E. Phillips, Baltimore, Md.

Lynona 36240.—Sire, Bombastic 9931; dam, Carlo's Pride 31371.

Butter, 16 lbs. 8 oz. Milk, 156 lbs. 4 oz.

Test made from Sept. 22 to 29, 1888; age, 3 yrs. 1 mo.; estimated weight, 800 lbs.; fed 56 qts. corn bran, 14 qts. cotton-seed meal and 14 qts. corn meal—poor pasture; property of A. J. Hawes & Son, Montgomery, Ala.

Mabel Crockett 18719.—Sire, Le Brocq's Pride 4871; dam, Lady Robin 1119.

Butter, 14 lbs. 9 oz. Milk, 307 lbs. 8 oz.

Test made from May 25 to 31, 1884; age, 2 yrs. 5 mos.; property of C. C. Crockett, Richmond, Ind.

Ma Belle 4942.—Sire, Euclid 520; dam, Bellflower 59.

Butter, 15 lbs.

Test made from May 4 to 10, 1884; age, 8 yrs.; property of C. E. Grosvenor, Canterbury, N. B., Can.

Mabel's Jewel 6251.—Sire, Rex 1330; dam, Mabel 1092.

Butter, 17 lbs. 13 oz. Milk, 305 lbs.

Test made from May 4 to 10, 1885; age, 7 yrs. 8 mos.; property of William Crozier, Northport, N. Y.

Mab of Deerfoot 3d 15345.—Sire, Deerfoot Boy 1926; dam, Mab of Deerfoot 8589.

Butter, 14 lbs. ½ oz.   Milk, 231 lbs. 3 oz.

Test made from Apr. 9 to 15, 1887; age, 6 yrs. 5 mos.; fed 12 lbs. wheat bran and mid-dlings, 5½ lbs. corn and oat meal and 2½ lbs. oil meal, daily; property of N. N. Palmer, Brodhead, Wis. .

Mab of Eau Claire 120020.—Sire, Melia Ann's Son 22041; dam, Callie of Eau Claire 94451.

Butter, 14 lbs. 10 oz.   Milk, 222 lbs. 4 oz.

Test made from June 1 to 8, 1897; age, 2 yrs. 2 mos.; estimated weight, 750 lbs.; fed 4 qts. ground corn and oats, 4 qts. wheat bran, 1 qt. oil meal and 1 qt. cotton-seed meal, daily—orchard pasture; property of E. I. Van Deusen, Ashley Falls, Mass.

Macie Pogis 103539.—Sire, Sig Pogis 22806; dam, Fanny Mason 45992.

Butter, 15 lbs. 1⅛ oz.   Milk, 197 lbs. 6 oz.

Test made from Jan. 10 to 17, 1898; age, 3 yrs. 1 mo.; estimated weight, 750 lbs.; fed 112 qts. cotton-seed hulls, 56 qts. cotton-seed, 56 qts. bran, 24 qts. oat chop, 24 qts. pea meal and 24 qts. corn meal—sorghum hay *ad lib.*; property of Gebhart & Kauffman, Dallas, Tex.

Maculac 24277.—Sire, Thalma 4288; dam, Junon 11269.

Butter, 15 lbs. 3 oz.   Milk, 149 lbs. 13 oz.

Test made from June 2 to 8, 1885; age, 2 yrs. 2 mos.; property of R. Peters, Atlanta, Ga.

Madame Argyle 19476.—Sire, Spangle's Boy 6229; dam, Louise of Lawnfield 14151.

Butter, 14 lbs. 6 oz.   Milk, 196 lbs. 7 oz.

Test made from Aug. 7 to 15, 1885; age, 3 yrs. 6 mos.; property of E. S. Henry, Rockville, Conn.

Madame Melrosa 103185.—Sire, Ida's Rioter's Prince 32355; dam, Madame Pogis 74452.

Butter, 16 lbs. 3 oz.   Milk, 179 lbs. 15 oz.

Test made from Dec. 14 to 21, 1896; age, 2 yrs. 5 mos.; estimated weight, 700 lbs.; fed 28 qts. cotton-seed, 56 qts. ground corn and oats and 56 qts. wheat bran—dry pasture; property of Mrs. E. M. Mirick, Cleburne, Tex.

Madame Pogis 74452.—Sire, Beauty's Stoke Pogis 14175; dam, Romping Lassie 25173.

Butter, 14 lbs. 4 oz.   Milk, 255 lbs. 6½ oz.

Test made from June 7 to 14, 1895; age, 5 yrs. 5 mos.; estimated weight, 775 lbs.; fed 21 qts. cotton-seed meal, 84 qts. bran and 84 qts. chopped corn—good Bermuda pasture; property of W. S. May, Ruston, La.

Madam Zophee 31679.—Sire, Signalda 4027; dam, Celia Belle 5865.

Butter, 16 lbs. 2½ oz.   Milk, 226 lbs. 6 oz.

Test made from Jan. 4 to 11, 1889; age, 4 yrs.; estimated weight, 800 lbs.; fed 22 lbs. per day of ground oats and corn, in equal parts, mixed with 4 bucketfuls cut timothy, wetted—timothy hay *ad lib.*; property of Matt. M. Gardner, Nashville, Tenn.

Madge 18904, imp.

Butter, 19 lbs. 2 oz.   Milk, 245 lbs. 8 oz.

Test made from Feb. 10 to 17, 1886; age, 6 yrs. 6 mos.; property of William H. Burr, Redding Ridge, Conn.

Madge Berg 24165.—Sire, Bergen 5567; dam, Madge H. 14441.

Butter, 14 lbs. Milk, 273 lbs. 12 oz.

Test made from Nov. 14 to 21, 1887; age, 4 yrs. 7 mos.; estimated weight, 900 lbs.; fed 21 qts. corn and cob meal, 21 qts. middlings and 21 qts. shorts; property of J. H. Walker, Worcester, Mass.

Madge of Glen Rouge 112455.—Sire, One Hundred per Cent. 16590; dam, Molina of St. Lambert 09454.

Butter, 15 lbs. 2 oz. Milk, 250 lbs. 6 oz.

Test made from May 13 to 20, 1896; age, 1 yr. 11 mos.; estimated weight, 700 lbs.; fed 56 lbs. oats and peas, chopped—hay *ad lib.* at night; clover and timothy pasture; property of William Rolph, Markham, Ont., Can.

Madolina H. 12327.—Sire, Homer H. 3683; dam, Bessie Spaulding 2780.

Butter, 19 lbs. 4 oz. Milk, 400 lbs. 2 oz.

Test made from Sept. 7 to 13, 1885; age, 6 yrs. 7 mos.; property of William H. Hayden, Albany, Vt.

Madora Gold 67302.—Sire, Gold of St. Lambert 16744; dam, Miss Madora 2d 39243.

Butter, 16 lbs. 7 oz. Milk, 277 lbs. 4 oz.

Test made from Mar. 3 to 10, 1896; age, 6 yrs. 5 mos.; estimated weight, 900 lbs.; fed 112 lbs. ground corn and oats, 14 lbs. old process oil meal and 231 lbs. ensilage—hay and corn stover *ad lib.*; property of N. N. Palmer & Son, Brodhead, Wis.

Madora Gold 67302.—Sire, Gold of St. Lambert 16744; dam, Miss Madora 2d 39243.

Butter, 17 lbs. 1 oz. Milk, 296 lbs. 11 oz.

Test made from May 20 to 27, 1897; age, 7 yrs. 7 mos.; estimated weight, 900 lbs.; fed 98 lbs. corn and oats and 14 lbs. oil meal—pasture; property of N. N. Palmer & Son, Brodhead, Wis.

Maggie C. 12216.—Sire, Champion of America 1567; dam, Maggie May 3255.

Butter, 14 lbs. 6 oz. Milk, 175 lbs.

Test made from Nov. 15 to 21, 1884; age, 4 yrs. 10 mos.; property of W. B. Montgomery, Starkville, Miss.

Maggie Cyrene 53462.—Sire, Chendado 12955; dam, Myra Cyrene 46590.

Butter, 15 lbs. 15 oz. Milk, 191 lbs.

Test made from Dec. 21 to 28, 1895; age, 9 yrs. 7 mos.; estimated weight, 1000 lbs.; fed 28 qts. ground oats, 28 qts. bran, 14 qts. corn meal and 7 qts. cotton-seed meal—hay *ad lib.* at night, blue-grass pasture; property of D. P. Carter, Lynnville, Tenn.

Maggie McM. 14073.—Sire, Thorndale 2582; dam, Pandora of Staatsburgh 3d 6497.

Butter, 19 lbs. 9 oz. Milk, 211 lbs.

Test made from May 5 to 11, 1885; age, 4 yrs.; property of R. McMichael, Lexington, Ky.

Maggie Martin 9562.—Sire, Much Ado 2405; dam, Lucette Martin 7770.

Butter, 17 lbs. 6 oz.

Test made from Dec. 25, 1885, to Jan. 1, 1886; age, 6 yrs. 6 mos.; property of H. L. Muldrow, Starkville, Miss.

Maggie May 3255.—Sire, Nelusko 479; dam, Lucky Belle 2214.

Butter, 14 lbs. 2½ oz.   Milk, 235 lbs. 8 oz.

Test made from Mar. 21 to 27, 1882; age, 8 yrs. 2 mos.; property of W. B. Montgomery, Starkville, Miss.

Maggie May 2d 12926.—Sire, Champion of America 1567; dam, Maggie May 3255.

Butter, 14 lbs. 6 oz.   Milk, 181 lbs.

Test made from Dec. 1 to 7, 1884; age, 4 yrs.; property of W. B. Montgomery, Starkville, Miss.

Maggie May of Tupelo 71280.—Sire, Duke of Coonewah 14548; dam, Irelia 21320.

Butter, 17 lbs. 6 oz.   Milk, 206 lbs. 11 oz.

Test made from Nov. 5 to 12, 1893; age, 7 yrs. 11 mos.; estimated weight, 900 lbs.; fed 12 qts. wheat bran and 12 qts. ground oats and corn meal, daily—prairie hay *ad lib.*; property of Terrell & Harris, Terrell, Tex.

Maggie May of Tupelo 71280.—Sire, Duke of Coonewah 14548; dam, Irelia 21320.

Butter, 20 lbs. 7 oz.   Milk, 257 lbs. 1 oz.

Test made from Mar. 15 to 22, 1897; age, 11 yrs. 3 mos.; actual weight, 840 lbs.; fed 28 lbs. cotton-seed meal, 72 lbs. corn meal, 63 lbs. wheat bran and 35 lbs. ground oats —Johnson grass, Johnson grass and Bermuda pasture; property of S. B. Hopkins, Dallas, Tex.

Maggie of St. Lambert 9776.—Sire, Stoke Pogis 3d 2238; dam, Ophelie 493.

Butter, 16 lbs. 3 oz.   Milk, 278 lbs.

Test made from Apr. 1 to 6, 1883; age, 3 yrs. 4 mos.; fed 8 lbs. meal, 4 lbs. bran and 1 pk. carrots, daily—hay *ad lib.*; property of W. D. Reesor, Markham, Ont., Can.

Maggie of Springvale 15931.—Sire, Duke of Springvale 4290; dam, Jessie Lee of Labyrinth 5290.

Butter, 18 lbs. 5 oz.   Milk, 223 lbs. 2 oz.

Test made from May 13 to 20, 1887; age, 5 yrs. 2 mos.; estimated weight, 850 lbs.; fed 70 qts. corn meal, 42 qts. bran, 42 qts. cotton-seed meal and 210 lbs. ensilage— plenty of hay and good clover pasture; property of James Crook, Jacksonville, Ala.

Maggie Rex 28623.—Sire, Cash Boy 2248; dam, Maggie 2d 2055.

—   Butter, 17 lbs. 2½ oz.   Milk, 298 lbs.

Test made from Mar. 16 to 22, 1885; age, 5 yrs. 7 mos.; property of Mrs. E. M. Jones, Brockville, Ont., Can.

Maggie Rule 31940.—Sire, Yoko 3951; dam, Niantic's Palestine 14519.

Butter, 21 lbs. 5 oz.   Milk, 184 lbs. 3 oz.

Test made from Sept. 17 to 24, 1890; age, 7 yrs. 4 mos.; actual weight, 960 lbs.; fed, daily, 6 lbs. corn meal, 6 lbs. wheat bran, 4 lbs. barley meal and 2 lbs. linseed meal; property of M. L. Frink, Oxford, Mich.

Maggie Sheldon 23583.—Sire, Sheldon 5250; dam, Maggie of St. Lambert 9776.

Butter, 21 lbs. 5 oz.   Milk, 288 lbs. 8 oz.

Test made from Feb. 14 to 20, 1886; age, 3 yrs. 11 mos.; fed 32 lbs. peas and oats, with clover hay, and 30 lbs. mangel-wurzels, daily; property of William Rolph, Markham, Ont., Can.

Magna 2238, imp.

Butter, 19 lbs. 1 oz., unsalted.

Test made from Oct. 19 to 26, 1882; age, 14 yrs.; property of W. B. Dinsmore, Staatsburg, N. Y.

Magnibel 7976.—Sire, Magnetic 1428; dam, Hinnibel 4040.

Butter, 14 lbs. 12 oz.

Test made from June 24 to July 1, 1883; age, 4 yrs. 6 mos.; property of Robert S. Taylor, Fort Wayne, Ind.

Magnibel 7976.—Sire, Magnetic 1428; dam, Hinnibel 4040.

Butter, 15 lbs. 2 oz. Milk, 281 lbs. 2 oz.

Test made from June 22 to 28, 1885; age, 6 yrs. 5 mos.; property of R. S. Taylor, Fort Wayne, Ind.

Magnolia Ridgely 17269.—Sire, Orange Skin 1216; dam, Belle Ridgely 6785.

Butter, 14 lbs. 8 oz. Milk, 181 lbs.

Test made from July 13 to 20, 1885; age, 8 yrs. 9 mos.; property of W. M. Bell, Miami, Mo.

Magyarland 86326.—Sire, Trial (P. S. 1187 J. H. B.); dam, Magyarland 4th (P. S. 3242 J. H. B.)

Butter, 17 lbs. 8 oz. Milk, 311 lbs. 4 oz.

Test made from Nov. 8 to 15, 1895; age, 5 yrs. 4 mos.; estimated weight, 900 lbs.; fed 1½ pts. cotton-seed meal, 1½ pts. oil meal, 2½ qts. corn meal, 10 qts. ground oats, 5 qts. middlings and 8 qts. bran, daily; property of R. A. Sibley, Rochester, N. Y.

Mahoning Bess 35834.—Sire, Prince of St. Lambert 5287; dam, Mercurenalia 14874.

Butter, 15 lbs. 8¾ oz. Milk, 322 lbs. 7 oz.

Test made from June 13 to 20, 1894; age, 8 yrs. 10 mos.; actual weight, 834 lbs.; fed 12 qts. daily of chop and wheat bran, the chop consisting of one part each of wheat and corn and two parts of oats—hay, and some cut grass; property of H. H. Hahn, Youngstown, Ohio.

Maiden of Jersey 2736, imp.

Butter, 14 lbs. 11 oz.

Test made from June 1 to 7, 1878; age, 8 yrs.; property of James A. Hayt, Patterson, N. Y.

Maid of Amboy 2929.—Sire. Standard 553; dam, Sukey 2d 1224.

Butter, 16 lbs. 1 oz.

Test made from May 1 to 7, 1880; property of J. W. Burke, Jacksonville, Ala.

Maid of Avranches 6959.—Sire, Tommy, on I. of J.; dam, Grey Queen (F. S. 571 J. H. B.)

Butter, 15 lbs., unsalted. Milk, 172 lbs.

Test made from Aug. 20 to 26, 1883; fed 5 qts. corn meal, daily—poor blue-grass pasture; property of Thomas H. Malone, Nashville, Tenn.

Maid of Berlin 12746.—Sire, Don Pedro of Binghamton 2974; dam, Belle of Bayside 1457.

 Butter, 14 lbs. 8 oz. Milk, 157 lbs. 10 oz.

Test made from Apr. 14 to 21, 1887; age, 7 yrs. 1 mo.; estimated weight, 800 lbs.; fed, daily, 8 lbs. corn meal, 4 lbs. bran, 2 lbs. middlings and 2 lbs. oil meal; property of James Stillman, Sing Sing, N. Y.

Maid of Fernwood 2d 29010.—Sire, Uproar 4th 5954; dam, Maid of Fernwood 10939.

 Butter, 17 lbs. 11 oz. Milk, 221 lbs. 9 oz.

Test made from June 2 to 9, 1888; age, 5 yrs.; estimated weight, 900 lbs.; fed, daily, 5 qts. corn meal, 1 qt. oil meal and 6 qts. bran; property of D. W. Voyles, Crandall, Ind.

Maid of Five Oaks 7178.—Sire, Butterstamp (P. S. 101 J. H. B.); dam, Fancy (F. S. 1528 J. H. B.)

 Butter, 15 lbs. 4 oz. Milk, 273 lbs. 9 oz.

Test made from May 7 to 13, 1883; age, 7 yrs.; property of Houghton Farm, Mountainville, N. Y.

Maid of Saragossa 9086.—Sire, Balboa 1244; dam, Prize Maid 3835.

 Butter, 16 lbs. 6 oz. Milk, 190 lbs. 5 oz.

Test made from Aug. 5 to 12, 1885; age, 7 yrs. 7 mos.; property of David W. Voyles, Crandall, Ind.

Maid of Tanglewood 51606.—Sire, Marjoram's Stoke Pogis 17203; dam, Maid of the Elms 18932.

 Butter, 16 lbs. Milk, 185 lbs. 4 oz.

Test made from Apr. 8 to 15, 1891; age, 3 yrs.; estimated weight, 800 lbs.; fed 9 lbs. corn meal, 9 lbs. crushed oats, 3 lbs. bran and 10 lbs. hay, daily; property of Ayer & McKinney, Philadelphia, Pa.

Maid of the Elms 6960.—Sire on I. of J.; dam, Grey Queen (F. S. 571 J. H. B.)

 Butter, 16 lbs., unsalted. Milk, 214 lbs. 8 oz.

Test made from May 1 to 8, 1881; age, about 5 yrs.; property of Thomas H. Malone, Nashville, Tenn.

Maid of the Elms 18932, imp.

 Butter, 16 lbs. 14½ oz. Milk, 239 lbs. 6½ oz.

Test made from May 3 to 9, 1885; age, 4 yrs. 11 mos.; property of D. A. Givens, Cynthiana, Ky.

Maid of the Elms 3d 15554.—Sire, Lord Lawrence 1414; dam, Maid of the Elms 6960.

 Butter, 16 lbs. 4 oz. Milk, 172 lbs. 8 oz.

Test made from July 21 to 27, 1885; age, 3 yrs. 4 mos.; property of Thomas H. Malone, Nashville, Tenn.

Maid of Yokun 2d 86350.—Sire, Tennessee's Tormentor 23498; dam, Maid of Yokun 82415.

 Butter, 14 lbs. 4 oz. Milk, 155 lbs. 4 oz.

Test made from Jan. 28 to Feb. 4, 1895; age, 2 yrs. 6 mos.; estimated weight, 700 lbs.; fed 56 qts. corn meal, 56 qts. ground oats and 28 qts. cooked cotton-seed—clover and millet hay; property of Morgan & Brown, Columbia, Tenn.

Malope 2d 11923.—Sire, Brother Jack 4042; dam, Malope 5872.

Butter, 15 lbs. 10 oz. Milk, 193 lbs. 8 oz.

Test made from Apr. 20 to 26, 1884; age, 3 yrs. 8 mos.; property of William Simpson, New York, N. Y.

Mamelle 20804.—Sire, Gold Basis 4038; dam, Jazel's Maid 11011.

Butter, 21 lbs. 8¼ oz. (official). Milk, 256 lbs. 1 oz.

Test made from June 5 to 12, 1885; age, 2 yrs. 2 mos.; fed 21 lbs. ground corn and oats and 1 qt. cotton-seed meal, mixed, daily—good blue-grass pasture; property of Thomas H. Malone, Nashville, Tenn.

Mamie Coburn 3798.—Sire, Ramchunder 718; dam, Bowley 1946.

Butter, 17 lbs. 8 oz.

Test made from May 2 to 9, 1878; age, 4 yrs.; property of Mrs. J. B. Ritzinger, Indianapolis, Ind.

Mamie H. Pogis 57050.—Sire, Buffer's Hugo Pogis 15315; dam, Lady Perfection 32116.

Butter, 16 lbs. 14 oz. Milk, 193 lbs. 8 oz.

Test made from Dec. 3 to 10, 1893; age, 4 yrs. 9 mos.; actual weight, 925 lbs.; fed 84 qts. ground oats, 21 qts. ground oil-cake, 42 qts. corn meal, 21 qts. wheat bran, 30 lbs. carrots and 105 lbs. boiled sweet corn—hay; property of Miller & Sibley, Franklin, Pa.

Mandana 3d 78778.—Sire, Aaron Stoke Pogis 21705; dam, Mandana 46198.

Butter, 19 lbs. 15¾ oz. Milk, 241 lbs. 8 oz.

Test made from Nov. 11 to 18, 1897; age, 6 yrs. 3 mos.; actual weight, 930 lbs.; fed 126 qts. ground oats, 42 qts. corn meal, 14 qts. wheat bran and 14 qts. ground oil-cake—clover hay; property of Miller & Sibley, Franklin, Pa.

Manoa 6340.—Sire, Duke of Lebanon 1880; dam, Winnowa 2614.

Butter, 16 lbs. 12 oz. Milk, 273 lbs. 1 oz.

Test made from Jan. 8 to 14, 1884; age, 6 yrs. 11 mos.; property of Mrs. L. M. Fair, Wallingford, Conn.

Manuscript 74881.—Sire, Combination 3d 17576; dam, Script 43308.

Butter, 15 lbs. 1 oz. Milk, 222 lbs. 8 oz.

Test made from Jan. 15 to 22, 1897; age, 5 yrs. 5 mos.; estimated weight, 750 lbs.; fed 63 lbs. corn meal, 49 lbs. ground oats, 14 lbs. cotton-seed meal, 7 lbs. linseed meal, 42 lbs. bran and 140 lbs. ensilage—some clover hay; property of B. G. Cox, Terre Haute, Ind.

Maple Glens Gold Lorna 84017.—Sire, Rioter of Maple Glens 21887; dam, Lorna of Maple Glens 48094.

Butter, 15 lbs. 1 oz. Milk, 225 lbs. 4 oz.

Test made from Jan. 6 to 13, 1898; age, 6 yrs. 8 mos.; estimated weight, 850 lbs.; fed 8 qts. wheat bran, 4 qts. corn and cob meal, 4 qts. ground oats, 1½ pts. oil meal and 1½ pts. cotton-seed meal, daily—hay and cut corn fodder *ad lib.*, 20 lbs. roots; property of J. E. Gray & Son, Youngstown, Ohio.

Maple Leaf 4768.—Sire, Ontario 865; dam, Echo 2223.

Butter, 14 lbs. 12 oz.

Property of John D. Wing, Millbrook, N. Y.

Maquilla 24043.—Sire, Rosy King 4433; dam, Maquita 7589.

Butter, 21 lbs. 1 oz.   Milk, 149 lbs. 10 oz.

Test made from Oct. 13 to 20, 1885; age, 2 yrs. 10 mos.; fed 8 qts. corn-hearts and 4 qts. oats, daily—poor pasture; property of Columbia Jersey Cattle Co., Columbia, Tenn.

Maquilla of Hood Farm 113067.—Sire, Fancy's Harry 9777; dam, Maquilla 24043.

Butter, 14 lbs. 13¾ oz.   Milk, 218 lbs. 4 oz.

Test made from Dec. 13 to 20, 1896; age. 2 yrs. 8 mos.; estimated weight, 700 lbs.; fed 42 lbs. bran, 33½ lbs. corn, 17½ lbs. linseed meal, 42 lbs. oats, 140 lbs. roots and 210 lbs. ensilage—hay *ad lib.*; property of C. I. Hood, Lowell, Mass.

Maquilla's Violet 69774.—Sire, Maquilla's Harry 22701; dam, Violet Denison 49997.

Butter, 14 lbs. 10½ oz.   Milk, 163 lbs.

Test made from May 15 to 22, 1894; age, 3 yrs. 9 mos.; estimated weight, 750 lbs.; fed 42 qts. corn meal, 28 qts. ground oats and 14 qts. cotton-seed meal—high, rocky pasture; property of H. A. Huntington, Higganum, Conn.

Maquilla's Violet 69774.—Sire, Maquilla's Harry 22701; dam, Violet Denison 49997.

Butter, 31 lbs. 1 oz.   Milk, 186 lbs. 8 oz.

Test made from Mar. 15 to 22, 1895; age, 4 yrs. 7 mos.; estimated weight, 900 lbs.; fed 111 lbs. provender, 32½ lbs. cotton-seed meal and 103 lbs. ensilage—hay; property of H. A. Huntington, Higganum, Conn.

Marburi 37991.—Sire, Niburn 7658; dam, Marchioness of Maplehurst 8018.

Butter, 16 lbs. 12 oz.   Milk, 192 lbs.

Test made from Sept. 2 to 9, 1889; age, 4 yrs. 2 mos.; actual weight, 860 lbs.; fed 12 lbs. wheat middlings daily; property of Mrs. Thomas Allen, Pittsfield, Mass.

Marchande 52258.—Sire, Southern Prince 10760; dam, Countess Potoka 7496.

Butter, 18 lbs. 12¼ oz.   Milk, 215 lbs.

Test made from Nov. 27 to Dec. 4, 1893; age, 5 yrs. 5 mos.; estimated weight, 875 lbs.; fed 112 lbs. corn chop and 56 lbs. ground oats—blue-grass pasture; property of Est. of Campbell Brown, Spring Hill, Tenn.

Marchande 52258.—Sire, Southern Prince 10760; dam, Countess Potoka 7496.

Butter, 23 lbs. 14¼ oz.   Milk, 175 lbs. 6½ oz.

Test made from Jan. 3 to 10, 1895; age, 6 yrs. 6 mos.; actual weight, 830 lbs.; fed 20 lbs. per day of ground oats and corn, in equal parts, mixed with 6 bucketfuls cut fodder corn, wetted—clover hay *ad lib.* at night; property of Matt. M. Gardner, Nashville, Tenn.

Marchande 2d 93823.—Sire. Bisson's Pogis 30781; dam, Marchande 52258.

Butter, 15 lbs. 9 oz.   Milk, 203 lbs. 13 oz.

Test made from Oct. 27 to Nov. 3, 1897; age, 4 yrs.; estimated weight, 750 lbs.; fed 19 lbs. per day of ground oats and corn, in equal parts, mixed with 4 bucketfuls cut sheaf oats, wetted—orchard and blue-grass pasture; property of Matt. M. Gardner, Nashville, Tenn.

Marchande 3d 133818.—Sire, Prince of Tennessee 20772; dam, Marchande 52258.

Butter, 15 lbs. 6¼ oz. Milk, 173 lbs. 5¾ oz.

Test made from Dec. 29, 1897, to Jan. 5, 1898; age, 2 yrs. 3 mos.; estimated weight, 600 lbs.; fed 18 lbs. per day of ground corn and oats, in equal parts, mixed with 4 bucketfuls cut sheaf oats, wetted—millet hay *ad lib.*; property of Matt. M. Gardner, Nashville, Tenn.

March Ida Pogis 101182.—Sire, Master Le Brocq 32047; dam, Persis Pogis 83919.

Butter, 47 lbs. 4½ oz. in 21 days. Milk, 479 lbs. 8 oz.

Test made from Feb. 18 to Mar. 10, 1898; age, 4 yrs. 11 mos.; estimated weight, 900 lbs.; fed 215 lbs. bran and 20 lbs. oil meal—2 bush. cut corn fodder and hay, daily; property of Peter Raab, Brightwood, Ind.

Marchioness of Maplehurst 8018.—Sire, Duke of Maplehurst 2290; dam, Clari 4266.

Butter, 14 lbs. 6 oz. Milk, 156 lbs. 8 oz.

Test made from Feb. 7 to 14, 1890; age, 11 yrs. 3 mos.; estimated weight, 900 lbs.; fed 12 lbs. wheat middlings daily; property of Mrs. Thomas Allen, Pittsfield, Mass.

Marcia G. 53211.—Sire, Osiris of Fernwood 16078; dam, Bounty Duchess 37955.

Butter, 14 lbs. 11 oz. Milk, 228 lbs. 13 oz.

Test made from June 10 to 17, 1892; age, 5 yrs. 3 mos.; estimated weight, 700 lbs.; fed 56 lbs. mill feed, 48 lbs. corn meal and 14 lbs. oil-cake meal—blue-grass and clover pasture; property of D. W. Voyles, Crandall, Ind.

Marea 10167.—Sire, Le Brocq's Prize 3350; dam, Mamie Coburn 3798.

Butter, 17 lbs. 10 oz. Milk, 203 lbs. 8 oz.

Test made from May 25 to 31, 1885; age, 5 yrs. 2 mos.; property of D. H. & S. S. Tripp, Peoria, Ill.

Margaret of Lagonda 35127.—Sire, Crystal 9324; dam, Queen of the Miamis 6793.

Butter, 14 lbs. 4 oz. Milk, 209 lbs. 1 oz.

Test made from June 11 to 18, 1892; age, 6 yrs. 9 mos.; estimated weight, 850 lbs.; fed 3 gals. corn, oats and barley chop and 1 qt. oil meal, daily—good pasture, but wet; property of Belmont Jersey Cattle Co., Columbus, Ohio.

Marian Holton 101478.—Sire, Sophomore 24253; dam, Paraphrie 73008.

Butter, 14 lbs. 12 oz. Milk, 222 lbs.

Test made from Mar. 31 to Apr. 7, 1898; age, 3 yrs. 11 mos.; estimated weight, 750 lbs.; fed 35 lbs. corn meal, 28 lbs. oats, 14 lbs. oil meal and 21 lbs. bran—corn ensilage; property of H. C. Taylor, Orfordville, Wis.

Maria of St. Lambert 64908.—Sire, Ismena's Prince 20041; dam, Maria of Allentown 2d 33443.

Butter, 20 lbs. 3½ oz. Milk, 216 lbs. 8 oz.

Test made from Nov. 27 to Dec. 4, 1894; age, 5 yrs. 4 mos.; actual weight, 850 lbs.; fed 17 lbs. cotton-seed, 8 lbs. old process linseed meal, 20 lbs. corn chop, 60 lbs. wheat bran, 98 lbs. corn-stalks ensilage and 42 lbs. hay; property of B. F. Fackenthal, Jr., Riegelsville, Pa.

Marie Bash 81441.—Sire, Beauty's Stoke Pogis 14175; dam, Southern Belle 18570.

Butter, 16 lbs. Milk, 266 lbs.

Test made from Feb. 20 to 27, 1898; age, 6 yrs.; actual weight, 751 lbs.; fed 6 lbs. bran, 6 lbs. corn and cob meal, 3 lbs. ground oats, 3 lbs. cotton-seed meal and about 112 lbs. peavine and prairie hay—oats pasture; property of J. R. Miller, Minden, La.

Marie C. Magnet 22903.—Sire, Champion Magnet 6480; dam, May Champion 14104.

Butter, 15 lbs. 8 oz. Milk, 205 lbs. 8 oz.

Test made from June 3 to 9, 1885; age, 2 yrs.; property of John Boyd, Elmhurst, Ill.

Marie S. 12043.—Sire, Farmer's Glory 5196; dam, Blossom 12013.

Butter, 15 lbs. 6 oz. Milk, 192 lbs. 15 oz.

Test made from May 15 to 21, 1884; age, 3 yrs. 2 mos.; property of William Simpson, New York, N. Y.

Marie's Maud 78467.—Sire, Nihilist of St. Lambert 18256; dam, Marie Rex 21423.

Butter, 18 lbs. 2 oz. Milk, 292 lbs.

Test made from Mar. 5 to 12, 1895; age, 5 yrs. 5 mos.; estimated weight, 850 lbs.; fed 42 lbs. wheat bran, 28 lbs. ground oats, 28 lbs. ground corn and 7 lbs. linseed oil-cake—hay *ad lib.*, 2 bush. roots; property of Albert Stracke, Warsaw, Ill.

Marilla of St. Lambert 31809.—Sire, Exile of St. Lambert 13657; dam, Duchess Lurlene 9048.

Butter, 21 lbs. 13½ oz. Milk, 258.9 lbs.

Test made from Nov. 13 to 20, 1892; age, 7 yrs. 7 mos.; actual weight, 978 lbs.; fed 42 qts. corn meal, 42 qts. bran, 42 qts. oats and peas, 42 qts. oil meal, 84 lbs. hay and 84 lbs. roots; property of H. L. S. Hall, Scottsville, N. Y.

Marion Earle 63908.—Sire, Matilda's Perfection 17100; dam, Marmaduke's Bella 42312.

Butter, 14 lbs. 8½ oz. Milk, 241 lbs. 12 oz.

Test made from June 11 to 18, 1894; age, 5 yrs. 2 mos.; estimated weight, 850 lbs.; fed 48 lbs. corn meal and 26 lbs. wheat bran—hill pasture, clover and timothy; property of Henry S. Redfield, Elmira, N. Y.

Marion Hill 40777.—Sire, Black Diamond V. 4970; dam, Silver Cloudlet 22710.

Butter, 15 lbs. 1 oz. Milk, 201 lbs. 8 oz.

Test made from June 13 to 20, 1891; age, 5 yrs. 11 mos.; estimated weight, 950 lbs.; fed 5 lbs. bran and 5 lbs. cotton-seed meal, daily—timothy, clover and blue-grass pasture; property of Bordwell & Cochran, Batavia, Ohio.

Marion of Eau Claire 120457.—Sire, Mella Ann's Son 22041; dam, Molly of Eau Claire 77583.

Butter, 21 lbs. 14 oz. Milk, 322 lbs. 8 oz.

Test made from Dec. 20 to 27, 1897; age, 3 yrs. 4 mos.; estimated weight, 900 lbs.; fed 3 qts. ground oats, 3 qts. corn meal, 4 qts. wheat middlings, 2 qts. oil meal and 2 qts. cotton-seed meal, daily—clover hay *ad lib.*; property of E. L. Van Deusen, Ashley Falls, Mass.

Marissa 65285.—Sire, Diploma 16219; dam, Damara 65001.

Butter, 16 lbs. 2 oz. Milk, 267 lbs. 1 oz.

Test made from June 1 to 8, 1893; age, 4 yrs. 1 mo.; estimated weight, 800 lbs.; fed about 15 lbs. daily of a mixture of corn meal, ground oats and oil meal—very little hay, clover pasture; property of Richardson Bros., Davenport, Iowa.

Maritana 12039.—Sire, Farmer's Glory 5196; dam, May Flower, on I. of J.

Butter, 16 lbs. 3½ oz. Milk, 254 lbs. 12 oz.

Test made from Nov. 10 to 16, 1884; age, 3 yrs. 5 mos.; property of William Simpson, New York, N. Y.

Marjoram 3239, imp.

Butter, 16 lbs.

Property of T. S. Cooper, Coopersburg, Pa.

Marjoram 2d 12805.—Sire, Stoke Pogis 1259; dam, Marjoram 3239.

Butter, 15 lbs.

Test made in June, 1881; age, 4 yrs. 3 mos.; property of Cooper, Maddux & Co., Reading, Ohio.

Marjoram of Glen Rouge 78420.—Sire, Canada's John Bull 5th 20092; dam, Leclair's Marjoram 36355.

Butter, 17 lbs. 3 oz. Milk, 245 lbs. 8 oz.

Test made from May 17 to 24, 1892; age, 2 yrs.; estimated weight, 700 lbs.; fed 32 lbs. chopped oats and peas and 30 lbs. ensilage, daily—clover hay, some grass; property of William Rolph, Markham, Ont., Can.

Marjoram of Glen Rouge 78420.—Sire, Canada's John Bull 5th 20092; dam, Leclair's Marjoram 36355.

Butter, 22 lbs. 12¼ oz. Milk, 255 lbs. 4 oz.

Test made from Apr. 11 to 18, 1897; age, 7 yrs.; actual weight, 1100 lbs.; fed 119 qts. ground oats, 77 qts. corn meal, 28 qts. ground oil-cake and 20 lbs. sugar beets—clover hay; property of Miller & Sibley, Franklin, Pa.

Marjoram of Glen Rouge 2d 123634.—Sire, One Hundred per Cent. 16590; dam, Marjoram of Glen Rouge 78420.

Butter, 17 lbs. 3 oz. Milk, 260 lbs.

Test made from Mar. 23 to 30, 1897; age, 2 yrs.; estimated weight, 750 lbs.; fed 18 qts. oats and peas, mixed and ground, and 2 lbs. oil-cake, per day—hay *ad lib.*, corn ensilage; property of William Rolph, Markham, Ont., Can.

Marjoram of Linden 43600.—Sire, Lorne 5248; dam, Marjoram 2d 12805.

Butter, 22 lbs. 12 oz. Milk, 285 lbs.

Test made from Dec. 10 to 17, 1893; age, 7 yrs. 9 mos.; estimated weight, 1000 lbs.; fed about 26 lbs. per day of a mixture of ground oats, middlings, wheat bran, corn meal and a little pea meal—clover hay and ensilage; property of T. S. Cooper, Coopersburg, Pa.

Marjoram's Delle 82954.—Sire, Rioter's Pride 11694; dam, Golden Delle 37698.

Butter, 19 lbs. 12 oz., unsalted. Milk, 280 lbs.

Test made from Jan 25 to Feb. 1, 1895; age, 2 yrs. 9 mos.; estimated weight, 1300 lbs.; . fed 84 lbs. daily of cut wet forage, 24 lbs. wheat middlings and ground oats and 4 lbs. wheat bran in a mash—hay *ad lib.*; property of Mrs. H. E. Tremain, Lake George, N. Y.

Marjoram's Matilda 77944.—Sire, Prospect's Rioter 9189; dam, Marjoram 2d 12805.

Butter, 19 lbs. 6 oz.   Milk, 256 lbs.

Test made from Jan. 15 to 22, 1895; age, 5 yrs. 10 mos.; estimated weight, 800 lbs.; fed 18 qts. to 20 qts. daily of a mixture of ground oats, wheat middlings, bran, corn meal and a little pea meal—clover hay and ensilage; property of T. S. Cooper, Coopersburg, Pa.

Marna 76108.—Sire, Sophie's Tormentor 20883; dam, Martona 58630.

Butter, 16 lbs. 12 oz.   Milk, 258 lbs.

Test made from Aug. 3 to 10, 1897; age, 6 yrs. 2 mos.; estimated weight, 700 lbs.; fed 28 lbs. bran, 28 lbs. corn meal, 21 lbs. ground oats and 7 lbs. oil meal—mixed clover and timothy pasture; property of C. I. Hood, Lowell, Mass.

Marno B. 56368.—Sire, Romp's Tormentor 7126; dam, Miss Flite of Bingo 31000.

Butter, 16 lbs. 10 oz.   Milk, 205 lbs. 8 oz.

Test made from June 24 to July 1, 1892; age, 5 yrs. 2 mos.; estimated weight, 800 lbs.; no grain fed—good pasture only; property of Mrs. J. R. Sherman Walsh, Pawlet, Vt.

Marpetra 10284.—Sire, Marpetro 3352; dam, Bettie D. 7677.

Butter, 14 lbs. 6 oz.   Milk, 261 lbs.

Test made from Jan. 24 to 30, 1883; age, 2 yrs. 10 mos.; property of C. E. Douglas, Crockett, Tex.

Marquette Tormentor 131897.—Sire, Oonan's Tormentor 22280; dam, Marchande 2d 93823.

Butter, 16 lbs. 14½ oz.   Milk, 200 lbs. 7 oz.

Test made from Jan. 29 to Feb. 5, 1898; age, 2 yrs. 4 mos.; estimated weight, 700 lbs.; fed 19 lbs. per day of ground oats and corn, in equal parts, mixed with 4 bucketfuls cut sheaf oats, wetted—millet hay at night *ad lib.*; property of Matt. M. Gardner, Nashville, Tenn.

Martha D. 66641.—Sire, Pride of New York 19787; dam, Lady Baush 11810.

Butter, 17 lbs. 2 oz.   Milk, 211 lbs. 14 oz.

Test made from Nov. 20 to 27, 1892; age, 4 yrs. 1 mo.; estimated weight, 900 lbs.; fed 20 qts. daily of equal parts corn meal, ground oats and wheat bran and a pint of oil meal—hay *ad lib.*; property of Walter W. Law, Whitson, N. Y.

Martha Hall 2d 57901.—Sire, Champion's Loyal 12673; dam, Martha Hall 40786.

Butter, 14 lbs. 10 oz.   Milk, 216 lbs.

Test made from June 13 to 20, 1891; age, 3 yrs. 3 mos.; estimated weight, 750 lbs.; fed 5 lbs. bran and 5 lbs. cotton-seed meal, daily—short pasture, timothy, clover and blue-grass; property of Bordwell & Cochran, Batavia, Ohio.

Martha Lafayette 17158.—Sire, Lord Harry 3445; dam, Mary Garnet 10371.

Butter, 17 lbs. 6 oz.   Milk, 220 lbs. 10 oz.

Test made from June 17 to 24, 1888; age, 7 yrs. 8 mos.; estimated weight, 850 lbs.; fed, daily, 8 lbs. bran, 5 lbs. corn meal, 2 lbs. cotton-seed meal and 2 lbs. pea meal; property of T. S. Webb, Knoxville, Tenn.

Martha of Eau Claire 120022.—Sire, Melia Ann's Son 22041; dam, Jennie Cream 2d 47466.

Butter, 14 lbs. 13 oz.   Milk, 220 lbs. 14 oz.

Test made from Mar. 10 to 17, 1897; age, 1 yr. 11 mos.; estimated weight, 850 lbs.; fed 42 qts. ground corn and oats, 42 qts. wheat bran, 14 qts. cotton-seed and oil meal and 140 lbs. carrots and beets—hay *ad lib.*; property of E. L. Van Deusen, Ashley Falls, Mass.

Marvel 13734.—Sire, Mercury 432; dam, Faustine 10354.

Butter, 15 lbs. 1 oz.   Milk, 182 lbs. 9 oz.

Test made from June 30 to July 6, 1884; age, 2 yrs. 8 mos.; property of William Simpson, New York, N. Y.

Mary Anne of St. Lambert 9770.—Sire, Stoke Pogis 3d 2238; dam, Lolly of St. Lambert 5480.

Butter, 27 lbs. 9¼ oz.   Milk, 251 lbs.

Test made from Sept. 23 to 29, 1883; age, 4 yrs. 6 mos.; fed first two days 3 qts. ground oats, 2 qts. pea meal and 1 qt. oil-cake, thrice daily, last five days the same five times daily; property of Valancey E. Fuller, Hamilton, Ont., Can.

(This test was conducted by appointees of the Canadian Jersey Breeders' Association, and its correctness certified under oath by them. It was the seventeenth week of her continued test.)

Mary Anne of St. Lambert 9770.—Sire, Stoke Pogis 3d 2238; dam, Lolly of St. Lambert 5480.

Butter, 36 lbs. 12¼ oz. (official).   Milk, 245 lbs.

Test made from Sept. 23 to 30, 1884; age, 5 yrs. 6 mos.; property of V. E. Fuller, Hamilton, Ont., Can.

Mary Anne of St. Lambert 9770.—Sire, Stoke Pogis 3d 2238; dam, Lolly of St. Lambert 5480.

Butter, 867 lbs. 14¾ oz. in one year.

Test made 1883-4; age, 4 yrs.; property of V. E. Fuller, Hamilton, Ont., Can.

Mary Brown 81213.—Sire, Beauty's Stoke Pogis 14175; dam, Idolette Melrose 61923.

Butter, 15 lbs. 11½ oz.   Milk, 308 lbs. 14 oz.

Test made from May 15 to 22, 1897; age, 7 yrs. 1 mo.; estimated weight, 875 lbs.; fed 14½ lbs. corn meal, 2 lbs. oil meal, 3 lbs. molasses feed and 6 lbs. ground oats, daily—rye and Bermuda pasture; property of William E. Oates & Son, Vicksburg, Miss.

Mary Clover 9998.—Sire, Beeswax 1931; dam, Betty Clover 3784.

Butter, 14 lbs. 15 oz.   Milk, 212 lbs. 8 oz.

Test made from May 21 to 28, 1883; age, 6 yrs. 4 mos.; property of C. J. Wemple, New Rochelle, N. Y.

Mary Fallon 80220.—Sire, Rival Rioter 11445; dam, Anna Neal 2d 7304.

Butter, 17 lbs. 1¼ oz.   Milk, 218 lbs. 8 oz.

Test made from Dec. 30, 1894, to Jan. 6, 1895; age, 4 yrs. 5 mos.; estimated weight, 800 lbs.; fed 42 qts. corn meal, 35 qts. ground oats, 35 qts. wheat bran and 14 qts. oil meal—hay; property of Bishop Clay, Lexington, Ky.

Mary Gordon 41429.—Sire, Lord Harry 3445; dam, Kate Gordon 8387.

Butter, 14 lbs. 6½ oz.   Milk, 121 lbs. 2 oz.

Test made from Mar. 12 to 19, 1889; age, 2 yrs. 9 mos.; estimated weight, 700 lbs.; fed, daily, 4 gals. corn meal, oats and wheat bran, in equal parts; property of Maury Jersey Farm, Columbia, Tenn.

Mary Gordon 41429.—Sire, Lord Harry 3445; dam, Kate Gordon 8387.

Butter, 18 lbs. 5¼ oz.   Milk, 237 lbs. 7 oz.

Test made from Jan. 1 to 8, 1895; age, 8 yrs. 7 mos.; actual weight, 751 lbs.; fed 45 lbs. wheat bran, 40 lbs. corn meal and 35 lbs. oil-cake meal—clover hay *ad lib.*; property of J. P. Bradbury, Pomeroy, Ohio.

Mary Hinman 17619.—Sire, Stoke Pogis 5th 5987; dam, Belle Hinman 3641.

Butter, 15 lbs. 11½ oz.   Milk, 225 lbs.

Test made from May 5 to 11, 1885; age, 2 yrs. 11 mos.; property of Miller & Sibley, Franklin, Pa.

Mary Idagold 88186.—Sire, Stoke Pogis 5th 5987; dam. Ida Marigold 32615.

Butter, 23 lbs. 9 oz.   Milk, 343 lbs. 5½ oz.

Test made from May 23 to 30, 1896; age, 3 yrs. 11 mos.; actual weight, 900 lbs.; fed 11 qts. ground oats, 7 qts. corn meal. 3 qts. oil meal and 6 qts. wheat bran, mixed, daily, in three rations—hay, fair pasture three hours daily; property of C. A. Sweet, East Aurora, N. Y.

Mary Jane of Belle Vue 6956.—Sire, Remarkable (F. S. 229 J. H. B.); dam, Nelly (F. S. 1509 J. H. B.)

Butter, 17 lbs. 7 oz.

Test made May, 1880; age, 4 yrs.; property of V. L. Kirkman, Nashville, Tenn.

Mary Jane of Belle Vue 6956.—Sire, Remarkable (F. S. 229 J. H. B.); dam, Nelly (F. S. 1509 J. H. B.)

Butter, 15 lbs. 2 oz. (official).   Milk, 238 lbs. 10 oz.

Test made from May 21 to 27, 1885; age, 9 yrs.; fed, daily, 11 lbs. ground oats, 5 lbs. bean meal, 2 lbs. pea meal, 2 lbs. cotton-seed meal and 1 lb. corn-hearts; property of Campbell Brown, Spring Hill, Tenn.

Mary Justice 37449.—Sire, Justice 9949; dam, Rose of Menard 13272.

Butter, 14 lbs. 9½ oz.   Milk, 260 lbs. 12 oz.

Test made from Mar. 3 to 10, 1888; age, 3 yrs. 10 mos.; weight, 925 lbs.; fed, daily, 3 lbs. oil meal and 15 lbs. middlings; property of J. B. Allen & Son, Delavan, Ill.

Mary M. Allison 6308.—Sire, Earl of Willow Glen 2043; dam, Traviata 3253.

Butter, 20 lbs. 14 oz., unsalted.   Milk, 254½ lbs.

Test made from Sept. 18 to 24, 1882; age, 5 yrs.; property of C. W. H. Eicke, West Monterey, Pa.

Mary M. Pogis 32455.—Sire, Stoke Pogis 5th 5987; dam, Mary M. Allison 6308.

Butter, 16 lbs. 4½ oz.   Milk, 204 lbs.

Test made from June 28 to July 5, 1891; age, 6 yrs. 2 mos.; estimated weight, 1060 lbs.; fed 16 qts. grain daily (equal parts corn meal. ground oats and oil meal)—good old meadow pasture; property of Miller & Sibley, Franklin, Pa.

Mary of Bear Lake 6171.—Sire, Prince of Warren 1512; dam, Warren's Duchess 4622.

Butter, 14 lbs. 7 oz.

Test made Dec., 1880; age, 4 yrs. 8 mos.; property of Chester Bordwell, Batavia, Ohio.

Mary of Bear Lake 6171.—Sire, Prince of Warren 1512; dam, Warren's Duchess 4622.

Butter, 15 lbs. 14 oz. Milk, 311 lbs. 1 oz.

Test made from June 25 to July 1, 1885; age, 9 yrs. 3 mos.; property of Bordwell & Cochran, Batavia, Ohio.

Mary of Gilderoy 11219.—Sire, Gilderoy 2107; dam, Contessa 3256.

Butter, 14 lbs. 4 oz., unsalted. Milk, 217 lbs. 8 oz.

Test made from June 29 to July 5, 1884; age, 4 yrs. 3 mos.; property of H. M. Howe, Bristol, R. I.

Mary of Glenoir 59940.—Sire, Leonette's Landseer 20399; dam, Lady of Four Pines 31737.

Butter, 18 lbs. 5 oz. Milk, 330 lbs. 8 oz.

Test made from Mar. 1 to 8, 1896; age, 7 yrs. 9 mos.; estimated weight, 850 lbs.; fed 6 lbs. corn chop, 6 lbs. wheat bran and 4 lbs. crushed oats, daily—hay *ad lib.*, poor pasture; property of J. E. McGuire, Gatesville, Tex.

Mary of Glenoir 59940.—Sire, Leonette's Landseer 20399; dam, Lady of Four Pines 31737.

Butter, 27 lbs. 3½ oz. Milk, 326 lbs. 5½ oz.

Test made from Jan. 26 to Feb. 2, 1898; age, 8 yrs. 9 mos.; estimated weight, 1000 lbs.; fed 1½ gals. wheat bran, 1 qt. corn chop, ½ gal. ground oats and 1 pt. cotton-seed meal, daily—hay *ad lib.*; property of J. M. Logan, Fort Worth, Tex.

Mary of Lodi 88517.—Sire, Sir Hugo of St. Lambert 13726; dam, Lady of Lodi 21784.

Butter, 15 lbs. 1 oz. Milk, 224 lbs. 8 oz.

Test made from Feb. 24 to Mar. 3, 1898; age, 5 yrs. 10 mos.; estimated weight, 800 lbs.; fed 5 lbs. ground oats, 5 lbs. wheat bran and 2 lbs. old process oil meal, daily, and about 20 lbs. ensilage—clover hay or corn stover *ad lib.*; property of N. N. Palmer & Son, Brodhead, Wis.

Mary of Pleasant View 13448.—Sire, William Devries 6064; dam, Lonette of Magnolia 9425.

Butter, 14 lbs. 6 oz. Milk, 246 lbs. 8 oz.

Test made from Apr. 18 to 24, 1885; age, 4 yrs. 2 mos.; property of Miller & Sibley, Franklin, Pa.

Mary Palestine 52439.—Sire, Jupiter Hugo Pogis 14342; dam, Lady Palestine 2d 28744.

Butter, 16 lbs. 7 oz. Milk, 220 lbs. 8 oz.

Test made from Oct. 17 to 24, 1894; age, 8 yrs.; estimated weight, 1050 lbs.; fed 21 lbs. coarse winter wheat bran, 28 lbs. corn meal, 21 lbs. ground oats and 14 lbs. cotton-seed meal—cut clover hay and corn stover, poor pasture; property of John C. McClintock, Meadville, Pa.

Mary's Nightingale 87263.—Sire, Nicklin 11161; dam, Fannie Jones 61654.

Butter, 14 lbs. 14½ oz. Milk, 201 lbs.

Test made from Sept. 10 to 17, 1897; age, 7 yrs. 1 mo.; estimated weight, 1000 lbs.; fed 6 qts. wheat bran, 4½ qts. wheat middlings and 3 pts. cotton-seed meal, daily—hay *ad lib.*, second crop clover pasture; property of Everett G. Campbell, Keister's, Pa.

Mary's Pet 40772.—Sire, Steletho 10505; dam, Mary's Silver Drop 14235.

Butter, 15 lbs. 7½ oz. Milk, 209 lbs. 8 oz.

Test made from June 13 to 20, 1891; age, 6 yrs. 2 mos.; estimated weight, 900 lbs.; fed 5 lbs. bran and 5 lbs. cotton-seed meal, daily—short pasture, timothy, clover and blue-grass; property of Bordwell & Cochran, Batavia, Ohio.

Mary's Silver Drop 14235.—Sire, Dudley 3999; dam, Mary of Bear Lake 6171.

Butter, 15 lbs. 4½ oz. Milk, 247 lbs. 10 oz.

Test made from May 28 to June 3, 1884; age, 3 yrs. 6 mos.; property of C. Bordwell, Batavia, Ohio.

Masher 64950.—Sire, Damascus 22222; dam, Merry Maiden 64949.

Butter, 16 lbs. 14½ oz. Milk, 228 lbs. 11 oz.

Test made from Oct. 28 to Nov. 4, 1897; age, 7 yrs. 5 mos.; estimated weight, 925 lbs.; fed 35 lbs. bran, 28 lbs. corn meal, 28 lbs. ground oats, 10½ lbs. oil meal and 70 lbs. roots—hay *ad lib.*, short clover and timothy pasture; property of C. I. Hood, Lowell, Mass.

Massena 25732.—Sire, Kago 1353; dam, Highland Mary 3d 19876.

Butter, 20 lbs. 7 oz. Milk, 210 lbs. 11 oz.

Test made from June 11 to 19, 1884; age, 8 yrs. 3 mos.; fed 24 lbs. middlings and bran in equal proportions by measure, daily—pasture of mixed grasses; property of P. P. Paddock, Malone, N. Y.

Massena 25732.—Sire, Kago 1353; dam, Highland Mary 3d 19876.

Butter, 145 lbs. 8½ oz. in 61 days. Milk, 1822 lbs. 8 oz.

Test made from Mar. 12 to May 11, 1891; age, 15 yrs.; fed, daily, 15 lbs. of a mixture consisting of equal parts of bran, ground oats, peas and corn, with a little oil meal, for the first eleven days, increased to 18 lbs. during the next nine days, again increased to 21 lbs. on the twenty-first day, and so continued to the forty-fourth day, when it was reduced to 15 lbs. for the remainder of the test; fed also good hay and one feed of roots per day until the forty-fifth day; property of Mrs. E. M. Jones, Brockville, Ont., Can.

Massena 25732.—Sire, Kago 1353; dam, Highland Mary 3d 19876.

Butter, 162 lbs. 3 oz. in 101 days. Milk, 1499 lbs. 10 oz.

Test made in 1885; age, 9 yrs.; property of P. P. Paddock, Malone, N. Y.

Massena 2d 31188.—Sire, Highland Kago 14349; dam, Massena 25732.

Butter, 17 lbs. 11 oz. Milk, 244 lbs.

Test made from Mar. 29 to Apr. 5, 1889; age, 9 yrs.; actual weight, 850 lbs.; fed, daily, about 14½ lbs. corn meal, 8 lbs. bran and 1¼ lbs. boiled flaxseed; property of P. P. Paddock, Malone, N. Y.

Massena 4th 47342.—Sire, Hugo Chief of St. Anne's 12070; dam, Massena 25732.

Butter, 14 lbs. Milk, 194 lbs. 8 oz.

Test made from May 8 to 15, 1896; age, 8 yrs. 7 mos.; estimated weight, 900 lbs.; fed 14 lbs. malt sprouts, 28 lbs. corn meal, 14 lbs. wheat bran, 14 lbs. oil meal and 140 lbs. ensilage—short pasture; property of Adelbert C. Davis, Deer Park, N. Y.

Massey Polo 67010.—Sire, Holyoke 8147; dam, Lily Polo 22218.

Butter, 22 lbs. 6 oz. Milk, 296 lbs.

Test made from Apr. 27 to May 4, 1893; age, 6 yrs. 1 mo.; actual weight, 860 lbs.; fed 6 qts. ground corn and oats, 3 qts. cotton-seed meal, 4 qts. Buffalo gluten feed, 5 qts. wheat bran and 45 lbs. corn ensilage, daily—hay *ad lib.*; property of E. L. Van Deusen, Ashley Falls, Mass.

Massey Polo 2d 89216.—Sire, Ramapose's Bachelor 25900; dam, Massey Polo 67010.

Butter, 16 lbs. 9 oz. Milk, 211 lbs. 8 oz.

Test made from May 22 to 29, 1894; age, 2 yrs. 9 mos.; actual weight, 705 lbs.; fed 2 qts. ground oats, 1 qt. cotton-seed meal, 2 qts. corn meal and 1 qt. oil meal, daily—upland pasture; property of E. L. Van Deusen, Ashley Falls, Mass.

Massey Polo 3d 99378.—Sire, Kindergarten's Duke 30065; dam, Massey Polo 67010.

Butter, 14 lbs. 13½ oz. Milk, 209 lbs.

Test made from Aug. 22 to 29, 1897; age, 3 yrs. 3 mos.; actual weight, 840 lbs.; fed 9 qts. ground oats, 12 qts. wheat bran, 3 qts. corn meal and 3 qts. oil meal, daily—hay, good clover pasture; property of C. A. Sweet, East Aurora, N. Y.

Matilda 3238, imp.

Butter, 17 lbs.

Property of T. S. Cooper, Coopersburg, Pa.

Matilda 4th 12816.—Sire, Stoke Pogis 1259; dam, Matilda 3238.

Butter, 21 lbs. 8½ oz. (official). Milk, 336 lbs.

Test made from July 7 to 13, 1885; age, 5 yrs. 3 mos.; fed about 31 lbs. sifted oats, pea meal, oil-cake and middlings, mixed, daily—poor pasture; property of Miller & Sibley, Franklin, Pa.

Matilda 4th 12816.—Sire, Stoke Pogis 1259; dam, Matilda 3238.

Butter, 279 lbs. 7½ oz. in 125 days.

Test made from Dec. 2, 1885, to Apr. 6, 1886; age, 5 yrs. at beginning of test; weight, 1000 lbs.; property of Miller & Sibley, Franklin, Pa.

(Above test was merely incidental to a test made for milk, in which Matilda 4th gave a total of 16,153¾ lbs. for the twelve months ending April 6, 1886, the cow being fed with special reference to flow of milk. Included in butter record given above is the month of January, her tenth month after calving, in which she produced 73 lbs. 1½ oz.)

Matilda 5th 18068.—Sire, Duke of Darlington 2460; dam, Matilda 3238.

Butter, 16 lbs. 4 oz. Milk, 146 lbs.

Test made from June 15 to 22, 1885; age, 2 yrs. 11 mos.; property of J. H. Walker, Worcester, Mass.

Matilda 6th 39290.—Sire, Lord Darlington 7285; dam, Matilda 3238.

> Butter, 20 lbs. 1 oz.   Milk, 202 lbs. 10 oz.

> Test made from Mar. 15 to 22, 1891; age, 4 yrs. 10 mos.; estimated weight, 950 lbs.; fed 28 lbs. oil meal, 28 lbs. pea meal, 42 lbs. corn meal, 56 lbs. oat meal, 56 lbs. bran, 70 lbs. hay and 1½ bush. roots; property of Ayer & McKinney, Philadelphia, Pa.

Matilda Ann Darling 105398.—Sire, Matilda's Lisgar 21374; dam, Independence of Nipsic 88307.

> Butter, 14 lbs. 1 oz.   Milk, 213 lbs. 8 oz.

> Test made from Mar. 4 to 11, 1897; age, 3 yrs. 7 mos.; estimated weight, 800 lbs.; fed 35 lbs. ensilage, 13 lbs. mixture of bran, gluten, corn and oats and 1 lb. oil meal, daily—hay *ad lib.*; property of W. J. Hussey, Mount Pleasant, Ohio.

Matilda Hilda 75346.—Sire, Matilda 4th's Son 20214; dam, The Colonel's Daughter 50230.

> Butter, 14 lbs. 4 oz.   Milk, 173 lbs. 2 oz.

> Test made from Dec. 18 to 25, 1892; age, 2 yrs. 5 mos.; estimated weight, 700 lbs.; fed 40 lbs. pea meal, 26 lbs. oil meal, 26 lbs. cotton-seed meal, 40 lbs. mixed feed and 140 lbs. ensilage; property of Ayer & McKinney, Philadelphia, Pa.

Matilda Mysinda 84487.—Sire, Matilda 4th's Son 20214; dam, Mysinda 2d 42250.

> Butter, 15 lbs. 3 oz.   Milk, 258 lbs.

> Test made from May 16 to 23, 1896; age, 5 yrs. 1 mo.; estimated weight, 950 lbs.; fed 42 lbs. corn and oats (half and half), 28 lbs. wheat bran and 14 lbs. cotton-seed meal—hay *ad lib.*; property of Ayer & McKinney, Philadelphia, Pa.

Matilda of Maplewood 49723.—Sire, Stoke Pogis Jr. 15300; dam, Bertha Maplewood 22944.

> Butter, 14 lbs. 15 oz.   Milk, 208 lbs. 12 oz.

> Test made from May 27 to June 3, 1891; age, 5 yrs. 2 mos.; estimated weight, 800 lbs.; fed 70 lbs. corn meal and 24½ lbs. oil meal—old pasture; property of R. A. Sibley, Spencer, Mass.

Matilda of Meridale 57176.—Sire, Ida's Rioter of St. L. 13656; dam, Matilda 4th 12816.

> Butter, 14 lbs. 14 oz.   Milk, 164 lbs. 13 oz.

> Test made from Nov. 21 to 28, 1892; age, 3 yrs. 7 mos.; estimated weight, 800 lbs.; fed 42 lbs. pea meal, 28 lbs. oil meal, 14 lbs. cotton-seed meal, 42 lbs. mixed feed and 140 lbs. ensilage; property of Ayer & McKinney, Philadelphia, Pa.

Matilda of Meridale 57176.—Sire, Ida's Rioter of St. L. 13656; dam, Matilda 4th 12816.

> Butter, 18 lbs. 3 oz.   Milk, 211 lbs.

> Test made from Feb. 23 to Mar. 2, 1895; age, 5 yrs. 10 mos.; actual weight, 950 lbs.; fed 42 qts. ground oats, 42 qts. corn meal, 24 qts. ground oil-cake, 28 qts. bran—15 lbs. cabbage per day; property of Miller & Sibley, Franklin, Pa.

Matilda of Meridale 57176.—Sire, Ida's Rioter of St. L. 13656; dam, Matilda 4th 12816.

> Butter, 23 lbs. 5¾ oz.   Milk, 307 lbs. 12 oz.

> Test made from June 21 to 28, 1897; age, 8 yrs. 2 mos.; actual weight, 940 lbs.; fed 84 qts. ground oats, 42 qts. corn meal and 21 qts. ground oil-cake—good old meadow pasture; property of Miller & Sibley, Franklin, Pa.

Matilda of the Ledges 97740.—Sire, Matilda 4th's Son 20214; dam, Flower of Meridale 64537.

Butter, 16 lbs. 9½ oz. Milk, 306 lbs. 5 oz.

Test made from May 29 to June 5, 1897; age, 4 yrs. 5 mos.; actual weight, 895 lbs.; fed 14 lbs. bran, 17½ lbs. ground oats, 28 lbs. corn meal and 8¾ lbs. oil meal—rather poor hay at night, good blue-grass and clover pasture; property of Mrs. M. E. McCalla, Columbus, Ohio.

Matilda's Matilda 75348.—Sire, Matilda 4th's Son 20214; dam, Matilda 5th 18068.

Butter, 14 lbs. 6 oz. Milk, 182 lbs.

Test made from Dec. 15 to 22, 1892; age, 2 yrs. 4 mos.; estimated weight, 750 lbs.; fed 40 lbs. pea meal, 27 lbs. oil meal, 28 lbs. cotton-seed meal, 41 lbs. mixed feed and 135 lbs. ensilage; property of Ayer & McKinney, Philadelphia, Pa.

Matilda's Matilda 75348.—Sire, Matilda 4th's Son 20214; dam, Matilda 5th 18068.

Butter, 17 lbs. 2½ oz. Milk, 281 lbs. 7 oz.

Test made from June 16 to 23, 1895; age, 4 yrs. 9 mos.; estimated weight, 850 lbs.; fed 16 qts. ground oats and corn, 4 qts. bran, 2 qts. middlings, 4 qts. cotton-seed meal and 1 qt. linseed meal, mixed, daily—hay *ad lib.*, woodland blue-grass pasture; property of E. A. Berauer, Waldron, Ind.

Matilda's Pansy 81311.—Sire, Matilda's Victor Hugo 20244; dam, Pogis' Pansy of Side View 59812.

Butter, 14 lbs. 7 oz. Milk, 189 lbs. 12 oz.

Test made from Dec. 5 to 12, 1897; age, 6 yrs. 10 mos.; estimated weight, 800 lbs.; fed 9 qts. corn and oats, ground together, and 4 qts. gluten, daily—stalks and hay *ad lib.*; property of George E. Nichols, Afton, N. Y.

Matilda's Queen 72663.—Sire, Matilda's Victor Hugo 20244; dam, Queen of Clinton 11500.

Butter, 16 lbs. Milk, 247 lbs. 12 oz.

Test made from Oct. 18 to 25, 1895; age, 5 yrs. 7 mos.; estimated weight, 850 lbs.; fed 7 qts. wheat bran, 4 qts. Buffalo gluten feed, 1 qt. cotton-seed meal, 1 bush. cut corn fodder and 8 qts. potatoes, daily—poor pasture; property of George E. Nichols, Afton, N. Y.

Matilda's Queen of St. L. 93746.—Sire, Ida's Rioter of St. L. 13656; dam, Matilda 4th 12816.

Butter, 18 lbs. 13¼ oz. Milk, 219 lbs.

Test made from Feb. 9 to 16, 1898; age, 4 yrs. 6 mos.; estimated weight, 875 lbs.; fed 84 qts. ground oats, 56 qts. corn meal, 7 qts. ground oil-cake, 14 qts. malt and 7 pks. carrots; property of Miller & Sibley, Franklin, Pa.

Matilda's Rose 81397.—Sire, Matilda's Victor Hugo 20244; dam, Rose of Side View 47275.

Butter, 15 lbs. 2 oz. Milk, 202 lbs. 8 oz.

Test made from Nov. 30 to Dec. 7, 1897; age, 7 yrs. 7 mos.; estimated weight, 750 lbs.; fed 9 qts. corn and oats, ground together, and 4 qts. gluten, daily—corn-stalks and hay *ad lib.*; property of George E. Nichols, Afton, N. Y.

Matin 7768.—Sire, Horace (P. S. 94 J. H. B.); dam, Cigarette (F. S. 1962 J. H. B.)

Butter, 17 lbs. 11 oz. Milk, 265 lbs. 4 oz.

Test made from June 1 to 7, 1884; age, 9 yrs.; property of Frederick Billings, Woodstock, Vt.

Matina of Riverside 51783.—Sire, Holyoke 8147; dam, Usilda's Pride 13921.

Butter, 20 lbs. 12 oz.   Milk, 338 lbs. 8 oz.

Test made from Nov. 12 to 19, 1893; age, 6 yrs. 4 mos.; estimated weight, 1000 lbs.; fed 42 qts. ground corn and oats, 21 qts. oil meal and cotton-seed meal, 49 qts. wheat bran, 14 qts. wheat middlings, 14 qts. Buffalo gluten feed and 350 lbs. corn ensilage —hay *ad lib.*; property of E. L. Van Deusen, Ashley Falls, Mass.

Matina of Riverside 51783.—Sire, Holyoke 8147; dam, Usilda's Pride 13921.

Butter, 27 lbs. 13 oz.   Milk, 426 lbs. 7 oz.

Test made from Apr. 29 to May 6, 1897; age, 9 yrs. 9 mos.; actual weight, 1030 lbs.; fed 9 qts. bran, 12 qts. ground oats, 9 qts. corn meal and 3 qts. oil meal, daily—hay *ad lib.*, ensilage; property of C. A. Sweet, East Aurora, N. Y.

Matlie 3d 9879.—Sire, Charlie Kittredge 1247; dam, Matlie 3213.

Butter, 14 lbs. 13 oz.   Milk, 161 lbs. 4 oz.

Test made from Nov. 26 to Dec. 2, 1885; age, 8 yrs. 3 mos.; property of C. W. Paris, Mount Healthy, Ohio.

Mattie Tormentor 62841.—Sire, Little Tormentor 17995; dam, Mattie Cole 47911.

Butter. 15 lbs. 14½ oz.   Milk, 224 lbs.

Test made from Mar. 30 to Apr. 6, 1893; age. 4 yrs. 4 mos.; estimated weight, 800 lbs.; fed 56 lbs. bran, 49 lbs. corn and cob meal, 17½ lbs. ground oats, 17½ lbs. cotton-seed meal, with 1 lb. cotton-seed hulls at each feed; property of James L. Cooper, Nashville, Tenn.

Mattie Tormentor 62841.—Sire, Little Tormentor 17995; dam, Mattie Cole 47911.

Butter, 14 lbs. 15 oz.   Milk, 251 lbs.

Test made from Apr. 27 to May 4, 1894; age, 5 yrs. 5 mos.; estimated weight, 900 lbs.; fed 59½ lbs. coarse corn meal, 38½ lbs. ground oats and 14 lbs. cotton-seed meal—fair blue-grass pasture; property of James L. Cooper, Nashville, Tenn.

Mattie Tormentor 62841.—Sire, Little Tormentor 17995; dam, Mattie Cole 47911.

Butter, 15 lbs. 8 oz.   Milk, 239 lbs.

Test made from July 17 to 24, 1895; age, 6 yrs. 7 mos.; estimated weight, 900 lbs.; fed 56 qts. ground oats, 42 qts. ground corn and 14 qts. cotton-seed meal—fair blue-grass pasture; property of James L. Cooper, Nashville, Tenn.

Mattie Tormentor 62841.—Sire, Little Tormentor 17995; dam, Mattie Cole 47911.

Butter, 16 lbs. 5½ oz.   Milk, 237 lbs.

Test made from July 24 to 31, 1895; age, 6 yrs. 7 mos.; estimated weight, 900 lbs.; fed 56 qts. ground oats, 42 qts. ground corn and 14 qts. cotton-seed meal—fair blue-grass pasture; property of James L. Cooper, Nashville, Tenn.

Mattie Tormentor 62841.—Sire, Little Tormentor 17995; dam, Mattie Cole 47911.

Butter, 31 lbs. 13½ oz. in 14 days.   Milk, 476 lbs.

Test made from July 17 to 31, 1895; age, 6 yrs. 7 mos.; estimated weight, 900 lbs.; fed 112 qts. ground oats, 84 qts. ground corn and 28 qts. cotton-seed meal—fair blue-grass pasture; property of James L. Cooper, Nashville, Tenn.

Maud Drew 38522.—Sire, Cattaraugus Chief 6026; dam, Pomona 2560.

Butter, 17 lbs. 12 oz.   Milk, 242 lbs. 4 oz.

Test made from Dec. 4 to 11, 1801; age, 6 yrs. 8 mos.; estimated weight, 1000 lbs.; fed 28 qts. corn meal, 14 qts. old process oil meal, 14 qts. ground oats, 250 lbs. ensilage (estimated), 70 lbs. hay (estimated) and 3½ bush. roots; property of Giltedge Farm Co., Syracuse, N. Y.

Maude Brown 85756.—Sire, Ida's Stoke Pogis 2d 16864; dam, Eva Daughtry 21951.

Butter, 15 lbs. 9 oz.   Milk, 265 lbs. 5 oz.

Test made from May 3 to 10, 1806; age, 4 yrs. 8 mos.; estimated weight, 775 lbs.; fed 56 lbs. bran, 56 lbs. ground oats and 91 lbs. crushed corn—prairie pasture; property of J. D. Gray, Terrell, Tex.

Maudie of St. Lambert 73603.—Sire, Count Rioter Pogis 21388; dam, Pogis' Twilight 53794.

Butter, 15 lbs. 7½ oz.   Milk, 242 lbs.

Test made from Apr. 3 to 10, 1894; age, 3 yrs. 8 mos.; estimated weight, 900 lbs.; fed 56 qts. corn meal, 56 qts. ground oats and 28 qts. cotton-seed meal—oats pasture; property of W. W. Lipscomb, Luling, Tex.

Maudie of White Lick 29682.—Sire, Kinsman 3338; dam, Io B. 3d 12881.

Butter, 15 lbs. 7 oz.   Milk, 189 lbs.

Test made from Dec. 30, 1888, to Jan. 6, 1889; age, 6 yrs. 2 mos.; estimated weight, 875 lbs.; fed, daily, about 18 lbs. corn and cob meal, ground oats and bran; property of S. H. Godman, Muncie, Ind.

Maudine of Elmwood 8718.—Sire, Hindoo 2282; dam, Maudine 3d 5646.

Butter, 16 lbs. 15 oz.   Milk, 227 lbs. 14 oz.

Test made from Sept. 19 to 26, 1883; age, 4 yrs. 7 mos.; property of W. B. Dugger, Carlinville, Ill.

Maud Lee 2416.—Sire, Cœur de Lion 318; dam, Matilda 2405.

Butter, 23 lbs.

Test made from Oct. 6 to 13, 1879; age, 9 yrs. 1 mo.; fed 4½ lbs. corn meal, daily—good rowen pasture; property of F. W. Tanner, West Stockbridge, Mass.

Maud Lee 2d 8839.—Sire, Wormwood 1746; dam, Maud Lee 2416.

Butter, 14 lbs. 9 oz.   Milk, 179 lbs. 8 oz.

Test made from Aug. 5 to 12, 1884; age, 5 yrs. 11 mos.; property of G. L. Foskett & Co., Winsted, Conn.

Maud Melinda 12126.—Sire, Mirza 1300; dam, Maud of Millbrook 2070.

Butter, 17 lbs. 8 oz.

Test made from Jan. 13 to 19, 1884; age, 5 yrs. 9 mos.; property of C. E. Grosvenor, Canterbury, N. B., Can.

Maud Miami Jr. 69131.—Sire, Little Miami 22076; dam, Maud Miami 57918.

Butter, 14 lbs. 8 oz.   Milk, 228 lbs.

Test made from May 17 to 24, 1896; age, 6 yrs. 2 mos.; estimated weight, 1000 lbs.; fed 6 qts. corn and oats twice daily—June grass on hills; property of E. G. Silvus, Athens, Ohio.

Maud of Glen Rouge 103781.—Sire, Canada's John Bull 5th 20092; dam, Cheerful Pogis 61748.

Butter, 22 lbs. 10 oz. Milk, 334 lbs. 8 oz.

Test made from Mar. 16 to 23, 1897; age, 4 yrs. 11 mos.; estimated weight, 700 lbs.; fed 250 lbs. peas and oats chop and 15 lbs. oil-cake—corn ensilage and hay *ad lib.*; property of William Rolph, Markham, Ont., Can.

Maud of Ingleside 74044.—Sire, Duke of Ingleside 14274; dam, Molly May 17202.

Butter, 18 lbs. 5½ oz. Milk, 284 lbs. 8 oz.

Test made from Aug. 3 to 10, 1895; age, 5 yrs. 8 mos.; estimated weight, 900 lbs.; fed 42 qts. ground oats, 42 qts. corn meal, 42 qts. ground oil-cake and 42 qts. wheat bran—good old meadow pasture; property of Miller & Sibley, Franklin, Pa.

Maud of Maple View 39511.—Sire, Nora's Longfellow 7427; dam, Maud Alice 17001.

Butter, 23 lbs. 7½ oz. Milk, 209 lbs. 10 oz.

Test made from Mar. 5 to 12, 1892; age, 5 yrs. 11 mos.; estimated weight, 900 lbs.; fed 6 lbs. wheat bran, 2 lbs. cotton-seed meal, 4 lbs. corn meal, 3 lbs. old process oil meal, 2 oz. salt and 30 lbs. ensilage, daily—126 lbs. hay; property of Mrs. Kate M. Busick, Wabash, Ind.

Maud Pogis 24240.—Sire, West Wind 4289; dam, Crocus of St. Lambert 8351.

Butter, 14 lbs. 12¾ oz. Milk, 207 lbs.

Test made from Aug. 2 to 8, 1886; age, 3 yrs. 3 mos.; estimated weight, 775 lbs.; fed, daily, 4½ lbs. corn meal, 3 lbs. bran, 4 lbs. oat meal and 2½ lbs. oil meal; property of Frederick Loeser, Somerville, N. J.

Maud's Crown Princess 90320.—Sire, Crown Prince of St. L. 20070; dam, Maud of Bois d'Arc 59639.

Butter, 17 lbs. Milk, 243 lbs. 10 oz.

Test made from Jan. 26 to Feb. 2, 1896; age, 4 yrs.; estimated weight, 900 lbs.; fed 72 lbs. corn and cob meal, 30 lbs. bran, 12 lbs. cotton-seed meal and 280 lbs. ensilage; property of Austin & Probert, Florence, Mich.

Maud's Crown Princess 90320.—Sire, Crown Prince of St. L. 20070; dam, Maud of Bois d'Arc 59639.

Butter, 66 lbs. 13 oz. in 30 days. Milk, 999 lbs. 4 oz.

Test made from Jan. 26 to Feb. 25, 1896; age, 4 yrs.; estimated weight, 900 lbs.; fed 308 lbs. corn and cob meal, 130 lbs. bran, 50 lbs. cotton-seed meal and 1250 lbs. ensilage; property of Austin & Probert, Florence, Mich.

Maud Signal 2d 68664.—Sire, Rioter of St. Lambert 16501; dam, Maud Signal 42187.

Butter, 22 lbs. 4 oz. Milk, 255 lbs. 14 oz.

Test made from July 25 to Aug. 1, 1896; age, 7 yrs. 4 mos.; estimated weight, 850 lbs.; fed 4 qts. ground corn and oats, 2 qts. corn meal, 8 qts. wheat bran and 2 qts. old process linseed meal, daily—blue-grass and timothy pasture; property of J. H. Kilbourne, Okemos, Mich.

Maud's Sultana 19518.—Sire, Ramapo 4679; dam, Princess Maude 7177.

Butter, 16 lbs. 4 oz. Milk, 18½ qts. per day.

Test made from June 25 to July 1, 1887; age, 4 yrs. 7 mos.; estimated weight, 800 lbs.; fed 3 qts. corn meal, 7 qts. oat meal, 2 qts. pea meal, 1 qt. oil meal and 3 qts. middlings, daily; property of D. F. Appleton, Ipswich, Mass.

Maumee Girl 118155.—Sire, Prince of Tennessee 20772; dam, Maumee Beauty 50836.

Butter, 15 lbs. 10¾ oz. Milk, 223 lbs.

Test made from June 29 to July 6, 1896; age, 4 yrs. 9 mos.; estimated weight, 750 lbs.; fed 24 lbs. per day of ground oats and corn, in equal parts—very good orchard and blue-grass pasture; property of Percy Kinnaird, Nashville, Tenn.

Mavourneen of St. Lambert 9777.—Sire, Stoke Pogis 3d 2238; dam, Amelia 484.

Butter, 15 lbs. 12 oz. Milk, 298 lbs.

Test made from Aug. 8 to 14, 1884; age, 4 yrs. 8 mos.; property of V. E. Fuller, Hamilton, Ont., Can.

Maxminta 85922.—Sire, Pogis Maximum 20761; dam, Erminta V. 53553.

Butter, 15 lbs. 13½ oz. Milk, 210 lbs. 6 oz.

Test made from Dec. 17 to 24, 1896; age, 4 yrs.; estimated weight, 800 lbs.; fed 35 lbs. wheat bran, 30 lbs. corn meal and 10 lbs. cotton-seed meal—about 80 lbs. hay; property of Ayer & McKinney, Philadelphia, Pa.

Max Polly 108994.—Sire, Pogis Maximum 20761; dam, Poolas 2d 53552.

Butter, 15 lbs. 8½ oz. Milk, 206 lbs. 8 oz.

Test made from Jan. 7 to 14, 1897; age, 3 yrs. 2 mos.; estimated weight, 800 lbs.; fed 50 lbs. wheat bran, 50 lbs. corn meal, 20 lbs. cotton-seed meal, 20 lbs. linseed meal and about 80 lbs. hay; property of Ayer & McKinney, Philadelphia, Pa.

Maybee H. 60620.—Sire, Tamy's Rioter 15288; dam, Iola H. 24225.

Butter, 18 lbs. 4 oz. Milk, 216 lbs.

Test made from Feb. 15 to 22, 1897; age, 9 yrs.; estimated weight, 900 lbs.; fed 15 lbs. per day of a mixture of peas, oats, corn meal and oil meal, and 25 lbs. to 30 lbs. ensilage—hay *ad lib.*; property of George W. Sisson, Jr., Potsdam, N. Y.

May Belle's Lady d'Or 74749.—Sire, Lady d'Or's Son 21048; dam, Le Brocq's May Belle 27980.

Butter, 15 lbs. 15 oz. Milk, 177 lbs. 8 oz.

Test made from Feb. 25 to Mar. 4, 1894; age, 2 yrs. 3 mos.; estimated weight, 750 lbs.; fed 28 qts. crushed corn and 56 qts. bran—hay; property of Abraham McNutt, Lewisburg, Ohio.

May Blossom 5657.—Sire, Litchfield 674; dam, Bessie Allen 3719.

Butter, 18 lbs. 11 oz. Milk, 228 lbs. 3 oz.

Test made from May 23 to 29, 1883; age, 6 yrs. 2 mos.; fed corn and oats, ground together, with a little oil meal—very good pasture; property of William Simpson, New York, N. Y.

May Bud of Monmouth 128864.—Sire, Happy Marquis 26801; dam, Bessie Day 82257.

Butter, 15 lbs. 8 oz. Milk, 278 lbs.

Test made from Jan. 2 to 9, 1898; age, 5 yrs.; actual weight, 810 lbs.; fed 5 lbs. wheat bran, 5 lbs. oats and corn, ground, 3 lbs. cotton-seed meal, 2 lbs. linseed meal, 20 lbs. hay and 10 lbs. carrots, daily; property of George M. Haynes, Monmouth, Me.

May Cuckoo M. 97417.—Sire, Prince of Tennessee 20772; dam, May Katydid 63503.

> Butter, 14 lbs. 8 oz. Milk, 203 lbs. 12½ oz.

Test made from Nov. 28 to Dec. 5, 1897; age, 6 yrs. 6 mos.; estimated weight, 700 lbs.; fed 19 lbs. ground oats and corn, in equal parts, mixed with 4 buckets cut sheaf oats, wetted, daily—millet hay *ad lib.*; property of J. E. Hart, Nashville, Tenn.

May Day 60451.—Sire, Exile of St. Lambert 23d 20712; dam, Marilla of St. Lambert 31809.

> Butter, 16 lbs. 3½ oz. Milk, 179.9 lbs.

Test made from June 14 to 21, 1892; age, 3 yrs. 1 mo.; actual weight, 870 lbs.; fed 56 qts. corn meal, 14 qts. wheat bran, 28 qts. ground oats, 12 qts. fine middlings and 14 lbs. hay—also green oats and peas; property of Mrs. L. B. Hall, Scottsville, N. Y.

May Day Stoke Pogis 28353.—Sire, Stoke Pogis 3d 2238; dam, May Day of St. Lambert 5109.

> Butter, 15 lbs. 3 oz. Milk, 284 lbs.

Test made from July 23 to 29, 1884; age, 4 yrs. 1 mo.; property of George Smith, Grimsby, Ont., Can.

May Day Stoke Pogis 28353.—Sire, Stoke Pogis 3d 2238; dam, May Day of St. Lambert 5109.

> Butter, 17 lbs. 7 oz. Milk, 298 lbs. 12 oz.

Test made from July 12 to 18, 1886; age, 6 yrs. 1 mo.; estimated weight, 1000 lbs.; fed, daily, 20 lbs. of corn, oats and bran; property of C. A. Reeser, Springfield, Ohio.

May Day Stoke Pogis 28353.—Sire, Stoke Pogis 3d 2238; dam, May Day of St. Lambert 5109.

> Butter, 19 lbs. ½ oz. Milk, 310 lbs. 8 oz.

Test made from Mar. 9 to 16, 1889; age, 8 yrs. 9 mos.; estimated weight, 1050 lbs.; fed, daily, 23 lbs. bran, corn chop and oat chop, mixed; property of W. B. von Richthofen, Montclair, Col.

May Dee 18058.—Sire, Eupidee 4097; dam, May of Lakeside 10826.

> Butter, 15 lbs. 10 oz. Milk, 206 lbs. 8 oz.

Test made from Sept. 29 to Oct. 5, 1886; age, 4 yrs. 6 mos.; estimated weight, 850 lbs.; fed 38 lbs. daily of mixed corn-hearts, oats, oil meal and middlings; property of Miller & Sibley, Franklin, Pa.

May Dee Pogis 36993.—Sire, Stoke Pogis 5th 5987; dam, May Dee 18058.

> Butter, 20 lbs. 5½ oz. Milk, 220 lbs.

Test made from Mar. 8 to 15, 1890; age, 4 yrs. 5 mos.; estimated weight, 1000 lbs.; fed, daily, 36 lbs. ground oats, oil meal and corn meal, mixed; property of Miller & Sibley, Franklin, Pa.

May Dee Pogis of C. H. 93568.—Sire, Ida's Rioter of St. L.'s Son 28869; dam, Ampelis 7th 29851.

> Butter, 15 lbs. 8 oz. Milk, 240 lbs. 14 oz.

Test made from Feb. 1 to 8, 1898; age, 4 yrs. 11 mos.; estimated weight, 900 lbs.; fed 28 qts. corn meal, 28 qts. bran, 7 qts. cotton-seed meal, 3½ qts. linseed meal and 14 qts. whole cotton-seed—little alfalfa daily; property of M. Lothrop, Marshall, Tex.

May Evening 15938.—Sire, Polonius 2513; dam, Pet's Beauty 15726.

Butter, 17 lbs. 13 oz. Milk, 289 lbs. 13 oz.

Test made from Aug. 19 to 26, 1887; age, 9 yrs. 2 mos.; fed 8 lbs. corn meal, 8 lbs. oat meal, 8 lbs. shorts and 8 lbs. oil meal, daily; estimated weight, 900 lbs.; property of J. Herbert Johnston, Plainfield, N. J.

May Fair 5184.—Sire, Guy Mannering 698; dam, Miss Fair 4135.

Butter, 16 lbs. 7 oz. Milk, 171 lbs.

Test made from Sept. 21 to 27, 1883; age, 7 yrs. 6 mos.; fed 6 qts. ground oats and 4 qts. bran, daily—good timothy, orchard grass and clover pasture; property of Columbia Jersey Cattle Co., Columbia, Tenn.

Mayhegtal 64454.—Sire, Romp's Torment 8789; dam, Nerviona 43678.

Butter, 16 lbs. 3 oz. Milk, 246 lbs. 5 oz.

Test made from June 4 to 11, 1893; age, 4 yrs. 4 mos.; estimated weight, 875 lbs.; fed 56 qts. bran, 28 qts. corn meal, 14 qts. oil meal—blue-grass pasture; property of Henry DuBois, Vigo, Ohio.

Mayhegtal 2d 101298.—Sire, Ethleel's Landseer 22341; dam, Mayhegtal 64454.

Butter, 17 lbs. 14 oz. Milk, 202 lbs. 9 oz.

Test made from Dec. 23 to 30, 1895; age, 2 yrs. 11 mos.; estimated weight, 750 lbs.; fed 28 qts. corn, 28 qts. oats and 56 qts. wheat bran, ground and mixed—hay *ad lib.*; property of Henry DuBois, Vigo, Ohio.

May Indian 16075.—Sire, Indian 1721; dam, May Leaf 10667.

Butter, 14 lbs. 5 oz. Milk, 110 lbs. 12 oz.

Test made from Nov. 20 to 27, 1890; age, 10 yrs. 6 mos.; estimated weight, 850 lbs.; fed, daily, 2½ gals. meal and a small quantity of bran; property of H. P. Webster, Columbia, Tenn.

May Keller 78874.—Sire, King of St. Lambert 15175; dam, Elsie Dee 2d 45133.

Butter, 19 lbs. 8 oz. Milk, 271 lbs. 11.2 oz.

Test made from Oct. 18 to 25, 1896; age, 6 yrs. 6 mos.; estimated weight, 850 lbs.; fed 49 lbs. bran, 49 lbs. hominy meal and 56 lbs. corn and oats—clover hay *ad lib.*; property of John E. Robbins, Jr., Greensburg, Ind.

May Koffee 45862.—Sire, Ona's Koffee 9745; dam, Quailona 31492.

Butter, 15 lbs. 5½ oz. Milk, 226 lbs. 8 oz.

Test made from July 3 to 10, 1892; age, 5 yrs. 2 mos.; estimated weight, 900 lbs.; fed 1 qt. cotton-seed meal, 3 qts. corn meal and 4 qts. bran, daily—hilly pasture; property of George Davis, East Montpelier, Vt.

May Lankton 15872.—Sire, Civil Rights 7223; dam, Chloe Lankton 15871.

Butter, 16 lbs. 1½ oz. Milk, 295 lbs.

Test made from May 9 to 15, 1885; age, 7 yrs.; property of Miller & Sibley, Franklin, Pa.

May Marjoram 123811.—Sire, Exile of Glen Rouge 37213; dam, Cassie Marjoram 120666.

Butter, 16 lbs. 5 oz. Milk, 241 lbs. 4 oz.

Test made from July 6 to 13, 1897; age, 2 yrs. 7 mos.; estimated weight, 750 lbs.; fed 20 lbs. daily of peas and oats, mixed, chopped—clover and timothy pasture; property of William Rolph, Markham, Ont., Can.

May Naomi 92277.—Sire, Golden Ray 10669; dam, Naomi B. 8066.

Butter, 14 lbs. 10½ oz.   Milk, 241 lbs. 4 oz.

Test made from Apr. 17 to 24, 1898; age, 9 yrs. 11 mos.; estimated weight, 900 lbs.; fed 42 lbs. bran, 35 lbs. corn meal, 21 lbs. oats, 17½ lbs. oil meal, 70 lbs. roots and 210 lbs. ensilage—hay *ad lib.*; property of C. I. Hood, Lowell, Mass.

May of Edgewood 74825.—Sire, King Coomassie 2d 19545; dam, Pridalia 17249.

Butter, 18 lbs. 2 oz.   Milk, 223 lbs. 2 oz.

Test made from June 7 to 14, 1893; age, 3 yrs. 1 mo.; actual weight, 705 lbs.; fed 50 lbs. corn meal and 25 lbs. bran—pasture of mixed grasses; property of Columbus Dixon, Gillespieville, Ohio.

May of Ingleside 56831.—Sire, Duke of Ingleside 14274; dam, Molly May 17202.

Butter, 21 lbs. 7 oz.   Milk, 315 lbs. 8 oz.

Test made from Oct. 6 to 13, 1894; age, 6 yrs. 7 mos.; estimated weight, 1100 lbs.; fed 28 qts. bran, 28 qts. corn meal, 28 qts. oats, 14 lbs. cotton-seed meal and 14 lbs. oil meal—pasture; property of John C. McClintock, Meadville, Pa.

May Pedro 50876.—Sire, King Peto 10541; dam, Maud Westfield 17555.

Butter, 15 lbs. 4 oz.   Milk, 229 lbs. 8 oz.

Test made from Sept. 30 to Oct. 7, 1890; age, 3 yrs. 11 mos.; actual weight, 810 lbs.; no grain fed—pasture only; property of Mrs. J. H. Martin, Granville, N. Y.

May's Grand Duchess 62607.—Sire, Clermont's Victor Pogis 20026; dam, May Koffee 45862.

Butter, 15 lbs. 3½ oz.   Milk, 226 lbs.

Test made from Mar. 8 to 15, 1895; age, 5 yrs. 1 mo.; estimated weight, 950 lbs.; fed 5 lbs. ground corn-cobs and oats, 4 lbs. bran and 3 lbs. cotton-seed, mixed, daily—hay *ad lib.*; property of George Davis, East Montpelier, Vt.

May Toltena 50405.—Sire, Toltec 6831; dam, Sadie 2d 3698.

Butter, 14 lbs. 5⅜ oz.   Milk, 121 lbs. 2 oz.

Test made from July 27 to Aug. 3, 1889; age, 3 yrs. 2 mos.; estimated weight, 775 lbs.; fed, daily, 4 gals. ground corn and oats, mixed in equal parts; property of Maury Jersey Farm, Columbia, Tenn.

May Violet 30250.—Sire, Lord Alfred 5215; dam, May Indian 16075.

Butter, 18 lbs. 11½ oz.   Milk, 196 lbs. 2 oz.

Test made from Sept. 9 to 16, 1890; age, 6 yrs. 5 mos.; estimated weight, 800 lbs.; fed 4 gals. meal, daily; property of H. P. Webster, Columbia, Tenn.

Medrena 3939.—Sire, Medway 717; dam, Renina 1431.

Butter, 18 lbs.   Milk, 298 lbs.

Test made from July 8 to 14, 1884; age, 9 yrs. 3 mos.; property of Beech Grove Farm, Beech Grove, Ind.

Meg Mitchell 4187.—Sire, Ishmael Hurd 1548; dam, Victoria Guelph 3898.

Butter, 15 lbs. 3 oz.

Test made from Sept. 2 to 9, 1879; age, 5 yrs. 7 mos.; property of C. S. Gladwin, East Haddam, Conn.

Meines 3d 7741.—Sire, St. Helier 45; dam, Meines 3559.

Butter, 20 lbs. 1 oz. Milk, 272 lbs.

Test made from July 6 to 12, 1883; age, 6 yrs. 3 mos.; fed 12 qts. daily of chopped oats and clover—clover hay in stable and clover pasture; property of V. E. Fuller, Hamilton, Ont., Can.

Meines St. Helier 29999.—Sire, Duke of St. Helier 7527; dam, Luella Berlin 16927.

Butter, 14 lbs. 1 oz. Milk, 152 lbs. 7 oz.

Test made from June 10 to 16, 1885; age, 1 yr. 8 mos.; property of A. J. Fish, Lima, Ohio.

Melbina 78875.—Sire, King of St. Lambert 15175; dam, Elsie Dee 18057.

Butter, 20 lbs. 2 oz. Milk, 295 lbs.

Test made from June 3 to 10, 1898; age, 8 yrs.; actual weight, 1000 lbs.; fed 16 qts. wheat bran, 6 qts. ground oats and 4 qts. corn meal, daily—good pasture; property of Mrs. Jennie E. Hall, Columbus, Ohio.

Melia Ann 5444.—Sire, Lord Aylmer 1067; dam, Amelia 2d 1730.

Butter, 18 lbs. ½ oz., unsalted. Milk, 245 lbs.

Test made from Aug. 2 to 9, 1883; age, 8 yrs. 2 mos.; property of J. H. Walker, Worcester, Mass.

Melia Ann 2d 47464.—Sire, Lord Darlington 7285; dam, Melia Ann 5444.

Butter, 15 lbs. 6 oz. Milk, 216 lbs. 15 oz.

Test made from June 17 to 24, 1889; age, 2 yrs. 11 mos.; estimated weight, 875 lbs.; fed, daily, 7 lbs. corn meal and coarse bran (ship stuff); property of Horace G. Westlake, Hillsdale, N. Y.

Melia Ann 2d 47464.—Sire, Lord Darlington 7285; dam, Melia Ann 5444.

Butter, 18 lbs. 6 oz. Milk, 243 lbs. 4 oz.

Test made from July 5 to 12, 1893; age, 6 yrs. 11 mos.; estimated weight, 900 lbs.; fed 42 qts. ship feed. 28 qts. corn meal and 21 qts. ground oats—good lowland pasture; property of H. G. Westlake, Hillsdale, N. Y.

Melia Ann 2d's Pogis 68757.—Sire, King of St. Lambert 2d 17823; dam, Melia Ann 2d 47464.

Butter, 17 lbs. 1 oz. Milk, 206 lbs. 12 oz.

Test made from Oct. 15 to 22, 1893; age, 3 yrs. 6 mos.; estimated weight, 900 lbs.; fed 35 qts. ship feed, 21 qts. corn meal. 14 qts. ground oats and about 7 bush. cut corn-stalks—upland timothy and clover pasture; property of H. G. Westlake, Hillsdale, N. Y.

Melia Ann 3d 68070.—Sire, Lucy's Stoke Pogis 11544; dam, Melia Ann 5444.

Butter, 21 lbs. Milk, 313 lbs. 4 oz.

Test made from May 18 to 25, 1896; age, 6 yrs. 8 mos.; estimated weight, 1000 lbs.; fed 4 qts. wheat bran, 1 qt. oil meal, 2 qts. ground oats and 2 qts. corn meal, daily—hay *ad lib.*, two hours daily on timothy and red-top pasture; property of C. N. Gilbert, Gt. Barrington, Mass.

Melia Ann 3d 68070.—Sire, Lucy's Stoke Pogis 11544; dam, Melia Ann 5444.

Butter, 28 lbs. 8 oz.   Milk, 375 lbs. 8 oz.

Test made from Feb. 17 to 24, 1897; age, 7 yrs. 5 mos.; actual weight, 1135 lbs.; fed 42 qts. corn-hearts, 14 qts. cotton-seed meal, 21 qts. oil meal, 42 qts. wheat middlings and 210 lbs. carrots and beets—hay *ad lib.*; property of Van Deusen & Bishop, Ashley Falls, Mass.

Melita of Hillcrest 7054.—Sire, Major Domo 2161; dam, Atlantis 5901.

Butter, 14 lbs. 2 oz.   Milk, 263 lbs.

Test made from May 7 to 13, 1884; age, 6 yrs. 3 mos.; property of Garretson Bros., Pendleton, Ind.

Mellie Argyle 20609.—Sire, Duke of Lorne 6510; dam, Louise of Lawnfield 14151.

Butter, 14 lbs. 2 oz.   Milk, 205 lbs. 8 oz.

Test made from June 22 to 29, 1885; age, 2 yrs. 7 mos.; property of E. Stevens Henry, Rockville, Conn.

Melody 2689.—Sire, Duke of Jersey 198; dam, Mignonne 693.

Butter, 14 lbs. 1 oz.   Milk, 300 lbs. 4 oz.

Test made from June 14 to 21, 1884; age, 12 yrs. 3 mos.; property of H. W. Douglass, Pevely, Mo.

Melrose Milly 61917.—Sire, Prince of Melrose 4819; dam, Melrose Princess 40343.

Butter, 16 lbs. 2½ oz.   Milk, 222 lbs. 2 oz.

Test made from May 27 to June 3, 1892; age, 4 yrs. 7 mos.; estimated weight, 700 lbs.; fed 42 lbs. bran, 14 lbs. cotton-seed meal, 14 lbs. cotton-seed and 49 lbs. corn, oats and peas, ground—good Bermuda and clover pasture; property of M. Lothrop, Marshall, Tex.

Melrose Princess 2d 62003.—Sire, Beauty's Stoke Pogis 14175; dam, Melrose Princess 40343.

Butter, 14 lbs. 11 oz.   Milk, 228 lbs. 3 oz.

Test made from Oct. 15 to 22, 1894; age, 5 yrs. 8 mos.; estimated weight, 1100 lbs.; fed 84 qts. bran, 56 qts. corn meal and 28 qts. cotton-seed meal—sorghum hay *ad lib.*; property of J. O. Lipscomb, Luling, Tex.

Melrose's Perfection 28895.—Sire, Duke of Melrose 5185; dam, Dewdrop Pansy 19736.

Butter, 24 lbs. 12 oz.   Milk, 252 lbs. 8 oz.

Test made from Nov. 26 to Dec. 3, 1892; age, 8 yrs. 11 mos.; estimated weight, 1000 lbs.; fed 56 qts. bran and shorts, 56 qts. corn meal, 28 qts. cotton-seed meal and 50 lbs. ensilage—hay, winter rye pasture; property of James Crook, Jacksonville, Ala.

Melrose's Signal 54830.—Sire, Prince of Melrose 4819; dam, Signal Beauty 22260.

Butter, 15 lbs. 5¾ oz.   Milk, 217 lbs. 3 oz.

Test made from Feb. 24 to Mar. 3, 1894; age, 8 yrs.; estimated weight, 1000 lbs.; fed 84 qts. wheat bran, 42 qts. ground corn, 28 qts. ground oats and 21 qts. cotton-seed meal—oats pasture; property of W. W. Lipscomb, Luling, Tex.

Même 34921.—Sire, Oonan's Rajah 8905; dam, Littalla 19314.

Butter, 18 lbs. 4 oz.   Milk, 207 lbs.

Test made from Apr. 2 to 9, 1893; age, 8 yrs. 3 mos.; actual weight, 930 lbs.; fed 14 qts. chop and 6 qts. wheat bran per day—clover hay, short pasture; property of Mrs. A. M. Halbeck, Columbus, Ohio.

Même 34921.—Sire, Oonan's Rajah 8965; dam, Littalla 19314.

Butter, 37 lbs. 5 oz. in 16 days. Milk, 453 lbs. 8 oz.

Test made from Mar. 26 to Apr. 11, 1893; age, 8 yrs. 3 mos.; actual weight, 930 lbs.; fed 12 qts. chop and 6 qts. wheat bran, daily, for first four days; 14 qts. chop and 6 qts. wheat bran, daily, for the remainder of the time—short pasture; property of Mrs. A. M. Hallock, Columbus, Ohio.

Même 2d 49946.—Sire, Alfonso 3013; dam, Même 34921.

Butter, 18 lbs. 12⅞ oz. Milk, 259 lbs. 12 oz.

Test made from Nov. 21 to 28, 1894; age, 7 yrs. 5 mos.; estimated weight, 1000 lbs.; fed 28 gals. cut hay and 41 gals. gluten, bran and corn chop, wet; property of Mrs. A. M. Hallock, Columbus, Ohio.

Même 2d 49946.—Sire, Alfonso 3013; dam, Même 34921.

Butter, 39 lbs. 10¼ oz. in 16 days. Milk, 576 lbs. 12 oz.

Test made from Nov. 19 to Dec. 5, 1894; age, 7 yrs. 5 mos.; estimated weight, 1000 lbs.; fed, first 8 days, 77 gals. wet chop and mill feed (gluten, bran and chop); last 8 days, 8 qts. chop and 4 qts. bran, daily, and hay and corn fodder *ad lib.*; property of Mrs. A. M. Hallock, Columbus, Ohio.

Memento 1913.—Sire, Lawrence 61; dam, Motto 80.

Butter, 14 lbs. 5 oz. Milk, 235 lbs.

Test made from Aug. 21 to 28, 1883; age, 11 yrs. 7 mos.; property of Thomas Beer, Bucyrus, Ohio.

Mendota 3d 26326.—Sire, Omaha 482; dam, Mendota 26324.

Butter, 15 lbs. 6 oz. Milk, 301 lbs.

Test made from June 15 to 21, 1885; age, 10 yrs. 1 mo.; property of T. L. Haecker & Co., Madison, Wis.

Méprise Bloomfield 50062.—Sire, Méprise's Tormentor 11554; dam, Signalda Bloomfield 30694.

Butter, 14 lbs. 14 oz. Milk, 137 lbs. 2 oz.

Test made from Nov. 11 to 18, 1891; age, 4 yrs.; estimated weight, 850 lbs.; fed 34 lbs. bran, 56 lbs. crushed oats and 56 lbs. corn meal, about 90 lbs. hay, 2 bush. roots and 126 lbs. ensilage; property of Ayer & McKinney, Philadelphia, Pa.

Mercedes H. 12326.—Sire, Homer H. 3683; dam, Gilda 2779.

Butter, 17 lbs. 12 oz. Milk, 356 lbs. 2 oz.

Test made from June 13 to 19, 1885; age, 7 yrs. 3 mos.; property of William H. Hayden, Albany, Vt.

Mercedes' Pet 20570.—Sire, Alaric of Lakeside 6621; dam, Gilda Mercedes H. 14451.

Butter, 17 lbs. 1½ oz. Milk, 351 lbs. 2 oz.

Test made from June 13 to 19, 1885; age, 2 yrs. 9 mos.; property of William H. Hayden, Albany, Vt.

Mercurina 64920.—Sire, Combination 4389; dam, Modita 16626.

Butter, 17 lbs. 10½ oz. Milk, 226 lbs. 15 oz.

Test made from Oct. 17 to 24, 1889; age, 5 yrs. 8 mos.; estimated weight, 900 lbs.; fed, daily, 12 qts. ground oats and corn meal; property of Richardson Bros., Davenport, Iowa.

Meridale Beauty Dee 97738.—Sire, Ida of St. Lambert's Bull 19169; dam, Beauty Dee 18065.

Butter, 14 lbs. 9 oz.   Milk, 211 lbs. 12 oz.

Test made from Jan. 9 to 16, 1897; age, 4 yrs. 3 mos.; estimated weight, 750 lbs.; fed 40 lbs. wheat bran, 40 lbs. corn meal, 20 lbs. cotton-seed meal and 10 lbs. linseed meal—about 75 lbs. hay and 1 bush. potatoes; property of Ayer & McKinney, Philadelphia, Pa.

Meridale Coquette 64535.—Sire, Ida of St. Lambert's Bull 19169; dam, St. Lambert's Coquette 41070.

Butter, 17 lbs.   Milk, 202 lbs. 2 oz.

Test made from Mar. 31 to Apr. 7, 1892; age, 2 yrs. 6 mos.; estimated weight, 700 lbs.; fed 16 lbs. oats, corn and bran and 2 lbs. oil meal, daily—75 lbs. hay; property of Ayer & McKinney, Philadelphia, Pa.

Meridale Oaklands Nora 64785.—Sire, Ida of St. Lambert's Bull 19169; dam, Oaklands Nora 14880.

Butter, 15 lbs. 1 oz.   Milk, 174 lbs. 6 oz.

Test made from Dec. 4 to 11, 1891; age, 2 yrs. 5 mos.; estimated weight, 750 lbs.; fed 49 lbs. oat meal, 49 lbs. corn meal, 28 lbs. bran, about 70 lbs. hay and 126 lbs. ensilage; property of Ayer & McKinney, Philadelphia, Pa.

Meridale Phinora 119398.—Sire, Garfield's Black Prince 24548; dam, Phinora's Beauty 96176.

Butter, 14 lbs. 9 oz.   Milk, 239 lbs. 5 oz.

Test made from Oct. 21 to 28, 1897; age, 3 yrs. 11 mos.; estimated weight, 900 lbs.; fed 42 lbs. wheat bran, 28 lbs. corn meal and 7 lbs. cotton-seed meal—pasture, timothy aftermath and some natural grass; property of Ayer & McKinney, Philadelphia, Pa.

Meridale Rioter Pink 64534.—Sire, Ida of St. Lambert's Bull 19169; dam, Rioter Pink of Berlin 23665.

Butter, 15 lbs. 13 oz.   Milk, 252 lbs. 9 oz.

Test made from Sept. 6 to 13, 1896; age, 7 yrs.; estimated weight, 1000 lbs.; fed 21 lbs. grain per day. consisting of two-fifths bran and three-fifths hominy and corn meal—pasture, timothy and natural grasses; property of Ayer & McKinney, Philadelphia, Pa.

Meridale Sweet Ida 94854.—Sire, Ida of St. Lambert's Bull 19169; dam, Rioter's Sweet Brier 30582.

Butter, 14 lbs. 5 oz.   Milk, 203 lbs.

Test made from June 24 to July 1, 1896; age, 3 yrs. 8 mos.; estimated weight, 800 lbs.; fed 10 lbs. cotton-seed meal, 44 lbs. ground oats and 16 lbs. corn meal—short pasture; property of W. R. Connal, Newport Centre, Vt.

Meridale Sweet Ida 94854.—Sire, Ida of St. Lambert's Bull 19169; dam, Rioter's Sweet Brier 30582.

Butter, 20 lbs. 8 oz.   Milk, 320 lbs. 7 oz.

Test made from May 23 to 30, 1897; age, 4 yrs. 7 mos.; estimated weight, 950 lbs.; fed 42 lbs. ground oats, 21 lbs. bran, 10½ lbs. corn meal and 10½ lbs. cotton-seed meal—pasture; property of W. R. Connal, Newport Centre, Vt.

Meridale Victoria 94848.—Sire, Ida of St. Lambert's Bull 19169; dam, Hugo's Victoria 29436.

Butter, 18 lbs. 7¼ oz. Milk, 315 lbs. 1 oz.

Test made from Jan. 29 to Feb. 5, 1897; age, 4 yrs. 9 mos.; estimated weight, 950 lbs.; fed 26 lbs. corn meal, 26 lbs. ground oats, 46 lbs. wheat bran, 20 lbs. cotton-seed meal, 12 lbs. linseed meal, 81 lbs. hay and 1 bush. carrots and potatoes; property of Ayer & McKinney, Philadelphia, Pa.

Merlette 4988.—Sire, Samson 1079; dam, Celestia 1898.

Butter, 16 lbs. Milk, 297 lbs. 13 oz.

Test made from July 2 to 8, 1882; age, 6 yrs. 4 mos.; property of J. D. Conner, Wabash, Ind.

Mermaid of St. Lambert 9771.—Sire, Stoke Pogis 3d 2238; dam, Pink of St. Lambert 5486.

Butter, 25 lbs. 13½ oz. (official). Milk, 307 lbs.

Test made from June 25 to July 2, 1884; age, 5 yrs. 2 mos.; property of V. E. Fuller, Hamilton, Ont., Can.

Merry Burlington 7600.—Sire, Merry Andrew 719; dam, Lady Burlington 2d 4082.

Butter, 15 lbs. 4 oz. Milk, 256 lbs.

Test made from May 22 to June 2, 1883; age, 5 yrs. 3 mos.; fed 12 lbs. of corn meal and 3 lbs. of cotton-seed meal per day—orchard grass and white clover pasture; property of J. T. & W. S. Shields, Bean's Station, Tenn.

Merry Maiden 64949.—Sire, Diploma 16219; dam, Costa Rica 64570.

Butter, 14 lbs. 1 oz. Milk, 170 lbs.

Test made from Oct. 3 to 10, 1892; age, 4 yrs.; estimated weight, 800 lbs.; fed about 4 gals. daily of a mixture consisting of ground corn. oats, bran and cut hay or cut corn fodder, also 1 qt. oil meal—pasture very short and dry; property of O. Graves, Maitland, Mo.

Merry Maiden 64949.—Sire, Diploma 16219; dam, Costa Rica 64570.

Butter, 66.695 lbs. in 30 days. Milk, 965 lbs.

Test made from Aug. 29 to Sept. 27, 1893, at World's Columbian Exposition; age, 5 yrs.; actual weight, 946 lbs.; fed 418 lbs. hay, 89.5 lbs. oil·meal, 112 lbs. ensilage, 179 lbs. bran, 90 lbs. oats, 210 lbs. corn-hearts, 45 lbs. cotton-seed, 30 lbs. middlings, 18 lbs. carrots and 6 lbs. old hay; property of C. I. Hood, Lowell, Mass.

Merry Maiden 64949.—Sire, Diploma 16219; dam, Costa Rica 64570.

Butter, 200.517 lbs. in 90 days. Milk, 3041.2 lbs.

Test made from May 31 to Aug. 28, 1893, at World's Columbian Exposition; age, 4 yrs. 8 mos. at beginning of test; actual weight, 936 lbs.; fed 322 lbs. old hay, 210 lbs. ensilage. 597.7 lbs. new hay, 180.5 lbs. oil meal, 494 lbs. bran, 152 lbs. oats, 535 lbs. corn-hearts, 105.75 lbs. cotton-seed, 293 lbs. middlings, 33 lbs. grano-gluten, 1053.4 lbs. clover and 10 lbs. swale-grass; property of O. Graves, Maitland, Mo.

Merty of Peach Hill 77774.—Sire, Young Weston 18439; dam, Golia of Peach Hill Farm 52931.

Butter, 14 lbs. 4½ oz. Milk, 188 lbs. 4 oz.

Test made from Sept. 10 to 17, 1897; age, 7 yrs. 5 mos.; estimated weight, 1000 lbs.; fed 6 qts. wheat bran. 4½ qts. wheat middlings and 3 pts. cotton-seed meal, daily, for first four days; 6 qts. wheat bran, 4½ qts. wheat middlings, daily, for last three days—hay *ad lib.*, second crop clover pasture; property of Everett G. Campbell, Keister's, Pa.

Metah's Baby 9710.—Sire, Omaha 482; dam, Metah 1295.

Butter, 14 lbs. 5 oz. Milk, 239 lbs.

Test made from June 10 to 16, 1885; age, 7 yrs. 2 mos.; property of George E. Bryant, Madison, Wis.

Metah's Queen 4886.—Sire, Omaha 482; dam, Metah 1295.

Butter, 17 lbs. 9 oz., unsalted. Milk, 258 lbs. 2 oz.

Test made from June 26 to July 2, 1881; age, 5 yrs.; property of George E. Bryant, Madison, Wis.

Metaphysics 100974.—Sire, Ida's Rioter of St. L. 13656; dam, Lady Anna Pogis 52734.

Butter, 23 lbs. 2¼ oz. Milk, 256 lbs.

Test made from Dec. 21 to 28, 1897; age, 3 yrs. 1 mo.; actual weight, 1068 lbs.; fed 21 qts. wheat bran, 63 qts. ground oats, 21 qts. oat meal, 42 qts. corn meal, 21 qts. ground oil-cake, 21 qts. malt and 140 lbs. sugar beets; property of Miller & Sibley, Franklin, Pa.

Mette D. 47433.—Sire, Little Harry 8808; dam, Cora of Linwood 12915.

Butter, 14 lbs. 12 oz. Milk, 190 lbs. 1 oz.

Test made from Apr. 10 to 17, 1891; age, 4 yrs. 10 mos.; estimated weight, 950 lbs.; fed 6 qts. corn meal, 6 qts. wheat bran, 4 qts. ground oats, 3 qts. linseed meal, mixed, on cut clover hay, daily—good pasture; property of Matthews & Humes, Huntsville, Ala.

Mhoon Lady 6560.—Sire, Ralph 957; dam, Mink 2d 3890.

Butter, 17 lbs. 3 oz., unsalted. Milk, 261 lbs. 8 oz.

Test made from May 29 to June 4, 1883; age, 5 yrs. 6 mos.; property of William B. Montgomery, Starkville, Miss.

Miaclara 48690.—Sire, Chief of the Miamis 2958; dam, Clara Barder 14311.

Butter, 14 lbs. 8 oz. Milk, 254 lbs.

Test made from Jan. 18 to 25, 1893; age, 5 yrs. 9 mos.; estimated weight, 900 lbs.; fed 8 qts. wheat middlings, 6 qts. ground oats, 3 qts. corn meal, 2 qts. oil meal and 30 lbs. ensilage, daily—hay *ad lib.*; property of George W. Sisson, Jr., Potsdam, N. Y.

Miamis Signaldini 43891.—Sire, Signaldini 8556; dam, Miamis Queen 23356.

Butter, 17 lbs. 14½ oz. Milk, 318 lbs. 3 oz.

Test made from Apr. 14 to 21, 1895; age, 8 yrs. 9 mos.; actual weight, 925 lbs.; fed 70 lbs. corn meal, 14½ lbs. oil meal, 14 lbs. cotton-seed meal, 29 lbs. oats and 29 lbs. bran—clover hay *ad lib.*, blue-grass pasture; property of M. A. Scovell, Lexington, Ky.

Miamotto 48691.—Sire, Chief of the Miamis 2958; dam, Sigmotto 34939.

Butter, 16 lbs. Milk, 276 lbs. 12 oz.

Test made from June 14 to 21, 1893; age, 6 yrs. 1 mo.; estimated weight, 950 lbs.; fed 8 qts. daily of a mixture of Buffalo gluten meal and wheat middlings—natural pasture; property of George W. Sisson, Jr., Potsdam, N. Y.

Midway 50000.—Sire, Diploma 16219; dam, Science 49561.

Butter, 16 lbs. 2 oz. Milk, 182 lbs.

Test made from Jan. 10 to 17, 1890; age, 2 yrs. 1 mo.; estimated weight, 700 lbs.; fed, daily, 16 qts. ground corn, oats and peas and 2 qts. shorts; property of Charles E. Hill, Denver, Col.

Mildred G. 45077.—Sire, Prince Pedro 10588; dam, Unique Ives 14384.

Butter, 17 lbs. 5 oz. Milk, 180 lbs. 2 oz.

Test made from Jan. 4 to 11, 1897; age, 9 yrs. 9 mos.; actual weight, 820 lbs.; fed 60 lbs. corn meal, 50 lbs. ground oats, 30 lbs. bran, 28 lbs. oil meal and 455 qts. finely-cut corn-stalks; property of L. C. Read & Son, East Pembroke, N. Y.

Mildred Montague 45476.—Sire, Lord Montague 12385; dam, Champion G.'s Duchess 40768.

Butter, 15 lbs. 5½ oz. Milk, 242 lbs.

Test made from May 23 to 30, 1891; age, 4 yrs.; estimated weight, 900 lbs.; fed 5 lbs. cotton-seed meal and 5 lbs. wheat bran, daily—timothy, clover and blue-grass pasture; property of Bordwell & Cochran, Batavia, Ohio.

Mildred of M. 15548.—Sire, Rabbi 2496; dam, Pauline 3d 8296.

Butter, 14 lbs. 2½ oz. Milk, 264 lbs.

Test made from May 19 to 25, 1886; age, 4 yrs. 8 mos.; estimated weight, 900 lbs.; fed 30 lbs. daily of mixed corn-hearts, oil meal, oats and middlings; property of Miller & Sibley, Franklin, Pa.

Milkgood 27828.—Sire, Lemon Peel of Francheville (P. S. 439 J. H. B.); dam, Tomboy 24348.

Butter, 14 lbs. 7½ oz. Milk, 210 lbs. 14 oz.

Test made from May 30 to June 6, 1888; age, 4 yrs.; estimated weight, 800 lbs.; fed, daily, 2 lbs. 6 oz. oil meal, 1 lb. 11 oz. pea meal, 2 lbs. 8 oz. bran, 4 lbs. 10 oz. ground oats; property of H. M. Baum, Frankfort, Ind.

Milkmaid Felch 12339.—Sire, Ike Felch 1292; dam, Milkmaid of Lake Forest 5010.

Butter, 16 lbs. 7½ oz. Milk, 238 lbs. 8 oz.

Test made from Apr. 19 to 25, 1884; age, 3 yrs. 11 mos.; property of C. R. C. Dye, Troy, Ohio.

Milkmaid of Burr Oaks 9035.—Sire, Ike Felch 1292; dam, Milkmaid of Lake Forest 5010.

Butter, 14 lbs. 5 oz. Milk, 236 lbs. 2 oz.

Test made from Aug. 26 to Sept. 1, 1883; age, 5 yrs. 4 mos.; property of T. Bacon, Wauconda, Ill.

Milk Weed 16402 (F. S. 1673 J. H. B.)

Butter, 14 lbs. 7 oz., unsalted. Milk, 189 lbs. 8 oz.

Test made from Jan. 28 to Feb. 3, 1883; age, 8 yrs. 7 mos.; fed 2 qts. ground corn and 4 qts. wheat bran, daily, with cut hay; property of Edward Worth, Wawa, Pa.

Milk Well 99301.—Sire, Spokane 23440; dam, Francesville 53367.

Butter, 14 lbs. 4 oz. Milk, 259 lbs. 8 oz.

Test made from May 20 to 27, 1897; age, 5 yrs. 4 mos.; estimated weight, 800 lbs.; fed 305 lbs. ensilage—blue-grass pasture; property of Baum & Hill, Frankfort, Ind.

Milky Way 18865 (F. S. 2607 J. H. B.)

Butter, 17 lbs. 8½ oz. Milk, 232 lbs. 15 oz.

Test made from Sept. 18 to 24, 1884; age, 6 yrs. 6 mos.; property of William Simpson, New York, N. Y.

Millerton Maid 74926.—Sire, Matilda's Perfection 17100; dam, Chemung Maid 34704.

Butter, 17 lbs. 3 oz.   Milk, 259 lbs. 8 oz.

Test made from May 6 to 13, 1897; age, 6 yrs. 1 mo.; actual· weight, 705 lbs.; fed 9 qts. bran, 9 qts. oats, 4½ qts. corn meal and 3 pts. oil meal, daily—hay *ad lib.*; property of C. A. Sweet, East Aurora, N. Y.

Millie of Eau Claire 120023.—Sire, Melia Ann's Son 22041; dam, Jennie Cream 2d 47466.

Butter, 14 lbs. 7 oz.   Milk, 215 lbs. 4 oz.

Test made from June 2 to 9, 1898; age, 1 yr. 11 mos.; estimated weight, 750 lbs.; fed 6 qts. ground corn and oats, 2 qts. oil meal and 6 qts. wheat bran, daily—clover pasture; property of E. L. Van Deusen, Ashley Falls, Mass.

Millplain Princess 78579.—Sire, Prince of Avon 17544; dam, Wilson's Beauty 27671.

Butter, 18 lbs. 2 oz.   Milk, 257 lbs.

Test made from Mar. 12 to 19, 1894; age, 4 yrs. 6 mos.; estimated weight, 800 lbs.; fed 10½ lbs. wheat bran, 31½ lbs. ground oats, 21 lbs. cotton-seed and 63 lbs. corn meal—hay, 28 lbs. potatoes; property of Charles H. Stevens, Canton Centre, Conn.

Milly Judd 85898.—Sire, Metropolitan 23888; dam, Ocla 58396.

Butter, 14 lbs. 3½ oz.   Milk, 237 lbs. 3 oz.

Test made from May 9 to 16, 1898; age, 7 yrs. 2 mos.; estimated weight, 850 lbs.; fed 38½ lbs. bran, 28 lbs. corn meal, 21 lbs. ground oats, 10½ lbs. oil meal, 7 lbs. cotton-seed meal and 70 lbs. roots—hay *ad lib.*, timothy and June grass pasture; property of C. I. Hood, Lowell, Mass.

Minette of St. Lambert 9774.—Sire, Stoke Pogis 3d 2238; dam, May Day of St. Lambert 5109.

Butter, 17 lbs. 4 oz.   Milk, 297 lbs.

Test made from July 25 to 31, 1883; age, 4 yrs. 2 mos.; property of William Rolph, Markham, Ont., Can.

Mink 2d 3890.—Sire, The Hub 1009; dam, Mink 2548.

Butter, 19 lbs. 11 oz.   Milk, 324 lbs.

Test made from July 14 to 20, 1882; age, 7 yrs. 7 mos.; fed steamed cotton-seed, wheat bran and corn meal—reasonably good pasture; property of W. B. Montgomery, Starkville, Miss.

Mink 3d 4868.—Sire, The Hub 1009; dam, Mink 2548.

Butter, 14 lbs. 9 oz.   Milk, 253 lbs. 8 oz.

Test made from Aug. 12 to 18, 1883; age, 7 yrs. 3 mos.; property of W. B. Montgomery, Starkville, Miss.

Minna Rajah 14847.—Sire, Marmaduke's King of Hearts 2949; dam, Nelly A. 1002.

Butter, 16 lbs. 8 oz.   Milk, 179 lbs.

Test made from June 3 to 9, 1885; age, 4 yrs.; property of Columbia Jersey Cattle Co., Columbia, Tenn.

Minnie C. 2d 48414.—Sire, Gen. U. S. Grant 14822; dam, Minnie C. 29003.

Butter, 16 lbs. 14 oz.   Milk, 213 lbs. 1 oz.

Test made from Sept. 5 to 12, 1891; age, 5 yrs. 4 mos.; estimated weight, 1000 lbs.; fed 40 lbs. mill feed (bran and shorts), 42 lbs. corn meal and 3½ gals. oil-cake—pasture of mixed grasses, blue, orchard, etc.; property of D. W. Voyles, Crandall, Ind.

Minnie Hauk 2d 99103.—Sire, Oonan's Pogis 17165; dam, Minnie Hauk 15630.

Butter, 16 lbs. 1 oz.   Milk, 177 lbs. 12 oz.

Test made from Jan. 10 to 17, 1897; age, 4 yrs. 3 mos.; estimated weight, 650 lbs.; fed 8 qts. of a mixture of corn, oats and bran, 1 qt. oil meal and 25 lbs. ensilage and hay, daily; property of George W. Sisson, Jr., Potsdam, N. Y.

Minnie Kersey 19895.—Sire, Watson 3451; dam, Cigarette 2849.

Butter, 14 lbs. 11 oz.   Milk, 224 lbs. 3 oz.

Test made from Nov. 14 to 20, 1885; age, 5 yrs. 5 mos.; property of W. B. Matthews & Son, Franklin, Tenn.

Minnie Lee 2d 12941.—Sire, Champion of America 1567; dam, Minnie Lee 6009.

Butter, 14 lbs. 3 oz.   Milk, 170 lbs.

Test made from Apr. 7 to 13, 1884; age, 3 yrs. 2 mos.; property of W. B. Montgomery, Starkville, Miss.

Minnie Mignon 90972.—Sire, Judge Watson 26286; dam, Marcella Maid's Pearl 23794.

Butter, 14 lbs. 12 oz.   Milk, 225 lbs. 1 oz.

Test made from Dec. 19 to 26, 1895; age, 4 yrs.; estimated weight, 800 lbs.; fed 42 lbs. bran and 28 lbs. corn meal—evergreen sweet corn fodder and mixed hay; property of A. D. McBain, Brockport, N. Y.

Minnie Montague 45474.—Sire, Lord Montague 12385; dam, Ida Imperial 2d 28710.

Butter, 15 lbs. 5½ oz.   Milk, 223 lbs. 4 oz.

Test made from May 23 to 30, 1891; age, 4 yrs. 1 mo.; estimated weight, 850 lbs.; fed 4 lbs. bran and 5 lbs. cotton-seed cake meal, daily—timothy, clover and blue-grass pasture; property of Bordwell & Cochran, Batavia, Ohio.

Minnie of Oxford 12806.—Sire, Stoke Pogis 2d 2414; dam, Matilda 2d 5471.

Butter, 16 lbs. 1 oz.

Test made from June 8 to 15, 1882; age, 5 yrs. 3 mos.; property of Horace G. West-lake, Hillsdale, N. Y.

Minnie of Oxford 12806.—Sire, Stoke Pogis 2d 2414; dam, Matilda 2d 5471.

Butter, 17 lbs.   Milk, 247 lbs.

Test made from Aug. 10 to 16, 1886; age, 9 yrs. 5 mos.; estimated weight, 750 lbs.; fed, daily, 8 qts. ground corn and oats and 2 qts. oil meal; property of Frederick Loeser, Somerville, N. J.

Minnie of Scituate 17829.—Sire, Duke of Scituate 3623; dam, Minnie 2d 17828.

Butter, 14 lbs. 4½ oz.   Milk, 251 lbs.

Test made from Dec. 24 to 31, 1883; age, 4 yrs. 7 mos.; fed 4 qts. bran, 2 qts. corn meal and 1 qt. cotton-seed meal, daily; property of Orestes Pierce, Baldwin, Me.

Minnie Parker 38718.—Sire, Gilderoy 5th 7426; dam, Respectability 16730.

Butter, 26 lbs. 2½ oz. Milk, 309 lbs.

Test made from July 18 to 25, 1896; age, 10 yrs. 5 mos.; estimated weight, 1000 lbs.; fed 350 lbs. cut corn and sorghum (green), 56 qts. bran, 28 qts. oats, 14 qts. cottonseed meal, 28 qts. corn meal and 14 qts. shorts—Bermuda grass pasture; property of T. G. Bush, Anniston, Ala.

Minnie Rioter 21336.—Sire, Greatorex 4218; dam, Minnie of Oxford 2d 12818.

Butter, 15 lbs. 10 oz. Milk, 249 lbs. 6 oz.

Test made from Apr. 30 to May 7, 1893; age, 9 yrs. 11 mos.; estimated weight, 900 lbs.; fed 10 qts. corn meal, 4 qts. oil meal, 12 qts. oat meal, 15 qts. bran, 60 lbs. ensilage and 1 bush. roots—clover and timothy pasture for 4 days; property of Mrs. E. F. Hawley, Pittsford, N. Y.

Minnie R. of St Lambert 61965.—Sire, Duke of St. Lambert 16160; dam, Minnie Rioter 21336.

Butter, 16 lbs. 9 oz. Milk, 235 lbs.

Test made from May 22 to 29, 1893; age, 5 yrs. 3 mos.; estimated weight, 900 lbs.; fed 2 qts. corn meal, 2 qts. oat meal and 2 qts. bran, daily—clover and timothy pasture; property of Mrs. E. F. Hawley, Pittsford, N. Y.

Minnie's Lady Allen 102925.—Sire, Royal Allen 28149; dam, Hawthorn Minnie 26687.

Butter, 17 lbs. 10 oz. Milk, 240 lbs.

Test made from Apr. 5 to 12, 1898; age, 3 yrs. 6 mos.; estimated weight, 750 lbs.; fed about 20 qts. daily of a mixture of ground oats, middlings, bran, corn meal and some pea meal—clover hay and ensilage; property of T. S. Cooper, Coopersburg, Pa.

Minnie's Ouida 33117.—Sire, Parole of Honor 4291; dam, Minnie of Orange 6016.

Butter, 14 lbs. Milk, 260 lbs. 7 oz.

Test made from Oct. 23 to 30, 1889; age, 5 yrs. 3 mos.; estimated weight, 950 lbs.; fed, daily, 16 lbs. ground corn, oats and wheat bran, one-third part of each; property of F. H. Bates, Hamburg, Ala.

Minnie's Pet Jersey 93119.—Sire, R. F. Eric 21330; dam, Minnie Montague 45474.

Butter, 15 lbs. 6 oz. Milk, 253 lbs. 6 oz.

Test made from July 5 to 12, 1897; age, 5 yrs. 4 mos.; estimated weight, 800 lbs.; fed 42 qts. corn and cob meal and 56 qts. wheat bran—good blue-grass pasture; property of Columbus Dixon, Gillespieville, Ohio.

Mintha 12812.—Sire, Rioter Vulcan 5380; dam, Matilda 3d 12808.

Butter, 15 lbs. Milk, 249 lbs. 8 oz.

Test made from Apr. 24 to 30, 1885; age, 5 yrs. 3 mos.; property of Campbell Brown, Spring Hill, Tenn.

Mintresse of Ingleside 103128.—Sire, Duke of Ingleside 14274; dam, Mintresse 20642.

Butter, 16 lbs. 10½ oz. Milk, 180 lbs. 9 oz.

Test made from July 7 to 14, 1895; age, 8 yrs. 1 mo.; estimated weight, 1000 lbs.; fed 28 lbs. coarse winter wheat bran, 28 lbs. ground oats and 28 lbs. corn meal—common pasture; property of John C. McClintock, Meadville, Pa.

Minty 2d of Hood Farm 112326.—Sire, Mint 23600; dam, Statue 56604.

Butter, 15 lbs. 3½ oz. Milk, 231 lbs. 8 oz.

Test made from Jan. 20 to 27, 1898; age, 3 yrs. 11 mos.; estimated weight, 750 lbs.; fed 49 lbs. bran, 28 lbs. corn meal, 28 lbs. ground oats, 10½ lbs. oil meal, 10½ lbs. cotton-seed meal, 210 lbs. ensilage and 35 lbs. roots—hay *ad lib.*; property of C. I. Hood, Lowell, Mass.

Minty 2d of Hood Farm 112326.—Sire, Mint 23600; dam, Statue 56604.

Butter, 23 lbs. 12 oz. in 11 days. Milk, 362 lbs. 5 oz.

Test made from Jan. 16 to 27, 1898; age, 3 yrs. 11 mos.; estimated weight, 750 lbs.; fed 77 lbs. bran, 44 lbs. corn meal, 44 lbs. ground oats, 18½ lbs. oil meal, 11 lbs. cotton-seed meal, 55 lbs. roots and 330 lbs. ensilage—hay *ad lib.*; property of C. I. Hood, Lowell, Mass.

Mirah Landseer 74827.—Sire, Leonette's Landseer 20399; dam, Lady Mirah 25266.

Butter, 17 lbs. 1 oz. Milk, 202 lbs. 1½ oz.

Test made from Mar. 8 to 15, 1896; age, 5 yrs. 8 mos.; estimated weight, 800 lbs.; fed 6 lbs. corn chop, 6 lbs. wheat bran and 4 lbs. crushed oats, daily—hay *ad lib.*, poor pasture; property of J. E. McGuire, Gatesville, Tex.

Miralba 65002.—Sire, North Pacific Prize 11459; dam, Laura Lee 8550.

Butter, 14 lbs. 7 oz. Milk, 248 lbs. 2 oz.

Test made from May 1 to 8, 1890; age, 3 yrs. 5 mos.; estimated weight, 800 lbs.; fed 15 qts. ground oats and corn, daily; property of Richardson Bros., Davenport, Iowa.

Miranda of Bolton 100226.—Sire, Viking of Bolton 24881; dam, Helena of Bolton 39109.

Butter, 21 lbs. 4 oz. Milk, 229 lbs. 3 oz.

Test made from Jan. 6 to 13, 1898; age, 4 yrs. 10 mos.; actual weight, 896 lbs.; fed 15½ qts. corn meal, 15½ qts. ground oats, 22½ qts. linseed meal, 21 lbs. corn fodder, 98 lbs. rowen, 175 lbs. roots (carrots and beets) and 31 qts. shorts; property of John A. Cunningham, Bolton, Mass.

Mirnie Potter 72670.—Sire, Paul Potter 12119; dam, Mocha's Pet's Rheta 37529.

Butter, 14 lbs. 10 oz. Milk, 273 lbs. 8 oz.

Test made from Mar. 29 to Apr. 5, 1895; age, 6 yrs.; estimated weight, 950 lbs.; fed 42 qts. bran, 42 qts. bruised oats, 21 qts. corn meal, 21 qts. oil meal, 168 lbs. hay and 70 lbs. roots; property of P. J. Cogswell, Rochester, N. Y.

Mirtha 3437.—Sire, Medway 717; dam, Mirth 92.

Butter, 17 lbs. 13½ oz. Milk, 314 lbs. 8 oz.

Test made from July 18 to 24, 1883; age, 8 yrs. 9 mos.; property of Campbell Brown, Spring Hill, Tenn.

Mischief Le Brocq 7680.—Sire, Pierrot (F. S. 143 J. H. B.); dam, Lady Jane Grey (F. S. 592 J. H. B.)

Butter, 15 lbs., unsalted.

Test made from June 7 to 14, 1876; age, 5 yrs.; property of A. E. Kapp, Northumberland, Pa.

Miss Albert 63138.—Sire, Eurotas' Victor Hugo 15689; dam. Albert's Fawn 34004.

Butter, 17 lbs. 9 oz.　Milk, 263 lbs.

Test made from Mar. 16 to 23, 1896; age, 6 yrs. 8 mos.; estimated weight, 900 lbs.; fed about 22 qts. daily of a mixture of ground oats, middlings, wheat bran, corn meal and some pea meal—clover hay and ensilage; property of T. S. Cooper, Coopersburg, Pa.

Miss Alexander 2d 26054.—Sire, Carlo 5559; dam, Miss Alexander 26041.

Butter, 15 lbs. 10 oz.　Milk, 177 lbs.

Test made from Mar. 8 to 15, 1890; age, 6 yrs. 3 mos.; actual weight, 960 lbs.; fed, daily, 10 qts. ground oats, 2 qts. oil meal and 2 qts. pea meal; property of T. R. Proctor, Utica, N. Y.

Miss Beauty 4053.—Sire, Apis 1206; dam, Miss Blossom 1986.

Butter, 16 lbs. 12 oz.　Milk, 251 lbs. 2 oz.

Test made from Mar. 6 to 13, 1886; age, 11 yrs. 1 mo.; property of Peter D. Hulst, East Penfield, N. Y.

Miss Belle 5083.—Sire, Apis 1206; dam, Miss Blossom 1986.

Butter, 14 lbs. 15 oz., unsalted.　Milk, 269 lbs.

Test made from May 23 to 29, 1882; age, 6 yrs. 5 mos.; property of F. Bronson, Greenfield Hill, Conn.

Miss Belle 5083.—Sire, Apis 1206; dam, Miss Blossom 1986.

Butter, 22 lbs. 9 oz. (official).　Milk, 276 lbs.

Test made from May 25 to 31, 1886; age, 10 yrs. 5 mos.; weight, 1100 lbs.; fed 87 lbs. crushed oats, 55 lbs. corn meal, 24 lbs. bran, 25 lbs. linseed oil cake meal and 25 lbs. pea meal; property of Frederic Bronson, Greenfield Hill, Conn.

Miss Bianca 12517.—Sire, Faust 503; dam, Miss Millie 12264.

Butter, 17 lbs. 14 oz.　Milk, 263 lbs. 3 oz.

Test made from Mar. 10 to 17, 1883; age, 5 yrs. 11 mos.; estimated weight, 850 lbs.; fed, daily, 18 qts. corn meal, oat meal and pea meal, mixed; property of Richardson Bros., Davenport, Iowa.

Miss Bingo 80096.—Sire, Bingo 2d 6749; dam, Mrs. Knickerbocker 19367.

Butter, 24 lbs. 12 oz.　Milk, 255 lbs. 4 oz.

Test made from Jan. 25 to Feb. 1, 1891; age, 8 yrs. 7 mos.; actual weight, 1050 lbs.; fed 24 qts. daily of a mixture composed of one part bran, two parts ground corn and three parts ground oats—bran slop daily, good hay *ad lib.*; property of L. L. Tozier, Batavia, N. Y.

Miss Blanche 2515.—Sire, Ringleader 392; dam, Zilpah 2510.

Butter, 20 lbs. 9 oz. in 10 days.　Milk, 357 lbs.

Test made from July 25 to Aug. 3, 1879; age, 8 yrs. 3 mos.; property of Z. C. Luse & Son, Iowa City, Iowa.

Miss Bligh 24561.—Sire, Berger (P. S. 278 J. H. B.); dam, Lady Bligh (F. S. 5621 J. H. B.)

Butter, 15 lbs. 2 oz.

Test made from Jan. 3 to 9, 1886; age, 4 yrs.; property of W. L. Conyngham, Wilkesbarre, Pa.

Miss Bluster 108231.—Sire, Bluster 22798; dam, Torlandseer's Pet 83371.

Butter, 15 lbs. 15⅝ oz. (confirmed); estimated butter on basis of 85 per cent. fat, 15 lbs. 8⅜ oz. Milk, 210.3 lbs.

Test made from Mar. 25 to 31, 1897; age, 2 yrs. 5 mos.; average weight, 824 lbs.; fed 11 lbs. corn meal, 4.5 lbs. oil meal, 10 lbs. ground oats and corn, 20 lbs. bran and 42 lbs. corn-hearts—clover hay *ad lib.*; property of J. Waller Rodes, Lexington, Ky.

Analyses of Butter.

First churning—Fat 81.21, water 14.66, ash 3.12. casein 1.01.
Second churning—Fat 80.69, water 15.73, ash 2.80, casein 0.78.

Miss Bluster 108231.—Sire, Bluster 22798; dam, Torlandseer's Pet 83371.

Butter, 50.95 lbs. in 24 days. Milk, 705.8 lbs.

Test made from Mar. 7 to 31, 1897; age, 2 yrs. 5 mos.; actual weight, 824 lbs.; fed 48 lbs. bran, 115 lbs. germ food, 4 lbs. cotton-seed meal, 18½ lbs. flaxseed meal, 8 lbs. ground oats, 14 lbs. corn meal and 112 lbs. mixed feed—hay *ad lib.*; property of J. Waller Rodes, Lexington, Ky.

Miss Bowris 12206.—Sire, Faust 503; dam, Bessie Ring 12175.

Butter, 17 lbs. 8½ oz. Milk, 276 lbs. 3 oz.

Test made from Nov. 20 to 27, 1883; age, 6 yrs. 7 mos.; estimated weight, 850 lbs.; fed, daily, 18 qts. of a mixture of meal, ground oats, corn and peas; property of Richardson Bros., Davenport, Iowa.

Miss Browney 7288.—Sire, Browny (P. S. 158 J. H. B.); dam, Miss De Carteret 7285.

Butter, 16 lbs. 13 oz. Milk, 336 lbs.

Test made from Feb. 9 to 15, 1885; age, 6 yrs. 10 mos.; property of William Crozier, Northport, N. Y.

Miss Clifford 27962.—Sire, Gold Rope 6139; dam, Jessie Benton 11183.

Butter, 17 lbs. 15½ oz. Milk, 256 lbs. 12 oz.

Test made from June 8 to 15, 1890; age, 6 yrs.; estimated weight, 900 lbs.; fed, daily, 5 lbs. corn meal, 3 lbs. coarse bran and 2 lbs. old process oil meal; property of A. S. Bell, London, Ohio.

Miss Clifford 2d 56149.—Sire, Casta's Champion 11794; dam, Miss Clifford 27962.

Butter, 16 lbs. 4½ oz. Milk, 270 lbs. 4 oz.

Test made from May 27 to June 3, 1890; age, 4 yrs. 1 mo.; estimated weight, 750 lbs.; fed, daily, 12 lbs. corn meal and coarse bran, mixed in equal parts, and 2 lbs. linseed meal; property of A. S. Bell, London, Ohio.

Miss Clifford 3d 63360.—Sire, Stoke Pogis Trio 15448; dam, Miss Clifford 27962.

Butter, 15 lbs. 2½ oz. Milk, 266 lbs. 2 oz.

Test made from June 20 to 27, 1893; age, 5 yrs. 2 mos.; estimated weight, 850 lbs.; fed 30 lbs. coarse wheat bran, 40 lbs. corn meal and 20 lbs. old process oil meal; property of A. S. Bell, London, Ohio.

Miss Clifford 5th 63364.—Sire, Stoke Pogis Trio 15448; dam, Miss Clifford 27962.

Butter, 14 lbs. 7 oz. Milk, 215 lbs. 1 oz.

Test made from June 11 to 18, 1893; age, 3 yrs. 8 mos.; estimated weight, 750 lbs.; fed 30 lbs. corn meal, 20 lbs. coarse bran and 15 lbs. old process oil meal; property of A. S. Bell, London, Ohio.

Miss D'Arcy 62596.—Sire, Signalda's Tormentor 9298; dam, Fair Geraldine 37690.

Butter, 15 lbs. Milk, 256 lbs. 8 oz.

Test made from Apr. 23 to 30, 1895; age, 5 yrs. 9 mos.; estimated weight, 700 lbs.; fed 63 lbs. corn meal, 63 lbs. wheat bran, 28 lbs. ground oats, 70 lbs. raw cotton-seed and 7 lbs. cotton-seed meal—mesquite grass pasture; property of S. N. Clark, Cleburne, Tex.

Miss Dora Deane 24505.—Sire, Argo 3737; dam, Eunice of Belle Vue 7704.

Butter, 16 lbs. 3½ oz. Milk, 226 lbs.

Test made from May 30 to June 5, 1886; age, 3 yrs. 11 mos.; property of Campbell Brown, Spring Hill, Tenn.

Miss Dora Deane 24505.—Sire, Argo 3737; dam, Eunice of Belle Vue 7704.

Butter, 15 lbs. 5 oz. (official). Milk, 230 lbs. 9 oz.

Test made from June 12 to 18, 1886; age, 3 yrs. 11 mos.; actual weight, 925 lbs.; fed, daily, about 15 lbs. corn meal and 4½ lbs. crushed oats; property of Campbell Brown, Spring Hill, Tenn.

Miss Dora Deane 24505.—Sire, Argo 3737; dam, Eunice of Belle Vue 7704.

Butter, 17 lbs. 9⅝ oz. Milk, 256 lbs. 8 oz.

Test made from May 11 to 18, 1887; age, 4 yrs. 10 mos.; estimated weight, 925 lbs.; well fed, but no record of feed kept; property of Campbell Brown, Spring Hill, Tenn.

Misselle 73488.—Sire, Kathy's Stoke Pogis 17566; dam, Cornie D. 73486.

Butter, 16 lbs. 5½ oz. Milk, 167 lbs. 10 oz.

Test made from May 17 to 24, 1894; age, 4 yrs. 3 mos.; estimated weight, 800 lbs.; fed 56 qts. corn-hearts, 28 qts. cotton-seed meal and 16 qts. whole oats, boiled, fed on cut green wheat—very poor pasture; property of William E. Matthews, Huntsville, Ala.

Miss Flora Brown 100532.—Sire, Plebian 31020; dam, Brown Flora 2d 96594.

Butter, 16 lbs. 2 oz. Milk, 238 lbs. 4 oz.

Test made from June 18 to 25, 1898; age, 3 yrs. 7 mos.; estimated weight, 800 lbs.; fed 28 lbs. corn meal, 28 lbs. oats and 17½ lbs. oil meal—mixed pasture; property of H. C. Taylor, Orfordville, Wis.

Miss Gracie 86495.—Sire, Oonan's Tormentor 22280; dam, Nancy's Grace 41421.

Butter, 37 lbs. 12 oz. in 14 days. Milk, 414 lbs. 12 oz.

Test made from Sept. 15 to 29, 1895; age, 3 yrs.; actual weight, 775 lbs.; fed 22 lbs. per day of ground oats and corn, in equal parts—extra good orchard and blue-grass pasture; property of Matt. M. Gardner, Nashville, Tenn.

Miss Helen Brice 88340.—Sire, Diploma 16219; dam, Treasure 79200.

Butter, 14 lbs. 3 oz. Milk, 252 lbs. 13 oz.

Test made from June 1 to 8, 1896; age, 4 yrs. 8 mos.; estimated weight, 825 lbs.; fed 35 lbs. bran, 42 lbs. oats, 11½ lbs. oil meal, 28 lbs. cotton-seed meal and 105 lbs. ensilage—hay *ad lib.*, timothy and clover pasture; property of C. I. Hood, Lowell, Mass.

Miss Helen Brice 88340.—Sire, Diploma 16219; dam, Treasure 79200.

Butter, 18 lbs. 1¼ oz. Milk, 304 lbs. 10 oz.

Test made from May 31 to June 7, 1897; age, 5 yrs. 8 mos.; estimated weight, 900 lbs.; fed 35 lbs. bran, 42 lbs. corn meal, 28 lbs. ground oats and 7 lbs. oil meal—cut grass in barn at night, mixed pasture; property of C. I. Hood, Lowell, Mass.

Miss Julia 26701.—Sire, Sir Alpha 6378; dam, Encore Lawrence 12688.

Butter, 21 lbs. 1½ oz. Milk, 265 lbs.

Test made from Aug. 25 to Sept. 1, 1889; age, 6 yrs. 5 mos.; estimated weight, 1000 lbs.; fed, daily, 5 qts. corn meal, 2 qts. ground oats, 2 qts. bran and 2 qts. flaxseed meal; property of Carson & Bro., Crab Orchard, Ky.

Miss Katie W. 47366.—Sire, Gilderoy 5th 7426; dam, Minkabel 27732.

Butter, 15 lbs. 11¾ oz. Milk, 197 lbs.

Test made from Aug. 20 to 27, 1896; age, 9 yrs. 4 mos.; estimated weight, 1100 lbs.; fed 63 qts. bran, 42 qts. oat meal, 42 qts. corn meal, 10½ qts. cotton-seed meal and 350 lbs. cut corn and sorghum—pasture in dry lot; property of T. G. Bush, Anniston, Ala.

Miss Le Dain 27472.—Sire, Nero 7266; dam, Fuchsia, on I. of J.

Butter, 14 lbs. 5¼ oz. Milk, 228 lbs. 8 oz.

Test made from Feb. 20 to 26, 1886; age, 3 yrs. 9 mos.; property of Joseph P. Gardner, Hamilton, Mass.

Miss Le Gros' Maid 83645.—Sire, Ida's Pogis 18000; dam, Koffee's Miss Le Gros 35107.

Butter, 18 lbs. 5¾ oz. Milk, 203 lbs. 4 oz.

Test made from Dec. 31, 1896, to Jan. 7, 1897; age, 5 yrs. 11 mos.; estimated weight, 825 lbs.; fed 65 qts. corn meal, 42 qts. ground oats, 21 qts. ground oil-cake, 10 lbs. clover hay and 12 lbs. carrots; property of Miller & Sibley, Franklin, Pa.

Miss Lofty 9718.—Sire, Duke of Mansfield 2277; dam, Belle of Brooklyn 9493.

Butter, 16 lbs. 10 oz. Milk, 222 lbs. 2 oz.

Test made from May 23 to 30, 1886; age, 8 yrs. 2 mos.; property of E. Stevens Henry, Rockville, Conn.

Miss Lorne 78132.—Sire, Lorne Victor Pogis 22214; dam, Ingleside Duchess 46268.

Butter, 15 lbs. 9 oz. Milk, 234 lbs. 8 oz.

Test made from June 6 to 13, 1896; age, 5 yrs. 3 mos.; estimated weight, 850 lbs.; fed 28 lbs. corn meal, 28 lbs. ground oats and 28 lbs. wheat middlings—fine June pasture; property of John C. McClintock, Meadville, Pa.

Miss Madrid 49419.—Sire, Combination 2d 16957; dam, Period 42640.

Butter, 14 lbs. 11 oz. Milk, 204 lbs. 13 oz.

Test made from June 1 to 8, 1889; age, 3 yrs. 2 mos.; estimated weight, 795 lbs.; fed 12 qts. corn and oat meal, daily—hay at night, orchard pasture; property of Richardson Bros., Davenport, Iowa.

Miss May of St. Lambert 37084.—Sire, Rubano 8806; dam, May Day Stoke Pogis 28353.

Butter, 15 lbs. 14 oz. Milk, 138 lbs. 4 oz.

Test made from July 7 to 14, 1887; age, 2 yrs. 2 mos.; estimated weight, 650 lbs.; fed, daily, 21 qts. mixed corn, bran and oats; property of C. A. Reeser, Springfield, Ohio.

Miss Milton 51851.—Sire, Chevalier 14748; dam, Carlo Belle 23091.

Butter, 24 lbs. 9½ oz. Milk, 241 lbs. 6 oz.

Test made from Nov. 28 to Dec. 5, 1897; age, 9 yrs. 11 mos.; actual weight, 900 lbs.; fed 5 qts. shorts, 2 qts. corn meal, 2 qts. ground oats, 2 qts. linseed, 16 lbs. hay and 5 lbs. corn fodder, daily; property of Mrs. J. A. Cunningham, Bolton, Mass.

Miss Montague 57927.—Sire, Lord Montague 12385; dam, Miss Loyal 40771.

Butter, 14 lbs. Milk, 202 lbs. 4 oz.

Test made from June 6 to 13, 1891; age, 2 yrs. 8 mos.; estimated weight, 700 lbs.; fed 5 lbs. bran and 5 lbs. cotton-seed cake meal, daily—timothy, clover and blue-grass pasture; property of Bordwell & Cochran, Batavia, Ohio.

Miss Morehouse 91703.—Sire, Pussie's Pogis 25935; dam, Patti Niobe 74649.

Butter, 15 lbs. 5½ oz. Milk, 316 lbs. 4 oz.

Test made from Feb. 7 to 14, 1897; age, 4 yrs.; actual weight, 800 lbs.; fed 28 lbs. corn meal, 14 lbs. oil meal, 49 lbs. bran, 21 lbs. oat meal, 7 lbs. pea meal and 49 lbs. carrots; property of Allen & Pease, Belmont, N. Y.

Miss Mouse 92512.—Sire, Prince of Tennessee 20772; dam, Gold Mosie 63502.

Butter, 19 lbs. 8¾ oz. Milk, 223 lbs. ¾ oz.

Test made from Dec. 18 to 25, 1895; age, 4 yrs. 6 mos.; estimated weight, 700 lbs.; fed 18 lbs. per day of ground oats and corn, in equal parts, mixed with 4 bucketfuls cut sheaf oats, wetted—millet hay *ad lib.*; property of Thomas H. Malone, Nashville, Tenn.

Miss Naomi 45098.—Sire, Molten Gold 12186; dam, Heathbell 20170.

Butter, 15 lbs. 3 oz. Milk, 224 lbs. 6 oz.

Test made from Aug. 24 to 31, 1890; age, 4 yrs. 9 mos.; actual weight, 800 lbs.; fed, daily, 12 lbs. cut oats, yellow meal and wheat bran, mixed, and 1 lb. oil meal; property of Holly Grove Farm, Plainfield, N. J.

Miss Nautila 49000.—Sire, Aella's Stoke Pogis 14769; dam, Nautila 14738.

Butter, 15 lbs. 5 oz. Milk, 215 lbs. 12 oz.

Test made from June 24 to July 1, 1896; age, 8 yrs. 6 mos.; estimated weight, 850 lbs.; fed 23 lbs. daily of a mixture of bran, oats, corn, oil meal and cotton-seed meal, with 20 per cent. pea meal—corn fodder and clover hay at night, fair pasture; property of R. F. Shannon, Pittsburg, Pa.

Miss Nellie Parker 24988.—Sire, Perrot (P. S. 342 J. H. B.); dam, Lass Nellie 24012.

Butter, 23 lbs. 11 oz. Milk, 240 lbs. 13 oz.

Test made from Apr. 18 to 25, 1889; age, 5 yrs. 4 mos.; estimated weight, 875 lbs.; fed 70 qts. corn meal, 42 qts. bran, 14 qts. pea meal, 14 qts. cotton-seed meal and 105 lbs. ensilage—hay, pasture of winter rye; property of James Crook, Jacksonville, Ala.

Miss Patti Rosa 73607.—Sire, St. Lambert's John Bull 16618; dam, Litta Pogis 30275.

Butter, 17 lbs. 15 oz. Milk, 226 lbs. 4 oz.

Test made from Feb. 14 to 21, 1898; age, 7 yrs. 4 mos.; estimated weight, 950 lbs.; fed 8 lbs. wheat bran, 11 lbs. corn meal, ½ lb. linseed meal, 7 lbs. cotton-seed meal and 3½ lbs. cotton-seed, daily—sorghum *ad lib.*, woods pasture; property of W. E. Johnson, Millican, Tex.

Miss Peacock 30871.—Sire, Oonan's Rajah 8965; dam, Little Torment 15581.

Butter, 16 lbs. 10 oz. Milk, 257 lbs. 1 oz.

Test made from Jan. 22 to 29, 1890; age, 5 yrs. 10 mos.; estimated weight, 900 lbs.; fed, daily, about 12 lbs. bran, 9 lbs. corn meal and 2½ lbs. cotton-seed meal; property of W. B. von Richthofen, Montclair, Col.

Miss Perrot 73801.—Sire, Ethleel's Harry 15962; dam, Perrot's Baroness 22508.

Butter, 16 lbs. Milk, 236 lbs.

Test made from May 11 to 18, 1896; age, 5 yrs. 1 mo.; estimated weight, 800 lbs.; fed 3 qts. wheat middlings, 3 qts. wheat bran and 3 qts. corn meal, daily—timothy and clover pasture; property of Alfred B. Darling, Ramsey's, N. J.

Miss Porter 20300.—Sire, Young Prince (P. S. 182 J. H. B.); dam, Mrs. Porter 16667.

Butter, 16 lbs. 6 oz. Milk, 180 lbs. 15 oz.

Test made from July 17 to 23, 1885; age, 6 yrs. 5 mos.; property of G. L. & A. C. Davis, Port Jefferson, N. Y.

Miss Puritan 34204.—Sire, Star Duke 8438; dam, Josephine Beacon 3306.

Butter, 18 lbs. 5 oz. Milk, 357 lbs. 11 oz.

Test made from Feb. 26 to Mar. 5, 1894; age, 8 yrs. 6 mos.; estimated weight, 900 lbs.; fed 28 qts. corn meal, 28 qts. pea meal, 28 qts. cotton-seed meal and 28 qts. wheat bran—pea straw *ad lib.*, two hours daily on green rye; property of W. L. Kennedy, Falling Creek, N. C.

Miss Rainy 66893.—Sire, Minnie's Stoke Pogis 17201; dam, Star's Fancy 55674.

Butter, 15 lbs. 2 oz. Milk, 230 lbs. 6 oz.

Test made from Dec. 14 to 22, 1896; age, 7 yrs. 5 mos.; actual weight, 1230 lbs.; fed 14 gals. shelled corn and oats, ground together (two-thirds corn and one-third oats) and 28 gals. wheat bran—clover hay at night, corn fodder during day; property of H. H. Anderson, Rockville, Ind.

Miss Remarkable 33446.—Sire, Gold Basis 4038; dam, Rosa of Belle Vue 6954.

Butter, 15 lbs. 2½ oz. Milk, 154 lbs. 11½ oz.

Test made from June 24 to July 1, 1895; age, 11 yrs. 2 mos.; estimated weight, 800 lbs.; fed 19 lbs. per day of ground oats and corn, in equal parts—good orchard and blue-grass pasture; property of Thomas H. Malone, Nashville, Tenn.

Miss Retta 105807.—Sire, Recorder 29239; dam, Belle Noble's Signal 78013.

Butter, 20 lbs. 1 oz. Milk, 235 lbs. 4 oz.

Test made from Mar. 30 to Apr. 6, 1897; age, 3 yrs. 2 mos.; estimated weight, 725 lbs.; fed 42 lbs. oats, 7 lbs. oil meal, 42 lbs. bran, 21 lbs. rye and 28 lbs. corn meal—45 lbs. ensilage daily, cut corn fodder and clover hay *ad lib.*; property of H. C. Taylor, Orfordville, Wis.

Miss Sassafras 70468.—Sire, Raleigh Lisgar 17981; dam, Arbutus of A. 42529.

Butter, 17 lbs. 2 oz. Milk, 298 lbs. 5 oz.

Test made from May 27 to June 3, 1896; age, 6 yrs. 1 mo.; actual weight, 1100 lbs.; fed 99 lbs. corn and oats, ground, in equal parts—timothy and clover pasture; property of D. C. Grunder, Angelica, N. Y.

Miss Satanella 31544.—Sire, Master Vermont 4304; dam, Satanella 8927.

Butter, 20 lbs. 6 oz. Milk, 275 lbs.

Test made from June 23 to 30, 1889; age, 5 yrs.; estimated weight, 900 lbs.; fed, daily, about 20 qts. beans, oats, peas and corn ground together, with a little oil meal; property of Mrs. E. M. Jones, Brockville, Ont., Can.

Miss Satanella 31544.—Sire, Master Vermont 4304; dam, Satanella 8927.

Butter, 78 lbs. 3½ oz. in 31 days. Milk, 1164 lbs. 4 oz.

Test made from May 12 to June 12, 1889; age, 4 yrs. 11 mos.; fed. daily,.18 lbs. ground oats, peas and wheat bran; good pasture; property of Mrs. E. M. Jones, Brockville, Ont., Can.

Miss Signal 20379.—Sire, Alpenoa 6062; dam, Euphorbia 11229.

Butter, 16 lbs. 4 oz. Milk, 223 lbs. 3 oz.

Test made from May 24 to 31, 1886; age, 4 yrs.; property of Peter D. Hulst, East Penfield, N. Y.

Miss Vermont 7698, imp.

Butter, 16 lbs. 5 oz.

Test made Dec., 1879; age, 6 yrs.; property of W. R. McCready, Saugatuck, Conn.

Miss Willie Jones 6918.—Sire, Mopsus 1165; dam, Bronx Pearl 3752.

Butter, 16 lbs. 4 oz. Milk, 316 lbs. 8 oz.

Test made from May 21 to 28, 1883; age, 7 yrs. 1 mo.; fed 4 qts. corn meal and 4 qts. bran, daily—good pasture; property of William S. Taylor, Burlington, N. J.

Misty Morn 88725.—Sire, Noremac 26958; dam, Vigo Princess 60806.

Butter, 17 lbs. 5 oz. Milk, 289 lbs. 12 oz.

Test made from May 24 to 31, 1897; age, 4 yrs. 1 mo.; estimated weight, 800 lbs.; fed 42 qts. corn and cob meal and 56 qts. wheat bran—good blue-grass pasture; property of Columbus Dixon, Gillespieville, Ohio.

Mitten 13368.—Sire, Niobe Duke 2304; dam, Martinet 6418.

Butter, 15 lbs. 11 oz. Milk, 173 lbs.

Test made from May 9 to 16, 1884; age, 4 yrs. 2 mos.; property of Edwin Thorne, Millbrook, N. Y.

Mittretta's Rose 56192.—Sire, Ida's Stoke Pogis 13658; dam, Mittretta 3d 42242.

Butter, 15 lbs. ½ oz. Milk, 156 lbs. 15 oz.

Test made from Nov. 17 to 24, 1894; age, 6 yrs.; actual weight, 780 lbs.; fed 17 lbs. per day of ground oats and corn, in equal parts, mixed with 4 bucketfuls of cut sheaf oats, wetted—clover hay *ad lib.* at night; property of Matt. M. Gardner, Nashville, Tenn.

Mocha's Pet's Rheta 37529.—Sire, Chippewa 5739; dam, Mocha's Pet 12985.

Butter, 14 lbs. 2½ oz. Milk, 325 lbs. 8 oz.

Test made from July 16 to 23, 1893; age, 8 yrs. 10 mos.; estimated weight, 800 lbs.; fed 6 qts. corn meal, 6 qts. ground oats, 3 qts. bran, ¾ qt. oil meal and 21 lbs. cut hay and green fodder, daily; property of Pierce J. Cogswell, Rochester, N. Y.

Modiste 74679.—Sire, Compass 16958; dam, Belboma's Beauty 65941.

Butter, 16 lbs. 11½ oz. Milk, 169 lbs. 6 oz.

Test made from Dec. 19 to 26, 1894; age, 4 yrs. 2 mos.; estimated weight, 750 lbs.; fed 6 qts. corn meal, 6 qts. oat meal, 12 qts. bran and 3 qts. oil meal, daily, with beets; property of C. A. Sweet, East Aurora, N. Y.

Modita 16626.—Sire, Tony Calypso 3354; dam, Laura Lee 8550.

Butter, 16 lbs. 8 oz. Milk, 243 lbs. 8 oz.

Test made from Dec. 3 to 10, 1891; age, 9 yrs. 11 mos.; estimated weight, 800 lbs.; fed about 5 lbs. corn meal, 6 lbs. oat meal, 6 lbs. bran and 2 lbs. oil meal, daily; property of Richardson Bros., Davenport, Iowa.

Moggy 59626.—Sire, Mythology 12448; dam, Scribe 53894.

Butter, 14 lbs. 8 oz. Milk, 230 lbs. 4 oz.

Test made from Apr. 3 to 10, 1895; age, 6 yrs. 3 mos.; estimated weight, 700 lbs.; fed 28 lbs. oats, 24 lbs. corn, 21 lbs. bran and 21 lbs. oil meal—corn ensilage and hay; property of H. C. Taylor, Orfordville, Wis.

Moggy Bright 25891.—Sire, Bright (F. S. 308 J. H. B.); dam on I. of J.

Butter, 16 lbs. 6 oz., unsalted. Milk, 136 lbs.

Test made from June 25 to July 1, 1884; age, 3 yrs.; property of W. R. McCready, Saugatuck, Conn.

Moggy Bright 25891.—Sire, Bright (F. S. 308 J. H. B.); dam on I. of J.

Butter, 21 lbs. 15 oz. Milk, 241 lbs. 4 oz.

Test made from May 31 to June 6, 1887; age, 5 yrs. 11 mos.; estimated weight, 1000 lbs.; fed, daily, 4 qts. ground corn and oats, 4 qts. bran and 1 qt. pea meal; property of M. Erskine Miller, Staunton, Va.

Molina of Glen Rouge 123635.—Sire, One Hundred per Cent. 16590; dam, Molina of St. Lambert 69454.

Butter, 16 lbs. 2½ oz. Milk, 225 lbs. 8 oz.

Test made from Mar. 19 to 26, 1897; age, 1 yr. 9 mos.; estimated weight, 750 lbs.; fed 12 qts. oats and peas, mixed, and 2 lbs. oil-cake, per day—hay and ensilage *ad lib.*; property of William Rolph, Markham, Ont., Can.

Mollie Davis 26378.—Sire, Killarney 5179; dam, Florrie Ringgold 10048.

Butter, 16 lbs. 12½ oz. Milk, 199 lbs. 1 oz.

Test made from Oct. 2 to 9, 1888; age, 6 yrs. 6 mos.; estimated weight, 800 lbs.; fed 26 lbs. per day of ground oats and corn in equal parts—orchard and blue-grass pasture; property of Matt. M. Gardner, Nashville, Tenn.

Mollie Garfield 12172.—Sire, Bel Caliph 1432; dam, Maple Dale 2907.

Butter, 18 lbs. 7 oz.

Property of F. S. Peer, East Palmyra, N. Y.

Mollie Garfield 12172.—Sire, Bel Caliph 1432; dam, Maple Dale 2907.

Butter, 22 lbs. 12 oz. Milk, 258 lbs. 3 oz.

Test made from June 7 to 14, 1883; age, 7 yrs. 2 mos.; fed 12 qts., half bran and half ground corn and oats, daily—soiled on oats and peas; property of F. S. Peer, East Palmyra, N. Y.

Mollie Garfield 12172.—Sire, Bel Caliph 1432; dam, Maple Dale 2907.

Butter, 163 lbs. in 62 days. Milk, 1877¾ lbs.

Test made July and August, 1882; age, 6 yrs. 5 mos. at end of test; property of F. S. Peer, East Palmyra, N. Y.

Mollie Garfield 2d 18662.—Sire, Judge Mayo 5149; dam, Mollie Garfield 12172.

Butter, 15 lbs. 14 oz. Milk, 162 lbs. 6 oz.

Test made from Aug. 16 to 23, 1883; age, 3 yrs. 4 mos.; property of F. S. Peer, East Palmyra, N. Y.

Mollie of Glen Rouge 129125.—Sire, Two Hundred per Cent. 33592; dam, Nora of Glen Rouge 69452.

Butter, 15 lbs. 12 oz. Milk, 200 lbs. 12 oz.

Test made from Mar. 2 to 9, 1898; age, 1 yr. 9 mos.; estimated weight, 600 lbs.; fed 15 lbs. peas and oats, chopped, 4 lbs. oil-cake and 30 lbs. corn ensilage, daily—hay *ad lib.*; property of William Rolph, Markham, Ont., Can.

Mollie of St. Lambert 34644.—Sire, Exile of St. Lambert 13657; dam, Mollie Garfield 2d 18662.

Butter, 15 lbs. 4 oz. Milk, 205 lbs.

Test made from July 16 to 23, 1891; age, 6 yrs.; estimated weight, 900 lbs.; fed 8 lbs. corn germs, 4 lbs. ground oats and 2 lbs. oil meal, daily—pasture, clover, timothy and brush sprouts; property of S. H. Evans, Tidioute, Pa.

Mollie St. Helier 2d 98389.—Sire, Warren's Duke 21380; dam, Mollie St. Helier 35717.

Butter, 14 lbs. 12 oz. Milk, 141 lbs.

Test made from Dec. 10 to 17, 1894; age, 2 yrs. 2 mos.; estimated weight, 600 lbs.; fed 42 qts. corn meal, 42 qts. wheat bran and 7 qts. cotton-seed meal—hay *ad lib.*; property of B. P. Holliday, Prairie Station, Miss.

Mollie's Fancy 69728.—Sire, Texas Pogis 20194; dam, Cousin May's Mollie 46382.

Butter, 14 lbs. 8 oz. Milk, 226 lbs. 9 oz.

Test made from Feb. 5 to 12, 1897; age, 6 yrs. 6 mos.; estimated weight, 800 lbs.; fed 42 lbs. oats and corn ground together and 116 lbs. ensilage—hay *ad lib.*; property of J. D. Gray, Terrell, Tex.

Moll Roberts 110201.—Sire, Leonette's Landseer 20399; dam, Champion's Caroline 4th 65300.

Butter, 14 lbs. 6½ oz. Milk, 199 lbs. 13 oz.

Test made from Mar. 2 to 9, 1897; age, 3 yrs.; estimated weight, 800 lbs.; fed 63 lbs. bran, 42 lbs. oats and corn ground together, 14 lbs. oil meal and 28 lbs. cotton-seed hulls—hay *ad lib.*, pastured on barley; property of J. D. Gray, Terrell, Tex.

Molly 3554.—Sire, Dick 1410; dam, Venus 112.

Butter, 15 lbs. 9 oz.

Test made from Sept. 5 to 12, 1871; age, 9 yrs. 2 mos.; property of William S. Lincoln, Worcester, Mass.

Molly May 17202.—Sire, Mingo 948; dam, Lottie May 17201.

Butter, 14 lbs. 15 oz. Milk, 246 lbs.

Test made from Apr. 18 to 24, 1885; age, 3 yrs. 11 mos.; property of M. M. Small, Cooperstown, Pa.

Molly May Pogis 102823.—Sire, Lorne Victor Pogis 22214; dam, Maud of Ingleside 74044.

Butter, 15 lbs. 12 oz. Milk, 228 lbs.

Test made from Sept. 5 to 12, 1896; age, 3 yrs. 5 mos.; estimated weight, 875 lbs.; fed 28 lbs. corn meal, 28 lbs. ground oats and 28 lbs. wheat middlings and bran, mixed in equal parts—common pasture; property of John C. McClintock, Meadville, Pa.

Mona 3d 23134.—Sire, Tom Kenelm 11023; dam, Mona 1461.

Butter, 14 lbs. 6 oz. Milk, 276 lbs.

Test made from May 12 to 18, 1884; age, 5 yrs. 2 mos.; property of R. Rowett, Carlinville, Ill.

Mona Caird 75776.—Sire, Aurora Chief 20039; dam, Daisy Harrison 69253.

Butter, 14 lbs. 11 oz. Milk, 297 lbs.

Test made from Apr. 17 to 24, 1895; age, 5 yrs. 2 mos.; actual weight, 900 lbs.; fed 21 qts. corn meal, 63 qts. bran, 10½ qts. oil meal, 21 qts. cotton-seed meal and 84 qts. ground oats; property of C. A. Sweet, East Aurora, N. Y.

Mona Caird 75776.—Sire, Aurora Chief 20039; dam, Daisy Harrison 69253.

Butter, 17 lbs. 10½ oz. Milk, 333 lbs.

Test made from Apr. 24 to May 1, 1895; age, 5 yrs. 2 mos.; actual weight, 900 lbs.; fed 21 qts. corn meal, 63 qts. wheat bran, 10½ qts. oil meal, 21 qts. cotton-seed meal and 84 qts. ground oats; property of C. A. Sweet, East Aurora, N. Y.

Mona Caird 75776.—Sire, Aurora Chief 20039; dam, Daisy Harrison 69253.

Butter, 21 lbs. 8½ oz. Milk, 346 lbs. 8 oz.

Test made from May 1 to 8, 1895; age, 5 yrs. 2 mos.; actual weight, 900 lbs.; fed 21 qts. corn meal, 63 qts. wheat bran, 10½ qts. oil meal, 21 qts. cotton-seed meal and 84 qts. ground oats; property of C. A. Sweet, East Aurora, N. Y.

Mona Caird 75776.—Sire, Aurora Chief 20039; dam, Daisy Harrison 69253.

Butter, 24 lbs. 1½ oz. Milk, 349 lbs. 12 oz.

Test made from May 8 to 15, 1895; age, 5 yrs. 2 mos.; actual weight, 900 lbs.; fed 21 qts. corn meal, 63 qts. wheat bran, 10½ qts. oil meal, 21 qts. cotton-seed meal and 84 qts. ground oats; property of C. A. Sweet, East Aurora, N. Y.

Mona Caird 75776.—Sire, Aurora Chief 20039; dam, Daisy Harrison 69253.

Butter, 23 lbs. 15 oz. Milk, 366 lbs. 4 oz.

Test made from May 15 to 22, 1895; age, 5 yrs. 2 mos.; actual weight, 900 lbs.; fed 21 qts. corn meal, 63 qts. wheat bran, 10½ qts. oil meal, 21 qts. cotton-seed meal and 84 qts. ground oats; property of C. A. Sweet, East Aurora, N. Y.

Mona Caird 75776.—Sire, Aurora Chief 20039; dam, Daisy Harrison 69253.

Butter, 25 lbs. 7¼ oz. Milk, 362 lbs. 4 oz.

Test made from May 22 to 29, 1895; age, 5 yrs. 3 mos.; actual weight, 900 lbs.; fed 21 qts. corn meal, 63 qts. wheat bran, 10½ qts. oil meal, 21 qts. cotton-seed meal and 84 qts. ground oats; property of C. A. Sweet, East Aurora, N. Y.

Mona Caird 75776.—Sire, Aurora Chief 20039; dam, Daisy Harrison 69253.

Butter, 127 lbs. 5¾ oz. in 42 days. Milk, 2054 lbs. 12 oz.

Test made from Apr. 17 to May 29, 1895; age, 5 yrs. 2 mos.; actual weight, 900 lbs.; fed 126 qts. corn meal, 378 qts. wheat bran, 63 qts. oil meal, 126 qts. cotton-seed meal and 504 qts. ground oats; property of C. A. Sweet, East Aurora, N. Y.

Monarch's Star 80232.—Sire, St. Lambert's Monarch 25802; dam, Susie Pogis 35425.

Butter, 18 lbs. 4 oz. Milk, 237 lbs.

Test made from Nov. 4 to 11, 1896; age, 4 yrs. 10 mos.; estimated weight, 850 lbs.; fed 8 qts. corn meal, 2 qts. pea meal, 4 qts. cotton-seed meal, 8 qts. cotton-seed, 12 qts. bran and 2 qts. oats, cooked together by steam, daily—hay *ad lib.*, wheat and Bermuda grass pasture; property of C. & J. Merzbacher, Marshall, Tex.

Mona's Beauty 50041.—Sire, Duke of Mona 10639; dam, Cæsarea 3d 22588.

Butter, 15 lbs. 4 oz. Milk, 171 lbs.

Test made from Apr. 11 to 18, 1892; age, 5 yrs. 3 mos.; estimated weight, 850 lbs.; fed 21 lbs. cotton-seed meal, 116 lbs. corn meal, 168 lbs. bran and fine feed, 128 lbs. hay and 35 lbs. roots; property of Pierce J. Cogswell, Rochester, N. Y.

Mona's Mab 31524.—Sire, Snedens 4882; dam, Mona of Woodstock 6914.

Butter, 15 lbs. 7 oz. Milk, 204 lbs. 12 oz.

Test made from Nov. 6 to 13, 1892; age, 8 yrs. 11 mos.; estimated weight, 875 lbs.; fed 42 qts. corn meal, 42 qts. shorts and 28 qts. ground oats—short pasture; property of C. H. Ellsworth, Worcester, Mass.

Monmouth Duchess 3895.—Sire, Hector 129; dam, Julia 3893.

Butter, 14 lbs. 7 oz., unsalted.

Test made from June 1 to 7, 1876; property of W. A. Conover, Hackettstown, N. J.

Monmouth Duchess 3d 4620.—Sire, Optimus 1607; dam, Monmouth Duchess 3895.

Butter, 14 lbs. 4 oz. Milk, 160 lbs. 10 oz.

Test made from Oct. 16 to 22, 1883; age, 10 yrs. 6 mos.; property of W. A. Conover, Hackettstown, N. J.

Monmouth Duchess 4th 7129.—Sire, Optimus 1607; dam, Monmouth Duchess 3895.

Butter, 18 lbs. Milk, 196 lbs.

Test made from June 20 to 26, 1883; age, 8 yrs. 2 mos.; property of W. A. Conover, Hackettstown, N. J.

Monocacy Dimple 9680.—Sire, Blennerhasset 3334; dam, Dimple 3248.

Butter, 14 lbs. 3 oz.

Test made from May 5 to 11, 1885; age, 5 yrs. 6 mos.; property of S. F. Gray, Indianapolis, Ind.

Mon Plaisir Lily 25969, imp.

Butter, 16 lbs. 2½ oz. Milk, 255 lbs.

Test made from May 5 to 12, 1889; age, 6 yrs. 1 mo.; estimated weight, 900 lbs.; fed 8 qts. ground oats, daily; property of J. K. Sheets, Troy, Ohio.

Mon Plaisir of Yerba Buena Ranch 12622, imp.

Butter, 18 lbs. 12 oz. Milk, 239 lbs. 8 oz.

Test made from Mar. 6 to 12, 1884; age, 6 yrs.; property of Henry Pierce, San Francisco, Cal.

Monte Sano Belle 77683.—Sire, Signalda's Torment 14673; dam, Genie Walker 38443.

Butter, 15 lbs. 15½ oz. Milk, 181 lbs. 4 oz.

Test made from Mar. 8 to 15, 1896; age, 6 yrs. 3 mos.; estimated weight, 800 lbs.; fed 38½ lbs. bran, 84 lbs. oats and corn and cob meal, mixed, 17½ lbs. oil meal, 21 lbs. cotton-seed meal, 105 lbs. roots and 210 lbs. ensilage—hay *ad lib.*; property of C. I. Hood, Lowell, Mass.

Mosly Pedro 77811.—Sire, Othick Pedro 20511; dam, May Pedro 50876.

Butter, 17 lbs. 2 oz. Milk, 252 lbs.

Test made from Sept. 4 to 11, 1895; age, 5 yrs. 11 mos.; estimated weight, 650 lbs.; fed 10 lbs. oats, peas and corn, ground together in equal parts. 2 lbs. oil meal and 6 lbs. wheat bran, daily—green corn; property of J. H. Martin, Granville, N. Y.

Moonah's Pet 7484.—Sire, Marmaduke 483; dam, Moonah 2d 7483.

Butter, 15 lbs. 6 oz. Milk, 222 lbs. 12 oz.

Test made from Apr. 1 to 7, 1885; age, 8 yrs. 11 mos.; property of V. Lowe, Palmyra, Wis.

Moorey of Lawn 68847.—Sire, Potiphar 19988; dam, Bettie Belle Nimmo 39960.

Butter, 21 lbs. 4 oz. Milk, 267 lbs. 8 oz.

Test made from Mar. 28 to Apr. 4, 1895; age, 4 yrs. 11 mos.; estimated weight, 850 lbs.; fed 14 lbs. cotton-seed meal, 42 lbs. corn meal, 42 lbs. wheat bran and 119 lbs. hay; property of T. H. Creager, Denison, Tex.

Moragina 26344.—Sire, Kapper's Victor 12340; dam, Belle Morgan 26219.

Butter, 17 lbs. 3 oz. Milk, 233 lbs.

Test made from May 15 to 22, 1889; age, 7 yrs. 1 mo.; estimated weight, 875 lbs.; fed 15 qts. corn and oat meal in three feeds, daily—hay *ad lib.*, orchard pasture; property of Richardson Bros., Davenport, Iowa.

Morganza Queen 77791.—Sire, Chieftain Pogis 25035; dam, Bristol Girl 66865.

Butter, 16 lbs. 2¾ oz. Milk, 247 lbs. 4 oz.

Test made from Mar. 17 to 24, 1897; age, 5 yrs. 8 mos.; estimated weight, 900 lbs.; fed 350 lbs. ensilage, 70 lbs. hay, 17½ lbs. ground corn, 17½ lbs. ground oats, 35 lbs. bran, 28 lbs. gluten and 21 lbs. oil meal; property of Pennsylvania Reform School, Morganza, Pa.

Morlacchi 2725.—Sire, Duke of Grayholdt 1035; dam, Caramel 2727.

Butter, 14 lbs.

Property of Tennessee Hospital for Insane, Nashville, Tenn.

Mosa 69691.—Sire, Lexicon 23110; dam, Moggy 59626.

Butter, 14 lbs. 6 oz. Milk, 255 lbs. 6 oz.

Test made from Apr. 27 to May 4, 1895; age, 4 yrs. 2 mos.; estimated weight, 700 lbs.; fed 21 lbs. corn meal, 14 lbs. oil meal, 28 lbs. oats and 21 lbs. bran—rye pasture; property of H. C. Taylor, Orfordville, Wis.

Moss Rose of Jackson 86709.—Sire, Moss Rose Victor 19123; dam, Rioter Lisgar's Rose 64497.

Butter, 16 lbs. 4 oz. Milk, 299 lbs. 8 oz.

Test made from June 28 to July 5, 1898; age, 5 yrs. 5 mos.; estimated weight, 800 lbs.; fed 4 lbs. bran, 4 lbs. gluten meal and 2 lbs. oil meal, daily—pasture during day, cut grass in barn at night; property of H. C. Vanderhoef, Belmont, N. Y.

Moss Rose of St. Lambert 5114.—Sire, Buffer 2055; dam, Camelia of St. Lambert 5106.

Butter, 14 lbs. ½ oz., unsalted. Milk, 208 lbs.

Test made from July 2 to 8, 1883; age, 8 yrs. 2 mos.; property of William Rolph, Markham, Ont., Can.

Moss Rose of Willow Farm 5194.—Sire, Young Concord 1406; dam, Molly 3554.

Butter, 23 lbs. 1 oz.   Milk, 228 lbs. 5 oz.

Test made from Nov. 26 to Dec. 2, 1885; age, 12 yrs.; fed 6 qts. corn meal and 2 qts. shorts, daily—hay *ad lib.*; property of Winslow S. Lincoln, Worcester, Mass.

Mother Cary 11746.—Sire, Ralph Guild 1917; dam, Argosy 4320.

Butter, 27 lbs. 1½ oz.   Milk, 149 lbs.

Test made from May 11 to 18, 1884; age, 4 yrs. 6 mos.; fed 2 qts. corn meal and 12 qts. wheat bran, daily—clover hay *ad lib.*, fair pasturage; property of Thomas L. Dodd, Nashville, Tenn.

Mother Hubbard 10331.—Sire, Duke of Darlington 2460; dam, Grace Darling 5574.

Butter, 24 lbs. 1½ oz.   Milk, 233 lbs. 8 oz.

Test made from Nov. 9 to 15, 1885; age, 5 yrs. 11 mos.; fed 4 qts. corn meal, 1 qt. pea meal, 1½ qts. oil meal, 3 qts. shorts, 2 qts. middlings, 4 qts. crushed oats, mixed, daily—a few roots and hay at night; property of F. C. Sayles, Pawtucket, R. I.

Moth of St. Lambert 9775.—Sire, Stoke Pogis 3d 2238; dam, Bessy of St. Lambert 5482.

Butter, 16 lbs. 2 oz.   Milk, 235 lbs. 8 oz.

Test made from June 13 to 19, 1883; age, 3 yrs. 10 mos.; fed 1 qt. barley meal in the morning and 3 pts. at night—old pasture; property of William Rolph, Markham, Ont., Can.

Mountain Lass 12921.—Sire, Ralph 957; dam, Starkville Beauty 4897.

Butter, 14 lbs. 9 oz.   Milk, 190 lbs.

Test made from May 10 to 16, 1884; age, 4 yrs. 5 mos.; property of W. B. Montgomery, Starkville, Miss.

Mousy 2d 14962.—Sire. Romulus (P. S. 181 J. H. B.); dam, Mousy (F. S. 1931 J. H. B.)

Butter, 17 lbs. 1 oz., unsalted.

Test made from Oct. 9 to 15, 1885; age, 6 yrs. 3 mos.; property of D. F. Appleton, Ipswich, Mass.

Moxie's Mode 97432.—Sire, Prince of Tennessee 20772; dam, Moxie G. 63504.

Butter, 16 lbs. 3¼ oz.   Milk, 199 lbs. 10½ oz.

Test made from Oct. 2 to 9, 1895; age, 2 yrs. 6 mos.; estimated weight, 650 lbs.; fed 15 lbs. per day of ground corn and oats, in equal parts, mixed, with three bucketfuls of cut sheaf oats, wetted—good orchard and blue-grass pasture; property of Matt. M. Gardner, Nashville, Tenn.

Moxie's Mode 2d 133253.—Sire. Maury's Tormentor 25113; dam, Moxie's Mode 97432.

Butter, 15 lbs. 9¼ oz.   Milk, 199 lbs. 11 oz.

Test made from Jan. 28 to Feb. 4, 1898; age, 2 yrs. 6 mos.; estimated weight, 625 lbs.; fed 17 lbs. per day of ground oats and corn, in equal parts, mixed with 4 bucketfuls of cut sheaf oats, wetted; property of Matt. M. Gardner, Nashville, Tenn.

Moyane 21595.—Sire, Progress 2d (P. S. 298 J. H. B.); dam, Tonic (F. S. 3804 J. H. B.)

Butter, 21 lbs. 12½ oz.   Milk, 276 lbs.

Test made from May 29 to June 5, 1889; age, 7 yrs. 2 mos.; estimated weight, 700 lbs.; fed, daily, about 3 gals. corn meal and 1 pt. cotton-seed meal; property of M. C. Campbell, Spring Hill, Tenn.

Moyane C. 86492.—Sire, Oonan's Tormentor 22280; dam, Moyane's Fancy 78452.

Butter, 37 lbs. 15¼ oz. in 14 days. Milk, 433 lbs. 7½ oz.

Test made from Dec. 18 to Jan. 1, 1896; age, 3 yrs. 6 mos.; actual weight, 750 lbs.; fed 22 lbs. per day of ground oats and corn, in equal parts, mixed with 4 bucketfuls of cut sheaf oats, wetted—millet hay *ad lib.* at night; property of Matt. M. Gardner, Nashville, Tenn.

Moyane's Pet 98115.—Sire, Moyane's Tormentor 21690; dam, Brenda St. Heller 20052.

Butter, 16 lbs. 12 oz. Milk, 259 lbs. 4 oz.

Test made from Jan. 2 to 9, 1898; age, 5 yrs. 2 mos.; estimated weight, 900 lbs.; fed 6 lbs. corn and cob meal, 6 lbs. Quaker oats feed, 3 lbs. bran, 1½ lbs. oil meal and 1½ lbs. cotton-seed meal, daily—32 lbs. corn and pea ensilage per day; property of Newton Frazier, Clark, Ky.

Mrs. Jack 50805.—Sire, Prince Pogis 10682; dam, Lorna Doone of P. 31335.

Butter, 15 lbs. 5 oz. Milk, 247 lbs. 12 oz.

Test made from May 4 to 11, 1893; age, 5 yrs. 6 mos.; actual weight, 740 lbs.; fed 28 qts. corn meal, 21 qts. ground oats, 14 qts. cotton-seed meal, 28 qts. wheat shorts, 40 lbs. hay and 350 lbs. corn ensilage; property of J. F. Gulliver, Andover, Mass.

Mrs. Knickerbocker 19367.—Sire, Bingo 1811; dam, Miss Beauty 4053.

Butter, 15 lbs. 8 oz. Milk, 305 lbs. 8 oz.

Test made from May 21 to 27, 1885; age, 6 yrs.; property of L. L. Cramer, Macedon, N. Y.

Mrs. Knickerbocker 19367.—Sire, Bingo 1811; dam, Miss Beauty 4053.

Butter, 16 lbs. 8½ oz. Milk, 190 lbs. 4 oz.

Test made from June 23 to 30, 1887; age, 8 yrs. 1 mo.; estimated weight, 850 lbs.; no grain fed during test—grass only; property of James Stillman, Sing Sing, N. Y.

Mulberry 22031.—Sire, Victor (P. S. 148 J. H. B.); dam, Unique (F. S. 1035 J. H. B.)

Butter, 14 lbs. 2 oz. Milk, 197 lbs.

Test made from July 10 to 17, 1887; age, 8 yrs.; estimated weight, 900 lbs.; fed, daily, about 10 qts. ground corn, oats and peas, with about 2 qts. bran; property of Mrs. E. M. Jones, Brockville, Ont., Can.

Mulford's Lizzie Walker 50213.—Sire, Broadsword 7149; dam, Lizzie Walker 10134.

Butter, 16 lbs. 5 oz. Milk, 229 lbs. 10 oz.

Test made from May 7 to 14, 1893; age, 6 yrs.; estimated weight, 850 lbs.; fed 25 qts. oat meal, 28 qts. corn meal, 10 qts. oil meal, 35 qts. bran and 1 bush. roots—clover and timothy pasture; property of Mrs. E. F. Hawley, Pittsford, N. Y.

Muriel 5th 19017.—Sire, Lenape 2732; dam, Muriel 3904.

Butter, 16 lbs. 12½ oz. Milk, 192 lbs. 4 oz.

Test made from Oct. 18 to 24, 1886; age, 3 yrs. 9 mos.; estimated weight, 900 lbs.; fed 4½ lbs. oats, 2½ lbs. oil meal, 3½ lbs. shorts, 3¾ lbs. bran, 3¾ lbs. corn meal, daily; property of C. W. H. Eicke, West Monterey, Pa.

Musicale of Greenwood 62932.—Sire, Profundo Basso 20321; dam, Miss Serenade 40301.

Butter, 15 lbs. 14 oz. Milk, 194 lbs. 6 oz.

Test made from July 5 to 12, 1894; age, 5 yrs. 9 mos.; estimated weight, 750 lbs.; fed 55 lbs. corn and 29 lbs. oats, ground together—very dry clover pasture; property of Houston & Zick, Beloit, Wis.

Musidove 25379.—Sire, Pilot of Bowker Farm 6085; dam, Dove 5th 14561.

Butter, 18 lbs. 2 oz. Milk, 267 lbs. 12 oz.

Test made from June 13 to 20, 1889; age, 6 yrs. 1 mo.; estimated weight, 875 lbs.; fed, daily, 4 qts. corn meal, 4 qts. shorts and 4 qts. ground oats; property of Terence Carrigan, Hopkinton, Mass.

Musidove 3d 61641.—Sire, Pilot of Bowker Farm 6085; dam, Musidove 25379.

Butter, 15 lbs. 1 oz. Milk, 217 lbs.

Test made from June 12 to 19, 1894; age, 6 yrs. 4 mos.; estimated weight, 850 lbs.; fed 12 lbs. per day of a mixture of bran, middlings, corn meal and oil meal—natural pasture; property of George W. Sisson, Jr., Potsdam, N. Y.

Musquita 28462.—Sire, Ramapo 4679; dam, Money Musk 13714.

Butter, 17 lbs. 3 oz. Milk, 227 lbs. 12 oz.

Test made from Aug. 27 to Sept. 3, 1889; age, 5 yrs. 6 mos.; estimated weight, 775 lbs.; fed, daily, 10 lbs. corn meal, 10 lbs. ground oats and 10 lbs. middlings; property of D. F. Appleton, Ipswich, Mass.

Muxio 97423.—Sire, Fancy's Pogis 21797; dam, Gold Mosie 63502.

Butter, 15 lbs. 2 oz. Milk, 214 lbs. 2 oz.

Test made from Jan. 15 to 22, 1897; age, 4 yrs. 9 mos.; estimated weight, 850 lbs.; fed 140 lbs. ensilage, 112 lbs. wheat bran, 112 lbs. corn and oats chop and 28 lbs. corn meal—alfalfa hay; property of Platter & Foster, Denison, Tex.

My Alpha's Beauty 56469.—Sire, Beauty's Stoke Pogis 14175; dam, My Alpha's Victory 43304.

Butter, 15 lbs. Milk, 164 lbs. 8 oz.

Test made from Aug. 3 to 10, 1894; age, 6 yrs. 6 mos.; estimated weight, 900 lbs.; fed 10 qts. chop, 4 qts. bran and 12 stalks green corn, daily—poor pasture; property of Mrs. A. M. Hallock, Columbus, Ohio.

My May Pogis 84236.—Sire, Odelle's Pogis 25075; dam, Myra C. 27837.

Butter, 14 lbs. 5 oz. Milk, 211 lbs.

Test made from Jan. 4 to 11, 1897; age, 4 yrs. 8 mos.; estimated weight, 850 lbs.; fed 56 lbs. bran, 91 lbs. corn and cob meal and 21 lbs. cotton-seed meal—small bundle sheaf oats daily, hay *ad lib.*; property of James L. Cooper, Nashville, Tenn.

My Queen 12614.—Sire, Young Duke 138; dam, Carrie 3894.

Butter, 15 lbs. 8 oz.

Property of John V. N. Willis, Marlborough, N. J.

Mylitta 15668.—Sire, Kapper 2033; dam, Clover Blossom 4057.

Butter, 16 lbs. 13 oz. Milk, 271 lbs. 5 oz.

Test made from Apr. 12 to 19, 1884; age, 4 yrs. 10 mos.; estimated weight, 850 lbs.; fed, daily, 18 qts. ground oats and corn and pea meal; property of Richardson Bros., Davenport, Iowa.

Myra Tormentor 84227.—Sire, Little Tormentor 17995; dam, Myra C. 27837.

Butter, 15 lbs. 11½ oz. Milk, 241 lbs. 8 oz.

Test made from May 17 to 24, 1895; age, 3 yrs. 11 mos.; estimated weight, 800 lbs.; fed 56 qts. ground oats, 56 qts. ground corn and 14 qts. cotton-seed meal—fair bluegrass pasture; property of J. L. Cooper, Nashville, Tenn.

Myretta 28231.—Sire, Myron 7157; dam, Ulalloa 14687.

Butter, 14 lbs. 11¼ oz. Milk, 183 lbs. 12 oz.

Test made from July 6 to 12, 1886; age, 2 yrs. 8 mos.; property of G. R. Hill, Oxford, Miss.

Myrrha 11299.—Sire, Top-Sawyer 1404; dam, Marietta 1813.

Butter, 16 lbs. 1 oz. Milk, 169 lbs. 4 oz.

Test made from Apr. 26 to May 2, 1886; age, 5 yrs. 6 mos.; property of Samuel McEwen, Columbia, Tenn.

Myrrha 3d 50406.—Sire, Fancy's Harry 9777; dam, Myrrha 11299.

Butter, 14 lbs. ½ oz. Milk, 119 lbs.

Test made from Jan. 14 to 21, 1894; age, 6 yrs. 9 mos.; estimated weight,.750 lbs.; fed 56 qts. ground oats and 56 qts. corn meal—clover hay; property of Morgan & Brown, Columbia, Tenn.

Myrtle 2d 211.—Sire, Blücher 2d 102; dam, Myrtle 208.

Butter, 15 lbs. 12 oz.

Test made from July 1 to 8, 1875; age, 6 yrs.; property of Thomas Fitch, New London, Conn.

Myrtle of Ridgewood 7858.—Sire, Ben Rajah 795; dam, Myrtle 2d 211.

Butter, 14 lbs. 1 oz. Milk, 184 lbs.

Test made from Nov. 6 to 12, 1882; age, 4 yrs. 2 mos.; property of Charles R. Christy, Stamford, Conn.

Myrtle Vance 84237.—Sire, Odelle's Pogis 25075; dam, Elsie Miller 59261.

Butter, 14 lbs. 3 oz. Milk, 212 lbs.

Test made from Feb. 7 to 14, 1897; age, 4 yrs. 8 mos.; estimated weight, 850 lbs.; fed 70 qts. corn and cob meal, 70 qts. bran and 14 qts. cotton-seed meal—hay *ad lib.*; property of James L. Cooper, Nashville, Tenn.

Myrtlise 64006.—Sire, Denise's Tormentor 11823; dam, Grace Pansy 2d 18764.

Butter, 16 lbs. 13½ oz. Milk, 176 lbs. 8 oz.

Test made from Nov. 11 to 18, 1893; age, 5 yrs. 2 mos.; estimated weight, 850 lbs.; fed 12 qts. ground corn and oats, 12 qts. wheat bran and 3 qts. oil meal, daily—timothy hay; property of Charles A. Sweet, Buffalo, N. Y.

Myrtlise 64006.—Sire, Denise's Tormentor 11823; dam, Grace Pansy 2d 18764.

Butter, 18 lbs. 9 oz. Milk, 165 lbs. 4 oz.

Test made from Nov. 18 to 25, 1893; age, 5 yrs. 2 mos.; estimated weight, 850 lbs.; fed 15 qts. ground corn and oats, 12 qts. wheat bran and 3 qts. oil meal, daily—timothy hay; property of Charles A. Sweet, Buffalo, N. Y.

Myrtlise 64006.—Sire, Denise's Tormentor 11823; dam, Grace Pansy 2d 18764.

Butter, 19 lbs. 9 oz. Milk, 204 lbs. 4 oz.

Test made from Nov. 25 to Dec. 2, 1893; age, 5 yrs. 2 mos.; estimated weight, 850 lbs.; fed 15 qts. ground corn and oats, 12 qts. wheat bran, 3 qts. oil meal, cooked, and 20 qts. beets, daily—timothy hay; property of Charles A. Sweet, Buffalo, N. Y.

Myrtlise 64006.—Sire, Denise's Tormentor 11823; dam, Grace Pansy 2d 18764.

> Butter, 54 lbs. 15½ oz. in 21 days. Milk, 546 lbs.

Test made from Nov. 11 to Dec. 2, 1893; age, 5 yrs. 2 mos.; estimated weight, 850 lbs.; fed 294 qts. ground corn and oats, 252 qts. wheat bran, 63 qts. oil meal and 140 qts. beets—timothy hay; property of Charles A. Sweet, Buffalo, N. Y.

Mysinda 2d 42250.—Sire, Ida's Stoke Pogis 13658; dam, Mysinda 27133.

> Butter, 18 lbs. 8 oz. Milk, 226 lbs. 4 oz.

Test made from May 11 to 18, 1891; age, 4 yrs. 8 mos.; estimated weight, 850 lbs.; fed 6 lbs. corn meal, 6 lbs. crushed oats, 3 lbs. oil meal, 8 lbs. pea meal and 3 lbs. bran, daily—65 lbs. hay; property of Ayer & McKinney, Philadelphia, Pa.

Myth 2837.—Sire, Count Bismarck 732; dam, Damsel 2d 1837.

> Butter, 14 lbs. 6 oz. Milk, 248 lbs.

Test made from May 22 to 28, 1881; age, 7 yrs.; property of Edward Worth, Wawa, Pa.

Naiad of St. Lambert 12965.—Sire, Stoke Pogis 3d 2238; dam, Lolly of St. Lambert 5480.

> Butter, 22 lbs. 2¼ oz. (official). Milk, 267 lbs.

Test made from June 5 to 12, 1884; age, 4 yrs. 4 mos.; property of V. E. Fuller, Hamilton, Ont., Can.

Nal Day 26553.—Sire, Signalda 2d 6748; dam, Nan Day 17192.

> Butter, 19 lbs. 13 oz. Milk, 210 lbs.

Test made from Apr. 28 to May 4, 1886; age, 2 yrs. 3 mos.; property of R. McMichael, Lexington, Ky.

Nanalda 32917.—Sire, Signalda 4027; dam, Oonan 1485.

> Butter, 14 lbs. 2⅛ oz. Milk, 137 lbs. 7 oz.

Test made from Dec. 22 to 29, 1889; age, 4 yrs. 11 mos.; estimated weight, 1000 lbs.; fed, daily, 4 gals. ground oats and corn meal, in equal parts; property of M. C. Campbell, Spring Hill, Tenn.

Nanalda 2d 97130.—Sire, Oonan's Pogis 17165; dam, Nanalda 32917.

> Butter, 16 lbs. 1 oz. Milk, 249 lbs. 11 oz.

Test made from June 8 to 15, 1897; age, 4 yrs.; estimated weight, 800 lbs.; fed 22 lbs. per day of ground oats and corn, in equal parts—good orchard and blue-grass pasture; property of Matt. M. Gardner, Nashville, Tenn.

Nancy Lee 7618.—Sire, Claimant (P. S. 84 J. H. B.); dam, Nonpareil (F. S. 1248 J. H. B.)

> Butter, 26 lbs. 8½ oz., unsalted. Milk, 360 lbs. 12 oz.

Test made from June 28 to July 4, 1883; age, 7 yrs. 2 mos.; fed 4 qts. corn meal and 6 qts. bran, daily—fair pasture; property of C. Easthope, Niles, Ohio.

Nancy Lovelock 15511.—Sire, Silver Mine 1658; dam, Empress Eugenie 13549.

> Butter, 17 lbs. 9 oz. Milk, 207 lbs. 8 oz.

Test made from Mar. 1 to 8, 1885; age, 4 yrs.; property of J. B. Wallace, Lexington Ky.

Nancy of St. Lambert 12964.—Sire, Stoke Pogis 3d 2238; dam, Lucy of St. Lambert 5116.

Butter, 14 lbs. 5 oz. Milk, 180 lbs. 8 oz.

Test made from Nov. 26 to Dec. 2, 1883; age, 3 yrs. 10 mos.; fed 8 lbs. bran and 8 lbs. crushed oats, daily—mangels and hay; property of V. E. Fuller, Hamilton, Ont., Can.

Nancy's Star 41423.—Sire, Nancy's Rioter 8390; dam, Gordonetta 19570.

Butter, 16 lbs. 1⅝ oz. Milk, 194 lbs. 10 oz.

Test made from Oct. 6 to 13, 1889; age, 4 yrs. 1 mo.; estimated weight, 800 lbs.; fed, daily, 4 gals. ground corn and oats, mixed in equal parts; property of Maury Jersey Farm, Columbia, Tenn.

Nan Day 17192.—Sire, Duke of Ouaquaga 2740; dam, Bonnie Jean 13116.

Butter, 20 lbs. 4 oz. Milk, 209 lbs.

Test made from Apr. 28 to May 4, 1885; age, 3 yrs.; fed 22 qts. oat, pea and oil meal and bran, daily—short blue-grass pasture; property of Robert McMichael, Lexington, Ky.

Nanette of Hood Farm 104447.—Sire, Pogis Whiting 28907; dam, Nanette Scituate 67183.

Butter, 15 lbs. 8½ oz. Milk, 241 lbs. 2 oz.

Test made from Mar. 29 to Apr. 5, 1897; age, 4 yrs. 1 mo.; estimated weight, 750 lbs.; fed 48 lbs. bran, 22½ lbs. corn meal, 28 lbs. oats, 7 lbs. oil meal, 105 lbs. roots and 280 lbs. ensilage—hay *ad lib.*; property of C. I. Hood, Lowell, Mass.

Nannaxie 18860.—Sire, Maxse 400; dam, Anne 412.

Butter, 17 lbs. 4 oz. Milk, 245 lbs. 6 oz.

Test made from Feb. 8 to 15, 1881; age, 7 yrs. 5 mos.; property of Cornelius Whitbeck, Craryville, N. Y.

Nannie of Hood Farm 108565.—Sire, Fancy's Harry 9777; dam, Nannie Lee Morgan 56863.

Butter, 14 lbs. 5¼ oz. Milk, 207 lbs. 11 oz.

Test made from Feb. 22 to Mar. 1, 1898; age, 4 yrs. 3 mos.; estimated weight, 800 lbs.; fed 42 lbs. bran, 28 lbs. corn meal, 21 lbs. ground oats, 14 lbs. oil meal, 210 lbs. ensilage and 70 lbs. roots—hay *ad lib.*; property of C. I. Hood, Lowell, Mass.

Nannie Pogis 53786.—Sire, Prince of Melrose 2d 11015; dam, Litta Oaks' Alphea 26676.

Butter, 16 lbs. 11 oz. Milk, 238 lbs. 14 oz.

Test made from Oct. 5 to 12, 1891; age, 3 yrs. 8 mos.; estimated weight, 850 lbs.; fed 28 qts. corn meal, 56 qts. bran, 28 qts. raw cotton-seed and 175 lbs. rich corn ensilage—sorghum hay, poor pasture; property of James Crook, Jacksonville, Ala.

Nan of Argyle 47561.—Sire, Baron Hugo 15208; dam, Belle of Argyle 18358.

Butter, 14 lbs. 5 oz. Milk, 280 lbs.

Test made from July 4 to 11, 1893; age, 4 yrs. 7 mos.; estimated weight, 900 lbs.; no grain fed during test—pasture, good new stocking, clover and mixed grasses; property of E. S. Henry, Rockville, Conn.

Nan of Ethelwyn 49502.—Sire, Champion of Ethelwyn 14302; dam, Purple Mulberry 3d 12100.

Butter, 15 lbs. 12½ oz.   Milk, 298 lbs. 8 oz.

Test made from Feb. 23 to Mar. 1, 1896; age, 9 yrs. 11 mos.; estimated weight, 900 lbs.; fed 98 lbs. of a mixture of bran, middlings, Chicago gluten and old process linseed meal, in equal parts—91 lbs. hay and corn fodder; property of H. G. Mansfield, Westborough, Mass.

Nan of St. Lambert 69453.—Sire, Canada's John Bull 5th 20092; dam, St. Lambert's Nancy 56503.

Butter, 15 lbs. 9 oz.   Milk, 222 lbs.

Test made from Aug. 18 to 25, 1891; age, 2 yrs.; estimated weight, 700 lbs.; fed 22 lbs. daily of chopped oats and peas—pasture, second crop clover; property of William Rolph, Markham, Ont., Can.

Nanoonan 98775.—Sire, Oonan's Tormentor 22280; dam, Ida's Nan 49957.

Butter, 14 lbs. 11⅝ oz. (confirmed); estimated butter, 85 per cent. fat, 14 lbs. 9¼ oz.   Milk, 276.5 lbs.

Test made from July 11 to 17, 1898; age, 4 yrs. 2 mos.; average weight, 860 lbs.; fed 56 lbs. corn meal and 70 lbs. bran—good pasture; property of T. R. Webber, Shelbyville, Ky.

Analyses of Butter.

First churning—Fat 81.15, water 15.28, ash 2.76, casein 0.81.
Second churning—Fat 81.17, water 14.75, ash 3.46, casein 0.62.

Naomi's Pride 16745.—Sire, Ridgely of Hampton 3046; dam, Naomi Ronsby 16742.

Butter, 15 lbs. 2 oz.   Milk, 190 lbs. 4 oz.

Test made from Aug. 30 to Sept. 5, 1883; age, 3 yrs. 3 mos.; property of A. C. Jennings, Urbana, Ohio.

Narda 38012.—Sire, Tormentor 3533; dam, Mattie Hunter 19798.

Butter, 15 lbs. 1 oz.   Milk, 172 lbs.

Test made from May 2 to 9, 1891; age, 6 yrs. 1 mo.; estimated weight, 1000 lbs.; fed 24 lbs. corn-hearts, daily; property of Campbell Brown, Spring Hill, Tenn.

Narmeoka H. 12323.—Sire, Bismarck (Fairbanks') 1277; dam, Lady Lightfoot 2745.

Butter, 18 lbs. 12 oz.   Milk, 381 lbs. 1 oz.

Test made from June 13 to 19, 1885; age, 8 yrs. 6 mos.; property of William H. Hayden, Albany, Vt.

Natalie Victor 73689.—Sire, D. of D.'s Victor 11726; dam, Pythoness 55199.

Butter, 17 lbs. 12 oz.   Milk, 324 lbs. 12 oz.

Test made from May 24 to 31, 1897; age, 5 yrs. 9 mos.; actual weight, 910 lbs.; fed 4 qts. ground corn and oats, in equal parts, and 6 qts. wheat bran, daily, fed dry—clover and timothy pasture; property of John A. Webber, Portland, Mich.

Natasqua 65598.—Sire, Fair Corinth 12455; dam, Queen Emma 32560.

Butter, 161.522 lbs. in 90 days.   Milk, 2463.9 lbs.

Test made from May 31 to Aug. 28, 1893, at World's Columbian Exposition; age, 5 yrs. 1 mo. at beginning of test; actual weight, 844 lbs.; fed 300 lbs. old hay, 123 lbs. ensilage, 486.6 lbs. new hay, 190.5 lbs. oil meal, 398 lbs. bran, 72 lbs. oats, 436.5 lbs. corn-hearts, 124.25 lbs. cotton-seed, 296 lbs. middlings, 34 lbs. grano-gluten, 1137.4 lbs. clover and 10 lbs. swale-grass; property of T. A. Havemeyer, Mahwah, N. J.

Nazell 2d 98271.—Sire, Little Harry 8808; dam, Nazell 80709.

Butter, 16 lbs. 8½ oz. Milk, 227 lbs. 5 oz.

Test made from June 17 to 24, 1898; age, 4 yrs. 2 mos.; estimated weight, 775 lbs.; fed 23 lbs. per day of ground oats and corn, in equal parts—very good orchard and blue-grass pasture; property of Matt. M. Gardner, Nashville, Tenn.

Nazli 10327.—Sire, Duke of Darlington 2460; dam, Grace Darlington 5574.

Butter, 15 lbs. 3½ oz., unsalted. Milk, 232 lbs. 8 oz.

Test made from Apr. 29 to May 5, 1883; age, 4 yrs. 6 mos.; fed 6 qts. corn, oats and bran, in equal parts, daily; property of A. B. Darling, Ramsey's, N. J.

Neitherme 64670.—Sire, Tormentor 2d 7124; dam, Lady Rareripe 23081.

Butter, 15 lbs. 1 oz. Milk, 258 lbs.

Test made from Mar. 6 to 13, 1897; age, 9 yrs. 2 mos.; estimated weight, 950 lbs.; fed 100 lbs. corn meal, 30½ lbs. bran, 15 lbs. oats and 8¾ lbs. oil meal—corn fodder *ad lib.*, woods pasture; property of W. E. Johnson, Millican, Tex.

Nelida 2d 8227.—Sire, Beauclerc 1882; dam, Nelida 5346.

Butter, 15 lbs. 2½ oz. Milk, 262 lbs. 2 oz.

Test made from Apr. 11 to 17, 1882; age, 3 yrs. 9 mos.; property of D. A. Givens, Cynthiana, Ky.

Nellie 1507.—Sire, Potomac 153; dam, Bell 1506.

Butter, 14 lbs. 2 oz.

Test made from June 26 to July 2, 1876; age, 9 yrs. 1 mo.; property of Clarke & Jones, Baltimore, Md.

Nellie Alphea 51274.—Sire, Champion of the West 14271; dam, Josie B. 32749.

Butter, 14 lbs. 1½ oz. Milk, 170 lbs. 8 oz.

Test made from Sept. 17 to 24, 1890; age, 4 yrs. 2 mos.; estimated weight, 700 lbs.; fed, daily, 24 lbs. corn fodder and 8 lbs. shorts; property of Mrs. M. B. Pack, Edinburg, Mo.

Nellie Bunker 52163.—Sire, Bunker 9025; dam, Zinnia 2d 20003.

Butter, 14 lbs. 4 oz. Milk, 170 lbs. 6 oz.

Test made from June 18 to 25, 1891; age, 5 yrs.; estimated weight, 750 lbs.: fed 6 qts. bran and 2 qts. corn meal, daily—fair pasture; property of T. R. Proctor, Utica, N. Y.

Nellie Darlington 5956.—Sire, Smith of Darlington 2458; dam, Grace Darlington 5574.

Butter, 15 lbs. 3 oz., unsalted. Milk, 194 lbs. 4 oz.

Test made from Mar. 20 to 26, 1883; fed 3 qts. corn and 3 qts. oats, daily; property of A. B. Darling, Ramsey's, N. J.

Nellie Hugo 53793.—Sire, St. Lambert's John Bull 16618; dam, Miss Nellie Parker 24988.

Butter, 20 lbs. Milk, 242 lbs. 6 oz.

Test made from May 8 to 15, 1892; age, 4 yrs. 1 mo.; estimated weight, 800 lbs.; fed 56 qts. pea meal, 56 qts. bran and shorts, 28 qts. cotton-seed meal and 35 lbs. corn ensilage—hay, good mixed pasture; property of James Crook, Jacksonville, Ala.

Nellie of St. Lambert 69458.—Sire, Canada's John Bull 5th 20092; dam, St. Lambert's Nora 42132.

Butter, 23 lbs. 12 oz. Milk, 318 lbs.

Test made from Dec. 6 to 13, 1897; age, 7 yrs. 9 mos.; estimated weight, 800 lbs.; fed 22 qts. chopped peas and oats, 4 qts. bran and 4 lbs. oil-cake, daily—hay *ad lib.*; property of William Rolph, Markham, Ont., Can.

Nelly 2402, imp.

Butter, 15 lbs. 14 oz.

Property of W. L. & W. Rutherford, Waddington, N. Y.

Nelly 6456.—Sire, Lemon (F. S. 170 J. H. B.); dam, Little Browny (P. S. 29 J. H. B.)

Butter, 21 lbs. Milk, 344 lbs.

Test made from May 27 to June 2, 1881; age, 9 yrs.; fed 6 qts. corn meal, 12 qts. bran and 1 qt. oil meal, daily—rye and blue-grass pasture; property of Samuel Stratton, Litchfield, Ill.

Nelly Gray of Clermont 10905.—Sire, Achmed 2115; dam, Signora 4723.

Butter, 14 lbs. 1 oz.

Test made from May 15 to 22, 1883; age, 5 yrs. 1 mo.; property of D. A. Smalley, Milford, Ohio.

Nelly Thorn 83344.—Sire, King Coomassie 2d 19545; dam, Thornleaf 39969.

Butter, 16 lbs. 7 oz. Milk, 234 lbs. 12 oz.

Test made from May 30 to June 6, 1894; age, 3 yrs. 5 mos.; actual weight, 820 lbs.; fed 42 qts. corn meal, 28 qts. oats and 56 qts. wheat bran—clover and blue-grass pasture; property of Columbus Dixon, Gillespieville, Ohio.

Nelly Torment 61903.—Sire, Romp's Torment 8789; dam, Glendale Beauty 27690.

Butter, 15 lbs. 7 oz. Milk, 233 lbs. 2 oz.

Test made from June 11 to 18, 1893; age, 5 yrs. 4 mos.; estimated weight, 800 lbs.; fed 56 qts. bran, 28 qts. corn meal and 14 qts. oil meal—blue-grass pasture; property of Henry DuBois, Vigo, Ohio.

Nelope Pedro 50879.—Sire, King Peto 10541; dam, Nellie of Cedar Glen 10949.

Butter, 15 lbs. 4 oz. Milk, 219 lbs. 12 oz.

Test made from June 24 to July 1, 1892; age, 5 yrs. 3 mos.; estimated weight, 850 lbs.; no grain fed—clover pasture only; property of Mrs. J. H. Martin, Granville, N. Y.

Nerbat 58664.—Sire, Nemonette's Rioter 16106; dam, Black Teat 32932.

Butter, 15 lbs. 7 oz. Milk, 212 lbs. 15½ oz.

Test made from Dec. 16 to 23, 1894; age, 6 yrs. 11 mos.; estimated weight, 850 lbs.; fed 70 lbs. crushed corn, 56 lbs. crushed oats and 5½ lbs. old process oil-cake—2 bush. roots, fall wheat pasture; property of J. M. Hund, Hund's Station, Kan.

Nerissa of Nyack 9692.—Sire, Rory O'More 3236; dam, Rosa of Salem 6476.

Butter, 15 lbs. 1 oz. Milk, 248 lbs. 8 oz.

Test made from Apr. 2 to 8, 1885; age, 5 yrs. 2 mos.; property of Miller & Sibley, Franklin, Pa.

Nerobella's Blossom 48051.—Sire, Toltec 2d 11073; dam, Nerobella 2d 29654.

Butter, 15 lbs. 1 oz. Milk, 225 lbs. 10 oz.

Test made from Apr. 8 to 15, 1894; age, 7 yrs. 7 mos.; estimated weight, 900 lbs.; fed 70 qts. ground corn, 14 qts. cotton-seed meal and 56 qts. ground oats—poor mesquite pasture; property of Samuel C. Bell, San Antonio, Tex.

Nervine 25932.—Sire, Tisquantum (P. S. 262 J. H. B.); dam, La Rouge (P. S. 45 J. H. B.)

Butter, 14 lbs. 1½ oz. Milk, 233 lbs. 10 oz.

Test made from July 14 to 20, 1884; age, 2 yrs. 4 mos.; property of F. P. Scearce, Lexington, Ky.

Nerviona 43678.—Sire, Catonoson 9667; dam, Nervine 25932.

Butter, 14 lbs. 15 oz. Milk, 215 lbs. 2 oz.

Test made from Dec. 19 to 26, 1891; age, 5 yrs. 7 mos.; estimated weight, 980 lbs.; fed 42 qts. bran, 28 qts. corn meal, 28 qts. ground oats and 14 qts. oil meal—clover and timothy hay *ad lib.*; property of Henry DuBois, Vigo, Ohio.

Nerviona 2d 98687.—Sire, Fancy's Harry 6th 24768; dam, Nerviona 43678.

Butter, 14 lbs. 3 oz. Milk, 208 lbs. 15 oz.

Test made from Oct. 19 to 26, 1895; age, 3 yrs.; estimated weight, 850 lbs.; fed 56 qts. ground oats, 42 qts. corn meal and 28 qts. wheat bran—blue-grass pasture; property of Henry DuBois, Vigo, Ohio.

Nestor's Beauty of St. L. 74884.—Sire, Nestor of St. Lambert 22385; dam, Miss Garda 40520.

Butter, 14 lbs. Milk, 119 lbs. 12 oz.

Test made from Aug. 21 to 28, 1894; age, 2 yrs. 11 mos.; estimated weight, 700 lbs.; fed 4 gals. cut corn fodder and bran, in equal parts, daily—pasture; property of Mrs. A. M. Hallock, Columbus, Ohio.

Nettie N. 58307.—Sire, Stoke Pogis Prince 11005; dam, Dora Norton 46293.

Butter, 15 lbs. Milk, 288 lbs.

Test made from June 16 to 23, 1894; age, 6 yrs.; actual weight, 810 lbs.; fed 28 qts. corn meal, 28 qts. ground oats and 28 qts. shorts—good clover and blue-grass pasture; property of T. H. Jones, Mount Sterling, Ohio.

Nettie Signal 37837.—Sire, Signal Duke 12661; dam, Printonette 26121.

Butter, 18 lbs. 8 oz. Milk, 249 lbs.

Test made from Sept. 20 to 27, 1892; age, 6 yrs. 7 mos.; estimated weight, 800 lbs.; fed 10 lbs. corn meal, 5 lbs. oat meal, 7 lbs. oil meal and 3 lbs. bran, daily—blue-grass pasture; property of John A. Middelton, Shelbyville, Ky.

Nettie's New Year 99295.—Sire, Stoke Pogis' Perfection 5984; dam, Nettie's Accident 61236.

Butter, 23 lbs. 3 oz. Milk, 372 lbs. 14.4 oz.

Test made from Feb. 28 to Mar. 7, 1897; age, 6 yrs. 2 mos.; estimated weight, 950 lbs.; fed 8 lbs. bran, 10 lbs. corn and oats, 2½ lbs. linseed meal, 2½ lbs. cotton-seed meal and 28 lbs. corn ensilage, daily—hay *ad lib.*; property of John E. Robbins, Greensburg, Ind.

Nettie's Picture 50960.—Sire, Eurotas' Victor Hugo 15689; dam, Nettie Artiste 31653.

Butter, 15 lbs. 8 oz.   Milk, 299 lbs. 4 oz.

Test made from May 27 to June 3, 1895; age, 7 yrs. 5 mos.; estimated weight, 930 lbs.; fed 1 qt. oil meal, 1 pt. cotton-seed meal and 6½ qts. of a mixture of corn meal and middlings, in equal parts, daily—timothy and clover pasture; property of A. B. Darling, Ramsey's, N. J.

Nettina of Winnikee 46406.—Sire, Hohokus 5569; dam, Phrynetine 16198.

Butter, 18 lbs. 15½ oz.   Milk, 290 lbs. 12 oz.

Test made from Dec. 24 to 31, 1889; age, 5 yrs. 4 mos.; estimated weight, 950 lbs.; fed, daily, 15 lbs. meal, 7½ lbs. bran and about 1 lb. oil meal; property of Rufus A. Sibley, Spencer, Mass.

Nettina Silverette 89997.—Sire, Ninetta's Son 29074; dam, Nutley Silverette 22410.

Butter, 18 lbs. 14½ oz.   Milk, 317 lbs. 4 oz.

Test made from Feb. 27 to Mar. 6, 1897; age, 4 yrs. 5 mos.; estimated weight, 900 lbs.; fed 63 qts. bran, 10½ qts. oil meal, 28 qts. corn meal, 168 lbs. ensilage and 136 lbs. hay; property of R. A. Sibley, Rochester, N. Y.

New London Gipsy 11667.—Sire, Pierrot 2d 1669; dam, Flirt 482.

Butter, 14 lbs. 8 oz.   Milk, 241 lbs.

Test made from Apr. 27 to May 3, 1882; age, 6 yrs. 2 mos.; fed one-half bush. of corn, boiled on the cob, daily—young blue-grass pasture; property of Garretson Bros., Pendleton, Ind.

New Maud Lee 16614.—Sire, Glue 3960; dam, Carrie Waite 10641.

Butter, 14 lbs. 10 oz.   Milk, 200 lbs.

Test made from July 2 to 8, 1885; age, 3 yrs. 8 mos.; property of A. J. Fish, Findlay, Ohio.

Nibbette 11625.—Sire, Fast Boy 2606; dam, Nibbie 6796.

Butter, 14 lbs. 7 oz.   Milk, 232 lbs. 12 oz.

Test made Oct., 1883; age, 4 yrs. 4 mos.; property of T. L. Haecker, Madison, Wis.

Nibbie 6796.—Sire, The Squire 1298; dam, Nimmie 968.

Butter, 14 lbs. 15 oz.   Milk, 220 lbs. 1 oz.

Test made from May 22 to 29, 1885; age, 7 yrs. 10 mos.; property of Wisconsin Agricultural Experiment Station.

Nice 68027.—Sire, Diploma 16219; dam, Nicomis 68026.

Butter, 14 lbs. 12 oz.   Milk, 184 lbs.

Test made from Dec. 16 to 23, 1894; age, 4 yrs. 7 mos.; estimated weight, 750 lbs.; fed 35 lbs. oats, 21 lbs. oil meal, 21 lbs. corn meal, 21 lbs. bran and 21 lbs. wheat—corn ensilage and clover hay; property of H. C. Taylor, Orfordville, Wis.

Nicomis 68026.—Sire, Upright 6147; dam, Laura Lee 8550.

Butter, 14 lbs. 3 oz.   Milk, 200 lbs. 2 oz.

Test made from June 1 to 8, 1890; age, 2 yrs. 5 mos.; estimated weight, 775 lbs.; fed, daily, 12 qts. ground corn and oats; property of Richardson Bros., Davenport, Iowa.

Nienchen 48698.—Sire, Chuffy Chief 17625; dam, Sigrianna 34941.

Butter, 14 lbs. 4 oz. Milk, 211 lbs.

Test made from Jan. 3 to 10, 1893; age, 5 yrs. 3 mos.; estimated weight, 800 lbs.; fed 8 qts. wheat middlings, 6 qts. ground oats, 2 qts. corn meal, 2 qts. oil meal, with 40 lbs. ensilage, daily—hay *ad lib.*; property of George W. Sisson, Jr., Potsdam, N. Y.

Nienchen 48698.—Sire, Chuffy Chief 17625; dam, Sigrianna 34941.

Butter, 17 lbs. 2 oz. Milk, 225 lbs.

Test made from Mar. 11 to 18, 1894; age, 6 yrs. 5 mos.; estimated weight, 875 lbs.; fed 12 lbs. to 14 lbs. per day of bran, middlings, corn meal and oil meal, mixed, and 30 lbs. ensilage per day—hay *ad lib.*; property of G. W. Sisson, Jr., Potsdam, N. Y.

Nienchen 48698.—Sire, Chuffy Chief 17625; dam, Sigrianna 34941.

Butter, 73 lbs. 13 oz. in 31 days. Milk, 987 lbs. 8 oz.

Test made from Feb. 26 to Mar. 29, 1894; age, 6 yrs. 5 mos.; estimated weight, 875 lbs.; fed 12 lbs. to 14 lbs. bran, middlings, corn meal and oil meal, and 30 lbs. ensilage per day—hay *ad lib.*; property of George W. Sisson, Jr., Potsdam, N. Y.

Nigella 7895.—Sire, Fast Boy 2606; dam, Nitella 4423.

Butter, 16 lbs. 3 oz. Milk, 215 lbs. 12 oz.

Test made from Apr. 4 to 11, 1887; age, 8 yrs. 9 mos.; estimated weight, 750 lbs.; fed, daily, 8 lbs. corn meal, 2 lbs. oil meal and 8 lbs. bran; property of James Stillman, Sing Sing, N. Y.

Nightingale K. 2d 19841.—Sire, Frank Warren 1490; dam, Nightingale K. 16056.

Butter, 16 lbs. 8 oz.

Test made from June 1 to 7, 1883; age, 9 yrs. 11 mos.; property of Henry Hamilton, Brookfield, Ohio.

Nightingale of Elmarch 8312.—Sire, Merry Boy, on I. of J.; dam, Nellie, on I. of J.

Butter, 14 lbs. 2 oz., unsalted. Milk, 196 lbs.

Test made from June 21 to 27, 1881; age, 7 yrs.; property of D. A. Givens, Cynthiana, Ky.

Nimble 22335.—Sire, Mercury 432; dam, Pyrrha 6100.

Butter, 14 lbs. 10 oz. Milk, 171 lbs. 5 oz.

Test made from Jan. 20 to 26, 1885; age, 1 yr. 9 mos.; property of William Simpson, New York, N. Y.

Niobe 99, imp.

Butter, 14 lbs.

Property of Sam. J. Sharpless, Philadelphia, Pa.

Niobe Gordon 3d 44041.—Sire, Pedro of the Valley 8750; dam, Niobe Gordon 13922.

Butter, 14 lbs. 3 oz. Milk, 234 lbs. 12 oz.

Test made from Oct. 6 to 13, 1889; age, 3 yrs. 9 mos.; estimated weight, 900 lbs.; fed, daily, 6 qts. provender with 2 qts. cotton-seed meal on each of the first three days, changed to 2 qts. old process oil meal on the remainder; property of M. W. Terrill, Middlefield, Conn.

Niobe Marigold 108363.—Sire, Stoke Pogis of Prospect 29121; dam, Niobe's Alpheanette 8d 39459.

Butter, 15 lbs. 8½ oz. Milk, 215 lbs. 4 oz.

Test made from June 27 to July 4, 1897; age, 2 yrs. 4 mos.; actual weight, 800 lbs.; fed 9 qts. bran, 9 qts. ground oats, 4½ qts. corn meal and 3 qts. oil meal, daily—plenty of good grass in barn; property of C. A. Sweet, Buffalo, N. Y. .

Niobe of Linwood 11134.—Sire, Prince Ridgely 3047; dam, Niobe of Beechwood 11131.

Butter, 14 lbs. 12 oz. Milk, 196 lbs. 8 oz.

Test made from July 19 to 26, 1885; age, 5 yrs. 5 mos.; property of W. M. Bell, Miami, Mo.

Niobe of St. Lambert 12969.—Sire, Stoke Pogis 3d 2238; dam, Estelle of St. Lambert 7011.

Butter, 21 lbs. 9¼ oz. (official). Milk, 280 lbs. 12 oz.

Test made from July 14 to 21, 1884; age, 4 yrs.; property of V. E. Fuller, Hamilton, Ont., Can.

Niobe's Alpheanette 23336.—Sire, Alpha of Clifton 1824; dam, Mary M. Allison 6308.

Butter, 22 lbs. 10½ oz. Milk, 183 lbs. 9½ oz.

Test made from Aug. 4 to 10, 1885; age, 2 yrs. 11 mos.; fed 3 lbs. corn meal, 3½ lbs. pea meal, 3 lbs. wheat middlings, 4 lbs. crushed oats, 3½ lbs. wheat bran and 1 lb. oil meal, mixed, and given as a warm mash twice a day—medium good pasture; property of C. W. H. Eicke, West Monterey, Pa.

Niobe's Alpheanette 3d 39459.—Sire, Yellow Boy 6381; dam, Niobe's Alpheanette 23336.

Butter, 19 lbs. 4½ oz. Milk, 216 lbs. 8 oz.

Test made from Feb. 3 to 10, 1894; age, 7 yrs. 6 mos.; estimated weight, 900 lbs.; fed 18 qts. bran, 3 qts. oil meal, 9 qts. ground corn and oats and 6 qts. wheat middlings, daily—hay; property of C. A. Sweet, East Aurora, N. Y.

Niobomba 26456.—Sire, Prince of Darlington 5089; dam, Blossom's Niobe 13855.

Butter, 14 lbs. 15 oz. Milk, 196 lbs.

Test made from Nov. 23 to 30, 1890; age, 7 yrs.; estimated weight, 1000 lbs.; fed 56 qts. corn bran and 14 qts. cotton-seed meal—Bermuda grass pasture; property of A. J. Hawes & Son, Montgomery, Ala.

Nitelis 102722.—Sire, Alney 7502; dam, Lady Walker of Creekside 21984.

Butter, 16 lbs. ½ oz. Milk, 329 lbs. 12 oz.

Test made from May 7 to 14, 1898; age, 6 yrs.; actual weight, 840 lbs.; fed 43 lbs. wheat bran, 46 lbs. corn meal, 28 lbs. cotton-seed meal and 44 lbs. cotton-seed hulls—pasture. crimson clover, oats, red clover and orchard grass; property of J. W. Hart, Clemson College, S. C.

Niva 7523.—Sire, Caen 2317; dam, Countess Gisela 2820.

Butter, 15 lbs. 8 oz. Milk, 201 lbs. 7 oz.

Test made from Apr. 14 to 20, 1883; age, 4 yrs. 9 mos.; property of C. A. Keefer, Sterling, Ill.

Noble Charity 67245.—Sire, Charity Boy 11765; dam, Calista Noble 36065.

Butter, 18 lbs. Milk, 259 lbs. 12 oz.

Test made from June 17 to 24, 1896; age, 7 yrs. 1 mo.; estimated weight, 800 lbs.; fed 8 lbs. ground oats and peas, 4 lbs. corn meal and 6 lbs. bran, daily—good mixed pasture; property of J. H. Martin, Granville, N. Y.

No Chemicals 71452.—Sire, Prince Pogis 10682; dam, Eulogy 30634.

Butter, 16 lbs. 9 oz. Milk, 236 lbs. 8 oz.

Test made from Feb. 21 to 28, 1893; age, 2 yrs. 8 mos.; estimated weight, 690 lbs.; fed 24 qts. wheat shorts, 13 qts. ground oats, 15½ qts. corn meal, 15½ qts. cotton-seed meal, 250 lbs. corn ensilage and about 50 lbs. hay; property of J. F. Gulliver, Andover, Mass.

Nonnie Pogis 41432.—Sire, Ida's Stoke Pogis 13658; dam, Tormentor's Cinderella 19564.

Butter, 18 lbs. 12 oz. Milk, 234 lbs. 4 oz.

Test made from Mar. 20 to 27, 1895; age, 8 yrs. 5 mos.; actual weight, 843 lbs.; fed 40 lbs. corn meal. 40 lbs. ground wheat, 20 lbs. wheat bran and 10 lbs. oil-cake meal, mixed with shredded corn-stover—clover hay *ad lib.*; property of J. P. Bradbury, Pomeroy, Ohio.

Nonsuch of Linwood 29028.—Sire, Gold Basis 4038; dam, Lady of Belle Vue 7705.

Butter, 14 lbs. 14½ oz. Milk, 209 lbs. 3 oz.

Test made from July 22 to 29, 1888; age, 3 yrs. 11 mos.; weight, 720 lbs.; fed, daily, 8 lbs. bran, 4 lbs. corn meal, 2 lbs. cotton-seed meal; property of Jacob L. Thomas, Knoxville, Tenn.

Nora Clinton 91938.—Sire, Fannie's Son 28014; dam, Nora Loadstone 2d 84482.

Butter, 18 lbs. 4 oz. Milk, 244 lbs. 10 oz.

Test made from Nov. 4 to 11, 1896; age, 4 yrs. 2 mos.; estimated weight, 1000 lbs.; fed 35 lbs. corn meal, 35 lbs. oats, 35 lbs. bran and 24 lbs. oil meal—frosted clover pasture; property of H. C. Taylor, Orfordville, Wis.

Nora of Glen Rouge 69452.—Sire, Canada's John Bull 5th 20092; dam, Rinora Pogis 40107.

Butter, 16 lbs. 13 oz. Milk, 217 lbs.

Test made from Jan. 18 to 25, 1892; age, 2 yrs. 8 mos.; estimated weight, 700 lbs.; fed 140 lbs. chopped oats and 320 lbs. corn ensilage—clover hay *ad lib.*; property of William Rolph, Markham, Ont., Can.

Nora of St. Lambert 12962.—Sire, Stoke Pogis 3d 2238; dam, Duchess of St. Lambert 5111.

Butter, 14 lbs. 7 oz. Milk, 243 lbs.

Test made from Dec. 6 to 12, 1882; age, 2 yrs. 11 mos.; property of W. D. Reesor, Markham, Ont., Can.

Nora of St. Lambert 12962.—Sire, Stoke Pogis 3d 2238; dam, Duchess of St. Lambert 5111.

Butter, 22 lbs. Milk, 289 lbs.

Test made from Sept. 2 to 8, 1885; age, 5 yrs. 8 mos.; fed 20 lbs. barley meal daily—clover and timothy hay; property of William Rolph, Markham, Ont., Can.

Nora Pogis 75345.—Sire, Ida of St. Lambert's Bull 19169; dam, Oaklands Nora 14880.

Butter, 14 lbs. 9 oz. Milk, 199 lbs. 6 oz.

Test made from Nov. 15 to 22, 1892; age, 2 yrs. 5 mos.; estimated weight, 900 lbs.; fed 42 lbs. pea meal, 28 lbs. oil meal, 28 lbs. cotton-seed meal, 42 lbs. mixed feed and 140 lbs. ensilage; property of Ayer & McKinney, Philadelphia, Pa.

Nora Pogis 75345.—Sire, Ida of St. Lambert's Bull 19169; dam, Oaklands Nora 14880.

Butter, 21 lbs. 14¾ oz. Milk, 283 lbs. 8 oz.

Test made from June 25 to July 2, 1895; age, 5 yrs.; actual weight, 920 lbs.; fed 56 qts. corn meal, 56 qts. ground oats and 28 qts. ground oil-cake—good old meadow pasture; property of Miller & Sibley, Franklin, Pa.

Nora Sheldon 43590.—Sire, Sheldon of St. Lambert 13831; dam, Pine Tree Nora 32799.

Butter, 17 lbs. 6 oz. Milk, 233 lbs. 3 oz.

Test made from Sept. 18 to 25, 1890; age, 4 yrs. 4 mos.; estimated weight, 750 lbs.; fed, daily, 8 lbs. corn meal, 8 lbs. middlings and 7 lbs. ground oats; property of D. F. Appleton, Ipswich, Mass.

Nora Stoke Pogis 34687.—Sire, Exile of St. Lambert 13657; dam, Eva Locust 21050.

Butter, 16 lbs. 1 oz. Milk, 248 lbs. 8 oz.

Test made from June 24 to July 1, 1888; age, 3 yrs. 2 mos.; estimated weight, 850 lbs.; fed, daily, 5½ lbs. corn meal, 5½ lbs. oil meal, 6¾ lbs. bran, 2¼ lbs. middlings and 3 lbs. oats; property of A. D. McBride, Rochester, N. Y.

Nora Tormentor of Lawn 99738.—Sire, Tormentor's Last 24930; dam, Chloe Tormentor of Lawn 71642.

Butter, 15 lbs. 12 oz. Milk, 226 lbs. 3 oz.

Test made from Dec. 28, 1895, to Jan. 4, 1896; age, 2 yrs. 10 mos.; estimated weight, 800 lbs.; fed 70 lbs. wheat bran, 70 lbs. ground corn and oats, 28 lbs. cotton-seed meal and 28 lbs. linseed meal—prairie hay *ad lib.*; property of Platter & Foster, Denison, Tex.

Nordheim Creamer 9758.—Sire, Eclipse 1449; dam, Hecuba 3155.

Butter, 14 lbs. Milk, 196¼ lbs.

Test made from Feb. 14 to 21, 1881; age, 4 yrs. 1 mo.; property of J. W. North, Jr., Augusta, Me.

Normanda 3914.—Sire, Normandy 1046; dam, Olive 4th 3018.

Butter, 14 lbs. 14 oz. Milk, 175 lbs. 10 oz.

Test made from Nov. 22 to 28, 1884; age, 9 yrs. 9 mos.; property of Samuel McKeen, Terre Haute, Ind.

Nuna 18669.—Sire, Southville Boy 4271; dam, Abbie Z. 4th 14743.

Butter, 21 lbs. Milk, 231 lbs. 12 oz.

Test made from Feb. 4 to 11, 1888; age, 5 yrs. 10 mos.; estimated weight, 850 lbs.; fed 28 lbs. corn meal, 28 lbs. wheat bran, 28 lbs. oat meal, 28 lbs. cotton-seed meal and 70 lbs. roots; property of J. H. Walker, Worcester, Mass.

Nuphar Houghton 36364.—Sire, Houghton 5886; dam, Nuphar 3d 20455.

Butter, 15 lbs. 3 oz. Milk, 185 lbs. 7 oz.

Test made from Nov. 3 to 10, 1890; age, 5 yrs. 6 mos.; estimated weight, 850 lbs.; fed, daily, about 5½ lbs. bran, 5½ lbs. oats, 5 lbs. corn meal and ¼ lb. oil meal; property of George W. Burchard, Fort Atkinson, Wis.

Nutley Darling 22412.—Sire, Duke of Darlington 2460; dam, Nutley Alma 13581.

Butter, 15 lbs. 3½ oz. Milk, 251 lbs. 11 oz.

Test made from Nov. 9 to 16, 1887; age, 4 yrs. 3 mos.; estimated weight, 800 lbs.; fed 11 lbs. corn meal, 9 lbs. ground oats, 10 lbs. middlings, 2 lbs. oil meal and 1 lb. pea meal per day; property of D. F. Appleton, Ipswich, Mass.

Nutley Dolores 13797.—Sire, Tommy (P. S. 249 J. H. B.); dam, Nutley Alma 13581.

Butter, 23 lbs. 2 oz. Milk, 234 lbs.

Test made from Mar. 24 to 31, 1890; age, 8 yrs. 3 mos.; actual weight, 700 lbs.; fed, daily, 12 qts. corn meal, 6 qts. oil meal and 12 qts. ground oats; property of T. R. Proctor, Utica, N. Y.

Nutley Dolores 2d 55901.—Sire, Bunker 9025; dam, Nutley Dolores 13797.

Butter, 15 lbs. 13 oz. Milk, 157 lbs.

Test made from Jan. 23 to 30, 1890; age, 2 yrs. 1 mo.; actual weight, 725 lbs.; fed, daily, 3 qts. corn meal, 8 qts. ground oats, 1 qt. oil meal and 2 qts. shorts; property of T. R. Proctor, Utica, N. Y.

Nutley St. Lambert 58469.—Sire, Exile of St. Lambert 13657; dam, Nutley Silverette 22410.

Butter, 16 lbs. 8 oz. Milk, 202 lbs.

Test made from Dec. 22 to 29, 1892; age, 4 yrs. 3 mos.; estimated weight, 1050 lbs.; fed 84 qts. corn meal, 147 qts. bran, 21 qts. oil meal, 84 lbs. hay and 105 lbs. roots; property of R. A. Sibley, Rochester, N. Y.

Nutley Silverette 22410.—Sire, Nutley's Favorite 5132; dam, Nutley Silver 13591.

Butter, 15 lbs. 12¼ oz. Milk, 195 lbs. 12 oz.

Test made from Sept. 13 to 19, 1885; age, 3 yrs. 4 mos.; property of Frederick Loeser, Somerville, N. J.

Nutley Silverette 22410.—Sire, Nutley's Favorite 5132; dam, Nutley Silver 13591.

Butter, 22 lbs. 7 oz. Milk, 281 lbs. 12 oz.

Test made from Nov. 4 to 11, 1890; age, 8 yrs. 5 mos.; estimated weight, 975 lbs.; fed, daily, 10½ lbs. corn meal, 4½ lbs. bran and 3 lbs. cotton-seed meal; property of R. A. Sibley, Rochester, N. Y.

Nutshell 2d 50962.—Sire, Eurotas' Victor Hugo 15689; dam, Nutshell 10326.

Butter, 19 lbs. 2 oz. Milk, 240 lbs. 1 oz.

Test made from Mar. 30 to Apr. 6, 1892; age, 4 yrs. 1 mo.; estimated weight, 700 lbs.; fed 10 qts. corn meal, 8 qts. ground oats, 9 qts. shorts and 1 qt. oil meal, daily— hay *ad lib.*, no pasture; property of D. F. Appleton, Ipswich, Mass.

Nymphæa 5141.—Sire, Mercury 432; dam, Phædra 2561.

Butter, 15 lbs. 14½ oz. Milk, 242 lbs. 8 oz.

Test made from Aug. 6 to 12, 1883; age, 7 yrs. 3 mos.; property of William Simpson, New York, N. Y.

Nymphæa 5141.—Sire, Mercury 432; dam, Phædra 2561.

Butter, 18 lbs. 7½ oz. Milk, 250 lbs. 9 oz.

Test made from June 30 to July 6, 1884; age, 8 yrs. 2 mos.; property of William Simpson, New York, N. Y.

Nymph of St. Lambert 12968.—Sire, Stoke Pogis 3d 2238; dam, Diana of St. Lambert 6636.

> Butter, 24 lbs. 14 oz. Milk, 280 lbs. 8 oz.

Test made from Nov. 23 to 29, 1885; age, 4 yrs. 5 mos.; fed 9 qts. oats, 8 qts. bran, a few mangels, ensilage and clover hay, daily—lucern pasture; property of Valancey E. Fuller, Hamilton, Ont., Can.

Oakland Girl 11103.—Sire, Thorndale 2582; dam, Clotilde 4575.

> Butter, 14 lbs. 10½ oz. Milk, 209 lbs. 8 oz.

Test made from Feb. 12 to 18, 1885; age, 4 yrs. 10 mos.; property of Miller & Sibley, Franklin, Pa.

Oaklands Cora 18853.—Sire, Jersey Boy (P. S. 92 J. H. B.); dam, Lively (F. S. 1401 J. H. B.)

> Butter, 19 lbs. 9½ oz., unsalted. Milk, 169 lbs.

Test made from Mar. 30 to Apr. 5, 1883; age, 6 yrs. 2 mos.; property of V. E. Fuller, Hamilton, Ont., Can.

Oaklands Lilly 14881.—Sire, Rambler of St. Lambert 5285; dam, Minette of St. Lambert 9774.

> Butter, 15 lbs. 4 oz. Milk, 247 lbs. 4 oz.

Test made from Feb. 12 to 18, 1888; age, 6 yrs. 10 mos.; weight, 1000 lbs.; fed 3 lbs. oats, 12 lbs. corn meal, 3¾ lbs. middlings and 2¼ lbs. oil meal, daily; property of Mrs. A. N. Martin, Summit, N. J.

Oaklands Nora 14880.—Sire, Lorne 5248; dam, Pet of St. Lambert 5123.

> Butter, 23 lbs. 5 oz. Milk, 153 lbs.

Test made from Sept. 20 to 26, 1885; age, 4 yrs. 7 mos.; fed 5 qts. pea meal, 3 qts. ground oats, 1 qt. bran, mixed, daily—good pasture; property of L. W. & H. W. Simonds, Berlin, Ont., Can.

Oaklands Nora 2d 45967.—Sire, Canada's John Bull Jr. 15693; dam, Oaklands Nora 14880.

> Butter, 15 lbs. 7 oz. Milk, 207 lbs. 9 oz.

Test made from June 2 to 9, 1890; age, 2 yrs. 9 mos.; estimated weight, 800 lbs.; fed, daily, 9 lbs. corn meal and ship stuff, in equal parts; property of H. G. Westlake, Hillsdale, N. Y.

Oaklands Nora 2d 45967.—Sire, Canada's John Bull Jr. 15693; dam, Oaklands Nora 14880.

> Butter, 19 lbs. 1 oz. Milk, 262 lbs. 13 oz.

Test made from Oct. 29 to Nov. 5, 1893; age, 6 yrs. 2 mos.; estimated weight, 900 lbs.; fed 42 qts. corn meal, 28 qts. ground oats, 14 qts. oil meal and 42 qts. ship feed —fair pasture; property of H. G. Westlake, Hillsdale, N. Y.

Oak Leaf 4769.—Sire, Devilshoof 866; dam, Echo 2223.

> Butter, 17 lbs. 10 oz.

Property of S. M. Burnham, Saugatuck, Conn.

Oak Leaf 4769.—Sire, Devilshoof 866; dam, Echo 2223.

> Butter, 63 lbs. 4 oz. in 31 days.

Property of John D. Wing, Millbrook, N. Y.

Odalette 76888.—Sire, Odelio 20223; dam, Roulette 50080.

Butter, 17 lbs. 3 oz. Milk, 281 lbs. 1½ oz.

Test made from May 9 to 16, 1895; age, 3 yrs. 4 mos.; estimated weight, 740 lbs.; fed 112 qts. wheat bran, 56 qts. corn meal, 21 qts. cotton-seed meal, 14 qts. linseed meal and 126 lbs. ensilage—green clover, short pasture; property of E. B. C. Hambley, Rockwell, N. C.

Odelle Sales 15564.—Sire, Tormentor 3533; dam, Oonan 1485.

Butter, 14 lbs. 5½ oz. Milk, 132 lbs. 8 oz.

Test made from Apr. 7 to 13, 1886; age, 5 yrs. 3 mos.; property of Campbell Brown, Spring Hill, Tenn.

Odelle Sales 15564.—Sire, Tormentor 3533; dam, Oonan 1485.

Butter, 16 lbs. 15⅝ oz. Milk, 157 lbs. 8 oz.

Test made from Apr. 14 to 21, 1887; age, 6 yrs. 8 mos.; estimated weight, 1000 lbs.; fed, daily, 6 lbs. cotton-seed meal, 6 lbs. bran and 12 lbs. corn meal, mixed; property of Campbell Brown, Spring Hill, Tenn.

Odelsign of Lynwood 114421.—Sire, Odelio 2d 30299; dam, Bell Ensign 56702.

Butter, 15 lbs. 8½ oz. Milk, 193 lbs. 14¾ oz.

Test made from Nov. 21 to 28, 1897; age, 3 yrs.; estimated weight, 750 lbs.; fed 20 lbs. per day of ground oats and corn, in equal parts, mixed with 4 bucketfuls of cut sheaf oats, wetted—orchard and blue-grass pasture; property of Matt. M. Gardner, Nashville, Tenn.

Œnone 8614.—Sire. Signal 1170; dam, Zina 2d 3082.

Butter, 15 lbs. 14 oz., unsalted. Milk, 185 lbs. 12 oz.

Test made from Oct. 10 to 17, 1883; age, 5 yrs. 6 mos.; property of M. M. Gardner, Nashville, Tenn.

Œnone 8614.—Sire, Signal 1170; dam, Zina 2d 3082.

Butter, 18 lbs. 15 oz. Milk, 139 lbs. 7 oz.

Test made from Dec. 29, 1885, to Jan. 4, 1886; age, 7 yrs. 8 mos.; property of William Morrow, Nashville, Tenn.

Ogden's Alice 77590.—Sire, Ethleel's Landseer 22341; dam, Alice Ogden 62684.

Butter, 17 lbs. 4 oz. Milk, 294 lbs. 6 oz.

Test made from June 10 to 17, 1896; age, 5 yrs. 3 mos.; estimated weight, 850 lbs.; fed 85 qts. wheat bran and 74 qts. corn meal—woods pasture; property of S. E. Hanna, McLean, Ohio.

Ogstree 90943.—Sire, Biveno of Ogston 13253; dam, Ogstveno 88458.

Butter, 14 lbs. 2 oz. Milk, 222 lbs. 8 oz.

Test made from Apr. 7 to 14, 1896; age, 4 yrs. 5 mos.; actual weight, 830 lbs.; fed 10½ lbs. oat meal, 10½ lbs. wheat bran, 21 lbs. corn meal and 19 lbs. oil meal; property of L. T. Becker, Galesburg, Mich.

Oitz 8649.—Sire, Clive Duke 1901; dam, Io A. 6284.

Butter, 15 lbs. 1 oz., unsalted. Milk, 293 lbs. 1 oz.

Test made from June 10 to 18, 1882; age, 3 yrs. 9 mos.; property of I. S. Earhart, Mulberry, Ind.

Oits 8649.—Sire, Clive Duke 1901; dam, Io A. 6284.

Butter, 15 lbs. 11 oz.   Milk, 319 lbs. 8 oz.

Test made from June 16 to 22, 1884; age, 5 yrs. 9 mos.; property of I. S. Earhart, Mulberry, Ind.

Oktibbeha Duchess 4422.—Sire, The Hub 1009; dam, Lucky Belle 2214.

Butter, 17 lbs. 4 oz.   Milk, 282 lbs. 8 oz.

Test made from Apr. 7 to 13, 1882; age, 7 yrs. 1 mo.; fed steamed cotton-seed, corn meal and wheat bran—good pasture; property of W. B. Montgomery, Starkville, Miss.

Old Noble's Heiress 99083.—Sire, Koffee's Noble 14631; dam, Brewer's Queen 68288.

Butter, 15 lbs. 6½ oz.   Milk, 231 lbs. 4 oz.

Test made from Feb. 20 to 27, 1897; age, 3 yrs. 1 mo.; actual weight, 681 lbs.; fed 9 qts. bran, 6 qts. ground oats, 3 qts. corn meal and 3 qts. oil meal, daily—hay; property of C. A. Sweet, East Aurora, N. Y.

Oléo 38475.—Sire, Seneca Chief 4098; dam, Pure Mocha 9186.

Butter, 18 lbs. 1 oz.   Milk, 256 lbs.

Test made from Aug. 4 to 11, 1888; age, 6 yrs. 4 mos.; estimated weight, 800 lbs.; fed, daily, about 18 qts. bran, 4 qts. oat meal, 2 qts. corn and oil meal; property of P. J. Cogswell, Rochester, N. Y.

Oleta 15625.—Sire, Romeo of Sacramento 3030; dam, Ione of Sacramento 5220.

Butter, 15 lbs. 12 oz.   Milk, 251 lbs. 8 oz.

Test made from Dec. 25, 1897, to Jan. 1, 1898; age, 16 yrs. 2 mos.; estimated weight, 750 lbs.; fed 30 lbs. bran and 30 lbs. middlings—hay; property of Peter J. Shields, Sacramento, Cal.

Olga H. 60619.—Sire, Tamy's Rioter 15288; dam, Olo H. 14455.

Butter, 15 lbs. 4 oz.   Milk, 195 lbs. 4 oz.

Test made from Nov. 30 to Dec. 7, 1893; age, 5 yrs. 9 mos.; estimated weight, 970 lbs.; fed, daily, 9 qts. corn and gluten meal, 4½ qts. oat meal and 3 pts. old process oil meal—hay, 8 qts. potatoes; property of E. C. Newton, Hudson, Ohio.

Olga H. 60619.—Sire, Tamy's Rioter 15288; dam, Olo H. 14455.

Butter, 61 lbs. 13 oz. in 30 days.   Milk, 841 lbs. 12 oz.

Test made from Nov. 16 to Dec. 16, 1893; age, 5 yrs. 9 mos.; estimated weight, 970 lbs.; fed 216 qts. corn meal, 108 qts. wheat and oats ground together and 36 qts. oil meal—4 qts. daily of cut, raw potatoes, no pasture; property of E. C. Newton, Hudson, Ohio.

Olie's Lady Teazle 12307.—Sire, Nero Chief 2d 4217; dam, Olie's May Bell 6567.

Butter, 16 lbs. 5 oz.   Milk, 275 lbs. 4 oz.

Test made from July 1 to 8, 1883; age, 3 yrs. 3 mos.; blue-grass pasture only; property of L. H. Smith, Lexington, Ill.

Olive Branch of Hillcrest 7447.—Sire, Jeweler 1385; dam, Olive Leaf 4257.

Butter, 15 lbs. 7 oz.   Milk, 224 lbs. 6 oz.

Test made from June 16 to 22, 1885; age, 7 yrs. 4 mos.; property of James Crook, Jacksonville, Ala.

Olive of One Pine 93449.—Sire, Osprey 17385; dam, Katy Barr 11066.

Butter, 17 lbs. 9 oz.   Milk, 282 lbs. 10 oz.

Test made from July 14 to 21, 1898; age, 6 yrs. 1 mo.; actual weight, 765 lbs.; fed 9 qts. bran, 9 qts. ground oats, 3 qts. corn meal and 3 pts. oil meal, mixed, wet, daily —good clover pasture; property of Charles A. Sweet, East Aurora, N. Y.

Olive Star 81102.—Sire, Lucky King 17350; dam, Olive Lass 52889.

Butter, 15 lbs. 7 oz.   Milk, 279 lbs. 6 oz.

Test made from May 24 to 31, 1895; age, 3 yrs. 3 mos.; estimated weight, 700 lbs.; fed 98 lbs. corn meal, 98 lbs. cracked oats, 49 lbs. Buffalo gluten, 28 lbs. bran and 49 lbs. linseed meal—hay *ad lib.*, fair blue-grass pasture; property of C. Delano, Mount Vernon, Ohio.

Olivia Albertine 3d 83438.—Sire, Aaron Stoke Pogis 21705; dam, Olivia Albertine 48007.

Butter, 15 lbs. 1¾ oz.   Milk, 194 lbs. 4 oz.

Test made from Oct. 21 to 28, 1897; age, 5 yrs. 2 mos.; actual weight, 925 lbs.; fed 42 qts. ground oats, 42 qts. corn meal and 28 qts. wheat bran—hay at night, pasture during day; property of Miller & Sibley, Franklin, Pa.

Olivia Marigold 108187.—Sire, Stoke Pogis of Prospect 29121; dam, Royal Bomba's Daisy 35996.

Butter, 18 lbs. ½ oz.   Milk, 254 lbs. 3 oz.

Test made from July 20 to 27, 1898; age, 3 yrs. 1 mo.; actual weight, 850 lbs.; fed 9 qts. wheat bran, 6 qts. ground oats, 3 qts. corn meal and 1½ qts. oil meal, mixed, fed wet, daily—green clover and green corn; property of Charles A. Sweet, Buffalo, N. Y.

Ollie Wilkes 77402.—Sire, In-and-In Bred S. P. Third 22595; dam, Ramblerotas 50036.

Butter, 14 lbs. 14 oz.   Milk, 288 lbs. 4 oz.

Test made from June 3 to 10, 1897; age, 6 yrs. 3 mos.; estimated weight, 900 lbs.; fed 77 lbs. corn and oat meal, in equal parts by weight—June grass, timothy and clover pasture; property of A. D. Lee, Moravia, N. Y.

Olymph 17957.—Sire, Vertumnus (P. S. 161 J. H. B.); dam, Butterfly (F. S. 1743 J. H. B.)

Butter, 15 lbs. 13 1-16 oz.   Milk, 242 lbs.

Test made from Apr. 15 to 22, 1885; age, 5 yrs. 7 mos.; property of D. M. McCullough, Troy, Ohio.

Ona 7840.—Sire, Khedive (P. S. 103 J. H. B.); dam, Écornée (F. S. 846 J. H. B.)

Butter, 20 lbs. 13 oz., unsalted.   Milk, 149 lbs. 6 oz.

Test made from June 20 to 27, 1883; age, 6 yrs. 4 mos.; fed 4 qts. bran, 4 qts. corn meal and 1 qt. oil meal, daily—June grass, timothy and red top pasture; property of S. M. Burnham, Saugatuck, Conn.

Ona 7840.—Sire, Khedive (P. S. 103 J. H. B.); dam, Écornée (F. S. 846 J. H. B.)

Butter, 22 lbs. 10½ oz.   Milk, 168¾ qts.

Test made from June 11 to 17, 1884; age, 7 yrs. 4 mos.; fed 8 qts. coarse corn meal and 8 qts. ground oats, mixed, daily—mixed grass pasture; property of S. M. Burnham, Saugatuck, Conn.

Oneida 42100.—Sire, Combination 4389; dam, Doe 3061.

Butter, 14 lbs. 2 oz.   Milk, 202 lbs. 12 oz.

Test made from Oct. 1 to 8, 1887; age, 2 yrs. 6 mos.; estimated weight, 800 lbs.; fed 15 qts. corn, oat and pea meal, daily—pastured in orchard; property of Richardson Bros., Davenport, Iowa.

Oneida 42100.—Sire, Combination 4389; dam, Doe 3061.

Butter, 16 lbs. 13 oz.   Milk, 338 lbs. 3 oz.

Test made from Nov. 22 to 29, 1896; age, 11 yrs. 8 mos.; estimated weight, 800 lbs.; fed 44½ lbs. bran, 52½ lbs. corn, 43½ lbs. oats, 21¼ lbs. linseed meal. 140 lbs. roots, 280 lbs. ensilage and hay *ad lib.*; property of C. I. Hood, Lowell, Mass.

Oneida 2d 43553.—Sire, Combination 3d 17576; dam, Oneida 42100.

Butter, 14 lbs. 14½ oz.   Milk, 208 lbs. 11 oz.

Test made from July 1 to 8, 1889; age, 2 yrs. 9 mos.; estimated weight, 750 lbs.; fed 15 qts. corn, oat and oil meal, daily—good pasture; property of Richardson Bros., Davenport, Iowa.

Oneida 2d 43553.—Sire, Combination 3d 17576; dam, Oneida 42100.

Butter, 17 lbs. 4½ oz.   Milk, 280 lbs. 1 oz.

Test made from Nov. 21 to 28, 1896; age, 10 yrs. 2 mos.; estimated weight, 1000 lbs.; fed 42 lbs. bran, 41¾ lbs. corn, 44 lbs. oats, 20½ lbs. linseed meal, 150 lbs. roots and 280 lbs. ensilage—hay *ad lib.*; property of C. I. Hood, Lowell, Mass.

Onnolee 23804.—Sire, Bingo 2d 6749; dam, Mrs. Bannister 23803.

Butter, 16 lbs. 4 oz.   Milk, 186 lbs. 4 oz.

Test made from Apr. 14 to 21, 1887; age, 6 yrs. 6 mos.; estimated weight, 800 lbs.; fed, daily, 8 lbs. corn meal, 4 lbs. bran, 2 lbs. middlings and 2 lbs. oil meal; property of James Stillman, Sing Sing, N. Y.

Onwa 59628.—Sire, Diploma 16219; dam, Oneida 2d 43553.

Butter, 16 lbs. 5 oz.   Milk, 247 lbs. 15 oz.

Test made from July 1 to 8, 1892; age, 3 yrs. 1 mo.; estimated weight, 775 lbs.; fed 12 qts. ground oats, corn and oil meal, daily, in three feeds—good pasture; property of Richardson Bros., Davenport, Iowa.

Onwa 59628.—Sire, Diploma 16219; dam, Oneida 2d 43553.

Butter, 18 lbs. 13½ oz.   Milk, 282 lbs. 7 oz.

Test made from May 13 to 20, 1898; age, 8 yrs. 11 mos.; estimated weight, 1000 lbs.; fed 35 lbs. bran, 42 lbs. corn meal, 21 lbs. ground oats, 10½ lbs. oil meal and 14 lbs. cotton-seed meal—clover and timothy pasture; property of C. I. Hood, Lowell, Mass.

Onyx of Argyle 76742.—Sire, Baron Hugo 15208; dam, Dassa Argyle 21180.

Butter, 18 lbs. 5 oz.   Milk, 280 lbs. 8 oz.

Test made from Jan. 27 to Feb. 3, 1894; age, 2 yrs. 10 mos.; estimated weight, 775 lbs.; fed 3 lbs. Spring wheat bran, 3 lbs. corn meal, 2 lbs. oats and rye provender, 2 lbs. cream gluten, 1 lb. cotton-seed meal, 30 lbs. corn ensilage and 10 lbs. cut clover hay, daily; property of E. Stevens Henry, Rockville, Conn.

Oonamistle 64469.—Sire, Oonan's Pogis 17165; dam, Mistle 26125.

Butter, 15 lbs. 4 oz. Milk, 232 lbs.

Test made from June 26 to July 3, 1893; age, 4 yrs. 9 mos.; estimated weight, 800 lbs.; fed 4 qts. each of Buffalo gluten meal and wheat middlings, mixed, per day—rough pasture, natural grasses; property of George W. Sisson, Jr., Potsdam, N. Y.

Oonan 1485.—Sire, Rajah 340; dam, Omoo 1247.

Butter, 22 lbs. 2½ oz. Milk, 205 lbs. 7 oz.

Test made from June 4 to 10, 1882; property of M. C. Campbell, Spring Hill, Tenn.

Oonan 2d 19569.—Sire, Lord Harry 3445; dam, Oonan 1485.

Butter, 18 lbs. 4¼ oz. Milk, 158 lbs. 11 oz.

Test made from July 23 to 30, 1888; age, 5 yrs. 6 mos.; estimated weight, 850 lbs.; fed 16 qts. corn, oats and barley, mixed in equal parts and ground, daily; property of M. C. Campbell, Spring Hill, Tenn.

Oonan 3d 49955.—Sire, Ida's Stoke Pogis 13658; dam, Oonan 1485.

Butter, 14 lbs. 15½ oz. Milk, 141 lbs. 12 oz.

Test made from Mar. 2 to 9, 1889; age, 1 yr. 9 mos.; estimated weight, 750 lbs.; fed, daily, 2 gals. oats and 2 gals. corn meal; property of M. C. Campbell, Spring Hill, Tenn.

Oonan of Riverside 69773.—Sire, Oonan's Tormentor 22280; dam, Hyacinth Pogis 56020.

Butter, 14 lbs. 9½ oz. Milk, 218 lbs.

Test made from May 16 to 23, 1893; age, 2 yrs. 9 mos.; estimated weight, 700 lbs.; fed 29 qts. middlings, 26 qts. ground oats, 23 qts. meal and 70 lbs. cut hay; property of H. A. Huntington, Higganum, Conn.

Oonan of Riverside 69773.—Sire, Oonan's Tormentor 22280; dam, Hyacinth Pogis 56020.

Butter, 34 lbs. 3 oz. Milk, 262 lbs. 8 oz.

Test made from Feb. 23 to Mar. 2, 1895; age, 4 yrs. 7 mos.; estimated weight, 700 lbs.; fed provender, ground oats, bran, cotton-seed meal and ensilage—no record of quantity kept; property of H. A. Huntington, Higganum, Conn.

Oonan of Riverside 69773.—Sire, Oonan's Tormentor 22280; dam, Hyacinth Pogis 56020.

Butter, 34 lbs. 3 oz. (official). Milk, 237 lbs.

Test made from Mar. 5 to 12, 1895; age, 4 yrs. 7 mos.; estimated weight, 700 lbs.; fed 20 lbs. corn meal, 22 lbs. ground oats, 18 lbs. cotton-seed meal, 38 lbs. provender, 5 lbs. bran, 144 lbs. ensilage—hay; property of H. A. Huntington, Higganum, Conn.

Analyses of Butter.

Moisture 12.78, fat 85.68, casein 1.05, ash 0.49.

Oonan Pogis 59835.—Sire, Kate's Tornado 20888; dam, Oonan 3d 49955.

Butter, 14 lbs. 6 oz. Milk, 215 lbs. 11.2 oz.

Test made from June 18 to 25, 1896; age, 7 yrs. 4 mos.; estimated weight, 950 lbs.; fed 56 lbs. bran, 35 lbs. corn and oats, ground together (two-thirds oats and one-third corn), 28 lbs. hominy meal and 14 lbs. cotton-seed meal—oats and clover pasture; property of J. E. Robbins, Jr., Greensburg, Ind.

Oonan's Alda 108337.—Sire, Oonan's Tormentor 22280; dam, Roonalda 3d 49958.

Butter, 20 lbs. 5¾ oz. Milk, 235 lbs. 14¼ oz.

Test made from Apr. 3 to 10, 1898; age, 3 yrs. 10 mos.; estimated weight, 750 lbs.; fed 21 lbs. per day of ground oats and corn, in equal parts—good orchard and blue-grass pasture; property of M. C. Campbell. Spring Hill, Tenn.

Oonan's Fancy 31887.—Sire, Toltec 6831; dam, Landseer's Fancy 2876.

Butter, 19 lbs. 10¾ oz. Milk, 285 lbs. 11 oz.

Test made from Aug. 20 to 27, 1894; age, 10 yrs.; actual weight, 1188 lbs.; fed 42 lbs. wheat bran, 38 lbs. old process oil meal and 48 lbs. corn meal, mixed—excellent blue-grass pasture; property of J. P. Bradbury, Pomeroy, Ohio.

Oonan's Grace 86489.—Sire, Oonan's Tormentor 22280; dam, Nancy's Grace 41421.

Butter, 24 lbs. 3 oz. Milk, 258 lbs. 12 oz.

Test made from June 24 to July 1, 1896; age, 4 yrs. 8 mos.; estimated weight. 850 lbs.; fed 21 lbs. daily of a mixture of bran, oats, corn, oil meal and cotton-seed meal, with 20 per cent. pea meal—green corn fodder and clover hay; property of R. F. Shannon, Pittsburg, Pa.

Oonan's Jeannette 45616.—Sire, Oonan's Rajah 8965; dam, Lady Jeannette of St. James 12054.

Butter, 20 lbs. 3 oz. Milk, 261 lbs. 12 oz.

Test made from Oct. 17 to 24, 1891; age, 5 yrs. 10 mos.; estimated weight, 750 lbs.; fed 24 lbs. daily of a mixture of bran and oats—rather poor fall pasture; property of George M. Jewett, Glenville, Md.

Oonan's Oonan 99574.—Sire, Oonan's Tormentor 22280; dam, Oonan 3d 49955.

Butter, 17 lbs. 12¾ oz. Milk, 221 lbs. 6½ oz.

Test made from July 15 to 22, 1898; age, 5 yrs. 3 mos.; estimated weight, 800 lbs.; fed 24 lbs. daily of ground oats and corn, in equal parts—orchard and blue-grass pasture; property of Samuel N. Warren, Spring Hill, Tenn.

Oonazeluss 77481.—Sire, Oonan's Pogis 17165; dam, Zélie de Lussan 48681.

Butter, 15 lbs. 2½ oz. Milk, 217 lbs.

Test made from Mar. 4 to 11, 1894; age, 3 yrs. 10 mos.; estimated weight, 750 lbs.; fed 6 qts. corn-hearts, 4 qts. ground oats, 4 qts. bran, 1 qt. cotton-seed meal, daily—clover hay and corn stover. poor pasture; property of Samuel N. Warren, Spring Hill, Tenn.

Opaline 7590.—Sire, Top-Sawyer 1404; dam, Silver Rose 4753.

Butter, 14 lbs. 10 oz. Milk, 223 lbs. 7 oz.

Test made from Aug. 15 to 22, 1883; age, 5 yrs. 4 mos.; property of Matt. M. Gardner, Nashville, Tenn.

Opaline 7590.—Sire, Top-Sawyer 1404; dam, Silver Rose 4753.

Butter, 15 lbs. 2 oz. Milk, 215 lbs.

Test made from Aug. 20 to 27, 1884; age, 6 yrs. 4 mos.; property of Matt. M. Gardner, Nashville, Tenn.

Ophi of St. Lambert 49004.—Sire, Exile of St. Lambert 13657; dam, Augerez Violet 28377.

Butter, 14 lbs. 10½ oz. Milk, 254 lbs. 8 oz.

Test made from July 23 to 30, 1893; age, 6 yrs. 2 mos.; estimated weight, 900 lbs.; fed 9 qts. corn meal, 9 qts. ground oats, 3 qts. bran, 1½ qts. oil meal and 21 lbs. cut hay and green fodder, daily; property of Pierce J. Cogswell, Rochester, N. Y.

Optima 6715.—Sire, Signal 1170; dam, Romp 1098.

Butter, 19 lbs. 2 oz. Milk, 191 lbs. 4 oz.

Test made from Jan. 24 to 31, 1885; age, 7 yrs. 9 mos.; property of L. J. & A. W. Hill, Atlanta, Ga.

Orange Extra 133951.—Sire, Hulee 18209; dam, Ada of Orange 2d 65246.

Butter, 18 lbs. Milk, 226 lbs.

Test made from Nov. 15 to 22, 1897; age, 4 yrs. 9 mos.; estimated weight, 850 lbs.; fed 16 lbs. corn meal and 8 lbs. wheat bran, daily—poor blue-grass pasture; property of Polk & Brown, Ashwood, Tenn.

Ora of Forester 54698.—Sire, Blackstone of Orange 8337; dam, Brier of Portland 28753.

Butter, 15 lbs. 12 oz. Milk, 207 lbs. 4 oz.

Test made from Mar. 2 to 9, 1896; age, 9 yrs. 2 mos.; estimated weight, 950 lbs.; fed 5 qts. corn and cob meal, 5 qts. oat meal and 2 qts. oat middlings, daily—few cut raw potatoes, cut corn fodder and hay *ad lib.*; property of E. C. Newton, Hudson, Ohio.

Ora Vibert 119585.—Sire, Vibert Stoke Pogis 25549; dam, Eliza Pogis 49203.

Butter, 17 lbs. 11 oz. Milk, 201 lbs.

Test made from June 16 to 23, 1898; age, 3 yrs. 5 mos.; estimated weight, 850 lbs.; fed 1 gal. corn and oat chop and 1 gal. bran, daily—good mixed pasture; property of H. A. Neff, Glencoe, Ohio.

Orie's Fancy 72039.—Sire, Fancy's Harry 6th 24768; dam, Romp's Orie 64456.

Butter, 15 lbs. 10½ oz. Milk, 196 lbs. 4 oz.

Test made from June 5 to 12, 1896; age, 4 yrs. 10 mos.; estimated weight, 800 lbs.; fed 28 qts. corn meal, 28 qts. coarse wheat bran and 14 qts. old process oil meal—good blue-grass and timothy pasture; property of A. S. Bell, London, Ohio.

Orie's Gem of St. Lambert 85472.—Sire, Rioter of St. Lambert 16501; dam, Orie Hugo Pogis 50210.

Butter, 18 lbs. 4 oz. Milk, 317 lbs. 3 oz.

Test made from May 24 to 31, 1896; age, 5 yrs. 4 mos.; estimated weight, 925 lbs.; fed 16 lbs. ground oats, 30 lbs. ground corn and oats, in equal parts, 20 lbs. bran and 12 lbs. oil meal—shady orchard pasture; property of D. S. Shourds, Macedon, N. Y.

Orphan Duchess 3d 21284.—Sire, Balboa 1244; dam, Orphan Duchess 4519.

Butter, 16 lbs. 3 oz. Milk, 231 lbs. 12 oz.

Test made from June 23 to 30, 1888; age, 6 yrs. 4 mos.; estimated weight, 850 lbs.; fed, daily, 4 qts. corn meal, 1 qt. oil meal and 6 qts. bran; property of D. W. Voyles, Crandall, Ind.

Orphean 4636.—Sire, Hurd's Ivanhoe 1522; dam, Hurd's Orpha 3346.

Butter, 15 lbs. 7 oz.   Milk, 83 qts. ¾ pt.

Test made from Nov. 6 to 13, 1883; age, 8 yrs. 6 mos.; property of S. W. Sterrett, Barnitz, Pa.

Otta of St. Lambert 49003.—Sire, Exile of St. Lambert 13657; dam, Chief's Guenn 29266.

Butter, 15 lbs. 8¾ oz. (official).   Milk, 219 lbs. 4 oz.

Test made from Feb. 13 to 20, 1893; age, 5 yrs. 9 mos.; estimated weight, 900 lbs.; fed 89¼ lbs. corn meal, 36¾ lbs. bran, 24½ lbs. cotton-seed meal, 14 lbs. hay and 119 lbs. ensilage; property of P. J. Cogswell, Rochester, N. Y.

Our Daisy Queen 94252.—Sire, Noble of Side View 28280; dam, Queen of Clinton 11500.

Butter, 15 lbs. 13 oz.   Milk, 240 lbs.

Test made from May 12 to 19, 1896; age, 4 yrs. 1 mo.; estimated weight, 650 lbs.; fed 5 qts. bran, 2½ qts. corn meal and 2 qts. cotton-seed meal, daily—new pasture; property of George E. Nichols, Afton, N. Y.

Oxalis 2d 15631.—Sire, St. Heller 45; dam, Oxalis 606.

Butter, 15 lbs.   Milk, 197 lbs. 11 oz.

Test made from June 1 to 8, 1877; age, 4 yrs.; no grain fed—grass only; property of Samuel F. Scofield, Stamford, Conn.

Oxford Catena 68905.—Sire, Oxford Catono 23192; dam, Allena 3d 14879.

Butter, 15 lbs. 11¼ oz.   Milk, 248 lbs.

Test made from May 20 to 27, 1898; age, 7 yrs. 4 mos.; estimated weight, 900 lbs.; fed 28 lbs. bran, 28 lbs. corn meal, 21 lbs. ground oats, 10½ lbs. oil meal and 3½ lbs. cotton-seed meal—mixed clover and timothy pasture; property of Mrs. C. De Nottbeck, New York, N. Y.

Oxford Kate 13646.—Sire, Pilot (P. S. 183 J. H. B.); dam, Verclut (F. S. 1846 J. H. B.)

Butter, 39 lbs. 12 oz. (official).   Milk, 248 lbs. 8 oz.

Test made from Apr. 1 to 8, 1885; age, 6 yrs.; fed 12 qts. pea meal, 16 qts. ground oats, 3 qts. linseed oil cake and 4 qts. wheat bran, daily—clover hay, beets and carrots; property of Andrew Banks, Baltimore, Md.

Palestina 4644.—Sire, Pierrot 2d 1669; dam, Palestine 3d 1104.

Butter, 15 lbs. 8 oz.

Test made from Apr. 30 to May 7, 1883; age, 8 yrs. 1 mo.; fed 4 qts. meal (half corn and half oats), 4 qts. wheat bran and 2 qts. oil meal, daily—hay first half of week and green rye second half; property of A. F. Mullin, Mount Holly Springs, Pa.

Palestine Bess 102970.—Sire, Banshee's Stoke Pogis 29261; dam, Mahogany Bess 68544.

Butter, 16 lbs. 12 oz.   Milk, 260 lbs. 12 oz.

Test made from Apr. 10 to 17, 1898; age, 3 yrs. 7 mos.; estimated weight, 850 lbs.; fed 18 qts. corn and cob meal, 9 qts. ground oats, 2 qts. oil meal and 25 lbs. ensilage, daily—clover hay *ad lib.*; property of Brooks & Pidgeon, Salem, Ohio.

Palestine of Oxford 42194.—Sire, Yoko 3951; dam, Niantic's Palestine 14519.

Butter, 21 lbs. Milk, 194 lbs. 11½ oz.

Test made from Nov. 25 to Dec. 2, 1890; age, 5 yrs. 6 mos.; actual weight, 810 lbs.; fed, daily, 5 lbs. wheat bran, 5½ lbs. corn meal, 3½ lbs. barley meal and 4¾ lbs. linseed meal; property of M. L. Frink, Oxford, Mich.

Palestine Pierrot 24099.—Sire, Pierrot 7th 1667; dam, Palestine's Last Daughter 12602.

Butter, 14 lbs. 6 oz. Milk, 163 lbs. 12 oz.

Test made from May 9 to 15, 1885; age, 5 yrs. 11 mos.; property of Webster & Weaver, Columbia, Tenn.

Palestine's Last Daughter 12602.—Sire, Pierrot 7th 1667; dam, Palestine 26.

Butter, 14 lbs. 6 oz.

Test made from June 1 to 8, 1881; age, 4 yrs.; property of Thomas Fitch, New London, Conn.

Paletta of Darlington 16255.—Sire, Duke of Darlington 2460; dam, Palestina 4644.

Butter, 27 lbs. 8 oz. Milk, 274 lbs. 4 oz.

Test made from June 1 to 8, 1888; age, 6 yrs. 2 mos.; weight, 950 lbs.; fed, daily, 17 lbs. chopped oats and corn and 7½ lbs. bran; property of W. A. & A. F. Mullin, Mount Holly Springs, Pa.

Paletta of Darlington 2d 75343.—Sire, Lord Lisgar Pogis 12149; dam, Paletta of Darlington 16255.

Butter, 14 lbs. 15½ oz. Milk, 214 lbs. 7 oz.

Test made from Sept. 6 to 13, 1896; age, 6 yrs. 4 mos.; estimated weight, 900 lbs.; fed 20 lbs. grain per day, consisting of two-fifths bran, two-fifths corn meal and hominy and one-fifth cotton-seed meal—pasture, timothy and natural grasses; property of Ayer & McKinney, Philadelphia, Pa.

Palm Rose 75110.—Sire, Khedive's Landseer 21926; dam, Bridal Rose 74883.

Butter, 15 lbs. 1 oz. Milk, 190 lbs. 8 oz.

Test made from June 23 to 30, 1897; age, 7 yrs. 3 mos.; estimated weight, 800 lbs.; fed 84 qts. bran, 28 qts. corn meal and 14 qts. oil meal—June grass pasture; property of Austin & Probert, Sandstone, Mich.

Panache 62055.—Sire, Romp's Torment 8789; dam, Lady Minnie's Lass 43794.

Butter, 16 lbs. 1 oz. Milk, 195 lbs. 12 oz.

Test made from Jan. 6 to 13, 1894; age, 4 yrs. 5 mos.; actual weight, 775 lbs.; fed 28 qts. corn meal, 28 qts. bran and 14 lbs. oil meal—good clover hay *ad lib.*; property of Henry DuBois, Vigo, Ohio.

Pandothra 22383.—Sire, Jethro 5218; dam, Pandora of Staatsburgh 3d 6497.

Butter, 17 lbs. 5 oz. Milk, 196 lbs.

Test made from Apr. 13 to 19, 1885; age, 2 yrs. 1 mo.; property of R. McMichael, Lexington, Ky.

Panola 85344.—Sire, Soapstone 14377; dam, Flora D. 11192.

Butter, 17 lbs. 3¾ oz. Milk, 278 lbs. 15 oz.

Test made from Feb. 20 to 27, 1898; age, 6 yrs. 4 mos.; estimated weight, 950 lbs.; fed 210 lbs. ensilage, 35 lbs. bran, 21 lbs. corn meal, 17½ lbs. ground oats, 10½ lbs. oil meal and 70 lbs. roots—hay *ad lib.*; property of C. I. Hood, Lowell, Mass.

Pansita 4th 18109.—Sire, Duke of Thornebrook 3832; dam, Pansita 7531.

Butter, 14 lbs. 6 oz. Milk, 248 lbs. 7 oz.

Test made from May 13 to 19, 1886; age, 4 yrs. 1 mo.; property of Mrs. E. B. Buchanan, Franklin, Tenn.

Pansy 1019.—Sire, Living Storm 173; dam, Dolly 2d 1020.

Butter, 574 lbs. 8 oz. in one year.

Property of J. H. Sutliff, Bristol, Conn.

Pansy Albert of Lawn 76634.—Sire, Pansy Pogis Rioter 14064; dam, Sadie's Daisy 40099.

Butter, 15 lbs. 14½ oz. Milk, 239 lbs. 11 oz.

Test made from Oct. 4 to 11, 1897; age, 6 yrs. 7 mos.; estimated weight, 1050 lbs.; fed 112 lbs. wheat bran, 112 lbs. corn and oat chop and 14 lbs. linseed meal—good crab-grass pasture; property of Platter & Foster, Denison, Tex.

Pansy Blossom 22413.—Sire, Kitty's Royal Rex 6176; dam, Dye's Pansy Buttercup 14914.

Butter, 14 lbs. 13½ oz. Milk, 315 lbs. 1 oz.

Test made from June 3 to 10, 1888; age, 5 yrs. 2 mos.; estimated weight, 800 lbs.; fed 1 gal. of mill feed daily; property of W. Gettys, Athens, Tenn.

Pansy Darling of Lawn 76637.—Sire, Darling's Prince 9513; dam, Pansy Page 13163.

Butter, 19 lbs. 2 oz. Milk, 233 lbs.

Test made from May 10 to 17, 1896; age, 5 yrs. 2 mos.; estimated weight, 850 lbs.; fed 28 lbs. cotton-seed meal, 35 lbs. corn meal and 112 lbs. wheat bran—wild grass pasture; property of Platter & Foster, Denison, Tex.

Pansy Dec. 99294.—Sire, Stoke Pogis' Perfection 5984; dam, Signaido's Pansy 53366.

Butter, 16 lbs. 4½ oz. Milk, 236 lbs. 14.4 oz.

Test made from Mar. 19 to 26, 1897; age, 6 yrs. 2 mos.; estimated weight, 900 lbs.; fed 6 lbs. bran, 8 lbs. corn and oats, 2 lbs. cotton-seed meal, 2 lbs. linseed meal and 24 lbs. corn ensilage, daily—hay *ad lib.*; property of John E. Robbins, Greensburg, Ind.

Pansy K. 23889.—Sire, Dickero 3542; dam, Pansy Hawes 2d 23752.

Butter, 14 lbs. 11½ oz.

Test made in June, 1884; age, 2 yrs. 1 mo.; property of J. C. Ritchie, Marissa, Ill.

Pansy Patterson 18612.—Sire, Black Diamond V. 4970; dam, Prudence of Bovina 3d 10747.

Butter, 15 lbs. 15 oz. Milk, 150 lbs. 12 oz.

Test made from May 27 to June 3, 1884; age, 2 yrs.; property of George L. Douglas, Lexington, Ky.

Pansy's Thoughts 69400.—Sire, Sophie's Tormentor 20883; dam, Ona's Pansy 34866.

Butter, 17 lbs. 6 oz. Milk, 248 lbs. 5 oz.

Test made from Oct. 3 to 10, 1897; age, 6 yrs. 9 mos.; estimated weight, 850 lbs.; fed 42 lbs. bran, 28 lbs. corn meal, 31½ lbs. ground oats and 10½ lbs. oil meal—green hay in barn at night, tethered on clover during day; property of C. I. Hood, Lowell, Mass.

Pansy's Thoughts 69400.—Sire, Sophie's Tormentor 20883; dam, Ona's Pansy 34866.

Butter, 33 lbs. in 14 days. Milk, 503 lbs. 14 oz.

Test made from Oct. 3 to 17, 1897; age, 6 yrs. 10 mos.; estimated weight, 850 lbs.; fed 84 lbs. bran, 63 lbs. ground oats, 59½ lbs. corn meal and 23 lbs. oil meal—hay *ad lib.*, short clover pasture; property of C. I. Hood, Lowell, Mass.

Paola Stoke Pogis 34691.—Sire, Exile of St. Lambert 13657; dam, St. John's Daisy 28388.

Butter, 19 lbs. 6¼ oz. Milk, 234 lbs.

Test made from Dec. 31, 1888, to Jan. 7, 1889; age, 3 yrs. 3 mos.; estimated weight, 800 lbs.; fed, daily, average of 29 lbs. corn meal, wheat bran, fine middlings and oil meal, mixed in equal parts; property of P. J. Cogswell, Rochester, N. Y.

Paola Stoke Pogis 34691.—Sire, Exile of St. Lambert 13657; dam, St. John's Daisy 28388.

Butter, 21 lbs. Milk, 279 lbs.

Test made from Jan. 11 to 18, 1891; age, 5 yrs. 3 mos.; estimated weight, 800 lbs.; fed 21 lbs. corn meal, 27 lbs. fine middlings, 78 lbs. bran, 10½ lbs. oil meal, 10½ lbs. pea meal, 6 lbs. roots and 126 lbs. ensilage—clover hay; property of P. J. Cogswell, Rochester, N. Y.

Paola Stoke Pogis 34691.—Sire, Exile of St. Lambert 13657; dam, St. John's Daisy 28388.

Butter, 22 lbs. 8 oz. Milk, 275 lbs. 8 oz.

Test made from Jan. 18 to 25, 1891; age, 5 yrs. 3 mos.; estimated weight, 800 lbs.; fed 133 lbs. grain (composed of 4 parts corn meal, 4 parts B wheat bran, 4 parts fine feed, 4 parts pea meal and 1 part oil meal), 42 lbs. roots and 84 lbs. ensilage—clover hay; property of P. J. Cogswell, Rochester, N. Y.

Paola Stoke Pogis 34691.—Sire, Exile of St. Lambert 13657; dam, St. John's Daisy 28388.

Butter, 23 lbs. Milk, 274 lbs.

Test made from Jan. 25 to Feb. 1, 1891; age, 5 yrs. 4 mos.; estimated weight, 800 lbs.; fed 35 lbs. corn meal, 21 lbs. pea meal, 42 lbs. oil meal, 42 lbs. bran, 21 lbs. fine feed, 14 lbs. ground oats, 42 lbs. roots and 84 lbs. ensilage—clover hay; property of P. J. Cogswell, Rochester, N. Y.

Paola Stoke Pogis 34691.—Sire, Exile of St. Lambert 13657; dam, St. John's Daisy 28388.

Butter, 21 lbs. 5¼ oz. Milk, 272 lbs.

Test made from Feb. 1 to 8, 1891; age, 5 yrs. 4 mos.; estimated weight, 800 lbs.; fed 35 lbs. corn meal, 21 lbs. pea meal, 21 lbs. fine feed, 28 lbs. oil meal, 14 lbs. ground oats, 28 lbs. bran and 100 lbs. roots—clover hay; property of P. J. Cogswell, Rochester, N. Y.

Paola Stoke Pogis 34691.—Sire, Exile of St. Lambert 13657; dam, St. John's Daisy 28388.

Butter, 21 lbs. 12¾ oz. Milk, 284 lbs. 8 oz.

Test made from Feb. 8 to 15, 1891; age, 5 yrs. 4 mos.; estimated weight, 800 lbs.; fed 35 lbs. corn meal, 14 lbs. oil meal, 42 lbs. fine middlings, 7 lbs. pea meal, 56 lbs. bran, 7 lbs. cotton-seed meal and 100 lbs. roots; property of P. J. Cogswell, Rochester, N. Y.

Paola Stoke Pogis 34691.—Sire, Exile of St. Lambert 13657; dam, St. John's Daisy 28388.

Butter, 109 lbs. 10 oz. in 35 days. Milk, 1385 lbs.

Test made from Jan. 11 to Feb. 15, 1891; age, 5 yrs. 4 mos. at end of test; estimated weight, 800 lbs.; fed 742 lbs. grain, 290 lbs. roots, 294 lbs. ensilage—clover hay; property of P. J. Cogswell, Rochester, N. Y.

Pa Pogis 72500.—Sire, Vibert Stoke Pogis 25549; dam, Patience Pogis 61110.

Butter, 16 lbs. 7 oz. Milk, 280 lbs. 11 oz.

Test made from May 19 to 26, 1897; age, 5 yrs. 11 mos.; actual weight, 1020 lbs.; fed 1 lb. corn, 2 lbs. oats, 2 lbs. bran, ½ lb. gluten and ½ lb. oil meal, daily, for first three days, 2 lbs. corn, 4 lbs. oats, 4 lbs. bran, 1 lb. gluten and 1 lb. oil meal, daily, during last four days—blue grass pasture; property of George E. Scott, Mount Pleasant, Ohio.

Pa Pogis 72500.—Sire, Vibert Stoke Pogis 25549; dam, Patience Pogis 61110.

Butter, 16 lbs. 8½ oz. Milk, 296 lbs. 15 oz.

Test made from May 26 to June 2, 1897; age, 5 yrs. 11 mos.; actual weight, 1020 lbs.; fed 2 lbs. corn, 4 lbs. oats, 4 lbs. bran, 1 lb. gluten and 1 lb. oil meal, daily—blue-grass pasture; property of George E. Scott, Mount Pleasant, Ohio.

Pa Pogis 72500.—Sire, Vibert Stoke Pogis 25549; dam, Patience Pogis 61110.

Butter, 32 lbs. 15½ oz. in 14 days. Milk, 577 lbs. 10 oz.

Test made from May 19 to June 2, 1897; age, 5 yrs. 11 mos.; actual weight, 1020 lbs.; fed 25 lbs. corn, 50 lbs. oats, 50 lbs. bran, 12½ lbs. gluten and 12½ lbs. oil meal—blue-grass pasture; property of George E. Scott, Mount Pleasant, Ohio.

Paradise 32082.—Sire, Combination 4389; dam, Goodbye 27366.

Butter, 17 lbs. 11 oz. Milk, 227 lbs. 8 oz.

Test made from May 10 to 17, 1887; age, 4 yrs. 10 mos.; property of Richardson Bros., Davenport, Iowa.

Paradise 2d 97112.—Sire, Toltec's Signal 29501; dam, Paradise 32082.

Butter, 18 lbs. 4 oz. Milk, 234 lbs. 8 oz.

Test made from June 19 to 26, 1897; age, 3 yrs. 8 mos.; estimated weight, 900 lbs.; fed 56 lbs. oats, 35 lbs. corn meal and 14 lbs. oil meal—mixed pasture; property of A. O. Auten, Jerseyville, Ill.

Paradox 65003.—Sire, Upright 6147; dam, Miss Madrid 49419.

Butter, 16 lbs. 9 oz. Milk, 234 lbs. 11 oz.

Test made from Nov. 3 to 10, 1892; age, 4 yrs. 9 mos.; estimated weight, 800 lbs.; fed 12 qts. ground oats, corn meal and oil meal, daily—hay and pasture; property of Richardson Bros., Davenport, Iowa.

Parloa 32357.—Sire, Duke of Darlington 3d 11096; dam, Cultured Cream 29196.

Butter, 14 lbs. 10 oz. Milk, 217 lbs. 14 oz.

Test made from Dec. 16 to 23, 1889; age, 4 yrs. 11 mos.; estimated weight, 850 lbs.; fed, daily, 6 lbs. corn meal, 4 lbs. ship stuff and 1 lb. oil meal; property of Horace G. Westlake, Hillsdale, N. Y.

Parole's Duchess 133953.—Sire, Parole 26114; dam, Duchess of Bloomfield 5th 64215.

Butter, 15 lbs. 11 oz. Milk, 261 lbs.

Test made from May 9 to 16, 1898; age, 3 yrs. 4 mos.; estimated weight, 900 lbs.; fed 12 lbs. coarse-ground corn, 6 lbs. wheat bran, 2 lbs. linseed meal and 2 lbs. cotton-seed meal, daily—good blue-grass pasture; property of George Campbell Brown, Spring Hill, Tenn.

Parthia 64925.—Sire, Damascus 22222; dam, Itura 61090.

Butter. 16 lbs. 11 oz. Milk, 253 lbs. 1 oz.

Test made from May 15 to 22, 1894; age. 4 yrs. 2 mos.; estimated weight, 750 lbs.; fed 12 qts. ground oats and corn, oil meal and pea meal, daily—some hay, good pasture; property of Richardson Bros., Davenport, Iowa.

Patina Pogis 110784.—Sire. Vibert Stoke Pogis 25549; dam, Patience Pogis 61110.

Butter, 15 lbs. 2½ oz. Milk, 186 lbs. 4 oz.

Test made from June 1 to 8, 1898; age, 4 yrs. 4 mos.; estimated weight, 950 lbs.; fed about 2 gals. bran, 1 lb. gluten and ½ lb. linseed, daily—good pasture of blue-grass, orchard grass, timothy and clover; property of H. A. Neff, Glencoe, Ohio.

Patona 55117.—Sire, Young Pedro 9033; dam, Saugus Lass 30542.

Butter, 16 lbs. 14½ oz. Milk, 227 lbs. 3 oz.

Test made from Jan. 23 to 30, 1893; age, 4 yrs. 5 mos.; estimated weight, 750 lbs.; fed 42 qts. corn meal. 56 qts. wheat bran and 56 qts. ground oats—hay *ad lib.*; property of D. F. Appleton, Ipswich, Mass.

Patrol 40490.—Sire, Young Combination 14550; dam, Lorita 33750.

Butter, 15 lbs. 4 oz. Milk, 242 lbs. 12 oz.

Test made from June 21 to 28, 1889; age, 2 yrs. 10 mos.; estimated weight, 850 lbs.; fed about 15 qts. chop feed, daily; property of Richardson Bros., Davenport, Iowa.

Patterson's Beauty 4760.—Sire, Bijou (F. S. 65 J. H. B.); dam, Arlene 1071.

Butter, 18 lbs.

Property of John Patterson, Philadelphia, Pa.

Patty Iris 101894.—Sire, Noble Paddy 31072; dam, Matilda's Lady Day 81310.

Butter, 14 lbs. 15 oz. Milk, 190 lbs. 8 oz.

Test made from June 11 to 18, 1897; age, 4 yrs. 2 mos.; estimated weight, 950 lbs.; no grain fed—clover pasture only; property of George E. Nichols, Afton, N. Y.

Patty Polonius 3d 38677.—Sire, Wosie 6802; dam, Patty Polonius 14016.

Butter, 16 lbs. 10 oz. Milk, 184 lbs. 8 oz.

Test made from Mar. 28 to Apr. 4, 1892; age, 6 yrs. 4 mos.; estimated weight, 1000 lbs.; fed 75 lbs. meal, 42 lbs. wheat bran, 13 lbs. cotton-seed meal and 70 lbs. hay; property of H. A. Huntington, Higganum, Conn.

Pauletta Pogis 39273.—Sire, Stoke Pogis 5th 5987; dam, Pauline 3d 8296.

Butter, 17 lbs. 12 oz. Milk, 218 lbs. 10 oz.

Test made from July 12 to 19, 1891; age, 5 yrs.; estimated weight, 1100 lbs. fed 3 lbs. corn meal, 3 lbs. oil meal, 3 lbs. pea meal, 9 lbs. oat meal and 4 lbs. bran, daily—good pasture; property of Ayer & McKinney, Philadelphia, Pa.

Pauline Hugo Pogis 37098.—Sire, Baron of St. Lambert 5286; dam, Atala 23238.

Butter, 14 lbs. 14 oz. Milk, 230 lbs.

Test made from Aug. 31 to Sept. 7, 1892; age, 6 yrs. 6 mos.; estimated weight, 875 lbs.; fed 96 qts. corn meal and wheat middlings, in equal parts—corn fodder, clover and timothy pasture; property of William Baylis, Bedford, N. Y.

Pauline Vivienne 11305.—Sire, Black Defiance 4014; dam, La Vivienne 2d 1324.

Butter, 16 lbs. 13 oz.

Test made from July 25 to 31, 1884; age, 4 yrs. 5 mos.; property of Andrew J. Fish, Toledo, Ohio.

Pavon 12485.—Sire, St. Helier 45; dam, Moss Rose of Willow Farm 5194.

Butter, 14 lbs. 8 oz., unsalted. Milk, 134 lbs. 8 oz.

Test made from Feb. 6 to 13, 1882; age, 2 yrs. 1 mo.; property of J. H. Walker, Worcester, Mass.

Pay Little 86189.—Sire, Colt's Pierrot 23528; dam, Little Gay 36475.

Butter, 15 lbs. 6¾ oz. Milk, 192 lbs. 1 oz.

Test made from Dec. 2 to 9, 1896; age, 4 yrs. 2 mos.; estimated weight, 750 lbs.; fed 4 qts. corn meal, 2 qts. cotton-seed meal, 8 qts. cooked cotton-seed and 12 qts. bran, daily—hay, wheat and Bermuda grass pasture; property of C. & J. Merzbacher, Marshall, Tex.

Pearl Armstrong 2670.—Sire, Young Baron 702; dam, Virtue Lass 1782.

Butter, 21 lbs. 10 oz. Milk, 317 lbs.

Test made from June 3 to 9, 1882; age, 10 yrs. 5 mos.; fed 4 lbs. corn meal and 4 lbs. wheat bran, daily—pasture; property of Woodside Farm Herd, Troy, N. Y.

Pearl Banks 85758.—Sire, Ida's Stoke Pogis 2d 16864; dam, Letitia 2d 29021.

Butter, 14 lbs. 11 oz. Milk, 255 lbs. 12 oz.

Test made from July 9 to 16, 1896; age, 4 yrs. 9 mos.; estimated weight, 800 lbs.; fed 84 lbs. bran and 98 lbs. chopped green corn—prairie pasture; property of J. D. Gray, Terrell, Tex.

Pearl Jenks 50834.—Sire, Albert's Royalist 10169; dam, Madam Jenks 34694.

Butter, 14 lbs. 14 oz. Milk, 246 lbs.

Test made from Aug. 23 to 30, 1895; age, 9 yrs. 1 mo.; estimated weight, 750 lbs.; fed 4 lbs. cotton-seed meal, 2 lbs. wheat shorts, 14 lbs. cotton-seed hulls and 2 lbs. peas, daily—lespedeza and water grass pasture; property of Mrs. Mattie A. Rubush, Meridian, Miss.

Pearl Jenks 2d 68908.—Sire, Albert's Royalist 10169; dam, Pearl Jenks 50834.

Butter, 18 lbs. 10 oz. Milk, 216 lbs.

Test made from Oct. 7 to 14, 1895; age, 7 yrs. 4 mos.; estimated weight, 750 lbs.; fed 12 lbs. cotton-seed hulls, 5 lbs. cotton-seed meal, 2 lbs. wheat shorts and 1 pk. sweet potatoes, daily—lespedeza and water grass pasture; property of Mrs. Mattie A. Rubush, Meridian, Miss.

Pearl Jenks 3d 84869.—Sire, Mississippi Rioter 20401; dam, Pearl Jenks 50834.

Butter, 15 lbs. 3 oz. Milk, 186 lbs.

Test made from Nov. 3 to 10, 1895; age, 4 yrs. 9 mos.; estimated weight, 900 lbs.; fed 15 lbs. cotton-seed hulls, 4 lbs. cotton-seed meal, 3 lbs. wheat bran, 2 lbs. oats and 1 pk. chopped turnips, daily—hay and oat straw *ad lib.*; property of Mrs. Mattie A. Rubush, Meridian, Miss.

Pearl Jenks 4th 84872.—Sire, Mississippi Rioter 20401; dam, Pearl Jenks 50834.

Butter, 14 lbs. 1½ oz. Milk, 195 lbs. 1½ oz.

Test made from Jan. 20 to 27, 1896; age, 3 yrs. 3 mos.; estimated weight, 750 lbs.; fed slops made of cotton-seed meal and hulls, sweet potatoes, oats and bran, with 5 lbs. cotton-seed meal—hay *ad lib.*; property of Mrs. Mattie A. Rubush, Meridian, Miss.

Pearl of Oakwood 37722.—Sire, Oonan's Signal 11586; dam, Majestic 24757.

Butter, 14 lbs. 1 oz. Milk, 186 lbs. 10 oz.

Test made from Aug. 15 to 22, 1888; age, 2 yrs. 6 mos.; weight, 762 lbs.; fed, daily, 4 lbs. bran, 5 lbs. corn meal and 6 lbs. oat meal; property of Jacob L. Thomas, Knoxville, Tenn.

Pearl of Riverside 55659.—Sire, Catono Koffee 18716; dam, Gladdis Roy 2d 55652.

Butter, 17 lbs. 5½ oz. Milk, 248 lbs.

Test made from May 31 to June 7, 1892; age, 3 yrs. 4 mos.; estimated weight, 800 lbs.; fed 6 qts. meal and 6 qts. No. 2 wheat feed, daily—plain and poor pasture; property of H. A. Huntington, Higganum, Conn.

Pearl of Riverside 55659.—Sire, Catono Koffee 18716; dam, Gladdis Roy 2d 55652.

Butter, 160.804 lbs. in 90 days. Milk, 2653.7 lbs.

Test made from May 31 to Aug. 28, 1893, at World's Columbian Exposition; age, 4 yrs. 3 mos. at beginning of test; actual weight, 1036 lbs.; fed 319 lbs. old hay, 210 lbs. ensilage. 527.8 lbs. new hay, 191 lbs. oil meal, 522 lbs. bran, 73 lbs. oats, 560.5 lbs. corn-hearts, 111.50 lbs. cotton-seed, 402 lbs. middlings, 34 lbs. grano-gluten, 1153.4 lbs. clover and 10 lbs. swale-grass; property of H. A. Huntington, Nashville, Tenn.

Pearl of St. Lambert 5527.—Sire, Buffer 2055; dam, Dot of St. Lambert 5525.

Butter, 14 lbs. 2 oz. Milk, 247 lbs. 8 oz.

Test made from Apr. 4 to 10, 1882; age, 6 yrs.; property of William Rolph, Markham, Ont., Can.

Pearl Pogis 38304.—Sire, Exile of St. Lambert 13657; dam, Laurette 2d 17148.

Butter, 16 lbs. 10¾ oz. Milk, 203 lbs. 8 oz.

Test made from Oct. 17 to 24, 1891; age, 5 yrs. 5 mos.; estimated weight, 825 lbs.; fed 12 qts. corn meal, 2 qts. oil meal, 4 qts. bran, 4 lbs. hay and 8 lbs. corn-stalks, daily; property of Pierce J. Cogswell, Rochester, N. Y.

Pearl's Lemon 41646.—Sire, Young Jupiter 12157; dam, Pearl of Warsaw 30543.

Butter, 17 lbs. 5¼ oz. Milk, 320 lbs.

Test made from Apr. 27 to May 4, 1890; age, 3 yrs. 11 mos.; estimated weight, 850 lbs.; fed, daily, 3 lbs. corn and oat meal, ground together, and 1 lb. ground oil-cake; property of Albert Stracke, Warsaw, Ill.

Pearl's Oonan 32288.—Sire, Oonan's Rajah 8965; dam, Pearl Button 10372.

Butter. 15 lbs. 5½ oz.   Milk, 180 lbs. 12 oz.

Test made from Mar. 17 to 24, 1895; age, 9 yrs.; estimated weight, 850 lbs.; fed 56 qts. ground oats and 56 qts. ground corn—clover hay, poor pasture; property of C. Morgan, Columbia, Tenn.

Pearl's Oonan 2d 86105.—Sire, Landseer's Pogis 15847; dam, Pearl's Oonan 32288.

Butter. 15 lbs. 1 oz.   Milk, 171 lbs. 8 oz.

Test made from June 16 to 23, 1893; age, 3 yrs.; estimated weight, 725 lbs.; fed 56 qts. ground oats, 56 qts. corn meal and 28 qts. wheat bran—a little clover hay, white clover and blue-grass pasture; property of C. Morgan, Columbia, Tenn.

Peasant 65401.—Sire, Lord Harry 3445; dam, Siva 24508.

Butter, 21 lbs. 15½ oz.   Milk, 220 lbs. 7 oz.

Test made from June 20 to 27, 1891; age, 2 yrs. 10 mos.; estimated weight, 1000 lbs.; fed about 125 lbs. of ground corn, oats and bran, in equal parts—blue-grass pasture; property of S. H. Godman, Muncie, Ind.

Pedro Girl 22400.—Sire, Pedro 2d 5419; dam, Frozine's Pansy 19926.

Butter, 18 lbs. 6 oz.   Milk, 233 lbs. 7 oz.

Test made from Mar. 9 to 16, 1886; age, 2 yrs. 8 mos.; property of J. C. Kiplinger & Co., Springfield, Ohio.

Pedroletta 26597.—Sire, Royalty 7210; dam, Romilly 14346.

Butter, 16 lbs. 6½ oz.   Milk, 269 lbs. 10 oz.

Test made from Mar. 26 to Apr. 2, 1888; age, 4 yrs.; estimated weight, 900 lbs.; fed, daily, 17 lbs. ground oats, 12 lbs. corn meal, 16 lbs. middlings and 2 lbs. oil meal; property of D. F. Appleton, Ipswich, Mass.

Pedronina 34803.—Sire, Young Pedro 9033; dam, Nina 2d 2557.

Butter, 21 lbs. 3 oz.   Milk, 248 lbs. 6 oz.

Test made from Oct. 14 to 21, 1891; age, 6 yrs. 6 mos.; estimated weight, 1000 lbs.; fed 8 qts. corn meal, 8 qts. oats and 5 qts. wheat middlings, daily—very good pasture; property of D. F. Appleton, Ipswich, Mass.

Pedro's Bonnie Rainbow 88612.—Sire, Pedro 3187; dam, Rainbow 16672.

Butter, 18 lbs. 5 oz.   Milk, 276 lbs.

Test made from Nov. 16 to 23, 1893; age, 4 yrs. 2 mos.; estimated weight, 950 lbs.; fed about 20 qts. daily of a mixture of ground oats, middlings, corn meal, bran and a little pea meal—clover hay and green corn, blue-grass and meadow pasture; property of T. S. Cooper, Coopersburg, Pa.

Pedro's Clara 96361.—Sire, Pedro Sheldon 22865; dam, Pine Tree Clara 55042.

Butter, 16 lbs. 13 oz.   Milk, 282 lbs.

Test made from Feb. 10 to 17, 1897; age, 4 yrs. 5 mos.; actual weight, 855 lbs.; fed 10 lbs. old process oil meal, 45½ lbs. bran, 49½ lbs. fine corn meal and 120 lbs. mixed clover hay; property of B. F. Carper, Rosemond, Ill.

Pedro's Dolly 77948.—Sire, Pedro 3187; dam, Edith Darlington 3d 26185.

Butter, 17 lbs. 10 oz.   Milk, 282 lbs.

Test made from Jan. 15 to 22, 1893; age, 2 yrs. 6 mos.; estimated weight, 850 lbs.; fed about 22 qts. daily of a mixture of ground oats, middlings, bran, corn meal and some pea meal—clover hay and ensilage; property of T. S. Cooper, Coopersburg, Pa.

Pedro's Duchess 29581.—Sire, Pedro 3187; dam, Great Happiness 18205.

Butter, 18 lbs. 7½ oz. Milk, 261 lbs.

Test made from Oct. 15 to 22, 1891; age, 7 yrs. 5 mos.; estimated weight, 900 lbs.; fed about 20 qts. daily of a mixture of ground oats, middlings, bran, corn meal and pea meal—green corn, clover hay, good pasture; property of T. S. Cooper, Coopersburg, Pa.

Pedro's Fair Marjoram 88489.—Sire, Pedro 3187; dam, Marjoram 2d 12805.

Butter, 18 lbs. 12 oz. Milk, 275 lbs.

Test made from Apr. 22 to 29, 1896; age, 4 yrs. 2 mos.; estimated weight, 1050 lbs.; fed about 22 qts. per day of a mixture of ground oats, wheat middlings, bran, corn meal and a little pea meal—clover hay and ensilage; property of T. S. Cooper, Coopersburg, Pa.

Pedro's Fame 33836.—Sire, Pedro 3187; dam, Marjoram 2d 12805.

Butter, 19 lbs. 12 oz. Milk, 278 lbs.

Test made from May 20 to 27, 1892; age, 7 yrs. 1 mo.; estimated weight, 850 lbs.; fed a mixture of ground oats, wheat middlings, bran, corn meal and a little pea meal, as much as she would eat up clean—clover hay *ad lib.*, some ensilage and mowed grass; property of T. S. Cooper, Coopersburg, Pa.

Pedro's Fancy 4th of H. F. 119842.—Sire, Pedro Signal Landseer 30212; dam, Julia Landseer 56268.

Butter, 14 lbs. 14 oz. Milk, 240 lbs. 14 oz.

Test made from May 30 to June 6, 1898; age, 3 yrs. 3 mos.; estimated weight, 750 lbs.; fed 42 lbs. bran, 35 lbs. corn meal, 21 lbs. oat meal, 7 lbs. oil meal and 10½ lbs. cotton-seed meal—clover and timothy pasture; property of C. I. Hood, Lowell, Mass.

Pedro's Gay Lady 88561.—Sire, Prospect's Rioter 9189; dam, Pedro's Lady 36715.

Butter, 16 lbs. 10 oz. Milk, 242 lbs.

Test made from Jan. 14 to 21, 1896; age, 3 yrs. 3 mos.; estimated weight, 750 lbs.; fed about 20 qts. daily of a mixture of ground oats, middlings, wheat bran, corn meal and some pea meal—clover hay and ensilage; property of T. S. Cooper, Coopersburg, Pa.

Pedro's Golden Beauty 110402.—Sire, Pedro Marjoram Albert 38326; dam, Pedro's Golden Harp 2d 103365.

Butter, 18 lbs. Milk, 231 lbs. 4 oz.

Test made from May 16 to 23, 1898; age, 3 yrs. 1 mo.; estimated weight, 700 lbs.; fed 14 qts. cotton-seed, 7 qts. corn meal, 21 qts. pea meal and 84 qts. wheat bran—some green corn, Bermuda grass pasture; property of Mrs. E. M. Mirick, Cleburne, Tex.

Pedro's Golden Harp 66098.—Sire, Pedro 3187; dam, Golden Harp 24410.

Butter, 15 lbs. 3 oz. Milk, 250 lbs.

Test made from May 4 to 11, 1892; age, 4 yrs. 5 mos.; estimated weight, 850 lbs.; fed about 20 qts. daily of a mixture of ground oats, corn meal, bran, middlings and some pea meal—clover hay and ensilage, meadow pasture; property of T. S. Cooper, Coopersburg, Pa.

Pedro's Golden Harp 2d 103365.—Sire, Pedro Royal Marjoram 28560; dam, Pedro's Golden Harp 66098.

Butter, 16 lbs. 9 oz. Milk, 236 lbs.

Test made from Feb. 18 to 25, 1897; age, 3 yrs. 9 mos.; estimated weight, 700 lbs.; fed about 18 qts. daily of a mixture of ground oats, middlings, wheat bran, corn meal and some pea meal—clover hay and ensilage; property of T. S. Cooper, Coopersburg, Pa.

Pedro's Hansome Phillis 96935.—Sire, Prospect's Rioter 9189; dam, Pretty Phillis Pogis 66100.

Butter, 17 lbs. 8 oz. Milk, 240 lbs.

Test made from Feb. 16 to 23, 1898; age, 3 yrs. 11 mos.; estimated weight, 850 lbs.; fed about 20 qts. daily of a mixture of ground oats, middlings, wheat bran, corn meal and some pea meal—clover hay and ensilage; property of T. S. Cooper, Coopersburg, Pa.

Pedro's Hansom Marjoram 96934.—Sire, Pedro 3187; dam, Marjoram's Matilda 77944.

Butter, 18 lbs. 6 oz. Milk, 254 lbs.

Test made from Mar. 10 to 17, 1897; age, 3 yrs. 5 mos.; estimated weight, 850 lbs.; fed 20 qts. daily of a mixture of ground oats, middlings, wheat bran, corn meal and some pea meal—clover hay and good ensilage; property of T. S. Cooper, Coopersburg, Pa.

Pedro's June 38790.—Sire, Pedro 3187; dam, Lady Lemon 22125.

Butter, 14 lbs. 6 oz. Milk, 180 lbs. 8 oz.

Test made from Feb. 28 to Mar. 7, 1890; age, 5 yrs. 9 mos.; estimated weight, 900 lbs.; fed, daily, 12 lbs. wheat middlings; property of Mrs. Thomas Allen, Pittsfield, Mass.

Pedro's Lady 36715.—Sire, Pedro 3187; dam, Gay Girl 18225.

Butter, 18 lbs. 12 oz. Milk, 282 lbs.

Test made from Jan. 15 to 22, 1892; age, 6 yrs.; estimated weight, 900 lbs.; fed a mixture of ground oats, middlings, bran, corn meal and little pea meal, quantity not known—clover hay ad lib.; property of T. S. Cooper, Coopersburg, Pa.

Pedro's Lady Phillis 88486.—Sire, Prospect's Rioter 9189; dam, Pedro's Phillis 66079.

Butter, 16 lbs. 4 oz. Milk, 252 lbs.

Test made from Oct. 17 to 24, 1894; age, 3 yrs.; estimated weight, 800 lbs.; fed about 18 qts. daily of a mixture of ground oats, middlings, wheat bran, corn meal and some pea meal—clover hay and green corn, meadow and blue-grass pasture; property of T. S. Cooper, Coopersburg, Pa.

Pedro's Laura B. 96363.—Sire, Pedro Sheldon 22865; dam, Laura Esther B. 49272.

Butter, 14 lbs. 10 oz. Milk, 232 lbs. 8 oz.

Test made from May 12 to 19, 1896; age, 3 yrs. 5 mos.; estimated weight, 700 lbs.; fed 42 lbs. bran, 28 lbs. corn and cob meal and 14 lbs. oil meal—timothy pasture; property of B. F. Carper, Rosemond, Ill.

Pedro's Little Wonder 72379.—Sire, Prospect's Rioter 9189; dam, Pedro's Wonder 66078.

Butter, 17 lbs. 14 oz. Milk, 287 lbs. 8 oz.

Test made from Mar. 1 to 8, 1896; age, 4 yrs. 9 mos.; estimated weight, 800 lbs.; fed 74 lbs. corn and cob meal, 27 lbs. bran, 34 lbs. oats and 16 lbs. oil meal—clover hay and corn fodder ad lib.; property of John F. Avery, Saline, Mich.

Pedro's Mab 40212.—Sire, Pedro 3187; dam, Lady Mabella 22138.

Butter, 16 lbs. 10 oz. Milk, 190 lbs. 8 oz.

Test made from Feb. 14 to 21, 1890; age, 4 yrs. 9 mos.; estimated weight, 650 lbs.; fed, daily, 12 lbs. wheat middlings; property of Mrs. Thomas Allen, Pittsfield, Mass.

Pedro's Maggie 102922.—Sire, Pedro 3187; dam, Pomona's Maggie 102921.

Butter, 16 lbs. 6 oz. Milk, 241 lbs.

Test made from Apr. 23 to 30, 1894; age, 4 yrs. 6 mos.; estimated weight, 750 lbs.; fed about 16 qts. daily of a mixture of ground oats, middlings, wheat bran and pea meal—clover hay and ensilage; property of T. S. Cooper, Coopersburg, Pa.

Pedro's Minette 66092.—Sire, Pedro 3187; dam, Minette Pogis 25210.

Butter, 19 lbs. 8 oz. Milk, 295 lbs.

Test made from Nov. 22 to 29, 1893; age, 6 yrs. 5 mos.; estimated weight, 1050 lbs.; fed about 22 qts. daily of a mixture of equal parts of ground oats, wheat middlings, bran, corn meal and a little pea meal—clover hay, green corn, good pasture; property of T. S. Cooper, Coopersburg, Pa.

Pedro's Pansy 33835.—Sire, Pedro 3187; dam, Phillis 2d 18198.

Butter, 19 lbs. 10 oz. Milk, 268 lbs.

Test made from May 3 to 10, 1891; age, 6 yrs. 2 mos.; estimated weight, 850 lbs.; fed about 22 qts. daily of a mixture of ground oats, wheat middlings, bran, corn chop and pea meal, in equal parts—clover hay, good ensilage, green clover; property of T. S. Cooper, Coopersburg, Pa.

Pedro's Phillis 66079.—Sire, Pedro 3187; dam, Phillis 2d 18198.

Butter, 17 lbs. 9 oz. Milk, 258 lbs.

Test made from Dec. 20 to 27, 1893; age, 7 yrs.; estimated weight, 850 lbs.; fed a mixture of ground oats, middlings, bran, corn chop and some pea meal, about 20 qts. daily—clover hay and ensilage; property of T. S. Cooper, Coopersburg, Pa.

Pedro's Polly Rex 99845.—Sire, Pedro's Royal Marjoram 28560; dam, Polly J. Rex 83278.

Butter, 21 lbs. 4 oz. Milk, 299 lbs.

Test made from June 15 to 22, 1897; age, 4 yrs. 2 mos.; estimated weight, 1000 lbs.; fed 4 qts. bran, 4 qts. middlings, 4 qts. ground oats and 2 qts. sugar feed, daily—clover hay *ad lib.*, meadow pasture at night; property of T. S. Cooper, Coopersburg, Pa.

Pedro's Pretty Belle 77946.—Sire, Pedro 3187; dam, Lorne's Belle Dame 66088.

Butter, 18 lbs. 8 oz. Milk, 261 lbs.

Test made from Mar. 18 to 25, 1895; age, 5 yrs. 4 mos.; estimated weight, 1200 lbs.; fed one pail bran mash and one bush. cut clover hay and corn fodder, night and morning; 10 lbs. wheat bran, middlings, cotton-seed meal, oil meal, ground oats and corn meal and one bush. turnips and potatoes, daily—84 lbs. hay; property of Mrs. H. E. Tremain, Lake George, N. Y.

Pedro's Pretty Dolly 102376.—Sire, Pedro's Silver Rioter 31320; dam, Pedro's Dolly 77948.

Butter, 18 lbs. 13 oz. Milk, 258 lbs.

Test made from Nov. 22 to 29, 1897; age, 3 yrs. 2 mos.; estimated weight, 750 lbs.; fed about 20 qts. daily of a mixture of crushed oats, middlings, wheat bran, corn meal and some pea meal—clover hay and green corn, meadow and blue-grass pasture; property of T. S. Cooper, Coopersburg, Pa.

Pedro's Pretty Duchess 88558.—Sire, Prospect's Rioter 9189; dam, Pedro's Duchess 29581.

       Butter, 16 lbs. 11 oz. Milk, 238 lbs.

Test made from Nov. 20 to 27, 1895; age, 4 yrs. 2 mos.; estimated weight, 850 lbs.; fed about 20 qts. daily of a mixture of ground oats, middlings, wheat bran, corn meal and some pea meal—clover hay and green corn, meadow and blue-grass pasture; property of T. S. Cooper, Coopersburg, Pa.

Pedro's Pretty Flower 88542.—Sire, Prospect's Rioter 9189; dam, Pedro's Pansy 33835.

       Butter, 18 lbs. 6 oz. Milk, 266 lbs.

Test made from Nov. 21 to 28, 1894; age, 3 yrs. 9 mos.; estimated weight, 900 lbs.; fed about 22 qts. daily of a mixture of ground oats, middlings, wheat bran, corn meal and some pea meal—clover hay and green corn, meadow and blue-grass pasture; property of T. S. Cooper, Coopersburg, Pa.

Pedro's Pretty Gem 69953.—Sire, Prospect's Rioter 9189; dam, Lorne's Gem 66080.

       Butter, 18 lbs. 8 oz. Milk, 243 lbs.

Test made from Oct. 6 to 13, 1895; age, 5 yrs. 11 mos.; estimated weight, 1100 lbs.; fed 84 lbs. grain (corn meal, cotton-seed meal, oil meal, wheat bran, ground oats, wheat middlings and buckwheat bran), 28 qts. potatoes, 28 qts. carrots, 10 qts. turnips and 84 lbs. hay; property of Mrs. H. E. Tremain, Lake George, N. Y.

Pedro's Pretty Girl 88488.—Sire, Prospect's Rioter 9189; dam, Pedro's Lady 36715.

       Butter, 17 lbs. 4 oz. Milk, 271 lbs.

Test made from Sept. 16 to 23, 1894; age, 2 yrs. 9 mos.; estimated weight, 900 lbs.; fed about 22 qts. daily of a mixture of ground oats, middlings, wheat bran, corn meal and some pea meal—clover hay and green corn, meadow and blue-grass pasture; property of T. S. Cooper, Coopersburg, Pa.

Pedro's Pretty Lady 102379.—Sire, Pedro's Silver Rioter 31320; dam, Pedro's Gay Lady 88561.

       Butter, 15 lbs. 12 oz. Milk, 256 lbs.

Test made from Nov. 22 to 29, 1897; age, 3 yrs. 1 mo.; estimated weight, 750 lbs.; fed about 18 qts. daily of a mixture of ground oats, middlings, wheat bran, corn meal and some pea meal—clover and green corn, meadow and blue-grass pasture; property of T. S. Cooper, Coopersburg, Pa.

Pedro's Pretty Lass 135067.—Sire, Pedro 3187; dam, Carlo's Lass 94405.

       Butter, 16 lbs. 6 oz. Milk, 248 lbs.

Test made from May 22 to 29, 1898; age, 1 yr. 11 mos.; estimated weight, 700 lbs.; fed about 18 qts. daily of a mixture of crushed oats, middlings, wheat bran, corn meal and some pea meal—clover hay and good ensilage, clover and meadow pasture; property of T. S. Cooper, Coopersburg, Pa.

Pedro's Pretty Maggie 102923.—Sire, Pedro 3187; dam, Pomona's Maggie 102921.

       Butter, 18 lbs. 9 oz. Milk, 260 lbs.

Test made from Jan. 21 to 28, 1896; age, 3 yrs. 11 mos.; estimated weight, 800 lbs.; fed about 18 qts. daily of a mixture of ground oats, wheat middlings, bran, corn meal and a little pea meal—clover hay and ensilage; property of T. S. Cooper, Coopersburg, Pa.

Pedro's Pretty Marjoram 77947.—Sire, Pedro 3187; dam, Marjoram 2d 12805.

Butter, 14 lbs. 10 oz. Milk, 244 lbs.

Test made from Dec. 4 to 11, 1892; age, 2 yrs. 7 mos.; estimated weight, 800 lbs.; fed about 18 qts. daily of a mixture of ground oats, wheat middlings, bran, corn meal and a little pea meal—clover hay and ensilage; property of T. S. Cooper, Coopersburg, Pa.

Pedro's Pretty Pansy 75275.—Sire, Prospect's Rioter 9189; dam, Pedro's Pansy 33835.

Butter, 18 lbs. 4 oz. Milk, 246 lbs.

Test made from Mar. 23 to 30, 1894; age, 3 yrs. 11 mos.; estimated weight, 850 lbs.; fed about 22 qts. daily of a mixture of ground oats, middlings, bran, corn meal and some pea meal—clover hay and ensilage; property of T. S. Cooper, Coopersburg, Pa.

Pedro's Pretty Phillis 102378.—Sire. Pedro 3187; dam, Pedro's Lady Phillis 88486.

Butter, 17 lbs. 4 oz. Milk, 248 lbs.

Test made from Nov. 20 to 27, 1896; age, 2 yrs. 2 mos.; actual weight, 750 lbs.; fed about 18 qts. daily of a mixture of ground oats, wheat middlings, bran, corn meal and a little pea meal—clover hay, green corn, good meadow pasture; property of T. S. Cooper, Coopersburg, Pa.

Pedro's Pretty Princess 102377.—Sire, Pedro's Silver Rioter 31320; dam, Pedro's Young Princess 72310. ·

Butter, 17 lbs. 6 oz. Milk, 248 lbs.

Test made from Nov. 23 to 30, 1897; age, 3 yrs. 2 mos.; estimated weight, 750 lbs.; fed about 20 qts. daily of a mixture of ground oats, middlings, wheat bran, corn meal and some pea meal—clover hay and green corn, meadow and blue-grass pasture; property of T. S. Cooper, Coopersburg, Pa.

Pedro's Princess of York 72378.—Sire, Prospect's Rioter 9189; dam, Pedro's Young Princess 72310.

Butter, 15 lbs. 4 oz. Milk, 159 lbs. 8 oz.

Test made from Nov. 15 to 22, 1894; age, 4 yrs. 6 mos.; estimated weight, 700 lbs.; fed sweet corn fodder—corn stubble in forenoon, grass pasture in afternoon; property of J. F. Avery, Saline, Mich.

Pedro's Priscilla 99413.—Sire, Pedro 3187; dam, Kedria 88645.

Butter, 17 lbs. 6 oz. Milk, 265 lbs.

Test made from Jan. 23 to 30, 1898; age, 3 yrs. 4 mos.; estimated weight, 775 lbs.; fed about 20 qts. daily of a mixture of ground oats, middlings, bran, corn meal and a little pea meal—clover hay and ensilage; property of T. S. Cooper, Coopersburg. Pa.

Pedro's Rainbow 66096.—Sire, Pedro 3187; dam, Rainbow 16672.

Butter, 18 lbs. 15 oz. Milk, 276 lbs.

Test made from Nov. 22 to 29, 1892; age, 5 yrs.; estimated weight, 850 lbs.; fed about 20 qts. daily of a mixture of ground oats, middlings, bran, corn meal and some pea meal—clover hay, green corn, blue-grass pasture; property of T. S. Cooper, Coopersburg, Pa.

Pedro's Rosebud 66086.—Sire, Pedro 3187; dam, Khedive's Rosebud 18173.

Butter, 19 lbs. 4 oz. Milk, 248 lbs.

Test made from Nov. 4 to 11, 1892; age, 5 yrs. 7 mos.; estimated weight, 850 lbs.; fed about 20 qts. daily of a mixture of ground oats, wheat middlings, wheat bran, corn meal and a little pea meal—clover hay, green corn, good pasture; property of T. S. Cooper, Coopersburg, Pa.

Pedro's Royal Princess 88485.—Sire, Pedro 3187; dam, Princess Lorne 68699.

Butter, 23 lbs. 14 oz. Milk, 295 lbs.

Test made from Sept. 11 to 18, 1894; age, 3 yrs. 8 mos.; estimated weight, 950 lbs.; fed about 26 qts. daily of a mixture of ground oats, bran, middlings, corn meal and a little pea meal—clover hay, blue-grass pasture; property of T. S. Cooper, Coopersburg, Pa.

Pedro's Silver Pearl 110401.—Sire, Pedro's Silver Rioter 31320; dam, Rioter's Pretty Pearl 75273.

Butter, 18 lbs. 6 oz. Milk, 253 lbs.

Test made from Jan. 23 to 30, 1898; age, 3 yrs.; estimated weight, 700 lbs.; fed about 20 qts. daily of a mixture of ground oats, middlings. wheat bran, corn meal and some pea meal—clover hay and ensilage; property of T. S. Cooper, Coopersburg, Pa.

Pedro's Very Pretty 110508.—Sire, Pedro 3187; dam, Pedro's Pretty Girl 88488.

Butter, 15 lbs. 4 oz. Milk, 235 lbs.

Test made from Nov. 23 to 30, 1897; age, 2 yrs. 4 mos.; estimated weight, 700 lbs.; fed about 15 qts. daily of a mixture of ground oats, middlings, wheat bran, corn meal and a little pea meal—clover hay, green corn, good blue-grass and meadow pasture; property of T. S. Cooper, Coopersburg, Pa.

Pedro's Young Princess 72310.—Sire, Pedro 3187; dam, Young Princess Pogis 29560.

Butter, 18 lbs. 4 oz. Milk, 252 lbs.

Test made from June 23 to 30, 1891; age, 2 yrs. 8 mos.; estimated weight, 850 lbs.; fed about 20 qts. daily of a mixture of ground oats, middlings, bran, corn meal and some pea meal—clover hay and ensilage; good blue-grass pasture; property of T. S. Cooper, Coopersburg, Pa.

Pedro's Zoe 95697.—Sire, Pedro Sheldon 22865; dam, Zoe Sheldon 48873.

Butter, 15 lbs. 10 oz. Milk, 240 lbs. 4 oz.

Test made from Feb. 18 to 25, 1896; age, 3 yrs. 8 mos.; estimated weight, 900 lbs.; fed 35 lbs. corn-cob meal, 35 lbs. crushed oats, 35 lbs. bran, 14 lbs. oil meal, 87½ lbs. clover hay and 87½ lbs. wheat straw; property of B. F. Carper, Rosemond, Ill.

Peggie Tormentor 98103.—Sire, Tormentor 5th 21962; dam, Mirodelle 76862.

Butter, 16 lbs. 2 oz. Milk, 217 lbs. 14 oz.

Test made from Nov. 10 to 17, 1897; age, 4 yrs. 11 mos.; actual weight, 910 lbs.; fed 48½ lbs. bran, 14¼ lbs. cotton-seed meal and 28¼ lbs. corn and cob meal—hay, ensilage and corn fodder; property of Biltmore Farms, Biltmore, N. C.

Peggy Ford 21713.—Sire, Palm Nut 4744; dam, Caddie Mackie 10411.

Butter, 14 lbs. 10 oz. Milk, 189 lbs.

Test made from June 21 to 27, 1885; age, 2 yrs. 11 mos.; test made on grass alone; property of Jacob Lusk, East Palmyra, N. Y.

Peggy of Staatsburgh 2342.—Sire, Orphan 891; dam, Princess 2d 2295.

Butter, 14 lbs. 1¼ oz., unsalted.

Test made from June 22 to 29, 1883; age, 10 yrs. 10 mos.; property of W. B. Dinsmore. Staatsburg, N. Y.

Pendule 2d 16709.—Sire, Cicero 7657; dam, Pendule 16673.

Butter, 16 lbs. 5 oz. Milk, 226 lbs. 5 oz.

Test made from Sept. 26 to Oct. 2, 1885; age, 3 yrs. 11 mos.; property of J. L. Shallcross, Anchorage, Ky.

Pera B. 86761.—Sire, Festive Florin 19967; dam, Perira 19433.

Butter, 17 lbs. 9 oz. Milk, 253 lbs. 8 oz.

Test made from May 25 to June 1, 1896; age, 4 yrs. 6 mos.; estimated weight. 875 lbs.; fed 56 lbs. wheat bran, 35 lbs. corn meal and 14 lbs. cotton-seed meal—dry prairie pasture; property of C. M. Bivins, Terrell, Tex.

Percie 14937.—Sire, Golden Lion 5239; dam, Bellita 2d 10311.

Butter, 14 lbs. 6½ oz. (official). Milk, 261 lbs. 4 oz.

Test made from June 17 to 23, 1884; age, 3 yrs.; estimated weight, 800 lbs.; fed, daily, 2 qts. corn meal, 3 qts. fine feed and 1 qt. linseed meal; property of Cornelius Wellington, East Lexington, Mass.

Percie 14937.—Sire, Golden Lion 5239; dam, Bellita 2d 10311.

Butter, 18 lbs. 10 oz. Milk, 234 lbs. 10 oz.

Test made from June 3 to 9, 1885; age, 4 yrs.; property of Cornelius Wellington, East Lexington, Mass.

Period 42640.—Sire, Combination 4389; dam, Coma 29330.

Butter, 16 lbs. 3 oz. Milk, 230 lbs. 11 oz.

Test made from Mar. 15 to 22, 1888; age, 3 yrs. 11 mos.; estimated weight, 800 lbs.; fed 18 qts. ground oats and corn, daily; property of Richardson Bros., Davenport, Iowa.

Perrot's Baroness 22508.—Sire, Perrot (P. S. 342 J. H. B.); dam, Lady Baronnella 22501.

Butter, 15 lbs. 14½ oz. Milk, 142 lbs. 6 oz.

Test made from Feb. 15 to 22, 1889; age, 5 yrs. 7 mos.; estimated weight, 700 lbs.; fed, daily, 4 gals. corn meal and ground oats, mixed; property of Maury Jersey Farm, Columbia, Tenn.

Perry Farm Golden Cloud 22872.—Sire, Rough (P. S. 239 J. H. B.); dam on I. of J.

Butter, 18 lbs. 9 oz. Milk, 208 lbs.

Test made from June 14 to 21, 1887; age, 6 yrs.; estimated weight, 850 lbs.; fed, daily, 4 qts. ground corn and oats, 4 qts. bran and 1 qt. pea meal; property of M. Erskine Miller, Staunton, Va.

Personia 90231.—Sire, Diploma 16219; dam, Modita 16626.

Butter, 16 lbs. 3¼ oz. Milk, 249 lbs. 10 oz.

Test made from Mar. 19 to 26, 1897; age, 5 yrs. 4 mos.; estimated weight, 750 lbs.; fed 49 lbs. bran, 48½ lbs. corn, 40 lbs. oats, 14 lbs. linseed meal, 140 lbs. roots and 245 lbs. ensilage—hay *ad lib.*; property of C. I. Hood, Lowell, Mass.

Pertinette 70348.—Sire, Kerchunkus 21111; dam, Pertinaxia 28300.

Butter, 14 lbs. Milk, 196 lbs.

Test made from July 22 to 29, 1891; age, 2 yrs. 3 mos.; estimated weight, 800 lbs.; fed 8 lbs. corn meal, 4 lbs. ground oats, 4 lbs. wheat bran and 8 lbs. clover hay, daily —white clover, orchard and blue-grass pasture; property of Mrs. Kate D. Nicholson, Knoxville, Tenn.

Perty W. 41721.—Sire, Great Ado 4227; dam, Perty J. 15648.

Butter, 14 lbs. 2 oz. Milk, 215 lbs. 12 oz.

Test made from Sept. 19 to 26, 1897; age, 12 yrs. 7 mos.; estimated weight, 875 lbs.; fed 61 lbs. wheat bran, 61 lbs. cotton-seed meal and 33 lbs. corn meal—short pasture during day, green clover at night; property of W. W. & F. B. Pike, Cornish, Me.

Pet Anna 1608.—Sire, Marmion 359; dam, Lillie Fair 1607.

Butter, 14 lbs., unsalted.

Test made in June, 1877; age, 6 yrs. 2 mos.; no feed—good pasture only; property of Beech Grove Farm, Beech Grove, Ind.

Pet Clover 14624.—Sire, Granite Rock 4391; dam, May Clover 12273.

Butter, 16 lbs. 8 oz., unsalted. Milk, 17 qts. per day.

Test made from Aug. 25 to Sept. 1, 1885; age, 5 yrs. 1 mo.; property of W. F. Daniell, Franklin, N. H.

Petite Mère 8516.—Sire, Caliph 1618; dam, Fannie G. 3829.

Butter, 15 lbs. 13 oz. Milk, 197 lbs.

Test made from Mar. 31 to Apr. 6, 1884; age, 6 yrs. 2 mos.; property of R. W. Miller, Lebanon, Tenn.

Pet Lee 7993.—Sire, Padisha 1623; dam, Julia Parks 3778.

Butter, 14 lbs. 12 oz.

Pet of Maplewood Farm 4854.—Sire, Victor 797; dam, Margeretta 1994.

Butter, 15 lbs. 2 oz. Milk, 259 lbs. 12 oz.

Test made from May 28 to June 3, 1883; property of Mrs. L. M. Fair, Wallingford, Conn.

Pet of Rose Lawn 11326.—Sire, Aventurier 4254; dam, Rose of Rose Lawn 9365.

Butter, 18 lbs. 1 oz. Milk, 258 lbs. 5 oz.

Test made from May 18 to 25, 1884; age, 4 yrs.; property of J. S. Rogers, Paterson, N. J.

Pet of Rose Lawn 11326.—Sire, Aventurier 4254; dam, Rose of Rose Lawn 9365.

Butter, 15 lbs. 8½ oz. (official). Milk, 246 lbs. 10 oz.

Test made from June 20 to 27, 1885; age, 5 yrs. 1 mo.; fed, daily, about 12 qts. ground oats and 3 qts. corn meal; property of Jacob S. Rogers, Paterson, N. J.

Petra 19267.—Sire, Le Brocq's Prize 3350; dam, Petrus 5563.

Butter, 16 lbs. 6 oz. Milk, 238 lbs.

Test made from Sept. 23 to 30, 1887; age, 5 yrs. 7 mos.; estimated weight, 850 lbs.; fed, daily, 8 qts. ground oats and corn and 8 qts. bran; property of H. M. Baum, Frankfort, Ind.

Petra's Lady 99791.—Sire, Stoke Pogis' Perfection 5984; dam, Petra 19267.

Butter, 15 lbs. 10½ oz. Milk, 213 lbs. 8 oz.

Test made from Mar. 6 to 13, 1897; age, 5 yrs. 5 mos.; estimated weight, 900 lbs.; fed 10½ lbs. oil meal, 84 lbs. corn meal, 42 lbs. bran, 35 lbs. shorts, 224 lbs. ensilage and 35 lbs. hay; property of Baum & Hill, Frankfort, Ind.

Petra's Signal 53376.—Sire, Signaldo Alexis 15051; dam, Petra 19267.

Butter, 14 lbs. 12 oz. Milk, 208 lbs.

Test made from Oct. 3 to 10, 1897; age, 9 yrs. 3 mos.; estimated weight, 800 lbs.; fed 245 lbs. ensilage—blue-grass pasture; property of Baum & Hill, Frankfort, Ind.

Pet Rex 20166.—Sire, Rex 1330; dam, Hybla 2991.

Butter, 14 lbs. 2½ oz. Milk, 161 lbs. 6 oz.

Test made from Jan. 20 to 26, 1885; age, 2 yrs.; property of William Simpson, New York, N. Y.

Pet's Fancy 80968.—Sire, Fancy Toltec 21167; dam, Buckeye Pet 19459.

Butter, 17 lbs. 4 oz. Milk, 244 lbs. 8 oz.

Test made from June 7 to 14, 1895; age, 3 yrs. 2 mos.; actual weight, 740 lbs.; fed 18 lbs. per day of a mixture, in equal parts by bulk, of ground corn, oats and bran—extra good clover and timothy pasture; property of J. W. L. Motherspaw, Newark, Ohio.

Pet's Torment 92255.—Sire, Tennessee's Tormentor 23498; dam, Fancy's Pet 36013.

Butter, 15 lbs. 14 oz. Milk, 162 lbs. 12 oz.

Test made from Apr. 26 to May 3, 1895; age, 3 yrs. 3 mos.; estimated weight, 700 lbs.; fed 56 qts. ground oats and 56 qts. corn meal—clover and millet hay, poor pasture; property of W. B. Gordon, Columbia, Tenn.

Pet Victor Pogis 77812.—Sire, Melia Ann's Victor Pogis 20697; dam, Pierrot's Beauty 48964.

Butter, 16 lbs. 4 oz. Milk, 243 lbs.

Test made from July 4 to 11, 1898; age, 8 yrs. 5 mos.; estimated weight, 1000 lbs.; fed 6 lbs. corn meal and 6 lbs. wheat bran, daily—mixed pasture; property of J. H. Martin, Granville, N. Y.

Phædra 2561.—Sire, Mercury 432; dam, Leda 799.

Butter, 19 lbs. 13 oz. Milk, 263 lbs. 12 oz.

Test made from Jan. 15 to 21, 1882; age, 9 yrs.; fed corn and oats, ground together, a little oil meal, roots, cut fine and sprinkled with bran, and plenty of good clover hay; property of William Simpson, New York, N. Y.

Philena of Briarcliff 68938.—Sire, American Rioter 13363; dam, Philine of Linden 2d 68319.

Butter, 15 lbs. 15½ oz. Milk, 219 lbs. 12 oz.

Test made from May 13 to 20, 1893; age, 5 yrs. 2 mos.; estimated weight, 850 lbs.; fed 15 qts. daily of corn meal, ground oats and wheat bran, in equal parts—hay *ad lib.* and a little green rye; property of Walter W. Law, Whitson, N. Y.

Philena S. 70375.—Sire, Sophie's Tormentor 20883; dam, Miss Allena 42174.

Butter, 18 lbs. 9½ oz. Milk, 265 lbs. 6 oz.

Test made from May 29 to June 5, 1898; age, 7 yrs. 1 mo.; estimated weight, 950 lbs.; fed 31½ lbs. bran, 31½ lbs. corn meal, 21 lbs. ground oats, 7 lbs. oil meal and 7 lbs. cotton-seed meal—clover and timothy pasture; property of C. I. Hood, Lowell, Mass.

Philena Victor Pogis 127138.—Sire, Melia Ann's Victor Pogis 20697; dam, Philena Polonius 24915.

Butter, 17 lbs. 3 oz. Milk, 289 lbs.

Test made from June 20 to 27, 1898; age, 4 yrs. 2 mos.; estimated weight, 725 lbs.; fed 4 lbs. corn meal and 4 lbs. wheat bran, daily—clover pasture; property of J. H. Martin, Granville, N. Y.

Phillis 3d 46681.—Sire, Stoke Pogis of Linden 10558; dam, Phillis 18162.

Butter, 18 lbs. 13 oz. Milk, 262 lbs.

Test made from Nov. 18 to 25, 1893; age, 8 yrs.; estimated weight, 900 lbs.; fed about 20 qts. daily of a mixture of ground oats, middlings, wheat bran, corn meal and some pea meal—clover hay and green corn, meadow and blue-grass pasture; property of T. S. Cooper, Coopersburg, Pa.

Phillis of St. Lambert 78867.—Sire, King of St. Lambert 15175; dam, Allie of St. Lambert 2d 43671.

Butter, 23 lbs. 12 oz. Milk, 320 lbs.

Test made from May 3 to 10, 1896; age, 7 yrs. 5 mos.; estimated weight, 1000 lbs.; fed 2 qts. bran, 2 qts. oats and peas, ground together, 2 qts. corn meal and 1 qt. oil meal, on ½ bush. cut clover hay, wet, night and morning; 2 qts. oats and peas and 2 qts. corn meal, on ½ bush. cut roots, at noon—hay *ad lib.*; property of George E. Peer, Rochester, N. Y.

Phlox 16399.—Sire, Guy Mannering 698; dam, Gazania 4513.

Butter, 16 lbs. 3½ oz. Milk, 175 lbs.

Test made from Aug. 19 to 25, 1883; fed 6 qts. daily of corn and oats, mixed in equal parts and ground together, fed with cut hay—timothy, orchard grass and blue-grass pasture; property of Columbia Jersey Cattle Co., Columbia, Tenn.

Phlox 16399.—Sire, Guy Mannering 698; dam, Gazania 4513.

Butter, 21 lbs. 11 oz. Milk, 233 lbs. 2 oz.

Test made from June 5 to 11, 1884; age, 7 yrs. 2 mos.; fed 10 qts. corn meal, 4 qts. bran, over cut hay, daily—pasture of timothy, orchard and blue-grass; property of Columbia Jersey Cattle Co., Columbia, Tenn.

Phlox 3d 31882.—Sire, Toltec 6831; dam, Phlox 16399.

Butter, 15 lbs. 6 oz. Milk, 100 lbs. 6 oz.

Test made from Jan. 5 to 11, 1886; age, 1 yr. 8 mos.; property of Columbia Jersey Cattle Co., Columbia, Tenn.

Phœbe N. 25401.—Sire, Samson Jr. 2723; dam, Phœbe 4th 2271.

Butter, 15 lbs. 3 oz. Milk, 240 lbs. 8 oz.

Test made from May 23 to 30, 1884; age, 4 yrs. 3 mos.; property of W. C. Norton, Aldenville, Pa.

Phœbe Rex 55084.—Sire, Col. Rex 12975; dam, Margaret of Lagonda 35127.

Butter, 16 lbs. ½ oz. Milk, 256 lbs. 5 oz.

Test made from Feb. 24 to Mar. 3, 1895; age, 7 yrs.; estimated weight, 900 lbs.; fed 35 lbs. bran. 30 lbs. shorts. 32 lbs. corn and oats, 8½ lbs. Blatchford's stock food and 210 lbs. ensilage—hay; property of Jud. Keller, Newark, Ohio.

Phœbe's Charity 59008.—Sire, Charity Boy 11765; dam, Phœbe Cross 38334.

Butter, 15 lbs. ½ oz. Milk, 256 lbs. 8 oz.

Test made from June 24 to July 1, 1896; age, 7 yrs. 4 mos.; estimated weight, 725 lbs.; fed 10 lbs. oats and corn, ground together, and 3 lbs. oil meal, daily—clover hay, mixed pasture; property of J. H. Martin, Granville, N. Y.

Phrynore of Allentown 3d 126321.—Sire, Fritz of Glynllyn 36579; dam, Phrynore of Allentown 45205.

Butter, 14 lbs. 8 oz. Milk, 233 lbs. 4 oz.

Test made from Feb. 23 to Mar. 2, 1898; age, 2 yrs. 8 mos.; estimated weight, 700 lbs.; fed 28 lbs. wheat middlings, 28 lbs. hominy meal, 12 lbs. oil meal and 7 lbs. cotton-seed meal—hay; property of J. A. Griffith, Ashley Falls, Mass.

Phyllis of Hillcrest 9067.—Sire, Date 2624; dam, Sukey 2d 1224.

Butter, 14 lbs. 12 oz.

Test made from May 28 to June 3, 1883; age, 3 yrs. 10 mos.; property of W. A. Mullin, Mount Holly Springs, Pa.

Phyllis of Hillcrest 9067.—Sire, Date 2624; dam, Sukey 2d 1224.

Butter, 16 lbs. Milk, 240 lbs.

Test made from June 11 to 18, 1884; age, 4 yrs. 10 mos.; property of W. A. & A. F. Mullin, Mount Holly Springs, Pa.

Picotu's Bess 86172.—Sire, Festive Florin 19967; dam, Picotu 24211.

Butter, 17 lbs. 8½ oz. Milk, 263 lbs. 12 oz.

Test made from Apr. 1 to 8, 1896; age, 4 yrs. 6 mos.; estimated weight, 900 lbs.; fed 14 lbs. cotton-seed meal, 56 lbs. corn meal and 70 lbs. wheat bran—prairie hay *ad lib.*; property of C. M. Bivins, Terrell, Tex.

Pierrot's Beauty 48964.—Sire, Pedroller 12412; dam, Island Beauty 2d 19885.

Butter, 15 lbs. 11 oz. Milk, 221 lbs. 12 oz.

Test made from Dec. 19 to 26, 1890; age, 3 yrs.; estimated weight, 750 lbs.; fed, daily, 5½ lbs. corn, ground in ear, and 2½ lbs. wheat bran; property of Mrs. J. H. Martin, Granville, N. Y.

Pierrot's Lady Bacon 12482.—Sire, Pierrot 7th 1667; dam, Princess of Mansfield 8070.

Butter, 16 lbs. 10 oz. Average milk, 4 gals. per day.

Test made from May 24 to 30, 1884; age, 6 yrs.; property of Joseph C. Kiplinger & Co., Eagle City, Ohio.

Pilot's Rose 17958.—Sire, Pilot (P. S. 183 J. H. B.); dam on I. of J.

Butter, 18 lbs. 3¾ oz. Milk, 205 lbs.

Test made from June 7 to 14, 1887; age, 7 yrs. 5 mos.; estimated weight, 850 lbs.; fed, daily, 4 qts. bran, 4 qts. ground corn and oats and 1 qt. pea meal; property of M. Erskine Miller, Staunton, Va.

Pilot's Veronica 18917.—Sire, Pilot (P. S. 183 J. H. B.); dam on I. of J.

Butter, 20 lbs. 2 oz. Milk, 234 lbs.

Test made from Oct. 9 to 16, 1885; age, 6 yrs. 1 mo.; fed 10 lbs. ground oats and corn and 6 lbs. cotton-seed meal, daily, mixed—good pasture; property of William H. Burr, Redding Ridge, Conn.

Pinafore 2d 15072.—Sire, Vermonter 5620; dam, Pinafore 8289.

Butter, 15 lbs. 8 oz. Milk, 182 lbs.

Test made from June 13 to 19, 1885; age, 3 yrs. 5 mos.; test made on grass alone; property of Jacob Lusk, East Palmyra, N. Y.

Pine Tree Clara 55042.—Sire, Sheldon of St. Lambert 13831; dam, Pine Tree Bessie 32798.

Butter, 17 lbs. 3 oz. Milk, 251 lbs. 12 oz.

Test made from Apr. 28 to May 5, 1896; age, 9 yrs. 11 mos.; estimated weight, 1000 lbs.; fed 56 lbs. bran, 28 lbs. corn-cob meal and 14 lbs. old process oil meal—timothy pasture; property of B. F. Carper, Rosemond, Ill.

Pinkey Spry 18316.—Sire, Tormentor 3533; dam, Varinka 3838.

Butter, 16 lbs. 12 oz. Milk, 207 lbs. 5 oz.

Test made from July 14 to 21, 1892; age, 11 yrs. 1 mo.; estimated weight, 800 lbs.; fed 2 gals. corn meal, 2 gals. bran, 3 gals. oat meal, mixed, fed on cut green corn; property of H. A. Huntington, Higganum, Conn.

Pinkey Spry 2d 92953.—Sire, Odello 20223; dam, Pinkey Spry 18316.

Butter, 14 lbs. 10½ oz. Milk, 143 lbs. 8 oz.

Test made from Nov. 23 to 30, 1894; age, 2 yrs. 6 mos.; estimated weight, 700 lbs.; fed 2 qts. corn-hearts, 2 qts. bran, 1 qt. cotton-seed meal and 2 qts. ground oats, daily—sheaf oats, cut clover hay, blue-grass pasture; property of Samuel N. Warren, Spring Hill, Tenn.

Pink of Meridale 114748.—Sire, Matilda 4th's Son 20214; dam, Meridale Rioter Pink 64584.

Butter, 16 lbs. ½ oz. Milk, 227 lbs. 14 oz.

Test made from Jan. 21 to 28, 1898; age, 3 yrs. 1 mo.; estimated weight, 875 lbs.; fed 56 lbs. wheat bran, 35 lbs. corn meal, 7 lbs. cotton-seed meal, 7 lbs. linseed meal, 140 lbs. ensilage and 1½ bush. beets—about 42 lbs. hay; property of Ayer & McKinney, Philadelphia, Pa.

Pink Ring 69394.—Sire, Sophie's Tormentor 20883; dam, Catono's Pink 2d 58626.

Butter, 16 lbs. 4½ oz. Milk, 239 lbs. 5 oz.

Test made from Apr. 6 to 13, 1898; age, 8 yrs.; estimated weight, 800 lbs.; fed 42 lbs. bran, 35 lbs. corn meal, 21 lbs. ground oats, 17½ lbs. oil meal, 210 lbs. ensilage and 70 lbs. roots—hay *ad lib.*; property of C. I. Hood, Lowell, Mass.

Pitapat 11136.—Sire, Tormentor 3533; dam, Ximena 5682.

Butter, 16 lbs. 5 oz. Milk, 197 lbs. 8 oz.

Test made from Nov. 5 to 12, 1888; age, 8 yrs. 8 mos.; actual weight, 878 lbs.; fed, daily, 20 lbs. mixed feed, consisting of equal parts of corn meal, oat meal and wheat bran; property of Jacob L. Thomas, Knoxville, Tenn.

Pitapat 2d 86907.—Sire, Fancy's Harry 9777; dam, Pitapat 11136.

Butter, 19 lbs. 3¼ oz. Milk, 218 lbs. 5 oz.

Test made from Jan. 14 to 21, 1897; age, 4 yrs. 10 mos.; estimated weight, 700 lbs.; fed 16 qts. daily of oats and corn, ground together, mixed with four bucketfuls of cut sheaf oats; property of J. E. Hart, Nashville, Tenn.

Pixie 4115.—Sire, Pertinatti 713; dam, Roxana 2d 2532.

Butter, 14 lbs. Milk, 184 lbs. 8 oz.

Test made from Dec. 7 to 13, 1881; age, 6 yrs. 8 mos.; property of Campbell Brown, Spring Hill, Tenn.

Platina Cary 73375.—Sire, Twice Signal 9582; dam, Arlette Cary 73372.

Butter, 17 lbs. 7¼ oz. Milk, 265 lbs.

Test made from May 14 to 21, 1895; age, 11 yrs. 1 mo.; estimated weight, 900 lbs.; fed 98 lbs. corn-hearts, 70 lbs. ground oats and 28 lbs. bran, wet—blue-grass pasture; property of Jeff. D. Clinard, Springfield, Tenn.

Pledge 59214.—Sire, Upright 6147; dam, Frankie's Lass 24900.

Butter, 17 lbs. 9½ oz. Milk, 256 lbs. 4 oz.

Test made from Dec. 17 to 24, 1894; age, 8 yrs. 9 mos.; estimated weight, 850 lbs.; fed 14 qts. daily of a mixture of corn meal, pea meal, oil meal and ground oats, for 5 days, then 16 lbs. daily—2 qts. roots daily, clover hay; property of Richardson Bros., Davenport, Iowa.

Plenty 950.—Sire, Comus 54; dam, Beck 463.

Butter, 14 lbs. 8 oz.

Test made June, 1873; age, 10 yrs.; property of Thomas T. Turner, St. Louis, Mo.

Plum 13228.—Sire, Gen. Rosser 4189; dam, Fannie Booth 12505.

Butter, 17 lbs. 4 oz. Milk, 229 lbs. 8 oz.

Test made from June 20 to 26, 1885; age, 4 yrs. 1 mo.; test made on grass alone; property of J. R. Gaston, Normal, Ill.

Plumage 53897.—Sire, Diploma 16219; dam, Paradise 32082.

Butter, 17 lbs. 5 oz. Milk, 266 lbs. 7 oz.

Test made from Nov. 23 to 30, 1890; age, 3 yrs. 8 mos.; estimated weight, 850 lbs.; fed, daily, 18 qts. ground corn and oats and peas; property of Richardson Bros., Davenport, Iowa.

Pluma of Jersey Lawn 89954.—Sire, Maud's Rioter Pogis 25345; dam, Pettie Fair Light 68207.

Butter, 14 lbs. Milk, 237 lbs.

Test made from May 26 to June 2, 1895; age, 3 yrs. 5 mos.; estimated weight, 900 lbs.; fed, first four days, warm mash of 10 lbs. bran and 8 lbs. ground corn and oats; last three days, 3 lbs. shorts added to bran, corn and oats, fed dry—mixed pasture; property of Mrs. Isa Bayler, Washington, Ill.

Plumida 23621.—Sire, Warpole 3500; dam, Mildrida 6743.

Butter, 15 lbs. ½ oz. Milk, 153 lbs. 8 oz.

Test made from May 13 to 20, 1887; age, 3 yrs. 9 mos.; estimated weight, 820 lbs.; blue-grass pasture only; property of Thos. C. Beer, Bucyrus, Ohio.

Pocia of Andalusia 62616.—Sire, Boss of the Cedars 17712; dam, Princess of Ashantee 4th 26307.

> Butter, 20 lbs. 15 oz. Milk, 328 lbs. 12 oz.

Test made from May 17 to 24, 1892; age, 3 yrs. 6 mos.; estimated weight, 900 lbs.; fed 6 qts. corn meal and 6 qts. ground oats, daily, in two feeds—pasture, timothy, common grass and some white clover; property of George Fox, Torresdale, Pa.

Pocia of Andalusia 2d 79548.—Sire, Daisy's Boss 24566; dam, Pocia of Andalusia 62616.

> Butter, 21 lbs. 7½ oz. Milk, 257 lbs.

Test made from Nov. 26 to Dec. 3, 1892; age, 2 yrs.; estimated weight, 700 lbs.; fed 5 qts. corn meal, 6 qts. ground oats, 4 qts. bran, 1 qt. cake meal and 1 pk. carrots, daily (estimated)—12 lbs. corn fodder and hay; property of George Fox, Torresdale, Pa.

Pogis' Dewdrop 40373.—Sire, Prince of Melrose 2d 11015; dam, Dewdrop Pansy 19736.

> Butter, 19 lbs. 12½ oz. Milk, 261 lbs. 6 oz.

Test made from Apr. 14 to 21, 1890; age, 3 yrs. 9 mos.; estimated weight, 900 lbs.; fed 56 qts. ground oats, 28 qts. pea meal, 28 qts. corn meal and 20 lbs. ensilage—hay, clover pasture; property of James Crook, Jacksonville, Ala.

Pogis' Ida May 50063.—Sire, Ida's Stoke Pogis 13058; dam, May Bud of St. Lambert 5105.

> Butter, 14 lbs. 11 oz. Milk, 192 lbs.

Test made from Feb. 13 to 20, 1892; age, 4 yrs. 10 mos.; estimated weight, 850 lbs.; fed 51 lbs. corn meal, 52 lbs. oat meal, 40 lbs. bran, 9 lbs. oil meal, 6 lbs. pea meal, 80 lbs. hay and 140 lbs. ensilage; property of Ayer & McKinney, Philadelphia, Pa.

Pogis' May 26950.—Sire, Stoke Pogis 5th 5987; dam, Lottie May 17201.

> Butter, 18 lbs. 2 oz. Milk, 219 lbs. 12 oz.

Test made from Dec. 20 to 27, 1890; age, 6 yrs. 7 mos.; estimated weight, 1000 lbs.; fed, daily, average of 9½ lbs. corn meal, 9½ lbs. crushed oats, 4¾ lbs. pea meal and 4¾ lbs. oil meal; property of Ayer & McKinney, Philadelphia, Pa.

Pogis' May's Matilda 75347.—Sire, Matilda 4th's Son 20214; dam, Pogis' May 26950.

> Butter, 15 lbs. 2 oz. Milk, 252 lbs. 12 oz.

Test made from Aug. 29 to Sept. 5, 1896; age, 6 yrs. 1 mo.; estimated weight, 950 lbs.; fed 30 lbs. grain per day, consisting of two-fifths bran, two-fifths hominy and corn meal and one-fifth cotton-seed meal—pasture, timothy and natural grasses; property of Ayer & McKinney, Philadelphia, Pa.

Pogis' May's Matilda 75347.—Sire, Matilda 4th's Son 20214; dam, Pogis' May 26950.

> Butter, 15 lbs. 11½ oz. Milk, 282 lbs. 8 oz.

Test made from Aug. 8 to 15, 1897; age, 7 yrs.; estimated weight, 1000 lbs.; fed 63 lbs. wheat bran, 30 lbs. corn meal and 30 lbs. cotton-seed meal—timothy and natural grass pasture; property of Ayer & McKinney, Philadelphia, Pa.

Pogis' May's Pride 58577.—Sire, Guy Mannering Pogis 16365; dam, Pogis' May 26950.

> Butter, 15 lbs. 2 oz. Milk, 159 lbs. 12 oz.

Test made from Sept. 30 to Oct. 7, 1891; age, 3 yrs. 11 mos.; estimated weight, 825 lbs.; fed 39 lbs. oat meal, 34 lbs. corn meal, 32 lbs. bran, 19 lbs. oil meal and about 70 lbs. hay—40 lbs. green cut corn and millet, daily; property of Ayer & McKinney, Philadelphia, Pa.

Pogis Oonan 29890.—Sire, Nancy's Rioter 8390; dam, Little Accident 15578.

Butter, 15 lbs. 8 oz. Milk, 204 lbs. 8 oz.

Test made from Nov. 30 to Dec. 7, 1889; age, 5 yrs. 1 mo.; estimated weight, 800 lbs.; fed, daily, 24 lbs. chopped corn and wheat bran, mixed; property of Algernon K. Johnston, Stapleton, N. Y.

Pogis Oonan 29890.—Sire, Nancy's Rioter 8390; dam, Little Accident 15578.

Butter, 17 lbs. 8 oz. Milk, 263 lbs. 12 oz.

Test made from May 17 to 24, 1894; age, 9 yrs. 6 mos.; estimated weight, 1000 lbs.; fed 12 lbs. to 14 lbs. per day of bran, middlings, corn meal and oil meal, mixed—natural pasture; property of George W. Sisson, Jr., Potsdam, N. Y.

Pogis Oonan 29890.—Sire, Nancy's Rioter 8390; dam, Little Accident 15578.

Butter, 73 lbs. 5 oz. in 31 days. Milk, 1128 lbs.

Test made from May 17 to June 17, 1894; age, 9 yrs. 6 mos.; estimated weight, 1000 lbs.; fed 12 to 14 lbs. per day of bran, middlings, corn meal and oil meal, mixed—natural pasture; property of George W. Sisson, Jr., Potsdam, N. Y.

Pogis' René 102854.—Sire, Sig Pogis 22806; dam, René of Hillside 12952.

Butter, 17 lbs. Milk, 204 lbs.

Test made from Apr. 16 to 23, 1896; age, 1 yr. 11 mos.; estimated weight, 550 lbs.; fed 28 qts. ground oats, 28 qts. crushed corn, 28 qts. wheat bran and 7 qts. cotton-seed meal—blue-grass pasture; property of D. P. Carter, Lynnville, Tenn.

Pogis Siglise 92889.—Sire, Denise's Tormentor 11823; dam, Duchess Sigletta 54519.

Butter, 14 lbs. 10 oz. Milk, 207 lbs.

Test made from Nov. 12 to 19, 1897; age, 5 yrs. 9 mos.; estimated weight, 850 lbs.; fed 60 qts. bran, corn meal, cotton-seed meal and oil meal, in proportions of 10 parts bran, two of corn meal, two of cotton-seed meal and one of oil meal—hay *ad lib.*; property of George V. Green, Hopkinsville, Ky.

Pogis' Twilight 53794.—Sire, St. Lambert's John Bull 16618; dam, Dewdrop Pansy 19736.

Butter, 19 lbs. 1 oz. Milk, 234 lbs. 9 oz.

Test made from Dec. 19 to 26, 1891; age, 3 yrs. 8 mos.; estimated weight, 875 lbs.; fed 56 qts. corn meal, 42 qts. ground oats, 7 gals. raw cotton-seed and 210 lbs. ensilage—hay, no pasture; property of James Crook, Jacksonville, Ala.

Point Lace 62471.—Sire, Matilda 4th's Son 20214; dam, Mary M. Pogis 32455.

Butter, 17 lbs. 8 oz. Milk, 232 lbs. 4 oz.

Test made from Apr. 22 to 29, 1892; age, 3 yrs. 4 mos.; actual weight, 920 lbs.; fed 12 qts. ground oats, 12 qts. corn meal and 4 qts. ground oil-cake, daily—clover hay, 70 lbs. roots, second crop clover and timothy pasture; property of Miller & Sibley, Franklin, Pa.

Polly Bonair 42849.—Sire, Romeo de Bonair 4091; dam, Polly of Scituate 13635.

Butter, 18 lbs. 7 oz. Milk, 182 lbs. 8 oz.

Test made from Sept. 16 to 23, 1894; age, 8 yrs. 8 mos.; actual weight, 1020 lbs.; fed 18.7 lbs. cotton-seed meal, 19.3 lbs. old process linseed meal, 37.4 lbs. wheat bran, 46.6 lbs. corn chop, 112 lbs. cut corn-stalks and 7 lbs. hay; property of B. F. Fackenthal, Jr., Riegelsville, Pa.

Polly Clover 7052.—Sire, Schinchon 1132; dam, Sally Clover 4468.

Butter, 16 lbs. 15 oz. Daily average of milk, 19 qts.

Test made from May 25 to 31, 1882; age, 6 yrs. 9 mos.; property of Warren F. Daniell, Franklin, N. H.

Polly Neptune 15214.—Sire, Neptune (P. S. 151 J. H. B.); dam, Polly (P. S. 69 J. H. B.)

Butter, 14 lbs. 7 oz. Milk, 232 lbs. 4 oz.

Test made from June 7 to 13, 1885; age, 6 yrs. 4 mos.; property of C. E. Rowley, Cortland, N. Y.

Polonia of Riverside 84036.—Sire, Maquilla's Harry 22701; dam, Patty Polonius 3d 88677.

Butter, 18 lbs. 10 oz. Milk, 149 lbs. 8 oz.

Test made from Jan. 3 to 10, 1896; age, 4 yrs.; estimated weight, 700 lbs.; fed 104 lbs. ground oats, 8 lbs. wheat bran, 28 lbs. cotton-seed meal, 5 lbs. corn meal and 8 lbs. provender—hay *ad lib.*, some roots; property of H. A. Huntington, Higganum, Conn.

Polyxia 10753.—Sire, Grand Duke Alexis 1040; dam, Zina 1434.

Butter, 16 lbs. 7 oz. Milk, 301 lbs. 1 oz.

Test made from May 2 to 8, 1884; age, 4 yrs. 9 mos.; property of I. D. Risher, Hope Church, Pa.

Pomona of Prospect 123426.—Sire, Ida's Rioter of St. L. 13656; dam, Gipsy's Berry Duchess 86124.

Butter, 20 lbs. 6¼ oz. Milk, 215 lbs.

Test made from Dec. 21 to 28, 1897; age, 2 yrs. 6 mos.; actual weight, 936 lbs.; fed 56 qts. ground oats, 14 lbs. wheat bran, 21 qts. oat meal, 35 qts. corn meal, 21 qts. oil-cake, about 30 qts. malt and 84 lbs. sugar beets; property of Miller & Sibley, Franklin, Pa.

Pomona Pogis 40371.—Sire, Prince of Melrose 2d 11015; dam, Maggie of Springvale 15931.

Butter, 17 lbs. 2½ oz. Milk, 229 lbs. 14 oz.

Test made from Nov. 11 to 18, 1890; age, 4 yrs. 4 mos.; estimated weight, 850 lbs.; fed 28 qts. pea meal, 28 qts. corn meal, 14 qts. cotton-seed meal, 28 qts. bran and 280 lbs. ensilage—hay, ordinary pasture; property of James Crook, Jacksonville, Ala.

Pomona's Coquette 52832.—Sire, Canada's John Bull 3d 17750; dam, St. Lambert's Coquette 41070.

Butter, 14 lbs. 10 oz. Milk, 182 lbs. 2 oz.

Test made from Sept. 4 to 11, 1891; age, 3 yrs. 1 mo.; estimated weight, 800 lbs.; fed 8 lbs. corn meal, 8 lbs. crushed oats and 6 lbs. bran, daily—millet, green oats and peas, also pasture; property of Ayer & McKinney, Philadelphia, Pa.

Pomona's Ida of St. L. 59565.—Sire, Stoke Pogis 5th 5987; dam, Ida of St. Lambert 24990.

Butter, 14 lbs. 14½ oz. Milk, 190 lbs. 10 oz.

Test made from Feb. 3 to 10, 1892; age, 4 yrs. 1 mo.; estimated weight, 900 lbs.; fed 49 lbs. corn meal, 21 lbs. bran, 14 lbs. pea meal, 49 lbs. oat meal, 7 lbs. oil meal, 80 lbs. hay and 140 lbs. ensilage; property of Ayer & McKinney, Philadelphia, Pa.

Pomona's May Day Pogis 56249.—Sire, King of St. Lambert 15175; dam, May Day Stoke Pogis 28353.

Butter, 15 lbs. ½ oz. Milk, 188 lbs. 10 oz.

Test made from Aug. 24 to 31, 1891; age, 2 yrs. 7 mos.; estimated weight, 750 lbs.; fed 14 lbs. oil meal, 42 lbs. corn meal, 42 lbs. oat meal and 38 lbs. bran—good pasture, green oats and peas; property of Ayer & McKinney, Philadelphia, Pa.

Pomona's Violetta Pogis 56246.—Sire, Diana's Rioter 10481; dam, Stoke Pogis' Violetta 32684.

Butter, 16 lbs. 7 oz. Milk, 208 lbs.

Test made from Jan. 1 to 8, 1892; age, 3 yrs. 5 mos.; estimated weight, 825 lbs.; fed 2 lbs. oil meal, 2 lbs. pea meal, 8 lbs. oat meal and 8 lbs. corn meal, daily—90 lbs. hay, 2 bush. roots and 140 lbs. ensilage during test; property of Ayer & McKinney, Philadelphia, Pa.

Pomona's Violetta Pogis 56246.—Sire, Diana's Rioter 10481; dam, Stoke Pogis' Violetta 32684.

Butter, 23 lbs. 2¾ oz. Milk, 267 lbs. 4 oz.

Test made from Apr. 11 to 18, 1897; age, 8 yrs. 8 mos.; actual weight, 1160 lbs.; fed 119 qts. ground oats, 77 qts. corn meal, 28 qts. ground oil-cake and 140 lbs. sugar beets—clover hay; property of Miller & Sibley, Franklin, Pa.

Pope's Flora 18697.—Sire, The Pope's Bull 2965; dam, Flora F. 5544.

Butter, 14 lbs. 2 oz. Milk, 210 lbs. 5 oz.

Test made from June 19 to 25, 1885; age, 3 yrs. 7 mos.; property of Columbia Jersey Cattle Co., Columbia, Tenn.

Poppy H. 34865.—Sire, Pedro of the Valley 8750; dam, Arawana Poppy 2d 9935.

Butter, 17 lbs. 5½ oz. Milk, 163 lbs. 9 oz.

Test made from Oct. 20 to 27, 1889; age, 4 yrs. 7 mos.; estimated weight, 750 lbs.; fed, daily, 5 gals. ground oats and corn, in equal parts; property of Maury Jersey Farm, Columbia, Tenn.

Portfolio 84633.—Sire, Bishop 28851; dam, Portrait 32592.

Butter, 15 lbs. 5½ oz. Milk, 253 lbs. 2 oz.

Test made from Mar. 22 to 29, 1897; age, 4 yrs. 8 mos.; estimated weight, 750 lbs.; fed 42 lbs. bran, 42 lbs. corn meal, 42 lbs. ground oats, 13½ lbs. oil meal, 105 lbs. roots and 245 lbs. ensilage—hay ad lib.; property of C. I. Hood, Lowell, Mass.

Portrait 32592.—Sire, Combination 4389; dam, Duchess of Jefferson 17144.

Butter, 15 lbs. 2½ oz. Milk, 243 lbs. 9 oz.

Test made from Aug. 22 to 29, 1888; age, 3 yrs. 5 mos.; estimated weight, 800 lbs.; fed 12 qts. corn and oat meal, daily; property of Richardson Bros., Davenport, Iowa.

Postscript 39366.—Sire, Golden Art 10998; dam, Transcript 31867.

Butter, 14 lbs. 1½ oz. Milk, 241 lbs. 9 oz.

Test made from Mar. 7 to 14, 1897; age, 10 yrs. 11 mos.; estimated weight, 750 lbs.; fed 42 lbs. bran, 38½ lbs. corn meal, 31½ lbs. oats, 10½ lbs. linseed meal, 280 lbs. ensilage and 105 lbs. roots—hay ad lib.; property of C. I. Hood, Lowell, Mass.

Praxitella 50072.—Sire, Ida's Stoke Pogis 13658; dam, Praxilla 31677.

    Butter, 14 lbs. 4 oz. Milk, 183 lbs. 12 oz.

Test made from Dec. 18 to 25, 1892; age, 5 yrs. 4 mos.; estimated weight, 850 lbs.; fed 40 lbs. pea meal. 27 lbs. oil meal. 26 lbs. cotton-seed, 40 lbs. mixed feed and 150 lbs. ensilage; property of Ayer & McKinney, Philadelphia, Pa.

Preference 26343.—Sire, Dominie 11135; dam, Buckwheat 13840.

    Butter, 15 lbs. 5 oz. Milk, 313 lbs. 2 oz.

Test made from Mar. 3 to 10, 1888; age, 4 yrs. 1 mo.; weight, 960 lbs.; fed, daily, 9 lbs. oats, 6 lbs. corn meal, 3 lbs. cotton-seed meal and 3 lbs. oil meal; property of Mrs. Mary A. Thomas, Bristol, Conn.

Pretty Flirt 28791.—Sire, Florian (P. S. 373 J. H. B.); dam, Flirt (F. S. 2250 J. H. B.)

    Butter, 15 lbs. Milk, 169 lbs.

Test made from May 1 to 7, 1886; age, 2 yrs. 7 mos.; property of John A. Middelton, Shelbyville, Ky.

Pretty Marjoram 77943.—Sire, Pedro 3187; dam, Marjoram of Linden 43600.

    Butter, 19 lbs. 6 oz. Milk, 274 lbs.

Test made from Dec. 15 to 22, 1892; age, 4 yrs.; estimated weight, 950 lbs.; fed 22 qts. daily of a mixture of ground oats, middlings, bran, corn meal and some pea meal—clover hay and ensilage; property of T. S. Cooper, Coopersburg, Pa.

Pretty Patty 44108.—Sire, Stoke Pogis 5th 5987; dam, Yellow Lass 32603.

    Butter, 15 lbs. 4½ oz. Milk, 225 lbs.

Test made from June 4 to 11, 1890; age, 3 yrs. 2 mos.; estimated weight, 800 lbs.; no record of feed kept—ordinary rations; property of Miller & Sibley, Franklin, Pa.

Pretty Patty 44108.—Sire, Stoke Pogis 5th 5987; dam, Yellow Lass 32603.

    Butter, 19 lbs. 1 oz. Milk, 231 lbs.

Test made from Apr. 5 to 12, 1891; age, 4 yrs.; estimated weight, 900 lbs.; fed 24 qts. of equal parts ground oats and corn meal, 30 lbs. sugar beets, 20 lbs. ensilage and 12 lbs. cooked clover hay, daily; property of Miller & Sibley, Franklin, Pa.

Pretty Phillis Pogis 66100.—Sire, Pedro 3187; dam, Phillis Pogis 29592.

    Butter, 17 lbs. 6 oz. Milk, 258 lbs.

Test made from May 6 to 13, 1893; age, 5 yrs. 3 mos.; estimated weight, 900 lbs.; fed about 22 qts. daily of a mixture of ground oats, bran, middlings, corn meal and some pea meal—clover hay and ensilage, meadow pasture; property of T. S. Cooper, Coopersburg, Pa.

Pretty Prattler 99080.—Sire, Fritz Pogis of Glynllyn 28042; dam, Prattler 55394.

    Butter, 21 lbs. 8 oz. Milk, 279 lbs. 1 oz.

Test made from May 3 to 10, 1897; age, 5 yrs. 2 mos.; estimated weight, 750 lbs.; fed 84 qts. ground oats, 72 qts. bran and 42 qts. corn meal—hay; property of D. Wolfe Bishop, Jr., Lenox, Mass.

Pridalia 17249.—Sire, Minx 3410; dam, Pride of the Hill 4877.

    Butter, 26 lbs. 4 oz. Milk, 289 lbs. 11 oz.

Test made from June 16 to 23, 1890; age, 8 yrs.; actual weight, 975 lbs.; fed, daily, 7 lbs. corn meal and 3 lbs. germ bran; property of C. Dixon, Gillespieville, Ohio.

Pride of Bay View 61997.—Sire, Chief 10663; dam, Allie L. 32073.

Butter, 16 lbs. 6 oz. Milk, 190 lbs. 6 oz.

Test made from Dec. 21 to 28, 1892; age, 4 yrs. 11 mos.; estimated weight, 525 lbs.; fed 245 lbs. corn refuse, 245 lbs. ensilage, 35 lbs. malt sprouts, 14 lbs. linseed meal and 21 lbs. hay; property of Charles Lautz, Buffalo, N. Y.

Pride of Bovina 8050.—Sire, Ben Butler of Bovina 2024; dam, Phœbe 4th 2271.

Butter, 16 lbs. 9 oz. Milk, 214 lbs. 11 oz.

Test made from Mar. 20 to 27, 1883; age, 6 yrs.; fed 5 lbs. daily of ground corn and wheat bran, mixed—good hay *ad lib.*; property of W. L. Rutherford, Franklin, N. Y.

Pride of Edgewood 64199.—Sire, Sir Signalier 12490; dam, Pridalia 17249.

Butter, 17 lbs. 2 oz. Milk, 202 lbs. 7 oz.

Test made from June 8 to 15, 1894; age, 5 yrs. 11 mos.; actual weight, 850 lbs.; fed 42 qts. corn meal, 42 qts. oats and 56 qts. wheat bran—good clover and blue-grass pasture; property of C. Dixon, Gillespieville, Ohio.

Pride of Ingleside 54545.—Sire, Duke of Ingleside 14274; dam, Gipsy's Pride 2d 29365.

Butter, 20 lbs. 4½ oz. Milk, 287 lbs. 8 oz.

Test made from June 20 to 27, 1893; age, 5 yrs. 8 mos.; actual weight, 1005 lbs.; fed 10 qts. corn meal, 4 qts. ground oats and 3 qts. ground oil-cake, daily—good meadow pasture; property of Miller & Sibley, Franklin, Pa.

Pride of Madelon 105711.—Sire, Recorder 29239; dam, Madelon 2d 51040.

Butter, 16 lbs. 6 oz. Milk, 260 lbs. 5 oz.

Test made from June 26 to July 3, 1896; age, 2 yrs. 3 mos.; estimated weight, 650 lbs.; fed 35 lbs. bran, 28 lbs. oats, 24½ lbs. corn and 17½ lbs. oil meal—mixed pasture; property of H. C. Taylor, Orfordville, Wis.

Pride of Mashamoquet Farm 6469.—Sire, Landseer 331; dam, Myrtle 2d 211.

Butter, 16 lbs. 1¾ oz. Milk, 110 lbs. 4 oz.

Test made from Dec. 28, 1885, to Jan. 3, 1886; age, 13 yrs. 8 mos.; property of William Morrow, Nashville, Tenn.

Pride of Rosemond 74449.—Sire, Tamaroa Rex 18604; dam, Prize's Bounty 67114.

Butter, 17 lbs. 14½ oz. Milk, 278 lbs. 4 oz.

Test made from May 5 to 12, 1896; age, 5 yrs. 2 mos.; estimated weight, 750 lbs.; fed 42 lbs. bran, 21 lbs. crushed oats, 21 lbs. corn meal and 14 lbs. oil meal—timothy pasture; property of B. F. Carper, Rosemond, Ill.

Pride of Springvale 20093.—Sire, Alpheus of Springvale 5645; dam, Alabama 7690.

Butter, 19 lbs. 2 oz. Milk, 245 lbs.

Test made from May 16 to 23, 1893; age, 10 yrs. 4 mos.; estimated weight, 900 lbs.; fed 56 qts. ground oats, 42 qts. corn meal and 28 qts. cotton-seed meal—hay and good winter rye pasture; property of James Crook, Jacksonville, Ala.

Pride of the Hill 4877.—Sire, Nutshell 729; dam, Guinevere 1484.

Butter, 14 lbs. 8 oz.

Test made from July 20 to 27, 1882; age, 7 yrs. 2 mos.; property of G. J. Shaw, Hartland, Me.

Pride of the Maner 22652, imp.

> Butter, 18 lbs. 10 oz. Milk, 276 lbs.

Test made from Feb. 16 to 23, 1884; age, 4 yrs.; property of William Whiting, Holyoke, Mass.

Pride of Winslow 2613, imp.

> Butter, 14 lbs. 3 oz.

Property of G. Dawson Coleman, Brickerville, Pa.

Pride's Lady Frances 39529.—Sire, Pride of the Island 5416; dam, Lady Leonora 11811.

> Butter, 16 lbs. 3½ oz. Milk, 225 lbs.

Test made from June 14 to 21, 1890; age, 5 yrs. 9 mos.; actual weight, 850 lbs.; fed, daily, 12 lbs., consisting of cut oats, yellow meal and wheat bran, mixed, and 1 lb. oil meal; property of Holly Grove Farm, Plainfield, N. J.

Pride's Lady Lillie 35181.—Sire, Pride of the Island 5416; dam, Lady Lillie 11813.

> Butter, 15 lbs. ¾ oz. Milk, 201 lbs. 1 oz.

Test made from June 7 to 14, 1892; age, 8 yrs. 5 mos.; estimated weight, 1000 lbs.; fed 8 qts. daily of equal parts corn meal, ground oats and wheat bran—good permanent pasture; property of Walter W. Law, Whitson, N. Y.

Pride's Lady Rushmore 35182.—Sire, Pride of the Island 5416; dam, Lady Rushmore 11812.

> Butter, 15 lbs. 1 oz. Milk, 210 lbs. 7 oz.

Test made from Dec. 1 to 8, 1892; age, 8 yrs. 10 mos.; estimated weight, 850 lbs.; fed 16 qts. equal parts of corn meal, ground oats and wheat bran, with 1 pt. oil meal, daily—hay *ad lib.* and 15 lbs. ensilage; property of Walter W. Law, Whitson, N. Y.

Pride's Lady Sarah 39537.—Sire, Pride of the Island 5416; dam, Lady Rushmore 11812.

> Butter, 15 lbs. 1½ oz. Milk, 232 lbs. 2 oz.

Test made from Mar. 22 to 29, 1891; age, 5 yrs.; actual weight, 920 lbs.; fed 6 lbs. yellow meal, 5 lbs. wheat bran, 1 lb. oil meal, 20 lbs. corn ensilage, 5 lbs. cut hay and 6 lbs. carrots, daily; property of S. K. Schwenk, Marconnier, N. J.

Pride's Olga 37186.—Sire, Rachel's Duke 7022; dam, Pride of the Grange 8914.

> Butter, 19 lbs. 12 oz. Milk, 191 lbs. 4 oz.

Test made from Nov. 20 to 27, 1896; age, 11 yrs. 7 mos.; estimated weight, 900 lbs.; fed 19 qts. corn meal, 46 qts. wheat bran, 25 qts. gluten meal, 13 qts. cotton-seed meal, 13½ qts. oil meal, 7 qts. oat hearts, 180 qts. carrots, 38 qts. turnips, 78 qts. beets, 14 qts. potatoes and 28 qts. apples—hay *ad lib.*; property of C. N. Gilbert, Great Barrington, Mass.

Pride's Olga 2d 79887.—Sire, Real Queen's Stoke Pogis 20535; dam. Pride's Olga 37186.

> Butter, 22 lbs. 12 oz. Milk, 241 lbs. 6 oz.

Test made from Feb. 11 to 18, 1896; age, 5 yrs. 3 mos.; estimated weight, 850 lbs.; fed 44 qts. wheat bran, 13 qts. Buffalo gluten meal, 13 qts. corn meal, 12 qts. oil meal, 8 qts. oat-hearts, 3½ bush. mixed carrots and beets, 7 bush. ensilage, 28 ears of corn and hay *ad lib.*; property of C. N. Gilbert, Great Barrington, Mass.

Pride's Olga 3d 79897.—Sire, Real Queen's Stoke Pogis 20535; dam, Pride's Olga 37186.

Butter, 18 lbs. 2 oz. Milk, 275 lbs. 2 oz.

Test made from Apr. 14 to 21, 1898; age, 6 yrs. 7 mos.; actual weight, 980 lbs.; fed 6 qts. bran, 6 qts. middlings, 3 qts. corn meal, 6 qts. ground oats and 3 qts. oil meal, wet, mixed, daily—hay and ensilage; property of C. A. Sweet, Buffalo, N. Y.

Pride's Olga 4th 96870.—Sire, Melia Ann's Stoke Pogis 22042; dam, Pride's Olga 37186.

Butter, 15 lbs. 2 oz. Milk, 242 lbs. 4 oz.

Test made from Apr. 9 to 16, 1896; age, 2 yrs. 8 mos.; estimated weight, 900 lbs.; fed 64 qts. grain mixture (consisting of 100 parts yellow corn meal, 100 parts winter wheat bran and 25 parts linseed meal), 31 qts. oat-hearts, 5 qts. Buffalo gluten meal, 7 bush. corn ensilage, 7 pks. potatoes and hay *ad lib.*; property of C. N. Gilbert, Great Barrington, Mass.

Pride's Olga 4th 96870.—Sire, Melia Ann's Stoke Pogis 22042; dam, Pride's Olga 37186.

Butter, 23 lbs. 3 oz. Milk, 368 lbs. 12 oz.

Test made from July 18 to 25, 1897; age, 3 yrs. 11 mos.; actual weight, 950 lbs.; fed 9 qts. bran, 9 qts. ground oats, 6 qts. corn meal and 3 qts. oil meal, daily—plenty of green corn fodder; property of C. A. Sweet, Buffalo, N. Y.

Pride's Queen 66442.—Sire, Pride of the Island 5416; dam, Dellia Regina 9474.

Butter, 15 lbs. 8½ oz. Milk, 231 lbs. 8 oz.

Test made from Sept. 3 to 10, 1890; age, 3 yrs. 6 mos.; actual weight, 810 lbs.; fed, daily, 12 lbs., consisting of cut oats, yellow meal and wheat bran, mixed, and 1 lb. oil meal; property of Holly Grove Farm, Plainfield, N. J.

Pride's St. Lambert Lady 82955.—Sire, Rioter's Pride 11694; dam, Edith Delle 66143.

Butter, 18 lbs. 12 oz., unsalted. Milk, 299 lbs. 14 oz.

Test made from Jan. 27 to Feb. 3, 1896; age, 3 yrs. 9 mos.; estimated weight, 1300 lbs.; fed 12 lbs. daily of a mixture of corn meal, bran, ground oats, oil meal, cotton-seed meal, wheat middlings and buckwheat meal, fed on 1 bush. cut corn fodder—84 lbs. hay and 4 bush. carrots; property of Mrs. H. E. Tremain, Lake George, N. Y.

Primrose 11956.—Sire, Tom (P. S. 77 J. H. B.); dam, Cherry (F. S. 1140 J. H. B.)

Butter, 21 lbs. 10 oz. Milk, 282 lbs. 8 oz.

Test made from June 21 to 28, 1885; age, 9 yrs. 4 mos.; fed 20 qts. ground oats, 8 qts. pea meal, 4 qts. bran, 4 qts. oil meal, daily—fair pasture; property of J. W. Johnson, Plainfield, N. J.

Primrose of Summerland 66059.—Sire, Primrose's Carlo 12174; dam, Marilla of St. Lambert 31800.

Butter, 15 lbs. 4½ oz. Milk, 188.6 lbs.

Test made from Aug. 10 to 17, 1894; age, 4 yrs. 1 mo.; actual weight, 826 lbs.; fed 40 qts. corn meal, 21 qts. oil meal, 56 qts. wheat bran, 12 qts. cotton-seed meal, 35 qts. ground oats and 84 lbs. cut clover hay, mixed with grain, wetted; property of H. L. S. Hall, Scottsville, N. Y.

Prince's Bloom 9729.—Sire, Prince of Croton 2490; dam, Bloom 2d 5927.

Butter, 14 lbs. 3 oz.

Property of E. J. Robbins, Wethersfield, Conn.

Prince's Christmas Gift 92513.—Sire, Prince of Tennessee 20772; dam, Cream's Christmas 63506.

Butter, 24 lbs. 12¼ oz.   Milk, 282 lbs. 15 oz.

Test made from June 5 to 12, 1896; age, 4 yrs. 11 mos.; estimated weight, 700 lbs.; fed 22 lbs. per day of ground oats and corn, in equal parts—extra good orchard and blue-grass pasture; property of Thomas H. Malone, Nashville, Tenn.

Prince's Nellie 23719.—Sire, Iowa Prince 2727; dam, Nellie Harrison 2d 23093.

Butter, 14 lbs. 6 oz.   Milk, 234 lbs. 12 oz.

Test made from May 24 to 31, 1888; age, 5 yrs. 1 mo.; estimated weight, 1000 lbs.; fed 13 lbs. oats and corn chop, daily; property of E. E. Harrison, West Liberty, Iowa.

Prince's Rubetta 92785.—Sire, Prince of Tennessee 20772; dam, Toltec's Ruby 54450.

Butter, 19 lbs. 4 oz.   Milk, 196 lbs. 12½ oz.

Test made from June 6 to 13, 1895; age, 5 yrs.; estimated weight, 775 lbs.; fed 17 lbs. per day of ground oats and corn, in equal parts—very good orchard and blue-grass pasture; property of Matt. M. Gardner, Nashville, Tenn.

Prince's Rubetta 2d 133914.—Sire, Maury's Tormentor 25113; dam, Prince's Rubetta 92785.

Butter, 14 lbs. 6¼ oz.   Milk, 140 lbs. 7 oz.

Test made from May 6 to 13, 1898; age, 3 yrs. 1 mo.; estimated weight, 675 lbs.; fed 19 lbs. per day of ground oats and corn, in equal parts—good orchard and blue-grass pasture; property of Matt. M. Gardner, Nashville, Tenn.

Princess 836, imp.

Butter, 14 lbs. 12 oz., unsalted.

Princess 2d 8046.—Sire, Khedive (P. S. 103 J. H. B.); dam, Princess (F. S. 452 J. H. B.)

Butter, 46 lbs. 12½ oz. (official).   Milk, 299 lbs. 8 oz.

Test made from Feb. 22 to Mar. 1, 1885; age, 8 yrs.; property of Mrs. S. M. Shoemaker, Baltimore, Md.

Princess Aurea Pogis 39266.—Sire, Stoke Pogis 5th 5987; dam, Gold Princess 8809.

Butter, 17 lbs. 7½ oz.   Milk, 208 lbs. 8 oz.

Test made from Aug. 26 to Sept. 2, 1890; age, 4 yrs. 4 mos.; estimated weight, 875 lbs.; no record of feed kept—ordinary testing feed; property of Miller & Sibley, Franklin, Pa.

Princess Bellwort 6801.—Sire, Rex 1330; dam, Belle of Middlefield 1516.

Butter, 15 lbs. 10½ oz.   Milk, 256 lbs. 4 oz.

Test made from June 16 to 22, 1883; age, 5 yrs. 6 mos.; property of John E. Phillips, Baltimore, Md.

Princess Bowen 9699.—Sire, Duke of Bloomfield 1544; dam, Lady Bowen 354.

Butter, 14 lbs. 12 oz.   Milk, 202 lbs. 14 oz.

Test made from Apr. 20 to 27, 1883; age, 6 yrs.; property of James Cloud & Son, Kennett Square, Pa.

Princess Carmen 58320.—Sire, Lanison 15283; dam, Clotaire's Beauty 28171.

Butter, 16 lbs. 14 oz. Milk, 217 lbs. 9 oz.

Test made from June 7 to 14, 1890; age, 3 yrs. 2 mos.; estimated weight, 900 lbs.; fed 12 qts. daily of a mixture of one part bran, two parts ground corn and three parts ground oats—bran slop daily, good hay, old pasture; property of L. L. Tozier, Batavia, N. Y.

Princess Chrysantha 34191.—Sire, Prince Pogis 10682; dam, Chrysantha 23465.

Butter, 14 lbs. 3½ oz. Milk, 232 lbs. 12 oz.

Test made from Dec. 29, 1889, to Jan. 5, 1890; age, 4 yrs. 3 mos.; estimated weight, 700 lbs.; fed, daily, 9 lbs. corn meal, 12 lbs. wheat middlings and 13 lbs. ground oats; property of D. F. Appleton, Ipswich, Mass.

Princess Chuck 45650.—Sire, Kansas Niobe Duke 7335; dam, Rosetta of Whiteland 6112.

Butter, 24 lbs. 14 oz. Milk, 242 lbs.

Test made from Aug. 26 to Sept. 2, 1890; age, 6 yrs. 4 mos.; estimated weight, 950 lbs.; fed, daily, about 33½ qts. bran and 4¼ qts. corn chop, with 5 qts. oil-cake and 7 qts. oat chop; property of G. F. Miller, Topeka, Kan.

Princess Corinne 48203.—Sire, Prince of Melrose 4819; dam, Corinne Moore 35748.

Butter, 19 lbs. 1 oz. Milk, 213 lbs. 6½ oz.

Test made from Jan. 4 to 11, 1894; age, 7 yrs. 8 mos.; estimated weight, 1000 lbs.; fed 56 qts. corn meal, 56 qts. corn and oats, ground together, and 42 qts. bran and 200 lbs. corn ensilage—35 lbs. hay, two hours per day pastured on oats; property of M. Lothrop, Marshall, Tex.

Princess Cross 2d 80649.—Sire, Charity Boy 11765; dam, Princess Cross 38329.

Butter, 20 lbs. 1 oz. Milk, 327 lbs.

Test made from Oct. 25 to Nov. 1, 1897; age, 7 yrs. 1 mo.; estimated weight, 800 lbs.; fed 8 lbs. corn meal, 4 lbs. wheat bran and 4 lbs. oil meal, daily—pasture; property of J. H. Martin, Granville, N. Y.

Princess Edith Hugo 45120.—Sire, Prince Pogis' Hugo 15660; dam, Edith Hanna 33146.

Butter, 15 lbs. 14½ oz. Milk, 246 lbs. 1 oz.

Test made from Feb. 19 to 26, 1893; age, 5 yrs. 10 mos.; actual weight, 800 lbs.; fed 8 lbs. each of barley, oats and corn, mixed and ground, daily—hay ad lib.; property of McGill Bros., Leavenworth, Kan.

Princess Haidee Signal 98652.—Sire, Croton's Signal 17614; dam, Haidee Signal 86448.

Butter, 16 lbs. 4 oz. Milk, 247 lbs.

Test made from June 5 to 12, 1896; age, 4 yrs. 11 mos.; estimated weight, 800 lbs.; fed 12 lbs. daily of a mixture of corn meal, ground oats, bran and oil meal—natural pasture; property of George W. Sisson, Jr., Potsdam, N. Y.

Princess Helene of St. L. 61972.—Sire, Duke of St. Lambert 16160; dam, Princess Helene 21505.

Butter, 16 lbs. 9 oz. Milk, 257 lbs. 4 oz.

Test made from Mar. 23 to 30, 1894; age, 5 yrs. 10 mos.; estimated weight, 950 lbs.; fed 12 qts. grain daily, consisting of three parts corn meal, one part ground wheat and one part oat meal; 2 lbs. daily old process oil meal, few cut potatoes, raw—good hay ad lib.; property of E. C. Newton, Hudson, Ohio.

Princess Helene of St. L. 61972.—Sire, Duke of St. Lambert 16160; dam, Princess Helene 21505.

Butter, 66 lbs. 6 oz. in 30 days. Milk, 1079 lbs. 10 oz.

Test made from Mar. 10 to Apr. 9, 1894; age, 5 yrs. 10 mos.; estimated weight, 950 lbs.; fed 360 qts. grain (three parts corn meal and one part each of wheat and oats ground together), 40 lbs. oil meal (old process), and 6 bush. potatoes—good hay *ad lib.*; property of E. C. Newton, Hudson, Ohio.

Princess Honoria 62548.—Sire, Garfield Stoke Pogis 15963; dam, Princess Honor 31805.

Butter, 16 lbs. 8 oz. Milk, 167 lbs. 8 oz.

Test made from Nov. 5 to 12, 1892; age, 2 yrs. 11 mos.; estimated weight, 845 lbs.; fed 56 lbs. cut corn fodder, 7 lbs. corn meal, 14 lbs. gluten meal, 14 lbs. bran, 7 lbs. linseed meal. 56 lbs. hay and 210 lbs. roots; property of Est. of Frederick Billings, Woodstock, Vt.

Princess Honoria 62548.—Sire, Garfield Stoke Pogis 15963; dam, Princess Honor 31805.

Butter, 159.447 lbs. in 90 days. Milk, 2690.4 lbs.

Test made from May 31 to Aug. 28, 1893, at World's Columbian Exposition; age, 3 yrs. 6 mos. at beginning of test; actual weight, 865 lbs.; fed 299 lbs. old hay, 111 lbs. ensilage, 511.4 lbs. new hay, 213.5 lbs. oil meal, 523 lbs. bran, 92 lbs. oats, 436.5 lbs. corn-hearts, 111.50 lbs. cotton-seed, 397 lbs. middlings, 34 lbs. grano-gluten, 1126.4 lbs. clover and 10 lbs. swale-grass; property of Est. of Frederick Billings, Woodstock, Vt.

Princess Inez 5402.—Sire, Hurd's Dandy 1627; dam, Kate Early 3802.

Butter, 14 lbs. 2 oz. Milk, 203 lbs. 14 oz.

Test made from June 8 to 15, 1885; age, 9 yrs.; property of Judson H. Clark, Elmira, N. Y.

Princess Jeannaise 53863.—Sire, Prince Jeannaise 17074; dam, Pilot's St. Jeannaise 25089.

Butter, 15 lbs. 12 oz. Milk, 201 lbs. 4 oz.

Test made from May 11 to 18, 1892; age, 5 yrs. 4 mos.; estimated weight, 750 lbs.; fed 8 qts. yellow corn meal daily—hay, good clover pasture; property of W. Cannon, Philadelphia, Tenn.

Princess Le Brocq 17261.—Sire, Le Brocq's Prize 3350; dam, Princess of St. Saviour's 7470.

Butter, 15 lbs. 8 oz.

Test made from June 12 to 19, 1884; age, 4 yrs. 9 mos.; property of O. Guitar, Columbia, Mo.

Princess Malita 66676.—Sire, Southern Prince 10760; dam, Rosa of Belle Vue 6954.

Butter, 20 lbs. 8¾ oz. Milk, 205 lbs. 8 oz.

Test made from Dec. 12 to 19, 1892; age, 3 yrs. 11 mos.; estimated weight, 800 lbs.; fed 24 lbs. mixed corn-hearts and wheat bran, daily—hay, short blue-grass pasture; property of Campbell Brown, Spring Hill, Tenn.

Princess Marie G. 120390.—Sire, Prince's Lambert 16490; dam, Princess Simmons 57520.

Butter, 15 lbs. 12⅛ oz. (confirmed); estimated butter on basis of 85 per cent. fat, 15 lbs. 10 oz. Milk, 251.9 lbs.

Test made from Feb. 8 to 14, 1897; age, 5 yrs. 8 mos.; average weight, 943 lbs.; fed 14 lbs. cotton-seed meal, 14 lbs. oil meal, 56 lbs. bran and 56 lbs. corn-hearts—clover hay *ad lib.*; property of W. F. Galbreath, Lexington, Ky.

Analyses of Butter.

First churning—Fat 78.90, water 15.72, ash 4.52, casein 0.86.
Second churning—Fat 80.31, water 14.48, ash 4.24, casein 0.97.
Third churning—Fat 83.66, water 12.55, ash 3.01, casein 0.78.

Princess Marie G. 120390.—Sire, Prince's Lambert 16490; dam, Princess Simmons 57520.

Butter, 14 lbs. 13½ oz. (confirmed); estimated butter on basis of 85 per cent. fat, 14 lbs. 9½ oz. Milk, 250.2 lbs.

Test made from Feb. 15 to 21, 1897; age, 5 yrs. 8 mos.; estimated weight, 926 lbs.; fed 14 lbs. cotton-seed meal, 14 lbs. oil meal, 56 lbs. bran and 56 lbs. corn-hearts—clover hay *ad lib.*; property of W. F. Galbreath, Lexington, Ky.

Analyses of Butter.

First churning—Fat 82.17, water 14.06, ash 2.82, casein 0.95.
Second churning—Fat 81.67, water 14.83, ash 2.53, casein 0.97.

Princess Marie G. 120390.—Sire, Prince's Lambert 16490; dam, Princess Simmons 57520.

Butter, 15 lbs. 2½ oz. (confirmed); estimated butter on basis of 85 per cent. fat, 15 lbs. 4⅝ oz. Milk, 259.7 lbs.

Test made from Feb. 22 to 28, 1897; age, 5 yrs. 8 mos.; estimated weight, 926 lbs.; fed 14 lbs. cotton-seed meal, 14 lbs. oil meal, 56 lbs. bran and 56 lbs. corn-hearts—clover hay *ad lib.*; property of W. F. Galbreath, Lexington, Ky.

Analyses of Butter.

First churning—Fat 82.81, water 13.06, ash 3.06, casein 1.07.
Second churning—Fat 83.31, water 14.03, ash 1.79, casein 0.87.

Princess Marie G. 120390.—Sire, Prince's Lambert 16490; dam, Princess Simmons 57520.

Butter, 15 lbs. 3½ oz. (confirmed); estimated butter on basis of 85 per cent. fat, 15 lbs. 3½ oz. Milk, 245.8 lbs.

Test made from Mar. 1 to 7, 1897; age, 5 yrs. 8 mos.; average weight, 910 lbs.; fed 14 lbs. cotton-seed meal, 14 lbs. oil meal. 56 lbs. bran and 56 lbs. corn-hearts—clover hay *ad lib.*; property of W. F. Galbreath, Lexington, Ky.

Analyses of Butter.

First churning—Fat 82.99, water 13.50, ash 2.45, casein 1.06.
Second churning—Fat 81.40, water 14.58, ash 3.12, casein 0.90.

Princess Marie G. 120390.—Sire, Prince's Lambert 16490; dam, Princess Simmons 57520.

Butter, 60.98 lbs. in 28 days (confirmed); estimated butter on basis of 85 per cent. fat, 60.69 lbs. Milk, 1007.6 lbs.

Test made from Feb. 8 to Mar. 7, 1897; age, 5 yrs. 8 mos.; average weight at end of test, 910 lbs.; fed 56 lbs. cotton-seed meal, 56 lbs. oil meal, 224 lbs. bran and 224 lbs. corn-hearts—clover hay *ad lib.*; property of W. F. Galbreath, Lexington, Ky.

(For analyses of butter see weekly tests above.)

Princess Marie G. 120390.—Sire, Prince's Lambert 16490; dam, Princess Simmons 57520.

Butter, 69 lbs. 8 oz. in 32 days. Milk, 1139.7 lbs.

Test made from Feb. 7 to Mar. 11, 1897; age, 5 yrs. 8 mos.; actual weight, 943 lbs.; fed 132 lbs. germ food, 224 lbs. bran, 81 lbs. cotton-seed meal, 64 lbs. flax-seed meal, 24 lbs. ground oats and 170 lbs. corn meal—clover hay *ad lib.*; property of W. F. Galbreath, Lexington, Ky.

Princess Mary of Woodlawn 11663.—Sire, Gilderoy 2107; dam, Queen Mary of Woodlawn 11659.

Butter, 14 lbs. 4 oz. Milk, 260 lbs. 15 oz.

Test made from Aug. 3 to 10, 1885; age, 4 yrs. 6 mos.; property of Frank S. Stevens, Swansea, Mass.

Princess Minette 24042.—Sire, Prince Boulivot 8757; dam, Minette of St. Lambert 9774.

Butter, 18 lbs. 6½ oz. Milk, 282 lbs. 6 oz.

Test made from June 12 to 19, 1892; age, 9 yrs. 10 mos.; estimated weight, 1000 lbs.; fed 14 lbs. oats and peas, chopped, daily—pasture of mixed grasses; property of William Rolph, Markham, Ont., Can.

Princess Mostar 9700.—Sire, Duke of Bloomfield 1544; dam, Mostar 6971.

Butter, 17 lbs. 3 oz. Milk, 232 lbs. 15 oz.

Test made from Apr. 22 to 29, 1882; age, 5 yrs.; property of James Cloud & Son, Kennett Square, Pa.

Princess of Ashantee 13467, imp.

Butter, 16 lbs. 5 oz., unsalted.

Test made from July 2 to 8, 1883; age, 4 yrs.; fed 4 qts. corn meal and 4 qts. bran, daily—June grass pasture; property of S. M. Burnham, Saugatuck, Conn.

Princess of Ashantee 13467, imp.

Butter, 16 lbs. 12 oz. Milk, 123 lbs. ⅞ oz.

Test made from June 18 to 24, 1884; age, 5 yrs.; property of S. M. Burnham, Saugatuck, Conn.

Princess of Clovernook 51598.—Sire, Dixie Prince 18379; dam, Sedate 2d 20389.

Butter, 16 lbs. 13 oz. Milk, 146 lbs. 1 oz.

Test made from June 16 to 23, 1890; age, 2 yrs. 4 mos.; estimated weight, 750 lbs.; fed 5 qts. corn chop and rye, daily; property of S. N. Warren, Spring Hill, Tenn.

Princess of Dorset 85902.—Sire, Melia Ann's Victor Pogis 20697; dam, Kitty Kringel 57120.

Butter, 35 lbs. 1 oz. Milk, 368 lbs.

Test made from May 27 to June 3, 1896; age, 5 yrs. 2 mos.; estimated weight, 800 lbs.; fed 10 lbs. peas, oats and corn, ground together, 5 lbs. wheat bran and 4 lbs. oil meal, daily—clover pasture; property of Mrs. J. H. Martin, Granville, N. Y.

Princess of Eau Claire 128219.—Sire, Melia Ann's Son 22041; dam, Dinah of Eau Claire 77581.

Butter, 17 lbs. 6 oz. Milk, 227 lbs. 8 oz.

Test made from Jan. 12 to 19, 1898; age, 4 yrs. 8 mos.; estimated weight, 750 lbs.; fed 25 lbs. wheat bran, 25 lbs. corn meal, 20 lbs. oil meal, 15 lbs. cotton-seed meal and 2 bush. roots; property of J. A. Griffith, Ashley Falls, Mass.

Princess of Hanover 46706.—Sire, Pogis Chief 3998; dam, Dia 13658.

Butter, 18 lbs. 12 oz. Milk, 268 lbs. 1 oz.

Test made from Aug. 22 to 29, 1893; age, 7 yrs.; estimated weight, 1000 lbs.; fed 112 qts. wheat bran, 28 qts. corn meal and 14 qts. cotton-seed meal—hay, Bermuda grass pasture; property of G. W. Talbot, Cleburne, Tex.

Princess of Lynwood 118458.—Sire, Prince of Tennessee 20772; dam, Maumee Beauty 50836.

Butter, 16 lbs. ½ oz. Milk, 218 lbs. 11½. oz.

Test made from Oct. 28 to Nov. 4, 1896; age, 6 yrs.; estimated weight, 700 lbs.; fed 19 lbs. per day of ground oats and corn, in equal parts, mixed with 4 bucketfuls cut sheaf oats, wetted—good orchard and blue-grass pasture; property of Matt. M. Gardner, Nashville, Tenn.

Princess of St. Saviour's 7470.—Sire, Noble (P. S. 93 J. H. B.); dam, Princess, on I. of J.

Butter, 15 lbs. 10½ oz. Milk, 174 lbs. 8 oz.

Test made from Oct. 27 to Nov. 2, 1885; age, 8 yrs. 5 mos.; property of T. W. Mc-Neely, Petersburg, Ill.

Princess of the Valley 22641.—Sire, Pretender (P. S. 187 J. H. B.); dam, Blush Rose (F. S. 453 J. H. B.)

Butter, 18 lbs. 12 oz. Milk, 263 lbs. 14 oz.

Test made from Feb. 19 to 25, 1885; age, 6 yrs.; property of William Whiting, Holyoke, Mass.

Princess of Trinity 23641, imp.

Butter, 14 lbs. 10 oz. Milk, 248 lbs.

Test made from June 24 to July 1, 1885; age, 6 yrs.; estimated weight, 1075 lbs.; no grain fed; property of T. R. Proctor, Utica, N. Y.

Princess of Yerba Buena Ranch 12626.—Sire, Napoléon (F. S. 249 J. H. B.); dam on I. of J.

Butter, 15 lbs. 10 oz. Milk, 221 lbs.

Test made from June 9 to 16, 1884; age, 5 yrs. 10 mos.; property of Henry Pierce, San Francisco, Cal.

Princess Oonan 84369.—Sire, Oonan's Signal 11586; dam, Romp's Princess 51185.

Butter, 15 lbs. 13½ oz. Milk, 225 lbs. 12 oz.

Test made from June 20 to 27, 1896; age, 5 yrs. 1 mo.; actual weight, 750 lbs.; fed 9 qts. bran, 9 qts. ground oats, 2¼ qts. corn meal and 3 qts. oil meal, daily—good timothy and clover pasture; property of C. A. Sweet, East Aurora, N. Y.

Princess Ora 66374.—Sire, Victor Hugo Signal 12135; dam, Princess Olga 26528.

Butter, 14 lbs. 5 oz. Milk, 231 lbs.

Test made from June 6 to 13, 1895; age, 7 yrs. 2 mos.; estimated weight, 850 lbs.; fed 3 qts. corn meal, 3 qts. cotton-seed meal and 8 qts. wheat bran, wet, daily—medium pasture; property of Jim Mitchell, Rusk, Tex.

Princess Rioter 29588.—Sire, Black Prince of Linden 9063; dam, Royal Princess 22013.

Butter, 17 lbs. 6 oz. Milk, 274 lbs.

Test made from Jan. 10 to 17, 1889; age. 4 yrs. 5 mos.; estimated weight, 900 lbs.; fed about 22 qts. daily of a mixture of whole oats, corn meal, middlings, wheat bran and pea meal—clover hay and ensilage; property of T. S. Cooper, Coopersburg, Pa.

Princess Rosalind 87288.—Sire, Prince of Amherst 25728; dam, Etta of Bay View 66028.

Butter, 15 lbs. 1 oz. Milk, 168 lbs. 13 oz.

Test made from Feb. 11 to 18, 1896; age, 3 yrs. 3 mos.; actual weight, 640 lbs.; fed 35 lbs. ground oats, 21 lbs. corn meal, 14 lbs. ground wheat, 14 lbs. oil meal and 140 lbs. corn ensilage; property of Charles Lautz, Buffalo, N. Y.

Princess Sheila 7297.—Sire, Champion of America 1567; dam, Elsie Burnside 5598.

Butter, 16 lbs. 4½ oz. Milk, 213 lbs. 12 oz.

Test made from Mar. 2 to 8, 1882; age, 4 yrs.; property of G. R. Dykeman, Shippensburg, Pa.

Princess Sesia 112753.—Sire, Tormentor's Rioter 21964; dam, Nellie Prince 58053.

Butter, 15 lbs. 8 oz. Milk, 199 lbs. 8 oz.

Test made from July 26 to Aug. 2, 1896; age, 1 yr. 9 mos.; estimated weight, 400 lbs.; fed 210 lbs. green sorghum, 42 qts. bran, 42 qts. oats, 14 qts. cotton-seed meal and 14 qts. corn meal—pasture in dry lot; property of T. G. Bush, Anniston, Ala.

Princess Victor Pogis 132724.—Sire, Melia Ann's Victor Pogis 20697; dam, Princess of Dorset 85902.

Butter, 16 lbs. 5 oz. Milk, 255 lbs. 4 oz.

Test made from May 18 to 25, 1898; age, 2 yrs. 11 mos.; estimated weight, 650 lbs.; fed 12 lbs. daily of corn meal and wheat bran, mixed in equal parts by measure—clover pasture; property of Mrs. J. H. Martin, Granville, N. Y.

Princess White Water 21137.—Sire, Finis Lawrence 4107; dam, Lady Mahopac 8613.

Butter, 14 lbs. 15 oz. Milk, 208 lbs. 15 oz.

Test made from June 13 to 20, 1889; age, 7 yrs.; estimated weight, 800 lbs.; fed, daily, 28 lbs. of a mixture consisting of ground corn, oats and bran; property of Mrs. L. C. Smith, White Water, Wis.

Princess Winnie 63508.—Sire, Prince of Tennessee 20772; dam, Winolo Gold 56826.

Butter, 21 lbs. 4¼ oz. Milk, 272 lbs. 10¾ oz.

Test made from June 29 to July 6, 1898; age, 8 yrs. 5 mos.; estimated weight, 760 lbs.; fed 26 lbs. per day of ground oats and corn, in equal parts—very good orchard and blue-grass pasture; property of Matt. M. Gardner, Nashville, Tenn.

Princess Zeroda 107242.—Sire, Prince of St. Lambert 5287; dam, Zeroda 2d 78630.

Butter, 14 lbs. Milk, 298 lbs.

Test made from June 1 to 7, 1897; age, 3 yrs. 9 mos.; estimated weight, 850 lbs.; fed 12 qts. bran and 6 qts. meal, daily—good orchard pasture; property of C. I. Kingsbury, Oswego, N. Y.

Printenette 26121.—Sire, Augerez' King (F. S. 380 J. H. B.); dam, Printonnière 26086.

Butter, 14 lbs. 8 oz. Milk, 181 lbs.

Test made from Feb. 9 to 16, 1886; age, 2 yrs.; estimated weight, 675 lbs.; fed 14 lbs. corn meal, 5 lbs. oat meal and 6 lbs. oil meal, daily—no pasture; property of John A. Middelton, Shelbyville, Ky.

Priscilla Pogis 39270.—Sire, Stoke Pogis 5th 5987; dam, Duchess of Darlington 13830.

Butter, 18 lbs. 6½ oz. Milk, 233 lbs. 12 oz.

Test made from Mar. 12 to 19, 1890; age, 3 yrs. 8 mos.; estimated weight, 850 lbs.; fed, daily, 7 lbs. 8 oz. corn meal, 5 lbs. cotton-seed, 3 lbs. 8 oz. oil meal and 6 lbs. crushed oats; property of Ayer & McKinney, Philadelphia, Pa.

Priscilla Pogis 39270.—Sire, Stoke Pogis 5th 5987; dam, Duchess of Darlington 13830.

Butter, 77 lbs. 12½ oz. in 31 days. Milk, 1042 lbs. 12 oz.

Test made from Feb. 23 to Mar. 26, 1890; age, 3 yrs. 9 mos.; fed, daily, 23 lbs. 8 oz. ensilage, 7 lbs. 8 oz. corn meal, 5 lbs. cotton-seed, 3 lbs. 8 oz. oil meal, 6 lbs. crushed oats and 8 lbs. hay; property of Ayer & McKinney, Philadelphia, Pa.

Pristine 126329.—Sire, Pasha (P. S. 1901 J. H. B.); dam, Mary Gold (P. S. 4188 J. H. B.)

Butter, 14 lbs. 4½ oz. Milk, 165 lbs. 12 oz.

Test made from Jan. 18 to 25, 1898; age, 2 yrs. 8 mos.; actual weight, 520 lbs.; fed 21 lbs. ground oats, 21 lbs. ground corn, 14 lbs. oil meal, 28 lbs. wheat bran, 200 lbs. corn ensilage and 49 lbs. hay; property of Charles Lautz, Buffalo, N. Y.

Prize Rose 16309.—Sire, Le Brocq's Prize 3350; dam, Violette 7471.

Butter, 15 lbs. 1 oz. Milk, 183 lbs. 4 oz.

Test made from Sept. 20 to 26, 1884; age, 2 yrs. 7 mos.; property of William Simpson, New York, N. Y.

Proctor's Alma Dolores 47107.—Sire, Bunker 9025; dam, Nutley Dolores 13797.

Butter, 17 lbs. 10 oz. Milk, 181 lbs.

Test made from Feb. 8 to 15, 1890; age, 3 yrs.; actual weight, 740 lbs.; fed, daily, 4 qts. corn meal, 8 qts. ground oats, 3 qts. oil meal and 2 qts. pea meal; property of T. R. Proctor, Utica, N. Y.

Proctor's Alma Dolores 47107.—Sire, Bunker 9025; dam, Nutley Dolores 13797.

Butter, 23 lbs. 9 oz. Milk, 270 lbs.

Test made from Dec. 30, 1891, to Jan. 6, 1892; age, 4 yrs. 11 mos.; actual weight, 795 lbs.; fed 1 qt. linseed meal, 4 qts. corn meal, 4 qts. ground oats. 6 qts. bran. 16 qts. beets, 15 lbs. ensilage and 10 lbs. hay, daily; property of T. R. Proctor, Utica, N. Y.

Proctor's Belle 47106.—Sire, Bunker 9025; dam, Stella's Pride 16855.

Butter, 16 lbs. 4 oz. Milk, 186 lbs.

Test made from Dec. 14 to 21, 1890; age, 3 yrs. 11 mos.; actual weight, 690 lbs.; fed 10 lbs. corn meal, 4 lbs. linseed meal, 5 lbs. ground oats, 10 lbs. hay, 15 lbs. ensilage and 6 qts. beets, daily; property of T. R. Proctor, Utica, N. Y.

Proctor's Dolores 38564.—Sire, Duke Coomassie 8409; dam, Nutley Dolores 13797.

Butter, 15 lbs. 10 oz. Milk, 188 lbs. 12 oz.

Test made from Dec. 14 to 21, 1890; age, 6 yrs.; estimated weight, 850 lbs.; fed, daily, 8 lbs. corn meal, 8 lbs. crushed oats, 4 lbs. oil meal and 4 lbs. pea meal; property of Ayer & McKinney, Philadelphia, Pa.

Proctor's Pansy 25688.—Sire, Lena's Lenox 6059; dam, La Pucelle 16829.

Butter, 15 lbs. 13 oz. Milk, 276 lbs. 15 oz.

Test made from July 10 to 17, 1887; age, 3 yrs. 4 mos.; estimated weight, 1050 lbs.; fed, daily, 8 qts. oats, middlings and shorts mixed, equal parts; property of T. R. Proctor, Utica, N. Y.

Prospect's Edwina 2d 86909.—Sire, Ethleel 2d's Jubilee 18249; dam, Prospect's Edwina 43185.

Butter, 18 lbs. 1 oz. Milk, 223 lbs.

Test made from Feb. 5 to 12, 1897; age, 4 yrs. 10 mos.; estimated weight, 800 lbs.; fed 8 lbs. ground peas and oats, 2 lbs. corn meal, 3 lbs. bran, 2 lbs. oil meal and 25 lbs. ensilage, daily—hay *ad lib.*; property of George W. Sisson, Jr., Potsdam, N. Y.

Prusa Pedro 50877.—Sire, King Peto 10541; dam, June Pansy 27317.

Butter, 15 lbs. 8 oz. Milk, 216 lbs. 12 oz.

Test made from Oct. 14 to 21, 1890; age, 3 yrs. 9 mos.; estimated weight, 700 lbs.; no grain fed—test made on grass alone; property of Mrs. J. H. Martin, Granville, N. Y.

Prusa Victor Pogis 134216.—Sire, Melia Ann's Victor Pogis 20697; dam, Prusa Pedro 50877.

Butter, 16 lbs. 8 oz. Milk, 271 lbs. 12 oz.

Test made from May 25 to June 1, 1898; age, 2 yrs. 6 mos.; estimated weight, 750 lbs.; no grain fed—clover and mixed grass pasture; property of Mrs. J. H. Martin, Granville, N. Y.

Purest 13730.—Sire, Mercury 432; dam, Nymphæa 5141.

Butter, 15 lbs. 4 oz. Milk, 206 lbs. 14 oz.

Test made from July 2 to 8, 1883; age, 2 yrs.; fed corn and oats, ground together, and bran mashes—extra good white clover pasture; property of William Simpson, New York, N. Y.

Pussy Stoke Pogis 62547.—Sire, Garfield Stoke Pogis 15963; dam, Pussy Baker 6994.

Butter, 20 lbs. 1 oz. Milk, 229 lbs. 10 oz.

Test made from May 22 to 29, 1898; age, 8 yrs. 7 mos.; estimated weight, 900 lbs.; fed 6 lbs. wheat bran, 3 lbs. Buffalo gluten, 3 lbs. new process linseed meal and 2 lbs. corn meal, daily—medium pasture; property of Charles H. Ellsworth, Worcester, Mass.

Pyrrha 6100.—Sire, Zeus 2634; dam, Themis 6076.

Butter, 16 lbs. 14½ oz. Milk, 243 lbs. 14 oz.

Test made from July 14 to 20, 1883; age, 6 yrs. 5 mos.; property of William Simpson, New York, N. Y.

Quachette 17091.—Sire, Waukesha 5330; dam, Suamico 11375.

Butter, 19 lbs. 11½ oz. Milk, 206 lbs. 12 oz.

Test made from Apr. 13 to 19, 1885; age, 2 yrs. 10 mos.; property of V. Lowe, Palmyra, Wis.

Quachette F. W. 83654.—Sire, Scituate F. W. 24624; dam, Rubano's Quachette 50556.

Butter, 15 lbs. 3 oz. Milk, 254 lbs. 12 oz.

Test made from Feb. 21 to 28, 1897; age, 6 yrs.; estimated weight, 850 lbs.; fed 21 lbs. bran, 14 lbs. corn meal, 14 lbs. oats, 7 lbs. oil meal and 280 lbs. ensilage—hay *ad lib.*; property of William Weis, Palmyra, Wis.

Quadruple Pogis 32359.—Sire, Marjoram's Rioter 5991; dam, Lily of Riverside 19599.

Butter, 14 lbs. 1½ oz. Milk, 140 lbs. 13 oz.

Test made from Feb. 8 to 15, 1888; age, 2 yrs. 11 mos.; estimated weight, 925 lbs.; fed, daily, 14 lbs. mixed ground corn, oats and ship stuff and 2 lbs. oil meal; property of H. G. Westlake, Hillsdale, N. Y.

Quail 3d 43568.—Sire, Pedro of the Valley 8750; dam, Quail 5895.

Butter, 16 lbs. 12 oz. Milk, 246 lbs. 5 oz.

Test made from Oct. 6 to 13, 1889; age, 3 yrs. 10 mos.; estimated weight, 900 lbs.; fed, daily, 6 qts. corn, ground oats and wheat bran, mixed, and 2 qts. oil meal; property of M. W. Terrill, Middlefield, Conn.

Queen Deeline 78869.—Sire, King of St. Lambert 15175; dam, Deeline 2d 55878.

Butter, 20 lbs. 13½ oz. Milk, 290 lbs. 14.4 oz.

Test made from May 24 to 31, 1897; age, 8 yrs. 2 mos.; estimated weight, 1000 lbs.; fed 8 lbs. bran, 8 lbs. corn and oats, 6 lbs. hominy feed, 2 lbs. oil meal, 2 lbs. cotton-seed meal and 28 lbs. corn ensilage, daily—hay *ad lib.*; property of John E. Robbins, Greensburg, Ind.

Queen Dodona 42375.—Sire, Gambetta M. 10012; dam, Dodona 4800.

Butter, 14 lbs. 2 oz. Milk, 258 lbs. 4 oz.

Test made from May 23 to 30, 1887; age, 2 yrs. 1 mo.; estimated weight, 700 lbs.; no grain fed during test, or previous thereto—clover pasture only; property of Jason Hotchkiss, Fauncetown, Pa.

Queen Fannie 10275.—Sire, Pierrot 2d 1669; dam, Fannie Landseer 1969.

Butter, 14 lbs. 2 oz. Milk, 220 lbs.

Test made from Jan. 2 to 8, 1881; age, 4 yrs. 10 mos.; property of Hoover & Co., Columbus, Ohio.

Queen Ione 113999.—Sire, Noble Paddy 31072; dam, Queen of Clinton 11500.

Butter, 16 lbs. 3 oz. Milk, 214 lbs. 8 oz.

Test made from May 20 to 27, 1897; age, 3 yrs. 2 mos.; estimated weight, 750 lbs.; fed 8 qts. daily bran and corn meal, in equal parts by weight—first class pasture; property of George E. Nichols, Afton, N. Y.

Queen Irene 101893.—Sire, Noble Paddy 31072; dam, Queen of Clinton 11500.

Butter, 15 lbs. 12 oz. Milk, 236 lbs. 4 oz.

Test made from Dec. 12 to 19, 1896; age, 3 yrs. 9 mos.; estimated weight, 750 lbs.; fed 8 qts. bran and corn and cob meal, 4 qts. Buffalo gluten and 2 qts. hominy, daily—some ensilage and hay; property of George E. Nichols, Afton, N. Y.

Queen Mary 6212.—Sire, Top-Sawyer 1404; dam, Duchess of Bloomfield 3653.

Butter, 14 lbs. 13½ oz. Milk, 122 lbs. 12 oz.

Test made from Apr. 3 to 9, 1886; age, 9 yrs. 2 mos.; property of Campbell Brown, Spring Hill, Tenn.

Queen Mary of Woodlawn 11659.—Sire, Gilderoy 2107; dam, Gold Proof 10860.

Butter, 22 lbs. 5 oz. Milk, 303 lbs. 4 oz.

Test made from July 15 to 22, 1882; age, 3 yrs. 5 mos.; fed 6 qts. meal (kind not stated) and 6 qts. shorts, mixed, daily—rye pasture; property of Raymond H. Perry, Bristol, R. I.

Queen May of St. Lambert 45770.—Sire, King of St. Lambert 15175; dam, Queen of the Miamis 6793.

Butter, 15 lbs. 12½ oz. Milk, 159 lbs. ½ oz.

Test made from Nov. 18 to 25, 1894; age, 7 yrs. 5 mos.; estimated weight, 750 lbs.; fed 17 lbs. per day of ground oats and corn, in equal parts, mixed with 4 bucketfuls of cut sheaf oats, wetted—clover hay *ad lib.*; property of Matt. M. Gardner, Nashville, Tenn.

Queen Neptune 15501.—Sire, Neptune 2d (P. S. 206 J. H. B.); dam, Lady Rushmore 11812.

Butter, 18 lbs. 13½ oz. Milk, 227 lbs. 12 oz.

Test made from Oct. 20 to 26, 1885; age, 4 yrs. 10 mos.; property of D. A. Givens, Cynthiana, Ky.

Queen of Ashantee 14554.—Sire, Guy Fawkes (F. S. 251 J. H. B.); dam, Gros Puits (F. S. 1454 J. H. B.)

Butter, 15 lbs. 2 oz., unsalted. Milk, 116 qts.

Test made from June 25 to July 2, 1883; age, 4 yrs. 4 mos.; property of S. M. Burnham, Saugatuck, Conn.

Queen of Beauty 17109.—Sire, Knight of St. Louis 3680; dam, Valentine of Trinity 7460.

Butter, 23 lbs. 14 oz. Milk, 221 lbs. 4 oz.

Test made from June 7 to 14, 1888; age, 6 yrs. 9 mos.; weight, 990 lbs.; fed, daily, 4¼ lbs. corn meal, 2 lbs. cotton-seed meal, 1 lb. pea meal and 9 lbs. bran; property of Mrs. Hunter Nicholson, Knoxville, Tenn.

Queen of Chenango 17771.—Sire, Tamerlane 4287; dam, Nerina 9183.

Butter, 14 lbs. 6 oz. Milk, 189 lbs. 8 oz.

Test made from June 21 to 27, 1884; age, 2 yrs. 4 mos.; property of J. V. N. Willis, Marlborough, N. J.

Queen of Clinton 11500.—Sire, Noble 901; dam, Queen of Sidney 11492.

Butter, 15 lbs. 8 oz. Milk, 217 lbs. 8 oz.

Test made from June 11 to 18, 1892; age, 12 yrs. 1 mo.; estimated weight, 850 lbs.; no grain fed—white clover, timothy and June grass pasture; property of G. E. Nichols, Afton, N. Y.

Queen of Corfu 83732.—Sire, St. Lambert Perrot 23689; dam, Tulia Perrot 46357.

Butter, 15 lbs. Milk, 190 lbs. 12 oz.

Test made from July 24 to 31, 1894; age, 2 yrs. 11 mos.; estimated weight, 750 lbs.; fed 12 qts. ground corn and oats, 12 qts. middlings and bran and 3 qts. oil meal, daily—oats and peas during day, clover pasture at night; property of C. A. Sweet, East Aurora, N. Y.

Queen of Corfu 83732.—Sire, St. Lambert Perrot 23689; dam, Tulla Perrot 46357.

Butter, 21 lbs. ½ oz. Milk, 272 lbs. 8 oz.

Test made from Jan. 15 to 22, 1896; age, 4 yrs. 5 mos.; actual weight, 860 lbs.; fed 12 qts. bran, 7½ qts. ground corn and oats and 3 qts. oil meal, daily—hay; property of C. A. Sweet, East Aurora, N. Y.

Queen of Cowes 46309.—Sire, Coomassie Welcome 9279; dam, May Colt 23491.

Butter, 18 lbs. 3 oz. Milk, 280 lbs.

Test made from June 23 to 30, 1898; age, 11 yrs. 2 mos.; estimated weight, 1000 lbs.; fed about 24 qts. daily of a mixture of ground oats, middlings, wheat bran, corn chop and some pea meal—clover hay and ensilage, clover and meadow pasture; property of T. S. Cooper, Coopersburg, Pa.

Queen of Delaware 17029.—Sire, Noble 901; dam, Jennie of Sidney 11488.

Butter, 18 lbs. 13 oz. Milk, 252 lbs.

Test made from June 24 to 30, 1882; age, 4 yrs. 1 mo.; grass alone—timothy, just heading; property of A. Baker, West Dryden, N. Y.

Queen of De Soto 12318.—Sire, Dode 3057; dam, Emiline 5185.

Butter, 14 lbs. 13 oz., unsalted.

Test made from Aug. 9 to 15, 1882; age, 2 yrs. 5 mos.; property of Edward Mayes, Oxford, Miss.

Queen of Nubbin Ridge 14528.—Sire, Sweepstakes Duke 1905; dam, Eastwood's Queen 14523.

Butter, 17 lbs. Milk, 241 lbs. 12 oz.

Test made from June 13 to 20, 1885; age, 4 yrs. 6 mos.; property of J. R. Kendall, Terre Haute, Ind.

Queen of Pomona 56250.—Sire, Ida of St. Lambert's Bull 19169; dam, St. Lambert's Violet 25278.

Butter, 16 lbs. 6 oz. Milk, 188 lbs. 8 oz.

Test made from Aug. 5 to 12, 1891; age, 2 yrs. 4 mos.; estimated weight, 760 lbs.; fed 36 lbs. corn meal, 48 lbs. oat meal, 28 lbs. bran, 19 lbs. pea meal and 19 lbs. oil meal—dry pasture and green oats and peas; property of Ayer & McKinney, Philadelphia, Pa.

Queen of Prospect 11997.—Sire, Shirley 1613; dam, Adeline 9115.

Butter, 14 lbs. 2 oz. Milk, 225 lbs. 8 oz.

Test made from June 21 to 27, 1883; age, 3 yrs. 3 mos.; no grain fed—timothy and clover pasture; property of R. S. Kingman, Sparta, Wis.

Queryan 42140.—Sire, Ryan 9891; dam, Metah's Queen 4886.

Butter, 16 lbs. 2 oz. Milk, 252 lbs.

Test made from Apr. 30 to May 7, 1895; age, 10 yrs.; estimated weight, 1000 lbs.; fed 18 lbs. oil meal, 40 lbs. oats and 35 lbs. wheat bran—hay and corn fodder *ad lib.*; property of Henry A. Egerton, Footville, Wis.

Quickstep 85969.—Sire, Perfect Pacific 16994; dam, Postscript 39366.

Butter, 17 lbs. 9 oz. Milk, 243 lbs.

Test made from July 13 to 20, 1895; age, 3 yrs. 10 mos.; estimated weight, 900 lbs.; fed 35 lbs. corn meal, 21 lbs. bran, 21 lbs. oats and 21 lbs. oil meal—mixed pasture; property of H. C. Taylor, Orfordville, Wis.

Quinnie Pogis 61924.—Sire, Beauty's Stoke Pogis 14175; dam, Ella Quin 35749.

Butter, 16 lbs. 10 oz. Milk, 237 lbs.

Test made from Apr. 27 to May 4, 1894; age, 5 yrs. 7 mos.; estimated weight, 950 lbs.; fed 56 qts. corn meal, 28 qts. bran and 28 qts. cotton-seed meal—pastured on oats and grass; property of W. W. Lipscomb, Luling, Tex.

Quintuple Pogis 44233.—Sire, Lucy's Stoke Pogis 11544; dam, Quadruple Pogis 32359.

Butter, 14 lbs. 6 oz. Milk, 192 lbs. 6 oz.

Test made from Oct. 11 to 18, 1889; age, 2 yrs. 9 mos.; estimated weight, 850 lbs.; fed, daily, 5 lbs. corn meal, 3 lbs. ship stuff and 2 lbs. oil meal; property of H. G. Westlake, Hillsdale, N. Y.

Quintuple Pogis' Fille 94975.—Sire, Lucy's Stoke Pogis 4th 26533; dam, Quintuple Pogis 44233.

Butter, 16 lbs. 8 oz. Milk, 244 lbs.

Test made from May 8 to 15, 1898; age, 4 yrs. 4 mos.; estimated weight, 900 lbs.; fed 12 lbs. daily of "Victor" corn and oat feed, bran and gluten, in equal parts—blue-grass pasture; property of W. J. Hussey, Mt. Pleasant, Ohio.

Quintuple's Daisy 86078.—Sire, Quintuple Victor Pogis 22343; dam, Daisy B. 4th 14277.

Butter, 14 lbs. 5 oz. Milk, 211 lbs. 2 oz.

Test made from June 5 to 12, 1896; age, 4 yrs. 8 mos.; estimated weight, 850 lbs.; fed 2 qts. wheat bran, 1 qt. corn meal and 1 qt. ground oats, daily—old orchard pasture; property of E. L. Van Deusen, Ashley Falls, Mass.

Quintuple's Florence 86076.—Sire, Quintuple Victor Pogis 22343; dam, Signolo's Florence 49430.

Butter, 15 lbs. 10 oz. Milk, 210 lbs. 4 oz.

Test made from Sept. 10 to 17, 1896; age, 5 yrs.; estimated weight, 900 lbs.; fed 2 qts. wheat bran, 1 qt. corn meal and 1 qt. ground oats, daily—lowland pasture; property of E. L. Van Deusen, Ashley Falls, Mass.

Quintuple's Wonder 86077.—Sire, Quintuple Victor Pogis 22343; dam, Gilderoy's Daisy 50023.

Butter, 15 lbs. 14 oz. Milk, 252 lbs. 12 oz.

Test made from June 5 to 12, 1896; age, 4 yrs. 8 mos.; estimated weight, 900 lbs.; fed 2 qts. wheat bran, 1 qt. corn meal and 1 qt. ground oats, daily—old orchard pasture; property of E. L. Van Deusen, Ashley Falls, Mass.

Rachel Spencer 50974.—Sire, Exile of St. Lambert 13657; dam, Eva Locust 21050.

Butter, 21 lbs. 1½ oz. Milk, 404 lbs. 2 oz.

Test made from Feb. 4 to 11, 1895; age, 6 yrs. 11 mos.; actual weight, 1000 lbs.; fed 49 qts. bran, 49 qts. ground oats, 24½ qts. oil meal, 14 qts. roots and 105 lbs. hay; property of P. J. Cogswell, Rochester, N. Y.

Rachel Spencer 50974.—Sire, Exile of St. Lambert 13657; dam, Eva Locust 21050.

Butter, 18 lbs. 15¾ oz. Milk, 385 lbs.

Test made from Feb. 11 to 18, 1895; age, 6 yrs. 11 mos.; actual weight, 1000 lbs.; fed 35 qts. corn meal, 56 qts. bran, 24½ qts. oil meal, 31½ qts. ground oats, 42 qts. roots and 105 lbs. hay; property of P. J. Cogswell, Rochester, N. Y.

Rachel Spencer 50974.—Sire, Exile of St. Lambert 13657; dam, Eva Locust 21050.

Butter, 20 lbs. 5½ oz. Milk, 413 lbs.

Test made from Feb. 18 to 25, 1895; age, 6 yrs. 11 mos.; actual weight, 1000 lbs.; fed 35 qts. corn meal, 56 qts. bran, 24½ qts. oil meal, 31½ qts. ground oats, 14 qts. cottonseed meal, 42 qts. roots and 105 lbs. hay; property of P. J. Cogswell, Rochester, N. Y.

Rachel Spencer 50974.—Sire, Exile of St. Lambert 13657; dam, Eva Locust 21050.

Butter, 22 lbs. ½ oz. Milk, 399 lbs. 12 oz.

Test made from Feb. 25 to Mar. 4, 1895; age, 6 yrs. 11 mos.; actual weight, 1000 lbs.; fed 56 qts. bran. 31½ qts. ground oats, 24½ qts. oil meal, 31½ qts. corn meal, 10½ qts. cotton-seed meal, 42 qts. roots and 105 lbs. hay; property of P. J. Cogswell, Rochester, N. Y.

Rachel Spencer 50974.—Sire, Exile of St. Lambert 13657; dam, Eva Locust 21050.

Butter, 23 lbs. 3¼ oz. Milk, 402 lbs. 12 oz.

Test made from Mar. 4 to 11, 1895; age, 6 yrs. 11 mos.; actual weight, 1000 lbs.; fed 33 qts. bran, 48½ qts. ground oats, 22½ qts. oil meal, 19 qts. corn meal, 19 qts. cotton-seed meal, 19 qts. pea meal, 42 qts. roots and 105 lbs. hay; property of P. J. Cogswell, Rochester, N. Y.

Rachel Spencer 50974.—Sire, Exile of St. Lambert 13657; dam, Eva Locust 21050.

Butter, 105 lbs. 10½ oz. in 35 days. Milk, 2004 lbs. 10 oz.

Test made from Feb. 4 to Mar. 11, 1895; age, 6 yrs. 11 mos.; actual weight, 1000 lbs.; fed 250 qts. bran, 192 qts. ground oats, 120½ qts. oil meal, 120½ qts. corn meal, 43½ qts. cotton-seed meal, 19 qts. pea meal, 182 qts. roots and 525 lbs. hay; property of P. J. Cogswell, Rochester, N. Y.

Rachel's Tamy 89987.—Sire, Lord Almer H. 24081; dam, Rachel Spencer 50974.

Butter, 14 lbs. 4½ oz. Milk, 208 lbs.

Test made from Nov. 28 to Dec. 5, 1895; age, 4 yrs. 2 mos.; actual weight, 930 lbs.; fed 28 qts. oil meal, 28 qts. corn meal, 42 qts. bran, 42 qts. ground oats, 140 lbs. hay and 160 lbs. roots; property of Pierce J. Cogswell, Rochester, N. Y.

Rainbow 2d 13962.—Sire, Doesticks 2387; dam, Rainbow 6493.

Butter, 21 lbs. 8 oz. Milk, 167 lbs.

Test made from June 8 to 14, 1888; age, 7 yrs. 5 mos.; estimated weight, 900 lbs.; fed 6¾ lbs. corn meal, 3¾ lbs. oats, 7 lbs. each of bran, middlings and oil meal, daily; property of A. D. McBride, Rochester, N. Y.

Ramapo's Pride 35497.—Sire, Ramapo 4679; dam, Pride of Franklin 25406.

Butter, 14 lbs. 4 oz. Milk, 225 lbs.

Test made from Apr. 10 to 17, 1889; age, 4 yrs. 10 mos.; estimated weight, 900 lbs.; fed, daily, 8 qts. cob meal, 2 qts. oil meal, average of 2½ qts. wheat bran; property of W. C. Norton, Agent, Aldenville, Pa.

Rarity 2d 7724.—Sire, Flash 2532; dam, Rarity 5923.

Butter, 14 lbs. 2 oz. Milk, 199 lbs.

Test made from Dec. 9 to 17, 1881; age, 4 yrs.; property of Louis Stracke, Warsaw, Ill.

Rayon d'Or's Chroma 114053.—Sire, Elm Place Rayon d'Or 17518; dam, Yankee's Chroma 62029.

Butter, 14 lbs. 15 oz. Milk, 251 lbs. 14 oz.

Test made from Mar. 29 to Apr. 5, 1897; age, 6 yrs. 7 mos.; estimated weight, 900 lbs.; fed 52½ lbs. corn meal, 63 lbs. ground oats, 42 lbs. bran, 24½ lbs. oil meal, 7 lbs. gluten meal and 112 lbs. roots—40 lbs. ensilage and a little clover hay, daily; property of L. D. Ely, Rochester, N. Y.

Reality 16537.—Sire, Mercury 432; dam, Result 11848.

Butter, 15 lbs. 3½ oz. Milk, 185 lbs. 13 oz.

Test made from June 30 to July 6, 1884; age, 2 yrs. 2 mos.; property of William Simpson, New York, N. Y.

Real· Queen 29198.—Sire, Rachel's Duke 7022; dam, Auraria 10688.

Butter, 18 lbs. 1 oz. Milk, 231 lbs. 1 oz.

Test made from Mar. 20 to 27, 1887; age, 3 yrs. 8 mos.; estimated weight, 1050 lbs.; fed, daily, 14 qts. mixed corn meal, oats and ship stuff and 3 lbs. oil meal; property of H. G. Westlake, Hillsdale, N. Y.

Reception 8557.—Sire, Jacques (P. S. 63 J. H. B.); dam on I. of J.

Butter, 19 lbs. 8 oz.

Property of W. R. McCready, Saugatuck, Conn.

Reception 8557.—Sire, Jacques (P. S. 63 J. H. B.); dam on I. of J.

Butter, 21 lbs. 4½ oz. Milk, 265 lbs.

Test made from Oct. 6 to 13, 1884; age, 10 yrs. 2 mos.; fed about 24 lbs. daily, two-thirds ground corn and oats (one part of corn to two of oats) and one-third old process oil meal—good pasture; property of George W. Farlee, Cresskill, N. J.

Redacta 26954.—Sire, Millennium 4791; dam, Hillsdale Gem 16640.

Butter, 14 lbs. 5 oz. Milk, 197 lbs. 11·oz.

Test made from Mar. 13 to 20, 1887; age, 3 yrs. 9 mos.; estimated weight, 975 lbs.; fed, daily, 13 qts. corn meal, oats and ship stuff and 2 lbs. oil meal; property of H. G. Westlake, Hillsdale, N. Y.

Regalia 64574.—Sire, Damascus 22222; dam, Diplomacy 53640.

Butter, 14 lbs. 12 oz. Milk, 218 lbs. 12 oz.

Test made from May 29 to June 5, 1895; age, 5 yrs.; estimated weight, 950 lbs.; fed 12 qts. ground oats, 12 qts. wheat bran, 6 qts. corn meal, 1½ qts. oil meal and 1½ qts. cotton-seed meal, daily; property of C. A. Sweet, Buffalo, N. Y.

Regalia 64574.—Sire, Damascus 22222; dam, Diplomacy 53640.

Butter, 24 lbs. Milk, 285 lbs. 14 oz.

Test made from Mar. 26 to Apr. 2, 1897; age, 6 yrs. 10 mos.; estimated weight, 1150 lbs.; fed 97 qts. oats, 42 qts. meal, 73 qts. bran and 140 lbs. roots; property of D. Wolfe Bishop, Jr., Lenox, Mass.

Regina 2d 2475.—Sire, Noble (F. S. 104 J. H. B.); dam, Regina (F. S. 32 J. H. B.)

Butter, 14 lbs. 8 oz.

Property of H. Borden-Bowen, Bristol, R. I.

Regina 4th 12732.—Sire, Pasha (P. S. 64 J. H. B.); dam, Regina (F. S. 32 J. H. B.)

Butter, 17 lbs. 13½ oz. Milk, 187 lbs. 8 oz.

Test made from Aug. 23 to 29, 1883; age, 7 yrs.; property of H. M. Howe, Bristol, R. I.

Regina Gilford 49891.—Sire, Sir Florian 11578; dam, Enna 33487.

Butter, 14 lbs. 11 oz. Milk, 167 lbs. 4 oz.

Test made from May 2 to 9, 1889; age, 3 yrs.; actual weight, 675 lbs.; fed, daily, 4 qts. oats, 4 qts. corn meal, 2 qts. cotton-seed meal and 8 qts. wheat bran; property of Charles E. Wheeler, Chesterville, Me.

Regina of Cedar Hill 111995.—Sire, Southern Duke 35777; dam, Fall Leaf's Heir-ess 54454.

Butter, 15 lbs. 8½ oz. Milk, 295 lbs. 13 oz.

Test made from May 16 to 23, 1898; age, 3 yrs.; actual weight, 760 lbs.; fed 42 qts. bran, 42 qts. corn meal, 12 qts. ground barley, 12 qts. boiled oats and 21 lbs. cotton-seed hulls—poor Bermuda grass pasture; property of J. W. Persohn, McKinney, Tex.

Regina Weston 70290.—Sire, Regina's Sir George 13569; dam, Weston's Paragon 33297.

Butter, 16 lbs. 9½ oz. Milk, 260 lbs. 12 oz.

Test made from July 14 to 21, 1895; age, 5 yrs. 6 mos.; estimated weight, 900 lbs.; fed 28 lbs. coarse wheat bran, 28 lbs. ground oats, 28 lbs. corn meal and 14 lbs. cotton-seed meal—common pasture; property of John C. McClintock, Meadville, Pa.

Reif 58398.—Sire, Diploma 16219; dam, Quarto 31828.

Butter, 19 lbs. 3 oz. Milk, 283 lbs. 12 oz.

Test made from July 7 to 14, 1895; age, 6 yrs. 2 mos.; estimated weight, 900 lbs.; fed 28 lbs. oats, 21 lbs. bran, 30 lbs. oil meal and 56 lbs. corn meal—mixed pasture; prop-erty of H. C. Taylor, Orfordville, Wis.

Relay 70353.—Sire, Diploma 16219; dam, Itura 61090.

Butter, 17 lbs. 2 oz. Milk, 277 lbs. 9 oz.

Test made from May 8 to 15, 1896; age, 5 yrs. 3 mos.; estimated weight, 1000 lbs.; fed 42 lbs. bran and oats, in equal parts by weight, 42 lbs. corn meal and 21 lbs. oil meal—mixed pasture; property of H. C. Taylor, Orfordville, Wis.

Renalba 4117.—Sire, Pertinatti 713; dam, Renébel 2772.

Butter, 17 lbs. 4½ oz. Milk, 267 lbs. 8 oz.

Test made from Feb. 20 to 26, 1882; age, 6 yrs. 10 mos.; fed wheat bran, cotton-seed meal, linseed meal, chopped corn and peas, mixed, 4 qts. per day—hay, winter pas-ture of blue-grass; property of Campbell Brown, Spring Hill, Tenn.

René of Hillside 12952.—Sire, Tormentor 3533; dam, Renébel 2772.

Butter, 18 lbs. 1 oz. Milk, 242 lbs.

Test made from June 20 to 27, 1896; age, 16 yrs. 10 mos.; estimated weight, 1000 lbs.; fed 28 qts. ground oats, 28 qts. crushed corn, 28 qts. wheat bran and 7 qts. cotton-seed meal—blue-grass pasture; property of D. P. Carter, Lynnville, Tenn.

René Ogden 1568.—Sire, Don 611; dam, René 2d 56.

Butter, 15 lbs.

Property of W. S. Taylor, Burlington, N. J.

Renini 9181.—Sire, Oxoli 1922; dam, Chroma 4572.

Butter, 14 lbs. 10½ oz. Milk, 206¾ lbs.

Test made from Oct. 17 to 23, 1881; age, 4 yrs. 4 mos.; property of Charles Keep, Lockport, N. Y.

Renown 13729.—Sire, Mercury 432; dam, Reserve 6051.

Butter, 14 lbs. 6 oz. Milk, 196 lbs.

Test made from June 30 to July 6, 1884; age, 3 yrs. 1 mo.; property of William Simpson, New York, N. Y.

Rescue of St. Lambert 48939.—Sire, Pauline's Stoke Pogis 11859; dam, Royalist's Daisy 19187.

Butter, 16 lbs. 8 oz. Milk, 181 lbs. 12 oz.

Test made from Nov. 23 to 30, 1893; age, 6 yrs. 10 mos.; estimated weight, 850 lbs.; fed 16 qts. bran, 8 qts. ground oats and corn, 2 qts. old process oil meal and 35 lbs. cut clover hay, daily; property of Mrs. Kate M. Busick, Wabash, Ind.

Retta Pogis 81132.—Sire, Sig Pogis 22806; dam, René of Hillside 12952.

Butter, 16 lbs. 3¼ oz. Milk, 143 lbs. 8 oz.

Test made from Mar. 24 to 31, 1895; age, 3 yrs. 4 mos.; estimated weight, 600 lbs.; fed 42 qts. ground oats, 7 qts. cotton-seed meal, 28 qts. crushed corn and 28 qts. bran —good rye pasture; property of D. P. Carter, Lynnville, Tenn.

Rêve 78944.—Sire, Khedive's Landseer 21926; dam, Ida's Dream 50070.

Butter, 14 lbs. ½ oz. Milk, 205 lbs. 8 oz.

Test made from Aug. 20 to 27, 1893; age, 2 yrs. 9 mos.; estimated weight, 750 lbs.; fed 112 lbs. ground oats and 56 lbs. corn-hearts—blue-grass pasture; property of Est. of Campbell Brown, Spring Hill, Tenn.

Rêve 78944.—Sire, Khedive's Landseer 21926; dam, Ida's Dream 50070.

Butter, 21 lbs. 12 oz. Milk, 273 lbs. 9 oz.

Test made from Oct. 10 to 17, 1897; age, 6 yrs. 10 mos.; estimated weight, 850 lbs.; fed 63 lbs. ground oats and 63 lbs. ground corn, mixed—second crop clover pasture; property of F. J. McDonald, Decatur, Ill.

Rexella 69413.—Sire, Farmer Rex 8898; dam, Clelie of Mapleton 27182.

Butter, 14 lbs. 4 oz. Milk, 211 lbs. 11 oz.

Test made from June 13 to 20, 1895; age, 6 yrs. 3 mos.; estimated weight, 800 lbs.; fed 16 lbs. gluten meal, 16 lbs. bran and 8 lbs. cotton-seed meal—clover and timothy pasture; property of George L. Ferris, Atwater, N. Y.

Rex's Queen of St. Lambert 59161.—Sire, Rioter of St. Lambert 16501; dam, Corinna's Rexena 33032.

Butter, 16 lbs. 2 oz. Milk, 155 lbs. 15 oz.

Test made from Jan. 3 to 10, 1897; age, 7 yrs. 8 mos.; estimated weight, 1050 lbs.; fed 56 lbs. corn meal, 42 lbs. crushed oats, 42 lbs. wheat bran, 21 lbs. old process oil meal and 56 lbs. roots; property of Ira A. Van Orsdal, Quincy, Mich.

Rheta of Glynllyn 74273.—Sire, Ramapose's Bachelor 25900; dam, Beechnut of Glynllyn 29462.

Butter, 20 lbs. 4½ oz. Milk, 318 lbs. 12 oz.

Test made from Aug. 3 to 10, 1896; age, 5 yrs. 1 mo.; actual weight, 1015 lbs.; fed 9 qts. bran, 9 qts. ground oats, 6 qts. corn meal and 3 qts. oil meal, daily—good pasture at night, green corn in stable day time; property of C. A. Sweet, East Aurora, N. Y.

Rho A. Pogis 39269.—Sire, Stoke Pogis 5th 5987; dam, Rho A. 16950.

Butter, 14 lbs. 15 oz. Milk, 174 lbs.

Test made from Nov. 7 to 14, 1890; age, 4 yrs. 4 mos.; estimated weight, 850 lbs.; fed, daily, 6 qts. bran, 8 qts. corn meal and 8 qts. ground oats; property of Miller & Sibley, Franklin, Pa.

Rhoda Hudson 48723.—Sire, Catono 3761; dam, Rosabel Hudson 5704.

Butter, 14 lbs. 4 oz. Milk, 211 lbs. 5 oz.

Test made from May 24 to 31, 1891; age, 5 yrs. 3 mos.; actual weight, 850 lbs.; fed 82 lbs. 3 oz. corn and oats ground together; property of Lyman A. Mills, Middlefield, Conn.

Ria Alexandra 83752.—Sire, St. Helier Berkshire 21030; dam, Vit 51519.

Butter, 14 lbs. 10 oz. Milk, 254 lbs. 12 oz.

Test made from Jan. 5 to 12, 1894; age, 3 yrs. 7 mos.; estimated weight, 900 lbs.; fed 6 qts. ground corn and oats, 1 qt. oil meal, 1 qt. cotton-seed meal, 4 qts. wheat bran and 35 lbs. corn ensilage, daily—hay *ad lib.*; property of E. L. Van Deusen, Ashley Falls, Mass.

Ribbon of St. Lambert 39450.—Sire, Ida's Rioter of St. L. 13656; dam, Ribbon's Fawn 26907.

Butter, 18 lbs. 2 oz. Milk, 206 lbs.

Test made from Oct. 17 to 24, 1893; age, 8 yrs. 6 mos.; actual weight, 960 lbs.; fed 84 qts. ground oats, 56 qts. corn meal and 28 qts. ground oil-cake—meadow pasture; property of Miller & Sibley, Franklin, Pa.

Ribbon's Gift 77375.—Sire, Stoke Pogis Bulletin 16619; dam, Ribbon of St. Lambert 39450.

Butter, 18 lbs. 1 oz. Milk, 304 lbs. 8 oz.

Test made from May 10 to 17, 1894; age, 4 yrs. 6 mos.; actual weight, 900 lbs.; fed 42 qts. ground oats, 28 qts. corn meal, 14 qts. ground oil-cake and 28 qts. wheat bran —good old meadow pasture; property of Miller & Sibley, Franklin, Pa.

Ribbon's Matilda 66109.—Sire, Matilda 4th's Son 20214; dam, Ribbon of St. Lambert 39450.

Butter, 16 lbs. 4 oz. Milk, 202 lbs. 8 oz.

Test made from Oct. 11 to 18, 1893; age, 4 yrs. 10 mos.; actual weight, 800 lbs.; fed 56 qts. ground oats, 42 qts. corn meal and 14 qts. ground oil-cake—meadow pasture; property of Miller & Sibley, Franklin, Pa.

Ribbon's Queen 66321.—Sire, Stoke Pogis Bulletin 16619; dam, Ribbon of St. Lambert 30450.

Butter, 16 lbs. 1 oz. Milk, 197 lbs.

Test made from Jan. 15 to 22, 1894; age, 7 yrs. 1 mo.; actual weight, 875 lbs.; fed 42 qts. corn meal, 56 qts. ground oats, 28 qts. bran, 14 qts. ground oil-cake and 20 lbs. roots; property of Miller & Sibley, Franklin, Pa.

Richland Belle 46920.—Sire, Southern Prince 10760; dam, Tette 20802.

Butter, 15 lbs. 5 oz. Milk, 171 lbs.

Test made from Nov. 24 to Dec. 1, 1896; age, 9 yrs. 10 mos.; estimated weight, 975 lbs.; fed 98 lbs. corn meal and 70 lbs. ground oats—extra good blue-grass pasture; property of George Campbell Brown, Spring Hill, Tenn.

Richness 16536.—Sire, Mercury 432; dam, Clytemnestra 2455.

Butter, 17 lbs. 5 oz. Milk, 219 lbs. 13 oz.

Test made from Dec. 3 to 9, 1885; age, 3 yrs. 7 mos.; property of William Simpson, New York, N. Y.

Rinora Pogis 40107.—Sire, Brier's Pogis 14163; dam, Rioter's Nora 21778.

Butter, 19 lbs. 1½ oz. Milk, 239 lbs. 12 oz.

Test made from Sept. 16 to 23, 1894; age, 9 yrs. 2 mos.; actual weight, 996 lbs.; fed 46 lbs. wheat bran, 50 lbs. corn meal, 24 lbs. old process oil meal and 7 lbs. cotton-seed meal—good blue-grass pasture; property of J. P. Bradbury, Pomeroy, Ohio.

Riotaletta 2d 34495.—Sire, Golden Ray 10669; dam, Riotaletta 29937.

Butter, 15 lbs. 15½ oz. Milk, 136 lbs. 4 oz.

Test made from June 8 to 14, 1887; age, 1 yr. 11 mos.; estimated weight, 600 lbs.; fed, daily, 4 qts. bran, 4 qts. corn and oats and 1 qt. pea meal; property of M. Erskine Miller, Staunton, Va.

Rioter Alphea 10091.—Sire, Jason Jr. 3270; dam, Chansonnette 5095.

Butter, 16 lbs. 7 oz., unsalted.

Test made from Sept. 2 to 8, 1884; age, 6 yrs. 3 mos.; property of D. F. Appleton, Ipswich, Mass.

Rioter Alphea 2d 29676.—Sire, Lord of Mountain Side 7111; dam, Rioter Alphea 10091.

Butter, 17 lbs. 15 oz. Milk, 238 lbs. 10 oz.

Test made from Aug. 13 to 20, 1890; age, 6 yrs.; estimated weight, 850 lbs.; fed, daily, 8 lbs. ground oats, 8 lbs. middlings and 7 lbs. corn meal; property of D. F. Appleton, Ipswich, Mass.

Rioter Alphea 3d 34073.—Sire, Rioter Hugo Pogis 13457; dam, Rioter Alphea 10091.

Butter, 17 lbs. 1½ oz. Milk, 233 lbs. 2 oz.

Test made from Oct. 28 to Nov. 4, 1888; age, 3 yrs. 3 mos.; estimated weight, 650 lbs.; fed, daily, 13 lbs. oats, 14 lbs. shorts, 10 lbs. corn meal and 2 lbs. oil meal; property of D. F. Appleton, Ipswich, Mass.

Rioter Carlotta 29667.—Sire, Lord of Mountain Side 7111; dam, Butterstamp Lass 19517.

Butter, 21 lbs. 2½ oz. Milk, 228 lbs. 6 oz.

Test made from May 24 to 31, 1888; age, 4 yrs. 1 mo.; estimated weight, 700 lbs.; fed 6 lbs. wheat middlings daily; property of James Stillman, Sing Sing, N. Y.

Rioter Flora Marjoram 111165.—Sire, Pedro Royal Marjoram 28560; dam, Hulabaloo's Flora 111164.

Butter, 16 lbs. 4 oz. Milk, 272 lbs.

Test made from July 6 to 13, 1897; age, 4 yrs. 1 mo.; estimated weight, 1050 lbs.; fed 4 qts. bran, 4 qts. middlings, 4 qts. crushed oats and 2 qts. sugar feed, daily— clover hay *ad lib.*, meadow pasture at night; property of T. S. Cooper, Coopersburg, Pa.

Rioter Pink of Berlin 23665.—Sire, Stoke Pogis 3d 2238; dam, Pink of St. Lambert 5486.

Butter, 19 lbs. 14 oz. (official). Milk, 290 lbs. 4 oz.

Test made from Aug. 8 to 15, 1884; age, 4 yrs. 3 mos.; property of L. W. & H. W. Simonds, Berlin, Ont., Can.

Rioter Rhea 10092.—Sire, Jason Jr. 3270; dam, Chansonnette 5695.

Butter, 19 lbs. 3½ oz. Milk, 18 qts. per day.

Test made from Aug. 7 to 13, 1886; age, 7 yrs. 4 mos.; estimated weight, 800 lbs.; fed 18 qts. grain per day; property of D. F. Appleton, Ipswich, Mass.

Rioter's Baby 106100.—Sire, Rioter of St. Lambert 16501; dam, Little Nellie Rex 50209.

Butter, 19 lbs. 12 oz. Milk, 298 lbs.

Test made from May 18 to 25, 1895; age, 5 yrs. 1 mo.; estimated weight, 750 lbs.; no grain fed—4 qts. potatoes night and morning, pasture of timothy, red-top, bluegrass and white clover; property of J. B. Briggs, Macedon, N. Y.

Rioter's Denise 3d 70238.—Sire, Fancy's Harry 9777; dam, Rioter's Denise 36057.

Butter, 17 lbs. 1½ oz. Milk, 196 lbs. 9 oz.

Test made from Nov. 11 to 18, 1894; age, 4 yrs.; estimated weight, 800 lbs.; fed 70 qts. crushed corn and barley, mixed, in equal parts—blue-grass pasture; property of John Woodard, Nashville, Tenn.

Rioter's Drosera 81056.—Sire, Belmont Rioter 22849; dam, Drosera's Beauty 67275.

Butter, 18 lbs. 4 oz. Milk, 231 lbs. 8 oz.

Test made from Jan. 25 to Feb. 1, 1896; age, 3 yrs. 11 mos.; estimated weight, 900 lbs.; fed 350 lbs. ensilage, 19½ lbs. old process oil meal, 42½ lbs. bran and 42½ lbs. Chicago maize gluten—hay *ad lib.*; property of Harmon Austin, Warren, Ohio.

Rioter's Esse 81901.—Sire, Rioter of St. Lambert 16501; dam, Esse Signal Alexis 42184.

Butter, 16 lbs. 7 oz. Milk, 159 lbs. 8 oz.

Test made from Dec. 6 to 13, 1896; age, 5 yrs. 5 mos.; estimated weight, 825 lbs.; fed 7 qts. chop, 2 qts. bran and 1 qt. kiln-dried malt sprouts, twice a day—corn fodder; property of Mrs. A. M. Hallock, Columbus, Ohio.

Rioter's Esse 81901.—Sire, Rioter of St. Lambert 16501; dam, Esse Signal Alexis 42184.

Butter, 31 lbs. 6¾ oz. in 14 days. Milk, 323 lbs. 8 oz.

Test made from Nov. 29 to Dec. 13, 1895; age, 5 yrs. 5 mos.; estimated weight, 825 lbs.; fed 5 qts. chop, 2 qts. bran and 1 qt. kiln-dried sprouts, twice a day from Nov. 30 to Dec. 4, then increased by 4 qts. chop per day—corn fodder; property of Mrs. A. M. Hallock, Columbus, Ohio.

Rioter's Fawn 69168.—Sire, Rioter of St. Lambert 16501; dam, Esse Signal Alexis 42184.

Butter, 20 lbs. 2 oz. Milk, 289 lbs.

Test made from Jan. 20 to 27, 1896; age, 6 yrs. 10 mos.; estimated weight, 900 lbs.; fed 70 qts. bran, 42 qts. corn meal, 7 qts. oil meal, 7 bush. cut hay and 3½ bush. cut roots, mixed—dry cut hay *ad lib.*; property of George E. Peer, Rochester, N. Y.

Rioter's Hortense 51499.—Sire, Nemonette's Rioter 16106; dam, Ruth Le Brocq 38861.

Butter, 15 lbs. 1½ oz. Milk, 222 lbs. 10 oz.

Test made from Jan. 14 to 21, 1895; age, 7 yrs. 6 mos.; estimated weight, 1000 lbs.; fed 8 qts. ground corn, 4 qts. bran, 3 lbs. cotton-seed meal, 2 lbs. oil-cake and 1 pk. mangels, mixed, daily—hay; property of J. M. Hund, Hund's Station, Kan.

Rioter's Letty 110450.—Sire, St. Lambert Boy 17408; dam, Letty Coles 2d 48128.

Butter, 21 lbs. 9½ oz. Milk, 296 lbs. 4 oz.

Test made from Mar. 31 to Apr. 7, 1898; age, 4 yrs.; actual weight, 863 lbs.; fed 44 lbs. corn meal, 30 lbs. wheat middlings, 30 lbs. wheat bran and 24 lbs. old process oil meal—corn ensilage and hay *ad lib.*; property of J. P. Bradbury, Pomeroy, Ohio.

Rioter's Lorne Primrose 88583.—Sire, Prospect's Rioter 9189; dam, Lorne's Primrose 66082.

Butter, 18 lbs. 15 oz. Milk, 258 lbs.

Test made from Nov. 21 to 28, 1896; age, 6 yrs. 9 mos.; estimated weight, 850 lbs.; fed about 20 qts. daily of a mixture of ground oats, middlings, wheat bran, corn meal and some pea meal—clover and green corn, meadow and blue-grass pasture; property of T. S. Cooper, Coopersburg, Pa.

Rioter's Nora 21778.—Sire, Sir George of St. Lambert 6036; dam, Nora of St. Lambert 12962.

Butter, 15 lbs. 9 oz. Milk, 223 lbs.

Test made from Oct. 14 to 20, 1884; age, 2 yrs. 10 mos.; property of William Rolph, Markham, Ont., Can.

Rioter's Normanda 102441.—Sire, Rioter of St. Lambert 16501; dam, Rubano's Normanda 68927.

Butter, 15 lbs. 8 oz. Milk, 210 lbs.

Test made from May 30 to June 6, 1897; age, 4 yrs. 2 mos.; estimated weight, 800 lbs.; fed 52½ lbs. bran, 35 lbs. corn meal, 7 lbs. linseed meal and 7 lbs. cotton-seed meal—blue-grass pasture; property of Samuel McKeen, Terre Haute, Ind.

Rioter's Pretty Belle 88543.—Sire, Pedro of Linden 26015; dam, Rioter's Pretty Pearl 75273.

Butter, 15 lbs. 10 oz. Milk, 245 lbs.

Test made from Nov. 22 to 29, 1896; age, 5 yrs.; estimated weight, 700 lbs.; fed about 20 qts. daily of a mixture of ground oats, middlings, bran, corn meal and some pea meal—clover hay and ensilage, meadow blue-grass pasture; property of T. S. Cooper, Coopersburg, Pa.

Rioter's Pretty Pearl 75273.—Sire, Prospect's Rioter 9189; dam, Lorne's Pretty Pearl 46683.

Butter, 15 lbs. 8 oz. Milk, 245 lbs.

Test made from Dec. 21 to 28, 1896; age, 6 yrs. 10 mos.; estimated weight, 800 lbs.; fed about 18 qts. daily of a mixture of ground oats, bran, middlings, corn meal and pea meal—clover hay and ensilage, meadow blue-grass pasture; property of T. S. Cooper, Coopersburg, Pa.

Rioter's Queen 14895.—Sire, Rambler of St. Lambert 5285; dam, Maud of St. Lambert 9772.

Butter, 17 lbs. 8 oz. Milk, 314 lbs.

Test made from June 20 to 27, 1885; age, 4 yrs.; estimated weight, 850 lbs.; fed, daily, about 10 qts. oats, corn and bran; property of Mrs. E. M. Jones, Brockville, Ont., Can.

Rioter's Roberta 80469.—Sire, Cherie's Rioter 14437; dam, Lady Roberta 29209.

Butter, 17 lbs. Milk, 294 lbs. 8 oz.

Test made from Jan. 12 to 19, 1895; age, 4 yrs. 10 mos.; estimated weight, 850 lbs.; fed 98 qts. crushed oats, 28 qts. corn meal, 3 qts. oil meal—3½ bush. carrots, hay; property of Allendale Farm, Honeoye, N. Y.

Rioter's Rosaline 35581.—Sire, Nancy's Rioter 8390; dam, Rosaline of Glenmore 3179.

Butter, 14 lbs. 14½ oz. Milk, 169 lbs. 8 oz.

Test made from Dec. 20 to 27, 1889; age, 4 yrs. 8 mos.; estimated weight, 800 lbs.; fed, daily, 20 lbs. chopped corn and wheat bran, mixed; property of Campbell Brown, Spring Hill, Tenn.

Rioter's Ruth 14882.—Sire, Rambler of St. Lambert 5285; dam, Moth of St. Lambert 9775.

Butter, 19 lbs. 6½ oz. Milk, 235 lbs. 8 oz.

Test made from Nov. 23 to 29, 1885; age, 5 yrs. 6 mos.; property of Valancey E. Fuller, Hamilton, Ont., Can.

Rioter's Sunshine 75108.—Sire, Rioter of St. Lambert 16501; dam, Farmer's Sunbeam 2d 55506.

Butter, 15 lbs. 10 oz. Milk, 244 lbs. 11 oz.

Test made from Mar. 24 to 31, 1896; age, 6 yrs.; estimated weight, 700 lbs.; fed 10½ lbs. gluten, 10½ lbs. linseed meal, 21 lbs. ground oats, 28 lbs. wheat bran, 3½ pcks. roots (beets or turnips), 49 lbs. hay and about 200 lbs. ensilage; property of Wilbur G. Davis, Concord, Mass.

Rioter's Sweet Brier 30582.—Sire, Diana's Rioter 10481; dam, Sweet Brier of St. Lambert 5481.

Butter, 17 lbs. 8½ oz. Milk, 254 lbs. 6 oz.

Test made from Mar. 13 to 20, 1890; age, 6 yrs. 9 mos.; estimated weight, 1050 lbs.; fed, daily, about 5½ lbs. bran, 6½ lbs. corn meal, 6 lbs. pea meal, 2 lbs. oats and 5½ lbs. oat meal; property of W. B. von Richthofen, Montclair, Col.

Rioter's Sweet Brier 30582.—Sire, Diana's Rioter 10481; dam, Sweet Brier of St. Lambert 5481.

Butter, 18 lbs. 9 oz. Milk, 212 lbs.

Test made from Feb. 11 to 18, 1891; age, 6 yrs. 8 mos.; estimated weight, 1100 lbs.; fed, daily, 6 lbs. pea meal, 12 lbs. corn meal and 16 lbs. crushed oats; property of Ayer & McKinney. Philadelphia, Pa.

Rioter's Violet 33774.—Sire, Oxford Rioter 5992; dam, Merlin's Violet 16430.

Butter, 14 lbs. 3 oz. Milk, 154 lbs.

Test made from Aug. 21 to 27, 1887; age, 2 yrs. 1 mo.; estimated weight, 650 lbs.; fed 6 qts. oats and 8 qts. bran, daily; property of Chas. E. Hill, Denver, Col.

Rioter's Violet Pogis 75342.—Sire, Ida's Rioter of St. L. 13656; dam, Little Pogis 32612.

Butter, 14 lbs. 15 oz. Milk, 190 lbs. 14 oz.

Test made from Dec. 18 to 25, 1892; age, 3 yrs.; estimated weight, 700 lbs.; fed 42 lbs. pea meal, 28 lbs. oil meal, 28 lbs. cotton-seed meal, 42 lbs. mixed feed and 145 lbs. ensilage; property of Ayer & McKinney, Philadelphia, Pa.

Rioter's Violetta 79391.—Sire, Violet 3d's Rioter 23470; dam, Stoke Pogis Violetta 2d 51873.

Butter, 14 lbs. 7½ oz. Milk, 271 lbs.

Test made from Feb. 8 to 15, 1897; age, 6 yrs. 8 mos.; estimated weight, 1100 lbs.; fed 56 qts. oats and 56 qts. corn, ground together, and 14 qts. oil meal—clover hay *ad lib.*; property of Hawkins & Garretson, Oskaloosa, Iowa.

Rioter's Zoe 19769.—Sire, Stoke Pogis 5th 5987; dam, Golden Zoe 3975.

Butter, 14 lbs. 12 oz. Milk, 239 lbs. 8 oz.

Test made from June 21 to 27, 1887; age, 4 yrs. 3 mos.; estimated weight, 800 lbs.; fed 25 lbs., daily, of mixed corn-hearts, oil meal, oats and middlings; property of Miller & Sibley, Franklin, Pa.

Riotress 3d 63166.—Sire, Young Pedro 9033; dam, Riotress 28062.

Butter, 17 lbs. 3½ oz. Milk, 252 lbs. 11 oz.

Test made from June 2 to 9, 1892; age, 3 yrs. 4 mos.; estimated weight, 700 lbs.; fed 6 qts. corn meal, 6 qts. ground oats and 6 qts. shorts, daily—good pasture; property of D. F. Appleton, Ipswich, Mass.

Riotress Signal 95430.—Sire, Croton's Torment 17616; dam, Letty Rioter 73475.

Butter, 21 lbs. 5½ oz. Milk, 297 lbs. 7 oz.

Test made from May 28 to June 4, 1896; age, 3 yrs. 10 mos.; actual weight, 898 lbs.; fed 45 lbs. wheat bran, 40 lbs. wheat middlings and 25 lbs. corn meal—little hay, excellent clover and timothy pasture; property of J. P. Bradbury, Pomeroy, Ohio.

Rissa 16014 (F. S. 2173 J. H. B.)

Butter, 19 lbs., unsalted. Milk, 210 lbs.

Test made from June 2 to 9, 1882; age, 4 yrs. 6 mos.; property of Nathan Brownell, Hubbardsville, N. Y.

Rita of Andalusia 29414.—Sire, Majestic Pedro 8043; dam, Muriel 4th 17146.

Butter, 24 lbs. 1 oz. Milk, 421 lbs. 5 oz.

Test made from May 28 to June 4, 1892; age, 8 yrs. 4 mos.; estimated weight, 1000 lbs.; fed 8 qts. corn meal and 6 qts. ground oats, daily—pasture, timothy, green grass and some white clover; property of George Fox, Torresdale, Pa.

Rival's Ochra 10172.—Sire, Rival 3762; dam, Ochra 4845.

Butter, 19 lbs. 10½ oz. Milk, 225 lbs. 14 oz.

Test made from May 29 to June 5, 1886; age, 6 yrs. 4 mos.; property of James N. Smith, Highland, Conn.

Robinella 63132.—Sire, Eurotas' Victor Hugo 15689; dam, Robinette 7114.

Butter, 15 lbs. 14 oz. Milk, 213 lbs.

Test made from Oct. 9 to 16, 1892; age, 3 yrs. 6 mos.; estimated weight, 800 lbs.; fed 16 qts. daily of equal parts corn meal, ground oats and wheat bran, and 1 pt. oil meal—hay; property of Walter W. Law, Whitson, N. Y.

Robinette 7114.—Sire, Domino of Darlington 2459; dam, Rachel Ray 1754.

Butter, 14 lbs. 1 oz., unsalted. Milk, 233 lbs. 4 oz.

Test made from Apr. 17 to 23, 1883; age, 5 yrs. 3 mos.; fed 3 qts. of corn and 3 qts. of bran, daily; property of A. B. Darling, Ramsey's, N. J.

Robinette Rioter 57206.—Sire, Mirama's Rioter 12611; dam, Alphea Gwendolen 32550.

Butter, 15 lbs. Milk, 268 lbs. 2 oz.

Test made from May 24 to 31, 1895; age, 7 yrs.; estimated weight, 950 lbs.; fed 21 qts. ground oats, 21 qts. crushed corn, 21 qts. ground wheat and 14 qts. flaxseed meal—good pasture; property of J. E. Carson, Crab Orchard, Ky.

Rochelle 15574.—Sire, Lord Harry 3445; dam, Roxana 2d 2532.

Butter, 15 lbs. 10 oz. Milk, 26 gals. 2 qts.

Test made from June 15 to 21, 1885; age, 3 yrs. 7 mos.; property of L. Q. C. Lamar, Oxford, Miss.

Rockwood Maid 8375.—Sire, Count Logan 1599; dam, Lady Clover 8024.

Butter, 15 lbs. 4½ oz. Milk, 273 lbs. 2½ oz.

Test made from June 21 to 28, 1885; age, 7 yrs. 5 mos.; property of E. D. Griswold, Orwell, Vt.

Rockys' Pogis 96272.—Sire, Michael Angelo Pogis 19901; dam, Lady of Rockys 96265.

Butter, 18 lbs. 3 oz. Milk, 276 lbs. 7 oz.

Test made from Mar. 28 to Apr. 4, 1898; age, 4 yrs. 1 mo.; actual weight, 915 lbs.; fed 56 lbs. wheat, oats and barley, in equal parts, ground together, 14 lbs. bran, 14 lbs. cotton-seed meal, 14 lbs. corn meal and about 245 lbs. hay—two hours on pasture daily; property of Charles E. Shattuck, Los Angeles, Cal.

Roland's Bonnie 2d 18054.—Sire, Star F. 8364; dam, Roland's Bonnie 18053.

Butter, 19 lbs. 2 oz., unsalted. Milk, 217 lbs.

Test made from Jan. 13 to 19, 1883; age, 4 yrs. 11 mos.; fed 9 qts. equal parts corn meal and shorts; property of J. H. Walker, Worcester, Mass.

Roll of Honor 13610.—Sire, Much Ado 2405; dam, Phryne 4289.

Butter, 14 lbs. 12 oz.

Test made from Jan. 2 to 8, 1883; age, 4 yrs. 4 mos.; property of James B. Wilder, Louisville, Ky.

Romantic 61626.—Sire, North Pacific Prize 11459; dam, Curfew 16498.

Butter, 15 lbs. 9 oz. Milk, 240 lbs. 11 oz.

Test made from June 1 to 8, 1889; age, 2 yrs. 11 mos.; estimated weight, 900 lbs.; fed 18 qts. meal, daily; property of Richardson Bros., Davenport, Iowa.

Romena of Hood Farm 114887.—Sire, Lord Landseer 24228; dam, Romena 72300.

Butter, 15 lbs. 11 oz. Milk, 262 lbs. 2 oz.

Test made from Aug. 18 to 25, 1897; age, 3 yrs. 5 mos.; estimated weight, 700 lbs.; fed 35 lbs. bran, 21 lbs. corn meal, 18 lbs. ground oats and 8 lbs. oil meal—green grass in barn at night, mixed clover and June grass pasture; property of C. I. Hood, Lowell, Mass.

Romping Lass 11021.—Sire, Tormentor 3533; dam, Romp Ogden 3d 5458.

Butter, 15 lbs. Milk, 226 lbs. 13½ oz.

Test made from Apr. 9 to 15, 1884; age, 3 yrs. 7 mos.; property of J. B. Wade, Edgewood, Ga.

Romping Lass 2d 51989.—Sire, Fancy's Harry 9777; dam, Romping Lass 11021.

Butter, 15 lbs. 14 oz. Milk, 148 lbs. 14 oz.

Test made from Mar. 29 to Apr. 5, 1891; age, 3 yrs. 1 mo.; estimated weight, 800 lbs.; fed 3 gals. corn-hearts, 2 gals. wheat bran and 1 gal. ground oats, daily—ensilage and clover hay; property of Matthews & Humes, Huntsville, Ala.

Romping Lass 2d 51989.—Sire, Fancy's Harry 9777; dam, Romping Lass 11021.

Butter, 21 lbs. 12 oz. Milk, 158 lbs. 12 oz.

Test made from Aug. 9 to 16, 1893; age, 5 yrs. 5 mos.; estimated weight, 1000 lbs.; fed 46 qts. wheat bran, 34 qts. corn and oats and 14 qts. cotton-seed meal, fed on cut green corn—very short and dry pasture; property of William E. Matthews, Huntsville, Ala.

Romping Mirth 64070.—Sire, Romp's Tormentor 7126; dam, Mirth of Bingo Farm 30999.

Butter, 17 lbs. ½ oz. Milk, 260 lbs. 8 oz.

Test made from May 24 to 31, 1894; age, 7 yrs. 3 mos.; estimated weight, 700 lbs.; fed 10 lbs. ground oats and corn, 6 lbs. bran and 2 lbs. oil meal, daily—good mixed pasture; property of Mrs. J. R. Sherman, Pawlet, Vt.

Romping Miss 54457.—Sire, Prince of Melrose 4819; dam, Romping Princess 40347.

Butter, 23 lbs. 1 oz. Milk, 295 lbs.

Test made from Aug. 31 to Sept. 7, 1893; age, 6 yrs. 4 mos.; estimated weight, 900 lbs.; fed 56 qts. corn meal, 28 qts. cotton-seed meal, 70 qts. crushed oats and 21 qts. ground peas—pastured on lespedeza, Bermuda and pea field; property of W. E. Oates, Vicksburg, Miss.

Romp Ogden 2d 4764.—Sire, Pertinatti 713; dam, Romp Ogden 1571.

Butter, 15 lbs. 5 oz. Milk, 265½ lbs.

Test made from May 20 to 26, 1882; age, 6 yrs. 3 mos.; property of William E. Oates, Vicksburg, Miss.

Romp Ogden 3d 5458.—Sire, Top-Sawyer 1404; dam, Romp Ogden 1571.

Butter, 14 lbs. 1 oz. Milk, 251 lbs. 8 oz.

Test made from July 9 to 15, 1881; age, 4 yrs. 6 mos.; property of Campbell Brown, Spring Hill, Tenn.

Romp Ogden 3d 5458.—Sire, Top-Sawyer 1404; dam, Romp Ogden 1571.

Butter, 15 lbs. 14 oz. Milk, 189 lbs.

Test made from Nov. 8 to 14, 1884; age, 7 yrs. 9 mos.; property of Baxter Smith, Nashville, Tenn.

Romp Ogden 5th 43181.—Sire, Fancy's Harry 9777; dam, Romp Ogden 1571.

Butter, 17 lbs. 9 oz. Milk, 240 lbs. 4 oz.

Test made from May 14 to 21, 1890; age, 4 yrs. 8 mos.; actual weight, 940 lbs.; fed, daily, between 16 and 24 lbs. corn and oats; property of Wm. Morrow & Son, Nashville, Tenn.

Romp's Nervilette 72282.—Sire, Romp's Torment 8789; dam, Nerviona 43678.

Butter, 17 lbs. 10 oz. Milk, 242 lbs. 14 oz.

Test made from June 18 to 25, 1893; age, 3 yrs. 5 mos.; estimated weight, 700 lbs.; fed 56 qts. bran, 28 qts. corn meal and 14 qts. oil meal—blue-grass pasture; property of Henry DuBois, Vigo, Ohio.

Romp's Princess 51185.—Sire, Romp's Torment 8789; dam, Grey Friar's Princess 27559.

Butter, 51.357 lbs. in 30 days. Milk, 704.7 lbs.

Test made from Aug. 29 to Sept. 27, 1893, at World's Columbian Exposition; age, 6 yrs. 5 mos.; actual weight, 785 lbs.; fed 340.7 lbs. hay, 66.5 lbs. oil meal, 106.5 lbs. ensilage, 14 lbs. corn meal, 214 lbs. bran, 73 lbs. oats, 84 lbs. corn-hearts, 28.5 lbs. cotton-seed, 4 lbs. middlings, 18 lbs. carrots and 2.5 lbs. old hay; property of C. A. Sweet, Buffalo, N. Y.

Romp's Princess 51185.—Sire, Romp's Torment 8789; dam, Grey Friar's Princess 27559.

Butter, 188.373 lbs. in 90 days. Milk, 2984.4 lbs.

Test made from May 31 to Aug. 28, 1893, at World's Columbian Exposition; age, 6 yrs. 2 mos. at beginning of test; actual weight, 780 lbs.; fed 304 lbs. old hay, 210 lbs. ensilage, 534.1 lbs. new hay, 191.5 lbs. oil meal, 450 lbs. bran, 103 lbs. oats, 545.5 lbs. corn-hearts, 126.75 lbs. cotton-seed, 385 lbs. middlings, 34 lbs. grano-gluten, 1110.4 lbs. clover and 10 lbs. swale-grass; property of C. A. Sweet, Buffalo, N. Y.

Roonan 5133.—Sire, Rupert 1456; dam, Oonan 1485.

Butter, 18 lbs. 2 oz. Milk, 284 lbs.

Test made from Apr. 4 to 10, 1883; property of M. C. Campbell, Spring Hill, Tenn.

Roonan 5133.—Sire, Rupert 1456; dam, Oonan 1485.

Butter, 20 lbs. 4 oz. Milk, 297 lbs. 7 oz.

Test made from June 10 to 16, 1884; age, 8 yrs. 6 mos.; fed 11 lbs. corn meal and wheat bran, mixed with 2 pts. cotton-seed meal—good pasture of blue and orchard grass, and an hour daily on an acre of oats; property of M. C. Campbell, Spring Hill, Tenn.

Rosabel Hudson 5704.—Sire, Oneco 918; dam, Arabel Hudson 2877.

Butter, 15 lbs. 12 oz.

Test made from June 15 to 21, 1884; age, 8 yrs. 2 mos.; property of Elbert Miller, Middlefield, Conn.

Rosa F. W. 13226.—Sire, Duke F. 6134; dam, Rose F. 2d 13224.

Butter, 29 lbs. 12 oz. Milk, 200 lbs. 8 oz.

Test made from Apr. 17 to 24, 1886; age, 9 yrs. 1 mo.; fed twice daily on 10 lbs. of fine feed, 7 lbs. of corn meal, 1 lb. pea meal and ½ lb. of Empire horse and cattle food— good hay twice daily and ensilage once; property of O. F. Fuller, Blackstone, Mass.

Rosa Hudson 2d 87631.—Sire, Tormentor Jr. 15005; dam, Rosa Hudson 34871.

Butter, 16 lbs. 14½ oz. Milk, 247 lbs.

Test made from July 25 to Aug. 1, 1897; age, 5 yrs. 2 mos.; actual weight, 820 lbs.; fed 42 qts. corn meal, 42 qts. ground oats, 14 qts. wheat bran and 14 qts. ground oil-cake—good old meadow pasture; property of Miller & Sibley, Franklin, Pa.

Rosaletta 9339.—Sire, Cushnoc Jr. 2358; dam, Fuchsia 2789.

Butter, 14 lbs.

Test made in June, 1883; age, 5 yrs. 6 mos.; property of Freeman Partridge, Prospect, Me.

Rosaline Collas 11187.—Sire, Browny (P. S. 158 J. H. B.); dam on I. of J.

Butter, 14 lbs. 15 oz. Milk, 225 lbs. 8 oz.

Test made from June 16 to 23, 1889; age, 11 yrs. 2 mos.; estimated weight, 1000 lbs.; fed, daily, 4 lbs. bran, 4 lbs. oat meal and 4 lbs. corn meal; property of H. M. Baum, Frankfort, Ind.

Rosaline of Glenmore 3179.—Sire, Saladin 447; dam, Elveta 2121.

Butter, 17 lbs. 10 oz. Milk, 183 lbs.

Test made from Dec. 19 to 25, 1881; age, 8 yrs. 9 mos.; property of Campbell Brown, Spring Hill, Tenn.

Rosaline of Glenmore 3179.—Sire, Saladin 447; dam, Elveta 2121.

Butter, 17 lbs. 12 oz. Milk, 206 lbs.

Test made from Feb. 25 to Mar. 2, 1884; age, 11 yrs.; property of Campbell Brown, Spring Hill, Tenn.

Rosa Miller 4333.—Sire, Albert 2d 1835; dam, Belle of Bloomfield 4331.

Butter, 17 lbs. 7 oz. Milk, 276 lbs. 8 oz.

Test made from Dec. 17 to 24, 1883; age, 9 yrs. 10 mos.; property of J. H. Walker, Worcester, Mass.

Rosa of Belle Vue 6954.—Sire, Remarkable (F. S. 229 J. H. B.); dam, Bye-Bye (F. S. 3180 J. H. B.)

Butter, 18 lbs. 7½ oz., unsalted. Milk, 205 lbs. 8 oz.

Test made from Aug. 3 to 9, 1882; age, 6 yrs.; fed 12 qts. corn meal daily—poor blue-grass pasture; property of Thomas H. Malone, Nashville, Tenn.

Rosa of Frisby Hill 36080.—Sire, Statesman Jr. 7328; dam, Minnie Sayre 16006.

Butter, 16 lbs. 3¼ oz. Milk, 216 lbs. 4 oz.

Test made from Jan. 26 to Feb. 2, 1891; age, 6 yrs. 10 mos.; actual weight, 1050 lbs.; fed 7 lbs. yellow meal, 4 lbs. wheat bran, 1 lb. oil meal, 25 lbs. corn ensilage, 5 lbs. cut hay and 5 lbs. roots, daily; property of S. K. Schwenk, Marconnier, N. J.

Rosa Oxford 66678.—Sire, Oxford Pride 19160; dam, Rosa of Belle Vue 6954.

Butter, 15 lbs. ½ oz. Milk, 203 lbs.

Test made from Dec. 16 to 23, 1893; age, 3 yrs. 9 mos.; estimated weight, 800 lbs.; fed 112 lbs. corn chop and 56 lbs. ground oats—blue-grass pasture; property of Est. of Campbell Brown, Spring Hill, Tenn.

Rosa Oxford 2d 127909.—Sire, Tormentor 5th 21962; dam, Rosa Oxford 66678.

Butter, 14 lbs. 8 oz. Milk, 171 lbs.

Test made from Nov. 13 to 20, 1896; age, 2 yrs. 1 mo.; estimated weight, 750 lbs.; fed 84 lbs. corn meal, 42 lbs. ground oats and 14 lbs. wheat bran—good blue-grass pasture; property of Geo. Campbell Brown, Spring Hill, Tenn.

Rosa Ramsey 25875.—Sire, Duke of Ramsey's 6947; dam, Rosa F. W. 18226.

Butter, 15 lbs. 1 oz. Milk, 224 lbs.

Test made from July 22 to 29, 1893; age, 9 yrs. 4 mos.; estimated weight, 900 lbs.; fed 4 qts. bran, 6 qts. ground oats, 4 qts. corn meal and 2 qts. oil meal, daily—pasture of natural grasses; property of George W. Sisson, Jr., Potsdam, N. Y.

Rosa Russell 81549.—Sire, Ida's Stoke Pogis 2d 16864; dam, Little L. 2d 54807.

Butter, 16 lbs. 1 oz. Milk, 256 lbs. 8 oz.

Test made from May 13 to 20, 1896; age, 5 yrs. 7 mos.; estimated weight, 900 lbs.; fed 56 lbs. wheat bran, 28 lbs. corn meal and 14 lbs. cotton-seed meal—prairie pasture; property of C. M. Blvins, Terrell, Tex.

Rose 240, imp.

Butter, 17 lbs.

Property of John T. Norton, Farmington, Conn.

Rosebud of Allerton 6352.—Sire, Fortunatus 1152; dam, Rhoda of Merion 5849.

Butter, 19 lbs. 12 oz., unsalted.

Age, 4 yrs. 6 mos.; property of Mrs. J. B. Turner, Winnsborough, S. C.

Rosebud of Belle Vue 7702.—Sire, Browny (P. S. 158 J. H. B.); dam, Rosa of Belle Vue 6954.

Butter, 14 lbs. 10½ oz. Milk, 207 lbs. 8 oz.

Test made from June 24 to 30, 1884; age, 6 yrs. 2 mos.; property of Campbell Brown, Spring Hill, Tenn.

Rose Cottage Beauty 99542.—Sire, Torlandseer 22102; dam, Anna Neal 2d 7304.

Butter, 14 lbs. 10½ oz. Milk, 145 lbs. 14 oz.

Test made from Jan. 20 to 27, 1895; age, 3 yrs. 4 mos.; estimated weight, 750 lbs.; fed 21 qts. corn meal, 35 qts. wheat bran, 21 qts. ground oats and 7 qts. oil meal—hay; property of Bishop Clay, Lexington, Ky.

Roselaine 7167.—Sire, Grand Duke Alexis 1040; dam, Rosa Gamp 2732.

Butter, 14 lbs. 5 oz. Milk, 236 lbs. 10 oz.

Test made from Apr. 29 to May 5, 1882; age, 4 yrs. 1 mo.; property of A. Garretson, Pendleton, Ind.

Roselaine 7167.—Sire, Grand Duke Alexis 1040; dam, Rosa Gamp 2732.

Butter, 15 lbs. 1 oz.

Test made from Dec. 5 to 11, 1883; age, 5 yrs. 8 mos.; property of Garretson Bros., Pendleton, Ind.

Rosella of Bay View 66032.—Sire, G. F. Train 18868; dam, Belladonna of Glen Dale 7384.

Butter, 14 lbs. 14 oz.   Milk, 172 lbs. 14 oz.

Test made from Mar. 22 to 29, 1893; age, 2 yrs. 9 mos.; estimated weight, 500 lbs.; fed 196 lbs. corn refuse, 224 lbs. ensilage, 14 lbs. oil meal, 21 lbs. middlings, 28 lbs. malt sprouts and 35 lbs. hay; property of Charles Lautz, Buffalo, N. Y.

Rosella of Bay View 66032.—Sire, G. F. Train 18868; dam, Belladonna of Glen Dale 7384.

Butter, 18 lbs. 10 oz.   Milk, 256 lbs. 2 oz.

Test made from Mar. 21 to 27, 1894; age, 3 yrs. 9 mos.; actual weight, 756 lbs.; fed 14 lbs. oil meal, 21 lbs. ground oats, 14 lbs. corn meal, 14 lbs. ground wheat, 7 lbs. rye bran and 170 lbs. corn ensilage—50 lbs. hay; property of Charles Lautz, Buffalo, N. Y.

Rosella of Bay View 66032.—Sire, G. F. Train 18868; dam, Belladonna of Glen Dale 7384.

Butter, 66 lbs. 2 oz. in 30 days.   Milk, 942 lbs.

Test made from Feb. 22 to Mar. 24, 1898; age, 7 yrs. 9 mos.; actual weight, 750 lbs.; fed 90 lbs. oil meal, 60 lbs. ground oats, 60 lbs. corn meal, 120 lbs. wheat bran, 1050 lbs. corn ensilage and 210 lbs. hay; property of Charles Lautz, Buffalo, N. Y.

Rose of Eden 13437.—Sire, Landseer (P. S. 162 J. H. B.); dam, Samarés (F. S. 944 J. H. B.)

Butter, 20 lbs. 1½ oz.   Milk, 296 lbs.

Test made from June 18 to 24, 1885; age, 7 yrs. 3 mos.; fed bran, pea meal and chopped oats—clover and timothy pasture; property of Valancey E. Fuller, Hamilton, Ont., Can.

Rose of Hillside 3866.—Sire, Ramchunder 718; dam, Mamie 1612.

Butter, 14 lbs. 3½ oz.   Milk, 152 lbs. 9 oz.

Test made from Dec. 23 to 30, 1883; age, 9 yrs. 6 mos.; property of Charles J. Reed, Fairfield, Iowa.

Rose of Oxford 13469.—Sire, Tormentor 3533; dam, Young Rose (P. S. 43 J. H. B.)

Butter, 15 lbs. 14½ oz.   Milk, 234 lbs. 10 oz.

Test made from Dec. 1 to 7, 1884; age, 5 yrs. 3 mos.; property of William Simpson, New York, N. Y.

Rose of Riverside 46953.—Sire, Denise's Tormentor 11823; dam, Whiting's Daisy 14858.

Butter, 19 lbs. 9 oz.   Milk, 254 lbs. 4 oz.

Test made from May 26 to June 2, 1894; age, 8 yrs. 2 mos.; estimated weight, 1000 lbs.; fed 12 qts. ground oats, 8 qts. corn meal, 5 qts. bran and 3 qts. oil meal, daily—hay, fair woods pasture; property of C. A. Sweet, East Aurora, N. Y.

Rose of Rose Lawn 9365.—Sire, Columbiad 2d 1515; dam, Rose Lawn 3690.

Butter, 14 lbs. 6 oz.   Milk, 225 lbs. 7 oz.

Test made from Aug. 13 to 20, 1883; age, 6 yrs. 1 mo.; property of J. S. Rogers, Paterson, N. J.

Rose of Rose Lawn 9365.—Sire, Columbiad 2d 1515; dam, Rose Lawn 3690.

Butter, 16 lbs. 3 oz. Milk, 275 lbs. 10 oz.

Test made from July 16 to 23, 1884; age, 7 yrs.; property of J. S. Rogers, Paterson, N. J.

Rose of St. Lambert 20426.—Sire, Rambler of St. Lambert 5285; dam, Estelle of St. Lambert 7011.

Butter, 21 lbs. 3½ oz. Milk, 217 lbs.

Test made from Apr. 1 to 7, 1885; age, 3 yrs. 11 mos.; property of Valancey E. Fuller, Hamilton, Ont., Can.

Rose Pogis 24626.—Sire, Stoke Pogis of Linden 10558; dam, Trintasia 3d 24622.

Butter, 15 lbs. 2½ oz. Milk, 233 lbs.

Test made from Apr. 9 to 16, 1889; age, 5 yrs. 3 mos.; estimated weight, 850 lbs.; fed, daily, 5 qts. bran (average), 2 qts. old process oil meal and 2 qts. corn meal; property of E. J. Packard, Santa Barbara, Cal.

Rose's Fancy 93047.—Sire, Daisy's Harry 21500; dam, Rose of Riverside 46953.

Butter, 14 lbs. 7 oz. Milk, 199 lbs. 8 oz.

Test made from Mar. 24 to 31, 1898; age, 6 yrs.; estimated weight, 850 lbs.; fed 56 qts. bran, 28 qts. corn meal, 28 qts. oat feed, 14 qts. cotton-seed meal and 7 qts. oil meal, fed with cotton-seed hulls—hay *ad lib.*, good pasture; property of George V. Green, Hopkinsville, Ky.

Rosetta Landseer 99379.—Sire, Landseer's Pogis 15847; dam, Rosetta Stone 75200.

Butter, 14 lbs. 4 oz. Milk, 207 lbs. 12 oz.

Test made from Dec. 26, 1897, to Jan. 2, 1898; age, 3 yrs. 7 mos.; actual weight, 900 lbs.; fed 9 qts. bran, 6 qts. ground oats, 3 qts. H. O. feed and 3 qts. oil meal, mixed, fed wet, daily—plenty good hay and ensilage; property of George H. Sweet, East Aurora, N. Y.

Rosetta of Whiteland 6112.—Sire, Rossman 1128; dam, Soekel 2d 2959.

Butter, 27 lbs. 2¾ oz. Milk, 201 lbs.

Test made from Aug. 26 to Sept. 2, 1890; age, 14 yrs. 4 mos.; estimated weight, 850 lbs.; fed, daily, about 29½ qts. bran, 4 qts. corn chop, 3½ qts. oil meal and 3½ qts. ground oats; property of G. F. Miller, Topeka, Kan.

Rosette M. 78677.—Sire, Khedive's Landseer 21926; dam, Lady Rex 3d 15391.

Butter, 14 lbs. 6.4 oz. Milk, 193 lbs. 12.8 oz.

Test made from Sept. 12 to 19, 1895; age, 4 yrs. 4 mos.; estimated weight, 850 lbs.; fed 119 lbs. bran and 49 lbs. cotton-seed meal—good red-top pasture; property of Hecla Jersey Cattle Co., Earlington, Ky.

Rosette of St. Lambert 5108.—Sire, Lord Lisgar 1066; dam, Victoria 411.

Butter, 14 lbs. 3½ oz. Milk, 199 lbs.

Test made from July 1 to 7, 1885; age, 11 yrs. 3 mos.; property of Valancey E. Fuller, Hamilton, Ont., Can.

Rose Woodmere 91092.—Sire, Kattie's Prince 22315; dam, Susie's Pertinattis 50499.

Butter, 19 lbs. 13 oz. Milk, 231 lbs. 13 oz.

Test made from May 5 to 12, 1896; age, 3 yrs.; estimated weight, 800 lbs.; fed 17 lbs. per day of ground oats and corn, in equal parts—fair orchard and blue-grass pasture; property of Matt. M. Gardner, Nashville, Tenn.

Rosona 12956.—Sire, Catono 3761; dam, Rosa Daniell 5441.

Butter, 16 lbs. 7 oz. Milk, 107 lbs. 12 oz.

Test made from June 11 to 17, 1885; age, 4 yrs.; property of Lyman A. Mills, Middlefield, Conn.

Rosora 87622.—Sire, Ethleel's Landseer 22341; dam, Rose of Riverside 46953.

Butter, 15 lbs. 8½ oz. Milk, 226 lbs.

Test made from Apr. 10 to 17, 1895; age, 4 yrs. 1 mo.; estimated weight, 800 lbs.; fed 22½ qts. ground oats and corn, 4½ qts. wheat bran and 4½ qts. corn meal, with cut hay, daily—short pasture; property of George V. Green, Hopkinsville, Ky.

Rosy Dream 9808.—Sire, Jubilee 2125; dam, Landseer's Fancy 2876.

Butter, 19 lbs. 1 oz. Milk, 141 lbs. 6 oz.

Test made from Aug. 12 to 18, 1884; age, 5 yrs. 2 mos.; property of Columbia Jersey Cattle Co., Columbia, Tenn.

Rosy Dream 2d 27127.—Sire, Toltec 6831; dam, Rosy Dream 9808.

Butter, 17 lbs. 11 oz. Milk, 163 lbs. 14 oz.

Test made from Sept. 6 to 13, 1890; age, 7 yrs. 2 mos.; estimated weight, 750 lbs.; fed 4 gals. ground corn, daily; property of Maury Jersey Farm, Columbia, Tenn.

Rosy Kate 10276.—Sire, Pierrot 2d 1669; dam, Rose of Mashamoquet Farm 6472.

Butter, 14 lbs. 4 oz. Milk, 224 lbs. 8 oz.

Test made from Feb. 16 to 22, 1881; age, 4 yrs. 8 mos.; property of Hoover & Co., Columbus, Ohio.

Roulette 50080.—Sire, Torrance 17872; dam, Miss Rollo 2d 35587.

Butter, 14 lbs. 5½ oz. Milk, 228 lbs.

Test made from May 31 to June 7, 1890; age, 2 yrs. 2 mos.; estimated weight, 650 lbs.; well fed, but no record kept of amount; property of Campbell Brown, Spring Hill, Tenn.

Rover's Swan 106934.—Sire, Rover of St. Lambert 28333; dam, Lad's Swan 73325.

Butter, 16 lbs. 3 oz. Milk, 251 lbs.

Test made from Apr. 27 to May 4, 1898; age, 5 yrs. 2 mos.; estimated weight, 850 lbs.; fed 8 lbs. wheat bran, 8 lbs. corn meal, 3 lbs. cotton-seed meal, ½ lb. linseed meal and 2½ lbs. cotton-seed, daily—sorghum *ad lib.*, woods pasture; property of W. E. Johnson, Millican, Tex.

Rowena of Elm Spring 54613.—Sire, Noble Cicero 13562; dam, Compeer's Rowena 22968.

Butter, 14 lbs. 4 oz. Milk, 195 lbs. 14½ oz.

Test made from Oct. 17 to 24, 1891; age, 4 yrs.; estimated weight, 750 lbs.; fed 69½ qts. corn, oats and bran, with oil meal—stubble pasture; property of Armstrong Bros., Clinton, Mo.

Rowena of St. Lambert 71526.—Sire, Ruby's Star of St. Lambert 20719; dam, Ona Coomassie 43569.

Butter, 19 lbs. 2¼ oz. Milk, 255 lbs. 12 oz.

Test made from Mar. 18 to 25, 1896; age, 6 yrs. 9 mos.; estimated weight, 900 lbs.; fed 20 lbs. per day of gluten, bran and chop, in equal parts, mixed with clover—a little clover hay; property of Mrs. A. M. Hallock, Columbus, Ohio.

Rowena of St. Lambert 71526.—Sire, Ruby's Star of St. Lambert 20719; dam, Ona Coomassie 43569.

Butter, 20 lbs. 4 oz. Milk, 272 lbs.

Test made from Mar. 25 to Apr. 1, 1896; age, 6 yrs. 9 mos.; estimated weight, 900 lbs.; fed 21 lbs. 8 oz. per day of gluten, wheat bran and corn and oats chop, mixed in equal parts, moistened; property of Mrs. A. M. Hallock, Columbus, Ohio.

Rowena of St. Lambert 71526.—Sire, Ruby's Star of St. Lambert 20719; dam, Ona Coomassie 43569.

Butter, 39 lbs. 6¼ oz. in 14 days. Milk, 527 lbs. 12 oz.

Test made from Mar. 18 to Apr. 1, 1896; age, 6 yrs. 9 mos.; estimated weight, 900 lbs.; fed 20 lbs. per day of gluten, bran and chop, for first 7 days; 21 lbs. 8 oz. per day of gluten, wheat bran and corn and oats chop, mixed, in equal parts, moistened, for last 7 days; property of Mrs. A. M. Hallock, Columbus, Ohio.

Rowena of St. Lambert 71526.—Sire, Ruby's Star of St. Lambert 20719; dam, Ona Coomassie 43569.

Butter, 108 lbs. 2½ oz. in 42 days. Milk, 1467 lbs. 12 oz.

Test made from Feb. 26 to Apr. 8, 1896; age, 6 yrs. 8 mos.; estimated weight, 900 lbs.; fed 10 gals. cut corn fodder and mill feed (one part gluten and two parts bran), in equal parts, daily, for first week; 11½ gals. of the same, daily, during the second week; 14 qts. mixed mill feed, daily, for third week; 20 lbs. mixed mill feed, daily, for fourth week; 21 lbs. 8 oz. mixed mill feed, daily, for fifth week, and 20 lbs. daily of the same for sixth week; property of Mrs. A. M. Hallock, Columbus, Ohio.

Roxena 64212.—Sire, Odelio 20223; dam, Roxana 2d 2532.

Butter, 14 lbs. 9¾ oz. Milk, 253 lbs. 8 oz.

Test made from May 9 to 16, 1895; age, 5 yrs. 11 mos.; estimated weight, 1000 lbs.; fed 56 qts. corn-hearts, 14 qts. cotton-seed meal and 14 qts. bran—good blue-grass pasture; property of Est. of Campbell Brown, Spring Hill, Tenn.

Roxena 64212.—Sire, Odelio 20223; dam, Roxana 2d 2532.

Butter, 15 lbs. 8½ oz. Milk, 214 lbs.

Test made from Nov. 25 to Dec. 2, 1897; age, 8 yrs. 5 mos.; estimated weight, 1050 lbs.; fed 12 lbs. coarse corn meal, 9 lbs. wheat bran, 1½ lbs. cotton-seed meal and 1½ lbs. oil meal, daily—fair blue-grass pasture; property of George Campbell Brown, Spring Hill, Tenn.

Roxie Landseer 74832.—Sire, Leonette's Landseer 20399; dam, Roy Clementine 2d 23106.

Butter, 14 lbs. 13 oz. Milk, 213 lbs. 10 oz.

Test made from Mar. 16 to 23, 1894; age, 4 yrs. 1 mo.; estimated weight, 800 lbs.; fed 14 gals. corn and oats chop and 14 gals. wheat bran—pasture; property of J. D. Gray, Terrell, Tex.

Royal Beauty 18908.—Sire, Browny (P. S. 158 J. H. B.); dam, Princess Royal (P. S. 240 J. H. B.)

Butter, 15 lbs. 2½ oz. Milk, 216 lbs. 4 oz.

Test made from July 16 to 22, 1885; age, 6 yrs. 4 mos.; property of J. V. & C. Ramsden, Morton, Pa.

Royal Bomba's Daisy 35996.—Sire, Bomba's Son 8372; dam, Royal Daisy 25214.

Butter, 21 lbs. 5½ oz. Milk, 193 lbs. 11 oz.

Test made from June 3 to 10, 1893; age, 7 yrs. 4 mos.; estimated weight, 950 lbs.; fed 12 qts. corn and oats, mixed, 6 qts. wheat bran and 3 qts. oil meal, daily—little hay. good clover and timothy pasture; property of Charles A. Sweet, East Aurora, N. Y.

Royal Bomba's Daisy 3d 99553.—Sire, Landseer's Pogis 15847; dam, Royal Bomba's Daisy 35996.

Butter, 15 lbs. 8 oz. Milk, 225 lbs. 8 oz.

Test made from July 4 to 11, 1897; age, 3 yrs. 3 mos.; actual weight, 805 lbs.; fed 9 qts. bran, 9 qts. ground oats, 4½ qts. corn meal and 3 qts. oil meal, daily—plenty of good grass in barn; property of C. A. Sweet, Buffalo, N. Y.

Royal Daisy 25214.—Sire, Yellow Boy 6381; dam, Mary M. Allison 6308.

Butter, 20 lbs. 9 oz. Milk, 129 lbs. 3 oz.

Test made from Mar. 22 to 29, 1886; age, 2 yrs. 3 mos.; fed three times a day on 18 lbs. of feed, mixed as follows: 4½ lbs. crushed oats, 3¾ lbs. wheat bran, 2½ lbs. oil meal, 3¾ lbs. corn meal, 3½ lbs. wheat shorts, prepared in warm mash, with hay and green cabbage in addition; property of C. W. H. Eicke, West Monterey, Pa.

Royalist's Daisy 19187.—Sire, Royalist 2d 3791; dam, Fairy of Forest City 4th 16260.

Butter, 21 lbs. 9½ oz. Milk, 236 lbs.

Test made from Sept. 22 to 29, 1890; age, 9 yrs. 5 mos.; actual weight, 650 lbs.; fed, daily, average of 6 lbs. bran and shorts, 1 2-7 lbs. cotton-seed meal and 1½ lbs. oil meal; property of Mrs. Kate M. Busick, Wabash, Ind.

Royal Lady 22078.—Sire, King (P. S. 238 J. H. B.); dam, Brunette (F. S. 2110 J. H. B.)

Butter, 14 lbs. 12 oz. Milk, 198 lbs. 12 oz.

Test made from Apr. 2 to 8, 1886; age, 5 yrs.; property of D. A. Givens, Cynthiana, Ky.

Royal Princess 2370.—Sire, Daniel Webster 403; dam, Princess Royal 2d 1005.

Butter, 17 lbs. 12 oz. Milk, 160 lbs.

Test made from Dec. 28, 1880, to Jan. 3, 1881; age, 8 yrs. 7 mos.; property of Mrs. A. R. Adams, Watertown, Mass.

Royal Queen 24428.—Sire, Nero 7266; dam, Question (F. S. 3132 J. H. B.)

'   Butter, 22 lbs. 6 oz. Milk, 224 lbs. 12 oz.

Test made from June 7 to 14, 1887; age, 5 yrs. 5 mos.; estimated weight, 950 lbs.; fed, daily, 4 qts. bran, 4 qts. ground oats and corn and 1 qt. pea meal; property of M. Erskine Miller, Staunton, Va.

Royal Rose 86791.—Sire, Chairman 29883; dam, Rosa Mosher 2d 69179.

Butter, 15 lbs. 11 oz. Milk, 250 lbs.

Test made from Feb. 28 to Mar. 6, 1896; age, 3 yrs. 6 mos.; actual weight, 1050 lbs.; fed 21 qts. oil meal, 35 qts. corn meal, 35 qts. bran, 35 qts. oat meal, 100 lbs. hay and 120 lbs. roots; property of Pierce J. Cogswell, Rochester, N. Y.

Royal Rose 86791.—Sire, Chairman 29883; dam, Rosa Mosher 2d 69179.

Butter, 20 lbs. 4 oz. Milk, 323 lbs. 8 oz.

Test made from Mar. 2 to 9, 1897; age, 4 yrs. 5 mos.; actual weight, 1000 lbs.; fed 56 qts. bran, 56 qts. oat meal, 70 qts. corn meal, 32 qts. oil meal, 175 lbs. ensilage, 105 lbs. hay and 70 qts. roots; property of P. J. Cogswell, Rochester, N. Y.

Royal Sister 12451.—Sire, Vesper's Royal Son 2946; dam, Home Matron 6707.

Butter, 14 lbs. 11 oz. Milk, 212 lbs. 8 oz.

Test made from Sept. 1 to 7, 1883; age, 3 yrs. 5 mos.; property of S. L. Hoover, Columbus, Ohio.

Rosel Lass 20268, imp.

Butter, 19 lbs. 9½ oz. Milk, 273 lbs. 14 oz.

Test made from June 9 to 15, 1885; age, 5 yrs. 2 mos.; property of Frederick Loeser, Somerville, N. J.

Ruba H. Pogis 81944.—Sire, Rubano's Stoke Pogis 23760; dam, Mamie H. Pogis 57050.

Butter, 18 lbs. 1½ oz. Milk, 270 lbs.

Test made from June 12 to 19, 1894; age, 3 yrs. 5 mos.; actual weight, 900 lbs.; fed 14 qts. oil-cake, 42 qts. corn meal, 42 qts. ground oats—good old meadow pasture; property of Miller & Sibley, Franklin, Pa.

Rubano's Normanda 68927.—Sire, Rubano 8806; dam, St. Albans' Normanda 41705.

Butter, 21 lbs. 4½ oz. Milk, 271 lbs.

Test made from May 17 to 24, 1895; age, 6 yrs. 1 mo.; estimated weight, 875 lbs.; fed 24 lbs. corn meal, 24 lbs. ground oats, 14 lbs. linseed meal, 14 lbs. cotton-seed meal, 42 lbs. bran and 210 lbs. ensilage—blue-grass pasture; property of Samuel McKeen, Terre Haute, Ind.

Rubano's Valentine 54703.—Sire, Rubano 8806; dam, Valentine of Trinity 7460.

Butter, 20 lbs. 3 oz. Milk, 291 lbs. 6 oz.

Test made from May 30 to June 6, 1893; age, 6 yrs. 3 mos.; estimated weight, 900 lbs.; fed 28 lbs. ground oats, 42 lbs. corn meal, 10½ lbs. linseed meal and 56 lbs. bran—good blue-grass pasture; property of Samuel McKeen, Terre Haute, Ind.

Ruby Jenks 56862.—Sire, Albert's Royalist 10169; dam, Madam Jenks 34694.

Butter, 18 lbs. 3 oz. Milk, 254 lbs.

Test made from July 26 to Aug. 2, 1895; age, 8 yrs. 1 mo.; estimated weight, 800 lbs.; fed 3 lbs. cotton-seed meal, 2 lbs. wheat shorts, 10 lbs. cotton-seed hulls, daily—sweet potato vines *ad lib.*, lespedeza and water grass pasture; property of Mrs. Mattie A. Rubush, Meridian, Miss.

Ruby Jenks 2d 76738.—Sire, Mississippi Rioter 20401; dam, Ruby Jenks 56862.

Butter, 23 lbs. 3½ oz. Milk, 290 lbs. 4 oz.

Test made from Dec. 18 to 25, 1895; age, 5 yrs. 10 mos.; estimated weight, 1000 lbs.; fed 6 qts. bran, 3 qts. cotton-seed meal, 12 qts. cotton-seed hulls and 2 qts. oats, daily—sweet potatoes, peas and bran in slop, hay *ad lib.*; property of Mrs. Mattie A. Rubush, Meridian, Miss.

Ruby Jenks 3d 84868.—Sire, Mississippi Rioter 20401; dam, Ruby Jenks 56862.

Butter, 23 lbs. 4½ oz.   Milk, 289 lbs. 9 oz.

Test made from Feb. 24 to Mar. 2, 1896; age, 5 yrs. 2 mos.; estimated weight, 1000 lbs.; fed slops, sweet potatoes, peas, oats, cotton-seed meal, shorts, corn meal, wheat bran, cotton-seed hulls and hay, as much as she would eat; property of Mrs. Mattie A. Rubush, Meridian, Miss.

Ruby Jenks 4th 84871.—Sire, Mississippi Rioter 20401; dam, Ruby Jenks 56862.

Butter, 18 lbs. 5 oz.   Milk, 236 lbs. 8 oz.

Test made from Nov. 19 to 26, 1895; age, 3 yrs. 2 mos.; estimated weight, 800 lbs.; fed 15 lbs. cotton-seed hulls, 3 lbs. wheat bran, 4 lbs. cotton-seed meal, 4 lbs. oats and ½ bush. chopped sweet potatoes, daily—hay and oat straw *ad lib.*; property of Mrs. Mattie A. Rubush, Meridian, Miss.

Ruby Love 16915.—Sire, Lymington 3767; dam, Will o' the Wisp 16131.

Butter, 14 lbs. 12 oz.   Milk, 216 lbs. 1 oz.

Test made from Apr. 14 to 20, 1884; age, 2 yrs. 11 mos.; property of V. Lowe, Palmyra, Wis.

Ruby Love 2d 32058.—Sire, Stamp 8009; dam, Ruby Love 16915.

Butter, 14 lbs. 1 oz.   Milk, 166 lbs. 5 oz.

Test made from Jan. 27 to Feb. 2, 1886; age, 1 yr. 11 mos.; property of V. Lowe, Palmyra, Wis.

Ruby of Springvale 14505.—Sire, Tunlaw Boy 2866; dam, Alabama 7690.

Butter, 14 lbs. 15 oz.   Milk, 225 lbs.

Test made from May 28 to June 4, 1885; age, 4 yrs. 4 mos.; property of James Crook, Jacksonville, Ala.

Ruby's Valentine 81780.—Sire, Ruby's Star of St. Lambert 20719; dam, St. Albans' Valentine 41707.

Butter, 14 lbs. 7 oz.   Milk, 221 lbs. 14 oz.

Test made from July 27 to Aug. 3, 1893; age, 2 yrs. 3 mos.; estimated weight, 450 lbs.; fed 28 lbs. corn meal, 14 lbs. oil meal, 28 lbs. bran and 20 lbs. ensilage; property of Samuel McKeen, Terre Haute, Ind.

Ruby Tiptoe 78960.—Sire, Ruby's Harry 15664; dam, Tiptoe 24957.

Butter, 16 lbs. 12½ oz.   Milk, 187 lbs. 8 oz.

Test made from Mar. 26 to Apr. 2, 1894; age, 4 yrs. 6 mos.; estimated weight, 700 lbs.; fed 5 qts. ground oats, 1 qt. cotton-seed meal, 2 qts. bran and 2 qts. corn meal, daily —clover hay, corn stover and poor blue-grass pasture; property of Samuel N. Warren, Spring Hill, Tenn.

Ruby Torment 32989.—Sire, Tormentor 3533; dam, Babette 3650.

Butter, 16 lbs.   Milk, 120 lbs. 10 oz.

Test made from Jan. 14 to 20, 1885; age, 2 yrs. 1 mo.; property of Columbia Jersey Cattle Co., Columbia, Tenn.

Runaway 87030.—Sire, Exile of St. Lambert 23d 20712; dam, Rainbow's Girl 4th 69199.

Butter, 16 lbs. 10 oz.   Milk, 205 lbs.

Test made from Jan. 2 to 9, 1897; age, 4 yrs. 2 mos.; actual weight, 925 lbs.; fed 56 qts. bran, 56 qts. oat meal, 28 qts. corn meal, 14 qts. oil meal, 105 lbs. ensilage, 84 lbs. hay and 56 qts. roots; property of P. J. Cogswell, Rochester, N. Y.

Rupertina 10409.—Sire, Cloverine 3510; dam, Signalana 7719.

Butter, 17 lbs. 1½ oz. Milk, average of 17 qts. a day.

Test made in June, 1885; age, 5 yrs. 3 mos.; property of George W. Farlee, Cresskill, N. J.

Rush of St. Lambert 61234.—Sire, King of St. Lambert Jr. 18808; dam, Belle of Woodstock 11658.

Butter, 14 lbs. 9¼ oz. Milk, 270 lbs. 12 oz.

Test made from June 19 to 26, 1894; age, 4 yrs. 9 mos.; estimated weight, 1100 lbs.; fed 8 qts. chop and 2 qts. bran, daily, the first three days; 16 qts. chop and 2 qts. bran, daily, the last four days—poor pasture; property of Mrs. A. M. Hallock, Columbus, Ohio.

Rush of St. Lambert 61234.—Sire, King of St. Lambert Jr. 18808; dam, Belle of Woodstock 11658.

Butter, 15 lbs. 8¾ oz. Milk, 259 lbs.

Test made from Mar. 11 to 18, 1897; age, 7 yrs. 6 mos.; estimated weight, 1200 lbs.; fed 5 gals. daily of soaked feed, consisting of 4 parts fine-cut Kaffir corn, 1 part gluten and 3 parts bran—corn fodder mornings, Kaffir corn nights; property of Mrs. A. M. Hallock, Columbus, Ohio.

Rush of St. Lambert 61234.—Sire, King of St. Lambert Jr. 18808; dam, Belle of Woodstock 11658.

Butter, 17 lbs. 6¾ oz. Milk, 252 lbs.

Test made from Mar. 18 to 25, 1897; age, 7 yrs. 6 mos.; estimated weight, 1200 lbs.; fed 2 qts. gluten, 2 qts. chop, 2 qts. bran, 4 qts. fine-cut Kaffir corn and ½ pt. oil meal, twice daily—a little hay; property of Mrs. A. M. Hallock, Columbus, Ohio.

Rush of St. Lambert 61234.—Sire, King of St. Lambert Jr. 18808; dam, Belle of Woodstock 11658.

Butter, 34 lbs. 12½ oz. in 15 days. Milk, 545 lbs.

Test made from Mar. 11 to 26, 1897; age, 7 yrs. 6 mos.; estimated weight, 1200 lbs.; fed 5 gals. daily of soaked feed, composed of 4 parts fine-cut Kaffir corn, 1 part gluten and 3 parts bran, for first seven days; 2 qts. gluten, 2 qts. chop, 2 qts. wheat bran and ½ pt. oil meal, twice daily, during last eight days—corn fodder first seven days, mixed hay during last eight days; property of Mrs. A. M. Hallock, Columbus, Ohio.

Ruthie Rex Pogis 100326.—Sire, Imperial Rex Rioter 24096; dam, Lily of Elm Spring 51474.

Butter, 15 lbs. ¾ oz. Milk, 181 lbs. 1 oz.

Test made from May 1 to 8, 1898; age, 5 yrs. 5 mos.; estimated weight, 800 lbs.; fed 4 qts. wheat bran, 3 qts. cotton-seed and 3 qts. corn meal, daily—sorghum hay *ad lib.*; common poor pasture; property of H. J. Mitchell, Winnsborough, Tex.

Ruthione 31514.—Sire, Johnson's Acme 3656; dam, Ruth Morgan 2d 24100.

Butter, 25 lbs. 12 oz. Milk, 263 lbs. 1 oz.

Test made from Oct. 9 to 16, 1890; age, 7 yrs. 4 mos.; estimated weight, 950 lbs.; fed, daily, 6 lbs. malt, 5 lbs. corn, oats and barley chop, 4 lbs. bran and 2 lbs. oil meal; property of Belmont Jersey Cattle Co., Columbus, Ohio.

Ruth L. 84578.—Sire, Clover Lad 21069; dam, Travlo's Roxy 52313.

Butter, 14 lbs. 12 oz. Milk, 193 lbs. 14 oz.

Test made from June 27 to July 4, 1896; age, 5 yrs.; estimated weight, 850 lbs.; fed 15 lbs. daily of oats, peas and corn, ground—upland pasture; property of Schuyler Bros., Cobleskill, N. Y.

Ruth Morgan 24098.—Sire, Pierrot 7th 1667; dam, Lady Morgan 11671.

Butter, 14 lbs. 6 oz. Milk, 168 lbs.

Test made from Aug. 8 to 15, 1887; age, 8 yrs. 4 mos.; estimated weight, 700 lbs.; fed 12 lbs. ground corn and oats, daily; property of John Moore, Jr., Columbia, Tenn.

Ruth Morgan 3d 57043.—Sire, Fancy's Harry 9777; dam, Ruth Morgan 24098.

Butter, 19 lbs. 6.4 oz. Milk, 232 lbs. 6.4 oz.

Test made from July 13 to 20, 1895; age, 9 yrs. 2 mos.; estimated weight, 850 lbs.; fed 56 lbs. corn and oats, ground together, 56 lbs. bran and 42 lbs. cotton-seed meal —clover pasture; property of Hecla Jersey Cattle Co., Earlington, Ky.

Sadie Le Pet 77100.—Sire, Founder 20928; dam, Teacher's Pet 60242.

Butter, 22 lbs. 5 oz. Milk, 269 lbs. 1 oz.

Test made from Apr. 28 to May 5, 1896; age, 4 yrs. 9 mos.; estimated weight, 750 lbs.; fed 35 lbs. middlings, 35 lbs. bran, 21 lbs. oil meal and 151 lbs. roots—woods pasture; property of Mrs. Adda F. Howie, Elm Grove, Wis.

Sadie R. 3d 83729.—Sire, Norton 31941; dam, Sadie R. 34813.

Butter, 15 lbs. 14 oz. Milk, 174 lbs. 6 oz.

Test made from July 18 to 25, 1893; age, 2 yrs. 4 mos.; estimated weight, 750 lbs.; fed 12 qts. ground corn and oats, 12 qts. wheat bran, 3 qts. oil meal, in three feeds, daily—green oats and peas, poor pasture; property of C. A. Sweet, East Aurora, N. Y.

Sadie R. 3d 83729.—Sire, Norton 31941; dam, Sadie R. 34813.

Butter, 20 lbs. 14¼ oz. Milk, 283 lbs. 12 oz.

Test made from May 9 to 16, 1894; age, 3 yrs. 2 mos.; estimated weight, 750 lbs.; fed 70 qts. ground oats, 42 qts. corn meal, 21 qts. bran and 14 qts. oil meal—hay, poor pasture; property of C. A. Sweet, East Aurora, N. Y.

Sadie's Delight 53392.—Sire, Nutley Duke 10706; dam, Philena's Pet 31296.

Butter, 14 lbs. 12 oz. Milk, 299 lbs. 4 oz.

Test made from Apr. 3 to 10, 1893; age, 5 yrs. 5 mos.; estimated weight, 925 lbs.; fed 42 lbs. wheat bran, 42 lbs. corn meal, about 56 lbs. hay, 105 lbs. roots and 250 lbs. ensilage; property of Vermont Experiment Station, Burlington, Vt.

Sadie T. 75896.—Sire, Pearleta's Scituate 17109; dam, Ramobina 31196.

Butter, 16 lbs. 6 oz. Milk, 268 lbs.

Test made from June 4 to 11, 1897; age, 5 yrs. 6 mos.; estimated weight, 850 lbs.; fed 35 lbs. corn and cob meal and 35 lbs. gluten feed—red and white clover, timothy and blue-grass pasture; property of Holloway & Plummer, Flushing, Ohio.

Safety 13463, imp.

Butter, 15 lbs. 8 oz. Milk, 255 lbs. 15 oz.

Test made from Feb. 8 to 14, 1886; age, 8 yrs.; property of Frederick Billings, Woodstock, Vt.

Safrano 4568.—Sire, St. Helier 45; dam, Kalmia 4561.

Butter, 14 lbs. 2½ oz. Milk, 180 lbs. 8 oz.

Test made from Oct. 24 to 30, 1881; age, 8 yrs. 7 mos.; property of Charles Keep, Lockport, N. Y.

St. Albans' Valentine 41707.—Sire, Duke of St. Albans 11234; dam, Valentine of Trinity 7460.

Butter, 22 lbs. 6 oz. Milk, 248 lbs. 2 oz.

Test made from Mar. 16 to 23, 1893; age, 7 yrs. 1 mo.; estimated weight, 850 lbs.; fed 84 lbs. bran, 50 lbs. corn meal, 50 lbs. ground oats, 12 lbs. oil meal, 240 lbs. ensilage and 30 lbs. hay; property of Samuel McKeen, Terre Haute, Ind.

St. Helier Gem 35438.—Sire, St. Helier Boy 11884; dam, Mother Carey of P. 14092.

Butter, 15 lbs. ½ oz. Milk, 218 lbs.

Test made from July 5 to 12, 1890; age, 5 yrs.; actual weight, 760 lbs.; fed, daily, 12 lbs. cut oats, yellow meal and wheat bran, mixed, and 1 lb. oil meal; property of Holly Grove Farm, Plainfield, N. J.

St. Helier's Matilda 52857.—Sire, L. D.'s Victor St. Helier 15286; dam, Matilda 5th 18068.

Butter, 14 lbs. 1 oz. Milk, 148 lbs. 4 oz.

Test made from Jan. 11 to 18, 1891; age, 2 yrs. 6 mos.; estimated weight, 750 lbs.; fed, daily, 6 lbs. crushed oats, 6 lbs. corn meal, 3 lbs. oil meal and 3 lbs. pea meal; property of Ayer & McKinney, Philadelphia, Pa.

St. Jeannaise 15789.—Sire, Happy (P. S. 211 J. H. B.); dam, Khedive's Daisy 18174.

Butter, 16 lbs. 4 oz., unsalted.

Test made in July, 1883; age, 3 yrs. 3 mos.; property of S. W. Robbins, Wethersfield, Conn.

St. Jeannaise 15789.—Sire, Happy (P. S. 211 J. H. B.); dam, Khedive's Daisy 18174.

Butter, 17 lbs. 8½ oz.

Test made in June, 1885; age, 5 yrs. 2 mos.; property of S. W. Robbins, Wethersfield, Conn.

St. Jeannaise 4th 57990.—Sire, Ida's Pogis 18000; dam, St. Jeannaise 15789.

Butter, 15 lbs. 11 oz. Milk, 304 lbs.

Test made from Jan. 19 to 26, 1895; age, 6 yrs. 1 mo.; estimated weight, 900 lbs.; fed 10½ lbs. corn and cob meal, 42 lbs. bran, 42 lbs. ground oats, 7 lbs. oil meal and 105 lbs. ensilage—84 lbs. hay, 105 lbs. roots; property of R. A. Sibley, Spencer, Mass.

St. Jeannaise 5th 96554.—Sire, Ida's Pogis 18000; dam, St. Jeannaise 15789.

Butter, 16 lbs. 12½ oz. Milk, 302 lbs. 4 oz.

Test made from Apr. 10 to 17, 1897; age, 5 yrs. 7 mos.; estimated weight, 900 lbs.; fed 70 qts. bran, 28 qts. ground oats, 28 qts. corn meal, 5¼ qts. cotton-seed meal, 10½ qts. oil meal, 168 lbs. ensilage, 84 lbs. hay and 70 qts. roots; property of R. A. Sibley, Spencer, Mass.

St. John's Daisy 28388.—Sire, St. John (P. S. 316 J. H. B.); dam, Letacq Bess.

Butter, 15 lbs. 4 oz. Milk, 196 lbs. 8 oz.

Test made from July 29 to Aug. 5, 1888; age, 5 yrs. 10 mos.; estimated weight, 750 lbs.; fed, daily, about 18 qts. bran, 4 qts. oat meal, 2 qts. corn and oil meal; property of P. J. Cogswell, Rochester, N. Y.

St. Lambert B. Kate 105205.—Sire, St. Lambert Boy 17408; dam, Emsie of St. Lambert 95428.

Butter, 15 lbs. Milk, 265 lbs. 14 oz.

Test made from Mar. 1 to 8, 1898; age, 4 yrs. 2 mos.; estimated weight, 850 lbs.; fed 44 lbs. ensilage, 5 lbs. bran, 2½ lbs. oats, 2½ lbs. corn, 1½ lbs. cotton-seed meal and about 10 lbs. hay, daily; property of Pennsylvania Reform School, Morganza, Pa.

St. Lambert Hebe 54069.—Sire, George of St. Lambert 17200; dam, Hebe of Canterbury 33917.

Butter, 15 lbs. 5 oz. Milk, 114 lbs.

Test made from Aug. 15 to 22, 1890; age, 1 yr. 10 mos.; estimated weight, 750 lbs.; fed, daily, 25 lbs. ground corn, oats and peas and 5 lbs. shorts; property of Charles E. Hill, Denver, Col.

St. Lambert's Alphea 74100.—Sire, St. Lambert's John Bull 16618; dam, Pride of Springvale 20093.

Butter, 20 lbs. ½ oz. Milk, 271 lbs. 8 oz.

Test made from Nov. 16 to 23, 1894; age, 4 yrs. 6 mos.; estimated weight, 1100 lbs.; fed 56 qts. boiled cotton-seed, 50 qts. corn meal, 42 qts. ground oats, 8 qts. pea meal and 56 qts. bran—hay *ad lib.*, green oats pasture; property of J. C. Munden, Marshall, Tex.

St. Lambert's Babette 99185.—Sire, St. Lambert Boy 17408; dam, Barberry 2d 46391.

Butter, 15 lbs. ¾ oz. Milk, 241 lbs. 9 oz.

Test made from Apr. 4 to 11, 1897; age, 3 yrs. 7 mos.; actual weight, 849 lbs.; fed 35 lbs. crushed corn and cob meal, 35 lbs. wheat bran, 25 lbs. wheat middlings and 14 lbs. old process oil meal—cut corn fodder and ensilage *ad lib.*; property of J. P. Bradbury, Pomeroy, Ohio.

St. Lambert's Bijou 78174.—Sire, Don of St. Lambert 17456; dam, St. Lambert's Jewel 53743.

Butter, 14 lbs. 15½ oz. Milk, 238 lbs. 4 oz.

Test made from Jan. 27 to Feb. 3, 1805; age, 4 yrs. 6 mos.; estimated weight, 850 lbs.; no record of feed kept—fed same as herd, with 4 qts. chop per day additional; property of Mrs. A. M. Hallock, Columbus, Ohio.

St. Lambert's Coquette 41070.—Sire, Diana's Rioter 10481; dam, Coquette of Glen Rouge 17559.

Butter, 16 lbs. 15 oz. Milk, 257 lbs. 8 oz.

Test made from Oct. 10 to 17, 1889; age, 4 yrs. 5 mos.; estimated weight, 850 lbs.; fed, daily, nearly 18 lbs. bran and corn chop, in equal parts; property of W. B. von Richthofen, Denver, Col.

St. Lambert's Crescent 39711.—Sire, Ida's Rioter of St. L. 13656; dam, Lottie May 17201.

Butter, 22 lbs. 10 oz. Milk, 204 lbs.

Test made from June 9 to 16, 1890; age, 5 yrs.; estimated weight, 950 lbs.; fed 56 qts. bran and 14 qts. feed (mixture of two-thirds corn and one-third oats)—very good pasture of mixed grasses; property of R. S. Hartley, Freehold, Pa.

St. Lambert's Duchess 56504.—Sire, Canada's John Bull 8388; dam, Duchess of St. Lambert 5111.

Butter, 15 lbs. 7½ oz. Milk, 214 lbs. 2 oz.

Test made from Dec. 13 to 20, 1890; age, 2 yrs. 7 mos.; estimated weight, 750 lbs.; fed 20 lbs. peas and oats, daily—clover hay at night, clover and timothy pasture; property of William Rolph, Markham, Ont., Can.

St. Lambert's Kate 61975.—Sire, Duke of St. Lambert 16100; dam, Kate Winslow 28055.

Butter, 16 lbs. 11 oz. Milk, 259 lbs. 2 oz.

Test made from Apr. 30 to May 7, 1893; age, 4 yrs. 10 mos.; estimated weight, 925 lbs.; fed 3 qts. corn meal, 2 qts. oil meal, 4 qts. oat meal, 5 qts. bran, daily, also 4 qts. roots—clover and timothy pasture; property of Mrs. E. F. Hawley, Pittsford, N. Y.

St. Lambert's Nancy 56503.—Sire, Canada's John Bull 8388; dam, Nancy of St. Lambert 12964.

Butter, 18 lbs. 7 oz. Milk, 200 lbs. 8 oz.

Test made from Aug. 8 to 15, 1890; age, 2 yrs. 7 mos.; estimated weight, 900 lbs.; fed 15 lbs. peas and oats, daily—clover and timothy pasture; property of William Rolph, Markham, Ont., Can.

St. Lambert's Nora 42132.—Sire, Brier's Pogis 14163; dam, Rioter's Nora 21778.

Butter, 18 lbs. 12 oz. Milk, 391 lbs.

Test made from May 23 to 30, 1893; age, 6 yrs. 9 mos.; estimated weight, 900 lbs.; fed 98 lbs. peas and oats chop and 14 lbs. oil-cake—clover and timothy pasture; property of William Rolph, Markham, Ont., Can.

St. Lambert's Nora 3d 111218.—Sire, One Hundred per Cent. 16590; dam, St. Lambert's Nora 42132.

Butter, 17 lbs. 5 oz. Milk, 256 lbs.

Test made from June 3 to 10, 1896; age, 3 yrs.; estimated weight, 650 lbs.; fed 70 qts. ground oats and peas—clover and timothy pasture; property of William Rolph, Markham, Ont., Can.

St. Lambert's Pride 53742.—Sire, Brier's Pogis 14163; dam, Beauty of Lee Farm 15694.

Butter, 16 lbs. 13½ oz. Milk, 174 lbs. 8 oz.

Test made from Nov. 1 to 8, 1894; age, 7 yrs. 7 mos.; estimated weight, 950 lbs.; fed first four days, 6 qts. chop and 4 qts. kiln-dried corn bran, over wet cut hay, daily; last three days, 2 qts. more chop per day added—corn fodder, poor pasture; property of Mrs. A. M. Hallock, Columbus, Ohio.

St. Lambert's Pride 53742.—Sire, Brier's Pogis 14163; dam, Beauty of Lee Farm 15694.

Butter, 32 lbs. ¼ oz. in 14 days. Milk, 339 lbs.

Test made from Oct. 26 to Nov. 9, 1894; age, 7 yrs. 7 mos.; estimated weight, 950 lbs.; fed 112 qts. chop and 56 qts. maizene, over cut hay—corn fodder, poor pasture; property of Mrs. A. M. Hallock, Columbus, Ohio.

St. Lambert's Rexina 113609.—Sire, St. Lambert Boy 17408; dam, Sanarex 85387.

Butter, 16 lbs. 2 oz. Milk, 257 lbs.

Test made from July 5 to 12, 1898; age, 4 yrs. 2 mos.; estimated weight, 800 lbs.; fed 6 qts. wheat bran, 2 qts. ground oats and 1½ qts. corn meal, daily—good pasture; property of Mrs. Jennie E. Hall, Columbus, Ohio.

St. Lambert's Riotress 106220.—Sire, St. Lambert Boy 17408; dam, Letty Rioter 73475.

Butter, 24 lbs. 6 oz. Milk, 337 lbs. 8 oz.

Test made from May 29 to June 5, 1898; age, 4 yrs. 11 mos.; actual weight, 871 lbs.; fed 40 lbs. corn meal, 40 lbs. wheat bran, 25 lbs. wheat middlings and 16 lbs. old process oil meal—some fresh cut clover, excellent blue-grass; property of J. P. Bradbury, Pomeroy, Ohio.

St. Lambert's Violet 25278.—Sire, Stunner 9679; dam, Lily of St. Lambert 5120.

Butter, 16 lbs. 12 oz. Milk, 283 lbs.

Test made from Nov. 7 to 14, 1888; age, 4 yrs. 9 mos.; estimated weight, 1100 lbs.; fed, daily, 5 lbs. oat chop, 2 lbs. corn chop, 2 lbs. oil meal and 7 lbs. bran; property of W. B. von Richthofen, Denver, Col.

St. L. Gazelle 78870.—Sire, King of St. Lambert 15175; dam, Gazelle Dee 2d 55880.

Butter, 22 lbs. 14 oz. Milk, 299 lbs. 14 oz.

Test made from Nov. 6 to 13, 1897; age, 8 yrs. 5 mos.; estimated weight, 1050 lbs.; fed 8 qts. corn meal, 8 qts. ground oats and 6 qts. bran, daily—hay *ad lib.* at night, blue-grass pasture during day; property of T. B. Kibler, Jenera, Ohio.

St. L.'s Silken Les Cateaux 61964.—Sire, Duke of St. Lambert 16160; dam, Silken Les Cateaux 23431.

Butter, 21 lbs. 12 oz. Milk, 262 lbs. 10 oz.

Test made from Apr. 18 to 25, 1893; age, 5 yrs. 2 mos.; actual weight, 800 lbs.; fed 40 qts. corn meal, 40 qts. oat meal, 14 qts. oil meal, 42 qts. wheat bran, 140 lbs. ensilage and 1 bush. roots; property of Mrs. E. F. Hawley, Pittsford, N. Y.

St. Nick's Flora 16195.—Sire, St. Nick 7224; dam, Flora of Hopelands 16194.

Butter, 14 lbs.

Property of G. C. Rowley, Blandford, Mass.

St. Perpetua 2d 5557.—Sire, Pine Cliff 1106; dam, St. Perpetua 3648.

Butter, 14 lbs. Milk, 200 lbs.

Test made from June 6 to 12, 1881; age, 4 yrs. 5 mos.; property of Thomas H. Malone, Nashville, Tenn.

Sallie Grimes 79122.—Sire, Sig-Nig 18659; dam, Imperial Signalla 38585.

Butter, 16 lbs. 2 oz. Milk, 248 lbs. 4 oz.

Test made from Dec. 12 to 19, 1894; age, 5 yrs. 9 mos.; estimated weight, 900 lbs.; fed 84 lbs. bran and 21 lbs. corn—hay *ad lib.*, short pasture; property of L. P. Burrier, South Elkhorn, Ky.

Sallie Guild 14524.—Sire, Ralph Guild 1917; dam, Wiczil 6222.

Butter, 16 lbs. Milk, 16 qts. daily.

Test made from Apr. 20 to 27, 1884; age, 5 yrs. 2 mos.; property of Thomas L. Dodd, Nashville, Tenn.

Sallie of the Ledges 87182.—Sire, Crocus' John Bull 21665; dam, Ethel of Shelburne 46285.

Butter, 16 lbs. 6 oz. Milk, 335.8 lbs.

Test made from May 24 to 31, 1896; age, 3 yrs. 6 mos.; estimated weight, 850 lbs.; fed 84 lbs. grain, consisting of one-quarter corn meal, one-quarter oil meal, one-quarter oats and one-quarter bran, besides 7 qts. bran, 3½ qts. oats and 3½ qts. oil meal, in slop—98 lbs. hay; property of James S. Wadsworth, Geneseo, N. Y.

Sallie Pomeroy 98806.—Sire, St. Lambert Boy 17408; dam, Princess of Pomeroy 52110.

Butter, 14 lbs. 10 oz. Milk, 222 lbs. 11 oz.

Test made from Apr. 6 to 13, 1897; age, 3 yrs. 7 mos.; actual weight, 891 lbs.; fed 30 lbs. wheat bran, 30 lbs. wheat middlings, 45 lbs. crushed corn and cob meal and 15 lbs. old process oil meal—corn ensilage and cut fodder *ad lib.*; property of J. P. Bradbury, Pomeroy, Ohio.

Sally Leavitt 2d 10471.—Sire, Statesman 2407; dam, Sally Leavitt 7258.

Butter, 15 lbs. 12 oz. Milk, 220 lbs.

Test made from Feb. 16 to 23, 1890; age, 10 yrs. 9 mos.; estimated weight, 975 lbs.; fed, daily, 15 qts. ground corn, oats and peas and 4 qts. bran; property of Charles E. Hill, Denver, Col.

Sally of St. Lambert 78873.—Sire, King of St. Lambert 15175; dam, Allie of St. Lambert 2d 43671.

Butter, 16 lbs. 5 oz. Milk, 224 lbs.

Test made from Oct. 22 to 29, 1896; age, 6 yrs. 8 mos.; estimated weight, 900 lbs.; fed 35 lbs. corn meal, 7 lbs. cotton-seed meal, 7 lbs. linseed meal, 70 lbs. bran and 210 lbs. ensilage—blue-grass pasture, very dry; property of Samuel McKeen, Terre Haute, Ind.

Salsoda 3721.—Sire, Sam 980; dam, Hattie 2d 2901.

Butter, 14 lbs. 7 oz.

Property of W. L. & W. Rutherford, Waddington, N. Y.

Salute 67682.—Sire, Damascus 22222; dam, Bermuda R. 67534.

Butter, 16 lbs. 8 oz. Milk, 219 lbs. 4 oz.

Test made from Aug. 16 to 23, 1894; age, 4 yrs. 5 mos.; estimated weight, 700 lbs.; fed 28 lbs. oats, 28 lbs. corn, 28 lbs. bran and 14 lbs. oil meal—mixed pasture; property of H. C. Taylor, Orfordville, Wis.

Santa Stoke Pogis 91680.—Sire, Letty's Snap 27796; dam, Lady Perfection 32116.

Butter, 19 lbs. 1½ oz. Milk, 274.6 lbs.

Test made from Jan. 7 to 14, 1808; age, 4 yrs. 11 mos.; estimated weight, 950 lbs.; fed 27 lbs. corn meal, oil meal and wheat bran, mixed, 15 lbs. hay and 30 lbs. ensilage, daily; property of F. W. Hart, Cleveland, Ohio.

Sapolio 61082.—Sire, Exile of St. Lambert 2d 17131; dam, Marcella Maid's Pearl 23794.

Butter, 14 lbs. 6 oz. Milk, 227 lbs. 5 oz.

Test made from Oct. 18 to 25, 1891; age, 3 yrs. 7 mos.; estimated weight, 850 lbs.; fed about 2 lbs. oil meal, 8 lbs. bran and middlings and 4 lbs. oats and fodder corn, daily—clover and timothy pasture; property of A. D. McBain, Brockport, N. Y.

Sasanda 50075.—Sire, Tormentor 3533; dam, Sirona 2d 38011.

> Butter, 16 lbs. 4 oz. Milk, 212 lbs.

Test made from June 10 to 17, 1891; age, 3 yrs. 8 mos.; estimated weight, 850 lbs.; fed 16 lbs. corn-hearts and 4 lbs. bran, mixed, daily—blue-grass pasture; property of Campbell Brown, Spring Hill, Tenn.

Sasanda 50075.—Sire, Tormentor 3533; dam, Sirona 2d 38011.

> Butter, 22 lbs. 1½ oz. Milk, 307 lbs. 3 oz.

Test made from Apr. 2 to 9, 1896; age, 8 yrs. 6 mos.; actual weight, 985 lbs.; fed 30 lbs. wheat bran, 30 lbs. wheat middlings, 30 lbs. corn meal, 28 lbs. old process oil meal and 10 lbs. cotton-seed meal—corn ensilage and clover hay *ad lib.*; property of J. P. Bradbury, Pomeroy, Ohio.

Sasco Belle 13601.—Sire, Daniel Deronda 2291; dam, Lady Louise 4339.

> Butter, 14 lbs.

Age, 4 yrs.; property of R. G. Skiff, Green's Farms, Conn.

Satala 50074.—Sire, Ida's Stoke Pogis 13658; dam, Little Butterstar 20926.

> Butter, 16 lbs. 2½ oz. Milk, 194 lbs. 3 oz.

Test made from Mar. 31 to Apr. 7, 1890; age, 2 yrs. 6 mos.; actual weight, 650 lbs.; fed 20 lbs. per day of ground oats and corn, in equal parts, mixed with 3 bucketfuls of cut sheaf oats, wetted—clover hay *ad lib.*, good orchard and blue-grass pasture; property of Matt. M. Gardner, Nashville, Tenn.

Sataline 64213.—Sire, Ida's Stoke Pogis 13658; dam, Little Butterstar 20926.

> Butter, 17 lbs. 6½ oz. Milk, 225 lbs.

Test made from Nov. 5 to 12, 1893; age, 4 yrs. 4 mos.; estimated weight, 850 lbs.; fed 112 lbs. corn chop and 84 lbs. ground oats—blue-grass pasture; property of Est. of Campbell Brown, Spring Hill, Tenn.

Sataline 2d 131898.—Sire, Oonan's Tormentor 22280; dam, Sataline 64213.

> Butter, 16 lbs. 7½ oz. Milk, 228 lbs. 10 oz.

Test made from Dec. 12 to 19, 1897; age, 2 yrs. 2 mos.; estimated weight, 725 lbs.; fed 22 lbs. per day of ground corn and oats, in equal parts, mixed with 4 bucketfuls of cut sheaf oats, wetted—millet hay *ad lib.*; property of Matt. M. Gardner, Nashville, Tenn.

Satin Bird 16380.—Sire, Hero (P. S. 90 J. H. B.); dam, Philippa (F. S. 1665 J. H. B.)

> Butter, 14 lbs. 15½ oz. Milk, 201 lbs. 8 oz.

Test made from Oct. 6 to 12, 1883; age, 6 yrs. 6 mos.; property of Valancey E. Fuller, Hamilton, Ont., Can.

Saucy Pogis 103535.—Sire, Sig Pogis 22806; dam, Betty Bluet 53464.

> Butter, 17 lbs. 2¼ oz. Milk, 220 lbs.

Test made from Apr. 23 to 30, 1896; age, 2 yrs. 3 mos.; estimated weight, 650 lbs.; fed 28 qts. ground oats, 28 qts. crushed corn, 28 qts. wheat bran and 3½ qts. cotton-seed meal—blue-grass pasture; property of Mrs. S. J. Carter, Lynnville, Tenn.

Saucy Princess 40350.—Sire, Prince of Melrose 4819; dam, Saucylip 18574.

> Butter, 15 lbs. 11 oz. Milk, 259 lbs. 7 oz.

Test made from Jan. 20 to 27, 1893; age, 7 yrs. 7 mos.; estimated weight, 950 lbs.; fed 1½ gals. corn meal, 1½ gals. bran, 1 qt. cotton-seed meal, 3 qts. corn-cob meal, 15 lbs. ensilage and 5 lbs. hay, daily; property of M. Lothrop, Marshall, Tex.

Saucy Sally 83059.—Sire, Matilda's Stoke Pogis 14832; dam, Saucy Ann S. 30573.

Butter, 16 lbs. 12½ oz. Milk, 270 lbs. 3 oz.

Test made from June 6 to 13, 1898; age, 6 yrs. 2 mos.; estimated weight, 900 lbs.; fed 30 lbs. corn chop, 32 lbs. bran, 27 lbs. oil meal, 15 lbs. cotton-seed meal, 27 lbs. gluten meal and 10 lbs. H. O. dairy feed—blue grass and white clover pasture; property of Filston Farm, Glencoe, Md.

Saugus Lass 30542.—Sire, Lord Darlington 7285; dam, Young Bosdet's Rose 20067.

Butter, 14 lbs. 9 oz. Milk, 238 lbs. 10 oz.

Test made from Oct. 4 to 11, 1888; age, 4 yrs. 3 mos.; estimated weight, 850 lbs.; fed, daily, 10 lbs. oats, 10 lbs. corn meal, 14 lbs. shorts and 2 lbs. oil meal; property of D. F. Appleton, Ipswich, Mass.

Sayda 3d 17317.—Sire, Prince of M. 2d 5507; dam, Sayda 4400.

Butter, 47.825 lbs. in 30 days. Milk, 843.6 lbs.

Test made from Aug. 29 to Sept. 27, 1893, at World's Columbian Exposition; age, 11 yrs. 6 mos.; actual weight, 990 lbs.; fed 349.6 lbs. hay, 70 lbs. oil meal, 110 lbs. silage, 11 lbs. corn meal, 189 lbs. bran, 97 lbs. oats, 120 lbs. corn-hearts, 35.5 lbs. cotton-seed, 49.5 lbs. middlings, 18 lbs. carrots and 4 lbs. old hay; property of Edgar Brewer, Hockanum, Conn.

Sayda 3d 17317.—Sire, Prince of M. 2d 5507; dam, Sayda 4400.

Butter, 170.094 lbs. in 90 days. Milk, 3043.1 lbs.

Test made from May 31 to Aug. 28, 1893, at World's Columbian Exposition; age, 11 yrs. 3 mos. at beginning of test; actual weight, 944 lbs.; fed 313.9 lbs. old hay, 210 lbs. silage, 478.2 lbs. new hay, 184.5 lbs. oil meal, 545 lbs. bran, 72.5 lbs. oats, 537 lbs. corn-hearts, 121.75 lbs. cotton-seed, 418 lbs. middlings, 34 lbs. grano-gluten, 1060.4 lbs. clover and 10 lbs. swale grass; property of Edgar Brewer, Hockanum, Conn.

Sayda M. 46195.—Sire, Torpedo 8671; dam, Sayda 3d 17317.

Butter 22 lbs. 11½ oz. Milk, 201 lbs. 13½ oz.

Test made from Sept. 27 to Oct. 4, 1893; age, 7 yrs. 11 mos.; estimated weight, 950 lbs.; fed 15 qts. ground corn and oats, 15 qts. wheat bran and 3 qts. oil meal, daily—clover hay and green corn; property of Charles A. Sweet, Buffalo, N. Y.

Sayda M. 3d 71134.—Sire, Pocatello 21953; dam, Sayda M. 46195.

Butter, 20 lbs. 10 oz. Milk, 288 lbs. 9 oz.

Test made from Nov. 2 to 9, 1895; age, 6 yrs. 3 mos.; actual weight, 810 lbs.; fed 9 qts. ground oats, 3 qts. corn meal, 12 qts. wheat bran and 3 qts. oil meal, daily—hay and ensilage; property of C. A. Sweet, East Aurora, N. Y.

Sayda of Aurora 86956.—Sire, Koffee's Noble 14631; dam, Sayda M. 46195.

Butter, 16 lbs. 11½ oz. Milk, 218 lbs. 8 oz.

Test made from Aug. 12 to 19, 1895; age, 2 yrs. 5 mos.; actual weight, 775 lbs.; fed 9 qts. wheat bran, 9 qts. ground corn and oats, 1½ qts. oil meal and 1½ qts. cotton-seed meal, daily—fair mixed pasture, green oats and peas in stable; property of C. A. Sweet, East Aurora, N. Y.

Scepter's Beauty 23234.—Sire, Scepter 5417; dam, Pierrot's Myrtle 10135.

Butter, 14 lbs. ½ oz. Milk, 107 lbs. 14 oz.

Test made from Nov. 15 to 22, 1888; age, 5 yrs. 6 mos.; estimated weight, 700 lbs.; fed, daily, 16 qts. corn and cob meal; property of Maury Jersey Farm, Columbia, Tenn.

School Marm 67264.—Sire, Mutiny 17028; dam, Comwa 35911.

    Butter, 18 lbs. 4 oz. Milk, 244 lbs.

    Test made from Nov. 4 to 11, 1894; age, 5 yrs. 2 mos.; estimated weight, 1000 lbs.; fed 35 lbs. oats, 28 lbs. wheat, 21 lbs. corn, 24 lbs. bran and 7 lbs. oil meal; property of H. C. Taylor, Orfordville, Wis.

Scroll 93498.—Sire, Major Appel Pogis 17861; dam, Lily of Shady Side 49194.

    Butter, 14 lbs. 15 oz. Milk, 201 lbs.

    Test made from Apr. 27 to May 4, 1894; age, 4 yrs. 3 mos.; actual weight, 875 lbs.; fed 56 qts. corn meal, 56 qts. ground oats and 28 qts. oil-cake—good second season pasture; property of Miller & Sibley, Franklin, Pa.

Sedate 11119.—Sire, Wanderer 3014; dam, Signalia 5303.

    Butter, 18 lbs. 13 oz. Milk, 223 lbs. 9 oz.

    Test made from Nov. 30 to Dec. 7, 1892; age, 12 yrs. 8 mos.; estimated weight, 850 lbs.; fed 70 lbs. corn meal, 42 lbs. ground oats, 42 lbs. wheat bran, 42 lbs. cotton-seed meal, 280 lbs. ensilage and 70 lbs. hay; property of William Hart Dexter, Springfield, Mass.

Sedate 2d 20389.—Sire, Master Vermont 4304; dam, Sedate 11119.

    Butter, 15 lbs. 8 oz. Milk, 160 lbs. 12 oz.

    Test made from Jan. 27 to Feb. 3, 1890; age, 7 yrs.; estimated weight, 700 lbs.; fed 6 qts. ground corn and barley, daily; property of S. N. Warren, Spring Hill, Tenn.

Sedate's Retta 86596.—Sire, Signoretta's Signal 18980; dam, Sedate 11119.

    Butter, 16 lbs. 8 oz. Milk, 153 lbs. 8 oz.

    Test made from July 31 to Aug. 7, 1896; age, 4 yrs. 1 mo.; estimated weight, 800 lbs.; fed 10 lbs. ground corn, oats and rye, 5 lbs. wheat bran and 2 lbs. oil meal, daily—green corn-stalks; property of Thomas Friant, Grand Rapids, Mich.

Segilinda 41741.—Sire, Holyoke 8147; dam, Arawana Iris 2d 10420.

    Butter, 14 lbs. 8½ oz. Milk, 193 lbs. 4 oz.

    Test made from Mar. 3 to 10, 1897; age, 11 yrs. 8 mos.; estimated weight, 950 lbs.; fed 9 qts. ground oats, 3 qts. corn meal, 9 qts. bran and 3 qts. oil meal, mixed, daily—hay; property of C. A. Sweet, East Aurora, N. Y.

Selita J. 32184.—Sire, Pertinax 1965; dam, Minnie Kersey 19895.

    Butter, 25 lbs. 2½ oz. Milk, 211 lbs. 12 oz.

    Test made from June 8 to 14, 1886; age, 4 yrs. 1 mo.; test made on grass alone—good blue-grass; property of W. B. Matthews & Son, Franklin, Tenn.

Seneca's Frankie 114052.—Sire, Elm Place Seneca 21939; dam, Elm Place Frankie 48283.

    Butter, 15 lbs. 7½ oz. Milk, 292 lbs. 9 oz.

    Test made from May 17 to 24, 1897; age, 7 yrs.; estimated weight, 800 lbs.; fed 35 lbs. corn meal, 35 lbs. ground oats, 22½ lbs. oil meal, 67 lbs. gluten meal and 98 lbs. potatoes; property of L. D. Ely, Rochester, N. Y.

Seneca's Mocha 23355.—Sire, Seneca Chief 4098; dam, Mocha 2d 4881.

    Butter, 20 lbs. 3 oz. Milk, 406 lbs. 9 oz.

    Test made from June 1 to 8, 1890; age, 6 yrs. 11 mos.; estimated weight, 950 lbs.; fed 10 qts. bran and 8 qts. corn meal, daily—fodder, clover and blue-grass pasture; property of John J. Mott, Statesville, N. C.

Señorita of St. Lambert 98665.—Sire, Tombigbee Cowboy 18372; dam, Donita 40049.

Butter, 15 lbs. 8½ oz. Milk, 223 lbs.

Test made from Apr. 10 to 17, 1897; age, 4 yrs. 6 mos.; estimated weight, 850 lbs.; fed 28 lbs. bran, 17½ lbs. linseed meal, 14 lbs. cotton-seed meal, 28 lbs. ground oats, 140 lbs. cotton-seed hulls, 31½ lbs. corn meal and 70 lbs. green cut oats; property of S. H. Butler, Como, Miss.

Señerita Pogis 53791.—Sire, St. Lambert's John Bull 16618; dam, Katie Putnam 40366.

Butter, 16 lbs. Milk, 211 lbs. 8 oz.

Test made from Mar. 3 to 10, 1892; age, 4 yrs.; estimated weight, 750 lbs.; fed 56 qts. bran, 28 qts. pea meal, 28 qts. corn meal and 175 lbs. ensilage—pea-vine hay, good rye pasture; property of James Crook, Jacksonville, Ala.

Sensa 87569.—Sire, Saulie 33223; dam, Sue Crittenden 87564.

Butter, 21 lbs. 3 oz. Milk, 320 lbs. 2 oz.

Test made from June 28 to July 5, 1896; age, 5 yrs. 7 mos.; actual weight, 825 lbs.; fed 56 lbs. corn meal, 56 lbs. oat meal and 21 lbs. oil meal—good blue-grass pasture; property of W. L. Scott, Scott's Station, Ky.

Seppie 58689.—Sire, Chendado 12955; dam, Sunlight's Fairy 38134.

Butter, 15 lbs. 6½ oz. Milk, 196 lbs.

Test made from June 27 to July 4, 1896; age, 7 yrs. 9 mos.; estimated weight, 900 lbs.; fed 28 qts. ground oats, 28 qts. wheat bran and 28 qts. crushed corn—short blue-grass pasture; property of Mrs. S. J. Carter, Lynnville, Tenn.

Seraph 72217.—Sire, Diploma 16219; dam, Clara's Belle 16582.

Butter, 18 lbs. 5 oz. Milk, 298 lbs.

Test made from Dec. 10 to 17, 1893; age, 4 yrs. 2 mos.; estimated weight, 850 lbs.; fed 14 gals. cotton-seed, 14 gals. chop, 7 gals. bran and 3½ gals. steamed oats—pastured on sorghum stubble; property of S. T. Howard, Quanah, Tex.

Seraphine 2d 37451.—Sire, Suzerain 8408; dam, Seraphine 19262.

Butter, 14 lbs. 6 oz. Milk, 212 lbs. 4 oz.

Test made from May 25 to June 1, 1887; age, 3 yrs.; weight, 780 lbs.; fed from 6 qts. to 10 qts. corn-hearts daily; property of J. B. Allen & Son, Delavan, Ill.

Seraphine 2d 37451.—Sire, Suzerain 8408; dam, Seraphine 19262.

Butter, 15 lbs. 12 oz. Milk, 284 lbs.

Test made from May 27 to June 3, 1889; age, 5 yrs.; estimated weight, 900 lbs.; fed 4½ lbs. hominy daily; property of J. B. Allen & Son, Delavan, Ill.

Serita 15520.—Sire, Solid South 4711; dam, Sallie Ward 7201.

Butter, 17 lbs. 2 oz. Milk, 158 lbs.

Test made from Jan. 6 to 13, 1888; age, 6 yrs. 4 mos.; estimated weight, 750 lbs.; fed, daily, 8 lbs. corn meal, 8 lbs. bran, 2 lbs. oil meal and 2 lbs. middlings; property of James Stillman, Sing Sing, N. Y.

Serona Hudson 84775.—Sire, Brie's Tristan 19249; dam, Star Serona 84773.

Butter, 15 lbs. 11 oz. Milk, 253 lbs.

Test made from June 13 to 20, 1896; age, 4 yrs.; estimated weight, 900 lbs.; fed 56 qts. corn and oats (two parts oats to one of corn) and 14 qts. oil meal—timothy and clover pasture; property of E. E. Runner, Plattville, Ill.

Shanenalawn 56892.—Sire, Albert W. 6355; dam, Very Pretty 29184.

Butter, 18 lbs. 5 oz. Milk, 260 lbs. 1 oz.

Test made from June 23 to 30, 1893; age, 6 yrs. 1 mo.; actual weight, 910 lbs.; fed 5 qts. corn meal, 6 qts. oat meal, 6 qts. bran and 1 qt. cotton-seed meal, daily—clover and timothy pasture; property of Mrs. E. F. Hawley, Pittsford, N. Y.

Sheba Rex 47429.—Sire, Windsor Rex 14493; dam, Lady Sheba 15479.

Butter, 57.511 lbs. in 30 days. Milk, 1004.2 lbs.

Test made from Aug. 29 to Sept. 27, 1893, at World's Columbian Exposition; age, 7 yrs. 9 mos.; actual weight, 1013 lbs.; fed 394.6 lbs. hay, 75 lbs. oil meal, 146 lbs. silage, 40 lbs. corn meal, 179 lbs. bran, 90.5 lbs. oats, 169 lbs. corn-hearts, 59.5 lbs. cotton-seed, 60 lbs. middlings, 18 lbs. carrots and 6 lbs. old hay; property of T. A. Havemeyer, Mahwah, N. J.

Sheba Rex 47429.—Sire, Windsor Rex 14493; dam, Lady Sheba 15479.

Butter, 190.617 lbs. in 90 days. Milk, 3283.3 lbs.

Test made from May 31 to Aug. 28, 1893, at World's Columbian Exposition; age, 7 yrs. 8 mos. at beginning of test; actual weight, 991 lbs.; fed 324 lbs. old hay, 123 lbs. silage, 603.7 lbs. new hay, 191.5 lbs. oil meal, 469 lbs. bran, 122.5 lbs. oats, 566 lbs. corn-hearts, 144.50 lbs. cotton-seed, 385 lbs. middlings, 34 lbs. grano-gluten, 1210.4 lbs. clover and 10 lbs. swale-grass; property of T. A. Havemeyer, Mahwah, N. J.

Sheldon's Daisy 30592.—Sire, Sheldon 5250; dam, Daisy of St. Lambert 25840.

Butter, 17 lbs. 12 oz. Milk, 214 lbs.

Test made from June 2 to 9, 1889; age, 5 yrs. 11 mos.; estimated weight, 900 lbs.; fed, daily, 10 qts. chopped barley and oats and 2 qts. oil-cake; property of Wm. Rolph, Markham, Ont., Can.

Sheldon's Daisy of St. L. 53047.—Sire, Canada's John Bull 8388; dam, Sheldon's Daisy 30592.

Butter, 20 lbs. 11 oz. Milk, 240 lbs. 3 oz.

Test made from Oct. 11 to 18, 1890; age, 3 yrs. 2 mos.; estimated weight, 800 lbs.; fed, daily, 16 qts. ground corn, oats and peas and 2 qts. shorts; property of Mrs. L. E. Hill, Denver, Col.

Sheldon's Daisy's Ida 101059.—Sire, Ida of St. Lambert's Bull 19169; dam, Sheldon's Daisy 30592.

Butter, 14 lbs. 2 oz. Milk, 195 lbs.

Test made from Dec. 21 to 28, 1896; age, 2 yrs. 1 mo.; estimated weight, 700 lbs.; fed 19 lbs. daily of a mixture of corn meal, ground oats, bran, pea meal, cotton-seed meal, and oil meal, 40 lbs. ensilage and about 10 lbs. hay; property of R. F. Shannon, Pittsburg, Pa.

Sheldon's Maud 86187.—Sire, Sheldon Pogis 22659; dam, Fair Corinth's Maud 54610.

Butter, 14 lbs. 15 oz. Milk, 241 lbs. 8 oz.

Test made from Dec. 7 to 14, 1897; age, 5 yrs. 4 mos.; estimated weight, 800 lbs.; fed 7 lbs. wheat bran, 6 lbs. corn meal, ½ lb. linseed meal, 2 lbs. cotton-seed meal and 2½ lbs. cotton-seed, daily—corn-tops *ad lib.*, woods pasture; property of W. E. Johnson, Millican, Tex.

Shelta F. 56437.—Sire, Signalda's Torment 14673; dam, Kate Grimball 56435.

Butter, 20 lbs. ½ oz. Milk, 275 lbs. 4 oz.

Test made from May 22 to 29, 1893; age, 5 yrs.; estimated weight, 750 lbs.; fed 100 qts. ground corn and oats (equal parts), 18 qts. bran, 7 qts. cotton-seed meal, fed on cut hay; property of W. R. Rison, Huntsville, Ala.

Shiloh Daughter 20378.—Sire, Chatelaine's Dayton 5311; dam, Amber of Cottage Grove 14225.

Butter, 14 lbs. 7½ oz. Milk, 238 lbs. 15 oz.

Test made from July 2 to 8, 1885; age, 3 yrs. 2 mos.; property of C. Easthope, Niles, Ohio.

Shiloh Daughter 20378.—Sire, Chatelaine's Dayton 5311; dam, Amber of Cottage Grove 14225.

Butter, 16 lbs. 7 oz. Milk, 254 lbs.

Test made from June 3 to 10, 1893; age, 11 yrs. 1 mo.; estimated weight, 900 lbs.; fed 70 lbs. bran, 64 lbs. corn meal and 40 lbs. oil meal—meadow and woods pasture; property of Cornelius Easthope, Niles, Ohio.

Shrine 59627.—Sire, Diploma 16219; dam, Pledge 59214.

Butter, 15 lbs. 1½ oz. Milk, 243 lbs.

Test made from Jan. 18 to 25, 1896; age, 6 yrs. 11 mos.; estimated weight, 925 lbs.; fed 49 lbs. bran, 94½ lbs. oats and corn and cob meal mixed, 10½ lbs. oil meal, 10½ lbs. cotton-seed meal, 140 lbs. roots and 245 lbs. ensilage—hay *ad lib.*; property of C. I. Hood, Lowell, Mass.

Sibby 87742.—Sire, Gridley's Duke 24997; dam, Riverside Belle 2d 14125.

Butter, 15 lbs. 4½ oz. Milk, 286 lbs. 12 oz.

Test made from May 31 to June 7, 1897; age, 5 yrs. 2 mos.; estimated weight, 850 lbs.; fed 10 qts. wheat bran and 6 qts. corn meal, daily—prairie hay *ad lib.*; property of J. D. Gray, Terrell, Tex.

Sibby Landseer 116107.—Sire, Leonette's Landseer 20399; dam, Sibby 87742.

Butter, 15 lbs. 2 oz. Milk, 217 lbs. 4 oz.

Test made from Feb. 15 to 22, 1898; age, 2 yrs. 11 mos.; estimated weight, 750 lbs.; fed 98 qts. sorghum ensilage, 112 qts. bran, 42 qts. oats and corn, ground together, and 14 lbs. cotton-seed meal—hay *ad lib.*; property of J. D. Gray, Terrell, Tex.

Sibyl's Beauty 25941.—Sire, Forget-me-not 6291; dam, Sibyl (P. S. 345 J. H. B.)

Butter, 16 lbs. 8½ oz. Milk, 170 lbs. 8 oz.

Test made from Mar. 1 to 7, 1886; age, 3 yrs. 10 mos.; property of Timothy J. Hennesy, Litchfield, Conn.

Sibyl's Beauty 25941.—Sire, Forget-me-not 6291; dam, Sibyl (P. S. 345 J. H. B.)

Butter, 18 lbs. Milk, 186 lbs. 4 oz.

Test made from Apr. 25 to May 1, 1887; age, 5 yrs.; fed 18 lbs. daily of mixed corn, oats and middlings; property of George E. Jones, Litchfield, Conn.

Sibyl's Fancy 25942.—Sire, Forget-me-not 6291; dam, Sibyl (P. S. 345 J. H. B.)

Butter, 17 lbs. Milk, 182 lbs. 12 oz.

Test made from Apr. 18 to 24, 1887; age, 5 yrs.; fed 18 lbs. daily of mixed corn, oats and middlings; property of George E. Jones, Litchfield, Conn.

Sicilienne 25010, imp.

Butter, 16 lbs. 11 oz.　Milk, 162 lbs. 8 oz.

Test made from June 14 to 21, 1887; age, 8 yrs. 2 mos.; estimated weight, 900 lbs.; fed, daily, 4 qts. corn and oats, 1 qt. pea meal and 4 qts. bran; property of M. Erskine Miller, Staunton, Va.

Siesta 37321.—Sire, Signalda 4027; dam, St. Perpetua's Princess 19072.

Butter, 15 lbs. 11 oz.　Milk, 207 lbs.

Test made from May 19 to 26, 1890; age, 5 yrs.; estimated weight, 800 lbs.; no record of feed kept; property of Campbell Brown, Spring Hill, Tenn.

Sig Eurota 64898.—Sire, Sig 9528; dam, Kittie Pedro 27720.

Butter, 15 lbs. 1½ oz.　Milk, 301 lbs. 11 oz.

Test made from May 14 to 21, 1891; age, 5 yrs. 4 mos.; estimated weight, 850 lbs.; fed 35 lbs. shelled oats, 35 lbs. corn meal, 42 lbs. bran and 2 lbs. flaxseed—rich blue-grass pasture; property of M. A. Scovell, Lexington, Ky.

Sigleonella 70091.—Sire, Leonidas 3010; dam, Tenella 6712.

Butter, 14 lbs. 8 oz.　Milk, 176 lbs.

Test made from Dec. 9 to 16, 1893; age, 2 yrs. 10 mos.; estimated weight, 775 lbs.; fed 35 lbs. wheat bran, 35 lbs. ground oats, 56 lbs. corn meal and 21 lbs. linseed meal—hay; property of William Hart Dexter, Springfield, Mass.

Sigleonella 70091.—Sire, Leonidas 3010; dam, Tenella 6712.

Butter, 21 lbs. 9¾ oz.　Milk, 282 lbs. 3 oz.

Test made from Mar. 23 to 30, 1896; age, 5 yrs. 1 mo.; estimated weight, 900 lbs.; fed 63 lbs. wheat bran, 126 lbs. corn and oats provender (one-fourth oats, three-fourths corn), 35 lbs. pea meal and 21 lbs. linseed meal—hay *ad lib.*; property of William Hart Dexter, Springfield, Mass.

Sigletta 32915.—Sire, Signalda 4027; dam, Kathletta 19567.

Butter, 16 lbs.　Milk, 182 lbs. 12 oz.

Test made from Apr. 10 to 17, 1889; age, 4 yrs. 8 mos.; estimated weight, 850 lbs.; fed 4 gals. corn meal and ground oats, daily; property of M. C. Campbell, Spring Hill, Tenn.

Sigletta 3d 108977.—Sire, Oonan's Tormentor 22280; dam, Sigletta 32915.

Butter, 20 lbs. 10½ oz.　Milk, 178 lbs. 4 oz.

Test made from Nov. 30 to Dec. 7, 1895; age, 3 yrs. 11 mos.; estimated weight, 750 lbs.; fed 140 qts. corn meal and ground oats, in equal parts—hay, barley pasture; property of M. C. Campbell, Spring Hill, Tenn.

Sigma 78676.—Sire, Ethleel 2d's Jubilee 18249; dam, Koffee's Grisette 30433.

Butter, 16 lbs. 7 oz.　Milk, 226 lbs.

Test made from May 7 to 14, 1896; age, 5 yrs.; estimated weight, 750 lbs.; fed bran, corn meal, hay and shucks, quantity not stated—pasture; property of J. E. Hart, Nashville, Tenn.

Sigma 78676.—Sire, Ethleel 2d's Jubilee 18249; dam, Koffee's Grisette 30433.

Butter, 18 lbs. 2¼ oz.　Milk, 230 lbs. 9 oz.

Test made from Apr. 22 to 29, 1897; age, 6 yrs.; estimated weight, 800 lbs.; fed 16 qts. daily ground oats and corn, in equal parts—orchard and blue-grass pasture; property of J. E. Hart, Nashville, Tenn.

Sigma 2d 89803.—Sire, Fill Pail's Duke 24650; dam, Sigma 78676.

Butter, 14 lbs. 15½ oz. Milk, 195 lbs. 14½ oz.

Test made from Oct. 18 to 25, 1896; age, 3 yrs. 8 mos.; estimated weight, 950 lbs.; fed 14 qts. ground oats and corn, in equal parts, mixed with 4 bucketfuls of cut sheaf oats, daily—blue-grass pasture; property of J. E. Hart, Nashville, Tenn.

Sigmanella 117023.—Sire, Signal Maximum 27286; dam, Sigleonella 70091.

Butter, 20 lbs. 2 oz. Milk, 276 lbs. 12 oz.

Test made from Apr. 4 to 11, 1898; age, 3 yrs. 3 mos.; estimated weight, 800 lbs.; fed 42 lbs. wheat bran, 84 lbs. corn meal, 56 lbs. ground oats and 28 lbs. linseed meal —hay *ad lib.*; property of William Hart Dexter, Springfield, Mass.

Signalana 7719.—Sire, Signal 1170; dam, Maiden of Jersey 2736.

Butter, 15 lbs. 4 oz. Milk (average), 15 qts. per day.

Test made in May, 1884; age, 6 yrs. 2 mos.; property of George W. Farlee, Cresskill, N. J.

Signalbianca 54948.—Sire, Signalbert 18840; dam, Signodelle 27132.

Butter, 15 lbs. 5 oz. Milk, 204 lbs. ½ oz.

Test made from May 24 to 31, 1891; age, 2 yrs. 8 mos.; actual weight, 725 lbs.; fed 21 lbs. per day of ground oats and corn, in equal parts—very good orchard and blue-grass pasture; property of Matt. M. Gardner, Nashville, Tenn.

Signalbianca 2d 119779.—Sire, Fancy's Harry 2d 20199; dam, Signalbianca 54948.

Butter, 17 lbs. ¼ oz. Milk, 249 lbs. 15 oz.

Test made from May 13 to 20, 1897; age, 4 yrs. 11 mos.; estimated weight, 725 lbs.; fed 17 lbs. per day of ground oats and corn, in equal parts—extra good orchard and blue-grass pasture; property of Matt. M. Gardner, Nashville, Tenn.

Signalda Bloomfield 30694.—Sire, Signalda 4027; dam, Duchess of Bloomfield 3d 15580.

Butter, 16 lbs. 3¾ oz. Milk, 226 lbs. 8 oz.

Test made from Dec. 9 to 16, 1887; age, 2 yrs. 11 mos.; estimated weight, 800 lbs.; no record of feed kept; property of Algernon K. Johnston, Stapleton, N. Y.

Signalda's Duchess 32173.—Sire, Signalda 4027; dam, Ashwood Maid 15570.

Butter, 14 lbs. 3 oz. Milk, 213 lbs.

Test made from Sept. 3 to 10, 1885; age, 2 yrs. 4 mos.; property of G. V. Green, Hopkinsville, Ky.

Signalda's Ethleel 77599.—Sire, Ethleel's Landseer 22341; dam, Signalda's Coomassie 55444.

Butter, 16 lbs. 11 oz. Milk, 217 lbs. 5 oz.

Test made from June 8 to 15, 1896; age, 4 yrs. 8 mos.; estimated weight, 900 lbs.; fed 140 lbs. wheat bran, 28 lbs. corn meal and 14 lbs. cotton-seed meal—wild grass pasture; property of Platter & Foster, Denison, Tex.

Signalda's Rosebud 35583.—Sire, Signalda 4027; dam, Rosebud of Belle Vue 7702.

Butter, 18 lbs. 10⅛ oz. Milk, 309 lbs.

Test made from Apr. 28 to May 5, 1889; age, 3 yrs. 10 mos.; estimated weight, 850 lbs.; fed, daily, 20 lbs. corn-hearts and bran, mixed; property of Campbell Brown, Spring Hill, Tenn.

Signaldella 24107.—Sire, Signalda 4027; dam, Jazella G. 14191.

Butter, 18 lbs. 1¾ oz.

Test made from June 5 to 11, 1885; age, 1 yr. 11 mos.; property of Thomas H. Malone, Nashville, Tenn.

Signaldo's Pansy 53366.—Sire, Signaldo Alexis 15051; dam, La France's Pansy Rex 49141.

Butter, 16 lbs. ½ oz.   Milk, 306 lbs.

Test made from May 29 to June 5, 1897; age, 10 yrs.; estimated weight, 1000 lbs.; fed 18 lbs. corn meal and 25 lbs. ensilage, daily—blue-grass pasture; property of Baum & Hill, Frankfort, Ind.

Signal Duke's Creamer 60812.—Sire, Signal Duke of Alfred 11927; dam, Signal Pomona 28398.

Butter, 14 lbs. 7 oz.   Milk, 201 lbs. 1 oz.

Test made from July 9 to 16, 1891; age, 2 yrs. 9 mos.; estimated weight, 775 lbs.; fed 14 lbs. corn and oats, mixed in equal parts by weight and ground, and 7 lbs. oil meal—potato parings, etc., pasture of white clover and native grasses; property of Giltedge Farm Co., Syracuse, N. Y.

Signal Fancy 30812.—Sire, Fancy's Harry 9777; dam, Lilly Signalda 23227.

Butter, 14 lbs. 8⅝ oz.   Milk, 106 lbs. 12 oz.

Test made from Feb. 12 to 19, 1888; age, 2 yrs. 8 mos.; estimated weight, 900 lbs.; fed, daily, 4 gals. corn and oats, equal parts, and 1 qt. cotton-seed meal; property of Webster & Morrow & Son, Nashville, Tenn.

Signal Fancy of Lawn 73540.—Sire, Tormentor's John 14715; dam, Signalda's Fancy 32724.

Butter, 15 lbs. 1 oz.   Milk, 241 lbs. 8 oz.

Test made from Feb. 17 to 24, 1895; age, 4 yrs. 7 mos.; estimated weight, 800 lbs.; fed 84 lbs. wheat bran, 84 lbs. ground oats and 56 lbs. corn meal—good prairie hay once a day; property of Platter & Foster, Denison, Tex.

Signal Hinman 44615.—Sire, Composite 10810; dam, Louisa Hinman 12802.

Butter, 15 lbs. 12¾ oz.   Milk, 233 lbs. 12 oz.

Test made from Dec. 13 to 20, 1890; age, 5 yrs.; actual weight, 1010 lbs.; fed, daily, 18 lbs. cut oats, bran and yellow meal, mixed; property of Holly Grove Farm, Plainfield, N. J.

Signalinda 27002.—Sire, Signalio 7649; dam, Miss Belinda 16638.

Butter, 18 lbs. 7 oz.   Milk, 234 lbs. 14½ oz.

Test made from Sept. 10 to 17, 1888; age, 4 yrs. 4 mos.; estimated weight, 850 lbs.; fed 24 lbs. per day of ground oats and corn, in equal parts, mixed with 4 bucketfuls of cut oats and rye, wetted—orchard and blue-grass pasture; property of Matt. M. Gardner, Nashville, Tenn.

Signalinda 2d 61142.—Sire, Cherokee's Gold Basis 16945; dam, Signalinda 27002.

Butter, 17 lbs. 12 oz.   Milk, 212 lbs. ½ oz.

Test made from Jan. 7 to 14, 1894; age, 4 yrs. 7 mos.; estimated weight, 800 lbs.; fed 56 qts. middlings, 35 qts. corn meal, 105 lbs. hay and 168 lbs. ensilage; property of H. A. Huntington, Higganum, Conn.

Signalinda 2d 61142.—Sire, Cherokee's Gold Basis 16945; dam, Signalinda 27002.

Butter, 21 lbs. 10 oz. Milk, 208 lbs. 1 oz.

Test made from Dec. 4 to 11, 1894; age, 5 yrs. 6 mos.; estimated weight, 800 lbs.; fed provender, bran, cotton-seed meal and oil meal—no record of quantity kept; property of H. A. Huntington, Higganum, Conn.

Signal Maid 19361.—Sire, Croton Maid's Duke 6658; dam, Eva Taylor 14875.

Butter, 19 lbs. 13 oz. Milk, 274 lbs. 8 oz.

Test made from Apr. 24 to 30, 1886; age, 3 yrs. 3 mos.; property of John A. Middelton, Shelbyville, Ky.

Signal Maid 4th 67492.—Sire, Signal Duke 12661; dam, Signal Maid 19361.

Butter, 15 lbs. 8 oz. Milk, 237 lbs.

Test made from July 14 to 21, 1895; age, 5 yrs. 2 mos.; actual weight, 980 lbs.; fed 170 lbs. of a mixture of three-fifths corn and oats and two-fifths bran, ground together—little hay, blue-grass and clover pasture; property of John A. Middelton, Shelbyville, Ky.

Signal Maiden 42793.—Sire, Signal Jr. 7166; dam, Maiden of Jersey 3d 21242.

Butter, 15 lbs. 3½ oz. Milk, 246 lbs. 8 oz.

Test made from May 25 to June 1, 1890; age, 4 yrs. 9 mos.; estimated weight, 900 lbs.; fed, daily, about 15 lbs. ground corn and oats and 4 lbs. bran; property of Henry M. Baum, Frankfort, Ind.

Signal Mattie 113952.—Sire, Tenella's Best 19736; dam, Mattie E. Royal 78728.

Butter, 15 lbs. 7½ oz. Milk, 215 lbs. 12 oz.

Test made from Mar. 3 to 10, 1898; age, 3 yrs. 11 mos.; estimated weight, 800 lbs.; fed 84 qts. bran, 28 qts. corn meal, 14 qts. cotton-seed meal, 84 qts. sorghum ensilage—prairie hay *ad lib.*, two hours per day on barley pasture; property of J. D. Gray, Terrell, Tex.

Signal Mine 2d 60789.—Sire, Fancy's Harry 9777; dam, Signal Mine 20832.

Butter, 17 lbs. 9½ oz. Milk, 267 lbs. 15 oz.

Test made from May 20 to 27, 1893; age, 3 yrs. 11 mos.; estimated weight, 700 lbs.; fed 5 qts. corn meal, 6 qts. ground oats, 7 qts. middlings and 1 pt. oil meal, daily—no pasture; property of D. F. Appleton, Ipswich, Mass.

Signal of Raceland 84299.—Sire, Wasp 15358; dam, Neatham's Rosa Signal 82309.

Butter, 18 lbs. 6 oz. Milk, 287 lbs. 4 oz.

Test made from May 8 to 15, 1896; age, 4 yrs. 5 mos.; actual weight, 860 lbs.; fed 50 lbs. bran, 56 lbs. corn and cob meal and 21 lbs. oil meal—timothy pasture; property of James Marsh, Reddington, Ind.

Signalona 72105.—Sire, Signal Jr. 7166; dam, Sirona 27128.

Butter, 14 lbs. 3 oz. Milk, 189 lbs. 12 oz.

Test made from June 9 to 16, 1893; age, 2 yrs.; estimated weight, 700 lbs.; fed 42 lbs. Chicago gluten meal, 42 lbs. ground oats, 42 lbs. wheat bran, 14 lbs. linseed meal—hay, no pasture; property of William Hart Dexter, Springfield, Mass.

Signalona's Gem 134219.—Sire, Signalona's Ben 32929; dam, Signal's New Geranium 134117.

Butter, 14 lbs. 4½ oz.   Milk, 214 lbs. 4 oz.

Test made from Mar. 28 to Apr. 4, 1898; age, 2 yrs. 3 mos.; estimated weight, 725 lbs.; fed 35 lbs. wheat bran, 56 lbs. corn meal, 49 lbs. ground oats and 21 lbs. linseed meal—hay *ad lib.*; property of William Hart Dexter, Springfield, Mass.

Signal Queen 30869.—Sire, Chief of the Miamis 2958; dam, Fairy Queen of St. Brelade's 7464.

Butter, 21 lbs. ½ oz.   Milk, 208 lbs.

Test made from Oct. 31 to Nov. 7, 1891; age, 8 yrs.; estimated weight, 850 lbs.; fed 4 lbs. linseed meal, 10 lbs. corn meal, 4 lbs. ground oats and 4 lbs. bran, daily, on cut hay; property of Wm. Hart Dexter, Springfield, Mass.

Signal Queen 30869.—Sire, Chief of the Miamis 2958; dam, Fairy Queen of St. Brelade's 7464.

Butter, 51.522 lbs. in 30 days.   Milk, 944.5 lbs.

Test made from Aug. 29 to Sept. 27, 1893, at World's Columbian Exposition; age, 9 yrs. 10 mos.; actual weight, 1094 lbs.; fed 374.2 lbs. hay, 75 lbs. oil meal, 112 lbs. ensilage, 20 lbs. corn meal, 206 lbs. bran, 120 lbs. oats, 188 lbs. corn-hearts, 45 lbs. cotton-seed, 60 lbs. middlings, 15 lbs. carrots and 5 lbs. old hay; property of Frank Eno, Pine Plains, N. Y.

Signal Queen 30869.—Sire, Chief of the Miamis 2958; dam, Fairy Queen of St. Brelade's 7464.

Butter, 165.601 lbs. in 90 days.   Milk, 3190.6 lbs.

Test made from May 31 to Aug. 28, 1893, at World's Columbian Exposition; age, 9 yrs. 7 mos. at beginning of test; actual weight, 1059 lbs.; fed 327 lbs. old hay, 210 lbs. ensilage, 559.9 lbs. new hay, 188.5 lbs. oil meal, 503 lbs. bran, 98 lbs. oats, 509.5 lbs. corn-hearts, 126.75 lbs. cotton-seed, 438 lbs. middlings, 34 lbs. granogluten, 1195.4 lbs. clover and 10 lbs. swale-grass; property of Frank Eno, Pine Plains, N. Y.

Signal's Bijou 60800.—Sire, Croton's Signal 17614; dam, The Maid's Gem 42750.

Butter, 15 lbs. 1½ oz.   Milk, 209 lbs. 4 oz.

Test made from Dec. 14 to 21, 1892; age, 3 yrs. 8 mos.; estimated weight, 750 lbs.; fed 280 lbs. ensilage, 70 lbs. hay, 42 lbs. corn meal, 42 lbs. ground oats, 42 lbs. wheat bran and 31¼ lbs. cotton-seed meal; property of William Hart Dexter, Springfield, Mass.

Signal's Crusta 97341.—Sire, Toltec's Signal 29501; dam, Crusta 29637.

Butter, 15 lbs. 3½ oz.   Milk, 210 lbs. 15 oz.

Test made from May 28 to June 4, 1896; age, 2 yrs. 6 mos.; estimated weight, 775 lbs.; fed 35 lbs. bran, 21 lbs. oats, 35 lbs. corn meal and 21 lbs. oil meal—mixed pasture; property of H. C. Taylor, Orfordville, Wis.

Signal's Empress 57767.—Sire, Signal Jr. 7166; dam, Maiden of Jersey 3d 21242.

Butter, 16 lbs. 6½ oz.   Milk, 318 lbs. 10 oz.

Test made from May 6 to 13, 1894; age, 5 yrs. 10 mos.; estimated weight, 900 lbs.; fed 10 lbs. per day of a mixture in equal parts of corn meal, ground oats and wheat bran—pasture of timothy and natural grasses; property of Austin Leonard, Troy, Pa.

Signal's Garland 116048.—Sire, Signal's Rebel 32928; dam, Garland Melrose 93858.

Butter. 15 lbs. 2 oz. Milk, 239 lbs.

Test made from Dec. 6 to 13, 1897; age, 2 yrs. 9 mos.; estimated weight, 700 lbs.; fed 7 lbs. bran, 5 lbs. corn meal, ½ lb. linseed meal, 2 lbs. cotton-seed meal and 2½ lbs. cotton-seed, daily—corn fodder, woods pasture; property of W. E. Johnson, Millican, Tex.

Signal's Gold Foil 90508.—Sire, Signal Maid's Duke 21742; dam, Fairy Lady 24748.

Butter, 15 lbs. 4 oz. Milk, 272 lbs.

Test made from July 5 to 12, 1898; age, 6 yrs. 8 mos.; estimated weight, 900 lbs.; fed 5 lbs. Quaker oats feed, 3 lbs. oil meal, 6 lbs. bran and 10 lbs. hominy meal, daily—blue-grass pasture; property of N. Frazier, Clark, Ky.

Signal's Lily Flagg 31035.—Sire, Georgian 6073; dam, Little Nan 15895.

Butter, 27 lbs. 3½ oz. (official). Milk, 189 lbs. 6 oz.

Test made from May 23 to 30, 1892; age, 8 yrs. 4 mos.; estimated weight, 1000 lbs.; fed 28 gals. corn meal. 28 gals. ground oats, 14 gals. bran and 9 gals. linseed meal—poor clover pasture; property of Matthews, Moore & Humes, Huntsville, Ala.

Signal's New Geranium 134117.—Sire, Geranium's Leo 29482; dam, Geranium Leaf 2d 56859.

Butter, 21 lbs. 15½ oz. Milk, 290 lbs. 8 oz.

Test made from Apr. 11 to 18, 1898; age, 5 yrs. 4 mos.; estimated weight, 900 lbs.; fed 56 lbs. wheat bran, 70 lbs. corn meal, 70 lbs. ground oats and 28 lbs. linseed meal—hay *ad lib.*; property of William Hart Dexter, Springfield, Mass.

Signal's Rhoda 98651.—Sire, Signal Jr. 7166; dam, Rhoda 2d 16097.

Butter, 14 lbs. 5½ oz. Milk, 186 lbs. 4 oz.

Test made from June 30 to July 7, 1895; age, 8 yrs. 8 mos.; estimated weight. 850 lbs.; fed 4 lbs. wheat middlings and 2 lbs. cotton-seed meal daily—natural grass pasture; property of George W. Sisson, Jr., Potsdam, N. Y.

Signal's Ripple 27093.—Sire, Allegany Signal 10178; dam, Gardiner's Ripple 11693.

Butter, 17 lbs. 5 oz. Milk, 236 lbs. 6 oz.

Test made from Aug. 8 to 15, 1887; age, 3 yrs. 4 mos.; estimated weight, 800 lbs.; grain ration not given—dry blue-grass pasture; property of John B. Wallace, Lexington, Ky.

Signal's Ripple 4th 64862.—Sire, Jupiter Pogis 18192; dam, Signal's Ripple 27093.

Butter, 16 lbs. 10¼ oz. Milk, 282 lbs. 8 oz.

Test made from May 23 to 30, 1892; age, 3 yrs. 11 mos.; estimated weight, 800 lbs.; fed 82 lbs. corn meal, 40 lbs. ground oats and 28 lbs. bran; property of R. McMichael, Lexington, Ky.

Signetilia 16333.—Sire, Signalda 4027; dam, Sadie's Choice 7979.

Butter, 18 lbs. 5½ oz. Milk, 233 lbs. 8 oz.

Test made from Dec. 28, 1885, to Jan. 3, 1886; age, 3 yrs. 11 mos.; property of Campbell Brown, Spring Hill, Tenn.

Signet's Fancy 58446.—Sire, Fancy's Harry 9777; dam, Signet of Linwood 40423.

Butter, 14 lbs. 11 oz. Milk, 176 lbs. 8 oz.

Test made from Nov. 12 to 19, 1892; age, 4 yrs.; estimated weight, 700 lbs.; fed 51 qts. corn meal, 33 qts. bran, 13 qts. cotton-seed meal and 112 lbs. hay; property of H. A. Huntington, Higganum, Conn.

Signet's Fancy 58446.—Sire, Fancy's Harry 9777; dam, Signet of Linwood 40423.

Butter, 18 lbs. 11½ oz. Milk, 189 lbs.

Test made from May 12 to 19, 1895; age, 6 yrs. 6 mos.; estimated weight, 900 lbs.; fed 73 lbs. provender and 14 lbs. cotton-seed meal—poor pasture, hay; property of H. A. Huntington, Higganum, Conn.

Signilla M. 65839.—Sire, Signalda's Torment 14673; dam, Signal's Lily Flagg 31035.

Butter, 21 lbs. 4 oz. Milk, 225 lbs. 2 oz.

Test made from Nov. 19 to 26, 1891; age, 3 yrs. 10 mos.; estimated weight, 700 lbs.; fed 3 gals. corn-hearts, 2 gals. bran and 1 gal. ground oats, daily, on ensilage—short pasture; property of Matthews & Moore, Huntsville, Ala.

Signodelle 3d 50067.—Sire, Ida's Stoke Pogis 13658; dam, Signodelle 27132.

Butter, 17 lbs. 1½ oz. Milk, 189 lbs. 8 oz.

Test made from Dec. 16 to 23, 1891; age, 4 yrs. 7 mos.; estimated weight, 900 lbs.; fed 42 lbs. bran, 63 lbs. oat meal, 63 lbs. corn meal, 80 lbs. hay and 140 lbs. ensilage; property of Ayer & McKinney, Philadelphia, Pa.

Signola M. 76122.—Sire, Signalda's Torment 14673; dam, Blizz 55835.

Butter, 14 lbs. 10 oz. Milk, 200 lbs. 3 oz.

Test made from May 5 to 12, 1892; age, 2 yrs. 8 mos.; estimated weight, 700 lbs.; fed 2 gals. corn meal, 3 gals. wheat bran and 2 qts. linseed meal, daily, on cut hay—good pasture; property of Matthews & Moore, Huntsville, Ala.

Signolo's Florence 2d 82446.—Sire, Quintuple Victor Pogis 22343; dam, Signolo's Florence 49430.

Butter, 16 lbs. Milk, 211 lbs.

Test made from Sept. 14 to 21, 1896; age, 5 yrs. 10 mos.; estimated weight, 900 lbs.; fed 2 qts. wheat bran, 1 qt. corn meal and 1 qt. ground oats, daily—lowland pasture; property of E. L. Van Deusen, Ashley Falls, Mass.

Signora Reid 27125.—Sire, Signalda 4027; dam, Ora Reid 6482.

Butter, 16 lbs. 8⅜ oz. Milk, 241 lbs. 8 oz.

Test made from July 13 to 20, 1887; age, 3 yrs. 9 mos.; estimated weight, 825 lbs.; fed chopped barley and wheat bran, not weighed; property of William Bruce, Spring Hill, Tenn.

Signora Reid 2d 93808.—Sire, Tormentor 5th 21962; dam, Signora Reid 27125.

Butter, 15 lbs. 10½ oz. Milk, 232 lbs. 9 oz.

Test made from June 15 to 22, 1897; age, 3 yrs. 9 mos.; estimated weight, 850 lbs.; fed 56 lbs. ground oats and 56 lbs. ground corn, mixed—good blue-grass pasture; property of F. J. McDonald, Decatur, Ill.

Signora Signal 39396.—Sire, Conklin's Signal 11050; dam, Magnorant 12509.

Butter, 15 lbs. Milk, 215 lbs. 4 oz.

Test made from Oct. 19 to 26, 1890; age, 4 yrs. 9 mos.; estimated weight, 800 lbs.; fed 84 lbs. corn meal—good fall pasture; property of George M. Jewett, Glenville, Md.

Signoretta 21546.—Sire, Signalio 7649; dam, Gazella 3d 9355.

Butter, 680 lbs. 6½ oz. in one year. Milk, 7621 lbs.

Age, 3 yrs. 2 mos. at beginning of test; estimated weight, 750 lbs.; fed about 8 lbs. of grain daily; property of Geo. W. Farlee, Cresskill, N. J.

Silene 4307.—Sire, St. Helier 45; dam, Lara 4306.

Butter, 14 lbs.

Property of William Simpson, New York, N. Y.

Silenta 17685.—Sire, Oxoli 1922; dam, Lara 4306.

Butter, 15 lbs. 10 oz.

Test made from Aug. 12 to 18, 1883; age, 6 yrs. 2 mos.; property of Charles W. Beardsley, Milford, Conn.

Silicen 25577.—Sire, Prince George 11571; dam, Sadie A. 25573.

Butter, 18 lbs. 13 oz. Milk, 202 lbs. 5 oz.

Test made from July 15 to 22, 1888; age, 4 yrs. 6 mos.; estimated weight, 675 lbs.; fed, daily, 8 lbs. bran and 4 lbs. corn meal; property of Jacob L. Thomas, Knoxville, Tenn.

Silistria 20264, imp.

Butter, 16 lbs. 1 oz. Milk, 197 lbs. 2 oz.

Test made from Dec. 28, 1883, to Jan. 3, 1884; age, 4 yrs. 9 mos.; property of C. E. Rowley, Cortland, N. Y.

Silkey Morrill 17058.—Sire, Hazel Eye 1514; dam, Pussy Cat 2731.

Butter, 14 lbs. 7 oz.

Test made from Apr. 28 to May 5, 1884; age, 6 yrs. 6 mos.; property of C. F. Cobb, South Vassalborough, Me.

Siloam 17623.—Sire, Silver Mine 1658; dam, Prunella 2d 5861.

Butter, 18 lbs. 9½ oz. Milk, 230 lbs. 12½ oz.

Test made from May 6 to 12, 1884; age, 3 yrs. 4 mos.; fed 16 qts. oats and bran daily—good blue-grass pasture; property of John B. Wallace, Lexington, Ky.

Silver Belle C. 90600.—Sire, Fritz Pogis of Glynllyn 28042; dam, Silver Hull 15758.

Butter, 16 lbs. 2½ oz. Milk, 249 lbs. 8 oz.

Test made from Mar. 26 to Apr. 2, 1898; age, 4 yrs. 11 mos.; estimated weight, 800 lbs.; fed 6½ qts. bran, 3 qts. ground oats, 4½ qts. corn meal, ¾ qt. cotton-seed meal, ¾ qt. oil meal and 21 lbs. ensilage, daily—clover hay *ad lib.*; property of R. A. Sibley, Spencer, Mass.

Silver Cloud 13461, imp.

Butter, 15 lbs. 2 oz., unsalted. Milk, 127 lbs. 8 oz.

Test made from Aug. 31 to Sept. 6, 1882; age, 6 yrs. 5 mos.; property of W. R. McCready, Saugatuck, Conn.

Silver Delle 40691.—Sire, Ellwood 13382; dam, Charlton Caroline 11724.

Butter, 55 lbs. in 29 days. Milk, 794 lbs. 8 oz.

Test made from May 22 to June 20, 1889; age, 3 yrs. 2 mos.; fed, daily, 15 lbs. ground oats, peas and wheat bran—pasture; property of Mrs. E. M. Jones, Brockville, Ont., Can.

Silver Delle 40691.—Sire, Ellwood 13382; dam, Charlton Caroline 11724.

Butter, 17 lbs. 4½ oz. Milk, 170 lbs. 8 oz.

Test made from Sept. 7 to 14, 1890; age, 4 yrs. 5 mos.; estimated weight, 800 lbs.; fed, daily, 14 qts. ground oats, peas and barley; property of Mrs. Eliza M. Jones, Brockville, Ont., Can.

Silver Delle 40691.—Sire, Ellwood 13382; dam, Charlton Caroline 11724.

Butter, 18 lbs. 14½ oz. Milk, 237 lbs.

Test made from May 1 to 8, 1892; age, 6 yrs. 1 mo.; estimated weight, 850 lbs.; fed 98 lbs. of a mixture of ground oats, ground peas, wheat bran and oil meal—short pasture; property of Mrs. E. M. Jones, Brockville, Ont., Can.

Silveretta 6852.—Sire, Champion of America 1567; dam, Mag 3351.

Butter, 16 lbs. 9 oz.

Test made from Nov. 20 to 27, 1881; age, 5 yrs. 6 mos.; property of R. J. Fair, Wallingford, Conn.

Silver Hair 4th 60449.—Sire, Climax of St. Lambert 17553; dam, Silver Hair 4665.

Butter, 15 lbs. 10¼ oz. Milk, 282 lbs. 8 oz.

Test made from May 27 to June 3, 1895; age, 5 yrs. 7 mos.; estimated weight, 1100 lbs.; fed 10 qts. chop and 5 qts. wheat bran, on cut hay, daily, May 28 to June 2; 12 qts. chop, 6 qts. wheat bran and 1 pt. oil meal, June 2 and 3—poor pasture; property of Mrs. A. M. Hallock, Columbus, Ohio.

Silver Hair 4th 60449.—Sire, Climax of St. Lambert 17553; dam, Silver Hair 4665.

Butter, 31 lbs. ¼ oz. in 14 days. Milk, 559 lbs. 4 oz.

Test made from June 9 to 23, 1895; age, 5 yrs. 7 mos.; estimated weight, 1100 lbs.; fed 12 qts. chop, 6 qts. wheat bran, 1½ qts. oil meal, daily—poor pasture; property of Mrs. A. M. Hallock, Columbus, Ohio.

Silver Hair 4th 60449.—Sire, Climax of St. Lambert 17553; dam, Silver Hair 4665.

Butter, 60 lbs. 8¾ oz. in 28 days. Milk, 1118 lbs. 8 oz.

Test made from May 26 to June 23, 1895; age, 5 yrs. 7 mos.; estimated weight, 1100 lbs.; fed 276 qts. chop, 110 qts. wheat bran and 22 qts. oil meal—poor pasture; property of Mrs. A. M. Hallock, Columbus, Ohio.

Silver Montague 45475.—Sire, Lord Montague 12385; dam, Silverhair of C. 19668.

Butter, 16 lbs. 10 oz. Milk, 268 lbs. 4 oz.

Test made from May 23 to 30, 1891; age, 4 yrs.; estimated weight, 950 lbs.; fed 5 lbs. bran and 5 lbs. cotton-seed cake meal, daily—short pasture of timothy, clover and blue-grass; property of Bordwell & Cochran, Batavia, Ohio.

Silver Rose 4753.—Sire, Pilot Jr. 141; dam, Rosa 122.

Butter, 16 lbs. 14 oz.   Milk, 264 lbs.

Test made from July 1 to 7, 1881; age, 5 yrs. 9 mos.; property of Campbell Brown, Spring Hill, Tenn.

Silver Sheen 26210.—Sire, Silver (P. S. 287 J. H. B.); dam on I. of J.

Butter, 19 lbs. 2 oz.   Milk, 282 lbs. 4 oz.

Test made from Apr. 18 to 25, 1894; age, 11 yrs. 4 mos.; estimated weight, 950 lbs.; fed 54 qts. corn, oats and barley chop, 22 qts. gluten feed, 64 qts. bran, 14 qts. oil meal, 140 lbs. clover hay, cut and steamed, and 75 lbs. loose hay; property of A. T. Dempsey, Columbus, Ohio.

Silver Steletho 40765.—Sire, Steletho 10505; dam, Silverhair of C. 19668.

Butter, 18 lbs. ½ oz.   Milk, 241 lbs. 8 oz.

Test made from June 6 to 13, 1891; age, 6 yrs. 5 mos.; estimated weight, 1000 lbs.; fed 5 lbs. bran and 5 lbs. cotton-seed meal, daily—short pasture of timothy, clover and blue-grass; property of Bordwell & Cochran, Batavia, Ohio.

Silver Venus 20270.—Sire, Silver Hair (F. S. 279 J. H. B.); dam, Lily 2d (P. S. 268 J. H. B.)

Butter, 16 lbs. 3¾ oz.   Milk, 205 lbs.

Test made from Mar. 27 to Apr. 2, 1886; age, 5 yrs. 11 mos.; property of Rollins & Day, Unionville, Conn.

Silver Wave 10844.—Sire, The Hub 1009; dam, Variella 6337.

Butter, 19 lbs.   Milk, 382 lbs. 8 oz.

Test made from July 6 to 13, 1885; age, 5 yrs. 8 mos.; property of William E. Oates, Vicksburg, Miss.

Silvia Baker 8793.—Sire, Silver Mine 1658; dam, Fannie Baker 4274.

Butter, 20 lbs. 10 oz.   Milk, 121 lbs. 12 oz.

Test made from Apr. 21 to 28, 1886; age, 6 yrs. 10 mos.; fed 6 qts. corn meal, 6 qts. shelled oats, 3 qts. oil meal and 3 qts. wheat bran, daily; property of John B. Wallace, Lexington, Ky.

Silvia of Linwood 61008.—Sire, Southern Prince 10760; dam, Silver Tresses 45412.

Butter, 16 lbs. 5.6 oz.   Milk, 183 lbs. 11.2 oz.

Test made from May 1 to 8, 1895; age, 6 yrs. 1 mo.; estimated weight, 800 lbs.; fed 42 lbs. ground corn and oats, 28 lbs. bran and 21 lbs. cotton-seed meal—rye pasture; property of Hecla Jersey Cattle Co., Earlington, Ky.

Silvia Torment 104668.—Sire, Noonday 12961; dam, Silvia of Linwood 61008.

Butter, 15 lbs. 11.2 oz.   Milk, 213 lbs. 5.6 oz.

Test made from Dec. 13 to 20, 1896; age, 3 yrs. 6 mos.; estimated weight, 775 lbs.; fed 13 lbs. cut hay and sorghum (mixed half and half), 12 lbs. bran, 4 lbs. cotton-seed meal, 28 lbs. beets and 3½ lbs. corn meal, daily, and a little hay—196 lbs. roots; property of Hecla Jersey Cattle Co., Earlington, Ky.

Sirona 27128.—Sire, Signalda 4027; dam, Signalinea 16336.

Butter, 21 lbs. 2 oz.   Milk, 179 lbs. 8 oz.

Test made from Oct. 28 to Nov. 4, 1890; age, 7 yrs.; estimated weight, 950 lbs.; fed, daily, about 15 lbs. corn meal, 5¼ lbs. wheat bran, 3½ lbs. ground oats and 3 lbs. linseed meal; property of William Hart Dexter, Thompsonville, Conn.

Sirona Croton 86597.—Sire, Croton's Signal 17614; dam, Sirona 27128.

> Butter, 18 lbs. 10 oz. Milk, 183 lbs. 1 oz.

Test made from Feb. 10 to 17, 1896; age, 3 yrs. 6 mos.; estimated weight, 800 lbs.; fed 70 lbs. corn and oats, 33 lbs. bran and 21 lbs. oil meal; property of Thomas Friant, Grand Rapids, Mich.

Sister Dorothy 2607.—Sire on I. of J.; dam, Sister 1427.

> Butter, 15 lbs.

Test made from Sept. 19 to 25, 1881; age, 10 yrs. 9 mos.; property of G. Dawson Coleman, Brickerville, Pa.

Sister of Charity 62453.—Sire, Stoke Pogis 5th 5987; dam, Nell Pogis 31991.

> Butter, 16 lbs. 5 oz. Milk, 174 lbs. 8 oz.

Test made from June 5 to 12, 1890; age, 2 yrs. 7 mos.; estimated weight, 750 lbs.; no record of feed kept—ordinary rations; property of Miller & Sibley, Franklin, Pa.

Sister of Charity 62453.—Sire, Stoke Pogis 5th 5987; dam, Nell Pogis 31991.

> Butter, 24 lbs. 14½ oz. Milk, 215 lbs. 5 oz.

Test made from Jan. 21 to 28, 1892; age, 4 yrs. 2 mos.; estimated weight, 800 lbs.; fed 14 lbs. ensilage, 3 gals. bran, 1½ gals. corn, oats and peas ground together, and 2 lbs. cotton-seed meal, daily—pastured on green oats one hour per day; property of M. Lothrop, Marshall, Tex.

Sister Rex 13194.—Sire, Cash Boy 2248; dam, Miss Seelock 6614.

> Butter, 16 lbs. 8 oz. Milk, 201 lbs.

Test made from June 22 to 28, 1885; age, 4 yrs. 9 mos.; property of John Boyd, Elmhurst, Ill.

Sister Sue 58447.—Sire, Romp's Tormentor 7126; dam, Stately Susan 32396.

> Butter, 19 lbs. 9 oz. Milk, 335 lbs.

Test made from Aug. 19 to 26, 1894; age, 6 yrs. 11 mos.; estimated weight, 850 lbs.; fed 6 lbs. corn meal and 8 lbs. bran, daily—green corn *ad lib.* morning and night, good mixed pasture; property of Mrs. J. R. Sherman, Pawlet, Vt.

Sitanda's Rose 64958.—Sire, Sitanda 20221; dam, Silver Rose 2d 18337.

> Butter, 24 lbs. 12 oz. Milk, 301 lbs.

Test made from May 11 to 18, 1895; age, 6 yrs. 2 mos.; estimated weight, 1000 lbs.; fed 4 qts. corn meal and 6 qts. bran, daily—blue-grass pasture; property of Boyce & Smell, Muncie, Ind.

Sitka M. 78680.—Sire, Fancy's Harry 9777; dam, Tenella Alexiana 26971.

> Butter, 19 lbs. 13 oz. Milk, 211 lbs. 11 oz.

Test made from Apr. 21 to 28, 1895; age, 3 yrs. 11 mos.; estimated weight, 850 lbs.; fed 14 qts. daily of oats and corn, ground together in equal parts—blue-grass and orchard pasture; property of T. C. Lipscomb, Shelbyville, Tenn.

Sitka's Star 104707.—Sire, Toltec of Bedford 34302; dam, Sitka M. 78680.

> Butter, 14 lbs. 9 oz. Milk, 208 lbs. 12¼ oz.

Test made from Aug. 8 to 15, 1897; age, 2 yrs. 6 mos.; estimated weight, 675 lbs.; fed 20 lbs. per day of ground oats and corn, in equal parts, mixed with 4 bucketfuls of cut sheaf oats, wetted—orchard and blue-grass pasture for 4 hours daily; property of Matt. M. Gardner, Nashville, Tenn.

Skipover 20217.—Sire, Pedro 3187; dam, Sylvie of St. Mary's 19469.

Butter, 16 lbs. 5½ oz. Milk, 17 qts. per day.

Test made from July 18 to 24, 1886; age, 3 yrs. 5 mos.; property of D. F. Appleton, Ipswich, Mass.

Smack 64575.—Sire, Diploma 16219; dam, Smax 48562.

Butter, 15 lbs. 12 oz. Milk, 229 lbs. 1 oz.

Test made from Nov. 1 to 8, 1894; age, 3 yrs. 5 mos.; estimated weight, 800 lbs.; fed 12 qts. of a mixture of pea meal, oil meal, corn meal and bran, daily—cut hay; property of Richardson Bros., Davenport, Iowa.

Smax 48562.—Sire, Combination 3d 17576; dam, Comwa 35911.

Butter, 16 lbs. 11½ oz. Milk, 243 lbs. 10 oz.

Test made from June 10 to 17, 1890; age, 3 yrs. 1 mo.; estimated weight, 800 lbs.; fed 15 qts. ground oats, corn and oil meal, daily; property of Richardson Bros., Davenport, Iowa.

Smoky 13733.—Sire, Mercury 432; dam, Colie 8309.

Butter, 14 lbs. 9 oz. Milk, 198 lbs. 12 oz.

Test made from June 30 to July 6, 1884; age, 2 yrs. 10 mos.; property of William Simpson, New York, N. Y.

Snap's Dainty 18958.—Sire, Snap (F. S. 301 J. H. B.); dam, Dainty (P. S. 404 J. H. B.)

Butter, 14 lbs. Milk, 203 lbs.

Test made from Aug. 19 to 26, 1886; age, 5 yrs. 5 mos.; fed, daily, 4 qts. corn meal, 8 qts. ground oats, 5 qts. shorts and 1 qt. oil meal; property of W. H. Haley, North Wilmington, Mass.

Snow Bank 75184.—Sire, Ida's Rioter of St. L. 13656; dam, Rho A. Pogis 39269.

Butter, 16 lbs. 7 oz. Milk, 198 lbs.

Test made from Aug. 6 to 13, 1893; age, 2 yrs. 11 mos.; actual weight, 900 lbs.; fed 56 qts. corn meal, 56 qts. ground oats and 28 qts. ground oil-cake; property of Miller & Sibley, Franklin, Pa.

Snowdrop F. W. 16948.—Sire, Duke F. 6134; dam, Chloe of Millwood 14083.

Butter, 14 lbs. 8 oz., unsalted. Milk, 168 lbs.

Test made from June 10 to 17, 1883; age, 8 yrs. 11 mos.; property of J. H. Walker, Worcester, Mass.

Sobriquet 68028.—Sire, Iowa's Pogis Chief 24060; dam, Connoisseur 65644.

Butter, 21 lbs. 3 oz. Milk, 278 lbs. 4 oz.

Test made from Nov. 8 to 15, 1897; age, 6 yrs. 11 mos.; estimated weight, 950 lbs.; fed 35 lbs. corn meal, 17½ lbs. oil meal, 35 lbs. bran and 35 lbs. oats—corn ensilage, frosted clover pasture; property of H. C. Taylor, Orfordville, Wis.

Soconee 76895.—Sire, Ida's Stoke Pogis 13658; dam, Harry's Duchess 60289.

Butter, 40 lbs. 1¾ oz. in 14 days. Milk, 466 lbs. ¼ oz.

Test made from Mar. 16 to 30, 1898; age, 6 yrs. 2 mos.; estimated weight, 800 lbs.; fed 19 lbs. ground oats and corn and 5 lbs. bran per day, mixed with 4 bucketfuls cut sheaf oats, wetted—millet hay *ad lib.*; property of Matt. M. Gardner, Nashville, Tenn.

Soconee 2d 114420.—Sire, Tormentor 5th 21962; dam, Soconee 76895.

Butter, 15 lbs. Milk, 159 lbs.

Test made from Dec. 30, 1896, to Jan. 6, 1897; age, 2 yrs. 1 mo.; estimated weight, 675 lbs.; fed 70 lbs. cotton-seed hulls, 22¾ lbs. cotton-seed meal, 17½ lbs. linseed meal, 21 lbs. bran, 31½ lbs. ground oats, 50¾ lbs. corn and cob meal and 35 lbs. hay —turf oats pasture; property of S. H. Butler, Como, Miss.

Soconee 3d 114422.—Sire, Oonan's Tormentor 22280; dam, Soconee 76895.

Butter, 20 lbs. 1½ oz. Milk, 233 lbs. 15 oz.

Test made from Feb. 9 to 16, 1898; age, 2 yrs. 3 mos.; estimated weight, 750 lbs.; fed 45½ qts. corn and 45½ qts. oats—hay *ad lib.*, orchard grass pasture; property of J. E. Hart, Nashville, Tenn.

Sœurette 52256.—Sire, Southern Prince 10760; dam, Tette 20802.

Butter, 17 lbs. 4¾ oz. Milk, 216 lbs.

Test made from Mar. 28 to Apr. 4, 1893; age, 5 yrs. 1 mo.; estimated weight, 850 lbs.; fed 24 lbs. daily of a mixture of corn feed, meal and wheat bran—short blue-grass pasture; property of Campbell Brown, Spring Hill, Tenn.

Sœurette 52256.—Sire, Southern Prince 10760; dam, Tette 20802.

Butter, 97 lbs. ½ oz. in 30 days. Milk, 1023 lbs. 15¾ oz.

Test made from Nov. 6 to Dec. 6, 1895; age, 7 yrs. 9 mos.; actual weight, 850 lbs.; fed 22 lbs. per day of ground oats and corn, in equal parts, mixed with 4 bucketfuls of cut sheaf oats, wetted—clover hay *ad lib.*; property of Matt. M. Gardner, Nashville, Tenn.

Sœurette 2d 86414.—Sire, Odello 20223; dam, Sœurette 52256.

Butter, 18 lbs. 1½ oz. Milk, 217 lbs. 13 oz.

Test made from June 23 to 30, 1896; age, 3 yrs. 5 mos.; estimated weight, 800 lbs.; fed 22 lbs. per day of ground corn and oats, in equal parts—good orchard and blue-grass pasture; property of Matt. M. Gardner, Nashville, Tenn.

Solava 20928.—Sire, Signalda 4027; dam, Odelle Sales 15564.

Butter, 15 lbs. 3⅝ oz. Milk, 210 lbs. 8 oz.

Test made from Oct. 28 to Nov. 4, 1887; age, 4 yrs. 5 mos.; estimated weight, 975 lbs.; fed, daily, 24 lbs. corn chop, wheat bran and cotton-seed meal, mixed; property of Campbell Brown, Spring Hill, Tenn.

Sombre 80796.—Sire, Diploma 16219; dam, Alexa 64924.

Butter, 15 lbs. 14 oz. Milk, 232 lbs. 11 oz.

Test made from Sept. 15 to 22, 1894; age, 3 yrs.; estimated weight, 800 lbs.; fed 15 qts. daily of a mixture of corn meal, ground oats, bran and oil meal—good pasture; property of Richardson Bros., Davenport, Iowa.

Sombre 80796.—Sire, Diploma 16219; dam, Alexa 64924.

Butter, 18 lbs. 14 oz. Milk, 282 lbs. 5 oz.

Test made from Oct. 10 to 17, 1895; age, 4 yrs. 1 mo.; estimated weight, 800 lbs.; fed 5 lbs. corn meal, 6 lbs. oat meal, 4 lbs. bran and 1½ lbs. oil meal, daily—some roots; property of Richardson Bros., Davenport, Iowa.

Son's Princess 118386.—Sire, Ida's Rioter of St. L.'s Son 28869; dam, Myrtle Melrose 114464.

Butter, 16 lbs. 15 oz. Milk, 231 lbs. 8 oz.

Test made from Jan. 24 to 31, 1898; age, 3 yrs. 5 mos.; estimated weight, 850 lbs.; fed 28 qts. corn meal, 28 qts. bran, 7 qts. cotton-seed meal, 3½ qts. linseed meal and 14 qts. whole cotton-seed—little alfalfa; property of M. Lothrop, Marshall, Tex.

Sophie 2d of Hood Farm 131236.—Sire, Sophie's Tormentor 20883; dam, Gipsy of Avon 79747.

Butter, 15 lbs. 8 oz. Milk, 262 lbs. 15 oz.

Test made from Mar. 21 to 28, 1898; age, 3 yrs. 3 mos.; estimated weight, 750 lbs.; fed 35 lbs. bran, 26½ lbs. corn meal, 16½ lbs. ground oats, 12½ lbs. oil meal and 70 lbs. roots—hay *ad lib.*; property of C. I. Hood, Lowell, Mass.

Sophie Hudson 76105.—Sire, Sophie's Tormentor 20883; dam, Rhoda Hudson 48723.

Butter, 19 lbs. 12½ oz. Milk, 335 lbs. 2 oz.

Test made from June 6 to 13, 1898; age, 7 yrs. 2 mos.; estimated weight, 800 lbs.; fed 28 lbs. bran, 28 lbs. corn meal, 21 lbs. ground oats, 7 lbs. oil meal and 7 lbs. cotton-seed meal—clover and timothy pasture; property of C. I. Hood, Lowell, Mass.

Sophie Hudson 76105.—Sire, Sophie's Tormentor 20883; dam, Rhoda Hudson 48723.

Butter, 28 lbs. ½ oz. in 10 days. Milk, 468 lbs. 2 oz.

Test made from June 6 to 16, 1898; age, 7 yrs. 2 mos.; estimated weight, 800 lbs.; fed 40 lbs. bran, 41½ lbs. corn, 30 lbs. ground oats, 10 lbs. oil meal and 10 lbs. cotton-seed meal—mixed clover and timothy pasture; property of C. I. Hood, Lowell, Mass.

Sophona 76110.—Sire, Sophie's Tormentor 20883; dam, Little Betsona 45344.

Butter, 16 lbs. ¾ oz. Milk, 247 lbs. 13 oz.

Test made from Oct. 18 to 25, 1897; age, 6 yrs. 2 mos.; actual weight, 664 lbs.; fed 28 lbs. bran, 7 lbs. corn meal, 21 lbs. ground oats and 7 lbs. oil meal—hay *ad lib.*, short clover pasture; property of C. I. Hood, Lowell, Mass.

Sophona 76110.—Sire, Sophie's Tormentor 20883; dam, Little Betsona 45344.

Butter, 18 lbs. 5¾ oz. Milk, 242 lbs. 4 oz.

Test made from May 17 to 24, 1898; age, 6 yrs. 9 mos.; estimated weight, 700 lbs.; fed 21 lbs. bran, 28 lbs. corn meal, 14 lbs. ground oats, 7 lbs. oil meal and 7 lbs. cotton-seed meal—clover and timothy pasture; property of C. I. Hood, Lowell, Mass.

Sophona 76110.—Sire, Sophie's Tormentor 20883; dam, Little Betsona 45344.

Butter, 22 lbs. 4¾ oz. in 10 days. Milk, 354 lbs. 12 oz.

Test made from Oct. 14 to 25, 1897; age, 6 yrs. 2 mos.; actual weight, 664 lbs.; fed 40 lbs. bran, 7 lbs. corn meal, 30 lbs. ground oats and 10 lbs. oil meal—hay *ad lib.*, short clover pasture; property of C. I. Hood, Lowell, Mass.

Sordella 76876.—Sire, Odelio 20223; dam, Sorina 42993.

Butter, 15 lbs. 9 oz. Milk, 233 lbs. 8 oz.

Test made from May 13 to 20, 1894; age, 3 yrs.; estimated weight, 800 lbs.; fed 140 lbs. coarsely-ground corn and 28 lbs. cut sheaf oats—blue-grass pasture; property of Est. of Campbell Brown, Spring Hill, Tenn.

Sorina 42993.—Sire, Fancy's Harry 9777; dam, Sirona 2d 38011.

Butter, 15 lbs. 11 oz.   Milk, 217 lbs. 8 oz.

Test made from June 2 to 9, 1891; age, 4 yrs. 6 mos.; estimated weight, 850 lbs.; fed 16 lbs. corn-hearts and 4 lbs. bran, mixed, daily—blue-grass pasture; property of Campbell Brown, Spring Hill, Tenn.

Sorina 42993.—Sire, Fancy's Harry 9777; dam, Sirona 2d 38011.

Butter, 17 lbs. 15½ oz.   Milk, 238 lbs.

Test made from May 7 to 14, 1894; age, 7 yrs. 6 mos.; estimated weight, 900 lbs.; fed 168 lbs. coarsely-ground corn chop or corn meal and 28 lbs. cut sheaf oats— blue-grass pasture; property of Est. of Campbell Brown, Spring Hill, Tenn.

Sosia 19198.—Sire, Mapleson 5951; dam, Glint 12896.

Butter, 17 lbs. 5½ oz.   Milk, 236 lbs. 11 oz.

Test made from Feb. 28 to Mar. 7, 1895; age, 12 yrs.; estimated weight, 1000 lbs.; fed 35 lbs. gluten meal, 35 lbs. bran, 35 lbs. cracked oats, 21 lbs. linseed meal and 14 lbs. corn meal—hay *ad lib.*; property of C. Delano, Mount Vernon, Ohio.

Southern Belle 18570.—Sire, Lord Longford 3997; dam, Romp Ogden 2d 4764.

Butter, 16 lbs.   Milk, 270 lbs.

Test made from May 1 to 7, 1885; age, 4 yrs. 11 mos.; property of William E. Oates, Vicksburg, Miss.

Southern Bell of C. H. 93565.—Sire, Ida's Rioter of St. L.'s Son 28869; dam, Melrose Milly 61917.

Butter, 14 lbs. 7½ oz.   Milk, 220 lbs. 14 oz.

Test made from Apr. 30 to May 5, 1897; age, 4 yrs. 3 mos.; estimated weight, 650 lbs.; fed 28 qts. corn meal, 28 qts. bran, 7 qts. cotton-seed meal, 3½ qts. linseed meal and 14 qts. whole cotton-seed—little alfalfa; property of M. Lothrop, Marshall, Tex.

Southern Daisy 38292.—Sire, Southern Prince 10760; dam, Myrrha 11209.

Butter, 14 lbs. 11 oz.   Milk, 108 lbs. 8 oz.

Test made from May 27 to June 3, 1888; age, 2 yrs. 3 mos.; estimated weight, 750 lbs.; fed 16 qts. oats and corn, daily; property of Morgan & Brown, Columbia, Tenn.

Southern Pet 26675.—Sire, Alpheus of Springvale 5645; dam, Lady Alice of Hillcrest 7450.

Butter, 18 lbs. 8 oz.   Milk, 238 lbs. 10 oz.

Test made from Sept. 4 to 11, 1888; age, 4 yrs. 8 mos.; estimated weight, 950 lbs.; fed 84 qts. corn meal, 42 qts. cotton-seed meal, 42 qts. bran and shorts and 2 pks. chopped pumpkins—hay, fair pasture; property of James Crook, Jacksonville, Ala.

Southern Princess 36320.—Sire, Southern Prince 10760; dam, Lorelle 12913.

Butter, 16 lbs. 4 oz.   Milk, 205 lbs.

Test made from Feb. 12 to 19, 1894; age, 8 yrs. 2 mos.; estimated weight, 800 lbs.; fed 4 qts. corn-hearts, 6 qts. ground oats, 2 qts. cotton-seed meal and 2 qts. bran, daily—clover hay and corn stover, poor pasture; property of Samuel N. Warren, Spring Hill, Tenn.

Sovereign's Elsie 15793.—Sire, Sovereign (F. S. 307 J. H. B.); dam, Salacia (F. S. 2197 J. H. B.)

Butter, 14 lbs. 3 oz.

Test made from July 1 to 7, 1884; age, 3 yrs. 5 mos.; property of James B. Wilder, Louisville, Ky.

Spangled Princess 117978.—Sire, Julia's Prince 17938; dam, Richland Pride 117977.

Butter, 15 lbs. 5 oz. Milk, 197 lbs. 10 oz.

Test made from July 26 to Aug. 2, 1897; age, 2 yrs. 7 mos.; estimated weight, 675 lbs.; fed 19 lbs. per day of ground oats and corn—orchard and blue-grass pasture; property of Matt. M. Gardner, Nashville, Tenn.

Spark 62689.—Sire, John Rioter 20695; dam, Daretta 62579.

Butter, 20 lbs. 3 oz. Milk, 231 lbs. 8 oz.

Test made from Feb. 7 to 14, 1895; age, 5 yrs. 4 mos.; estimated weight, 950 lbs.; fed 98 lbs. corn and oats, ground together in equal parts by bulk, 35 lbs. bran, 21 lbs. oil meal and 40 lbs. corn ensilage—little clover hay; property of H. C. Taylor, Orfordville, Wis.

Sparkle 56775.—Sire, Concert 16499; dam, Sparks 41041.

Butter, 17 lbs. 3 oz. Milk, 246 lbs. 4 oz.

Test made from Oct. 11 to 18, 1892; age, 3 yrs. 8 mos.; estimated weight, 800 lbs.; fed 12 qts. daily of ground oats, corn and oil meal—plenty of hay, pasture; property of Richardson Bros., Davenport, Iowa.

Sparkling 41042.—Sire, Citizen 13186; dam, Sparks 41041.

Butter, 14 lbs. 1 oz. Milk, 225 lbs. 4 oz.

Test made from Mar. 2 to 9, 1889; age, 3 yrs. 5 mos.; estimated weight, 825 lbs.; fed, daily, 12 qts. meal and 6 qts. bran; property of Richardson Bros., Davenport, Iowa.

Sparks 41041.—Sire, Combination 4389; dam, Romp Lawrence 13819.

Butter, 16 lbs. 4½ oz. Milk, 236 lbs. 8 oz.

Test made from Nov. 21 to 28, 1886; age, 3 yrs.; estimated weight, 900 lbs.; fed, daily, about 8 qts. of ground corn and oats; property of Richardson Bros., Davenport, Iowa.

Sparks' Maid 45575.—Sire, Sparks Dee 8365; dam, Chemung Maid 34704.

Butter, 16 lbs. 6 oz. Milk, 279 lbs. 8 oz.

Test made from June 14 to 21, 1895; age, 8 yrs. 4 mos.; estimated weight, 925 lbs.; fed 58 lbs. corn meal and 43 lbs. wheat bran—timothy and clover pasture; property of Henry S. Redfield, Elmira, N. Y.

Sparks' Maid 45575.—Sire, Sparks Dee 8365; dam, Chemung Maid 34704.

Butter, 19 lbs. 8 oz. Milk, 345 lbs. 11 oz.

Test made from Apr. 6 to 13, 1898; age, 11 yrs. 2 mos.; actual weight, 1020 lbs.; fed 6 qts. bran, 6 qts. middlings, 4½ qts. corn meal, 6 qts. ground oats and 3 qts. oil meal, mixed, wet, daily—ensilage and hay; property of C. A. Sweet, Buffalo, N. Y.

Sparta 75000.—Sire, Herotas 26500; dam, Spark 62689.

Butter, 21 lbs. 5 oz. Milk, 301 lbs. 7 oz.

Test made from June 3 to 10, 1896; age, 4 yrs. 8 mos.; estimated weight, 925 lbs.; fed 35 lbs. ground oats, 35 lbs. bran, 49 lbs. corn meal and 21 lbs. old process oil meal—mixed pasture; property of H. C. Taylor, Orfordville, Wis.

Spiræa 4th 20075.—Sire, Lenox Cash Boy 6804; dam, Spiræa 3915.

Butter, 18 lbs. 5½ oz. Milk, 187 lbs.

Test made from Nov. 27 to Dec. 3, 1885; age, 2 yrs. 7 mos.; property of Campbell Brown, Spring Hill, Tenn.

Spot Victor Pogis 85901.—Sire, Melia Ann's Victor Pogis 20697; dam, Carrie Pedrolier 48966.

Butter, 17 lbs. 3 oz. Milk, 226 lbs.

Test made from June 10 to 17, 1893; age, 2 yrs. 3 mos.; estimated weight, 650 lbs.; fed 49 lbs. grain—pasture of white clover and mixed grasses; property of Mrs. J. H. Martin, Granville, N. Y.

Stanstead Belle 4709.—Sire, Orleans 533; dam, Beauty 1319.

Butter, 15 lbs. 4 oz. Milk, 220 lbs. 4 oz.

Test made from July 7 to 13, 1885; age, 10 yrs. 1 mo.; property of J. C. Hatheway, Lancaster, N. B., Can.

Starkville Beauty 4897.—Sire, Aldine 1136; dam, Dolly Berry 2004.

Butter, 14 lbs., unsalted. Milk, 185 lbs.

Test made from July 10 to 16, 1883; age, 7 yrs. 5 mos.; property of W. B. Montgomery, Starkville, Miss.

Star of Edgewood 74823.—Sire, King Coomassie 2d 19545; dam, Pride of Edgewood 64199.

Butter, 16 lbs. 10 oz. Milk, 213 lbs. 1 oz.

Test made from Aug. 12 to 19, 1894; age, 4 yrs. 3 mos.; actual weight, 825 lbs.; fed 42 qts. corn meal and 84 qts. wheat bran—good clover and blue-grass pasture; property of Columbus Dixon, Gillespieville, Ohio.

Star of Mahoning 16579.—Sire, Lewis' Ringgold 6379; dam, Silver Queen of Mahoning 13839.

Butter, 14 lbs. 7 oz. Milk, 158 lbs. 8 oz.

Test made from June 2 to 8, 1885; age, 3 yrs.; property of A. R. Hill, North Washington, Pa.

Star's Valentine 81778.—Sire, Ruby's Star of St. Lambert 20719; dam, Rubano's Valentine 54703.

Butter, 16 lbs. 2 oz. Milk, 277 lbs. 7 oz.

Test made from Aug. 8 to 15, 1893; age, 2 yrs. 6 mos.; estimated weight, 500 lbs.; fed 42 lbs. corn meal, 21 lbs. linseed meal, 35 lbs. bran and 15 lbs. ensilage; property of Samuel McKeen, Terre Haute, Ind.

Startling 94942.—Sire, Herotas 26500; dam, Sparkle 56775.

Butter, 17 lbs. 14 oz. Milk, 233 lbs. 14 oz.

Test made from Oct. 11 to 18, 1896; age, 4 yrs. 4 mos.; estimated weight, 1000 lbs.; fed 49 lbs. bran, 28 lbs. oats, 28 lbs. corn meal, 21 lbs. oil meal and 200 lbs. pumpkins—corn stover; property of H. C. Taylor, Orfordville, Wis.

Stella of Edgewood 114473.—Sire, King Phil 33443; dam, Isa of Edgewood 96452.

Butter, 14 lbs. 12 oz. Milk, 191 lbs. 3 oz.

Test made from Nov. 23 to 30, 1896; age, 2 yrs. 6 mos.; actual weight, 750 lbs.; fed 42 qts. corn and cob meal and 56 qts. wheat bran—hay *ad lib.*, fair blue-grass pasture; property of Columbus Dixon, Gillespieville, Ohio.

Sterling Merit 62456.—Sire, Stoke Pogis 5th 5987; dam, Lady Appel 8612.

Butter, 15 lbs. 4 oz. Milk, 203 lbs.

Test made from Jan. 16 to 23, 1891; age, 2 yrs. 10 mos.; estimated weight, 800 lbs.; fed, daily, 32 lbs. ground oats and corn meal in equal parts; property of Miller & Sibley, Franklin, Pa.

Stoke Pogis' Regina 48309.—Sire, Stoke Pogis of Linden 10558; dam, Regina's Pride 24407.

Butter, 60.268 lbs. in 30 days. Milk, 1012.2 lbs.

Test made from Aug. 29 to Sept. 27, 1893, at World's Columbian Exposition; age, 8 yrs. 1 mo.; actual weight, 886 lbs.; fed 369.6 lbs. hay, 73 lbs. oil meal, 90 lbs. ensilage, 112 lbs. corn meal, 180 lbs. bran, 120 lbs. oats, 85 lbs. corn-hearts, 44 lbs. cotton-seed, 80 lbs. middlings, 9 lbs. carrots and 5 lbs. old hay; property of Est. of Frederick Billings, Woodstock, Vt.

Stoke's Tipsey 99380.—Sire, Stoke Pogis of Prospect 29121; dam, Tipsey Pogis 83492.

Butter, 14 lbs. 7½ oz. Milk, 214 lbs.

Test made from June 10 to 17, 1897; age, 3 yrs. 1 mo.; actual weight, 830 lbs.; fed 9 qts. bran, 9 qts. ground oats, 4½ qts. corn meal and 3 qts. oil meal, daily—plenty of good grass in barn; property of George H. Sweet, East Aurora, N. Y.

Stolen Kisses 16864, imp.

Butter, 16 lbs. 14 oz. Milk, 236 lbs. 6 oz.

Test made from Mar. 11 to 17, 1886; age, 5 yrs. 10 mos.; property of E. S. Henry, Rockville, Conn.

Storm-Beaten of Lawn 61375.—Sire, Lord Hugo 15830; dam, Storm-Beaten 36269.

Butter, 18 lbs. 3 oz. Milk, 254 lbs. 14 oz.

Test made from Jan. 21 to 28, 1895; age, 5 yrs. 3 mos.; estimated weight, 900 lbs.; fed 56 lbs. wheat bran, 56 lbs. ground oats, 28 lbs. corn meal, 14 lbs. cotton-seed meal and 84 lbs. hay; property of Platter & Foster, Denison, Tex.

Success of St. Lambert 28489.—Sire, Sir George of St. Lambert 6036; dam, Success 8782.

Butter, 16 lbs. 2 oz. Milk, 238 lbs. 4 oz.

Test made from May 6 to 13, 1889; age, 4 yrs. 8 mos.; estimated weight, 1000 lbs.; fed, daily, 9 qts. ground corn and oats, 6 qts. bran, 3 qts. middlings and 3 qts. oil meal; property of Peter D. Hulst, East Penfield, N. Y.

Successor 72503.—Sire, Odelio 20223; dam, Signalda Bloomfield 30694.

Butter, 17 lbs. 4 oz. Milk, 208 lbs. 12 oz.

Test made from Mar. 3 to 10, 1895; age, 4 yrs. 8 mos.; estimated weight, 900 lbs.; fed 6 lbs. bran, 6 lbs. middlings, 2 lbs. linseed meal and 2 lbs. cotton-seed meal, daily—40 lbs. ensilage per day, hay *ad lib.*; property of George W. Sisson, Jr., Potsdam, N. Y.

Sudie St. Helier 40235.—Sire, Yakout 6842; dam, Sudie Mac 24202.

Butter, 18 lbs. 9 oz.   Milk, 211 lbs. 8 oz.

Test made from Feb. 20 to 27, 1893; age, 6 yrs. 11 mos.; actual weight, 794 lbs.; fed 12 qts. chop and 6 qts. wheat bran, daily—clover hay; property of Mrs. A. M. Hallock, Columbus, Ohio.

Sue Gallagher 15945.—Sire, Meadow King of Farmington 4911; dam, Mabel Cowles 3879.

Butter, 23 lbs. 1½ oz.   Milk, 188 lbs. 11 oz.

Test made from July 26 to Aug. 1, 1885; age, 4 yrs. 3 mos.; fed 10 qts. corn meal, 5 qts. oil-cake, 5 qts. wheat bran, 10 qts. ground oats, mixed, daily—good blue-grass pasture; property of James S. Stoll, Lexington, Ky.

Sulphide 88038.—Sire, Prince of Mahaska 16159; dam, Crusta 29637.

Butter, 18 lbs. 5 oz.   Milk, 271 lbs. 6 oz.

Test made from June 1 to 8, 1895; age, 6 yrs. 3 mos.; estimated weight, 800 lbs.; fed 6 lbs. corn meal, 8 lbs. oat meal, 6 lbs. bran and 1 lb. pea meal, daily—clover hay, access to pasture; property of Richardson Bros., Davenport, Iowa.

Sultana 2d 11798.—Sire, Son of Rosa 663; dam, Sultana 403.

Butter, 15 lbs. 4 oz., unsalted.   Milk, 285 lbs. 4 oz.

Test made from June 12 to 18, 1884; age, 5 yrs. 9 mos.; property of Moulton Bros., West Randolph, Vt.

Sultan's Sultane 32854.—Sire, Sultan of St. Saviour's 5328; dam, Sultane Americaine 11874.

Butter, 15 lbs. 1 oz.   Milk, 186 lbs. 10 oz.

Test made from May 29 to June 5, 1888; age, 3 yrs. 2 mos.; estimated weight, 900 lbs.; fed, daily, 3 qts. oat meal and 3 qts. bran; property of M. H. Messchert, Douglassville, Pa.

Su Lu 4705.—Sire, Rioter 670; dam, Angela 1682.

Butter, 17 lbs. 15 oz. (official).   Milk, 241 lbs. 11 oz.

Test made from June 6 to 12, 1883; age, 7 yrs. 8 mos.; fed 7 lbs. chopped oats, 6 lbs. corn meal, 2 lbs. bean meal, 2 lbs. cotton-seed meal and 2 lbs. cut hay, mixed and dampened, daily; property of Campbell Brown, Spring Hill, Tenn.

Summerline 8001.—Sire, Bristol Chief 1476; dam, Salsoda 3721.

Butter, 18 lbs. 6 oz., unsalted.   Milk, 264 lbs. 6 oz.

Test made from Aug. 7 to 14, 1883; age, 5 yrs. 5 mos.; property of C. W. H. Eicke, West Monterey, Pa.

Summer Morning 75197.—Sire, June Pogis 19872; dam, Blanche of Castile 43793.

Butter, 16 lbs. 15 oz.   Milk, 162 lbs. 8 oz.

Test made from Dec. 10 to 17, 1894; age, 3 yrs. 7 mos.; estimated weight, 850 lbs.; fed 6 qts. corn meal, 12 qts. wheat bran, 12 qts. ground oats and 3 qts. oil meal, daily, with beets; property of C. A. Sweet, East Aurora, N. Y.

Summer Morning 75197.—Sire, June Pogis 19872; dam, Blanche of Castile 43793.

Butter, 20 lbs. ½ oz.   Milk, 253 lbs. 4 oz.

Test made from Aug. 25 to Sept. 1, 1895; age, 4 yrs. 4 mos.; actual weight, 1030 lbs.; fed 12 qts. ground oats, 3 qts. corn meal, 12 qts. bran and 3 qts. oil meal, daily—poor pasture, green oats and peas; property of C. A. Sweet, East Aurora, N. Y.

Sunbeam's Pet 2d 9112.—Sire, Lucifer 2696; dam, Sunbeam's Pet 9111.

Butter, 16 lbs. 4 oz. Milk, 312 lbs.

Test made from June 23 to 29, 1884; age, 5 yrs. 2 mos.; test made on grass alone; property of O. A. Gilman, Paris, Ky.

Sünderin 125092.—Sire, Prince of Tennessee 20772; dam, Lucile des Moulins 92726.

Butter, 15 lbs. ½ oz. Milk, 191 lbs. 10½ oz.

Test made from May 17 to 24, 1898; age, 2 yrs. 6 mos.; actual weight, 550 lbs.; fed 20 lbs. per day of ground corn and oats, in equal parts—very good orchard and blue-grass pasture; property of Thomas H. Malone, Nashville, Tenn.

Sunny Lass 6063.—Sire, Balboa 1244; dam, Twilight Lass 2698.

Butter, 14 lbs. 7 oz., unsalted. Milk, 192 lbs. 12½ oz.

Test made from Jan. 13 to 19, 1882; age, 4 yrs. 9 mos.; property of D. A. Givens, Cynthiana, Ky.

Sunny of Cedar Hill 90410.—Sire, Ida's Rioter Stoke Pogis 26496; dam, Sunny Pogis 61919.

Butter, 19 lbs. 5 oz. Milk, 298 lbs.

Test made from Jan. 11 to 18, 1898; age, 5 yrs. 11 mos.; actual weight, 858 lbs.; fed 28 lbs. corn and cob meal, 28 lbs. ground oats, 56 lbs. bran, 14 lbs. cotton-seed meal, 84 lbs. prairie hay, 28 lbs. corn fodder and 14 lbs. pea hay—pasture of winter oats; property of J. R. Miller, Minden, La.

Sunny Pogis 61919.—Sire, Prince of Melrose 4819; dam, Sunny Girl 26276.

Butter, 15 lbs. 3½ oz. Milk, 249 lbs. 2 oz.

Test made from Apr. 29 to May 6, 1893; age, 5 yrs. 3 mos.; estimated weight, 700 lbs.; fed 14 gals. corn meal, 17½ gals. bran, 3½ gals. cob meal and 3½ gals. cotton-seed meal—Bermuda grass pasture; property of M. Lothrop, Marshall, Tex.

Sunny Pogis 61919.—Sire, Prince of Melrose 4819; dam, Sunny Girl 26276.

Butter, 17 lbs. 9½ oz. Milk, 281 lbs. 12 oz.

Test made from Mar. 11 to 18, 1897; age, 9 yrs. 1 mo.; estimated weight, 800 lbs.; fed 35 lbs. bran, 70 lbs. corn meal, 14 lbs. linseed meal and 28 lbs. whole cotton-seed, boiled—Bermuda grass and Japanese clover pasture; property of M. Lothrop, Marshall, Tex.

Sunset 15130.—Sire, Prince of Woodstock 5030; dam, Noonday 15127.

Butter, 16 lbs. 2½ oz. Milk, 210 lbs.

Test made from Aug. 12 to 19, 1885; age, 3 yrs. 5 mos.; property of B. F. Rogers, Versailles, Ky.

Sunset of Pleasant View 13071.—Sire, Elsmore 4384; dam, Plenita 6143.

Butter, 15 lbs. 2 oz. Milk, 235 lbs.

Test made from Dec. 17 to 23, 1884; age, 5 yrs. 7 mos.; property of J. L. Shallcross, Anchorage, Ky.

Sunswick Dolly 91844.—Sire, Fritz Pogis of Glynllyn 28042; dam, Tavie 2d 40253.

Butter, 15 lbs. 12 oz. Milk, 246 lbs. 2 oz.

Test made from Apr. 18 to 25, 1897; age, 4 yrs.; estimated weight, 700 lbs.; fed 7 qts. oil meal, 42 qts. corn meal, 63 qts. bran, 63 qts. ground oats and 56 lbs. roots—hay; property of D. W. Bishop, Jr., Lenox, Mass.

Surprise Melrose 93562.—Sire, Ida's Rioter of St. L.'s Son 28869; dam, Surprised Princess 40349.

Butter, 21 lbs. 14⅝ oz.   Milk, 325 lbs. 3 oz.

Test made from Mar. 26 to Apr. 2, 1897; age, 4 yrs. 3 mos.; estimated weight, 850 lbs.; fed 4 qts. cotton-seed, 4 qts. ground corn, 4 qts. ground oats and 12 qts. wheat bran, daily—hay *ad lib.*, oats pasture; property of Mrs. E. M. Mirick, Cleburne, Tex.

Susa Drew 38521.—Sire, Cattaraugus Chief 6026; dam, Signal Pomona 28308.

Butter, 21 lbs. 1 oz.   Milk, 262 lbs. 5 oz.

Test made from June 4 to 11, 1892; age, 7 yrs. 2 mos.; estimated weight, 950 lbs.; fed 28 qts. bran, 28 qts. yellow corn meal and 7 lbs. linseed meal—cut corn fodder and pasture; property of Giltedge Farm Co., Syracuse, N. Y.

Susan Bell 112146.—Sire, Odello 20223; dam, Mina B. 28947.

Butter, 14 lbs. 13 oz.   Milk, 225 lbs.

Test made from May 31 to June 7, 1898; age, 3 yrs. 1 mo.; actual weight, 790 lbs.; fed 8 lbs. corn and cob meal, 8 lbs. ground oats and 2 lbs. cotton-seed meal, daily—pea-vine hay and corn fodder *ad lib.*, Bermuda grass and oats stubble pasture; property of J. R. Miller, Minden, La.

Susanna Frank 18261.—Sire, Frank 309; dam, Buttercup 311.

Butter, 14 lbs.

Test made from Mar. 21 to 28, 1882; age, 12 yrs.; property of Frederick Barnard, Lombard, Ill.

Susa's Darling 44915.—Sire, Signal Duke of Alfred 11927; dam, Susa Drew 38521.

Butter, 15 lbs. 12 oz.   Milk, 225 lbs. 3 oz.

Test made from June 4 to 11, 1892; age, 5 yrs. 6 mos.; estimated weight, 900 lbs.; fed 28 qts. corn meal, 28 qts. bran and 7 lbs. linseed meal—good white clover and native grass pasture; property of Giltedge Farm Co., Syracuse, N. Y.

Susette C. 18602.—Sire, Delaware Darling 3461; dam, Lila Bailey 18601.

Butter, 14 lbs. 8 oz.   Milk, 140 lbs.

Test made from June 3 to 9, 1886; age, 3 yrs. 5 mos.; property of William Heyser, Chambersburg, Pa.

Susie B. of Springvale 24986.—Sire, Jersey King 9458; dam, Atlanta 2d 24011.

Butter, 17 lbs. 5 oz.   Milk, 227 lbs. 2 oz.

Test made from July 6 to 13, 1888; age, 4 yrs. 7 mos.; estimated weight, 800 lbs.; fed 42 qts. corn meal, 70 qts. shorts and 14 qts. bran—hay, mixed pasture; property of James Crook, Jacksonville, Ala.

Susie Pogis 35425.—Sire, Prince of Melrose 2d 11015; dam, Susie B. of Springvale 24986.

Butter, 16 lbs. 9 oz.   Milk, 245 lbs. 15 oz.

Test made from Mar. 2 to 9, 1890; age, 4 yrs. 6 mos.; estimated weight, 800 lbs.; fed 28 qts. corn meal, 56 qts. bran and shorts, 14 qts. cotton-seed meal and 210 lbs. ensilage—hay, winter rye pasture; property of James Crook, Jacksonville, Ala.

Susie V. 54855.—Sire, Charley Willis 14866; dam, Bertha's Star 24902.

Butter, 14 lbs. 14 oz. Milk, 194 lbs.

Test made from Oct. 1 to 8, 1890; age, 2 yrs. 11 mos.; estimated weight, 700 lbs.; fed, daily, 4 qts. bran and 4 qts. ground oats and corn; property of C. W. Paris, Mount Healthy, Ohio.

Süsschen 125091.—Sire, Prince of Tennessee 20772; dam, Petite Lamballe 92260.

Butter, 15 lbs. 6¾ oz. Milk, 178 lbs. 11¾ oz.

Test made from Mar. 18 to 25, 1898; age, 2 yrs. 7 mos.; estimated weight, 575 lbs.; fed 17 lbs. per day of ground oats and corn, in equal parts, mixed with 3 bucketfuls of cut sheaf oats, wetted—very good orchard and blue-grass pasture; property of Thomas H. Malone, Nashville, Tenn.

Sweet Blossom Pogis 36995.—Sire, Stoke Pogis 5th 5987; dam, Blossom of Prospect 18627.

Butter, 16 lbs. 1½ oz. Milk, 177 lbs.

Test made from Mar. 1 to 8, 1890; age, 4 yrs. 2 mos.; estimated weight, 950 lbs.; fed, daily, 36 lbs. grain, 21 lbs. sugar beets and 10 lbs. cooked hay; property of Miller & Sibley, Franklin, Pa.

Sweet Brier of St. Lambert 5481.—Sire, Lord Lisgar 1066; dam, Lily of St. Lambert 5120.

Butter, 14 lbs. 3 oz. Milk, 267 lbs.

Test made from June 6 to 12, 1883; age, 7 yrs. 1 mo.; property of David Reesor, Toronto, Ont., Can.

Sweet Brier of St. Lambert 5481.—Sire, Lord Lisgar 1066; dam, Lily of St. Lambert 5120.

Butter, 22 lbs. 10 oz. Milk, 266 lbs.

Test made from June 17 to 23, 1884; age, 8 yrs. 2 mos.; fed 16 lbs. ground oats and barley, 2 lbs. linseed oil-cake and 8 lbs. bran, mixed, daily—good pasture; property of W. S. Reesor, Toronto, Ont., Can.

Sweet Leona B. 21934.—Sire, Stoke Pogis 5th 5987; dam, Lady of Oakland 11101.

Butter, 14 lbs. 1½ oz. Milk, 248 lbs. 8 oz.

Test made from Aug. 1 to 7, 1887; age, almost 4 yrs.; estimated weight, 800 lbs.; fed 30 lbs. daily of mixed corn-hearts, oil meal, oats and middlings; property of Miller & Sibley, Franklin, Pa.

Sweet Lily of Meridale 94267.—Sire, Matilda 4th's Son 20214; dam, Coquette of Glen Rouge 17559.

Butter, 14 lbs. 1½ oz. Milk, 242 lbs.

Test made from Mar. 3 to 10, 1896; age, 3 yrs. 3 mos.; estimated weight, 850 lbs.; fed 14 gals. bran, 7 gals. corn and cob meal, 7 gals. oats, 2 gals. oil meal, 14 bush. cut corn stover, 7 gals. turnips and 7 gals. mangels—clover hay at noon; property of J. E. Gray & Son, Youngstown, Ohio.

Sweet Lily of Meridale 94267.—Sire, Matilda 4th's Son 20214; dam, Coquette of Glen Rouge 17559.

Butter, 14 lbs. 15 oz. Milk, 245 lbs. 12 oz.

Test made from Mar. 10 to 17, 1896; age, 3 yrs. 3 mos.; estimated weight, 850 lbs.; fed 14 gals. bran, 7 gals. corn and cob meal, 7 gals. oats, 2 gals. oil meal, 14 bush. cut corn stover, 7 gals. turnips and 7 gals. mangels—clover hay at noon; property of J. E. Gray & Son, Youngstown, Ohio.

Sweet Lily of Meridale 94267.—Sire, Matilda 4th's Son 20214; dam, Coquette of Glen Rouge 17559.

Butter, 15 lbs. 8½ oz. Milk, 253 lbs. 4 oz.

Test made from Dec. 19 to 26, 1897; age, 5 yrs.; estimated weight, 900 lbs.; fed 8 qts. wheat bran, 4 qts. corn and cob meal, 4 qts. ground oats, 1½ pts. oil meal and 1½ pts. cotton-seed meal, daily—clover and cut corn fodder *ad lib.*, 20 lbs. roots; property of J. E. Gray & Son, Youngstown, Ohio.

Sweet Lily of Meridale 94267.—Sire, Matilda 4th's Son 20214; dam, Coquette of Glen Rouge 17559.

Butter, 29 lbs. ½ oz. in 14 days. Milk, 487 lbs. 12 oz.

Test made from Mar. 3 to 17, 1896; age, 3 yrs. 3 mos.; estimated weight, 850 lbs.; fed 28 gals. bran, 14 gals. corn and cob meal, 14 gals. oats, 4 gals. oil meal, 28 bush. cut corn stover, 14 gals. turnips and 14 gals. mangels—clover hay at noon; property of J. E. Gray & Son, Youngstown, Ohio.

Sweet Perrot 100749.—Sire, Stoke Pogis of Prospect 29121; dam, Tulla Perrot 46357.

Butter, 17 lbs. 2 oz. Milk, 246 lbs. 12 oz.

Test made from Aug. 5 to 12, 1897; age, 3 yrs.; actual weight, 960 lbs.; fed 9 qts. ground oats, 12 qts. wheat bran, 3 qts. corn meal and 3 qts. oil meal, mixed, daily—hay, good clover and timothy pasture; property of C. A. Sweet, East Aurora, N. Y.

Sweet Rock 2d 18256.—Sire, Jazel 3501; dam, Sweet Rock 20442.

Butter, 14 lbs. 11½ oz., unsalted.

Test made from May 1 to 8, 1883; age, 4 yrs.; property of W. B. Dinsmore, Staatsburg, N. Y.

Sweet Sixteen 10682.—Sire, Gilderoy 2107; dam, Lily Lenape 2d 8760.

Butter, 14 lbs. 15 oz. Milk, 127 lbs.

Test made from July 4 to 10, 1883; age, 3 yrs.; property of H. M. Howe, Bristol, R. I.

Sylvia 687.—Sire, Abe Lincoln 268; dam, Millie 690.

Butter, 15 lbs. 8 oz.

Test made from June 10 to 16, 1876; age, 11 yrs. 1 mo.; property of Fairbanks & Co., St. Johnsbury, Vt.

Sylvia Marigold 108183.—Sire, Stoke Pogis of Prospect 29121; dam, Sylvia of Aurora 86954.

Butter, 17 lbs. 8 oz. Milk, 223 lbs.

Test made from June 17 to 24, 1897; age, 2 yrs. 6 mos.; actual weight, 935 lbs.; fed 9 qts. bran, 9 qts. ground oats, 4½ qts. corn meal and 3 qts. oil meal, daily—plenty of good grass in barn; property of C. A. Sweet, East Aurora, N. Y.

Sylvia Perrot 30963.—Sire, Perrot (P. S. 342 J. H. B.); dam, Chestnut Farm Sylvia 25022.

Butter, 16 lbs. ½ oz. Milk, 182 lbs. 13 oz.

Test made from Nov. 21 to 28, 1892; age, 9 yrs. 8 mos.; estimated weight, 950 lbs.; fed 22 qts. daily of a mixture composed of one-third corn meal and two-thirds ground oats—also 147 lbs. hay; property of Charles A. Sweet, East Aurora, N. Y.

Syren V. 14619.—Sire, Kisber 4377; dam, Syren (F. S. 371 J. H. B.)

Butter, 17 lbs. 5 oz. Milk, 215 lbs. 12 oz.

Test made from Feb. 14 to 21, 1885; age, 5 yrs. 8 mos.; property of Watts & Seth, Baltimore, Md.

Taglioni 9182.—Sire, Oxoli 1922; dam, Auria 4567.

Butter, 14 lbs. 1 oz. Milk, 237 lbs. 1 oz.

Test made from May 15 to 21, 1884; age, 6 yrs. 9 mos.; property of Mrs. Brewster R. Burr, Commack, N. Y.

Tait's Beauty 121221.—Sire, Van Amburgh 22416; dam, Jennie Graham 2d 121220.

Butter, 18 lbs. 4 oz. Milk, 285 lbs.

Test made from Apr. 3 to 10, 1898; age, 6 yrs. 11 mos.; actual weight, 838 lbs.; fed 6 lbs. corn and cob meal, 6 lbs. bran, 4 lbs. ground oats, 2 lbs. cotton-seed meal and about 6 lbs. green cut oats and 10 lbs. pea vines and prairie hay, daily—Bermuda grass pasture; property of J. R. Miller, Minden, La.

Talaney 69361.—Sire, Texas Wanderer 14453; dam, Talasse 41315.

Butter, 15 lbs. 3 oz. Milk, 253 lbs. 14 oz.

Test made from Mar. 23 to 30, 1893; age, 3 yrs. 4 mos.; estimated weight, 750 lbs.; fed 56 lbs. cotton-seed, 30 lbs. ground oats and 87 lbs. corn meal—pasture; property of Neal Coldwell, Centre Point, Tex.

Tamarack 69900.—Sire, Authority 27197; dam, Ability 69897.

Butter, 14 lbs. 3 oz. Milk, 186 lbs. 4 oz.

Test made from Nov. 22 to 29, 1895; age, 5 yrs. 2 mos.; estimated weight, 950 lbs.; fed 21 lbs. oil meal, 21 lbs. oats and 28 lbs. corn—corn ensilage, frosted clover pasture; property of H. C. Taylor, Orfordville, Wis.

Tamasese Naomi 128774.—Sire, Tamasese of Bay View 39903; dam, Mildred G. 45077.

Butter, 14 lbs. 15 oz. Milk, 175 lbs. 12 oz.

Test made from Feb. 4 to 11, 1898; age, 2 yrs. 1 mo.; actual weight, 750 lbs.; fed 12 qts. daily of corn meal, bran, ground oats and gluten feed, in equal parts, and ½ bush. cut corn-stalks—hay; property of L. C. Read & Son, East Pembroke, N. Y.

Tancreda Signal 27789.—Sire, Signal Jr. 7166; dam, Tancreda 7938.

Butter, 21 lbs. 8½ oz. Milk, 310 lbs. 8 oz.

Test made from July 12 to 19, 1892; age, 8 yrs. 4 mos.; estimated weight, 900 lbs.; fed 2 qts. flaxseed meal, 4 qts. ground oats and 4 qts. corn meal, daily—short bluegrass, timothy and clover pasture; property of J. E. Carson, Crab Orchard, Ky.

Tancreda Torment 114090.—Sire, Tancreda's Signal King 35742; dam, Ethio's Princess 2d 83990.

Butter, 15 lbs. 1 oz. Milk, 164 lbs. 7 oz.

Test made from June 21 to 28, 1896; age, 2 yrs. 1 mo.; estimated weight, 700 lbs.; fed 84 qts. of a mixture of corn and cob meal, ground wheat and ground oats, in equal parts by measure, and 14 qts. oil-cake meal—timothy and blue-grass pasture; property of H. C. Schlegel, Daleville, Ind.

Tapestry 56607.—Sire, Compass 16958; dam, Sarita Victoria 56451.

Butter, 23 lbs. 12 oz. Milk, 313 lbs. 5 oz.

Test made from June 16 to 23, 1896; age, 7 yrs. 2 mos.; estimated weight, 900 lbs.; fed 35 lbs. oats, 35 lbs. bran, 42 lbs. corn and 21 lbs. oil meal—mixed pasture; property of H. C. Taylor, Orfordville, Wis.

Teacher's Pet 60242.—Sire, Teacher 15982; dam, Dame Dew Drop 38947.

Butter, 18 lbs. 12 oz. Milk, 268 lbs.

Test made from Apr. 7 to 14, 1895; age, 6 yrs. 6 mos.; estimated weight, 900 lbs.; fed 35 lbs. oats, 24½ lbs. corn meal, 24 lbs. bran and 24 lbs. oil meal—corn ensilage and clover hay; property of H. C. Taylor, Orfordville, Wis.

Teasel 75358.—Sire, Combination 3d 17576; dam, Brown Bessie 74997.

Butter, 20 lbs. 4 oz. Milk, 294 lbs. 4 oz.

Test made from June 6 to 13, 1896; age, 4 yrs. 9 mos.; estimated weight, 1050 lbs.; fed 42 lbs. oats, 28 lbs. rye, 42 lbs. corn meal and 7 lbs. oil meal—mixed pasture; property of H. C. Taylor, Orfordville, Wis.

Tease of Clovernook 92954.—Sire, Oonan's Tormentor 22280; dam, Torment of Clovernook 51969.

Butter, 15 lbs. 2 oz. Milk, 211 lbs. 11¼ oz.

Test made from June 23 to 30, 1896; age, 4 yrs.; estimated weight, 725 lbs.; fed 21 lbs. per day of ground oats and corn, in equal parts—extra good orchard and blue-grass pasture; property of S. H. Butler, Como, Miss.

Telka 8037.—Sire on I. of J.; dam, Duchess (F. S. 935 J. H. B.)

Butter, 14 lbs. 3 oz.

Property of W. R. McCready, Saugatuck, Conn.

Temalema 44761.—Sire, Fancy's Harry 9777; dam, Pearl's Oonan 32288.

Butter, 14 lbs. 10 oz. Milk, 219 lbs. 2 oz.

Test made from Sept. 27 to Oct. 4, 1889; age, 3 yrs. 1 mo.; estimated weight, 650 lbs.; fed, daily, 4 gals. ground oats and corn, mixed equally; property of Morgan & Brown, Columbia, Tenn.

Temalema 2d 82411.—Sire, Landseer's Pogis 15847; dam, Temalema 44761.

Butter, 16 lbs. 4 oz. Milk, 218 lbs. 8 oz.

Test made from Apr. 12 to 19, 1893; age, 3 yrs. 7 mos.; estimated weight, 750 lbs.; fed 8 qts. ground oats, 8 qts. corn meal and 4 qts. wheat bran, daily—ordinary pasture; property of C. Morgan, Columbia, Tenn.

Temisia of Winnikee 46604.—Sire, Rienzi of Winnikee 18948; dam, Tesina of Winnikee 46402.

Butter, 15 lbs. 3½ oz. Milk, 205 lbs. 12 oz.

Test made from Nov. 24 to Dec. 1, 1891; age, 5 yrs. 2 mos.; estimated weight, 900 lbs.; fed 7 qts. corn meal, 15 qts. oat meal, 30 qts. bran, 5 qts. cotton-seed meal, 3½ qts. oil meal and 7 bush. ensilage; property of R. A. Sibley, Spencer, Mass.

Tenella 6712.—Sire, Signal 1170; dam, Alda 3873.

Butter, 22 lbs. 1½ oz.

Test made from Nov. 18 to 25, 1881; age, 4 yrs. 6 mos.; property of J. B. Wade, Atlanta, Ga.

Tenella 2d 19521.—Sire, Sharpshooter of Atlanta 3011; dam, Tenella 6712.

Butter, 18 lbs. 12 oz. Milk, 314 lbs.

Test made from Apr. 8 to 14, 1884; age, 3 yrs. 8 mos.; property of J. B. Wade, Atlanta, Ga.

Tessia H. 94223.—Sire, A. C. L. Pearl H. 27011; dam, Dett H. 2d 74514.

Butter, 15 lbs. 15 oz. Milk, 216 lbs. 12 oz.

Test made from Mar. 2 to 9, 1896; age, 2 yrs. 11 mos.; estimated weight, 800 lbs.; fed 8 qts. corn and cob meal and oat meal, in equal parts, and 2 qts. oat middlings, daily—a few potatoes, cut corn fodder and hay *ad lib.*; property of J. T. Newton, Hudson, Ohio.

Tette 20802.—Sire, Gold Basis 4038; dam, Syringa 3d 6778.

Butter, 17 lbs. 6 oz. (official). Milk, 188 lbs. 4 oz.

Test made from June 8 to 14, 1885; age, 2 yrs. 7 mos.; property of Thomas H. Malone, Nashville, Tenn.

Thaley 14299.—Sire, Guy Fawkes (F. S. 251 J. H. B.); dam, Fauvette (F. S. 1972 J. H. B.)

Butter, 16 lbs. 13 oz. Milk, 264 lbs.

Test made from Feb. 1 to 7, 1885; age, 6 yrs.; property of William Rolph, Markham, Ont., Can.

The Colonel's Daughter 50230.—Sire, Colonel Stoke Pogis 18756; dam, Hildmella 22315.

Butter, 15 lbs. 4½ oz. Milk, 285 lbs. 2 oz.

Test made from July 17 to 24, 1893; age, 5 yrs. 3 mos.; estimated weight, 900 lbs.; fed 70 lbs. ground oats and corn, in equal parts, 14 lbs. old process linseed meal and 21 lbs. cotton-seed meal—short pasture; property of Ayer & McKinney, Philadelphia, Pa.

Theda H. 20567.—Sire, Clusius H. 5781; dam, Dancy H. 14446.

Butter, 17 lbs. 1 oz. Milk, 409 lbs. 11 oz.

Test made from July 7 to 13, 1885; age, 3 yrs. 2 mos.; property of William H. Hayden, Albany, Vt.

Theda's Belle 106891.—Sire, Theda's Son 31202; dam, Belle of Passaic 85846.

Butter, 16 lbs. 2 oz. Milk, 251 lbs.

Test made from Mar. 23 to 30, 1897; age, 3 yrs. 2 mos.; estimated weight, 900 lbs.; fed 70 qts. bran, 28 qts. ground oats, 28 qts. corn meal, 10½ qts. oil meal, 168 lbs. ensilage and 84 lbs. hay; property of R. A. Sibley, Rochester, N. Y.

Thekla of Clover Nook 33445.—Sire, Gold Basis 4038; dam, Jazel's Maid 11011.

Butter, 15 lbs. 9 oz. Milk, 191 lbs. 10 oz.

Test made from May 11 to 18, 1890; age, 6 yrs. 1 mo.; estimated weight, 800 lbs.; no grain fed—pasture only; property of S. N. Warren, Spring Hill, Tenn.

The Queen's Gift 75186.—Sire, Ida's Rioter of St. L. 13656; dam, Matilda 4th 12816.

Butter, 17 lbs. 11 oz. Milk, 152 lbs. 9 oz.

Test made from Sept. 28 to Oct. 5, 1893; age, 3 yrs.; estimated weight, 850 lbs.; fed 12 qts. corn and oats, 15 qts. bran and 3 qts. oil meal, daily—green corn and clover hay; property of Charles A. Sweet, East Aurora, N. Y.

The Queen's Gift 75186.—Sire, Ida's Rioter of St. L. 13656; dam, Matilda 4th 12816.

Butter, 22 lbs. 9 oz.   Milk, 323 lbs. 12 oz.

Test made from Apr. 29 to May 6, 1897; age, 6 yrs. 7 mos.; actual weight, 985 lbs.; fed 9 qts. bran, 12 qts. ground oats, 9 qts. corn meal and 3 qts. oil meal, daily—hay and ensilage; property of C. A. Sweet, East Aurora, N. Y.

Theresa H. 14447.—Sire, Homer H. 3683; dam, Lady Gray Hayden 3512.

Butter, 15 lbs. 3 oz.   Milk, 315 lbs. 4 oz.

Test made from June 22 to 28, 1885; age, 7 yrs. 3 mos.; property of William H. Hayden, Albany, Vt.

Therese G. 16390.—Sire, Champion of Indiana 3075; dam, Miss Thérèse 7327.

Butter, 17 lbs. 2 oz.   Milk, 212 lbs. 10 oz.

Test made from June 6 to 13, 1887; age, 4 yrs. 11 mos.; estimated weight, 850 lbs.; fed, daily, 20 to 30 qts. corn, oats and wheat bran, in equal parts; property of G. V. Green, Hopkinsville, Ky.

Therese M. 8364.—Sire, Champion of America 1567; dam, Julia Evelyn 6007.

Butter, 14 lbs. 2 oz.

Property of W. B. Montgomery, Starkville, Miss.

Théroigne 92259.—Sire, Julia's Prince 17938; dam, Miss Remarkable 33446.

Butter, 18 lbs. 11 oz.   Milk, 228 lbs. 11¼ oz.

Test made from June 29 to July 6, 1898; age, 6 yrs. 10 mos.; estimated weight, 775 lbs.; fed 22 lbs. per day of ground oats and corn, in equal parts—very good orchard and blue-grass pasture; property of Thomas H. Malone, Nashville, Tenn.

The Sister 21135.—Sire, Pierrot's Telephone 5440; dam, Palestine's Last Daughter 12602.

Butter, 14 lbs. 4½ oz.   Milk, 134 lbs. 8 oz.

Test made from Apr. 24 to May 1, 1889; age, 7 yrs. 11 mos.; estimated weight, 825 lbs.; fed, daily, 4 gals. ground corn and oats (two-thirds corn and one-third oats); property of Maury Jersey Farm, Columbia, Tenn.

The Widow's Daughter 11507.—Sire, Cargo 5370; dam, The Young Widow 11505.

Butter, 19 lbs. 8½ oz.   Milk, 323 lbs. 6 oz.

Test made from June 8 to 14, 1885; age, 4 yrs. 10 mos.; property of U. F. Shalter, Terre Haute, Ind.

Thisbe 2d 2201.—Sire, Cliff 176; dam, Thisbe 607.

Butter, 19 lbs. 1½ oz.   Milk, 372 lbs. 14 oz.

Test made from Apr. 19 to 26, 1882; age, 11 yrs. 6 mos.; fed raw cotton-seed and dry bran—good pasture; property of John E. Stiles, Artesia, Miss.

Thorndale Belle 5265.—Sire, Barney 1491; dam, Lena Lewis 3735.

Butter, 14 lbs. 8 oz., unsalted.   Milk, 172 lbs.

Test made from May 4 to 10, 1880; age, 6 yrs. 6 mos.; property of Edwin Thorne, Millbrook, N. Y.

Thorndale Belle 3d 10459.—Sire, Niobe Duke 2364; dam, Thorndale Belle 5265.

Butter, 15 lbs. 15 oz. Milk, 176 lbs.

Test made from May 6 to 13, 1884; age, 6 yrs. 2 mos.; property of Edwin Thorne, Millbrook, N. Y.

Thorndale Belle 6th 44885.—Sire, Dido's Duke 4678; dam, Thorndale Belle 5265.

Butter, 15 lbs. Milk, 226 lbs. 6 oz.

Test made from Nov. 28 to Dec. 5, 1891; age, 7 yrs. 1 mo.; estimated weight, 800 lbs.; fed 64 lbs. dried brewer's grains, 56 lbs. bran and 54 lbs. corn meal—hay and cut corn-stalks; property of Oakleigh Thorne, Millbrook, N. Y.

Thornleaf 39969.—Sire, Silver Duke 7125; dam, Tillie Thorn 24082.

Butter, 15 lbs. 12 oz. Milk, 214 lbs. 5 oz.

Test made from Oct. 16 to 23, 1892; age, 6 yrs. 9 mos.; estimated weight, 875 lbs.; fed 105 lbs. ground corn and cob and 60 lbs. wheat bran—pasture of mixed grasses; property of Columbus Dixon, Gillespieville, Ohio.

Tib's Jewel 6808.—Sire, Jeweler 1385; dam, Tib 4312.

Butter, 14 lbs. 4 oz. Milk, 228 lbs. 8 oz.

Test made from Mar. 31 to Apr. 6, 1885; age, 8 yrs.; property of Garretson Bros., Pendleton, Ind.

Tidy of St. Lambert 31114.—Sire, Lord Monck 304; dam, Victoria 411.

Butter, 14 lbs. 2 oz. Milk, 194 lbs. 4 oz.

Test made from Mar. 25 to 31, 1885; age, 12 yrs. 1 mo.; property of George G. Creamer and Frank Morgan, Hamilton, Mass.

Tiger Lill 69771.—Sire, Sherbet's Koffee 18512; dam, Gladdis Roy 18566.

Butter, 15 lbs. Milk, 200 lbs. 8 oz.

Test made from Dec. 28, 1894, to Jan. 4, 1895; age, 4 yrs. 8 mos.; estimated weight, 800 lbs.; fed provender, bran, cotton-seed meal and oil meal, no record of quantity kept; property of H. A. Huntington, Higganum, Conn.

Tilda Ives 17707.—Sire, Ned Dean 5387; dam, Tilda of Brookside 10871.

Butter, 15 lbs. Milk, 285 lbs.

Test made from June 11 to 18, 1893; age, 11 yrs. 4 mos.; estimated weight, 1000 lbs.; fed 4 qts. Buffalo gluten meal and 4 qts. middlings, mixed, daily—natural pasture; property of George W. Sisson, Jr., Potsdam, N. Y.

Tillie Texas 62840.—Sire, Little Tormentor 17995; dam, Tillie Meysenburg 23148.

Butter, 15 lbs. 13 oz. Milk, 206 lbs.

Test made from Feb. 21 to 28, 1897; age, 8 yrs. 4 mos.; estimated weight, 900 lbs.; fed 70 qts. corn and cob meal, 70 qts. wheat bran and 14 lbs. cotton-seed meal—hay *ad lib.*; property of James L. Cooper, Nashville, Tenn.

Tinney 2d 26253.—Sire, Hazel Eye 1514; dam, Tinney 18015.

Butter, 17 lbs. 12 oz. Milk, 257 lbs. 4 oz.

Test made from May 6 to 12, 1886; age, 5 yrs. 7 mos.; property of Archer N. Martin, Summit, N. J.

Tobira 8400.—Sire, Champion of America 1567; dam, Merry 4814.

Butter, 15 lbs. 13 oz. Milk, 235 lbs. 8 oz.

Test made from Apr. 16 to 22, 1882; age, 3 yrs. 5 mos.; property of W. B. Montgomery, Starkville, Miss.

Tolonga 76868.—Sire, Toltec 6831; dam, Lady Longfield 23524.

Butter, 15 lbs. 5½ oz. Milk, 220 lbs. 8 oz.

Test made from May 18 to 25, 1894; age, 3 yrs. 5 mos.; estimated weight, 875 lbs.; fed 140 lbs. corn chop and 21 lbs. cut sheaf oats—good blue-grass pasture; property of Est. of Campbell Brown, Spring Hill, Tenn.

Tolteca 44762.—Sire, Toltec 6831; dam, Duke's Signal Queen 32323.

Butter, 14 lbs. 11½ oz. Milk, 218 lbs. 9 oz.

Test made from July 26 to Aug. 2, 1890; age, 4 yrs.; estimated weight, 900 lbs.; fed, daily, 4 gals. ground oats and corn, mixed equally; property of Morgan & Brown, Columbia, Tenn.

Tolteca 3d 86344.—Sire, Landseer's Pogis 15847; dam, Tolteca 44762.

Butter, 14 lbs. 11½ oz. Milk, 196 lbs.

Test made from Apr. 26 to May 3, 1894; age, 2 yrs. 9 mos.; estimated weight, 650 lbs.; fed 56 qts. ground oats and 56 qts. ground corn—clover hay, poor pasture; property of Morgan & Brown, Columbia, Tenn.

Toltecida 2d 96030.—Sire, Vexer 20889; dam, Toltecida 70200.

Butter, 16 lbs. 6 oz. Milk, 188 lbs. ½ oz.

Test made from May 1 to 8, 1896; age, 3 yrs. 2 mos.; estimated weight, 725 lbs.; fed 56 qts. corn and cob meal, 28 qts. bran, 28 qts. middlings and 14 qts. old process oil-cake meal—clover hay, blue-grass pasture; property of H. C. Schlegel, Daleville, Ind.

Toltec's Alice 31885.—Sire, Toltec 6831; dam, Sweet Alice 6402.

Butter, 17 lbs. 10 oz. Milk, 153 lbs. 12 oz.

Test made from Jan. 28 to Feb. 4, 1890; age, 5 yrs. 7 mos.; estimated weight, 750 lbs.; fed, daily, 6 qts. ground corn and barley; property of S. N. Warren, Spring Hill, Tenn.

Toltec's Fancy 27172.—Sire, Toltec 6831; dam, Landseer's Fancy 2876.

Butter, 17 lbs. 6 oz. Milk, 124 lbs. 11 oz.

Test made from July 19 to 25, 1885; age, 1 yr. 11 mos.; property of Columbia Jersey Cattle Co., Columbia, Tenn.

Toltec's Fancy 27172.—Sire, Toltec 6831; dam, Landseer's Fancy 2876.

Butter, 27 lbs. 5½ oz. Milk, 130 lbs. 4 oz.

Test made from Jan. 8 to 15, 1889; age, 5 yrs. 4 mos.; estimated weight, 800 lbs.; fed 4 gals. corn and oats, daily; property of Maury Jersey Farm, Columbia, Tenn.

Toltec's Nan 70199.—Sire, Toltec's Prince 21961; dam, Tormentor's Nan 42251.

Butter, 14 lbs. Milk, 194 lbs. 8 oz.

Test made from May 19 to 26, 1893; age, 2 yrs. 7 mos.; estimated weight, 725 lbs.; fed 14 lbs. corn-hearts, 4 lbs. cotton-seed meal and 2 lbs. cut sheaf oats, daily—blue-grass pasture; property of Campbell Brown, Spring Hill, Tenn.

Toltec's Roonan 64564.—Sire, Toltec 6831; dam, Daisy's Roonan 38482.

Butter, 19 lbs. 7½ oz. Milk, 278 lbs. 11.2 oz.

Test made from Mar. 9 to 16, 1897; age, 7 yrs. 10 mos.; estimated weight, 850 lbs.; fed 7 lbs. bran, 8 lbs. corn and oats (two-thirds oats and one-third corn, ground together), 2 lbs. hominy meal, 2 lbs. linseed meal, 2 lbs. cotton-seed meal and 24 lbs. corn ensilage, daily—hay *ad lib.*; property of John E. Robbins, Greensburg, Ind.

Toltec's Sally 94461.—Sire, Toltec 6th 14507; dam, Alice Landseer 2d 76815.

Butter, 17 lbs. 2¼ oz. Milk, 248 lbs. 9½ oz.

Test made from Feb. 12 to 19, 1897; age, 5 yrs.; actual weight, 856 lbs.; fed, for first two days, 16 qts. corn meal, 16 qts. ground oats, 12 qts. wheat bran, 4 qts. cotton-seed meal; for remaining five days, 80 qts. corn and oats, ground together, half and half, 30 qts. wheat bran and 10 qts. cotton-seed meal; 70 lbs. corn fodder and 35 lbs. hay during test; property of W. H. Webb, Minden, La.

Toltec's Silvia 86911.—Sire, Toltec 6831; dam, Silvia Baker 8793.

Butter, 17 lbs. 1¼ oz. Milk, 209 lbs. 4¾ oz.

Test made from Jan. 5 to 12, 1897; age, 4 yrs. 9 mos.; estimated weight, 700 lbs.; fed 16 qts. per day of oats and corn, ground together, mixed with four bucketfuls of cut sheaf oats—some hay; property of J. E. Hart, Nashville, Tenn.

Toma May 74836.—Sire, Tombeau 25347; dam, May of Four Pines 52020.

Butter, 14 lbs. 13 oz. Milk, 217 lbs. 8 oz.

Test made from Jan. 17 to 24, 1897; age, 5 yrs. 9 mos.; estimated weight, 900 lbs.; fed 24 lbs. oats, 21 lbs. bran, 15 lbs. ground oats and corn, 7 lbs. 10½ oz. oil meal and 26 lbs. 4 oz. corn meal—corn fodder *ad lib.*, short rye pasture; property of W. E. Johnson, Millican, Tex.

Tomassee 76875.—Sire, Tormentor 5th 21962; dam, Cara Mia 64224.

Butter, 14 lbs. 12½ oz. Milk, 213 lbs. 8 oz.

Test made from June 9 to 16, 1893; age, 2 yrs. 1 mo.; estimated weight, 700 lbs.; fed 12 lbs. corn-hearts and 4 lbs. cut sheaf oats—blue-grass pasture; property of Campbell Brown, Spring Hill, Tenn.

Tomassee 76875.—Sire, Tormentor 5th 21962; dam, Cara Mia 64224.

Butter, 17 lbs. 5 oz. Milk, 267 lbs.

Test made from May 23 to 30, 1894; age, 3 yrs.; estimated weight, 800 lbs.; fed 168 lbs. corn chop and 21 lbs. sheaf oats—blue-grass pasture; property of Est. of Campbell Brown, Spring Hill, Tenn.

Tomassee 76875.—Sire, Tormentor 5th 21962; dam, Cara Mia 64224.

Butter, 41 lbs. 8¼ oz. in 14 days. Milk, 582 lbs. 8 oz.

Test made from May 19 to June 2, 1895; age, 4 yrs.; actual weight, 825 lbs.; fed 22 lbs. per day of ground oats and corn, in equal parts—extra good orchard and blue-grass pasture; property of Matt. M. Gardner, Nashville, Tenn.

Tones 46306.—Sire, Coomassie Welcome 9279; dam, Emma's Favorite 14078.

Butter, 14 lbs. 9 oz. Milk, 224 lbs. 4 oz.

Test made from Apr. 21 to 28, 1890; age, 3 yrs. 6 mos.; estimated weight, 850 lbs.; fed, daily, 14 lbs. corn meal, 8 lbs. bran and 2½ lbs. oil-cake meal; property of R. A. Sibley, Spencer, Mass.

Tonnage Girl 103331.—Sire, Tonnage 20000; dam, Kissing Girl 67725.

Butter, 15 lbs. 1½ oz. Milk, 242 lbs. 5 oz.

Test made from Mar. 18 to 25, 1898; age, 4 yrs. 11 mos.; estimated weight, 800 lbs.; fed 42 lbs. bran, 28 lbs. corn meal, 21 lbs. ground oats, 14 lbs. oil meal, 210 lbs. ensilage and 70 lbs. roots—hay *ad lib.*; property of C. I. Hood, Lowell, Mass.

Topaz of Woodlawn 11661.—Sire, Gilderoy 2107; dam, Gold Proof 10860.

Butter, 16 lbs. 4 oz. Milk, 214 lbs. 14 oz.

Test made from Aug. 3 to 10, 1885; age, 5 yrs. 1 mo.; property of Spencer Borden, Fall River, Mass.

Topsey Doe 25378.—Sire, Pilot of Bowker Farm 6085; dam, Topsey of Lakeside 3d 12138.

Butter, 15 lbs. 4 oz. Milk, 219 lbs.

Test made from June 22 to 29, 1889; age, 6 yrs. 1 mo.; estimated weight, 900 lbs.; fed, daily, 4 qts. corn meal, 4 qts. shorts and 4 qts. ground oats; property of Terence Carrigan, Hopkinton, Mass.

Topsey K. 22769.—Sire, Napoleon 2d 527; dam, Cora K. 22768.

Butter, 14 lbs.

Property of Nathan Brinley, Tyngsborough, Mass.

Topsy of Malone 49478.—Sire, Hugo Chief of St. Anne's 12070; dam, Malone 4986.

Butter, 14 lbs. 4 oz. Milk, 192 lbs.

Test made from June 14 to 21, 1889; age, 2 yrs. 2 mos.; estimated weight, 800 lbs.; fed, daily, 12 to 15 qts. oats, peas, corn and bran; property of Mrs. E. M. Jones, Brockville, Ont., Can.

Topsy of Malone 49478.—Sire, Hugo Chief of St. Anne's 12070; dam, Malone 4986.

Butter, 55 lbs. 12 oz. in 31 days. Milk, 790 lbs. 8 oz.

Test made from May 31 to July 1, 1889; age, 2 yrs. 1 mo.; fed, daily, 15 lbs. ground oats, peas and wheat bran—good pasture; property of Mrs. E. M. Jones, Brockville, Ont., Can.

Topsy's Oonan 97132.—Sire, Oonan's Pogis 17165; dam, Tormentor's Topsy 60737.

Butter, 15 lbs. 9½ oz. Milk, 194 lbs. 7½ oz.

Test made from Dec. 29, 1896, to Jan. 5, 1897; age, 3 yrs. 4 mos.; estimated weight, 675 lbs.; fed 20 lbs. per day of ground oats and corn, in equal parts, mixed with 3 bucketfuls cut sheaf oats, wetted—millet hay *ad lib.*; property of Matt. M. Gardner, Nashville, Tenn.

Torfrida 3596.—Sire, Rioter 2d 469; dam, Europa 176.

Butter, 17 lbs. 6½ oz. Milk, 256 lbs. 3½ oz.

Test made from July 14 to 20, 1883; age, 9 yrs.; property of William Simpson, New York, N. Y.

Toridex 76873.—Sire, Tormentor 5th 21962; dam, Tormentor's Idex 64209.

Butter, 14 lbs. 13 oz. Milk, 228 lbs.

Test made from June 3 to 10, 1897; age, 6 yrs. 4 mos.; estimated weight, 800 lbs.; fed 8 lbs. wheat bran, 13 lbs. ground oats and 4 lbs. linseed meal, daily—excellent blue-grass pasture; property of George Campbell Brown, Spring Hill, Tenn.

Torlona 76872.—Sire, Tormentor 7th 22911; dam, Lemon Twig 42028.

Butter, 14 lbs. 2 oz. Milk, 177 lbs.

Test made from June 16 to 23, 1893; age, 2 yrs. 4 mos.; estimated weight, 725 lbs.; fed 12 lbs. corn-hearts and 4 lbs. cut sheaf oats—blue-grass pasture; property of Campbell Brown, Spring Hill, Tenn.

Torlona 76872.—Sire, Tormentor 7th 22911; dam, Lemon Twig 42028.

Butter, 17 lbs. 1½ oz. Milk, 172 lbs. 8 oz.

Test made from Oct. 23 to 30, 1896; age, 5 yrs. 9 mos.; estimated weight, 875 lbs.; fed 98 lbs. coarse corn meal and 70 lbs. ground oats—fair blue-grass pasture; property of Est. of Campbell Brown, Spring Hill, Tenn.

Tormendova 74885.—Sire, Odelio 20223; dam, Tormentor's Bloomfield 55530.

Butter, 16 lbs. 8 oz. Milk, 206 lbs.

Test made from July 15 to 22, 1893; age, 2 yrs. 2 mos.; estimated weight, 650 lbs.; fed 16 lbs. corn-hearts and 12 lbs. ground oats, mixed, daily; property of Campbell Brown, Spring Hill, Tenn.

Tormendova 74885.—Sire, Odelio 20223; dam, Tormentor's Bloomfield 55530.

Butter, 19 lbs. 12 oz. Milk, 204 lbs.

Test made from Oct. 29 to Nov. 5, 1895; age, 4 yrs. 6 mos.; estimated weight, 950 lbs.; fed 70 lbs. corn meal, 84 lbs. ground oats and 42 lbs. cotton-seed meal—blue-grass pasture; property of Est. of Campbell Brown, Spring Hill, Tenn.

Tormendova 74885.—Sire, Odelio 20223; dam, Tormentor's Bloomfield 55530.

Butter, 37 lbs. 11 oz. in 14 days. Milk, 389 lbs.

Test made from Oct. 28 to Nov. 11, 1895; age, 4 yrs. 6 mos.; estimated weight, 950 lbs.; fed 140 lbs. corn meal, 168 lbs. ground oats and 84 lbs. cotton-seed meal—short blue-grass pasture; property of Est. of Campbell Brown, Spring Hill, Tenn.

Tormendova 2d 118161.—Sire, Tormentor 5th 21962; dam, Tormendova 74885.

Butter, 15 lbs. 9 oz. Milk, 207 lbs.

Test made from May 16 to 23, 1898; age, 3 yrs. 7 mos.; estimated weight, 850 lbs.; fed 8 lbs. ground corn, 4 lbs. wheat bran, 2 lbs. cotton-seed meal and 2 lbs. oil meal, daily—blue-grass pasture; property of George Campbell Brown, Spring Hill, Tenn.

Torment 15579.—Sire, Tormentor 3533; dam, Irene of Short Hills 5137.

Butter, 17 lbs. 8½ oz. Milk, 17½ qts. per day.

Test made from July 10 to 16, 1886; age, 4 yrs. 5 mos.; property of D. F. Appleton, Ipswich, Mass.

Torment 4th 70660.—Sire, Fancy's Harry 9777; dam, Torment 15579.

Butter, 15 lbs. 1¼ oz. Milk, 208 lbs. 6½ oz.

Test made from June 20 to 27, 1897; age, 6 yrs. 9 mos.; estimated weight, 700 lbs.; fed 20 lbs. per day of ground oats and corn, in equal parts—good orchard and blue-grass pasture; property of Matt. M. Gardner, Nashville, Tenn.

Tormento Coomassie 110447.—Sire. Tormentor Stoke Pogis 20485; dam, Croton Maid 4th 26729.

   Butter, 22 lbs. 2 oz. Milk, 203 lbs. 12 oz.

Test made from Oct. 29 to Nov. 5, 1897; age, 3 yrs. 9 mos.; actual weight, 830 lbs.; fed 35 lbs. corn meal, 35 lbs. wheat middlings, 35 lbs. wheat bran and 15 lbs. old process oil meal—sweet corn fodder *ad lib.*, excellent blue-grass pasture; property of J. P. Bradbury, Pomeroy, Ohio.

Tormentor's Barbara 98801.—Sire. Tormentor Stoke Pogis 20485; dam, Barberry 2d 46391.

   Butter, 17 lbs. 3 oz. Milk, 206 lbs. 15 oz.

Test made from Sept. 6 to 13, 1894; age, 1 yr. 11 mos.; actual weight, 746 lbs.; fed 30 lbs. corn meal, 30 lbs. old process oil meal and 30 lbs. bran, mixed—fine blue-grass pasture; property of J. P. Bradbury, Pomeroy, Ohio.

Tormentor's Bloomfield 55530.—Sire, Tormentor 3533; dam, Signalda Bloomfield 30694.

   Butter, 22 lbs. 5½ oz. Milk, 302 lbs. 13 oz.

Test made from May 5 to 12, 1896; age, 7 yrs. 7 mos.; actual weight, 967 lbs.; fed 16 lbs. oil-cake meal, 40 lbs. ground wheat, 40 lbs. corn meal, 16 lbs. cotton-seed meal, 20 lbs. bran and 35 lbs. clover hay—excellent blue-grass pasture; property of J. P. Bradbury, Pomeroy, Ohio.

Tormentor's Browny 73079.—Sire, Tormentor Stoke Pogis 20485; dam, Anise Burns 44635.

   Butter, 14 lbs. 14 oz. Milk, 218 lbs. 4 oz.

Test made from July 5 to 12, 1892; age, 3 yrs. 1 mo.; estimated weight, 650 lbs.; fed 56 lbs. mill feed, 42 lbs. corn-mids and 14 lbs. oil meal—blue-grass and clover pasture; property of D. W. Voyles, Crandall, Ind.

Tormentor's Cinderella 19564.—Sire, Tormentor 3533; dam, Paraph 6134.

   Butter, 15 lbs. 8½ oz. Milk, 230 lbs. 4 oz.

Test made from Dec. 24 to 30, 1885; age, 5 yrs. 9 mos.; property of M. C. Campbell, Spring Hill, Tenn.

Tormentor's Content 119129.—Sire, Denise's Tormentor 11823; dam, Landseer's Content 56985.

   Butter, 14 lbs. 1 oz. Milk. 157 lbs. 8 oz.

Test made from Apr. 26 to May 3, 1897; age, 2 yrs. 1 mo.; estimated weight, 700 lbs.; fed 70 qts. wheat bran, 14 qts. corn-hearts, 14 qts. cotton-seed meal and 7 qts. oil meal—good pasture; property of Geo. V. Green, Hopkinsville, Ky.

Tormentor's Daisy 52496.—Sire, Denise's Tormentor 11823; dam, Dolly of Riverside 32944.

   Butter, 15 lbs. 4 oz. Milk, 209 lbs. 8 oz.

Test made from Apr. 17 to 24, 1890; age, 3 yrs. 5 mos.; estimated weight, 800 lbs.; fed, daily, from 12 to 20 qts. corn, bran and oats, in equal parts; property of George V. Green, Hopkinsville, Ky.

Tormentor's Daisy 52496.—Sire, Denise's Tormentor 11823; dam, Dolly of Riverside 32944.

   Butter, 20 lbs. 8½ oz. Milk, 250 lbs.

Test made from Dec. 7 to 14, 1893; age, 7 yrs. 1 mo.; estimated weight, 900 lbs.; fed 12 lbs. wheat bran, 19½ lbs. corn and oats, 5½ lbs. oil meal and 28 lbs. beets, daily —timothy hay; property of Charles A. Sweet, Buffalo, N. Y.

Tormentor's Fairy 108510.—Sire, Moyane's Tormentor 21690; dam, Fairy Lady 24748.

Butter, 22 lbs. Milk, 363 lbs.

Test made from May 22 to 29, 1898; age, 4 yrs. 7 mos.; estimated weight, 900 lbs.; fed 5 lbs. Quaker oats feed, 10 lbs. hominy meal, 3 lbs. oil meal and 6 lbs. wheat bran, daily—blue-grass pasture; property of N. Frazier, Clark, Ky.

Tormentor's Fancy Wax 73212.—Sire, Tormentor 3533; dam, Fancy Wax 37159.

Butter, 15 lbs. 4½ oz. Milk, 248 lbs. 8 oz.

Test made from Jan. 9 to 16, 1897; age, 7 yrs. 3 mos.; estimated weight, 850 lbs.; fed 62 lbs. bran, 44 lbs. corn, 56 lbs. oats, 14 lbs. linseed meal, 210 lbs. ensilage and 140 lbs. roots—hay *ad lib.*; property of C. I. Hood, Lowell, Mass.

Tormentor's Favorite 36873.—Sire, Tormentor 3533; dam, Nilsson's Favorite 17680.

Butter, 19 lbs. 2 oz. Milk, 142 lbs.

Test made from Dec. 2 to 9, 1890; age, 5 yrs. 1 mo.; estimated weight, 825 lbs.; fed 5 gals. meal daily; property of E. H. Hatcher, Columbia, Tenn.

Tormentor's Fawn 2d 94485.—Sire, Sir Oonan Pogis 26521; dam, Tormentor's Fawn 82224.

Butter, 15 lbs. 4 oz. Milk, 216 lbs. 8 oz.

Test made from Mar. 2 to 9, 1897; age, 4 yrs. 9 mos.; estimated weight, 1000 lbs.; fed 42 lbs. corn meal, 35 lbs. bran, 35 lbs. ground oats and 14 lbs. oil meal, mixed with cut oats—clover hay; property of J. W. Bradrick, Farmland, Ind.

Tormentor's Fawn 2d 94485.—Sire, Sir Oonan Pogis 26521; dam, Tormentor's Fawn 82224.

Butter, 18 lbs. 8 oz. Milk, 267 lbs. 8 oz.

Test made from Jan. 26 to Feb. 2, 1898; age, 5 yrs. 8 mos.; estimated weight, 1000 lbs.; fed 56 lbs. corn meal, 42 lbs. bran and 42 lbs. ground oats, with cut sheaf oats, scalded—clover hay *ad lib.*; property of J. W. Bradrick, Farmland, Ind.

Tormentor's Idex 64209.—Sire, Tormentor 3533; dam, Gilderoy's Idex 42027.

Butter, 14 lbs. 9 oz. Milk, 216 lbs.

Test made from July 18 to 25, 1895; age, 6 yrs. 3 mos.; estimated weight, 800 lbs.; fed 70 lbs. corn-hearts and 42 lbs. cotton-seed meal—blue-grass pasture; property of Est. of Campbell Brown, Spring Hill, Tenn.

Tormentor's Isis 73350.—Sire, Tormentor Stoke Pogis 20485; dam, Sara Iris 53217.

Butter, 19 lbs. 7 oz. Milk, 292 lbs. ¾ oz.

Test made from May 7 to 14, 1898; age, 7 yrs. 10 mos.; estimated weight, 850 lbs.; fed about 70 qts. wheat bran, 21 qts. corn and cob meal and 7 qts. oil meal—very fine orchard and blue-grass pasture; property of C. W. Brubeck, Georgetown, Ind.

Tormentor's Laura 98346.—Sire, Tormentor Stoke Pogis 20485; dam, Katie Laurie 2d 46392.

Butter, 16 lbs. 9½ oz. Milk, 231 lbs. 9 oz.

Test made from Mar. 29 to Apr. 5, 1895; age, 2 yrs. 7 mos.; actual weight, 791 lbs.; fed 16 lbs. old process oil meal, 42 lbs. corn meal, 42 lbs. wheat bran and 21 lbs. ground oats—clover hay, corn stover and oat straw *ad lib.*; property of J. P. Bradbury, Pomeroy, Ohio.

Tormentor's Oona Toltec 105860.—Sire, Ethol 19104; dam, Bridal Rose 74883.

Butter, 14 lbs. 12 oz. Milk, 205 lbs. 12 oz.

Test made from Jan. 29 to Feb. 5, 1898; age, 2 yrs. 9 mos.; estimated weight, 800 lbs.; fed 18 qts. bran, 12 qts. corn meal and 3 qts. oil meal, daily—hay *ad lib.*; property of Mrs. E. A. Berauer, Waldron, Ind.

Tormentor's Ravalita 97909.—Sire, Tormentor Stoke Pogis 20485; dam, Ravalita 46390.

Butter, 15 lbs. 11¾ oz. Milk, 206 lbs. 8½ oz.

Test made from Mar. 29 to Apr. 5, 1895; age, 2 yrs. 7 mos.; actual weight, 753 lbs.; fed 6 lbs. cotton-seed meal, 6 lbs. old process oil meal, 55 lbs. corn meal and 55 lbs. ground wheat—clover hay *ad lib.*; property of J. P. Bradbury, Pomeroy, Ohio.

Tormentor's Rexea 38906.—Sire, Denise's Tormentor 11823; dam, Gilt Edge Rexea 32942.

Butter, 15 lbs. 14 oz. Milk, 176 lbs. 8 oz.

Test made from June 6 to 13, 1888; age, 2 yrs. 11 mos.; estimated weight, 750 lbs.; fed, daily, 30 lbs., one-half ground oats and one-half corn-hearts; property of Matt. M. Gardner, Nashville, Tenn.

Tormentor's Rosebud 50069.—Sire, Tormentor 3533; dam, Rosebud of Belle Vue 7702.

Butter, 14 lbs. 2 oz. Milk, 155 lbs. 8 oz.

Test made from Nov. 14 to 21, 1890; age, 2 yrs. 5 mos.; estimated weight. 700 lbs.; well fed, but no record kept of amount; property of Campbell Brown, Spring Hill, Tenn.

Tormentor's Silver 36058.—Sire, Tormentor 3533; dam, Silver Rose 4753.

Butter, 14 lbs. 7½ oz. Milk, 201 lbs. 4 oz.

Test made from May 15 to 22, 1889; age, 3 yrs. 7 mos.; actual weight, 940 lbs.; fed, daily, between 16 and 24 lbs. ground corn and oats; property of Wm. Morrow & Son, Nashville, Tenn.

Tormentor's Spiræa 42248.—Sire, Tormentor 3533; dam, Spiræa 4th 20075.

Butter, 14 lbs. Milk, 147 lbs. 8 oz.

Test made from Dec. 1 to 8, 1888; age, 2 yrs. 3 mos.; estimated weight, 850 lbs.; well fed, but no record of feed kept; property of Campbell Brown, Spring Hill, Tenn.

Tormentor's Su Lu 54943.—Sire, Tormentor 3533; dam, Su Lu 4705.

Butter, 14 lbs. 13 oz. Milk, 180 lbs. 8 oz.

Test made from May 12 to 19, 1891; age, 3 yrs. 1 mo.; estimated weight. 750 lbs.; fed 10 lbs. corn-hearts and 2 lbs. wheat bran, mixed, daily—blue-grass pasture; property of Campbell Brown, Spring Hill, Tenn.

Tormentor Victor Pogis 133021.—Sire, Melia Ann's Victor Pogis 20697; dam, Belle of Rydal Grange 41787.

Butter, 14 lbs. 7 oz. Milk, 237 lbs.

Test made from May 31 to June 7, 1898; age, 2 yrs.; estimated weight, 700 lbs.; no grain fed—clover pasture; property of J. H. Martin, Granville, N. Y.

Tormentris 49659.—Sire, Denise's Tormentor 11823; dam, Grace Pansy 2d 18764.

Butter, 18 lbs. 14½ oz. Milk, 204 lbs. 2 oz.

Test made from Nov. 6 to 13, 1893; age, 7 yrs. 3 mos.; estimated weight, 1000 lbs.; fed 15 qts. ground corn and oats, 15 qts. wheat bran and 3 qts. oil meal, daily—hay; property of Charles A. Sweet, East Aurora, N. Y.

Tormentris of Aurora 93948.—Sire, Alf. Signal 31184; dam, Tormentris 49659.

Butter, 15 lbs. 1½ oz. Milk, 230 lbs. 8 oz.

Test made from Mar. 28 to Apr. 4, 1896; age, 2 yrs. 6 mos.; actual weight, 900 lbs.; fed 12 qts. wheat bran, 3 qts. corn meal, 6 qts. ground oats and 3 qts. oil meal, daily—hay; property of Charles A. Sweet, East Aurora, N. Y.

Tormintha 37327.—Sire, Tormentor 3533; dam, Mintha 12812.

Butter, 16 lbs. 10¼ oz. Milk, 225 lbs. 8 oz.

Test made from May 10 to 17, 1894; age, 8 yrs. 1 mo.; actual weight, 850 lbs.; fed 35 lbs. bran, 35 lbs. corn meal and 30 lbs. old process oil meal, mixed with chopped clover hay—excellent blue-grass pasture; property of J. P. Bradbury, Pomeroy, Ohio.

Tornella 78453.—Sire, Oonan's Tormentor 22280; dam, Tormentor's Lass 59832.

Butter, 14 lbs. 7¼ oz. Milk, 246 lbs. 15 oz.

Test made from Jan. 9 to 16, 1895; age, 4 yrs. 3 mos.; estimated weight, 800 lbs.; fed 42 lbs. wheat bran, 28 lbs. corn meal, 21 lbs. cotton-seed meal, 14 lbs. linseed meal, 42 lbs. corn fodder and 280 lbs. ensilage; property of E. B. C. Hambley, Rockwell, N. C.

Tornella 78453.—Sire, Oonan's Tormentor 22280; dam, Tormentor's Lass 59832.

Butter, 15 lbs. 12 oz. Milk, 251 lbs. 1½ oz.

Test made from Dec. 24 to 31, 1895; age, 5 yrs. 2 mos.; estimated weight, 750 lbs.; fed 42 lbs. corn meal, 42 lbs. wheat bran, 14 lbs. ground oats and 245 lbs. ensilage; property of E. B. C. Hambley, Rockwell, N. C.

Torpedo Exile 105287.—Sire, Wells' Exile 20170; dam, Jenny Brewster Jr. 73133.

Butter, 16 lbs. 4½ oz. Milk, 206 lbs.

Test made from Sept. 15 to 22, 1896; age, 4 yrs. 5 mos.; estimated weight, 700 lbs.; fed 56 qts. bran, 14 qts. oil-cake meal and 1½ bush. nubbin corn—hill blue-grass pasture; property of E. G. Silvus, Athens, Ohio.

Torraline 54945.—Sire, Torrance 17872; dam, Rosaline Pogis 42253.

Butter, 17 lbs. 6¾ oz. Milk, 216 lbs. 6¼ oz.

Test made from June 28 to July 5, 1893; age, 5 yrs. 2 mos.; actual weight, 920 lbs.; fed 22 lbs. per day of ground oats and corn, in equal parts—very good orchard and blue-grass pasture; property of Matt. M. Gardner, Nashville, Tenn.

Torrancella 59827.—Sire, Torrance 17872; dam, Idarella 41433.

Butter, 17 lbs. 8½ oz. Milk, 216 lbs. 1 oz.

Test made from May 23 to 30, 1892; age, 4 yrs.; actual weight, 1120 lbs.; fed 24 lbs. per day of ground oats and corn, in equal parts—extra good orchard and blue-grass pasture; property of Matt. M. Gardner, Nashville, Tenn.

Tortilla 55528.—Sire, Tormentor 3533; dam, Solava 20928.

Butter, 15 lbs. 3 oz. Milk, 175 lbs. 8 oz.

Test made from May 2 to 9, 1893; age, 4 yrs. 9 mos.; estimated weight, 900 lbs.; fed 20 lbs. corn-hearts and 4 lbs. cut sheaf oats, daily—blue-grass pasture; property of Campbell Brown, Spring Hill, Tenn.

Tortilla 55528.—Sire, Tormentor 3533; dam, Solava 20928.

Butter, 18 lbs. 4½ oz. Milk, 280 lbs. 14½ oz.

Test made from Dec. 23 to 30, 1895; age, 7 yrs. 4 mos.; actual weight, 819 lbs.; fed 24 lbs. old process oil meal, 40 lbs. ground wheat and 60 lbs. corn meal—shredded corn-stover and clover hay *ad lib.*; property of J. P. Bradbury, Pomeroy, Ohio.

Tower View Princess 2d 87452.—Sire, Hilarious 7651; dam, Tower View Princess 22595.

Butter, 17 lbs. 12½ oz. Milk, 195 lbs. 9 oz.

Test made from June 3 to 10, 1893; age, 4 yrs. 4 mos.; estimated weight, 700 lbs.; fed 84 qts. corn and oats (ground half and half), 28 qts. wheat bran, 7 qts. cottonseed meal and cut clover hay, dampened—short mixed pasture; property of W. R. Rison, Huntsville, Ala.

Toy 20604.—Sire, Duke of Belmont Jr. 7794; dam, Queen Fan 18013.

Butter, 17 lbs. 8 oz. Milk, 258 lbs. 8 oz.

Test made from Oct. 24 to 31, 1890; age, 7 yrs. 8 mos.; estimated weight, 1000 lbs.; fed, daily, 4 lbs. oil meal, 2 lbs. bran and 20 lbs. corn sortings boiled on cob; property of Garretson Bros., Pendleton, Ind.

Transcript 31867.—Sire, Combination 4389; dam, Chronicle 21625.

Butter, 17 lbs. 7 oz. Milk, 239 lbs.

Test made from June 1 to 7, 1887; age, 4 yrs. 1 mo.; property of Richardson Bros., Davenport, Iowa.

Treasure 79200.—Sire, Upright 6147; dam, Mercurina 64920.

Butter, 17 lbs. 5 oz. Milk, 231 lbs.

Test made from Oct. 1 to 8, 1892; age, 5 yrs.; estimated weight, 800 lbs.; fed 36 qts. ground oats, corn and oil meal, daily—hay, orchard pasture; property of Richardson Bros., Davenport, Iowa.

Tremona 93017.—Sire, Chromo 26113; dam, Josephina 64921.

Butter, 15 lbs. 2½ oz. Milk, 217 lbs. 7 oz.

Test made from July 12 to 19, 1897; age, 4 yrs.; estimated weight, 700 lbs.; fed 35 lbs. bran, 28 lbs. corn meal, 31½ lbs. ground oats and 10½ lbs. oil meal—mixed clover and June grass pasture; property of C. I. Hood, Lowell, Mass.

Tremona 93017.—Sire, Chromo 26113; dam, Josephina 64921.

Butter, 17 lbs. 2 oz. Milk, 273 lbs. 12 oz.

Test made from June 20 to 27, 1898; age, 5 yrs.; estimated weight, 850 lbs.; fed 28 lbs. bran, 30 lbs. corn meal, 17½ lbs. ground oats, 7 lbs. oil meal and 7 lbs. cottonseed meal—mixed clover and timothy pasture; property of C. I. Hood, Lowell, Mass.

Trenie 17770.—Sire, Tamerlane 4287; dam, Cathiella 16947.

Butter, 14 lbs. 10 oz. Milk, 202 lbs. 4 oz.

Test made from July 9 to 15, 1884; age, 2 yrs. 10 mos.; property of J. V. N. Willis, Marlborough, N. J.

Trident 88647.—Sire. Stoke Pogis of Prospect 29121; dam, Elopement 65578.

Butter, 17 lbs. 4½ oz. Milk, 283 lbs. 4 oz.

Test made from July 3 to 10, 1898; age, 5 yrs. 4 mos.; actual weight, 1050 lbs.; fed 9 qts. wheat bran, 9 qts. ground oats, 3 qts. corn meal and 3 pts. oil meal, mixed, wet, daily--green clover, clover pasture; property of Charles A. Sweet, East Aurora, N. Y.

Trilby M. 103547.—Sire, Kathy's Stoke Pogis 17566; dam, Frushia Wilson 47432.

Butter, 14 lbs. 4 oz. Milk, 240 lbs. 8 oz.

Test made from Apr. 24 to May 1, 1898; age, 3 yrs. 7 mos.; estimated weight, 800 lbs.; fed 42 lbs. wheat bran, 28 lbs. corn chop, 14 lbs. whole oats and 14 lbs. cotton-seed meal—native grass pasture; property of W. A. Ponder, Denton, Tex.

Tritoma Pogis 44952.—Sire, Prince Pogis' Hugo 15660; dam, Fannie Keene 40097.

Butter, 16 lbs. 13 oz. Milk, 179 lbs. 15 oz.

Test made from June 20 to 27, 1893; age, 6 yrs. 3 mos.; estimated weight, 950 lbs.; fed 4 lbs. ground oats and 6 lbs. ground corn, daily—poor blue-grass pasture; property of D. L. Heinsheimer, Glenwood, Iowa.

Trixket 16292.—Sire, Berlin Prince 3360; dam, Gold Trinket 9518.

Butter, 14 lbs. 2½ oz. Milk, 188 lbs.

Test made from Mar. 31 to Apr. 7, 1889; age, 7 yrs. 8 mos.; estimated weight, 600 lbs.; fed ground corn, oats, barley and hay, quantity unknown—blue-grass pasture; property of Samuel N. Warren, Spring Hill, Tenn.

Troth 6139.—Sire, Hornbeam 2123; dam, Blondette 1817.

Butter, 16 lbs. 5 oz. Milk, 104 qts.

Test made from June 19 to 26, 1881; age, 4 yrs.; property of Thos. J. Hand, Sing Sing, N. Y.

Trudie 2d 4084.—Sire, Bismarck 1423; dam, Trudie 277.

Butter, 15 lbs.

Test made from May 10 to 17, 1882; age, 6 yrs. 11 mos.; property of Thomas H. Faile, New York, N. Y.

True Rex Pogis 105467.—Sire. Imperial Rex Rioter 24096; dam, Daisy Staunton 46592.

Butter, 15 lbs. 11¼ oz. Milk, 170 lbs. 1¾ oz.

Test made from July 2 to 9, 1898; age, 4 yrs. 6 mos.; estimated weight, 800 lbs.; fed 8 qts. wheat bran, 8 qts. corn meal and 8 qts. cotton-seed, daily—common pasture; property of H. J. Mitchell, Winnsborough, Tex.

Trut 8090.—Sire, Lord Francis 1857; dam, Tyke 8089.

Butter, 14 lbs. 3 oz. Milk, 270 lbs.

Test made from May 17 to 24, 1894; age, 10 yrs. 8 mos.; estimated weight, 950 lbs.; fed 101 lbs. of a mixture consisting of corn meal, No. 2 wheat feed, oats and corn, in equal parts, ground—poor pasture; property of James T. Inglis, Middlefield, Conn.

Tuäa Perrot 46357.—Sire, Thotmes 17157; dam, Sylvia Perrot 30963.

Butter, 19 lbs. 5½ oz. Milk, 207 lbs. 12½ oz.

Test made from July 16 to 23, 1893; age, 6 yrs. 1 mo.; estimated weight, 850 lbs.; fed 15 qts. ground corn and oats, 12 qts. wheat bran and 3 qts. oil meal, daily—green oats and peas; property of Charles A. Sweet, East Aurora, N. Y.

Tuäa St. Lambert 59992.—Sire, Menephtah 18943; dam, Tuäa Perrot 46357.

Butter, 17 lbs. 7½ oz. Milk, 187 lbs. 15½ oz.

Test made from July 25 to Aug. 1, 1893; age, 3 yrs. 11 mos.; estimated weight, 850 lbs.; fed 15 qts. ground corn and oats (equal parts), 12 qts. wheat bran and 3 qts. oil meal, daily—green oats and peas; property of Charles A. Sweet, East Aurora, N. Y.

Tuäa St. Lambert 59992.—Sire, Menephtah 18943; dam, Tuäa Perrot 46357.

Butter, 23 lbs. 1 oz. Milk, 306 lbs. 4 oz.

Test made from Dec. 15 to 22, 1895; age, 6 yrs. 3 mos.; actual weight, 850 lbs.; fed 12 qts. ground corn and oats, 12 qts. wheat bran and 3 qts. oil meal, daily—plenty of good hay; property of Charles A. Sweet, East Aurora, N. Y.

Tuäa St. L. of Aurora 89388.—Sire, Stoke Pogis of Prospect 29121; dam, Tuäa St. Lambert 59992.

Butter, 17 lbs. 1 oz. Milk, 211 lbs. 12 oz.

Test made from Aug. 5 to 12, 1895; age, 2 yrs. 1 mo.; actual weight, 800 lbs.; fed 9 qts. wheat bran, 9 qts. ground corn and oats, 1½ qts. oil meal and 1½ qts. cotton-seed meal, daily—poor pasture, green oats and peas in stable; property of C. A. Sweet, East Aurora, N. Y.

Tuleema Pogis 100089.—Sire, Oracle 30340; dam, Tuleema 43160.

Butter, 15 lbs. 4 oz. Milk, 252 lbs.

Test made from Apr. 27 to May 4, 1898; age, 5 yrs.; estimated weight, 900 lbs.; fed 8 lbs. ground corn, 4 lbs. bran and 2 lbs. oil meal, daily—spear-grass pasture; property of C. L. Boyer, Finleyville, Pa.

Turquoise 1129.—Sire, Lawrence 61; dam, Topaz 75.

Butter, 14 lbs. 3 oz.

Property of John D. Wing, Millbrook, N. Y.

Tweedledee of Nipsic 60427.—Sire, Dandelion's Augustus 19867; dam, Dame Dee 18060.

Butter, 14 lbs. 1 oz. Milk, 219 lbs. 8 oz.

Test made from June 11 to 18, 1895; age, 7 yrs. 6 mos.; estimated weight, 750 lbs.; fed 9 qts. gluten and bran and 1 qt. oil meal, daily—hay *ad lib.*, poor blue-grass pasture; property of W. J. Hussey, Mount Pleasant, Ohio.

Tylesholm Blythesomie 101347.—Sire, Baron of East View 17630; dam, Duke's Blythesomie 101346.

Butter, 16 lbs. Milk, 301 lbs. 8 oz.

Test made from June 20 to 27, 1898; age, 7 yrs. 2 mos.; estimated weight, 800 lbs.; fed 49 qts. grain (wheat bran and corn meal, mixed in proportion of 3 qts. bran to 1 qt. corn meal)—mixed pasture; property of Jonathan F. Gould, East Hampton, N. Y.

Typha 5870.—Sire, Asgard 1379; dam, Torfrida 3596.

Butter, 16 lbs. 11 oz. Milk, 234 lbs. 9 oz.

Test made from July 16 to 22, 1884; age, 7 yrs. 4 mos.; property of William Simpson, New York, N. Y.

Uinta 5743.—Sire, Baronet 2240; dam, Lucy Dale 5129.

Butter, 14 lbs. 10 oz. Milk, 234 lbs. 4 oz.

Test made from May 23 to 29, 1883; age, 6 yrs. 1 mo.; property of M. L. Frazier, Hudson, Mich.

Ulricalla 22225.—Sire, Wessex 3638; dam, Rosa Thornton 12233.

Butter, 18 lbs. 14 oz. Milk, 195 lbs. 10 oz.

Test made from Feb. 24 to Mar. 2, 1887; age, 3 yrs. 5 mos.; estimated weight, 800 lbs.; fed, daily, 4 lbs. crushed oats, 4 lbs. bran, 1 lb. oil meal, 1½ lbs. corn meal, 1½ lbs. pea meal and 2 lbs. shorts; property of C. W. H. Eicke, West Monterey, Pa.

Ultima 14456.—Sire, Phil Bree 2957; dam, Maggie 4th 3224.

Butter, 15 lbs. 12 oz. Milk, 228 lbs.

Test made from Oct. 26 to Nov. 2, 1885; age, 5 yrs. 3 mos.; property of William R. Goodspeed, East Haddam, Conn.

Una of St. Lambert 80117.—Sire, Canada's Sir George 18290; dam, Miss Uarda 40473.

Butter, 16 lbs. 15½ oz. Milk, 297 lbs. 4 oz.

Test made from Sept. 8 to 15, 1896; age, 4 yrs. 5 mos.; estimated weight, 1000 lbs.; fed 28 lbs. ground corn, bran and oil meal, daily—green corn fodder, poor pasture; property of R. F. Shannon, Pittsburg, Pa.

Undeniah Pogis 85990.—Sire, Count Rioter Pogis 21388; dam, Litta Pogis 30275.

Butter, 14 lbs. 2 oz. Milk, 174 lbs. 12 oz.

Test made from Aug. 12 to 19, 1897; age, 6 yrs.; estimated weight, 1000 lbs.; fed 27 lbs. 11 oz. cotton-seed meal, 27 lbs. 11 oz. wheat shorts, 25 lbs. 11 oz. wheat bran, 11 lbs. 11 oz. corn meal and 9 lbs. 3 oz. ground oats—Bermuda, lespedeza and Japan clover pasture; property of J. H. Wright, Meridian, Miss.

Upper Ten 37603.—Sire, Citizen 13186; dam, Income 19472.

Butter, 14 lbs. 7 oz. Milk, 247 lbs. 13 oz.

Test made from Sept. 3 to 10, 1888; age, 2 yrs. 10 mos.; estimated weight, 800 lbs.; fed 15 qts. meal daily; property of Richardson Bros., Davenport, Iowa.

Upright's Brownie 103755.—Sire, Upright 6147; dam, Laura Lee 8550.

Butter, 15 lbs. 3 oz. Milk, 210 lbs. 12 oz.

Test made from Nov. 29 to Dec. 6, 1895; age, 6 yrs. 10 mos.; estimated weight, 800 lbs.; fed 48 lbs. oats, 21 lbs. oil meal, 28 lbs. corn meal and 21 lbs. bran—corn ensilage; property of H. C. Taylor, Orfordville, Wis.

Upright's Malden 105616.—Sire, Diploma 3d 36376; dam, Upright's Brownie 103755.

Butter, 14 lbs. 8 oz. Milk, 200 lbs. 4 oz.

Test made from Nov. 24 to Dec. 1, 1896; age, 1 yr. 9 mos.; estimated weight, 675 lbs.; fed 28 lbs. bran, 28 lbs. corn meal, 31½ lbs. oats and 17½ lbs. oil meal—cut green corn fodder, rape and hay *ad lib.*; property of H. C. Taylor, Orfordville, Wis.

Urbana 5597.—Sire, Duke Glen Dale 1819; dam, La Belle Desreaux 2d 5096.

Butter, 16 lbs.

Test made from June 1 to 7, 1883; age, 6 yrs.; property of J. C. Johnson, Marion, Ohio.

Utilla 11215.—Sire, Frankfort 2990; dam, Metela 2d 3242.

Butter, 16 lbs. 2 oz., unsalted.

Test made from May 21 to 27, 1884; age, 4 yrs.; property of James R. Crane, Washington, Ill.

Valentine of Trinity 7460.—Sire, Duke (P. S. 76 J. H. B.); dam, Milkmaid 3d (P. S. 4 J. H. B.)

Butter, 19 lbs. 4½ oz. Milk, 277 lbs. 3 oz.

Test made from May 25 to 31, 1885; age, 9 yrs. 3 mos.; property of Samuel McKeen, Terre Haute, Ind.

Valerie 6044.—Sire, Butter Print 1863; dam, Oktibbeha Duchess 4422.

Butter, 15 lbs. 13 oz. Milk, 284 lbs. 8 oz.

Test made from June 9 to 15, 1882; age, 5 yrs.; property of W. B. Montgomery, Starkville, Miss.

Valhalla 5300.—Sire, Signal 1170; dam, Azelda 3872.

Butter, 16 lbs.

Property of C. P. Markle & Sons, West Newton, Pa.

Valhalla 5300.—Sire, Signal 1170; dam, Azelda 3872.

Butter, 34 lbs. in 14 days.

Test made in 1882; age, 6 yrs.; property of C. P. Markle & Sons, West Newton, Pa.

Valita's Fancy 115842.—Sire, Harry Torment's Signal 39846; dam, Valita 105202.

Butter, 17 lbs. 1¼ oz. Milk, 255 lbs. 1 oz.

Test made from Aug. 1 to 8, 1897; age, 3 yrs.; actual weight, 816 lbs.; fed 10 lbs. old process oil meal, 40 lbs. corn meal, 40 lbs. wheat middlings and 18 lbs. wheat bran—blue-grass pasture; property of J. P. Bradbury, Pomeroy, Ohio.

Valma Hoffman 4500.—Sire, Orange Peel 864; dam, Valma 2192.

Butter, 21 lbs. 9 oz.

Test made from July 22 to 29, 1883; age, 9 yrs. 4 mos.; property of Samuel T. Earle, Centreville, Md.

Value 2d 6844.—Sire, Hurrah 2814; dam, Value 5433.

Butter, 25 lbs. 2 11-12 oz., unsalted (official). Milk, 327 lbs.

Test made from June 19 to 25, 1883; age, 7 yrs. 4 mos.; property of Watts & Seth, Baltimore, Md.

Value Belle 33080.—Sire, Champion of Kansas 4585; dam, Gertie Rex 23946.

Butter, 14 lbs. 3 oz. Milk, 266 lbs. 4 oz.

Test made from Apr. 15 to 22, 1892; age, 7 yrs. 3 mos.; estimated weight, 1000 lbs.; fed 5 lbs. linseed meal (old process), 2 lbs. bran and 8 lbs. corn meal, daily—140 lbs. hay; property of E. B. Jones, Mount Pleasant, Ohio.

Value Belle 33080.—Sire, Champion of Kansas 4585; dam, Gertie Rex 23946.

Butter, 16 lbs. 6 oz. Milk, 276 lbs. 12 oz.

Test made from Mar. 13 to 20, 1895; age, 10 yrs. 2 mos.; estimated weight, 1000 lbs.; fed 42 lbs. bran, 14 lbs. gluten meal, 14 lbs. old process linseed meal and 56 lbs. hay —40 lbs. ensilage per day; property of E. B. Jones, Mount Pleasant, Ohio.

Vaniah 6597.—Sire, Iron Bank 1120; dam, Matchless 1277.

Butter, 15 lbs. 9½ oz. Milk, 216 lbs. 4 oz.

Test made from Jan. 28 to Feb. 3, 1883; property of T. F. Shotwell, Bucyrus, Ohio.

Vaniah of St. Lambert 71374.—Sire, Diana's Rioter 10481; dam, Vaniah 6th 48126.

Butter, 15 lbs. Milk, 205 lbs. 13 oz.

Test made from Dec. 18 to 25, 1893; age, 3 yrs. 2 mos.; estimated weight, 700 lbs.; fed 8 qts. bran, 12 qts. ground oats and corn, 2 qts. old process linseed meal and 60 lbs. clover hay; property of Mrs. Kate M. Busick, Wabash, Ind.

Variella of Linwood 10954.—Sire, Mirabeau 3800; dam, Vixen 7591.

Butter, 14 lbs. 1 oz., unsalted. Milk, 228 lbs. 2 oz.

Test made from Aug. 17 to 24, 1883; age, 3 yrs. 3 mos.; property of M. M. Gardner, Nashville, Tenn.

Velveteen 7703, imp.

Butter, 14 lbs. 13½ oz., unsalted. Milk, 212 lbs.

Test made from May 1 to 7, 1882; age, about 4 yrs.; no grain—fine blue-grass pasture; property of Thomas H. Malone, Nashville, Tenn.

Velvet Mostar 17533.—Sire, Clifton Monarch 3546; dam, Princess Mostar 9700.

Butter, 15 lbs. 3 oz. Milk, 202 lbs. 14 oz.

Test made from Apr. 7 to 14, 1884; age, 5 yrs. 1 mo.; property of James Cloud & Son, Kennett Square, Pa.

Velvet Queen 112795.—Sire, Rioter of St. Lambert 16501; dam, Corinna Belle 21051.

Butter, 15 lbs. 8 oz. Milk, 257 lbs. 8 oz.

Test made from July 4 to 11, 1897; age, 7 yrs. 2 mos.; actual weight, 900 lbs.; fed 12 qts. bran, 9 qts. oat meal, 6 qts. corn meal and 3 qts. oil meal, daily—plenty of good grass in barn; property of C. A. Sweet, East Aurora, N. Y.

Vemicou 79322.—Sire, Vexer 20889; dam, Miacou 48697.

Butter, 14 lbs. 7 oz. Milk, 224 lbs. 4 oz.

Test made from May 22 to 29, 1896; age, 4 yrs. 7 mos.; actual weight, 900 lbs.; fed 28 qts. ground corn and 56 qts. wheat bran—good blue-grass pasture; property of Columbus Dixon, Gillespieville, Ohio.

Vemist 79323.—Sire, Vexer 20889; dam, Mist 26087.

Butter, 14 lbs. 2 oz. Milk, 198 lbs. 8 oz.

Test made from May 19 to 26, 1898; age, 6 yrs. 6 mos.; estimated weight, 700 lbs.; fed 4 lbs. Quaker oats feed, 8 lbs. hominy meal, 2 lbs. oil meal and 6 lbs. bran, daily —blue-grass and clover pasture; property of Frazier & Byars, Clark, Ky.

Venice 18192, imp.

Butter, 14 lbs. 4 oz. Milk, 250 lbs. 11 oz.

Test made from June 10 to 17, 1886; age, 7 yrs. 4 mos.; estimated weight, 1075 lbs.; fed 6 qts. shorts, daily; property of T. R. Proctor, Utica, N. Y.

Venna's Zoka 26670.—Sire, Count Coomassie 7542; dam, Venna 9525.

Butter, 14 lbs. ½ oz. Milk, 146 lbs. 8 oz.

Test made from Feb. 10 to 17, 1888; age, 4 yrs. 3 mos.; estimated weight, 700 lbs.; fed, daily, about 3 gals. ground corn and oats; property of Webster & Morrow & Son, Nashville, Tenn.

Ventnor Beauty 46308.—Sire, Coomassie Welcome 9279; dam, Golden Lena 15396.

Butter, 15 lbs. 11 oz. Milk, 155 lbs. 8 oz.

Test made from June 20 to 27, 1890; age, 3 yrs. 5 mos.; estimated weight, 700 lbs.; fed, daily, 10 lbs. corn meal, 4 lbs. shorts and 1½ lbs. oil-cake meal; property of R. A. Sibley, Spencer, Mass.

Ventnor Beauty 46308.—Sire, Coomassie Welcome 9279; dam, Golden Lena 15396.

Butter, 17 lbs. 5½ oz. Milk, 240 lbs. 4 oz.

Test made from May 29 to June 5, 1894; age, 7 yrs. 4 mos.; estimated weight, 900 lbs.; fed 14 qts. corn meal, 56 qts. bran, 42 qts. ground oats and 14 lbs. cotton-seed meal—short pasture; property of R. A. Sibley, Spencer, Mass.

Venus 112.—Sire, Typhoon 77; dam, Brenda 789.

Butter, 14 lbs. 11 oz.

Test made from July 20 to 26, 1863; age, 9 yrs. 11 mos.; property of W. S. Lincoln, Worcester, Mass.

Vera McMillan 66520.—Sire, Angela's Rioter 20120; dam, Miss Susy McMillan 50386.

Butter, 17 lbs. ½ oz. Milk, 270 lbs.

Test made from Mar. 13 to 20, 1895; age, 4 yrs. 9 mos.; estimated weight, 950 lbs.; fed 49 lbs. wheat bran, 49 lbs. corn and cob meal, 32 lbs. old process oil meal, mixed —steamed cut corn stover and mixed hay *ad lib.*; property of B. F. Gibbons & Son, Warren, Ohio.

Vera of Briarcliff 28687.—Sire, Young Garenne's Duke 6863; dam, Lady Horton 2d 15499.

Butter, 15 lbs. 1 oz. Milk, 129 lbs. 8 oz.

Test made from Apr. 30 to May 7, 1887; age, 2 yrs. 4 mos.; estimated weight, 750 lbs.; fed, daily, 12 lbs. corn meal, 3 lbs. oil meal, 4 lbs. middlings and 4 lbs. bran; property of James Stillman, Sing Sing, N. Y.

Verbena of Fernwood 9088.—Sire, Balboa 1244; dam, Vanilla 3834.

Butter, 15 lbs. Milk, 196 lbs. 13 oz.

Test made from Jan. 6 to 13, 1883; age, 4 yrs. 7 mos.; property of Harrison Leib, Cincinnati, Ohio.

Vernon Dolly 52284.—Sire, Dalton 20117; dam, Vernon Beauty 50687.

Butter, 21 lbs. Milk, 308 lbs.

Test made from May 27 to June 3, 1893; age, 5 yrs. 2 mos.; actual weight, 1040 lbs.; clover pasture; property of Theodore L. Flood, Meadville, Pa.

Verona 10766.—Sire, Orawapum 2833; dam, Veronica 6684.

Butter, 15 lbs. 1½ oz. Milk, 227 lbs.

Test made from Oct. 25 to 31, 1884; age, 4 yrs. 2 mos.; property of Simon Beery, Urbana, Ohio.

Very Pretty 29184.—Sire, Lemon Peel's Duke (P. S. 479 J. H. B.); dam, Rather Pretty (F. S. 5933 J. H. B.)

Butter, 15 lbs. 2 oz. Milk, 236 lbs. 4 oz.

Test made from June 29 to July 6, 1892; age, 8 yrs. 1 mo.; estimated weight, 800 lbs.; fed 42 qts. corn meal, 42 qts. ground oats, 3½ qts. oil meal and 21 pks. cut clover hay; property of Mrs. Flora A. Howe, Rochester, N. Y.

Vesper 1395, imp.

Butter, 14 lbs.

Property of G. Dawson Coleman, Brickerville, Pa.

Vespucia 17455.—Sire, Indiaman 2071; dam, Fenelle 9095.

Butter, 14 lbs. 4 oz.

Test made from Oct. 11 to 18, 1883; age, 3 yrs. 7 mos.; fed slop of 2 gals. wheat bran and cut sheaf oats, twice daily—clover and timothy pasture; property of A. J. Fish, Lima, Ohio.

Vestina 2458.—Sire, Mercury 432; dam, Vesta 1235.

Butter, 14 lbs. 2 oz. Milk, 201 lbs. 1 oz.

Test made from Sept. 20 to 26, 1884; age, 12 yrs. 6 mos.; property of William Simpson, New York, N. Y.

Vexanella 79306.—Sire, Vexer 20889; dam, Joanella 19063.

Butter, 16 lbs. 5 oz. Milk, 218 lbs. 4 oz.

Test made from June 9 to 16, 1895; age, 4 yrs. 4 mos.; estimated weight, 900 lbs.; fed 7 lbs. per day of a mixture of one part cotton-seed meal to two parts wheat middlings—pasture of natural grasses; property of George W. Sisson, Jr., Potsdam, N. Y.

Victor Hugo's Maggie 30274.—Sire, Prince of Melrose 2d 11015; dam, Maggie of Springvale 15931.

Butter, 18 lbs. 14 oz. Milk, 242 lbs. 8 oz.

Test made from May 13 to 20, 1888; age, 3 yrs. 6 mos.; estimated weight, 800 lbs.; fed 70 qts. corn meal, 42 qts. bran and shorts, 42 qts. cotton-seed meal and 140 lbs. ensilage—hay, good clover pasture; property of James Crook, Jacksonville, Ala.

Victoria 3175.—Sire, Ned 523; dam, Jessie 3207.

Butter, 16 lbs. 1 oz.

Property of W. L. & W. Rutherford, Waddington, N. Y.

Victor's Lady 52691.—Sire, Victor F. W. 16548; dam, Lady Livingston 33374.

Butter, 15 lbs. 14 oz. Milk, 203 lbs. 9 oz.

Test made from June 23 to 30, 1892; age, 3 yrs. 10 mos.; estimated weight, 1000 lbs.; fed 17½ qts. corn meal, bran and oil meal, daily, mixed with 1½ bush. cut clover hay; property of George E. Peer, Rochester, N. Y.

Victor's Matilda 46834.—Sire, L. D.'s Victor St. Helier 15286; dam, Rioter's Daisy M. 35374.

Butter, 14 lbs. 6 oz. Milk, 237 lbs. 4 oz.

Test made from July 28 to Aug. 4, 1893; age, 5 yrs. 9 mos.; estimated weight, 825 lbs.; fed 6 qts. mixed feed, 4 qts. Buffalo gluten, 4 qts. coarse linseed meal and 4 qts. oat feed, daily—green oats and peas; property of C. H. Ellsworth, Worcester, Mass.

Victor's May 97958.—Sire, Victor Scituate 2d 26330; dam, May Bijou 63911.

Butter, 14 lbs. 8 oz. Milk, 235 lbs. 12 oz.

Test made from May 28 to June 4, 1897; age, 3 yrs. 1 mo.; estimated weight, 700 lbs.; fed 25 lbs. 2 oz. ground corn and oats (one-third corn and two-thirds oats), 19 lbs. bran, 7 lbs. corn meal and 5 lbs. oil meal—timothy and clover pasture; property of W. W. Bontecon, Spring Valley, Minn.

Victory 16379, imp.

Butter, 15 lbs. 4½ oz., unsalted. Milk, 277 lbs.

Test made from Apr. 16 to 23, 1883; age, 6 yrs.; property of V. E. Fuller, Hamilton, Ont., Can.

Vidalia 101047.—Sire, Sophomore 24253; dam, Vidalinne 50101.

Butter, 14 lbs. 8 oz. Milk, 217 lbs.

Test made from Aug. 26 to Sept. 2, 1897; age, 3 yrs. 6 mos.; estimated weight, 750 lbs.; fed 28 lbs. corn meal, 21 lbs. oats, 28 lbs. bran and 14 lbs. oil meal—mixed pasture; property of H. C. Taylor, Orfordville, Wis.

Vidalinne 50101.—Sire, Young Combination 14550; dam, Violette Foster 20405.

Butter, 14 lbs. 2 oz. Milk, 211 lbs.

Test made from May 30 to June 6, 1897; age, 10 yrs. 8 mos.; estimated weight, 875 lbs.; fed 42 lbs. corn meal, 14 lbs. oil meal, 21 lbs. rye and 48 lbs. oats—mixed pasture; property of H. C. Taylor, Orfordville, Wis.

Vida's Crown Princess 76010.—Sire, Crown Prince of St. L. 20070; dam, Vida Pogis of Bois d'Arc 60205.

Butter, 16 lbs. 9 oz. Milk, 230 lbs. 7 oz.

Test made from Dec. 8 to 15, 1895; age, 4 yrs. 9 mos.; estimated weight, 800 lbs.; fed 72 lbs. corn and cob meal, 28 lbs. bran, 20 lbs. cotton-seed meal and 315 lbs. ensilage; property of Austin & Probert, Florence, Mich.

Vida's Crown Princess 76010.—Sire, Crown Prince of St. L. 20070; dam, Vida Pogis of Bois d'Arc 60205.

Butter, 67 lbs. 13½ oz. in 30 days. Milk, 969 lbs. 13 oz.

Test made from Dec. 8, 1895, to Jan. 7, 1896; age, 4 yrs. 9 mos.; estimated weight, 800 lbs.; fed 371 lbs. corn and cob meal, 143 lbs. bran, 86 lbs. cotton-seed meal and 1350 lbs. ensilage; property of Austin & Probert, Florence, Mich.

Vieva 3d 7642.—Sire, Sydney 3262; dam, Vieva 2117.

Butter, 16 lbs. 5 oz. Milk, 311 lbs. 12 oz.

Test made from July 15 to 21, 1883; age, 4 yrs. 10 mos.; property of John E. Phillips, Baltimore, Md.

Vigo Princess 60806.—Sire, Vigo King 19238; dam, Lady Bobita 2d 43571.

Butter, 18 lbs. Milk, 350 lbs.

Test made from May 21 to 28, 1893; age, 4 yrs. 1 mo.; estimated weight, 900 lbs.; fed 420 lbs. mash and 35 lbs. crushed feed—poor blue-grass pasture; property of Jacob Warner, Chillicothe, Ohio.

Vilify 50107.—Sire, Young Combination 14550; dam, Violette Foster 20405.

Butter, 14 lbs. 3 oz. Milk, 227 lbs. 14 oz.

Test made from Jan. 2 to 9, 1891; age, 3 yrs.; estimated weight, 800 lbs.; fed, daily, 18 qts. ground corn, oats and peas; property of Richardson Bros., Davenport, Iowa.

Vinala H. 61835.—Sire, Tamy's Rioter 15288; dam, Ardell H. 12330.

Butter, 16 lbs. Milk, 196 lbs. 4 oz.

Test made from Oct. 1 to 8, 1895; age, 6 yrs. 9 mos.; estimated weight, 800 lbs.; fed 8 qts. H. O. dairy feed and 4 qts. corn meal, daily—natural pasture; property of George W. Sisson, Jr., Potsdam, N. Y.

Vina Marigold 108186.—Sire, Stoke Pogis of Prospect 29121; dam, Regalia 64574.

Butter, 14 lbs. 8½ oz. Milk, 211 lbs. 4 oz.

Test made from Aug. 22 to 29, 1897; age, 2 yrs. 5 mos.; actual weight, 860 lbs.; fed 9 qts. ground oats, 12 qts. wheat bran, 3 qts. corn meal and 3 qts. oil meal, daily—hay, good clover pasture; property of C. A. Sweet, East Aurora, N. Y.

Vintage 53335.—Sire, Diploma 16219; dam, Belna 41043.

Butter, 14 lbs. 12 oz. Milk, 274 lbs. 3 oz.

Test made from June 22 to 29, 1897; age, 8 yrs. 11 mos.; estimated weight, 900 lbs.; fed 49 lbs. bran, 17½ lbs. corn meal, 21 lbs. ground oats, 7 lbs. oil meal and 10½ lbs. cotton-seed meal—mixed clover, timothy and June grass pasture; property of C. I. Hood, Lowell, Mass.

Vintonette 62931.—Sire, Combine 20238; dam, Vidalinne 50101.

Butter, 19 lbs. 4 oz. Milk, 251 lbs.

Test made from Jan. 14 to 21, 1894; age, 5 yrs. 5 mos.; estimated weight, 1000 lbs.; fed 28 lbs. oats, 28 lbs. corn, 35 lbs. bran and 21 lbs. oil meal—corn ensilage; property of H. C. Taylor, Orfordville, Wis.

Viola of Briarcliff 37617.—Sire, Young Garenne's Duke 6863; dam, Violet of Briarcliff 24186.

Butter, 14 lbs. 8 oz. Milk, 152 lbs. 9 oz.

Test made from May 31 to June 7, 1888; age, 2 yrs. 2 mos.; estimated weight, 650 lbs.; fed, daily, 8 lbs. corn meal and 4 lbs. middlings; property of James Stillman, Sing Sing, N. Y.

Violente 28819.—Sire, Prince George 11571; dam, Violente (F. S. 4566 J. H. B.)

Butter, 17 lbs. 10½ oz. Milk, 284 lbs.

Test made from June 21 to 28, 1890; age, 6 yrs. 4 mos.; estimated weight, 700 lbs.; fed, daily, 7 lbs. ground oats, 1 lb. oil meal and average of 4½ lbs. corn meal; property of James T. Henderson, Auvergne, Ark.

Violet 3d 3240, imp.

Butter, 15 lbs. 8 oz.

Property of T. S. Cooper, Coopersburg, Pa.

Violet Denison 49997.—Sire, Lord Harry 3445; dam, Moyane 21595.

Butter, 14 lbs. 13½ oz. Milk, 250 lbs.

Test made from May 27 to June 3, 1893; age, 6 yrs.; estimated weight, 800 lbs.; fed 28 qts. meal, 28 qts. middlings, 28 qts. ground oats and 70 lbs. hay—poor pasture; property of H. A. Huntington, Higganum, Conn.

Violet of Glencairn 10221.—Sire, Fortunatus 1152; dam, Mollie Brown 7831.

      Butter, 14 lbs. 4 oz.   Milk, 219 lbs.

Test made from July 13 to 20, 1883; age, 3 yrs. 4 mos.; fed 12 qts. ground oats, daily —clover pasture; property of V. E. Fuller, Hamilton, Ont., Can.

Violet of Maplewood 49724.—Sire, Stoke Pogis Jr. 15300; dam, Belle of Maplewood 22942.

      Butter, 14 lbs. 10¼ oz.   Milk, 192 lbs. 4 oz.

Test made from June 4 to 11, 1889; age, 3 yrs. 2 mos.; estimated weight, 800 lbs.; fed, daily, 7 lbs. corn meal, 2 lbs. cotton-seed meal and 3 lbs. shorts; property of Rufus A. Sibley, Spencer, Mass.

Vionetta Pogis 91978.—Sire, Manifold Rioter Pogis 26158; dam, Viozetta Pogis 56294.

      Butter, 17 lbs. 2 oz.   Milk, 250 lbs.

Test made from May 26 to June 2, 1898; age, 6 yrs. 5 mos.; estimated weight, 1000 lbs.; fed 10 lbs. daily of corn and oats chop, mixed with gluten and bran in equal parts by weight—blue-grass pasture; property of E. B. Jones, Mount Pleasant, Ohio.

Viozetta Pogis 56294.—Sire, Violator 17940; dam, Spot's Grisette 25905.

      Butter, 14 lbs. 5½ oz.   Milk, 258 lbs. 8 oz.

Test made from Feb. 28 to Mar. 7, 1895; age, 6 yrs. 1 mo.; estimated weight, 1100 lbs.; fed 42 lbs. bran, 14 lbs. gluten meal, 14 lbs. old process linseed meal and 56 lbs. hay—40 lbs. ensilage per day; property of E. B. Jones, Mount Pleasant, Ohio.

Virgie Landseer 74831.—Sire, Leonette's Landseer 20399; dam, Mercy Blossom 26453.

      Butter, 14 lbs. 15 oz.   Milk, 242 lbs.

Test made from June 29 to July 6, 1897; age, 7 yrs. 5 mos.; estimated weight, 900 lbs.; fed 42 lbs. wheat bran and 80½ lbs. corn meal—woods pasture; property of W. E. Johnson, Millican, Tex.

Virgina 80904.—Sire, Jersey King 9458; dam, Virginia Louise 42402.

      Butter, 16 lbs. 11½ oz.   Milk, 280 lbs. 13½ oz.

Test made from May 4 to 11, 1895; age, 4 yrs. 2 mos.; estimated weight, 775 lbs.; fed 84 qts. wheat bran, 42 qts. corn meal, 21 qts. cotton-seed meal, 14 qts. linseed meal and 126 lbs. ensilage—little green clover, short pasture; property of E. B. C. Hambley, Rockwell, N. C.

Virginia's Oonan 58107.—Sire, Landseer's Pogis 15847; dam, Pearl's Oonan 32288.

      Butter, 15 lbs. 3 oz.   Milk, 127 lbs. 6 oz.

Test made from Aug. 21 to 28, 1890; age, 2 yrs.; estimated weight, 600 lbs.; fed, daily, 3½ gals. ground corn and 1 gal. oats; property of Miss Virginia Webster, Columbia, Tenn.

Virginia Taylor 117964.—Sire, Oonan's Tormentor Pogis 30505; dam, Torette 75487.

      Butter, 17 lbs. 2¾ oz. (confirmed); estimated butter on basis of 85 per cent. fat, 16 lbs. 13¾ oz.   Milk, 278.4 lbs.

Test made from Apr. 2 to 8, 1898; age, 3 yrs. 2 mos.; average weight, 829 lbs.; fed 56 lbs. corn meal, 28 lbs. cotton-seed meal, 14 lbs. oil meal, 42 lbs. bran and 70 lbs. beets; property of Mrs. J. W. Sayre, Lexington, Ky.

Analyses of Butter.

First churning—Fat 81.46, water 12.28, ash 5.39, casein 0.87.
Second churning—Fat 83.02, water 13.54, ash 2.53, casein 0.91.

Viva Le Brocq 13702.—Sire, Le Brocq's Prize 3350; dam, Violette 7471.

Butter, 18 lbs. 3 oz. Milk, 206 lbs. 6 oz.

Test made from Aug. 31 to Sept. 6, 1885; age, 4 yrs. 5 mos.; property of G. B. & C. S. Smith, Eagle, Mich.

Vivian 15813.—Sire, Black Defiance 4014; dam, Juno W. 8553.

Butter, 16 lbs. Milk, 193 lbs. 8 oz.

Test made from Nov. 18 to 25, 1886; age, 4 yrs. 6 mos.; fed, daily, 6 qts. corn meal, 6 qts. oats and 6 qts. shorts; property of W. H. Haley, North Wilmington, Mass.

Vivienne Pogis 41547.—Sire, Sir George of St. Lambert 6036; dam, Maude Vivienne 2d 30755.

Butter, 15 lbs. 1 oz. Milk, 314 lbs.

Test made from Jan. 29 to Feb. 5, 1893; age, 6 yrs. 1 mo.; estimated weight, 1150 lbs.; fed 2 bush. cut hay, 6½ lbs. ground oats, 4 lbs. shorts, 6½ lbs. corn meal, 2 lbs. linseed meal, 6 qts. carrots and mangels, mixed, and 4 lbs. loose hay, daily; property of Horatio J. Gilbert, Milton, Mass.

Vixen 7591.—Sire, Top-Sawyer 1404; dam, Roxana 1761.

Butter, 17 lbs. 6 oz., unsalted. Milk, 196 lbs. 12 oz.

Test made from Dec. 12 to 19, 1883; age, 5 yrs. 6 mos.; property of M. M. Gardner, Nashville, Tenn.

V. K.'s May 41929.—Sire, Victor's King 12678; dam, May of Lakeside 10826.

Butter, 14 lbs. 8 oz. Milk, 204 lbs. 15 oz.

Test made from Oct. 19 to 26, 1892; age, 6 yrs.; estimated weight, 900 lbs.; fed 42 qts. corn meal, 42 qts. shorts and 14 qts. pea meal—short pasture; property of C. H. Ellsworth, Worcester, Mass.

Volie 19465.—Sire, Oxoli 1922; dam, Safrano 4568.

Butter, 18 lbs. 1 oz., unsalted.

Test made from June 12 to 18, 1883; age, 6 yrs. 2 mos.; no feed—very good old pasture; property of Henry Sanford, Bridgeport, Conn.

Volie of Sennett 2d 78638.—Sire, Jalisco of Sennett 21581; dam, Volie of Sennett 49059.

Butter, 14 lbs. 13 oz. Milk, 242 lbs.

Test made from Apr. 13 to 20, 1897; age, 7 yrs.; actual weight, 870 lbs.; fed 42 qts. bran, 42 qts. oat meal, 35 qts. corn meal, 21 qts. oil meal, 105 lbs. ensilage, 105 lbs. hay and 56 qts. roots; property of P. J. Cogswell, Rochester, N. Y.

Vone 89138.—Sire, Favorite's Prize 33042; dam, Levon 17319.

Butter, 14 lbs. 3 oz. Milk, 238 lbs. 13 oz.

Test made from May 26 to June 2, 1897; age, 4 yrs.; actual weight, 790 lbs.; fed 2 lbs. corn, 3 lbs. bran, 3 lbs. oats, 1 lb. gluten and 1 lb. oil meal, daily—blue-grass pasture; property of George E. Scott, Mount Pleasant, Ohio.

Waiter Girl 12776.—Sire, Elkornah 4401; dam, Millie Waite 10646.

Butter, 16 lbs. 7 oz. Milk, 228 lbs.

Test made from May 2 to 9, 1886; age, 5 yrs. 8 mos.; property of McBride & Cogswell, Rochester, N. Y.

Waiter Girl 2d 29265.—Sire, Mercury Boy 5721; dam, Waiter Girl 12776.

Butter, 18 lbs. 14½ oz. Milk, 225 lbs. 8 oz.

Test made from June 7 to 14, 1888; age, 5 yrs. 10 mos.; estimated weight, 900 lbs.; fed, daily, 6¾ lbs. corn meal, 7 lbs. each of bran, middlings and oil meal, and 3¾ lbs. oats; property of A. D. McBride, Rochester, N. Y.

Walkyrie 5708.—Sire, Ginx 1536; dam, Regina 2d 2475.

Butter, 14 lbs. 1 oz. Milk, 169 lbs. 12 oz.

Test made in Sept., 1883; age, 6 yrs. 5 mos.; property of H. M. Howe, Bristol, R. I.

Wamla 32083.—Sire, Combination 4389; dam, Wamba 30641.

Butter, 17 lbs. 5½ oz. Milk, 262 lbs. 14 oz.

Test made from Oct. 1 to 8, 1888; age, 4 yrs. 1 mo.; estimated weight. 800 lbs.; fed, daily, 24 qts. meal and bran, with a little oil meal; property of Richardson Bros., Davenport, Iowa.

Wanderer's Shiloh 65942.—Sire, Azelda's Wanderer 12997; dam, Shiloh Daughter 20378.

Butter, 15 lbs. Milk, 215 lbs.

Test made from Mar. 14 to 21, 1893; age, 5 yrs. 11 mos.; estimated weight, 900 lbs.; fed 70 lbs. bran, 60 lbs. corn meal and 45 lbs. oil meal; property of Cornelius Easthope, Niles, Ohio.

Wandrina 2d 82420.—Sire, Landseer's Pogis 15847; dam, Wandrina 44939.

Butter, 15 lbs. 9½ oz. Milk, 230 lbs.

Test made from Apr. 22 to 29, 1893; age, 3 yrs. 7 mos.; estimated weight, 750 lbs.; fed 8 qts. crushed oats, 8 qts. corn meal and 4 qts. wheat bran, daily—ordinary pasture; property of Morgan & Brown, Columbia, Tenn.

Wapello 56044.—Sire, Sindbad 12294; dam, Liberty Ann 56029.

Butter, 14 lbs. 12 oz. Milk, 204 lbs. 8 oz.

Test made from May 13 to 20, 1890; age, 4 yrs. 1 mo.; estimated weight, 700 lbs.; fed. daily, 14 lbs. wheat bran and 3 lbs. oil meal; property of J. L. Comstock, Sac City, Iowa.

Wardalia 2d 33970.—Sire, Oxford Rioter 5992; dam, Wardalia 12669.

Butter, 24 lbs. 6 oz. Milk, 293 lbs.

Test made from Jan. 10 to 17, 1897; age, 11 yrs. 8 mos.; estimated weight, 1000 lbs.; fed about 24 qts. daily of a mixture of ground oats, middlings, bran, corn meal and some pea meal—clover hay and ensilage; property of T. S. Cooper, Coopersburg, Pa.

Warren's Duchess 4622.—Sire, Optimus 1607; dam, Countess of Warren 3896.

Butter, 16 lbs. 1 oz. Milk, 250 lbs. 3 oz.

Test made from Aug. 20 to 26, 1882; age, 8 yrs. 6 mos.; property of C. Bordwell, Batavia, Ohio.

Warren's Duchess 3d 26836.—Sire, Champion's Son 3286; dam, Warren's Duchess 4622.

Butter, 17 lbs. 2½ oz. Milk, 210 lbs. 14 oz.

Test made from June 1 to 7, 1886; age, 3 yrs.; property of Bordwell & Cochran, Batavia, Ohio.

Waukesra 19721.—Sire, Tom Brown 2940; dam, Gipsy Countess 9188.

Butter, 16 lbs. Milk, 314 lbs. 8 oz.

Test made from June 22 to 28, 1883; age, 3 yrs. 2 mos.; property of A. P. Foster, Plainview, Minn.

Waxie 19706.—Sire, Beaconsfield 3416; dam, Beeswax 9807.

Butter, 20 lbs. 2½ oz. Milk, 218 lbs.

Test made from Apr. 18 to 25, 1889; age, 7 yrs. 3 mos.; estimated weight, 850 lbs.; fed, daily, 5 gals. of a mixture consisting of one-half corn meal, one-quarter wheat bran and one-quarter ground oats; property of Morgan & Brown, Columbia, Tenn.

Waxie 2d 83358.—Sire, Landseer's Pogis 15847; dam, Waxie 19706.

Butter, 14 lbs. 8½ oz. Milk, 132 lbs.

Test made from Mar. 5 to 12, 1895; age, 4 yrs.; estimated weight, 650 lbs.; fed 56 qts. ground oats and 84 qts. corn meal—clover and millet hay; property of Morgan & Brown, Columbia, Tenn.

Waxie of Aurora 86955.—Sire, Bisson's Landseer 27520; dam, Waxie 19706.

Butter, 17 lbs. 4 oz. Milk, 218 lbs.

Test made from Sept. 2 to 9, 1895; age, 2 yrs. 7 mos.; actual weight, 855 lbs.; fed 9 qts. wheat bran, 6 qts. ground oats, 3 qts. corn meal and 1½ qts. oil meal, mixed, daily—poor pasture, supplemented with green corn; property of C. A. Sweet, East Aurora, N. Y.

Webster Pet 4103.—Sire, Champion of America 1567; dam, Churchill Betsey 4105.

Butter, 14 lbs. 2 oz. Milk, 197 lbs.

Test made from Jan. 15 to 22, 1881; age, 6 yrs.; fed 5 lbs. oat meal and 2 lbs. corn meal, daily—hay; property of W. H. Walrath, La Fargeville, N. Y.

Well Done 25987.—Sire, Forget-me-not 6291; dam, Village Lass, on I of J.

Butter, 19 lbs. 4 oz. Milk, 260 lbs.

Test made in June, 1885; age, 3 yrs. 4 mos.; property of Jewett M. Richmond, Buffalo, N. Y.

Well-to-do 37724.—Sire, Disturbance 9684; dam, Well Met 24571.

Butter, 14 lbs. 3½ oz. Milk, 179 lbs. 8 oz.

Test made from Jan. 26 to Feb. 2, 1890; age, 3 yrs. 11 mos.; estimated weight, 900 lbs.; well fed, but no record of feed kept; property of Campbell Brown, Spring Hill, Tenn.

Welma 5942.—Sire, Marius 760; dam, Annie Wells 1947.

Butter, 17 lbs. 8 oz., unsalted. Milk, 252 lbs. 1 oz.

Test made from Oct. 2 to 8, 1882; property of Ariel Low, Jr., Warrensville, Ill.

Westphalia 24384.—Sire, Kisber 4377; dam, Gold Mark 24361.

Butter, 24 lbs. 9½ oz.

Age, 5 yrs.; property of Mrs. S. M. Shoemaker, Baltimore, Md.

Whisk of Meridale `131023.—Sire, Ben D. 24164; dam, Miss Whisk 38640.

Butter, 15 lbs. Milk, 204 lbs. 11 oz.

Test made from Jan. 25 to Feb. 1, 1898; age, 7 yrs. 3 mos.; actual weight, 960 lbs.; fed 50 lbs. wheat bran, 42 lbs. corn meal, 7 lbs. linseed meal, 7 lbs. cotton-seed meal and 2 bush. carrots and beets—about 200 lbs. ensilage and 50 lbs. hay; property of Ayer & McKinney, Philadelphia, Pa.

White Frost 17431, imp.

Butter, 16 lbs. 2 oz. Milk, 295 lbs. 8 oz.

Test made from Jan. 9 to 15, 1885; age, 4 yrs. 11 mos.; property of William Crozier, Northport, N. Y.

Whiting's Daisy 14858.—Sire, Brisk 2077; dam, Brewer's Daisy 7333.

Butter, 14 lbs. 10½ oz. Milk, 229 lbs. 2 oz.

Test made from June 25 to July 1, 1886; age, 4 yrs. 9 mos.; property of Columbia Jersey Cattle Co., Columbia, Tenn.

Wild Crocus 92486.—Sire, Rab 22937; dam, Young Idea 71458.

Butter, 18 lbs. 2¾ oz. Milk, 263 lbs. 8 oz.

Test made from May 14 to 21, 1897; age, 4 yrs. 7 mos.; actual weight, 820 lbs.; fed 56 qts. ground oats, 21 qts. ground oil-cake and 28 qts. corn meal—good old meadow pasture; property of Miller & Sibley, Franklin, Pa.

Wild Rose Bloom 110583.—Sire, Sig Pogis 22806; dam, Helen Walker 53465.

Butter, 17 lbs. 1¼ oz. Milk, 248 lbs.

Test made from Apr. 17 to 24, 1898; age, 3 yrs. 2 mos.; estimated weight, 700 lbs.; fed 56 qts. ground oats and corn, mixed, and 28 qts. bran—rye pasture; property of D. P. Carter, Lynnville, Tenn.

Willimenia H. 12325.—Sire, Homer H. 3683; dam, Lucy Gray 2746.

Butter, 20 lbs. 14 oz. Milk, 404 lbs. 14 oz.

Test made from June 22 to 28, 1885; age, 7 yrs. 4 mos.; fed 8 qts. corn and oats, mixed, daily—hill pasture; property of William H. Hayden, Albany, Vt.

Willis 2d 4461.—Sire, Iron Bank 1120; dam, Willis 3573.

Butter, 16 lbs. 3 oz. Milk, average of 24 qts. daily.

Test made from Oct. 10 to 16, 1880; age, 5 yrs. 4 mos.; property of G. Dawson Coleman, Brickerville, Pa.

Winder Gem 90447.—Sire, Winder 29955; dam, Little Germ 19965.

Butter, 14 lbs. Milk, 193 lbs. 3 oz.

Test made from Feb. 12 to 19, 1898; age, 4 yrs. 4 mos.; estimated weight, 950 lbs.; fed 5 lbs. ground oats, 5 lbs. wheat bran, 1½ lbs. old process oil meal and 20 lbs. ensilage, daily—corn stover and clover hay *ad lib.*; property of N. N. Palmer & Son, Brodhead, Wis.

Wine 15739.—Sire, Orawapum 2833; dam, Corn 10504.

Butter, 16 lbs. 4 oz., unsalted. Milk, 232 lbs. 5 oz.

Test made from May 14 to 21, 1886; age, 4 yrs.; property of W. H. McCurdy, Cleveland, Ohio.

Winetka 33669.—Sire, L'Allegro 7127; dam, Alice Donnal 12726.

Butter, 20 lbs., unsalted. Milk, 194 lbs. 8 oz.

Test made from June 8 to 14, 1886; age, 2 yrs. 1 mo.; fed 2 gals. corn meal, 1½ gals. wheat bran and ½ gal. oil meal, mixed, daily—poor pasture, green cut clover *ad lib.*; property of D. R. Nelson, Loudon, Tenn.

Winifred Victor Pogis 91658.—Sire, Melia Ann's Victor Pogis 20697; dam, Pet Victor Pogis 77812.

Butter, 14 lbs. 1¾ oz. Milk, 221 lbs. 4 oz.

Test made from June 1 to 8, 1894; age, 2 yrs. 5 mos.; estimated weight, 500 lbs.; fed about 105 lbs. grain—good mixed pasture, on clover night and morning; property of J. H. Martin, Granville, N. Y.

Winifred Victor Pogis 91658.—Sire, Melia Ann's Victor Pogis 20697; dam, Pet Victor Pogis 77812.

Butter, 22 lbs. 9 oz. Milk, 319 lbs. 4 oz.

Test made from June 18 to 20, 1898; age, 6 yrs. 6 mos.; estimated weight, 800 lbs.; fed 8 lbs. corn meal and 5 lbs. wheat bran, daily—pasture of mixed grasses; property of Mrs. J. H. Martin, Granville, N. Y.

Winksette 99280.—Sire, Lord Landseer 24228; dam, Romena 72300.

Butter, 23 lbs. Milk, 276 lbs. 8 oz.

Test made from Sept. 29 to Oct. 6, 1896; age, 3 yrs. 9 mos.; estimated weight, 975 lbs.; fed 70 lbs. of a mixture of equal parts of cotton-seed, wheat bran, crushed oats and corn meal—Kaffir corn stalks; property of S. T. Howard, Quanah, Tex.

Winsome of Ipswich 9213.—Sire, Duke Jr. 2465; dam, Amy of Ipswich 5583.

Butter, 14 lbs. 7 oz.

Test made in 1884; age, 5 yrs.; property of J. H. Towle, Westfield, Mass.

Witch-Hazel 1360.—Sire, Southampton 117; dam, Hazel 91.

Butter, 14 lbs. Milk, a fraction over 10 qts. daily.

Test made Dec., 1879; age, 9½ yrs.; property of Thos. J. Hand, Sing Sing, N. Y.

Witch-Hazel 4th 6131.—Sire, Lord Lawrence 1414; dam, Witch-Hazel 1360.

Butter, 15 lbs. 5½ oz. Milk, 260 lbs. 8 oz.

Test made from May 28 to June 3, 1882; age, 5 yrs. 6 mos.; property of Campbell Brown, Spring Hill, Tenn.

Witch-Hazel 4th 6131.—Sire, Lord Lawrence 1414; dam, Witch-Hazel 1360.

Butter, 16 lbs. 5 oz. Milk, 210 lbs. 8 oz.

Test made from Mar. 28 to Apr. 8, 1886; age, 9 yrs. 4 mos.; property of Campbell Brown, Spring Hill, Tenn.

Witch-Hazel 4th 6131.—Sire, Lord Lawrence 1414; dam, Witch-Hazel 1360.

Butter, 18 lbs. 2⅞ oz. Milk, 260 lbs. 8 oz.

Test made from Apr. 21 to 28, 1887; age, 10 yrs. 5 mos.; estimated weight, 1000 lbs.; no record of feed kept, but well fed; property of Campbell Brown, Spring Hill, Tenn.

Witching 42252.—Sire, Southern Prince 10760; dam, Little Witch 35813.

Butter, 17 lbs. 14½ oz. Milk, 192 lbs.

Test made from Dec. 29, 1889, to Jan. 5, 1890; age, 3 yrs. 3 mos.; estimated weight, 900 lbs.; fed, daily, 20 lbs. bran and corn chop; property of Campbell Brown, Spring Hill, Tenn.

Witching Hour 54947.—Sire, Odelio 20223; dam, Witching 42252.

Butter, 16 lbs. 13½ oz. Milk, 196 lbs.

Test made from Dec. 9 to 16, 1894; age, 6 yrs. 3 mos.; estimated weight, 700 lbs.; fed 2 qts. corn-hearts, 2 qts. bran, 1 qt. cotton-seed meal and 2 qts. ground oats, daily—sheaf oats, cut clover hay and blue-grass pasture; property of Samuel N. Warren, Spring Hill, Tenn.

Woodland Maid 8156.—Sire, Browser 1624; dam, Clarencia 3484.

Butter, 14 lbs. 2 oz. Milk, 258 lbs. 6 oz.

Test made from June 15 to 21, 1884; age, 6 yrs. 6 mos.; property of C. Bordwell, Batavia, Ohio.

Woodland Maid 2d 19666.—Sire, Cazique 3103; dam, Woodland Maid 8156.

Butter, 14 lbs. Milk, 294 lbs.

Test made from June 3 to 10, 1886; age, 4 yrs. 6 mos.; property of Bordwell & Cochran, Batavia, Ohio.

Woodland Rose 68381.—Sire, Jersey King 9458; dam, Virginia Louise 2d 60838.

Butter, 14 lbs. 6½ oz. Milk, 247 lbs. 2 oz.

Test made from Feb. 27 to Mar. 5, 1896; age, 5 yrs. 4 mos.; estimated weight, 750 lbs.; fed 42 lbs. wheat bran, 42 lbs. corn meal, 14 lbs. cotton-seed meal, 7 lbs. linseed meal, 245 lbs. ensilage and 42 lbs. hay; property of Egbert B. C. Hambley, Rockwell, N. C.

Woodstock Lady 80619.—Sire, Woodstock Chief 21172; dam, Lady Jane of St. Peter's 7475.

Butter, 23.027 lbs. in 21 days. Milk, 398.1 lbs.

Test made from Sept. 30 to Oct. 20, 1893, at World's Columbian Exposition; age, 2 yrs. 2 mos.; actual weight, 691 lbs.; fed 201 lbs. hay, 6 lbs. ensilage, 42 lbs. oil meal, 84 lbs. corn meal, 126 lbs. bran, 63 lbs. oats, 42 lbs. cotton-seed and 42 lbs. middlings; property of Est. of Frederick Billings, Woodstock, Vt.

Woodstock Mystery 77746.—Sire, Woodstock Chief 21172; dam, Alrajah's Mystery 48466.

Butter, 26.223 lbs. in 21 days. Milk, 388.6 lbs.

Test made from Sept. 30 to Oct. 20, 1893, at World's Columbian Exposition; age, 2 yrs. 2 mos.; actual weight, 660 lbs.; fed 176 lbs. hay, 6 lbs. ensilage, 38 lbs. oil meal, 61 lbs. corn meal, 114 lbs. bran, 57 lbs. oats, 24.5 lbs. cotton-seed and 19 lbs. mids; property of Est. of Frederick Billings, Woodstock, Vt.

Worthy Noble 19629.—Sire, Noble 4th 6246; dam, Pride of Sidney 19210.

Butter, 15 lbs. 9½ oz. Milk, 166 lbs. 8 oz.

Test made from Apr. 11 to 18, 1889; age, 7 yrs.; estimated weight, 950 lbs.; fed, daily, about 26 lbs. (one-third corn meal and one-fourth bran, the remainder fine feed), with a very little oil meal; property of P. J. Cogswell, Rochester, N. Y.

Wybie 595.—Sire, Charleston 1; dam, Princess 836.

Butter, 17 lbs. 4 oz.

Test made in June, 1875; property of O. S. Hubbell, Stratford, Conn.

Wyoming 48849.—Sire, Butternut of Home Farm 6965; dam, Duchess of Green Hill 15804.

Butter, 15 lbs. 5 oz. Milk, 234 lbs. 7 oz.

Test made from May 14 to 21, 1893; age, 5 yrs. 10 mos.; estimated weight, 900 lbs.; fed 8 qts. corn meal, 8 qts. ground oats, 4 qts. wheat middlings, 2 qts. oil meal, daily—pasture; property of George Fox, Torresdale, Pa.

Xarama 76887.—Sire, Tormentor 5th 21962; dam, Xeres 64235.

Butter, 14 lbs. 15¾ oz. Milk, 230 lbs. 14½ oz.

Test made from Dec. 18 to 25, 1894; age, 3 yrs.; estimated weight, 750 lbs.; fed 42 lbs. bran, 28 lbs. corn meal, 21 lbs. cotton-seed meal, 14 lbs. linseed meal, 28 lbs. corn fodder and 280 lbs. ensilage; property of E. B. C. Hambley, Rockwell, N. C.

Xarama 76887.—Sire, Tormentor 5th 21962; dam, Xeres 64235.

Butter, 15 lbs. 4½ oz. **Milk, 286** lbs. 8½ oz.

Test made from Dec. 25, 1894, to Jan. 1, 1895; age, 3 yrs.; estimated weight, 750 lbs.; fed 42 lbs. bran, 28 lbs. corn meal, 21 lbs. cotton-seed meal, 14 lbs. linseed meal, 28 lbs. corn fodder and 280 lbs. ensilage; property of E. B. C. Hambley, Rockwell, N. C.

Xarama 76887.—Sire, Tormentor 5th 21962; dam, Xeres 64235.

Butter, 15 lbs. 6¼ oz. Milk, 231 lbs. 11 oz.

Test made from Jan. 1 to 8, 1895; age, 3 yrs.; estimated weight, 750 lbs.; fed 42 lbs. bran, 28 lbs. corn meal, 21 lbs. cotton-seed meal, 14 lbs. linseed meal, 28 lbs. corn fodder and 280 lbs. ensilage; property of E. B. C. Hambley, Rockwell, N. C.

Xarama 76887.—Sire, Tormentor 5th 21962; dam, Xeres 64235.

Butter, 17 lbs. 7¼ oz. Milk, 254 lbs. 1 oz.

Test made from Dec. 26, 1895, to Jan. 2, 1896; age, 4 yrs.; estimated weight, 700 lbs.; fed 42 lbs. corn meal, 42 lbs. wheat bran, 14 lbs. ground oats and 245 lbs. corn ensilage; property of E. B. C. Hambley, Rockwell, N. C.

Xarama 76887.—Sire, Tormentor 5th 21962; dam, Xeres 64235.

Butter, 45 lbs. 10½ oz. in 21 days. Milk, 699 lbs. 2 oz.

Test made from Dec. 18, 1894, to Jan. 8, 1895; age, 3 yrs.; estimated weight, 750 lbs.; fed 126 lbs. bran, 84 lbs. corn meal, 63 lbs. cotton-seed meal, 42 lbs. linseed meal, 84 lbs. corn fodder and 840 lbs. ensilage; property of E. B. C. Hambley, Rockwell, N. C.

Xarifina 34176.—Sire, Trailer 7160; dam, Kosi 5th 10870.

Butter, 14 lbs. 9 oz. Milk, 206 lbs. 8 oz.

Test made from May 24 to 31, 1893; age, 8 yrs.; estimated weight, 750 lbs.; fed 12 qts. daily of a mixture of corn meal, middlings and coarse linseed meal, in equal parts—pasture; property of C. H. Ellsworth, Worcester, Mass.

Xyst 101247.—Sire, Gipsy's Lorne Pogis 20346; dam, Dorothy Smith 92095.

Butter, 23 lbs. 6½ oz. Milk, 342 lbs. 6 oz.

Test made from Feb. 9 to 16, 1898; age, 4 yrs. 5 mos.; actual weight, 900 lbs.; fed 10 lbs. ground oats, 10 lbs. wheat bran and 12 lbs. corn meal, daily; property of J. Gerow Dutcher, Pawling, N. Y.

Yankee Girl 2d 29261.—Sire, Catoctin 2692; dam, Yankee Girl 24832.

Butter, 14 lbs. 7 oz. Milk, 225 lbs. 8 oz.

Test made from June 15 to 22, 1886; age, 6 yrs. 2 mos.; property of P. J. Cogswell, Rochester, N. Y.

Yellow Belle 69780.—Sire, Yellow Star 13472; dam, Miss Belle Pogis 47038.

Butter, 14 lbs. 9½ oz. Milk, 287 lbs. 7 oz.

Test made from May 8 to 15, 1896; age, 6 yrs. 3 mos.; estimated weight, 950 lbs.; fed 100 lbs. corn and cob and oats, ground together, 21 lbs. bran and 26 lbs. oil meal, mixed—good natural pasture; property of W. H. O. Golst, Girard, Ohio.

Yellow Locust 10679.—Sire, Gilderoy 2107; dam, Idex 2d 5429.

Butter, 14 lbs. 10½ oz. Milk, 163 lbs.

Test made from July 6 to 12, 1883; age, 3 yrs. 3 mos.; property of H. M. Howe, Bristol, R. I.

Yokun Maid 19073.—Sire, Yokun Chief 4399; dam, Briseis 3875.

Butter, 16 lbs. 6 oz. Milk, 223 lbs. 9 oz.

Test made from May 17 to 23, 1886; age, 4 yrs. 3 mos.; property of S. W. Taliaferro, Guthrie, Ky.

Yokun Maid 2d 86563.—Sire, Landseer's Pogis 15847; dam, Yokun Maid 19073.

Butter, 14 lbs. 9½ oz. Milk, 204 lbs. 4 oz.

Test made from Apr. 13 to 20, 1895; age, 2 yrs. 11 mos.; estimated weight, 700 lbs.; fed 56 qts. ground oats and 56 qts. corn meal—clover and millet hay, short woodland pasture; property of Morgan & Brown, Columbia, Tenn.

Yokun's Princess 35643.—Sire, King Yokun 7068; dam, Isolda's Princess 12474.

Butter, 15 lbs. 10¼ oz. (official). Milk, 300 lbs. 9 oz.

Test made from June 17 to 24, 1895; age, 9 yrs. 11 mos.; estimated weight, 850 lbs.; fed 31 qts. corn meal, 15 qts. cotton-seed meal and 55 qts. wheat bran—timothy, red-top and clover pasture; property of O. N. Gilbert, Great Barrington, Mass.

Analyses of Butter.

Water 14.65, fat 82.65, curd 0.99, salt 1.71.

Young Anne Lee 31668.—Sire, Mahkeenac 3290; dam, Lady Anerly 10595.

Butter, 14 lbs. 7 oz. Milk, 212 lbs. 8 oz.

Test made from July 5 to 11, 1887; age, 4 yrs. 2 mos.; estimated weight, 700 lbs.; fed, daily, 4 lbs. corn meal, 2 lbs. oil meal and 8 lbs. bran; property of A. H. Cooley, Little Britain, N. Y.

Young Cherry 3d 14980.—Sire, Rough (P. S. 239 J. H. B.); dam, Young Cherry 14945.

Butter, 14 lbs. 7½ oz. Milk, 201 lbs. 8½ oz.

Test made from Mar. 23 to 29, 1886; age, 4 yrs. 4 mos.; property of D. A. Givens, Cynthiana, Ky.

Young Duchess 497.—Sire on I. of J.; dam, Duchess 101.

Butter, 15 lbs. 8 oz. Milk, 133½ qts.

Test made from June 30 to July 6, 1874; age, 6 yrs. 2 mos.; fed 3 qts. bran, 2 qts. corn meal and 1 qt. linseed oil cake, daily; property of H. S. Parker, Bay Side, N. Y.

Young Fanny 9032.—Sire, Jimmy, on I. of J.; dam, Lily, on I. of J.

Butter, 17 lbs. Milk, 144 qts.

Test made from May 26 to June 1, 1880; age, 3 yrs. 2 mos.; property of Isaac W. Stokes, Medford, N. J.

Young Floribundus 28172.—Sire, Theron 5722; dam, Floribundus 2d 14949.

Butter, 17 lbs. 12¾ oz. Milk, 200 lbs. 14 oz.

Test made from Dec. 24 to 31, 1890; age, 6 yrs. 7 mos.; estimated weight, 1050 lbs.; fed 16 qts. daily of a mixture of one part bran, two parts ground corn and three parts ground oats—bran mash daily, good hay *ad lib.*; property of L. L. Tozier, Batavia, N. Y.

Young Floribundus 3d 59312.—Sire, Lanison 15283; dam, Young Floribundus 28172.

Butter, 19 lbs. 14¾ oz. Milk, 230 lbs. 1 oz.

Test made from Nov. 1 to 8, 1890; age, 3 yrs. 6 mos.; estimated weight, 1050 lbs.; fed 16 qts. daily of a mixture of one part bran, two parts ground corn and three parts ground oats—bran slop daily, good hay *ad lib.*; property of L. L. Tozier, Batavia, N. Y.

Young Floribundus 4th 59314.—Sire, Ko Ko of St. Lambert 16527; dam, Young Floribundus 28172.

Butter, 15 lbs. 14 oz. Milk, 225 lbs.

Test made from May 18 to 25, 1891; age, 3 yrs. 1 mo.; estimated weight, 950 lbs.; fed 16 qts. daily of a mixture of one part bran, two parts ground corn and three parts ground oats—bran slop daily; property of L. L. Tozier, Batavia, N. Y.

Young Garenne 3d 13648.—Sire, Sir George 7656; dam, Young Garenne 13641.

Butter, 16 lbs. 3½ oz.

Test made in June, 1885; age, 4 yrs. 5 mos.; property of S. W. Robbins, Wethersfield, Conn.

Young Gazelle 19837.—Sire, Signal (F. S. 278 J. H. B.); dam, Gazelle 15666.

Butter, 14 lbs. 11 oz.

Test made in June, 1885; age, 4 yrs. 5 mos.; property of S. W. Robbins, Wethersfield, Conn.

Young Jersey Maiden 42794.—Sire, Signal Jr. 7166; dam, Maiden of Jersey 3d 21242.

Butter, 15 lbs. 10½ oz. Milk, 218 lbs.

Test made from Aug. 24 to 31, 1896; age, 10 yrs.; estimated weight, 1000 lbs.; fed 98 lbs. bran and 70 lbs. hominy meal—blue-grass pasture; property of J. E. Robbins, Jr., Greensburg, Ind.

Young Lass 67471.—Sire, Young Pedro 9033; dam, Saugus Lass 30542.

Butter, 17 lbs. 3½ oz. Milk, 252 lbs. 9 oz.

Test made from Aug. 2 to 9, 1893; age, 2 yrs. 11 mos.; estimated weight, 725 lbs.; fed 42 qts. corn meal, 42 qts. ground oats, 42 qts. shorts and 7 pts. oil meal—good pasture; property of D. F. Appleton, Ipswich, Mass.

Young Miss 99631.—Sire, Odelle's Pogis 25075; dam, Old Miss 84233.

Butter, 16 lbs. Milk, 200 lbs. 8 oz.

Test made from Apr. 22 to 29, 1898; age, 4 yrs. 6 mos.; estimated weight, 800 lbs.; fed 63 qts. wheat bran, 56 qts. corn and cob meal and 10½ qts. cotton-seed meal—hay *ad lib.*, good pasture; property of James L. Cooper, Nashville, Tenn.

Yuba Stoke Pogis 37294.—Sire, Exile of St. Lambert 13657; dam, Walter Girl 2d 29265.

Butter, 15 lbs. 14¾ oz. Milk, 231 lbs.

Test made from May 1 to 8, 1889; age, 3 yrs. 3 mos.; estimated weight, 750 lbs.; fed, daily, 26 lbs. on the first three days, and 28 lbs. on the last four days, of a mixture of one-fourth corn meal, one-fourth fine middlings and one-half bran; property of P. J. Cogswell, Rochester, N. Y.

Yum Yum of St. Lambert 39152.—Sire, Exile of St. Lambert 13657; dam, Floribundus 2d 14949.

Butter, 21 lbs. 4 oz. Milk, 219 lbs. 11 oz.

Test made from Jan. 8 to 15, 1891; age, 4 yrs. 9 mos.; estimated weight, 950 lbs.; fed 16 qts. of a mixture of one part bran, two parts ground corn and three parts ground oats, daily—bran mash daily, good hay; property of L. L. Tozier, Batavia, N. Y.

Yum Yum Signal 40688.—Sire, Signal's Laddie 11100; dam, Odette 13668.

Butter, 16 lbs. 7 oz. Milk, 198 lbs. 12 oz.

Test made from Nov. 8 to 15, 1891; age, 6 yrs. 3 mos.; estimated weight, 900 lbs.; fed 94 lbs. corn meal, 70 lbs. bran and 70 lbs. oats—clover, timothy and blue-grass pasture; property of Hall & Harris, Rushville, Ind.

Zalma 8778.—Sire, Mercury 432; dam, Hybla 2991.

Butter, 15 lbs. 5 oz. Milk, 221 lbs. 8 oz.

Test made from Feb. 15 to 21, 1883; age, 4 yrs. 11 mos.; fed corn and oats, ground together, a little oil meal, best clover hay and some roots; property of William Simpson, New York, N. Y.

Zélie de Lussan 48681.—Sire, Signally Signal 8875; dam, Printonnière 26086.

Butter, 20 lbs. 1 oz. Milk, 324 lbs.

Test made from June 3 to 10, 1893; age, 7 yrs.; estimated weight, 1000 lbs.; fed 4 qts. Buffalo gluten meal and 8 qts. mixed corn meal, oats, middlings and oil meal, daily—natural pasture; property of George W. Sisson, Jr., Potsdam, N. Y.

Zélie de Lussan 48681.—Sire, Signally Signal 8875; dam, Printonnière 26086.

Butter, 20 lbs. 7 oz. Milk, 326 lbs. 8 oz.

Test made from June 27 to July 4, 1893; age, 7 yrs. 1 mo.; estimated weight, 1000 lbs.; fed 8 qts. Buffalo gluten meal, 8 qts. mixed corn and oats, ground, and some oil meal, daily—natural pasture; property of George W. Sisson, Jr., Potsdam, N. Y.

Zélie de Lussan 48681.—Sire, Signally Signal 8875; dam, Printonnière 26086.

Butter, 86 lbs. 13 oz. in 31 days. Milk, 1446 lbs. 8 oz.

Test made from June 4 to July 5, 1893; age, 7 yrs. 1 mo.; estimated weight, 1000 lbs.; fed from 14 qts. to 18 qts. mixed grain per day—pasture of natural grasses; property of George W. Sisson, Jr., Potsdam, N. Y.

Zélie de Lussan 48681.—Sire, Signally Signal 8875; dam, Printonnière 26086.

Butter, 450 lbs. 5 oz. in 6 mos. Milk, 6864 lbs. 8 oz.

Test made from June 4 to Dec. 5, 1893; age, 7 yrs. 6 mos. at end of test; estimated weight, 1000 lbs.; fed 14 to 18 lbs. per day of a mixture of bran, corn meal and oil meal, with pasture until Nov. 1, then 40 lbs. ensilage per day, with hay *ad lib.*; property of George W. Sisson, Jr., Potsdam, N. Y.

Zélie's Zilpah 109344.—Sire, Zélie's Boy 30249; dam, Zilpah S. 108026.

Butter, 17 lbs. Milk, 226 lbs.

Test made from May 31 to June 7, 1898; age, 3 yrs. 7 mos.; estimated weight, 700 lbs.; fed 12 lbs. daily of a mixture of corn, oats, bran and pea meal—natural pasture; property of George W. Sisson, Jr., Potsdam, N. Y.

Zenie Lisgar 2d 87453.—Sire, Hilarious 7651; dam, Zenie Lisgar 31290.

Butter, 20 lbs. 14 oz. Milk, 270 lbs. 12 oz.

Test made from June 12 to 19, 1895; age, 5 yrs. 4 mos.; estimated weight, 900 lbs.; fed 14 gals. oat meal, 17½ gals. wheat bran, 7 gals. corn meal and 7 gals. cotton-seed meal—poor pasture; property of Milton Humes, Huntsville, Ala.

Zenitsa 19190.—Sire, Tormentor 3533; dam, Neata 4748.

Butter, 17 lbs. 5½ oz. Milk, 216 lbs. 4 oz.

Test made from July 1 to 8, 1888; age, 5 yrs. 10 mos.; weight, 1035 lbs.; fed, daily, 10 lbs. bran, 8 lbs. corn meal, 1 lb. cotton-seed meal and 1 lb. pea meal; property of Jacob L. Thomas, Knoxville, Tenn.

Zenobia Pogis 35421.—Sire, Prince of Melrose 2d 11015; dam, Zenobia of Springvale 20094.

Butter, 15 lbs. 10 oz. Milk, 183 lbs.

Test made from July 8 to 15, 1888; age, 3 yrs. 1 mo.; estimated weight, 800 lbs.; fed 56 qts. bran, 14 qts. corn meal and 14 qts. cotton-seed meal—hay, good mixed pasture; property of James Crook, Jacksonville, Ala.

Zenobia Stoke Pogis 37292.—Sire, Exile of St. Lambert 13657; dam, Minto Fay 27705.

Butter, 15 lbs. 8 oz. Milk, 162 lbs. 8 oz.

Test made from Apr. 23 to 30, 1889; age, 3 yrs. 3 mos.; estimated weight, 800 lbs.; fed, daily, 25 lbs. (one-fourth corn meal, one-fourth fine feed and one-half bran); property of P. J. Cogswell, Rochester, N. Y.

Zeroda 43723.—Sire, Alfonso 3013; dam, Prudence Hunt 22614.

Butter, 16 lbs. 7 oz. Milk, 219 lbs.

Test made from Oct. 16 to 23, 1889; age, 3 yrs. 9 mos.; estimated weight, 850 lbs.; fed, daily, 4 qts. corn meal, 2 qts. ground oats, 2 qts. rye meal and 2 qts. flaxseed meal; property of Carson & Bro., Crab Orchard, Ky.

Zillia 2d 1693.—Sire, Beacon Comet 675; dam, Zillia 1692.

Butter, 15 lbs. 2 oz. Milk, 249 lbs. 8 oz.

Test made from Mar. 9 to 15, 1885; age, 15 yrs. 10 mos.; property of William Crozier, Northport, N. Y.

Zitella 2d 11922.—Sire, Brother Jack 4042; dam, Zitella 8780.

Butter, 17 lbs. 8½ oz. Milk, 206 lbs. 3 oz.

Test made from Apr. 20 to 26, 1884; age, 3 yrs. 9 mos.; property of William Simpson, New York, N. Y.

Zoe Henry 6693.—Sire, Greysteel 1294; dam, Duchess 3d 1661.

Butter, 15 lbs. 14¾ oz. Milk, 202 lbs.

Test made from June 13 to 19, 1884; age, 9 yrs. 6 mos.; property of Patrick Henry, Clarksville, Tenn.

Zophar H. 12329.—Sire, Homer H. 3683; dam, Jenny Gray 3511.

Butter, 21 lbs. 3 oz. Milk, 422 lbs. 4 oz.

Test made from June 13 to 19, 1885; age, 6 yrs. 3 mos.; fed 8 qts. corn and oats, daily —pasture; property of William H. Hayden, Albany, Vt.

Zulisca Pogis 78777.—Sire, Aaron Stoke Pogis 21705; dam, Zulisca 2d 49025.

Butter, 18 lbs. 1¼ oz. Milk, 251 lbs.

Test made from May 15 to 22, 1897; age, 5 yrs. 10 mos.; estimated weight, 1200 lbs.; fed 70 qts. ground oats, 28 qts. ground oil-cake and 42 qts. corn meal—good old meadow pasture; property of Miller & Sibley, Franklin, Pa.

Zunilda 2d 93811.—Sire, Tormentor 5th 21962; dam, Zunilda 65701.

Butter, 15 lbs. 14 oz. Milk, 196 lbs.

Test made from Jan. 26 to Feb. 2, 1896; age, 3 yrs. 6 mos.; estimated weight, 800 lbs.; fed 28 lbs. cotton-seed meal, 28 lbs. bran, 56 lbs. corn meal, about 35 lbs. hay and 42 lbs. roots—pastured on blue-grass and winter oats; property of George Campbell Brown, Spring Hill, Tenn.

TESTED COWS

ARRANGED IN

ORDER OF MERIT.

	BUTTER.	MILK.	PAGE.
Princess 2d 8046	46 lbs. 12½ oz....	299 lbs. 8 oz.....	368
Oxford Kate 13646	39 lbs. 12 oz....	248 lbs. 8 oz.....	338
Mary Anne of St. Lambert 9770	36 lbs. 12¼ oz....	245 lbs. 0 oz.....	277
Princess of Dorset 85902	35 lbs. 1 oz....	368 lbs. 0 oz.....	372
Little Goldie 38671	34 lbs. 8½ oz....	251 lbs. 5 oz.....	254
Oonan of Riverside 69773	34 lbs. 3 oz....	237 lbs. 0 oz.....	335
Exile's Belle 40524	32 lbs. 6 oz....	230 lbs. 8 oz.....	138
Maquilla's Violet 69774	31 lbs. 1 oz....	186 lbs. 8 oz.....	272
Ethleel 2d 32291	30 lbs. 15 oz....	155 lbs. 11 oz.....	131
Insie of Riverside 23825	30 lbs. 5¾ oz....	289 lbs. 12 oz.....	200
Ida of St. Lambert 24990	30 lbs. 2½ oz....	290 lbs. 8 oz.....	194
Cornwall Maid 19024	29 lbs. 12 oz....	19½ qts. per day...	82
Rosa F. W. 13226	29 lbs. 12 oz....	200 lbs. 8 oz.....	393
Bess Pogis of Prospect 87045	29 lbs. 1¾ oz....	408 lbs. 0 oz.....	38
Landseer's Fancy 2876	29 lbs. ½ oz....	152 lbs. 8 oz.....	237
Daphne of Arcadia 21710	28 lbs. 12 oz....	274 lbs. 0 oz.....	99
Bisson's Belle 31144	28 lbs. 10 oz....	257 lbs. 12 oz.....	41
Melia Ann 3d 68070	28 lbs. 8 oz....	375 lbs. 8 oz.....	292
Gipsy's Berry Duchess 86124	28 lbs. 6¼ oz....	328 lbs. 0 oz.....	169
Aldarine 3d 27482	28 lbs. 1 oz....	248 lbs. 12 oz.....	10
Matina of Riverside 51783	27 lbs. 13 oz....	426 lbs. 7 oz.....	284
Gipsy's Pride 2d 29365	27 lbs. 12 oz....	295 lbs. 8 oz.....	170
Dove Dee 18059	27 lbs. 9 oz....	197 lbs. 11 oz.....	115
Paletta of Darlington 16255	27 lbs. 8 oz....	274 lbs. 4 oz.....	339
Lady Cloud 19358	27 lbs. 6 oz....	235 lbs. 8 oz.....	226
Toltec's Fancy 27172	27 lbs. 5½ oz....	130 lbs. 4 oz.....	444
Mary of Glenoir 59940	27 lbs. 3½ oz....	326 lbs. 5½ oz.....	279
Signal's Lily Flagg 31035	27 lbs. 3½ oz....	189 lbs. 6 oz.....	421
Rosetta of Whiteland 6112	27 lbs. 2¾ oz....	201 lbs. 0 oz.....	397
Eurotisama 29668	27 lbs. 1½ oz....	267 lbs. 10 oz.....	136
Mother Cary 11746	27 lbs. 1½ oz....	149 lbs. 0 oz.....	314
Canata 10523	27 lbs. 0 oz....	248 lbs. 1 oz.....	55
Eastwood Clearwater 30445	27 lbs. 0 oz....	263 lbs. 0 oz.....	120
Hilda Glenn 81195	27 lbs. 0 oz....	239 lbs. 14 oz.....	188
Allie of St. Lambert 24991	26 lbs. 12 oz....	329 lbs. 0 oz.....	12
Flower of Meridale 64537	26 lbs. 11½ oz....	356 lbs. 8 oz.....	159
Nancy Lee 7618	26 lbs. 8½ oz....	360 lbs. 12 oz.....	318
Chilioness Queen 82973	26 lbs. 8 oz....	323 lbs. 4 oz.....	66
Fable 62520	26 lbs. 5¼ oz....	304 lbs. 0 oz.....	145

	BUTTER.	MILK.	PAGE.
Geranium 2d 7838	26 lbs. 4¾ oz	244 lbs. 12½ oz	166
Pridalia 17249	26 lbs. 4 oz	280 lbs. 11 oz	364
Minnie Parker 38718	26 lbs. 2½ oz	309 lbs. 0 oz	300
Fill Pail 2d 24388	26 lbs. 2 oz	207 lbs. 14 oz	156
Inez of Riverside 51781	26 lbs. 1½ oz	463 lbs. 12½ oz	200
Ida's Alpha 64536	25 lbs. 15¾ oz	280 lbs. 12 oz	195
Mermaid of St. Lambert 9771	25 lbs. 13½ oz	307 lbs. 0 oz	295
Daisy Morrison 14035	25 lbs. 12½ oz	254 lbs. 0 oz	96
Ruthione 31514	25 lbs. 12 oz	263 lbs. 1 oz	403
Duchess Noble 27090	25 lbs. 8 oz	268 lbs. 8 oz	116
Mona Caird 75776	25 lbs. 7¼ oz	362 lbs. 4 oz	311
Lady Grace of Upholme 39569	25 lbs. 5½ oz	319 lbs. 12 oz	228
Gazelle's Fawn 93704	25 lbs. 4½ oz	333.1 lbs	165
Jersey Belle of Scituate 7828	25 lbs. 3 oz	44 to 45 lbs. daily	206
Value 2d 6844	25 lbs. 2 11-12 oz	327 lbs. 0 oz	456
Ida Marigold 32615	25 lbs. 2½ oz	244 lbs. 8 oz	193
Selita J. 32184	25 lbs. 2½ oz	211 lbs. 12 oz	412
Dora 4th 3936	25 lbs. 2 oz		113
Daisy Harrison 69253	25 lbs. 1 oz	303 lbs. 4 oz	95
Chromatess 51578	25 lbs. ½ oz	314 lbs. 0 oz	67
Sister of Charity 62453	24 lbs. 14½ oz	215 lbs. 5 oz	426
Nymph of St. Lambert 12968	24 lbs. 14 oz	280 lbs. 8 oz	330
Princess Chuck 45650	24 lbs. 14 oz	242 lbs. 0 oz	369
Prince's Christmas Gift 92513	24 lbs. 12¼ oz	282 lbs. 15 oz	368
Melrose's Perfection 28895	24 lbs. 12 oz	252 lbs. 8 oz	292
Miss Bingo 80096	24 lbs. 12 oz	255 lbs. 4 oz	302
Sitanda's Rose 64958	24 lbs. 12 oz	301 lbs. 0 oz	426
Hazen's Bess 7329	24 lbs. 11 oz	344 lbs. 13½ oz	184
Daisy Hinman 61537	24 lbs. 10 oz	296 lbs. 4 oz	95
Lily Scituate 12665	24 lbs. 9½ oz	366 lbs. 8 oz	252
Miss Milton 51851	24 lbs. 9½ oz	241 lbs. 6 oz	305
Westphalia 24384	24 lbs. 9½ oz		465
Hallie Rice 75777	24 lbs. 8½ oz	290 lbs. 12 oz	179
Eugénie Chouteau 6186	24 lbs. 8 oz	262 lbs. 0 oz	134
Ethlo's Jean 71954	24 lbs. 6 oz	261 lbs. 13½ oz	132
Lady K. of St. Lambert 112294	24 lbs. 6 oz	333.4 lbs	229
St. Lambert's Riotress 106220	24 lbs. 6 oz	337 lbs. 8 oz	408
Wardalia 2d 33970	24 lbs. 6 oz	293 lbs. 0 oz	464
Count's Fillpail 30975	24 lbs. 5 oz	170 lbs. 12 oz	86
Hugo's Joie de Vie 40372	24 lbs. 5 oz	243 lbs. 8 oz	191
King's Princess 30948	24 lbs. 5 oz	232 lbs. 4 oz	218
Duchess of Bolton 98150	24 lbs. 4 oz	212 lbs. 8 oz	117
Oonan's Grace 86489	24 lbs. 3 oz	258 lbs. 12 oz	336
Letty Rioter 73475	24 lbs. 2 oz	318 lbs. 14 oz	248
Mother Hubbard 10331	24 lbs. 1½ oz	233 lbs. 8 oz	314
Rita of Andalusia 29414	24 lbs. 1 oz	421 lbs. 5 oz	390
Alteration 56436	24 lbs. ½ oz	264 lbs. 3 oz	15
Regalia 64574	24 lbs. 0 oz	285 lbs. 14 oz	382
Flower of Glen Rouge 17560	23 lbs. 14¾ oz	262 lbs. 0 oz	159
Marchande 52258	23 lbs. 14¼ oz	175 lbs. 6½ oz	272
Pedro's Royal Princess 88485	23 lbs. 14 oz	295 lbs. 0 oz	352
Queen of Beauty 17109	23 lbs. 14 oz	221 lbs. 4 oz	378
Golden Dewdrop 96551	23 lbs. 13¾ oz	269 lbs. 8 oz	172
Nellie of St. Lambert 69458	23 lbs. 12 oz	318 lbs. 0 oz	322

	BUTTER.	MILK.	PAGE.
Phillis of St. Lambert 78867	23 lbs. 12 oz.	320 lbs. 0 oz.	356
Tapestry 56607	23 lbs. 12 oz.	313 lbs. 5 oz.	440
Calculus 75185	23 lbs. 11½ oz.	347 lbs. 12 oz.	54
Miss Nellie Parker 24988	23 lbs. 11 oz.	240 lbs. 13 oz.	306
Chief's Charella 61999	23 lbs. 10½ oz.	213 lbs. 0 oz.	65
Cherokee Rose 20921	23 lbs. 10 oz.	227 lbs. 0 oz.	64
Lady Panalphrex 17400	23 lbs. 9 oz.	21 qts. per day	233
Mary Idagold 88186	22 lbs. 9 oz.	343 lbs. 5½ oz.	278
Proctor's Alma Dolores 47107	23 lbs. 9 oz.	270 lbs. 0 oz.	375
Bridal Rose 74883	23 lbs. 8 oz.	316 lbs. 5½ oz.	46
Maud of Maple View 39511	23 lbs. 7½ oz.	209 lbs. 10 oz.	286
Xyst 101247	23 lbs. 6½ oz.	342 lbs. 6 oz.	470
Euphorbia 11229	23 lbs. 6¼ oz.	131 lbs. 12 oz.	134
Matilda of Meridale 57176	23 lbs. 5¾ oz.	307 lbs. 12 oz.	282
Hilda 2d 5447	23 lbs. 5 oz.	289 lbs. 0 oz.	187
Louisa of Riverside 69769	23 lbs. 5 oz.	220 lbs. 8 oz.	260
Oaklands Nora 14880	23 lbs. 5 oz.	153 lbs. 0 oz.	330
Clem Pogis 49932	23 lbs. 4½ oz.	290 lbs. 5 oz.	72
Ruby Jenks 3d 84868	23 lbs. 4½ oz.	289 lbs. 9 oz.	402
Golightly 25597	23 lbs. 4 oz.	275 lbs. 11 oz.	174
Koffee's Lady Pogis 65219	23 lbs. 4 oz.	253 lbs. 11 oz.	222
Lady Golddust 2d 19861	23 lbs. 4 oz.	21 qts. per day	228
Ruby Jenks 2d 76738	23 lbs. 3½ oz.	290 lbs. 4 oz.	401
Rachel Spencer 50974	23 lbs. 3¼ oz.	402 lbs. 12 oz.	381
Beauty Dee 18065	23 lbs. 3 oz.	221 lbs. 8 oz.	26
Nettie's New Year 99295	23 lbs. 3 oz.	372 lbs. 14.4 oz.	323
Pride's Olga 4th 96870	23 lbs. 3 oz.	368 lbs. 12 oz.	367
Pomona's Violetta Pogis 56246	23 lbs. 2¾ oz.	267 lbs. 4 oz.	363
Ida Twinkle 36994	23 lbs. 2½ oz.	266 lbs. 4 oz.	197
Little Torment 15581	23 lbs 2½ oz.	224 lbs. 8 oz.	255
Metaphysics 100974	23 lbs. 2¼ oz.	256 lbs. 0 oz.	296
Alphea Rajah 20605	23 lbs. 2 oz.	247 lbs. 7 oz.	14
Nutley Dolores 13797	23 lbs. 2 oz.	234 lbs. 0 oz.	329
Sue Gallagher 15945	23 lbs. 1½ oz.	188 lbs. 11 oz.	434
Callie Nan 2d 64207	23 lbs. 1 oz.	302 lbs. 0 oz.	54
Diana Victor Pogis 100763	23 lbs. 1 oz.	364 lbs. 0 oz.	104
Moss Rose of Willow Farm 5194	23 lbs. 1 oz.	228 lbs. 5 oz.	314
Romping Miss 54457	23 lbs. 1 oz.	295 lbs. 0 oz.	392
Tuiia St. Lambert 59992	23 lbs. 1 oz.	306 lbs. 4 oz.	454
Garella 62541	23 lbs. 0 oz.	301 lbs. 8 oz.	163
Maud Lee 2416	23 lbs. 0 oz.		285
Paola Stoke Pogis 34691	23 lbs. 0 oz.	274 lbs. 0 oz.	341
Winksette 99280	23 lbs. 0 oz.	276 lbs. 8 oz.	467
Coquette of Glen Rouge 17559	22 lbs. 14 oz.	232 lbs. 10 oz.	79
St. L. Gazelle 78870	22 lbs. 14 oz.	299 lbs. 14 oz.	408
Etta of Bay View 66028	22 lbs. 13 oz.	245 lbs. 7 oz.	133
Experience 62458	22 lbs. 13 oz.	276 lbs. 4 oz.	145
Kathletta 19567	22 lbs. 12½ oz.	174 lbs. 6 oz.	214
Marjoram of Glen Rouge 78420	22 lbs. 12¼ oz.	255 lbs. 4 oz.	275
Dacie's Lena 100796	22 lbs. 12 oz.	280 lbs. 6 oz.	92
Faith of Eau Claire 120456	22 lbs. 12 oz.	307 lbs. 9 oz.	148
Lady Louise D. 58056	22 lbs. 12 oz.	278 lbs. 0 oz.	230
Marjoram of Linden 43600	22 lbs. 12 oz.	285 lbs. 0 oz.	275
Mollie Garfield 12172	22 lbs. 12 oz.	258 lbs. 3 oz.	309

	BUTTER.	MILK.	PAGE.
Pride's Olga 2d 79887.....................	22 lbs. 12 oz....	241 lbs. 6 oz.....	366
Sayda M. 46195..........................	22 lbs. 11½ oz....	201 lbs. 13½ oz....	411
Harry's Pet 52838........................	22 lbs. 11 oz....	157 lbs. 10 oz.....	182
Leonora of Canada 94906.................	22 lbs. 11 oz....	284 lbs. 8 oz.....	247
Gipsy Bess of Prospect 96928.............	22 lbs. 10½ oz....	322 lbs. 0 oz.....	169
Niobe's Alpheanette 23336................	22 lbs. 10½ oz....	183 lbs. 9½ oz.....	326
Ona 7840................................	22 lbs. 10½ oz....	168⅞ qts..........	333
Fena Eric 69137.........................	22 lbs. 10 oz....	323 lbs. 8 oz.....	155
Maud of Glen Rouge 103781..............	22 lbs. 10 oz....	334 lbs. 8 oz.....	286
St. Lambert's Crescent 39711.............	22 lbs. 10 oz....	294 lbs. 0 oz.....	406
Sweet Brier of St. Lambert 5481..........	22 lbs. 10 oz....	266 lbs. 0 oz.....	437
Chinqua 27384...........................	22 lbs. 9½ oz....	270 lbs. 0 oz.....	66
Miss Belle 5083..........................	22 lbs. 9 oz....	276 lbs. 0 oz.....	302
The Queen's Gift 75186...................	22 lbs. 9 oz....	323 lbs. 12 oz.....	442
Winifred Victor Pogis 91658..............	22 lbs. 9 oz....	319 lbs. 4 oz.....	467
Fadette of Verna 3d 11122...............	22 lbs. 8½ oz....	22 qts. per day....	146
Belle Smith 65807........................	22 lbs. 8¼ oz....	300 lbs. 0 oz.....	33
Allie Victor Pogis 126907,...............	22 lbs. 8 oz....	300 lbs. 8 oz.....	12
Angelo's Torment 39390..................	22 lbs. 8 oz....	237 lbs. 1 oz.....	17
Fill Pail's Countess 24462...............	22 lbs. 8 oz....	223 lbs. 8 oz.....	156
Lady Alice Pogis 30273..................	22 lbs. 8 oz....	251 lbs. 8 oz.....	224
Clothilde of St. Lambert 72271...........	22 lbs. 7½ oz....	201 lbs. 1 oz.....	72
Eurotas 2454............................	22 lbs. 7 oz....	216 lbs. 8 oz.....	135
Geranium Leaf 2d 56859.................	22 lbs. 7 oz....	261 lbs. 0 oz.....	166
Nutley Silverette 22410..................	22 lbs. 7 oz....	281 lbs. 12 oz.....	329
Massey Polo 67010.......................	22 lbs. 6 oz....	296 lbs. 0 oz.....	281
Royal Queen 24428......................	22 lbs. 6 oz....	224 lbs. 12 oz.....	400
St. Albans' Valentine 41707.............	22 lbs. 6 oz....	248 lbs. 2 oz.....	405
Grace Davy 8292........................	22 lbs. 5½ oz....	242 lbs. 11 oz.....	175
Tormentor's Bloomfield 55530.............	22 lbs. 5½ oz....	302 lbs. 13 oz.....	448
Attractive Maid 16925...................	22 lbs. 5 oz....	254 lbs. 0 oz.....	21
Cleodoxa 59311..........................	22 lbs. 5 oz....	235 lbs. 2 oz.....	72
Queen Mary of Woodlawn 11659..........	22 lbs. 5 oz....	303 lbs. 4 oz.....	378
Sadie Le Pet 77100......................	22 lbs. 5 oz....	269 lbs. 1 oz.....	404
Maud Signal 2d 68664...................	22 lbs. 4 oz....	255 lbs. 14 oz.....	286
Annabel Lee 55833.......................	22 lbs. 3 oz....	279 lbs. 4 oz.....	18
Oonan 1485..............................	22 lbs. 2½ oz....	205 lbs. 7 oz.....	335
Naiad of St. Lambert 12965...............	22 lbs. 2¼ oz....	267 lbs. 0 oz.....	318
Beulah De Gruchy 13480.................	22 lbs. 2 oz....	308 lbs. 0 oz.....	39
Cheerful of St. Lambert 2d 43745........	22 lbs. 2 oz....	228 lbs. 8 oz.....	63
Tormento Coomassie 110447..............	22 lbs. 2 oz....	293 lbs. 12 oz.....	448
Blanche of Castile 43793.................	22 lbs. 1½ oz....	299 lbs. 0 oz.....	43
Edith Haley 92643.......................	22 lbs. 1½ oz....	304 lbs. 14 oz.....	121
Sasanda 50075...........................	22 lbs. 1½ oz....	307 lbs. 3 oz.....	410
Tenella 6712............................	22 lbs. 1½ oz....	....	440
Biederkeit 52257.........................	22 lbs. 1¼ oz....	178 lbs. 0 oz.....	39
Garfield's Princess 62544.................	22 lbs. 0 oz....	283 lbs. 0 oz.....	163
Grey Friar's Bess 25128..................	22 lbs. 0 oz....	280 lbs. 8 oz.....	178
Nora of St. Lambert 12962...............	22 lbs. 0 oz....	289 lbs. 0 oz.....	327
Tormentor's Fairy 108510.................	22 lbs. 0 oz....	363 lbs. 0 oz.....	449
Lizzie of Glynllyn 74274.................	21 lbs. 15½ oz....	322 lbs. 4 oz.....	256
Peasant 65401...........................	21 lbs. 15½ oz....	220 lbs. 7 oz.....	346
Signal's New Geranium 134117...........	21 lbs. 15½ oz....	290 lbs. 8 oz.....	421
Moggy Bright 25891.....................	21 lbs. 15 oz....	241 lbs. 4 oz.....	309

	BUTTER.	MILK.	PAGE.
Nora Pogis 75345	21 lbs. 14¾ oz	283 lbs. 8 oz	328
Surprise Melrose 93562	21 lbs. 14⅝ oz	325 lbs. 3 oz	436
Marion of Eau Claire 120457	21 lbs. 14 oz	322 lbs. 8 oz	274
Giulietta Cooke 32193	21 lbs. 13½ oz	257 lbs. 4 oz	170
Marilla of St. Lambert 31809	21 lbs. 13½ oz	258.9 lbs	274
Eva of Verna 15228	21 lbs. 13 oz	192 lbs. 8 oz	136
Lady Dartmouth 23159	21 lbs. 13 oz	243 lbs. 5 oz	226
Moyane 21595	21 lbs. 12½ oz	276 lbs. 0 oz	314
Brown Elsie 96595	21 lbs. 12 oz	272 lbs. 8 oz	50
Duke's Pierrot Lady 31023	21 lbs. 12 oz	266 lbs. 12 oz	118
Rêve 78944	21 lbs. 12 oz	278 lbs. 9 oz	384
Romping Lass 2d 51989	21 lbs. 12 oz	158 lbs. 12 oz	392
St. L.'s Silken Les Cateaux 61964	21 lbs. 12 oz	262 lbs. 10 oz	408
Bomba 10330	21 lbs. 11½ oz	205 lbs. 6 oz	45
Croton Maid 5305	21 lbs. 11½ oz	254 lbs. 6 oz	90
Edith of St. Lambert 2d 67327	21 lbs. 11 oz	308 lbs. 7 oz	121
Ellenetta of Bay View 48921	21 lbs. 11 oz	211 lbs. 4 oz	124
Lady Marjoram 123887	21 lbs. 11 oz	355 lbs. 0 oz	230
Phlox 16399	21 lbs. 11 oz	233 lbs. 2 oz	356
Pearl Armstrong 2670	21 lbs. 10 oz	317 lbs. 0 oz	344
Primrose 11956	21 lbs. 10 oz	282 lbs. 8 oz	367
Signalinda 2d 61142	21 lbs. 10 oz	208 lbs. 1 oz	419
Sigleonella 70091	21 lbs. 9¾ oz	282 lbs. 3 oz	416
Rioter's Letty 110450	21 lbs. 9½ oz	296 lbs. 4 oz	388
Royalist's Daisy 19187	21 lbs. 9½ oz	236 lbs. 0 oz	400
Niobe of St. Lambert 12969	21 lbs. 9¼ oz	280 lbs. 12 oz	326
Lady Matilda Pogis 36270	21 lbs. 9 oz	282 lbs. 5 oz	231
Valma Hoffman 4500	21 lbs. 9 oz		456
Matilda 4th 12816	21 lbs. 8½ oz	336 lbs. 0 oz	281
Tancreda Signal 27789	21 lbs. 8½ oz	310 lbs. 8 oz	439
Mamelle 20804	21 lbs. 8¼ oz	256 lbs. 1 oz	271
Jenny Dodo H. 14448	21 lbs. 8 oz		205
Khelula 17970	21 lbs. 8 oz	184 lbs. 13 oz	216
Letty Coles 2d 48128	21 lbs. 8 oz	300 lbs. 15 oz	248
Pretty Prattler 99080	21 lbs. 8 oz	279 lbs. 1 oz	364
Rainbow 2d 13962	21 lbs. 8 oz	167 lbs. 0 oz	381
Pocia of Andalusia 2d 79548	21 lbs. 7½ oz	257 lbs. 0 oz	360
Limited Pedro 56382	21 lbs. 7 oz	288 lbs. 0 oz	253
May of Ingleside 56831	21 lbs. 7 oz	315 lbs. 8 oz	290
Cora of St. Lambert 8347	21 lbs. 6¾ oz	239 lbs. 8 oz	80
Costa Rica 64570	21 lbs. 6½ oz	282 lbs. 5 oz	82
Lily Niobe 55765	21 lbs. 6½ oz	225 lbs. 0 oz	251
Bronzie 9368	21 lbs. 6 oz	138 lbs. 8 oz	48
Gift 39901	21 lbs. 6 oz	249 lbs. 7 oz	167
Lady Antoinette 24391	21 lbs. 6 oz	272 lbs. 0 oz	224
Riotress Signal 95430	21 lbs. 5½ oz	297 lbs. 7 oz	390
Royal Bomba's Daisy 35996	21 lbs. 5½ oz	198 lbs. 11 oz	400
Dura 26001	21 lbs. 5 oz	282 lbs. 2 oz	119
Maggie Rule 31940	21 lbs. 5 oz	184 lbs. 3 oz	268
Maggie Sheldon 23583	21 lbs. 5 oz	288 lbs. 8 oz	268
Sparta 75000	21 lbs. 5 oz	301 lbs. 7 oz	432
Cheerful Pogis 61748	21 lbs. 4½ oz	286 lbs. 6 oz	63
Edith Campbell 23011	21 lbs. 4½ oz	256 lbs. 8 oz	120
Judy of Riverside 16495	21 lbs. 4½ oz	284 lbs. 0 oz	210

	BUTTER.	MILK.	PAGE.
Reception 8557	21 lbs. 4½ oz	265 lbs. 0 oz	382
Rubano's Normanda 68927	21 lbs. 4½ oz	271 lbs. 0 oz	401
Princess Winnie 63508	21 lbs. 4¼ oz	272 lbs. 10¾ oz	374
Albert's Comtesse 111718	21 lbs. 4 oz	121 lbs. 12 oz	9
Emma of Eau Claire 127826	21 lbs. 4 oz	267 lbs. 4 oz	128
Gipsy W. 132879	21 lbs. 4 oz	290 lbs. 0 oz	170
Guarder 90825	21 lbs. 4 oz	279 lbs. 10 oz	178
Miranda of Bolton 100226	21 lbs. 4 oz	229 lbs. 3 oz	301
Moorey of Lawn 68847	21 lbs. 4 oz	267 lbs. 8 oz	313
Pedro's Polly Rex 99845	21 lbs. 4 oz	299 lbs. 0 oz	349
Signilla M. 55839	21 lbs. 4 oz	225 lbs. 2 oz	422
Yum Yum of St. Lambert 39152	21 lbs. 4 oz	219 lbs. 11 oz	472
Rose of St. Lambert 20426	21 lbs. 3½ oz	217 lbs. 0 oz	397
Atlanta's Beauty 12949	21 lbs. 3 oz	146 lbs. 0 oz	21
Baron's Lorna 101623	21 lbs. 3 oz	260 lbs. 8 oz	25
Beauty of Seekonk 14651	21 lbs. 3 oz	251 lbs. 0 oz	27
Island Star 11876	21 lbs. 3 oz	150⅞ qts	201
Pedronina 34803	21 lbs. 3 oz	248 lbs. 6 oz	346
Sensa 87569	21 lbs. 3 oz	320 lbs. 2 oz	413
Sobriquet 68028	21 lbs. 3 oz	278 lbs. 4 oz	427
Zophar H. 12329	21 lbs. 3 oz	422 lbs. 4 oz	474
Lady Larkspur 56051	21 lbs. 2¾ oz	234 lbs. 8 oz	230
Gertie of Glynllyn 74474	21 lbs. 2½ oz	359 lbs. 0 oz	167
Hilda D. 6683	21 lbs. 2½ oz	249 lbs. 8 oz	188
Rioter Carlotta 29667	21 lbs. 2½ oz	228 lbs. 6 oz	386
Clara Oonan 78454	21 lbs. 2 oz	145 lbs. 0 oz	70
Sirona 27128	21 lbs. 2 oz	179 lbs. 8 oz	425
Daisette's Tormentress 106532	21 lbs. 1⅞ oz	289 lbs. 12 oz	94
Ida's Rose of St. Lambert 71555	21 lbs. 1½ oz	313 lbs. 12 oz	197
Miss Julia 26701	21 lbs. 1½ oz	265 lbs. 0 oz	305
Busy Melrose 54456	21 lbs. 1¼ oz	289 lbs. 0 oz	52
Bright Tass 46173	21 lbs. 1 oz	249 lbs. 9 oz	47
Gold Lace 10726	21 lbs. 1 oz	285 lbs. 6½ oz	173
Maquilla 24043	21 lbs. 1 oz	149 lbs. 10 oz	272
Susa Drew 38521	21 lbs. 1 oz	262 lbs. 5 oz	436
Granny's Gem 30406	21 lbs. 13-16 oz	179 lbs. 4 oz	177
La France's Pogis 99986	21 lbs. ½ oz	320 lbs. 11.2 oz	236
Queen of Corfu 83732	21 lbs. ½ oz	272 lbs. 8 oz	379
Signal Queen 30860	21 lbs. ½ oz	208 lbs. 0 oz	420
Elsie Bonner 78864	21 lbs. 0 oz	299 lbs. 12 oz	125
Gem of Hope 17102	21 lbs. 0 oz	253 lbs. 8 oz	165
Handsome Myra 14244	21 lbs. 0 oz	276 lbs. 0 oz	179
Lampedo 46204	21 lbs. 0 oz	156 lbs. 8 oz	237
Luani 37989	21 lbs. 0 oz	190 lbs. 0 oz	261
Nelly 6456	21 lbs. 0 oz	344 lbs. 0 oz	322
Nuna 18669	21 lbs. 0 oz	231 lbs. 12 oz	328
Palestine of Oxford 42194	21 lbs. 0 oz	194 lbs. 11½ oz	339
Vernon Dolly 52284	21 lbs. 0 oz	308 lbs. 0 oz	458
Lady's Blossom 18491	20 lbs. 15¾ oz	148 lbs. 8 oz	234
Pocia of Andalusia 62616	20 lbs. 15 oz	328 lbs. 12 oz	360
Fadette Signal 39472	20 lbs. 14½ oz	234 lbs. 0 oz	146
Sadie R. 3d 83729	20 lbs. 14¼ oz	283 lbs. 12 oz	404
Jocal 95182	20 lbs. 14 oz	268 lbs. 8 oz	209
Mary M. Allison 6308	20 lbs. 14 oz	254 lbs. 8 oz	278

	BUTTER.	MILK.	PAGE.
Willimenia H. 12325	20 lbs. 14 oz.	404 lbs. 14 oz.	466
Zenie Lisgar 2d 87453	20 lbs. 14 oz.	270 lbs. 12 oz.	473
Queen Deeline 78869	20 lbs. 13½ oz.	290 lbs. 14.4 oz.	377
Connoisseur 65644	20 lbs. 13 oz.	304 lbs. 0 oz.	78
Lady Longfield 23524	20 lbs. 13 oz.	227 lbs. 8 oz.	230
Beauty of Lee Farm 15694	20 lbs. 12¾ oz.	233 lbs. 12 oz.	27
Crown 75198	20 lbs. 12 oz.	310 lbs. 4 oz.	91
Cultured Cream 29196	20 lbs. 12 oz.	241 lbs. 5 oz.	91
Flight of Willow Farm 24783	20 lbs. 12 oz.	199 lbs. 0 oz.	157
Alberta Signal 18611	20 lbs. 11 oz.	224 lbs. 15 oz.	8
Daisy Delafield 59230	20 lbs. 11 oz.	266 lbs. 7 oz.	94
Denise's Pearl 95757	20 lbs. 11 oz.	272 lbs. 15 oz.	103
Harry's Signaline 2d 82428	20 lbs. 11 oz.	234 lbs. 12 oz.	182
Sheldon's Daisy of St. L. 53047	20 lbs. 11 oz.	240 lbs. 3 oz.	414
La Petite Pogis 28757	20 lbs. 10¾ oz.	233 lbs. 0 oz.	240
Sigletta 3d 108977	20 lbs. 10½ oz.	178 lbs. 4 oz.	416
Cassia 2d 21370	20 lbs. 10¼ oz.	283 lbs. 8 oz.	59
Chrome Skin 7881	20 lbs. 10 oz.	226 lbs. 8 oz.	67
Comely of St. Lambert 2d 41177	20 lbs. 10 oz.	266 lbs. 8 oz.	76
Cynthia A. 43721	20 lbs. 10 oz.	246 lbs. 0 oz.	92
Hallie's Jewel 17113	20 lbs. 10 oz.	197 lbs. 8 oz.	179
King of St. L.'s Allie 86797	20 lbs. 10 oz.	291 lbs. 11 oz.	217
Sayda M. 3d 71134	20 lbs. 10 oz.	288 lbs. 9 oz.	411
Silvia Baker 8793	20 lbs. 10 oz.	121 lbs. 12 oz.	425
Dolly of Edgewood 83345	20 lbs. 9½ oz.	289 lbs. 14.4 oz.	111
Lady St. Helier of S. 51380	20 lbs. 9½ oz.	271 lbs. 12 oz.	234
Dicta 55221	20 lbs. 9 oz.	255 lbs. 2 oz.	104
Royal Daisy 25214	20 lbs. 9 oz.	129 lbs. 3 oz.	400
Princess Malita 66676	20 lbs. 8¾ oz.	205 lbs. 8 oz.	370
Tormentor's Daisy 52496	20 lbs. 8½ oz.	250 lbs. 0 oz.	448
Brenda of Elmhurst 10762	20 lbs. 8 oz.	201 lbs. 0 oz.	46
Brown Bessie 74997	20 lbs. 8 oz.	283 lbs. 2 oz.	48
Cheerful of St. Lambert 8348	20 lbs. 8 oz.	239 lbs. 8 oz.	63
Clem Pedro 73961	20 lbs. 8 oz.	259 lbs. 9 oz.	71
Guess Not 57142	20 lbs. 8 oz.	266 lbs. 8½ oz.	178
Meridale Sweet Ida 94854	20 lbs. 8 oz.	320 lbs. 7 oz.	294
Analysis 2d 55834	20 lbs. 7½ oz.	242 lbs. 11 oz.	17
Honey Belle 25824	20 lbs. 7½ oz.	257 lbs. 3 oz.	190
Lady of Union Valley 101733	20 lbs. 7½ oz.	264 lbs. 0 oz.	233
Fedora of Greenwood 46892	20 lbs. 7 oz.	280 lbs. 0 oz.	154
Jazella G. 14191	20 lbs. 7 oz.		203
Le Roy's Pet 94106	20 lbs. 7 oz.	233 lbs. 0 oz.	247
Maggie May of Tupelo 71280	20 lbs. 7 oz.	257 lbs. 1 oz.	268
Massena 25732	20 lbs. 7 oz.	210 lbs. 11 oz.	280
Zélie de Lussan 48681	20 lbs. 7 oz.	326 lbs. 8 oz.	472
Pomona of Prospect 123426	20 lbs. 6¼ oz.	215 lbs. 0 oz.	362
Bess of Edgewood 71178	20 lbs. 6 oz.	276 lbs. 2 oz.	37
Calla Europa 77810	20 lbs. 6 oz.	346 lbs. 14 oz.	54
Chroma 4572	20 lbs. 6 oz.		67
L. D.'s Melia 29087	20 lbs. 6 oz.	222 lbs. 0 oz.	242
Miss Satanella 31544	20 lbs. 6 oz.	275 lbs. 0 oz.	307
Oonan's Alda 108337	20 lbs. 5¾ oz.	235 lbs. 14¼ oz.	336
Chief's Beauty 39391	20 lbs. 5½ oz.	305 lbs. 3 oz.	65
Daisy of St. Peter's 18175	20 lbs. 5½ oz.	346 lbs. 1 oz.	97

	BUTTER.	MILK.	PAGE.
May Dee Pogis 36993	20 lbs. 5½ oz	220 lbs. 0 oz	288
Honeymoon of St. Lambert 11221	20 lbs. 5¼ oz	250 lbs. 8 oz	190
Etta Lyne 83298	20 lbs. 5 oz	301 lbs. 6 oz	133
Hazen's Nora 4791	20 lbs. 5 oz	299 lbs. 0 oz	184
Jennie Cream 2d 47466	20 lbs. 5 oz	309 lbs. 10 oz	204
Koffee's May Belle 65220	20 lbs. 5 oz	242 lbs. 12 oz	223
Lady Grace Pogis of S. 79462	20 lbs. 5 oz	248 lbs. 0 oz	228
Lanison's Belle 84986	20 lbs. 5 oz	285 lbs. 8 oz	239
Pride of Ingleside 54545	20 lbs. 4½ oz	287 lbs. 8 oz	365
Rheta of Glynllyn 74273	20 lbs. 4½ oz	318 lbs. 12 oz	385
Bonnie Wherry 115088	20 lbs. 4 oz	245 lbs. 0 oz	45
Clotaire's Beauty 28171	20 lbs. 4 oz	262 lbs. 1½ oz	72
Duke's Minnie 42785	20 lbs. 4 oz	190 lbs. 0 oz	118
Lady Anna Pogis 52734	20 lbs. 4 oz	286 lbs. 8 oz	224
Nan Day 17192	20 lbs. 4 oz	209 lbs. 0 oz	319
Roonan 5133	20 lbs. 4 oz	297 lbs. 7 oz	393
Rowena of St. Lambert 71526	20 lbs. 4 oz	272 lbs. 0 oz	399
Royal Rose 86791	20 lbs. 4 oz	323 lbs. 8 oz	401
Teasel 75358	20 lbs. 4 oz	294 lbs. 4 oz	440
Fairy of Verna 2d 10973	20 lbs. 3¾ oz	280 lbs. 0 oz	147
Fancy Wax 3d 86345	20 lbs. 3¾ oz	206 lbs. 11½ oz	151
Maria of St. Lambert 64908	20 lbs. 3½ oz	216 lbs. 8 oz	273
Brown Bessie's Bonita 109265	20 lbs. 3 oz	241 lbs. 4 oz	49
Camelia 2d 11188	20 lbs. 3 oz	296 lbs. 0 oz	55
Chief's Queen 69967	20 lbs. 3 oz	246 lbs. 15 oz	65
Daisy Glen 71976	20 lbs. 3 oz	222 lbs. 0 oz	95
Dorine's Brunette 29300	20 lbs. 3 oz	289 lbs. 8 oz	114
Lamba 79039	20 lbs. 3 oz	272 lbs. 10 oz	237
Oonan's Jeannette 45616	20 lbs. 3 oz	261 lbs. 12 oz	336
Rubano's Valentine 54703	20 lbs. 3 oz	291 lbs. 6 oz	401
Seneca's Mocha 23355	20 lbs. 3 oz	406 lbs. 9 oz	412
Spark 62689	20 lbs. 3 oz	231 lbs. 8 oz	431
Edna of Verna 34537	20 lbs. 2½ oz	180 lbs. 2 oz	121
Harry's Fancita 96721	20 lbs. 2½ oz	266 lbs. 8 oz	181
Waxie 19706	20 lbs. 2½ oz	218 lbs. 0 oz	465
Bosnia 87508	20 lbs. 2 oz	271 lbs. 8 oz	46
Brown Flora 2d 96594	20 lbs. 2 oz	269 lbs. 10 oz	50
Fawn Tormentor of Lawn 71639	20 lbs. 2 oz	294 lbs. 10 oz	154
Grace Harribee 91836	20 lbs. 2 oz	260 lbs. 0 oz	176
Haidee Signal 86448	20 lbs. 2 oz	304 lbs. 0 oz	179
Lass Signal 16308	20 lbs. 2 oz	212 lbs. 0 oz	241
Melbina 78875	20 lbs. 2 oz	295 lbs. 0 oz	291
Pilot's Veronica 18917	20 lbs. 2 oz	234 lbs. 0 oz	358
Rioter's Fawn 69168	20 lbs. 2 oz	289 lbs. 0 oz	388
Sigmanella 117023	20 lbs. 2 oz	276 lbs. 12 oz	417
Golden Plover 22388	20 lbs. 1¾ oz	309 lbs. 0 oz	172
Helene of St. Lambert 73601	20 lbs. 1½ oz	207 lbs. 7 oz	185
Rose of Eden 13437	20 lbs. 1½ oz	296 lbs. 0 oz	396
Soconee 3d 114422	20 lbs. 1½ oz	233 lbs. 15 oz	428
Cretesia 13657	20 lbs. 1 oz	269 lbs. 0 oz	88
Grace Marigold 99377	20 lbs. 1 oz	257 lbs. 10 oz	176
Lalla Rookh of Sugar Grove 15882	20 lbs. 1 oz	296 lbs. 3 oz	236
Matilda 6th 39290	20 lbs. 1 oz	202 lbs. 10 oz	282
Melnes 3d 7741	20 lbs. 1 oz	272 lbs. 0 oz	291

	BUTTER.	MILK.	PAGE.
Miss Retta 105807	20 lbs. 1 oz.	235 lbs. 4 oz.	307
Princess Cross 2d 80649	20 lbs. 1 oz.	327 lbs. 0 oz.	369
Pussy Stoke Pogis 62547	20 lbs. 1 oz.	229 lbs. 10 oz.	376
Dora Neptune 20318	20 lbs. ½ oz.	324 lbs. 4 oz.	113
Duchess of Bloomfield 3653	20 lbs. ½ oz.	317 lbs. 8 oz.	116
St. Lambert's Alphea 74100	20 lbs. ½ oz.	271 lbs. 8 oz.	406
Shelta F. 56437	20 lbs. ½ oz.	275 lbs. 4 oz.	415
Summer Morning 75197	20 lbs. ½ oz.	253 lbs. 4 oz.	434
Allie's Emblem 45620	20 lbs. 0 oz.	221 lbs. 0 oz.	12
Benefit's Daughter 63932	20 lbs. 0 oz.	370 lbs. 0 oz.	34
Bride of Evergreen 72022	20 lbs. 0 oz.	288 lbs. 7 oz.	47
Hilda A. 2d 11120	20 lbs. 0 oz.	280 lbs. 0 oz.	187
Nellie Hugo 53793	20 lbs. 0 oz.	242 lbs. 6 oz.	321
Winetka 33669	20 lbs. 0 oz.	194 lbs. 8 oz.	467
Baron's Sophie 17615	19 lbs. 15⅞ oz.	271 lbs. 8 oz.	25
Mandana 3d 78778	19 lbs. 15¾ oz.	241 lbs. 8 oz.	271
Lady Mary of Prospect 19768	19 lbs. 15½ oz.	261 lbs. 8 oz.	231
Ida's Rioter's Elopement 98412	19 lbs. 15¼ oz.	263 lbs. 0 oz.	196
Young Floribundus 3d 59312	19 lbs. 14¾ oz.	230 lbs. 1 oz.	471
Ethleel 18724	19 lbs. 14 oz.	163 lbs. 9 oz.	131
John Bull's Princess 49670	19 lbs. 14 oz.	233 lbs. 0 oz.	209
Rioter Pink of Berlin 23665	19 lbs. 14 oz.	290 lbs. 4 oz.	387
Hypathia 2d 14774	19 lbs. 13½ oz.	208 lbs. 8 oz.	192
Exile's Signal Belle 2d 106097	19 lbs. 13 oz.	197 lbs. 2½ oz.	145
Nal Day 26553	19 lbs. 13 oz.	210 lbs. 0 oz.	318
Phædra 2561	19 lbs. 13 oz.	263 lbs. 12 oz.	355
Rose Woodmere 91092	19 lbs. 13 oz.	231 lbs. 13 oz.	397
Signal Maid 19361	19 lbs. 13 oz.	274 lbs. 8 oz.	419
Sitka M. 78680	19 lbs. 13 oz.	211 lbs. 11 oz.	426
Lora Nicholson 27185	19 lbs. 12¾ oz.	236 lbs. 0 oz.	257
Ella of Endegeest 87376	19 lbs. 12½ oz.	309 lbs. 5 oz.	123
Gardiner's Ripple 11693	19 lbs. 12½ oz.	264 lbs. 12 oz.	163
Pogis' Dewdrop 40373	19 lbs. 12½ oz.	261 lbs. 6 oz.	360
Sophie Hudson 76105	19 lbs. 12½ oz.	335 lbs. 2 oz.	429
Chirp 65551	19 lbs. 12 oz.	258 lbs. 8 oz.	66
Hulla 7898	19 lbs. 12 oz.	226 lbs. 0 oz.	191
Josephine O. 103635	19 lbs. 12 oz.	234 lbs. 0 oz.	209
Kitty Victor Pogis 85904	19 lbs. 12 oz.	311 lbs. 4 oz.	222
Lille Bonne 8108	19 lbs. 12 oz.	275 lbs. 0 oz.	250
Marjoram's Delle 82954	19 lbs. 12 oz.	280 lbs. 0 oz.	275
Pedro's Fame 33836	19 lbs. 12 oz.	278 lbs. 0 oz.	347
Pride's Olga 37186	19 lbs. 12 oz.	191 lbs. 4 oz.	366
Rioter's Baby 106100	19 lbs. 12 oz.	298 lbs. 0 oz.	387
Rosebud of Allerton 6352	19 lbs. 12 oz.		395
Tormendova 74885	19 lbs. 12 oz.	204 lbs. 0 oz.	447
Annie's Prize 86548	19 lbs. 11¾ oz.	261 lbs. 0 oz.	19
Clotaire's Annie 58319	19 lbs. 11½ oz.	212 lbs. 11 oz.	72
Quachette 17091	19 lbs. 11½ oz.	206 lbs. 12 oz.	376
Alpheon's Belle 27194	19 lbs. 11 oz.	217 lbs. 12 oz.	14
Countess Matilda 74928	19 lbs. 11 oz.	270 lbs. 3 oz.	84
Mink 2d 3890	19 lbs. 11 oz.	324 lbs. 0 oz.	298
Oonan's Fancy 31887	19 lbs. 10¾ oz.	285 lbs. 11 oz.	336
Belle Williamson 8386	19 lbs. 10½ oz.	157 lbs. 6½ oz.	33
Bess of Ingleside 2d 59648	19 lbs. 10½ oz.	240 lbs. 8 oz.	37

	BUTTER.	MILK.	PAGE.
Evelina of Verna 10971..................	19 lbs. 10½ oz. ...	264 lbs. 13 oz.	136
Helen Walker 53465......................	19 lbs. 10½ oz. ...	241 lbs. 4 oz.	185
Rival's Ochra 10172.....................	19 lbs. 10½ oz. ...	225 lbs. 14 oz.	390
Le Gros' Lily of the Valley 2d 13386.......	19 lbs. 10¼ oz. ...	158 lbs. 4 oz.	243
Lady Fair 22103.........................	19 lbs. 10 oz.		227
Pedro's Pansy 33835.....................	19 lbs. 10 oz. ...	268 lbs. 0 oz.	349
Koffee's Clarinda 65216..................	19 lbs. 9½ oz. ...	259 lbs. 14 oz.	222
Oaklands Cora 18853.....................	19 lbs. 9½ oz. ...	169 lbs. 0 oz.	330
Rozel Lass 20268........................	19 lbs. 9½ oz. ...	273 lbs. 14 oz.	401
Maggie McM. 14073......................	19 lbs. 9 oz. ...	211 lbs. 0 oz.	267
Myrtlise 64006..........................	19 lbs. 9 oz. ...	204 lbs. 4 oz.	317
Rose of Riverside 46953..................	19 lbs. 9 oz. ...	254 lbs. 4 oz.	396
Sister Sue 58447........................	19 lbs. 9 oz. ...	335 lbs. 0 oz	426
Miss Mouse 92512.......................	19 lbs. 8¾ oz. ...	223 lbs. ¾ oz.	306
The Widow's Daughter 11507.............	19 lbs. 8½ oz. ...	323 lbs. 6 oz.	442
Corinne Moore 35748....................	19 lbs. 8 oz. ...	270 lbs. 9 oz.	81
Dewdrop Pansy 19736...................	19 lbs. 8 oz. ...	289 lbs. 0 oz.	103
Fannie's Fairy 67304....................	19 lbs. 8 oz. ...	254 lbs. 8 oz.	152
Lily of Elm Spring 51474................	19 lbs. 8 oz. ...	223 lbs. 10 oz.	252
May Keller 78874.......................	19 lbs. 8 oz. ...	271 lbs. 11.2 oz.	289
Pedro's Minette 66092...................	19 lbs. 8 oz. ...	295 lbs. 0 oz.	349
Sparks' Maid 45575.....................	19 lbs. 8 oz. ...	345 lbs. 11 oz.	431
Exile's Dolly 67452.....................	19 lbs. 7½ oz. ...	327 lbs. 4 oz.	139
Toltec's Roonan 64564...................	19 lbs. 7½ oz. ...	278 lbs. 11.2 oz.	445
Fairy Queen of St. Brelade's 7464........	19 lbs. 7¼ oz. ...	222 lbs. 7 oz.	147
Alphea Vaudrada 17468.................	19 lbs. 7 oz. ...	212 lbs. 6 oz.	14
Christmas Nannie 4075..................	19 lbs. 7 oz. ...	....	67
Countess of Lakeside 12135.............	19 lbs. 7 oz. ...	251 lbs. 0 oz.	85
Croton Maid 4th 26729..................	19 lbs. 7 oz. ...	281 lbs. ½ oz.	90
Katie of St. Lambert 54776.............	19 lbs. 7 oz. ...	235 lbs. 9 oz.	215
Tormentor's Isis 73350..................	19 lbs. 7 oz. ...	292 lbs. ¾ oz.	449
Lady Oonan Pogis 119154...............	19 lbs. 6¾ oz. ...	231 lbs. 3 oz.	233
Rioter's Ruth 14882.....................	19 lbs. 6½ oz. ...	235 lbs. 8 oz.	389
Ruth Morgan 3d 57043..................	19 lbs. 6.4 oz. ...	232 lbs. 6.4 oz.	404
Cream Custard 97429...................	19 lbs. 6¼ oz. ...	229 lbs. 6½ oz.	87
Bertha Morgan 4770....................	19 lbs. 6 oz. ...	294 lbs. 0 oz.	35
Bright Eyes 2d 2290....................	19 lbs. 6 oz. ...	....	47
Ella Victor Pogis 100762................	19 lbs. 6 oz. ...	325 lbs. 8 oz.	124
Lora Martin 88381......................	19 lbs. 6 oz. ...	254 lbs. 0 oz.	257
Lorne's Pretty Pearl 46683..............	19 lbs. 6 oz. ...	278 lbs. 0 oz.	259
Marjoram's Matilda 77944...............	19 lbs. 6 oz. ...	256 lbs. 0 oz.	276
Pretty Marjoram 77943..................	19 lbs. 6 oz. ...	274 lbs. 0 oz.	364
Tulia Perrot 46357......................	19 lbs. 5½ oz. ...	207 lbs. 12½ oz.	453
Good Advice 62476.....................	19 lbs. 5¼ oz. ...	215 lbs. 0 oz.	175
Alluring 5541...........................	19 lbs. 5 oz. ...	....	13
Belle Noble's Bessie 105710..............	19 lbs. 5 oz. ...	285 lbs. 0 oz.	30
Bright Eyes St. Lambert 2d 72699........	19 lbs. 5 oz. ...	317 lbs. 0 oz.	47
Christel 6565............................	19 lbs. 5 oz. ...	205 lbs. 0 oz.	67
Editha Westray 89755...................	19 lbs. 5 oz. ...	385 lbs. 6 oz.	120
Happy Blossom 18218...................	19 lbs. 5 oz. ...	247 lbs. 8 oz.	180
Sunny of Cedar Hill 90410..............	19 lbs. 5 oz. ...	298 lbs. 0 oz.	435
Kitty of Prospect 111121................	19 lbs. 4¾ oz. ...	230 lbs. 8 oz.	220
Jane Neptune 52345.....................	19 lbs. 4½ oz. ...	308 lbs. 0 oz.	202
Niobe's Alpheanette 3d 39459............	19 lbs. 4½ oz. ...	216 lbs. 8 oz.	326

	BUTTER.	MILK.	PAGE.
Valentine of Trinity 7460.................	19 lbs. 4½ oz....	277 lbs. 3 oz.....	456
Cherry of Glynllyn 81145.................	19 lbs. 4 oz....	214 lbs. 4 oz.....	64
Dot of Bear Lake 6170...................	19 lbs. 4 oz....	213 lbs. 0 oz.....	115
Elena of Oakdale 84162..................	19 lbs. 4 oz....	302 lbs. 13 oz.....	122
Gray Beauty 3d 19845...................	19 lbs. 4 oz....	376 lbs. 0 oz.....	177
Madolina H. 12327......................	19 lbs. 4 oz....	400 lbs. 2 oz.....	267
Pedro's Rosebud 66086..................	19 lbs. 4 oz....	248 lbs. 0 oz.....	352
Prince's Rubetta 92785..................	19 lbs. 4 oz....	196 lbs. 12½ oz.....	368
Vintonette 62931.......................	19 lbs. 4 oz....	251 lbs. 0 oz.....	461
Well Done 25987.......................	19 lbs. 4 oz....	260 lbs. 0 oz.....	465
Fancy Wax 37159.......................	19 lbs. 3½ oz....	116 lbs. 7 oz.....	150
King's Stevia 98737.....................	19 lbs. 3½ oz....	191 lbs. 0 oz.....	218
Rioter Rhea 10092......................	19 lbs. 3½ oz....	18 qts. per day....	387
Pitapat 2d 86907.......................	19 lbs. 3¼ oz....	218 lbs. 5 oz.....	359
Fan's Grouville Beauty 10079............	19 lbs. 3 oz....	234 lbs. 6 oz.....	152
Grace G. Parks 29263...................	19 lbs. 3 oz....	257 lbs. 8 oz.....	175
Harmony 2d 17118......................	19 lbs. 3 oz....	273 lbs. 3 oz.....	180
Jonquil of Bingo Farm 41786............	19 lbs. 3 oz....	289 lbs. 12 oz.....	209
Reif 58398.............................	19 lbs. 3 oz....	283 lbs. 12 oz.....	383
Annie Dale's Princess 12664.............	19 lbs. 2½ oz....	335 lbs. 1 oz.....	19
Hildah of Clovernook 51597.............	19 lbs. 2½ oz....	221 lbs. 0 oz.....	188
Beauty of Jersey 7850..................	19 lbs. 2 oz....	280 lbs. 15 oz.....	26
Madge 18904...........................	19 lbs. 2 oz....	245 lbs. 8 oz.....	266
Nutshell 2d 50962......................	19 lbs. 2 oz....	240 lbs. 1 oz.....	329
Optima 6715...........................	19 lbs. 2 oz....	191 lbs. 4 oz.....	337
Pansy Darling of Lawn 76637............	19 lbs. 2 oz....	233 lbs. 0 oz.....	340
Pride of Springvale 20093...............	19 lbs. 2 oz....	245 lbs. 0 oz.....	365
Roland's Bonnie 2d 18054...............	19 lbs. 2 oz....	217 lbs. 0 oz.....	391
Silver Sheen 26210.....................	19 lbs. 2 oz....	282 lbs. 4 oz.....	425
Tormentor's Favorite 36873..............	19 lbs. 2 oz....	142 lbs. 0 oz.....	449
Rinora Pogis 40107.....................	19 lbs. 1½ oz....	239 lbs. 12 oz.....	386
Santa Stoke Pogis 91680................	19 lbs. 1½ oz....	274.6 lbs...........	409
Thisbe 2d 2201.........................	19 lbs. 1½ oz....	372 lbs. 14 oz.....	442
Edith Golding 63709....................	19 lbs. 1¼ oz....	250 lbs. 0 oz.....	121
Adonia 64699..........................	19 lbs. 1 oz....	264 lbs. 4 oz.....	7
Columbine of St. Lambert 8350..........	19 lbs. 1 oz....	263 lbs. 8 oz.....	75
Cora Belmont 2d 48868.................	19 lbs. 1 oz....	308 lbs. 0 oz.....	79
Cosmetic 75357........................	19 lbs. 1 oz....	239 lbs. 8 oz.....	82
Eoline of St. Lambert 104819............	19 lbs. 1 oz....	296 lbs. 12 oz.....	128
Fantine of Guilford 64480...............	19 lbs. 1 oz....	266 lbs. 3 oz.....	153
Magna 2238............................	19 lbs. 1 oz....	....	269
Oaklands Nora 2d 45967................	19 lbs. 1 oz....	262 lbs. 13 oz.....	330
Pogis' Twilight 53794...................	19 lbs. 1 oz....	234 lbs. 9 oz.....	361
Pretty Patty 44108.....................	19 lbs. 1 oz....	231 lbs. 0 oz.....	364
Princess Corinne 48203.................	19 lbs. 1 oz....	213 lbs. 6½ oz.....	369
Rosy Dream 9808.......................	19 lbs. 1 oz....	141 lbs. 6 oz.....	398
May Day Stoke Pogis 28353..............	19 lbs. ½ oz....	310 lbs. 8 oz.....	288
Belle of Prospect 2d 14326..............	19 lbs. 0 oz....	204 lbs. 0 oz.....	31
Durex 66320...........................	19 lbs. 0 oz....	344 lbs. 8 oz.....	119
Fair Lady 6723........................	19 lbs. 0 oz....	233 lbs. 0 oz.....	146
Gilderoy's Idex 42027..................	19 lbs. 0 oz....	265 lbs. 8 oz.....	168
Lorne's Diamond 66087.................	19 lbs. 0 oz....	318 lbs. 0 oz.....	258
Rissa 16014............................	19 lbs. 0 oz....	210 lbs. 0 oz.....	390
Silver Wave 10844.....................	19 lbs. 0 oz....	382 lbs. 8 oz.....	425

	BUTTER.	MILK.	PAGE.
Cordelia Signal 33452	18 lbs. 15½ oz.	199 lbs. 8 oz.	80
Nettina of Winnikee 46406	18 lbs. 15½ oz.	290 lbs. 12 oz.	324
Bijou Ogston 8210	18 lbs. 15 oz.	217 lbs. 4 oz.	40
Countess Potoka 7496	18 lbs. 15 oz.	220 lbs. 9 oz.	85
Lady Phillis 18240	18 lbs. 15 oz.	245 lbs. 8 oz.	234
Œnone 8614	18 lbs. 15 oz.	139 lbs. 7 oz.	331
Pedro's Rainbow 60096	18 lbs. 15 oz.	276 lbs. 0 oz.	351
Rioter's Lorne Primrose 88583	18 lbs. 15 oz.	258 lbs. 0 oz.	388
Nettina Silverette 89997	18 lbs. 14½ oz.	317 lbs. 4 oz.	324
Silver Delle 40691	18 lbs. 14½ oz.	237 lbs. 0 oz.	424
Tormentris 49659	18 lbs. 14½ oz.	204 lbs. 2 oz.	450
Walter Girl 2d 29265	18 lbs. 14½ oz.	225 lbs. 8 oz.	464
Amanda of St. Lambert 64955	18 lbs. 14 oz.	197 lbs. 0 oz.	15
Clover of Bay View 65424	18 lbs. 14 oz.	188 lbs. 0 oz.	73
Derjava 59830	18 lbs. 14 oz.	204 lbs. 4 oz.	103
Golden Princess 4557	18 lbs. 14 oz.	108 lbs. 8 oz.	172
Josephine Noble 66441	18 lbs. 14 oz.	261 lbs. 4 oz.	209
Sombre 80796	18 lbs. 14 oz.	282 lbs. 5 oz.	428
Ulricalla 22225	18 lbs. 14 oz.	195 lbs. 10 oz.	455
Victor Hugo's Maggie 30274	18 lbs. 14 oz.	242 lbs. 8 oz.	459
Hearty Pogis 81297	18 lbs. 13¾ oz.	220 lbs. 0 oz.	184
Flower of Glen Rouge 2d 55559	18 lbs. 13½ oz.	257 lbs. 4 oz.	159
Onwa 59628	18 lbs. 13½ oz.	282 lbs. 7 oz.	334
Queen Neptune 15501	18 lbs. 13½ oz.	227 lbs. 12 oz.	378
Matilda's Queen of St. L. 93746	18 lbs. 13¼ oz.	219 lbs. 0 oz.	283
Angelina Pogis 40367	18 lbs. 13 oz.	218 lbs. 0 oz.	17
Birdie Nicholson 31076	18 lbs. 13 oz.	214 lbs. 10 oz.	40
Happy Winn 35358	18 lbs. 13 oz.	282 lbs. 0 oz.	180
Imogene Pogis 44907	18 lbs. 13 oz.	223 lbs. 14 oz.	198
Pedro's Pretty Dolly 102376	18 lbs. 13 oz.	258 lbs. 0 oz.	349
Phillis 3d 46681	18 lbs. 13 oz.	262 lbs. 0 oz.	356
Queen of Delaware 17029	18 lbs. 13 oz.	252 lbs. 0 oz.	379
Sedate 11119	18 lbs. 13 oz.	223 lbs. 9 oz.	412
Silicon 25577	18 lbs. 13 oz.	202 lbs. 5 oz.	423
Même 2d 49046	18 lbs. 12⅞ oz.	259 lbs. 12 oz.	293
Beauty of Meridale 114744	18 lbs. 12½ oz.	305 lbs. 2 oz.	27
Exile's Bessie 49985	18 lbs. 12½ oz.	198 lbs. 8 oz.	139
Baroness Argyle 40498	18 lbs. 12 oz.	310 lbs. 0 oz.	24
Belle Dame 2d 22043	18 lbs. 12 oz.	143 lbs. 0 oz.	29
Belmeda 6229	18 lbs. 12 oz.	202 lbs. 12 oz.	33
Delia Martin 92358	18 lbs. 12 oz.	274 lbs. 12 oz.	101
Duke's Rowena 2d's Queen 79894	18 lbs. 12 oz.	275 lbs. 14 oz.	118
Gladdis Roy 2d 55652	18 lbs. 12 oz.	256 lbs. 0 oz.	171
Hallie Signal 61726	18 lbs. 12 oz.	251 lbs. 0 oz.	179
Hawthorn Minnie 26687	18 lbs. 12 oz.	278 lbs. 0 oz.	183
Lady Gray of Hilltop 6850	18 lbs. 12 oz.	42 lbs. per day	228
Mon Plaisir of Yerba Buena Ranch 12622	18 lbs. 12 oz.	239 lbs. 8 oz.	312
Narmeoka H. 12323	18 lbs. 12 oz.	381 lbs. 1 oz.	320
Nonnie Pogis 41432	18 lbs. 12 oz.	234 lbs. 4 oz.	327
Pedro's Fair Marjoram 88489	18 lbs. 12 oz.	275 lbs. 0 oz.	347
Pedro's Lady 36715	18 lbs. 12 oz.	282 lbs. 0 oz.	348
Pride's St. Lambert Lady 82955	18 lbs. 12 oz.	299 lbs. 14 oz.	367
Princess of Hanover 46706	18 lbs. 12 oz.	268 lbs. 1 oz.	373
Princess of the Valley 22641	18 lbs. 12 oz.	263 lbs. 14 oz.	373

	BUTTER.	MILK.	PAGE.
St. Lambert's Nora 42132	18 lbs. 12 oz.	391 lbs. 0 oz.	407
Teacher's Pet 60242	18 lbs. 12 oz.	268 lbs. 0 oz.	440
Tenella 2d 19521	18 lbs. 12 oz.	314 lbs. 0 oz.	441
Carrie Lena 3d 20077	18 lbs. 11¾ oz.	245 lbs. 0 oz.	58
Holyoke's Leda 52812	18 lbs. 11½ oz.	265 lbs. 12 oz.	189
May Violet 30250	18 lbs. 11½ oz.	196 lbs. 2 oz.	290
Signet's Fancy 58446	18 lbs. 11½ oz.	189 lbs. 0 oz.	422
Bet Arlington 8970	18 lbs. 11 oz.	236 lbs. 0 oz.	38
Lady Wellington of Milton 12234	18 lbs. 11 oz.	253 lbs. 0 oz.	235
May Blossom 5657	18 lbs. 11 oz.	228 lbs. 3 oz.	287
Théroigne 92259	18 lbs. 11 oz.	228 lbs. 11¼ oz.	442
Signalda's Rosebud 35583	18 lbs. 10⅛ oz.	309 lbs. 0 oz.	417
B. D. Exile 105286	18 lbs. 10 oz.	240 lbs. 0 oz.	25
Chancery 37987	18 lbs. 10 oz.	140 lbs. 8 oz.	62
Composite 58774	18 lbs. 10 oz.	288 lbs. 10 oz.	77
Hazel St. L. 100035	18 lbs. 10 oz.	314 lbs. 8 oz.	184
Pearl Jenks 2d 68908	18 lbs. 10 oz.	216 lbs. 0 oz.	344
Percle 14937	18 lbs. 10 oz.	234 lbs. 10 oz.	353
Polonia of Riverside 84036	18 lbs. 10 oz.	149 lbs. 8 oz.	362
Pride of the Manor 22652	18 lbs. 10 oz.	276 lbs. 0 oz.	366
Rosella of Bay View 66032	18 lbs. 10 oz.	256 lbs. 2 oz.	396
Sirona Croton 86597	18 lbs. 10 oz.	283 lbs. 1 oz.	426
Gingerbread 5th 37081	18 lbs. 9¾ oz.	248 lbs. 8 oz.	169
Philena S. 70375	18 lbs. 9½ oz.	265 lbs. 6 oz.	356
Siloam 17623	18 lbs. 9½ oz.	230 lbs. 12½ oz.	423
Between 86664	18 lbs. 9¼ oz.	317 lbs. 8 oz.	39
Amna 2d 98200	18 lbs. 9 oz.	283 lbs. 8 oz.	16
Beauty Romeril 29090	18 lbs. 9 oz.	219 lbs. 8 oz.	27
Ethel of Glen Rouge 131019	18 lbs. 9 oz.	273 lbs. 0 oz.	130
Pedro's Pretty Maggie 102923	18 lbs. 9 oz.	260 lbs. 0 oz.	350
Perry Farm Golden Cloud 22872	18 lbs. 9 oz.	208 lbs. 0 oz.	353
Rioter's Sweet Brier 30582	18 lbs. 9 oz.	212 lbs. 0 oz.	389
Sudie St. Helier 40235	18 lbs. 9 oz.	211 lbs. 8 oz.	434
Kittura of St. Lambert 78344	18 lbs. 8¾ oz.	217 lbs. 4 oz.	219
Elsie Harribee 91834	18 lbs. 8½ oz.	309 lbs. 12 oz.	126
Belle Grinnell 4073	18 lbs. 8 oz.		29
Brown Bonnie 102406	18 lbs. 8 oz.	215 lbs. 8 oz.	49
Brunette Hammond 7284	18 lbs. 8 oz.	254 lbs. 0 oz.	50
Calcina 80702	18 lbs. 8 oz.	273 lbs. 4¾ oz.	53
Cleora of Cornwall 63408	18 lbs. 8 oz.	261 lbs. 5 oz.	72
Crotonia Signal 89308	18 lbs. 8 oz.	212 lbs. 4 oz.	90
Doña Marina 7049	18 lbs. 8 oz.	171 lbs. 4 oz.	112
Floribundus 2d 14949	18 lbs. 8 oz.		158
King's Lucy Pogis 111380	18 lbs. 8 oz.	249 lbs. 8 oz.	218
Kitty's Ida 97736	18 lbs. 8 oz.	319 lbs. 6 oz.	221
Lady Phillis 2d 35629	18 lbs. 8 oz.	200 lbs. 12 oz.	234
Lena Stoke Pogis 39756	18 lbs. 8 oz.	259 lbs. 0 oz.	245
Loessin Fancy 91590	18 lbs. 8 oz.	229 lbs. 12 oz.	256
Mysinda 2d 42250	18 lbs. 8 oz.	226 lbs. 4 oz.	318
Nettie Signal 37837	18 lbs. 8 oz.	249 lbs. 0 oz.	323
Pedro's Pretty Belle 77946	18 lbs. 8 oz.	261 lbs. 0 oz.	349
Pedro's Pretty Gem 69953	18 lbs. 8 oz.	243 lbs. 0 oz.	350
Southern Pet 26675	18 lbs. 8 oz.	238 lbs. 10 oz.	430
Tormentor's Fawn 2d 94485	18 lbs. 8 oz.	267 lbs. 8 oz.	449

	BUTTER.	MILK.	PAGE.
Nymphæa 5141	18 lbs. 7½ oz.	250 lbs. 9 oz.	329
Pedro's Duchess 29581	18 lbs. 7½ oz.	261 lbs. 0 oz.	347
Rosa of Belle Vue 6954	18 lbs. 7½ oz.	205 lbs. 8 oz.	394
Amanda Pogis 79412	18 lbs. 7¼ oz.	294 lbs. 4½ oz.	15
Meridale Victoria 94848	18 lbs. 7¼ oz.	315 lbs. 1 oz.	295
Cretesia's Temisia 80996	18 lbs. 7 oz.	294 lbs. 8 oz.	89
Gipsy Pogis 35422	18 lbs. 7 oz.	199 lbs. 8 oz.	169
Helen Barry 55840	18 lbs. 7 oz.	225 lbs. 15 oz.	185
High Spirits of Lawn 61371	18 lbs. 7 oz.	237 lbs. 12 oz.	187
Jodie Jenks 76739	18 lbs. 7 oz.	244 lbs. 10 oz.	209
Leoni 11868	18 lbs. 7 oz.	224 lbs. 0 oz.	246
Polly Bonair 42849	18 lbs. 7 oz.	182 lbs. 8 oz.	361
St. Lambert's Nancy 56503	18 lbs. 7 oz.	200 lbs. 8 oz.	407
Signalinda 27002	18 lbs. 7 oz.	234 lbs. 14½ oz.	418
Annie Hines 84038	18 lbs. 6¾ oz.	284 lbs. 3 oz.	19
Harriet Neptune 52344	18 lbs. 6½ oz.	315 lbs. 0 oz.	180
Lucy Cobb 15133	18 lbs. 6½ oz.	229 lbs. 0 oz.	262
Princess Minette 24042	18 lbs. 6½ oz.	282 lbs. 6 oz.	372
Priscilla Pogis 39270	18 lbs. 6½ oz.	233 lbs. 12 oz.	375
Ben's Bijou 86599	18 lbs. 6 oz.	180 lbs. 7 oz.	34
Bloomer 2d 2321	18 lbs. 6 oz.	312 lbs. 3 oz.	43
Columbiana 74998	18 lbs. 6 oz.	216 lbs. 4 oz.	75
Eveline of Jersey 6781	18 lbs. 6 oz.	18 qts. per day	136
Herpa S. 127829	18 lbs. 6 oz.	279 lbs. 2 oz.	186
Ida Eva Pogis 39271	18 lbs. 6 oz.	245 lbs. 8 oz.	192
Lady Alice of Hillcrest 7450	18 lbs. 6 oz.	215 lbs. 0 oz.	224
Le Brocq's Pansy Rex 23789	18 lbs. 6 oz.	280 lbs. 0 oz.	243
Litta Pogis 30275	18 lbs. 6 oz.	236 lbs. 8 oz.	253
Melia Ann 2d 47464	18 lbs. 6 oz.	243 lbs. 4 oz.	291
Pedro Girl 22400	18 lbs. 6 oz.	233 lbs. 7 oz.	346
Pedro's Hansom Marjoram 96934	18 lbs. 6 oz.	254 lbs. 0 oz.	348
Pedro's Pretty Flower 88542	18 lbs. 6 oz.	266 lbs. 0 oz.	350
Pedro's Silver Pearl 110401	18 lbs. 6 oz.	253 lbs. 0 oz.	352
Signal of Raceland 84299	18 lbs. 6 oz.	287 lbs. 4 oz.	419
Summerline 8001	18 lbs. 6 oz.	264 lbs. 6 oz.	434
Miss Le Gros' Maid 83645	18 lbs. 5¾ oz.	203 lbs. 4 oz.	305
Sophona 76110	18 lbs. 5¾ oz.	242 lbs. 4 oz.	429
Alaric's Millicent 35363	18 lbs. 5½ oz.	163 lbs. 8 oz.	8
Maud of Ingleside 74044	18 lbs. 5½ oz.	284 lbs. 8 oz.	286
Signetilla 16333	18 lbs. 5½ oz.	233 lbs. 8 oz.	421
Spiræa 4th 20075	18 lbs. 5½ oz.	187 lbs. 0 oz.	432
Fanny Mason 45992	18 lbs. 5¼ oz.	209 lbs. 0 oz.	152
Mary Gordon 41429	18 lbs. 5¼ oz.	237 lbs. 7 oz.	278
Cobweb 3d 21325	18 lbs. 5 oz.	214 lbs. 0 oz.	74
Daisy of Malone 36505	18 lbs. 5 oz.	259 lbs. 0 oz.	97
Denise Landseer of Lawn 84411	18 lbs. 5 oz.	226 lbs. 10 oz.	102
Dora Victor Pogis 100921	18 lbs. 5 oz.	281 lbs. 12 oz.	114
Express 69574	18 lbs. 5 oz.	270 lbs. 12 oz.	145
King's Lassie 79103	18 lbs. 5 oz.	282 lbs. 0 oz.	218
Kitty Potter 9893	18 lbs. 5 oz.		221
Maggie of Springvale 15931	18 lbs. 5 oz.	223 lbs. 2 oz.	268
Miss Puritan 34204	18 lbs. 5 oz.	357 lbs. 11 oz.	307
Onyx of Argyle 76742	18 lbs. 5 oz.	280 lbs. 8 oz.	334
Pedro's Bonnie Rainbow 88612	18 lbs. 5 oz.	276 lbs. 0 oz.	346

	BUTTER.	MILK.	PAGE.
Ruby Jenks 4th 84871	18 lbs. 5 oz.	236 lbs. 8 oz.	402
Seraph 72217	18 lbs. 5 oz.	298 lbs. 0 oz.	413
Shanenalawn 56892	18 lbs. 5 oz.	260 lbs. 1 oz.	414
Sulphide 88038	18 lbs. 5 oz.	271 lbs. 6 oz.	434
Butterstar 7799	18 lbs. 4½ oz.	316 lbs. 8 oz.	53
Tortilla 55528	18 lbs. 4½ oz.	280 lbs. 14½ oz.	452
Clover of St. Lambert 60945	18 lbs. 4¼ oz.	227 lbs. 8 oz.	73
Oonan 2d 19569	18 lbs. 4¼ oz.	158 lbs. 11 oz.	335
Chestnut's Beauty 21576	18 lbs. 4⅛ oz.	273 lbs. 0 oz.	65
Bonnie Signaldina 108681	18 lbs. 4 oz.	244 lbs. 1½ oz.	45
Brier of St. Lambert 61750	18 lbs. 4 oz.	267 lbs. 1 oz.	47
Catchfly 25405	18 lbs. 4 oz.	207 lbs. 0 oz.	59
Colle 8309	18 lbs. 4 oz.	241 lbs. 14 oz.	75
Exile's Dewdrop 106104	18 lbs. 4 oz.	326 lbs. 0 oz.	139
Frankness 62451	18 lbs. 4 oz.	198 lbs. 0 oz.	161
Imperial Riotress 30259	18 lbs. 4 oz.	292 lbs. 6 oz.	199
Judith Coleman 11391	18 lbs. 4 oz.	256 lbs. 10 oz.	210
Maybee H. 60620	18 lbs. 4 oz.	216 lbs. 0 oz.	287
Même 34921	18 lbs. 4 oz.	207 lbs. 0 oz.	292
Monarch's Star 80232	18 lbs. 4 oz.	237 lbs. 0 oz.	311
Nora Clinton 91938	18 lbs. 4 oz.	244 lbs. 10 oz.	327
Orie's Gem of St. Lambert 85472	18 lbs. 4 oz.	317 lbs. 3 oz.	337
Paradise 2d 97112	18 lbs. 4 oz.	234 lbs. 8 oz.	342
Pedro's Pretty Pansy 75275	18 lbs. 4 oz.	246 lbs. 0 oz.	351
Pedro's Young Princess 72310	18 lbs. 4 oz.	252 lbs. 0 oz.	352
Rioter's Drosera 81056	18 lbs. 4 oz.	231 lbs. 8 oz.	387
School Marm 67264	18 lbs. 4 oz.	244 lbs. 0 oz.	412
Tait's Beauty 121221	18 lbs. 4 oz.	285 lbs. 0 oz.	439
Exile's Penelope 77182	18 lbs. 3¾ oz.	268 lbs. 8 oz.	144
Pilot's Rose 17958	18 lbs. 3¾ oz.	205 lbs. 0 oz.	357
Bunker's Pride 38571	18 lbs. 3½ oz.	264 lbs. 0 oz.	51
Belle Dawson 8270	18 lbs. 3 oz.	278 lbs. 0 oz.	29
Cicero's Jolie 18246	18 lbs. 3 oz.	233 lbs. 7 oz.	69
Cinxia's Pavonia 64441	18 lbs. 3 oz.	321 lbs. 8 oz.	70
Countess Queen 13519	18 lbs. 3 oz.		85
Dolly Bond 82662	18 lbs. 3 oz.	244 lbs. 8 oz.	111
Elsie Lofty 61451	18 lbs. 3 oz.	303 lbs. 0 oz.	126
Fandango 12908	18 lbs. 3 oz.	260 lbs. 3 oz.	151
Fill Pail's Nancy 95435	18 lbs. 3 oz.	238 lbs. 0 oz.	156
Georgiadear 11508	18 lbs. 3 oz.	242 lbs. 12 oz.	166
Jim's Pride 31376	18 lbs. 3 oz.	178 lbs. 8 oz.	208
Kate Crecraft 2d 75461	18 lbs. 3 oz.	225 lbs. 10 oz.	213
Lady Appel 8612	18 lbs. 3 oz.	288 lbs. 0 oz.	224
Lady Julia G. 16199	18 lbs. 3 oz.	210 lbs. 12 oz.	229
Queen of Cowes 46309	18 lbs. 3 oz.	280 lbs. 0 oz.	379
Rockys' Pogis 96272	18 lbs. 3 oz.	276 lbs. 7 oz.	391
Ruby Jenks 56862	18 lbs. 3 oz.	254 lbs. 0 oz.	401
Storm-Beaten of Lawn 61375	18 lbs. 3 oz.	254 lbs. 14 oz.	433
Viva Le Brocq 13702	18 lbs. 3 oz.	206 lbs. 6 oz.	463
Witch-Hazel 4th 6131	18 lbs. 2⅞ oz.	260 lbs. 8 oz.	467
Wild Crocus 92486	18 lbs. 2¾ oz.	263 lbs. 8 oz.	466
Belle Lyman 3d 86346	18 lbs. 2½ oz.	282 lbs. 15 oz.	30
Garibaldi's Kate 2d 56478	18 lbs. 2½ oz.	311 lbs. 0 oz.	163
Khelula of Briarcliff 46799	18 lbs. 2½ oz.	241 lbs. 6 oz.	217

	BUTTER.	MILK.	PAGE.
Sigma 78676............................	18 lbs. 2¼ oz....	230 lbs. 9 oz.....	416
Bonnie Yost 7943.......................	18 lbs. 2 oz....	226 lbs. 6 oz.....	45
Country Girl 4th 51877..................	18 lbs. 2 oz....	232 lbs. 13 oz.....	86
Fancy Work 116422......................	18 lbs. 2 oz....	168 lbs. 0 oz.....	151
Flora Rex Carlo 2d 102140..............	18 lbs. 2 oz....	236 lbs. 0 oz.....	158
Gold Ear 2d 3592.......................	18 lbs. 2 oz....	....	172
Jewel of Argyle 68752..................	18 lbs. 2 oz....	295 lbs. 8 oz.....	208
Marie's Maud 78467....................	18 lbs. 2 oz....	292 lbs. 0 oz.....	274
May of Edgewood 74825.................	18 lbs. 2 oz....	223 lbs. 2 oz.....	290
Millplain Princess 78579...............	18 lbs. 2 oz....	257 lbs. 0 oz.....	298
Musidove 25379........................	18 lbs. 2 oz....	267 lbs. 12 oz.....	316
Pogis' May 26950......................	18 lbs. 2 oz....	219 lbs. 12 oz.....	360
Pride's Olga 3d 79897..................	18 lbs. 2 oz....	275 lbs. 2 oz.....	367
Ribbon of St. Lambert 39450...........	18 lbs. 2 oz....	206 lbs. 0 oz.....	385
Signaldella 24107......................	18 lbs. 1¾ oz....	....	418
Aleph Judea 11389.....................	18 lbs. 1½ oz....	314 lbs. 13 oz.....	10
Diploma's Loreda 101528...............	18 lbs. 1½ oz....	240 lbs. 14 oz.....	105
Highland Ida 38427....................	18 lbs. 1½ oz....	279 lbs. 8 oz.....	186
Kitty Sands 77432.....................	18 lbs. 1½ oz....	230 lbs. 0 oz.....	221
Ruba H. Pogis 81944...................	18 lbs. 1½ oz....	270 lbs. 0 oz.....	401
Sœurette 2d 86414....................	18 lbs. 1½ oz....	217 lbs. 13 oz.....	428
Dollie's Valentine 105049..............	18 lbs. 1⅜ oz....	273.6 lbs..........	107
Miss Helen Brice 88340................	18 lbs. 1¼ oz....	304 lbs. 10 oz.....	304
Zulisca Pogis 78777...................	18 lbs. 1¼ oz....	251 lbs. 0 oz.....	474
Albert's Lobelia 111902................	18 lbs. 1 oz....	106 lbs. 9 oz.....	9
Bertha's Brunette 82336...............	18 lbs. 1 oz....	279 lbs. 12 oz.....	35
Daisy Leto 72404.....................	18 lbs. 1 oz....	221 lbs. 8 oz.....	96
Gay Orphan 25985....................	18 lbs. 1 oz....	269 lbs. 2 oz.....	164
Harry's Nervilette 86262..............	18 lbs. 1 oz....	213 lbs. 8 oz.....	182
Hettie of Briarcliff 26621..............	18 lbs. 1 oz....	185 lbs. 0 oz.....	186
Jennie Myrtle 22977..................	18 lbs. 1 oz....	....	205
Lady Crouse 69336....................	18 lbs. 1 oz....	305 lbs. 0 oz.....	226
Oléo 38475...........................	18 lbs. 1 oz....	256 lbs. 0 oz.....	332
Pet of Rose Lawn 11326...............	18 lbs. 1 oz....	258 lbs. 5 oz.....	354
Prospect's Edwina 2d 86909............	18 lbs. 1 oz....	223 lbs. 0 oz.....	376
Real Queen 29198.....................	18 lbs. 1 oz....	231 lbs. 1 oz.....	382
René of Hillside 12952.................	18 lbs. 1 oz....	242 lbs. 0 oz.....	383
Ribbon's Gift 77375...................	18 lbs. 1 oz....	304 lbs. 8 oz.....	385
Volle 19465...........................	18 lbs. 1 oz....	....	463
Conover's Beauty 2d 25315.............	18 lbs. ¾ oz....	252 lbs. 0 oz.....	78
Belle Mardi 18362.....................	18 lbs. ½ oz....	215 lbs. 12 oz.....	30
Champion Azuline 20542...............	18 lbs. ½ oz....	168 lbs. 5½ oz.....	61
Dubenna 2d 74920....................	18 lbs. ½ oz....	319 lbs. 4 oz.....	116
Ida Longfield 64227...................	18 lbs. ½ oz....	241 lbs. 1¼ oz.....	193
Melia Ann 5444.......................	18 lbs. ½ oz....	245 lbs. 0 oz.....	291
Olivia Marigold 108187................	18 lbs. ½ oz....	254 lbs. 3 oz.....	333
Silver Steletho 40765..................	18 lbs. ½ oz....	241 lbs. 8 oz.....	425
Bessie's Mabel 40939..................	18 lbs. 0 oz....	281 lbs. 0 oz.....	37
Bisson's Wandrina 100409.............	18 lbs. 0 oz....	227 lbs. 15 oz.....	42
Bobby's Lily 22055....................	18 lbs. 0 oz....	202 lbs. 0 oz.....	44
Clotaire's Daisy 45392.................	18 lbs. 0 oz....	231 lbs. 15 oz.....	72
Duke's Medusa 43511..................	18 lbs. 0 oz....	235 lbs. 8 oz.....	118
Eric's Ida 69134......................	18 lbs. 0 oz....	223 lbs. 0 oz.....	129

	BUTTER.	MILK.	PAGE.
Ida Bashan 4725	18 lbs. 0 oz.	126 qts.	192
King's Trust 18946	18 lbs. 0 oz.	137⅝ qts.	219
Lady Elfled 40757	18 lbs. 0 oz.	213 lbs. 0 oz.	227
Laura of Eau Claire 110739	18 lbs. 0 oz.	268 lbs. 8 oz.	241
Medrena 3939	18 lbs. 0 oz.	298 lbs. 0 oz.	290
Monmouth Duchess 4th 7129	18 lbs. 0 oz.	196 lbs. 0 oz.	312
Noble Charity 67245	18 lbs. 0 oz.	259 lbs. 12 oz.	327
Orange Extra 133951	18 lbs. 0 oz.	226 lbs. 0 oz.	337
Patterson's Beauty 4760	18 lbs. 0 oz.		343
Pedro's Golden Beauty 110402	18 lbs. 0 oz.	231 lbs. 4 oz.	347
Sibyl's Beauty 25941	18 lbs. 0 oz.	186 lbs. 4 oz.	415
Vigo Princess 60806	18 lbs. 0 oz.	350 lbs. 0 oz.	460
Denise's Ida 54942	17 lbs. 15½ oz.	309 lbs. 0 oz.	102
Helpmeet 62459	17 lbs. 15½ oz.	202 lbs. 4 oz.	185
Miss Clifford 27962	17 lbs. 15½ oz.	256 lbs. 12 oz.	303
Sorina 42993	17 lbs. 15½ oz.	238 lbs. 0 oz.	430
Annie L. 12934	17 lbs. 15¼ oz.	223 lbs. 4 oz.	19
Beckie Y. Pogis 63069	17 lbs. 15¼ oz.	205 lbs. 8 oz.	28
Celia of Whitehall 68004	17 lbs. 15 oz.	196 lbs. 15 oz.	60
Flashy Jessie 9722	17 lbs. 15 oz.		156
Glen's Festina 25851	17 lbs. 15 oz.	254 lbs. 10½ oz.	171
Grace Pansy 2d 18764	17 lbs. 15 oz.	236 lbs. 15 oz.	176
Jenny Gray 3511	17 lbs. 15 oz.	368 lbs. 5 oz.	205
Miss Patti Rosa 73607	17 lbs. 15 oz.	226 lbs. 4 oz.	306
Rioter Alphea 2d 29676	17 lbs. 15 oz.	238 lbs. 10 oz.	386
Su Lu 4705	17 lbs. 15 oz.	241 lbs. 11 oz.	434
Acacia of Home Farm 3d 78203	17 lbs. 14½ oz.	271 lbs. 8 oz.	5
Miamis Signaldini 43891	17 lbs. 14½ oz.	318 lbs. 3 oz.	296
Pride of Rosemond 74449	17 lbs. 14½ oz.	278 lbs. 4 oz.	365
Witching 42252	17 lbs. 14½ oz.	192 lbs. 0 oz.	468
Belle of Lynwood 18364	17 lbs. 14 oz.	220 lbs. 12 oz.	31
Coming Girl 3d 71794	17 lbs. 14 oz.	227 lbs. 10 oz.	76
Lydia Darrach 4903	17 lbs. 14 oz.	238 lbs. 8 oz.	265
Mayhegtal 2d 101298	17 lbs. 14 oz.	202 lbs. 9 oz.	289
Miss Bianca 12517	17 lbs. 14 oz.	263 lbs. 3 oz.	302
Pedro's Little Wonder 72379	17 lbs. 14 oz.	287 lbs. 8 oz.	348
Startling 94942	17 lbs. 14 oz.	233 lbs. 14 oz.	432
Elma Marigold 100630	17 lbs. 13½ oz.	257 lbs. 0 oz.	124
Exile's Myrtle 2d 94446	17 lbs. 13½ oz.	255 lbs. 8 oz.	143
Fleece 18568	17 lbs. 13½ oz.	284 lbs. 8 oz.	156
Gladdis Roy 18566	17 lbs. 13½ oz.	265 lbs. 8 oz.	171
Mirtha 3437	17 lbs. 13½ oz.	314 lbs. 8 oz.	301
Regina 4th 12732	17 lbs. 13½ oz.	187 lbs. 8 oz.	383
Early Morn 50661	17 lbs. 13¼ oz.	248.4 lbs.	119
Lady Robin 1119	17 lbs. 13¼ oz.	344 lbs. 9 oz.	234
Candelabrum 75187	17 lbs. 13 oz.	275 lbs. 0 oz.	55
Exile's Pansy 79800	17 lbs. 13 oz.	278 lbs. 12 oz.	144
Italian 4096	17 lbs. 13 oz.	274 lbs. 12 oz.	202
Loreda 19801	17 lbs. 13 oz.	286 lbs. 10 oz.	258
Mabel's Jewel 6251	17 lbs. 13 oz.	305 lbs. 0 oz.	265
May Evening 15938	17 lbs. 13 oz.	289 lbs. 13 oz.	289
Oonan's Oonan 99574	17 lbs. 12¾ oz.	221 lbs. 6½ oz.	336
Young Floribundus 28172	17 lbs. 12¾ oz.	200 lbs. 14 oz.	471

	BUTTER.	MILK.	PAGE.
Belle of Riverview 21464................	17 lbs. 12½ oz....	283 lbs. 4 oz.....	32
Tower View Princess 2d 87452...........	17 lbs. 12½ oz....	195 lbs. 9 oz.....	452
Chromia E. 97511.....................	17 lbs. 12 oz....	278 lbs. 4 oz.....	68
Countess Europa 35820..................	17 lbs. 12 oz....	264 lbs. 5 oz.....	84
Cowslip of St. Lambert 8349.............	17 lbs. 12 oz....	229 lbs. 8 oz.....	87
Crocus of St. Lambert 8351.............	17 lbs. 12 oz....	282 lbs. 0 oz.....	90
Daisy Pilotta 68852....................	17 lbs. 12 oz....	274 lbs. 0 oz.....	98
Elsie Coomassie 98081..................	17 lbs. 12 oz....	231 lbs. 13 oz.....	125
Ethleel's Nerviona 101300..............	17 lbs. 12 oz....	211 lbs. 3 oz.....	132
Gold Baby 105745......................	17 lbs. 12 oz....	233 lbs. 8 oz.....	171
Homestead Lucy 96325..................	17 lbs. 12 oz....	294 lbs. 8 oz.....	189
Iuka of Riverview 2d 61271.............	17 lbs. 12 oz....	211 lbs. 0 oz.....	202
Jubilethleel 99697.....................	17 lbs. 12 oz....	199 lbs. 6.4 oz.....	210
Louise Obella 37060....................	17 lbs. 12 oz....	265 lbs. 14 oz.....	260
Maud Drew 38522......................	17 lbs. 12 oz....	242 lbs. 4 oz.....	285
Mercedes H. 12326....................	17 lbs. 12 oz....	356 lbs. 2 oz.....	293
Natalie Victor 73689..................	17 lbs. 12 oz....	324 lbs. 12 oz.....	320
Pauletta Pogis 39273..................	17 lbs. 12 oz....	218 lbs. 10 oz.....	343
Rosaline of Glenmore 3179..............	17 lbs. 12 oz....	206 lbs. 0 oz.....	394
Royal Princess 2370...................	17 lbs. 12 oz....	160 lbs. 0 oz.....	400
Sheldon's Daisy 30592..................	17 lbs. 12 oz....	214 lbs. 0 oz.....	414
Tinney 2d 26253.......................	17 lbs. 12 oz....	257 lbs. 4 oz.....	443
Belle of Passaic 85846..................	17 lbs. 11 oz....	305 lbs. 4 oz.....	31
Big Wyuka 58345......................	17 lbs. 11 oz....	245 lbs. 0 oz.....	40
Bisson's Idex 133952..................	17 lbs. 11 oz....	270 lbs. 0 oz.....	42
Country Girl 3d 12538.................	17 lbs. 11 oz....	219 lbs. 15 oz.....	86
Daisy Hudson 24647...................	17 lbs. 11 oz....	304 lbs. 4 oz.....	95
Fresh 91786..........................	17 lbs. 11 oz....	231 lbs. 0 oz.....	161
Maid of Fernwood 2d 29010.............	17 lbs. 11 oz....	221 lbs. 9 oz.....	270
Massena 2d 31188.....................	17 lbs. 11 oz....	244 lbs. 0 oz.....	280
Matin 7768..........................	17 lbs. 11 oz....	265 lbs. 4 oz.....	283
Ora Vibert 119585....................	17 lbs. 11 oz....	201 lbs. 0 oz.....	337
Paradise 32082.......................	17 lbs. 11 oz....	227 lbs. 8 oz.....	342
Rosy Dream 2d 27127..................	17 lbs. 11 oz....	163 lbs. 14 oz.....	398
Mercurina 64920......................	17 lbs. 10½ oz....	226 lbs. 15 oz.....	293
Violente 28819........................	17 lbs. 10½ oz....	284 lbs. 0 oz.....	461
Brunette Star 27270...................	17 lbs. 10 oz....	279 lbs. 10 oz.....	51
Exile's Lady Sarah 107938.............	17 lbs. 10 oz....	252 lbs. 8 oz.....	142
Lady Thurlow 12410...................	17 lbs. 10 oz....	256 lbs. 2 oz.....	235
La Septième 13432....................	17 lbs. 10 oz....	166 lbs. 2 oz.....	240
Lilly Signalda 23227...................	17 lbs. 10 oz....	158 lbs. 14 oz.....	251
Marea 10167..........................	17 lbs. 10 oz....	203 lbs. 8 oz.....	273
Minnie's Lady Allen 102925.............	17 lbs. 10 oz....	240 lbs. 0 oz.....	300
Oak Leaf 4769........................	17 lbs. 10 oz....	....	330
Pedro's Dolly 77948...................	17 lbs. 10 oz....	282 lbs. 0 oz.....	346
Romp's Nervilette 72282...............	17 lbs. 10 oz....	242 lbs. 14 oz.....	393
Toltec's Alice 31885..................	17 lbs. 10 oz....	153 lbs. 12 oz.....	444
Empress 6th 3203.....................	17 lbs. 9¾ oz....	....	128
Miss Dora Deane 24505................	17 lbs. 9⅝ oz....	256 lbs. 8 oz.....	304
Belle of Wayne 16457.................	17 lbs. 9½ oz....	230 lbs. 8 oz.....	32
Clara's Belle 16582...................	17 lbs. 9½ oz....	219 lbs. 2 oz.....	70
Little Texas 84229....................	17 lbs. 9½ oz....	177 lbs. 0 oz.....	255
Pledge 59214.........................	17 lbs. 9½ oz....	256 lbs. 4 oz.....	359

	BUTTER.	MILK.	PAGE.
Signal Mine 2d 60789	17 lbs. 9½ oz.	267 lbs. 15 oz.	419
Sunny Pogis 61919	17 lbs. 9½ oz.	281 lbs. 12 oz.	435
Alberta Hugo 35424	17 lbs. 9 oz.	228 lbs. 0 oz.	8
Cinxia St. Helier 46730	17 lbs. 9 oz.	238 lbs. 0 oz.	69
Cordelia Baker 8814	17 lbs. 9 oz.		80
Dora of Glynllyn 95993	17 lbs. 9 oz.	166 lbs. 9 oz.	113
Fawn of Ingleside 59049	17 lbs. 9 oz.	231 lbs. 0 oz.	154
Goldthread 4945	17 lbs. 9 oz.		174
Hugo's Golden Sheen 61804	17 lbs. 9 oz.	241 lbs. 7 oz.	191
Lady Vain 18573	17 lbs. 9 oz.	302 lbs. 8 oz.	235
Lorna 2d 33634	17 lbs. 9 oz.	203 lbs. 4 oz.	258
Metah's Queen 4886	17 lbs. 9 oz.	258 lbs. 2 oz.	296
Miss Albert 63138	17 lbs. 9 oz.	263 lbs. 0 oz.	302
Nancy Lovelock 15511	17 lbs. 9 oz.	207 lbs. 8 oz.	318
Olive of One Pine 93449	17 lbs. 9 oz.	282 lbs. 10 oz.	333
Pedro's Phillis 66079	17 lbs. 9 oz.	258 lbs. 0 oz.	349
Pera B. 86761	17 lbs. 9 oz.	253 lbs. 8 oz.	353
Quickstep 85969	17 lbs. 9 oz.	243 lbs. 0 oz.	379
Romp Ogden 5th 43181	17 lbs. 9 oz.	240 lbs. 4 oz.	393
Callinette Pogis 99700	17 lbs. 8.8 oz.	209 lbs. 15.2 oz.	54
Milky Way 18865	17 lbs. 8½ oz.	232 lbs. 15 oz.	297
Miss Bowris 12206	17 lbs. 8½ oz.	276 lbs. 3 oz.	303
Picotu's Bess 86172	17 lbs. 8½ oz.	263 lbs. 12 oz.	357
St. Jeannaise 15789	17 lbs. 8½ oz.		405
Torment 15579	17 lbs. 8½ oz.	17½ qts. per day	447
Torrancella 59827	17 lbs. 8½ oz.	216 lbs. 1 oz.	451
Zitella 2d 11922	17 lbs. 8½ oz.	206 lbs. 3 oz.	473
Beauty Montague 57923	17 lbs. 8 oz.	257 lbs. 0 oz.	26
Cerita of Meadow Brook 5056	17 lbs. 8 oz.	119 qts.	61
Coomassie Fawkes 122445	17 lbs. 8 oz.	377 lbs. 0 oz.	78
Cordelia Signal 3d 57808	17 lbs. 8 oz.	194 lbs. 8 oz.	81
Countess Lowndes 26874	17 lbs. 8 oz.	210 lbs. 0 oz.	84
Denise 8281	17 lbs. 8 oz.	218 lbs. 0 oz.	102
Diploma's Queen 98151	17 lbs. 8 oz.	263 lbs. 7 oz.	106
Duke's Signal Queen 2d 82418	17 lbs. 8 oz.	313 lbs. 15 oz.	118
Embla 4799	17 lbs. 8 oz.		127
Fanny of Cream Brook 13930	17 lbs. 8 oz.	259 lbs. 8 oz.	152
Fermentor 2d's Pet 114503	17 lbs. 8 oz.	183 lbs. 1 oz.	155
Founder's Pet 94107	17 lbs. 8 oz.	292 lbs. 2 oz.	160
Gabriel Champion 14102	17 lbs. 8 oz.	178 lbs. 8 oz.	163
Gipsy's Riotess 62482	17 lbs. 8 oz.	245 lbs. 4 oz.	170
Helen Stoke Pogis 31947	17 lbs. 8 oz.	236 lbs. 8 oz.	185
Homestead Matilda 96448	17 lbs. 8 oz.	220 lbs. 8 oz.	189
Ilinka 91418	17 lbs. 8 oz.	267 lbs. 2 oz.	198
Io 5th 280	17 lbs. 8 oz.		200
Julia of Eau Claire 120018	17 lbs. 8 oz.	231 lbs. 10 oz.	210
Lady Armington 7610	17 lbs. 8 oz.	200 lbs. 0 oz.	224
Magyarland 86326	17 lbs. 8 oz.	311 lbs. 4 oz.	269
Mamie Coburn 3798	17 lbs. 8 oz.		271
Maud Melinda 12126	17 lbs. 8 oz.		285
Pedro's Hansome Phillis 96935	17 lbs. 8 oz.	240 lbs. 0 oz.	348
Pogis Oonan 29890	17 lbs. 8 oz.	263 lbs. 12 oz.	361
Point Lace 62471	17 lbs. 8 oz.	232 lbs. 4 oz.	361

	BUTTER.			MILK.			PAGE.
Rioter's Queen 14895......................	17 lbs.	8	oz....	314 lbs.	0	oz.....	389
Sylvia Marigold 108183...................	17 lbs.	8	oz....	223 lbs.	0	oz.....	438
Toy 20604...............................	17 lbs.	8	oz....	258 lbs.	8	oz.....	452
Welma 5942.............................	17 lbs.	8	oz....	252 lbs.	1	oz.....	465
Favorite of Avon 13438..................	17 lbs.	7¾	oz....	287 lbs.	4	oz.....	153
Fair Starlight 7745......................	17 lbs.	7½	oz....	250 lbs.	4	oz.....	147
Fairy Oxford 66679......................	17 lbs.	7½	oz....	236 lbs.	8	oz.....	147
Hara of St. Lambert 42128..............	17 lbs.	7½	oz....	332 lbs.	4	oz.....	180
Jennie of the Vale 9553.................	17 lbs.	7½	oz....	250 lbs.	0	oz.....	205
Princess Aurea Pogis 39266.............	17 lbs.	7½	oz....	208 lbs.	8	oz.....	368
Platina Cary 73375......................	17 lbs.	7¼	oz....	265 lbs.	0	oz.....	359
Xarama 76887...........................	17 lbs.	7¼	oz....	254 lbs.	1	oz.....	469
Cottage Lass 5332......................	17 lbs.	7	oz....	201 lbs.	0	oz.....	83
Ethleel Koffee of Lawn 116757..........	17 lbs.	7	oz....	212 lbs.	9	oz.....	132
Jean Ingelow 42515.....................	17 lbs.	7	oz....	221 lbs.	14	oz.....	203
Mary Jane of Belle Vue 6956............	17 lbs.	7	oz....	....			278
Rosa Miller 4333........................	17 lbs.	7	oz....	276 lbs.	8	oz.....	394
Transcript 31867........................	17 lbs.	7	oz....	239 lbs.	0	oz.....	452
Kathletta's Fancy 60738.................	17 lbs.	6¾	oz....	344 lbs.	9	oz.....	214
Rush of St. Lambert 61234..............	17 lbs.	6¾	oz....	252 lbs.	0	oz.....	403
Torraline 54945.........................	17 lbs.	6¾	oz....	216 lbs.	6¼	oz.....	451
Daisy Brown 12213......................	17 lbs.	6½	oz....	237 lbs.	8	oz.....	94
Exile's Success 49986...................	17 lbs.	6½	oz....	283 lbs.	4	oz.....	145
Fancy Nenita 62702.....................	17 lbs.	6½	oz....	231 lbs.	13.4	oz.....	149
High Tea 65577.........................	17 lbs.	6½	oz....	202 lbs.	0	oz.....	187
Lady Creamly 4th 19077.................	17 lbs.	6½	oz....	215 lbs.	12	oz.....	226
Leila of Briarcliff 24184.................	17 lbs.	6½	oz....	171 lbs.	6	oz.....	243
Sataline 64213..........................	17 lbs.	6½	oz....	225 lbs.	0	oz.....	410
Torfrida 3596...........................	17 lbs.	6½	oz....	256 lbs.	3½	oz.....	446
Adonia 2d 103302.......................	17 lbs.	6	oz....	296 lbs.	4	oz.....	7
Braw Lassie 36221......................	17 lbs.	6	oz....	174 lbs.	0	oz.....	46
Caressa 88474..........................	17 lbs.	6	oz....	229 lbs.	1	oz.....	56
Carlo's Last Polly 94628................	17 lbs.	6	oz....	273 lbs.	0	oz.....	56
Caro of St. Lambert 90875..............	17 lbs.	6	oz....	174 lbs.	8	oz.....	57
Clara Tormentor 2d 97279..............	17 lbs.	6	oz....	215 lbs.	12½	oz.....	71
Commotion 52960.......................	17 lbs.	6	oz....	227 lbs.	11	oz.....	76
Cordelia Signal 2d 44489...............	17 lbs.	6	oz....	237 lbs.	0	oz.....	81
Ethleel's Pet 85428.....................	17 lbs.	6	oz....	209 lbs.	0	oz.....	132
Floret 9959.............................	17 lbs.	6	oz....	255 lbs.	8	oz.....	158
Golden Susette 96552...................	17 lbs.	6	oz....	282 lbs.	8	oz.....	173
Infanta Pedro Marjoram 88584..........	17 lbs.	6	oz....	285 lbs.	0	oz.....	200
Maggie Martin 9562.....................	17 lbs.	6	oz....	....			267
Martha Lafayette 17158.................	17 lbs.	6	oz....	220 lbs.	10	oz.....	276
Nora Sheldon 43590.....................	17 lbs.	6	oz....	233 lbs.	3	oz.....	328
Pansy's Thoughts 69400.................	17 lbs.	6	oz....	248 lbs.	5	oz.....	340
Pedro's Pretty Princess 102377..........	17 lbs.	6	oz....	248 lbs.	0	oz.....	351
Pedro's Priscilla 99413..................	17 lbs.	6	oz....	265 lbs.	0	oz.....	351
Pretty Phillis Pogis 66100..............	17 lbs.	6	oz....	258 lbs.	0	oz.....	364
Princess of Eau Claire 128219...........	17 lbs.	6	oz....	227 lbs.	8	oz.....	373
Princess Rioter 29588...................	17 lbs.	6	oz....	274 lbs.	0	oz.....	374
Tette 20802.............................	17 lbs.	6	oz....	188 lbs.	4	oz.....	441
Vixen 7591.............................	17 lbs.	6	oz....	196 lbs.	12	oz.....	463
Faultless 12018.........................	17 lbs.	5½	oz....	234 lbs.	13	oz.....	153

	BUTTER.	MILK.	PAGE.
Ida Bellman 48273......................	17 lbs. 5½ oz....	211 lbs. 8 oz.....	192
Lady Ramaposa 26232..................	17 lbs. 5½ oz....	245 lbs. 3 oz.....	234
Pearl of Riverside 55659..................	17 lbs. 5½ oz....	248 lbs. 0 oz.....	345
Poppy H. 34865...........................	17 lbs. 5½ oz....	163 lbs. 9 oz.....	363
Sosia 19198...............................	17 lbs. 5½ oz....	236 lbs. 11 oz.....	430
Ventnor Beauty 46308...................	17 lbs. 5½ oz....	240 lbs. 4 oz.....	458
Wamla 32083............................	17 lbs. 5½ oz....	262 lbs. 14 oz.....	464
Zenitza 19190............................	17 lbs. 5½ oz....	216 lbs. 4 oz.....	473
Pearl's Lemon 41646....................	17 lbs. 5¼ oz....	320 lbs. 0 oz.....	345
Baronet's Victory 2d 39176..............	17 lbs. 5 oz....	231 lbs. 0 oz.....	24
Beeswax 9807............................	17 lbs. 5 oz....	187 lbs. 13 oz.....	28
Easter Ona 2d 97158.....................	17 lbs. 5 oz....	283 lbs. 4 oz.....	120
Letitia Hunter 35575....................	17 lbs. 5 oz....	211 lbs. 4 oz.....	248
Little Lass 28512........................	17 lbs. 5 oz....	163 lbs. 8 oz.....	254
Mildred G. 45077........................	17 lbs. 5 oz....	180 lbs. 2 oz.....	297
Misty Morn 88725.......................	17 lbs. 5 oz....	289 lbs. 12 oz.....	308
Pandothra 22383.........................	17 lbs. 5 oz....	196 lbs. 0 oz.....	339
Plumage 53897...........................	17 lbs. 5 oz....	266 lbs. 7 oz.....	359
Richness 16536..........................	17 lbs. 5 oz....	219 lbs. 13 oz.....	386
St. Lambert's Nora 3d 111218............	17 lbs. 5 oz....	256 lbs. 0 oz.....	407
Signal's Ripple 27093....................	17 lbs. 5 oz....	236 lbs. 6 oz.....	421
Susie B. of Springvale 24986.............	17 lbs. 5 oz....	227 lbs. 2 oz.....	436
Syren V. 14619..........................	17 lbs. 5 oz....	215 lbs. 12 oz.....	439
Tomassee 76875..........................	17 lbs. 5 oz....	267 lbs. 0 oz.....	445
Treasure 79200..........................	17 lbs. 5 oz....	231 lbs. 0 oz.....	452
Sœurette 52256..........................	17 lbs. 4¾ oz....	216 lbs. 0 oz.....	428
Oneida 2d 43553.........................	17 lbs. 4½ oz....	280 lbs. 1 oz.....	334
Renalba 4117............................	17 lbs. 4½ oz....	267 lbs. 8 oz.....	383
Trident 88647............................	17 lbs. 4½ oz....	283 lbs. 4 oz.....	453
Duchess of Bloomfield 2d 12250.........	17 lbs. 4 7-16 oz..	111 lbs. 10 oz.....	116
Daisy Harris 92471......................	17 lbs. 4¼ oz....	235 lbs. 0 oz.....	95
Ida's Lucille 90400......................	17 lbs. 4¼ oz....	197 lbs. 0 oz.....	196
Ida's St. Jeannaise 2d 106316.............	17 lbs. 4¼ oz....	295 lbs. 11 oz.....	197
Ada Fy 45479............................	17 lbs. 4 oz....	268 lbs. 4 oz.....	6
Anita H. 12334..........................	17 lbs. 4 oz....	331 lbs. 4 oz.....	18
Bonny of Woodford 58718...............	17 lbs. 4 oz....	296 lbs. 11 oz.....	45
Catchfly 4th 47682.......................	17 lbs. 4 oz....	247 lbs. 0 oz.....	60
Chloe 4th 4612...........................	17 lbs. 4 oz....	207 lbs. 0 oz.....	66
Diploma's Princess 104002...............	17 lbs. 4 oz....	263 lbs. 7 oz.....	106
Esclarmonde 59509......................	17 lbs. 4 oz....	251 lbs. 8 oz.....	129
Faith of Oaklands 19696.................	17 lbs. 4 oz....	267 lbs. 0 oz.....	148
Fennel of Glynllyn 109755...............	17 lbs. 4 oz....	226 lbs. 5 oz.....	155
Jennie Cream 30622.....................	17 lbs. 4 oz....	237 lbs. 10 oz.....	204
Jessie Lee Pogis 44960..................	17 lbs. 4 oz....	180 lbs. 11 oz.....	207
Leila Stratton 25198.....................	17 lbs. 4 oz....	222 lbs. 3 oz.....	243
Lemon Twig 42028.......................	17 lbs. 4 oz....	253 lbs. 8 oz.....	244
Lulu Pedro 71902........................	17 lbs. 4 oz....	328 lbs. 0 oz.....	263
Minette of St. Lambert 9774.............	17 lbs. 4 oz....	297 lbs. 0 oz.....	298
Nannaxie 18860..........................	17 lbs. 4 oz....	245 lbs. 6 oz.....	319
Ogden's Alice 77590.....................	17 lbs. 4 oz....	294 lbs. 6 oz.....	331
Oktibbeha Duchess 4422.................	17 lbs. 4 oz....	282 lbs. 8 oz.....	332
Pedro's Pretty Girl 88488................	17 lbs. 4 oz....	271 lbs. 0 oz.....	350
Pedro's Pretty Phillis 102378.............	17 lbs. 4 oz....	248 lbs. 0 oz.....	351

	BUTTER.	MILK.	PAGE.
Pet's Fancy 80968	17 lbs. 4 oz.	244 lbs. 8 oz.	355
Plum 13228	17 lbs. 4 oz.	229 lbs. 8 oz.	359
Successor 72503	17 lbs. 4 oz.	208 lbs. 12 oz.	433
Waxie of Aurora 86955	17 lbs. 4 oz.	218 lbs. 0 oz.	465
Wyble 595	17 lbs. 4 oz.		469
Almond Blossom 59709	17 lbs. 3¾ oz.	163 lbs. 8 oz.	13
Panola 85344	17 lbs. 3¾ oz.	278 lbs. 15 oz.	339
Coomassie of Ingleside 74045	17 lbs. 3½ oz.	266 lbs. 8 oz.	78
Frankie's Lass 24900	17 lbs. 3½ oz.	217 lbs. 8 oz.	161
Lelia Koffee of Lawn 120982	17 lbs. 3½ oz.	211 lbs. 8 oz.	244
Riotress 3d 63166	17 lbs. 3½ oz.	252 lbs. 11 oz.	390
Young Lass 67471	17 lbs. 3½ oz.	252 lbs. 9 oz.	471
Belnina 19189	17 lbs. 3¼ oz.	214 lbs. 0 oz.	34
Ada of Orange 2d 65246	17 lbs. 3 oz.	292 lbs. 0 oz.	6
Coomassie Veda 55838	17 lbs. 3 oz.	237 lbs. 11 oz.	79
Cretesia's Silverette 74807	17 lbs. 3 oz.	251 lbs. 0 oz.	89
Fear Not 6059	17 lbs. 3 oz.	255 lbs. 0 oz.	154
Friendship 62460	17 lbs. 3 oz.	276 lbs. 12 oz.	161
Marjoram of Glen Rouge 2d 123634	17 lbs. 3 oz.	260 lbs. 0 oz.	275
Mhoon Lady 6560	17 lbs. 3 oz.	261 lbs. 8 oz.	296
Millerton Maid 74926	17 lbs. 3 oz.	259 lbs. 8 oz.	298
Moragina 26344	17 lbs. 3 oz.	233 lbs. 0 oz.	313
Musquita 28462	17 lbs. 3 oz.	227 lbs. 12 oz.	316
Odalette 76888	17 lbs. 3 oz.	281 lbs. 1½ oz.	331
Philena Victor Pogis 127138	17 lbs. 3 oz.	289 lbs. 0 oz.	356
Pine Tree Clara 55042	17 lbs. 3 oz.	251 lbs. 12 oz.	358
Princess Mostar 9700	17 lbs. 3 oz.	232 lbs. 15 oz.	372
Sparkle 56775	17 lbs. 3 oz.	246 lbs. 4 oz.	431
Spot Victor Pogis 85901	17 lbs. 3 oz.	226 lbs. 0 oz.	432
Tormentor's Barbara 98801	17 lbs. 3 oz.	206 lbs. 15 oz.	448
Frugal 14925	17 lbs. 2¾ oz.	229 lbs. 4 oz.	162
Virginia Taylor 117964	17 lbs. 2¾ oz.	278.4 lbs.	462
Cetewayo's Silver Bell 18952	17 lbs. 2½ oz.	214 lbs. 15 oz.	61
Chromo's Pansy 114416	17 lbs. 2½ oz.	274 lbs. 8 oz.	68
Colt's La Biche 6399	17 lbs. 2½ oz.	282 lbs. 6 oz.	75
Cream of Sidney 17028	17 lbs. 2½ oz.	245 lbs. 14 oz.	87
Design 18481	17 lbs. 2½ oz.	250 lbs. 4 oz.	103
Fireside 75194	17 lbs. 2½ oz.	250 lbs. 4 oz.	156
Jessie Stoke Pogis 37291	17 lbs. 2½ oz.	187.8 lbs.	207
Lette Signal 26823	17 lbs. 2½ oz.	224 lbs. 8 oz.	248
Maggie Rex 28623	17 lbs. 2½ oz.	298 lbs. 0 oz.	268
Matilda's Matilda 75348	17 lbs. 2½ oz.	281 lbs. 7 oz.	283
Pomona Pogis 40371	17 lbs. 2½ oz.	229 lbs. 14 oz.	362
Warren's Duchess 3d 26836	17 lbs. 2½ oz.	210 lbs. 14 oz.	464
Saucy Pogis 103535	17 lbs. 2¼ oz.	220 lbs. 0 oz.	410
Toltec's Sally 94461	17 lbs. 2¼ oz.	248 lbs. 9½ oz.	445
Bella Marigold 94124	17 lbs. 2 oz.	223 lbs. 8 oz.	29
Bellita 4553	17 lbs. 2 oz.	16 qts. per day	33
Christina Pogis 35426	17 lbs. 2 oz.	211 lbs. 1 oz.	67
Cicero's Juno 16726	17 lbs. 2 oz.	20 qts. per day	69
Cream's Pride 47570	17 lbs. 2 oz.	310 lbs. 0 oz.	88
Cupid's Cloud 74040	17 lbs. 2 oz.	264 lbs. 0 oz.	91
English Elm 3d 62906	17 lbs. 2 oz.	233 lbs. 0 oz.	128

	BUTTER.	MILK.	PAGE.
Gipsy 5th 2252	17 lbs. 2 oz....	....	169
Girletta 61472	17 lbs. 2 oz....	262 lbs. 12 oz....	170
Islithanks 90544	17 lbs. 2 oz....	285 lbs. 8 oz....	202
Lady Velvetine 15771	17 lbs. 2 oz....	132 lbs. 8 oz....	235
Martha D. 66641	17 lbs. 2 oz....	211 lbs. 14 oz....	276
Miss Sassafras 70468	17 lbs. 2 oz....	208 lbs. 5 oz....	307
Mooly Pedro 77811	17 lbs. 2 oz....	252 lbs. 0 oz....	313
Nienchen 48698	17 lbs. 2 oz....	225 lbs. 0 oz....	325
Pride of Edgewood 64199	17 lbs. 2 oz....	202 lbs. 7 oz....	365
Relay 70353	17 lbs. 2 oz....	277 lbs. 9 oz....	383
Serita 15520	17 lbs. 2 oz....	158 lbs. 0 oz....	413
Sweet Perrot 100749	17 lbs. 2 oz....	246 lbs. 12 oz....	438
Therese G. 16390	17 lbs. 2 oz....	212 lbs. 10 oz....	442
Tremona 93017	17 lbs. 2 oz....	273 lbs. 12 oz....	452
Vionetta Pogis 91978	17 lbs. 2 oz....	250 lbs. 0 oz....	462
Busy Princess 48202	17 lbs. 1¾ oz....	257 lbs. 2 oz....	52
Belle Melrose 61920	17 lbs. 1½ oz....	260 lbs. 4 oz....	30
Bisson's Fancy Pogis 106415	17 lbs. 1½ oz....	240 lbs. 12 oz....	41
Frisky Elsie 81590	17 lbs. 1½ oz....	238 lbs. 0 oz....	162
Lactine 10680	17 lbs. 1½ oz....	197 lbs. 0 oz....	223
Mercedes' Pet 20570	17 lbs. 1½ oz....	351 lbs. 2 oz....	293
Rioter Alphea 3d 34073	17 lbs. 1½ oz....	233 lbs. 2 oz....	386
Rioter's Denise 3d 70238	17 lbs. 1½ oz....	196 lbs. 9 oz....	387
Rupertina 10409	17 lbs. 1½ oz....	17 qts. per day....	403
Signodelle 3d 50067	17 lbs. 1½ oz....	189 lbs. 8 oz....	422
Torlona 76872	17 lbs. 1½ oz....	172 lbs. 8 oz....	447
Donney Pogis 2d 82426	17 lbs. 1¼ oz....	280 lbs. 3 oz....	112
Mary Fallon 80220	17 lbs. 1¼ oz....	218 lbs. 8 oz....	277
Toltec's Silvia 86911	17 lbs. 1¼ oz....	209 lbs. 4¾ oz....	445
Valita's Fancy 115842	17 lbs. 1¼ oz....	255 lbs. 1 oz....	456
Wild Rose Bloom 110583	17 lbs. 1¼ oz....	248 lbs. 0 oz....	466
Chenda 4599	17 lbs. 1⅛ oz....	251 lbs. 0 oz....	64
Beauty of Oakland 45009	17 lbs. 1 oz....	223 lbs. 1 oz....	27
Belle's Charity 82217	17 lbs. 1 oz....	314 lbs. 8 oz....	32
Bernice of Clovernook 72805	17 lbs. 1 oz....	189 lbs. 12 oz....	34
Bride of Evergreen 3d 90473	17 lbs. 1 oz....	249 lbs. 8 oz....	47
Countess Micawber 1759	17 lbs. 1 oz....	152 lbs. 14 oz....	85
Cyrene 4th 480	17 lbs. 1 oz....	112 qts..........	92
Daisette 64004	17 lbs. 1 oz....	264 lbs. 15 oz....	93
Gold Violet 85078	17 lbs. 1 oz....	321 lbs. 8 oz....	174
Gordonetta 19570	17 lbs. 1 oz....	200 lbs. 8 oz....	175
Hilda A. 3d 16636	17 lbs. 1 oz....	246 lbs. 15 oz....	187
Indulgence 50105	17 lbs. 1 oz....	271 lbs. 8 oz....	199
Jetty d'Or 2d 62403	17 lbs. 1 oz....	235 lbs. 6 oz....	208
Jewel of Ingleside 54544	17 lbs. 1 oz....	294 lbs. 0 oz....	208
Koffee's Duchess 2d 57988	17 lbs. 1 oz....	362 lbs. 0 oz....	222
Lady Cleveland 30251	17 lbs. 1 oz....	178 lbs. 1 oz....	226
Lionette 18038	17 lbs. 1 oz....	19½ qts. per day...	253
Lucilla 3d 9786	17 lbs. 1 oz....	222 lbs. 0 oz....	261
Madora Gold 67302	17 lbs. 1 oz....	296 lbs. 11 oz....	267
Melia Ann 2d's Pogis 68757	17 lbs. 1 oz....	296 lbs. 12 oz....	291
Mirah Landseer 74827	17 lbs. 1 oz....	202 lbs. 1½ oz....	301
Mousy 2d 14962	17 lbs. 1 oz....	....	314

	BUTTER.	MILK.	PAGE.
Theda H. 20567........................	17 lbs. 1 oz.	409 lbs. 11 oz.	441
Tuia St. L. of Aurora 89388..............	17 lbs. 1 oz.	211 lbs. 12 oz.	454
Foolish 64069............................	17 lbs. ½ oz.	306 lbs. 8 oz.	160
Lilium Excelsum 24945..................	17 lbs. ½ oz.	231 lbs. 0 oz.	249
Romping Mirth 64070....................	17 lbs. ½ oz.	260 lbs. 8 oz.	392
Vera McMillan 66520...................	17 lbs. ½ oz.	270 lbs. 0 oz.	458
Clarintha 96919.........................	17 lbs. ¼ oz.	203 lbs. 13 oz.	71
Harry's Rovalti 115843..................	17 lbs. ¼ oz.	241 lbs. 0 oz.	182
Signalbianca 2d 119779..................	17 lbs. ¼ oz.	249 lbs. 15 oz.	417
Abbie Z. 3d 14742.......................	17 lbs. 0 oz.	288 lbs. 0 oz.	5
Bertha Stewart 99882...................	17 lbs. 0 oz.	262 lbs. 13 oz.	35
Brown Bessie's Brownie 114433..........	17 lbs. 0 oz.	234 lbs. 4 oz.	49
Brunette of Scarsdale 13276.............	17 lbs. 0 oz.	240 lbs. 0 oz.	51
Butterfly of Riverview 3d 85499..........	17 lbs. 0 oz.	265 lbs. 0 oz.	52
Cetewayo's Lily 18950..................	17 lbs. 0 oz.	261 lbs. 4 oz.	61
Frolic's Pride 31667.....................	17 lbs. 0 oz.	201 lbs. 12 oz.	162
Gem's Louisa 39564.....................	17 lbs. 0 oz.	194 lbs. 10 oz.	165
Gildercream 39480......................	17 lbs. 0 oz.	216 lbs. 0 oz.	168
Jersey Cream 3151......................	17 lbs. 0 oz.	266 lbs. 8 oz.	206
Landseer's Hilda 49788.................	17 lbs. 0 oz.	229 lbs. 12 oz.	238
Landseer's Regina 85988.................	17 lbs. 0 oz.	140 lbs. 0 oz.	238
Lisgar's Ella 24992.....................	17 lbs. 0 oz.	314 lbs. 8 oz.	253
Lulu of Riverside 19175.................	17 lbs. 0 oz.	198 lbs. 8 oz.	263
Matilda 3238............................	17 lbs. 0 oz.		281
Maud's Crown Princess 90320...........	17 lbs. 0 oz.	243 lbs. 10 oz.	286
Meridale Coquette 64535................	17 lbs. 0 oz.	202 lbs. 2 oz.	294
Minnie of Oxford 12806.................	17 lbs. 0 oz.	247 lbs. 0 oz.	299
Pogis' René 102854.....................	17 lbs. 0 oz.	204 lbs. 0 oz.	361
Queen of Nubbin Ridge 14528..........	17 lbs. 0 oz.	241 lbs. 12 oz.	379
Rioter's Roberta 80469..................	17 lbs. 0 oz.	204 lbs. 8 oz.	389
Rose 240................................	17 lbs. 0 oz.		395
Sibyl's Fancy 25942.....................	17 lbs. 0 oz.	182 lbs. 12 oz.	415
Young Fanny 9632.......................	17 lbs. 0 oz.	144 qts.	471
Zélie's Zilpah 109344....................	17 lbs. 0 oz.	226 lbs. 0 oz.	473
Exile's Gretchen 79245..................	16 lbs. 15½ oz.	230 lbs. 8 oz.	140
Una of St. Lambert 80117..............	16 lbs. 15½ oz.	297 lbs. 4 oz.	455
Odelle Sales 15564......................	16 lbs. 15⅜ oz.	157 lbs. 8 oz.	331
Calcium 79733..........................	16 lbs. 15 oz.	212 lbs. 5 oz.	53
Carlo's Lass 94405......................	16 lbs. 15 oz.	265 lbs. 0 oz.	56
Creole Maid 11017......................	16 lbs. 15 oz.	229 lbs. 0 oz.	88
Dora Lowndes 64136....................	16 lbs. 15 oz.	243 lbs. 4 oz.	113
Dudu of Linwood 8336..................	16 lbs. 15 oz.	233 lbs. 8 oz.	117
Effie of Hillside 1521..................	16 lbs. 15 oz.	290 lbs. 0 oz.	122
Exile's Pauline 37530...................	16 lbs. 15 oz.	186 lbs. 0 oz.	144
Fair Marigold 89387.....................	16 lbs. 15 oz.	230 lbs. 6 oz.	147
Forest Queen 12229.....................	16 lbs. 15 oz.	210 lbs. 0 oz.	160
Herberta 8811...........................	16 lbs. 15 oz.	231 lbs. 15 oz.	186
Jessie Lorne 25213......................	16 lbs. 15 oz.	270 lbs. 8 oz.	207
Lena Hogan 41506.......................	16 lbs. 15 oz.	212 lbs. 4 oz.	245
Lizzie D. 10408.........................	16 lbs. 15 oz.	207 lbs. 9 oz.	256
Maudine of Elmwood 8718...............	16 lbs. 15 oz.	227 lbs. 14 oz.	285
Polly Clover 7052.......................	16 lbs. 15 oz.	19 qts. per day.	362
St. Lambert's Coquette 41070............	16 lbs. 15 oz.	257 lbs. 8 oz.	406
Son's Princess 118386....................	16 lbs. 15 oz.	231 lbs. 8 oz.	429

	BUTTER.	MILK.	PAGE.
Gold Maid 98787	16 lbs. 14¾ oz	176 lbs. 0 oz	174
Baronetti 8425	16 lbs. 14½ oz	258 lbs. 0 oz	25
Maid of the Elms 18932	16 lbs. 14½ oz	239 lbs. 6½ oz	270
Marquette Tormentor 131897	16 lbs. 14½ oz	200 lbs. 7 oz	276
Masher 64950	16 lbs. 14½ oz	228 lbs. 11 oz	280
Patona 55117	16 lbs. 14½ oz	227 lbs. 3 oz	343
Pyrrha 6100	16 lbs. 14½ oz	243 lbs. 14 oz	376
Rosa Hudson 2d 87631	16 lbs. 14½ oz	247 lbs. 0 oz	394
Almah of Oakland 11102	16 lbs. 14 oz	253 lbs. 9 oz	13
Alma of Springvale 26951	16 lbs. 14 oz	229 lbs. 5 oz	13
Amelia Pogis 53796	16 lbs. 14 oz	238 lbs. 11 oz	16
Bertha Black 26275	16 lbs. 14 oz	256 lbs. 0 oz	34
Bertie Don 2d 74088	16 lbs. 14 oz	146 lbs. 8 oz	36
Chrissy 2d 7720	16 lbs. 14 oz	18 qts. per day	67
Clover Bud 4th 18992	16 lbs. 14 oz	124 lbs. 2 oz	73
Corinne Melrose 54455	16 lbs. 14 oz	190 lbs. 8 oz	81
Daisy Staunton 46592	16 lbs. 14 oz	224 lbs. 4 oz	98
Dassa Argyle 21180	16 lbs. 14 oz	275 lbs. 0 oz	100
Koffmadge 2d 77790	16 lbs. 14 oz	253 lbs. 4 oz	223
Louisa D. 90959	16 lbs. 14 oz	227 lbs. 12 oz	260
Lucky Belle 2d 6037	16 lbs. 14 oz	276 lbs. 8 oz	262
Mamie H. Pogis 57050	16 lbs. 14 oz	193 lbs. 8 oz	271
Minnie C. 2d 48414	16 lbs. 14 oz	213 lbs. 1 oz	299
Princess Carmen 58320	16 lbs. 14 oz	217 lbs. 9 oz	369
Silver Rose 4753	16 lbs. 14 oz	264 lbs. 0 oz	425
Stolen Kisses 16864	16 lbs. 14 oz	236 lbs. 6 oz	433
Allie St. Heller 45794	16 lbs. 13½ oz	220 lbs. 8 oz	12
Armon 10862	16 lbs. 13½ oz	227 lbs. 5 oz	20
Fleeter 61921	16 lbs. 13½ oz	234 lbs. 15½ oz	157
Heiress of St. Lambert 56507	16 lbs. 13½ oz	209 lbs. 0 oz	185
Kathy Torment 32910	16 lbs. 13½ oz	134 lbs. 0 oz	214
St. Lambert's Pride 53742	16 lbs. 13½ oz	174 lbs. 8 oz	407
Witching Hour 54947	16 lbs. 13½ oz	196 lbs. 0 oz	468
Brunette Landseer 100904	16 lbs. 13¼ oz	202 lbs. 0 oz	50
Belle Yakout 38920	16 lbs. 13 oz	303 lbs. 8 oz	33
Brunhilde Regina 41742	16 lbs. 13 oz	282 lbs. 4 oz	51
Celestia 2d 29482	16 lbs. 13 oz	177 lbs. 8 oz	60
Dove 5th 14561	16 lbs. 13 oz	253 lbs. 0 oz	115
Goodbye 27366	16 lbs. 13 oz	223 lbs. 0 oz	175
King's Beauty 24397	16 lbs. 13 oz	196 lbs. 8 oz	217
Miss Browney 7288	16 lbs. 13 oz	336 lbs. 0 oz	303
Mylitta 15668	16 lbs. 13 oz	271 lbs. 5 oz	316
Nora of Glen Rouge 69452	16 lbs. 13 oz	217 lbs. 0 oz	327
Oneida 42100	16 lbs. 13 oz	338 lbs. 3 oz	334
Pauline Vivienne 11305	16 lbs. 13 oz		344
Pedro's Clara 96361	16 lbs. 13 oz	282 lbs. 0 oz	346
Princess of Clovernook 51598	16 lbs. 13 oz	146 lbs. 1 oz	372
Thaley 14299	16 lbs. 13 oz	264 lbs. 0 oz	441
Tritoma Pogis 44952	16 lbs. 13 oz	179 lbs. 15 oz	453
Lucy T. 48200	16 lbs. 12.7 oz	194 lbs. 3 oz	263
Bonnie Grisette 2d 19526	16 lbs. 12½ oz	191 lbs. 12 oz	45
Lady Fawn of St. Anne's 10920	16 lbs. 12½ oz	263 lbs. 4 oz	227
Lottie Dorey 21578	16 lbs. 12½ oz	175 lbs. 8 oz	259

	BUTTER.	MILK.	PAGE.
Mollie Davis 26378	16 lbs. 12½ oz	199 lbs. 1 oz	309
Muriel 5th 19017	16 lbs. 12½ oz	192 lbs. 4 oz	315
Ruby Tiptoe 78900	16 lbs. 12½ oz	187 lbs. 8 oz	402
St. Jeannaise 5th 96554	16 lbs. 12½ oz	302 lbs. 4 oz	405
Saucy Sally 83059	16 lbs. 12½ oz	270 lbs. 3 oz	411
Carlo's Polly Rex 94627	16 lbs. 12 oz	265 lbs. 0 oz	56
Down of Chatsworth 15251	16 lbs. 12 oz	222 lbs. 14 oz	115
Ettagem 80258	16 lbs. 12 oz	215 lbs. 6 oz	133
Exile's Anna 46881	16 lbs. 12 oz	219 lbs. 8 oz	137
Exile's Flossie 116608	16 lbs. 12 oz	224 lbs. 8 oz	140
Fancy Mème 79748	16 lbs. 12 oz	204 lbs. 12 oz	148
Florence of St. Lambert 95192	16 lbs. 12 oz	190 lbs. 8 oz	158
Lady Cloud 2d 26825	16 lbs. 12 oz	229 lbs. 8 oz	226
Lady of Bay View 66029	16 lbs. 12 oz	213 lbs. 4 oz	232
Lorne's Primrose 66082	16 lbs. 12 oz	254 lbs. 0 oz	259
Manoa 6340	16 lbs. 12 oz	273 lbs. 1 oz	271
Marburi 37991	16 lbs. 12 oz	192 lbs. 0 oz	272
Marna 76108	16 lbs. 12 oz	258 lbs. 0 oz	276
Miss Beauty 4053	16 lbs. 12 oz	251 lbs. 2 oz	302
Moyane's Pet 98115	16 lbs. 12 oz	259 lbs. 4 oz	315
Palestine Bess 102970	16 lbs. 12 oz	260 lbs. 12 oz	338
Pinkey Spry 18316	16 lbs. 12 oz	207 lbs. 5 oz	358
Princess of Ashantee 13467	16 lbs. 12 oz	123 lbs. ⅞ oz	372
Quail 3d 43568	16 lbs. 12 oz	246 lbs. 5 oz	377
St. Lambert's Violet 25278	16 lbs. 12 oz	283 lbs. 0 oz	408
Lisgar's Rose 2d 56976	16 lbs. 11¾ oz	213 lbs. 11 oz	253
Fleecy Princess 40342	16 lbs. 11½ oz	243 lbs. 13 oz	157
Modiste 74679	16 lbs. 11½ oz	169 lbs. 6 oz	308
Sayda of Aurora 86956	16 lbs. 11½ oz	218 lbs. 8 oz	411
Smax 48562	16 lbs. 11½ oz	243 lbs. 10 oz	427
Virgina 80904	16 lbs. 11½ oz	280 lbs. 13½ oz	462
Guess Not 2d 133252	16 lbs. 11¼ oz	224 lbs. 11¾ oz	179
Ashwood Pet 2d 84408	16 lbs. 11⅛ oz	244 lbs. 0 oz	21
Black Teat 32932	16 lbs. 11 oz	239 lbs. 9 oz	43
Butterstamp Lass 19517	16 lbs. 11 oz	19 qts. per day	52
Coomassie Darlton 29519	16 lbs. 11 oz	221 lbs. 1 oz	78
Dimple 3248	16 lbs. 11 oz	20 qts. per day	105
Ethel Marjoram 87266	16 lbs. 11 oz	248 lbs. 11 oz	130
Ida's Mysinda 64868	16 lbs. 11 oz	213 lbs. 8 oz	196
Kitty of St. Lambert 6637	16 lbs. 11 oz	267 lbs. 8 oz	220
Lotta Pogis 40368	16 lbs. 11 oz	212 lbs. 1 oz	259
Nannie Pogis 53786	16 lbs. 11 oz	238 lbs. 14 oz	319
Parthia 64925	16 lbs. 11 oz	253 lbs. 1 oz	343
Pedro's Pretty Duchess 88558	16 lbs. 11 oz	238 lbs. 0 oz	350
St. Lambert's Kate 61975	16 lbs. 11 oz	250 lbs. 2 oz	407
Sicilienne 25010	16 lbs. 11 oz	162 lbs. 8 oz	416
Signalda's Ethleel 77599	16 lbs. 11 oz	217 lbs. 5 oz	417
Typha 5870	16 lbs. 11 oz	234 lbs. 9 oz	454
Dodona A. 88102	16 lbs. 10¾ oz	249 lbs. 0 oz	106
Lass Rex Alphea 16965	16 lbs. 10¾ oz	218 lbs. 0 oz	241
Pearl Pogis 38304	16 lbs. 10¾ oz	203 lbs. 8 oz	345
Chief's Stevia 122187	16 lbs. 10½ oz	149 lbs. 12 oz	65
Countess Gilderine 29027	16 lbs. 10½ oz	233 lbs. 12 oz	84

	BUTTER.	MILK.	PAGE.
Greatæna 85823	16 lbs. 10½ oz.	247 lbs. 0 oz.	178
Little Accident 15578	16 lbs. 10½ oz.	176 lbs. 8 oz.	254
Mintresse of Ingleside 103128	16 lbs. 10½ oz.	180 lbs. 9 oz.	300
Dosia Pogis 89049	16 lbs. 10¼ oz.	195 lbs. 8 oz.	114
Signal's Ripple 4th 64862	16 lbs. 10¼ oz.	282 lbs. 8 oz.	421
Tormintha 37327	16 lbs. 10¼ oz.	225 lbs. 8 oz.	451
Belle Steuben 20115	16 lbs. 10 oz.	230 lbs. 0 oz.	33
Bloom of Amherst Villa 70507	16 lbs. 10 oz.	184 lbs. 10 oz.	44
Carria H. 14454	16 lbs. 10 oz.	358 lbs. 10 oz.	57
Chamomilla 7552	16 lbs. 10 oz.	223 lbs. 8 oz.	61
Chromoline of Argyle 103709	16 lbs. 10 oz.	227 lbs. 0 oz.	68
Compass' Pansy 105706	16 lbs. 10 oz.	267 lbs. 10 oz.	77
Countess Coomassie 19339	16 lbs. 10 oz.	260 lbs. 12 oz.	83
Crusta 29637	16 lbs. 10 oz.	236 lbs. 10 oz.	91
Duenna's Duchess 5508	16 lbs. 10 oz.	239 lbs. 0 oz.	118
Dusky 2525	16 lbs. 10 oz.		119
Gem of Ellasleigh 38532	16 lbs. 10 oz.	221 lbs. 0 oz.	165
Grinnell Lass 11859	16 lbs. 10 oz.	252 lbs. 8 oz.	178
Ianthe 4562	16 lbs. 10 oz.		192
Marno B. 56368	16 lbs. 10 oz.	205 lbs. 8 oz.	276
Miss Lofty 9718	16 lbs. 10 oz.	222 lbs. 2 oz.	305
Miss Peacock 30871	16 lbs. 10 oz.	257 lbs. 1 oz.	306
Patty Polonius 3d 38677	16 lbs. 10 oz.	184 lbs. 8 oz.	343
Pedro's Gay Lady 88561	16 lbs. 10 oz.	242 lbs. 0 oz.	347
Pedro's Mab 40212	16 lbs. 10 oz.	190 lbs. 8 oz.	349
Pierrot's Lady Bacon 12482	16 lbs. 10 oz.	4 gals. per day	357
Quinnie Pogis 61924	16 lbs. 10 oz.	237 lbs. 0 oz.	380
Runaway 87030	16 lbs. 10 oz.	205 lbs. 0 oz.	402
Silver Montague 45475	16 lbs. 10 oz.	268 lbs. 4 oz.	424
Star of Edgewood 74823	16 lbs. 10 oz.	213 lbs. 1 oz.	432
Agnes of Clovernook 72553	16 lbs. 9½ oz.	192 lbs. 3½ oz.	7
Blanche Pomeroy 105013	16 lbs. 9½ oz.	245 lbs. 8 oz.	43
Heclatette 119403	16 lbs. 9½ oz.	221 lbs. 3.2 oz.	185
Isadore Stoke Pogis 37297	16 lbs. 9½ oz.	200 lbs. 0 oz.	201
Lady Musa 98730	16 lbs. 9½ oz.	175 lbs. 3 oz.	232
Matilda of the Ledges 97740	16 lbs. 9½ oz.	306 lbs. 5 oz.	283
Regina Weston 70290	16 lbs. 9½ oz.	260 lbs. 12 oz.	383
Tormentor's Laura 98346	16 lbs. 9½ oz.	231 lbs. 9 oz.	449
Ada S. 18366	16 lbs. 9 oz.	214 lbs. 1 oz.	6
Arawana Queen 5368	16 lbs. 9 oz.	367 lbs. 0 oz.	20
Chansonnette 2d 29672	16 lbs. 9 oz.	237 lbs. 14 oz.	62
Dandelion 2521	16 lbs. 9 oz.	345 lbs. 6½ oz.	99
Dariole 64106	16 lbs. 9 oz.	248 lbs. 6 oz.	99
Dark and Fair 24468	16 lbs. 9 oz.	167 lbs. 8 oz.	99
Duchess of Jefferson 17144	16 lbs. 9 oz.	241 lbs. 10 oz.	117
Eudora of Barre 12419	16 lbs. 9 oz.	249 lbs. 15 oz.	134
Income 19472	16 lbs. 9 oz.	251 lbs. 9 oz.	109
La Petite Mère 3d 12814	16 lbs. 9 oz.	227 lbs. 6 oz.	240
No Chemicals 71452	16 lbs. 9 oz.	236 lbs. 8 oz.	327
Massey Polo 2d 89216	16 lbs. 9 oz.	211 lbs. 8 oz.	281
Minnie R. of St. Lambert 61965	16 lbs. 9 oz.	235 lbs. 0 oz.	300
Paradox 65003	16 lbs. 9 oz.	234 lbs. 11 oz.	342
Pedro's Golden Harp 2d 103365	16 lbs. 9 oz.	236 lbs. 0 oz.	348

	BUTTER.	MILK.	PAGE.
Pride of Bovina 8050	16 lbs. 9 oz.	214 lbs. 11 oz.	365
Princess Helene of St. L. 61972	16 lbs. 9 oz.	257 lbs. 4 oz.	369
Silveretta 6852	16 lbs. 9 oz.		424
Susie Pogis 35425	16 lbs. 9 oz.	245 lbs. 15 oz.	436
Vida's Crown Princess 76010	16 lbs. 9 oz.	230 lbs. 7 oz.	460
Chief's Malita 25186	16 lbs. 8½ oz.	190 lbs. 4 oz.	65
Cocotte 11958	16 lbs. 8½ oz.	172 lbs. 15 oz.	75
Dainty Doris 85497	16 lbs. 8½ oz.	195 lbs. 0 oz.	93
Gambetta's Topsy 42376	16 lbs. 8½ oz.	263 lbs. 0 oz.	163
Mrs. Knickerbocker 19367	16 lbs. 8½ oz.	190 lbs. 4 oz.	315
Nazell 2d 98271	16 lbs. 8½ oz.	227 lbs. 5 oz.	321
Pa Pogis 72500	16 lbs. 8½ oz.	296 lbs. 15 oz.	342
Signora Reid 27125	16 lbs. 8⅜ oz.	241 lbs. 8 oz.	422
Analysis 29162	16 lbs. 8 oz.	234 lbs. 7 oz.	16
Brenhilda 7649	16 lbs. 8 oz.	145 qts.	46
Broadway 50883	16 lbs. 8 oz.	237 lbs. 3 oz.	48
Caroline Pauncefote 62919	16 lbs. 8 oz.	149 lbs. 14 oz.	57
Carrie 3894	16 lbs. 8 oz.		58
Carrie Gold 56348	16 lbs. 8 oz.	237 lbs. 8 oz.	58
Chrissy 1448	16 lbs. 8 oz.		66
Cretesia's Rachel 89998	16 lbs. 8 oz.	279 lbs. 8 oz.	89
Crust 4775	16 lbs. 8 oz.		91
Dagmar of Florence 120863	16 lbs. 8 oz.	185 lbs. 10 oz.	93
Daisy of Belhurst 3114	16 lbs. 8 oz.		96
Daretta 62579	16 lbs. 8 oz.	217 lbs. 5 oz.	99
Diana of St. Lambert 6636	16 lbs. 8 oz.	298 lbs. 0 oz.	104
Dixie Princess 55658	16 lbs. 8 oz.	164 lbs. 0 oz.	106
Edy Signal 19430	16 lbs. 8 oz.	250 lbs. 8 oz.	122
Elsie of Shadyside 85500	16 lbs. 8 oz.	200 lbs. 4 oz.	126
Empress of Ely 2d 6771	16 lbs. 8 oz.		128
Golden Skin 10861	16 lbs. 8 oz.	254 lbs. 0 oz.	172
Katy Signal 2d 44673	16 lbs. 8 oz.	244 lbs. 8 oz.	216
Kisberine Rex 31436	16 lbs. 8 oz.	257 lbs. 12 oz.	219
Kitty Kringel 57120	16 lbs. 8 oz.	229 lbs. 12 oz.	220
Lady Alexis 26916	16 lbs. 8 oz.	244 lbs. 14 oz.	223
Lady Bowen 354	16 lbs. 8 oz.	223 lbs. 0 oz.	225
Lady Love 2d 2212	16 lbs. 8 oz.		230
Lady Negelia 98735	16 lbs. 8 oz.	204 lbs. 0 oz.	232
Landseer's Content 56985	16 lbs. 8 oz.	232 lbs. 2 oz.	237
Landseer's Signal Queen 66363	16 lbs. 8 oz.	224 lbs. 8 oz.	239
Leonice 2d 8342	16 lbs. 8 oz.		246
Lida Mullin 9198	16 lbs. 8 oz.	269 lbs. 4 oz.	249
Lily Martin 49954	16 lbs. 8 oz.	256 lbs. 0 oz.	251
Little Myers 104650	16 lbs. 8 oz.	196 lbs. 9 oz.	255
Lynona 36240	16 lbs. 8 oz.	156 lbs. 4 oz.	265
Minna Rajah 14847	16 lbs. 8 oz.	179 lbs. 0 oz.	298
Modita 16626	16 lbs. 8 oz.	243 lbs. 8 oz.	309
Nightingale K. 2d 19841	16 lbs. 8 oz.		325
Nutley St. Lambert 58469	16 lbs. 8 oz.	202 lbs. 0 oz.	329
Pet Clover 14624	16 lbs. 8 oz.	17 qts. per day	354
Princess Honoria 62548	16 lbs. 8 oz.	167 lbs. 8 oz.	370
Prusa Victor Pogis 134216	16 lbs. 8 oz.	271 lbs. 12 oz.	376
Quintuple Pogis' Fille 94975	16 lbs. 8 oz.	244 lbs. 0 oz.	380

	BUTTER.	MILK.	PAGE.
Rescue of St. Lambert 48989	16 lbs. 8 oz.	181 lbs. 12 oz.	384
Salute 67682	16 lbs. 8 oz.	219 lbs. 4 oz.	409
Sedate's Retta 86596	16 lbs. 8 oz.	153 lbs. 8 oz.	412
Sister Rex 13194	16 lbs. 8 oz.	201 lbs. 0 oz.	426
Bettie Pogis 44542	16 lbs. 7½ oz.	263 lbs. 4 oz.	38
Blossom's Niobe 13855	16 lbs. 7½ oz.	243 lbs. 13 oz.	44
Carrie Rex 10271	16 lbs. 7½ oz.	245 lbs. 12 oz.	58
Chendado's Maid 45991	16 lbs. 7½ oz.	189 lbs. 0 oz.	64
Dame Ida 76878	16 lbs. 7½ oz.	224 lbs. 12 oz.	98
Homestead Rowena Pogis 89591	16 lbs. 7½ oz.	225 lbs. 8 oz.	189
Lady Mary Lenox 105574	16 lbs. 7½ oz.	252 lbs. 8 oz.	231
Milkmaid Felch 12339	16 lbs. 7½ oz.	238 lbs. 8 oz.	297
Sataline 2d 131898	16 lbs. 7½ oz.	228 lbs. 10 oz.	410
Ashantee's Brunette 65946	16 lbs. 7 oz.	185 lbs. 8 oz.	20
Chromo's Modita 121439	16 lbs. 7 oz.	250 lbs. 8 oz.	68
Corinna 2d 6594	16 lbs. 7 oz.		81
Cream Calla 40233	16 lbs. 7 oz.	262 lbs. 12 oz.	87
Damara 65001	16 lbs. 7 oz.	240 lbs. 2 oz.	98
Dew 62474	16 lbs. 7 oz.	181 lbs. 0 oz.	103
Gala 1375	16 lbs. 7 oz.		168
Gladys of Belle Vue 9569	16 lbs. 7 oz.		171
Gold Badger 67301	16 lbs. 7 oz.	226 lbs. 12 oz.	172
Harmintha 96920	16 lbs. 7 oz.	209 lbs. 0 oz.	180
Kamaretta 80479	16 lbs. 7 oz.	266 lbs. 8 oz.	212
Lady Hugo 29430	16 lbs. 7 oz.	239 lbs. 8 oz.	229
La Petite Mère 2d 12810	16 lbs. 7 oz.	346 lbs. 0 oz.	240
Lotchen 19823	16 lbs. 7 oz.	248 lbs. 10 oz.	259
Mary Palestine 52439	16 lbs. 7 oz.	220 lbs. 8 oz.	279
May Fair 5184	16 lbs. 7 oz.	171 lbs. 0 oz.	289
Nelly Thorn 83344	16 lbs. 7 oz.	234 lbs. 12 oz.	322
Polynia 10753	16 lbs. 7 oz.	301 lbs. 1 oz.	362
Rioter Alphea 10091	16 lbs. 7 oz.		386
Rioter's Esse 81901	16 lbs. 7 oz.	159 lbs. 8 oz.	387
Rosona 12956	16 lbs. 7 oz.	107 lbs. 12 oz.	398
Shiloh Daughter 20378	16 lbs. 7 oz.	254 lbs. 0 oz.	415
Snow Bank 75184	16 lbs. 7 oz.	198 lbs. 0 oz.	427
Walter Girl 12776	16 lbs. 7 oz.	228 lbs. 0 oz.	463
Yum Yum Signal 40688	16 lbs. 7 oz.	198 lbs. 12 oz.	472
Zeroda 43723	16 lbs. 7 oz.	219 lbs. 0 oz.	473
Pedroletta 26597	16 lbs. 6½ oz.	269 lbs. 10 oz.	346
Signal's Empress 57767	16 lbs. 6½ oz.	318 lbs. 10 oz.	420
Exile's Harriet 100716	16 lbs. 6¼ oz.	221 lbs. 8 oz.	140
Alnora 112132	16 lbs. 6 oz.	242 lbs. 1 oz.	13
Aurated Pogis 2d 65666	16 lbs. 6 oz.	280 lbs. 5 oz.	22
Baby Mar 115089	16 lbs. 6 oz.	237.3 lbs.	23
Belle of Patterson 5664	16 lbs. 6 oz.	241 lbs. 10 oz.	31
Belle of Rydal Grange 41787	16 lbs. 6 oz.	232 lbs. 0 oz.	32
Cill of Glen Rouge 13818	16 lbs. 6 oz.	259 lbs. 0 oz.	69
Donnona's Nectarine 40093	16 lbs. 6 oz.	277 lbs. 2 oz.	113
Fantine 1271	16 lbs. 6 oz.	252 lbs. 4 oz.	152
Francesville 53367	16 lbs. 6 oz.	232 lbs. 8 oz.	161
Gazellathirdella 89407	16 lbs. 6 oz.	350 lbs. 0 oz.	164
Jessaline 26099	16 lbs. 6 oz.	264 lbs. 8 oz.	206

	BUTTER.	MILK.	PAGE.
Lolly Darling 78005......................	16 lbs. 6 oz....	255 lbs. 5 oz....	257
Lucilla Kent 8892........................	16 lbs. 6 oz....	149 lbs. 8 oz....	262
Maid of Saragossa 9086..................	16 lbs. 6 oz....	190 lbs. 5 oz....	270
Miss Porter 20300.......................	16 lbs. 6 oz....	180 lbs. 15 oz....	307
Pedro's Maggie 102922...................	16 lbs. 6 oz....	241 lbs. 0 oz....	349
Pedro's Pretty Lass 135067..............	16 lbs. 6 oz....	248 lbs. 0 oz....	350
Petra 19267.............................	16 lbs. 6 oz....	238 lbs. 0 oz....	354
Pride of Bay View 61997.................	16 lbs. 6 oz....	190 lbs. 6 oz....	365
Pride of Madelon 105711.................	16 lbs. 6 oz....	260 lbs. 5 oz....	365
Queen of Pomona 56250..................	16 lbs. 6 oz....	188 lbs. 8 oz....	379
Sadie T. 75896..........................	16 lbs. 6 oz....	268 lbs. 0 oz....	404
Sallie of the Ledges 87182...............	16 lbs. 6 oz....,	335.8 lbs...........	409
Toltecida 2d 96030......................	16 lbs. 6 oz....	188 lbs. ½ oz....	444
Value Belle 33080.......................	16 lbs. 6 oz....	276 lbs. 12 oz....	456
Yokun Maid 19073.......................	16 lbs. 6 oz....	223 lbs. 9 oz....	470
Silvia of Linwood 61008..................	16 lbs. 5.6 oz....	183 lbs. 11.2 oz....	425
Couch's Lily 3237.......................	16 lbs. 5½ oz....	....	83
Mattie Tormentor 62841.................	16 lbs. 5½ oz....	237 lbs. 0 oz....	284
Misselle 73488..........................	16 lbs. 5½ oz....	167 lbs. 10 oz....	304
Skipover 20217..........................	16 lbs. 5½ oz....	17 qts. per day....	427
Benetia 21511...........................	16 lbs. 5⅓ oz....	162 lbs. 12 oz....	34
Adelaide Saufley 96284..................	16 lbs. 5 oz....	243 lbs. 15 oz....	6
Beautiful Blondine 82051................	16 lbs. 5 oz....	241 lbs. 4 oz....	25
Bertha Thorne 74824....................	16 lbs. 5 oz....	210 lbs. 4 oz....	35
Blanche Ferry 108501....................	16 lbs. 5 oz....	244 lbs. 4 oz....	43
Buttercup 17285.........................	16 lbs. 5 oz....	295 lbs. 0 oz....	52
Cassie Marjoram 120666.................	16 lbs. 5 oz....	250 lbs. 12 oz....	59
Diploma's Marissa 95590.................	16 lbs. 5 oz....	218 lbs. 4 oz....	106
Dorchester Maid 101385.................	16 lbs. 5 oz....	225 lbs. 12 oz....	114
Ettie Palmer 4th 47481..................	16 lbs. 5 oz....	241 lbs. 13¾ oz....	133
Flora Lee of Tennessee 7694.............	16 lbs. 5 oz....	334 lbs. 0 oz....	157
Flora of St. Peter's 8622.................	16 lbs. 5 oz....	....	157
Handy's Daisy 74017.....................	16 lbs. 5 oz....	251 lbs. 12¾ oz....	180
Hattie Douglass 24960...................	16 lbs. 5 oz....	262 lbs. 0 oz....	183
Jersey Cream 3d 8521...................	16 lbs. 5 oz....	239 lbs. 12 oz....	206
Kate Landseer of Lawn 84415............	16 lbs. 5 oz....	216 lbs. 11 oz....	213
Kitty of St. Lambert 2d 40106...........	16 lbs. 5 oz....	204 lbs. 0 oz....	220
Lady Superior 22865....................	16 lbs. 5 oz....	264 lbs. 0 oz....	234
Lord Harry's Lucy 86091.................	16 lbs. 5 oz....	220 lbs. 0 oz....	257
May Marjoram 123811...................	16 lbs. 5 oz....	241 lbs. 4 oz....	289
Miss Vermont 7698......................	16 lbs. 5 oz....	....	308
Mulford's Lizzie Walker 50213............	16 lbs. 5 oz....	229 lbs. 10 oz....	315
Olie's Lady Teazle 12307.................	16 lbs. 5 oz....	275 lbs. 4 oz....	332
Pendule 2d 16709.......................	16 lbs. 5 oz....	226 lbs. 5 oz....	353
Pitapat 11136...........................	16 lbs. 5 oz....	197 lbs. 8 oz....	358
Princess Victor Pogis 132724............	16 lbs. 5 oz....	255 lbs. 4 oz....	374
Sally of St. Lambert 78873...............	16 lbs. 5 oz....	224 lbs. 0 oz....	409
Troth 6130..............................	16 lbs. 5 oz....	104 qts...........	453
Vexanella 79306.........................	16 lbs. 5 oz....	218 lbs. 4 oz....	459
Viera 3d 7642...........................	16 lbs. 5 oz....	311 lbs. 12 oz....	460
Elphie May 69396.......................	16 lbs. 4½ oz....	227 lbs. 8 oz....	125
Jersey Jane 38308.......................	16 lbs. 4½ oz....	222 lbs. 8 oz....	206
Jessaline 2d 38295.......................	16 lbs. 4½ oz....	270 lbs. 5½ oz....	206

	BUTTER.	MILK.	PAGE.
Lorne's Marjoram 43327	16 lbs. 4½ oz		259
Mary M. Pogis 32455	16 lbs. 4½ oz	204 lbs. 0 oz	278
Miss Clifford 2d 56149	16 lbs. 4½ oz	270 lbs. 4 oz	303
Pansy Dec. 99294	16 lbs. 4½ oz	236 lbs. 14.4 oz	340
Pink Ring 69394	16 lbs. 4½ oz	239 lbs. 5 oz	358
Princess Sheila 7297	16 lbs. 4½ oz	213 lbs. 12 oz	374
Sparks 41041	16 lbs. 4½ oz	236 lbs. 8 oz	431
Torpedo Exile 105287	16 lbs. 4½ oz	206 lbs. 0 oz	451
Alfieda 6744	16 lbs. 4 oz	250 lbs. 4 oz	10
Bisson's Beauty 95023	16 lbs. 4 oz	166 lbs. 0 oz	41
Bramballetta 10451	16 lbs. 4 oz		46
Busy Bee 6336	16 lbs. 4 oz	277 lbs. 0 oz	52
Cetewayo's Daisy 18230	16 lbs. 4 oz	188 lbs. 8 oz	61
Chansonnette 5695	16 lbs. 4 oz	16 qts. per day	62
Daisy of Dryden 77778	16 lbs. 4 oz	196 lbs. 8 oz	97
Daisy Queen 9619	16 lbs. 4 oz		98
Diamond's Lass 81767	16 lbs. 4 oz	172 lbs. 13 oz	104
Elorita 101742	16 lbs. 4 oz	234 lbs. 12 oz	125
Eltekeh 28266	16 lbs. 4 oz	213 lbs. 11 oz	126
Emily of Hillside 8073	16 lbs. 4 oz	229 lbs. 10 oz	127
Exile's Daisy 46885	16 lbs. 4 oz	244 lbs. 0 oz	139
Fancy's Gem 2d 86342	16 lbs. 4 oz	215 lbs. 8 oz	149
Fancy's Pet 36013	16 lbs. 4 oz	165 lbs. 0 oz	150
Favorite of the Elms 1656	16 lbs. 4 oz	314 lbs. 0 oz	153
Harry's Duchess 60289	16 lbs. 4 oz	206 lbs. 8 oz	181
Kate Kyle 64434	16 lbs. 4 oz	244 lbs. 8 oz	213
Katie Putnam 40366	16 lbs. 4 oz	207 lbs. 6 oz	215
Laura Lee 2d 128023	16 lbs. 4 oz	244 lbs. 10 oz	241
Lydia Darrach 3d 10662	16 lbs. 4 oz		265
Maid of the Elms 3d 15554	16 lbs. 4 oz	172 lbs. 8 oz	270
Matilda 5th 18068	16 lbs. 4 oz	146 lbs. 0 oz	281
Maud's Sultana 19518	16 lbs. 4 oz	18½ qts. per day	286
Miss Signal 20379	16 lbs. 4 oz	223 lbs. 3 oz	308
Miss Willie Jones 6918	16 lbs. 4 oz	316 lbs. 8 oz	308
Moss Rose of Jackson 86709	16 lbs. 4 oz	299 lbs. 8 oz	313
Onnolee 23804	16 lbs. 4 oz	186 lbs. 4 oz	334
Pedro's Lady Phillis 88486	16 lbs. 4 oz	252 lbs. 0 oz	348
Pet Victor Pogis 77812	16 lbs. 4 oz	243 lbs. 0 oz	355
Princess Haidee Signal 98652	16 lbs. 4 oz	247 lbs. 0 oz	369
Proctor's Belle 47106	16 lbs. 4 oz	186 lbs. 0 oz	375
Ribbon's Matilda 66109	16 lbs. 4 oz	202 lbs. 8 oz	385
Rioter Flora Marjoram 111165	16 lbs. 4 oz	272 lbs. 0 oz	387
Southern Princess 36320	16 lbs. 4 oz	205 lbs. 0 oz	430
Sunbeam's Pet 2d 9112	16 lbs. 4 oz	312 lbs. 0 oz	435
Temalema 2d 82411	16 lbs. 4 oz	218 lbs. 8 oz	440
Topaz of Woodlawn 11661	16 lbs. 4 oz	214 lbs. 14 oz	446
Wine 15739	16 lbs. 4 oz	232 lbs. 5 oz	466
Signalda Bloomfield 30694	16 lbs. 3¾ oz	226 lbs. 8 oz	417
Silver Venus 20270	16 lbs. 3¾ oz	205 lbs. 0 oz	425
Beech Leaf 99381	16 lbs. 3½ oz	221 lbs. 0 oz	28
Belle Bonair 33960	16 lbs. 3½ oz	257 lbs. 7 oz	29
Belle Lyman 28522	16 lbs. 3½ oz	220 lbs. 3 oz	30
Blossie Reynolds 6082	16 lbs. 3½ oz		44

	BUTTER.	MILK.	PAGE.
Hermosa Pogis 40369................	16 lbs. 3½ oz....	166 lbs. 14 oz.....	186
Maritana 12039.....................	16 lbs. 3½ oz....	254 lbs. 12 oz.....	275
May Day 60451.....................	16 lbs. 3½ oz....	179.9 lbs.............	288
Pride's Lady Frances 39529..........	16 lbs. 3½ oz....	225 lbs. 0 oz.....	366
Young Garenne 3d 13648.............	16 lbs. 3½ oz....	....	471
Damask Rose 22065..................	16 lbs. 3¼ oz....	218 lbs. 2 oz.....	98
Hollie Pogissa 110490...............	16 lbs. 3¼ oz....	213 lbs. 0 oz.....	188
King's Erica 22000.................	16 lbs. 3¼ oz....	248 lbs. 3¾ oz.....	217
Moxie's Mode 97432.................	16 lbs. 3¼ oz....	199 lbs. 10½ oz.....	314
Personia 90231....................	16 lbs. 3¼ oz....	249 lbs. 10 oz.....	353
Retta Pogis 81132.................	16 lbs. 3¼ oz....	143 lbs. 8 oz.....	384
Rosa of Frisby Hill 36080...........	16 lbs. 3¼ oz....	216 lbs. 4 oz.....	394
Alhena 15995.....................	16 lbs. 3 oz....	....	11
Allie's Leaf 45617.................	16 lbs. 3 oz....	228 lbs. 12 oz.....	12
Bicora 2d 91188..................	16 lbs. 3 oz....	218 lbs. 8 oz.....	39
Bonnie Molf 95616.................	16 lbs. 3 oz....	245 lbs. 13 oz.....	45
Comanca 19380....................	16 lbs. 3 oz....	218 lbs. 0 oz.....	76
Compact 52705....................	16 lbs. 3 oz....	226 lbs. 8 oz.....	77
Countess Stoke Pogis 62540..........	16 lbs. 3 oz....	194 lbs. 8 oz.....	86
Diploma's Pansy 112871.............	16 lbs. 3 oz....	230 lbs. 8 oz.....	106
Elmwood Daisy 2d 34193.............	16 lbs. 3 oz....	227 lbs. 0 oz.....	125
Exile's Beauty 61220................	16 lbs. 3 oz....	199 lbs. 0 oz.....	138
Fadette Signal's Queen 109281........	16 lbs. 3 oz....	206 lbs. 14 oz.....	146
Felsticea 17083...................	16 lbs. 3 oz....	245 lbs. 14 oz.....	155
Gazella 3d 9355..................	16 lbs. 3 oz....	258 lbs. 8 oz.....	164
Gertie Victor Pogis 100022..........	16 lbs. 3 oz....	232 lbs. 12 oz.....	167
Golden Zoe 3975...................	16 lbs. 3 oz....	282 lbs. 8 oz.....	173
Julia Rex of Ingleside 74046.........	16 lbs. 3 oz....	210 lbs. 0 oz.....	211
Kate Oxford 58102.................	16 lbs. 3 oz....	153 lbs. 8 oz.....	214
Lady of Dryden 27642..............	16 lbs. 3 oz....	328 lbs. 0 oz.....	232
Lenore H. 2d 29767................	16 lbs. 3 oz....	294 lbs. 8 oz.....	245
Leslie 9179......................	16 lbs. 3 oz....	187 lbs. 8 oz.....	247
Letta of Eau Claire 120021..........	16 lbs. 3 oz....	260 lbs. 6 oz.....	248
Lily of Maple Grove 5079...........	16 lbs. 3 oz....	204 lbs. 4 oz.....	252
Madame Melrosa 103185.............	16 lbs. 3 oz....	179 lbs. 15 oz.....	266
Maggie of St. Lambert 9776..........	16 lbs. 3 oz....	278 lbs. 0 oz.....	268
Mayhegtal 64454..................	16 lbs. 3 oz....	246 lbs. 5 oz.....	289
Nigella 7895.....................	16 lbs. 3 oz....	215 lbs. 12 oz.....	325
Orphan Duchess 3d 21284............	16 lbs. 3 oz....	231 lbs. 12 oz.....	337
Period 42640.....................	16 lbs. 3 oz....	230 lbs. 11 oz.....	353
Queen Ione 113999.................	16 lbs. 3 oz....	214 lbs. 8 oz.....	377
Rose of Rose Lawn 9365.............	16 lbs. 3 oz....	275 lbs. 10 oz.....	397
Rover's Swan 106934...............	16 lbs. 3 oz....	251 lbs. 0 oz.....	398
Willis 2d 4461...................	16 lbs. 3 oz....	24 qts. per day....	466
Morganza Queen 77791..............	16 lbs. 2¾ oz....	247 lbs. 4 oz.....	313
Bertha's Royal Daisy 104901.........	16 lbs. 2½ oz....	230 lbs. 0 oz.....	35
Conger's M. 32751.................	16 lbs. 2½ oz....	183 lbs. 8 oz.....	78
Dame Miller 59260.................	16 lbs. 2½ oz....	203 lbs. 8 oz.....	99
Exile's Kosi 107505................	16 lbs. 2½ oz....	250 lbs. 8 oz.....	141
Ida Zoe Pogis 38685...............	16 lbs. 2½ oz....	172 lbs. 6 oz.....	197
Madam Zophee 31679................	16 lbs. 2½ oz....	226 lbs. 6 oz.....	266
Melrose Milly 61917................	16 lbs. 2½ oz....	222 lbs. 2 oz.....	292
Molina of Glen Rouge 123635.........	16 lbs. 2½ oz....	225 lbs. 8 oz.....	309

	BUTTER.	MILK.	PAGE.
Mon Plaisir Lily 25969	16 lbs. 2½ oz.	255 lbs. 0 oz.	312
Satala 50074	16 lbs. 2½ oz.	194 lbs. 3 oz.	410
Silver Belle C. 90600	16 lbs. 2½ oz.	249 lbs. 8 oz.	423
Sunset 15130	16 lbs. 2½ oz.	210 lbs. 0 oz.	435
Cetewayo's Dorcas 20287	16 lbs. 2¼ oz.	223 lbs. 4 oz.	61
Donney Pogis 38907	16 lbs. 2⅛ oz.	143 lbs. 4 oz.	112
Albert's Cretesia 22881	16 lbs. 2 oz.		9
Amelia S. King 112862	16 lbs. 2 oz.	243 lbs. 14 oz.	16
Callie Nan 7959	16 lbs. 2 oz.	252 lbs. 8 oz.	54
Catchfly 5th 55483	16 lbs. 2 oz.	274 lbs. 8 oz.	60
Corn 10504	16 lbs. 2 oz.		82
Cream Caroline 17357	16 lbs. 2 oz.	252 lbs. 8 oz.	87
Crensa 74203	16 lbs. 2 oz.	276 lbs. 8 oz.	88
Domett 88179	16 lbs. 2 oz.	205 lbs. 12 oz.	112
Dott Buttercup 16358	16 lbs. 2 oz.	167 lbs. 6 oz.	115
Edessa 21844	16 lbs. 2 oz.	218 lbs. 5. oz.	120
Emma Hudson 12469	16 lbs. 2 oz.		127
Eudora 1863	16 lbs. 2 oz.	268 lbs. 8 oz.	134
Exile's Lulu 49984	16 lbs. 2 oz.	328 lbs. 8 oz.	142
Fall Leaf's Heiress 54454	16 lbs. 2 oz.	210 lbs. 10 oz.	148
Fancy's Nienchen 104377	16 lbs. 2 oz.	205 lbs. 0 oz.	150
Fear Not 2d 6061	16 lbs. 2 oz.	216 lbs. 0 oz.	154
Goldie Pogis 45423	16 lbs. 2 oz.	206 lbs. 9 oz.	173
Gold Trinket 9518	16 lbs. 2 oz.	240 lbs. 12 oz.	174
Jennette Darling 10702	16 lbs. 2 oz.	215 lbs. 7 oz.	204
La Vivienne 2d 1324	16 lbs. 2 oz.	95 qts.	242
Lora of Clovernook 51599	16 lbs. 2 oz.	179 lbs. 8 oz.	257
Lord Fife's Joliette 30591	16 lbs. 2 oz.	172 lbs. 10 oz.	257
Lorilla 87505	16 lbs. 2 oz.	227 lbs. 4 oz.	258
Marissa 65285	16 lbs. 2 oz.	207 lbs. 1 oz.	274
Midway 50000	16 lbs. 2 oz.	182 lbs. 0 oz.	296
Miss Flora Brown 100532	16 lbs. 2 oz.	238 lbs. 4 oz.	304
Moth of St. Lambert 9775	16 lbs. 2 oz.	235 lbs. 8 oz.	314
Peggie Tormentor 98103	16 lbs. 2 oz.	217 lbs. 14 oz.	352
Queryan 42140	16 lbs. 2 oz.	252 lbs. 0 oz.	379
Rex's Queen of St. Lambert 59161	16 lbs. 2 oz.	155 lbs. 15 oz.	384
St. Lambert's Rexina 113609	16 lbs. 2 oz.	257 lbs. 0 oz.	408
Sallie Grimes 79122	16 lbs. 2 oz.	248 lbs. 4 oz.	408
Star's Valentine 81778	16 lbs. 2 oz.	277 lbs. 7 oz.	432
Success of St. Lambert 28489	16 lbs. 2 oz.	238 lbs. 4 oz.	433
Theda's Belle 106891	16 lbs. 2 oz.	251 lbs. 0 oz.	441
Utilla 11215	16 lbs. 2 oz.		456
White Frost 17431	16 lbs. 2 oz.	295 lbs. 8 oz.	466
Pride of Mashamoquet Farm 6469	16 lbs. 1¾ oz.	110 lbs. 4 oz.	365
Nancy's Star 41423	16 lbs. 1⅝ oz.	194 lbs. 10 oz.	319
Fadette of Canaan 14807	16 lbs. 1½ oz.	119 lbs. 12 oz.	146
Hazel Lee 76864	16 lbs. 1½ oz.	237 lbs. 0 oz.	184
Ida's Fawn 56018	16 lbs. 1½ oz.	244 lbs. 13 oz.	195
May Lankton 15872	16 lbs. 1½ oz.	295 lbs. 0 oz.	289
Sweet Blossom Pogis 36995	16 lbs. 1½ oz.	177 lbs. 0 oz.	437
Gladdys Gaily 20878	16 lbs. 1⅜ oz.	102 lbs. 1 oz.	171
Bessie Pope St. Lambert 80352	16 lbs. 1¼ oz.	234 lbs. 0 oz.	36
Ivy Tormento 121700	16 lbs. 1¼ oz.	234 lbs. 13 oz.	202

	BUTTER.	MILK.	PAGE.
Amy Landseer 74826	16 lbs. 1 oz....	219 lbs. 12 oz.....	16
Badger Girl 21463	16 lbs. 1 oz....	260 lbs. 4½ oz.....	23
Bettie's Pet 22550	16 lbs. 1 oz....	180 lbs. 8 oz.....	39
Chuffy's Ethel 102195	16 lbs. 1 oz....	250 lbs. 5 oz.....	68
Clematis 3d 6653	16 lbs. 1 oz....	209 lbs. 2 oz.....	71
Cotton Patchwork 85926	16 lbs. 1 oz....	211 lbs. 11 oz.....	83
Donna Signal 29407	16 lbs. 1 oz....	243 lbs. 0 oz.....	112
Harry's May Fair 37499	16 lbs. 1 oz....	131 lbs. 4 oz.....	182
Joan Montague 40787	16 lbs. 1 oz....	246 lbs. 0 oz.....	208
Lady Rareripe 23081	16 lbs. 1 oz....	199 lbs. 0 oz.....	234
La Jolie Lizette 16426	16 lbs. 1 oz....	....	236
Little Pogis 32612	16 lbs. 1 oz....	200 lbs. 12 oz.....	255
Maid of Amboy 2929	16 lbs. 1 oz....	....	269
Minnie Hauk 2d 99103	16 lbs. 1 oz....	177 lbs. 12 oz.....	299
Myrrha 11299	16 lbs. 1 oz....	169 lbs. 4 oz.....	317
Nanalda 2d 97130	16 lbs. 1 oz....	249 lbs. 11 oz.....	318
Nora Stoke Pogis 34687	16 lbs. 1 oz....	248 lbs. 8 oz.....	328
Panache 62055	16 lbs. 1 oz....	195 lbs. 12 oz.....	339
Ribbon's Queen 66321	16 lbs. 1 oz....	197 lbs. 0 oz.....	385
Rosa Russell 81549	16 lbs. 1 oz....	256 lbs. 8 oz.....	395
Silistria 20264	16 lbs. 1 oz....	197 lbs. 2 oz.....	423
Victoria 3175	16 lbs. 1 oz....	....	459
Warren's Duchess 4622	16 lbs. 1 oz....	250 lbs. 3 oz.....	464
Donney Pogis 3d 100199	16 lbs. ¾ oz....	198 lbs. 8 oz.....	113
Catono's Kate 59976	16 lbs. ½ oz....	206 lbs. 0 oz.....	60
Euphonia 6783	16 lbs. ½ oz....	261 lbs. 6 oz.....	134
Geneva M.'s Fancy 104917	16 lbs. ½ oz....	195 lbs. 15 oz.....	166
Ida's Dream 50070	16 lbs. ½ oz....	136 lbs. 14 oz.....	195
Nitelis 102722	16 lbs. ½ oz....	329 lbs. 12 oz.....	326
Phœbe Rex 55084	16 lbs. ½ oz....	256 lbs. 5 oz.....	357
Pink of Meridale 114748	16 lbs. ½ oz....	227 lbs. 14 oz.....	358
Princess of Lynwood 118458	16 lbs. ½ oz....	218 lbs. 11½ oz.....	373
Signaldo's Pansy 53306	16 lbs. ½ oz....	306 lbs. 0 oz.....	418
Sylvia Perrot 30963	16 lbs. ½ oz....	182 lbs. 13 oz.....	438
Exile's Lady Star 63943	16 lbs. ¼ oz....	217 lbs. 12 oz.....	142
Ajax Riotress 113709	16 lbs. 0 oz....	266 lbs. 7 oz.....	7
Alarm of Highland 41864	16 lbs. 0 oz....	252 lbs. 4 oz.....	8
Ashantee's Lady 35951	16 lbs. 0 oz....	129 lbs. 12 oz.....	21
Belle of Scituate 7977	16 lbs. 0 oz....	....	32
Chloe of Bolton 81340	16 lbs. 0 oz....	259 lbs. 15 oz.....	66
Countess 114	16 lbs. 0 oz....	....	83
Cream Lily 24015	16 lbs. 0 oz....	221 lbs. 0 oz.....	87
Dairy Pride 4th 21681	16 lbs. 0 oz....	302 lbs. 4 oz.....	93
Daisy of Hillside Farm 6025	16 lbs. 0 oz....	274 lbs. 0 oz.....	97
Dolly's Daughter 31392	16 lbs. 0 oz....	197 lbs. 10 oz.....	111
Dom Pedro's Julian 8631	16 lbs. 0 oz....	....	112
Durwood's Lass 19710	16 lbs. 0 oz....	176 lbs. 14 oz.....	119
Ettarock 81103	16 lbs. 0 oz....	233 lbs. 10 oz.....	133
Fanfare 49927	16 lbs. 0 oz....	225 lbs. 10½ oz.....	151
Fleurette of Linwood 12918	16 lbs. 0 oz....	151 lbs. 12 oz.....	157
Frolic of Chestnutwood 19405	16 lbs. 0 oz....	251 lbs. 4 oz.....	162
Ida of Bear Lake 6169	16 lbs. 0 oz....	12 qts. daily......	193
Idessimus 50508	16 lbs. 0 oz....	213 lbs. 8 oz.....	198

	BUTTER.	MILK.	PAGE.
Katy Scituate's Beauty 100259............	16 lbs. 0 oz....	257 lbs. 10 oz.....	215
Lady Emma C. 57041....................	16 lbs. 0 oz....	272 lbs. 8 oz.....	227
Lady of Malone 25734..................	16 lbs. 0 oz....	173 lbs. 6 oz.....	233
Leoni Landseer 65317..................	16 lbs. 0 oz....	192 lbs. 12 oz.....	246
Letitia of St. Lambert 91054.............	16 lbs. 0 oz....	174 lbs. 8 oz.....	248
Lily of Glynllyn 31340..................	16 lbs. 0 oz....	227 lbs. 8 oz.....	252
Lydia Darrach 2d 8056.................	16 lbs. 0 oz....	....	265
Maid of Tanglewood 51606..............	16 lbs. 0 oz....	185 lbs. 4 oz.....	270
Maid of the Elms 6960..................	16 lbs. 0 oz....	214 lbs. 8 oz.....	270
Marie Bash 81441......................	16 lbs. 0 oz....	266 lbs. 0 oz.....	274
Marjoram 3239........................	16 lbs. 0 oz....	....	275
Matilda's Queen 72663.................	16 lbs. 0 oz....	247 lbs. 12 oz.....	283
Merlette 4988.........................	16 lbs. 0 oz....	297 lbs. 13 oz.....	295
Miamotto 48691.......................	16 lbs. 0 oz....	276 lbs. 12 oz.....	296
Miss Perrot 73801.....................	16 lbs. 0 oz....	236 lbs. 0 oz.....	307
Phyllis of Hillcrest 9067...............	16 lbs. 0 oz....	240 lbs. 0 oz.....	357
Ruby Torment 32989...................	16 lbs. 0 oz....	120 lbs. 10 oz.....	402
Sallie Guild 14524....................	16 lbs. 0 oz....	16 qts. daily.....	408
Señorita Pogis 53791..................	16 lbs. 0 oz....	211 lbs. 8 oz.....	413
Sigletta 32915........................	16 lbs. 0 oz....	182 lbs. 12 oz.....	416
Signolo's Florence 2d 82446............	16 lbs. 0 oz....	211 lbs. 0 oz.....	422
Southern Belle 18570..................	16 lbs. 0 oz....	270 lbs. 0 oz.....	430
Tylesholm Blythesomie 101347..........	16 lbs. 0 oz....	301 lbs. 8 oz.....	454
Urbana 5597..........................	16 lbs. 0 oz....	....	455
Valhalla 5300.........................	16 lbs. 0 oz....	....	456
Vinala H. 61835......................	16 lbs. 0 oz....	196 lbs. 4 oz.....	461
Vivian 15813.........................	16 lbs. 0 oz....	193 lbs. 8 oz.....	463
Waukesra 19721.......................	16 lbs. 0 oz....	314 lbs. 8 oz.....	465
Young Miss 90631.....................	16 lbs. 0 oz....	200 lbs. 8 oz.....	472
Miss Bluster 108231...................	15 lbs. 15⅝ oz....	210.3 lbs...........	303
Geneva S. 57170......................	15 lbs. 15½ oz....	202 lbs. 9 oz.....	166
Julia Evelyn 6007.....................	15 lbs. 15½ oz....	233 lbs. 8 oz.....	210
Monte Sano Belle 77683...............	15 lbs. 15½ oz....	181 lbs. 4 oz.....	312
Philena of Briarcliff 68938.............	15 lbs. 15½ oz....	219 lbs. 12 oz.....	355
Riotaletta 2d 34495...................	15 lbs. 15½ oz....	136 lbs. 4 oz.....	386
Beauty's Crescent 26733...............	15 lbs. 15⅜ oz....	126 lbs. 0 oz.....	28
Idabel Landseer 88399.................	15 lbs. 15¼ oz....	299 lbs. 8 oz.....	192
Beauty of Hamilton 55827..............	15 lbs. 15 oz....	227 lbs. 8 oz.....	26
Brunette Le Gros 9755.................	15 lbs. 15 oz....	....	50
Bunker's Utica Belle 38567.............	15 lbs. 15 oz....	236 lbs. 0 oz.....	51
Championa Pogis M. 75926.............	15 lbs. 15 oz....	255 lbs. 5½ oz.....	61
Clara Signal of Lawn 85586............	15 lbs. 15 oz....	183 lbs. 14 oz.....	70
Cretesia's Matilda 90000...............	15 lbs. 15 oz....	284 lbs. 0 oz.....	89
Heritage 50077.......................	15 lbs. 15 oz....	240 lbs. 10 oz.....	186
Kate Gordon 8387.....................	15 lbs. 15 oz....	194 lbs. 7 oz.....	213
Maggie Cyrene 53462..................	15 lbs. 15 oz....	191 lbs. 0 oz.....	267
May Belle's Lady d'Or 74749...........	15 lbs. 15 oz....	177 lbs. 8 oz.....	287
Pansy Patterson 18612.................	15 lbs. 15 oz....	150 lbs. 12 oz.....	340
Tessia H. 94223.......................	15 lbs. 15 oz....	216 lbs. 12 oz.....	441
Thorndale Belle 3d 10459...............	15 lbs. 15 oz....	176 lbs. 0 oz.....	443
Yuba Stoke Pogis 37294...............	15 lbs. 14¾ oz....	231 lbs. 0 oz.....	472
Zoe Henry 6693.......................	15 lbs. 14¾ oz....	202 lbs. 0 oz.....	474
Dark Rose 101892.....................	15 lbs. 14½ oz....	255 lbs. 8 oz.....	100

	BUTTER.	MILK.	PAGE.
Exile's Alphea 45393	15 lbs. 14½ oz.	200 lbs. 6½ oz.	137
Hattie N. 45433	15 lbs. 14½ oz.	150 lbs. 9½ oz.	183
Landseer's Fancy 3d 70653	15 lbs. 14½ oz.	115 lbs. 1½ oz.	238
Pansy Albert of Lawn 76634	15 lbs. 14½ oz.	239 lbs. 11 oz.	340
Perrot's Baroness 22508	15 lbs. 14½ oz.	142 lbs. 6 oz.	353
Princess Edith Hugo 45120	15 lbs. 14½ oz.	246 lbs. 1 oz.	369
Rose of Oxford 13469	15 lbs. 14½ oz.	234 lbs. 10 oz.	396
Kitty Better 32911	15 lbs. 14⅜ oz.	221 lbs. 4 oz.	219
Lucy S. King 106584	15 lbs. 14¼ oz.	212 lbs. 8 oz.	263
Allie of Glen Rouge 129123	15 lbs. 14 oz.	206 lbs. 8 oz.	12
Alvoretta 93817	15 lbs. 14 oz.	178 lbs. 8 oz.	15
Belle of Vermillion 8798	15 lbs. 14 oz.	252 lbs. 0 oz.	32
Cricket's Minnie 26270	15 lbs. 14 oz.	157 lbs. 9 oz.	90
Fancy's Maybee 104383	15 lbs. 14 oz.	215 lbs. 8 oz.	150
Ida of Clinton 25419	15 lbs. 14 oz.	236 lbs. 0 oz.	194
Kaleta 2d 38810	15 lbs. 14 oz.	188 lbs. 4 oz.	212
Lady Alice of the Wilderness 12207	15 lbs. 14 oz.	266 lbs. 0 oz.	224
Lanison's Queen 2d 128773	15 lbs. 14 oz.	222 lbs. 0 oz.	239
Lass of Scituate 9555	15 lbs. 14 oz.	196 lbs. 12 oz.	241
Lois A. 107594	15 lbs. 14 oz.	223 lbs. 1 oz.	256
Mary of Bear Lake 6171	15 lbs. 14 oz.	311 lbs. 1 oz.	279
Miss May of St. Lambert 37084	15 lbs. 14 oz.	138 lbs. 4 oz.	305
Mollie Garfield 2d 18662	15 lbs. 14 oz.	162 lbs. 6 oz.	310
Musicale of Greenwood 62932	15 lbs. 14 oz.	194 lbs. 6 oz.	315
Nelly 2402	15 lbs. 14 oz.		322
Pet's Torment 92255	15 lbs. 14 oz.	162 lbs. 12 oz.	355
Quintuple's Wonder 86077	15 lbs. 14 oz.	252 lbs. 12 oz.	380
Robinella 63132	15 lbs. 14 oz.	213 lbs. 0 oz.	391
Romp Ogden 3d 5458	15 lbs. 14 oz.	189 lbs. 0 oz.	392
Tormentor's Rexea 38906	15 lbs. 14 oz.	176 lbs. 8 oz.	450
Victor's Lady 52691	15 lbs. 14 oz.	203 lbs. 9 oz.	459
Young Floribundus 4th 59314	15 lbs. 14 oz.	225 lbs. 0 oz.	471
Zunilda 2d 93811	15 lbs. 14 oz.	196 lbs. 0 oz.	474
Bell Ensign 56702	15 lbs. 13½ oz.	233 lbs. 7 oz.	30
Dia 13658	15 lbs. 13½ oz.	166 lbs. 10 oz.	103
Etta Bartlett 80254	15 lbs. 13½ oz.	239 lbs. 4 oz.	133
Glory of Elmarch 21521	15 lbs. 13½ oz.	204 lbs. 12 oz.	171
Jolie of St. Lambert 5126	15 lbs. 13½ oz.	298 lbs. 0 oz.	209
Maxminta 85922	15 lbs. 13½ oz.	210 lbs. 6 oz.	287
Princess Oonan 84369	15 lbs. 13½ oz.	225 lbs. 12 oz.	373
Olymph 17957	15 lbs. 13 1-16 oz.	242 lbs. 0 oz.	333
Badger Girl 4th 87488	15 lbs. 13 oz.	208 lbs. 4 oz.	23
Berylla 2d 15117	15 lbs. 13 oz.	237 lbs. 0 oz.	36
Edwina 6713	15 lbs. 13 oz.		121
Ethel McM. 100267	15 lbs. 13 oz.	270 lbs. 14 oz.	130
Exile's Hazel 85181	15 lbs. 13 oz.	197 lbs. 6 oz.	141
Hulabaloo's Flora 111164	15 lbs. 13 oz.	138 qts. 1 pt.	191
Lily of Burr Oaks 11001	15 lbs. 13 oz.	192 lbs. 15 oz.	251
Lucy Gray 2746	15 lbs. 13 oz.		262
Meridale Rioter Pink 64534	15 lbs. 13 oz.	252 lbs. 9 oz.	294
Nutley Dolores 2d 55901	15 lbs. 13 oz.	157 lbs. 0 oz.	329
Our Daisy Queen 94252	15 lbs. 13 oz.	240 lbs. 0 oz.	338
Petite Mère 8516	15 lbs. 13 oz.	197 lbs. 0 oz.	354

	BUTTER.	MILK.	PAGE.
Proctor's Pansy 25088....................	15 lbs. 13 oz....	276 lbs. 15 oz.....	376
Tillie Texas 62840.......................	15 lbs. 13 oz....	206 lbs. 0 oz.....	443
Tobira 8400..............................	15 lbs. 13 oz....	235 lbs. 8 oz.....	444
Valerie 6044.............................	15 lbs. 13 oz....	284 lbs. 8 oz.....	456
Signal Hinman 44615.....................	15 lbs. 12¾ oz....	233 lbs. 12 oz.....	418
Diploma's Clara 103867..................	15 lbs. 12½ oz....	233 lbs. 8 oz.....	105
Fanfare 2d 119780.......................	15 lbs. 12½ oz....	229 lbs. 15¾ oz.....	151
Kalma of Milford 52742.................	15 lbs. 12½ oz....	217 lbs. 12 oz.....	212
Kate McCollum 107306..................	15 lbs. 12½ oz....	206 lbs. 0 oz.....	213
King's Rosella 80767....................	15 lbs. 12½ oz....	205 lbs. 14 oz.....	218
Nan of Ethelwyn 49502..................	15 lbs. 12½ oz....	298 lbs. 8 oz.....	320
Queen May of St. Lambert 45770.........	15 lbs. 12½ oz....	159 lbs. ½ oz.....	378
Princess Marie G. 120390...............	15 lbs. 12⅛ oz....	251.9 lbs..........	371
Almeda Rex 10416.......................	15 lbs. 12 oz....	290 lbs. 4 oz.....	13
Angela's Pet 69776......................	15 lbs. 12 oz....	228 lbs. 0 oz.....	17
Bobby's Magnolia 24401..................	15 lbs. 12 oz....	244 lbs. 0 oz.....	44
Chief's Iona 111704.....................	15 lbs. 12 oz....	142 lbs. 2 oz.....	65
Countess of Croton 5307.................	15 lbs. 12 oz....	277 lbs. 4 oz.....	85
Country Girl 3515.......................	15 lbs. 12 oz....	174 lbs. 15 oz.....	86
Daisy Ida Torment 121824...............	15 lbs. 12 oz....	224 lbs. 0 oz.....	96
Daisy of Denison 53137..................	15 lbs. 12 oz....	223 lbs. 12 oz.....	97
Dora's Fawnie Exile 108696.............	15 lbs. 12 oz....	229 lbs. 0 oz.....	113
Fanny Taylor 6714......................	15 lbs. 12 oz....	223 lbs. 0 oz.....	152
Gold Gem 72024........................	15 lbs. 12 oz....	202 lbs. 6 oz.....	173
Julia Walker 10133.....................	15 lbs. 12 oz....	....	211
Lady Bidwell 10303.....................	15 lbs. 12 oz....	265 lbs. 4 oz.....	225
Lady McDowell 37414...................	15 lbs. 12 oz....	159 lbs. 2 oz.....	230
Lerna 3634.............................	15 lbs. 12 oz....	242 lbs. 9 oz.....	247
Mavourneen of St. Lambert 9777.........	15 lbs. 12 oz....	208 lbs. 0 oz.....	287
Mollie of Glen Rouge 129125.............	15 lbs. 12 oz....	200 lbs. 12 oz.....	310
Molly May Pogis 102823.................	15 lbs. 12 oz....	228 lbs. 0 oz.....	310
Myrtle 2d 211..........................	15 lbs. 12 oz....	....	317
Nora Tormentor of Lawn 99738..........	15 lbs. 12 oz....	226 lbs. 3 oz.....	328
Oleta 15625............................	15 lbs. 12 oz....	251 lbs. 8 oz.....	332
Ora of Forester 54698..................	15 lbs. 12 oz....	207 lbs. 4 oz.....	337
Pedro's Pretty Lady 102379.............	15 lbs. 12 oz....	256 lbs. 0 oz.....	350
Princess Jeannaise 53863................	15 lbs. 12 oz....	201 lbs. 4 oz.....	370
Queen Irene 101893....................	15 lbs. 12 oz....	236 lbs. 4 oz.....	377
Rosabel Hudson 5704...................	15 lbs. 12 oz....	....	393
Sally Leavitt 2d 10471.................	15 lbs. 12 oz....	220 lbs. 0 oz.....	409
Seraphine 2d 37451.....................	15 lbs. 12 oz....	284 lbs. 0 oz.....	413
Smack 64575...........................	15 lbs. 12 oz....	229 lbs. 1 oz.....	427
Sunswick Dolly 91844..................	15 lbs. 12 oz....	246 lbs. 2 oz.....	435
Susa's Darling 44915...................	15 lbs. 12 oz....	225 lbs. 3 oz.....	436
Thornleaf 39960........................	15 lbs. 12 oz....	214 lbs. 5 oz.....	443
Tornella 78453.........................	15 lbs. 12 oz....	251 lbs. 1½ oz.....	451
Ultima 14456...........................	15 lbs. 12 oz....	228 lbs. 0 oz.....	455
Miss Katie W. 47366...................	15 lbs. 11¾ oz....	197 lbs. 0 oz.....	305
Tormentor's Ravalita 97909.............	15 lbs. 11¾ oz....	206 lbs. 8½ oz.....	450
Adelpha Marigold 133768................	15 lbs. 11½ oz....	236 lbs. 10 oz.....	6
Exile's Nina 40522.....................	15 lbs. 11½ oz....	232 lbs. 0 oz.....	143
Mary Brown 81213......................	15 lbs. 11½ oz....	308 lbs. 14 oz.....	277
Mary H'nman 17619....................	15 lbs. 11½ oz....	225 lbs. 0 oz.....	278

	BUTTER.	MILK.	PAGE.
Myra Tormentor 84227	15 lbs. 11½ oz.	241 lbs. 8 oz.	316
Pogis' May's Matilda 75347	15 lbs. 11½ oz.	282 lbs. 8 oz.	360
Oxford Catena 68905	15 lbs. 11¼ oz.	248 lbs. 0 oz.	338
True Rex Pogis 105467	15 lbs. 11¼ oz.	170 lbs. 1¾ oz.	453
Silvia Torment 104668	15 lbs. 11.2 oz.	213 lbs. 5.6 oz.	425
Agnes of Glen Duart 82475	15 lbs. 11 oz.	283 lbs. 10 oz.	7
Alabama 7690	15 lbs. 11 oz.	209 lbs. 1 oz.	8
Barbara Lambert 105208	15 lbs. 11 oz.	264 lbs. 0 oz.	24
Countess Gisela of Belle Vue 9571	15 lbs. 11 oz.	213 lbs. 10 oz.	84
Duchess of St. Lambert 5111	15 lbs. 11 oz.	216 lbs. 0 oz.	117
Exile's Lady Angela 46882	15 lbs. 11 oz.	263 lbs. 14 oz.	141
Fillpail 16530	15 lbs. 11 oz.	183 lbs. 13 oz.	156
Geneva 13220	15 lbs. 11 oz.	193 lbs. 10 oz.	166
June Sweets 2d 87967	15 lbs. 11 oz.	283 lbs. 6 oz.	211
Lady of Belle Vue 7705	15 lbs. 11 oz.	169 lbs. 8 oz.	232
Lady Petite 100562	15 lbs. 11 oz.	196 lbs. 9 oz.	234
Mitten 13368	15 lbs. 11 oz.	173 lbs. 0 oz.	308
Oitz 8649	15 lbs. 11 oz.	319 lbs. 8 oz.	332
Parole's Duchess 133953	15 lbs. 11 oz.	261 lbs. 0 oz.	343
Pierrot's Beauty 48964	15 lbs. 11 oz.	221 lbs. 12 oz.	357
Romena of Hood Farm 114887	15 lbs. 11 oz.	262 lbs. 2 oz.	392
St. Jeannaise 4th 57990	15 lbs. 11 oz.	304 lbs. 0 oz.	405
Saucy Princess 40350	15 lbs. 11 oz.	259 lbs. 7 oz.	410
Serona Hudson 84775	15 lbs. 11 oz.	253 lbs. 0 oz.	414
Siesta 37321	15 lbs. 11 oz.	207 lbs. 0 oz.	416
Flora Pansy 2d 78678	15 lbs. 10.8 oz.	215 lbs. 6.4 oz.	158
Jennie Stoke Pogis 32010	15 lbs. 10¾ oz.	206 lbs. 8 oz.	205
Maumee Girl 118155	15 lbs. 10¾ oz.	223 lbs. 0 oz.	287
Belle France 22697	15 lbs. 10½ oz.		29
Daisy Brown 2d 37325	15 lbs. 10½ oz.	255 lbs. 8 oz.	94
Fashion Plate 62457	15 lbs. 10½ oz.	268 lbs. 12 oz.	153
Orie's Fancy 72039	15 lbs. 10½ oz.	196 lbs. 4 oz.	337
Petra's Lady 99791	15 lbs. 10½ oz.	213 lbs. 8 oz.	355
Princess Bellwort 6801	15 lbs. 10½ oz.	256 lbs. 4 oz.	368
Princess of St. Saviour's 7470	15 lbs. 10½ oz.	174 lbs. 8 oz.	373
Signora Reid 2d 93808	15 lbs. 10½ oz.	232 lbs. 9 oz.	422
Young Jersey Maiden 42794	15 lbs. 10½ oz.	218 lbs. 0 oz.	471
Baroness Hugo 47563	15 lbs. 10¼ oz.	237 lbs. 1½ oz.	24
Hattie Swett 109709	15 lbs. 10¼ oz.	155 lbs. 12½ oz.	183
Silver Hair 4th 60449	15 lbs. 10¼ oz.	282 lbs. 8 oz.	424
Yokun's Princess 35643	15 lbs. 10¼ oz.	300 lbs. 9 oz.	470
Albert's Ideal 111719	15 lbs. 10 oz.	125 lbs. 15 oz.	9
Auraria 10688	15 lbs. 10 oz.	184 lbs. 8 oz.	22
Belle of River's Bank 48023	15 lbs. 10 oz.	231 lbs. 3 oz.	31
Bertha 18912	15 lbs. 10 oz.	261 lbs. 8 oz.	34
Brunette Lass 1780	15 lbs. 10 oz.		50
Cabinet 22662	15 lbs. 10 oz.	149 lbs. 13 oz.	53
Clover of Argyle 76745	15 lbs. 10 oz.	213 lbs. 4 oz.	73
Corinne of St. Lambert 76579	15 lbs. 10 oz.	205 lbs. 0 oz.	81
Cream of Springvale 16621	15 lbs. 10 oz.	229 lbs. 0 oz.	88
Czarina of Spring Hill 47568	15 lbs. 10 oz.	278 lbs. 8 oz.	92
Dainty Rioter of St. L. 81168	15 lbs. 10 oz.	165 lbs. 8 oz.	93
Dame Lofty 47567	15 lbs. 10 oz.	296 lbs. 0 oz.	99

	BUTTER.		MILK.			PAGE.
Denise's Fancy 92500	15 lbs. 10	oz	210 lbs. 5	oz		102
Dilwa 30515	15 lbs. 10	oz	179 lbs. 8	oz		105
Diploma's Eureka 103865	15 lbs. 10	oz	227 lbs. 2	oz		105
Donna Argyle 61456	15 lbs. 10	oz	231 lbs. 8	oz		112
Dorothy Taylor 99485	15 lbs. 10	oz	280 lbs. 3	oz		114
Exile's Fawn 40979	15 lbs. 10	oz	185 lbs. 0	oz		140
Fancy Juno 6086	15 lbs. 10	oz				148
Francesville 2d 99307	15 lbs. 10	oz	278 lbs. 8	oz		161
Ida's Signaline 108819	15 lbs. 10	oz	153 lbs. 0	oz		197
Jessie's Sayda 105697	15 lbs. 10	oz	207 lbs. 4	oz		207
King's Fosie 80768	15 lbs. 10	oz	145 lbs. 12	oz		217
King's Gesnera 98727	15 lbs. 10	oz	158 lbs. 10	oz		218
Lady Kingscote 26085	15 lbs. 10	oz	239 lbs. 0	oz		229
Lalla Rookh's Queen 65848	15 lbs. 10	oz	226 lbs. 0	oz		237
Lisetta Johnson 5321	15 lbs. 10	oz	219 lbs. 12	oz		253
Malope 2d 11923	15 lbs. 10	oz	193 lbs. 8	oz		271
May Dee 18058	15 lbs. 10	oz	206 lbs. 8	oz		288
Minnie Rioter 21336	15 lbs. 10	oz	249 lbs. 6	oz		300
Miss Alexander 2d 26054	15 lbs. 10	oz	177 lbs. 0	oz		302
Pedro's Zoe 95697	15 lbs. 10	oz	240 lbs. 4	oz		352
Princess of Yerba Buena Ranch 12626	15 lbs. 10	oz	221 lbs. 0	oz		373
Proctor's Dolores 38564	15 lbs. 10	oz	188 lbs. 12	oz		376
Quintuple's Florence 86076	15 lbs. 10	oz	210 lbs. 4	oz		380
Rioter's Pretty Belle 88543	15 lbs. 10	oz	245 lbs. 0	oz		388
Rioter's Sunshine 75108	15 lbs. 10	oz	244 lbs. 11	oz		389
Rochelle 15574	15 lbs. 10	oz	26 gals. 2 qts			391
Silenta 17685	15 lbs. 10	oz				423
Zenobia Pogis 35421	15 lbs. 10	oz	183 lbs. 0	oz		473
Exile's Echo 111276	15 lbs. 9¾	oz	273 lbs. 0	oz		140
Belle's Esperanza 12053	15 lbs. 9½	oz	154 lbs. 12	oz		32
King's Charella 80765	15 lbs. 9½	oz	160 lbs. 11	oz		217
Kitty Colt 2213	15 lbs. 9½	oz	214 lbs. 8	oz		220
Topsy's Oonan 97132	15 lbs. 9½	oz	194 lbs. 7½	oz		446
Vaniah 6597	15 lbs. 9½	oz	216 lbs. 4	oz		457
Wandrina 2d 82420	15 lbs. 9½	oz	230 lbs. 0	oz		464
Worthy Noble 19629	15 lbs. 9½	oz	166 lbs. 8	oz		468
Cara Mia 64224	15 lbs. 9¼	oz	216 lbs. 0	oz		55
Moxie's Mode 2d 133253	15 lbs. 9¼	oz	190 lbs. 11	oz		314
Aurora Pogis 53798	15 lbs. 9	oz	166 lbs. 14	oz		22
Beauty Messenger 17450	15 lbs. 9	oz	245 lbs. 0	oz		26
Calista of Newark 13296	15 lbs. 9	oz	195 lbs. 12	oz		54
Carrie Pogis 22568	15 lbs. 9	oz	219 lbs. 0	oz		58
Cassie of Glen Rouge 112456	15 lbs. 9	oz	244 lbs. 12	oz		59
Cora's Beauty 92179	15 lbs. 9	oz	237 lbs. 5	oz		80
Delesline 44474	15 lbs. 9	oz	254 lbs. 12	oz		101
Diana Doon 61539	15 lbs. 9	oz	236 lbs. 0	oz		104
Dido Pogis 61260	15 lbs. 9	oz	202 lbs. 2	oz		104
Fannie Landseer 1969	15 lbs. 9	oz	97 lbs. 3	oz		151
Gay Riotress 105999	15 lbs. 9	oz	189 lbs. 1	oz		164
Joy of Argyle 40495	15 lbs. 9	oz	281 lbs. 0	oz		210
Julippa 69000	15 lbs. 9	oz	248 lbs. 2	oz		211
Kitty Pride Pogis 78134	15 lbs. 9	oz	240 lbs. 8	oz		221
La Reine Centennielle 4999	15 lbs. 9	oz	209 lbs. 8	oz		240

	BUTTER.		MILK.		PAGE.
Leoni 2d 29750............................	15 lbs. 9 oz....	148 lbs. 1 oz.....	246
Lily Darling 11713.......................	15 lbs. 9 oz....	197 lbs. 8 oz.....	251
Marchande 2d 93823......................	15 lbs. 9 oz....	203 lbs. 13 oz.....	272
Maude Brown 85756......................	15 lbs. 9 oz....	265 lbs. 5 oz.....	285
Miss Lorne 78132.........................	15 lbs. 9 oz....	234 lbs. 8 oz.....	305
Molly 3554...............................	15 lbs. 9 oz....	....	310
Nan of St. Lambert 69453.................	15 lbs. 9 oz....	222 lbs. 0 oz.....	320
Rioter's Nora 21778......................	15 lbs. 9 oz....	223 lbs. 0 oz.....	388
Romantic 61626...........................	15 lbs. 9 oz....	240 lbs. 11 oz.....	391
Sordella 76876...........................	15 lbs. 9 oz....	233 lbs. 8 oz.....	429
Thekla of Clover Nook 33445..............	15 lbs. 9 oz....	191 lbs. 10 oz.....	441
Tormendova 2d 118161....................	15 lbs. 9 oz....	207 lbs. 0 oz.....	447
Ko Ko's Allie 59317......................	15 lbs. 8¾ oz....	191 lbs. 8 oz.....	223
Mahoning Bess 35834.....................	15 lbs. 8¾ oz....	322 lbs. 7 oz.....	269
Otta of St. Lambert 49003................	15 lbs. 8¾ oz....	219 lbs. 4 oz.....	338
Comtesse de St. Ouen 22600..............	15 lbs. 8½ oz....	228 lbs. 3 oz.....	77
Edith Hugo Pogis 105045.................	15 lbs. 8½ oz....	290 lbs. 10 oz.....	121
Girletta 4th 109264......................	15 lbs. 8½ oz....	249 lbs. 6 oz.....	170
Granny's Baby 5th 95182.................	15 lbs. 8½ oz....	214 lbs. 8 oz.....	177
Idalene 11841............................	15 lbs. 8½ oz....	234 lbs. 4 oz.....	192
Katie Herr 39970.........................	15 lbs. 8½ oz....	222 lbs. 10 oz.....	215
Les Marais' Dell 20314...................	15 lbs. 8½ oz....	189 lbs. 4 oz.....	247
Lustre 2062..............................	15 lbs. 8½ oz....	286 lbs. 0 oz.....	264
Max Polly 108994........................	15 lbs. 8½ oz....	206 lbs. 8 oz.....	287
Nanette of Hood Farm 104447.............	15 lbs. 8½ oz....	241 lbs. 2 oz.....	319
Niobe Marigold 108363...................	15 lbs. 8½ oz....	215 lbs. 4 oz.....	326
Odelsign of Lynwood 114421..............	15 lbs. 8½ oz....	193 lbs. 14¾ oz.....	331
Pride's Queen 66442......................	15 lbs. 8½ oz....	231 lbs. 8 oz.....	367
Regina of Cedar Hill 111995..............	15 lbs. 8½ oz....	295 lbs. 13 oz.....	383
Rosora 87622............................	15 lbs. 8½ oz....	226 lbs. 0 oz.....	398
Roxena 64212............................	15 lbs. 8½ oz....	214 lbs. 0 oz.....	399
Señorita of St. Lambert 98665............	15 lbs. 8½ oz....	223 lbs. 0 oz.....	413
Sweet Lily of Meridale 94267.............	15 lbs. 8½ oz....	253 lbs. 4 oz.....	438
Tormentor's Cinderella 19564.............	15 lbs. 8½ oz....	230 lbs. 4 oz.....	448
Belle of Corfu 83731.....................	15 lbs. 8 oz....	209 lbs. 4 oz.....	30
Busy Bee 2d 25166.......................	15 lbs. 8 oz....	211 lbs. 8 oz.....	52
Carlo's Rosebud 18223...................	15 lbs. 8 oz....	184 lbs. 8 oz.....	57
Caucus Maid 69525......................	15 lbs. 8 oz....	229 lbs. 8 oz.....	60
Cretesia's Charma 81685.................	15 lbs. 8 oz....	215 lbs. 15 oz.....	88
Daisy 2d 15761..........................	15 lbs. 8 oz....	231 lbs. 8 oz.....	94
Delight's Darling 45700..................	15 lbs. 8 oz....	282 lbs. 0 oz.....	101
Duchess Caroline 3d 6039................	15 lbs. 8 oz....	224 lbs. 0 oz.....	116
Emblem of Del Valle 102384..............	15 lbs. 8 oz....	317 lbs. 8 oz.....	127
Ethleel Khedive 99698...................	15 lbs. 8 oz....	182 lbs. 11 1-5 oz....	132
Etiquette 4300...........................	15 lbs. 8 oz....	....	132
Eureen of Lawn 61372...................	15 lbs. 8 oz....	222 lbs. 7 oz.....	135
Fancy Bee 37496.........................	15 lbs. 8 oz....	116 lbs. 8 oz.....	148
Forget-me-not 5809......................	15 lbs. 8 oz....	....	160
Fulda Stoke Pogis 44992.................	15 lbs. 8 oz....	207 lbs. 8 oz.....	162
Hilda B-B-B 17640.......................	15 lbs. 8 oz....	246 lbs. 8 oz.....	187
Jeanne Le Bas 2476......................	15 lbs. 8 oz....	....	203
King's Lady Natica 80769................	15 lbs. 8 oz....	144 lbs. 10 oz.....	218
La Belle Petite 5472.....................	15 lbs. 8 oz....	....	223

	BUTTER.	MILK.	PAGE.
La Pucelle 16829......................	15 lbs. 8 oz....	271 lbs. 2 oz.....	240
Little Ida Hilda 97739...................	15 lbs. 8 oz....	239 lbs. 0 oz.....	254
Marie C. Magnet 22003..................	15 lbs. 8 oz....	205 lbs. 8 oz.....	274
May Bud of Monmouth 128864...........	15 lbs. 8 oz....	278 lbs. 0 oz.....	287
May Dee Pogis of C. H. 93568.............	15 lbs. 8 oz....	240 lbs. 14 oz.....	288
My Queen 12614.......................	15 lbs. 8 oz....	....	316
Nettie's Picture 50960..................	15 lbs. 8 oz....	209 lbs. 4 oz.....	324
Niva 7523.............................	15 lbs. 8 oz....	201 lbs. 7 oz.....	326
Palestina 4644.........................	15 lbs. 8 oz....	....	338
Pinafore 2d 15072......................	15 lbs. 8 oz....	182 lbs. 0 oz.....	358
Princess Le Brocq 17261................	15 lbs. 8 oz....	....	370
Princess Sosia 112753...................	15 lbs. 8 oz....	199 lbs. 8 oz.....	374
Prusa Pedro 50877......................	15 lbs. 8 oz....	216 lbs. 12 oz.....	376
Queen of Clinton 11500.................	15 lbs. 8 oz....	217 lbs. 8 oz.....	378
Rioter's Normanda 102441...............	15 lbs. 8 oz....	210 lbs. 0 oz.....	388
Rioter's Pretty Pearl 75273...............	15 lbs. 8 oz....	245 lbs. 0 oz.....	389
Royal Bomba's Daisy 3d 99553...........	15 lbs. 8 oz....	225 lbs. 8 oz.....	400
Safety 13463...........................	15 lbs. 8 oz....	255 lbs. 15 oz.....	404
Sedate 2d 20389........................	15 lbs. 8 oz....	160 lbs. 12 oz.....	412
Signal Maid 4th 67492..................	15 lbs. 8 oz....	237 lbs. 0 oz.....	419
Sophie 2d of Hood Farm 131236..........	15 lbs. 8 oz....	262 lbs. 15 oz.....	429
Sylvia 687.............................	15 lbs. 8 oz....	....	438
Velvet Queen 112795....................	15 lbs. 8 oz....	257 lbs. 8 oz.....	457
Violet 3d 3240.........................	15 lbs. 8 oz....	....	461
Young Duchess 407.....................	15 lbs. 8 oz....	133½ qts..........	471
Zenobia Stoke Pogis 37292..............	15 lbs. 8 oz....	162 lbs. 8 oz.....	473
Exile's Lucy 46883.....................	15 lbs. 7¾ oz....	231 lbs. 8 oz.....	142
Euphorbia 3d 45359....................	15 lbs. 7½ oz....	217 lbs. 12½ oz....	135
Mary's Pet 40772......................	15 lbs. 7½ oz....	209 lbs. 8 oz.....	280
Maudie of St. Lambert 73603.............	15 lbs. 7½ oz....	242 lbs. 0 oz.....	285
St. Lambert's Duchess 56504.............	15 lbs. 7½ oz....	214 lbs. 2 oz.....	407
Seneca's Frankie 114052................	15 lbs. 7½ oz....	292 lbs. 9 oz.....	412
Signal Mattie 113952...................	15 lbs. 7½ oz....	215 lbs. 12 oz.....	419
Chestnuts 63449.......................	15 lbs. 7¼ oz....	260 lbs. 0 oz.....	64
Angela of Coal Hill 79380...............	15 lbs. 7 oz....	243 lbs. 8 oz.....	17
Beauty 2076...........................	15 lbs. 7 oz....	205 lbs. 6 oz.....	25
Betsy Marjoram 65145..................	15 lbs. 7 oz....	301 lbs. 0 oz.....	38
Brown Lassie 92950.....................	15 lbs. 7 oz....	223 lbs. 5 oz.....	50
Catchfly 3d 35499......................	15 lbs. 7 oz....	280 lbs. 8 oz.....	59
Chenda 3d 43246.......................	15 lbs. 7 oz....	197 lbs. 8 oz.....	64
Copper 1979...........................	15 lbs. 7 oz....	....	79
Cora of Hillside 25253..................	15 lbs. 7 oz....	300 lbs. 7 oz.....	80
Countess Godiva 10820..................	15 lbs. 7 oz....	210 lbs. 0 oz.....	84
Deerlick Pojoram 99988.................	15 lbs. 7 oz....	238 lbs. 12 oz.....	101
Dorizella 20525........................	15 lbs. 7 oz....	230 lbs. 3 oz.....	114
Ette 10315.............................	15 lbs. 7 oz....	15 qts. per day....	133
False Step 69647.......................	15 lbs. 7 oz....	199 lbs. 2 oz.....	148
Friz Cam 14655........................	15 lbs. 7 oz....	217 lbs. 0 oz.....	162
Gertie of Eau Claire 120019.............	15 lbs. 7 oz....	228 lbs. 10 oz.....	166
Gilderine of Linwood 24488.............	15 lbs. 7 oz....	163 lbs. 12 oz.....	168
Grace's Pride 61018....................	15 lbs. 7 oz....	211 lbs. 14 oz.....	177
Jemele 2d 101437......................	15 lbs. 7 oz....	247 lbs. 0 oz.....	204
Khedive's Fairy 66849..................	15 lbs. 7 oz....	186 lbs. 8 oz.....	216

	BUTTER.	MILK.	PAGE.
Koffee's Grisette 30433	15 lbs. 7 oz.	147 lbs. 8 oz.	222
Lady Dorlena 36105	15 lbs. 7 oz.	249 lbs. 0 oz.	227
Le Broemer 10670	15 lbs. 7 oz.	277 lbs. 8 oz.	243
Lily's Crown Princess 76011	15 lbs. 7 oz.	214 lbs. 7 oz.	252
Maudie of White Lick 29682	15 lbs. 7 oz.	189 lbs. 0 oz.	285
Mona's Mab 31524	15 lbs. 7 oz.	204 lbs. 12 oz.	312
Nelly Torment 61903	15 lbs. 7 oz.	233 lbs. 2 oz.	322
Nerbat 58664	15 lbs. 7 oz.	212 lbs. 15½ oz.	322
Olive Branch of Hillcrest 7447	15 lbs. 7 oz.	224 lbs. 6 oz.	332
Olive Star 81102	15 lbs. 7 oz.	279 lbs. 6 oz.	333
Orphean 4636	15 lbs. 7 oz.	83 qts. ¾ pt.	338
Pay Little 86189	15 lbs. 6¾ oz.	192 lbs. 1 oz.	344
Süsschen 125091	15 lbs. 6¾ oz.	178 lbs. 11¾ oz.	437
Alice Rex 3d 63907	15 lbs. 6½ oz.	291 lbs. 2 oz.	11
Anita Lawrence 63909	15 lbs. 6½ oz.	232 lbs. 5 oz.	18
Clifty Beauty 30249	15 lbs. 6½ oz.	173 lbs. 0 oz.	72
Dixie Landseer 88681	15 lbs. 6½ oz.	151 lbs. 0 oz.	106
Ethleel 5th 70652	15 lbs. 6½ oz.	178 lbs. 15 oz.	131
Harry's Ernestine 66232	15 lbs. 6½ oz.	199 lbs. 0 oz.	181
Lauison's Nora 59316	15 lbs. 6½ oz.	226 lbs. 11 oz.	239
Lorita 33750	15 lbs. 6½ oz.	231 lbs. 10 oz.	258
Old Noble's Heiress 99083	15 lbs. 6½ oz.	231 lbs. 4 oz.	332
Seppie 58689	15 lbs. 6½ oz.	196 lbs. 0 oz.	413
Kitty of Seguin 74347	15 lbs. 6¼ oz.	208 lbs. 5 oz.	221
Marchande 3d 133818	15 lbs. 6¼ oz.	173 lbs. 5¾ oz.	273
Abbia H. 24221	15 lbs. 6 oz.	326 lbs. 1 oz.	5
Alchesleda 103710	15 lbs. 6 oz.	235 lbs. 4 oz.	9
Alphea 171	15 lbs. 6 oz.		13
Anna Smith 10324	15 lbs. 6 oz.	278 lbs. 4 oz.	18
Annie Lane 45345	15 lbs. 6 oz.	259 lbs. 0 oz.	19
Belle of Milford 7445	15 lbs. 6 oz.	228 lbs. 8 oz.	31
Blossom's Belle 97433	15 lbs. 6 oz.	219 lbs. 10 oz.	44
Carida 37322	15 lbs. 6 oz.	190 lbs. 8 oz.	56
Edna Hayes 90901	15 lbs. 6 oz.	227 lbs. 8 oz.	121
Enigma 5360	15 lbs. 6 oz.	234 lbs. 4 oz.	128
Epple Pogis 44972	15 lbs. 6 oz.	226 lbs. 4 oz.	129
Fanella 2d 64578	15 lbs. 6 oz.	235 lbs. 11 oz.	151
Gem Nehushta 73258	15 lbs. 6 oz.	204 lbs. 8 oz.	165
Griselda Marjoram 111166	15 lbs. 6 oz.	230 lbs. 0 oz.	178
Hyacinth Pogis 56020	15 lbs. 6 oz.	187 lbs. 8 oz.	191
Ida's Nan 49957	15 lbs. 6 oz.	239 lbs. 0 oz.	196
Jersey 3260	15 lbs. 6 oz.	212 lbs. 0 oz.	205
Lula M. 32643	15 lbs. 6 oz.	198 lbs. 0 oz.	263
Marie S. 12043	15 lbs. 6 oz.	192 lbs. 15 oz.	274
Mendota 3d 26326	15 lbs. 6 oz.	301 lbs. 0 oz.	293
Minnie's Pet Jersey 93119	15 lbs. 6 oz.	253 lbs. 6 oz.	300
Moonah's Pet 7484	15 lbs. 6 oz.	222 lbs. 12 oz.	313
Phlox 3d 31882	15 lbs. 6 oz.	100 lbs. 6 oz.	356
Douskha 63439	15 lbs. 5¾ oz.	227 lbs. 14 oz.	115
Melrose's Signal 54830	15 lbs. 5¾ oz.	217 lbs. 3 oz.	292
Cara Mia 2d 93827	15 lbs. 5½ oz.	178 lbs. 0 oz.	56
Champion's Chloe 12255	15 lbs. 5½ oz.	247 lbs. 0 oz.	62
City Belle 16539	15 lbs. 5½ oz.	245 lbs. 0 oz.	70

	BUTTER.	MILK.	PAGE.
Dairy 2d 3891	15 lbs. 5½ oz.	226 lbs. 0 oz.	93
Elodie 30222	15 lbs. 5½ oz.	243 lbs. 8 oz.	125
Fawn of St. Lambert 27942	15 lbs. 5½ oz.	353 lbs. 8 oz.	154
Gretchen of Pevely 46325	15 lbs. 5½ oz.	204 lbs. 13 oz.	178
Jennie Fordyce 57169	15 lbs. 5½ oz.	218 lbs. 1 oz.	204
Kate Gordon 2d 83662	15 lbs. 5½ oz.	251 lbs. 5 oz.	213
Leah Darlington 13836	15 lbs. 5½ oz.	187 lbs. 14 oz.	242
Litta Oaks 19734	15 lbs. 5½ oz.	220 lbs. 0 oz.	253
May Koffee 45862	15 lbs. 5½ oz.	226 lbs. 8 oz.	289
Mildred Montague 45476	15 lbs. 5½ oz.	242 lbs. 0 oz.	297
Minnie Montague 45474	15 lbs. 5½ oz.	223 lbs. 4 oz.	299
Miss Morehouse 91703	15 lbs. 5½ oz.	316 lbs. 4 oz.	306
Pearl's Oonan 32288	15 lbs. 5½ oz.	180 lbs. 12 oz.	346
Portfolio 84633	15 lbs. 5½ oz.	253 lbs. 2 oz.	363
Tolonga 76868	15 lbs. 5½ oz.	220 lbs. 8 oz.	444
Katie's Kind 97434	15 lbs. 5¼ oz.	199 lbs. 10 oz.	215
La Financière 11970	15 lbs. 5¼ oz.	238 lbs. 0 oz.	236
Arawana Buttercup 6052	15 lbs. 5 oz.	343 lbs. 0 oz.	19
Bessie of Montverde 18496	15 lbs. 5 oz.	233 lbs. 8 oz.	36
Daffy Wilcox 2d 18317	15 lbs. 5 oz.	256 lbs. 7 oz.	92
Daisy Mekeel 70400	15 lbs. 5 oz.	213 lbs. 0 oz.	96
Etta Allman 12185	15 lbs. 5 oz.	278 lbs. 7 oz.	133
Eurotas' Blossom 63134	15 lbs. 5 oz.	200 lbs. 4 oz.	135
Ex Presage 128529	15 lbs. 5 oz.	212 lbs. 8 oz.	145
Gilded Pansy 26552	15 lbs. 5 oz.	163 lbs. 0 oz.	168
Iliana 44594	15 lbs. 5 oz.	232 lbs. 0 oz.	198
Irene of St. Lambert 2d 75759	15 lbs. 5 oz.	259 lbs. 8 oz.	200
Landseer's Rosamond 56142	15 lbs. 5 oz.	189 lbs. 8 oz.	239
Leonette 29752	15 lbs. 5 oz.	130 lbs. 8 oz.	246
Letitia 3977	15 lbs. 5 oz.	209 lbs. 4 oz.	248
Little Meysie 84230	15 lbs. 5 oz.	191 lbs. 8 oz.	254
Little Tipsey 79398	15 lbs. 5 oz.	264 lbs. 8 oz.	255
Miss Nautila 49000	15 lbs. 5 oz.	215 lbs. 12 oz.	306
Mrs. Jack 50805	15 lbs. 5 oz.	247 lbs. 12 oz.	315
Preference 26343	15 lbs. 5 oz.	313 lbs. 2 oz.	364
Richland Belle 46920	15 lbs. 5 oz.	171 lbs. 0 oz.	386
Romp Ogden 2d 4764	15 lbs. 5 oz.	265 lbs. 8 oz.	392
St. Lambert Hebe 54069	15 lbs. 5 oz.	114 lbs. 0 oz.	406
Signalbianca 54948	15 lbs. 5 oz.	204 lbs. ½ oz.	417
Spangled Princess 117978	15 lbs. 5 oz.	197 lbs. 10 oz.	431
Wyoming 48840	15 lbs. 5 oz.	234 lbs. 7 oz.	469
Zalma 8778	15 lbs. 5 oz.	221 lbs. 8 oz.	472
Calypris 5943	15 lbs. 4½ oz.	234 lbs. 0 oz.	55
Cenie Wallace 2d 6557	15 lbs. 4½ oz.	239 lbs. 0 oz.	61
Cora of Sandusky 42038	15 lbs. 4½ oz.	188 lbs. 0 oz.	80
Dolly Doyle 22788	15 lbs. 4½ oz.	153 lbs. 15 oz.	111
Leta of Llorac 70220	15 lbs. 4½ oz.	242 lbs. 0 oz.	247
Lula Gordon 2d 90067	15 lbs. 4½ oz.	222 lbs. 8 oz.	263
Mary's Silver Drop 14235	15 lbs. 4½ oz.	241 lbs. 10 oz.	280
Primrose of Summerland 66059	15 lbs. 4½ oz.	188.6 lbs.	367
Rockwood Maid 8375	15 lbs. 4½ oz.	273 lbs. 2½ oz.	391
Sibby 87742	15 lbs. 4½ oz.	286 lbs. 12 oz.	415
The Colonel's Daughter 50230	15 lbs. 4½ oz.	285 lbs. 2 oz.	441

	BUTTER.	MILK.	PAGE.
Tormentor's Fancy Wax 73212............	15 lbs. 4½ oz....	248 lbs. 8 oz.....	449
Victory 16379.........................	15 lbs. 4½ oz....	277 lbs. 0 oz.....	460
Exile's May 52765.....................	15 lbs. 4¼ oz....	157 lbs. 8 oz.....	143
Idalette Pogis 64220...................	15 lbs. 4¼ oz....	223 lbs. 13½ oz.....	192
Lady Thekla 98270.....................	15 lbs. 4¼ oz....	204 lbs. 3½ oz.....	235
Augetta 19404.........................	15 lbs. 4 oz....	275 lbs. 8 oz.....	18
Baby Stanton 72397....................	15 lbs. 4 oz....	226 lbs. 0 oz.....	23
Baron's Rosette 25988.................	15 lbs. 4 oz....	181 lbs. 13 oz.....	25
Bijou of St. Lambert 5112.............	15 lbs. 4 oz....	150 lbs. 10 oz.....	40
Charming of St. Lambert 69077..........	15 lbs. 4 oz....	137 lbs. 0 oz.....	63
Clark's Gift 91708....................	15 lbs. 4 oz....	244 lbs. 0 oz.....	71
Coix 85258............................	15 lbs. 4 oz....	223 lbs. 0 oz.....	75
Corinne's Dollie Rex 110987............	15 lbs. 4 oz....	233 lbs. 4 oz.....	81
Cowslip 5th 849.......................	15 lbs. 4 oz....	....	87
Dorothy of Bovina 9373.................	15 lbs. 4 oz....	205 lbs. 0 oz.....	114
El Mora Mostar 15955..................	15 lbs. 4 oz....	231 lbs. 3 oz.....	124
Elsie Lane 13302......................	15 lbs. 4 oz....	118⅞ qts.........	126
Ethelka 2d 14128......................	15 lbs. 4 oz....	244 lbs. 8 oz.....	130
Eupidee's Perfection 20175.............	15 lbs. 4 oz....	174 lbs. 0 oz.....	135
Evri 5282.............................	15 lbs. 4 oz....	203 lbs. 8 oz.....	137
Exile's Butterball 65152...............	15 lbs. 4 oz....	154 lbs. 0 oz.....	139
Exile's Moss Rose 101155..............	15 lbs. 4 oz....	231 lbs. 0 oz.....	143
Farmer's Pride 12284..................	15 lbs. 4 oz....	252 lbs. 0 oz.....	153
Favorite Rajah Rex 16153..............	15 lbs. 4 oz....	108 lbs. 4 oz.....	153
First Prize 95323.....................	15 lbs. 4 oz....	249 lbs. 12 oz.....	156
Forget Me Not O 10564.................	15 lbs. 4 oz....	220 lbs. 8 oz.....	160
Gavotte 64205.........................	15 lbs. 4 oz....	235 lbs. 5½ oz.....	164
Gracie A. 54524.......................	15 lbs. 4 oz....	235 lbs. 0 oz.....	177
Hilda 18178...........................	15 lbs. 4 oz....	196 lbs. 9 oz.....	187
Janie Caruth 73311....................	15 lbs. 4 oz....	240 lbs. 1 oz.....	203
J. S. Exile 105495....................	15 lbs. 4 oz....	255 lbs. 0 oz.....	210
Kammerfrau 125093.....................	15 lbs. 4 oz....	172 lbs. 0 oz.....	212
Khedive's Princess 38568..............	15 lbs. 4 oz....	172 lbs. 0 oz.....	216
Kitty Pedrothorn 2d 84562.............	15 lbs. 4 oz....	198 lbs. 0 oz.....	221
Lady Bingo 24160......................	15 lbs. 4 oz....	172 lbs. 8 oz.....	225
La Fantine 24489......................	15 lbs. 4 oz....	120 lbs. 11 oz.....	236
Maid of Five Oaks 7178................	15 lbs. 4 oz....	273 lbs. 9 oz.....	270
May Pedro 50876.......................	15 lbs. 4 oz....	229 lbs. 8 oz.....	290
Merry Burlington 7600.................	15 lbs. 4 oz....	256 lbs. 0 oz.....	295
Mollie of St. Lambert 34644...........	15 lbs. 4 oz....	205 lbs. 0 oz.....	310
Mona's Beauty 50041...................	15 lbs. 4 oz....	171 lbs. 0 oz.....	312
Nelope Pedro 50879....................	15 lbs. 4 oz....	219 lbs. 12 oz.....	322
Oaklands Lilly 14881..................	15 lbs. 4 oz....	247 lbs. 4 oz.....	330
Olga H. 60619.........................	15 lbs. 4 oz....	195 lbs. 4 oz.....	332
Oonamistle 64469......................	15 lbs. 4 oz....	232 lbs. 0 oz.....	335
Patrol 40490..........................	15 lbs. 4 oz....	242 lbs. 12 oz.....	343
Pedro's Princess of York 72378........	15 lbs. 4 oz....	159 lbs. 8 oz.....	351
Pedro's Very Pretty 110508............	15 lbs. 4 oz....	235 lbs. 0 oz.....	352
Purest 13730..........................	15 lbs. 4 oz....	206 lbs. 14 oz.....	376
St. John's Daisy 28388................	15 lbs. 4 oz....	196 lbs. 8 oz.....	405
Signalana 7719........................	15 lbs. 4 oz....	15 qts. per day....	417
Signal's Gold Foil 90508..............	15 lbs. 4 oz....	272 lbs. 0 oz.....	421
Stanstead Belle 4709..................	15 lbs. 4 oz....	220 lbs. 4 oz.....	432

	BUTTER.	MILK.	PAGE.
Sterling Merit 62456......................	15 lbs. 4 oz....	203 lbs. 0 oz.....	433
Sultana 2d 11798.........................	15 lbs. 4 oz....	285 lbs. 4 oz.....	434
Topsey Doe 25378........................	15 lbs. 4 oz....	219 lbs. 0 oz.....	446
Tuleema Pogis 100089....................	15 lbs. 4 oz....	252 lbs. 0 oz.....	454
Solava 20928............................	15 lbs. 3⅝ oz....	210 lbs. 8 oz.....	428
Angelina D. 65514.......................	15 lbs. 3½ oz....	232 lbs. 13 oz.....	17
Beauty of Ninon 2d 18444...............	15 lbs. 3½ oz....	308 lbs. 8 oz.....	27
Belle Williamson 3d 104290.............	15 lbs. 3½ oz....	205 lbs. 10 oz.....	33
Bristol Bella 15697......................	15 lbs. 3½ oz....	257 lbs. 8 oz.....	48
Clytemnestra 2455.......................	15 lbs. 3½ oz....	216 lbs. 4 oz.....	74
Dark Cloud 9364.........................	15 lbs. 3½ oz....	256 lbs. 2 oz.....	99
Ella of Ingleside 76414..................	15 lbs. 3½ oz....	167 lbs. 3 oz.....	123
Howdy 55110............................	15 lbs. 3½ oz....	237 lbs. 11 oz.....	190
Koffee's Lily 25515......................	15 lbs. 3½ oz....	134 lbs. 8 oz.....	222
May's Grand Duchess 62607..............	15 lbs. 3½ oz....	226 lbs. 0 oz.....	290
Minty 2d of Hood Farm 112326...........	15 lbs. 3½ oz....	231 lbs. 8 oz.....	301
Nazli 10327.............................	15 lbs. 3½ oz....	232 lbs. 8 oz.....	321
Nutley Darling 22412....................	15 lbs. 3½ oz....	251 lbs. 11 oz.....	329
Reality 16537...........................	15 lbs. 3½ oz....	185 lbs. 13 oz.....	382
Signal Maiden 42793.....................	15 lbs. 3½ oz....	246 lbs. 8 oz.....	419
Signal's Crusta 97341....................	15 lbs. 3½ oz....	210 lbs. 15 oz.....	420
Temisia of Winnikee 46604..............	15 lbs. 3½ oz....	205 lbs. 12 oz.....	440
Eastern Sultane 20298...................	15 lbs. 3⅛ oz....	222 lbs. 14 oz.....	119
Adaphine 101425.........................	15 lbs. 3 oz....	227 lbs. 12 oz.....	6
Alfritha 13673..........................	15 lbs. 3 oz....	194 lbs. 4 oz.....	10
Alice of Clovernook 77532..............	15 lbs. 3 oz....	138 lbs. 0 oz.....	11
Arnold's Lulu 7328......................	15 lbs. 3 oz....	273 lbs. 0 oz.....	20
Atricia 6029............................	15 lbs. 3 oz....	....	21
Belle Garner 23682......................	15 lbs. 3 oz....	172 lbs. 4 oz.....	29
Ben's Nettie 118137.....................	15 lbs. 3 oz....	239 lbs. 15 oz.....	34
Brown Bessie's Modita 128259...........	15 lbs. 3 oz....	252 lbs. 0 oz.....	49
Complexia 56774........................	15 lbs. 3 oz....	249 lbs. 1 oz.....	77
Daisy Stillson 28174....................	15 lbs. 3 oz....	225 lbs. 0 oz.....	98
Darling of Neatham 20086...............	15 lbs. 3 oz....	194 lbs. 0 oz.....	100
Easter Pogis 52371......................	15 lbs. 3 oz....	217 lbs. 14 oz.....	120
Edna F. of Lawn 73545..................	15 lbs. 3 oz....	209 lbs. 10 oz.....	121
Fernside 113278.........................	15 lbs. 3 oz....	226 lbs. 5 oz.....	155
Fragrance 4059..........................	15 lbs. 3 oz....	....	160
Gilderoy P. 99308.......................	15 lbs. 3 oz....	297 lbs. 4 oz.....	168
Golden Gem of St. L. 81167.............	15 lbs. 3 oz....	187 lbs. 0 oz.....	172
Khedive's Fancy 18180..................	15 lbs. 3 oz....	219 lbs. 2 oz.....	216
Lady Adams 2d 6529....................	15 lbs. 3 oz....	....	223
Lady Monmouth 15173..................	15 lbs. 3 oz....	258 lbs. 4 oz.....	232
Leclair's Marjoram 36355...............	15 lbs. 3 oz....	110 lbs. 8 oz.....	243
Lou Tormentor of Lawn 84413...........	15 lbs. 3 oz....	210 lbs. 0 oz.....	260
Lydia Libby 11698......................	15 lbs. 3 oz....	266 lbs. 4 oz.....	265
Maculac 24277..........................	15 lbs. 3 oz....	149 lbs. 13 oz.....	266
Matilda Mysinda 84487.................	15 lbs. 3 oz....	258 lbs. 0 oz.....	282
Meg Mitchell 4187......................	15 lbs. 3 oz....	....	290
Miss Naomi 45098.......................	15 lbs. 3 oz....	224 lbs. 6 oz.....	306
Nellie Darlington 5956..................	15 lbs. 3 oz....	194 lbs. 4 oz.....	321
Nuphar Houghton 36364.................	15 lbs. 3 oz....	185 lbs. 7 oz.....	328
Pearl Jenks 3d 84869....................	15 lbs. 3 oz....	186 lbs. 0 oz.....	345

	BUTTER.	MILK.	PAGE.
Sibby Landseer 116107	15 lbs. 2 oz....	217 lbs. 4 oz	415
Signal's Garland 116048	15 lbs. 2 oz....	239 lbs. 0 oz	421
Silver Cloud 13461	15 lbs. 2 oz....	127 lbs. 8 oz	424
Sunset of Pleasant View 13071	15 lbs. 2 oz....	235 lbs. 0 oz	435
Tease of Clovernook 92954	15 lbs. 2 oz....	211 lbs. 11¼ oz	440
Very Pretty 29184	15 lbs. 2 oz....	236 lbs. 4 oz	459
Zillia 2d 1693	15 lbs. 2 oz....	249 lbs. 8 oz	473
Bisson's Rosa 97830	15 lbs. 1¾ oz....	176 lbs. 7½ oz	42
Olivia Albertine 3d 83438	15 lbs. 1¾ oz....	194 lbs. 4 oz	333
Aldarine 5301	15 lbs. 1½ oz....	232 lbs. 15 oz	9
Bessie Russ 2d 14649	15 lbs. 1½ oz....	196 lbs. 15 oz	87
Exile's Onnalinda 65189	15 lbs. 1½ oz....	158.6 lbs	144
Florine Value 2d 80910	15 lbs. 1½ oz....	248 lbs. 0 oz	159
Lassie 1134	15 lbs. 1½ oz....	122 lbs. 2 oz	240
Lemon Spray 44532	15 lbs. 1½ oz....	234 lbs. 8 oz	244
Leonetta H. 67514	15 lbs. 1½ oz....	196 lbs. 12 oz	245
Pride's Lady Sarah 39537	15 lbs. 1½ oz....	232 lbs. 2 oz	366
Rioter's Hortense 51499	15 lbs. 1½ oz....	222 lbs. 10 oz	388
Shrine 59627	15 lbs. 1½ oz....	243 lbs. 0 oz	415
Sig Eurota 64898	15 lbs. 1½ oz....	301 lbs. 11 oz	416
Signal's Bijou 60800	15 lbs. 1½ oz....	209 lbs. 4 oz	420
Tonnage Girl 103331	15 lbs. 1½ oz....	242 lbs. 5 oz	446
Tormentris of Aurora 93948	15 lbs. 1½ oz....	230 lbs. 8 oz	451
Verona 10766	15 lbs. 1½ oz....	227 lbs. 0 oz	458
Careta Pogis 42784	15 lbs. 1¼ oz....	295 lbs. 2 oz	56
Torment 4th 70660	15 lbs. 1¼ oz....	208 lbs. 6½ oz	447
Macie Pogis 103539	15 lbs. 1⅛ oz....	197 lbs. 6 oz	266
Bessie Landseer 58103	15 lbs. 1 oz....	147 lbs. 8 oz	36
Bessie Wolcott 91417	15 lbs. 1 oz....	241 lbs. 7 oz	37
Bronze Leaf 14902	15 lbs. 1 oz....	18 qts. per day	48
Cannie M. 10698	15 lbs. 1 oz....	226 lbs. 8 oz	55
Columbia Beauty 30263	15 lbs. 1 oz....	179 lbs. 5 oz	75
Coomassie L. 73273	15 lbs. 1 oz....	296 lbs. 10 oz	78
Cretesia's Dawson 89995	15 lbs. 1 oz....	247 lbs. 4 oz	89
Daisy of Sunnyside 16338	15 lbs. 1 oz....	267 lbs. 4 oz	97
Darling Pansy 44830	15 lbs. 1 oz....	223 lbs. 4 oz	100
Dot of Rocky Ford 40763	15 lbs. 1 oz....	246 lbs. 8 oz	115
Duchess of Bloomfield 3d 15580	15 lbs. 1 oz....	159 lbs. 0 oz	117
Empress Pedro 50878	15 lbs. 1 oz....	194 lbs. 12 oz	128
Forsaken 7520	15 lbs. 1 oz....		160
Kate Pansy 15177	15 lbs. 1 oz....	204 lbs. 6 oz	214
King's Antoinette 40456	15 lbs. 1 oz....	102 lbs. 0 oz	217
Lady Cecilia 24821	15 lbs. 1 oz....	257 lbs. 8 oz	225
Lemon Fern 22140	15 lbs. 1 oz....	223 lbs. 0 oz	244
Manuscript 74881	15 lbs. 1 oz....	222 lbs. 8 oz	271
Maple Glens Gold Lorna 84017	15 lbs. 1 oz....	225 lbs. 4 oz	271
Marion Hill 40777	15 lbs. 1 oz....	201 lbs. 8 oz	274
Marvel 13734	15 lbs. 1 oz....	182 lbs. 9 oz	277
Mary of Lodi 88517	15 lbs. 1 oz....	224 lbs. 8 oz	279
Meridale Oaklands Nora 64785	15 lbs. 1 oz....	174 lbs. 6 oz	294
Musidove 3d 61641	15 lbs. 1 oz....	217 lbs. 0 oz	316
Narda 38012	15 lbs. 1 oz....	172 lbs. 0 oz	320
Neitherme 64670	15 lbs. 1 oz....	258 lbs. 0 oz	321

	BUTTER.		MILK.		PAGE.
Nerissa of Nyack 9692	15 lbs. 1 oz		248 lbs. 8 oz		322
Nerobella's Blossom 48051	15 lbs. 1 oz		225 lbs. 10 oz		323
Palm Rose 75110	15 lbs. 1 oz		190 lbs. 8 oz		339
Pearl's Oonan 2d 86105	15 lbs. 1 oz		171 lbs. 8 oz		346
Pride's Lady Rushmore 35182	15 lbs. 1 oz		210 lbs. 7 oz		366
Princess Rosalind 87288	15 lbs. 1 oz		168 lbs. 13 oz		374
Prize Rose 16309	15 lbs. 1 oz		183 lbs. 4 oz		375
Rosa Ramsey 25875	15 lbs. 1 oz		224 lbs. 0 oz		395
Roselaine 7167	15 lbs. 1 oz				395
Signal Fancy of Lawn 73540	15 lbs. 1 oz		241 lbs. 8 oz		418
Sultan's Sultane 32854	15 lbs. 1 oz		186 lbs. 10 oz		434
Tancreda Torment 114090	15 lbs. 1 oz		164 lbs. 7 oz		439
Vera of Briarcliff 28687	15 lbs. 1 oz		129 lbs. 8 oz		458
Vivienne Pogis 41547	15 lbs. 1 oz		314 lbs. 0 oz		463
Pride's Lady Lillie 35181	15 lbs. ¾ oz		201 lbs. 1 oz		366
Ruthie Rex Pogis 100326	15 lbs. ¾ oz		181 lbs. 1 qz		403
St. Lambert's Babette 99185	15 lbs. ¾ oz		241 lbs. 9 oz		406
Buttery 2d 27138	15 lbs. ⅝ oz		146 lbs. 0 oz		53
Belle of St. John's 2d 29829	15 lbs. ½ oz		175 lbs. 0 oz		32
Bessie Signalda 3d 86349	15 lbs. ½ oz		236 lbs. 12 oz		37
Dairy C. 12227	15 lbs. ½ oz		217 lbs. 0 oz		93
Euphorbia 2d 27094	15 lbs. ½ oz		258 lbs. 8 oz		135
Fancy Ona 34867	15 lbs. ½ oz		224 lbs. 12 oz		149
Fancy's Moyane 122270	15 lbs. ½ oz		195 lbs. 10¾ oz		150
Ida of Coal Hill 12542	15 lbs. ½ oz		232 lbs. 0 oz		194
Ida of Oakland 2d 72061	15 lbs. ½ oz		185 lbs. 6 oz		194
Landseer's Fancy 2d 43184	15 lbs. ½ oz		103 lbs. 3 oz		238
Mittretta's Rose 56192	15 lbs. ½ oz		156 lbs. 15 oz		308
Phœbe's Charity 59008	15 lbs. ½ oz		256 lbs. 8 oz		357
Plumida 23621	15 lbs. ½ oz		153 lbs. 8 oz		359
Pomona's May Day Pogis 56249	15 lbs. ½ oz		188 lbs. 10 oz		363
Rosa Oxford 66678	15 lbs. ½ oz		203 lbs. 0 oz		394
St. Helier Gem 35438	15 lbs. ½ oz		218 lbs. 0 oz		405
Sünderin 125092	15 lbs. ½ oz		191 lbs. 10½ oz		435
Leneca 17178	15 lbs. ¼ oz				245
Lucy of Wabash 9741	15 lbs. ¼ oz				262
Idarella 41433	15 lbs. ⅛ oz		204 lbs. 0 oz		194
Amalthea 11319	15 lbs. 0 oz		241 lbs. 14 oz		15
Ampelis 5th 17548	15 lbs. 0 oz		194 lbs. 12 oz		16
Archie 1112	15 lbs. 0 oz				20
Arietta 5115	15 lbs. 0 oz				20
Arthur's Frolic 4438	15 lbs. 0 oz		241 lbs. 0 oz		20
Beauty 17414	15 lbs. 0 oz		194 lbs. 0 oz		26
Bettie Dixon 4527	15 lbs. 0 oz		281 lbs. 8 oz		38
Canossa 65243	15 lbs. 0 oz		183 lbs. 8 oz		55
Cora of Arcadia 16151	15 lbs. 0 oz		200 lbs. 4 oz		79
Crescent's Rachael 74738	15 lbs. 0 oz		278 lbs. 11 oz		88
Dairy Dolly 100605	15 lbs. 0 oz		253 lbs. 14 oz		93
Daisy Grant 1445	15 lbs. 0 oz				95
Dido B. 112926	15 lbs. 0 oz		262 lbs. 3 oz		104
Dora Doon 12909	15 lbs. 0 oz		198 lbs. 0 oz		113
Easter Prize Rex 102166	15 lbs. 0 oz		246 lbs. 5 oz		120
Ella Rosewood's Daisy 32787	15 lbs. 0 oz		280 lbs. 0 oz		124

	BUTTER.	MILK.	PAGE.
Ethleel 4th 54275	15 lbs. 0 oz	156 lbs. 11½ oz	131
Fan of Grouville 7458	15 lbs. 0 oz	263 lbs. 12 oz	152
Fringe 16875	15 lbs. 0 oz	298 lbs. 14 oz	161
Girlie's Glory 46305	15 lbs. 0 oz	181 lbs. 12 oz	170
Golden Violet 30136	15 lbs. 0 oz	170 lbs. 12 oz	173
Grace Felch 8291	15 lbs. 0 oz	192 lbs. 5 oz	175
Jersey Lily 14044	15 lbs. 0 oz	246 lbs. 12 oz	206
Kittie Green 88379	15 lbs. 0 oz	217 lbs. 0 oz	219
Lady Jane of St. Peter's 7475	15 lbs. 0 oz	207 lbs. 0 oz	229
Lady Louise 4339	15 lbs. 0 oz		230
Landseer's Ethel 74315	15 lbs. 0 oz	206 lbs. 12 oz	237
Landseer's Pogene 85743	15 lbs. 0 oz	227 lbs. 12 oz	238
Leonette's Orange 108521	15 lbs. 0 oz	212 lbs. 4 oz	246
Luna Dean 3d 84309	15 lbs. 0 oz	235 lbs. 12 oz	264
Lydia Darrach 5th 16577	15 lbs. 0 oz		265
Ma Belle 4942	15 lbs. 0 oz		265
Maid of Avranches 6959	15 lbs. 0 oz	172 lbs. 0 oz	269
Marjoram 2d 12805	15 lbs. 0 oz		275
Mintha 12812	15 lbs. 0 oz	249 lbs. 8 oz	300
Mischief Le Brocq 7680	15 lbs. 0 oz		301
Miss D'Arcy 62596	15 lbs. 0 oz	256 lbs. 8 oz	304
My Alpha's Beauty 56469	15 lbs. 0 oz	164 lbs. 8 oz	316
Nettle N. 58307	15 lbs. 0 oz	288 lbs. 0 oz	323
Oxalis 2d 15631	15 lbs. 0 oz	197 lbs. 11 oz	338
Pretty Flirt 28791	15 lbs. 0 oz	169 lbs. 0 oz	364
René Ogden 1568	15 lbs. 0 oz		383
Robinette Rioter 57206	15 lbs. 0 oz	208 lbs. 2 oz	391
Romping Lass 11021	15 lbs. 0 oz	226 lbs. 13½ oz	392
St. Lambert B. Kate 105205	15 lbs. 0 oz	265 lbs. 14 oz	406
Signora Signal 39396	15 lbs. 0 oz	215 lbs. 4 oz	423
Sister Dorothy 2607	15 lbs. 0 oz		426
Soconee 2d 114420	15 lbs. 0 oz	159 lbs. 0 oz	428
Thorndale Belle 6th 44885	15 lbs. 0 oz	226 lbs. 6 oz	443
Tiger Lill 69771	15 lbs. 0 oz	200 lbs. 8 oz	443
Tilda Ives 17707	15 lbs. 0 oz	285 lbs. 0 oz	443
Trudie 2d 4084	15 lbs. 0 oz		453
Vaniah of St. Lambert 71374	15 lbs. 0 oz	205 lbs. 13 oz	457
Verbena of Fernwood 9088	15 lbs. 0 oz	196 lbs. 13 oz	458
Wanderer's Shiloh 65942	15 lbs. 0 oz	215 lbs. 0 oz	464
Whisk of Meridale 131023	15 lbs. 0 oz	204 lbs. 11 oz	466
Azuline 2d 3888	14 lbs. 15¾ oz	238 lbs. 0 oz	23
Amanda May 83011	14 lbs. 15½ oz	261 lbs. 0 oz	15
Bertha Stoke Pogis 35206	14 lbs. 15½ oz	142 lbs. 0 oz	35
Deletta 21305	14 lbs. 15½ oz	145 lbs. 8 oz	101
Ella Europa 61062	14 lbs. 15½ oz	208 lbs. 2 oz	123
Happy May Girl 75588	14 lbs. 15½ oz	240 lbs. 12½ oz	180
Harry's May Bud 111817	14 lbs. 15½ oz	250 lbs. 3 oz	182
Lady Bountiful 17946	14 lbs. 15½ oz	229 lbs. 14 oz	225
Landseer's Croton Maid 82535	14 lbs. 15½ oz	176 lbs. 12 oz	237
Lucy Lee H. 58106	14 lbs. 15½ oz	229 lbs. 8 oz	262
Oonan 3d 49955	14 lbs. 15½ oz	141 lbs. 12 oz	335
Paletta of Darlington 2d 75343	14 lbs. 15½ oz	214 lbs. 7 oz	339
St. Lambert's Bijou 78174	14 lbs. 15½ oz	238 lbs. 4 oz	406

	BUTTER.	MILK.	PAGE.
Satin Bird 16380	14 lbs. 15½ oz	201 lbs. 8 oz	410
Sigma 2d 89803	14 lbs. 15½ oz	195 lbs. 14½ oz	417
Justa Pogis 64863	14 lbs. 15¼ oz	262 lbs. 10½ oz	211
Alexa 64924	14 lbs. 15 oz	307 lbs. 1 oz	10
Casta's Belle 56148	14 lbs. 15 oz	244 lbs. 0 oz	59
Chemung Maid 34704	14 lbs. 15 oz	286 lbs. 10 oz	63
Christine of Side View 2d 94253	14 lbs. 15 oz	211 lbs. 8 oz	67
Ethleel 6th 89802	14 lbs. 15 oz	241 lbs. 1 oz	132
Fancy's Gem 41797	14 lbs. 15 oz	171 lbs. 4 oz	149
Hatita 34538	14 lbs. 15 oz	197 lbs. 8 oz	183
Iryl Dane 3d 38375	14 lbs. 15 oz	223 lbs. 12 oz	201
Kitty Clover 2d 16099	14 lbs. 15 oz	231 lbs. 12 oz	220
Le Brocq's May Belle 27980	14 lbs. 15 oz	220 lbs. 4 oz	242
Louisa Deming 23469	14 lbs. 15 oz	230 lbs. 4 oz	260
Mary Clover 9998	14 lbs. 15 oz	212 lbs. 8 oz	277
Matilda of Maplewood 49723	14 lbs. 15 oz	208 lbs. 12 oz	282
Molly May 17202	14 lbs. 15 oz	246 lbs. 0 oz	310
Nerviona 43678	14 lbs. 15 oz	215 lbs. 2 oz	323
Nibble 6796	14 lbs. 15 oz	220 lbs. 1 oz	324
Niobomba 26456	14 lbs. 15 oz	196 lbs. 0 oz	326
Patty Iris 101894	14 lbs. 15 oz	190 lbs. 8 oz	343
Princess White Water 21137	14 lbs. 15 oz	203 lbs. 15 oz	374
Rayon d'Or's Chroma 114053	14 lbs. 15 oz	251 lbs. 14 oz	382
Rho A. Pogis 39269	14 lbs. 15 oz	174 lbs. 0 oz	385
Rioter's Violet Pogis 75342	14 lbs. 15 oz	190 lbs. 14 oz	390
Rosaline Collas 11187	14 lbs. 15 oz	225 lbs. 8 oz	394
Ruby of Springvale 14505	14 lbs. 15 oz	225 lbs. 0 oz	402
Scroll 93498	14 lbs. 15 oz	201 lbs. 0 oz	412
Sheldon's Maud 86187	14 lbs. 15 oz	241 lbs. 8 oz	414
Sweet Sixteen 10682	14 lbs. 15 oz	127 lbs. 0 oz	438
Tamasese Naomi 128774	14 lbs. 15 oz	175 lbs. 12 oz	439
Virgie Landseer 74831	14 lbs. 15 oz	242 lbs. 0 oz	462
Aunt Lera 77855	14 lbs. 14⅞ oz	273 lbs. 4 oz	22
Lady White 29213	14 lbs. 14¾ oz	197 lbs. 8 oz	235
Beauty Wynn 45482	14 lbs. 14½ oz	204 lbs. 0 oz	28
Belle of Echo 12432	14 lbs. 14½ oz	222 lbs. 0 oz	31
Faustine 10354	14 lbs. 14½ oz	225 lbs. 9 oz	153
Mary's Nightingale 87263	14 lbs. 14½ oz	201 lbs. 0 oz	280
Nonsuch of Linwood 29028	14 lbs. 14½ oz	209 lbs. 3 oz	327
Pomona's Ida of St. L. 59565	14 lbs. 14½ oz	190 lbs. 10 oz	362
Rioter's Rosaline 35581	14 lbs. 14½ oz	169 lbs. 8 oz	389
Alberta d'Or 76452	14 lbs. 14 oz	262 lbs. 2 oz	8
Carola's Crown Princess 90321	14 lbs. 14 oz	246 lbs. 14 oz	57
Chrysippe 53373	14 lbs. 14 oz	316 lbs. 0 oz	68
Countess of Lorne 20822	14 lbs. 14 oz	207 lbs. 10 oz	85
Daltonia Pogis 95203	14 lbs. 14 oz	284 lbs. 0 oz	98
Etta M. 2d 30820	14 lbs. 14 oz	287 lbs. 2 oz	133
Exile's Lady Palestine 107626	14 lbs. 14 oz	247 lbs. 0 oz	142
Fairy of Eau Claire 128596	14 lbs. 14 oz	240 lbs. 8 oz	147
Fancy's Harry's Sadie 77202	14 lbs. 14 oz	135 lbs. estimated	149
Fanny of Yerba Buena 2d 10689	14 lbs. 14 oz	244 lbs. 8 oz	152
Florry Keep 6556	14 lbs. 14 oz	231 lbs. 8 oz	159
Gold Mark 10727	14 lbs. 14 oz	197 lbs. 0 oz	174

	BUTTER.	MILK.	PAGE.
Honeysuckle of St. Anne's 18674..........	14 lbs. 14 oz....	285 lbs. 0 oz.....	190
Ida of Maple Glens 97785.................	14 lbs. 14 oz....	219 lbs. 4 oz.....	194
Imperial's Omega 2d 19664.................	14 lbs. 14 oz....	207 lbs. 12 oz.....	199
Island Chrissie 12007.....................	14 lbs. 14 oz....	247 lbs. 12 oz.....	201
Itura 61090..............................	14 lbs. 14 oz....	193 lbs. 14 oz.....	202
Jenny Le Brocq 9757.....................	14 lbs. 14 oz....	....	205
Lady Gilmore 12136.....................	14 lbs. 14 oz....	226 lbs. 8 oz.....	227
Lanison's Cora 59315....................	14 lbs. 14 oz....	171 lbs. 2 oz.....	239
Lass Pogis 40740........................	14 lbs. 14 oz....	258 lbs. 12 oz.....	241
Lorne's Pretty Marjoram 77942...........	14 lbs. 14 oz....	265 lbs. 0 oz.....	259
Lucy's Love 20568.......................	14 lbs. 14 oz....	341 lbs. 7 oz.....	263
Méprise Bloomfield 50062.................	14 lbs. 14 oz....	137 lbs. 2 oz.....	293
Normanda 3914..........................	14 lbs. 14 oz....	175 lbs. 10 oz.....	328
Ollie Wilkes 77402.......................	14 lbs. 14 oz....	288 lbs. 4 oz.....	333
Pauline Hugo Pogis 37098...............	14 lbs. 14 oz....	230 lbs. 0 oz.....	344
Pearl Jenks 50834.......................	14 lbs. 14 oz....	246 lbs. 0 oz.....	344
Pedro's Fancy 4th of H. F. 119842........	14 lbs. 14 oz....	240 lbs. 14 oz.....	347
Susie V. 54855..........................	14 lbs. 14 oz....	194 lbs. 0 oz.....	437
Tormentor's Browny 73079..............	14 lbs. 14 oz....	218 lbs. 4 oz.....	448
Maquilla of Hood Farm 113067...........	14 lbs. 13¾ oz....	218 lbs. 4 oz.....	272
Arietta 3d 14274........................	14 lbs. 13½ oz....	295 lbs. 12 oz.....	20
Chloe Tormentor of Lawn 71642..........	14 lbs. 13½ oz....	199 lbs. 14 oz.....	66
Classic 21402...........................	14 lbs. 13½ oz....	216 lbs. 0 oz.....	71
Colza Melrose 93859.....................	14 lbs. 13½ oz....	218 lbs. 12 oz.....	75
Compressa 55756........................	14 lbs. 13½ oz....	248 lbs. 3 oz.....	77
Cowslip of Rockwell 115354..............	14 lbs. 13½ oz....	233 lbs. 12 oz.....	87
Frances C. Magnet 22904.................	14 lbs. 13½ oz....	199 lbs. 4 oz.....	160
Homestead Prue 82203...................	14 lbs. 13½ oz....	215 lbs. 6 oz.....	189
Massey Polo 3d 99378...................	14 lbs. 13½ oz....	209 lbs. 0 oz.....	281
Pansy Blossom 22413....................	14 lbs. 13½ oz....	315 lbs. 1 oz.....	340
Queen Mary 6212........................	14 lbs. 13½ oz....	122 lbs. 12 oz.....	377
Velveteen 7703..........................	14 lbs. 13½ oz....	212 lbs. 0 oz.....	457
Violet Denison 49997....................	14 lbs. 13½ oz....	250 lbs. 0 oz.....	461
Chelka 79033............................	14 lbs. 13¼ oz....	170 lbs. 0 oz.....	63
L. Toltec's First 130688..................	14 lbs. 13¼ oz....	154 lbs. 0 oz.....	261
Alberta d'Or 2d 124269..................	14 lbs. 13 oz....	206 lbs. 4 oz.....	8
Alta B. 73003...........................	14 lbs. 13 oz....	215 lbs. 6 oz.....	14
Chump 81289............................	14 lbs. 13 oz....	149 lbs. 4 oz.....	68
Cymbelina 67201........................	14 lbs. 13 oz....	217 lbs. 0 oz.....	92
Duchess of Argyle 3758..................	14 lbs. 13 oz....	108 lbs. 12 oz.....	116
Florence Billot 7849....................	14 lbs. 13 oz....	205 lbs. 6 oz.....	158
Jefferson Albina 12196..................	14 lbs. 13 oz....	186 lbs. 0 oz.....	204
Kobe's Marcella 4597....................	14 lbs. 13 oz....	234 lbs. 10 oz.....	222
Lizentris 81821..........................	14 lbs. 13 oz....	207 lbs. 0 oz.....	256
Louvie 3d 6159...........................	14 lbs. 13 oz....	....	261
Martha of Eau Claire 120022.............	14 lbs. 13 oz....	220 lbs. 14 oz.....	277
Matlie 3d 9879..........................	14 lbs. 13 oz....	161 lbs. 4 oz.....	284
Queen of De Soto 12318..................	14 lbs. 13 oz....	....	379
Roxie Landseer 74832....................	14 lbs. 13 oz....	213 lbs. 10 oz.....	399
Susan Bell 112146.......................	14 lbs. 13 oz....	225 lbs. 0 oz.....	436
Toma May 74836.........................	14 lbs. 13 oz....	217 lbs. 8 oz.....	445
Toridex 76873...........................	14 lbs. 13 oz....	228 lbs. 0 oz.....	446
Tormentor's Su Lu 54943.................	14 lbs. 13 oz....	180 lbs. 8 oz.....	450

	BUTTER.	MILK.	PAGE.
Volie of Sennett 2d 78638................	14 lbs. 13 oz....	242 lbs. 0 oz.....	463
Maud Pogis 24240......................	14 lbs. 12¾ oz....	207 lbs. 0 oz.....	286
Amna 35588...........................	14 lbs. 12½ oz....	229 lbs. 8 oz.....	16
Die Königin 124952....................	14 lbs. 12½ oz....	194 lbs. 7⅜ oz.....	105
Harry's Fancy Belle 2d 82416...........	14 lbs. 12½ oz....	157 lbs. 0 oz.....	181
Hilda 2d 14967........................	14 lbs. 12½ oz....	115 lbs. 6 oz.....	187
Ideal 11842...........................	14 lbs. 12½ oz....	204 lbs. 0 oz.....	198
Lady Bloomfield 4704..................	14 lbs. 12½ oz....	....	225
Little Butterstar 20926.................	14 lbs. 12½ oz....	127 lbs. 0 oz.....	254
Lady Mitchell 29214...................	14 lbs. 12¼ oz....	252.6 lbs..........	231
Agnes Clover 86378....................	14 lbs. 12 oz....	200 lbs. 11 oz.....	7
Alice of the Meadows 20748.............	14 lbs. 12 oz....	243 lbs. 13 oz.....	11
Bloomfield Lady 6912..................	14 lbs. 12 oz....	....	44
Bright Lady 5938......................	14 lbs. 12 oz....	14 qts. per day.....	47
Cabbie Oak 90061.....................	14 lbs. 12 oz....	197 lbs. 0 oz.....	53
Charmer 4771.........................	14 lbs. 12 oz....	257 lbs. 11 oz.....	62
Content of Linwood 6950..............	14 lbs. 12 oz....	158 lbs. 0 oz.....	78
Copper Queen 53659...................	14 lbs. 12 oz....	266.7 lbs..........	79
Cowles' Nonesuch 6199................	14 lbs. 12 oz....	255 lbs. 4 oz.....	86
Duchess of Argyle 4th 7571.............	14 lbs. 12 oz....	154 lbs. 0 oz.....	116
Ella L. 84123.........................	14 lbs. 12 oz....	261 lbs. 3 oz.....	123
Elma's Star 102275....................	14 lbs. 12 oz....	202 lbs. 10 oz.....	124
Erie Maid of River View 94307..........	14 lbs. 12 oz....	207 lbs. 12 oz.....	129
Estrella 2831..........................	14 lbs. 12 oz....	....	130
Gamma 4147..........................	14 lbs. 12 oz....	200 lbs. 13 oz.....	163
Gold Princess 8809....................	14 lbs. 12 oz....	....	174
Goldstraw 3d 14724...................	14 lbs. 12 oz....	277 lbs. 0 oz.....	174
Good Friday 20081....................	14 lbs. 12 oz....	197 lbs. 4 oz.....	175
Harry's Florrie 86263..................	14 lbs. 12 oz....	198 lbs. 3 oz.....	181
Hertha 26102.........................	14 lbs. 12 oz....	274 lbs. 10 oz.....	186
Inez of Ingleside 28977................	14 lbs. 12 oz....	155 lbs. 8 oz.....	199
Jersey Cream 2d 8519..................	14 lbs. 12 oz....	....	206
Katydidn't 2734.......................	14 lbs. 12 oz....	....	215
Khedive's Pearl 71056.................	14 lbs. 12 oz....	199 lbs. 12 oz.....	216
King's Queen 18712...................	14 lbs. 12 oz....	223 lbs. 8 oz.....	218
Lady Brown 4th 6911..................	14 lbs. 12 oz....	....	225
Lady Gray of Hilltop 2d 14641..........	14 lbs. 12 oz....	243 lbs. 15 oz.....	228
Lady May Jefferson 90791..............	14 lbs. 12 oz....	235 lbs. 7 oz.....	231
Lady Thalma 54701...................	14 lbs. 12 oz....	243 lbs. 8 oz.....	235
Maple Leaf 4768......................	14 lbs. 12 oz....	....	271
Marian Holton 101478.................	14 lbs. 12 oz....	222 lbs. 0 oz.....	273
Mette D. 47433.......................	14 lbs. 12 oz....	190 lbs. 1 oz.....	296
Minnie Mignon 90672.................	14 lbs. 12 oz....	225 lbs. 1 oz.....	299
Mollie St. Helier 2d 98389.............	14 lbs. 12 oz....	141 lbs. 0 oz.....	310
Nice 68027...........................	14 lbs. 12 oz....	184 lbs. 0 oz.....	324
Niobe of Linwood 11134...............	14 lbs. 12 oz....	196 lbs. 8 oz.....	326
Pet Lee 7993..........................	14 lbs. 12 oz....	....	354
Petra's Signal 53376...................	14 lbs. 12 oz....	203 lbs. 0 oz.....	355
Princess 836..........................	14 lbs. 12 oz....	....	368
Princess Bowen 9699..................	14 lbs. 12 oz....	202 lbs. 14 oz.....	368
Rioter's Zoe 19769....................	14 lbs. 12 oz....	239 lbs. 8 oz.....	390
Roll of Honor 13610..................	14 lbs. 12 oz....	....	391
Royal Lady 22078.....................	14 lbs. 12 oz....	198 lbs. 12 oz.....	400

	BUTTER.		MILK.		PAGE.
Ruby Love 16915	14 lbs. 12	oz....	216 lbs. 1	oz.....	402
Ruth L. 84578	14 lbs. 12	oz....	193 lbs. 14	oz.....	404
Sadie's Delight 53392	14 lbs. 12	oz....	209 lbs. 4	oz.....	404
Stella of Edgewood 114473	14 lbs. 12	oz....	191 lbs. 3	oz.....	433
Tormentor's Oona Toltec 105860	14 lbs. 12	oz....	205 lbs. 12	oz.....	450
Vintage 53335	14 lbs. 12	oz....	274 lbs. 3	oz.....	461
Wapello 56044	14 lbs. 12	oz....	204 lbs. 8	oz.....	464
Landseer's Signaline 58101	14 lbs. 11⅞ oz....		126 lbs. 2	oz.....	239
Bisson's Ethleel 80021	14 lbs. 11¾ oz....		211 lbs. 8	oz.....	41
Blanche Duncan 2d 73333	14 lbs. 11¾ oz....		221 lbs. 9½ oz.....		43
Alfaretta C. 34928	14 lbs. 11½ oz....		288 lbs. 3	oz.....	10
Bessie Ridgely 8293	14 lbs. 11½ oz....		....		36
Bonnie 2d 5742	14 lbs. 11½ oz....		200 lbs. 2	oz.....	45
Bunker's Kitty 47103	14 lbs. 11½ oz....		187 lbs. 0	oz.....	51
Fancy Alice 106402	14 lbs. 11½ oz....		183 lbs. 15½ oz.....		148
Louise of Lawnfield 14151	14 lbs. 11½ oz....		268 lbs. 0	oz.....	260
Pansy K. 23889	14 lbs. 11½ oz....		....		340
Sweet Rock 2d 18256	14 lbs. 11½ oz....		....		438
Tolteca 44762	14 lbs. 11½ oz....		218 lbs. 9	oz.....	444
Tolteca 3d 86344	14 lbs. 11½ oz....		196 lbs. 0	oz.....	444
Nanoonan 98775	14 lbs. 11⅜ oz....		276.5 lbs........		320
Myretta 28231	14 lbs. 11¼ oz....		183 lbs. 12	oz.....	317
Abbie Z. 14002	14 lbs. 11	oz....	329 lbs. 0	oz.....	5
Ashantee's Rosy 59258	14 lbs. 11	oz....	160 lbs. 0	oz.....	21
Belle Thorne 13369	14 lbs. 11	oz....	215 lbs. 0	oz.....	33
Bissonise 110476	14 lbs. 11	oz....	212 lbs. 12	oz.....	40
Chautauqua Queen 26403	14 lbs. 11	oz....	225 lbs. 13	oz.....	63
Chief's Fosie 61998	14 lbs. 11	oz....	216 lbs. 6	oz.....	65
Chrissy 3d 15615	14 lbs. 11	oz....	212 lbs. 0	oz.....	67
Chromo's Princess 121906	14 lbs. 11	oz....	233 lbs. 4	oz.....	68
Clara C. Magnet 31563	14 lbs. 11	oz....	190 lbs. 4	oz.....	70
Daisy Morrison 4th 46830	14 lbs. 11	oz....	216 lbs. 0	oz.....	96
Duchess of Darlington 13830	14 lbs. 11	oz....	274 lbs. 0	oz.....	117
Exile's Dumbella 79452	14 lbs. 11	oz....	277 lbs. 4	oz.....	140
Hermione C. 90598	14 lbs. 11	oz....	222 lbs. 0	oz.....	186
Homestead Pogis 44231	14 lbs. 11	oz....	228 lbs. 9	oz.....	189
Ida's Myrrha 97939	14 lbs. 11	oz....	216 lbs. 8	oz.....	196
Idlewild Pogis 80271	14 lbs. 11	oz....	218 lbs. 10	oz.....	198
Kitty Ona 24209	14 lbs. 11	oz....	233 lbs. 4	oz.....	221
Leurona 43271	14 lbs. 11	oz....	262 lbs. 8	oz.....	249
Light Foot Pogis 57531	14 lbs. 11	oz....	233 lbs. 4	oz.....	249
Lizzette's Mary 12723	14 lbs. 11	oz....	....		256
Maiden of Jersey 2736	14 lbs. 11	oz....	....		269
Marcia G. 53211	14 lbs. 11	oz....	228 lbs. 13	oz.....	273
Melrose Princess 2d 62003	14 lbs. 11	oz....	228 lbs. 3	oz.....	292
Minnie Kersey 19895	14 lbs. 11	oz....	224 lbs. 3	oz.....	299
Miss Madrid 49419	14 lbs. 11	oz....	204 lbs. 13	oz.....	305
Pearl Banks 85758	14 lbs. 11	oz....	255 lbs. 12	oz.....	344
Pogis' Ida May 50063	14 lbs. 11	oz....	192 lbs. 0	oz.....	360
Regina Gilford 49891	14 lbs. 11	oz....	167 lbs. 4	oz.....	383
Royal Sister 12451	14 lbs. 11	oz....	212 lbs. 8	oz.....	401
Southern Daisy 38292	14 lbs. 11	oz....	108 lbs. 8	oz.....	430
Venus 112	14 lbs. 11	oz....	....		458
Young Gazelle 19837	14 lbs. 11	oz....	....		471

	Butter.	Milk.	Page.
Idelle's Idea 92510	14 lbs. 10¾ oz.	185 lbs. 12 oz.	198
Betsy of Rochester 23562	14 lbs. 10½ oz.	220 lbs. 9 oz.	38
Grace Torment 87957	14 lbs. 10½ oz.	232 lbs. 8 oz.	177
La France's Pansy Rex 49141	14 lbs. 10½ oz.	214 lbs. 12 oz.	236
May Naomi 92277	14 lbs. 10½ oz.	241 lbs. 4 oz.	290
Oakland Girl 11103	14 lbs. 10½ oz.	209 lbs. 8 oz.	330
Ophi of St. Lambert 49004	14 lbs. 10½ oz.	254 lbs. 8 oz.	337
Pinkey Spry 2d 92953	14 lbs. 10½ oz.	143 lbs. 8 oz.	358
Renini 9181	14 lbs. 10½ oz.	206¾ lbs.	384
Rosebud of Belle Vue 7702	14 lbs. 10½ oz.	207 lbs. 8 oz.	395
Rose Cottage Beauty 99542	14 lbs. 10½ oz.	145 lbs. 14 oz.	395
Whiting's Daisy 14858	14 lbs. 10½ oz.	229 lbs. 2 oz.	466
Yellow Locust 10679	14 lbs. 10½ oz.	163 lbs. 0 oz.	470
Effie Baal 134222	14 lbs. 10¼ oz.	165 lbs. 10¼ oz.	122
Exile's Aggie 52764	14 lbs. 10¼ oz.	141 lbs. 8 oz.	137
Exile's Dell 67533	14 lbs. 10¼ oz.	191 lbs. 8 oz.	139
Violet of Maplewood 49724	14 lbs. 10¼ oz.	192 lbs. 4 oz.	462
Balmoral 2d 13574	14 lbs. 10 oz.	191 lbs. 10 oz.	24
Bell Rex 11700	14 lbs. 10 oz.	201 lbs. 8 oz.	33
Buckmentor 62846	14 lbs. 10 oz.	195 lbs. 4 oz.	51
Charity of Argyle 35036	14 lbs. 10 oz.	189 lbs. 10 oz.	62
Diploma's Coma 54070	14 lbs. 10 oz.	137 lbs. 6 oz.	105
Dolly Landseer 70201	14 lbs. 10 oz.	173 lbs. 0 oz.	111
Grace of Glynllyn 67759	14 lbs. 10 oz.	186 lbs. 14 oz.	176
Idalco 117140	14 lbs. 10 oz.	173 lbs. 4 oz.	192
Idana 27974	14 lbs. 10 oz.	201 lbs. 7 oz.	193
Ida of Elma 92316	14 lbs. 10 oz.	210 lbs. 0 oz.	194
Lady Aurora 98734	14 lbs. 10 oz.	156 lbs. 8 oz.	224
Lady Dorcas 74348	14 lbs. 10 oz.	243 lbs. 4 oz.	227
Lucia Rex 78326	14 lbs. 10 oz.	190 lbs. 8 oz.	261
Mab of Eau Claire 120020	14 lbs. 10 oz.	222 lbs. 4 oz.	266
Martha Hall 2d 57901	14 lbs. 10 oz.	216 lbs. 0 oz.	276
Mirnie Potter 72670	14 lbs. 10 oz.	273 lbs. 8 oz.	301
New Maud Lee 16614	14 lbs. 10 oz.	200 lbs. 0 oz.	324
Nimble 22335	14 lbs. 10 oz.	171 lbs. 5 oz.	325
Parloa 32357	14 lbs. 10 oz.	217 lbs. 14 oz.	342
Pedro's Laura B. 96363	14 lbs. 10 oz.	232 lbs. 8 oz.	348
Pedro's Pretty Marjoram 77947	14 lbs. 10 oz.	244 lbs. 0 oz.	351
Peggy Ford 21713	14 lbs. 10 oz.	189 lbs. 0 oz.	352
Pogis Siglise 92889	14 lbs. 10 oz.	207 lbs. 0 oz.	361
Pomona's Coquette 52832	14 lbs. 10 oz.	182 lbs. 2 oz.	362
Princess of Trinity 23641	14 lbs. 10 oz.	248 lbs. 0 oz.	373
Ria Alexandra 83752	14 lbs. 10 oz.	254 lbs. 12 oz.	385
Sallie Pomeroy 98806	14 lbs. 10 oz.	222 lbs. 11 oz.	409
Signola M. 76122	14 lbs. 10 oz.	200 lbs. 3 oz.	422
Temalema 44761	14 lbs. 10 oz.	219 lbs. 2 oz.	440
Trenie 17770	14 lbs. 10 oz.	202 lbs. 4 oz.	452
Uinta 5743	14 lbs. 10 oz.	234 lbs. 4 oz.	455
Cinderella's Oonan 108764	14 lbs. 9¾ oz.	145 lbs. 3¼ oz.	69
Ella's Tina 68408	14 lbs. 9¾ oz.	242 lbs. 14 oz.	124
Lydia D. 2d 97860	14 lbs. 9¾ oz.	246 lbs. 0 oz.	264
Bisson's Perfection 103628	14 lbs. 9½ oz.	201 lbs. 11 oz.	42
Blizz Pogis 82873	14 lbs. 9½ oz.	276 lbs. 12 oz.	43
Carlo's Daisy 16702	14 lbs. 9½ oz.	153 lbs. 14 oz.	56

	BUTTER.	MILK.	PAGE.
Deerlick Petra Pogis 99987	14 lbs. 9½ oz.	241 lbs. 12 oz.	101
Emma Landseer 4th 109233	14 lbs. 9½ oz.	150 lbs. 8 oz.	127
Gilded Queen 42029	14 lbs. 9½ oz.	178 lbs. 8 oz.	168
Grace Sheen 2d 79788	14 lbs. 9½ oz.	229 lbs. 13 oz.	176
Lena's Florence 2d 78688	14 lbs. 9½ oz.	210 lbs. 6¼ oz.	245
Mary Justice 37449	14 lbs. 9½ oz.	260 lbs. 12 oz.	278
Yellow Belle 69780	14 lbs. 9½ oz.	287 lbs. 7 oz.	470
Yokun Maid 2d 86563	14 lbs. 9½ oz.	204 lbs. 4 oz.	470
Ida's Lassie 77315	14 lbs. 9¼ oz.	238 lbs. 0 oz.	195
Justa Pogis 2d 115494	14 lbs. 9⅛ oz.	256 lbs. 8 oz.	212
Alpheus Mostar 20037	14 lbs. 9 oz.	240 lbs. 5 oz.	14
Arawana Chèvre Feuille 15097	14 lbs. 9 oz.	214 lbs. 4 oz.	20
Auntybel 12582	14 lbs. 9 oz.	120 lbs. 0 oz.	22
Belbonnie 29630	14 lbs. 9 oz.	203 lbs. 0 oz.	28
Belle of Dorset 85905	14 lbs. 9 oz.	232 lbs. 0 oz.	31
Bisson's Oonan 108518	14 lbs. 9 oz.	163 lbs. 8 oz.	42
Brown Princess 30041	14 lbs. 9 oz.	12 to 14 qts. per day	50
Brunette Le Gros 3d 23326	14 lbs. 9 oz.		50
Clover Mel 16159	14 lbs. 9 oz.	190 lbs. 8 oz.	73
Cretesia's Violet A. 90051	14 lbs. 9 oz.	270 lbs. 8 oz.	90
Daisy Dell 14562	14 lbs. 9 oz.	179 lbs. 0 oz.	94
Elsie's Lady 81493	14 lbs. 9 oz.	204 lbs. 4 oz.	126
Etelka Tolemy 93807	14 lbs. 9 oz.	157 lbs. 8 oz.	130
Fauchon 2d 18200	14 lbs. 9 oz.	234 lbs. 9 oz.	153
Golden Trudie 34535	14 lbs. 9 oz.	157 lbs. 10 oz.	173
Guinevere Sinclair 11167	14 lbs. 9 oz.	234 lbs. 4 oz.	179
Hebe's Fancy 84314	14 lbs. 9 oz.	230 lbs. 13 oz.	184
Hoey Rex 82364	14 lbs. 9 oz.	220 lbs. 1 oz.	188
Island Dots 17003	14 lbs. 9 oz.	202 lbs. 8½ oz.	201
Jennie 766	14 lbs. 9 oz.	105 qts.	204
Lady Pedrona 63173	14 lbs. 9 oz.	206 lbs. 5 oz.	233
Lady Torenia 98736	14 lbs. 9 oz.	168 lbs. 10 oz.	235
Laundress 2d 24649	14 lbs. 9 oz.	138 lbs. 0 oz.	241
Lena Lowndes 23202	14 lbs. 9 oz.	168 lbs. 0 oz.	245
Lilley Rex Pogis 62545	14 lbs. 9 oz.	183 lbs. 12 oz.	250
Lilly of Leeds 17059	14 lbs. 9 oz.		250
Lily of Riverside 19599	14 lbs. 9 oz.	214 lbs. 0 oz.	252
Lois of Pittsford Farms 88848	14 lbs. 9 oz.	249 lbs. 10 oz.	257
Mabel Crockett 18719	14 lbs. 9 oz.	307 lbs. 8 oz.	265
Maud Lee 2d 8839	14 lbs. 9 oz.	179 lbs. 8 oz.	285
Meridale Beauty Dee 97738	14 lbs. 9 oz.	211 lbs. 12 oz.	294
Meridale Phinora 119398	14 lbs. 9 oz.	239 lbs. 5 oz.	294
Mink 3d 4868	14 lbs. 9 oz.	253 lbs. 8 oz.	298
Mountain Lass 12921	14 lbs. 9 oz.	190 lbs. 0 oz.	314
Saugus Lass 30542	14 lbs. 9 oz.	238 lbs. 10 oz.	411
Sitka's Star 104707	14 lbs. 9 oz.	208 lbs. 12¼ oz.	426
Smoky 13733	14 lbs. 9 oz.	198 lbs. 12 oz.	427
Tones 46306	14 lbs. 9 oz.	224 lbs. 4 oz.	445
Tormentor's Idex 64209	14 lbs. 9 oz.	216 lbs. 0 oz.	449
Xarifina 34176	14 lbs. 9 oz.	206 lbs. 8 oz.	469
Signal Fancy 30812	14 lbs. 8⅝ oz.	106 lbs. 12 oz.	418
Duchess of Bloomfield 4th 35584	14 lbs. 8½ oz.	168 lbs. 4 oz.	117
Exile's Lizzie 61477	14 lbs. 8½ oz.	168 lbs. 8 oz.	142
Fancy Torment 56273	14 lbs. 8½ oz.	139 lbs. 4 oz.	150

	BUTTER.		MILK.		PAGE.
Gem of St. Cloud 7342	14 lbs.	8½ oz....	200 lbs.	0 oz.....	165
Harry's Signaline 82419	14 lbs.	8½ oz....	163 lbs.	4 oz.....	182
June Sweets 59978	14 lbs.	8½ oz....	188 lbs.	8 oz.....	211
Kattie 41431	14 lbs.	8½ oz....	159 lbs.	8 oz.....	215
King's Sultana 76057	14 lbs.	8½ oz....	192 lbs.	0 oz.....	219
Lelia Talbot 96922	14 lbs.	8½ oz....	205 lbs.	0 oz.....	244
Marion Earle 63908	14 lbs.	8½ oz....	241 lbs.	12 oz.....	274
Segilinda 41741	14 lbs.	8½ oz....	193 lbs.	4 oz.....	412
Vina Marigold 108186	14 lbs.	8½ oz....	211 lbs.	4 oz.....	461
Waxie 2d 83358	14 lbs.	8½ oz....	132 lbs.	0 oz.....	465
Alice of Salem 5053	14 lbs.	8 oz....			11
Belle Cotton 96205	14 lbs.	8 oz....	226 lbs.	2 oz.....	29
Brown Coomassie 20322	14 lbs.	8 oz....	163 lbs.	8 oz.....	49
Bryant 4193	14 lbs.	8 oz....	235 lbs.	0 oz.....	51
Butterstamp Pedro 56383	14 lbs.	8 oz....	232 lbs.	6 oz.....	52
Caroline 12019	14 lbs.	8 oz....	218 lbs.	12 oz.....	57
Charm 3d's St. Lambert 96872	14 lbs.	8 oz....	240 lbs.	2 oz.....	62
Chloe Beach 3931	14 lbs.	8 oz....			66
Cretesia's Tamy 89994	14 lbs.	8 oz....	245 lbs.	8 oz.....	89
Daisy Morrison 3d 40300	14 lbs.	8 oz....	187 lbs.	8 oz.....	96
Davy's Dot 41693	14 lbs.	8 oz....	183 lbs.	8 oz.....	100
Deborana 4718	14 lbs.	8 oz....			101
Dinah Alexis 69345	14 lbs.	8 oz....	237 lbs.	4 oz.....	105
Dolly of Lakeside 10824	14 lbs.	8 oz....	204 lbs.	8 oz.....	111
Duchess of Argyle 7th 47569	14 lbs.	8 oz....	229 lbs.	4 oz.....	116
Duke's Rowena 37190	14 lbs.	8 oz....	190 lbs.	12 oz.....	118
Ernestina 2d 77682	14 lbs.	8 oz....	222 lbs.	9 oz.....	129
Ethel of Shelburne 46285	14 lbs.	8 oz....	270 lbs.	6 oz.....	131
Exile's Agnes 79796	14 lbs.	8 oz....	302 lbs.	0 oz.....	137
Fair Dairy-maid 29839	14 lbs.	8 oz....	182 lbs.	4 oz.....	146
Fall Leaf 8587	14 lbs.	8 oz....	256 lbs.	0 oz.....	148
Fides 2d 1576	14 lbs.	8 oz....			155
Florrie May Baker 10728	14 lbs.	8 oz....	207 lbs.	2 oz.....	159
Gilda Mercedes H. 14451	14 lbs.	8 oz....	336 lbs.	3 oz.....	167
Goddess of Staatsburgh 5252	14 lbs.	8 oz....			171
Harry's Fancy Belle 44764	14 lbs.	8 oz....	197 lbs.	0 oz.....	181
Hartwick Belle 7722	14 lbs.	8 oz....	189 lbs.	13 oz.....	183
Jessie Wilson 76121	14 lbs.	8 oz....	197 lbs.	7 oz.....	207
Lady Ives 3d 6740	14 lbs.	8 oz....			229
Lady Kate Rex 91751	14 lbs.	8 oz....	239 lbs.	0 oz.....	229
La Pera 2d 13404	14 lbs.	8 oz....	210 lbs.	8 oz.....	240
Lena's Florence 42460	14 lbs.	8 oz....	134 lbs.	7 oz.....	245
Lily's Daffodil 89469	14 lbs.	8 oz....	190 lbs.	8 oz.....	253
Lorraine 1435	14 lbs.	8 oz....			259
Lucie of Bay View 66027	14 lbs.	8 oz....	243 lbs.	0 oz.....	261
Magnolia Ridgely 17269	14 lbs.	8 oz....	181 lbs.	0 oz.....	269
Maid of Berlin 12746	14 lbs.	8 oz....	157 lbs.	10 oz.....	270
Maud Miami Jr. 69131	14 lbs.	8 oz....	228 lbs.	0 oz.....	285
May Cuckoo M. 97417	14 lbs.	8 oz....	203 lbs.	12½ oz.....	288
Minclara 48690	14 lbs.	8 oz....	254 lbs.	0 oz.....	296
Moggy 59626	14 lbs.	8 oz....	230 lbs.	4 oz.....	309
Mollie's Fancy 69728	14 lbs.	8 oz....	226 lbs.	9 oz.....	310
New London Gipsy 11667	14 lbs.	8 oz....	241 lbs.	0 oz.....	324
Pavon 12485	14 lbs.	8 oz....	134 lbs.	8 oz.....	344

	Butter.	Milk.	Page.
Phrynore of Allentown 3d 126321	14 lbs. 8 oz.	233 lbs. 4 oz.	357
Plenty 950	14 lbs. 8 oz.		359
Pride of the Hill 4877	14 lbs. 8 oz.		365
Printonette 26121	14 lbs. 8 oz.	181 lbs. 0 oz.	375
Regina 2d 2475	14 lbs. 8 oz.		382
Rosa Oxford 2d 127909	14 lbs. 8 oz.	171 lbs. 0 oz.	395
Snowdrop F. W. 16948	14 lbs. 8 oz.	168 lbs. 0 oz.	427
Susette C. 18602	14 lbs. 8 oz.	140 lbs. 0 oz.	436
Thorndale Belle 5265	14 lbs. 8 oz.	172 lbs. 0 oz.	442
Upright's Maiden 105616	14 lbs. 8 oz.	200 lbs. 4 oz.	455
Victor's May 97958	14 lbs. 8 oz.	235 lbs. 12 oz.	460
Vidalia 101047	14 lbs. 8 oz.	217 lbs. 0 oz.	460
Viola of Briarcliff 37617	14 lbs. 8 oz.	152 lbs. 9 oz.	461
V. K.'s May 41929	14 lbs. 8 oz.	204 lbs. 15 oz.	463
Carrie's Daffodil 15232	14 lbs. 7¾ oz.		59
Calma of Briarcliff 52599	14 lbs. 7½ oz.	200 lbs. 1 oz.	54
Carrie's Beauty 14601	14 lbs. 7½ oz.		59
Curfew 16498	14 lbs. 7½ oz.	210 lbs. 11 oz.	92
Desda Pogis 58344	14 lbs. 7½ oz.	247 lbs. 2 oz.	103
Dubenna 65024	14 lbs. 7½ oz.	184 lbs. 0 oz.	116
Dutsy Gordon 2d 108503	14 lbs. 7½ oz.	160 lbs. 4 oz.	119
Enid 2d 10782	14 lbs. 7½ oz.	187 lbs. 8 oz.	128
Esther Thorne 35545	14 lbs. 7½ oz.	250 lbs. 8 oz.	130
Exile's Antoinette 111279	14 lbs. 7½ oz.	243 lbs. 8 oz.	138
Milkgood 27828	14 lbs. 7½ oz.	210 lbs. 14 oz.	297
Rioter's Violetta 79391	14 lbs. 7½ oz.	271 lbs. 0 oz.	390
Southern Bell of C. H. 93565	14 lbs. 7½ oz.	220 lbs. 14 oz.	430
Stoke's Tipsey 99380	14 lbs. 7½ oz.	214 lbs. 0 oz.	433
Tormentor's Silver 36058	14 lbs. 7½ oz.	201 lbs. 4 oz.	450
Young Cherry 3d 14980	14 lbs. 7½ oz.	201 lbs. 8¼ oz.	470
Exile's Elf 111811	14 lbs. 7¼ oz.	206 lbs. 8 oz.	140
Ida's Dream 3d 131890	14 lbs. 7¼ oz.	199 lbs. 13¼ oz.	195
Kate Ritchey 54283	14 lbs. 7¼ oz.	244 lbs. 6 oz.	214
Ann Victor Pogis 77813	14 lbs. 7 oz.	245 lbs. 12 oz.	19
Aspirante 9272	14 lbs. 7 oz.	238 lbs. 0 oz.	21
Barbara of Melrose 90486	14 lbs. 7 oz.	183 lbs. 15 oz.	24
Bertha Scituate 75977	14 lbs. 7 oz.	202 lbs. 4 oz.	35
Clubs 2d 16725	14 lbs. 7 oz.	208 lbs. 8 oz.	74
Coronilla 8367	14 lbs. 7 oz.	208 lbs. 0 oz.	82
Czaretta 17358	14 lbs. 7 oz.	199 lbs. 7 oz.	92
Daffy of St. Lambert 69894	14 lbs. 7 oz.	211 lbs. 0 oz.	92
Daisy of Chenango 18582	14 lbs. 7 oz.	252 lbs. 12 oz.	96
Darlymora 37771	14 lbs. 7 oz.	211 lbs. 12 oz.	100
Diamond Rose 25944	14 lbs. 7 oz.	190 lbs. 0 oz.	103
Doris of Mt. Pleasant 73330	14 lbs. 7 oz.	235 lbs. 8 oz.	114
Duchess of Jefferson 2d 118168	14 lbs. 7 oz.	227 lbs. 0 oz.	117
English Elm 17600	14 lbs. 7 oz.	114 lbs. 3 oz.	128
Eurus 60801	14 lbs. 7 oz.	244 lbs. 3 oz.	136
Exile's Nancy 92030	14 lbs. 7 oz.	288 lbs. 0 oz.	143
Flory of the Oaks 8141	14 lbs. 7 oz.		159
Fragrance 4th 16509	14 lbs. 7 oz.		160
Gilt Lady 2d 33969	14 lbs. 7 oz.	199 lbs. 0 oz.	169
Hindoo Rose 14602	14 lbs. 7 oz.		188
Hinds' Tormentor of Lawn 71640	14 lbs. 7 oz.	203 lbs. 14 oz.	188

	BUTTER.	MILK.	PAGE.
Ida's Little Wonder 104961...............	14 lbs. 7 oz....	209 lbs. 4 oz.....	196
Jap Pogis 69122.........................,	14 lbs. 7 oz....	264 lbs. 15 oz.....	203
Jessie Lee of Labyrinth 5290..............	14 lbs. 7 oz....	....	207
Katie Bisson 100200......................	14 lbs. 7 oz....	137 lbs. 0 oz.....	215
Kosi 3431...............................	14 lbs. 7 oz....	198 lbs. 3 oz.....	223
Lady Charella 66187......................	14 lbs. 7 oz....	220 lbs. 8 oz.....	226
Lady Oxford 2d 15101.....................	14 lbs. 7 oz....	202 lbs. 13 oz.....	233
Landseer's Gilderine 68048................	14 lbs. 7 oz....	177 lbs. 8 oz.....	238
La Violette 2d 52604......................	14 lbs. 7 oz....	158 lbs. 8 oz......	241
Lilley Rex 9852..........................	14 lbs. 7 oz....	257 lbs. 4 oz.....	250
Lorelle 12913............................	14 lbs. 7 oz....	148 lbs. 8 oz.....	258
Matilda's Pansy 81311....................	14 lbs. 7 oz....	189 lbs. 12 oz.....	283
Milk Weed 16402.........................	14 lbs. 7 oz....	189 lbs. 8 oz.....	297
Millie of Eau Claire 120023..............	14 lbs. 7 oz....	215 lbs. 4 oz.....	298
Miralba 65002...........................	14 lbs. 7 oz....	248 lbs. 2 oz.....	301
Miss Clifford 5th 63364...................	14 lbs. 7 oz....	215 lbs. 1 oz.....	303
Monmouth Duchess 3895..................	14 lbs. 7 oz....	....	312
Nibbette 11625...........................	14 lbs. 7 oz....	232 lbs. 12 oz.....	324
Polly Neptune 15214......................	14 lbs. 7 oz....	232 lbs. 4 oz.....	362
Rose's Fancy 93047......................	14 lbs. 7 oz....	199 lbs. 8 oz.....	397
Ruby's Valentine 81780...................	14 lbs. 7 oz....	221 lbs. 14 oz.,...	402
Salsoda 3721............................	14 lbs. 7 oz....	....	409
Signal Duke's Creamer 60812....:.......	14 lbs. 7 oz....	201 lbs. 1 oz.....	418
Silkey Morrill 17058.....................:.	14 lbs. 7 oz....	....	423
Star of Mahoning 16579...................	14 lbs. 7 oz....	158 lbs. 8 oz.....	432
Sunny Lass 6033.........................	14 lbs. 7 oz....	192 lbs. 12½ oz.....	435
Tormentor Victor Pogis 133021............	14 lbs. 7 oz....	237 lbs. 0 oz.....	450
Upper Ten 37603.........................	14 lbs. 7 oz....	247 lbs. 13 oz.....	455
Vemicou 79322..........................	14 lbs. 7 oz....	224 lbs. 4 oz.....	457
Winsome of Ipswich 9213.................	14 lbs. 7 oz....	....	467
Yankee Girl 2d 29261.....................	14 lbs. 7 oz....	225 lbs. 8 oz.....	470
Young Anne Lee 31668...................	14 lbs. 7 oz....	212 lbs. 8 oz.....	470
Imperial Pansy Pogis 49750..............	14.42 lbs........	283.1 lbs...........	199
Bluster's Pip 101165.....................	14.41 lbs........	210.2 lbs...........	44
Allie Minka 2982........................	14 lbs. 6½ oz....	208 lbs. 8 oz.....	11
Beulah of Baltimore 3270.................	14 lbs. 6½ oz....	....	39
Harry's Fancy 39114....................	14 lbs. 6½ oz....	233 lbs. 12 oz.....	181
Irene of Short Hills 5137.................	14 lbs. 6½ oz....	237 lbs. 8 oz.....	201
Liberty 2d 16717.........................	14 lbs. 6½ oz....	186 lbs. 10 oz.....	249
Lily Martin 2d 89080.....................	14 lbs. 6½ oz....	208 lbs. 0 oz.....	251
Moll Roberts 110201.....................	14 lbs. 6½ oz....	199 lbs. 13 oz.....	310
Woodland Rose 68381....................	14 lbs. 6½ oz....	247 lbs. 2 oz.....	468
Rosette M. 78677........................	14 lbs. 6.4 oz....	193 lbs. 12.8 oz.....	397
Prince's Rubetta 2d 133914..............	14 lbs. 6¼ oz....	140 lbs. 7 oz.....	368
Aloha's Tormentor 90239.................	14 lbs. 6 oz....	224 lbs. 0 oz.....	13
Anklet 93500............................	14 lbs. 6 oz....	219 lbs. 8 oz.....	18
Athlete 28137.........................:..	14 lbs. 6 oz....	213 lbs. 10 oz.....	21
Augerez Girl 17015......................	14 lbs. 6 oz....	....	22
Champion Flower 26887..................	14 lbs. 6 oz....	226 lbs. 8 oz.....	62
Charity Victor Pogis 134217.............	14 lbs. 6 oz....	215 lbs. 8 oz.....	62
Chlora 8325.............................	14 lbs. 6 oz....	172 lbs. 0 oz.....	66
Como of Briarcliff 35849.................	14 lbs. 6 oz....	155 lbs. 0 oz.....	77
Countess of Scarsdale 18633.............	14 lbs. 6 oz....	230 lbs. 0 oz.....	85
Cupid of Lee Farm 5997.................	14 lbs. 6 oz....	199 lbs. 8 oz.....	91

	BUTTER.	MILK.	PAGE.
Daisy of Guilford 18583	14 lbs. 6 oz	243 lbs. 8 oz	97
Dempsey's Beauty 62808	14 lbs. 6 oz	112 lbs. 10 oz	102
Donna Le Brocq 99305	14 lbs. 6 oz	211 lbs. 0 oz	112
Edessa 3d 52043	14 lbs. 6 oz	173 lbs. 1 oz	120
Effie of Verna 8928	14 lbs. 6 oz	216 lbs. 8¼ oz	122
Esther Lofty 68754	14 lbs. 6 oz	226 lbs. 2 oz	129
Eveline Rose 80512	14 lbs. 6 oz	156 lbs. 0 oz	137
Exile's Lucile 65151	14 lbs. 6 oz	283 lbs. 8 oz	142
Extract 3d 85261	14 lbs. 6 oz	212 lbs. 0 oz	145
Gilda 2779	14 lbs. 6 oz		167
Ideal Alphea 18755	14 lbs. 6 oz	182 lbs. 10 oz	198
Jaquenetta 10958	14 lbs. 6 oz	141 lbs. 12 oz	203
Jazel's Maid 11011	14 lbs. 6 oz	250 lbs. 11 oz	203
Jersey Belle of Belmont 68337	14 lbs. 6 oz	243 lbs. 10 oz	205
Julia Tyre 45187	14 lbs. 6 oz	228 lbs. 6 oz	211
Kate Crecraft 40783	14 lbs. 6 oz	173 lbs. 4 oz	212
Khedive's Fancy 3d 70282	14 lbs. 6 oz	211 lbs. 0 oz	216
Lady Greville 12930	14 lbs. 6 oz	30 gals. 1 pt	228
Lady Nolia 87972	14 lbs. 6 oz	178 lbs. 0 oz	232
L. D.'s Mabel 29368	14 lbs. 6 oz	157 lbs. 6 oz	242
Leonie of Linwood 25319	14 lbs. 6 oz	152 lbs. 0 oz	246
Lobelia 2d 6650	14 lbs. 6 oz		256
Luella Berlin 16927	14 lbs. 6 oz		263
Lutdefy 46313	14 lbs. 6 oz	179 lbs. 0 oz	264
Madame Argyle 19476	14 lbs. 6 oz	196 lbs. 7 oz	266
Maggie C. 12216	14 lbs. 6 oz	175 lbs. 0 oz	267
Maggie May 2d 12026	14 lbs. 6 oz	181 lbs. 0 oz	268
Marchioness of Maplehurst 8018	14 lbs. 6 oz	156 lbs. 8 oz	273
Marpetra 10284	14 lbs. 6 oz	261 lbs. 0 oz	276
Mary of Pleasant View 13448	14 lbs. 6 oz	246 lbs. 8 oz	279
Mona 3d 23134	14 lbs. 6 oz	276 lbs. 0 oz	311
Mosa 69691	14 lbs. 6 oz	255 lbs. 6 oz	313
Myth 2837	14 lbs. 6 oz	248 lbs. 0 oz	318
Oonan Pogis 59835	14 lbs. 6 oz	215 lbs. 11.2 oz	335
Palestine Pierrot 24099	14 lbs. 6 oz	163 lbs. 12 oz	339
Palestine's Last Daughter 12602	14 lbs. 6 oz		339
Pansita 4th 18109	14 lbs. 6 oz	248 lbs. 7 oz	340
Pedro's June 38790	14 lbs. 6 oz	180 lbs. 8 oz	348
Prince's Nellie 23719	14 lbs. 6 oz	234 lbs. 12 oz	368
Queen of Chenango 17771	14 lbs. 6 oz	189 lbs. 8 oz	378
Quintuple Pogis 44233	14 lbs. 6 oz	192 lbs. 6 oz	380
Renown 13729	14 lbs. 6 oz	196 lbs. 0 oz	384
Ruth Morgan 24098	14 lbs. 6 oz	168 lbs. 0 oz	404
Sapollo 61082	14 lbs. 6 oz	227 lbs. 5 oz	409
Victor's Matilda 46834	14 lbs. 6 oz	237 lbs. 4 oz	459
Beech Grove Elsie 67758	14 lbs. 5½ oz	188 lbs. 3 oz	28
Fancy Wax 2d 83357	14 lbs. 5½ oz	158 lbs. 12 oz	150
Lady Clarendon 3d 17578	14 lbs. 5½ oz		226
Laurie Pansy Champion 20396	14 lbs. 5½ oz	160 lbs. 8 oz	241
Leaf Pogis 130184	14 lbs. 5½ oz	207 lbs. 1 oz	242
Roulette 50080	14 lbs. 5½ oz	228 lbs. 0 oz	398
Signal's Rhoda 98651	14 lbs. 5½ oz	186 lbs. 4 oz	421
Viozetta Pogis 56204	14 lbs. 5½ oz	258 lbs. 8 oz	462
May Toltena 50405	14 lbs. 5⅝ oz	121 lbs. 2 oz	290

	BUTTER.	MILK.	PAGE.
Eve of Christmas 9473..................	14 lbs. 5¼ oz....	257 lbs. 6 oz.	137
Miss Le Dain 27472.....................	14 lbs. 5¼ oz....	228 lbs. 8 oz.	305
Nannie of Hood Farm 108565............	14 lbs. 5¼ oz....	207 lbs. 11 oz.	319
Carrie Lena 2d 19074..................	14 lbs. 5⅛ oz....	219 lbs. 0 oz.	58
Acme's Bonnie Bess 31509...............	14 lbs. 5 oz....	221 lbs. 10 oz.	5
Alphea Marigold 117632.................	14 lbs. 5 oz....	223 lbs. 3 oz.	14
Azelda's Valhalla 45234................	14 lbs. 5 oz....	187 lbs. 0 oz.	23
Brown Bessie 7th of H. F. 119841.........	14 lbs. 5 oz....	226 lbs. 13 oz.	49
Carlotta Tom 76228....................	14 lbs. 5 oz....	229 lbs. 0 oz.	57
Catono's Rosebud 23674................	14 lbs. 5 oz....	132 lbs. 3 oz.	60
Cordelia Signal 5th 83097..............	14 lbs. 5 oz....	181 lbs. 8 oz.	81
Coupon Diamond 103559................	14 lbs. 5 oz....	226 lbs. 14 oz.	86
Daisy Clementaise 77341...............	14 lbs. 5 oz....	183 lbs. 0 oz.	94
Elmwood Melissa 4th 82578.............	14 lbs. 5 oz....	241 lbs. 10 oz.	125
Elsie Rowls of Lawn 71633.............	14 lbs. 5 oz....	211 lbs. 11 oz.	126
Exile's Claribel 107507................	14 lbs. 5 oz....	224 lbs. 8 oz.	139
Fruitland May 93002..................	14 lbs. 5 oz....	248 lbs. 10 oz.	162
Jetsam's May 62530....................	14 lbs. 5 oz....	150 lbs. 8 oz.	208
Lady Jane of St. Peter's 2d 62536........	14 lbs. 5 oz....	148 lbs. 12 oz.	229
Lady Mary Hampton 4861..............	14 lbs. 5 oz....	219 lbs. 8 oz.	231
Lady Palestine 2769...................	14 lbs. 5 oz....	....	233
Lilium Excelsum 2d 66040..............	14 lbs. 5 oz....	210 lbs. 8 oz.	249
Lillie Pope 8589......................	14 lbs. 5 oz....	213 lbs. 15 oz.	250
May Indian 16075.....................	14 lbs. 5 oz....	110 lbs. 12 oz.	289
Memento 1913.........................	14 lbs. 5 oz....	235 lbs. 0 oz.	293
Metah's Baby 9710....................	14 lbs. 5 oz....	239 lbs. 0 oz.	296
Milkmaid of Burr Oaks 9035............	14 lbs. 5 oz....	236 lbs. 2 oz.	297
My May Pogis 84236...................	14 lbs. 5 oz....	211 lbs. 0 oz.	316
Nancy of St. Lambert 12964.............	14 lbs. 5 oz....	180 lbs. 8 oz.	319
Nan of Argyle 47561...................	14 lbs. 5 oz....	280 lbs. 0 oz.	319
Princess Ora 66374....................	14 lbs. 5 oz....	231 lbs. 0 oz.	374
Quintuple's Daisy 86078................	14 lbs. 5 oz....	211 lbs. 2 oz.	380
Redacta 26954.........................	14 lbs. 5 oz....	197 lbs. 11 oz.	382
Exile's Lady Alexis 79112..............	14 lbs. 4¾ oz....	208 lbs. 8 oz.	141
Alphea Star 16532.....................	14 lbs. 4½ oz....	180 lbs. 4 oz.	14
Bertie Briggs 5213....................	14 lbs. 4½ oz....	234 lbs. 0 oz.	35
Eve of Bolton 63665...................	14 lbs. 4½ oz....	177 lbs. 4 oz.	137
Exile's Helen 61251...................	14 lbs. 4½ oz....	198 lbs. 0 oz.	141
Frushia Wilson 47432..................	14 lbs. 4½ oz....	181 lbs. 12 oz.	162
Golden Sheen 25561....................	14 lbs. 4½ oz....	205 lbs. 0 oz.	172
Italica 58775.........................	14 lbs. 4½ oz....	208 lbs. 13 oz.	202
L. D.'s Helita 29366...................	14 lbs. 4½ oz....	199 lbs. 8 oz.	242
Merty of Peach Hill 77774.............	14 lbs. 4½ oz....	188 lbs. 4 oz.	295
Minnie of Scituate 17829...............	14 lbs. 4½ oz....	251 lbs. 0 oz.	299
Pristine 126329........................	14 lbs. 4½ oz....	165 lbs. 12 oz.	375
Rachel's Tamy 89087...................	14 lbs. 4½ oz....	208 lbs. 0 oz.	381
Signalona's Gem 134219................	14 lbs. 4½ oz....	214 lbs. 4 oz.	420
The Sister 21135......................	14 lbs. 4½ oz....	134 lbs. 8 oz.	442
Eterna 27134.........................	14 lbs. 4¼ oz....	169 lbs. 8 oz.	130
Exile's Arcadia 66201.................	14 lbs. 4¼ oz....	243 lbs. 8 oz.	138
Abbie K. 13227........................	14 lbs. 4 oz....	224 lbs. 0 oz.	5
Adina 1942...........................	14 lbs. 4 oz....	234 lbs. 0 oz.	7
Blonde 2d 9268........................	14 lbs. 4 oz....	204 lbs. 0 oz.	43
Brown Bessie's Princess 120166..........	14 lbs. 4 oz....	224 lbs. 4 oz.	49

	BUTTER.	MILK.	PAGE.
Buckeye Lass 10355......................	14 lbs. 4 oz....	208 lbs. 15 oz.....	51
Calisida 76889..........................	14 lbs. 4 oz....	176 lbs. 13 oz.....	54
Cigarette 2849..........................	14 lbs. 4 oz....	182 lbs. 7 oz.....	69
Countess Bee 32630......................	14 lbs. 4 oz....	160 lbs. 0 oz.....	83
Daisey Dixie 9469.......................	14 lbs. 4 oz....	205 lbs. 7 oz.....	94
Dame of Eau Claire 127862...............	14 lbs. 4 oz....	202 lbs. 10 oz.....	99
Davy's Dot 2d 85744.....................	14 lbs. 4 oz....	168 lbs. 4 oz.....	100
Della of Grouville 16428................	14 lbs. 4 oz....	....	101
Fay Signal 2d 134018....................	14 lbs. 4 oz....	201 lbs. 8 oz.....	154
Gilderoy's Enid 32924...................	14 lbs. 4 oz....	235 lbs. 8 oz.....	168
Gowan 12272.............................	14 lbs. 4 oz....	....	175
Gray Bessie 96596.......................	14 lbs. 4 oz....	215 lbs. 4 oz.....	177
Grey Jessamine 17444....................	14 lbs. 4 oz....	265 lbs. 8 oz.....	178
Hattie Landseer 72719...................	14 lbs. 4 oz....	246 lbs. 14 oz.....	183
Ida's Landseer's Gold 121641............	14 lbs. 4 oz....	167 lbs. 0 oz.....	195
Jeannette Geranium 61023................	14 lbs. 4 oz....	121 lbs. 4 oz.....	203
Jeannie Platt 6005......................	14 lbs. 4 oz....	120⅝ qts..........	204
Jim's Glory 60527.......................	14 lbs. 4 oz....	181 lbs. 6 oz.....	208
Juliette Gulon 13143....................	14 lbs. 4 oz....	187 lbs. 8 oz.....	211
Kate Daisy 8204.........................	14 lbs. 4 oz....	276 lbs. 4 oz.....	213
Kate Oxford 2d 69270....................	14 lbs. 4 oz....	188 lbs. 12 oz.....	214
Lady Delphine 28460.....................	14 lbs. 4 oz....	277 lbs. 0 oz.....	227
Lady Emily Kilduff 101053...............	14 lbs. 4 oz....	234 lbs. 8 oz.....	227
Lebanon Daughter 6106...................	14 lbs. 4 oz....	16 to 18 qts. per day	242
Leoline 2d 18315........................	14 lbs. 4 oz....	173 lbs. 8 oz.....	245
Lillie Victor Pogis 85471...............	14 lbs. 4 oz....	206 lbs. 8 oz.....	250
Lottie Rex 18757........................	14 lbs. 4 oz....	164 lbs. 10 oz.....	259
Lucile of Pearl Hill 91151..............	14 lbs. 4 oz....	256 lbs. 4 oz.....	261
Madame Pogis 74452......................	14 lbs. 4 oz....	255 lbs. 6½ oz.....	266
Maid of Yokun 2d 86350.................	14 lbs. 4 oz....	155 lbs. 4 oz.....	270
Margaret of Lagonda 35127...............	14 lbs. 4 oz....	209 lbs. 1 oz.....	273
Mary of Gilderoy 11219..................	14 lbs. 4 oz....	217 lbs. 8 oz.....	279
Matilda Hilda 75346.....................	14 lbs. 4 oz....	173 lbs. 2 oz.....	282
Milk Well 90301.........................	14 lbs. 4 oz....	259 lbs. 8 oz.....	297
Monmouth Duchess 3d 4620................	14 lbs. 4 oz....	160 lbs. 10 oz.....	312
Nellie Bunker 52163.....................	14 lbs. 4 oz....	179 lbs. 6 oz.....	321
Praxitella 50072........................	14 lbs. 4 oz....	183 lbs. 12 oz.....	364
Princess Mary of Woodlawn 11663........	14 lbs. 4 oz....	260 lbs. 15 oz.....	372
Ramapo's Pride 35497....................	14 lbs. 4 oz....	225 lbs. 0 oz.....	381
Rexella 69413...........................	14 lbs. 4 oz....	211 lbs. 11 oz.....	384
Rhoda Hudson 48723......................	14 lbs. 4 oz....	211 lbs. 5 oz.....	385
Rosetta Landseer 90379..................	14 lbs. 4 oz....	207 lbs. 12 oz.....	397
Rosy Kate 10276.........................	14 lbs. 4 oz....	224 lbs. 8 oz.....	398
Rowena of Elm Spring 54613..............	14 lbs. 4 oz....	195 lbs. 14½ oz.....	398
Tib's Jewel 6808........................	14 lbs. 4 oz....	228 lbs. 8 oz.....	443
Topsy of Malone 49478...................	14 lbs. 4 oz....	192 lbs. 0 oz.....	446
Trilby M. 103547........................	14 lbs. 4 oz....	240 lbs. 8 oz.....	453
Venice 18192............................	14 lbs. 4 oz....	250 lbs. 11 oz.....	457
Vespucia 17455..........................	14 lbs. 4 oz....	....	459
Violet of Glencairn 10221...............	14 lbs. 4 oz....	219 lbs. 0 oz.....	462
Exile's Ona 106212......................	14 lbs. 3¾ oz....	224 lbs. 8 oz.....	144
Bianca Lass 14997.......................	14 lbs. 3½ oz....	223 lbs. 8 oz.....	39
Bintana 9837............................	14 lbs. 3½ oz....	183 lbs. 15 oz.....	40
Effie's Girl 103452.....................	14 lbs. 3½ oz....	135 lbs. estimated..	122

	BUTTER.	MILK.	PAGE.
Elainette's Odello 102115..................	14 lbs. 3½ oz....	206 lbs. 8 oz.....	122
Fancy Millicent 71059.....................	14 lbs. 3½ oz....	178 lbs. 8 oz.....	149
Gem of Sassafras 8434....................	14 lbs. 3½ oz....	253 lbs. 12 oz.....	165
Gilt Edge C. 12223.......................	14 lbs. 3½ oz....	....	169
Halsie McCurdy 12379...................	14 lbs. 3½ oz....	212 lbs. 8 oz.....	179
Hennette 11624..........................	14 lbs. 3½ oz....	144 lbs. 8 oz.....	185
Lottie Signal 90039......................	14 lbs. 3½ oz....	203 lbs. 4 oz.....	260
Lucky Olive Germ 80256.................	14 lbs. 3½ oz....	236 lbs. 4 oz.....	262
Milly Judd 85898........................	14 lbs. 3½ oz....	237 lbs. 3 oz.....	298
Princess Chrysantha 34191..............	14 lbs. 3½ oz....	232 lbs. 12 oz.....	369
Rose of Hillside 3866....................	14 lbs. 3½ oz....	152 lbs. 9 oz.....	396
Rosette of St. Lambert 5108.............	14 lbs. 3½ oz....	199 lbs. 0 oz.....	397
Well-to-do 37724........................	14 lbs. 3½ oz....	179 lbs. 8 oz.....	465
Addais 92923............................	14 lbs. 3 oz....	210 lbs. 4 oz.....	6
Adora 18569.............................	14 lbs. 3 oz....	294 lbs. 8 oz.....	7
Austie May 7185.........................	14 lbs. 3 oz....	....	22
Beauty Blücher 308......................	14 lbs. 3 oz....	232 lbs. 5 oz.....	26
Betsona 16776...........................	14 lbs. 3 oz....	105½ qts..........	38
Bisson's Emma 100408...................	14 lbs. 3 oz....	160 lbs. 0 oz.....	41
Bisson's Signaline 93844................	14 lbs. 3 oz....	153 lbs. 8 oz.....	42
Box 2d 4297.............................	14 lbs. 3 oz....	190 lbs. 4 oz.....	46
Celebrity 52011..........................	14 lbs. 3 oz....	212 lbs. 2 oz.....	60
Celia Belle 5865.........................	14 lbs. 3 oz....	246 lbs. 4 oz.....	60
Clematis of St. Lambert 5478............	14 lbs. 3 oz....	285 lbs. 0 oz.....	71
Countess Buttercup 13505...............	14 lbs. 3 oz....	175 lbs. 0 oz.....	83
Countess Hebe 2d 45899.................	14 lbs. 3 oz....	241 lbs. 0 oz.....	84
Deoine 6343.............................	14 lbs. 3 oz....	....	103
Duke of St. L.'s Vail's Pet 61970.........	14 lbs. 3 oz....	230 lbs. 14 oz.....	118
Elegantie 11943..........................	14 lbs. 3 oz....	253 lbs. 8 oz.....	122
Embla Brick 15690.......................	14 lbs. 3 oz....	199 lbs. 0 oz.....	127
Epple Lady Pogis 110638.................	14 lbs. 3 oz....	263 lbs. 4 oz.....	129
Eufielda 62546...........................	14 lbs. 3 oz....	151 lbs. 8 oz.....	134
Fairy Le Brocq 15283....................	14 lbs. 3 oz....	34 to 36 lbs. per day	147
Fancy Phlox 49652.......................	14 lbs. 3 oz....	166 lbs. 9 oz.....	149
Irondequoit Belle 74696..................	14 lbs. 3 oz....	214 lbs. 1 oz.....	201
Kitty of Jefferson 2d 12313..............	14 lbs. 3 oz....	196 lbs. 6 oz.....	220
Kitty's Matilda 75349....................	14 lbs. 3 oz....	199 lbs. 3 oz.....	221
Lady Brown 2d 2348.....................	14 lbs. 3 oz....	195 lbs. 15 oz.....	225
Lillian Mostar 10364.....................	14 lbs. 3 oz....	219 lbs. 12 oz.....	250
Lilly Cross 13796........................	14 lbs. 3 oz....	287 lbs. 8 oz.....	250
Little Mistiss 62849.....................	14 lbs. 3 oz....	220 lbs. 12 oz.....	254
Litza 6338..............................	14 lbs. 3 oz....	268 lbs. 4 oz.....	256
Lucetta 6856............................	14 lbs. 3 oz....	160 lbs. 8 oz.....	261
Lucy McClung 20368.....................	14 lbs. 3 oz....	204 lbs. 13 oz.....	262
Minnie Lee 2d 12941.....................	14 lbs. 3 oz....	170 lbs. 0 oz.....	299
Monocacy Dimple 9680...................	14 lbs. 3 oz....	....	312
Myrtle Vance 84237......................	14 lbs. 3 oz....	212 lbs. 0 oz.....	317
Nerviona 2d 98687.......................	14 lbs. 3 oz....	208 lbs. 15 oz.....	323
Nicomis 68026...........................	14 lbs. 3 oz....	200 lbs. 2 oz.....	324
Niobe Gordon 3d 44041..................	14 lbs. 3 oz....	234 lbs. 12 oz.....	325
Pride of Winslow 2613...................	14 lbs. 3 oz....	....	366
Prince's Bloom 9729.....................	14 lbs. 3 oz....	....	367
Rioter's Violet 33774....................	14 lbs. 3 oz....	154 lbs. 0 oz.....	390
Signalda's Duchess 32173................	14 lbs. 3 oz....	213 lbs. 0 oz.....	417

	BUTTER.	MILK.	PAGE.
Signalona 72105	14 lbs. 3 oz.	189 lbs. 12 oz.	419
Sovereign's Elsie 15793	14 lbs. 3 oz.		431
Tamarack 69900	14 lbs. 3 oz.	186 lbs. 4 oz.	439
Telka 8037	14 lbs. 3 oz.		440
Trut 8090	14 lbs. 3 oz.	270 lbs. 0 oz.	453
Turquoise 1129	14 lbs. 3 oz.		454
Villfy 50107	14 lbs. 3 oz.	227 lbs. 14 oz.	461
Vono 89138	14 lbs. 3 oz.	238 lbs. 13 oz.	463
Antigna 109426	14 lbs. 2¾ oz.	203 lbs. 12 oz.	19
Delphina 105895	14 lbs. 2¾ oz.	164 lbs. 6 oz.	102
Emma Landseer 2d 69220	14 lbs. 2¾ oz.	186 lbs. 0 oz.	127
Bee Princess 40345	14 lbs. 2⅝ oz.	201 lbs. 7 oz.	28
Aggie of St. Lambert 37085	14 lbs. 2½ oz.	134 lbs. 4 oz.	7
Alphetta 16531	14 lbs. 2½ oz.	186 lbs. 5 oz.	14
Bisson's Southern Daisy 112992	14 lbs. 2½ oz.	177 lbs. 10 oz.	42
Carrie's Wonder 15233	14 lbs. 2½ oz.		59
Combination of Lawn 76635	14 lbs. 2½ oz.	218 lbs. 3 oz.	76
Emma Landseer 58104	14 lbs. 2½ oz.	162 lbs. 6 oz.	127
Friar's Lady 36119	14 lbs. 2½ oz.	155 lbs. 8 oz.	161
Gerty of Peach Hill 77773	14 lbs. 2½ oz.	224 lbs. 8 oz.	167
Gilfilla Pogis 38688	14 lbs. 2½ oz.	258 lbs. 4 oz.	168
Grildde 22564	14 lbs. 2½ oz.	187 lbs. 8 oz.	177
Julia Landseer 56268	14 lbs. 2½ oz.	91 lbs. 0 oz.	210
Kaembe's Maud Eurota 91066	14 lbs. 2½ oz.	171 lbs. 5 oz.	212
Maggie May 3255	14 lbs. 2½ oz.	235 lbs. 8 oz.	268
Mildred of M. 15548	14 lbs. 2½ oz.	264 lbs. 0 oz.	297
Mocha's Pet's Rheta 37529	14 lbs. 2½ oz.	325 lbs. 8 oz.	308
Pet Rex 20166	14 lbs. 2½ oz.	161 lbs. 6 oz.	355
Safrano 4568	14 lbs. 2½ oz.	180 lbs. 8 oz.	405
Trixket 16292	14 lbs. 2½ oz.	188 lbs. 0 oz.	453
Lily of Staatsburgh 5427	14 lbs. 2¼ oz.		252
Nanalda 32917	14 lbs. 2⅛ oz.	137 lbs. 7 oz.	318
Ada of Elmwood 98728	14 lbs. 2 oz.	166 lbs. 2 oz.	6
Albert's Alba 111721	14 lbs. 2 oz.	117 lbs. 4 oz.	8
Amrah of Ingleside 56832	14 lbs. 2 oz.	156 lbs. 2 oz.	10
Angela 1682	14 lbs. 2 oz.		17
Augustella 63018	14 lbs. 2 oz.	126 lbs. 0 oz.	22
Bay View Queen 60429	14 lbs. 2 oz.	212 lbs. 8 oz.	25
Becky Z. 71311	14 lbs. 2 oz.	223 lbs. 12 oz.	28
Bella Delaine 10356	14 lbs. 2 oz.	222 lbs. 1 oz.	29
Belle Grinnell 3d 16503	14 lbs. 2 oz.		29
Bessie Bradford 7269	14 lbs. 2 oz.		36
Bessie of Clover Leaf 13033	14 lbs. 2 oz.	207 lbs. 7½ oz.	36
Bettie of Edgewood 31162	14 lbs. 2 oz.	217 lbs. 8 oz.	38
Birdsey's Surprise 48326	14 lbs. 2 oz.	255 lbs. 7 oz.	40
Bo-Peep's Maid 100538	14 lbs. 2 oz.	275 lbs. 12 oz.	46
Bronzie of St. Lambert 49001	14 lbs. 2 oz.	234 lbs. 8 oz.	48
Clarina B. 98430	14 lbs. 2 oz.	238 lbs. 0 oz.	71
Combonnie 40260	14 lbs. 2 oz.	227 lbs. 1 oz.	76
Comedia 38885	14 lbs. 2 oz.	223 lbs. 8 oz.	76
Coquette's Molly 23841	14 lbs. 2 oz.	192 lbs. 13 oz.	79
Countess Alberta 42922	14 lbs. 2 oz.	245 lbs. 3 oz.	83
D. D.'s Coomassie 63673	14 lbs. 2 oz.	195 lbs. 11 oz.	100
Edith Darby 6246	14 lbs. 2 oz.	233 lbs. 8 oz.	120

	BUTTER.	MILK.	PAGE.
Elite 4299...............................	14 lbs. 2 oz....	....	123
Eugenie 2d 12733.......................	14 lbs. 2 oz....	256 lbs. 0 oz.....	134
Euro Polono 2d 101052..................	14 lbs. 2 oz....	222 lbs. 0 oz.....	135
Faith of Cloverdale 29277................	14 lbs. 2 oz....	128 lbs. 5½ oz.....	147
Flora B. 12446..........................	14 lbs. 2 oz....	201 lbs. 13 oz.....	157
Gazania 4513...........................	14 lbs. 2 oz....	198 lbs. 0 oz.....	164
Grace's Nightingale 19855..............	14 lbs. 2 oz....	186 lbs. 8 oz.....	176
Ida Mary Pogis 56126...................	14 lbs. 2 oz....	150 lbs. 14 oz.....	193
Jessie Belle B. 100285..................	14 lbs. 2 oz....	197 lbs. 2 oz.....	207
Jessie Leavenworth 8248................	14 lbs. 2 oz....	....	207
Lady Gray of Hilltop 3d 14642...........	14 lbs. 2 oz....	253 lbs. 4 oz.....	228
Lady Woodbine 26803...................	14 lbs. 2 oz....	107 lbs. 8 oz.....	236
Landseer's Lily 87294..................	14 lbs. 2 oz....	177 lbs. 0 oz.....	238
Lebanon Lass 6108......................	14 lbs. 2 oz....	18 qts. per day.....	242
Leila of Clovernook 72554..............	14 lbs. 2 oz....	206 lbs. 7½ oz.....	243
Lightning of St. Lambert 91392..........	14 lbs. 2 oz....	271 lbs. 8 oz.....	249
Lilly Mac 22976........................	14 lbs. 2 oz....	....	250
Little Princess Potoka 79285............	14 lbs. 2 oz....	221 lbs. 8 oz.....	255
Lois B. 107595.........................	14 lbs. 2 oz....	228 lbs. 10 oz.....	256
Melita of Hillcrest 7054................	14 lbs. 2 oz....	203 lbs. 0 oz.....	292
Mellie Argyle 20609....................	14 lbs. 2 oz....	205 lbs. 8 oz.....	292
Mulberry 22031.........................	14 lbs. 2 oz....	197 lbs. 0 oz.....	315
Nellie 1507............................	14 lbs. 2 oz....	....	321
Nightingale of Elmarch 8312............	14 lbs. 2 oz....	196 lbs. 0 oz.....	325
Ogstree 90943..........................	14 lbs. 2 oz....	222 lbs. 8 oz.....	331
Pearl of St. Lambert 5527..............	14 lbs. 2 oz....	247 lbs. 8 oz.....	345
Perty W. 41721.........................	14 lbs. 2 oz....	215 lbs. 12 oz.....	354
Pope's Flora 18697.....................	14 lbs. 2 oz....	210 lbs. 5 oz.....	363
Princess Inez 5402.....................	14 lbs. 2 oz....	203 lbs. 14 oz.....	370
Queen Dodona 42375....................	14 lbs. 2 oz....	258 lbs. 4 oz.....	377
Queen Fannie 10275....................	14 lbs. 2 oz....	220 lbs. 0 oz.....	377
Queen of Prospect 11997................	14 lbs. 2 oz....	225 lbs. 8 oz.....	379
Rarity 2d 7724.........................	14 lbs. 2 oz....	199 lbs. 0 oz.....	381
Sheldon's Daisy's Ida 101059............	14 lbs. 2 oz....	195 lbs. 0 oz.....	414
Therese M. 8364........................	14 lbs. 2 oz....	....	442
Tidy of St. Lambert 31114..............	14 lbs. 2 oz....	194 lbs. 4 oz.....	443
Tormentor's Rosebud 50069..............	14 lbs. 2 oz....	155 lbs. 8 oz.....	450
Undeniah Pogis 85990...................	14 lbs. 2 oz....	174 lbs. 12 oz.....	455
Vemist 79323...........................	14 lbs. 2 oz....	198 lbs. 8 oz.....	457
Vestina 2458...........................	14 lbs. 2 oz....	201 lbs. 1 oz.....	459
Vidalinne 50101........................	14 lbs. 2 oz....	211 lbs. 0 oz.....	460
Webster Pet 4103.......................	14 lbs. 2 oz....	197 lbs. 0 oz.....	465
Woodland Maid 8156....................	14 lbs. 2 oz....	258 lbs. 6 oz.....	468
Exile's Jessamine 111104...............	14 lbs. 1¾ oz....	211 lbs. 0 oz.....	141
Hugo's Lass 107656.....................	14 lbs. 1¾ oz....	282 lbs. 4¼ oz.....	191
Lemon's Flossie 71214..................	14 lbs. 1¾ oz....	250 lbs. 8 oz.....	244
Ethleel D. 83006........................	14 lbs. 1 3-5 oz..	222 lbs. 10 oz.....	132
Harry's Ruth Morgan 107773............	14 lbs. 1.6 oz....	213 lbs. 8 oz.....	182
Cicero's Ruby 29040....................	14 lbs. 1½ oz....	174 lbs. 7 oz.....	69
Goldie La V. 2d 26319..................	14 lbs. 1½ oz....	157 lbs. 0 oz.....	173
Hurrah Pansy 12153....................	14 lbs. 1½ oz....	166 lbs. 0 oz.....	191
Lass of Burrwood 37437................	14 lbs. 1½ oz....	233 lbs. 8 oz.....	240
Lernella 22322.........................	14 lbs. 1½ oz....	191 lbs. 11 oz.....	247
Luta Pogis 69501.......................	14 lbs. 1½ oz....	229 lbs. 0 oz.....	264

	BUTTER.	MILK.	PAGE.
Nellie Alphea 51274	14 lbs. 1½ oz....	170 lbs. 8 oz.....	321
Nervine 25932	14 lbs. 1½ oz....	233 lbs. 10 oz.....	323
Pearl Jenks 4th 84872	14 lbs. 1½ oz....	195 lbs. 1½ oz.....	345
Postscript 39366	14 lbs. 1½ oz....	241 lbs. 9 oz.....	363
Quadruple Pogis 32359	14 lbs. 1½ oz....	140 lbs. 13 oz.....	377
Sweet Leona B. 21934	14 lbs. 1½ oz....	248 lbs. 8 oz.....	437
Lady Maud of Cinque Park 2d 71548	14 lbs. 1¼ oz....	265 lbs. 0 oz.....	231
Peggy of Staatsburgh 2342	14 lbs. 1¼ oz....	....	353
Alicealda 43604	14 lbs. 1 oz....	248 lbs. 8 oz.....	11
Analysis 3d 82870	14 lbs. 1 oz....	155 lbs. 6 oz.....	17
Anselma 37798	14 lbs. 1 oz....	212 lbs. 5 oz.....	22
Ballet Girl 18750	14 lbs. 1 oz....	186 lbs. 13 oz.....	23
Beauty Bismarck 4967	14 lbs. 1 oz....	....	26
Bermuda R. 67534	14 lbs. 1 oz....	197 lbs. 14 oz.....	34
Betsy's Glory 45604	14 lbs. 1 oz....	219 lbs. 13 oz.....	38
Buttery 3502	14 lbs. 1 oz....	203 lbs. 0 oz.....	53
Carrie Franklin 25635	14 lbs. 1 oz....	218 lbs. 0 oz.....	58
Cowry 4432	14 lbs. 1 oz....	188 lbs. 15 oz.....	86
Creamer 2467	14 lbs. 1 oz....	167 lbs. 13 oz.....	87
Creole Maid 2d 31881	14 lbs. 1 oz....	77 lbs. 11 oz.....	88
Daisy-Seeker 43099	14 lbs. 1 oz....	156 lbs. 0 oz.....	98
Dot Badger Girl 101417	14 lbs. 1 oz....	201 lbs. 10 oz.....	115
Esther B. 24834	14 lbs. 1 oz....	207 lbs. 12 oz.....	129
Eva of Snipsic 17650	14 lbs. 1 oz....	232 lbs. 0 oz.....	136
Fancy's Pet 2d 85912	14 lbs. 1 oz....	173 lbs. 12 oz.....	150
Flora Lee 13294	14 lbs. 1 oz....	199 lbs. 12 oz.....	157
Hazelwood 52239	14 lbs. 1 oz....	257 lbs. 0 oz.....	184
Howland Nightingale 33291	14 lbs. 1 oz....	159 lbs. 0 oz.....	190
Josephina 64921	14 lbs. 1 oz....	218 lbs. 8 oz.....	209
Kate Durwood 2d 82417	14 lbs. 1 oz....	152 lbs. 4 oz.....	213
Leila Victor 85433	14 lbs. 1 oz....	203 lbs. 4 oz.....	244
Lord Hallock's Princess 96924	14 lbs. 1 oz....	207 lbs. 2 oz...,..	257
Matilda Ann Darling 105308	14 lbs. 1 oz....	213 lbs. 8 oz.....	282
Meines St. Helier 29999	14 lbs. 1 oz....	152 lbs. 7 oz.....	291
Melody 2689	14 lbs. 1 oz....	300 lbs. 4 oz.....	292
Merry Maiden 64949	14 lbs. 1 oz....	170 lbs. 0 oz.....	295
Myrtle of Ridgewood 7858	14 lbs. 1 oz....	184 lbs. 0 oz.....	317
Nelly Gray of Clermont 10905	14 lbs. 1 oz....	....	322
Pearl of Oakwood 37722	14 lbs. 1 oz....	186 lbs. 10 oz.....	345
Robinette 7114	14 lbs. 1 oz....	233 lbs. 4 oz.....	391
Ruby Love 2d 32058	14 lbs. 1 oz....	166 lbs. 5 oz.....	402
St. Helier's Matilda 52857	14 lbs. 1 oz....	148 lbs. 4 oz.....	405
Sparkling 41042	14 lbs. 1 oz....	225 lbs. 4 oz.....	431
Taglioni 9182	14 lbs. 1 oz....	237 lbs. 1 oz.....	439
Tormentor's Content 119129	14 lbs. 1 oz....	157 lbs. 8 oz.....	448
Tweedledee of Nipsic 60427	14 lbs. 1 oz....	219 lbs. 8 oz.....	454
Variella of Linwood 10954	14 lbs. 1 oz....	228 lbs. 2 oz.....	457
Walkyrie 5708	14 lbs. 1 oz....	169 lbs. 12 oz.....	464
Anna Paris 67989	14 lbs. ¾ oz....	237 lbs. 14 oz.....	18
Amelia Davenport 81044	14 lbs. ⅝ oz....	250 lbs. 0 oz.....	15
Bisson's Pride 108517	14 lbs. ½ oz....	157 lbs. 8 oz.....	42
Comtesse D'Espagne 10308	14 lbs. ½ oz....	204 lbs. 0 oz.....	77
Cupid of St. Lambert 5104	14 lbs. ½ oz....	185 lbs. 8 oz.....	91
Exile's Rosa 66552	14 lbs. ½ oz....	159 lbs. 8 oz.....	144

	BUTTER.	MILK.	PAGE.
Honeydrop 10033.......................	14 lbs. ½ oz....	236 lbs. 12 oz.....	190
Ida's Grace 56019......................	14 lbs. ½ oz....	186 lbs. 1 oz.....	195
Kitty Black Prince 104472...............	14 lbs. ½ oz....	194 lbs. 12 oz.....	219
Mab of Deerfoot 3d 15345...............	14 lbs. ½ oz....	231 lbs. 3 oz.....	266
Moss Rose of St. Lambert 5114..........	14 lbs. ½ oz....	208 lbs. 0 oz.....	313
Myrrha 3d 50406.......................	14 lbs. ½ oz....	119 lbs. 0 oz.....	317
Scepter's Beauty 23234.................	14 lbs. ½ oz....	107 lbs. 14 oz.....	411
Venna's Zeka 26670....................	14 lbs. ½ oz....	146 lbs. 8 oz.....	458
Elinor Wells 2d 72071..................	14 lbs. 1-16 oz....	198 lbs. 8 oz.....	123
Actress 2311..........................	14 lbs. 0 oz....	....	6
Aldarine 2d 14196.....................	14 lbs. 0 oz....	168 lbs. 8 oz.....	10
Alice Donnal 12726....................	14 lbs. 0 oz....	283 lbs. 11 oz.....	11
Alphea Jewel 22331....................	14 lbs. 0 oz....	187 lbs. 6 oz.....	13
Baby Princess 57493...................	14 lbs. 0 oz....	229 lbs. 4 oz.....	23
Belle of Ogden Farm 1570..............	14 lbs. 0 oz....	....	31
Birdie Le Brocq 17263.................	14 lbs. 0 oz....	167 lbs. 8 oz.....	40
Bounty 1606..........................	14 lbs. 0 oz....	....	46
Clover Bloom 2d 12736.................	14 lbs. 0 oz....	236 lbs. 0 oz.....	73
Clover Bud 5th 27743..................	14 lbs. 0 oz....	134 lbs. 12½ oz.....	73
Clyde Landseer 74834..................	14 lbs. 0 oz....	198 lbs. 0 oz.....	74
Comlassa 40671.......................	14 lbs. 0 oz....	237 lbs. 10 oz.....	76
Cora Scituate 2d 94979................	14 lbs. 0 oz....	245 lbs. 12 oz.....	80
Countess of Warren 3896...............	14 lbs. 0 oz....	....	85
Daisy of Clermont 3492................	14 lbs. 0 oz....	....	96
Elinor Wells 12068....................	14 lbs. 0 oz....	183 lbs. 0 oz.....	122
Ella of Briarcliff 68930...............	14 lbs. 0 oz....	201 lbs. 5 oz.....	123
Elmwood Duchess 8716.................	14 lbs. 0 oz....	206 lbs. 4 oz.....	125
Eugenie 2d 1623......................	14 lbs. 0 oz....	....	134
Eureka McHenry 8341..................	14 lbs. 0 oz....	....	135
Evergreen Maid 11536.................	14 lbs. 0 oz....	....	137
Fancy's Gem 4th 86348................	14 lbs. 0 oz....	191 lbs. 4 oz.....	149
Fenella of Bolton 2d 116768...........	14 lbs. 0 oz....	198 lbs. 12 oz.....	155
Fidelia 5817..........................	14 lbs. 0 oz....	....	155
Frisky Miss 18572.....................	14 lbs. 0 oz....	194 lbs. 0 oz.....	162
Gazelle 15961.........................	14 lbs. 0 oz....	109⅝ qts..........	165
Gazelle of Mobile 1735................	14 lbs. 0 oz....	....	165
Gentle of Glastonbury 4651............	14 lbs. 0 oz....	....	166
Gilt Edge 2d 4420....................	14 lbs. 0 oz....	....	169
Harry's Ethel 63970...................	14 lbs. 0 oz....	177 lbs. 4 oz.....	181
Hattie Adams 80909...................	14 lbs. 0 oz....	226 lbs. 15 oz.....	183
Hazalena's Butterfly 10123.............	14 lbs. 0 oz....	....	184
Home Matron 6707....................	14 lbs. 0 oz....	153 lbs. 0 oz.....	189
Kitty Clover 1113.....................	14 lbs. 0 oz....	....	219
Lady Aspinwall 8374..................	14 lbs. 0 oz....	190 lbs. 0 oz.....	224
Lady Brown 433.......................	14 lbs. 0 oz....	....	225
Lady Caroline of St. Aubin's 11372........	14 lbs. 0 oz....	16 qts. daily.......	225
Lady Ona 50642.......................	14 lbs. 0 oz....	216 lbs. 0 oz.....	233
Le Rosa 10078........................	14 lbs. 0 oz....	222 lbs. 8 oz.....	247
Lily of St. Lambert 5120...............	14 lbs. 0 oz....	....	252
Litty 8017............................	14 lbs. 0 oz....	258 lbs. 0 oz.....	255
Lorna of Maple Glens 48094...........	14 lbs. 0 oz....	242 lbs. 12 oz.....	258
Louisa Hinman 12802..................	14 lbs. 0 oz....	....	260
Lucilla 2735..........................	14 lbs. 0 oz....	....	261
Lucy Gaines' Buttercup 5058...........	14 lbs. 0 oz....	229 lbs. 0 oz.....	262

	BUTTER.	MILK.	PAGE.
Madge Berg 24165........................	14 lbs. 0 oz....	273 lbs. 12 oz.....	267
Massena 4th 47342......................	14 lbs. 0 oz....	194 lbs. 8 oz.....	281
Minnie's Ouida 33117....................	14 lbs. 0 oz....	260 lbs. 7 oz.....	300
Miss Montague 57927....................	14 lbs. 0 oz....	202 lbs. 4 oz.....	306
Morlacchi 2725...........................	14 lbs. 0 oz....	....	313
Nestor's Beauty of St. L. 74884...........	14 lbs. 0 oz....	110 lbs. 12 oz.....	323
Niobe 90.................................	14 lbs. 0 oz....	....	325
Nordheim Creamer 9758...................	14 lbs. 0 oz....	196¼ lbs...........	328
Pertinette 70348..........................	14 lbs. 0 oz....	196 lbs. 0 oz.....	354
Pet Anna 1608............................	14 lbs. 0 oz....	....	354
Pixie 4115...............................	14 lbs. 0 oz....	184 lbs. 8 oz.....	359
Pluma of Jersey Lawn 89954..............	14 lbs. 0 oz....	237 lbs. 0 oz.....	359
Princess Zeroda 107242..................	14 lbs. 0 oz....	298 lbs. 0 oz.....	375
Rosaletta 9339............................	14 lbs. 0 oz....	....	394
St. Nick s Flora 16195...................	14 lbs. 0 oz....	....	408
St. Perpetua 2d 5557.....................	14 lbs. 0 oz....	200 lbs. 0 oz.....	408
Sasco Belle 13601.......................	14 lbs. 0 oz....	....	410
Silene 4307..............................	14 lbs. 0 oz....	....	423
Snap s Dainty 18958.....................	14 lbs. 0 oz....	203 lbs. 0 oz.....	427
Starkville Beauty 4897...................	14 lbs. 0 oz....	185 lbs. 0 oz.....	432
Susanna Frank 18261.....................	14 lbs. 0 oz....	....	436
Toltec's Nan 70190.......................	14 lbs. 0 oz....	194 lbs. 8 oz.....	444
Topsey K. 22709..........................	14 lbs. 0 oz....	....	446
Tormentor's Spiræa 42248................	14 lbs. 0 oz....	147 lbs. 8 oz.....	450
Vesper 1395..............................	14 lbs. 0 oz....	....	459
Winder Gem 90447.......................	14 lbs. 0 oz....	193 lbs. 3 oz.....	466
Witch-Hazel 1360........................	14 lbs. 0 oz....	10 qts. daily......	467
Woodland Maid 2d 19666.................	14 lbs. 0 oz....	294 lbs. 0 oz.....	468

Tests for Longer Periods than Seven Days.

EIGHT DAYS.

	BUTTER.	MILK.	PAGE.
Lady Josephine 11560....................	19 lbs. 2 oz....	....	229

TEN DAYS.

	BUTTER.	MILK.	PAGE.
Sophie Hudson 76105.....................	28 lbs. ½ oz....	468 lbs. 2 oz.....	429
Sophona 76110.........................	22 lbs. 4¾ oz....	354 lbs. 12 oz.....	429
Lady Oxford 4860......................	22 lbs. 2 oz....	310 lbs. 0 oz.....	233
Miss Blanche 2515.....................	20 lbs. 9 oz....	357 lbs. 0 oz.....	302
Birdie 2611............................	20 lbs. 0 oz....	....	40

ELEVEN DAYS.

	BUTTER.	MILK.	PAGE.
Minty 2d of Hood Farm 112326..........	23 lbs. 12 oz....	362 lbs. 5 oz.....	301

THIRTEEN DAYS.

	BUTTER.	MILK.	PAGE.
Lulu Pedro 71902.......................	31 lbs. 4 oz....	611 lbs. 8 oz.....	263

FOURTEEN DAYS.

	BUTTER.	MILK.	PAGE.
Fable 62520............................	52 lbs. 7 oz....	611 lbs. 0 oz.....	146
Tomassee 76875........................	41 lbs. 8¼ oz....	582 lbs. 8 oz.....	445
Ida's Dream 50070......................	40 lbs. 8½ oz....	570 lbs. 6½ oz.....	195
Ida's Rose of St. Lambert 71555...........	40 lbs. 2½ oz....	607 lbs. 0 oz.....	197
Soconee 76895.........................	40 lbs. 1¾ oz....	466 lbs. ¼ oz.....	427
Rowena of St. Lambert 71526............	39 lbs. 6¼ oz....	527 lbs. 12 oz.....	399
Moyane C. 86492......................	37 lbs. 15¼ oz....	433 lbs. 7½ oz.....	315
Miss Gracie 86495.....................	37 lbs. 12 oz....	414 lbs. 12 oz.....	304
Tormendova 74885.....................	37 lbs. 11 oz....	389 lbs. 0 oz.....	447
Emma Landseer 3d 110513..............	37 lbs. 2¼ oz....	572 lbs. 14½ oz.....	127
Valhalla 5300..........................	34 lbs. 0 oz....	....	456
Pansy's Thoughts 69400.................	33 lbs. 0 oz....	503 lbs. 14 oz.....	341
Pa Pogis 72500........................	32 lbs. 15½ oz....	577 lbs. 10 oz.....	342
St. Lambert's Pride 53742.................	32 lbs. ¼ oz....	339 lbs. 0 oz.....	407
Mattie Tormentor 62841.................	31 lbs. 13½ oz....	476 lbs. 0 oz.....	284
Exile's Harriet 100716..................	31 lbs. 7½ oz....	448 lbs. 4 oz.....	141
Rioter's Esse 81901.....................	31 lbs. 6¾ oz....	323 lbs. 8 oz.....	387
Exile's Nina 40522.....................	31 lbs. 6½ oz....	469 lbs. 8 oz.....	144
Silver Hair 4th 60449...................	31 lbs. ¼ oz....	550 lbs. 4 oz.....	424
Sweet Lily of Meridale 94267.............	29 lbs. ½ oz....	487 lbs. 12 oz.....	438
Lydia D. 2d 97860......................	29 lbs. 0 oz....	505.9 lbs...........	264
Light Foot Pogis 57531..................	28 lbs. 15 oz....	459 lbs. 15 oz.....	249
Ethel of Shelburne 46285................	28 lbs. 13½ oz....	523 lbs. 14 oz.....	131

FIFTEEN DAYS.

	BUTTER.	MILK.	PAGE.
Clover of St. Lambert 60945	37 lbs. 13¼ oz	465 lbs. 12 oz	74
King's Trust 18946	36 lbs. 6¼ oz		219
Rush of St. Lambert 61234	34 lbs. 12½ oz	545 lbs. 0 oz	403

SIXTEEN DAYS.

Même 2d 49946	39 lbs. 10¼ oz	576 lbs. 12 oz	293
Même 34921	37 lbs. 5 oz	453 lbs. 8 oz	293
Exile's Lady Star 63943	35 lbs. 7½ oz	517 lbs. 8 oz	142

TWENTY-ONE DAYS.

Inez of Riverside 51781	71 lbs. ½ oz	1309 lbs. 4½ oz	200
Myrtlise 64006	54 lbs. 15½ oz	546 lbs. 0 oz	318
March Ida Pogis 101182	47 lbs. 4½ oz	479 lbs. 8 oz	273
Xarama 76887	45 lbs. 10½ oz	699 lbs. 2 oz	469
Lily Garfield 79819	37.488 lbs	562.7 lbs	251
Jeannette of Pittsford 73185	29.973 lbs	501.8 lbs	204
Campania 88475	28.127 lbs	556.7 lbs	55
Woodstock Mystery 77746	26.223 lbs	388.6 lbs	468
Iola F. 85529	25.251 lbs	465.3 lbs	200
Elturia 80701	24.137 lbs	483.4 lbs	126
Woodstock Lady 80619	23.027 lbs	398.1 lbs	468

TWENTY-FOUR DAYS.

Miss Bluster 108231	50.95 lbs	705.8 lbs	303
Golden Gem of St. L. 81167	49 lbs. 4¼ oz	629 lbs. 8 oz	172

TWENTY-EIGHT DAYS.

Princess Marie G. 120390	60.98 lbs	1007.6 lbs	371
Silver Hair 4th 60449	60 lbs. 8¾ oz	1118 lbs. 8 oz	424

TWENTY-NINE DAYS.

Silver Delle 40691	55 lbs. 0 oz	794 lbs. 8 oz	424

THIRTY DAYS.

	BUTTER.	MILK.	PAGE.
Exile's Belle 40524	122 lbs. 6½ oz.	902 lbs. 8 oz.	139
Sœurette 52256	97 lbs. ½ oz.	1023 lbs. 15¾ oz.	428
Denise's Ida 54942	90 lbs. 6¼ oz.	1243 lbs. 6½ oz.	102
Costa Rica 64570	87 lbs. 14¼ oz.	1225 lbs. 9 oz.	82
Countess Matilda 74928	82 lbs. 15 oz.	532 lbs. 14 oz.	85
Brown Bessie 74997	72.235 lbs.	1134.6 lbs.	48
Vida's Crown Princess 76010	67 lbs. 13½ oz.	969 lbs. 13 oz.	460
Maud's Crown Princess 90320	66 lbs. 13 oz.	999 lbs. 4 oz.	286
Merry Maiden 64949	66.695 lbs.	905 lbs.	295
Princess Helene of St. L. 61972	66 lbs. 6 oz.	1079 lbs. 10 oz.	370
Rosella of Bay View 66032	66 lbs. 2 oz.	942 lbs. 0 oz.	396
Lily's Crown Princess 76011	64 lbs. 12 oz.	879 lbs. 5 oz.	252
Leonetta H. 67514	62 lbs. 3 oz.	814 lbs. 12 oz.	246
Olga H. 60619	61 lbs. 13 oz.	841 lbs. 12 oz.	332
Abbie Z. 14002	61 lbs. 2 oz.	1406 lbs. 0 oz.	5
Stoke Pogis' Regina 48309	60.268 lbs.	1012.2 lbs.	433
Ida Marigold 32615	59.367 lbs.	985.8 lbs.	193
Sheba Rex 47429	57.511 lbs.	1004.2 lbs.	414
Baroness Argyle 40498	56.215 lbs.	925.5 lbs.	24
Cupid's Jersey Maid 35040	55.163 lbs.	1028.7 lbs.	91
Flora Temple 3d 40086	55.058 lbs.	923.6 lbs.	158
Katherine of Pittsford 73169	54.107 lbs.	1062.3 lbs.	214
Exile's Lulu 49984	54.017 lbs.	988.4 lbs.	143
Signal Queen 30869	51.522 lbs.	944.5 lbs.	420
Romp's Princess 51185	51.357 lbs.	704.7 lbs.	393
Hugo's Countess 68394	48.172 lbs.	684.2 lbs.	190
Sayda 3d 17317	47.825 lbs.	843.6 lbs.	411
Islip Lenox 51703	47.699 lbs.	714.6 lbs.	201

THIRTY-ONE DAYS. •

	BUTTER.	MILK.	PAGE.
Daisy Hinman 61537	102 lbs. 0 oz.	1308 lbs. 14 oz.	95
Coquette of Glen Rouge 17559	93 lbs. 4 oz.	1014 lbs. 2 oz.	79
Costa Rica 64570	90 lbs. 11¼ oz.	1264 lbs. 9 oz.	82
Bomba 10330	89 lbs. 14 oz.		45
Zélie de Lussan 48681	86 lbs. 13 oz.	1446 lbs. 8 oz.	472
Busy Melrose 54456	85 lbs. 2¾ oz.	1311 lbs. 0 oz.	52
Haidee Signal 86448	79 lbs. 11 oz.	1264 lbs. 4 oz.	179
Miss Satanella 31544	78 lbs. 3½ oz.	1164 lbs. 4 oz.	308
Priscilla Pogis 39270	77 lbs. 12½ oz.	1042 lbs. 12 oz.	375
Nienchen 48698	73 lbs. 13 oz.	987 lbs. 8 oz.	325
Pogis Oonan 29890	73 lbs. 5 oz.	1128 lbs. 0 oz.	361
Couch's Lily 3237	71 lbs. 0 oz.		83
Clover of St. Lambert 60945	70 lbs. 3 oz.	1028 lbs. 4 oz.	74
Oak Leaf 4769	63 lbs. 4 oz.		330
Topsy of Malone 49478	55 lbs. 12 oz.	790 lbs. 8 oz.	446

THIRTY-TWO DAYS.

	BUTTER.	MILK.	PAGE.
Princess Marie G. 120390	69 lbs. 8 oz.	1139.7 lbs.	372

THIRTY-FIVE DAYS.

	BUTTER.	MILK.	PAGE.
Hilda Glenn 81195....................	121 lbs. 7½ oz.	..1181 lbs. 13 oz.	 188
Paola Stoke Pogis 34691................	109 lbs. 10 oz.	..1385 lbs. 0 oz.	 342
Rachel Spencer 50974..................	105 lbs. 10½ oz.	..2004 lbs. 10 oz.	 381

FORTY-TWO DAYS.

Mona Caird 75776......................	127 lbs. 5¾ oz.	..2054 lbs. 12 oz.	 311
Rowena of St. Lambert 71526...........	108 lbs. 2½ oz.	..1467 lbs. 12 oz.	 399
Clover of St. Lambert 60945.............	94 lbs. 6¼ oz.	..1169 lbs. 6½ oz.	 74

SIXTY-ONE DAYS.

Lady Mel 2d 1795.....................	183 lbs. 0 oz.	.. 18 qts. per day	... 231
Massena 25732........................	145 lbs. 8½ oz.	..1822 lbs. 8 oz.	 280

SIXTY-TWO DAYS.

Mollie Garfield 12172..................	163 lbs. 0 oz.	..1877¾ lbs.	 309

NINETY DAYS.

Countess Matilda 74928................	220 lbs. 6 oz.	....3171 lbs.	 85
Brown Bessie 74997....................	216.640 lbs.	....3634 lbs.	 49
Merry Maiden 64949...................	200.517 lbs.	....3041.2 lbs.	 295
Ida Marigold 32615....................	199.756 lbs.	....3448.3 lbs.	 193
Baroness Argyle 40498.................	194.400 lbs.	....3266.2 lbs.	 24
Hugo's Countess 68394.................	191.894 lbs.	....3542.9 lbs.	 191
Sheba Rex 47429......................	190.617 lbs.	....3283.3 lbs.	 414
Romp's Princess 51185.................	188.373 lbs.	....2984.4 lbs.	 393
Alteration 56436......................	179.879 lbs.	....3115.7 lbs.	 15
Islip Lenox 31703.....................	178.066 lbs.	....3070 lbs.	 202
Flora Temple 3d 40086.................	176.751 lbs.	....3038.2 lbs.	 158
Little Goldie 38671...................	176.394 lbs.	....3284.1 lbs.	 254
Sayda 3d 17317.......................	170.094 lbs.	....3043.1 lbs.	 411
Exile's Lulu 49984....................	168.538 lbs.	....3224.5 lbs.	 148
Albert's Gem 34006...................	165.777 lbs.	....2666.4 lbs.	 9
Signal Queen 30869...................	165.601 lbs.	....3190.6 lbs.	 420
Lily Martin 49954....................	164.227 lbs.	....3520.2 lbs.	 251
Natasqua 65598.......................	161.522 lbs.	....2463.9 lbs.	 320
Pearl of Riverside 55659...............	160.804 lbs.	....2653.7 lbs.	 345
Princess Honoria 62548................	159.447 lbs.	....2690.4 lbs.	 370
Justa Pogis 64863....................	157.697 lbs.	....2745.3 lbs.	 212
Daisy Hinman 61537..................	155.131 lbs.	....2677.8 lbs.	 95
Grace Pansy 2d 18764.................	147.009 lbs.	....2344.4 lbs.	 176
Lorita 33750.........................	146.619 lbs.	....2320.3 lbs.	 258
Gay Orphan 25985...................	138.973 lbs.	....2175.9 lbs.	 164
Annice Magnet 60256..................	119.284 lbs.	....2064 lbs.	 18

THIRTEEN WEEKS.

ONE HUNDRED AND ONE DAYS.

ONE HUNDRED AND TWENTY-FIVE DAYS.

TWENTY-ONE WEEKS.

SIX MONTHS.

FIFTY WEEKS.

ONE YEAR.

Sires With Their Tested Daughters.

Aaron Stoke Pogis 21705.
Mandana 3d 78778............. 19 lbs. 15¾ oz.
Zulisca Pogis 78777........... 18 lbs. 1¼ oz.
Olivia Albertine 3d 83438...... 15 lbs. 1¾ oz.

Abe Lincoln 268.
Sylvia 687..................... 15 lbs. 8 oz.

Achmed 2115.
Nelly Gray of Clermont 10905. 14 lbs. 1 oz.

A. C. L. Pearl H. 27011.
Tessia H. 94223............... 15 lbs. 15 oz.

Admiral Farragut 2666.
Beauty of Seekonk 14651...... 21 lbs. 3 oz.
Cobweb 3d 21325.............. 18 lbs. 5 oz.

Aella's Stoke Pogis 14769.
Miss Nautila 49000............ 15 lbs. 5 oz.

Ajax of St. Lambert 19295.
Edith of St. Lambert 2d 67327. 21 lbs. 11 oz.

Alaric of Lakeside 6621.
Alaric's Milicent 35363........ 18 lbs. 5½ oz.
Mercedes' Pet 20570........... 17 lbs. 1½ oz.
Abbia H. 24221................ 15 lbs. 6 oz.

Albert 44.
Lady Love 2d 2212............ 16 lbs. 8 oz.
Couch's Lily 3237.............. 16 lbs. 5½ oz.
Kitty Colt 2213................. 15 lbs. 9½ oz.
Fragrance 4069................. 15 lbs. 3 oz.
Lady Brown 4th 6911.......... 14 lbs. 12 oz.
Lady Brown 2d 2348.......... 14 lbs. 3 oz.
Couch's Lily 3237 (31 days).... 71 lbs.
Lady Mel 2d 1795 (61 days).... 183 lbs.

Albert 2d 1835.
Rosa Miller 4333............... 17 lbs. 7 oz.
Bloomfield Lady 6912.......... 14 lbs. 12 oz.

Alberta's Fancy Harry 30472.
Alberta d'Or 2d 124269........ 14 lbs. 13 oz.

Albert Fair 15920.
Ada of Orange 2d 65246....... 17 lbs. 3 oz.

Albert Johnson 29637.
Katy Scituate's Beauty 100259. 16 lbs.

Albert of Amherst Villa 33165.
Albert's Comtesse 111718...... 21 lbs. 4 oz.
Albert's Lobelia 111902........ 18 lbs. 1 oz.
Albert's Ideal 111719.......... 15 lbs. 10 oz.
Albert's Alba 111721.......... 14 lbs. 2 oz.

Albert Rex 7724.
Albert's Cretesia 22881........ 16 lbs. 2 oz.
Albert's Lilley 19489.......... 15 lbs. 2 oz.
Albert's Gem 34006 (90 days).. 165.777 lbs.

Albert's Royalist 10169.
Pearl Jenks 2d 68908.......... 18 lbs. 10 oz.
Ruby Jenks 56862............. 18 lbs. 3 oz.
Pearl Jenks 50834............. 14 lbs. 14 oz.

Albert W. 6355.
Shanenalawn 56892........... 18 lbs. 5 oz.

Aldine 1136.
Lucky Belle 2d 6037.......... 16 lbs. 14 oz.
Julia Evelyn 6007............. 15 lbs. 15½ oz.
Duchess Caroline 3d 6039..... 15 lbs. 8 oz.
Bettie Dixon 4527............. 15 lbs.
Starkville Beauty 4897........ 14 lbs.

Alexis of M. 7963.
Good Friday 20081............ 14 lbs. 12 oz.

Alfonso 3013.
Cynthia A. 43721.............. 20 lbs. 10 oz.
Harmony 2d 17118............ 19 lbs. 3 oz.
Même 2d 49946............... 18 lbs. 12¾ oz.
Zeroda 43723................. 16 lbs. 7 oz.
Alfaretta C. 34928............ 14 lbs. 11½ oz.
Cordelia Signal 5th 83097..... 14 lbs. 5 oz.
Lottie Signal 90039........... 14 lbs. 3½ oz.
Même 2d 49946 (16 days)....... 39 lbs. 10¼ oz.

Alf. Signal 31184.
Tormentris of Aurora 93948... 15 lbs. 1½ oz.

Allegany Chief 2918.
Euphorbia 11229.............. 23 lbs. 6¼ oz.
Gardiner's Ripple 11693....... 19 lbs. 12½ oz.
Augustella 63018.............. 14 lbs. 2 oz.

Allegany Signal 10178.

Signal's Ripple 27093......... 17 lbs. 5 oz.
Euphorbia 2d 27094.......... 15 lbs. ½ oz.

Almont 2789.

Kitty Potter 9893............. 18 lbs. 5 oz.

Alney 7502.

Nitelis 102722.................. 16 lbs. ½ oz.

Alpenoa 6062.

Miss Signal 20379............. 16 lbs. 4 oz.

Alphadron 16460.

Adonia 64699.................. 19 lbs. 1 oz.

Alpha of Clifton 1824.

Niobe's Alpheanette 23336..... 22 lbs. 10½ oz.

Alphea Star 7487.

Conger's M. 32751............. 16 lbs. 2½ oz.

Alpheon 6082.

Alpheon's Belle 27194......... 19 lbs. 11 oz.

Alpheus 1168.

Crust 4775.................... 16 lbs. 8 oz.

Alpheus of Springvale 5645.

Pride of Springvale 20093...... 19 lbs. 2 oz.
Southern Pet 26675........... 18 lbs. 8 oz.
Alma of Springvale 26951...... 16 lbs. 14 oz.

Altaica's Perrot 21441.

Daisy Mekeel 70400........... 15 lbs. 5 oz.

American Rioter 13363.

Philena of Briarcliff 68938.... 15 lbs. 15½ oz.
Ella of Briarcliff 68939........ 14 lbs.

Angela's Rioter 20120.

Vera McMillan 66520.......... 17 lbs. ½ oz.
Angela's Pet 69776............ 15 lbs. 12 oz.

Antoinette's Koffee 26010.

Ethleel Koffee of Lawn 116757. 17 lbs. 7 oz.
Lelia Koffee of Lawn 120982.. 17 lbs. 3½ oz.

Antrascine 4747.

Bettie of Edgewood 31162..... 14 lbs. 2 oz.

Apis 1206.

Miss Belle 5083............... 22 lbs. 9 oz.
Ida Bashan 4725.............. 18 lbs.
Miss Beauty 4053............. 16 lbs. 12 oz.

Aquila 1208.

La Reine Centennielle 4999.... 15 lbs. 9 oz.

Arawana's Glory 8001.

Betsy's Glory 45604........... 14 lbs. 1 oz.

Argo 3737.

Miss Dora Deane 24505....... 17 lbs. 9¾ oz.

Arnold's Bronx 3309.

Herberta 8811................. 16 lbs. 15 oz.

Asgard 1379.

Typha 5870................... 16 lbs. 11 oz.

Asterisk's Stoke Pogis 29182.

Fernside 113278............... 15 lbs. 3 oz.

Astronomer 4150.

Brunette of Scarsdale 13276... 17 lbs.

Atamasco 2731.

Arawana Chèvre Feuille 15097 14 lbs. 9 oz.

Augerez' King (F. S. 380 J. H. B.)

Printonette 26121............. 14 lbs. 8 oz.

Augustus Davy 15583.

Davy's Dot 41693............. 14 lbs. 8 oz.

Auroraboreëllis 2408.

Goodbye 27366................ 16 lbs. 13 oz.
Income 19472................. 16 lbs. 9 oz.

Aurora Chief 20039.

Mona Caird 75776............. 25 lbs. 7¼ oz.
Hallie Rice 75777............. 24 lbs. 8½ oz.
Mona Caird 75776 (42 days)... 127 lbs. 5¾ oz.

Authority 27197.

Tamarack 69900............... 14 lbs. 3 oz.

Autocrat 1065.

Lady Wellington of Milton
12234 .... 18 lbs. 11 oz.

Aventurier 4254.

Pet of Rose Lawn 11326....... 18 lbs. 1 oz.

Azelda's Wanderer 12997.

Wanderer's Shiloh 65942....... 15 lbs.
Azelda's Valhalla 45234........ 14 lbs. 5 oz.

Babylon 4723.

Countess Lowndes 26874....... 17 lbs. 8 oz.

Bachelor of St. Lambert 4558.

Fawn of St. Lambert 27942.... 15 lbs. 5¼ oz.
Fawn of St. Lambert 27942
(one year).................. 571 lbs. 6 oz.

Balboa 1244.
Maid of Saragossa 9086........ 16 lbs. 6 oz.
Orphan Duchess 3d 21284...... 16 lbs. 3 oz.
Verbena of Fernwood 9088.... 15 lbs.
Sunny Lass 6033.............. 14 lbs. 7 oz.

Baldwin's Frolic 1384.
Lady Delphine 28460.......... 14 lbs. 4 oz.

Ballad 20239.
Connoisseur 65644............ 20 lbs. 13 oz.

Baltimore Boy 837.
Lady Oaks 2d 5246............ 15 lbs. 2 oz.

Banshee's Stoke Pogis 29261.
Palestine Bess 102970......... 16 lbs. 12 oz.

Baritone 1075.
Empress of Ely 2d 6771....... 16 lbs. 8 oz.

Barker's Dandy 3758.
Kate Daisy 8264............... 14 lbs. 4 oz.

Barney 1491.
Thorndale Belle 5265......... 14 lbs. 8 oz.

Baron (P. S. 289 J. H. B.)
Baron's Sophie 17615.......... 19 lbs. 15⅜ oz.
Baron's Rosette 25983........ 15 lbs. 4 oz.

Baronet 2240.
Baronetti 8425................. 16 lbs. 14½ oz.
Chamomilla 7552.............. 16 lbs. 10 oz.
Bonnie 2d 5742................ 14 lbs. 11½ oz.
Uinta 5743.................... 14 lbs. 10 oz.

Baron Hugo 15208.
Baroness Argyle 40498........ 18 lbs. 12 oz.
Onyx of Argyle 76742......... 18 lbs. 5 oz.
Elsie Lofty 61451.............. 18 lbs. 3 oz.
Jewel of Argyle 68752........ 18 lbs. 2 oz.
Cream's Pride 47570.......... 17 lbs. 2 oz.
Baroness Hugo 47563......... 15 lbs. 10¼ oz.
Clover of Argyle 76745....... 15 lbs. 10 oz.
Czarina of Spring Hill 47568.. 15 lbs. 10 oz.
Dame Lofty 47567............ 15 lbs. 10 oz.
Duchess of Argyle 7th 47569... 14 lbs. 8 oz.
Esther Lofty 68754............ 14 lbs. 6 oz.
Nan of Argyle 47561.......... 14 lbs. 5 oz.
Addais 92923.................. 14 lbs. 3 oz.
Baroness Argyle 40498 (30
days) ....56.215 lbs.
Baroness Argyle 40498 (90
days) .... 194.400 lbs.

Baron Neptune 10090.
Harriet Neptune 52344........ 18 lbs. 6½ oz.

Baron of East View 17630.
Tylesholm Blythesomie 101347 16 lbs.

Baron of Riverside 23793.
Lightning of St. Lambert 91392 14 lbs. 2 oz.

Baron of St. Lambert 5286.
Pauline Hugo Pogis 37098..... 14 lbs. 14 oz.

Baron of the Isles 9146.
Lady Elfied 40757............. 18 lbs.

Baron's Hugo Pogis 14008.
Baron's Lorna 101623.......... 21 lbs. 3 oz.

Barry St. Helier 22065.
Bertie Don 2d 74088........... 16 lbs. 14 oz.

Bay View Chief 23853.
Lady of Bay View 66029....... 16 lbs. 12 oz.

Beacon Comet 675.
Zillia 2d 1693.................. 15 lbs. 2 oz.

Beaconsfield 3416.
Cherokee Rose 20921........... 23 lbs. 10 oz.
Waxie 19706................... 20 lbs. 2½ oz.

Beauclerc 1882.
Nelida 2d 8227................. 15 lbs. 2½ oz.

Beauty Boy of Jefferson 5753.
Duchess of Jefferson 17144.... 16 lbs. 9 oz.

Beauty's Stoke Pogis 14175.
Quinnie Pogis 61924........... 16 lbs. 10 oz.
Marie Bash 81441.............. 16 lbs.
Mary Brown 81213............. 15 lbs. 11½ oz.
My Alpha's Beauty 56469...... 15 lbs.
Melrose Princess 2d 62003.... 14 lbs. 11 oz.
Madame Pogis 74452........... 14 lbs. 4 oz.
Luta Pogis 69501.............. 14 lbs. 1½ oz.

Beeswax 1931.
Cordelia Baker 8814........... 17 lbs. 9 oz.
Mary Clover 9998.............. 14 lbs. 15 oz.

Bee's Wing (P. S. 59 J. H. B.)
Daisy 2d 15761................. 15 lbs. 8 oz.

Bel Caliph 1432.
Mollie Garfield 12172......... 22 lbs. 12 oz.
Mollie Garfield 12172 (62 days). 163 lbs.

Bellman's Boy 14003.
Ida Bellman 48273............. 17 lbs. 5½ oz.

Belmont Rioter 22849.
Rioter's Drosera 81056........ 18 lbs. 4 oz.

Bem 18867.
Alta B. 73003................. 14 lbs. 13 oz.

Ben Butler of Bovina 2024.
Pride of Bovina 8050......... 16 lbs. 9 oz.
Dorothy of Bovina 9373....... 15 lbs. 4 oz.

Ben D. 24164.
Whisk of Meridale 131023...... 15 lbs.

Ben Nora 2262.
Emma Hudson 12469......... 16 lbs. 2 oz.

Ben Rajah 795.
Myrtle of Ridgewood 7858..... 14 lbs. 1 oz.

Ben Weston 3111.
Comtesse D'Espagne 10308.... 14 lbs. ½ oz.

Bergen 5567.
Madge Berg 24165............. 14 lbs.

Berger (P. S. 278 J. H. B.)
Miss Bligh 24561.............. 15 lbs. 2 oz.

Berlin 4507.
Luella Berlin 16927........... 14 lbs. 6 oz.

Berlin Prince 3360.
Guinevere Sinclair 11167...... 14 lbs. 9 oz.
Trixket 16292................. 14 lbs. 2½ oz.

Berry 26432.
Gipsy's Berry Duchess 86124.. 28 lbs. 6¼ oz.

Bertha's Pogis 17993.
Lady Grace Pogis of S. 79462.. 20 lbs. 5 oz.

Bertha's Pogis 2d 27065.
Bertha's Royal Daisy 104901.. 16 lbs. 2½ oz.

Bertram 1883.
Lady Josephine 11560 (8 days). 19 lbs. 2 oz.

Bessie's Toltec 19859.
Lula Gordon 2d 90067......... 15 lbs. 4½ oz.
Dutsy Gordon 2d 108503....... 14 lbs. 7½ oz.

Bijou (F. S. 65 J. H. B.)
Patterson's Beauty 4760....... 18 lbs.

Billy Russell 26384.
Between 86664................. 18 lbs. 9¼ oz.

Bingo 1811.
Mrs. Knickerbocker 19367..... 16 lbs. 8½ oz.

Bingo 2d 6749.
Miss Bingo 80096.............. 24 lbs. 12 oz.
Onnolee 23804................. 16 lbs. 4 oz.
Lady Bingo 24160.............. 15 lbs. 4 oz.

Bishop 28851.
Portfolio 84633................ 15 lbs. 5½ oz.

Bismarck 1423.
Trudie 2d 4084................ 15 lbs.

Bismarck (Fairbanks') 1277.
Narmeoka H. 12323........... 18 lbs. 12 oz.

Bisson's Landseer 27520.
Bisson's Wandrina 100409..... 18 lbs.
Waxie of Aurora 86955........ 17 lbs. 4 oz.
Bisson's Fancy Pogis 106415.. 17 lbs. 1½ oz.
Bisson's Beauty 95023........ 16 lbs. 4 oz.
Donney Pogis 3d 100199....... 16 lbs. ¾ oz.
Bisson's Ethleel 80921........ 14 lbs. 11¾ oz.
Bissonise 110476.............. 14 lbs. 11 oz.
Bisson's Perfection 103628.... 14 lbs. 9½ oz.
Emma Landseer 4th 109233.... 14 lbs. 9½ oz.
Bisson's Oonan 108518........ 14 lbs. 9 oz.
Katie Bisson 100200........... 14 lbs. 7 oz.
Bisson's Emma 100408........ 14 lbs. 3 oz.
Bisson's Signaline 93844....... 14 lbs. 3 oz.
Bisson's Southern Daisy 112992 14 lbs. 2½ oz.
Bisson's Pride 108517......... 14 lbs. ½ oz.

Bisson's Pogis 30781.
Bisson's Idex 133952.......... 17 lbs. 11 oz.
Marchande 2d 93823........... 15 lbs. 9 oz.
Bisson's Rosa 97830........... 15 lbs. 1¾ oz.

Biveno of Ogston 13253.
Bicora 2d 91188................ 16 lbs. 3 oz.
Ogstree 90943................. 14 lbs. 2 oz.

Black Defiance 4014.
Pauline Vivienne 11305....... 16 lbs. 13 oz.
Vivian 15813.................. 16 lbs.

Black Diamond V. 4970.
Alberta Signal 18611.......... 20 lbs. 11 oz.
Alphea Vaudrada 17468....... 19 lbs. 7 oz.
Letitia Hunter 35575.......... 17 lbs. 5 oz.
Pansy Patterson 18612........ 15 lbs. 15 oz.
Marion Hill 40777............. 15 lbs. 1 oz.

Black Prince of Linden 9063.
Princess Rioter 29588......... 17 lbs. 6 oz.

Blackstone of Orange 8337.
Ora of Forester 54698......... 15 lbs. 12 oz.

Blennerhasset 3334.
Monocacy Dimple 9680........ 14 lbs. 3 oz.

Blondin 3d 1935.
Daisy of Hillside Farm 6025.. 16 lbs.
Atricia 6029.................. 15 lbs. 3 oz.

Bloomfield's Pogis 17209.
Douskha 63439................ 15 lbs. 5¾ oz.

Blossom's Tennessee 6060.

Lady Creamly 4th 19077...... 17 lbs. 6½ oz.
Lily of Riverside 19599........ 14 lbs. 9 oz.

Blücher 49.

Beauty Blücher 308........... 14 lbs. 3 oz.

Blücher 2d 102.

Myrtle 2d 211................. 15 lbs. 12 oz.

Bluetooth 1821.

Bessie Bradford 2d 7271....... 15 lbs. 2 oz.

Bluster 22798.

Domett 88179................. 16 lbs. 2 oz.
Miss Bluster 108231.......... 15 lbs. 15⅝ oz.
Caucus Maid 69525............ 15 lbs. 8 oz.
Bluster's Pip 101165.......... 14.41 lbs.
Miss Bluster 108231 (24 days).. 50.95 lbs.

Bobby, on I. of J.

Fear Not 2d 6061.............. 16 lbs. 3 oz.

Bobby (P. S. 208 J. H. B.)

Bobby's Lily 22055............ 18 lbs.
Damask Rose 22065............ 16 lbs. 3¼ oz.
Bobby's Magnolia 24401....... 15 lbs. 12 oz.

Bobby Welcome Pogis 12514.

Janie Caruth 73311............ 15 lbs. 4 oz.

Bob Dawson 18069.

Durex 66320.................... 19 lbs.

Bomba's Son 8372.

Royal Bomba's Daisy 35996... 21 lbs. 5½ oz.

Bombastic 9931.

Lynona 36240.................. 16 lbs. 8 oz.

Bonair St. Lambert 17600.

Elsie Rowls of Lawn 71633.... 14 lbs. 5 oz.

Bonnie Blue Pogis 26285.

Bonnie Moll 95616............. 16 lbs. 3 oz.

Bonnie's Polonius 10138.

Julia Tyre 45187.............. 14 lbs. 6 oz.

Boss of the Cedars 17712.

Pocia of Andalusia 62616...... 20 lbs. 15 oz.
Garibaldi's Kate 2d 56473...... 18 lbs. 2½ oz.

Bradley Pogis 22891.

Jap Pogis 69122................ 14 lbs. 7 oz.

Bravy 1923.

Belle of Riverview 21464...... 17 lbs. 12½ oz.

Braxton 1715.

Creole Maid 11017............. 16 lbs. 15 oz.

Brie 6591.

Lady Grace of Upholme 39569 25 lbs. 5½ oz.

Brier's Pogis 14163.

Cheerful Pogis 61748.......... 21 lbs. 4½ oz.
Comely of St. Lambert 2d 41177 20 lbs. 10 oz.
Rinora Pogis 40107............ 19 lbs. 1½ oz.
St. Lambert's Nora 42132..... 18 lbs. 12 oz.
St. Lambert's Pride 53742..... 16 lbs. 13½ oz.
Kitty of St. Lambert 2d 40106. 16 lbs. 5 oz.
St. Lambert's Pride 53742 (14
days) .... 32 lbs. ¼ oz.

Brie's Tristan 19249.

Serona Hudson 84775.......... 15 lbs. 11 oz.

Bright (F. S. 308 J. H. B.)

Moggy Bright 25891............ 21 lbs. 15 oz.

Bright Lad 11035.

Bright Tass 46173.............. 21 lbs. 1 oz.

Brisk 2077.

Whiting's Daisy 14858......... 14 lbs. 10½ oz.

Bristol Chief 1496.

Summerline 8001............... 18 lbs. 6 oz.

Broadsword 7149.

Mulford's Lizzie Walker 50213 16 lbs. 5 oz.

Broker 873.

Christmas Nannie 4075........ 19 lbs. 7 oz.

Brookside 1104.

Gold Ear 2d 3592.............. 18 lbs. 2 oz.

Brother Jack 4042.

Zitella 2d 11922................ 17 lbs. 8½ oz.
Malope 2d 11923............... 15 lbs. 10 oz.
City Belle 16539............... 15 lbs. 5½ oz.

Brown Bessie's Duke 42166.

Brown Bessie's Modita 128259. 15 lbs. 3 oz.

Brown Bessie's Son 34550.

Brown Bessie's Bonita 109265. 20 lbs. 3 oz.
Brown Bonnie 102406......... 18 lbs. 8 oz.
Brown Bessie's Brownie 114433 17 lbs.
Brown Lassie 92950........... 15 lbs. 7 oz.
Brown Bessie 7th of H. F.
119841 .... 14 lbs. 5 oz.
Brown Bessie's Princess 120166 14 lbs. 4 oz.

Browny (P. S. 158 J. H. B.)

Miss Browney 7288............ 16 lbs. 13 oz.
Royal Beauty 18908........... 15 lbs. 2½ oz.
Beauty 17414.................. 15 lbs.
Rosaline Collas 11187......... 14 lbs. 15 oz.
Rosebud of Belle Vue 7702.... 14 lbs. 10½ oz.

Browser 1624.
Woodland Maid 8156.......... 14 lbs. 2 oz.

Brunette's Prince 7115.
Dorine's Brunette 29309....... 20 lbs. 3 oz.

Buckstone 26035.
Kate McCollum 107306........ 15 lbs. 12½ oz.

Buffer 2055.
Pearl of St. Lambert 5527..... 14 lbs. 2 oz.
Moss Rose of St. Lambert 5114 14 lbs. ½ oz.

Buffer's Hugo Pogis 15315.
Mamie H. Pogis 57050......... 16 lbs. 14 oz.

Bugler 2127.
Fannie Bugler 19962........... 15 lbs. 2 oz.

Bullion 2d 5246.
Bessie Russ 2d 14649......... 15 lbs. 1½ oz.

Bunker 9025.
Proctor's Alma Dolores 47107. 23 lbs. 9 oz.
Cora Belmont 2d 48868....... 19 lbs. 1 oz.
Bunker's Pride 38571.......... 18 lbs. 3½ oz.
Proctor's Belle 47106.......... 16 lbs. 4 oz.
Bunker's Utica Belle 38567.... 15 lbs. 15 oz.
Nutley Dolores 2d 55901...... 15 lbs. 13 oz.
Khedive's Princess 38568..... 15 lbs. 4 oz.
Bunker's Kitty 47103......... 14 lbs. 11½ oz.
Nellie Bunker 52163........... 14 lbs. 4 oz.

Burnside 1234.
Lobella 2d 6650............... 14 lbs. 6 oz.

Butterbred 18270.
Leta of Llorac 70220.......... 15 lbs. 4½ oz.

Butter Maker 3098.
Armon 10862.................. 16 lbs. 13½ oz.

Butternut of Home Farm 6965.
Wyoming 48849............... 15 lbs. 5 oz.

Butter Print 1863.
Valerie 6044.................. 15 lbs. 13 oz.

Butterstamp (P. S. 101 J. H. B.)
Butterstar 7799............... 18 lbs. 4½ oz.
Maid of Five Oaks 7178....... 15 lbs. 4 oz.

Cadet 23958.
Cosmetic 75357............... 19 lbs. 1 oz.

Caen 2317.
Niva 7523..................... 15 lbs. 8 oz.

Caen Jr. 6522.
Badger Girl 21463............. 16 lbs. 1 oz.

Caliph 1618.
Petite Mère 8516.............. 15 lbs. 13 oz.

Callie's Lad 21437.
Daisy Glen 71976.............. 20 lbs. 3 oz.

Callis 1696.
Callie Nan 7959............... 16 lbs. 2 oz.

Camerlengo 3012.
Friz Cam 14655............... 15 lbs. 7 oz.
Lady Clarendon 3d 17578..... 14 lbs. 5½ oz.

Campbell's Rex 6812.
Dassa Argyle 21180........... 16 lbs. 14 oz.

Canada's John Bull 8388.
Cheerful of St. Lambert 2d
43745 .... 22 lbs. 2 oz.
Sheldon's Daisy of St. L. 53047 20 lbs. 11 oz.
John Bull's Princess 49670.... 19 lbs. 14 oz.
St. Lambert's Nancy 56503.... 18 lbs. 7 oz.
Lisgar's Rose 2d 56976....... 16 lbs. 11¾ oz.
St. Lambert's Duchess 56504.. 15 lbs. 7½ oz.

Canada's John Bull Jr. 15693.
Oaklands Nora 2d 45967....... 19 lbs. 1 oz.

Canada's John Bull 3d 17750.
Flower of Glen Rouge 2d 55559 18 lbs. 13½ oz.
Brier of St. Lambert 61750.... 18 lbs. 4 oz.
Pomona's Coquette 52832..... 14 lbs. 10 oz.

Canada's John Bull 5th 20092.
Nellie of St. Lambert 69458... 23 lbs. 12 oz.
Marjoram of Glen Rouge 78420 22 lbs. 12¼ oz.
Leonora of Canada 94906...... 22 lbs. 11 oz.
Maud of Glen Rouge 103781... 22 lbs. 10 oz.
Lady Marjoram 123887........ 21 lbs. 11 oz.
Nora of Glen Rouge 69452.... 16 lbs. 13 oz.
Nan of St. Lambert 69453..... 15 lbs. 9 oz.
Daffy of St. Lambert 69894.... 14 lbs. 7 oz.

Canada's Sir George 18290.
Una of St. Lambert 80117..... 16 lbs. 15½ oz.

Captain Fox 23942.
Jersey Belle of Belmont 68337. 14 lbs. 6 oz.

Cardinal of Rosland 3335.
Grace M. 10697................ 15 lbs. 2 oz.
Cannie M. 10698.............. 15 lbs. 1 oz.

Cargo 5370.
The Widow's Daughter 11507.. 19 lbs. 8½ oz.
Georgiadear 11508............. 18 lbs. 3 oz.

Carlo 5559.
Belle Dame 2d 22043.......... 18 lbs. 12 oz.
Miss Alexander 2d 26054...... 15 lbs. 10 oz.
Carlo's Rosebud 18223........ 15 lbs. 8 oz.
Carlo's Daisy 16702........... 14 lbs. 9½ oz.

Carlos, on I. of J.
Bisson's Belle 31144......... 28 lbs. 10 oz.
Bisson's Belle 31144 (1 yr.).. 1028 lbs. 15⅝ oz.

Carnival 5110.
Busy Bee 2d 25166............ 15 lbs. 8 oz.

Cash Boy 2248.
Maggie Rex 28623.............. 17 lbs. 2½ oz.
Sister Rex 13194.............. 16 lbs. 8 oz.
Eve of Christmas 9473........ 14 lbs. 5¼ oz.

Casta's Champion 11794.
Miss Clifford 2d 56149........ 16 lbs. 4½ oz.
Casta's Belle 56148............ 14 lbs. 15 oz.

Catoctin 2692.
Betsy of Rochester 23562...... 14 lbs. 10½ oz.
Yankee Girl 2d 29261......... 14 lbs. 7 oz.

Catono 3761.
Gladdis Roy 18566............. 17 lbs. 13½ oz.
Daisy Hudson 24647........... 17 lbs. 11 oz.
Rosona 12956................. 16 lbs. 7 oz.
Annie Lane 45345............. 15 lbs. 6 oz.
Elsie Lane 13302............. 15 lbs. 4 oz.
Kitty Ona 24209.............. 14 lbs. 11 oz.
Leurona 43271................ 14 lbs. 11 oz.
Catono's Rosebud 23674...... 14 lbs. 5 oz.
Rhoda Hudson 48723........... 14 lbs. 4 oz.
Betsona 16776................ 14 lbs. 3 oz.
Grädde 22564................. 14 lbs. 2½ oz.
Lady Ona 50642............... 14 lbs.

Catono Koffee 18716.
Pearl of Riverside 55659....... 17 lbs. 5½ oz.
Pearl of Riverside 55659 (90
days) .... 160.804 lbs.

Catono's Harry 20096.
Hilda Glenn 81195............. 27 lbs.
Angelina D. 65514............. 15 lbs. 3½ oz.
Hilda Glenn 81195 (35 days)... 121 lbs. 7½ oz.

Catonoson 9667.
Nerviona 43678............... 14 lbs. 15 oz.

Cattaraugus Chief 6026.
Susa Drew 38521.............. 21 lbs. 1 oz.
Maud Drew 38522............. 17 lbs. 12 oz.

Cazique 3103.
Woodland Maid 2d 19666....... 14 lbs.

Cecco 1673.
Idalene 11841................. 15 lbs. 8½ oz.
Ideal 11842................... 14 lbs. 12½ oz.

Cetewayo (P. S. 224 J. H. B.)
Cetewayo's Silver Bell 18952.. 17 lbs. 2½ oz.
Cetewayo's Lily 18950........ 17 lbs.
Cetewayo's Daisy 18230....... 16 lbs. 4 oz.
Cetewayo's Dorcas 20287...... 16 lbs. 2¼ oz.

Chairman 29883.
Royal Rose 86791............. 20 lbs. 4 oz.

Champion Albert 6023.
Donnona's Nectarine 40093.... 16 lbs. 6 oz.
Countess Alberta 42922....... 14 lbs. 2 oz.

Champion Magnet 6480.
Marie C. Magnet 22903........ 15 lbs. 8 oz.
Frances C. Magnet 22904...... 14 lbs. 13½ oz.
Clara C. Magnet 31563........ 14 lbs. 11 oz.
Annice Magnet 60256 (90 days) 119.234 lbs.

Champion of America 1567.
Champion Azuline 20542...... 18 lbs. ½ oz.
Annie L. 12934................ 17 lbs. 15¼ oz.
Gabriel Champion 14102....... 17 lbs. 8 oz.
Forest Queen 12229........... 16 lbs. 15 oz.
Silveretta 6852................ 16 lbs. 9 oz.
Princess Sheila 7297.......... 16 lbs. 4½ oz.
Tobira 8400................... 15 lbs. 13 oz.
Champion's Chloe 12255....... 15 lbs. 5¼ oz.
Gilded Pansy 20552........... 15 lbs. 5 oz.
Dairy C. 12227................ 15 lbs. ½ oz.
Coronilla 8367................ 14 lbs. 7 oz.
Champion Flower 20887....... 14 lbs. 6 oz.
Lady Greville 12930........... 14 lbs. 6 oz.
Maggie C. 12216.............. 14 lbs. 6 oz.
Maggie May 2d 12926......... 14 lbs. 6 oz.
Gilt Edge C. 12223........... 14 lbs. 3½ oz.
Minnie Lee 2d 12941......... 14 lbs. 3 oz.
Jessie Leavenworth 8248...... 14 lbs. 2 oz.
Therese M. 8364.............. 14 lbs. 2 oz.
Webster Pet 4103............. 14 lbs. 2 oz.

Champion of Ethelwyn 14302.
Nan of Ethelwyn 49502........ 15 lbs. 12½ oz.

Champion of Hilltop 1839.
Lady Gray of Hilltop 2d 14641. 14 lbs. 12 oz.
Lady Gray of Hilltop 3d 14642. 14 lbs. 2 oz.

Champion of Indiana 3075.
Therese G. 16390.............. 17 lbs. 2 oz.

Champion of Kansas 4585.
Value Belle 33080............. 16 lbs. 6 oz.

Champion of Silver Bluff 7378.
Carlotta Tom 76228........... 14 lbs. 5 oz.

Champion of the West 14271.
Nellie Alphea 51274........... 14 lbs. 1½ oz.

Champion Royal 7983.
Lena Hogan 41506............. 16 lbs. 15 oz.

Champion's Ensign 9642.
Bell Ensign 56702............. 15 lbs. 13½ oz.

Champion's Loyal 12673.
Martha Hall 2d 57901......... 14 lbs. 10 oz.
Kate Crecraft 40783........... 14 lbs. 6 oz.

Champion's Son 3286.
Warren's Duchess 3d 26836.... 17 lbs. 2½ oz.
Laurie Pansy Champion 20396 14 lbs. 5½ oz.

Charity Boy 11765.

Princess Cross 2d 80649........	20 lbs. 1 oz.
Noble Charity 67245..........	18 lbs.
Belle's Charity 82217..........	17 lbs. 1 oz.
Phœbe's Charity 59008........	15 lbs. ½ oz.

Charleston 1.

Wyble 595....................	17 lbs. 4 oz.

Charley Willis 14866.

Susie V. 54855................	14 lbs. 14 oz.

Charlie Kittredge 1247.

Matlie 3d 9879................	14 lbs. 13 oz.
Gold Princess 8809...........	14 lbs. 12 oz.

Chatelaine's Dayton 5311.

Shiloh Daughter 20378.......	16 lbs. 7 oz.

Chelten Duke 924.

Louisa Hinman 12802.........	14 lbs.

Chendado 12955.

Helen Walker 53465..........	19 lbs. 10½ oz.
Fanny Mason 45992...........	18 lbs. 5¾ oz.
Chendado's Maid 45991.......	16 lbs. 7½ oz.
Maggie Cyrene 53462.........	15 lbs. 15 oz.
Sepple 58689.................	15 lbs. 6½ oz.

Cherie's Rioter 14437.

Rioter's Roberta 80469........	17 lbs.

Cherokee's Gold Basis 16945.

Signalinda 2d 61142...........	21 lbs. 10 oz.

Chevalier 14748.

Miss Milton 51851.............	24 lbs. 9½ oz.
Eve of Bolton 63665..........	14 lbs. 4½ oz.

Chief 10663.

Chief's Charella 61999........	23 lbs. 10½ oz.
Ellenetta of Bay View 48921..	21 lbs. 11 oz.
Pride of Bay View 61997......	16 lbs. 6 oz.
Chief's Fosie 61998..........	14 lbs. 11 oz.
Lady Charella 66187..........	14 lbs. 7 oz.
Bay View Queen 60429........	14 lbs. 2 oz.

Chief Baron 2984.

Jaquenetta 10958..............	14 lbs. 6 oz.

Chief Justice 2d 1643.

Hilda 2d 5447.................	23 lbs. 5 oz.
Hilda D. 6683.................	21 lbs. 2½ oz.

Chief Modoc 1718.

Countess Europa 35820.......	17 lbs. 12 oz.

Chief of Glen Duart 24340.

Agnes of Glen Duart 82475....	15 lbs. 11 oz.

Chief of St. Lambert 18863.

Irondequoit Belle 74696.......	14 lbs. 8 oz.

Chief of the Miamis 2958.

Signal Queen 30869............	21 lbs. ½ oz.
Fadette Signal 39472..........	20 lbs. 14½ oz.
Chief's Beauty 39391.........	20 lbs. 5½ oz.
Lass Signal 16308.............	20 lbs. 2 oz.
Jean Ingelow 42515...........	17 lbs. 7 oz.
Miamotto 48691................	16 lbs.
Miaclara 48690...............	14 lbs. 8 oz.
Signal Queen 30869 (30 days)..	51.522 lbs.
Signal Queen 30869 (90 days)..	165.601 lbs.

Chieftain Pogis 25035.

Bertha Stewart 99882..........	17 lbs.
Koffmadge 2d 77790...........	16 lbs. 14 oz.
Morganza Queen 77791........	16 lbs. 2¾ oz.

Chieftain Rex 29707.

Corinne's Dollie Rex 110987...	15 lbs. 4 oz.

Chippewa 5739.

Mocha's Pet's Rheta 37529....	14 lbs. 2½ oz.

Chromo 26113.

Chromia E. 97511.............	17 lbs. 12 oz.
Chromo's Pansy 114416........	17 lbs. 2½ oz.
Tremona 93017................	17 lbs. 2 oz.
Chromoline of Argyle 103709..	16 lbs. 10 oz.
Chromo's Modita 121439.......	16 lbs. 7 oz.
Laura Lee 2d 128023..........	16 lbs. 4 oz.
Chromo's Princess 121906.....	14 lbs. 11 oz.

Chuffy Chief 17625.

Nienchen 48698................	17 lbs. 2 oz.
Nienchen 48698 (31 days)......	73 lbs. 13 oz.

Cicero 7657.

Lady Longfield 23524.........	20 lbs. 13 oz.
Lady Fair 22103...............	19 lbs. 10 oz.
Cicero's Jolie 18246..........	18 lbs. 3 oz.
Brunette Star 27270..........	17 lbs. 10 oz.
Cicero's Juno 16726...........	17 lbs. 2 oz.
Lionette 18038................	17 lbs. 1 oz.
Pendule 2d 16709.............	16 lbs. 5 oz.
Cicero's Mabel 18238..........	15 lbs. 2 oz.

Cill's Stoke Pogis 14149.

Homestead Pogis 44231.......	14 lbs. 11 oz.

Citizen 13186.

Upper Ten 37603	14 lbs. 7 oz.
Sparkling 41042...............	14 lbs. 1 oz.

Civil Rights 7223.

May Lankton 15872...........	16 lbs. 1½ oz.

Claimant (P. S. 84 J. H. B.)

Nancy Lee 7618...............	26 lbs. 8½ oz.

Clara's Rioter Pogis 18317.

Daisy of Denison 53137.......	15 lbs. 12 oz.

Clement 115.

Alice of Salem 5053...........	14 lbs. 8 oz.

Clermont's Victor Pogis 20026.
May's Grand Duchess 62607... 15 lbs. 3½ oz.

Cliff 176.
Thisbe 2d 2201................ 19 lbs. 1½ oz.

Clifton Hero 6759.
Alpheus Mostar 20037........ 14 lbs. 9 oz.

Clifton Monarch 3546.
El Mora Mostar 15955........ 15 lbs. 4 oz.
Velvet Mostar 17533.......... 15 lbs. 3 oz.

Climax of St. Lambert 17553.
Silver Hair 4th 60449......... 15 lbs. 10¼ oz.
Silver Hair 4th 60449 (14 days) 31 lbs. ¼ oz.
Silver Hair 4th 60449 (28 days) 60 lbs. 8¾ oz.

Clive Duke 1901.
Oltz 8649..................... 15 lbs. 11 oz.
Halsie McCurdy 12379......... 14 lbs. 3½ oz.

Clotaire 9884.
Cleodoxa 59311................ 22 lbs. 5 oz.
Clotaire's Beauty 28171........ 20 lbs. 4 oz.
Clotaire's Annie 58319........ 19 lbs. 11½ oz.
Clotaire's Daisy 45392......... 18 lbs.

Cloverine 3510.
Rupertina 10409.............. 17 lbs. 1½ oz.

Clover Lad 21069.
Ruth L. 84578................. 14 lbs. 12 oz.

Clusius H. 5781.
Theda H. 20567............... 17 lbs. 1 oz.
Lucy's Love 20568............. 14 lbs. 14 oz.
Esther B. 24834............... 14 lbs. 1 oz.

Cobden 1871.
Frankie's Lass 24900.......... 17 lbs. 3½ oz.

Cobonon 15744.
Beech Grove Elsie 67758....... 14 lbs. 5½ oz.

Cockade 1979.
Forsaken 7520................. 15 lbs. 1 oz.

Cœur de Lion 318.
Maud Lee 2416................ 23 lbs.

Cœur de Lion (P. S. 140 J. H. B.)
Brown Princess 30941........ 14 lbs. 9 oz.

Col. Rex 12975.
Phœbe Rex 55084............. 16 lbs. ½ oz.

Colonel Stoke Pogis 18756.
The Colonel's Daughter 50230. 15 lbs. 4½ oz.

Colonsay 3591.
Carrie Lena 2d 19074......... 14 lbs. 5½ oz.

Color Blind 16861.
Flora Temple 3d 40086 (30 days) .... 55.058 lbs.
Flora Temple 3d 40086 (90 days) .... 176.751 lbs.

Colt Jr. 825.
Chloe Beach 3931............. 14 lbs. 8 oz.

Colt's Pierrot 23528.
Pay Little 86189.............. 15 lbs. 6¾ oz.

Columbiad 2d 1515.
Alluring 5541................. 19 lbs. 5 oz.
Rose of Rose Lawn 9365...... 16 lbs. 3 oz.
Aspirante 9272................ 14 lbs. 7 oz.
Deoine 6343.................. 14 lbs. 3 oz.

Column 9622.
Lorna of Maple Glens 48094... 14 lbs.

Combination 4389.
Paradise 32082............... 17 lbs. 11 oz.
Mercurina 64920.............. 17 lbs. 10½ oz.
Transcript 31867............. 17 lbs. 7 oz.
Commotion 52960............. 17 lbs. 6 oz.
Wamla 32083................. 17 lbs. 5½ oz.
Oneida 42100................. 16 lbs. 13 oz.
Crusta 29637................. 16 lbs. 10 oz.
Lady Hugo 29430............. 16 lbs. 7 oz.
Sparks 41041................. 16 lbs. 4½ oz.
Comanca 19389............... 16 lbs. 3 oz.
Period 42640................. 16 lbs. 3 oz.
Lorita 33750................. 15 lbs. 6½ oz.
Complexia 56774.............. 15 lbs. 3 oz.
Coma 29330.................. 15 lbs. 2½ oz.
Portrait 32592............... 15 lbs. 2½ oz.
Classic 21402................. 14 lbs. 13½ oz.
Belbonnie 29630.............. 14 lbs. 9 oz.
Curfew 16498................. 14 lbs. 7½ oz.
Combonnie 40260............. 14 lbs. 2 oz.
Comedia 38885................ 14 lbs. 2 oz.
Comlassa 40671............... 14 lbs.
Lorita 33750 (90 days)........ 146.619 lbs.

Combination 2d 16957.
Miss Madrid 49419............ 14 lbs. 11 oz.

Combination 3d 17576.
Brown Elsie 96595............ 21 lbs. 12 oz.
Teasel 76358.................. 20 lbs. 4 oz.
Oneida 2d 43553.............. 17 lbs. 4½ oz.
Smax 48562.................. 16 lbs. 11½ oz.
Compact 52705................ 16 lbs. 3 oz.
Manuscript 74881............. 15 lbs. 1 oz.
Cymbelina 67201.............. 14 lbs. 13 oz.
Gray Bessie 96596............. 14 lbs. 4 oz.
Celebrity 52011............... 14 lbs. 3 oz.

Combine 20238.
Vintonette 62931.............. 19 lbs. 4 oz.
Lady Mary Lenox 105574...... 16 lbs. 7½ oz.

Comet 130.
Abbie Z. 14002................ 14 lbs. 11 oz.
Abbie Z. 14002 (30 days)....... 61 lbs. 2 oz.

Commodore Roxbury 1586.
Belle Dawson 8270............ 18 lbs. 3 oz.

Compass 16958.
Tapestry 56607................. 23 lbs. 12 oz.
Modiste 74679 .... 16 lbs. 11½ oz.
Compass' Pansy 105706....... 16 lbs. 10 oz.
False Step 69647.............. 15 lbs. 7 oz.
Italica 58775................... 14 lbs. 4½ oz.

Compeer 2367.
Forget Me Not O 10564....... 15 lbs. 4 oz.

Compo Boy 2830.
Gazella 3d 9355............... 16 lbs. 3 oz.
Gazella 3d 9355 (1 yr.)........ 751 lbs. 6 oz.

Composite 10810.
Signal Hinman 44615.......... 15 lbs. 12¾ oz.

Comus 54.
Plenty 950.................... 14 lbs. 8 oz.

Comus Rex 26804.
Easter Ona 2d 97158.......... 17 lbs. 5 oz.
Easter Prize Rex 102166...... 15 lbs.
Lucia Rex 78326.............. 14 lbs. 10 oz.

Concert 16499.
Sparkle 56775................. 17 lbs. 3 oz.

Concord B. 29699.
Clarina B. 98430.............. 14 lbs. 2 oz.

Conklin's Signal 11050.
Signora Signal 39396.......... 15 lbs.

Conqueror (P. S. 89 J. H. B.)
Lucilla Kent 8892............. 16 lbs. 6 oz.

Coomassie 1484.
Deborana 4718................ 14 lbs. 8 oz.

Coomassie Dick 9815.
Baby Stanton 72397.......... 15 lbs. 4 oz.

Coomassie's Gilderoy 14672.
Lemon Spray 44532........... 15 lbs. 1½ oz.

Coomassie's Triple King 37939.
Ex Presage 128529............ 15 lbs. 5 oz.

Coomassie Welcome 9279.
Queen of Cowes 46309......... 18 lbs. 3 oz.
Ventnor Beauty 46308......... 17 lbs. 5½ oz.
Girlie's Glory 46305............ 15 lbs.
Tones 46306................... 14 lbs. 9 oz.

Copake 1387.
Country Girl 3d 12538........ 17 lbs. 11 oz.

Cornelius 4204.
Fragrance 4th 16509.......... 14 lbs. 7 oz.

Corporal St. Helier 13397.
Allie St. Helier 45794......... 16 lbs. 13½ oz.

Councillor 4468.
Deletta 21305................. 14 lbs. 15½ oz.

Count Bismarck 732.
Myth 2837.................... 14 lbs. 6 oz.

Count Cicero (F. S. 398 J. H. B.)
Count's Fillpail 30975........ 24 lbs. 5 oz.
Cicero's Ruby 29040.......... 14 lbs. 1½ oz.

Count Coomassie 7542.
Elodie 30222.................. 15 lbs. 5½ oz.
Venna's Zeka 26670........... 14 lbs. ½ oz.

Count Logan 1599.
Rockwood Maid 8375......... 15 lbs. 4½ oz.
Lady Aspinwall 8374.......... 14 lbs.

Count Potoka 9831.
Dame Miller 59260............ 16 lbs. 2½ oz.

Count Rioter Pogis 21388.
Amanda Pogis 79412.......... 18 lbs. 7¼ oz.
Maudie of St. Lambert 73603.. 15 lbs. 7½ oz.
Undeniah Pogis 85990........ 14 lbs. 2 oz.

Count St. George 8403.
Fill Pail's Countess 24462..... 22 lbs. 8 oz.

Coventry Boy 5847.
Lulu of Riverside 19175....... 17 lbs.

Creamer's Sir George 8721.
Lady White 29213............. 14 lbs. 14¾ oz.

Cretesia's Albert Pogis 21730.
Cretesia's Temisia 89996...... 18 lbs. 7 oz.
Cretesia's Silverette 74807..... 17 lbs. 3 oz.
Cretesia's Rachel 89998....... '16 lbs. 8 oz.
Cretesia's Matilda 90000....... 15 lbs. 15 oz.
Cretesia's Charma 81685....... 15 lbs. 8 oz.
Cretesia's Dawson 89995...... 15 lbs. 1 oz.
Cretesia's Violet A. 90061..... 14 lbs. 9 oz.
Cretesia's Tamy 89994........ 14 lbs. 8 oz.

Critic 540.
Gilda 2779.................... 14 lbs. 6 oz.

Crocker 19563.
Girletta 61472................. 17 lbs. 2 oz.

Crocus' John Bull 21665.
Sallie of the Ledges 37182..... 16 lbs. 6 oz.

Croton Maid's Duke 6658.
Signal Maid 19361............. 19 lbs. 13 oz.
Lette Signal 26823............ 17 lbs. 2½ oz.
Lady Cloud 2d 26825......... 16 lbs. 12 oz.

Croton's Signal 17614.
Sirona Croton 86597.......... 18 lbs. 10 oz.
Princess Haidee Signal 98652. 16 lbs. 4 oz.
Fadette Signal's Queen 109231. 16 lbs. 3 oz.
Fricka 86450................. 15 lbs. 2½ oz.
Signal's Bijou 60800.......... 15 lbs. 1½ oz.

Croton's Torment 17616.
Riotress Signal 95430........ 21 lbs. 5½ oz.

Crown Prince Melrose 21790.
Beautiful Blondine 82051...... 16 lbs. 5 oz.

Crown Prince of St. L. 20070.
Maud's Crown Princess 90320. 17 lbs.
Vida's Crown Princess 76010.. 16 lbs. 9 oz.
Lily's Crown Princess 76011.. 15 lbs. 7 oz.
Carola's Crown Princess 90321 14 lbs. 14 oz.
Vida's Crown Princess 76010
(30 days)..................... 67 lbs. 13½ oz.
Maud's Crown Princess 90320
(30 days).................... 66 lbs. 13 oz.
Lily's Crown Princess 76011
(30 days)................... 64 lbs. 12 oz.

Crystal 9324.
Darling Pansy 44830........... 15 lbs. 1 oz.
Margaret of Lagonda 35127... 14 lbs. 4 oz.

Cushnoc Jr. 2358.
Evergreen Maid 11536......... 14 lbs.
Rosaletta 9339................ 14 lbs.

Czar 251.
Adina 1942.................... 14 lbs. 4 oz.

Czar of Lenox 6875.
Islip Lenox 31703 (30 days).... 47.699 lbs.
Islip Lenox 31703 (90 days).... 178.066 lbs.

Czar of New York 4049.
Bertha Black 26275............ 16 lbs. 14 oz.
Czaretta 17358................ 14 lbs. 7 oz.

Dainty Boy 2955.
Hattie Douglass 24960......... 16 lbs. 5 oz.
Alabama 7690................. 15 lbs. 11 oz.

Daisy's Boss 24566.
Pocia of Andalusia 2d 79548... 21 lbs. 7½ oz.

Daisy's Harry 21500.
Denise Landseer of Lawn
84411 .... 18 lbs. 5 oz.
Kate Landseer of Lawn 84415. 16 lbs. 5 oz.
Rose s Fancy 93047............ 14 lbs. 7 oz.

Daisy's Signal 12954.
Fanfare 49927................. 16 lbs.

Dalton 20117.
Belle Smith 65807............. 22 lbs. 8¼ oz.
Vernon Dolly 52284........... 21 lbs.
Fresh 91786................... 17 lbs. 11 oz.
Esclarmonde 59509........... 17 lbs. 4 oz.
Daltonia Pogis 95203.......... 14 lbs. 14 oz.

Damascus 22222.
Regalia 64574................. 24 lbs.
Masher 64950................. 16 lbs. 14½ oz.
Parthia 64925................. 16 lbs. 11 oz.
Salute 67682................. 16 lbs. 8 oz.

Dana 3620.
Lucilla 3d 9786............... 17 lbs. 1 oz.

Dande 1556.
Flora Lee of Tennessee 7694.. 16 lbs. 5 oz.

Dandelion's Augustus 19867.
Tweedlcdee of Nipsic 60427... 14 lbs. 1 oz.

Daniel Deronda 2291.
Sasco Belle 13601............. 14 lbs.

Daniel Webster 403.
Royal Princess 2370.......... 17 lbs. 12 oz.

Darling Scituate 7499.
Ella Rosewood's Daisy 32787.. 15 lbs.

Darling's Prince 9513.
Pansy Darling of Lawn 76637. 19 lbs. 2 oz.

Date 2624.
Phyllis of Hillcrest 9067...... 16 lbs.

Day's Romp 24344.
Dubenna 2d 74920............. 18 lbs. ½ oz.

Deerfoot Boy 1926.
Abbie Z. 3d 14742............ 17 lbs.
Mab of Deerfoot 3d 15345...... 14 lbs. ½ oz.

Deerfoot Boy of Somerset 6944.
Grace G. Parks 29263......... 19 lbs. 3 oz.

De Gruchy 3652.
Ampelis 5th 17548............. 15 lbs.

Delaware Darling 3461.
Darling of Neatham 20086..... 15 lbs. 3 oz.
Susette C. 18602.............. 14 lbs. 8 oz.

Denise's Tormentor 11823.

Denise's Pearl 95757	20 lbs. 11 oz.
Tormentor's Daisy 52496	20 lbs. 8½ oz.
Myrtlise 64006	19 lbs. 9 oz.
Rose of Riverside 46953	19 lbs. 9 oz.
Tormentris 49659	18 lbs. 14½ oz.
Daisette 64004	17 lbs. 1 oz.
Jessaline 2d 38295	16 lbs. 4½ oz.
Lady Emma C. 57041	16 lbs.
Tormentor's Rexea 38906	15 lbs. 14 oz.
Denise's Fancy 92500	15 lbs. 10 oz.
Lou Tormentor of Lawn 84413	15 lbs. 3 oz.
Kittie Green 88379	15 lbs.
Pogis Siglise 92889	14 lbs. 10 oz.
Lady Nolia 87972	14 lbs. 6 oz.
Tormentor's Content 119129	14 lbs. 1 oz.
Myrtlise 64006 (21 days)	54 lbs. 15½ oz.

Devilshoof 866.

Oak Leaf 4769	17 lbs. 10 oz.
Oak Leaf 4769 (31 days)	63 lbs. 4 oz.

Dexter Montague 17294.

Beauty Wynn 45482	14 lbs. 14½ oz.

Dexter of Jefferson 5681.

Clara's Belle 16582	17 lbs. 9½ oz.
Kitty of Jefferson 2d 12313	14 lbs. 3 oz.

Diamond Earl 3116.

Belle of Vermillion 8798	15 lbs. 14 oz.
Lillian Mostar 10364	14 lbs. 3 oz.
Bessie of Clover Leaf 13033	14 lbs. 2 oz.

Diana's Rioter 10481.

Letty Rioter 73475	24 lbs. 2 oz.
Pomona's Violetta Pogis 56246	23 lbs. 2¾ oz.
Letty Coles 2d 48128	21 lbs. 8 oz.
Rioter's Sweet Brier 30582	18 lbs. 9 oz.
St. Lambert's Coquette 41070	16 lbs. 15 oz.
Vaniah of St. Lambert 71374	15 lbs.
Dinah Alexis 69345	14 lbs. 8 oz.

Diana's Stoke Pogis 13686.

Diana Doon 61539	15 lbs. 9 oz.

Diana's Top-Sawyer 21310.

Emblem of Del Valle 102384	15 lbs. 8 oz.

Dick (F. S. 171 J. H. B.)

Buckeye Lass 10355	14 lbs. 4 oz.

Dick (F. S. 223 J. H. B.)

Beauty of Jersey 7850	19 lbs. 2 oz.

Dick 1410.

Molly 3554	15 lbs. 9 oz.

Dickero 3542.

Pansy K. 23889	14 lbs. 11½ oz.

Dickinson's Sea Gull 2d 7373.

Clifty Beauty 30249	15 lbs. 6½ oz.

Dick Swiveller Jr. 276.

Countess of Lakeside 12135	19 lbs. 7 oz.
Jersey 3260	15 lbs. 6 oz.

Dido's Duke 4678.

Thorndale Belle 6th 44885	15 lbs.

Diploma 16219.

Bosnia 87508	20 lbs. 2 oz.
Reif 58398	19 lbs. 3 oz.
Sombre 80796	18 lbs. 14 oz.
Onwa 59628	18 lbs. 13½ oz.
Composite 58774	18 lbs. 10 oz.
Columbiana 74998	18 lbs. 6 oz.
Seraph 72217	18 lbs. 5 oz.
Diploma's Loreda 101528	18 lbs. 1½ oz.
Miss Helen Brice 88340	18 lbs. 1¼ oz.
Diploma's Queen 98151	17 lbs. 8 oz.
Plumage 53897	17 lbs. 5 oz.
Diploma's Princess 104002	17 lbs. 4 oz.
Relay 70353	17 lbs. 2 oz.
Dariole 64106	16 lbs. 9 oz.
Diploma's Marissa 95590	16 lbs. 5 oz.
Personia 90231	16 lbs. 3¼ oz.
Diploma's Pansy 112871	16 lbs. 3 oz.
Marissa 65285	16 lbs. 2 oz.
Midway 50000	16 lbs. 2 oz.
Diploma's Clara 103867	15 lbs. 12½ oz.
Smack 64575	15 lbs. 12 oz.
Diploma's Eureka 103865	15 lbs. 10 oz.
Julippa 69000	15 lbs. 9 oz.
Shrine 59627	15 lbs. 1½ oz.
Alexa 64924	14 lbs. 15 oz.
Itura 61090	14 lbs. 14 oz.
Nice 68027	14 lbs. 12 oz.
Vintage 53335	14 lbs. 12 oz.
Diploma's Coma 54070	14 lbs. 10 oz.
Eurus 60801	14 lbs. 7 oz.
Bermuda R. 67534	14 lbs. 1 oz.
Josephina 64921	14 lbs. 1 oz.
Merry Maiden 64949	14 lbs. 1 oz.
Campania 88475 (21 days)	28.127 lbs.
Elturia 80701 (21 days)	24.137 lbs.
Merry Maiden 64949 (30 days)	66.695 lbs.
Merry Maiden 64949 (90 days)	200.517 lbs.

Diploma 3d 36376.

Upright's Maiden 105616	14 lbs. 8 oz.

Discard H. 5763.

Carria H. 14454	16 lbs. 10 oz.
Gilda Mercedes H. 14451	14 lbs. 8 oz.

Disturbance 9684.

Well-to-do 37724	14 lbs. 8½ oz.

Dixie Prince 18379.

Hildah of Clovernook 51597	19 lbs. 2½ oz.
Princess of Clovernook 51598	16 lbs. 13 oz.
Dixie Princess 55658	16 lbs. 8 oz.
Lora of Clovernook 51599	16 lbs. 2 oz.
Alice of Clovernook 77532	15 lbs. 3 oz.

Dr. Ben Franklin 2d 7265.
Carrie Franklin 25635........ 14 lbs. 1 oz.

Doctor H. 2132.
Lydia Darrach 4903........... 17 lbs. 14 oz.

Dode 3057.
Queen of De Soto 12318....... 14 lbs. 13 oz.

Doesticks 2387.
Rainbow 2d 13962............ 21 lbs. 8 oz.

D. of D.'s Victor 11726.
Natalie Victor 73689.......... 17 lbs. 12 oz.

Doge of St. Lambert 17889.
Jeannette of Pittsford 73185
(21 days).................... 29.973 lbs.

Delphin St. Lambert 16431.
Bonny of Woodford 58718..... 17 lbs. 4 oz.

Dominie 11135.
Preference 26343.............. 15 lbs. 5 oz.

Domino of Darlington 2459.
Hettie of Briarcliff 26621...... 18 lbs. 1 oz.
Leila of Briarcliff 24184....... 17 lbs. 6½ oz.
Como of Briarcliff 35849....... 14 lbs. 6 oz.
Robinette 7114................ 14 lbs. 1 oz.

Dom Pedro 2092.
Dom Pedro's Julian 8631...... 16 lbs.

Don 611.
René Ogden 1568............. 15 lbs.
Belle of Ogden Farm 1570.... 14 lbs.

Don Hugo 19402.
Donna Argyle 61456.......... 15 lbs. 10 oz.

Don of St. Lambert 17456.
St. Lambert's Bijou 78174..... 14 lbs. 15½ oz.

Don Pedro of Binghamton 2974.
Attractive Maid 16925........ 22 lbs. 5 oz.
Maid of Berlin 12746.......... 14 lbs. 8 oz.

Dudley 3999.
Mary's Silver Drop 14235..... 15 lbs. 4½ oz.

Duffee 2083.
Blossom's Niobe 13855........ 16 lbs. 7½ oz.

Duke (P. S. 76 J. H. B.)
(Sweepstakes Duke 1905 A. J. C. C.)
Duke's Pierrot Lady 31023.... 21 lbs. 12 oz.
Fairy Queen of St. Brelade's
7464 .... 19 lbs. 7¼ oz.
Valentine of Trinity 7460..... 19 lbs. 4½ oz.
Bijou Ogston 8210............ 18 lbs. 15 oz.
Queen of Nubbin Ridge 14528. 17 lbs.
Forget-me-not 5809........... 15 lbs. 8 oz.
Florence Billot 7849.......... 14 lbs. 13 oz.

Duke Coomassie 8409.
Proctor's Dolores 38564...... 15 lbs. 10 oz.

Duke F. 6134.
Rosa F. W. 13226............. 29 lbs. 12 oz.
Snowdrop F. W. 16948........ 14 lbs. 8 oz.
Abbie K. 13227............... 14 lbs. 4 oz.

Duke Glen Dale 1819.
Urbana 5597.................. 16 lbs.

Duke Jr. 2465.
Winsome of Ipswich 9213..... 14 lbs. 7 oz.

Duke of Albany 3899.
Lady Alexis 26916............ 16 lbs. 8 oz.

Duke of Argyle 1517.
Duchess of Argyle 4th 7571... 14 lbs. 12 oz.
Louise of Lawnfield 14151.... 14 lbs. 11½ oz.

Duke of Baldwin 7137.
Daisy Staunton 46592........ 16 lbs. 14 oz.

Duke of Belmont Jr. 7794.
Toy 20604.................... 17 lbs. 8 oz.

Duke of Bloomfield 1544.
Princess Mostar 9700......... 17 lbs. 3 oz.
Princess Bowen 9699......... 14 lbs. 12 oz.

Duke of Brandywine 2213.
Lydia Darrach 3d 10662....... 16 lbs. 4 oz.
Lydia Darrach 2d 8056........ 16 lbs.
Lydia Darrach 5th 16577...... 15 lbs.

Duke of Burlington 1639.
Island Chrissie 12007........ 14 lbs. 14 oz.

Duke of Cloverdale 6994.
Duke's Medusa 43511.......... 18 lbs.
Faith of Cloverdale 29277.... 14 lbs. 2 oz.

Duke of Qoonewah 14548.
Maggie May of Tupelo 71280.. 20 lbs. 7 oz.

Duke of Daffodil 1662.
Alice of the Meadows 20748... 14 lbs. 12 oz.
Elmwood Duchess 8716........ 14 lbs.

Duke of Darlington 2460.
Paletta of Darlington 16255... 27 lbs. 8 oz.
Mother Hubbard 10331........ 24 lbs. 1½ oz.
Lady Golddust 2d 19861...... 23 lbs. 4 oz.
Bomba 10330.................. 21 lbs. 11½ oz.
Honey Belle 25824............ 20 lbs. 7½ oz.
Bonnie Grisette 2d 19526...... 16 lbs. 12½ oz.
Matilda 5th 18068............ 16 lbs. 4 oz.
Dolly's Daughter 31392........ 16 lbs.
Leah Darlington 13836........ 15 lbs. 5½ oz.
Nazli 10327.................. 15 lbs. 3½ oz.
Nutley Darling 22412.......... 15 lbs. 3½ os.
Duchess of Darlington 13830.. 14 lbs. 11 oz.
Bomba 10330 (31 days)....... 89 lbs. 14 oz.

Duke of Darlington 2d 6948.
Coomassie Darlton 29519...... 16 lbs. 11 oz.
Darlymora 37771............. 14 lbs. 7 oz.

Duke of Darlington 3d 11096.
Parloa 32357................. 14 lbs. 10 oz.

Duke of Grayholdt 1035.
Duenna's Duchess 5508....... 16 lbs. 10 oz.
Gazania 4513................. 14 lbs. 2 oz.
Morlacchi 2725............... 14 lbs.

Duke of Hamburg 5527.
Bessie's Mabel 40939......... 18 lbs.

Duke of Hill View 22179.
Coomassie Fawkes 122445..... 17 lbs. 8 oz.

Duke of Ingleside 14274.
May of Ingleside 56831....... 21 lbs. 7 oz.
Pride of Ingleside 54545.... 20 lbs. 4½ oz.
Bess of Ingleside 2d 59648... 19 lbs. 10½ oz.
Maud of Ingleside 74044...... 18 lbs. 5½ oz.
Fawn of Ingleside 59649...... 17 lbs. 9 oz.
Gipsy's Riotess 62482......... 17 lbs. 8 oz.
Coomassie of Ingleside 74045. 17 lbs. 3½ oz.
Jewel of Ingleside 54544...... 17 lbs. 1 oz.
Mintresse of Ingleside 103128. 16 lbs. 10½ oz.
Julia Rex of Ingleside 74046.. 16 lbs. 3 oz.
Gracie A. 54524............... 15 lbs. 4 oz.
Ella of Ingleside 76414........ 15 lbs. 3½ oz.
Amrah of Ingleside 56832..... 14 lbs. 2 oz.

Duke of Jersey 198.
Melody 2689................... 14 lbs. 1 oz.

Duke of La Grange 27682.
Ilinka 91418................... 17 lbs. 8 oz.
Bessie Wolcott 91417........ 15 lbs. 1 oz.

Duke of Lebanon 1880.
Manoa 6340................... 16 lbs. 12 oz.
Blossie Reynolds 6082........ 16 lbs. 3½ oz.
Home Matron 6707........... 14 lbs.

Duke of Lincoln 15475.
Chloe Tormentor of Lawn
71642 .... 14 lbs. 13½ oz.

Duke of Lorne 6510.
Mellie Argyle 20609........... 14 lbs. 2 oz.

Duke of Magnolia 2826.
Ette 10315.................... 15 lbs. 7 oz.

Duke of Mansfield 2277.
Miss Lofty 9718............... 16 lbs. 10 oz.

Duke of Mansfield's Pierrot 6261.
Insie of Riverside 23825....... 30 lbs. 5¾ oz.
Eva of Snipsic 17650......... 14 lbs. 1 oz.

Duke of Maplehurst 2290.
Marchioness of Maplehurst
8018 .... 14 lbs. 6 oz.
Litty 8017.................... 14 lbs.

Duke of Melrose 5185.
Melrose's Perfection 28895.... 24 lbs. 12 oz.

Duke of Milford 3820.
Benetia 21511................ 16 lbs. 5 1-3 oz.

Duke of Mona 10639.
Mona's Beauty 50041......... 15 lbs. 4 oz.

Duke of Normandy 3446.
Belle of St. John's 2d 29829 .. 15 lbs. ½ oz.

Duke of Oakland 1984.
Enid 2d 10782................. 14 lbs. 7½ oz.

Duke of Ouaquaga 2740.
Nan Day 17192................ 20 lbs. 4 oz.

Duke of Portage 1270.
Ida of Coal Hill 12542........ 15 lbs. ½ oz.
Celia Belle 5865.............. 14 lbs. 3 oz.

Duke of Ramsey's 6947.
Rosa Ramsey 25875.......... 15 lbs. 1 oz.

Duke of St. Albans 11234.
St. Albans' Valentine 41707... 22 lbs. 6 oz.
Esther Thorne 35545.......... 14 lbs. 7½ oz.

Duke of St. Helier 7527.
Meines St. Helier 29999....... 14 lbs. 1 oz.

Duke of St. Lambert 16160.
St. L.'s Silken Les Cateaux
61964 .... 21 lbs. 12 oz.
Duke's Minnie 42785.......... 20 lbs. 4 oz.
Katie of St. Lambert 54776... 19 lbs. 7 oz.
St. Lambert's Kate 61975..... 16 lbs. 11 oz.
Minnie R. of St. Lambert
61965 .... 16 lbs. 9 oz.
Princess Helene of St. L.
61972 .... 16 lbs. 9 oz.
Duke of St. L.'s Vail's Pet
61970 .... 14 lbs. 3 oz.
Princess Helene of St. L. 61972
(30 days)................... 66 lbs. 6 oz.
Katherine of Pittsford 73169
(30 days) .... 54.107 lbs.

Duke of Sassafras 2431.
Gem of Sassafras 8434........ 14 lbs. 3½ oz.

Duke of Scituate 3623.
Minnie of Scituate 17829...... 14 lbs. 4½ oz.

Duke of Springvale 4290.
Maggie of Springvale 15931.... 18 lbs. 5 oz.

Duke of Thornebrook 3832.
Hindoo Rose 14602............ 14 lbs. 7 oz.
Pansita 4th 18109............. 14 lbs. 6 oz.
Carrie's Wonder 15233........ 14 lbs. 2½ oz.

Duke of Wellington 35.
Lassie 1134.................... 15 lbs. 1½ oz.

Duke of Wellington 608.
Bounty 1606................... 14 lbs.

Duke of Woodlawn 4160.
Hypathia 2d 14774............. 19 lbs. 13½ oz.

Duke Thor 17323.
Grace of Glynllyn 67759....... 14 lbs. 10 oz.

Duke Thorne 7020.
Lenore H. 2d 29767............ 16 lbs. 3 oz.
Joy of Argyle 46495........... 15 lbs. 9 oz.
Charity of Argyle 35036....... 14 lbs. 10 oz.

Dunraven 7950.
Donna Signal 29407............ 16 lbs. 1 oz.

Durwood 4527.
Durwood's Lass 19710......... 16 lbs.

Earl of Eastmore 18595.
Almond Blossom 59709........ 17 lbs. 3¾ oz.

Earl of Russellville 11935.
Kate Ritchey 54283............ 14 lbs. 7¼ oz.

Earl of Willow Glen 2043.
Mary M. Allison 6308......... 20 lbs. 14 oz.

Easter Boy 3032.
Golden Plover 22388........... 20 lbs. 1¾ oz.
Kisberine Rex 31436.......... 16 lbs. 8 oz.

Easter Boy 2d 5310.
Lady Panalphrex 17400........ 23 lbs. 9 oz.

Eastern Chief (P. S. 171 J. H. B.)
Eastern Sultane 20298........ 15 lbs. 3 1-3 oz.

Eastwood's Pogis 15279.
Easter Pogis 52371............ 15 lbs. 3 oz.

Eclipse 1449.
Nordheim Creamer 9758....... 14 lbs.

Eddystone 23413.
Luna Dean 3d 84309.......... 15 lbs.
Alberta d'Or 76452............ 14 lbs. 14 oz.

Edward Earle 10462.
Beauty's Crescent 26733....... 15 lbs. 15⅔ oz.

Elkornah 4401.
Walter Girl 12776............. 16 lbs. 7 oz.

Ellenetta's Chief 34896.
Chief's Stevia 122187.......... 16 lbs. 10½ oz.
Chief's Iona 111704........... 15 lbs. 12 oz.

Ellwood 13382.
Silver Delle 40691............. 18 lbs. 14½ oz.
Silver Delle 40691 (29 days)... 55 lbs.

Elm Place Rayon d'Or 17518.
Rayon d'Or's Chroma 114053.. 14 lbs. 15 oz.

Elm Place Seneca 21939.
Seneca's Frankie 114052....... 15 lbs. 7½ oz.

Elmwood 2322.
English Elm 17600............ 14 lbs. 7 oz.

Elmwood Carrie's Pogis 19494.
Becky Z. 71311................ 14 lbs. 2 oz.

Elmwood Nona's Pogis 24986.
Dorothy Taylor 99485......... 15 lbs. 10 oz.

Elmwood Stoke Pogis 26552.
Elmwood Melissa 4th 82578.... 14 lbs. 5 oz.

Elsmore 4384.
Sunset of Pleasant View 13071 15 lbs. 2 oz.

Ely Fy 18244.
Ada Fy 45479................. 17 lbs. 4 oz.

Eminent (P. S. 1842 J. H. B.)
Effie Baal 134222.............. 14 lbs. 10¼ oz.

Emperor 287.
Actress 2311.................. 14 lbs.

Eric J. 33246.
Josephine O. 103635.......... 19 lbs. 12 oz.

Erie Chief 13438.
Daisy Harrison 69253......... 25 lbs. 1 oz.

Ethleel 2d's Jubilee 18249.
Sigma 78676................... 18 lbs. 2¼ oz.
Prospect's Edwina 2d 86909... 18 lbs. 1 oz.
Jubilethleel 99697............. 17 lbs. 12 oz.
Elorita 101742................. 16 lbs. 4 oz.
Ethleel Khedive 99698......... 15 lbs. 8 oz.
Ethleel D. 83006.............. 14 lbs. 1 3-5oz.

Ethleel 2d's Son 25885.
Ethleel's Pet 85428........... 17 lbs. 6 oz.

Ethleel's Harry 15962.

Miss Perrot 73801 .... 16 lbs.
Harry's Ethel 63970 .... 14 lbs.

Ethleel's Landseer 22341.

Mayhegtal 2d 101298 .... 17 lbs. 14 oz.
Ethleel's Nerviona 101300 .... 17 lbs. 12 oz.
Ogden's Alice 77590 .... 17 lbs. 4 oz.
Signalda's Ethleel 77599 .... 16 lbs. 11 oz.
Chuffy's Ethel 102195 .... 16 lbs. 1 oz.
Rosora 87622 .... 15 lbs. 8½ oz.

Ethlo's Prince 24119.

Guarder 90825 .... 21 lbs. 4 oz.

Ethlo Tormentor 19189.

Ethlo's Jean 71954 .... 24 lbs. 6 oz.
Ethlo's Belle 69053 .... 15 lbs. 2 oz.

Ethol 19104.

Belle Williamson 3d 104290 .... 15 lbs. 3½ oz.
Tormentor's Oona Toltec 105860 .... 14 lbs. 12 oz.

Ettamarius 17762.

Ettagem 80258 .... 16 lbs. 12 oz.
Kamaretta 80479 .... 16 lbs. 7 oz.
Ettarock 81103 .... 16 lbs.
Etta Bartlett 80254 .... 15 lbs. 13½ oz.

Etzel 3593.

Bianca Lass 14997 .... 14 lbs. 3½ oz.

Euclid 520.

Ma Belle 4942 .... 15 lbs.

Eupidee 4097.

Dove Dee 19059 .... 27 lbs. 9 oz.
Beauty Dee 19065 .... 23 lbs. 3 oz.
May Dee 18058 .... 15 lbs. 10 oz.
Eupidee's Perfection 20175 .... 15 lbs. 4 oz.

Euroclydon 4789.

Lizzette's Mary 12723 .... 14 lbs. 11 oz.

Europa's Duke 4832.

Cream Calla 40233 .... 16 lbs. 7 oz.

Eurotas' Victor Hugo 15689.

Nutshell 2d 56062 .... 19 lbs. 2 oz.
Belle of Passaic 85846 .... 17 lbs. 11 oz.
Miss Albert 63138 .... 17 lbs. 9 oz.
Robinelia 63132 .... 15 lbs. 14 oz.
Nettie's Picture 56060 .... 15 lbs. 8 oz.
Eurotas' Blossom 63134 .... 15 lbs. 5 oz.
Leila Victor 85433 .... 14 lbs. 1 oz.

Euroto Pogis Le Brocq 19503.

Kaembe's Maud Eurota 91066 .... 14 lbs. 2½ oz.

Evangelist 25700.

Alnora 112132 .... 16 lbs. 6 oz.

Excelsior of Jersey 949.

Lady Oxford 4860 (10 days) .... 22 lbs. 2 oz.

Exile of Glen Rouge 37213.

Ethel of Glen Rouge 131019 .... 18 lbs. 9 oz.
May Marjoram 123811 .... 16 lbs. 5 oz.

Exile of Jersey Lawn 29421.

Dora's Fawnie Exile 108696 .... 15 lbs. 12 oz.

Exile of St. Lambert 13657.

Exile's Belle 40524 .... 32 lbs. 6 oz.
Rachel Spencer 50974 .... 23 lbs. 3¼ oz.
Paola Stoke Pogis 34691 .... 23 lbs.
Marilla of St. Lambert 31809 .... 21 lbs. 13½ oz.
Yum Yum of St. Lambert 39152 .... 21 lbs. 4 oz.
Exile's Dolly 67452 .... 19 lbs. 7½ oz.
Exile's Bessie 49985 .... 18 lbs. 12½ oz.
Lena Stoke Pogis 39736 .... 18 lbs. 8 oz.
Exile's Dewdrop 106104 .... 18 lbs. 4 oz.
Exile's Penelope 77182 .... 18 lbs. 3¾ oz.
Exile's Myrtle 2d 94446 .... 17 lbs. 13½ oz.
Helen Stoke Pogis 31947 .... 17 lbs. 8 oz.
Exile's Success 49986 .... 17 lbs. 6½ oz.
Jessie Stoke Pogis 37291 .... 17 lbs. 2½ oz.
Exile's Gretchen 79245 .... 16 lbs. 15½ oz.
Exile's Pauline 37530 .... 16 lbs. 15 oz.
Exile's Anna 46881 .... 16 lbs. 12 oz.
Exile's Flossie 116908 .... 16 lbs. 12 oz.
Pearl Pogis 38904 .... 16 lbs. 10¾ oz.
Isadore Stoke Pogis 37297 .... 16 lbs. 9½ oz.
Nutley St. Lambert 58469 .... 16 lbs. 8 oz.
Exile's Harriet 100716 .... 16 lbs. 6¼ oz.
Exile's Daisy 46885 .... 16 lbs. 4 oz.
Exile's Kosi 107306 .... 16 lbs. 2½ oz.
Exile's Lulu 49984 .... 16 lbs. 2 oz.
Nora Stoke Pogis 34657 .... 16 lbs. 1 oz.
Exile's Lady Star 65943 .... 16 lbs. ¼ oz.
Yuba Stoke Pogis 37294 .... 15 lbs. 14¾ oz.
Exile's Alphea 45393 .... 15 lbs. 14½ oz.
Exile's Hazel 85181 .... 15 lbs. 13 oz.
Exile's Nina 40622 .... 15 lbs. 11½ oz.
Exile's Lady Angela 46882 .... 15 lbs. 11 oz.
Jennie Stoke Pogis 32010 .... 15 lbs. 10¾ oz.
Exile's Fawn 40979 .... 15 lbs. 10 oz.
Exile's Echo 111276 .... 15 lbs. 9¾ oz.
Otta of St. Lambert 49603 .... 15 lbs. 8¾ oz.
Zenobia Stoke Pogis 37292 .... 15 lbs. 8 oz.
Exile's Lucy 46883 .... 15 lbs. 7¼ oz.
Exile's May 52765 .... 15 lbs. 4¼ oz.
Exile's Butterball 65152 .... 15 lbs. 4 oz.
Exile's Moss Rose 101155 .... 15 lbs. 4 oz.
Mollie of St. Lambert 34644 .... 15 lbs. 4 oz.
Exile's Effie 40626 .... 15 lbs. 2 oz.
Exile's Onnalinda 65189 .... 15 lbs. 1½ oz.
Bertha Stoke Pogis 35536 .... 14 lbs. 15½ oz.
Exile's Lady Palestine 107626 .... 14 lbs. 14 oz.
Exile's Dumbella 79452 .... 14 lbs. 11 oz.
Orphi of St. Lambert 49004 .... 14 lbs. 10½ oz.
Exile's Aggie 52764 .... 14 lbs. 10¼ oz.
Exile's Dell 67733 .... 14 lbs. 10¼ oz.
Exile's Lizzie 61477 .... 14 lbs. 8½ oz.
Exile's Antoinette 111279 .... 14 lbs. 7½ oz.
Exile's Elf 111811 .... 14 lbs. 7¼ oz.
Exile's Nancy 92030 .... 14 lbs. 7 oz.
Exile's Lucile 65151 .... 14 lbs. 6 oz.

Exile's Claribel 107507........ 14 lbs. 5 oz.
Exile's Lady Alexis 79112..... 14 lbs. 4¾ oz.
Exile's Helen 61251........... 14 lbs. 4½ oz.
Exile's Arcadia 66201.......... 14 lbs. 4¼ oz.
Exile's Ona 106212............ 14 lbs. 3¾ oz.
Bronzie of St. Lambert 49001.. 14 lbs. 2 oz.
Exile's Jessamine 111104...... 14 lbs. 1¾ oz.
Daisy-Seeker 43099............ 14 lbs. 1 oz.
Exile's Rosa 66552............ 14 lbs. ½ oz.
Exile's Harriet 100716 (14 days) 31 lbs. 7½ oz.
Exile's Nina 40522 (14 days)... 31 lbs. 6½ oz.
Exile's Lady Star 63943 (16
    days) .... 35 lbs. 7½ oz.
Exile's Belle 40524 (30 days).. 122 lbs. 6½ oz.
Exile's Lulu 49984 (30 days).. 54.017 lbs.
Paola Stoke Pogis 34691 (35
    days) .... 109 lbs. 10 oz.
Rachel Spencer 50974 (35 days) 105 lbs. 10½ oz.
Exile's Lulu 49984 (90 days).. 168.538 lbs.

Exile of St. Lambert 2d 17131.
Sapolio 61082................. 14 lbs. 6 oz.

Exile of St. Lambert 4th 17196.
Exile's Pansy 79800........... 17 lbs. 13 oz.
Exile's Agnes 79796........... 14 lbs. 8 oz.

Exile of St. Lambert 16th 19155.
Lois of Pittsford Farms 85848 14 lbs. 9 oz.

Exile of St. Lambert 23d 20712.
Jocal 95182................... 20 lbs. 14 oz.
Runaway 87030................. 16 lbs. 10 oz.
May Day 60451................. 16 lbs. 3½ oz.

Exile of Scituate 19235.
Bertha Scituate 75977........ 14 lbs. 7 oz.

Exile's Duke 19795.
Exile's Beauty 61220.......... 16 lbs. 3 oz.

Exile's Rebel 28928.
Hattie Swett 109709........... 15 lbs. 10¼ oz.
Delphina 105895............... 14 lbs. 2¾ oz.

Exile's St. John 20202.
Exile's Lady Sarah 107938.... 17 lbs. 10 oz.

Fadette's Son 16301.
Etta Lyne 83298............... 20 lbs. 5 oz.

Fair Corinth 12455.
Natasqua 65598 (90 days)...... 161.522 lbs.

Fanchon's King 2637.
Countess Potoka 7496........ 18 lbs. 15 oz.

Fancy Ethleel's Pogis 30025.
Fancy's Nienchen 104377...... 16 lbs. 2 oz.
Fancy's Maybee 104383........ 15 lbs. 14 oz.

Fancy Farmer 12181.
Flora Rex Carlo 2d 102140.... 18 lbs. 2 oz.

Fancy's Harry 9777.
Harry's Pet 52838.............. 22 lbs. 11 oz.
Romping Lass 2d 51989........ 21 lbs. 12 oz.
Harry's Fancita 96721......... 20 lbs. 2½ oz.
Sitka M. 78680................ 19 lbs. 13 oz.
Ruth Morgan 3d 57043......... 19 lbs. 6.4 oz.
Fancy Wax 37159.............. 19 lbs. 3½ oz.
Pitapat 2d 86907.............. 19 lbs. 3¼ oz.
Signet's Fancy 58446......... 18 lbs. 11½ oz.
Sorina 42993................. 17 lbs. 15½ oz.
Signal Mine 2d 60789......... 17 lbs. 9½ oz.
Romp Ogden 5th 43181......... 17 lbs. 9 oz.
Fancy Nenita 62702........... 17 lbs. 6½ oz.
English Elm 3d 62906......... 17 lbs. 2 oz.
Rioter's Denise 3d 70238..... 17 lbs. 1½ oz.
Clarintha 96919.............. 17 lbs. ¼ oz.
Harmintha 96920.............. 16 lbs. 7 oz.
Harry's Duchess 60289........ 16 lbs. 4 oz.
Harry's May Fair 37499....... 16 lbs. 1 oz.
Leoni Landseer 65317......... 16 lbs.
Fancy Bee 37496.............. 15 lbs. 8 oz.
Harry's Ernestine 66232...... 15 lbs. 6½ oz.
Torment 4th 70660............ 15 lbs. 1¼ oz.
Ethleel 4th 54275............ 15 lbs.
Ethleel 6th 89802............ 14 lbs. 15 oz.
Fancy's Harry's Sadie 77202.. 14 lbs. 14 oz.
Maquilla of Hood Farm 113067 14 lbs. 13¾ oz.
Temalema 44761............... 14 lbs. 10 oz.
Signal Fancy 39812........... 14 lbs. 8⅓ oz.
Harry's Signaline 82419...... 14 lbs. 8½ oz.
Harry's Fancy Belle 44764.... 14 lbs. 8 oz.
Lena's Florence 42460........ 14 lbs. 8 oz.
Harry's Fancy 39114.......... 14 lbs. 6½ oz.
Nannie of Hood Farm 108565.. 14 lbs. 5¼ oz.
Fancy Phlox 49652............ 14 lbs. 3 oz.
Myrrha 3d 50406.............. 14 lbs. ½ oz.

Fancy's Harry 2d 20199.
Signalbianca 2d 119779........ 17 lbs. ¼ oz.
Fanfare 2d 119780............ 15 lbs. 12½ oz.

Fancy's Harry 6th 24768.
Harry's Nervilette 86262....... 18 lbs. 1 oz.
Fancy Même 79748............. 16 lbs. 12 oz.
Orie's Fancy 72039........... 15 lbs. 10½ oz.
Harry's Florrie 86263........ 14 lbs. 12 oz.
Nerviona 2d 98687............ 14 lbs. 3 oz.

Fancy's Jubilee 25205.
Bonnie Signaldina 108681...... 18 lbs. 4 oz.
Adelaide Saufley 96284....... 16 lbs. 5 oz.
Eveline Rose 80512........... 14 lbs. 6 oz.

Fancy's Landseer 12996.
Landseer's Content 56985...... 16 lbs. 8 oz.
Fancy's Pet 36013............. 16 lbs. 4 oz.
Landseer's Rosamond 56142... 15 lbs. 6 oz.
Fancy's Gem 41797............ 14 lbs. 15 oz.
Fancy Torment 56273........ 14 lbs. 8½ oz.
Fancy Millcent 71059......... 14 lbs. 3½ oz.

Fancy's Pogis 21797.
Muzio 97423.................. 15 lbs. 2 oz.

Fancy Toltec 21167.
Pet's Fancy 80968............. 17 lbs. 4 oz.
Dempsey's Beauty 62808...... 14 lbs. 6 oz.

Fannie's Son 28014.
Nora Clinton 91938............ 18 lbs. 4 oz.

Farmer Rex 8898.
Rexella 69413................... 14 lbs. 4 oz.

Farmer's Glory 5196.
Beulah De Gruchy 13480...... 22 lbs. 2 oz.
Maritana 12039................. 16 lbs. 3½ oz.
Geneva 13220................... 15 lbs. 11 oz.
Marie S. 12043................. 15 lbs. 6 oz.
Farmer's Pride 12284......... 15 lbs. 4 oz.
Hilda 2d 14967................. 14 lbs. 12½ oz.

Fast Boy 2606.
Nigella 7895................... 16 lbs. 3 oz.
Nibbette 11625.................. 14 lbs. 7 oz.
Hennette 11624................. 14 lbs. 3½ oz.

Faust 503.
Miss Bianca 12517............. 17 lbs. 14 oz.
Miss Bowris 12206............. 17 lbs. 8½ oz.

Favori (P. S. 185 J. H. B.)
Litta Oaks 19734.............. 15 lbs. 5½ oz.

Favorite's Prize 33042.
Vono 89138.................... 14 lbs. 3 oz.

Fayette Boy 4737.
Lady Cloud 19358.............. 27 lbs. 6 oz.

Fearnaught 1854.
Aldarine 2d 14196............. 14 lbs.

Fenwick 26451.
Kitty Sands 77432............. 18 lbs. 1½ oz.

Festive Florin 19967.
Pera B. 86761.................. 17 lbs. 9 oz.
Picotu's Bess 86172........... 17 lbs. 8½ oz.

Fill Pail's Duke 24650.
Fill Pail's Nancy 95435....... 18 lbs. 3 oz.
Sigma 2d 89803................ 14 lbs. 15½ oz.

Finis Lawrence 4107.
Design 18481.................. 17 lbs. 2½ oz.
Princess White Water 21137.. 14 lbs. 15 oz.

Fitz 1988.
Bet Arlington 8970............ 18 lbs. 11 oz.

Flash 2532.
Flashy Jessie 9722............ 17 lbs. 15 oz.
Rarity 2d 7724................. 14 lbs. 2 oz.

Flora Lee's Victor Hugo 17675.
Iuka of Riverview 2d 61271.... 17 lbs. 12 oz.

Florian (P. S. 373 J. H. B.)
Pretty Flirt 28791............. 15 lbs.

Florinde's Duke 4368.
Celestia 2d 29482.............. 16 lbs. 13 oz.

Footstep 5163.
Hilda A. 3d 16636............. 17 lbs. 1 oz.
Eltekeh 28266................. 16 lbs. 4 oz.
Edessa 21844.................. 16 lbs. 2 oz.
Hatita 34538.................. 14 lbs. 15 oz.

Forget-me-not 6291.
Well Done 25987.............. 19 lbs. 4 oz.
Lady Phillis 18240............ 18 lbs. 15 oz.
Sibyl's Beauty 25941.......... 18 lbs.
Sibyl's Fancy 25942........... 17 lbs.
Fringe 16875.................. 15 lbs.

Fortunatus 1152.
Rosebud of Allerton 6352..... 19 lbs. 12 oz.
Violet of Glencairn 10221...... 14 lbs. 4 oz.

Founder 20928.
Sadie Le Pet 77100........... 22 lbs. 5 oz.
Le Roy's Pet 94106........... 20 lbs. 7 oz.
Founder's Pet 94107.......... 17 lbs. 8 oz.

Frank 309.
Susanna Frank 18261......... 14 lbs.

Frankfort 2990.
Utilla 11215................... 16 lbs. 2 oz.

Frank Warren 1490.
Nightingale K. 2d 19841....... 16 lbs. 8 oz.

Fritz of Glynllyn 36579.
Phrynore of Allentown 3d
126321 .... 14 lbs. 8 oz.

Fritz Pogis of Glynllyn 28042.
Pretty Prattler 99080.......... 21 lbs. 8 oz.
Silver Belle C. 90600.......... 16 lbs. 2½ oz.
Sunswick Dolly 91844......... 15 lbs. 12 oz.
Hermione C. 90598............ 14 lbs. 11 oz.

Gambetta M. 10012.
Gambetta's Topsy 42376....... 16 lbs. 8½ oz.
Queen Dodona 42375.......... 14 lbs. 2 oz.

Garfield's Black Prince 24548.
Meridale Phinora 119398....... 14 lbs. 9 oz.

Garfield Stoke Pogis 15963.
Garella 62541.................. 23 lbs.
Garfield's Princess 62544...... 22 lbs.
Pussy Stoke Pogis 62547...... 20 lbs. 1 oz.

Princess Honoria 62548........ 16 lbs. 8 oz.
Countess Stoke Pogis 62540... 16 lbs. 3 oz.
Lilley Rex Pogis 62545....... 14 lbs. 9 oz.
La Violette 2d 52604......... 14 lbs. 7 oz.
Lady Jane of St. Peter's 2d
  62536 .... 14 lbs. 5 oz.
Eufielda 62546............... 14 lbs. 3 oz.
Lily Garfield 79819 (21 days).. 37.488 lbs.
Princess Honoria 62548 (90
  days) .... 159.447 lbs.

Garibaldi (P. S. 242 J. H. B.)
Lady Antoinette 24391........ 21 lbs. 6 oz.

Garibaldi H. 7106.
Lady Livingston 33374........ 15 lbs. 2 oz.

Gazella's Signal 15600.
Gazellathirdella 89407........ 16 lbs. 6 oz.

Gen. James A. Garfield 14874.
Kitty Livingston 34303....... 15 lbs. 2 oz.

General Lyman 6631.
Belle Lyman 28522............ 16 lbs. 3½ oz.

Gen. Marion 2298.
Etta Allman 12185............ 15 lbs. 5 oz.

Gen. Rosser 4189.
Countess Queen 13519........ 18 lbs. 3 oz.
Plum 13228................... 17 lbs. 4 oz.

Gen. U. S. Grant 14822.
Minnie C. 2d 48414............ 16 lbs. 14 oz.

George of St. Lambert 17200.
St. Lambert Hebe 54069....... 15 lbs. 5 oz.

Georgia 10606.
Daisy of Dryden 77778......... 16 lbs. 4 oz.
Daisy Clementaise 77341...... 14 lbs. 5 oz.

Georgian 6073.
Signal's Lily Flagg 31035...... 27 lbs. 3½ oz.

Gerald 895.
Daisy of Clermont 3492....... 14 lbs.

Geranium's Chief 18607.
Chief's Queen 69967.......... 20 lbs. 3 oz.
Jeannette Geranium 61023..... 14 lbs. 4 oz.

Geranium's Leo 29482.
Signal's New Geranium 134117 21 lbs. 15½ oz.

Geranium's Rajah 18605.
Grace's Pride 61018........... 15 lbs. 7 oz.

G. F. Train 18868.
Etta of Bay View 66028........ 22 lbs. 13 oz.
Rosella of Bay View 66032.... 18 lbs. 10 oz.
Bloom of Amherst Villa 70507 16 lbs. 10 oz.
Lucie of Bay View 66027....... 14 lbs. 8 oz.
Rosella of Bay View 66032 (30
  days) .... 66 lbs. 2 oz.

Gil Blas 1193.
Leonice 2d 8342............... 16 lbs. 8 oz.

Gilderoy 2107.
Queen Mary of Woodlawn
  11659 .... 22 lbs. 5 oz.
Chrome Skin 7881............ 20 lbs. 10 oz.
Gilderoy's Idex 42027.......... 19 lbs.
Lemon Twig 42028............. 17 lbs. 4 oz.
Lactine 10680................. 17 lbs. 1½ oz.
Gildercream 39480............ 17 lbs.
Topaz of Woodlawn 11661..... 16 lbs. 4 oz.
Lady Alice of the Wilderness
  12207 .... 15 lbs. 14 oz.
Sweet Sixteen 10682.......... 14 lbs. 15 oz.
Gold Mark 10727.............. 14 lbs. 14 oz.
Yellow Locust 10679.......... 14 lbs. 10½ oz.
Gilded Queen 42029.......... 14 lbs. 9½ oz.
Clover Mel 16159............. 14 lbs. 9 oz.
Gilderoy's Enid 3.924........ 14 lbs. 4 oz.
Mary of Gilderoy 11219...... 14 lbs. 4 oz.
Princess Mary of Woodlawn
  11663 .... 14 lbs. 4 oz.
Eugenie 2d 12733............. 14 lbs. 2 oz.

Gilderoy 5th 7426.
Minnie Parker 38718.......... 26 lbs. 2½ oz.
Miss Katie W. 47366......... 15 lbs. 11¾ oz.

Gillo, on I. of J.
Jessaline 26099................ 16 lbs. 6 oz.

Gilroy 1653.
Euphonia 6783................ 16 lbs. ½ oz.

Ginx 1536.
Walkyrie 5708................. 14 lbs. 1 oz.

Gipsy's Lorne Pogis 20346.
Bess Pogis of Prospect 87045.. 29 lbs. 1¾ oz.
Xyst 101247................... 23 lbs. 6½ oz.
Gipsy Bess of Prospect 96923.. 22 lbs. 10½ oz.

Glorene's King 22813.
Lolly Darling 78005............ 16 lbs. 6 oz.

Glue 3960.
Loreda 19801.................. 17 lbs. 13 oz.
New Maud Lee 16614.......... 14 lbs. 10 oz.

Glynllyn Boy 22396.
Girietta 4th 109264............ 15 lbs. 8½ oz.

Gold Base Boy 27116.
Gold Maid 98787............... 16 lbs. 14¾ oz.

Gold Basis 4038.

Mamelle 20804	21 lbs. 8¼ oz.
Jazella G. 14191	20 lbs. 7 oz.
Lady Julia G. 16199	18 lbs. 8 oz.
Tette 20802	17 lbs. 6 oz.
Countess Gilderine 29027	16 lbs. 10½ oz.
Thekla of Clover Nook 33445	15 lbs. 9 oz.
Gilderine of Linwood 24488	15 lbs. 7 oz.
Miss Remarkable 33446	15 lbs. 2½ oz.
Nonsuch of Linwood 29028	14 lbs. 14½ oz.
Leonie of Linwood 25319	14 lbs. 6 oz.

Gold Basis 5th 15725.

Gladdis Roy 2d 55652	18 lbs. 12 oz.
Catono's Kate 59976	16 lbs. ½ oz.
June Sweets 59978	14 lbs. 8½ oz.

Gold Bugler 20036.

Gold Baby 105745	17 lbs. 12 oz.

Gold Coast Jr 9217.

Golden Dewdrop 96551	23 lbs. 13¾ oz.
Golden Susette 96552	17 lbs. 6 oz.
Dorchester Maid 101385	16 lbs. 5 oz.

Golden Art 10998.

Postscript 39366	14 lbs. 1½ oz.

Golden Ear 1025.

Golden Zoe 3975	16 lbs. 3 oz.

Golden Lion 5239.

Percie 14937	18 lbs. 10 oz.
Berylla 2d 15117	15 lbs. 13 oz.

Golden Ray 10669.

Riotaletta 2d 34495	15 lbs. 15½ oz.
Copper Queen 53659	14 lbs. 12 oz.
May Naomi 92277	14 lbs. 10½ oz.

Golden Rule 4785.

Golden Violet 30136	15 lbs.

Gold Finder 2225.

Golden Trudie 34535	14 lbs. 9 oz.

Gold Mine 7272.

Etta M. 2d 30820	14 lbs. 14 oz.

Gold of St. Lambert 16744.

Bride of Evergreen 72022	20 lbs.
Fannie's Fairy 67304	19 lbs. 8 oz.
Madora Gold 67302	17 lbs. 1 oz.
Carrie Gold 56348	16 lbs. 8 oz.
Gold Badger 67301	16 lbs. 7 oz.
Badger Girl 4th 87488	15 lbs. 13 oz.
Gold Gem 72024	15 lbs. 12 oz.
Iola F. 85529 (21 days)	25.251 lbs.

Gold Plate 10733.

Lampedo 46204	21 lbs.

Gold Ridge 18951.

Express 69574	18 lbs. 5 oz.

Gold Rope 6189.

Miss Clifford 27962	17 lbs. 15½ oz.

Grace's Duke of Darlington 8289.

Delight's Darling 45700	15 lbs. 8 oz.

Grand Duke Alexis 1040.

Chrissy 2d 7720	16 lbs. 14 oz.
Polynia 10753	16 lbs. 7 oz.
Countess of Croton 5307	15 lbs. 12 oz.
Roselaine 7167	15 lbs. 1 oz.
Kitty Clover 2d 16099	14 lbs. 15 oz.
Chrissy 3d 15615	14 lbs. 11 oz.
Hartwick Belle 7722	14 lbs. 8 oz.

Granger 1591.

Brenhilda 7649	16 lbs. 8 oz.

Granite Rock 4391.

Pet Clover 14624	16 lbs. 8 oz.

Gray Beauty's Rioter 27131.

Lucile of Pearl Hill 91151	14 lbs. 4 oz.

Graycoat 1105.

Corinna 2d 6594	16 lbs. 7 oz.
Lady Louise 4339	15 lbs.

Graylock 740.

Daisy of Belhurst 3114	16 lbs. 8 oz.

Great Ado 4227.

Perty W. 41721	14 lbs. 2 oz.

Great Combination 20326.

Editha Westray 89755	19 lbs. 5 oz.

Greatorex 4218.

Minnie Rioter 21336	15 lbs. 10 oz.

Grey (F. S. 244 J. H. B.)

Handsome Myra 14244	21 lbs.
Lady Superior 22865	16 lbs. 5 oz.

Grey Friar 7627.

Grey Friar's Bess 25128	22 lbs.
Friar's Lady 36119	14 lbs. 2¼ oz.

Grey King (P. S. 169 J. H. B.)

Camelia 2d 11188	20 lbs. 3 oz.
Favorite of Avon 13438	17 lbs. 7¾ oz.
La Financière 11970	15 lbs. 5¼ oz.
Jersey Lily 14044	15 lbs.
Lady Jane of St. Peter's 7475	15 lbs.

Grey of the West (F. S. 317 J. H. B.)

Grey Jessamine 17444	14 lbs. 4 oz.

Grey Prince (F. S. 168 J. H. B.)
Eveline of Jersey 6781........ 18 lbs. 6 oz.

Greysteel 1294.
Zoe Henry 6693............... 15 lbs. 14¾ oz.

Gridley's Duke 24997.
Sibby 87742.................... 15 lbs. 4½ oz.

Grisette's Koffee 26008.
Grace Harribee 91836......... 20 lbs. 2 oz.
Lady Emily Kilduff 101053.... 14 lbs. 4 oz.

Guy Fawkes (F. S. 251 J. H. B.)
Island Star 11876.............. 21 lbs. 3 oz.
Thaley 14299.................. 16 lbs. 13 oz.
Queen of Ashantee 14554...... 15 lbs. 2 oz.
Auntybel 12582................ 14 lbs. 9 oz.

Guy Mannering 698.
Phlox 16399................... 21 lbs. 11 oz.
Fair Lady 6723................ 19 lbs.
Cottage Lass 5332............. 17 lbs. 7 oz.
May Fair 5184................. 16 lbs. 7 oz.

Guy Mannering Pogis 16365.
Pogis' May's Pride 58577...... 15 lbs. 2 oz.
Crescent's Rachael 74738...... 15 lbs.

Guy Warwick 1450.
Honeydrop 10033............... 14 lbs. ½ oz.

Halo 10517.
Edna of Verna 34537.......... 20 lbs. 2½ oz.
Edessa 3d 52043............... 14 lbs. 6 oz.

Hamilton 1074.
Bellita 4553.................... 17 lbs. 2 oz.
Lisetta Johnson 5321.......... 15 lbs. 10 oz.
Etiquette 4300................. 15 lbs. 8 oz.
Élite 4299..................... 14 lbs. 2 oz.

Handy's Stoke Pogis 18810.
Handy's Daisy 74017.......... 16 lbs. 5 oz.

Hannibal 618.
Fides 2d 1576................. 14 lbs. 8 oz.

Hans of Amherst Villa 27451.
Lady Musa 98730.............. 16 lbs. 9½ oz.
Lady Negella 98735........... 16 lbs. 8 oz.
Lady Aurora 98734........... 14 lbs. 10 oz.
Lady Torenia 98736.......... 14 lbs. 9 oz.

Happy (P. S. 211 J. H. B.)
Happy Blossom 18218......... 19 lbs. 5 oz.
St. Jeannaise 15789........... 17 lbs. 8½ oz.

Happy Boy 9846.
Auselma 37798................ 14 lbs. 1 oz.

Happy Cicero 10601.
Golden Sheen 25561............ 14 lbs. 4½ oz.

Happy Coomassie 12142.
D. D.'s Coomassie 63673....... 14 lbs. 2 oz.

Happy Marquis 26801.
May Bud of Monmouth 128864. 15 lbs. 8 oz.
Happy May Girl 75588........ 14 lbs. 15½ oz.

Hard Trials 5050.
Lady's Blossom 18491........ 20 lbs. 15¾ oz.

Harpado 1859.
Alfieda 6744.................. 16 lbs. 4 oz.

Harry Torment's Signal 39846.
Valita's Fancy 115842......... 17 lbs. 1½ oz.
Harry's Rovalti 115843....... 17 lbs. ¼ oz.
Ivy Tormento 121700.......... 16 lbs. 1¼ oz.

Hazel Eye 1514.
Tinney 2d 26253............... 17 lbs. 12 oz.
Silkey Morrill 17058.......... 14 lbs. 7 oz.

Hazel Eye 2d 7822.
Lilly of Leeds 17059.......... 14 lbs. 9 oz.

Hazel's Rex 7443.
Hoey Rex 82364............... 14 lbs. 9 oz.

Hector 129.
Monmouth Duchess 3895...... 14 lbs. 7 oz.
Countess of Warren 3896..... 14 lbs.

Hector 791.
Box 2d 4297................... 14 lbs. 3 oz.
Lilly Cross 13796.............. 14 lbs. 3 oz.

Hector 3814.
Eureka McHenry 8341........ 14 lbs.

Hector 2d 2837.
Hulla 7898.................... 19 lbs. 12 oz.

Hector of Plymouth Rock 886.
Bloomer 2d 2321.............. 18 lbs. 6 oz.

Heir of Signal 8080.
Euphorbia 3d 45359........... 15 lbs. 7½ oz.

Henry Ward Beecher 2297.
Daisey Dixie 9469............. 14 lbs. 4 oz.

Hero (P. S. 90 J. H. B.)
Daisy of St. Peter's 18175..... 20 lbs. 5½ oz.
Cocotte 11958................. 16 lbs. 8½ oz.
Satin Bird 16380.............. 14 lbs. 15½ oz.
Jenny Le Brocq 9757......... 14 lbs. 14 oz.

Herotas 26500.

Sparta 75000	21 lbs. 5 oz.
Calcina 80702	18 lbs. 8 oz.
Startling 94942	17 lbs. 14 oz.

Highland Blade 2164.

Jennie of the Vale 9553	17 lbs. 7½ oz.

Highland Kago 14349.

Massena 2d 31188	17 lbs. 11 oz.

Highland Pogis 22146.

Kitty Pedrothorn 2d 84562	15 lbs. 4 oz.

Hilarious 7651.

Zenie Lisgar 2d 87453	20 lbs. 14 oz.
Analysis 2d 55834	20 lbs. 7½ oz.
Tower View Princess 2d 87452	17 lbs. 12½ oz.

Hindoo 2282.

Maudine of Elmwood 8718	16 lbs. 15 oz.

Hockanum 792.

Copper 1979	15 lbs. 7 oz.

Hohokus 5569.

Nettina of Winnikee 46406	18 lbs. 15½ oz.

Holyoke 8147.

Matina of Riverside 51783	27 lbs. 13 oz.
Inez of Riverside 51781	26 lbs. 1½ oz.
Massey Polo 67010	22 lbs. 6 oz.
Holyoke's Leda 52812	18 lbs. 11½ oz.
Brunhilde Regina 41742	16 lbs. 13 oz.
Lily of Glynllyn 31340	16 lbs.
Segilinda 41741	14 lbs. 8½ oz.
Inez of Riverside 51781 (21 days)	71 lbs. ½ oz.

Homer H. 3683.

Jenny Dodo H. 14448	21 lbs. 8 oz.
Zophar H. 12329	21 lbs. 3 oz.
Willimenia H. 12325	20 lbs. 14 oz.
Madolina H. 12327	19 lbs. 4 oz.
Mercedes H. 12326	17 lbs. 12 oz.
Anita H. 12334	17 lbs. 4 oz.
Theresa H. 14447	15 lbs. 3 oz.

Homestead Victor Pogis 26535.

Hazel St. L. 100035	18 lbs. 10 oz.

Horace (P. S. 94 J. H. B.)

Matin 7768	17 lbs. 11 oz.

Hornbeam 2123.

Troth 6139	16 lbs. 5 oz.

Houghton 5886.

Nuphar Houghton 36364	15 lbs. 3 oz.

Hughes 954.

Gentle of Glastonbury 4651	14 lbs.

Hugo Chief of St. Anne's 12070.

Daisy of Malone 36505	18 lbs. 5 oz.
Topsy of Malone 49478	14 lbs. 4 oz.
Massena 4th 47342	14 lbs.
Topsy of Malone 49478 (31 days)	55 lbs. 12 oz.

Hugo Lofty 29176.

Alchesleda 103710	15 lbs. 6 oz.

Hugo Pogis of Elmarch 16318.

Imperial Pansy Pogis 49750	14.42 lbs.

Hugo Pogis of St. Anne's 19316.

Jemele 2d 101437	15 lbs. 7 oz.

Hulabaloo 6788.

Hulabaloo's Flora 111164	15 lbs. 13 oz.

Hulda's Duke 23810.

Ella's Tina 68408	14 lbs. 9¾ oz.

Hulce 18209.

Orange Extra 133951	18 lbs.
Hazel Lee 76864	16 lbs. 1½ oz.

Hurd's Dandy 1627.

Princess Inez 5402	14 lbs. 2 oz.

Hurd's Ivanhoe 1522.

Orphean 4636	15 lbs. 7 oz.

Hurrah 2814.

Value 2d 6844	25 lbs. 2 11-12 oz.
Goldie La V. 2d 26319	14 lbs. 1½ oz.
Hurrah Pansy 12153	14 lbs. 1½ oz.

Ida of St. Lambert's Bull 19169.

Flower of Meridale 64537	26 lbs. 11½ oz.
Ida's Alpha 64536	25 lbs. 15¾ oz.
Nora Pogis 75345	21 lbs. 14¾ oz.
Ida's Rose of St. Lambert 71555	21 lbs. 1½ oz.
Meridale Sweet Ida 94854	20 lbs. 8 oz.
Beauty of Meridale 114744	18 lbs. 12½ oz.
Meridale Victoria 94848	18 lbs. 7¼ oz.
Meridale Coquette 64535	17 lbs.
Ida's Mysinda 64868	16 lbs. 11 oz.
Queen of Pomona 56250	16 lbs. 6 oz.
Meridale Rioter Pink 64534	15 lbs. 13 oz.
Meridale Oaklands Nora 64785	15 lbs. 1 oz.
Meridale Beauty Dee 97738	14 lbs. 9 oz.
Ida Mary Pogis 56126	14 lbs. 2 oz.
Sheldon's Daisy's Ida 101059	14 lbs. 2 oz.
Ida's Rose of St. Lambert 71555 (14 days)	40 lbs. 2½ oz.

Ida of St. L.'s Last Son 23601.

Kitty's Ida 97736	18 lbs. 8 oz.
Little Ida Hilda 97739	15 lbs. 8 oz.

Ida's Harry Tormenter 27836.

Ida of Maple Glens 97785	14 lbs. 14 oz.

Ida's Landseer 17745.

Kathletta's Fancy 60738	17 lbs. 6¾ oz.
Idabel Landseer 88399	15 lbs. 15¼ oz.
Ida's Signaline 108819	15 lbs. 10 oz.
Ida's Myrrha 97939	14 lbs. 11 oz.
Dolly Landseer 70201	14 lbs. 10 oz.
Landseer's Gilderine 68048	14 lbs. 7 oz.
Ida's Landseer's Gold 121641	14 lbs. 4 oz.
Emma Landseer 3d 110513 (14 days)	37 lbs. 2¼ oz.

Ida's Pogis 18000.

Miss Le Gros' Maid 83645	18 lbs. 5¾ oz.
Koffee's Duchess 2d 57988	17 lbs. 1 oz.
St. Jeannaise 5th 96554	16 lbs. 12½ oz.
St. Jeannaise 4th 57990	15 lbs. 11 oz.
Ida's Lassie 77315	14 lbs. 9¼ oz.

Ida's Rioter of Coal Hill 19661.

Angela of Coal Hill 79380	15 lbs. 7 oz.

Ida's Rioter of St. L. 13656.

Fable 62520	26 lbs. 5¼ oz.
Ida Marigold 32615	25 lbs. 2½ oz.
Daisy Hinman 61537	24 lbs. 10 oz.
Calculus 75185	23 lbs. 11½ oz.
Matilda of Meridale 57176	23 lbs. 5¾ oz.
Ida Twinkle 36994	23 lbs. 2½ oz.
Metaphysics 100974	23 lbs. 2¼ oz.
Experience 62458	22 lbs. 13 oz.
St. Lambert's Crescent 39711	22 lbs. 10 oz.
The Queen's Gift 75186	22 lbs. 9 oz.
Crown 75198	20 lbs. 12 oz.
Pomona of Prospect 123426	20 lbs. 6¼ oz.
Ida's Rioter's Elopement 98412	19 lbs. 15¼ oz.
Good Advice 62476	19 lbs. 5¼ oz.
Matilda's Queen of St. L. 93746	18 lbs. 13¼ oz.
Ida Eva Pogis 39271	18 lbs. 6 oz.
Fancy Work 116422	18 lbs. 2 oz.
Ribbon of St. Lambert 39450	18 lbs. 2 oz.
Highland Ida 38427	18 lbs. 1½ oz.
Helpmeet 62459	17 lbs. 15¼ oz.
Friendship 62460	17 lbs. 3 oz.
Dew 62474	16 lbs. 7 oz.
Snow Bank 75184	16 lbs. 7 oz.
Ida Zoe Pogis 38685	16 lbs. 2½ oz.
Rioter's Violet Pogis 75342	14 lbs. 15 oz.
Fable 62520 (14 days)	52 lbs. 7 oz.
Ida Marigold 32615 (30 days)	59.367 lbs.
Daisy Hinman 61537 (31 days)	102 lbs.
Ida Marigold 32615 (90 days)	199.756 lbs.
Daisy Hinman 61537 (90 days)	155.131 lbs.

Ida's Rioter of St. L.'s Son 28869.

Surprise Melrose 93562	21 lbs. 14⅝ oz.
Son's Princess 118386	16 lbs. 15 oz.
May Dee Pogis of C. H. 93568	15 lbs. 8 oz.
Colza Melrose 93859	14 lbs. 13½ oz.
Southern Bell of C. H. 93565	14 lbs. 7½ oz.

Ida's Rioter's Prince 32355.

Madame Melrosa 103185	16 lbs. 3 oz.

Ida's Rioter Stoke Pogis 26496.

Sunny of Cedar Hill 90410	19 lbs. 5 oz.

Ida's Spot of St. L. 15442.

Ida of Oakland 2d 72061	15 lbs. ½ oz.

Ida's Stoke Pogis 13658.

Edith Golding 63709	19 lbs. 1¼ oz.
Nonnie Pogis 41432	18 lbs. 12 oz.
Mysinda 2d 42250	18 lbs. 8 oz.
Ida Longfield 64227	18 lbs. ½ oz.
Denise's Ida 54942	17 lbs. 15½ oz.
Sataline 64213	17 lbs. 6½ oz.
Signodelle 3d 50067	17 lbs. 1½ oz.
Lily Martin 49954	16 lbs. 8 oz.
Dame Ida 76878	16 lbs. 7½ oz.
Satala 50064	16 lbs. 2½ oz.
Ida's Fawn 56018	16 lbs. 1½ oz.
Ida's Dream 50070	16 lbs. ½ oz.
Heritage 50077	15 lbs. 15 oz.
Daisy Brown 2d 37325	15 lbs. 10½ oz.
Carida 37322	15 lbs. 6 oz.
Hyacinth Pogis 56020	15 lbs. 6 oz.
Ida's Nan 49957	15 lbs. 6 oz.
Idalette Pogis 64220	15 lbs. 4¼ oz.
Gavotte 64205	15 lbs. 4 oz.
Mittretta's Rose 56192	15 lbs. ½ oz.
Idarella 41433	15 lbs. ⅛ oz.
Oonan 3d 49955	14 lbs. 15½ oz.
Pogis' Ida May 50063	14 lbs. 11 oz.
Calisida 76889	14 lbs. 4 oz.
Praxitella 50072	14 lbs. 4 oz.
Ida's Grace 56019	14 lbs. ½ oz.
Ida's Dream 50070 (14 days)	40 lbs. 8½ oz.
Soconee 76895 (14 days)	40 lbs. 1¾ oz.
Denise's Ida 54942 (30 days)	90 lbs. 6¼ oz.
Lily Martin 49954 (90 days)	164.227 lbs.

Ida's Stoke Pogis 2d 16864.

Rosa Russell 81549	16 lbs. 1 oz.
Maude Brown 85756	15 lbs. 9 oz.
Pearl Banks 85758	14 lbs. 11 oz.

Ike Felch 1292.

Milkmaid Felch 12339	16 lbs. 7½ oz.
Lily of Burr Oaks 11001	15 lbs. 13 oz.
Grace Felch 8291	15 lbs.
Milkmaid of Burr Oaks 9035	14 lbs. 5 oz.

Imperial 2d 5254.

Imperial's Omega 2d 19664	14 lbs. 14 oz.

Imperial Rex Rioter 24096.

True Rex Pogis 105467	15 lbs. 11¼ oz.
Gay Riotress 105999	15 lbs. 9 oz.
Ruthie Rex Pogis 100326	15 lbs. ¾ oz.

In-and-In Bred S. P. Third 22595.

Ollie Wilkes 77402	14 lbs. 14 oz.

Indiaman 2071.

Vespucia 17455	14 lbs. 4 oz.

Indian 1721.

May Indian 16075	14 lbs. 5 oz.

Intrepid 5511.

Fillpail 16530	15 lbs. 11 oz.

Iowa Prince 2727.
Prince's Nellie 23719......... 14 lbs. 6 oz.

Iowa's Pogis Chief 24060.
Sobriquet 68028............... 21 lbs. 3 oz.

Iron Bank 1120.
Willis 2d 4461.................. 16 lbs. 3 oz.
Vaniah 6597.................... 15 lbs. 9½ oz.
Lebanon Daughter 6106....... 14 lbs. 4 oz.
Lebanon Lass 6108............ 14 lbs. 2 oz.

Isaac B. 1951.
Lily of Maple Grove 5079..... 16 lbs. 3 oz.

Ishmael Hurd 1548.
Meg Mitchell 4187............. 15 lbs. 3 oz.

Ismena's Prince 20041.
Maria of St. Lambert 64908... 20 lbs. 3½ oz.

Jack Dasher 932.
Duchess of Argyle 3758........ 14 lbs. 13 oz.

Jack Frost of St. Lambert 2419.
Coquette of Glen Rouge 17559. 22 lbs. 14 oz.
Cill of Glen Rouge 13818...... 16 lbs. 6 oz.
Honeysuckle of St. Anne's
18674 .... 14 lbs. 14 oz.
Coquette of Glen Rouge 17559
(31 days).................... 93 lbs. 4 oz.

Jack Libby 3307.
Lydia Libby 11698............. 15 lbs. 3 oz.

Jacques (P. S. 63 J. H. B.)
Reception 8557................ 21 lbs. 4½ oz.

Jalisco of Sennett 21581.
Volie of Sennett 2d 78638..... 14 lbs. 13 oz.

Jason Jr. 3270.
Rioter Rhea 10092............. 19 lbs. 3½ oz.
Rioter Alphea 10091........... 16 lbs. 7 oz.

Jazel 3501.
Sweet Rock 2d 18256......... 14 lbs. 11½ oz.
Jazel's Maid 11011............. 14 lbs. 6 oz.

Jersey Boy (P. S. 92 J. H. B.)
Oaklands Cora 18853.......... 19 lbs. 9½ oz.

Jersey Boy Thalma 12511.
Lady Thalma 54701........... 14 lbs. 12 oz.

Jersey Express 5771.
Jersey Jane 38308............. 16 lbs. 4½ oz.
Dilwa 30515................... 15 lbs. 10 oz.

Jersey King 879.
Bright Lady 5938.............. 14 lbs. 12 oz.

Jersey King 9458.
Susie B. of Springvale 24986.. 17 lbs. 5 oz.
Virgina 80904.................. 16 lbs. 11½ oz.
Florine Value 2d 80910....... 15 lbs. 1½ oz.
Amanda May 83011........... 14 lbs. 15½ oz.
Woodland Rose 68331......... 14 lbs. 6½ oz.
Hattie Adams 80909........... 14 lbs.

Jersey Monarch 14084.
Greatæna 85823............... 16 lbs. 10½ oz.

Jersey Pride 2355.
Jefferson Albina 12196........ 14 lbs. 13 oz.

Jersey Prince 1062.
Lady Palestine 2769.......... 14 lbs. 5 oz.

Jethro 5218.
Pandothra 22383.............. 17 lbs. 5 oz.

Jeweler 1385.
Lady Alice of Hillcrest 7450.. 18 lbs. 6 oz.
Olive Branch of Hillcrest 7447 15 lbs. 7 oz.
Tib's Jewel 6808.............. 14 lbs. 4 oz.

Jim (P. S. 346 J. H. B.)
Jim's Pride 31376............. 18 lbs. 3 oz.

Jimmy, on I. of J.
Young Fanny 9032............ 17 lbs.

Joe Lozer 7538.
Lady Mitchell 29214........... 14 lbs. 12¼ oz.

John Gilpin 2199.
Emily of Hillside 8073........ 16 lbs. 4 oz.
Effie of Verna 8928........... 14 lbs. 6 oz.

John Knox 3289.
Judith Coleman 11391........ 18 lbs. 4 oz.
Aleph Judea 11389........... 18 lbs. 1½ oz.

John Le Brocq 5424.
Brunette Le Gros 3d 23326.... 14 lbs. 9 oz.

John Morrison 4551.
Daisy Morrison 14035........ 25 lbs. 12¼ oz.

John Rex 2761.
Belle of Echo 12432........... 14 lbs. 14¼ oz.

John Rex 6th 4579.
Jim's Glory 60527............. 14 lbs. 4 oz.

John Ridgely 3045.
Bessie Ridgely 8293.......... 14 lbs. 11½ oz.

John Rieter 20695.
Spark 62689.................. 20 lbs. 3 oz.

Johnson's Acme 3656.

Ruthione 31514	25 lbs. 12 oz.
Acme's Bonnie Bess 31509	14 lbs. 5 oz.

John Street 6156.

Countess of Scarsdale 18633	14 lbs. 6 oz.

Jo Jones 3441.

Alice Donnal 12726	14 lbs.

Jubilee 2125.

Rosy Dream 9808	19 lbs. 1 oz.

Jubilee of Bois d'Arc 29041.

Dagmar of Florence 120863	16 lbs. 8 oz.

Judge Mayo 5149.

Mollie Garfield 2d 18662	15 lbs. 14 oz.

Judge Watson 26286.

Minnie Mignon 90972	14 lbs. 12 oz.

Julia's Prince 17938.

Théroigne 92259	18 lbs. 11 oz.
Spangled Princess 117978	15 lbs. 5 oz.

June Pogis 19872.

Summer Morning 75197	20 lbs. ½ oz.
Fireside 75194	17 lbs. 2½ oz.

Jupiter Hugo Pogis 14342.

Mary Palestine 52439	16 lbs. 7 oz.

Jupiter Pogis 18192.

Signal's Ripple 4th 64862	16 lbs. 10¼ oz.
Justa Pogis 64863	14 lbs. 15¼ oz.
Justa Pogis 64863 (90 days)	157.697 lbs.

Justice 9949.

Mary Justice 37449	14 lbs. 9½ oz.

Kago 1353.

Massena 25732	20 lbs. 7 oz.
Massena 25732 (61 days)	145 lbs. 8½ oz.
Massena 25732 (101 days)	162 lbs. 3 oz.

Kahela 2859.

La Pera 2d 13404	14 lbs. 8 oz.

Kansas Niobe Duke 7335.

Princess Chuck 45650	24 lbs. 14 oz.

Kapper 2033.

Mylitta 15668	16 lbs. 13 oz.

Kapper's Victor 12340.

Moragina 26344	17 lbs. 3 oz.

Kate's Tornado 20888.

Oonan Pogis 59835	14 lbs. 6 oz.

Kathy's Stoke Pogis 17566.

Coomassie Veda 55838	17 lbs. 3 oz.
Misselle 73488	16 lbs. 5½ oz.
Idlewild Pogis 80271	14 lbs. 11 oz.
Blizz Pogis 82873	14 lbs. 9¼ oz.
Ernestina 2d 77682	14 lbs. 8 oz.
Trilby M. 103547	14 lbs. 4 oz.

Kattie's Prince 22315.

Rose Woodmere 91092	19 lbs. 13 oz.

Kenilworth 8091.

Fair Dairy-maid 29839	14 lbs. 8 oz.

Kerchunkus 21111.

Pertinette 70348	14 lbs.

Khedive (P. S. 103 J. H. B.)

Princess 2d 8046	46 lbs. 12½ oz.
Ona 7840	22 lbs. 10¼ oz.
Daisy Queen 9619	16 lbs. 4 oz.
Khedive's Fancy 18180	15 lbs. 3 oz.
Blonde 2d 9268	14 lbs. 4 oz.

Khedive's Landseer 21926.

Rêve 78944	21 lbs. 12 oz.
Khedive's Fairy 66849	15 lbs. 7 oz.
Palm Rose 75110	15 lbs. 1 oz.
Landseer's Croton Maid 82535	14 lbs. 15¼ oz.
Khedive's Pearl 71056	14 lbs. 12 oz.
Rosette M. 78677	14 lbs. 6.4 oz.

Killarney 5179.

Mollie Davis 26378	16 lbs. 12½ oz.

Kindergarten's Duke 30065.

Blanche Ferry 108501	16 lbs. 5 oz.
Massey Polo 3d 99378	14 lbs. 13½ oz.

King (P. S. 238 J. H. B.)

Fill Pail 2d 24388	26 lbs. 2 oz.
King's Princess 30918	24 lbs. 5 oz.
Granny's Gem 30406	21 lbs. 13-16 oz.
Khelula 17970	21 lbs. 8 oz.
King's Trust 15946	18 lbs.
King's Beauty 24397	16 lbs. 13 oz.
King's Erica 22096	16 lbs. 3¼ oz.
Royal Lady 22078	14 lbs. 12 oz.
Brown Coomassie 20322	14 lbs. 8 oz.
King's Trust 18946 (15 days)	36 lbs. 6¼ oz.

King 2d 11570.

King's Antoinette 40456	15 lbs. 1 oz.

King Albert's Rioter 14533.

Aunt Lera 77855	14 lbs. 14⅞ oz.

King Allie 14290.

Allie's Emblem 45620	20 lbs.
Allie's Leaf 45617	16 lbs. 3 oz.
Allie's Queen 45618	15 lbs. 2 oz.

King Arthur 2170.

Doña Marina 7049 .... 18 lbs. 8 oz.

King Coomassie, on I. of J.

Inez of Ingleside 28977 .... 14 lbs. 12 oz.

King Coomassie 2d 19545.

Dolly of Edgewood 83345 ...... 20 lbs. 9½ oz.
Bess of Edgewood 71178 .... 20 lbs. 6 oz.
May of Edgewood 74825 .... 18 lbs. 2 oz.
Elsie Coomassie 98081 .... 17 lbs. 12 oz.
Star of Edgewood 74823 .... 16 lbs. 10 oz.
Nelly Thorn 83344 .... 16 lbs. 7 oz.
Bertha Thorne 74824 .... 16 lbs. 5 oz.
Ethel McM. 100267 .... 15 lbs. 13 oz.

King Cotton 25391.

Cotton Patchwork 85926 .... 16 lbs. 1 oz.
Belle Cotton 96205 .... 14 lbs. 8 oz.

King Jim 24346.

Hebe's Fancy 84314 .... 14 lbs. 9 oz.

King Koffee 5522.

Koffee's Grisette 30433 .... 15 lbs. 7 oz.
Koffee's Lily 25515 .... 15 lbs. 3½ oz.

King Koffee Jr. 12317.

King's Sultana 76657 .... 14 lbs. 8½ oz.
Effie's Girl 103452 .... 14 lbs. 3½ oz.

King of Amherst 25902.

King's Stevia 98737 .... 19 lbs. 3½ oz.
King's Rosella 80767 .... 15 lbs. 12½ oz.
King's Fosie 80768 .... 15 lbs. 10 oz.
King's Gesnera 98727 .... 15 lbs. 10 oz.
King's Charella 80765 .... 15 lbs. 9½ oz.
King's Lady Natica 80769 ..... 15 lbs. 8 oz.

King of Ashantee 6677.

Ashantee's Brunette 65946 .... 16 lbs. 7 oz.
Ashantee's Lady 35951 .... 16 lbs.
Laila Rookh's Queen 65848 ... 15 lbs. 10 oz.
Ashantee's Rosy 59258 .... 14 lbs. 11 oz.

King of Fairview 778.

Fair Starlight 7745 .... 17 lbs. 7½ oz.
Jersey Cream 2d 8519 .... 14 lbs. 12 oz.

King of St. Lambert 15175.

Gazelle's Fawn 93704 .... 25 lbs. 4½ oz.
Phillis of St. Lambert 78867 ... 23 lbs. 12 oz.
St. L. Gazelle 78870 .... 22 lbs. 14 oz.
Elsie Bonner 78864 .... 21 lbs.
Queen Deeline 78869 .... 20 lbs. 13½ oz.
King of St. L.'s Allie 86797 ... 20 lbs. 10 oz.
Melbina 78875 .... 20 lbs. 2 oz.
May Keller 78874 .... 19 lbs. 8 oz.
King's Lassie 79103 .... 18 lbs. 5 oz.
Butterfly of Riverview 3d 85499 17 lbs.
Dainty Doris 85497 .... 16 lbs. 8½ oz.
Elsie of Shadyside 85500 ...... 16 lbs. 8 oz.
Sally of St. Lambert 78873 .... 16 lbs. 5 oz.
Queen May of St. Lambert
   45770 .... 15 lbs. 12½ oz.
Pomona's May Day Pogis 56249 15 lbs. ½ oz.

King of St. Lambert 2d 17823.

Melia Ann 2d's Pogis 68757 .... 17 lbs. 1 oz.
Aurated Pogis 2d 65666 .... 16 lbs. 6 oz.
Homestead Prue 82203 .... 14 lbs. 13½ oz.

King of St. Lambert Jr. 18808.

Rush of St. Lambert 61234 .... 17 lbs. 6¾ oz.
Grace Sheen 2d 79788 .... 14 lbs. 9½ oz.
Epple Lady Pogis 110638 ...... 14 lbs. 3 oz.
Rush of St. Lambert 61234 (15
   days) .... 34 lbs. 12½ oz.

King of St. Lambert's King 30752.

Lady K. of St. Lambert 112294 24 lbs. 6 oz.
Gipsy W. 132879 .... 21 lbs. 4 oz.

King Pete 10541.

Celia of Whitehall 68004 ...... 17 lbs. 15 oz.
Prusa Pedro 50877 .... 15 lbs. 8 oz.
May Pedro 50876 .... 15 lbs. 4 oz.
Nelope Pedro 50879 .... 15 lbs. 4 oz.
Empress Pedro 50873 .... 15 lbs. 1 oz.

King Phil 33443.

Stella of Edgewood 114473 ..... 14 lbs. 12 oz.

King Philip of Mt. Hope 2399.

Christel 6565 .... 19 lbs. 5 oz.

King Pin 1878.

Bessie Bradford 7269 .... 14 lbs. 2 oz.

King Signal 12491.

Beauty of Oakland 45009 ...... 17 lbs. 1 oz.

King Torment 18609.

King's Queen 98712 .... 14 lbs. 12 oz.
Grace Torment 87957 .... 14 lbs. 10½ oz.

King Yokun 7068.

Yokun's Princess 35643 .... 15 lbs. 10¼ oz.

Kinsman 3338.

Maudie of White Lick 29682 .. 15 lbs. 7 oz.

Kinsman of Greenwood 12881.

Fedora of Greenwood 46892 ... 20 lbs. 7 oz.

Kisber 4377.

Westphalia 24384 .... 24 lbs. 9½ oz.
Syren V. 14619 .... 17 lbs. 5 oz.

Kit Carson 1772.

Austle May 7185 .... 14 lbs. 3 oz.

Kitty's Royal Rex 6176.

Pansy Blossom 22413 .... 14 lbs. 13¼ oz.

Knave 1856.

Colt's La Biche 6399 .... 17 lbs. 2½ oz.

Knight of St. Louis 3680.
Queen of Beauty 17109....... 23 lbs. 14 oz.
Gem of Hope 17102............ 21 lbs.
Hallie's Jewel 17113.......... 20 lbs. 10 oz.

Know Nothing 11638.
Lady Crouse 69336............ 18 lbs. 1 oz.

Koffee (F. S. 233 J. H. B.)
Gazelle 15961.................. 14 lbs.

Koffee of Ridgeside 11659.
Lady Phillis 2d 35629......... 18 lbs. 8 oz.

Koffee's Noble 14631.
Sayda of Aurora 86956........ 16 lbs. 11½ oz.
Old Noble's Heiress 99083..... 15 lbs. 6½ oz.

Ko Ko of St. Lambert 16527.
Young Floribundus 4th 59314... 15 lbs. 14 oz.
Ko Ko's Allie 59317........... 15 lbs. 8¾ oz.

Lady d'Or's Son 21048.
May Belle's Lady d'Or 74749.. 15 lbs. 15 oz.

Lady Mary's Chelten Duke 6002.
Lady Dorlena 36105........... 15 lbs. 7 oz.

L'Allegro 7127.
Winetka 33669................. 20 lbs.

Landseer 331.
Landseer's Fancy 2876........ 29 lbs. ½ oz.
Pride of Mashamoquet Farm 6469 .... 16 lbs. 1¾ oz.
Fannie Landseer 1969........ 15 lbs. 9 oz.
Landseer's Fancy 2876 (1 yr.) 936 lbs. 14¾ oz.

Landseer (P. S. 162 J. H. B.)
Rose of Eden 13437............ 20 lbs. 1½ oz.

Landseer's Major 25759.
Landseer's Lily 87294........ 14 lbs. 2 oz.

Landseer's Pogis 15847.
Fancy Wax 3d 86345.......... 20 lbs. 3¾ oz.
Belle Lyman 3d 86346........ 18 lbs. 2½ oz.
Duke's Signal Queen 2d 82418. 17 lbs. 8 oz.
Landseer's Hilda 49788....... 17 lbs.
Landseer's Signal Queen 66363 16 lbs. 8 oz.
Fancy's Gem 2d 86342......... 16 lbs. 4 oz.
Temalema 2d 82411........... 16 lbs. 4 oz.
Kate Oxford 58102............ 16 lbs. 3 oz.
Wandrina 2d 82420........... 15 lbs. 9½ oz.
Royal Bomba's Daisy 3d 99553 15 lbs. 8 oz.
Virginia's Oonan 58107....... 15 lbs. 3 oz.
Bessie Signalda 2d 86341...... 15 lbs. 2½ oz.
Bessie Landseer 58103........ 15 lbs. 1 oz.
Pearl's Oonan 2d 86105....... 15 lbs. 1 oz.
Bessie Signalda 3d 86349..... 15 lbs. ½ oz.
Landseer's Pogene 85743...... 15 lbs.
Lucy Lee H. 58106............ 14 lbs. 15½ oz

Harry's Fancy Belle 2d 82416. 14 lbs. 12½ oz.
Landseer's Signaline 58101.... 14 lbs. 11⅞ oz.
Tolteca 3d 86344.............. 14 lbs. 11½ oz.
Yokun Maid 2d 86563......... 14 lbs. 9½ oz.
Waxie 2d 83358............... 14 lbs. 8½ oz.
Fancy Wax 2d 83357.......... 14 lbs. 5½ oz.
Rosetta Landseer 99379....... 14 lbs. 4 oz.
Emma Landseer 58104........ 14 lbs. 2½ oz.
Julia Landseer 56268......... 14 lbs. 2½ oz.
Fancy's Pet 2d 85912......... 14 lbs. 1 oz.
Kate Durwood 2d 82417....... 14 lbs. 1 oz.
Fancy's Gem 4th 86348....... 14 lbs.

Landseer's Toltec 25426.
Landseer's Ethel 74315....... 15 lbs.

Lanison 15283.
Lanison's Belle 84986......... 20 lbs. 5 oz.
Young Floribundus 3d 59312.. 19 lbs. 14¾ oz.
Princess Carmen 58320....... 16 lbs. 14 oz.
Lanison's Queen 2d 128773.... 15 lbs. 14 oz.
Lanison's Nora 59316......... 15 lbs. 6½ oz.
Lanison's Cora 59315......... 14 lbs. 14 oz.

Laval 506.
Cupid of St. Lambert 5104.... 14 lbs. ½ oz.
Lily of St. Lambert 5120..... 14 lbs.

Lawrence 61.
Memento 1913................. 14 lbs. 5 oz.
Turquoise 1129............... 14 lbs. 3 oz.

L. D.'s Hugo 17374.
Hugo's Lass 107656........... 14 lbs. 1¾ oz.

L. D.'s Victor St. Helier 15286.
Cinxia's Pavonia 64441....... 18 lbs. 3 oz.
Kate Kyle 64434.............. 16 lbs. 4 oz.
Daisy Morrison 4th 46830..... 14 lbs. 11 oz.
Victor's Matilda 46834........ 14 lbs. 6 oz.
St. Helier's Matilda 52857..... 14 lbs. 1 oz.

Leander 1184.
Lady Thurlow 12410.......... 17 lbs. 10 oz.

Le Brocq's Pride 4871.
Mabel Crockett 18719......... 14 lbs. 9 oz.

Le Brocq's Prize 3350.
(Snap F. S. 254 J. H. B.)
Fan's Grouville Beauty 10079. 19 lbs. 3 oz.
Le Brocq's Pansy Rex 23769.. 18 lbs. 6 oz.
Viva Le Brocq 13702.......... 18 lbs. 3 oz.
Marea 10167.................. 17 lbs. 10 oz.
Petra 19267.................. 16 lbs. 6 oz.
Princess Le Brocq 17261....... 15 lbs. 8 oz.
Le Broemer 10670............. 15 lbs. 7 oz.
Prize Rose 16309.............. 15 lbs. 1 oz.
Fairy Le Brocq 15283......... 14 lbs. 3 oz.
Birdie Le Brocq 17263........ 14 lbs.
Elinor Wells 12068........... 14 lbs.
Le Rosa 10678................ 14 lbs.

Lemon (F. S. 170 J. H. B.)
Nelly 6456...................... 21 lbs.

Lemon Peel (P. S. 320 J. H. B.)
Lemon Fern 22140............. 15 lbs. 1 oz.

Lemon Peel of Francheville (P. S. 439 J. H. B.)
Milkgood 27828................ 14 lbs. 7½ oz.

Lemon Peel's Duke (P. S. 479 J. H. B.)
Very Pretty 29184............. 15 lbs. 2 oz.

Lenape 2732.
Muriel 5th 19017............... 16 lbs. 12½ oz.
Dorizella 20525................. 15 lbs. 7 oz.
Gowan 12272................... 14 lbs. 4 oz.

Lena's Lenox 6059.
Proctor's Pansy 25688........ 15 lbs. 13 oz.

Lenox Cash Boy 6804.
Carrie Lena 3d 20077......... 18 lbs. 11¾ oz.
Spiræa 4th 20075.............. 18 lbs. 5½ oz.

Lenox Pogis 27365.
Dacie's Lena 100796........... 22 lbs. 12 oz.

Leonette's Landseer 20399.
Mary of Glenoir 59940........ 27 lbs. 3½ oz.
Mirah Landseer 74827........ 17 lbs. 1 oz.
Amy Landseer 74826.......... 16 lbs. 1 oz.
Sibby Landseer 116107........ 15 lbs. 2 oz.
Leonette's Orange 108521...... 15 lbs.
Virgie Landseer 74831........ 14 lbs. 15 oz.
Roxie Landseer 74832........ 14 lbs. 13 oz.
Moll Roberts 110201........... 14 lbs. 6½ oz.
Clyde Landseer 74834.......... 14 lbs.

Leonidas 3010.
Sigleonella 70091.............. 21 lbs. 9¾ oz.

Leonidas' Germ 14289.
Geranium Leaf 2d 56859....... 22 lbs. 7 oz.

Lee of Glynllyn 32015.
Little Myers 104650............ 16 lbs. 8 oz.
Lady Petite 100562............ 15 lbs. 11 oz.

Leo Pogis 10785.
Lady Anna Pogis 52734........ 20 lbs. 4 oz.

Letty's Snap 27796.
Santa Stoke Pogis 91680....... 19 lbs. 1½ oz.

Lewis' Ringgold 6379.
Star of Mahoning 16579....... 14 lbs. 7 oz.

Lexicon 23110.
Mosa 69691..................... 14 lbs. 6 oz.

Lex Linmore 18623.
Lady of Union Valley 101733.. 20 lbs. 7½ oz.

Lisgar's Pride 14534.
Broadway 50883................ 16 lbs. 8 oz.

Lisgar's Victor 25887.
Loessin Fancy 91599.......... 18 lbs. 8 oz.

Litchfield 674.
May Blossom 5657............. 18 lbs. 11 oz.
Beauty Bismarck 4967........ 14 lbs. 1 oz.

Litchfield 15th 5802.
Belle Garner 23682............ 15 lbs. 8 oz.

Little Harry 8808.
Little Goldie 38671............ 34 lbs. 8½ oz.
Alteration 56436................ 24 lbs. ½ oz.
Helen Barry 55840............ 18 lbs. 7 oz.
Nazell 2d 98271............... 16 lbs. 8½ oz.
Lady Thekla 98270............ 15 lbs. 4¼ oz.
Mette D. ʋ433................. 14 lbs. 12 oz.
Frushia Wilson 47432.......... 14 lbs. 4½ oz.
Alteration 56436 (90 days)..... 179.879 lbs.
Little Goldie 38671 (90 days).. 176.394 lbs.

Little Harry's Wonder 35289.
Harry's May Bud 111817....... 14 lbs. 15½ oz.

Little Miami 22076.
Maud Miami Jr. 69131........ 14 lbs. 8 oz.

Little Toltec 24445.
L. Toltec's First 130688....... 14 lbs. 13¼ oz.

Little Tormentor 17995.
Little Texas 84229............. 17 lbs. 9¼ oz.
Mattie Tormentor 62841....... 16 lbs. 5¼ oz.
Tillie Texas 62840............ 15 lbs. 13 oz.
Myra Tormentor 84227........ 15 lbs. 11½ oz.
Little Meysie 84230........... 15 lbs. 5 oz.
Little Tipsey 79398............ 15 lbs. 5 oz.
Buckmentor 62846............. 14 lbs. 10 oz.
Little Mistiss 62849............ 14 lbs. 3 oz.
Little Princess Potoka 79285.. 14 lbs. 2 oz.
Mattie Tormentor 62841 (14 days) .... 31 lbs. 13½ oz.

Live Oak of Brushy 16155.
Cabbie Oak 90061.............. 14 lbs. 12 oz.

Living Storm 173.
Pansy 1019 (one year)........ 574 lbs. 8 oz.

Lofty 2856.
Coquette's Molly 23841........ 14 lbs. 2 oz.

Lopez 313.
Bertha Morgan 4770........... 19 lbs. 6 oz.

Lord Alfred 5215.

May Violet 30250............... 18 lbs. 11½ oz.
Lady Cleveland 30251.......... 17 lbs. 1 oz.

Lord Allison 16402.

Lady St. Heller of S. 51380.... 20 lbs. 9½ oz.

Lord Almer H. 24081.

Rachel's Tamy 89987.......... 14 lbs. 4½ oz.

Lord Anglesea 4537.

Belle Grinnell 3d 16503........ 14 lbs. 2 oz.

Lord Aylmer 1067.

Melia Ann 5444................ 18 lbs. ½ oz.

Lord Bronx 2d 1730.

Hazen's Bess 7329............. 24 lbs. 11 oz.
Arnold's Lulu 7328............ 15 lbs. 3 oz.

Lord Clive 3313.

Clover Bloom 2d 12736......... 14 lbs.

Lord Darlington 7285.

L. D.'s Melia 29087............ 20 lbs. 6 oz.
Matilda 6th ᴗᴗᴗ90.............. 20 lbs. 1 oz.
Melia Ann 2d 47464........... 18 lbs. 6 oz.
Saugus Lass 30542............. 14 lbs. 9 oz.
Gilt Lady 2d 33969............. 14 lbs. 7 oz.
L. D.'s Mabel 29368............ 14 lbs. 6 oz.
L. D.'s Helita 29366........... 14 lbs. 4½ oz.

Lord Dartmouth 6302.

Lady Dartmouth 23159........ 21 lbs. 13 oz.
Cretesia 13657................ 20 lbs. 1 oz.
Dia 13658..................... 15 lbs. 13½ oz.

Lord Ducie 2500.

Bessie of Montverde 18496..... 15 lbs. 5 oz.

Lord Fife 5148.

Lord Fife's Joliette 30591..... 16 lbs. 2 oz.

Lord Francis 1857.

Trut 8090..................... 14 lbs. 3 oz.

Lord Gordon 24901.

Lydia D. 2d 97860............. 14 lbs. 9¾ oz.
Lydia D. 2d 97860 (14 days)... 29 lbs.

Lord Hallock 29589.

Lord Hallock's Princess 96924 14 lbs. 1 oz.

Lord Harry 3445.

Ethleel 2d 32291.............. 30 lbs. 15 oz.
Louisa of Riverside 69769..... 23 lbs. 5 oz.
Kathletta 19567............... 22 lbs. 12½ oz.
Peasant 65401................. 21 lbs. 15½ oz.
Derjava 59830................. 18 lbs. 14 oz.
Mary Gordon 41429............ 18 lbs. 5¼ oz.
Oonan 2d 19569................ 18 lbs. 4¼ oz.

Martha Lafayette 17158........ 17 lbs. 6 oz.
Gordonetta 19570.............. 17 lbs. 1 oz.
Clover Bud 4th 18992.......... 16 lbs. 14 oz.
Kitty Better 32911............. 15 lbs. 14⅜ oz.
Kaleta 2d 38810................ 15 lbs. 14 oz.
Rochelle 15574................. 15 lbs. 10 oz.
Ethleel 5th 70652.............. 15 lbs. 6½ oz.
Duchess of Bloomfield 3d 15580 15 lbs. 1 oz.
Buttery 2d 27138.............. 15 lbs. ⅝ oz.
Violet Denison 49997.......... 14 lbs. 13½ oz.

Lord Harry 3d 27964.

Lord Harry's Lucy 86091...... 16 lbs. 5 oz.

Lord Harry 5th 29752.

Ida's Lucille 99400............. 17 lbs. 4¼ oz.

Lord Hugo 15830.

High Spirits of Lawn 61371... 18 lbs. 7 oz.
Storm-Beaten of Lawn 61375.. 18 lbs. 3 oz.
Eureen of Lawn 61372........ 15 lbs. 8 oz.

Lord Hugo Pogis 23673.

Ella of Endegeest 87376........ 19 lbs. 12½ oz.

Lord Landseer 24228.

Winksette 99280............... 23 lbs.
Landseer's Regina 85988....... 17 lbs.
Romena of Hood Farm 114887 15 lbs. 11 oz.

Lord Lawrence 1414.

Witch-Hazel 4th 6131.......... 18 lbs. 2⅞ oz.
Leila Stratton 25198........... 17 lbs. 4 oz.
Gladys of Belle Vue 9569...... 16 lbs. 7 oz.
Maid of the Elms 3d 15554.... 16 lbs. 4 oz.
Countess Gisela of Belle Vue
9571 .... 15 lbs. 11 oz.
Lady of Belle Vue 7705........ 15 lbs. 11 oz.
Fall Leaf 8587................. 14 lbs. 8 oz.
Lorelle 12913.................. 14 lbs. 7 oz.

Lord Lisgar 1066.

Sweet Brier of St. Lambert
5481 .... 22 lbs. 10 oz.
Jolie of St. Lambert 5126...... 15 lbs. 13½ oz.
Duchess of St. Lambert 5111.. 15 lbs. 11 oz.
Bijou of St. Lambert 5112..... 15 lbs. 4 oz.
Rosette of St. Lambert 5108.. 14 lbs. 3½ oz.
Clematis of St. Lambert 5478. 14 lbs. 3 oz.

Lord Lisgar Pogis 12149.

Paletta of Darlington 2d 75343 14 lbs. 15½ oz.

Lord Longford 3997.

Lady Vain 18573............... 17 lbs. 9 oz.
Southern Belle 18570.......... 16 lbs.
Frisky Miss 18572............. 14 lbs.

Lord McDuff 5147.

Flower of Glen Rouge 17560... 23 lbs. 14¾ oz.

Lord Monck 304.

Tidy of St. Lambert 31114..... 14 lbs. 2 oz.

Lord Montague 12385.

Kate Crecraft 2d 75461........	18 lbs. 3 oz.
Beauty Montague 57923........	17 lbs. 8 oz.
Silver Montague 45475.........	16 lbs. 10 oz.
Joan Montague 40787..........	16 lbs. 1 oz.
Mildred Montague 45476......	15 lbs. 5½ oz.
Minnie Montague 45474.......	15 lbs. 5½ oz.
Miss Montague 57927.........	14 lbs.

Lord of Mountain Side 7111.

Rioter Carlotta 29667.........	21 lbs. 2½ oz.
Rioter Alphea 2d 29676.......	17 lbs. 15 oz.

Lord Rex 4113.

Down of Chatsworth 15251....	16 lbs. 12 oz.

Lord St. Helier 15071.

Canossa 65243................	15 lbs.

Lord Tormentor Pogis 29753.

Daisy Ida Torment 121824.....	15 lbs. 12 oz.

Lorne 5248.

Oaklands Nora 14880.........	23 lbs. 5 oz.
Marjoram of Linden 43600.....	22 lbs. 12 oz.
Lorne's Pretty Pearl 46683....	19 lbs. 6 oz.
Lorne's Diamond 66087.......	19 lbs.
Jessie Lorne 25213............	16 lbs. 15 oz.
Lorne's Primrose 66082.......	16 lbs. 12 oz.
Lorne's Marjoram 43327.......	16 lbs. 4½ oz.

Lorne Victor Pogis 22214.

Molly May Pogis 102823.......	15 lbs. 12 oz.
Kitty Pride Pogis 78134.......	15 lbs. 9 oz.
Miss Lorne 78132..............	15 lbs. 9 oz.

Louis of Highland 4854.

Daisy of Sunnyside 16338......	15 lbs. 1 oz.

Louis of St. Lambert 11013.

Lady Matilda Pogis 36270.....	21 lbs. 9 oz.
Louise Obella 37060...........	17 lbs. 12 oz.
Gem's Louisa 39564...........	17 lbs.

Lucifer 2696.

Sunbeam's Pet 2d 9112........	16 lbs. 4 oz.
Beatrice of Elmarch 11367.....	15 lbs. 2 oz.

Lucky King 17350.

Olive Star 81102..............	15 lbs. 7 oz.
Lucky Olive Germ 80256......	14 lbs. 3½ oz.

Lucullus 2695.

Cora of Hillside 25253.........	15 lbs. 7 oz.

Lucy's Stoke Pogis 11544.

Melia Ann 3d 68070...........	28 lbs. 8 oz.
Jennie Cream 2d 47466.......	20 lbs. 5 oz.
Quintuple Pogis 44233.........	14 lbs. 6 oz.

Lucy's Stoke Pogis 4th 26533.

King's Lucy Pogis 111380.....	18 lbs. 8 oz.
Homestead Lucy 96325.......	17 lbs. 12 oz.
Homestead Matilda 96448......	17 lbs. 8 oz.
Louisa D. 90959..............	16 lbs. 14 oz.
Quintuple Pogis' Fille 94975..	16 lbs. 8 oz.
Homestead Rowena Pogis 89591	16 lbs. 7½ oz.

Lymington 3767.

Ruby Love 16915..............	14 lbs. 12 oz.

McClellan 4th 85.

Lady Brown 433..............	14 lbs.

Macgregor 2178.

Lucy Gaines' Buttercup 5058.	14 lbs.

McHenry 5890.

Cream Caroline 17357.........	16 lbs. 2 oz.

Mack 722.

Buttery 3502..................	14 lbs. 1 oz.

Magnetic 1428.

Magnibel 7976................	15 lbs. 2 oz.

Magog 1868.

Country Girl 4th 51877.........	18 lbs. 2 oz.

Maha Rajah 794.

Italian 4096..................	17 lbs. 13 oz.

Mahkeenac 3290.

Frolic's Pride 31667...........	17 lbs.
Frolic of Chestnutwood 19405.	16 lbs.
Angetta 19404................	15 lbs. 4 oz.
Young Anne Lee 31668........	14 lbs. 7 oz.

Majestic Pedro 8043.

Rita of Andalusia 29414.......	24 lbs. 1 oz.

Majestic Pedro's Son 9492.

Juno of Milton 71528.........	15 lbs. 2 oz.

Major Adams 1044.

Jenny Gray 3511..............	17 lbs. 15 oz.
Lucy Gray 2746...............	15 lbs. 13 oz.

Major Appel Pogis 17861.

Beckie Y. Pogis 63069.........	17 lbs. 15¼ oz.
High Tea 65577...............	17 lbs. 6½ oz.
Fair Marigold 89387...........	16 lbs. 15 oz.
Scroll 93493..................	14 lbs. 15 oz.
Anklet 93500..................	14 lbs. 6 oz.

Major Domo 2161.

Melita of Hillcrest 7054........	14 lbs. 2 oz.

Major Dott 7101.

Dott Buttercup 16358.........	17 lbs. 2 oz.

Major Hamilton 30164.
Dairy Dolly 100605............ 15 lbs.

Major Marion 6922.
Le Brocq's May Belle 27980... 14 lbs. 15 oz.

Manchester's Prospect 2817.
Jersey Cream 3d 8521......... 16 lbs. 5 oz.
Auraria 10688................... ·15 lbs. 10 oz.

Manifold Rioter Pogis 26158.
Vionetta Pogis 91978........... 17 lbs. 2 oz.
Elma's Star 102275............. 14 lbs. 12 oz.

Man of Ipswich 1510.
Jennette Darling 10702........ 16 lbs. 2 oz.

Mapleson 5951.
Sosia 19198.................... 17 lbs. 5½ oz.

Maquilla's Harry 22701.
Maquilla's Violet 69774....... 31 lbs. 1 oz.
Polonia of Riverside 84036.... 18 lbs. 10 oz.
Annie Hines 84038............. 18 lbs. 6¾ oz.

Marcot 726.
Golden Princess 4557.......... 18 lbs. 14 oz.

Mardi Gras 2927.
Belle Mardi 18362.............. 18 lbs. ½ oz.
Belle of Lynwood 18364........ 17 lbs. 14 oz.
Ada S. 18366.................... 16 lbs. 9 oz.

Marius 760.
Welma 5942...:................ 17 lbs. 8 oz.
Chenda 4599.................... 17 lbs. 1¼ oz.
Calypris 5943.................. 15 lbs. 4½ oz.
Evri 5282...................... 15 lbs. 4 oz.

Marjoram's Hugo Pogis 18311.
Khedive's Fancy 3d 702>2..... 14 lbs. 6 oz.

Marjoram's Rioter 5991.
Quadruple Pogis 32359........ 14 lbs. 1½ oz.

Marjoram's Signal 11183.
Betsy Marjoram 65145......... 15 lbs. 7 oz.

Marjoram's Stoke Pogis 17203.
Maid of Tanglewood 51606..... 16 lbs.

Marmaduke 483.
Moonah's Pet 7484............. 15 lbs. 6 oz.

Marmaduke's King of Hearts 2949.
Minna Rajah 14847............. 16 lbs. 8 oz.

Marmion 359.
Pet Anna 1608.................. 14 lbs.

Marpetro 3352.
Marpetra 10284................. 14 lbs. 6 oz.

Marquis of Lessie 3512.
Lalla Rookh of Sugar Grove
15882 .... 20 lbs. 1 oz.

Master 723.
Eugénie Chouteau 6186........ 24 lbs. 8 oz.
Lillie Pope 8589............... 14 lbs. 5 oz.

Master Le Brocq 32047.
March Ida Pogis 101182 (21
days) .... 47 lbs. 4½ oz.

Master Richmond 7429.
Catchfly 3d 35499.............. 15 lbs. 7 oz.

Master Vermont 4304.
Miss Satanella 31544.......... 20 lbs. 6 oz.
Sedate 2d 20389............... 15 lbs. 8 oz.
Miss Satanella 31544 (31 days) 78 lbs. 3½ oz.

Matilda 4th's Son 20214.
Point Lace 62471.............. 17 lbs. 8 oz.
Matilda's Matilda 75348....... 17 lbs. 2½ oz.
Matilda of the Ledges 97740... 16 lbs. 9½ oz.
Ribbon's Matilda 66109........ 16 lbs. 4 oz.
Pink of Meridale 114748....... 16 lbs. ½ oz.
Pogis' May's Matilda 75347... 15 lbs. 11½ oz.
Sweet Lily of Meridale 94267.. 15 lbs. 8½ oz.
Matilda Mysinda 84487........ 15 lbs. 3 oz.
Matilda Hilda 75346........... 14 lbs. 4 oz.
Kitty's Matilda 75349......... 14 lbs. 3 oz.
Sweet Lily of Meridale 94267
(14 days).................. 29 lbs. ½ oz.

Matilda's Lisgar 21374.
Matilda Ann Darling 105398... 14 lbs. 1 oz.

Matilda's Perfection 17100.
Countess Matilda 74928........ 19 lbs. 11 oz.
Millerton Maid 74926.......... 17 lbs. 3 oz.
Alice Rex 3d 63907............. 15 lbs. 6½ oz.
Anita Lawrence 63909......... 15 lbs. 6½ oz.
Marion Earle 63908............ 14 lbs. 8½ oz.
Countess Matilda 74928 (30
days) .... 82 lbs. 15 oz.
Countess Matilda 74928 (90
days) .... 220 lbs. 6 oz.

Matilda's Stoke Pogis 14832.
Saucy Sally 83059............. 16 lbs. 12½ oz.

Matilda's Victor Hugo 20244.
Matilda's Queen 72663........ 16 lbs.
Matilda's Rose 81397.......... 15 lbs. 2 oz.
Matilda's Pansy 81311........ 14 lbs. 7 oz.

Mattituck's Bachelor 20920.
Bertha's Brunette 82336....... 18 lbs. 1 oz.

Maud's Rioter Pogis 25345.
Pluma of Jersey Lawn 89954.. 14 lbs.

Maury's Tormentor 25113.
Moxie's Mode 2d 133253....... 15 lbs. 9¼ oz.
Prince's Rubetta 2d 133914.... 14 lbs. 6¼ oz.

Maximum Rioter 18539.
Lady Pedrona 63173.......... 14 lbs. 9 oz.

Maxse 400.
Nannaxie 18860................ 17 lbs. 4 oz.

May's Pedro 8588.
Glen's Festina 25851.......... 17 lbs. 15 oz.

Meadow King of Farmington 4911.
Sue Gallagher 15945.......... 23 lbs. 1½ oz.

Medway 717.
Medrena 3939.................. 18 lbs.
Mirtha 3437................... 17 lbs. 13½ oz.

Meg's Lord Bronx 11537.
Idessimus 50508............... 16 lbs.
Bronx Clover Blossom 38676.. 15 lbs. 2½ oz.

Melia Ann's Son 22041.
Faith of Eau Claire 120456.... 22 lbs. 12 oz.
Marion of Eau Claire 120457... 21 lbs. 14 oz.
Emma of Eau Claire 127826... 21 lbs. 4 oz.
Herpa S. 127829............... 18 lbs. 6 oz.
Laura of Eau Claire 110739.... 18 lbs.
Julia of Eau Claire 120018.... 17 lbs. 8 oz.
Princess of Eau Claire 128219. 17 lbs. 6 oz.
Letta of Eau Claire 120021.... 16 lbs. 3 oz.
Gertie of Eau Claire 120019... 15 lbs. 7 oz.
Fairy of Eau Claire 128596.... 14 lbs. 14 oz.
Martha of Eau Claire 120022... 14 lbs. 13 oz.
Mab of Eau Claire 120020..... 14 lbs. 10 oz.
Millie of Eau Claire 120023... 14 lbs. 7 oz.
Dame of Eau Claire 127862.... 14 lbs. 4 oz.

Melia Ann's Stoke Pogis 22042.
Pride's Olga 4th 96870........ 23 lbs. 3 oz.
Charm 3d's St. Lambert 96872 14 lbs. 8 oz.
Cora Scituate 2d 94979........ 14 lbs.

Melia Ann's Victor Pogis 20697.
Princess of Dorset 85902....... 35 lbs. 1 oz.
Diana Victor Pogis 100763..... 23 lbs. 1 oz.
Winifred Victor Pogis 91658.. 22 lbs. 9 oz.
Allie Victor Pogis 126907...... 22 lbs. 8 oz.
Kitty Victor Pogis 85904....... 19 lbs. 12 oz.
Ella Victor Pog's 100762...... 19 lbs. 6 oz.
Dora Victor Pogis 100921..... 18 lbs. 5 oz.
Philena Victor Pogis 127138... 17 lbs. 3 oz.
Spot Victor Pogis 85901....... 17 lbs. 3 oz.
Prusa Victor Pogis 134216.... 16 lbs. 8 oz.
Princess Victor Pogis 132724.. 16 lbs. 5 oz.
Pet Victor Pogis 77812........ 16 lbs. 4 oz.
Gertie Victor Pogis 100922.... 16 lbs. 3 oz.
Belle of Dorset 85905......... 14 lbs. 9 oz.
Ann Victor Pogis 77813....... 14 lbs. 7 oz.
Tormentor Victor Pogis 133021 14 lbs. 7 oz.
Charity Victor Pogis 134217... 14 lbs. 6 oz.
Lillie Victor Pogis 85471...... 14 lbs. 4 oz.

Menephtah 18943.
Tuäa St. Lambert 59992....... 23 lbs. 1 oz.

Méprise's Tormentor 11554.
Méprise Bloomfield 50062..... 14 lbs. 14 oz.

Mercury 432.
Phædra 2561................... 19 lbs. 13 oz.
Nymphæa 5141................. 18 lbs. 7½ oz.
Richness 16536................. 17 lbs. 5 oz.
Lerna 3634.................... 15 lbs. 12 oz.
Zalma 8778.................... 15 lbs. 5 oz.
Purest 13730.................. 15 lbs. 4 oz.
Clytemnestra 2455............. 15 lbs. 3½ oz.
Reality 16537................. 15 lbs. 3½ oz.
Iola 4627..................... 15 lbs. 2½ oz.
Marvel 13734.................. 15 lbs. 1 oz.
Nimble 22335................. 14 lbs. 10 oz.
Smoky 13733.................. 14 lbs. 9 oz.
Ideal Alphea 18755............ 14 lbs. 6 oz.
Renown 13729................. 14 lbs. 6 oz.
Alphea Star 16532............. 14 lbs. 4½ oz.
Alphetta 16531.......... .....14 lbs. 2¼ oz.
Vestina 2458.................. 14 lbs. 2 oz.
Lernella 22322................ 14 lbs. 1½ oz.
Ballet Girl 18750............. 14 lbs. 1 oz.
Alphea Jewel 22331........... 14 lbs.

Mercury Boy 5721.
Walter Girl 2d 29265......... 18 lbs. 14½ oz.

Mercutio 4591.
La Pucelle 16829.............. 15 lbs. 8 oz.

Meridale Pauline Pogis 25133.
Caro of St. Lambert 90875..... 17 lbs. 6 oz.
Florence of St. Lambert 95192 16 lbs. 12 oz.
Letitia of St. Lambert 91054.. 16 lbs.

Merry Andrew 719.
Merry Burlington 7600........ 15 lbs. 4 oz.

Merry Boy, on I. of J.
Nightingale of Elmarch 8312.. 14 lbs. 2 oz.

Messenger (P. S. 223 J. H. B.)
Beauty Messenger 17450...... 15 lbs. 9 oz.

Meteor 453.
Estrella 2831.................. 14 lbs. 12 oz.

Metropolitan 23888.
Milly Judd 85898.............. 14 lbs. 3½ oz.

Mhoon Stoke Pogis 19193.
Championa Pogis M. 75926.... 15 lbs. 15 oz.

Micawber 4796.
Dove 5th 14561................ 16 lbs. 13 oz.
Daisy Dell 14562.............. 14 lbs. 9 oz.
Dolly of Lakeside 10824....... 14 lbs. 8 oz.

Michael Angelo 10116.
Angelo's Torment 39390....... 22 lbs. 8 oz.
Fashion Plate 62457.......... 15 lbs. 10½ oz.

Michael Angelo Pogis 19901.
Rockys' Pogis 96272.......... 18 lbs. 3 oz.

Midas of Oxford 5986.
La Petite Mère 3d 12814....... 16 lbs. 9 oz.
Ethel of Shelburne 46285...... 14 lbs. 8 oz.
Ethel of Shelburne 46285 (14
days) .... 28 lbs. 13½ oz.

Millennium 4791.
Redacta 26954................. 14 lbs. 5 oz.

Millionaire Pogis 29688.
Ida's St. Jeannaise 2d 106316.. 17 lbs. 4¼ oz.

Milo 590.
Gamma 4147.................... 14 lbs. 12 oz.
Allie Minka 2982.............. 14 lbs. 6½ oz.
Cigarette 2849................ 14 lbs. 4 oz.

Mingo 948.
Molly May 17202............... 14 lbs. 15 oz.

Minnie's Duke of Darlington 6934.
Hawthorn Minnie 26687........ 18 lbs. 12 oz.

Minnie's Stoke Pogis 17201.
Miss Rainy 66893.............. 15 lbs. 2 oz.

Mint 23600.
Adonia 2d 103302.............. 17 lbs. 6 oz.
Minty 2d of Hood Farm 112326 15 lbs. 3½ oz
Adaphine 101425............... 15 lbs. 3 oz.
Minty 2d of Hood Farm 112326
(11 days).................... 23 lbs. 12 oz.

Minx 3410.
Pridalia 17249................ 26 lbs. 4 oz.

Mirabeau 3800.
Variella of Linwood 10954.... 14 lbs. 1 oz.

Mirama's Rioter 12611.
Robinette Rioter 57206........ 15 lbs.

Miramon 1451.
Amalthea 11319................ 15 lbs.

Mirza 1300.
Maud Melinda 12126........... 17 lbs. 8 oz.

Mississippi Rioter 20401.
Ruby Jenks 3d 84868.......... 23 lbs. 4½ oz.
Ruby Jenks 2d ...o168......... 23 lbs. 3½ oz.
Jodie Jenks 76739............. 18 lbs. 7 oz.
Ruby Jenks 4th 84871.......... 18 lbs. 5 oz.
Pearl Jenks 3d 84869.......... 15 lbs. 3 oz.
Pearl Jenks 4th 84872......... 14 lbs. 1½ oz.

Mistletoe's Carlo 12166.
Carlo's Last Polly 94623....... 17 lbs. 6 oz.
Carlo's Lass 94405............ 16 lbs. 15 oz.
Carlo's Polly Rex 94627....... 16 lbs. 12 oz.

Moidore 15911.
Lady Louise D. 58056......... 22 lbs. 12 oz.
Josephine Noble 66441........ 18 lbs. 14 oz.

Molten Gold 12186.
Miss Naomi 45098............. 15 lbs. 3 oz.

Monarch of New Jersey 5199.
Goldstraw 3d 14724............ 14 lbs. 12 oz.

Money Pogis 16674.
Lady May Jefferson 96591..... 14 lbs. 12 oz.

Monitor 878.
Belle Grinnell 4073........... 18 lbs. 8 oz.

Monmouth 210.
Cyrene 4th 480................ 17 lbs. 1 oz.

Monmouth Cyrene 6835.
Lucy McClung 20368........... 14 lbs. 3 oz.

Monmouth Laddie 6996.
Carrie's Daffodil 15232........ 14 lbs. 7¾ oz.

Monsieur 1723.
Azelda 2d 7022................ 15 lbs. 2 oz.

Mopsus 1165.
Miss Willie Jones 6918........ 16 lbs. 4 oz.
Faustine 10354................ 14 lbs. 14½ oz.

Morgan Victor Hugo 15330.
Fantine of Guilford 64480..... 19 lbs. 1 oz.

Moscow 2303.
Lady Adams 2d 6529........... 15 lbs. 3 oz.

Moss Rose Victor 19123.
Moss Rose of Jackson 86709.. 16 lbs. 4 oz.

Moyane's Tormentor 21690.
Tormentor's Fairy 108510...... 22 lbs.
Moyane's Pet 98115........... 16 lbs. 12 oz.
Fay Signal 2d 134018.......... 14 lbs. 4 oz.

Mr. Guppy 993.
Hazalena's Butterfly 10123.... 14 lbs.

Mr. Micawber 556.
Countess Micawber 1759....... 17 lbs. 1 oz.

Much Ado 2405.
Maggie Martin 9562........... 17 lbs. 6 oz.
Roll of Honor 13610.......... 14 lbs. 12 oz.

Muggy 22004.
Diamond's Lass 81767........ 16 lbs. 4 oz.
Lily's Daffodil 89469......... 14 lbs. 8 oz.

Mulberry Lad 3954.
Jennie Myrtle 22977........... 18 lbs. 1 oz.
Lilly Mac 22976............... 14 lbs. 2 oz.

Music (P. S. 118 J. H. B.)
Bella Delaine 10356........... 14 lbs. 2 oz.

Mutiny 17028.
School Marm 67264........... 18 lbs. 4 oz.

Myron 7157.
Myretta 28231................. 14 lbs. 11¼ oz.

Mythology 12448.
Moggy 59626.................. 14 lbs. 8 oz.

Naiad's St. Lambert King 30645.
Amelia S. King 112862........ 16 lbs. 2 oz.
Lucy S. King 106584.......... 15 lbs. 14¼ oz.

Nancy's Rioter 8390.
Pogis Oonan 29890............ 17 lbs. 8 oz.
Nancy's Star 41423............ 16 lbs. 1⅝ oz.
Rioter's Rosaline 35581....... 14 lbs. 14½ oz.
Amna 35588................... 14 lbs. 12¼ oz.
Duchess of Bloomfield 4th
  35584 .... 14 lbs. 8½ oz.
Pogis Oonan 29890 (31 days).. 73 lbs. 5 oz.

Napier (F. S. 275 J. H. B.)
Chestnut's Beauty 21576....... 18 lbs. 4¼ oz.

Napier 2d (P. S. 244 J. H. B.)
Balmoral 2d 13574............. 14 lbs. 10 oz.

Napoleon 291.
Bright Eyes 2d 2290........... 19 lbs. 6 oz.
Gipsy 5th 2252................ 17 lbs. 2 oz.

Napoléon (F. S. 249 J. H. B.)
Princess of Yerba Buena
  Ranch 12626................ 15 lbs. 10 oz.

Napoleon 2d 527.
Topsey K. 22769............... 14 lbs.

Ned 523.
Victoria 3175................. 16 lbs. 1 oz.

Ned Dean 5387.
Tilda Ives 17707............... 15 lbs.
Iryl Dane 3d 38375........... 14 lbs. 15 oz.

Nell's John Bull 21921.
Kitty of Prospect 111121...... 19 lbs. 4¾ oz.
Erie Maid of River View 94307 14 lbs. 12 oz.

Nelusko 479.
Belle of Milford 7445.......... 15 lbs. 6 oz.
Maggie May 3255.............. 14 lbs. 2½ oz.
Gilt Edge 2d 4420............. 14 lbs.

Nemonette's Rioter 16106.
Nerbat 58664.................. 15 lbs. 7 oz.
Rioter's Hortense 51499....... 15 lbs. 1½ oz.

Neptune 5368.
Dora Neptune 20318........... 20 lbs. ½ oz.

Neptune (P. S. 151 J. H. B.)
Polly Neptune 15214.......... 14 lbs. 7 oz.

Neptune 2d (P. S. 206 J. H. B.)
Queen Neptune 15501......... 18 lbs. 13½ oz.

Nero 13.
Effie of Hillside 1521......... 16 lbs. 15 oz.

Nero 7266.
Royal Queen 24428............ 22 lbs. 6 oz.
Dark and Fair 24468.......... 16 lbs. 9 oz.
Lotchen 19823................. 16 lbs. 7 oz.
Miss Le Dain 27472........... 14 lbs. 5¼ oz.

Nero Chief 2d 4217.
Olie's Lady Teazle 12307...... 16 lbs. 5 oz.

Nerontes 9222.
Gray Beauty 3d 19843........ 19 lbs. 4 oz.

Nestor of St. Lambert 22385.
Dainty Rioter of St. L. 81168.. 15 lbs. 10 oz.
Golden Gem of St. L. 81167... 15 lbs. 3 oz.
Nestor's Beauty of St. L.
  74884 .... 14 lbs.
Golden Gem of St. L. 81167 (24
  days) .... 49 lbs. 4¼ oz.

Newton Prince 15221.
Barbara of Melrose 90496...... 14 lbs. 7 oz.

New Year's 4352.
Arletta 3d 14274............... 14 lbs. 13½ oz.

Niburn 7658.
Luani 37989................... 21 lbs.
Chancery 37987............... 18 lbs. 10 oz.
Marburl 37991................. 16 lbs. 12 oz.

Nicklin 11161.
Mary's Nightingale 87263..... 14 lbs. 14½ oz.

Nightingale's Rioter 14935.
Elsie's Lady 81493............ 14 lbs. 9 oz.

Nihilist of St. Lambert 18256.
Marie's Maud 78467........... 18 lbs. 2 oz.
Lemon's Flossie 71214......... 14 lbs. 1¾ oz.

Ninetta's Son 29074.
Nettina Silverette 89997....... 18 lbs. 14½ oz.

Niobe Duke 2364.
Thorndale Belle 3d 10459...... 15 lbs. 15 oz.
Mitten 13368.................. 15 lbs. 11 oz.
Belle Thorne 13369........... 14 lbs. 11 oz.

Niobe Grand Duke 4510.
Alfritha 13673................. 15 lbs. 3 oz.

Niobe's Stoke Pogis 13448.
Lily Niobe 55765............. 21 lbs. 6½ oz.

Noble 901.
Queen of Delaware 17029...... 18 lbs. 13 oz.
Cream of Sidney 17028........ 17 lbs. 2½ oz.
Queen of Clinton 11500........ 15 lbs. 8 oz.

Noble (F. S. 71 J. H. B.)
Jeanne Le Bas 2476........... 15 lbs. 8 oz.

Noble (F. S. 104 J. H. B.)
Regina 2d 2475................ 14 lbs. 8 oz.

Noble (P. S. 93 J. H. B.)
Princess of St. Saviour's 7470. 15 lbs. 10½ oz.

Noble 4th 6246.
Duchess Noble 27090......... 25 lbs. 8 oz.
Worthy Noble 19629.......... 15 lbs. 9½ oz.

Noble Cicero 13562.
Rowena of Elm Spring 54613.. 14 lbs. 4 oz.

Noble of Side View 28280.
Our Daisy Queen 94252........ 15 lbs. 13 oz.
Christine of Side View 2d
94253 .... 14 lbs. 15 oz.

Noble of Sidney 5369.
Ida of Clinton 25419........... 15 lbs. 14 oz.

Noble Paddy 31072.
Queen Ione 113999............ 16 lbs. 3 oz.
Dark Rose 101892............. 15 lbs. 14½ oz.
Queen Irene 101893........... 15 lbs. 12 oz.
Patty Iris 101894.............. 14 lbs. 15 oz.
Idalco 117140.................. 14 lbs. 10 oz.

Noble Rex 5174.
Chautauqua Queen 26403...... 14 lbs. 11 oz.

No License 16030.
Cherry of Glynllyn 81145...... 19 lbs. 4 oz.

Noonday 12961.
Callinette Pogis 99700......... 17 lbs. 84-5 oz.
Silvia Torment 104668......... 15 lbs. 11.2 oz.
Harry's Ruth Morgan 107773.. 14 lbs. 1.6 oz.

Norajah 812.
Hazen's Nora 4791............. 20 lbs. 5 oz.
Arawana Buttercup 6052...... 15 lbs. 5 oz.
Arawana Poppy 6053......... 15 lbs. 2 oz.

Nora's Champion 20739.
Delia Martin 92358........... 18 lbs. 12 oz.

Nora's Longfellow 7427.
Maud of Maple View 39511.... 23 lbs. 7½ oz.

Noremac 26958.
Misty Morn 88725............. 17 lbs. 5 oz.

Norman B. 7001.
Edith Campbell 23611.......... 21 lbs. 4½ oz.

Normandy 1046.
Normanda 3914................. 14 lbs. 14 oz.

North Pacific Prize 11459.
Damara 65001................. 16 lbs. 7 oz.
Romantic 61626............... 15 lbs. 9 oz.
Miralba 65002................. 14 lbs. 7 oz.

Norton 31941.
Sadie R. 3d 83729............. 20 lbs. 14¼ oz.

Norwood 1077.
Goldthread 4945............... 17 lbs. 9 oz.

Nutley Duke 10706.
Sadie's Delight 53392.......... 14 lbs. 12 oz.

Nutley's Favorite 5132.
Nutley Silverette 22410........ 22 lbs. 7 oz.

Nutshell 729.
Pride of the Hill 4877........ 14 lbs. 8 oz.

Odelio 20223.
Tormendova 74885............. 19 lbs. 12 oz.
Sœurette 2d 86414............. 18 lbs. 1½ oz.
Successor 72503............... 17 lbs. 4 oz.
Odalette 76888................. 17 lbs. 3 oz.
Bernice of Clovernook 72805.. 17 lbs. 1 oz.
Witching Hour 54947.......... 16 lbs. 13½ oz.
Sordella 76876................. 15 lbs. 9 oz.
Roxena 64212................. 15 lbs. 8½ oz.
Susan Bell 112146............. 14 lbs. 13 oz.
Pinkey Spry 2d 92953.......... 14 lbs. 10½ oz.
Elainette's Odelio 102115..... 14 lbs. 3½ oz.
Tormendova 74885 (14 days).. 37 lbs. 11 oz.

Odelio 2d 30299.
Guess Not 2d 133252.......... 16 lbs. 11¼ oz.
Odelsign of Lynwood 114421... 15 lbs. 8½ oz.

Odelle's Pogis 25075.
Young Miss 99631............. 16 lbs.
My May Pogis 84236.......... 14 lbs. 5 oz.
Myrtle Vance 84237............ 14 lbs. 3 oz.

Okubo 1876.
Bronze Leaf 14902............. 15 lbs. 1 oz.
Irene of Short Hills 5137...... 14 lbs. 6½ oz.

Old Hickory 2d 9224.
Grace's Nightingale 19855..... 14 lbs. 2 oz.

Old Tenella's Signal 34802.
Heclatette 119403............. 16 lbs. 9½ oz.
Fancy Alice 106402............ 14 lbs. 11½ oz.

Olymph's Koffee 20087.
Koffee's Lady Pogis 65219..... 23 lbs. 4 oz.
Koffee's May Belle 65220..... 20 lbs. 5 oz.
Koffee's Clarinda 65216....... 19 lbs. 9½ oz.

Omaha 482.
Metah's Queen 4886............ 17 lbs. 9 oz.
Mendota 3d 26326.............. 15 lbs. 6 oz.
Bryant 4193................... 14 lbs. 8 oz.
Metah's Baby 9710............. 14 lbs. 5 oz.

O'Malley 6441.
Flight of Willow Farm 24783.. 20 lbs. 12 oz.
Cinxia St. Helier 46730........ 17 lbs. 9 oz.

Ona's Koffee 9745.
May Koffee 45862.............. 15 lbs. 5½ oz.

Ona's Koffee 2d 17609.
Annie Linn 82892............. 15 lbs. 2 oz.

Oneco 918.
Rosabel Hudson 5704......... 15 lbs. 12 oz.

One Hundred per Cent. 16590.
Annie's Prize 86548............ 19 lbs. 11¾ oz.
St. Lambert's Nora 3d 111218.. 17 lbs. 5 oz. ,
Marjoram of Glen Rouge 2d
  123634 .... 17 lbs. 3 oz.
Heiress of St. Lambert 56507.. 16 lbs. 13¼ oz.
Ethel Marjoram 87266......... 16 lbs. 11 oz.
Cassie Marjoram 120666....... 16 lbs. 5 oz.
Molina of Glen Rouge 123635.. 16 lbs. 2½ oz.
Cassie of Glen Rouge 112456.. 15 lbs. 9 oz.
Madge of Glen Rouge 112455.. 15 lbs. 2 oz.
Jetsam's May 62530........... 14 lbs. 5 oz.

Ontario 865.
Maple Leaf 4768.............. 14 lbs. 12 oz.

Oonan's Harry 25877.
Exile's Signal Belle 2d 106097. 19 lbs. 13 oz.

Oonan's Pogis 17165.
Minnie Hauk 2d 99103........ 16 lbs. 1 oz.
Nanalda 2d 97130.............. 16 lbs. 1 oz.
Topsy's Oonan 97132.......... 15 lbs. 9½ oz.
Oonamistle 64469............. 15 lbs. 4 oz.
Oonazeluss 77481.............. 15 lbs. 2½ oz.
Fancy's Moyane 122270....... 15 lbs. ½ oz.

Oonan's Rajah 8965.
Oonan's Jeannette 45616...... 20 lbs. 3 oz.
Même 34921................... 18 lbs. 4 oz.
Miss Peacock 30871........... 16 lbs. 10 oz.
Pearl's Oonan 32288.......... 15 lbs. 5½ oz.
Même 34921 (16 days)......... 37 lbs. 5 oz.

Oonan's Signal 11586.
Princess Oonan 84369......... 15 lbs. 13½ oz.
Iliana 44594.................. 15 lbs. 5 oz.
Pearl of Oakwood 37722....... 14 lbs. 1 oz.

Oonan's Tormentor 22280.
Oonan of Riverside 69773..... 34 lbs. 3 oz.
Oonan's Grace 86489.......... 24 lbs. 3 oz.
Clara Oonan 78454............ 21 lbs. 2 oz.
Sigletta 3d 108977............ 20 lbs. 10½ oz.
Oonan's Alda 108337.......... 20 lbs. 5¾ oz.
Soconee 3d 114422............ 20 lbs. 1½ oz.
Lady Oonan Pogis 119154..... 19 lbs. 6¾ oz.
Oonan's Oonan 99574......... 17 lbs. 12¾ oz.
Clara Tormentor 2d 97279..... 17 lbs. 6 oz.
Marquette Tormentor 131897.. 16 lbs. 14½ oz.
Sataline 2d 131898............ 16 lbs. 7½ oz.
Alvoretta 93817................ 15 lbs. 14 oz.
Tornella 78453................ 15 lbs. 12 oz.
Kate Gordon 2d 83662......... 15 lbs. 5½ oz.
Tease of Clovernook 92954.... 15 lbs. 2 oz.
Nanoonan 98775............... 14 lbs. 11¾ oz.
Cinderella's Oonan 108764..... 14 lbs. 9¾ oz.
Ida's Dream 3d 131899........ 14 lbs. 7¼ oz.
Lily Martin 2d 89080.......... 14 lbs. 6½ oz.
Moyane C. 86492 (14 days).... 37 lbs. 15¼ oz.
Miss Gracie 86495 (14 days).... 37 lbs. 12 oz.

Oonan's Tormentor Pogis 30505.
Dollie's Valentine 105049...... 18 lbs. 1¾ oz.
Virginia Taylor 117964........ 17 lbs. 2¾ oz.
Justa Pogis 2d 115494......... 14 lbs. 9½ oz.
Dollie's Valentine 105049 (21
  weeks) .... 345 lbs. 11¾ oz.

Optimus 1607.
Monmouth Duchess 4th 7129.. 18 lbs.
Warren's Duchess 4622........ 16 lbs. 1 oz.
Monmouth Duchess 3d 4620... 14 lbs. 4 oz.

Oracle 30340.
Tuleema Pogis 100069......... 15 lbs. 4 oz.

Orangepeel 502.
Lustre 2062................... 15 lbs. 8½ oz.

Orange Peel 864.
Valma Hoffman 4500......... 21 lbs. 9 oz.
Lady Mary Hampton 4861.... 14 lbs. 5 oz.

Orange Skin 1216.
Gold Trinket 9518............. 16 lbs. 2 oz.
Magnolia Ridgely 17269....... 14 lbs. 8 oz.

Orawapum 2833.
Wine 15739................... 16 lbs. 4 oz.
Corn 10504................... 16 lbs. 2 oz.
Verona 10766................. 15 lbs. 1½ oz.

Orleans 533.
Stanstead Belle 4709......... 15 lbs. 4 oz.

Orloff 3143.
Lisgar's Ella 24992........... 17 lbs.
Carrie Pogis 22568............. 15 lbs. 9 oz.

Orphan 891.
Peggy of Staatsburgh 2342.... 14 lbs. 1¼ oz.

Orphan (P. S. 256 J. H. B.)
Gay Orphan 25985............. 18 lbs. 1 oz.
Gay Orphan 25985 (90 days)... 138.973 lbs.

Osiris 3792.
Countess Godiva 10820........ 15 lbs. 7 oz.

Osiris of Fernwood 16078.
Marcia G. 53211................ 14 lbs. 11 oz.

Osprey 17385.
Olive of One Pine 93449....... 17 lbs. 9 oz.

Othello 1114.
Dark Cloud 9364.............. 15 lbs. 3½ oz.

Othello of Elmarch 3780.
Cream of Springvale 16621.... 15 lbs. 10 oz.

Othick Pedro 20511.
Calla Europa 77810........... 20 lbs. 6 oz.
Mooly Pedro 77811............. 17 lbs. 2 oz.

Owen St. Lambert of Lawn 24992.
Edna F. of Lawn 73545........ 15 lbs. 3 oz.

Oxford Catono 23192.
Oxford Catena 68905.......... 15 lbs. 11¼ oz.

Oxford Pride 19160.
Fairy Oxford 66679........... 17 lbs. 7½ oz.
Rosa Oxford 66678............. 15 lbs. ½ oz.

Oxford Rioter 5992.
Wardalia 2d 33970............. 24 lbs. 6 oz.
Rioter's Violet 33774.......... 14 lbs. 3 oz.

Oxeli 1922.
Volie 19465.................... 18 lbs. 1 oz.
Lesbie 9179.................... 16 lbs. 3 oz.
Silenta 17685.................. 15 lbs. 10 oz.
Renini 9181.................... 14 lbs. 10½ oz.
Bintana 9837.................. 14 lbs. 3½ oz.
Taglioni 9182.................. 14 lbs. 1 oz.

Paddy Wilson 3084.
Lady Bidwell 10303............ 15 lbs. 12 oz.

Padisha 1623.
Pet Lee 7993................... 14 lbs. 12 oz.

Palm Nut 4744.
Peggy Ford 21713............. 14 lbs. 10 oz.

Pansy Pogis Rioter 14064.
Pansy Albert of Lawn 76634... 15 lbs. 14½ oz.
Combination of Lawn 76635... 14 lbs. 2½ oz.

Pansy's Double Rex 9997.
Lass of Burrwood 37437....... 14 lbs. 1½ oz.

Parade 30875.
Duchess of Jefferson 2d 113168 14 lbs. 7 oz.

Parole 26114.
Parole's Duchess 133953....... 15 lbs. 11 oz.

Parole of Honor 4291.
Minnie's Ouida 33117.......... 14 lbs.

Pascarrelle 16603.
Blanche of Castile 43793....... 22 lbs. 1½ oz.

Pasha (P. S. 64 J. H. B.)
Regina 4th 12732.............. 17 lbs. 13½ oz.
Faultless 12018................ 17 lbs. 5½ oz.
Frugal 14925.................. 17 lbs. 2¾ oz.

Pasha (P. S. 1901 J. H. B.)
Pristine 126329................ 14 lbs. 4½ oz.

Paterson 11.
Cowslip 5th 849............... 15 lbs. 4 oz.

Pauline's Stoke Pogis 11859.
Katy Signal 2d 44673.......... 16 lbs. 8 oz.
Rescue of St. Lambert 48939... 16 lbs. 8 oz.
Goldie Pogis 45423............. 16 lbs. 2 oz.

Paul Potter 12119.
Mirnie Potter 72670........... 14 lbs. 10 oz.

Pearleta's Scituate 17109.
Sadie T. 75896................. 16 lbs. 6 oz.

Pearl Rex 4438.
Beauty of Ninon 2d 18444..... 15 lbs. 3½ oz.

Pedro 3187.

Pedro's Royal Princess 88485.	23 lbs. 14 oz.
Golightly 25597	23 lbs. 4 oz.
Pedro's Fame 33836	19 lbs. 12 oz.
Pedro's Pansy 33835	19 lbs. 10 oz.
Pedro's Minette 66092	19 lbs. 8 oz.
Pretty Marjoram 77943	19 lbs. 6 oz.
Pedro's Rosebud 66086	19 lbs. 4 oz.
Pedro's Rainbow 66096	18 lbs. 15 oz.
Pedro's Fair Marjoram 88489	18 lbs. 12 oz.
Pedro's Lady 36715	18 lbs. 12 oz.
Pedro's Pretty Maggie 102923	18 lbs. 9 oz.
Pedro's Pretty Belle 77946	18 lbs. 8 oz.
Pedro's Duchess 29581	18 lbs. 7½ oz.
Pedro's Hansom Marjoram 96934	18 lbs. 6 oz.
Pedro's Bonnie Rainbow 88612	18 lbs. 5 oz.
Pedro's Young Princess 72310	18 lbs. 4 oz.
Pedro's Dolly 77948	17 lbs. 10 oz.
Pedro's Phillis 66079	17 lbs. 9 oz.
Pedro's Priscilla 99413	17 lbs. 6 oz.
Pretty Phillis Pogis 66100	17 lbs. 6 oz.
Pedro's Pretty Phillis 102378	17 lbs. 4 oz.
Pedro's Mab 40212	16 lbs. 10 oz.
Pedro's Maggie 102922	16 lbs. 6 oz.
Pedro's Pretty Lass 135067	16 lbs. 6 oz.
Skipover 20217	16 lbs. 5½ oz.
Pedro's Very Pretty 110508	15 lbs. 4 oz.
Pedro's Golden Harp 66098	15 lbs. 3 oz.
Lorne's Pretty Marjoram 77942	14 lbs. 14 oz.
Pedro's Pretty Marjoram 77947	14 lbs. 10 oz.
Pedro's June 38790	14 lbs. 6 oz.

Pedro 2d 5419.

Pedro Girl 22400	18 lbs. 6 oz.

Pedro 3d 12583.

Benefit's Daughter 63932	20 lbs.

Pedrolier 12412.

Pierrot's Beauty 48964	15 lbs. 11 oz.

Pedro Marjoram Albert 38326.

Pedro's Golden Beauty 110402.	18 lbs.

Pedro of Linden 26015.

Rioter's Pretty Belle 88543	15 lbs. 10 oz.

Pedro of the Valley 8750.

Braw Lassie 36221	17 lbs. 6 oz.
Poppy H. 34865	17 lbs. 5½ oz.
Quail 3d 43568	16 lbs. 12 oz.
Fancy Ona 34867	15 lbs. ½ oz.
Niobe Gordon 3d 44041	14 lbs. 3 oz.
Birdsey's Surprise 48326	14 lbs. 2 oz.

Pedro Royal Marjoram 28560.

Pedro's Polly Rex 99845	21 lbs. 4 oz.
Infanta Pedro Marjoram 88584	17 lbs. 6 oz.
Pedro's Golden Harp 2d 103365	16 lbs. 9 oz.
Rioter Flora Marjoram 111165	16 lbs. 4 oz.
Griselda Marjoram 111166	15 lbs. 6 oz.

Pedro Sheldon 22865.

Pedro's Clara 96361	16 lbs. 13 oz.
Pedro's Zoe 95697	15 lbs. 10 oz.
Pedro's Laura B. 96363	14 lbs. 10 oz.

Pedro Signal Landseer 30212.

Pedro's Fancy 4th of H. F. 119842	14 lbs. 14 oz.

Pedro's Silver Rioter 31320.

Pedro's Pretty Dolly 102376	18 lbs. 13 oz.
Pedro's Silver Pearl 110401	18 lbs. 6 oz.
Pedro's Pretty Princess 102377	17 lbs. 6 oz.
Pedro's Pretty Lady 102379	15 lbs. 12 oz.

Peggy's Royal Signal 22613.

Lady Kate Rex 91751	14 lbs. 8 oz.

Perfect Pacific 16994.

Quickstep 85969	17 lbs. 9 oz.

Perrot (P. S. 342 J. H. B.)

Eastwood Clearwater 30445	27 lbs.
Miss Nellie Parker 24988	23 lbs. 11 oz.
Sylvia Perrot 30963	16 lbs. ½ oz.
Perrot's Baroness 22508	15 lbs. 14½ oz.

Pertinatti 713.

Renalba 4117	17 lbs. 4½ oz.
Romp Ogden 2d 4764	15 lbs. 5 oz.
Pixie 4115	14 lbs.

Pertinax 1965.

Selita J. 32184	25 lbs. 2½ oz.
Belle Williamson 8386	19 lbs. 10½ oz.
Kate Gordon 8387	15 lbs. 15 oz.

Peter 835.

Beauty 2076	15 lbs. 7 oz.

Petit Victor 24788.

Coix 85258	15 lbs. 4 oz.
Extract 3d 85261	14 lbs. 6 oz.

Pet's Landseer 29392.

Davy's Dot 2d 85744	14 lbs. 4 oz.

Pharos 3552.

Belle of Scituate 7977	16 lbs.

Pharos Jr. 3621.

Lass of Scituate 9555	15 lbs. 14 oz.

Phil Bree 2957.

Ultima 14456	15 lbs. 12 oz.

Philidor (P. S. 276 J. H. B.)

Glory of Elmarch 21521	15 lbs. 13½ oz.

Pickwick 2d 8915.

Jonquil of Bingo Farm 41786.	19 lbs. 3 oz.

Pierrot (F. S. 143 J. H. B.)
Mischief Le Brocq 7680....... 15 lbs.

Pierrot 2d 1669.
Julia Walker 10133............ 15 lbs. 12 oz.
Palestina 4644................ 15 lbs. 8 oz.
New London Gipsy 11667...... 14 lbs. 8 oz.
Rosy Kate 10276.............. 14 lbs. 4 oz.
Queen Fannie 10275.......... 14 lbs. 2 oz.

Pierrot 7th 1667.
Pierrot's Lady Bacon 12482... 16 lbs. 10 oz.
Palestine Pierrot 24099....... 14 lbs. 6 oz.
Palestine's Last Daughter
12602 .... 14 lbs. 6 oz.
Ruth Morgan 24098........... 14 lbs. 6 oz.

Pierrot's Telephone 5440.
The Sister 21135.............. 14 lbs. 4½ oz.

Pilot 3549.
Lady Gilmore 12136.......... 14 lbs. 14 oz.

Pilot (P. S. 183 J. H. B.)
Oxford Kate 13646............ 39 lbs. 12 oz.
Pilot's Veronica 18917........ 20 lbs. 2 oz.
Pilot's Rose 17958............. 18 lbs. 3¾ oz.
Cabinet 22662................. 15 lbs. 10 oz.
Fauchon 2d 18200............. 14 lbs. 9 oz.

Pilot Jr. 141.
Silver Rose 4753.............. 16 lbs. 14 oz.

Pilot of Bowker Farm 6085.
Musidove 25379............... 18 lbs. 2 oz.
Daisy Pilotta 68852........... 17 lbs. 12 oz.
Topsey Doe 25378............. 15 lbs. 4 oz.
Musidove 3d 61641............ 15 lbs. 1 oz.

Pine Cliff 1106.
St. Perpetua 2d 5557.......... 14 lbs.

Planet 1130.
Chlora 8325................... 14 lbs. 6 oz.

Platon (F. S. 310 J. H. B.)
Dewdrop Pansy 19736........ 19 lbs. 8 oz.

Plebeian 31020.
Miss Flora Brown 100532...... 16 lbs. 2 oz.

Plough Boy (P. S. 102 J. H. B.)
Lille Bonne 8108.............. 19 lbs. 12 oz.

Plymouth Champion 27227.
Agnes Clover 86378........... 14 lbs. 12 oz.

Pocatello 21953.
Sayda M. 3d 71134............ 20 lbs. 10 oz.

Pogis' Boom 18295.
Corinne of St. Lambert 76579.. 15 lbs. 10 oz.

Pogis Chief 3998.
Princess of Hanover 46706.... 18 lbs. 12 oz.
Donney Pogis 38907........... 16 lbs. 2¼ oz.

Pogis Maximum 20761.
Maxminta 85922.............. 15 lbs. 13½ oz.
Max Polly 108994............. 15 lbs. 8½ oz.

Pogis of Shadow Brook 27926.
Fennel of Glynllyn 109755..... 17 lbs. 4 oz.

Pogis Whiting 28907.
Nanette of Hood Farm 104447. 15 lbs. 8½ oz.

Polonius 2513.
May Evening 15938........... 17 lbs. 13 oz.

Polonius De Pansy 6625.
Countess Bee 32630........... 14 lbs. 4 oz.

Pompus 2881.
Bristol Bella 15697............ 15 lbs. 3½ oz.

Ponterrin's Duke 14999.
Belle of River's Bank 48023.. 15 lbs. 10 oz.

Pope St. Lambert 19850.
Bessie Pope St. Lambert 80352 16 lbs. 1¼ oz.

Potiphar 19988.
Moorey of Lawn 68847........ 21 lbs. 4 oz.

Potomac 153.
Nellie 1507................... 14 lbs. 2 oz.

Pouch of Gold 4121.
Leoline 2d 18315.............. 14 lbs. 4 oz.

Pretender (P. S. 187 J. H. B.)
Princess of the Valley 22641.. 18 lbs. 12 oz.

Pride of New York 19787.
Martha D. 66641.............. 17 lbs. 2 oz.

Pride of the Island 5416.
Pride's Lady Frances 39529... 16 lbs. 3½ oz.
Pride's Queen 66442.......... 15 lbs. 8½ oz.
Pride's Lady Sarah 39537..... 15 lbs. 1½ oz.
Pride's Lady Rushmore 35182. 15 lbs. 1 oz.
Pride's Lady Lillie 35181..... 15 lbs. ¾ oz.
Island Dots 17003............. 14 lbs. 9 oz.

Pride's Landseer 18590.
Geneva M.'s Fancy 104917.... 16 lbs. ½ oz.
Hattie Landseer 72719......... 14 lbs. 4 oz.

Pride's Pogis 23459.
Anna Paris 67989.............. 14 lbs. ⅜ oz.

Primrose's Carlo 12174.
Cora of Sandusky 42038....... 15 lbs. 4½ oz.
Primrose of Summerland
66059 .... 15 lbs. 4½ oz.

Prince, on I. of J.
Flora of St. Peter's 8622...... 16 lbs. 5 oz.

Prince Boulivot 8757.
Princess Minette 24042........ 18 lbs. 6½ oz.

Prince George 11571.
Silicon 25577.................... 18 lbs. 13 oz.
Violente 28819................. 17 lbs. 10½ oz.

Prince Harry 5176.
Cricket's Minnie 26270......... 15 lbs. 14 oz.

Prince Jeannaise 17074.
Princess Jeannaise 53863...... 15 lbs. 12 oz.

Prince Milan 2318.
Beauty of Lee Farm 15694.... 20 lbs. 12¾ oz.

Prince of Amherst 25728.
Princess Rosalind 87288....... 15 lbs. 1 oz.

Prince of Avon 17544.
Millplain Princess 78579....... 18 lbs. 2 oz.
Baby Princess 57493.......... 14 lbs.

Prince of Beech Grove 16234.
Lelia Talbot 96922............. 14 lbs. 8½ oz.

Prince of Bridgewater 2d 4489.
Hilda B-B-B 17640............. 15 lbs. 8 oz.

Prince of Broadmoor 8005.
Hazelwood 52239............. 14 lbs. 1 oz.

Prince of Croton 2490.
Prince's Bloom 9729.......... 14 lbs. 3 oz.

Prince of Darlington 5089.
Niobomba 26456............... 14 lbs. 15 oz.

Prince of Elmwood 20994.
Ada of Elmwood 98728........ 14 lbs. 2 oz.

Prince of Ferncliff 24294.
Granny's Baby 5th 95132...... 15 lbs. 8½ oz.

Prince of M. 2811.
Dura 26001.................... 21 lbs. 5 oz.
Lilley Rex 9852................ 14 lbs. 7 oz.

Prince of M. 2d 5507.
Sayda 3d 17317 (30 days)...... 47.825 lbs.
Sayda 3d 17317 (90 days)...... 170.094 lbs.

Prince of Mahaska 16159.
Sulphide 88088................ 18 lbs. 5 oz.

Prince of Maple Lane 14361.
Baronet's Victory 2d 39176.... 17 lbs. 5 oz.

Prince of Melrose 4819.
Romping Miss 54457.......... 23 lbs. 1 oz.
Busy Melrose 54456........... 21 lbs. 1¼ oz.
Princess Corinne 48203....... 19 lbs. 1 oz.
Sunny Pogis 61919............ 17 lbs. 9½ oz.
Busy Princess 48202.......... 17 lbs. 1¾ oz.
Belle Melrose 61920........... 17 lbs. 1½ oz.
Corinne Melrose 54455........ 16 lbs. 14 oz.
Fleeter 61921.................. 16 lbs. 13½ oz.
Fleecy Princess 40342........ 16 lbs. 11½ oz.
Melrose Milly 61917.......... 16 lbs. 2½ oz.
Fall Leaf's Heiress 54454..... 16 lbs. 2 oz.
Saucy Princess 40350......... 15 lbs. 11 oz.
Delesline 44474............... 15 lbs. 9 oz.
Melrose's Signal 54830........ 15 lbs. 5⅝ oz.
Bee Princess 40345............ 14 lbs. 2⅝ oz.
Busy Melrose 54456 (31 days).. 85 lbs. 2¾ oz.

Prince of Melrose 2d 11015.
Hugo's Joie de Vie 40372...... 24 lbs. 5 oz.
Lady Alice Pogis 30273........ 22 lbs. 8 oz.
Clothilde of St. Lambert 72271 22 lbs. 7½ oz.
Pogis' Dewdrop 40373......... 19 lbs. 12½ oz.
Amanda of St. Lambert 64955. 18 lbs. 14 oz.
Victor Hugo's Maggie 30274.. 18 lbs. 14 oz.
Angelina Pogis 40367.......... 18 lbs. 13 oz.
Imogene Pogis 44967.......... 18 lbs. 13 oz.
Gipsy Pogis 35422............. 18 lbs. 7 oz.
Litta Pogis 30275.............. 18 lbs. 6 oz.
Alberta Hugo 35424........... 17 lbs. 9 oz.
Jessie Lee Pogis 44969........ 17 lbs. 4 oz.
Pomona Pogis 40371.......... 17 lbs. 2½ oz.
Christina Pogis 35426......... 17 lbs. 2 oz.
Lotta Pogis 40368............. 16 lbs. 11 oz.
Nannie Pogis 53786........... 16 lbs. 11 oz.
Susie Pogis 35425............. 16 lbs. 9 oz.
Katie Putnam 40366.......... 16 lbs. 4 oz.
Hermosa Pogis 40369......... 16 lbs. 3½ oz.
Zenobia Pogis 35421.......... 15 lbs. 10 oz.
Epple Pogis 44972............. 15 lbs. 6 oz.

Prince of Milford 7951.
Kalma of Milford 52742....... 15 lbs. 12½ oz.

Prince of Montverde 2d 19975.
Coupon Diamond 103759....... 14 lbs. 5 oz.

Prince of St. Lambert 5287.
Hara of St. Lambert 42128..... 17 lbs. 7½ oz.
Mahoning Bess 35834.......... 15 lbs. 8¾ oz.
Princess Zeroda 107242........ 14 lbs.

Prince of Scituate 3888.
Lily Scituate 12665............ 24 lbs. 9½ oz.
Annie Dale's Princess 12664... 19 lbs. 2½ oz.

Prince of Sidney 2665.

Daisy of Chenango 18582	14 lbs.	7 oz.
Daisy of Guilford 18583	14 lbs.	6 oz.

Prince of Tennessee 20772.

Prince's Christmas Gift 92513	24 lbs.	12¼ oz.
Princess Winnie 63508	21 lbs.	4¼ oz.
Miss Mouse 92512	19 lbs.	8¾ oz.
Cream Custard 97429	19 lbs.	6¼ oz.
Prince's Rubetta 92785	19 lbs.	4 oz.
Moxie's Mode 97432	16 lbs.	3¼ oz.
Princess of Lynwood 118458	16 lbs.	½ oz.
Maumee Girl 118155	15 lbs.	10¾ oz.
Süsschen 125091	15 lbs.	6¾ oz.
Marchande 3d 133818	15 lbs.	6¼ oz.
Blossom's Belle 97433	15 lbs.	6 oz.
Katie's Kind 97434	15 lbs.	5¼ oz.
Kammerfrau 125093	15 lbs.	4 oz.
Sünderin 125092	15 lbs.	½ oz.
Die Königin 124952	14 lbs.	12½ oz.
Idelle's Idea 92510	14 lbs.	10¾ oz.
May Cuckoo M. 97417	14 lbs.	8 oz.

Prince of the Realm 7842.

Bettie's Pet 22550	16 lbs.	1 oz.

Prince of the Valley (P. S. 88 J. H. B.)

Faith of Oaklands 19696	17 lbs.	4 oz.

Prince of Warren 1512.

Canata 10523	27 lbs.	
Dot of Bear Lake 6170	19 lbs.	4 oz.
Conover's Beauty 2d 25315	18 lbs.	¾ oz.
Ida of Bear Lake 6169	16 lbs.	
Mary of Bear Lake 6171	15 lbs.	14 oz.
Lady Monmouth 15173	15 lbs.	3 oz.
Lena Lowndes 23202	14 lbs.	9 oz.

Prince of Woodstock 5030.

Sunset 15130	16 lbs.	2½ oz.

Prince Pedro 10588.

Mildred G. 45077	17 lbs.	5 oz.

Prince Pogis 10682.

Lady Larkspur 56051	21 lbs.	2¾ oz.
No Chemicals 71452	16 lbs.	9 oz.
Mrs. Jack 50805	15 lbs.	5 oz.
Princess Chrysantha 34191	14 lbs.	3½ oz.

Prince Pogis' Hugo 15660.

Tritoma Pogis 44952	16 lbs.	13 oz.
Princess Edith Hugo 45120	15 lbs.	14½ oz.
Hugo's Countess 68394 (30 days)	48.172 lbs.	
Hugo's Countess 68394 (90 days)	191.894 lbs.	

Prince Ridgely 3047.

Niobe of Linwood 11134	14 lbs.	12 oz.

Prince's Lambert 16490.

Princess Marie G. 120390	15 lbs.	12½ oz.
Princess Marie G. 120390 (28 days)	60.98 lbs.	
Princess Marie G. 120390 (32 days)	69 lbs.	8 oz.

Prince Victor Pogis 14257.

Bettie Pogis 44542	16 lbs.	7½ oz.

Prize La France 8162.

La France's Pansy Rex 49141	14 lbs.	10½ oz.

Producer 9367.

Gretchen of Pevely 46325	15 lbs.	5½ oz.

Profundo Basso 20321.

Musicale of Greenwood 62932	15 lbs.	14 oz.

Progress (F. S. 286 J. H. B.)

Beauty Romeril 26090	18 lbs.	9 oz.
La Jolie Lizette 16426	16 lbs.	1 oz.
Delia of Grouville 16428	14 lbs.	4 oz.

Progress 2d (P. S. 298 J. H. B.)

Moyane 21595	21 lbs.	12½ oz.

Prospect 2047.

Edith Darby 6246	14 lbs.	2 oz.

Prospect's Rioter 9189.

Marjoram's Matilda 77944	19 lbs.	6 oz.
Rioter's Lorne Primrose 88583	18 lbs.	15 oz.
Pedro's Pretty Gem 69953	18 lbs.	8 oz.
Pedro's Pretty Flower 88542	18 lbs.	6 oz.
Pedro's Pretty Pansy 75275	18 lbs.	4 oz.
Pedro's Little Wonder 72379	17 lbs.	14 oz.
Pedro's Hansome Phillis 96935	17 lbs.	8 oz.
Pedro's Pretty Girl 88488	17 lbs.	4 oz.
Pedro's Pretty Duchess 88558	16 lbs.	11 oz.
Pedro's Gay Lady 88561	16 lbs.	10 oz.
Pedro's Lady Phillis 88486	16 lbs.	4 oz.
Rioter's Pretty Pearl 75273	15 lbs.	8 oz.
Pedro's Princess of York 72378	15 lbs.	4 oz.

Proven Pogis 19594.

Dido Pogis 67260	15 lbs.	9 oz.

Pulaski 1932.

Cowles' Nonesuch 6199	14 lbs.	12 oz.

Pussie's Pogis 25935.

Miss Morehouse 91703	15 lbs.	5½ oz.
Clark's Gift 91708	15 lbs.	4 oz.

Quaker 887.

Country Girl 3515	15 lbs.	12 oz.

Queen's Legacy 29652.

Dido B. 112926	15 lbs.	

Quintuple Victor Pogis 22343.

Signolo's Florence 2d 82446...	16 lbs.
Quintuple's Wonder 86077.....	15 lbs. 14 oz.
Quintuple's Florence 86076....	15 lbs. 10 oz.
Quintuple's Daisy 86078......	14 lbs. 5 oz.

Rab 22937.

Wild Crocus 92486...........	18 lbs. 2¾ oz.

Rabbi 2496.

Mildred of M. 15548...........	14 lbs. 2½ oz.

Rachel's Duke 7022.

Cultured Cream 29196........	20 lbs. 12 oz.
Pride's Olga 37186...........	19 lbs. 12 oz.
Real Queen 29198.............	18 lbs. 1 oz.
Jennie Cream 30622...........	17 lbs. 4 oz.
Duke's Rowena 37190.........	14 lbs. 8 oz.

Rajah 340.

Oonan 1485	22 lbs. 2½ oz.
Fantine 1271	16 lbs. 6 oz.

Rajah Stoke Pogis 25276.

Dubenna 65024................	14 lbs. 7½ oz.

Raleigh Lisgar 17981.

Miss Sassafras 70463.........	17 lbs. 2 oz.

Ralph 957.

Mhoon Lady 6560.............	17 lbs. 3 oz.
Cenie Wallace 2d 6557........	15 lbs. 4½ oz.
Florry Keep 6556.............	14 lbs. 14 oz.
Mountain Lass 12921..........	14 lbs. 9 oz.

Ralph Guild 1917.

Mother Cary 11746...........	27 lbs. 1½ oz.
Sallie Guild 14524.............	16 lbs.

Ramapo 4679.

Cornwall Maid 19024..........	29 lbs. 12 oz.
Lady Ramaposa 26232........	17 lbs. 5½ oz.
Musquita 28482................	17 lbs. 3 oz.
Butterstamp Lass 19517......	16 lbs, 11 oz.
Maud's Sultana 19518........	16 lbs. 4 oz.
Ramapo's Pride 35497........	14 lbs. 4 oz.

Ramapo Chieftain 15324.

Catchfly 4th 47682............	17 lbs. 4 oz.
Catchfly 6th 55483............	16 lbs. 2 oz.

Ramapogis 3d 26443.

Lois A. 107594................	15 lbs. 14 oz.
Lois B. 107595................	14 lbs. 2 oz.

Ramapose's Bachelor 25900.

Lizzie of Glynllyn 74274......	21 lbs. 15½ oz.
Gertie of Glynllyn 74474......	21 lbs. 2½ oz.
Rheta of Glynllyn 74273......	20 lbs. 4½ oz.
Massey Polo 2d 89216.........	16 lbs. 9 oz.

Rambler of St. Lambert 5285.

Rose of St. Lambert 20426....	21 lbs. 3½ oz.
Rioter's Ruth 14882...........	19 lbs. 6½ oz.
Rioter's Queen 14895.........	17 lbs. 8 oz.
Oaklands Lilly 14881.........	15 lbs. 4 oz.

Ramchunder 718.

Mamie Coburn 3798...........	17 lbs. 8 oz.
Rose of Hillside 3866.........	14 lbs. 3½ oz.

Rayon d'Or 7516.

Jetty d'Or 39845..............	15 lbs. 2 oz.

Ready Money Jack 2986.

Clematis 3d 6653..............	16 lbs. 1 oz.

Real Queen's Stoke Pogis 20535.

Chilioness Queen 82973........	26 lbs. 8 oz.
Pride's Olga 2d 79887..........	22 lbs. 12 oz.
Duke's Rowena 2d's Queen 79894	18 lbs. 12 oz.
Pride's Olga 3d 79897.........	18 lbs. 2 oz.

Record 22801.

Crensa 74203.................	18 lbs. 2 oz.

Recorder 29239.

Miss Retta 105807.............	20 lbs. 1 oz.
Belle Noble's Bessie 105710...	19 lbs. 5 oz.
Pride of Madelon 105711......	16 lbs. 6 oz.

Rector 1458.

Leoni 11868...................	18 lbs. 7 oz.
Bonnie Yost 7943.............	18 lbs. 2 oz.
Dudu of Linwood 8336.......	16 lbs. 15 oz.
Lucetta 6856..................	14 lbs. 3 oz.

Red Cloud 2d 2260.

Fancy Juno 6086..............	15 lbs. 10 oz.

Regina's Sir George 13569.

Big Wyuka 58345..............	17 lbs. 11 oz.
Regina Weston 70290.........	16 lbs. 9½ oz.

Remarkable (F. S. 229 J. H. B.)

Rosa of Belle Vue 6954........	18 lbs. 7½ oz.
Mary Jane of Belle Vue 6956..	17 lbs. 7 oz.
Caroline 12019................	14 lbs. 8 oz.

Revenue 1744.

Bramballetta 10451...........	16 lbs. 4 oz.

Rex 1330.

Mabel's Jewel 6251............	17 lbs. 13 oz.
Arawana Queen 5368.........	16 lbs. 9 oz.
Carrie Rex 10271..............	16 lbs. 7½ oz.
Almeda Rex 10416............	15 lbs. 12 oz.
Princess Bellwort 6801........	15 lbs. 10½ oz.
Favorite Rajah Rex 16153.....	15 lbs. 4 oz.
Louvie 3d 6159................	14 lbs. 13 oz.
Bell Rex 11700...............	14 lbs. 10 oz.
Jeannie Platt 6005............	14 lbs. 4 oz.
Lottie Rex 18757..............	14 lbs. 4 oz.
Pet Rex 20166.................	14 lbs. 2½ oz.

Rex Alphea 4509.
Lass Rex Alphea 16965........ 16 lbs. 10¾ oz.

Rex Alphea's Rioter 9190.
Lily of Elm Spring 51474...... 19 lbs. 8 oz.

Rex of Jersey 12878.
Countess Hebe 2d 45899........ 14 lbs. 3 oz.

R. F. Eric 21330.
Fena Eric 69137................ 22 lbs. 10 oz.
Eric's Ida 69134................ 18 lbs.
Minnie's Pet Jersey 93119..... 15 lbs. 6 oz.

Ridgely of Hampton 3046.
Naomi's Pride 16745.......... 15 lbs. 2 oz.

Rienzi of Winnikee 18948.
Temisia of Winnikee 46604.... 15 lbs. 3½ oz.

Ringleader 392.
Miss Blanche 2515 (10 days).. 20 lbs. 9 oz.

Riot Act 20520.
Chirp 65551..................... 19 lbs. 12 oz.

Rioter 670.
Duchess of Bloomfield 3653.. 20 lbs. ½ oz.
Su Lu 4705..................... 17 lbs. 15 oz.
Letitia 3977.................... 15 lts. 5 oz.
Lady Bloomfield 4704.......... 14 lbs. 12½ oz.

Rioter 2d 469.
Eurotas 2454.................. 22 lbs. 7 oz.
Torfrida 3596................. 17 lbs. 6½ oz.
Eurotas 2454 (one year)....... 7⅞ lbs. 1 oz.

Rioter Hugo Pogis 13457.
Gift 39901..................... 21 lbs. 6 oz.
Rioter Alpha 3d 34073........ 17 lbs. 1½ oz.

Rioter of Brookside 18702.
Gem Nehushta 73258.......... 15 lbs. 6 oz.

Rioter of Maple Glens 21887.
Maple Glens Gold Lorna
84017 .... 15 lbs. 1 oz.

Rioter of Rocky Farm 18183.
Lilium Excelsum 2d 66040..... 14 lbs. 5 oz.

Rioter of St. Lambert 16501.
Maud Signal 2d 68664......... 22 lbs. 4 oz.
Rioter's Fawn 69168........... 20 lbs. 2 oz.
Rioter's Baby 106100.......... 19 lbs. 12 oz.
Bright Eyes St. Lambert 2d
72699 .... 19 lbs. 5 oz.
Orie's Gem of St. Lambert
85472 .... 18 lbs. 4 oz.
Rioter's Esse 81901............ 16 lbs. 7 oz.

Rex's Queen of St. Lambert
59161 .... 16 lbs. 2 oz.
Rioter's Sunshine 75108........ 15 lbs. 10 oz.
Rioter's . Normanda 102441.... 15 lbs. 8 oz.
Velvet Queen 112795........... 15 lbs. 8 oz.
Rioter's Esse 81901 (14 days).. 31 lbs. 6¾ oz.

Rioter's Combination 10363.
Cleora of Cornwall 63408...... 18 lbs. 8 oz.
Alarm of Highland 41564..... 16 lbs.
Careta Pogis 42784............ 15 lbs. 1¼ oz.
Calma of Briarcliff 52599...... 14 lbs. 7½ oz.

Rioter's Pride 11694.
Marjoram's Delle 82954....... 19 lbs. 12 oz.
Pride's St. Lambert Lady
82955 .... 18 lbs. 12 oz.
Charming of St. Lambert
69077 .... 15 lbs. 4 oz.

Rioter's Victory 16446.
Ella Europa 61062............. 14 lbs. 15½ oz.

Rioter Vulcan 5380.
Mintha 12812.................. 15 lbs.
ʟɪɪ d's Jersey Maid 35040 (30
days) .... 55.163 lbs.

Rival 3762.
Rival's Ochra 10172........... 19 lbs. 10½ oz

Rival Rioter 11445.
Mary Fallon 80220............. 17 lbs. 1¼ oz.

Riverside Stoke Pogis 29235.
June Sweets 2d 87967......... 15 lbs. 11 oz.

Roanoke 1448.
Gold Lace 10726............... 21 lbs. 1 oz.
Golden Skin 10861............. 16 lbs. 8 oz.

Robbins 953.
Dusky 2525.................... 16 lbs. 10 oz.

Robert of Green Lawn 27007.
Fruitland May 93002.......... 14 lbs. 5 oz.

Rob Roy 17.
Eugenie 2d 1623............... 14 lbs.

Rocco 4517.
Catchfly 25405................ 18 lbs. 4 oz.
Cora of Arcadia 16151........ 15 lbs.

Rodney 1941.
Lady Oxford 2d 15101......... 14 lbs. 7 oz.

Roger 121.
Io 5th 280..................... 17 lbs. 8 oz.

Roi Rene 23730.
Lady Dorcas 74348.......... 14 lbs. 10 oz.

Romeo de Bonair 4091.
Polly Bonair 42849........... 18 lbs. 7 oz.
Belle Bonair 33960............ 16 lbs. 3½ oz.

Romeo of Sacramento 3030.
Oleta 15625.................... 15 lbs. 12 oz.

Romp's Torment 8789.
Romp's Nervilette 72282...... 17 lbs. 10 oz.
Mayhegtal 64454.............. 16 lbs. 3 oz.
Panache 62065................. 16 lbs. 1 oz.
Nelly Torment 61903.......... 15 lbs. 7 oz.
Romp's Princess 51185 (30 days) .... 51.357 lbs.
Romp's Princess 51185 (90 days) .... 188.373 lbs.

Romp's Tormentor 7126.
Sister Sue 58447............... 19 lbs. 9 oz.
Foolish 64069.................. 17 lbs. ½ oz.
Romping Mirth 64070......... 17 lbs. ½ oz.
Marno B. 56368................ 16 lbs. 10 oz.
Belle of Rydal Grange 41787.. 16 lbs. 6 oz.

Romulus (P. S. 181 J. H. B.)
Mousy 2d 14962............... 17 lbs. 1 oz.

Rory O'More 3236.
Nerissa of Nyack 9692........ 15 lbs. 1 oz.

Rosalie's Pogis 27654.
Daisy Harris 92471........... 17 lbs. 4¼ oz.

Rossman 1128.
Rosetta of Whiteland 6112.... 27 lbs. 2¾ oz.

Rosy King 4433.
Maquilla 24043................ 21 lbs. 1 oz.

Rough (P. S. 239 J. H. B.)
Perry Farm Golden Cloud 22872 .... 18 lbs. 9 oz.
Young Cherry 3d 14980........ 14 lbs. 7½ oz.

Rover of St. Lambert 28333.
Rover's Swan 106934.......... 16 lbs. 3 oz.

Roxana's Gilderoy 11585.
Hattie N. 45433............... 15 lbs. 14½ oz.

Roxbury 247.
Angela 1682.................... 14 lbs. 2 oz.

Royal Allen 28149.
Minnie's Lady Allen 102925... 17 lbs. 10 oz.

Royalist 2d 3791.
Royalist's Daisy 19187......... 21 lbs. 9¼ oz.

Royalist 3d 4500.
Elmwood Daisy 2d 34193...... 16 lbs. 3 oz.
Kate Pansy 15177............. 15 lbs. 1 oz.

Royalist 6th 4977.
Black Teat 32932.............. 16 lbs. 11 oz.

Royalty 7210.
Pedroletta 26597............... 16 lbs. 6½ oz.

Roy Brick 5637.
Embla Brick 15690............ 14 lbs. 3 oz.

Rubano 8806.
Rubano's Normanda 68927.... 21 lbs. 4½ oz.
Rubano's Valentine 54703..... 20 lbs. 3 oz.
Miss May of St. Lambert 37084 .... 15 lbs. 14 oz.
Aggie of St. Lambert 37085... 14 lbs. 2¼ oz.

Rubano's Stoke Pogis 23760.
Ruba H. Pogis 81944........... 18 lbs. 1¼ oz.

Ruby's Harry 15664.
Guess Not 57142................ 20 lbs. 8 oz.
Early Morn 50661.............. 17 lbs. 13¼ oz.
Ruby Tiptoe 78960............. 16 lbs. 12½ oz.

Ruby's Star of St. Lambert 20719.
Rowena of St. Lambert 71526. 20 lbs. 4 oz.
Kittura of St. Lambert 78344.. 18 lbs. 8¾ oz.
Clover of St. Lambert 6 945... 18 lbs. 4¼ oz.
Star's Valentine 81778........ 16 lbs. 2 oz.
Ruby's Valentine 81780....... 14 lbs. 7 oz.
Rowena of St. Lambert 71526 (14 days) .... 39 lbs. 6¼ oz.
Clover of St. Lambert 60945 (15 days) .... 37 lbs. 13¼ oz.
Clover of St. Lambert 60945 (31 days) .... 70 lbs. 3 oz.
Rowena of St. Lambert 71526 (42 days) .... 108 lbs. 2½ oz.
Clover of St. Lambert 60945 (42 days) .... 94 lbs. 6¼ oz.

Rupert 1456.
Roonan 5133................... 20 lbs. 4 oz.

Ryan 9891.
Queryan 42140................. 16 lbs. 2 oz.

St. Helier 45.
Chroma 4572................... 20 lbs. 6 oz.
Meines 3d 7741................ 20 lbs. 1 oz.
Ianthe 4562.................... 16 lbs. 10 oz.
Oxalis 2d 15631................ 15 lbs.
Pavon 12485................... 14 lbs. 8 oz.
Safrano 4568.................. 14 lbs. 2½ oz.
Silene 4307.................... 14 lbs.

St. Helier Berkshire 21030.
Ria Alexandra 83752.......... 14 lbs. 10 oz.

St. Helier Boy 11884.
St. Helier Gem 35438......... 15 lbs. ½ oz.

St. Helier Brave 17343.
Daisy Leto 72404.............. 18 lbs. 1 oz.

St. Helier of Idleside 12200.
Chromatess 51578............. 25 lbs. ½ oz.

St. Ion 18876.
Elena of Oakdale 84162....... 19 lbs. 4 oz.

St. Jacob 6438.
. Gingerbread 5th 37681........ 18 lbs. 9¾ oz.
Idana 27974.................... 14 lbs. 10 oz.

St. John (P. S. 316 J. H. B.)
St. John's Daisy 28388........ 15 lbs. 4 oz.

St. Lambert Boy 17408.
St. Lambert's Riotress 106220.. 24 lbs. 6 oz.
Rioter's Letty 110450......... 21 lbs. 9½ oz.
Lamba 79039.................... 20 lbs. 3 oz.
Eoline of St. Lambert 104819.. 19 lbs. 1 oz.
Blanche Pomeroy 105013....... 16 lbs. 9½ oz.
St. Lambert's Rexina 113609.. 16 lbs. 2 oz.
Ajax Riotress 113709.......... 16 lbs.
Barbara Lambert 105208....... 15 lbs. 11 oz.
St. Lambert's Babette 99185.. 15 lbs. ¾ oz.
St. Lambert B. Kate 105205... 15 lbs.
Sallie Pomeroy 98806......... 14 lbs. 10 oz.
Elinor Wells 2d 72071......... 14 lbs. 1-16 oz.

St. Lambert Gilderoy 20210.
Kitty of Seguin 74347........ 15 lbs. 6¼ oz.

St. Lambert Perrot 23689.
Queen of Corfu 83732......... 21 lbs. ½ oz.
Belle of Corfu 83731.......... 15 lbs. 8 oz.

St. Lambert's Hugo 17457.
Coming Girl 3d 71794......... 17 lbs. 14 oz.

St. Lambert's John Bull 16618.
Helene of St. Lambert 73601.. 20 lbs. 1½ oz.
St. Lambert's Alphea 74100... 20 lbs. ½ oz.
Nellie Hugo 53793............. 20 lbs.
Pogis' Twilight 53794.......... 19 lbs. 1 oz.
Miss Patti Rosa 73607......... 17 lbs. 15 oz.
Hugo's Golden Sheen 61804... 17 lbs. 9 oz.
Amelia Pogis 53796............ 16 lbs. 14 oz.
Señorita Pogis 53791.......... 16 lbs.
Aurora Pogis 53798........... 15 lbs. 9 oz.

St. Lambert's Monarch 25802.
Monarch's Star 80232.......... 18 lbs. 4 oz.

St. Martin 1482.
Enigma 5360................... 15 lbs. 6 oz.

St. Nick 7224.
St. Nick's Flora 16195........ 14 lbs.

St. Valentine 2251.
Colie 8309..................... 18 lbs. 4 oz.

Saladin 447.
Rosaline of Glenmore 3179.... 17 lbs. 12 oz.
Embla 4799.................... 17 lbs. 8 oz.

Sam 980.
Salsoda 3721.................. 14 lbs. 7 oz.

Samson 1079.
Bronzie 9368.................. 21 lbs. 6 oz.
Merlette 4988................. 16 lbs.

Samson Jr. 2723.
Calista of Newark 13296....... 15 lbs. 9 oz.
Phœbe N. 25401............... 15 lbs. 3 oz.
Flora Lee 13294............... 14 lbs. 1 oz.

Sans-Peur (F. S. 201 J. H. B.)
Fear Not 6059................. 17 lbs. 3 oz.
Buttercup 17285............... 16 lbs. 5 oz.
Fan of Grouville 7458.......... 15 lbs.

Saturn 94.
Alphea 171.................... 15 lbs 6 oz.

Saulie 33223.
Sensa 87569................... 21 lbs. 3 oz.

Sayda's Prince 28501.
Jessie's Sayda 105697......... 15 lbs. 10 oz.

Scales' Stoke Pogis 30098.
Ida's Little Wonder 104961.... 14 lbs. 7 oz.

Scepter 5417.
Scepter's Beauty 23234........ 14 lbs. ½ oz.

Schinchon 1132.
Polly Clover 7052............. 16 lbs. 15 oz.

Scion 1033.
Charmer 4771................. 14 lbs. 12 oz.

Scituate F. W. 24624.
Quachette F. W. 83654........ 15 lbs. 3 oz.

Scituate's Prize 7206.
Athlete 28137................. 14 lbs. 6 oz.

Secretary 4074.
Daffy Wilcox 2d 18317......... 15 lbs. 5 oz.

Seneca Chief 4098.
Giulietta Cooke 32193......... 21 lbs. 13½ oz.
Seneca's Mocha 23355......... 20 lbs. 3 oz.
Oléo 38475.................... 18 lbs. 1 oz.
Daisy Stillson 28174.......... 15 lbs. 3 oz.

Sharpshooter of Atlanta 3011.
Tenella 2d 19521.............. 18 lbs. 12 oz.

Sheldon 5250.
Maggie Sheldon 23583........ 21 lbs. 5 oz.
Sheldon's Daisy 30592........ 17 lbs. 12 oz.

Sheldon of St. Lambert 13831.
Nora Sheldon 43590........... 17 lbs. 6 oz.
Pine Tree Clara 55042......... 17 lbs. 3 oz.

Sheldon Pogis 22659.
Sheldon's Maud 86187........ 14 lbs. 15 oz.

Sherbet's Coomassie 24115.
Coomassie L. 73273........... 15 lbs. 1 oz.

Sherbet's Koffee 18512.
Agnes of Clovernook 72553.... 16 lbs. 9½ oz.
Tiger Lill 69771................ 15 lbs.
Leila of Clovernook 72554..... 14 lbs. 2 oz.

Shirley 1613.
Belle of Prospect 2d 14326..... 19 lbs.
Queen of Prospect 11997....... 14 lbs. 2 oz.

Sig 9528.
Sig Eurota 64898.............. 15 lbs. 1½ oz.

Signal 1170.
Geranium 2d 7838.............. 26 lbs. 4¾ oz.
Tenella 6712.................... 22 lbs. 1½ oz.
Croton Maid 5305.............. 21 lbs. 11½ oz.
Optima 6715.................... 19 lbs. 2 oz.
Œnone 8614.................... 18 lbs. 15 oz.
Belle of Patterson 5664........ 16 lbs. 6 oz.
Valhalla 5300................... 16 lbs.
Edwina 6713.................... 15 lbs. 13 oz.
Fanny Taylor 6714............. 15 lbs. 12 oz.
Signalana 7719................. 15 lbs. 4 oz.
Aldarine 5301.................. 15 lbs. 1½ oz.
Valhalla 5300 (14 days)........ 34 lbs.

Signal (F. S. 278 J. H. B.)
Lottie Dorey 21578............. 16 lbs. 12½ oz.
Young Gazelle 19837........... 14 lbs. 11 oz.

Signal Ben Signal 27186.
Ben's Bijou 86599............. 18 lbs. 6 oz.
Ben's Nettie 118137............ 15 lbs. 3 oz.

Signalbert 18840.
Signalbianca 54948............. 15 lbs. 5 oz.

Signalda 4027.
Sirona 27128.................... 21 lbs. 2 oz.
Birdie Nicholson 31676........ 18 lbs. 13 oz.
Signalda's Rosebud 35583...... 18 lbs. 10½ oz.
Signetilla 16333................. 18 lbs. 5½ oz.
Signaldella 24107............... 18 lbs. 1¾ oz.
Lilly Signalda 23227........... 17 lbs. 10 oz.
Signora Reid 27125............ 16 lbs. 8⅜ oz.

Signalda Bloomfield 30694..... 16 lbs. 3¾ oz.
Madam Zophee 31679.......... 16 lbs. 2½ oz.
Sigletta 32915.................. 16 lbs.
Siesta 37321.................... 15 lbs. 11 oz.
Solava 20928................... 15 lbs. 3⅝ oz.
Eterna 27134................... 14 lbs. 4¼ oz.
Signalda's Duchess 32173..... 14 lbs. 3 oz.
Nanalda 32917................. 14 lbs. 2½ oz.
Clover Bud 5th 27743.......... 14 lbs.

Signalda 2d 6748.
Nal Day 26553................. 19 lbs. 13 oz.

Signalda's Torment 14673.
Annabel Lee 55833............. 22 lbs. 3 oz.
Signilla M. 55839.............. 21 lbs. 4 oz.
Shelta F. 56437................ 20 lbs. ½ oz.
Monte Sano Belle 77683....... 16 lbs. 15½ oz
Jennie Fordyce 57169......... 15 lbs. 5½ oz.
Chump 81289................... 14 lbs. 13 oz.
Signola M. 76122.............. 14 lbs. 10 oz.
Jessie Wilson 76121........... 14 lbs. 8 oz.
Analysis 3d 82870............. 14 lbs. 1 oz.

Signalda's Tormentor 9298.
Miss D'Arcy 62596............ 15 lbs.

Signaldini 8556.
Miamis Signaldini 43891....... 17 lbs. 14½ oz.
Gem of Ellasleigh 38532...... 16 lbs. 10 oz.

Signaldo Alexis 15051.
Francesville 53367............. 16 lbs. 6 oz.
Signaldo's Pansy 53366........ 16 lbs. ½ oz.
Chrysippe 53373............... 14 lbs. 14 oz.
Petra's Signal 53376........... 14 lbs. 12 oz.

Signal Duke 12661.
Hallie Signal 61726........... 18 lbs. 12 oz.
Nettie Signal 37837............ 18 lbs. 8 oz.
Cupid's Cloud 74040.......... 17 lbs. 2 oz.
Signal Maid 4th 67492......... 15 lbs. 8 oz.

Signal Duke of Alfred 11927.
Susa's Darling 44915.......... 15 lbs. 12 oz.
Signal Duke's Creamer 60812.. 14 lbs. 7 oz.

Signalio 7649.
Signalinda 27002............... 18 lbs. 7 oz.
Signoretta 21546 (one year).... 680 lbs. 6½ oz.

Signal Jr. 7166.
Tancreda Signal 27789........ 21 lbs. 8½ oz.
Haidee Signal 86448........... 20 lbs. 2 oz.
Crotonia Signal 89308......... 18 lbs. 8 oz.
Edy Signal 19430.............. 16 lbs. 8 oz.
Signal's Empress 57767........ 16 lbs. 6½ oz.
Young Jersey Maiden 42794.. 15 lbs. 10½ oz.
Signal Maiden 42793.......... 15 lbs. 3½ oz.
Signal's Rhoda 98651.......... 14 lbs. 5½ oz.
Signalona 72105................ 14 lbs. 3 oz.
Haidee Signal 86448 (31 days). 79 lbs. 11 oz.

Signal Lad 12199.

Cordelia Signal 3d 57808....... 17 lbs. 8 oz.
Cordelia Signal 2d 44489...... 17 lbs. 6 oz.

Signally Signal 8875.

Zélie de Lussan 48681......... 20 lbs. 7 oz.
Zélie de Lussan 48681 (31 days) 86 lbs. 13 oz.
Zélie de Lussan 48681 (6 mos.) 450 lbs. 5 oz.

Signal Maid's Duke 21742.

Signal's Gold Foil 90508....... 15 lbs. 4 oz.

Signal Maximum 27286.

Sigmanella 117023............. 20 lts. 2 oz.

Signal of Springwood 28545.

Clara Signal of Lawn 85586... 15 lbs. 15 oz.

Signalona's Ben 32929.

Signalona's Gem 134219...... 14 lbs. 4½ oz.

Signal Sawyer 16786.

Lady Maud of Cinque Park 2d
71548 .... 14 lbs. 1¼ oz.

Signal's Laddie 11100.

Yum Yum Signal 40688........ 16 lbs. 7 oz.

Signal's Rebel 32928.

Signal's Garland 116048....... 15 lbs. 2 oz.

Sig-Nig 18659.

Sallie Grimes 79122........... 16 lbs. 2 oz.

Signoretta's Signal 18980.

Sedate's Retta 86596.......... 16 lbs. 8 oz.

Sig Pogis 22806.

Hearty Pogis 81297............ 18 lbs. 13¾ oz.
Saucy Pogis 103535........... 17 lbs. 2¼ oz.
Wild Rose Bloom 110583....... 17 lbs. 1¼ oz.
Pogis' René 102854............ 17 lbs.
Dosia Pogis 89049............. 16 lbs. 10¼ oz.
Hollie Pogissa 110490......... 16 lbs. 3¼ oz.
Retta Pogis 81132............. 16 lbs. 3¼ oz.
Macie Pogis 103539........... 15 lbs. 1⅛ oz.

Silver (P. S. 287 J. H. B.)

Silver Sheen 26210............ 19 lbs. 2 oz.
Cream Lily 24015.............. 16 lbs.
Les Marais' Dell 20314........ 15 lbs. 8½ oz.

Silver Duke 7125.

Thornleaf 39969............... 15 lbs. 12 oz.
Katie Herr 39970.............. 15 lbs. 8½ oz.

Silver Hair (F. S. 279 J. H. B.)

Silver Venus 20270............ 16 lbs. 3¾ oz.

Silver Mine 1658.

Silvia Baker 8793............. 20 lbs. 10 oz.
Siloam 17623.................. 18 lbs. 9½ oz.
Nancy Lovelock 15511......... 17 lbs. 9 oz.
Countess Coomassie 19339..... 16 lbs. 10 oz.

Sindbad 12294.

Wapello 56044................. 14 lbs. 12 oz.

Sir Alpha 6378.

Miss Julia 26701.............. 21 lbs. 1½ oz.

Sir Charles 131.

Carrie 3894................... 16 lbs. 8 oz.

Sir Davy 84.

Beulah of Baltimore 3270..... 14 lbs. 6½ oz.

Sir Farmington 3316.

Belle of Wayne 16457......... 17 lbs. 9½ oz.

Sir Florian 11578.

Regina Gilford 49891......... 14 lbs. 11 oz.
Doris of Mt. Pleasant 73330... 14 lbs. 7 oz.

Sir George 7656.

Young Garenne 3d 13648...... 16 lbs. 3½ oz.
Liberty 2d 16717.............. 14 lbs. 6½ oz.

Sir George of St. Lambert 6036.

Success of St. Lambert 28489.. 16 lbs. 2 oz.
Rioter's Nora 21778........... 15 lbs. 9 oz.
Irene of St. Lambert 2d 75759. 15 lbs. 5 oz.
Vivienne Pogis 41547......... 15 lbs. 1 oz.
Lass Pogis 40740.............. 14 lbs. 14 oz.

Sir Harry (F. S. 314 J. H. B.)

Comtesse de St. Ouen 22600... 15 lbs. 8½ oz.

Sir Hugo of St. Lambert 13726.

Mary of Lodi 88517........... 15 lbs. 1 oz.

Sir Joseph Peek 4978.

Laundress 2d 24649........... 14 lbs. 9 oz.

Sir Julian Pauncefote 22149.

Caroline Pauncefote 62919..... 16 lbs. 8 oz.

Sir Lawrence Angelo 19628.

Dodona A. 88102.............. 16 lbs. 10¾ oz.

Sir Oonan Pogis 26521.

Tormentor's Fawn 2d 94485... 18 lbs. 8 oz.

Sir Pedro Pogis 16010.

Kitty Kringel 57120........... 16 lbs. 8 oz.

Sir Roger's T. 3938.

Lucy T. 48200................. 16 lbs. 12.7 oz.

Sir Samuel Cunard 2231.
Albena 15995................... 16 lbs. 3 oz.

Sir Signal 3018.
Corinne Moore 35748.......... 19 lbs. 8 oz.

Sir Signalier 12490.
Pride of Edgewood 64199...... 17 lbs. 2 oz.

Sitanda 20221.
Sitanda's Rose 64958.......... 24 lbs. 12 oz.
Callie Nan 2d 64207............ 23 lbs. 1 oz.
Cara Mia 64224................. 15 lbs. 9¼ oz.

Smith of Darlington 2458.
Anna Smith 10324.............. 15 lbs. 6 oz.
Nellie Darlington 5956........ 15 lbs. 3 oz.

Smith's St. Lambert 30332.
Edith Haley 92643............. 22 lbs. 1½ oz.
Edith Hugo Pogis 105045..... 15 lbs. 8½ oz.
Bo-Peep's Maid 100538......... 14 lbs. 2 oz.

Smoke Smith 4844.
Atlanta's Beauty 12949........ 21 lbs. 3 oz.
Lucy Cobb 15133.............. 18 lbs. 6¼ oz.
Grace Pansy 2d 18764.......... 17 lbs. 15 oz.
Grace Pansy 2d 18764 (90 days) 147.000 lbs.

Snap (F. S. 254 J. H. B.)
(Le Brocq's Prize 3350 A. J. C. C.)
Fan's Grouville Beauty 10079. 19 lbs. 3 oz.
Le Brocq's Pansy Rex 23789.. 18 lbs. 6 oz.
Viva Le Brocq 13702........... 18 lbs. 3 oz.
Marea 10167.................... 17 lbs. 10 oz.
Petra 19267..................... 16 lbs. 6 oz.
Princess Le Brocq 17261....... 15 lbs. 8 oz.
Le Broemer 10670............. 15 lbs. 7 oz.
Prize Rose 16309.............. 15 lbs. 1 oz.
Fairy Le Brocq 15283.......... 14 lbs. 8 oz.
Birdie Le Brocq 17263........ 14 lbs.
Elinor Wells 12068............. 14 lbs.
Le Rosa 10078.................. 14 lbs.

Snap (F. S. 301 J. H. B.)
Snap's Dainty 15958............ 14 lbs.

Snedens 4882.
Mona's Mab 31524............ 15 lbs. 7 oz.
Alice McClellan 25237.......... 15 lbs. 2 oz.

Soapstone 14377.
Panola 85344................... 17 lbs. 3¾ oz.

Solid South 4711.
Serita 15520................... 17 lbs. 2 oz.

Son of Rosa 663.
Sultana 2d 11798.............. 15 lbs. 4 oz.

Sophie's Tormentor 20883.
Sophie Hudson 76105......... 19 lbs. 12½ oz.
Philena S. 70375.............. 18 lbs. 9½ oz.
Sophona 76110................. 18 lbs. 5¾ oz.
Pansy's Thoughts 69400....... 17 lbs. 6 oz.
Marna 76108.................... 16 lbs. 12 oz.
Elphie May 69396.............. 16 lbs. 4½ oz.
Pink Ring 69394.............. 16 lbs. 4½ oz.
Sophie 2d of Hood Farm 131236 15 lbs. 8 oz.
Sophie Hudson 76105 (10 days) 28 lbs. ½ oz.
Sophona 76110 (10 days)....... 22 lbs. 4¾ oz.
Pansy's Thoughts 69400 (14
days) .... 33 lbs.

Sophomore 24253.
Marian Holton 101478......... 14 lbs. 12 oz.
Vidalia 101047................. 14 lbs. 8 oz.

Sorona's Prince Pogis 16319.
Jessie Belle B. 100285.......... 14 lbs. 2 oz.

Southampton 117.
Witch-Hazel 1360.............. 14 lbs.

Southern Duke 35777.
Regina of Cedar Hill 111995.. 15 lbs. 8½ oz.

Southern Prince 10760.
Marchande 52258.............. 23 lbs. 14¼ oz.
Biederkeit 52257................ 22 lbs. 1¼ oz.
Princess Malita 66676......... 20 lbs. 8¾ oz.
Witching 42252................. 17 lbs. 14½ oz.
Sœurette 52256................. 17 lbs. 4¾ oz.
Silvia of Linwood 61008....... 16 lbs. 5.6 oz.
Southern Princess 36320....... 16 lbs. 4 oz.
Richland Belle 46920.......... 15 lbs. 5 oz.
Southern Daisy 38292.......... 14 lbs. 11 oz.
Sœurette 52256 (30 days)...... 97 lbs. ½ oz.

Southville Boy 4271.
Nuna 18669.................... 21 lbs.

Sovereign (F. S. 307 J. H. B.)
Sovereign's Elsie 15793........ 14 lbs. 3 oz.

Spangle's Boy 6229.
Countess of Lorne 20822....... 14 lbs. 14 oz.
Madame Argyle 19476......... 14 lbs. 6 oz.

Sparks Dee 8365.
Sparks' Maid 45575............ 19 lbs. 8 oz.

Spokane 23440.
Francesville 2d 99307.......... 15 lbs. 10 oz.
First Prize 95313.............. 15 lbs. 4 oz.
Donna Le Brocq 99305........ 14 lbs. 6 oz.
Milk Well 99301............... 14 lbs. 4 oz.

Spotless 1860.
Eva Horner 3d 24663.......... 15 lbs. 2 oz.

Squire Deerfoot 7835.
Felsticea 17083................. 16 lbs. 3 oz.

Stamp 8609.
Ruby Love 2d 32058............ 14 lbs. 1 oz.

Standard 553.
Maid of Amboy 2929......... 16 lbs. 1 oz.

Stand Point 7th 7929.
Lady Woodbine 26803......... 14 lbs. 2 o:.

Star Duke 8438.
Miss Puritan 34204............ 18 lbs. 5 oz.

Star F. 8364.
Roland's Bonnie 2d 18054...... 19 lbs. 2 oz.

Star Neal 1495.
Kobe's Marcella 4597......... 14 lbs. 13 oz.

Star of Glynllyn 31097.
Dora of Glynllyn 95993........ 17 lbs. 9 oz.

Statesman 2407.
Sally Leavitt 2d 10471.......... 15 lbs. 12 oz.

Statesman Jr. 7328.
Rosa of Frisby Hill 36080..... 16 lbs. 3¼ oz.

Steamboat 4422.
Leneca 17178.................. 15 lbs. ¼ oz.

Steletho 10505.
Silver Steletho 40765.......... 18 lbs. ½ oz.
Mary's Pet 40772.............. 15 lbs. 7½ oz.
Dot of Rocky Ford 40763...... 15 lbs. 1 oz.

Steuben 4751.
Belle Steuben 20115.......... 16 lbs. 10 oz.

Stiletto 8320.
Lorna 2d 33634................. 17 lbs. 9 oz.

Stockholder 7214.
Lora Nicholson 27135.......... 19 lbs. 12¾ oz.

Stockholder 2d 15616.
Daisy Delafield 59230......... 20 lbs. 11 oz.

Stokenter 25422.
Lizentris 81821................ 14 lbs. 13 oz.

Stoke Pogis 1259.
Matilda 4th 12816............. 21 lbs. 8½ oz.
La Petite Mère 2d 12810...... 16 lbs. 7 oz.
Leclair's Marjoram 36355...... 15 lbs. 3 oz.
Marjoram 2d 12805........... 15 lbs.
Matilda 4th 12816 (125 days)... 279 lbs. 7½ oz.
La Petite Mère 2d 12810 (1 yr.) 660 lbs. 4 oz.

Stoke Pogis 2d 2414.
Minnie of Oxford 12806........ 17 lbs.

Stoke Pogis 3d 2238.
Mary Anne of St. Lambert 9770 .... 36 lbs. 12¼ oz.
Ida of St. Lambert 24990...... 30 lbs. 2½ oz.
Allie of St. Lambert 24991.... 26 lbs. 12 oz.
Mermaid of St. Lambert 9771. 25 lbs. 13½ oz.
Nymph of St. Lambert 12968.. 24 lbs. 14 oz.
Naiad of St. Lambert 12965.... 22 lbs. 2¼ oz.
Nora of St. Lambert 12962.... 22 lbs.
Niobe of St. Lambert 12969... 21 lbs. 9¼ oz.
Cora of St. Lambert 8347..... 21 lbs. 6¾ oz.
Brenda of Elmhurst 10762.... 20 lbs. 8 oz.
Cheerful of St. Lambert 8348. 20 lbs. 8 oz.
Honeymoon of St. Lambert 11221 .... 20 lbs. 5¼ oz.
Rioter Pink of Berlin 23665... 19 lbs. 14 oz.
Columbine of St. Lambert 8350 19 lbs. 1 oz.
May Day Stoke Pogis 28353... 19 lbs. ½ oz.
Cowslip of St. Lambert 8349.. 17 lbs. 12 oz.
Crocus of St. Lambert 8351... 17 lbs. 12 oz.
Minette of St. Lambert 9774.. 17 lbs. 4 oz.
Kitty of St. Lambert 6637..... 16 lbs. 11 oz.
Diana of St. Lambert 6636..... 16 lbs. 8 oz.
Maggie of St. Lambert 9776... 16 lbs. 3 oz.
Moth of St. Lambert 9775..... 16 lbs. 2 oz.
Mavourneen of St. Lambert 9777 .... 15 lbs. 12 oz.
La Belle Petite 5472........... 15 lbs. 8 oz.
Cupid of Lee Farm 5997....... 14 lbs. 6 oz.
Nancy of St. Lambert 12964... 14 lbs. 5 oz.
Mary Anne of St. Lambert 9770 (one year)............. 867 lbs. 14¾ oz.

Stoke Pogis 5th 5987.
Sister of Charity 62453........ 24 lbs. 14½ oz.
Mary Idagold 88186............ 23 lbs. 9 oz.
May Dee Pogis 36993......... 20 lbs. 5½ oz.
Lady Mary of Prospect 19768.. 19 lbs. 15½ oz.
Pretty Patty 44108............. 19 lbs. 1 oz.
Priscilla Pogis 39270........... 18 lbs. 6½ oz.
Frankness 62451............... 18 lbs. 4 oz.
Pogis' May 26950.............. 18 lbs. 2 oz.
Candelabrum 75187............. 17 lbs. 13 oz.
Pauletta Pogis 39273........... 17 lbs. 12 oz.
Princess Aurea Pogis 39266... 17 lbs. 7½ oz.
Mary M. Pogis 32455......... 16 lbs. 4½ oz.
Sweet Blossom Pogis 36995... 16 lbs. 1½ oz.
Little Pogis 32612.............. 16 lbs. 1 oz.
Mary Hinman 17619.......... 15 lbs. 11½ oz.
Sterling Merit 62456........... 15 lbs. 4 oz.
Rho A. Pogis 39269........... 14 lbs. 15 oz.
Pomona's Ida of St. L. 59565.. 14 lbs. 14½ oz.
Rioter's Zoe 19769............. 14 lbs. 12 oz.
Gilfilia Pogis 38688........... 14 lbs. 2½ oz.
Sweet Leona B. 21934......... 14 lbs. 1½ oz.
Priscilla Pogis 39270 (31 days) 77 lbs. 12½ oz.

Stoke Pogis Bulletin 18619.
Ribbon's Gift 77375........... 18 lbs. 1 oz.
Ribbon's Queen 66321........ 16 lbs. 1 oz.

Stoke Pogis Jr. 15300.
Matilda of Maplewood 49723.. 14 lbs. 15 oz.
Violet of Maplewood 49724.... 14 lbs. 10¼ oz.

Stoke Pogis of Linden 10558.

Phillis 3d 46681	18 lbs. 13 oz.
Rose Pogis 24626	15 lbs. 2½ oz.
Stoke Pogis' Regina 48309 (30 days)	60.268 lbs.

Stoke Pogis of Prospect 29121.

Grace Marigold 99377	20 lbs. 1 oz.
Olivia Marigold 108187	18 lbs. ½ oz.
Elma Marigold 100630	17 lbs. 13½ oz.
Sylvia Marigold 108183	17 lbs. 8 oz.
Trident 88647	17 lbs. 4½ oz.
Bella Marigold 94124	17 lbs. 2 oz.
Sweet Perrot 100749	17 lbs. 2 oz.
Tuäa St. L. of Aurora 89388..	17 lbs. 1 oz.
Beech Leaf 99381	16 lbs. 3½ oz.
Adelpha Marigold 133768	15 lbs. 11½ oz.
Niobe Marigold 108363	15 lbs. 8½ oz.
Ida of Elma 92316	14 lbs. 10 oz.
Vina Marigold 108186	14 lbs. 8½ oz.
Stoke's Tipsey 99380	14 lbs. 7½ oz.
Leaf Pogis 130184	14 lbs. 5½ oz.
Alphea Marigold 117632	14 lbs. 5 oz.

Stoke Pogis' Perfection 5984.

Nettie's New Year 99295	23 lbs. 3 oz.
La France's Pogis 99986	21 lbs. ½ oz.
Pansy Dec. 99294	16 lbs. 4½ oz.
Petra's Lady 99791	15 lbs. 10½ oz.
Deerlick Pojoram 99988	15 lbs. 7 oz.
Gilderoy P. 99308	15 lbs. 3 oz.
Deerlick Petra Pogis 99987....	14 lbs. 9½ oz.

Stoke Pogis Prince 11005.

Nettie N. 58307	15 lbs.
Desda Pogis 58344	14 lbs. 7½ oz.

Stoke Pogis Trio 15448.

Miss Clifford 3d 63360	15 lbs. 2½ oz.
Miss Clifford 5th 63364	14 lbs. 7 oz.

Stunner 9679.

St. Lambert's Violet 25278	16 lbs. 12 oz.

Success 2097.

Lady Ives 3d 6740	14 lbs. 8 oz.

Success of Pa. 23374.

Elsie Harribee 91834	18 lbs. 8½ oz.

Sue's Signal 14467.

Geneva S. 57170	15 lbs. 15½ oz.

Sultan Chief 3216.

Chief's Malita 25186	16 lbs. 8½ oz.

Sultan of New York 6186.

Lady of Dryden 27042	16 lbs. 3 oz.

Sultan of St. Saviour's 5328.

Sultan's Sultane 32854	15 lbs. 1 oz.

Sumach 5249.

Dolly Bond 82662	18 lbs. 3 oz.

Superb 1956.

Belmeda 6229	18 lbs. 12 oz.
Floret 9959	17 lbs. 6 oz.
Lizzie D. 10408	16 lbs. 15 oz.
Lida Mullin 9198	16 lbs. 8 oz.

Surpry 2d (P. S. 459 J. H. B.)

Analysis 29162	16 lbs. 8 oz.

Suzerain 8408.

Seraphine 2d 37451	15 lbs. 12 oz.

Sweepstakes Duke 1905.

(Duke P. S. 76 J. H. B.)

Duke's Pierrot Lady 31023....	21 lbs. 12 oz.
Fairy Queen of St. Brelade's 7464	19 lbs. 7¼ oz.
Valentine of Trinity 7460	19 lbs. 4½ oz.
Bijou Ogston 8210	18 lbs. 15 oz.
Queen of Nubbin Ridge 14528.	17 lbs.
Forget-me-not 5809	15 lbs. 8 oz.
Florence Billot 7849	14 lbs. 13 oz.

Sydney 3262.

Vieva 3d 7642	16 lbs. 5 oz.

Talmadge 3992.

Daphne of Arcadia 21710	28 lbs. 12 oz.

Tamaroa Rex 18604.

Pride of Rosemond 74449	17 lbs. 14½ oz.
Frisky Elsie 81590	17 lbs. 1½ oz.

Tamasese of Bay View 39903.

Tamasese Naomi 128774	14 lbs. 15 oz.

Tamerlane 4287.

Trenie 17770	14 lbs. 10 oz.
Queen of Chenango 17771	14 lbs. 6 oz.

Tamy's Rioter 15288.

Maybee H. 60620	18 lbs. 4 oz.
Vinala H. 61835	16 lbs.
Olga H. 60619	15 lbs. 4 oz.
Leonetta H. 67514	15 lbs. 1½ oz.
Leonetta H. 67514 (30 days)...	62 lbs. 3 oz.
Olga H. 60619 (30 days)	61 lbs. 13 oz.

Tancreda's Signal King 35742.

Fermentor 2d's Pet 114503	17 lbs. 8 oz.
Tancreda Torment 114090	15 lbs. 1 oz.

Tarquin 3d 8203.

Lady of Malone 25734	16 lbs.

Tasso of Mount Waite 2334.

Eudora of Barre 12419	16 lbs. 9 oz.

Taureau d'Or 11743.

Jetty d'Or 2d 62403	17 lbs. 1 oz.

Teacher 15952.

Teacher's Pet 60242	18 lbs. 12 oz.

Telegraph 9457.
Chinqua 27384................. 22 lbs. 9½ oz.

Tenella's Alexis 13170.
Bridal Rose 74883............. 23 lbs. 8 oz.

Tenella's Best 19736.
Signal Mattie 113952......... 15 lbs. 7½ oz.

Tenella's Tormentor 22387.
Leta May 82171............... 15 lbs. 2 oz.

Tennessee of Bolton 27508.
Fenella of Bolton 2d 116768... 14 lbs.

Tennessee's Landseer 23015.
Donney Pogis 2d 82426........ 17 lbs. 1¼ oz.
Kate Oxford 2d 69270.......... 14 lbs. 4 oz.
Emma Landseer 2d 69220..... 14 lbs. 2¾ oz.

Tennessee's Tormentor 23498.
Harry's Signaline 2d 82428.... 20 lbs. 11 oz.
Pet's Torment 92255........... 15 lbs. 14 oz.
Maid of Yokun 2d 86350....... 14 lbs. 4 oz.
Antigna 109426................ 14 lbs. 2¾ oz.

Texas Pogis 20194.
Mollie's Fancy 69728.......... 14 lbs. 8 oz.

Texas Wanderer 14453.
Talaney 69861................. 15 lbs. 3 oz.

Thalma 4288.
Maculac 24277................. 15 lbs. 3 oz.

Theda's Son 31202.
Theda's Belle 106891.......... 16 lbs. 2 oz.

The Hub 1009.
Mink 2d 3890.................. 19 lbs. 11 oz.
Silver Wave 10844............. 19 lbs.
Fleece 18568.................. 17 lbs. 13½ oz.
Oktibbeha Duchess 4422....... 17 lbs. 4 oz.
Dairy 2d 3891................. 15 lbs. 5½ oz.
Azuline 2d 3888............... 14 lbs. 15¾ oz.
Mink 3d 4863.................. 14 lbs. 9 oz.
Adora 18569................... 14 lbs. 3 oz.

The Marquis 3805.
Grinnell Lass 1009............ 16 lbs. 10 oz.

The Pope's Bull 2965.
Pope's Flora 18697............ 14 lbs. 2 oz.

Theron 5722.
Young Floribundus 28172...... 17 lbs. 12¾ oz.

The Squire 1298.
Nibble 6796................... 14 lbs. 15 oz.
Cowry 4432.................... 14 lbs. 1 oz.

Thomaston 4600.
Ethelka 2d 14128.............. 15 lbs. 4 oz.

Thorndale 2582.
Maggie McM. 14073........... 19 lbs. 9 oz.
Almah of Oakland 11102...... 16 lbs. 14 oz.
Oakland Girl 11103........... 14 lbs. 10½ oz.
Florrie May Baker 10728...... 14 lbs. 8 oz.

Thotmes 17157.
Tuäa Perrot 46357............. 19 lbs. 5½ oz.

Tisquantum (P. S. 262 J. H. B.)
Nervine 25932................. 14 lbs. 1½ oz.

Titan 8548.
La Petite Pogis 28757........ 20 lbs. 10¾ oz.
Luteefy 46313................. 14 lbs. 6 oz.

Toltec 6831.
Toltec's Fancy 27172......... 27 lbs. 5½ oz.
Oonan's Fancy 31887.......... 19 lbs. 10¾ oz.
Toltec's Roonan 64564........ 19 lbs. 7½ oz.
Rosy Dream 2d 27127......... 17 lbs. 11 oz.
Toltec's Alice 31885........... 17 lbs. 10 oz.
Toltec's Silvia 36911........... 17 lbs. 1¾ oz.
Little Accident 15578......... 16 lbs. 10½ oz.
Landseer's Fancy 3d 70653.... 15 lbs. 14½ oz.
Leoni 2d 29750................ 15 lbs. 9 oz.
Phlox 3d 31882................ 15 lbs. 6 oz.
Tolonga 76868................. 15 lbs. 5½ oz.
Leonette 29752................ 15 lbs. 5 oz.
Columbia Beauty 30263....... 15 lbs. 1 oz.
Landseer's Fancy 2d 43184.... 15 lbs. ½ oz.
Tolteca 44762................. 14 lbs. 11¼ oz.
Lena's Florence 2d 78688..... 14 lbs. 9½ oz.
May Toltena 50405........... 14 lbs. 5¾ oz.
Creole Maid 2d 31881......... 14 lbs. 1 oz.

Toltec 2d 11073.
Nerobella's Blossom 48051..... 15 lbs. 1 oz.

Toltec 6th 14507.
Toltec's Sally 94461........... 17 lbs. 2¼ oz.
Brunette Landseer 109904..... 16 lbs. 13¼ oz.

Toltec of Bedford 34302.
Sitka's Star 104707........... 14 lbs. 9 oz.

Toltec's Prince 21961.
Ashwood Pet 2d 84408........ 16 lbs. 11½ oz.
Toltec's Nan 70199............ 14 lbs.

Toltec's Signal 29501.
Paradise 2d 97112............. 18 lbs. 4 oz.
Lorilla 87505.................. 16 lbs. 2 oz.
Signal's Crusta 97341.......... 15 lbs. 3½ oz.

Tom (P. S. 77 J. H. B.)
Primrose 11956................ 21 lbs. 10 oz.
Brunette Le Gros 9755........ 15 lbs. 15 oz.

Tombeau 25347.
Toma May 74836.............. 14 lbs. 13 oz.

Tombigbee Cowboy 18372.

Señorita of St. Lambert 98665.	15 lbs.	8½ oz.
Light Foot Pogis 57531........	14 lbs.	11 oz.
Light Foot Pogis 57531 (14 days)	23 lbs.	15 oz.

Tom Brown 2940.

Waukesra 19721................	16 lbs.	

Tom Dasher 420.

Jersey Cream 3151.............	17 lbs.	
Creamer 2467...................	14 lbs.	1 oz.

Tom Doyle 4695.

Dolly Doyle 22788.............	15 lbs.	4½ oz.

Tom Kenelm 11023.

Mona 3d 23134.................	14 lbs.	6 oz.

Tom McGreevy 1692.

Le Gros' Lily of the Valley 2d 13386	19 lbs.	10¼ oz.

Tommy, on L. of J.

Maid of Avranches 6959.......	15 lbs.	

Tommy (P. S. 249 J. H. B.)

Nutley Dolores 13797..........	23 lbs.	2 oz.

Tonnage 20000.

Tonnage Girl 103331..........	15 lbs.	1½ oz.

Tony Calypso 3354.

Modita 16626...................	16 lbs.	8 oz.

Tony Lumpkin 31154.

Lora Martin 88381.............	19 lbs.	6 oz.

Topris 4135.

Alphea Rajah 20605...........	23 lbs.	2 oz.

Top-Sawyer 1404.

Fandango 12908................	18 lbs.	3 oz.
Denise 8281...................	17 lbs.	8 oz.
Vixen 7591....................	17 lbs.	6 oz.
Beeswax 9807..................	17 lbs.	5 oz.
Busy Bee 6336.................	16 lbs.	4 oz.
Myrrha 11299..................	16 lbs.	1 oz.
Fleurette of Linwood 12918...	16 lbs.	
Romp Ogden 3d 5458...........	15 lbs.	14 oz.
La Fantine 24489.............	15 lbs.	4 oz.
Opaline 7590..................	15 lbs.	2 oz.
Dora Doon 12909..............	15 lbs.	
Queen Mary 6212..............	14 lbs.	13½ oz.
Litza 6338....................	14 lbs.	3 oz.

Torlandseer 22102.

Rose Cottage Beauty 99542....	14 lbs.	10½ oz.

Tormenter 3533.

Little Torment 15581.........	23 lbs.	2½ oz.
Tormentor's Bloomfield 55530.	22 lbs.	5½ oz.
Sasanda 50075.................	22 lbs.	1½ oz.
Ethleel 18724.................	19 lbs.	14 oz.
Croton Maid 4th 26729........	19 lbs.	7 oz.
Tormentor's Favorite 36873...	19 lbs.	2 oz.
Tortilla 55528...............	'18 lbs.	4½ oz.
René of Hillside 12952.......	18 lbs.	1 oz.
Torment 15579................	17 lbs.	8½ oz.
Daisy Brown 12213............	17 lbs.	6½ oz.
Zenitza 19190................	17 lbs.	5½ oz.
Little Lass 28512............	17 lbs.	5 oz.
Duchess of Bloomfield 2d 12250	17 lbs.	4 7-16 oz.
Belnina 19189................	17 lbs.	3¼ oz.
Odelle Sales 15564...........	16 lbs.	15¾ oz.
Kathy Torment 32910..........	16 lbs.	13½ oz.
Pinkey Spry 18316............	16 lbs.	12 oz.
Tormintha 37327..............	16 lbs.	10¼ oz.
Pitapat 11136................	16 lbs.	5 oz.
Gladdys Gaily 20878..........	16 lbs.	1¾ oz.
Ruby Torment 32989...........	16 lbs.	
Rose of Oxford 13469.........	15 lbs.	14½ oz.
Tormentor's Cinderella 19564.	15 lbs.	8½ oz.
Chenda 3d 43246..............	15 lbs.	7 oz.
Tormentor's Fancy Wax 73212	15 lbs.	4½ oz.
Narda 38012..................	15 lbs.	1 oz.
Romping Lass 11021...........	15 lbs.	
Tormentor's Su Lu 54943......	14 lbs.	13 oz.
Little Butterstar 20926......	14 lbs.	12½ oz.
Tormentor's Idex 64209.......	14 lbs.	9 oz.
Kattie 41431.................	14 lbs.	8½ oz.
Tormentor's Silver 36058.....	14 lbs.	7½ oz.
Tormentor's Rosebud 50069....	14 lbs.	2 oz.
Tormentor's Spiræa 42248.....	14 lbs.	
Little Torment 15581 (13 wks.).	2-7 lbs.	11-32 oz.

Tormentor 2d 7124.

Lady Rareripe 23081..........	16 lbs.	1 oz.
Neitherme 64670..............	15 lbs.	1 oz.

Tormentor 5th 21962.

Xarama 76887.................	17 lbs.	7¼ oz.
Tomassee 76875...............	17 lbs.	5 oz.
Peggie Tormentor 98103.......	16 lbs.	2 oz.
Zunilda 2d 93811.............	15 lbs.	14 oz.
Signora Reid 2d 93808	15 lbs.	10½ oz.
Tormendova 2d 118161........	15 lbs.	9 oz.
Dixie Landseer 88681.........	15 lbs.	6½ oz.
Cara Mia 2d 93827............	15 lbs.	5½ oz.
Soconee 2d 114420............	15 lbs.	
Chelka 79033.................	14 lbs.	13¼ oz.
Toridex 76873................	14 lbs.	13 oz.
Etelka Tolemy 93807..........	14 lbs.	9 oz.
Rosa Oxford 2d 127909........	14 lbs.	8 oz.
Tomassee 76875 (14 days).....	41 lbs.	8¼ oz.
Xarama 76887 (21 days).......	45 lbs.	10½ oz.

Tormentor 7th 22911.

Torlona 76872................	17 lbs.	1½ oz.

Tormenter Cateno 17004.

Bonnie Wherry 115088........	20 lbs.	4 oz.
Baby Mar 115089.............	16 lbs.	6 oz.

Tormentor Jr. 15005.
Rosa Hudson 2d 87631........ 16 lbs. 14½ oz.
Flora Pansy 2d 78678......... 15 lbs. 10.8 oz.

Tormentor's Jersey King 26928.
Aloha's Tormentor 90239...... 14 lbs. 6 oz.

Tormentor's John 14715.
Fawn Tormentor of Lawn
71639 .... 20 lbs. 2 oz.
Signal Fancy of Lawn 73540.. 15 lbs. 1 oz.
Hinds' Tormentor of Lawn
71640 .... 14 lbs. 7 oz.

Tormentor's Last 24930.
Nora Tormentor of Lawn 99738 15 lbs. 12 oz.

Tormentor's Rioter 21964.
Princess Sosia 112753.......... 15 lbs. 8 oz.
Edna Hayes 90901.............. 15 lbs. 6 oz.

Tormentor Stoke Pogis 20485.
Tormento Coomassie 110447... 22 lbs. 2 oz.
Daisette's Tormentress 106532 21 lbs. 1⅞ oz.
Tormentor's Isis 73350........ 19 lbs. 7 oz.
Tormentor's Barbara 98801.... 17 lbs. 3 oz.
Tormentor's Laura 98346..... 16 lbs. 9½ oz.
Tormentor's Ravalita 97909... 15 lbs. 11¾ oz.
Tormentor's Browny 73079.... 14 lbs. 14 oz.
Blanche Duncan 2d 73333..... 14 lbs. 11¾ oz.

Torpedo 8671.
Sayda M. 46195.............. 22 lbs. 11½ oz.

Torrance 17872.
Torrancella 59827.............. 17 lbs. 8½ oz.
Torraline 54945................. 17 lbs. 6¾ oz.
Roulette 50080................. 14 lbs. 5½ oz.

Trailer 7160.
Xariñna 34176................. 14 lbs. 9 oz.

Traveller (P. S. 280 J. H. B.)
Clubs 2d 16725................. 14 lbs. 7 oz.

Trial (P. S. 1187 J. H. B.)
Magyarland 86326............. 17 lbs. 8 oz.

Tristram Shandy 1767.
Gem of St. Cloud 7342........ 14 lbs. 8½ oz.

Troubadour 481.
Cerita of Meadow Brook 5056. 17 lbs. 8 oz.

Troy's Duke 13003.
Ettie Palmer 4th 47481........ 16 lbs. 5 oz.

Trump 22048.
Islithanks 90544............... 17 lbs. 2 oz.

Truxton 4654.
Lady McDowell 37414......... 15 lbs. 12 oz.

Tug Wilson 9680.
Imperial Riotress 30259....... 18 lbs. 4 oz.
Chemung Maid 34704......... 14 lbs. 15 oz.

Tunlaw Boy 2866.
Ruby of Springvale 14505..... 14 lbs. 15 oz.

Tunxes Chief 3705.
Louisa Deming 23469......... 14 lbs. 15 oz.

Tuscarora of Clermont 7232.
Gipsy's Pride 2d 29365........ 27 lbs. 12 oz.

Twice Signal 9582.
Platina Cary 73375............ 17 lbs. 7¼ oz.

Two Hundred per Cent. 33592.
Allie of Glen Rouge 129123.... 15 lbs. 14 oz.
Mollie of Glen Rouge 129125.. 15 lbs. 12 oz.

Tycoon Jr. 1212.
Jessie Lee of Labyrinth 5290.. 14 lbs. 7 oz.

Typhoon 77.
Venus 112..................... 14 lbs. 11 oz.

Uncle Toby Brown 5814.
Flora B. 12446................. 14 lbs. 2 oz.

Upright 6147.
Costa Rica 64570.............. 21 lbs. 6½ oz.
Pledge 59214.................. 17 lbs. 9½ oz.
Caressa 88474.................. 17 lbs. 6 oz.
Treasure 79200................. 17 lbs. 5 oz.
Calcium 79733................. 16 lbs. 15 oz.
Paradox 65003................. 16 lbs. 9 oz.
Daretta 62579.................. 16 lbs. 8 oz.
Chestnuts 63449............... 15 lbs. 7¼ oz.
Upright's Brownie 103755..... 15 lbs. 3 oz.
Compressa 55756.............. 14 lbs. 13½ oz.
Nicomis 68026................. 14 lbs. 3 oz.
Costa Rica 64570 (30 days).... 87 lbs. 14¼ oz.
Costa Rica 64570 (31 days).... 90 lbs. 11¼ oz.

Uproar 4609.
Belle's Esperanza 12053........ 15 lbs. 9½ oz.

Uproar 4th 5954.
Maid of Fernwood 2d 29010.... 17 lbs. 11 oz.

Vailey King 19752.
Belle Miller 60488.............. 15 lbs. 2¾ oz.

Van Amburgh 22416.
Tait's Beauty 121221.......... 18 lbs. 4 oz.

Velpeau 2146.
Lilium Excelsum 24945........ 17 lbs. ½ oz.

Vermont 893.
Dora 4th 3936.................. 25 lbs. 2 oz.
Empress 6th 3203............. 17 lbs. 9¾ oz.
Goddess of Staatsburgh 5252.. 14 lbs. 8 oz.

Vermonter 5620.
Pinafore 2d 15072............. 15 lbs. 8 oz.

Verte's Chief 16276.
Clover of Bay View 65424..... 18 lbs. 14 oz.

Vertumnus (P. S. 161 J. H. B.)
Lady Velvetine 15771........ 17 lbs. 2 oz.
Olymph 17957................. 15 lbs. 13 1-16 oz.
Lady Kingscote 26085........ 15 lbs. 10 oz.

Vesper's Royal Son 2946.
Royal Sister 12451............. 14 lbs. 11 oz.

Vexer 20889.
Toltecida 2d 96030............. 16 lbs. 6 oz.
Vexanella 79306................ 16 lbs. 5 oz.
Vemicou 79322................. 14 lbs. 7 oz.
Vemist 79323.................. 14 lbs. 2 oz.

Vibert Stoke Pogis 25549.
Ora Vibert 119585............. 17 lbs. 11 oz.
Pa Pogis 72500................ 16 lbs. 8½ oz.
Patina Pogis 110784........... 15 lbs. 2½ oz.
Pa Pogis 72500 (14 days)...... 32 lbs. 15½ oz.

Victor 797.
Pet of Maplewood Farm 4854.. 15 lbs. 2 oz.

Victor 3550.
Jersey Belle of Scituate 7828.. 25 lbs. 3 oz.
Jersey Belle of Scituate 7828
(one year).................... 705 lbs.

Victor (P. S. 148 J. H. B.)
Floribundus 2d 14949......... 18 lbs. 8 oz.
Dairy Pride 4th 21681......... 16 lbs.
Mulberry 22031................ 14 lbs. 2 oz.

Victor F. W. 16548.
Victor's Lady 52691........... 15 lbs. 14 oz.
Cora's Beauty 92179.......... 15 lbs. 9 oz.

Victor Hugo 197.
Lady Fawn of St. Anne's 10920 16 lbs. 12½ oz.

Victor Hugo Signal 12135.
Princess Ora 66374............ 14 lbs. 5 oz.

Victor of Yerba Buena 3809.
Fanny of Yerba Buena 2d 10689 14 lbs. 14 oz.

Victor Pogis 15592.
Clem Pogis 49932............. 23 lbs. 4½ oz.

Victor Scituate 2d 26330.
Victor's May 97958............ 14 lbs. 8 oz.

Victor's King 12678.
V. K.'s May 41929............. 14 lbs. 8 oz.

Victory 23051.
Acacia of Home Farm 3d 78203 17 lbs. 14½ oz.

Vigo King 19238.
Vigo Princess 60806........... 18 lbs.

Viking of Bolton 24881.
Duchess of Bolton 98150....... 24 lbs. 4 oz.
Miranda of Bolton 100226...... 21 lbs. 4 oz.
Chloe of Bolton 81340......... 16 lbs.

Violator 17940.
Viozetta Pogis 56294......... 14 lbs. 5½ oz.

Violet 3d's Rioter 23470.
Rioter's Violetta 79391........ 14 lbs. 7½ oz.

Violetta's Hugo Pogis 15740.
Gold Violet 85078............. 17 lbs. 1 oz.

Volco 7890.
Brown Bessie 74997........... 20 lbs. 8 oz.
Brown Flora 2d 96594......... 20 lbs. 2 oz.
Brown Bessie 74997 (30 days).. 72.235 lbs.
Brown Bessie 74997 (90 days).. 216.640 lbs.

Von Bismarck 3d 1780.
Lady Armington 7610......... 17 lbs. 8 oz.

Wallace Barnes 1264.
Dimple 3248................... 16 lbs. 11 oz.

Wallenstein 2261.
Elegantie 11943................ 14 lbs. 3 oz.

Walnut Chief 3130.
Juliette Guion 13143........... 14 lbs. 4 oz.

Wanderer 3014.
Aldarine 3d 27482.............. 28 lbs. 1 oz.
Fadette of Verna 3d 11122..... 22 lbs. 8½ oz.
Eva of Verna 15228........... 21 lbs. 13 oz.
Fairy of Verna 2d 10973....... 20 lbs. 3¾ oz.
Hilda A. 2d 11120.............. 20 lbs.
Evelina of Verna 10971........ 19 lbs. 10½ oz.
Cordelia Signal 33452......... 18 lbs. 15½ oz.
Sedate 11119.................. 18 lbs. 13 oz.
Fadette of Canaan 14807...... 16 lbs. 1½ oz.

Warpole 3500.
Plumida 23621................. 15 lbs. ½ oz.

Warren Lowndes 17591.
Dora Lowndes 64136.......... 16 lbs. 15 oz.

Warren's Duke 21380.
Mollie St. Helier 2d 98389..... 14 lbs. 12 oz.

Warren's Prince 27973.
Cowslip of Rockwell 115354.... 14 lbs. 13½ oz.

Wasp 15358.
Signal of Raceland 84299...... 18 lbs. 6 oz.
Amelia Davenport 81044....... 14 lbs. ⅝ oz.

Watson 3451.
Minnie Kersey 19895.......... 14 lbs. 11 oz.

SIRES WITH THEIR TESTED DAUGHTERS. 601

Waukesha 5330.
Quachette 17091................ 19 lbs. 11½ oz.

Waxy 2116.
Fanny of Cream Brook 13930. 17 lbs. 8 oz.

Wells' Exile 20170.
B. D. Exile 105286............. 18 lbs. 10 oz.
Torpedo Exile 105287......... 16 lbs. 4½ oz.
J. S. Exile 105495.............. 15 lbs. 4 oz.

Wessex 3638.
Judy of Riverside 16495....... 21 lbs. 4½ oz.
Ulricalla 22225................. 18 lbs. 14 oz.

Westchester 1266.
Chansonnette 5695............. 16 lbs. 4 oz.

Weston 6285.
Howland Nightingale 33291.... 14 lbs. 1 oz.

West Wind 4289.
Maud Pogis 24240............. 14 lbs. 12¾ oz.

Wethersfield 966.
Lady Gray of Hilltop 6850.... 18 lbs. 12 oz.

Whip 2638.
Lady Cecilia 24821............. 15 lbs. 1 oz.

William Devries 6064.
Mary of Pleasant View 13448.. 14 lbs. 6 oz.

Winder 29955.
Bride of Evergreen 3d 90473.. 17 lbs. 1 oz.
Dot Badger Girl 101417........ 14 lbs. 1 oz.
Winder Gem 90447............. 14 lbs.

Windsor Rex 14493.
Sheba Rex 47429 (30 days)..... 57.511 lbs.
Sheba Rex 47429 (90 days)..... 190.617 lbs.

Wine 29954.
Euro Polono 2d 101052........ 14 lbs. 2 oz.
Kitty Black Prince 104472..... 14 lbs. ½ oz.

Winner 5572.
Happy Winn 35358............. 18 lbs. 13 oz.

Winner's Lisgar 11557.
Daisy Morrison 3d 40300...... 14 lbs. 8 oz.

Witch's Token 9966.
Alicealda 43604................ 14 lbs. 1 oz.

Woodstock Chief 21172.
Woodstock Mystery 77746 (21
days) .... 26.223 lbs.
Woodstock Lady 80619 (21
days) .... 23.027 lbs.

Wormwood 1746.
Maud Lee 2d 8839.............. 14 lbs. 9 oz.

Wosie 6802.
Patty Polonius 3d 38677....... 16 lbs. 10 oz.

Wrangler 20881.
Amna 2d 98269................ 18 lbs. 9 oz.

Yakout 6842.
Sudie St. Helier 40235......... 18 lbs. 9 oz.
Belle Yakout 38020............ 16 lbs. 13 oz.

Yankee 1003.
Cassia 2d 21370................ 20 lbs. 10¼ oz.
Chloe 4th 4612................ 17 lbs. 4 oz.

Yankee Stoke Pogis 15904.
Fulda Stoke Pogis 44992....... 15 lbs. 8 oz.

Yellow Boy 6381.
Royal Daisy 25214............. 20 lbs. 9 oz.
Niobe's Alpheanette 3d 39459.. 19 lbs. 4½ oz.

Yellow Star 13472.
Yellow Belle 69780............. 14 lbs. 9½ oz.

Yoko 3951.
Maggie Rule 31940............. 21 lbs. 5 oz.
Palestine of Oxford 42194..... 21 lbs.

Yokun Chief 4399.
Yokun Maid 19073............. 16 lbs. 6 oz.

Young Baltimore Boy 2048.
Countess Buttercup 13505..... 14 lbs. 3 oz.

Young Baron 702.
Pearl Armstrong 2670......... 21 lbs. 10 oz.
Arietta 5115.................... 15 lbs.
Bertie Briggs 5213............. 14 lbs. 4½ oz.

Young Combination 14550.
Indulgence 50105.............. 17 lbs. 1 oz.
Patrol 40490.................. 15 lbs. 4 oz.
Vilify 50107................... 14 lbs. 3 oz.
Vidalinne 50101............... 14 lbs. 2 oz.

Young Concord 1406.
Moss Rose of Willow Farm
5194 .... 23 lbs. 1 oz.

Young Duke 138.
My Queen 12614............... 15 lbs. 8 oz.
Carrie's Beauty 14601.......... 14 lbs. 7½ oz.

Young Garenne's Duke 6863.
Khelula of Briarcliff 46799.... 18 lbs. 2½ oz.
Vera of Briarcliff 28687........ 15 lbs. 1 oz.
Viola of Briarcliff 37617....... 14 lbs. 8 oz.

Young Jupiter 12157.
Pearl's Lemon 41646.......... 17 lbs. 5¼ oz.

Young Neptune 17656.
Jane Neptune 52345.......... 19 lbs. 4½ oz.

Young Pedro 9033.
Eurotisama 29668............. 27 lbs. 1½ oz.
Pedronina 34803............... 21 lbs. 3 oz.
Dicta 55221.................... 20 lbs. 9 oz.
Riotress 3d 63166............. 17 lbs. 3½ oz.
Young Lass 67471............. 17 lbs. 3½ oz.
Patona 55117.................. 16 lbs. 14½ oz.
Chansonnette 2d 29672........ 16 lbs. 9 oz.
Beauty of Hamilton 55827.... 15 lbs. 15 oz.
Fanella 2d 64578.............. 15 lbs. 6 oz.
Howdy 55119.................. 15 lbs. 3½ oz.
Eurotisama 29668 (one year).. 945 lbs. 9 oz.

Young Pedro 2d 15011.
Limited Pedro 56382......... 21 lbs. 7 oz.
Clem Pedro 73961............. 20 lbs. 8 oz.
Lulu Pedro 71902............. 17 lbs. 4 oz.
Butterstamp Pedro 56383...... 14 lbs. 8 oz.
Lulu Pedro 71902 (13 days).... 31 lbs. 4 oz.

Young Prince (P. S. 182 J. H. B.)
Miss Porter 20300............. 16 lbs. 6 oz.

Young St. Martin 2219.
Lady Appel 8612.............. 18 lbs. 3 oz.

Young Sir Davy 3034.
Grace Davy 8292.............. 22 lbs. 5½ oz.

Young Weston 18439.
Merty of Peach Hill 77774..... 14 lbs. 4½ oz.
Gerty of Peach Hill 77773..... 14 lbs. 2¼ oz.

Yumroar 19656.
Ella L. 84123................. 14 lbs. 12 oz.

Zalma's Mercury 6983.
Lula M. 32643................. 15 lbs. 6 oz.

Zélie's Boy 30249.
Zélie's Zilpah 109344.......... 17 lbs.

Zeus 2634.
Pyrrha 6100................... 16 lbs. 14½ oz.

Dams With Their Tested Daughters.

Abbie Z. 14002.
Abbie Z. 3d 14742............... 17 lbs.

Abbie Z. 4th 14743.
Nuna 18669..................... 21 lbs.

Ability 69897.
Tamarack 69900................ 14 lbs. 3 oz.

Acacia of Home Farm 12094.
Acacia of Home Farm 3d 78203 17 lbs. 14½ oz.

Ada of Orange 16749.
Ada of Orange 2d 65246........ 17 lbs. 3 oz.

Ada of Orange 2d 65246.
Orange Extra 133961........... 18 lbs.

Addie Tormentor 78464.
Cowslip of Rockwell 115354... 14 lbs. 13½ oz.

Adeline 9115.
Queen of Prospect 11997....... 14 lbs. 2 oz.

Adonia 64699.
Adonia 2d 103302............. 17 lbs. 6 oz.
Adaphine 101425................ 15 lbs. 3 oz.

Adopted Girl 12767.
Loreda 19801.................. 17 lbs. 13 oz.

Aggie of St. Lambert 37085.
Exile's Aggie 52764............ 14 lbs. 10¼ oz.

Aggie of St. Lambert 2d 77184.
Exile's Jessamine 111104...... 14 lbs. 1¾ oz.

Alabama 7690.
Pride of Springvale 20093..... 19 lbs. 2 oz.
Christina Pogis 35426......... 17 lbs. 2 oz.
Alma of Springvale 26951..... 16 lbs. 14 oz.
Aurora Pogis 53798............ 15 lbs. 9 oz.
Ruby of Springvale 14505...... 14 lbs. 15 oz.

Alaric's Millicent 2d 63588.
Fancy Millicent 71059.......... 14 lbs. 3½ oz.

Alberta d'Or 76452.
Alberta d'Or 2d 124269......... 14 lbs. 13 oz.

Albert's Fawn 34004.
Miss Albert 63138....., ... 17 lbs. 9 oz.

Albert's Susette 47727.
Golden Susette 96552........... 17 lbs. 6 oz.

Alda 3873.
Tenella 6712.................... 22 lbs. 1½ oz.
Aldarine 5301.................. 15 lbs. 1½ oz.

Aldabella 3928.
Bianca Lass 14997............. 14 lbs. 3½ oz.

Aldarine 5301.
Aldarine 3d 27482............. 28 lbs. 1 oz.
Aldarine 2d 14196............. 14 lbs.

Alexa 64924.
Sombre 80796.................. 18 lbs. 14 oz.

Alexina 3836.
Grace M. 10697................ 15 lbs. 2 oz.

Alfieda 6744.
Alfritha 13673................. 15 lbs. 3 oz.

Alfritha 13673.
Lady Elfied 40757............. 18 lbs.

Alice 540.
Alice of the Meadows 20748... 14 lbs. 12 oz.

Alice Brand 17836.
Little Goldie 38671............ 34 lbs. 8½ oz.
Little Goldie 38671 (90 days).. 176.394 lbs.

Alice Donnal 12726.
Winetka 33669................. 20 lbs.

Alice Fairfax 83297.
Etta Lyne 83298............... 20 lbs. 5 oz.

Alice Landseer 2d 76815.
Toltec's Sally 94461.......... 17 lbs. 3¼ oz.
Brunette Landseer 109904..... 16 lbs. 13¼ oz.

Alice Ogden 62684.
Ogden's Alice 77590........... 17 lbs. 4 oz.

DAMS WITH THEIR TESTED DAUGHTERS.

Alice Rex 33535.
Alice Rex 3d 63907............ 15 lbs. 6½ oz.

Allena 3d 14879.
Oxford Catena 68905......... 15 lbs. 11¼ oz.

Allie L. 32073.
Bloom of Amherst Villa 70507. 16 lbs. 10 oz.
Pride of Bay View 61997...... 16 lbs. 6 oz.

Allie Minka 3d 27137.
Amna 35588................... 14 lbs. 12½ oz.

Allie of St. Lambert 24991.
Fawn of St. Lambert 27942.... 15 lbs. 5½ oz.
Fawn of St. Lambert 27942
(one year)................... 571 lbs. 6 oz.

Allie of St. Lambert 2d 43671.
Phillis of St. Lambert 78867... 23 lbs. 12 oz.
Sally of St. Lambert 78873.... 16 lbs. 5 oz.

Allie's Queen 45618.
Chief's Queen 69967.......... 20 lbs. 3 oz.

Alma of Springvale 26951.
Lotta Pogis 40368............. 16 lbs. 11 oz.

Almeda 3842.
Almeda Rex 10416............. 15 lbs. 12 oz.

Almond 11747.
Almond Blossom 59709........ 17 lbs. 3¾ oz.

Alnora 112132.
Brown Bessie's Brownie 114433 17 lbs.

Aloha of St. Lambert 63445.
Aloha's Tormentor 90239...... 14 lbs. 6 oz.

Alphalia 13820.
Athlete 28137.................. 14 lbs. 6 oz.

Alphea Alexia 15951.
Countess Bee 32630............ 14 lbs. 4 oz.

Alphea Gwendolen 32550.
Robinette Rioter 57206........ 15 lbs.

Alphea Hortense 14904.
Gem of Ellasleigh 38532....... 16 lbs. 10 oz.

Alphea Rajah 20605.
Clotaire's Daisy 45392......... 18 lbs.

Alrajah's Mystery 48466.
Woodstock Mystery 77746 (21
days) .... 26.223 lbs.

Amaranth 6200.
Alfieda 6744................... 16 lbs. 4 oz.

Amber of Cottage Grove 14225.
Shiloh Daughter 20378......... 16 lbs. 7 oz.

Amelia 484.
Mavourneen of St. Lambert
9777 .... 15 lbs. 12 oz.
Cupid of St. Lambert 5104..... 14 lbs. ½ oz.

Amelia 2d 1730.
Melia Ann 5444............... 18 lbs. ½ oz.
Gem of St. Cloud 7342........ 14 lbs. 8½ oz.

Amité 18877.
Eurotisama 29668............. 27 lbs. 1½ oz.
Eurotisama 29668 (one year).. 945 lbs. 9 oz.

Amna 35588.
Amna 2d 98269................ 18 lbs. 9 oz.

Ampelis 5622.
Ampelis 5th 17548............. 15 lbs.

Ampelis 7th 29851.
Amanda Pogis 79412.......... 18 lbs. 7¼ oz.
May Dee Pogis of C. H. 93568 15 lbs. 8 oz.

Amrah J. 23284.
Julia Rex of Ingleside 74046.. 16 lbs. 3 oz.
Amrah of Ingleside 56832..... 14 lbs. 2 oz.

Amy of Glenoir 47047.
Amy Landseer 74826.......... 16 lbs. 1 oz.

Amy of Ipswich 5583.
Winsome of Ipswich 9213...... 14 lbs. 7 oz.

Amy Valentine 68378.
Amanda May 83011............ 14 lbs. 15½ oz.

Analysis 29162.
Alteration 56436............... 24 lbs. ½ oz.
Analysis 2d 55834............. 20 lbs. 7½ oz.
Analysis 3d 82870............. 14 lbs. 1 oz.
Alteration 56436 (90 days)..... 179.879 lbs.

Andrews' Ethel 4108.
Eve of Christmas 9473......... 14 lbs. 5¼ oz.

Angela 1682.
Duchess of Bloomfield 3653... 20 lbs. ½ oz.
Su Lu 4705.................... 17 lbs. 15 oz.
Letitia 3977................... 15 lbs. 5 oz.

Angeline Pomeroy 2d 94684.
Fairy of Eau Claire 128596.... 14 lbs. 14 oz.

Angelo's Annie 58888.
Exile's Pansy 79800........... 17 lbs. 13 oz.
Exile's Agnes 79796.......... 14 lbs. 8 oz.

Anise Burns 44635.
Tormentor's Browny 73079.... 14 lbs. 14 oz.

Annabel Lee 55833.
Idlewild Pogis 80271.......... 14 lbs. 11 oz.

Anna Gold Ear 21053.
Exile's Anna 46881............. 16 lbs. 12 oz.

Anna Neal 2d 7304.
Mary Fallon 80220............ 17 lbs. 1¼ oz.
Rose Cottage Beauty 99542... 14 lbs. 10½ oz.

Anna of Mountain Side 15544.
Lady Anna Pogis 52734........ 20 lbs. 4 oz.

Anne 412.
Nannaxie 18860................ 17 lbs. 4 oz.

Annie 2d 776.
Actress 2311.................... 14 lbs.

Annie Bropho 39052.
Annice Magnet 60256 (90 days) 119.284 lbs.

Annie Dale 12662.
Lily Scituate 12665............. 24 lbs. 9½ oz.
Annie Dale's Princess 12664... 19 lbs. 2¼ oz.

Annie Landers 2d 7670.
Annie L. 12934................ 17 lbs. 15¼ oz.

Annie Laurie 3d 13236.
Laurie Pansy Champion 20396 14 lbs. 5½ oz.

Annie Wells 1947.
Welma 5942.................... 17 lbs. 8 oz.
Elinor Wells 12068............. 14 lbs.

Annorette 20178.
Alfaretta C. 34928.............. 14 lbs. 11½ oz.

Anolialla 19298.
Adonia 64699.................. 19 lbs. 1 oz.

Antianira 2457.
Faustine 10354................ 14 lbs. 14½ oz.

April Flower 4421.
Champion Flower 20887....... 14 lbs. 6 oz.

Arabel Hudson 2877.
Rosabel Hudson 5704......... 15 lbs. 12 oz.

Arawana Iris 2d 10420.
Lizzie of Glynllyn 74274...... 21 lbs. 15½ oz.
Segilinda 41741................ 14 lbs. 8½ oz.

Arawana Marigold 9380.
Ida Marigold 32615............ 25 lbs. 2½ oz.
Ida Marigold 32615 (30 days).. 59.367 lbs.
Ida Marigold 32615 (90 days).. 199.756 lbs.

Arawana Poppy 2d 9935.
Poppy H. 34865................ 17 lbs. 5½ oz.

Arawana Rose 3810.
Arawana Queen 5368......... 16 lbs. 9 oz.

Arbutus of A. 42529.
Miss Sassafras 70468.......... 17 lbs. 2 oz.

Archie Signal of Croton 73831.
Fricka 86450................... 15 lbs. 2½ oz.

Ardell H. 12330.
Vinala H. 61835............... 16 lbs.
Abbia H. 24221................ 15 lbs. 6 oz.

Argosy 4320.
Mother Cary 11746............ 27 lbs. 1½ oz.

Ariadne 2d 1135.
Estrella 2831.................. 14 lbs. 12 oz.

Ariella 9178.
Lesbie 9179................... 16 lbs. 3 oz.

Ariene 1071.
Patterson's Beauty 4160....... 18 lbs.

Arletta 14264.
Arletta 3d 14274............... 14 lbs. 13½ oz.

Arlette Cary 73372.
Platina Cary 73375............ 17 lbs. 7¼ oz.

Arthur's Frolic 4438.
Frolic's Pride 31667........... 17 lbs.
Frolic of Chestnutwood 19405. 16 lbs.

Ashwood Maid 15570.
Signalda's Duchess 32173..... 14 lbs. 3 oz.

Ashwood Pet 59355.
Ashwood Pet 2d 84408......... 16 lbs. 11¼ oz.

Astel 2d 10792.
Grace of Glynllyn 67759....... 14 lbs. 10 oz.

Atala 23238.
Pauline Hugo Pogis 37098..... 14 lbs. 14 oz.

Atlanta 2d 24011.
Susie B. of Springvale 24986.. 17 lbs. 5 oz.

Atlantis 5901.
Melita of Hillcrest 7054........ 14 lbs. 2 oz.

Augeres Georgiana 24421.
Coomassie Darlton 29519...... 16 lbs. 11 oz.

Augeres Violet 28377.
Ophi of St. Lambert 49004.... 14 lbs. 10½ oz.

Aunty 16648.
Auntybel 12582................. 14 lbs. 9 oz.

Auraria 10688.
Real Queen 29198............ 18 lbs. 1 oz.

Aurated Pogis 48220.
Aurated Pogis 2d 65666....... 16 lbs. 6 oz.

Aureola 8617.
Alberta Signal 18611.......... 20 lbs. 11 oz.

Auria 4567.
Taglioni 9182.................. 14 lbs. 1 oz.

Ava (F. S. 6217 J. H. B.)
Analysis 29162................ 16 lbs. 8 oz.

Azelda 3872.
Belle of Patterson 5664....... 16 lbs. 6 oz.
Gold Trinket 9518............. 16 lbs. 2 oz.
Valhalla 5300.................. 16 lbs.
Azelda 2d 7022................ 15 lbs. 2 oz.
Valhalla 5300 (14 days)........ 34 lbs.

Azuline 3360.
Azuline 2d 3888............... 14 lbs. 15¾ oz.

Azuline 2d 3888.
Champion Azuline 20542...... 18 lbs. ½ oz.

Bab 3d 40258.
Hermione C. 90598............ 14 lbs. 11 oz.

Babette 3650.
Ruby Torment 32989.......... 16 lbs.

Baby Ryan 42139.
Gold Baby 105745............. 17 lbs. 12 oz.

Bachelor's Effie 75678.
Effie's Girl 103452............. 14 lbs. 3½ oz.

Badger Girl 21463.
Badger Girl 4th 87488......... 15 lbs. 13 oz.

Badger Girl 2d 49717.
Gold Badger 67301............ 16 lbs. 7 oz.

Balmoral 13464.
Balmoral 2d 13574............. 14 lbs. 10 oz.

Barberry 2d 46391.
Tormentor's Barbara 98801.... 17 lbs. 3 oz.
St. Lambert's Babette 99185... 15 lbs. ¾ oz.

Barberry 2d's Princess 69073.
Barbara Lambert 105208....... 15 lbs. 11 oz.

Baroness Argyle 40498.
Chromoline of Argyle 103709.. 16 lbs. 10 oz.
Donna Argyle 61456.......... 15 lbs. 10 oz.

Baronet's Victory 20740.
Baronet's Victory 2d 39176..... 17 lbs. 5 oz.

Baronetti 8425.
Lucy McClung 20368.......... 14 lbs. 3 oz.

Beatrice Cenci 16629.
Alpheon's Belle 27194......... 19 lbs. 11 oz.

Beautiful 5981.
Brunette of Scarsdale 13276... 17 lbs.
Countess of Scarsdale 18633... 14 lbs. 6 oz.

Beauty 309.
Beauty Blücher 308........... 14 lbs. 3 oz.

Beauty 1319.
Beauty of Lee Farm 15694.... 20 lbs. 12¾ oz.
Stanstead Belle 4709.......... 15 lbs. 4 oz.

Beauty (F. S. 1159 J. H. B.)
Beauty Messenger 17450....... 15 lbs. 9 oz.

Beauty Dee 18065.
Beauty of Meridale 114744.... 18 lbs. 12½ oz.
Meridale Beauty Dee 97738.... 14 lbs. 9 oz.

Beauty Huge 63143.
Ethleel's Pet 85428............ 17 lbs. 6 oz.

Beauty of Darlington 5736.
Bomba 10330................... 21 lbs. 11½ oz.
Bomba 10330 (31 days)........ 89 lbs. 14 oz.

Beauty of Edgeworth 51361.
Grace Harribee 91836......... 20 lbs. 2 oz.
Lady Emily Kilduff 101053.... 14 lbs. 4 oz.

Beauty of Four Pines 31738.
Clyde Landseer 74834......... 14 lbs.

Beauty of Lakeside 2d 11322.
Beauty Dee 18065............. 23 lbs. 3 oz.

Beauty of Lee Farm 15694.
St. Lambert's Pride 53742.... 16 lbs. 13½ oz.
St. Lambert's Pride 53742 (14
days) .... 32 lbs. ¼ oz.

Beauty of Ninon 9691.
Beauty of Ninon 2d 18444..... 15 lbs. 3½ oz.

Beauty of Oakland 45009.
Bess of Edgewood 71178....... 20 lbs. 6 oz.

Beauty of Shady Side 49193.
Anklet 93500.................. 14 lbs. 6 oz.

Beauty of Snipsic 22909.
Edith Campbell 23011.......... 21 lbs. 4½ oz.

Beauty of Springvale 26677.
Corinne of St. Lambert 76579. 15 lbs. 10 oz.

Beauty's Baby 92511.
Kammerfrau 125093............ 15 lbs. 4 oz.

Beauvine 1593.
Gazania 4513.................. 14 lbs. 2 oz.

Beck 463.
Plenty 950.................... 14 lbs. 8 oz.

Beckie of Shady Side 39464.
Beckie Y. Pogis 63069........ 17 lbs. 15¼ oz.

Bee 448.
Bounty 1606................... 14 lbs.

Beech Grove 42709.
Gilderoy P. 99308............. 15 lbs. 3 oz.

Beech Grove Elsie 67758.
Elsie Coomassie 98081......... 17 lbs. 12 oz.

Beechnut of Glynllyn 29462.
Rheta of Glynllyn 74273....... 20 lbs. 4½ oz.

Beeswax 9807.
Waxie 19706................... 20 lbs. 2½ oz.
Ethleel 18724................. 19 lbs. 14 oz.
Fancy Wax 37159............... 19 lbs. 3½ oz.
Fancy Bee 37496............... 15 lbs. 8 oz.

Belboma's Beauty 65941.
Modiste 74679................. 16 lbs. 11½ oz.

Belbennie 29630.
Combonnie 40260............... 14 lbs. 2 oz.

Beletrice 7687.
Comtesse D'Espagne 10308.... 14 lbs. ½ oz.

Belinda R. 13150.
Curfew 16498.................. 14 lbs. 7½ oz.

Bell 1506.
Edith Darby 6246.............. 14 lbs. 2 oz.
Nellie 1507................... 14 lbs. 2 oz.

Belladonna of Glen Dale 7384.
Rosella of Bay View 66032..... 18 lbs. 10 oz.
Rosella of Bay View 66032 (30
days) .... 66 lbs. 2 oz.

Belle 53.
Belle of Ogden Farm 1570..... 14 lbs.

Belle (F. S. 302 J. H. B.)
Cocotte 11958................. 16 lbs. 8½ oz.

Belle Atwood 5907.
Belle of Echo 12432........... 14 lbs. 14½ oz.

Belle Clarendon 36334.
High Tea 65577................ 17 lbs. 6½ oz.

Belle Dame 11951.
Belle Dame 2d 22043........... 18 lbs. 12 oz.

Belle Dawson 8270.
Belle Yakout 38020............ 16 lbs. 13 oz.

Bell Earle 26732.
Bell Ensign 56702............. 15 lbs. 13½ oz.

Bell Ensign 56702.
Odelsign of Lynwood 114421... 15 lbs. 8½ oz.

Belle Grinnell 4073.
Belle Grinnell 3d 16503....... 14 lbs. 2 oz.

Belle Grinnell 2d 9009.
Grinnell Lass 11859........... 16 lbs. 10 oz.

Belle Hinman 3641.
Mary Hinman 17619............ 15 lbs. 11½ oz.

Belle Lyman 28522.
Belle Lyman 3d 86346......... 18 lbs. 2½ oz.
Harry's Fancy Belle 44764.... 14 lbs. 8 oz.
Emma Landseer 58104......... 14 lbs. 2½ oz.

Belle Lyman 2d 82424.
Bisson's Beauty 95023......... 16 lbs. 4 oz.

Belle Marquise 34859.
Lightning of St. Lambert 91392 14 lbs. 2 oz.

Belle Morgan 26219.
Moragina 26344 .... 17 lbs. 3 oz.

Belle Noble 2d 77233.
Belle Noble's Bessie 105710...   19 lbs. 5 oz.

Belle Noble's Signal 78013.
Miss Retta 105807.............   20 lbs. 1 oz.

Belle of Allentown 83636.
Kitty Black Prince 104472....'..   14 lbs. ½ oz.

Belle of Argyle 18358.
Nan of Argyle 47561..........   14 lbs. 5 oz.

Belle of Aurora 67665.
Hallie Rice 75777.............   24 lbs. 8½ oz.

Belle of Bayside 1457.
Maid of Berlin 12746.........   14 lbs. 8 oz.

Belle of Beaumont 2d 9269.
Leneca 17178..................   15 lbs. ¼ oz.

Belle of Bloomfield 4331.
Rosa Miller 4333..............   17 lbs. 7 oz.

Belle of Brooklyn 9493.
Miss Lofty 9718...............   16 lbs. 10 oz.

Belle of Canandaigua 10139.
Belle of Wayne 16457.........   17 lbs. 9½ oz.

Belle of Cherry Valley 15837.
Dura 26001....................   21 lbs. 5 oz.

Belle of Chester 4442.
Celia Belle 5865..............   14 lbs. 3 oz.

Belle of Cork 17403.
Letitia Hunter 35575..........   17 lbs. 5 oz.

Belle of Jessamine 79456.
Jessie Belle B. 100285.........   14 lbs. 2 oz.

Belle of Kennett 17085.
Belle's Charity 82217..........   17 lbs. 1 oz.

Belle of Kent 4371.
Gem of Sassafras 5434........   14 lbs. 3½ oz.

Belle of Maple Grove 11334.
Lady Alexis 26916............   16 lbs. 8 oz.

Belle of Maplewood 22942.
Violet of Maplewood 49724....   14 lbs. 10¼ oz.

Belle of Middlefield 1516.
Princess Bellwort 6801........   15 lbs. 10½ oz.

Belle of New York 6963.
Lady Cleveland 30251.........   17 lbs. 1 oz.

Belle of New York 2d 81731.
Idabel Landseer 88399.......   15 lbs. 15¼ oz.

Belle of Passaic 85846.
Theda's Belle 106891.........   16 lbs. 2 oz.

Belle of Patterson 5664.
Guinevere Sinclair 11167......   14 lbs. 9 oz.

Belle of Prospect 6627.
Belle of Prospect 2d 14326.....   19 lbs.

Belle of Rydal Grange 41787.
Belle of Dorset 85905.........   14 lbs. 9 oz.
Tormentor Victor Pogis 133021   14 lbs. 7 oz.

Belle of St. John's 8078.
Belle of St. John's 2d 29829...   15 lbs. ½ oz.

Belle of Tennessee 9573.
Lilly Signalda 23227..........   17 lbs. 10 oz.
Belnina 19189................   17 lbs. 3¼ oz.

Belle of Utica 11356.
Bunker's Utica Belle 38567...   15 lbs. 15 oz.

Belle of Wolcott 8938.
Belle Lyman 28522...........   16 lbs. 3½ oz.

Belle of Woodstock 11658.
Rush of St. Lambert 61234....   17 lbs. 6¾ oz.
Rush of St. Lambert 61234 (15
days) ....   34 lbs. 12¼ oz.

Belle Ridgely 6785.
Magnolia Ridgely 17269.......   14 lbs. 8 oz.

Belle Rosy 27291.
Ashantee's Rosy 59258.........   14 lbs. 11 oz.

Belle's Charity 82217.
Charity Victor Pogis 134217...   14 lbs. 6 oz.

Belle Warren 7978.
Belle's Esperanza 12053......?   15 lbs. 9½ oz.

Belle Williamson 8386.
Durwood's Lass 19710........   16 lbs.
Belle Williamson 3d 104290....   15 lbs. 3½ oz.

Bellflower 59.
Ma Belle 4942.................   15 lbs.

Bellis 38284.
Beauty of Hamilton 55827.....   15 lbs. 15 oz.

Bellita 2d 10311.
Percie 14937...................   18 lbs. 10 oz.

Belna 41043.
Vintage 53335.................. 14 lbs. 12 oz.

Bemona 52994.
Anna Paris 67989............. 14 lbs. ¾ oz.

Benefit 15218.
Benefit's Daughter 63932...... 20 lbs.

Benefit's Daughter 63932.
Coomassie Fawkes 122445..... 17 lbs. 8 oz.

Bennie Hinman 7166.
Louisa Hinman 12802.......... 14 lbs.

Berlin Daisy 3759.
Duchess of Argyle 3758........ 14 lbs. 13 oz.

Bermuda R. 67534.
Salute 67682.................... 16 lbs. 8 oz.

Bertha Maplewood 22944.
Matilda of Maplewood 49723... 14 lbs. 15 oz.

Bertha Morgan 4770.
Lydia Darrach 4903........... 17 lbs. 14 oz.

Bertha Polonius 24914.
Duchess of Bolton 98150....... 24 lbs. 4 oz.

Bertha's Star 24902.
Susie V. 54855.................. 14 lbs. 14 oz.

Bertie Don 32071.
Bertie Don 2d 74088........... 16 lbs. 14 oz.

Bertie Don 2d 74088.
Ada of Elmwood 98728........ 14 lbs. 2 oz.

Bertina B. 81715.
Clarina B. 98430............... 14 lbs. 2 oz.

Berylla 8957.
Berylla 2d 15117............... 15 lbs. 13 oz.

Bessie (P. S. 232 J. H. B.)
Lucilla Kent 8892.............. 16 lbs. 6 oz.

Bessie Allen 3719.
May Blossom 5657............. 18 lbs. 11 oz.

Bessie Bradford 7269.
Bessie Bradford 2d 7271....... 15 lbs. 2 oz.

Bessie Day 82257.
May Bud of Monmouth 128864 15 lbs. 8 oz.

Bessie Dunedin 73665.
B. D. Exile 105286............. 18 lbs. 10 oz.

Bessie Morris 26730.
Bessie Pope St. Lambert 80352 16 lbs. 1¼ oz.

Bessie Niobe 55763.
Clark's Gift 91708............. 15 lbs. 4 oz.

Bessie of Greenwood 9752.
Hattie N. 45433................ 15 lbs. 14½ oz.

Bessie Ring 12175.
Miss Bowris 12206............. 17 lbs. 8½ oz.

Bessie Russ 14648.
Bessie Russ 2d 14649.......... 15 lbs. 1½ oz.

Bessie Russ 2d 14649.
Bessie Landseer 58103........ 15 lbs. 1 oz.

Bessie Signalda 38178.
Bessie Signalda 2d 86341...... 15 lbs. 2½ oz.
Bessie Signalda 3d 86349....... 15 lbs. ½ oz.
Landseer's Signaline 58101.... 14 lbs. 11⅞ oz.

Bessie Signalda 2d 86341.
Ida's Signaline 108819........ 15 lbs. 10 oz.

Bessie Spaulding 2780.
Madolina H. 12327............ 19 lbs. 4 oz.

Bess Lena 3349.
Lady Gray of Hilltop 6850.... 18 lbs. 12 oz.

Bess of Edgewood 71178.
Belle Cotton 96205............. 14 lbs. 8 oz.

Bess of Ingleside 42172.
Bess Pogis of Prospect 87045.. 29 lbs. 1¾ oz.
Gipsy Bess of Prospect 96928. 22 lbs. 10½ oz.
Bess of Ingleside 2d 59648..... 19 lbs. 10½ oz.

Bessy of St. Lambert 5482.
Moth of St. Lambert 9775.... 16 lbs. 2 oz.

Betsy 3d 8162.
Betsy's Glory 45604.......... 14 lbs. 1 oz.

Betsy B. 9413.
Betsona 16776.................. 14 lbs. 3 oz.

Betsy Rain 37454.
Betsy Marjoram 65145........ 15 lbs. 7 oz.

Bettie Belle Nimmo 39960.
Moorey of Lawn 68847........ 21 lbs. 4 oz.

Bettie D. 7677.
Marpetra 10284................ 14 lbs. 6 oz.

Bettie of Hillside 3864.
Bettie of Edgewood 31162..... 14 lbs. 2 oz.

Betty 683.
Countess of Lakeside 12135.... 19 lbs. 7 oz.

Betty Bluet 53464.
Saucy Pogis 103535............ 17 lbs. 2¼ oz.

Betty Clover 3784.
Mary Clover 9998............. 14 lbs. 15 oz.

Beulah of Baltimore 3270.
Jessie Lee of Labyrinth 5290.. 14 lbs. 7 oz.

Bicora 88461.
Bicora 2d 91188............... 16 lbs. 3 oz.

Biddy of Bovina 6655.
Calista of Newark 13296...... 15 lbs. 9 oz.

Bijou of St. Lambert 5112.
Honeymoon of St. Lambert
11221 .... 20 lbs. 5¼ oz.
Columbine of St. Lambert
8350 .... 19 lbs. 1 oz.

Bina of Eau Claire 77585.
Emma of Eau Claire 127826.. 21 lbs. 4 oz.

Birdie Birchette 66773.
Cabbie Oak 90061.............. 14 lbs. 12 oz.

Bisma 1669.
Lula M. 32643................. 15 lbs. 6 oz.

Bisma 3d 1870.
Beeswax 9807.................. 17 lbs. 5 oz.
Little Lass 28512.............. 17 lbs. 5 oz.
Busy Bee 6336................. 16 lbs. 4 oz.

Black Diamond's Queen 11865.
Alphea Vaudrada 17468........ 19 lbs. 7 oz.

Black Teat 32932.
Nerbat 58064.................. 15 lbs. 7 oz.

Blanche 594.
Ianthe 4562................... 16 lbs. 10 oz.

Blanche Amory 7592.
Jaquenetta 10958.............. 14 lbs. 6 oz.

Blanche Duncan 73077.
Blanche Duncan 2d 73333..... 14 lbs. 11¾ oz.

Blanche Duncan 2d 73333.
Blanche Pomeroy 105013...... 16 lbs. 9½ oz.

Blanche of Castile 43793.
Summer Morning 75197....... 20 lbs. ½ oz.

Bliss 15219.
Lorne's Pretty Pearl 46683.... 19 lbs. 6 oz.

Blizz 55835.
Signola M. 76122.............. 14 lbs. 10 oz.
Blizz Pogis 82873.............. 14 lbs. 9½ oz.

Blonde (F. S. 214 J. H. B.)
Blonde 2d 9268................. 14 lbs. 4 oz.

Blondette 1817.
Troth 6139.................... 16 lbs. 5 oz.

Bloom 2d 5927.
Prince's Bloom 9729.......... 14 lbs. 3 oz.

Bloomer 2262.
Bloomer 2d 2321............... 18 lbs. 6 oz.

Bloomfield Fairy 2d 3210.
Lady Bloomfield 4704......... 14 lbs. 12½ oz.

Bloom of Amherst Villa 70507.
Lady Musa 98730.............. 16 lbs. 9½ oz.

Blossom 12013.
Marie S. 12043................ 15 lbs. 6 oz.

Blossom of Hanover 13655.
Dia 13658..................... 15 lbs. 13½ oz.

Blossom of Prospect 18627.
Sweet Blossom Pogis 36995... 16 lbs. 1¼ oz.

Blossom's Niobe 13855.
Niobomba 26456............... 14 lbs. 15 oz.

Blush Rose (F. S. 453 J. H. B.)
Princess of the Valley 22641... 18 lbs. 12 oz.

Bombazine 2d 49327.
Editha Westray 89755......... 19 lbs. 5 oz.

Bonanza 3965.
Fanny Taylor 6714........... 15 lbs. 12 oz.

Bonnie 491.
Bonnie 2d 5742................ 14 lbs. 11½ oz.

Bonnie Eloise 15244.
Down of Chatsworth 15251.... 16 lbs. 12 oz.

Bonnie Gladesia 47792.
Bonnie Wherry 115088........ 20 lbs. 4 oz.

Bonnie Grisette 6979.
Bonnie Grisette 2d 19526....... 16 lbs. 12½ oz.

Bonnie Grisette 2d 19526.
Koffee's Grisette 30433........ 15 lbs. 7 oz.

Bonnie Jean 13116.
Nan Day 17192................. 20 lbs. 4 oz.

Bonnie Katrina 80509.
Bonnie Signaldina 108681..... 18 lbs. 4 oz.

Bo Peep 2850.
Bonnie Yost 7943.............. 18 lbs. 2 oz.

Bornou (F. S. 4338 J. H. B.)
King's Beauty 24397.......... 16 lbs. 13 oz.

Bother 25595.
Golightly 25597................ 23 lbs. 4 oz.

Bounty Duchess 37955.
Marcia G. 53211............... 14 lbs. 11 oz.

Bowley 1946.
Mamie Coburn 3798........... 17 lbs. 8 oz.

Box 1982.
Box 2d 4297................... 14 lbs. 3 oz.

Brenda 789.
Venus 112..................... 14 lbs. 11 oz.

Brenda St. Helier 20052.
Moyane's Pet 98115............ 16 lbs. 12 oz.

Brenda Yorke 34460.
Dicta 55221................... 20 lbs. 9 oz.

Brewer's Daisy 7333.
Whiting's Daisy 14858......... 14 lbs. 10½ oz.

Brewer's Queen 68288.
Old Noble's Heiress 99083.... 15 lbs. 6½ oz.

Bridal Rose 74883.
Palm Rose 75110.............. 15 lbs. 1 oz.
Tormentor's Oona Toltec
105860 .... 14 lbs. 12 oz.

Bride of Evergreen 72022.
Bride of Evergreen 3d 90473... 17 lbs. 1 oz.

Bride of Lakeside 44139.
Bride of Evergreen 72022..... 20 lbs.

Brier of Portland 28753.
Ora of Forester 54698......... 15 lbs. 12 oz.

Brighteyes 1517.
Chloe Beach 3931.............. 14 lbs. 8 oz.

Bright Eyes 1537.
Bright Eyes 2d 2290.......... 19 lbs. 6 oz.

Bright Eyes St. Lambert 53720.
Bright Eyes St. Lambert 2d
72699 .... 19 lbs. 5 oz.

Briseis 3875.
Yokun Maid 19073............ 16 lbs. 6 oz.

Bristol Bella 15697.
Bella Marigold 94124.......... 17 lbs. 2 oz.

Bristol Girl 66865.
Morganza Queen 77791....... 16 lbs. 2¾ oz.

Bronx Pearl 3752.
Miss Willie Jones 6918........ 16 lbs. 4 oz.

Bronzie 9368.
Bronzie of St. Lambert 49001. 14 lbs. 2 oz.

Brown Bessie 74997.
Teasel 75358.................. 20 lbs. 4 oz.

Brown Flora 74996.
Brown Bessie 74997........... 20 lbs. 8 oz.
Brown Flora 2d 96594......... 20 lbs. 2 oz.
Brown Bessie 74997 (30 days). 72.235 lbs.
Brown Bessie 74997 (90 days). 216.640 lbs.

Brown Flora 2d 96594.
Brown Elsie 96595............. 21 lbs. 12 oz.
Miss Flora Brown 100532...... 16 lbs. 2 oz.
Gray Bessie 96596............. 14 lbs. 4 oz.

Brown Lily 32770.
Laura of Eau Claire 110739.... 18 lbs.

Brunette (F. S. 1256 J. H. B.)
Fan of Grouville 7458......... 15 lbs.

Brunette (F. S. 2110 J. H. B.)
Royal Lady 22078............. 14 lbs. 12 oz.

Brunette (P. S. 101 J. H. B.)
Brunette Star 27270........... 17 lbs. 10 oz.

Brunette Balliet 6733.
Dom Pedro's Julian 8631...... 16 lbs.

Brunette Lass 1780.
Leoni 11868................... 18 lbs. 7 oz.

Brunette Le Gros 9755.
Brunette Le Gros 3d 23326.... 14 lbs. 9 oz.

Brunette Star 27270.
Ashantee's Brunette 65946.... 16 lbs. 7 oz.

Buckeye Pet 19459.
Pet's Fancy 80968 .... 17 lbs. 4 oz.

Buckwheat 13840.
Preference 26343 .... 15 lbs. 5 oz.

Buff Nepte 35641.
Cherry of Glynllyn 81145 ...... 19 lbs. 4 oz.

Bull's Kitty Clover 3897.
Jessie Leavenworth 8248 ...... 14 lbs. 2 oz.

Burckel's Clover Blossom 15609.
Bronx Clover Blossom 38676 .. 15 lbs. 2½ oz.

Busy Bee 6336.
Busy Melrose 54456 .... 21 lbs. 1¼ oz.
Fleece 18568 .... 17 lbs. 13½ oz.
Busy Bee 2d 25166 .... 15 lbs. 8 oz.
Busy Melrose 54456 (31 days) .. 85 lbs. 2¾ oz.

Busy Bee 2d 25166.
Busy Princess 48202 .... 17 lbs. 1¾ oz.
Bee Princess 40345 .... 14 lbs. 2⅝ oz.

Buttercup 311.
Susanna Frank 18261 .... 14 lbs.

Buttercup 3d 10253.
Dott Buttercup 16358 .... 16 lbs. 2 oz.

Buttercup Darling 28020.
Lolly Darling 78005 .... 16 lbs. 6 oz.

Buttercup of Riverview 19952.
Belle of Riverview 21464 ...... 17 lbs. 12½ oz.

Butterfly 12020.
Ballet Girl 18750 .... 14 lbs. 1 oz.

Butterfly (F. S. 1743 J. H. B.)
Olymph 17957 .... 15 lbs. 13 1-16 oz.

Butterfly of Riverview 21466.
Butterfly of Riverview 3d
85499 .... 17 lbs.

Butterstamp Lass 19517.
Rioter Carlotta 29667 .... 21 lbs. 2½ oz.

Butterstar 7799.
Little Butterstar 20926 .... 14 lbs. 12½ oz.

Buttery 3502.
Buttery 2d 27138 .... 15 lbs. ⅝ oz.

Bye-Bye (F. S. 3180 J. H. B.)
Rosa of Belle Vue 6954 .... 18 lbs. 7½ oz.

Caddie Mackie 10411.
Peggy Ford 21713 .... 14 lbs. 10 oz.

Cad Lee 4th 35721.
Allie St. Helier 45794 .... 16 lbs. 13½ oz.

Cæsarea 3d 22588.
Mona's Beauty 50041 .... 15 lbs. 4 oz.

Calcium 79733.
Calcina 80702 .... 18 lbs. 8 oz.

Caleri 7016.
Clara C. Magnet 31563 .... 14 lbs. 11 oz.

Calista Appel 10760.
Countess Godiva 10820 .... 15 lbs. 7 oz.

Calista Noble 36065.
Noble Charity 67245 .... 18 lbs.

Callie Nan 7959.
Callie Nan 2d 64207 .... 23 lbs. 1 oz.

Callie Nan 2d 64207.
Calisida 76889 .... 14 lbs. 4 oz.

Callie of Eau Claire 94451.
Mab of Eau Claire 120020 ...... 14 lbs. 10 oz.

Callie Rex 21672.
Addals 92923 .... 14 lbs. 3 oz.

Callinette 2d 38908.
Callinette Pogis 99700 .... 17 lbs. 8 4-5 oz.
Ethleel D. 83006 .... 14 lbs. 1 3-5 oz.

Calpurnia 13267.
Lady Hugo 29430 .... 16 lbs. 7 oz.

Camelia (F. S. 1687 J. H. B.)
Camelia 2d 11188 .... 20 lbs. 3 oz.
Favorite of Avon 13438 .... 17 lbs. 7¾ oz.

Camelia of St. Lambert 5106.
Moss Rose of St. Lambert
5114 .... 14 lbs. ½ oz.

Camilla 4th 2023.
Bramballetta 10451 .... 16 lbs. 4 oz.

Caramel 2727.
Morlacchi 2725 .... 14 lbs.

Cara Mia 64224.
Tomassee 76875 .... 17 lbs. 5 oz.
Cara Mia 2d 93827 .... 15 lbs. 5½ oz.
Tomassee 76875 (14 days) ..... 41 lbs. 8¼ oz.

Caressa 88474.
Campania 88475 (21 days)..... 28.127 lbs.

Careta 19092.
Careta Pogis 42784............ 15 lbs. 1¼ oz.

Carida 37322.
Cara Mia 64224................. 15 lbs. 9¼ oz.

Carlo Belle 23091.
Miss Milton 51851............. 24 lbs. 9½ oz.

Carlo's Lass 94405.
Pedro's Pretty Lass 135067.... 16 lbs. 6 oz.

Carlo's Polly Rex 94627.
Griselda Marjoram 111166..... 15 lbs. 6 oz.

Carlo's Pride 31371.
Lynona 36240................... 16 lbs. 8 oz.

Carma 6361.
Le Broemer 10670.............. 15 lbs. 7 oz.
Lass of Burrwood 37437...... 14 lbs. 1½ oz.

Carola of Bois d'Arc 40198.
Carola's Crown Princess 90321 14 lbs. 14 oz.

Carrie 3894.
My Queen 12614................ 15 lbs. 8 oz.
Carrie's Beauty 14601........ 14 lbs. 7½ oz.

Carrie Clark 6584.
Carrie Rex 10271............. 16 lbs. 7½ oz.

Carrie H. 12182.
Dolly Doyle 22788.............. 15 lbs. 4½ oz.

Carrie Lena 3348.
Countess Potoka 7496......... 18 lbs. 15 oz.
Carrie Lena 3d 20077.......... 18 lbs. 11¾ oz.
Carrie Lena 2d 19074.......... 14 lbs. 5⅓ oz.

Carrie Lena 3d 20077.
Carida 37322................... 15 lbs. 6 oz.

Carrie Pedrolier 48966.
Spot Victor Pogis 85901....... 17 lbs. 3 oz.
Ann Victor Pogis 77813....... 14 lbs. 7 oz.

Carrie Ryan 47413.
Carrie Gold 56348............. 16 lbs. 8 oz.

Carrie's Beauty 14601.
Carrie's Daffodil 15232....... 14 lbs. 7¾ oz.
Hindoo Rose 14602........... 14 lbs. 7 oz.
Carrie's Wonder 15233....... 14 lbs. 2½ oz.

Carrie Waite 10641.
New Maud Lee 16614........ 14 lbs. 10 oz.

Cassia 3615.
Cassia 2d 21370............... 20 lbs. 10¼ oz.

Cassia Catoctin 23561.
Betsy of Rochester 23562..... 14 lbs. 10½ oz.

Cassie Marjoram 120666.
May Marjoram 123811......... 16 lbs. 5 oz.

Casta Diva 8154.
Casta's Belle 56148........... 14 lbs. 15 oz.

Castaledes (F. S. 2876 J. H. B.)
Lady Antoinette 24391........ 21 lbs. 6 oz.

Castara 2d 46853.
Combination of Lawn 76635.. 14 lbs. 2½ oz.

Castelara 2d 9859.
Lady Panalphrex 17400........ 23 lbs. 9 oz.

Catchfly 25405.
Catchfly 4th 47682............ 17 lbs. 4 oz.
Catchfly 5th 55483............ 16 lbs. 2 oz.
Catchfly 3d 35499............ 15 lbs. 7 oz.

Catchfly 4th 47682.
Colx 85258.................... 15 lbs. 4 oz.

Cathiella 16947.
Trenie 17770................... 14 lbs. 10 oz.

Catoctin's Daughter 29262.
Exile's Lucy 46883............ 15 lbs. 7¾ oz.

Catono's Pink 2d 58626.
Pink Ring 69394............... 16 lbs. 4½ oz.

Caucus 33462.
Caucus Maid 69525........... 15 lbs. 8 oz.

Cauline 15298.
Crensa 74203.................. 16 lbs. 2 oz.

Celestia 1898.
Celestia 2d 29482.............. 16 lbs. 13 oz.
Merlette 4988.................. 16 lbs.

Celia Belle 5865.
Madam Zophee 31679......... 16 lbs. 2½ oz.
Ida of Coal Hill 12542........ 15 lbs. ½ oz.

Cenie Wallace 2663.
Cenie Wallace 2d 6557....... 15 lbs. 4½ oz.
Bettie Dixon 4527............. 15 lbs.

Cérise (F. S. 737 J. H. B.)
Faultless 12018................ 17 lbs. 5½ oz.

Cetewayo's Lily 18950.
Calma of Briarcliff 52599...... 14 lbs. 7½ oz.

Chaffinch 16524.
Chancery 37987................ 18 lbs. 10 oz.

Champion G.'s Duchess 40768.
Fena Eric 69137................ 22 lbs. 10 oz.
Mildred Montague 45476...... 15 lbs. 5½ oz.

Champion Regina 21227.
Championa Pogis M. 75926.... 15 lbs. 15 oz.

Champion's Caroline 4th 65300.
Moll Roberts 110201.......... 14 lbs. 6½ oz.

Chansonnette 5695.
Rioter Rhea 10092........... 19 lbs. 3½ oz.
Chansonnette 2d 29672....... 16 lbs. 9 oz.
Rioter Alphea 10091.......... 16 lbs. 7 oz.

Charella 12017.
Lady Charella 66187.......... 14 lbs. 7 oz.

Charella of Bay View 48923.
Chief's Charella 61999........ 23 lbs. 10½ oz.

Charlotte Brooks 29580.
Helen Barry 55840............ 18 lbs. 7 oz.

Charlton Caroline 11724.
Silver Delle 40691............ 18 lbs. 14½ oz.
Cream Caroline 17357......... 16 lbs. 2 oz.
Silver Delle 40691 (29 days)... 55 lbs.

Charm 3d 79876.
Charm 3d's St. Lambert 96872 14 lbs. 8 oz.

Charmante (F. S. 1866 J. H. B.)
Lady Fair 22103............... 19 lbs. 10 oz.

Chatelaine 1916.
Chenda 4599................... 17 lbs. 1½ oz.

Cheerful of St. Lambert 8348.
Cheerful of St. Lambert 2d
43745 .... 22 lbs. 2 oz.

Cheerful of St. Lambert 2d 43745.
Cheerful Pogis 61748......... 21 lbs. 4½ oz.

Cheerful Pogis 61748.
Maud of Glen Rouge 103781... 22 lbs. 10 oz.

Chemung Maid 34704.
Sparks' Maid 45575........... 19 lbs. 8 oz.
Millerton Maid 74926......... 17 lbs. 3 oz.

Chenda 4599.
Chenda 3d 43246............... 15 lbs. 7 oz.

Cherry (F. S. 1140 J. H. B.)
Primrose 11956................ 21 lbs. 10 oz.

Chestnut (F. S. 3125 J. H. B.)
Chestnut's Beauty 21576...... 18 lbs. 4¼ oz.

Chestnut Farm Sylvia 25022.
Sylvia Perrot 30963........... 16 lbs. ½ oz.

Chief's Charella 61999.
King's Stevia 98737........... 19 lbs. 3½ oz.

Chief's Fosie 61998.
King's Fosie 80763............ 15 lbs. 10 oz.

Chief's Guenn 29266.
Otta of St. Lambert 49003..... 15 lbs. 8¾ oz.

Chief's Nattie 38469.
Clover of Bay View 65424..... 18 lbs. 14 oz.

Chilioness 44696.
Chilioness Queen 82973........ 26 lbs. 8 oz.

Chinquapin 4501.
Chinqua 27384................. 22 lbs. 9½ oz.

Chloe 129.
Champion's Chloe 12255....... 15 lbs. 5½ oz.

Chloe 1540.
Chloe 4th 4612................ 17 lbs. 4 oz.

Chloe Lankton 15871.
May Lankton 15872............ 16 lbs. 1¼ oz.

Chloe of Millwood 14083.
Snowdrop F. W. 16948........ 14 lbs. 8 oz.

Chloe Tormentor of Lawn 71642.
Nora Tormentor of Lawn
99738 .... 15 lbs. 12 oz.

Chrissy 1448.
Chrissy 2d 7720............... 16 lbs. 14 oz.
Countess of Croton 5307....... 15 lbs. 12 oz.
Chrissy 3d 15615.............. 14 lbs. 11 oz.

Chrissy Signal 27790.
Cynthia A. 43721.............. 20 lbs. 10 oz.

Christel 6565.
Gildercream 39480............. 17 lbs.

Christine of Side View 47278.
Christine of Side View 2d
94253 .... 14 lbs. 15 oz.

Chroma 4572.
Renini 9181................... 14 lbs. 10½ oz.

Chromo's Modita 121439.
Brown Bessie's Modita 128259   15 lbs. 3 oz.

Chronicle 21625.
Transcript 31867............... 17 lbs. 7 oz.

Chrysantha 23465.
Princess Chrysantha 34191.... 14 lbs. 3½ oz.

Chuffy Mianella 70915.
Chuffy's Ethel 102195.......... 16 lbs. 1 oz.

Churchill Betsey 4105.
Webster Pet 4103.............. 14 lbs. 2 oz.

Cicely 997.
Cowry 4432.................... 14 lbs. 1 oz.

Cigarette 2849.
Minnie Kersey 19895.......... 14 lbs. 11 oz.

Cigarette (F. S. 1962 J. H. B.)
Matin 7768.................... 17 lbs. 11 oz.

Cinderella St. Helier 27241.
Cinxia St. Helier 46730....... 17 lbs. 9 oz.

Cintra Cowles 6198.
Cowles' Nonesuch 6199........ 14 lbs. 12 oz.

Cinxia St. Helier 46730.
Cinxia's Pavonia 64441........ 18 lbs. 3 oz.

Clara 4th 1541.
Country Girl 3515............. 15 lbs. 12 oz.

Clara Barder 14311.
Miaclara 48690................ 14 lbs. 8 oz.

Clara of Jefferson 12311.
Clara's Belle 16582........... 17 lbs. 9½ oz.

Clara of Riverside 26301.
Inez of Riverside 51781....... 26 lbs. 1½ oz.
Inez of Riverside 51781 (21
  days) .... 71 lbs. ½ oz.

Clara's Belle 16582.
Seraph 72217.................. 18 lbs. 5 oz.
Diploma's Clara 103867........ 15 lbs. 12½ oz.

Clara Tormentor 41427.
Clara Oonan 78454............ 21 lbs. 2 oz.
Clara Tormentor 2d 97279...... 17 lbs. 6 oz.

Clarencia 3484.
Woodland Maid 8156.......... 14 lbs. 2 oz.

Clari 4266.
Marchioness of Maplehurst
  8018 .... 14 lbs. 6 oz.

Clarinda 2d 51037.
Koffee's Clarinda 65216....... 19 lbs. 9½ oz.

Clearwater 24382.
Eastwood Clearwater 30445... 27 lbs.

Clelie of Mapleton 27182.
Rexella 69413................. 14 lbs. 4 oz.

Clematis 3174.
Clematis 3d 6653.............. 16 lbs. 1 oz.

Clematis 3d 6653.
Baroness Hugo 47563......... 15 lbs. 10¼ oz.

Clematis of St. Lambert 5478.
Flower of Glen Rouge 17560... 23 lbs. 14⅝ oz.
Honeysuckle of St. Anne's
  18674 .... 14 lbs. 14 oz.

Clem Pogis 49932.
Clem Pedro 73961............. 20 lbs. 8 oz.

Clem's Lucy 26694.
Clem Pogis 49932............. 23 lbs. 4½ oz.

Clio 2d 1248.
Charmer 4771.................. 14 lbs. 12 oz.

Clio of Staatsburgh 2d 12540.
Cora of Arcadia 16151......... 15 lbs.

Clochette d'Or 5696.
Chansonnette 5695............. 16 lbs. 4 oz.

Clotaire's Alphea 32502.
Lanison's Nora 59316.......... 15 lbs. 6½ oz.

Clotaire's Annie 58319.
Cleodoxa 59311................ 22 lbs. 5 oz.

Clotaire's Beauty 28171.
Princess Carmen 58320........ 16 lbs. 14 oz.

Clotaire's Belle 28841.
Lanison's Belle 84986......... 20 lbs. 5 oz.
Exile's Alphea 45393.......... 15 lbs. 14½ oz.

Cloth of Gold 18490.
Gold Violet 85078............. 17 lbs. 1 oz.

Clotilde 4575.
Oakland Girl 11103............ 14 lbs. 10½ oz.

Clover Bloom 9788.
Clover Mel 16159............... 14 lbs. 9 oz.
Clover Bloom 2d 12736........ 14 lbs.

Clover Blossom 4057.
Mylitta 15968.................. 16 lbs. 13 oz.

Clover Bud 4074.
Clover Bud 4th 18992......... 16 lbs. 14 oz.
Clover Bud 5th 27743......... 14 lbs.

Clover Bud 5th 27743.
Alicealda 43694............... 14 lbs. 1 oz.

Clover of Bay View 65424.
Albert's Lobelia 111902........ 18 lbs. 1 oz.

Clubs 16721.
Clubs 2d 16725................. 14 lbs. 7 oz.

Clytemnestra 2455.
Richness 16536................ 17 lbs. 5 oz.
Hartwick Belle 7722.......... 14 lbs. 8 oz.

Clytie Lass 3395.
Cottage Lass 5332............. 17 lbs. 7 oz.

Cobweb 5006.
Cobweb 3d 21325.............. 18 lbs. 5 oz.

Cocette, on I. of J.
Happy Blossom 18218.......... 19 lbs. 5 oz.

Cocoa Butter 64983.
Between 86964................. 18 lbs. 9¼ oz.

Cocotte 7392.
Belle of Lynwood 18364....... 17 lbs. 14 oz.

Cocotte 11958.
Como of Briarcliff 35849...... 14 lbs. 6 oz.

Cocotte (P. S. 38 J. H. B.)
Lady Kingscote 26085......... 15 lbs. 10 oz.

Coe's Stella 3930.
Arawana Buttercup 6052...... 15 lbs. 5 oz.
Arawana   Chèvre   Feuille
15597 .... 14 lbs. 9 oz.

Cœur d'Alene 49802.
Deerlick Pojoram 99988........ 15 lbs. 7 oz.

Colie 8309.
Smoky 13733.................. 14 lbs. 9 oz.

Columbine (F. S. 131 J. H. B.)
Jessaline 26099................ 16 lbs. 6 oz.

Coma 29330.
Commotion 52960............. 17 lbs. 6 oz.
Period 42640.................. 16 lbs. 3 oz.
Diploma's Coma 54070........ 14 lbs. 10 oz.

Comanca 19389.
Bosnia 87598.................. 20 lbs. 2 oz.

Comballina 52701.
Cymbelina 67291............... 14 lbs. 13 oz.

Comely of St. Lambert 6639.
Comely of St. Lambert 2d
41177 .... 20 lbs. 10 oz.

Comille 86371.
Elma's Star 102275............ 14 lbs. 12 oz.

Coming Girl 12772.
Coming Girl 3d 71794......... 17 lbs. 14 oz.

Commilla 79614.
Diploma's Queen 98151........ 17 lbs. 8 oz.
Diploma's Princess 104002.... 17 lbs. 4 oz.
Calcium 79733................. 16 lbs. 15 oz.
Brown Lassie 92950........... 15 lbs. 7 oz.

Commotion 52960.
Complexia 56774............... 15 lbs. 3 oz.
Compressa 55756.............. 14 lbs. 13½ oz.

Compeer's Rowena 22968.
Rowena of Elm Spring 54613.. 14 lbs. 4 oz.

Complexia 56774.
Eurus 60801................... 14 lbs. 7 oz.

Comwa 35911.
School Marm 67264........... 18 lbs. 4 oz.
Smax 48562................... 16 lbs. 11½ oz.

Connoisseur 65644.
Sobriquet 68628............... 21 lbs. 3 oz.

Conover's Beauty 12650.
Conover's Beauty 2d 25315..... 18 lbs. ¾ oz.

Content of Linwood 6950.
Landseer's Content 56985..... 16 lbs. 8 oz.

Contessa 3256.
Mary of Gilderoy 11219........ 14 lbs. 4 oz.

Coomassie Darlton 29519.
Coomassie Veda 55838......... 17 lbs. 3 oz.

Coomassie Dawson 52710.
Cretesia's Dawson 89995...... 15 lbs. 1 oz.

Coomassiella 2d 9860.
Coomassie of Ingleside 74045.. 17 lbs. 3½ oz.
Ella of Ingleside 76414........ 15 lbs. 3½ oz.

Copper 2d 2960.
Copper Queen 53659........... 14 lbs. 12 oz.

Coquette 2d 1933.
Coquette's Molly 23841........ 14 lbs. 2 oz.

Coquette of Glen Rouge 17559.
St. Lambert's Coquette 41070. 16 lbs. 15 oz.
Sweet Lily of Meridale 94267.. 15 lbs. 8½ oz.
Sweet Lily of Meridale 94267
(14 days) .... 29 lbs. ½ oz.

Cora Belmont 18869.
Cora Belmont 2d 48868........ 19 lbs. 1 oz.

Cora K. 22768.
Topsey K. 22769.............. 14 lbs.

Coralie 4446.
English Elm 17600............. 14 lbs. 7 oz.

Coral's Calico 91825.
Hollie Pogissa 110490.......... 16 lbs. 3¼ oz.

Cora of Arcadia 16151.
Exile's Arcadia 66201.......... 14 lbs. 4¼ oz.

Cora of Ashland 57500.
Cora's Beauty 92179........... 15 lbs. 9 oz.

Cora of Eau Claire 77584.
Gertie of Eau Claire 120019... 15 lbs. 7 oz.

Cora of Lebanon 11637.
Cora of Hillside 25253......... 15 lbs. 7 oz.

Cora of Linwood 12915.
Mette D. 47433................ 14 lbs. 12 oz.

Cora Scituate 47524.
Cora Scituate 2d 94979........ 14 lbs.

Cordelia Signal 33452.
Cordelia Signal 3d 57808...... 17 lbs. 8 oz.
Cordelia Signal 2d 44489...... 17 lbs. 6 oz.
Cordelia Signal 5th 83097...... 14 lbs. 5 oz.

Coreopsis 4188.
Arnold's Lulu 7328............ 15 lbs. 3 oz.

Corinna 3907.
Corinna 2d 6594.............. 16 lbs. 7 oz.

Corinna Belle 21051.
Velvet Queen 112795........... 15 lbs. 8 oz.

Corinna's Rexena 33032.
Rex's Queen of St. Lambert
59161 .... 16 lbs. 2 oz.

Corinne Bright 66074.
Corinne's Dollie Rex 110987... 15 lbs. 4 oz.

Corinne Moore 35748.
Princess Corinne 48203........ 19 lbs. 1 oz.
Corinne Melrose 54455......... 16 lbs. 14 oz.

Corn 10504.
Wine 15739.................... 16 lbs. 4 oz.

Cornie D. 73486.
Misselle 73488................. 16 lbs. 5½ oz.

Cosette of Scituate 14726.
Chemung Maid 34704.......... 14 lbs. 15 oz.

Costa Rica 64570.
Chirp 65551................... 19 lbs. 12 oz.
Merry Maiden 64949........... 14 lbs. 1 oz.
Merry Maiden 64949 (30 days) 66.695 lbs.
Merry Maiden 64949 (90 days) 200.517 lbs.

Countess (F. S. 1302 J. H. B.)
Bijou Ogston 8210............. 18 lbs. 15 oz.

Countess 3d 990.
Countess Micawber 1759...... 17 lbs. 1 oz.

Countess Coomassie 19339.
Countess Alberta 42922....... 14 lbs. 2 oz.
Hugo's Countess 68394 (30
days) .... 48.172 lbs.
Hugo's Countess 68394 (90
days) .... 191.894 lbs.

Countess Dee 18061.
Countess Matilda 74928........ 19 lbs. 11 oz.
Countess Matilda 74928 (30
days) .... 82 lbs. 15 oz.
Countess Matilda 74928 (90
days) .... 220 lbs. 6 oz.

Countess Gazelle 4th 47729.
Golden Dewdrop 96551........ 23 lbs. 13¾ oz.

Countess Gilderine 29027.
Guess Not 57142.............. 20 lbs. 8 oz.

Countess Gisela 2820.
Niva 7523..................... 15 lbs. 8 oz.

Countess Gisela of Belle Vue 9571.
Countess Gilderine 29027...... 16 lbs. 10½ oz.
Gilderine of Linwood 24488... 15 lbs. 7 oz.

Countess Hebe 8102.
Alberta d'Or 76152............. 14 lbs. 14 oz.
Countess Hebe 2d 45899....... 14 lbs. 3 oz.

Countess of Lakeside 12135.
Eupidee's Perfection 20175.... 15 lbs. 4 oz.

Countess of Lorne 20822.
Alchesieda 103710............. 15 lbs. 6 oz.

Countess of Warren 3896.
Warren's Duchess 4622....... 16 lbs. 1 oz.
Ida of Bear Lake 6169........ 16 lbs.

Countess of Windsor 2024.
Crust 4775..................... 16 lbs. 8 oz.

Countess Potoka 7496.
Marchande 52258.............. 23 lbs. 14¼ oz.

Countess Snap Pogis 36807.
Countess Stoke Pogis 62540... 16 lbs. 3 oz.

Country Girl 3515.
Country Girl 4th 51877....... 18 lbs. 2 oz.
Country Girl 3d 12538........ 17 lbs. 11 oz.

Cousin May's Mollie 46382.
Mollie's Fancy 69728......... 14 lbs. 8 oz.

Cowles' Eirene 3926.
Gladdis Roy 18566............. 17 lbs. 13½ oz.

Cowslip 893.
Cowslip 5th 849................ 15 lbs. 4 oz.

Cowslip 1773.
Arietta 5115.................... 15 lbs.

Cowslip of St. Lambert 8349.
Carrie Pogis 22568............ 15 lbs. 9 oz.

Cream Calla 40233.
Calla Europa 77810............ 20 lbs. 6 oz.

Cream Caroline 17357.
Cream's Pride 47570.......... 17 lbs. 2 oz.

Creamlie Westfield 17818.
Cream Calla 40233............. 16 lbs. 7 oz.

Cream of Springvale 16621.
Alberta Hugo 35424........... 17 lbs. 9 oz.

Creampot 460.
Jersey Cream 3151............. 17 lbs.
Creamer 2467.................. 14 lbs. 1 oz.

Cream's Christmas 63506.
Prince's Christmas Gift 92513. 24 lbs. 12¼ oz.

Creole Maid 11017.
Creole Maid 2d 31881.......... 14 lbs. 1 oz.

Creole Maid 2d 31881.
Fancy Nenita 62702........... 17 lbs. 6½ oz.

Cretesia 13657.
Albert's Cretesia 22881........ 16 lbs. 2 oz.

Cricket of Belle Vue 9570.
Cricket's Minnie 26270........ 15 lbs. 14 oz.

Crocus of St. Lambert 8351.
Maud Pogis 24240............. 14 lbs. 12¾ oz.

Croton Maid 5305.
Croton Maid 4th 26729........ 19 lbs. 7 oz.

Croton Maid 3d 26726.
Landseer's Croton Maid 82535 14 lbs. 15½ oz.

Croton Maid 4th 26729.
Tormento Coomassie 110447... 22 lbs. 2 oz.

Crusta 29637.
Sulphide 88038................ 18 lbs. 5 oz.
Signal's Crusta 97341......... 15 lbs. 3½ oz.

Cultured Cream 29196.
Parlon 32357................... 14 lbs. 10 oz.

Cupid of Collingwood 6867.
Cupid's Jersey Maid 35040 (30
days) .... 55.163 lbs.

Cupid of Lee Farm 5997.
Beatrice of Elmarch 11367.... 15 lbs. 2 oz.

Cupid of St. Lambert 5104.
Bijou of St. Lambert 5112.... 15 lbs. 4 oz.
Cupid of Lee Farm 5997....... 14 lbs. 6 oz.

Curfew 16498.
Romantic 61626................ 15 lbs. 9 oz.

Custard 321.
Eureka McHenry 8341......... 14 lbs.

Cymelia 18706.
Lady Mitchell 29214........... 14 lbs. 12¼ oz.

Cyrene 137.
Cyrene 4th 480................. 17 lbs. 1 oz.

Cyrene 4th 480.
Cerita of Meadow Brook 5056. 17 lbs. 8 oz.

Czaretta 17358.
Czarina of Spring Hill 47568.. 15 lbs. 10 oz.

Dacie St. Helier 86972.
Dacie's Lena 100796........... 22 lbs. 12 oz.

Daffodil 307.
Golden Princess 4557......... 18 lbs. 14 oz.

Daffodil of Maplewood Farm 4853.
Albena 15995................. 16 lbs. 3 oz.

Daffy Wilcox 4046.
Daffy Wilcox 2d 18317........ 15 lbs. 5 oz.

Dahlia 22005.
Della of Grouville 16428....... 14 lbs. 4 oz.

Dainty (P. S. 404 J. H. B.)
Snap's Dainty 18958.......... 14 lbs.

Dainty Bessie 31475.
Dainty Rioter of St. L. 81168.. 15 lbs. 10 oz.

Dairy 2861.
Dairy 2d 3891................. 15 lbs. 5½ oz.

Dairy 2d 3891.
Dairy C. 12227............... 15 lbs. ½ oz.

Dairymaid 2d 13651.
Lamba 79039................. 20 lbs. 3 oz.

Dairy Pride (F. S. 348 J. H. B.)
Jeanne Le Bas 2476.......... 15 lbs. 8 oz.

Dairy Pride 2d (P. S. 37 J. H. B.)
Dairy Pride 4th 21681......... 16 lbs.

Daisette 64004.
Daisette's Tormentress 106532 21 lbs. 1¾ oz.

Daisy 684.
Daisy of Chenango 18582...... 14 lbs. 7 oz.
Daisy of Guilford 18583........ 14 lbs. 6 oz.

Daisy (F. S. 742 J. H. B.)
Bella Delaine 10356........... 14 lbs. 2 oz.

Daisy (F. S. 1260 J. H. B.)
Daisy 2d 15761................ 15 lbs. 8 oz.

Daisy (F. S. 1355 J. H. B.)
Eveline of Jersey 6781......... 18 lbs. 6 oz.

Daisy, on I. of J.
Flora of St. Peter's 8622...... 16 lbs. 5 oz.

Daisy (P. S. 92 J. H. B.)
Brunette Le Gros 9755........ 15 lbs. 15 oz.

Daisy (P. S. 218 J. H. B.)
Handsome Myra 14244........ 21 lbs.

Daisy B. 4th 14277.
Quintuple's Daisy 86078...... 14 lbs. 5 oz.

Daisy Bismarck 2697.
Beauty Bismarck 4967........ 14 lbs. 1 oz.

Daisy Brown 12213.
Daisy Brown 2d 37325........ 15 lbs. 10½ oz.

Daisy Darling 6386.
Cretesia 13657................ 20 lbs. 1 oz.

Daisy Dean 6855.
Dora Doon 12909.............. 15 lbs.

Daisy Deane of St. L. 35962.
Clover of St. Lambert 60945.. 18 lbs. 4¼ oz.
Clover of St. Lambert 60945
(15 days)................... 37 lbs. 13¼ oz.
Clover of St. Lambert 60945
(31 days)................... 70 lbs. 3 oz.
Clover of St. Lambert 60945
(42 days)................... 94 lbs. 6¼ oz.

Daisy Dell 14562.
Daisy Pilotta 68852........... 17 lbs. 12 oz.

Daisy Diamond 103558.
Coupon Diamond 103559....... 14 lbs. 5 oz.

Daisy Fawn 34854.
Fawn Tormentor of Lawn
71639 .... 20 lbs. 2 oz.

Daisy Gray 12893.
Handy's Daisy 74017.......... 16 lbs. 5 oz.

Daisy Harrison 69253.
Mona Caird 75776............. 26 lbs. 7¼ oz.
Mona Caird 75776 (42 days).... 127 lbs. 5¾ oz.

Daisy Ida 95190.
Daisy Ida Torment 121824..... 15 lbs. 12 oz.

Daisy Morrison 14035.
Daisy Morrison 4th 46830..... 14 lbs. 11 oz.
Daisy Morrison 3d 40300...... 14 lbs. 8 oz.

Daisy of Chenango 18582.
Daisy of Dryden 77778........ 16 lbs. 4 oz.

Daisy of Denison 53137.
Edna F. of Lawn 73545........ 15 lbs. 3 oz.

Daisy of Jersey 4th 4585.
Daisy of Hillside Farm 6025.. 16 lbs.

Daisy of St. John's 18170.
Cetewayo's Daisy 18230....... 16 lbs. 4 oz.

Daisy of St. Lambert 25840.
Sheldon's Daisy 30592........ 17 lbs. 12 oz.
Lord Fife's Joliette 30591..... 16 lbs. 2 oz.

Daisy Orton 16221.
Luella Berlin 16927........... 14 lbs. 6 oz.

Daisy S. of Amherst Villa 70508.
Lady Negelia 98735........... 16 lbs. 8 oz.

Daisy Staunton 46592.
True Rex Pogis 105467....... 15 lbs. 11¼ oz.

Daisy Stillson 28174.
Exile's Daisy 46885........... 16 lbs. 4 oz.

Daisy's Last 46376.
Daisy Clementaise 77341...... 14 lbs. 5 oz.

Daisy's Roonan 38482.
Toltec's Roonan 64564........ 19 lbs. 7½ oz.

Daisy Workman 19990.
Belle Garner 23682........... 15 lbs. 3 oz.

Damara 65001.
Marissa 65285................. 16 lbs. 2 oz.

Damask 10259.
Design 18481................. 17 lbs. 2½ oz.

Damask Lawrence 14850.
Composite 58774.............. 18 lbs. 10 oz.

Dame Dee 18060.
Tweedledee of Nipsic 60427... 14 lbs. 1 oz.

Dame Dew Drop 38947.
Teacher's Pet 60242........... 18 lbs. 12 oz.

Dame Miller 59260.
Dame Ida 76878............... 16 lbs. 7½ oz.

Damsel 2d 1837.
Myth 2837.................... 14 lbs. 6 oz.

Dancy H. 14446.
Theda H. 20567............... 17 lbs. 1 oz.

Dandelion 2521.
Gentle of Glastonbury 4651... 14 lbs.

Daphne of Staatsburgh 2d 3027.
Daphne of Arcadia 21710...... 28 lbs. 12 oz.
Dorothy of Bovina 9373....... 15 lbs. 4 oz.

Dappled (F. S. 2773 J. H. B.)
Lady Longfield 23524......... 20 lbs. 13 oz.

Daretta 62579.
Spark 62689................... 20 lbs. 3 oz.

Dassa Argyle 21180.
Onyx of Argyle 76742......... 18 lbs. 5 oz.
Jewel of Argyle 68752......... 18 lbs. 2 oz.

Davy's Dot 41693.
Landseer's Pogene 85743...... 15 lbs.
Davy's Dot 2d 85744.......... 14 lbs. 4 oz.

Deborah 2536.
Deborana 4718................ 14 lbs. 8 oz.

Deeline 2d 55878.
Queen Deeline 78869.......... 20 lbs. 13½ oz.

Delia Lawrence 61064.
Delia Martin 92358........... 18 lbs. 12 oz.

Delle 3789.
Bertha Black 26275........... 16 lbs. 14 oz.

Delle 2d 17737.
Czaretta 17358................ 14 lbs. 7 oz.

Dellia Regina 9474.
Pride's Queen 66442.......... 15 lbs. 8½ oz.

Dell's Daughter 48990.
Annie Linn 82892............. 15 lbs. 2 oz.

Delpha 2d 10713.
Gold Lace 10726.............. 21 lbs. 1 oz.
Golden Skin 10861............ 16 lbs. 8 oz.

Denise 8281.
Denise's Ida 54942............ 17 lbs. 15½ oz.
Denise's Ida 54942 (30 days)... 90 lbs. 6¼ oz.

Denison's Pet 72901.
Clara Signal of Lawn 85586... 15 lbs. 15 oz.

Derwood's Christine 41044.
Ida Bellman 48273............. 17 lbs. 5¼ oz.

Desdemona Belle 19366.
Ethlo's Belle 69053........... 15 lbs. 2 oz.
Desda Pogis 58344............. 14 lbs. 7½ oz.

Dett H. 2d 74514.
Tessia H. 94223................ 15 lbs. 15 oz.

Dewdrop Pansy 19736. ·
Melrose's Perfection 28895.... 24 lbs. 12 oz.
Pogis' Dewdrop 40373......... 19 lbs. 12½ oz.
Pogis' Twilight 53794......... 19 lbs. 1 oz.
Eppie Pogis 44972............. 15 lbs. 6 oz.

Dewolf's Picture 85462.
Little Myers 104650........... 16 lbs. 8 oz.

Dia 13658.
Princess of Hanover 46706.... 18 lbs. 12 oz.

Diana 2d 1718.
Beauty 2076.................... 15 lbs. 7 oz.

Diana 3d 263.
Halsie McCurdy 12379......... 14 lbs. 3½ oz.

Diana Doon 61539.
Landseer's Regina 85988...... 17 lbs.

Diana Franklin 12445.
Flora B. 12446.................. 14 lbs. 2 oz.

Diana of St. Lambert 6636.
Nymph of St. Lambert 12968.. 24 lbs. 14 oz.

Diana's Gold 56141.
Ida's Landseer's Gold 121641.. 14 lbs. 4 oz.

Dice 18182.
Jean Ingelow 42515........... 17 lbs. 7 oz.

Dimple 3248.
Monocacy Dimple 9680......... 14 lbs. 3 oz.

Dinah of Camp Oaks 58291.
Kitty of Seguin 74347......... 15 lbs. 6¼ oz.

Dinah of Eau Claire 77581.
Princess of Eau Claire 128219 17 lbs. 6 oz.

Dioneince 19607.
Carrie Franklin 25635......... 14 lbs. 1 oz.

Diplomacy 53640.
Regalia 64574.................. 24 lbs..

Diploma's Pansy 112871.
Chromo's Pansy 114416....... 17 lbs. 2½ oz.

Diploma's Princess 104002.
Brown Bessie's Princess
120166 .... 14 lbs. 4 oz.

Diplomate 2d (P. S. 3243 J. H. B.)
Effie Baal 134222.............. 14 lbs. 10¼ oz.

Dixie 5341.
Daisey Dixie 9469............. 14 lbs. 4 oz.

Dodona 4800.
Gambetta's Topsy 42376....... 16 lbs. 8½ oz.
Queen Dodona 42375.......... 14 lbs. 2 oz.

Dodona's Bee 88101.
Daisy Harris 92471............ 17 lbs. 4¼ oz.

Doe 3061.
Oneida 42100.................. 16 lbs. 13 oz.

Dollie Argyle 26987.
Baroness Argyle 40498......... 18 lbs. 12 oz.
Baroness Argyle 40498 (30
'.. ,.' .... 56.215 lbs.
Baroness Argyle 40498 (90
days) .... 194.400 lbs.

Dollie Fay 105047.
Dollie's Valentine 105049...... 18 lbs. 1⅜ oz.
Dollie's Valentine 105049 (21
weeks) .... 345 lbs. 11⅜ oz.

Dolly 2d 1020.
Pansy 1019 (one year)........ 574 lbs. 8 oz.

Dolly Berry 2004.
Starkville Beauty 4897........ 14 lbs.

Dolly Bloomfield 7842.
Le Brocq's May Belle 27980... 14 lbs. 15 oz.

Dolly Daisy 2d 12005.
Chautauqua Queen 26403...... 14 lbs. 11 oz.

Dolly Darlo 32200.
Fresh 91786.................... 17 lbs. 11 oz.

Dolly Doyle 2d 58043.
Dolly Landseer 70201.......... 14 lbs. 10 oz.

Dolly Dunn 18298.
Dolly's Daughter 31392........ 16 lbs.

Dolly Landseer 70201.
Dixie Landseer 88681.......... 15 lbs. 6½ oz.

Dolly of Riverside 32944.
Tormentor's Daisy 52496...... 20 lbs. 8½ oz.

Dolly P. 10129.
Bristol Bella 15697........... 15 lbs. 3½ oz.

Donita 40049.
Señorita of St. Lambert 98665 15 lbs. 8½ oz.

Donna Fay 6294.
Donna Signal 29407........... 16 lbs. 1 oz.

Donney Pogis 38907.
Donney Pogis 2d 82426........ 17 lbs. 1¼ oz.
Donney Pogis 3d 100199....... 16 lbs. ¾ oz.

Donney Pogis 2d 82426.
Bisson's Fancy Pogis 106415.. 17 lbs. 1½ oz.

Donnona 10120.
Donnona's Nectarine 40093.... 16 lbs. 6 oz.

Don's Fanchonella 22175.
Donney Pogis 38907........... 16 lbs. 2⅛ oz.

Dora 1550.
Dora 4th 3936................. 25 lbs. 2 oz.

Dora (P. S. 155 J. H. B.)
Dora Neptune 20318.......... 20 lbs. ½ oz.

Dora Doon 12909.
Diana Doon 61539............. 15 lbs. 9 oz.

Dora Norton 46293.
Nettie N. 58367................. 15 lbs.

Dora Signal 2d 73628.
Dora of Glynllyn 95993........ 17 lbs. 9 oz.

Dorcas (F. S. 1851 J. H. B.)
Cetewayo's Dorcas 20297...... 16 lbs. 2¼ oz.

Dorcas 4th 3533.
Jennie Myrtle 22977........... 18 lbs. 1 oz.

Dorcas of Wayne 2d 60557.
Exile's Moss Rose 101155..... 15 lbs. 4 oz.

Dorchester Lass 18297.
Dorchester Maid 101385....... 16 lbs. 5 oz.
Ida's Lassie 77315............. 14 lbs. 9¼ oz.

Dorine 7456.
Dorine's Brunette 29309....... 20 lbs. 3 oz.

Doris C. 33491.
Doris of Mt. Pleasant 73330.... 14 lbs. 7 oz.

Dorothea Day 43238.
Etta Bartlett 80254............ 15 lbs. 13½ oz.

Dorothy Smith 92095.
Xyst 101247................... 23 lbs. 6½ oz.

Dot 2d 621.
Davy's Dot 41693............. 14 lbs. 8 oz.

Dot of St. Lambert 5525.
Pearl of St. Lambert 5527..... 14 lbs. 2 oz.

Dot's Lily 49220.
Lily's Daffodil 89469.......... 14 lbs. 8 oz.

Dottie 15407.
Glory of Elmarch 21521........ 15 lbs. 13½ oz.

Dove 7824.
Dove 5th 14561................ 16 lbs. 13 oz.

Dove 2d 8742.
Daisy Dell 14562.............. 14 lbs. 9 oz.
Dolly of Lakeside 10824....... 14 lbs. 8 oz.

Dove 4th 10945.
Dove Dee 18059............... 27 lbs. 9 oz.

Dove 5th 14561.
Musidove 25379............... 18 lbs. 2 oz.

Dovie of Linwood 6857.
Fancy's Pet 36013............. 16 lbs. 4 oz.
Fancy's Gem 41797............ 14 lbs. 15 oz.

Draxy Leto 47889.
Daisy Leto 72404.............. 18 lbs. 1 oz.

Drosera's Beauty 67275.
Rioter's Drosera 81056........ 18 lbs. 4 oz.

Drosera's Pet 42427.
Angela's Pet 69776............ 15 lbs. 12 oz.

Dubenna 65024.
Dubenna 2d 74920............. 18 lbs. ½ oz.

Duchess 101.
Young Duchess 497........... 15 lbs. 8 oz.

Duchess (F. S. 935 J. H. B.)
Telka 8037.................... 14 lbs. 3 oz.

Duchess 3d 1661.
Zoe Henry 6693............... 15 lbs. 14¾ oz.

Duchess Caroline 2022.
Duchess Caroline 3d 6039...... 15 lbs. 8 oz.

Duchess Cicero 41426.
Dixie Princess 55658.......... 16 lbs. 8 oz.

Duchess Corona 2d 8598.
Favorite Rajah Rex 16153..... 15 lbs. 4 oz.

Duchess Lurlene 9048.
Marilla of St. Lambert 31809.. 21 lbs. 13½ oz

Duchess of Argyle 3758.
Duchess of Argyle 4th 7571... 14 lbs. 12 oz.
Duchess of Argyle 7th 47569.. 14 lbs. 8 oz.

Duchess of Argyle 3d 7569.
Louise of Lawnfield 14151.... 14 lbs. 11½ oz.

Duchess of Argyle 3d 7569.
Dassa Argyle 21180........... 16 lbs. 14 oz.

Duchess of Argyle 4th 7571.
Joy of Argyle 40495........... 15 lbs. 9 oz.
Charity of Argyle 35036....... 14 lbs. 10 oz.

Duchess of Bloomfield 3653.
Duchess of Bloomfield 2d 12250   17 lbs. 4 7-16 oz.
Duchess of Bloomfield 3d 15580   15 lbs. 1 oz.
ueen Mary 6212..............   14 lbs. 13½ oz.
Duchess of Bloomfield 4th
35584 ....   14 lbs. 8½ oz.

Duchess of Bloomfield 2d 12250.
Harry's Duchess 60289........   16 lbs. 4 oz.

Duchess of Bloomfield 3d 15580.
Signalda Bloomfield 30694.....   16 lbs. 3¾ oz.

Duchess of Bloomfield 5th 64215.
Parole's Duchess 133953......   15 lbs. 11 oz.

Duchess of Cherwell (F. S. 2955 J. H. B.)
Bobby's Lily 22055............   18 lbs.

Duchess of Croton 6717.
Crotonia Signal 89308.........   18 lbs. 8 oz.

Duchess of Darlington 13830.
Priscilla Pogis 39270.........   18 lbs. 6½ oz.
Priscilla Pogis 39270 (31 days)   77 lbs. 12½ oz.

Duchess of Green Hill 15804.
Wyoming 48849...............   15 lbs. 5 oz.

Duchess of Jefferson 17144.
Portrait 32592................   15 lbs. 2½ oz.
Duchess of Jefferson 2d 113163   14 lbs. 7 oz.

Duchess of St. Lambert 5111.
Nora of St. Lambert 12962....   22 lbs.
St. Lambert's Duchess 56504..   15 lbs. 7½ oz.

Duchess Sigletta 54519.
Pogis Siglise 92889............   14 lbs. 10 oz.

Duenna 2716.
Duenna's Duchess 5508........   16 lbs. 10 oz.

Duke's Blossom 36669.
Eurotas' Blossom 63134.......   15 lbs. 5 oz.

Duke's Blythesomie 101346.
Tylesholm Blythesomie 101347   16 lbs.

Duke's Del 37702.
D. D.'s Coomassie 63673.......   14 lbs. 2 oz.

Duke's Dream 42783.
Jeannette of Pittsford 73185
(21 days)....................   29.973 lbs.

Duke's Flo 79109.
Exile's Elf 111811.............   14 lbs. 7¼ oz.

Duke's Naomi 51629.
Belle Smith 65807.............   22 lbs. 8¼ oz.

Duke's Rowena 2d 79877.
Duke's Rowena 2d's Queen
79894 ....   18 lbs. 12 oz.

Duke's Signal Queen 32323.
Duke's Signal Queen 2d 82418   17 lbs. 8 oz.
Landseer's Signal Queen 66363   16 lbs. 8 oz.
Tolteca 44762..................   14 lbs. 11½ oz.

Dumbella 3d 11819.
Exile's Dumbella 79452........   14 lbs. 11 oz.

Dusky Le Brocq 22967.
Holyoke's Leda 52812.........   18 lbs. 11½ oz.
Ella L. 84123..................   14 lbs. 12 oz.

Dutsy Gordon 108502.
Dutsy Gordon 2d 108593.......   14 lbs. 7½ oz.

Dye's Pansy Buttercup 14914.
Pansy Blossom 22413.........   14 lbs. 13½ oz.

Early Rosette 32070.
Bay View Queen 60429........   14 lbs. 2 oz.

Easter Belle 20806.
Easter Pogis 52371...........   15 lbs. 3 oz.

Easter Ona 68721.
Easter Ona 2d 97158..........   17 lbs. 5 oz.

Easter Prize 40324.
Easter Prize Rex 102166......   15 lbs.
Lucia Rex 78326...............   14 lbs. 10 oz.

Eastwood's Queen 14523.
Queen of Nubbin Ridge 14528.   17 lbs.

Echo 2223.
Oak Leaf 4769.................   17 lbs. 10 oz.
Maple Leaf 4768..............   14 lbs. 12 oz.
Oak Leaf 4769 (31 days).......   63 lbs. 4 oz.

Écornée (F. S. 846 J. H. B,)
Ona 7840......................   22 lbs. 10½ oz.

Edessa 21844.
Edna of Verna 34537..........   20 lbs. 2½ oz.
Edessa 3d 52043...............   14 lbs. 6 oz.

Edith 4th 817.
Bessie Bradford 7269.........   14 lbs. 2 oz.

Edith Darlington 3d 26185.
Pedro's Dolly 77948...........   17 lbs. 10 oz.

Edith Delle 66143.
Pride's St. Lambert Lady
82955 .... 18 lbs. 12 oz.

Edith Hanna 33146.
Princess Edith Hugo 45120.... 15 lbs. 14½ oz.

Edith of St. Lambert 27054.
Edith of St. Lambert 2d 67327 21 lbs. 11 oz.

Edith of St. Lambert 2d 67327.
Edith Haley 92643............. 22 lbs. 1½ oz.
Edith Hugo Pogis 105045...... 15 lbs. 8½ oz.

Edna Browning 12292.
Azelda's Valhalla 45234....... 14 lbs. 5 oz.

Edy Bashan 1032.
Ida Bashan 4725............... 18 lbs.

Edy Bashan 2d 16098.
Edy Signal 19430............. 16 lbs. 8 oz.

Efficious 2d 15204.
Esther B. 24834................. 14 lbs. 1 oz.

Effie of Hillside 1521.
Eva of Verna 15228........... 21 lbs. 13 oz.
Evelina of Verna 10971........ 19 lbs. 10¼ oz.
Eltekeh 28266.................. 16 lbs. 4 oz.
Effie of Verna 8928........... 14 lbs. 6 oz.

Effie of Verna 8928.
Edessa 21844.................... 16 lbs. 2 oz.

Elainette 18314.
Bessie of Montverde 18496.... 15 lbs. 5 oz.

Elainette's Myra 28233.
Elainette's Odelio 102115...... 14 lbs. 3½ oz.

Elfieda of Maple Dale 17554.
Frisky Elsie 81590............. 17 lbs. 1½ oz.

Elfrida of Hillside 4056.
Emily of Hillside 8073........ 16 lbs. 4 oz.

Elinor Wells 12068.
Elinor Wells 2d 72071......... 14 lbs. 1-16 oz.

Elise Fosdick 65515.
Lora Martin 88381............. 19 lbs. 6 oz.

Eliza A. Perkins 14604.
Eva of Snipsic 17650.......... 14 lbs. 1 oz.

Eliza Pogis 49203.
Ora Vibert 119585............. 17 lbs. 11 oz.

Ella Esmond 35225.
Ella Europa 61062............. 14 lbs. 15½ oz.

Ella Quin 35749.
Quinnie Pogis 61924........... 16 lbs. 10 oz.

Ella Rosewood 2d 23602.
Ella Rosewood's Daisy 32787.. 15 lbs.

Ella Smith 69540.
Loessin Fancy 91599........... 18 lbs. 8 oz.

Ellenetta 9446.
Ellenetta of Bay View 48921.. 21 lbs. 11 oz.

Ellenetta of Bay View 48921.
Lucie of Bay View 66027..... 14 lbs. 8 oz.

Ellen of Collingwood 8285.
Bessie of Clover Leaf 13033... 14 lbs. 2 oz.

Elmanetta 2d 37945.
Exile's Kosi 107505........... 16 lbs. 2½ oz.

Elm Place Frankie 48283.
Seneca's Frankie 114052....... 15 lbs. 7½ oz.

Elmwood Daisy 18402.
Elmwood Daisy 2d 34193...... 16 lbs. 3 oz.

Elmwood Melissa 71275.
Elmwood Melissa 4th 82578... 14 lbs. 5 oz.

Eloise 735.
Cigarette 2849................. 14 lbs. 4 oz.

Elopement 65578.
Ida's Rioter's Elopement 98412 19 lbs. 15¼ oz.
Trident 86647................... 17 lbs. 4½ oz.

Elsie Brown 4026.
Jennie of the Vale 9553....... 17 lbs. 7½ oz.

Elsie Burnside 5598.
Princess Sheila 7297......... 16 lbs. 4½ oz.

Elsie Dee 18057.
Elsie Bonner 78864........... 21 lbs.
Melbina 78875.................. 20 lbs. 2 oz.
Elsie of Shadyside 85500...... 16 lbs. 8 oz.

Elsie Dee 2d 45133.
May Keller 78874............. 19 lbs. 8 oz.

Elsie Dinsmore 5834.
Belle Steuben 20115........... 16 lbs. 10 oz.

Elsie Lane 13302.
Annie Lane 45345............. 15 lbs. 6 oz.

Elsie Lane 2d 58629.
Elphie May 69396............. 16 lbs. 4½ oz.

Elsie Miller 59261.
Myrtle Vance 84237........... 14 lbs. 3 oz.

Elveta 2121.
Rosaline of Glenmore 3179.... 17 lbs. 12 oz.
Embla 4799.................... 17 lbs. 8 oz.

Embla 4799.
Embla Brick 15690............. 14 lbs. 3 oz.

Emiline 5185.
Queen of De Soto 12318........ 14 lbs. 13 oz.

Emily Le Brocq 36610.
Beech Grove Elsie 67758...... 14 lbs. 5½ oz.

Emma Bowling 23969.
Kate Ritchey 54253........... 14 lbs. 7¼ oz.

Emma Hudson 12469.
Fancy Ona 34867.............. 15 lbs. ½ oz.

Emma Landseer 58104.
Emma Landseer 4th 109233.... 14 lbs. 9½ oz.
Bisson's Emma 100408........ 14 lbs. 3 oz.
Emma Landseer 2d 69220..... 14 lbs. 2¾ oz.
Emma Landseer 3d 110513 (14 days) .... 37 lbs. 2¼ oz.

Emma's Favorite 14078.
Tones 46306.................. 14 lbs. 9 oz.

Emmie's Pet 3568.
Lebanon Daughter 6106....... 14 lbs. 4 oz.

Empress 1552.
Empress 6th 3203............. 17 lbs. 9¾ oz.

Empress 6th 3203.
Empress Pedro 50878......... 15 lbs. 1 oz.

Empress Eugenie 13549.
Nancy Lovelock 15511........ 17 lbs. 9 oz.

Empress of Ely 4823.
Empress of Ely 2d 6771....... 16 lbs. 8 oz.

Empress of St. Lambert 56502.
Daffy of St. Lambert 69894.... 14 lbs. 7 oz.

Empress Pedro 50878.
Ella Victor Pogis 100762...... 19 lbs. 6 oz.

Emsie of St. Lambert 95428.
St. Lambert B. Kate 105205... 15 lbs.

Encore Lawrence 12688.
Miss Julia 26701.............. 21 lbs. 1½ oz.

English Elm 17600.
English Elm 3d 62906........ 17 lbs. 2 oz.

Enid 1482.
Enid 2d 10782................. 14 lbs. 7½ oz.

Enid 3d 19582.
Gilderoy's Enid 32924........ 14 lbs. 4 oz.

Enna 33487.
Regina Gilford 49891.......... 14 lbs. 11 oz.

Ephrella 20651.
Braw Lassie 36221............. 17 lbs. 6 oz.

Epicure 33345.
Founder's Pet 94107.......... 17 lbs. 8 oz.

Erica 25005.
King's Erica 22096............. 16 lbs. 3¼ oz.

Erminta V. 53553.
Maxminta 85922............... 15 lbs. 13½ oz.

Ernestina 38444.
Ernestina 2d 77682............ 14 lbs. 8 oz.

Ernestine Mc. 10686.
Harry's Ernestine 66232...... 15 lbs. 6½ oz.

Ernestine's Annie 85645.
Annie's Prize 86548........... 19 lbs. 11¾ oz.

Esse Signal Alexis 42184.
Rioter's Fawn 69168......... 20 lbs. 2 oz.
Rioter's Esse 81901........... 16 lbs. 7 oz.
Rioter's Esse 81901 (14 days).. 31 lbs. 6¾ oz.

Estella Parks 15435.
Grace G. Parks 29263......... 19 lbs. 3 oz.

Estelle of St. Lambert 7011.
Niobe of St. Lambert 12969... 21 lbs. 9¼ oz.
Rose of St. Lambert 20426.... 21 lbs. 3½ oz.

Esther 889.
Élite 4299.................... 14 lbs. 2 oz.

Esther Lee 38131.
Fanny Mason 45992.......... 18 lbs. 5¼ oz.

Eterna 27134.
Idalette Pogis 64220.......... 15 lbs. 4¼ oz.

Ethel Hugo 30164.
Ethel of Shelburne 46285...... 14 lbs. 8 oz.
Ethel of Shelburne 46285 (14 days) .... 28 lbs. 13½ oz.

Ethelka 9515.
Ethelka 2d 14128.............. 15 lbs. 4 oz.

Ethel Marjoram 87266.
Lady Marjoram 123387......... 21 lbs. 11 oz.
Ethel of Glen Rouge 131019... 18 lbs. 9 oz.

Ethel of Shelburne 46285.
Sallie of the Ledges 87182..... 16 lbs. 6 oz.

Ethel's Nelly 19097.
Jessie Stoke Pogis 37291....... 17 lbs. 2½ oz.

Ethel Wanda 56144.
Landseer's Ethel 74315........ 15 lbs.
Landseer's Lily 87294......... 14 lbs. 2 oz.

Ethleel 18724.
Ethleel 2d 32291................ 30 lbs. 15 oz.
Ethleel 5th 70652.............. 15 lbs. 6½ oz.
Ethleel 4th 54275.............. 15 lbs.
Ethleel 6th 89802.............. 14 lbs. 15 oz.

Ethleel's Ida 66512.
Bisson's Ethleel 80921........ 14 lbs. 11¾ oz.

Ethlo's Princess 2d 83990.
Tancreda Torment 114090..... 15 lbs. 1 oz.

Etta 1756.
Etiquette 4300................. 15 lbs. 8 oz.

Etta Blanche 80253.
Ettarock 81103................. 16 lbs.

Etta M. 15901.
Etta M. 2d 30820.............. 14 lbs. 14 oz.

Etta of Bay View 66028.
Princess Rosalind 87288....... 15 lbs. 1 oz.
Lady Torenia 98736........... 14 lbs. 9 oz.

Etta of Briarcliff 68320.
Ella of Briarcliff 68939....... 14 lbs.

Ettie Palmer 14281.
Ettie Palmer 4th 47481........ 16 lbs. 5 oz.

Eudora 1863.
Daisy of Belhurst 3114....... 16 lbs. 8 oz.
Euphonia 6783................. 16 lbs. ½ oz.

Eufielda 62546.
Dido B. 112926................. 15 lbs.

Eugenie 498.
Eugenie 2d 12733.............. 14 lbs. 2 oz.

Eugenie 792.
Eugenie 2d 1623................ 14 lbs.

Eugenie 5th 6573.
Lady Cecilia 24821............ 15 lbs. 1 oz.

Eulogy 30634.
No Chemicals 71452.......... 16 lbs. 9 oz.

Eunice of Belle Vue 7704.
Miss Dora Deane 24505....... 17 lbs. 9⅜ oz.

Euphonia 5th 39124.
Eufielda 62546................. 14 lbs. 3 oz.

Euphorbia 11229.
Miss Signal 20379............. 16 lbs. 4 oz.
Euphorbia 3d 45359........... 15 lbs. 7½ oz.
Euphorbia 2d 27094........... 15 lbs. ½ oz.

Eureen 40312.
Eureen of Lawn 61372......... 15 lbs. 8 oz.

Europa 121.
Angela 1682................... 14 lbs. 2 oz.

Europa 176.
Eurotas 2454.................. 22 lbs. 7 oz.
Torfrida 3596................. 17 lbs. 6½ oz.
Eurotas 2454 (one year)....... 778 lbs. 1 oz.

Euro Polono 31962.
Elsie Harribee 91834.......... 18 lbs. 8½ oz.
Euro Polono 2d 101052........ 14 lbs. 2 oz.

Eutoga 58027.
Express 69574................. 18 lbs. 5 oz.

Eva Daughtry 21951.
Maude Brown 85756........... 15 lbs. 9 oz.

Eva Horner 2644.
Eva Horner 3d 24663.......... 15 lbs. 2 oz.

Eva Locust 21050.
Rachel Spencer 50974.......... 23 lbs. 3¼ oz.
Nora Stoke Pogis 34687....... 16 lbs. 1 oz.
Exilie's Lady Angela 46882.... 15 lbs. 11 oz.
Rachel   Spencer   50974 (35
   days) .... 105 lbs. 10½ oz.

Eva of Snipsic 17650.
Ida Eva Pogis 39271.......... 18 lbs. 6 oz.

Eva's Belle 20491.
Daisy Mekeel 70400............ 15 lbs. 5 oz.

Eva Taylor 14875.
Signal Maid 19361............. 19 lbs. 13 oz.

Eve 456.
Evri 5282...................... 15 lbs. 4 oz.

Evelina 446.
Effie of Hillside 1521......... 16 lbs. 15 oz.

Exhibit 23245.
Crusta 29637................... 16 lbs. 10 oz.

Exile's Alphea 45393.
Lanison's Cora 59315.......... 14 lbs. 14 oz.

Exile's Beauty 61220.
Lucy S. King 106584........... 16 lbs. 14¼ oz.

Exile's Belle 40524.
Irondequoit Belle 74696....... 14 lbs. 3 oz.

Exile's Dora Lowndes 80523.
Dora's Fawnie Exile 108696.. 15 lbs. 12 oz.

Exile's Lucy 46883.
Exile's Gretchen 79245........ 16 lbs. 15½ oz.
Lord Harry's Lucy 86091...... 16 lbs. 5 oz.
Exile's Beauty 61220.......... 16 lbs. 3 oz.
Exile's Lucile 65151:......... 14 lbs. 6 oz.

Exile's Majel 112861.
Amelia S. King 112862....... 16 lbs. 2 oz.

Exile's Myrtle 51351.
Exile's Myrtle 2d 94446....... 17 lbs. 13½ oz.
Exile's Butterball 65152....... 15 lbs. 4 oz.

Exile's Nina 40522.
Exile's Lady Palestine 107626. 14 lbs. 14 oz.

Exile's Phyllis 79111.
Exile's Lady Sarah 107938.... 17 lbs. 10 oz.

Exile's Signal Belle 82363.
Exile's Signal Belle 2d 106097. 19 lbs. 13 oz.

Exile's Success 49986.
Exile's Harriet 100716........ 16 lbs. 6¼ oz.
Exile's Harriet 100716 (14
days) .... 31 lbs. 7½ oz.

Extract 58032.
Extract 3d 85261.............. 14 lbs. 6 oz.

Fadette of Canaan 14807.
Fadette Signal 39472........... 20 lbs. 14½ oz.

Fadette of Verna 6814.
Fadette of Verna 3d 11122..... 22 lbs. 8½ oz.

Fadette of Verna 2d 9024.
Fadette of Canaan 14807....... 16 lbs. 1½ oz.

Fadette Signal 39472.
Fadette Signal's Queen 109281 16 lbs. 3 oz.

Fair Corinth's Maud 54610.
Sheldon's Maud 86187......... 14 lbs. 15 oz.

Fair Daisy (F. S. 2591 J. H. B.)
Carlo's Daisy 16702........... 14 lbs. 9½ oz.

Fair Geraldine 37690.
Miss D'Arcy 62596............. 15 lbs.

Fair Leaf 81293.
Beech Leaf 99381.............. 16 lbs. 3½ oz.
Leaf Pogis 130184............. 14 lbs. 5½ oz.

Fair Maid of Perth 13705.
Fair Dairy-maid 29839......... 14 lbs. 8 oz.

Fair Rosamond 11135.
Landseer's Rosamond 56142... 15 lbs. 5 oz.

Fairy (F. S. 964 J. H. B.)
Buckeye Lass 10355............ 14 lbs. 4 oz.

Fairy Lady 24748.
Tormentor's Fairy 108510..... 22 lbs.
Signal's Gold Foil 90508....... 15 lbs. 4 oz.

Fairy of Forest City 4th 16260.
Royalist's Daisy 19187......... 21 lbs. 9½ oz.

Fairy of Verna 6813.
Fairy of Verna 2d 10973....... 20 lbs. 3¾ oz.

Fairy Queen of St. Brelade's 7464.
Signal Queen 30869............ 21 lbs. ½ oz.
Allie's Queen 45618........... 15 lbs. 2 oz.
Signal Queen 30869 (30 days).. 51.522 lbs.
Signal Queen 30869 (90 days).. 165.601 lbs.

Faith of Argyle 29765.
Clover of Argyle 76745........ 15 lbs. 10 oz.

Fall Leaf 8587.
Corinne Moore 35748.......... 19 lbs. 8 oz.

Fall Leaf 2d 25171.
Fall Leaf's Heiress 54454..... 16 lbs. 2 oz.
Delphina 105895............... 14 lbs. 2¾ oz.

Fall Leaf's Heiress 54454.
Regina of Cedar Hill 111995.. 15 lbs. 8½ oz.

Fancy 9.
Fantine 1271.................. 16 lbs. 6 oz.

Fancy (F. S. 1528 J. H. B.)
Maid of Five Oaks 7178....... 15 lbs. 4 oz.

Fancy Fair 2858.
Fancy Juno 6086.............. 15 lbs. 10 oz.

Fancy Farmer's Flora 111163.
Hulabaloo's Flora 111164...... 15 lbs. 13 oz.

Fancy of Bovina 8055.
Flora Lee 13294............... 14 lbs. 1 oz.

Fancy's Gem 41797.
Fancy's Gem 2d 86342........ 16 lbs. 4 oz.
Fancy's Gem 4th 86348....... 14 lbs.

Fancy's Pet 36013.
Pet's Torment 92255.......... 15 lbs. 14 oz.
Fancy's Pet 2d 85912.......... 14 lbs. 1 oz.

Fancy Wax 37159.
Fancy Wax 3d 86345.......... 20 lbs. 3¾ oz.
Tormentor's Fancy Wax 73212 15 lbs. 4½ oz.
Fancy Wax 2d 83357........... 14 lbs. 5½ oz.

Fandango 12908.
Fanfare 49927................. 16 lbs.

Fanella 8006.
Fanella 2d 64578.............. 15 lbs. 6 oz.

Fanfare 49927.
Fanfare 2d 119780............. 15 lbs. 12½ oz.

Fannette 3070.
Fandango 12908............... 18 lbs. 3 oz.

Fannie Baker 4274.
Silvia Baker 8793............. 20 lbs. 10 oz.
Florrie May Baker 10728...... 14 lbs. 8 oz.

Fannie Booth 12505.
Plum 13228................... 17 lbs. 4 oz.

Fannie Bugler 19962.
Fannie's Fairy 67304......... 19 lbs. 8 oz.
Badger Girl 21463............. 16 lbs. 1 oz.

Fannie G. 3829.
Fleurette of Linwood 12918.... 16 lbs.
Petite Mère 8516.............. 15 lbs. 13 oz.

Fannie Jones 61654.
Mary's Nightingale 87263..... 14 lbs. 14½ oz.

Fannie Keene 40097.
Tritoma Pogis 44952.......... 16 lbs. 13 oz.

Fannie Landseer 1969.
Julia Walker 10133............ 15 lbs. 12 oz.
Queen Fannie 10275........... 14 lbs. 2 oz.

Fanny Fair 4136.
Fair Lady 6723................ 19 lbs.

Fanny Mason 45992.
Macie Pogis 103539........... 15 lbs. 1⅛ oz.

Fanny Micawber 1804.
Fanny of Cream Brook 13930. 17 lbs. 8 oz.

Fanny of Yerba Buena 8151.
Fanny of Yerba Buena 2d
10689 .... 14 lbs. 14 oz.

Fanny White 9804.
Harry's Ethel 63970........... 14 lbs.

Fan of Grouville 7458.
Fan's Grouville Beauty 10079. 19 lbs. 3 oz.

Fan's Grouville Beauty 10079.
Chief's Beauty 39391......... 20 lbs. 5½ oz.

Farmer's Sunbeam 2d 55506.
Rioter's Sunshine 75108....... 15 lbs. 10 oz.

Fauchon 22001.
Fauchon 2d 18200.............. 14 lbs. 9 oz.

Faultless 12018.
Fillpail 16530................. 15 lbs. 11 oz.

Fausetta 29334.
False Step 69647.............. 15 lbs. 7 oz.

Faustine 10354.
Marvel 13734................. 15 lbs. 1 oz.

Fauvette (F. S. 1972 J. H. B.)
Thaley 14299................. 16 lbs. 13 oz.

Favorite of Avon 13438.
Jane Neptune 52345.......... 19 lbs. 4½ oz.

Fawnette of Woodstock 3710.
Lydia Libby 11698............. 15 lbs. 3 oz.

Fawn of River View 78299.
Erie Maid of River View 94307 14 lbs. 12 oz.

Fay Signal 82159.
Fay Signal 2d 134018......... 14 lbs. 4 oz.

Fear Not 6059.
Fear Not 2d 6061.............. 16 lbs. 2 oz.

Fenella of Bolton 49572.
Fenella of Bolton 2d 116768.... 14 lbs.

Fenelle 9095.
Vespucia 17455................ 14 lbs. 4 oz.

Fermentor 2d 74267.
Fermentor 2d's Pet 114503..... 17 lbs. 8 oz.

Fides 51.
Fides 2d 1576................. 14 lbs. 8 oz.

Fidette of Woodruff 40536.
Fedora of Greenwood 46892.. 20 lbs. 7 oz.

Fille de l'Air (F. S. 3548 J. H. B.)
Count's Fillpail 30975......... 24 lbs. 5 oz.

Fill Pail 24341.
Fill Pail 2d 24388............. 26 lbs. 2 oz.

Fill Pail 2d 24388.
Fill Pail's Countess 24462..... 22 lbs. 8 oz.

Fir Shade Bonnibel 27235.
Bonny of Woodford 58718..... 17 lbs. 4 oz.

Flashy Jessie 2d 23332.
Birdsey's Surprise 48326...... 14 lbs. 2 oz.

Fleece 18568.
Fleecy Princess 40342......... 16 lbs. 11½ oz.

Fleetling 18576.
Fleeter 61921................... 16 lbs. 13½ oz.

Fleta 3859.
Creole Maid 11017.............. 16 lbs. 15 oz.

Flirt 482.
New London Gipsy 11667..... 14 lbs. 8 oz.

Flirt (F. S. 2250 J. H. B.)
Pretty Flirt 28791............. 15 lbs.

Flo 3d 14754.
Faith of Cloverdale 29277..... 14 lbs. 2 oz.

Flora Brown 75356.
Cosmetic 75357................. 19 lbs. 1 oz.

Flora D. 11192.
Panola 85344................... 17 lbs. 3¾ oz.

Flora du Coin 39492.
Lady Kate Rex 91751......... 14 lbs. 8 oz.

Flora F. 5544.
Pope's Flora 18697............. 14 lbs. 2 oz.

Flora G. 38119.
Jersey Belle of Belmont 68337 14 lbs. 6 oz.

Flora Lee of Tennessee 7694.
Alice Donnal 12726............ 14 lbs.

Floralia 6230.
Floret 9959.................... 17 lbs. 6 oz.

Flora of Hopelands 16194.
St. Nick's Flora 16195......... 14 lbs.

Flora Pansy 19935.
Flora Pansy 2d 78678......... 15 lbs. 10.8 oz.

Flora Rex Carlo 94407.
Flora Rex Carlo 2d 102140.... 18 lbs. 2 oz.

Flora Temple 3768.
Flora Temple 3d 40086 (30 days) .... 55.058 lbs.
Flora Temple 3d 40086 (90 days) .... 176.751 lbs.

Florence 1043.
Attractive Maid 16925......... 22 lbs. 5 oz.

Florence Pond 30980.
Chromatess 51578............. 25 lbs. ½ oz.

Floribel 2460.
Flight of Willow Farm 24783. 20 lbs. 12 oz.

Floribel Pogis 28759.
Fashion Plate 62457........... 15 lbs. 10½ oz.

Floribundus (F. S. 659 J. H. B.)
Floribundus 2d 14949......... 18 lbs. 8 oz.

Floribundus 2d 14949.
Yum Yum of St. Lambert 39152 .... 21 lbs. 4 oz.
Clotaire's Annie 58319......... 19 lbs. 11½ oz.
Young Floribundus 28172..... 17 lbs. 12¾ oz.

Florine Value 58278.
Florine Value 2d 80910........ 15 lbs. 1½ oz.

Floriole 13069.
Blanche of Castile 43793...... 22 lbs. 1½ oz.

Florrie Ringgold 10048.
Mollie Davis 26378............ 16 lbs. 12½ oz.

Floss of Lawnfield 16085.
Judy of Riverside 16495....... 21 lbs. 4½ oz.

Flower of Glen Rouge 17560.
Flower of Meridale 64537..... 26 lbs. 11½ oz.
Flower of Glen Rouge 2d 55559 18 lbs. 13½ oz.

Flower of Meridale 64537.
Matilda of the Ledges 97740.. 16 lbs. 9½ oz.

Floy Le Premier 12468.
Frances C. Magnet 22904..... 14 lbs. 13½ oz.

Forest Blossom 63501.
Blossom's Belle 97433........ 15 lbs. 6 oz.

Forest Q. Pogis 63567.
Emblem of Del Valle 102384.. 15 lbs. 8 oz.

Forget-me-not 5809.
Forsaken 7520................ 15 lbs. 1 oz.

Fosie 20255.
Chief's Fosie 61998........... 14 lbs. 11 oz.

Fragrance 4059.
Fragrance 4th 16509........... 14 lbs. 7 oz.

Francesville 53367.
Francesville 2d 99307......... 15 lbs. 10 oz.
Milk Well 99301.............. 14 lbs. 4 oz.

Frankie 4th 2253.
Duchess of Jefferson 17144... 16 lbs. 9 oz.

Frankie 5th 3542.
Frankie's Lass 24900.......... 17 lbs. 3½ oz.

Frankie's Dido 22424.
Dido Pogis 67260.............. 15 lbs. 9 oz.

Frankie's Lass 24900.
Pledge 59214.................. 17 lbs. 9½ oz.
Goodbye 27366................. 16 lbs. 13 oz.
Comlassa 40671................ 14 lbs.

Friendship 62460.
Fireside 75194................ 17 lbs. 2½ oz.

Friga's Delight 26649.
Delight's Darling 45700....... 15 lbs. 8 oz.

Froxine's Pansy 19926.
Pedro Girl 22400............... 18 lbs. 6 oz.

Frushia Wilson 47432.
Jessie Wilson 76121........... 14 lbs. 8 oz.
Trilby M. 103547.............. 14 lbs. 4 oz.

Fuchsia 2789.
Evergreen Maid 11536......... 14 lbs.
Rosaletta 9339................ 14 lbs.

Fuchsia, on I. of J.
Miss Le Dain 27472........... 14 lbs. 5¼ oz.

Fulda 25687.
Fulda Stoke Pogis 44992...... 15 lbs. 8 oz.

Gardiner's Ripple 11693.
Signal's Ripple 27093......... 17 lbs. 5 oz.

Gardiner's Thistle 11338.
Gardiner's Ripple 11693....... 19 lbs. 12½ oz.

Garibaldi's Kate 25114.
Garibaldi's Kate 2d 56478..... 18 lbs. 2½ oz.

Garland 851.
La Reine Centennielle 4999... 15 lbs. 9 oz.

Garland Melrose 93858.
Signal's Garland 116048....... 15 lbs. 2 oz.

Gay Girl 18225.
Pedro's Lady 36715........... 18 lbs. 12 oz.

Gazania 4513.
Phlox 16399................... 21 lbs. 11 oz.

Gazella 1880.
Gazella 3d 9355............... 16 lbs. 3 oz.
Gazella 3d 9355 (one year)..... 751 lbs. 6 oz.

Gazella 3d 9355.
Signoretta 21546 (one year)... 680 lbs. 6½ oz.

Gazelle 15666.
Young Gazelle 19837........... 14 lbs. 11 oz.

Gazelle 15961.
Island Star 11876............. 21 lbs. 3 oz.

Gazelle Dee 18056.
Gazelle's Fawn 93704.......... 25 lbs. 4½ oz.
King's Lassie 79103........... 18 lbs. 5 oz.

Gazelle Dee 2d 55880.
St. L. Gazelle 78870.......... 22 lbs. 14 oz.

Gemmi 2d 7314.
Gowan 12272.................. 14 lbs. 4 oz.

Gem of Ellasleigh 38532.
Ettagem 80258................ 16 lbs. 12 oz.

Gem of Oakland 2d 29756.
Gem's Louisa 39564........... 17 lbs.

Geneva M. 62904.
Geneva M.'s Fancy 104917.... 16 lbs. ¼ oz.

Genie Walker 38443.
Monte Sano Belle 77683....... 15 lbs. 15½ oz.
Chump 81289.................. 14 lbs. 13 oz.

Georgetta 93.
Lady Mary Hampton 4861..... 14 lbs. 5 oz.

Geranium 3963.
Geranium 2d 7838............. 26 lbs. 4¾ oz.

Geranium Leaf 30875.
Geranium Leaf 2d 56859....... 22 lbs. 7 oz.
Allie's Leaf 45617............ 16 lbs. 3 oz.

Geranium Leaf 2d 56859.
Signal's New Geranium 134117 21 lbs. 15½ oz.

Gertie Cowles 10736.
Esther Thorne 35545.......... 14 lbs. 7½ oz.

Gertie McM. 64200.
Ethel McM. 100267............ 15 lbs. 13 oz.

Gertie Rex 23946.
Value Belle 33080.............. 16 lbs. 6 oz.

Gigia 4447.
Giulietta Cooke 32193......... 21 lbs. 13½ oz.

Gilda 2779.
Mercedes H. 12326............. 17 lbs. 12 oz.
Anita H. 12334................ 17 lbs. 4 oz.

Gilda Mercedes H. 14451.
Mercedes' Pet 20570.......... 17 lbs. 1½ oz.

Gildercream 39480.
Gavotte 64205.................. 15 lbs. 4 oz.

Gilderine of Linwood 24488.
Landseer's Gilderine 68048.... 14 lbs. 7 oz.

Gilderoy's Daisy 50023.
Quintuple's Wonder 86077..... 15 lbs. 14 oz.

Gilderoy's Girlie 24131.
Girlie's Glory 46305............ 15 lbs.

Gilderoy's Idex 42027.
Tormentor's Idex 64209....... 14 lbs. 9 oz.

Gilderoy's Mary 53004.
Louisa D. 90959............... 16 lbs. 14 oz.

Gilfilia 2d 15990.
Gilfilia Pogis 38688............ 14 lbs. 2½ oz.

Gilt Edge 2662.
Gilt Edge 2d 4420.............. 14 lbs.

Gilt Edge 3d 6041.
Gabriel Champion 14102...... 17 lbs. 8 oz.
Gilded Pansy 20552........... 15 lbs. 5 oz.
Gilt Edge C. 12223............ 14 lbs. 3½ oz.

Gilt Edge Rexea 32942.
Tormentor's Rexea 38906..... 15 lbs. 14 oz.

Gilt Lady 14981.
Gilt Lady 2d 33969............ 14 lbs. 7 oz.

Gingerbread 7922.
Gingerbread 5th 37681........ 18 lbs. 9¾ oz.

Gipsy 319.
Gipsy 5th 2252................ 17 lbs. 2 oz.

Gipsy 4th 782.
Goddess of Staatsburgh 5252.. 14 lbs. 8 oz.

Gipsy Countess 9188.
Waukesra 19721................ 16 lbs.

Gipsy F. Eric 75462.
Guarder 90825.................. 21 lbs. 4 oz.

Gipsy of Avon 79747.
Sophie 2d of Hood Farm
131236 .... 15 lbs. 8 oz.

Gipsy's Berry Duchess 86124.
Pomona of Prospect 123426... 20 lbs. 6¼ oz.

Gipsy's Bo-Peep 90295.
Bo-Peep's Maid 100538........ 14 lbs. 2 oz.

Gipsy's Duchess 56833.
Gipsy's Berry Duchess 86124.. 28 lbs. 6¼ oz.

Gipsy's Pride 25290.
Gipsy's Pride 2d 29365........ 27 lbs. 12 oz.
Gipsy's Riotess 62482.......... 17 lbs. 8 oz.

Gipsy's Pride 2d 29365.
Pride of Ingleside 54545....... 20 lbs. 4½ oz.

Girletta 61472.
Gertie of Glynllyn 74474...... 21 lbs. 2½ oz.
Girletta 4th 109264............. 15 lbs. 8½ oz.

Girl of St. Lambert 20423.
Charming of St. Lambert
69077 .... 15 lbs. 4 oz.

Gladdis Roy 18566.
Gladdis Roy 2d 55652.......... 18 lbs. 12 oz.
Catono's Kate 59976........... 16 lbs. ½ oz.
Tiger Lill 69771................ 15 lbs.

Gladdis Roy 2d 55652.
Pearl of Riverside 55659...... 17 lbs. 5½ oz.
Pearl of Riverside 55659 (90
days) .... 160.804 lbs.

Glen Belma 17243.
Glen's Festina 25351.......... 17 lbs. 15 oz.

Glendale Beauty 27690.
Nelly Torment 61903........... 15 lbs. 7 oz.

Glenida 2d 5770.
Augustella 63018............... 14 lbs. 2 oz.

Glenmore Belle 4801.
Countess Buttercup 13505..... 14 lbs. 3 oz.

Glenn Forest Queen 4809.
Forest Queen 12229.......... 16 lbs. 15 oz.

Glint 12896.
Gift 39901...................... 21 lbs. 6 oz.
Sosia 19198...................... 17 lbs. 5½ oz.

Gold Badger Girl 74902.
Dot Badger Girl 101417....... 14 lbs. 1 oz.

Gold Ear 2200.
Gold Ear 2d 3592.............. 18 lbs. 2 oz.
Lady Louise 4339.............. 15 lbs.

Golden Delle 37698.
Marjoram's Delle 82954....... 19 lbs. 12 oz.

Golden Harp 24410.'
Pedro's Golden Harp 66098.... 15 lbs. 3 oz.

Golden Lena 15396.
Ventnor Beauty 46308......... 17 lbs. 5½ oz.

Golden Sheen 25561.
Iliana 44594.................... 15 lbs. 5 oz.

Golden Trudie 2d 38082.
Golden Gem of St. L. 81167... 15 lbs. 3 oz.
Golden Gem of St. L. 81167 (24
days) .... 49 lbs. 4¼ oz.

Golden Zoe 3975.
Rioter's Zoe 19769........... 14 lbs. 12 oz.

Goldie C. 8104.
Lass Signal 16308.............. 20 lbs. 2 oz.
Goldie Pogis 45423.............. 16 lbs. 2 oz.
Gold Princess 8809............. 14 lbs. 12 oz.

Goldie La V. 6983.
Goldie La V. 2d 26319......... 14 lbs. 1¼ oz.

Gold Lace 10726.
Lady Alice of the Wilderness
12207 .... 15 lbs. 14 oz.
Gold Mark 10727.............. 14 lbs. 14 oz.

Gold Leaf 2d 5860.
Bronze Leaf 14902............. 15 lbs. 1 oz.

Gold Mark 24361.
Westphalia 24384............... 24 lbs. 9½ oz.

Gold Mosie 63502.
Miss Mouse 92512.............. 19 lbs. 8¾ oz.
Muzio 97423................... 15 lbs. 2 oz.

Gold Princess 8809.
Princess Aurea Pogis 39266.... 17 lbs. 7½ oz.

Gold Proof 10860.
Queen Mary of Woodlawn
11659 .... 22 lbs. 5 oz.
Topaz of Woodlawn 11661..... 16 lbs. 4 oz.

Goldstraw 85.
Goldstraw 3d 14724............ 14 lbs. 12 oz.

Gold Trinket 9518.
Trixket 16292.................. 14 lbs. 2½ oz.

Golia of Peach Hill Farm 52931.
Merty of Peach Hill 77774..... 14 lbs. 4½ oz.
Gerty of Peach Hill 77773..... 14 lbs. 2½ oz.

Goodbye 27366.
Paradise 32082................. 17 lbs. 11 oz.
Celebrity 52011................ 14 lbs. 3 oz.

Gordonetta. 19570.
Nancy's Star 41423........... 16 lbs. 1⅝ oz.

Grace Carpenter 30934.
Hazelwood 52239.............. 14 lbs. 1 oz.

Grace Darling 2d 304.
Grace Davy 8292.............. 22 lbs. 5¼ oz.
Grace Felch 8291.............. 15 lbs.

Grace Darlington 5574.
Mother Hubbard 10331........ 24 lbs. 1½ oz.
Nazli 10327.................... 15 lbs. 3½ oz.
Nellie Darlington 5956......... 15 lbs. 3 oz.

Grace Davy 8292.
Bessie Ridgely 8293........... 14 lbs. 11½ oz.

Grace G. Parks 29263.
Isadore Stoke Pogis 37297.... 16 lbs. 9½ oz.

Grace Pansy 12156.
Grace Pansy 2d 18764......... 17 lbs. 15 oz.
Grace Pansy 2d 18764 (90 days) 147.009 lbs.

Grace Pansy 2d 18764.
Myrtlise 64006................. 19 lbs. 9 oz.
Tormentris 49659.............. 18 lbs. 14½ oz.
Lady Nolia 87972.............. 14 lbs. 6 oz.
Myrtlise 64006 (21 days)....... 54 lbs. 15¼ oz.

Grace Sheen 62319.
Grace Sheen 2d 79788......... 14 lbs. 9½ oz.

Grace's Nightingale 19855.
Howland Nightingale 33291... 14 lbs. 1 oz.

Gracie Washington 95045.
Daltonia Pogis 95203........... 14 lbs. 14 oz.

Granda 92755.
Alnora 112132.................. 16 lbs. 6 oz.

Granny (P. S. 495 J. H. B.)
Granny's Gem 30406........... 21 lbs. 13-16 oz.

Granny's Baby 30478.
Granny's Baby 5th 95132...... 15 lbs. 8½ oz.

Gray Beauty 16053.
Gray Beauty 3d 19845........ 19 lbs. 4 oz.

Gray Diamond 30946.
Lorne's Diamond 66087....... 19 lbs.

Gray Therese 5322.
Lady Ramaposa 26232........ 17 lbs. 5½ oz.

Great Happiness 18205.
Pedro's Duchess 29581........ 18 lbs. 7½ oz.

Gretna Coomassie 71177.
Cotton Patchwork 85926...... 16 lbs. 1 oz.

Grey Dame 114000.
Ex Presage 128529............ 15 lbs. 5 oz.

Grey Friar's Princess 27559.
Romp's Princess 51185 (30
days) .... 51.357 lbs.
Romp's Princess 51185 (90
days) .... 188.373 lbs.

Grey Queen (F. S. 571 J, H. B.)
Maid of the Elms 6960........ 16 lbs.
Maid of Avranches 6959....... 15 lbs.

Grinnella 3d 2209.
Belle Grinnell 4073............ 18 lbs. 8 oz.

Grise, on I. of J.
King's Princess 30948........ 24 lbs. 5 oz.

Gros Puits (F. S. 1454 J. H. B.)
Queen of Ashantee 14554...... 15 lbs. 2 oz.

Guenn 25284.
Frushia Wilson 47432........ 14 lbs. 4½ oz.

Guess Not 57142.
Guess Not 2d 133252.......... 16 lbs. 11¼ oz.

Guinevere 1484.
Pride of the Hill 4877........ 14 lbs. 8 oz.

Gunilda 662.
Brenhilda 7649................ 16 lbs. 8 oz.
Alice of Salem 5053............ 14 lbs. 8 oz.

Gussie Richards 1673.
Lerna 3634.................... 15 lbs. 12 oz.
Iola 4627..................... 15 lbs. 2½ oz.

Haidee Signal 86448.
Princess Haidee Signal 98652.. 16 lbs. 4 oz.

Hallie Carter 5329.
Hallie's Jewel 17113.......... 20 lbs. 10 oz.

Hallie's Jewel 17113.
Hallie Signal 61726........... 18 lbs. 12 oz.

Hamra 36522.
Luteefy 46313................. 14 lbs. 6 oz.

Happy H. 20566.
Happy Winn 35358............ 18 lbs. 13 oz.

Hara Rex 19005.
Hara of St. Lambert 42128.... 17 lbs. 7½ oz.

Harmony 4148.
Harmony 2d 17118............ 19 lbs. 3 oz.

Harrella 32912.
Idarella 41433................ 15 lbs. ⅛ oz.

Harry's Duchess 60289.
Soconee 76895 (14 days)....... 40 lbs. 1¾ oz.

Harry's Fancy Belle 44764.
Harry's Fancy Belle 2d 82416. 14 lbs. 12½ oz.

Harry's Signaline 82419.
Harry's Signaline 2d 82428.... 20 lbs. 11 oz.
Bisson's Perfection 103628.... 14 lbs. 9½ oz.

Hartwick Belle 7722.
Alphea Jewel 22331............ 14 lbs.

Hattie 795.
Hattie Douglass 24960........ 16 lbs. 5 oz.

Hattie 2d 2901.
Salsoda 3721.................. 14 lbs. 7 oz.

Hattie D. Rex 38772.
Hattie Landseer 72719........ 14 lbs. 4 oz.

Hattie Parks 3776.
Bet Arlington 8970............ 18 lbs. 11 oz.

Hawthorn Minnie 26687.
Minnie's Lady Allen 102925.... 17 lbs. 10 oz.

Hazalena 3275.
Hazalena's Butterfly 10123.... 14 lbs.

Hazel 91.
Witch-Hazel 1360............. 14 lbs.

Hazel Stoke Pogis 51793.
Exile's Hazel 85181........... 15 lbs. 13 oz.

Hazen's Bess 7329.
Herberta 8811................. 16 lbs. 15 oz.

Heathbell 20170.
Miss Naomi 45098............ 15 lbs. 3 oz.

Hebe of Canterbury 33917.
St. Lambert Hebe 54069....... 15 lbs. 5 oz.

Hebe's Pride 76453.
Hebe's Fancy 84314.......... 14 lbs. 9 oz.

Hecuba 3155.
Nordheim Creamer 9755....... 14 lbs.

Heedless 9522.
Daisy Brown 12213........... 17 lbs. 6½ oz.

Helena of Bolton 39109.
Miranda of Bolton 100226..... 21 lbs. 4 oz.

Helen of Bay View 66030.
Albert's Alba 111721.......... 14 lbs. 2 oz.

Helen of Maple Glens 56339.
Ida of Maple Glens 97785...... 14 lbs. 14 oz.

Helen Stoke Pogis 31947.
Exile's Helen 61251........... 14 lbs. 4½ oz.

Helen Walker 53465.
Hearty Pogis 81297........... 18 lbs. 13¾ oz.
Wild Rose Bloom 110583...... 17 lbs. 1¼ oz.

Helfre 15445.
Belle of Passaic 85846........ 17 lbs. 11 oz.

Heliotrope of Linwood 6952.
Columbia Beauty 30263........ 15 lbs. 1 oz.

Helita 22113.
L. D.'s Helita 29366........... 14 lbs. 4½ oz.

Hennette 11624.
Hettie of Briarcliff 26621...... 18 lbs. 1 oz.

Hennie 3335.
Hennette 11624................ 14 lbs. 3½ oz.

Herpa 54293.
Herpa S. 127829............... 18 lbs. 6 oz.

Hester D. 2283.
Bertie Briggs 5213............ 14 lbs. 4½ oz.

Highland Mary 3d 19876.
Massena 25732................. 20 lbs. 7 oz.
Massena 25732 (61 days)..... 145 lbs. 8½ oz.
Massena 25732 (101 days)..... 162 lbs. 3 oz.

High Life 3259.
Jersey 3260.................... 15 lbs. 6 oz.

High Spirits 35390.
High Spirits of Lawn 61371... 18 lbs. 7 oz.

High Tea 65577.
Fancy Work 116422........... 18 lbs. 2 oz.

Hild 46148.
Alta B. 73003................. 14 lbs. 13 oz.

Hilda 942.
Hilda 2d 5447................. 23 lbs. 5 oz.

Hilda 18178.
Hilda 2d 14967................ 14 lbs. 12½ oz.

Hilda 2d 14967.
Landseer's Hilda 49788........ 17 lbs.

Hilda A. 3951.
Hilda A. 2d 11120............. 20 lbs.
Hilda A. 3d 16636............. 17 lbs. 1 oz.

Hilda B-B 8599.
Hilda B-B-B 17640............. 15 lbs. 8 oz.

Hilda C. 3869.
Hilda D. 6683................. 21 lbs. 2½ oz.

Hilda D. 6683.
Hatita 34538.................. 14 lbs. 15 oz.

Hildah of Clovernook 51597.
Agnes of Clovernook 72553.... 16 lbs. 9½ oz.

Hildmella 22315.
The Colonel's Daughter 50230. 15 lbs. 4½ oz.

Hillsdale Gem 16640.
Redacta 26954................. 14 lbs. 5 oz.

Hilpah 5879.
Hulla 7898.................... 19 lbs. 12 oz.

Hinnibel 4040.
Magnibel 7976................. 15 lbs. 2 oz.

Hoey 6th 5315.
Hoey Rex 82364............... 14 lbs. 9 oz.

Holyoke's Belle 30453.
Bettie Pogis 44542............ 16 lbs. 7½ oz.

Home Matron 6707.
Royal Sister 12451............ 14 lbs. 11 oz.

Homestead Pogis 2d 57451.
Homestead Rowena Pogis
89591 .... 16 lbs. 7½ oz.
Homestead Prue 82203........ 14 lbs. 13½ oz.

Homestead Prue 82203.
King's Lucy Pogis 111380..... 18 lbs. 8 oz.

Homestead St. Lambert 56199.
Homestead Lucy 96325....... 17 lbs. 12 oz.

Honeydrop 10033.
Honey Belle 25824............. 20 lbs. 7½ oz.

Honeymoon of St. Lambert 11221.
Lady Larkspur 56051......... 21 lbs. 2¾ oz.

Honeysuckle 1313.
Kitty Potter 9893............. 18 lbs. 5 oz.

Hope of Pittsford 73186.
Lois of Pittsford Farms 88848 14 lbs. 9 oz.

Houri 2842.
Lady Aspinwall 8374......... 14 lbs.

Hugo's Charm 107655.
Hugo's Lass 107656.......... 14 lbs. 1¾ oz.

Hugo's Victoria 29436.
Meridale Victoria 94848....... 18 lbs. 7¼ oz.

Hulabaloo's Flora 111164.
Rioter Flora Marjoram 111165 16 lbs. 4 oz.

Hulabaloo's Lass 34680.
Carlo's Lass 94405............. 16 lbs. 15 oz.

Hulda R. 16497.
Belbonnie 29630............... 14 lbs. 9 oz.

Hurd's Orpha 3346.
Orphean 4636................... 15 lbs. 7 oz.

Hyacinth Pogis 56020.
Oonan of Riverside 69773...... 34 lbs. 3 oz.

Hybla 2991.
Zalma 8778.................... 15 lbs. 5 oz.
Pet Rex 20166................. 14 lbs. 2½ oz.

Hypathia 13358.
Hypathia 2d 14774............. 19 lbs. 13½ oz.

Ianthe 4562.
Chroma 4572................... 20 lbs. 6 oz.

Ida 9th 6528.
Idana 27974................... 14 lbs. 10 oz.

Ida Imperial 2d 28710.
Minnie Montague 45474....... 15 lbs. 5½ oz.

Idalette Pogis 64220.
Chelka 79033.................. 14 lbs. 13¼ oz.

Ida Marigold 32615.
Mary Idagold 88186........... 23 lbs. 9 oz.
Fair Marigold 89387.............16 lbs. 15 oz.

Ida of Clinton 25419.
Idalco 117140.................. 14 lbs. 10 oz.

Ida of Oakland 35798.
Ida of Oakland 2d 72061....... 15 lbs. ½ oz.

Ida of Oakland 2d 72061.
Elma Marigold 100630........ 17 lbs. 13½ oz.
Adelpha Marigold 133768...... 15 lbs. 11½ oz.
Ida of Elma 92316............. 14 lbs. 10 oz.

Ida of Rocky Ford 40762.
Eric's Ida 69134............... 18 lbs.

Ida of St. Lambert 24990.
Pomona's Ida of St. L. 59565.. 14 lbs. 14½ oz.

Ida Regina 5204.
Brunhilde Regina 41742....... 16 lbs. 13 oz.

Idarella 41433.
Torrancella 59827............. 17 lbs. 8½ oz.

Ida's Dream 50070.
Rêve 78944.................... 21 lbs. 12 oz.
Ida's Dream 3d 131899........ 14 lbs. 7¼ oz.

Ida's Nan 49957.
Nanoonan 98775................ 14 lbs. 11⅞ oz.

Ida's Own 49959.
Ida's Lucille 99400............ 17 lbs. 4¼ oz.

Ida's St. Jeannaise 83646.
Ida's St. Jeannaise 2d 106316.. 17 lbs. 4¼ oz.

Ida Twinkle 36994.
Candelabrum 75187........... 17 lbs. 13 oz.

Ideal 11842.
Ideal Alphea 18755............ 14 lbs. 6 oz.
Alphetta 16531................ 14 lbs. 2½ oz.

Idelle Pogis 56953.
Idelle's Idea 92510............ 14 lbs. 10¾ oz.

Ides 14097.
Idessimus 50508............... 16 lbs.

Idex 2d 5429.
Christel 6565.................. 19 lbs. 5 oz.
Gilderoy's Idex 42027......... 19 lbs.
Yellow Locust 10679.......... 14 lbs. 10½ oz.

Idolette Melrose 61923.
Mary Brown 81213............ 15 lbs. 11½ oz.

Imogene 4552.
Bronzie 9368.................. 21 lbs. 6 oz.

Imperial Pansy 26830.
Imperial Pansy Pogis 49750... 14.42 lbs.

Imperial Signalla 38585.
Sallie Grimes 79122............ 16 lbs. 2 oz.

Imperial's Omega 13241.
Dot of Rocky Ford 40763...... 15 lbs. 1 oz.
Imperial's Omega 2d 19664... 14 lbs. 14 oz.
Kate Crecraft 40783............ 14 lbs. 6 oz.

Imperial's Omega 2d 19664.
Ada Fy 45479.................. 17 lbs. 4 oz.

Ina Kyle 71806.
Leta May 82171............... 15 lbs. 2 oz.

Income 19472.
Classic 21402.................. 14 lbs. 13½ oz.
Upper Ten 37603.............. 14 lbs. 7 oz.
Italica 58775.................. 14 lbs. 4½ oz.

Independence of Nipsic 88307.
Matilda Ann Darling 105398... 14 lbs. 1 oz.

Ingleside Duchess 46268.
Miss Lorne 78132.............. 15 lbs. 9 oz.

Io 3d 245.
Io 5th 280..................... 17 lbs. 8 oz.

Io A. 6284.
Oitz 8649..................... 15 lbs. 11 oz.

Io B. 3d 12881.
Maudie of White Lick 29682.. 15 lbs. 7 oz.

Iola H. 24225.
Maybee H. 60620.............. 18 lbs. 4 oz.

Ione of Sacramento 5220.
Oleta 15625.................... 15 lbs. 12 oz.

Irelia 21320.
Maggie May of Tupelo 71280.. 20 lbs. 7 oz.

Irene of St. Lambert 27056.
Irene of St. Lambert 2d 75759 15 lbs. 5 oz.

Irene of Short Hills 5137.
Torment 15579................ 17 lbs. 8½ oz.

Irma 1298.
Adina 1942.................... 14 lbs. 4 oz.

Iryl Dane 22842.
Iryl Dane 3d 38375............ 14 lbs. 15 oz.

Isa of Edgewood 96452.
Stella of Edgewood 114473.... 14 lbs. 12 oz.

Island Beauty 2d 19885.
Pierrot's Beauty 48964....... 15 lbs. 11 oz.

Island Lily 6969.
Island Chrissie 12007......... 14 lbs. 14 oz.

Islianne 19545.
Islithanks 90544............... 17 lbs. 2 oz.

Islip 4th 1884.
Islip Lenox 31703 (30 days).... 47.699 lbs.
Islip Lenox 31703 (90 days).... 178.066 lbs.

Islip Lenox 31703.
Indulgence 50105.............. 17 lbs. 1 oz.

Isolda 5900.
Lady Appel 8612.............. 18 lbs. 3 oz.

Isolda's Princess 12474.
Yokun's Princess 35643....... 15 lbs. 10¼ oz.

Itura 61090.
Relay 70353................... 17 lbs. 2 oz.
Parthia 64925................. 16 lbs. 11 oz.

Iuka of Riverview 40648.
Iuka of Riverview 2d 61271... 17 lbs. 12 oz.

Iversnaid 8978.
Iola F. 85529 (21 days)........ 25.251 lbs.

Jane Glenn 56947.
Hilda Glenn 81195............ 27 lbs.
Hilda Glenn 81195 (35 days).. 121 lbs. 7½ oz.

Jane Riley 11455.
Jersey Jane 38308............. 16 lbs. 4½ oz.

Janie Moore 17514.
Janie Caruth 73311............ 15 lbs. 4 oz.

Jarto 1902.
Irene of Short Hills 5137..... 14 lbs. 6½ oz.

Jazella G. 14191.
Signaldella 24107.............. 18 lbs. 1¾ oz.

Jazel's Maid 11011.
Mamelle 20804................ 21 lbs. 8¼ oz.
Jazella G. 14191.............. 20 lbs. 7 oz.
Thekla of Clover Nook 33445.. 15 lbs. 9 oz.

Jemele 36524.
Jemele 2d 101437.............. 15 lbs. 7 oz.

Jennie 5th 2269.
Daisy of Clermont 3492....... 14 lbs.

Jennie Cream 30622.
Jennie Cream 2d 47466....... 20 lbs. 5 oz.

Jennie Cream 2d 47466.
Martha of Eau Claire 120022.. 14 lbs. 13 oz.
Millie of Eau Claire 120023.... 14 lbs. 7 oz.

Jennie Graham 2d 121220.
Tait's Beauty 121221.......... 18 lbs. 4 oz.

Jennie of Sidney 11488.
Queen of Delaware 17029...... 18 lbs. 13 oz.

Jennie's Pet 26728.
Durex 66320..........:......... 19 lbs.

Jennie Stoke Pogis 32010.
Exile's Dolly 67452............ 19 lbs. 7½ oz.

Jenny 7827.
Jersey Belle of Scituate 7828.. 25 lbs. 3 oz.
Jersey Belle of Scituate 7828
(one year).................... 705 lbs.

Jenny B. 4190.
Jennette Darling 10702....... 16 lbs. 2 oz.

Jenny Brewster Jr. 73133.
Torpedo Exile 105287.......... 16 lbs. 4½ oz.

Jenny Gray 3511.
Jenny Dodo H. 14448......... 21 lbs. 8 oz.
Zophar H. 12329.............. 21 lbs. 3 oz.
Carria H. 14454................ 16 lbs. 10 oz.

Jenny Justa 43731.
Justa Pogis 64863............. 14 lbs. 15¼ oz.
Justa Pogis 64863 (90 days)... 157.197 lbs.

Jenny Sutliff 2d 73131.
J. S. Exile 105495............. 15 lbs. 4 oz.

Jersey Belle of Scituate 7828.
Belle of Scituate 7977.......... 16 lbs.
Lass of Scituate 9555......... 15 lbs. 14 oz.

Jersey Cream 3151.
Jersey Cream 3d 8521.......... 16 lbs. 5 oz.
Jersey Cream 2d 8519.......... 14 lbs. 12 oz.

Jersey Cream 2d 8519.
Cultured Cream 29196......... 20 lbs. 12 oz.
Jennie Cream 30622............ 17 lbs. 4 oz.
Auraria 10688.................. 15 lbs. 10 oz.

Jersey Lily 14044.
Leila of Briarcliff 24184....... 17 lbs. 6½ oz.

Jessaline 26099.
Jessaline 2d 38295............ 16 lbs. 4½ oz.

Jessamine (F. S. 2188 J. H. B.)
Grey Jessamine 17444......... 14 lbs. 4 oz.

Jessamine of St. Lambert 5125.
Cheerful of St. Lambert 8348.. 20 lbs. 8 oz.

Jessica Fries 2d 60836.
Hattie Adams 80909........... 14 lbs.

Jessie 3207.
Victoria 3175.................. 16 lbs. 1 oz.

Jessie 2d 1301.
Flashy Jessie 9722............ 17 lbs. 15 oz.

Jessie Benton 11183.
Miss Clifford 27962........... 17 lbs. 15¼ oz.

Jessie Leavenworth 8248.
Bell Rex 11700................. 14 lbs. 10 oz.

Jessie Lee of Labyrinth 5290.
Maggie of Springvale 15931... 18 lbs. 5 oz.

Jessie S. 59822.
Jessie's Sayda 105697.......... 15 lbs. 10 oz.

Jetsam 32893.
Jetsam's May 62530........... 14 lbs. 5 oz.

Jetty d'Or 39845.
Jetty d'Or 2d 62403........... 17 lbs. 1 oz.

Jetty Lee 7988.
Jetty d'Or 39845.............. 15 lbs. 2 oz.

Jeune Jenny (F. S. 706 J. H. B.)
Jenny Le Brocq 9757.......... 14 lbs. 14 oz.

Jewel of Ridley 17231.
Jewel of Ingleside 54544....... 17 lbs. 1 oz.

Jim's Pride 31376.
Jim's Glory 60527............. 14 lbs. 4 oz.

Joanella 19063.
Vexanella 79306................ 16 lbs. 5 oz.

Joey of Prospect 27022.
Fable 62520.................... 26 lbs. 5¼ oz.
Fable 62520 (14 days)......... 52 lbs. 7 oz.

John's Fancy 58256.
Denise's Fancy 92500........ 15 lbs. 10 oz.

Johnson's Daisy 8391.
Hurrah Pansy 12153........... 14 lbs. 1½ oz.

Jolie (F. S. 1453 J. H. B.)
Cicero's Jolie 18246........... 18 lbs. 3 oz.

Jolie of Linden 16695.
  Julia of Eau Claire 120018.... 17 lbs. 8 oz.

Josephina 64921.
  Tremona 93017.................. 17 lbs. 2 oz.

Josephine Beacon 3306.
  Miss Puritan 34204........... 18 lbs. 5 oz.
  Lady Josephine 11560 (8 days) 19 lbs. 2 oz.

Josey Brown 7533.
  Belle Williamson 8386........ 19 lbs. 10½ oz.

Josie B. 32749.
  Nellie Alphea 51274........... 14 lbs. 1½ oz.

Judy Fagan 3d 37653.
  Jocal 95182................... 20 lbs. 14 oz.

Jule 3640.
  Jeannie Platt 6005............ 14 lbs. 4 oz.

Julia 3893.
  Monmouth Duchess 3.95....... 14 lbs. 7 oz.
  Countess of Warren 3896...... 14 lbs.

Julia Belle 13233.
  Beauty Montague 57923....... 17 lbs. 8 oz.

Julia Evelyn 6007.
  Therese M. 8364.............. 14 lbs. 2 oz.

Julia Landseer 56268.
  Pedro's Fancy 4th of H. F.
  119842 .... 14 l s. 14 oz.

Julia Parks 3778.
  Pet Lee 7993................. 14 lbs. 12 oz.

Julia Walker 10133.
  Julia Landseer 56268......... 14 lbs. 2½ oz.

Juliet 3d 5163.
  Belle of Milford 7445......... 15 lbs. 6 oz.

Juliette of St. Lambert 5483.
  Judith Coleman 11391......... 18 lbs. 4 oz.
  Aleph Judea 11389............ 18 lbs. 1½ oz.

June 3d 11644.
  Amelia Davenport 81044....... 14 lbs. ⅝ oz.

June Pansy 27317.
  Prusa Pedro 50877............ 15 lbs. 8 oz.

June Sweets 59978.
  June Sweets 2d 87967......... 15 lbs. 11 oz.

Juno Grey 16722.
  Cicero's Juno 16726........... 17 lbs. 2 oz.

Junon 11269.
  Maculac 24277................. 15 lbs. 3 oz.

June W. 8553.
  Vivian 15813.................. 16 lbs.

Justa Pogis 64863.
  Justa Pogis 2d 115494......... 14 lbs. 9¼ oz.

Kaleta 19711.
  Kaleta 2d 38810.............. 15 lbs. 14 oz.

Kalmia 4561.
  Bintana 9837................. 14 lbs. 3½ oz.
  Safrano 4568................. 14 lbs. 2¼ oz.

Kamris 5941.
  Kamaretta 80479.............. 16 lbs. 7 oz.

Kate Clover 16952.
  Agnes Clover 86378........... 14 lbs. 12 oz.

Kate Crecraft 40783.
  Kate Crecraft 2d 75461........ 18 lbs. 3 oz.

Kate Durwood 48971.
  Kate Durwood 2d 82417....... 14 lbs. 1 oz.

Kate Durwood 2d 82417.
  Katie Bisson 100200........... 14 lbs. 7 oz.

Kate Early 3802.
  Princess Inez 5402............ 14 lbs. 2 oz.

Kate Gordon 8387.
  Kathletta 19567............... 22 lbs. 12½ oz.
  Mary Gordon 41429........... 18 lbs. 5¼ oz.
  Gordonetta 19570............. 17 lbs. 1 oz.
  Kitty Better 32911............ 15 lbs. 14⅝ oz.
  Kate Gordon 2d 83662........ 15 lbs. 5½ oz.

Kate Grimball 56435.
  Shelta F. 56437............... 20 lbs. ½ oz.

Kate Linn 77176.
  Kate Landseer of Lawn 84415. 16 lbs. 5 oz.

Kate Oxford 58102.
  Kate Oxford 2d 69270......... 14 lbs. 4 oz.
  Bisson's Southern Daisy 112992 14 lbs. 2½ oz.

Kate Weston 47286.
  Kate McCollum 107306........ 15 lbs. 12¼ oz.

Kate Winslow 28055.
  Katie of St. Lambert 54776.... 19 lbs. 7 oz.
  St. Lambert's Kate 61975..... 16 lbs. 11 oz.
  Katherine of Pittsford 73169
    (30 days).................... 54.107 lbs.

Kathleen of Glen Duart 59982.
Agnes of Glen Duart 82475.... 15 lbs. 11 oz.

Kathleen of St. Lambert 5122.
Ida of St. Lambert 24990...... 30 lbs. 2½ oz.
Allie of St. Lambert 24991.... 26 lbs. 12 oz.

Kathletta 19567.
Kathletta's Fancy 60738....... 17 lbs. 6¾ oz.
Kathy Torment 32910......... 16 lbs. 13½ oz.
Sigletta 32915.................. 16 lbs.
Kattle 41431.................... 14 lbs. 8½ oz.

Katie Laurie 2d 46392.
Tormentor's Laura 98346..... 16 lbs. 9½ oz.

Katie Putnam 40366.
Señorita Pogis 53791.......... 16 lbs.

Katy Barr 11066.
Olive of One Pine 93449...... 17 lbs. 9 oz.

Katy Scituate 29300.
Katy Scituate's Beauty 100259 16 lbs.

Katy Signal 24690.
Katy Signal 2d 44673......... 16 lbs. 8 oz.

Kedria 88645.
Pedro's Priscilla 99413........ 17 lbs. 6 oz.

Kerima 7242.
Dorizella 20525................ 15 lbs. 7 oz.

Khedive's Daisy 18174.
St. Jeannaise 15789........... 17 lbs. 8½ oz.

Khedive's Fancy 18180.
Khedive's Fancy 3d 70282.... 14 lbs. 6 oz.

Khedive's Rosebud 18173.
Pedro's Rosebud 66086........ 19 lbs. 4 oz.
Carlo's Rosebud 18223........ 15 lbs. 8 oz.

Khelula 17970.
Khelula of Briarcliff 46799.... 18 lbs. 2½ oz.

King's Bravy 78868.
King of St. L.'s Allie 86797.. 20 lbs. 10 oz.

King's Florrie 2d 64455.
Harry's Florrie 86263.......... 14 lbs. 12 oz.

King's Fosie 80768.
Chief's Iona 111704........... 15 lbs. 12 oz.

King's Lady Natica 80769.
Lady Aurora 98734........... 14 lbs. 10 oz.

King's Stevia 98737.
Chief's Stevia 122187......... 16 lbs. 10½ oz.

Kisberine 23480.
Kisberine Rex 31436.......... 16 lbs. 8 oz.

Kismet 2d 22895.
Dorothy Taylor 99485........ 15 lbs. 10 oz.

Kissing Girl 67725.
Tonnage Girl 103331........... 15 lbs. 1½ oz.

Kit 34376.
Kitty of Prospect 111121...... 19 lbs. 4¾ oz.

Kitarene 4267.
Litty 8017..................... 14 lbs.

Kittie Dolan 66834.
Kittie Green 88379............ 15 lbs.

Kittie Lightfoet 2d 1300.
Kitty Colt 2213............... 15 lbs. 9½ oz.

Kittie Pedro 27720.
Sig Eurota 64898.............. 15 lbs. 1½ oz.

Kitty Better 32911.
Hyacinth Pogis 56020......... 15 lbs. 6 oz.

Kitty Clover 1113.
Chrissy 1448.................... 16 lbs. 8 oz.
Kitty Clover 2d 16099........ 14 lbs. 15 oz.

Kitty Clover of Burlington 11788.
Lady Wellington of Milton
12234 .... 18 lbs. 11 oz.

Kitty Floos 11539.
Kitty Ona 24209............... 14 lbs. 11 oz.

Kitty Kringel 57120.
Princess of Dorset 85902...... 35 lbs. 1 oz.

Kitty of Cedar Glen 10948.
Kitty Kringel 57120........... 16 lbs. 8 oz.

Kitty of Jefferson 12193.
Kitty of Jefferson 2d 12313.... 14 lbs. 3 oz.

Kitty of Jefferson 2d 12313.
Kitty's Ida 97736.............. 18 lbs. 8 oz.
Bunker's Kitty 47103.......... 14 lbs. 11½ oz.
Kitty's Matilda 75349......... 14 lbs. 3 oz.

Kitty of St. Lambert 6637.
Leonora of Canada 94906...... 22 lbs. 11 oz.
Kitty of St. Lambert 2d 40106. 16 lbs. 5 oz.

Kitty Pedrothorn 49581.
Kitty Pedrothorn 2d 84562.... 15 lbs. 4 oz.

Kobe 3055.
Kobe's Marcella 4597......... 14 lbs. 13 oz.

Koffee's Duchess 35113.
Koffee's Duchess 2d 57988..... 17 lbs. 1 oz.

Koffee's Duchess 2d 57988.
Exile's Ona 106212............ 14 lbs. 3¾ oz.

Koffee's Grisette 30433.
Sigma 78676................... 18 lbs. 2¼ oz.

Koffee's Medusa 26306.
Duke's Medusa 43511......... 18 lbs.

Koffee's Miss Le Gros 35107.
Miss Le Gros' Maid 83645..... 18 lbs. 5¾ oz.

Koffmadge 63735.
Koffmadge 2d 77790............ 16 lbs. 14 oz.

Kosi 5th 10870.
Xarlfina 34176................. 14 lbs. 9 oz.

La Belle Desreaux 2d 5096.
Urbana 5597................... 16 lbs.

La Biche 2d 4023.
Colt's La Biche 6399......... 17 lbs. 2½ oz.

Lad's Swan 73325.
Rover's Swan 106934........... 16 lbs. 3 oz.

Lady (P. S. 214 J. H. B.)
Beauty Romeril 26090......... 18 lbs. 9 oz.

Lady Adams 4919.
Lady Adams 2d 6529........... 15 lbs. 3 oz.

Lady Adams 4th 11464.
Lady Crouse 69336............. 18 lbs. 1 oz.

Lady Ajax 2d 89637.
Ajax Riotress 113709.......... 16 lbs.

Lady Alexis 26916.
Exile's Onnalinda 65189....... 15 lbs. 1½ oz.
Exile's Lady Alexis 79112..... 14 lbs. 4¾ oz.

Lady Alice of Hillcrest 7450.
Lady Alice Pogis 30273........ 22 lbs. 8 oz.
Southern Pet 26675............ 18 lbs. 8 oz.

Lady Anerly 10595.
Angetta 19404.................. 15 lbs. 4 oz.
Young Anne Lee 31668......... 14 lbs. 7 oz.

Lady Anna Pogis 52734.
Metaphysics 100974............ 23 lbs. 2¼ oz.

Lady Antoinette 24391.
King's Antoinette 40456...... 15 lbs. 1 oz.

Lady Appel 8612.
Dew 62474..................... 16 lbs. 7 oz.
Sterling Merit 62456......... 15 lbs. 4 oz.

Lady Aurora 98734.
Albert's Ideal 111719.......... 15 lbs. 10 oz.

Lady Baronnella 22501.
Perrot's Baroness 22508....... 15 lbs. 14½ oz.

Lady Baush 11810.
Martha D. 66641............... 17 lbs. 2 oz.

Lady Bingo 24160.
Clotaire's Beauty 28171....... 20 lbs. 4 oz.

Lady Bligh (F. S. 5621 J. H. B.)
Miss Bligh 24561.............. 15 lbs. 2 oz.

Lady Bobita 2d 43571.
Vigo Princess 60806........... 18 lbs.

Lady Bountiful 17946.
Ashantee's Lady 35951........ 16 lbs.

Lady Bowen 354.
Princess Bowen 9699......... 14 lbs. 12 oz.

Lady Brown 433.
Lady Brown 4th 6911........ 14 lbs. 12 oz.
Lady Brown 2d 2348.......... 14 lbs. 3 oz.

Lady Brown 4th 6911.
Bloomfield Lady 6912......... 14 lbs. 12 oz.

Lady Bullard 24648.
Bertha Stoke Pogis 35206...... 14 lbs. 15½ oz.

Lady Burlington 1713.
Lady of Belle Vue 7705........ 15 lbs. 11 oz.

Lady Burlington 2d 4032.
Merry Burlington 7600........ 15 lbs. 4 oz.

Lady Charella 66187.
King's Charella 80765......... 15 lbs. 9½ oz.

Lady Clarendon 7030.
Lady Clarendon 3d 17578...... 14 lbs. 5½ oz.

Lady Cloud 19358.
Lady Cloud 2d 26825.......... 16 lbs. 12 oz.

Lady Cloud 2d 26825.
Cupid's Cloud 74040.......... 17 lbs. 2 oz.

Lady Clover 8024.
Rockwood Maid 8375.......... 15 lbs. 4½ oz.

Lady Cornwall 7179.
Cornwall Maid 19024......... 29 lbs. 12 oz.

Lady Creamly 1975.
Lady Creamly 4th 19077........ 17 lbs. 6½ oz.

Lady Delphine 28460.
Helen Stoke Pogis 31947...... 17 lbs. 8 oz.

Lady Dove 4418.
Lady Rareripe 23081.......... 16 lbs. 1 oz.

Lady Elfied 40757.
Lady Dorcas 74348........... 14 lbs. 10 oz.

Lady Elkhorn 44407.
Elsie Rowls of Lawn 71633.... 14 lbs. 5 oz.

Lady Ellen 11660.
Lady's Blossom 18491.......... 20 lbs. 15¾ oz.

Lady Elsie 33142.
Elsie's Lady 81493............ 14 lbs. 9 oz.

Lady Eppie 72375.
Eppie Lady Pogis 110638...... 14 lbs. 3 oz.

Lady Feronia 17607.
Edith Golding 63709.......... 19 lbs. 1¼ oz.

Lady Gilmore 2d 12140.
Darlymora 37771............... 14 lbs. 7 oz.

Lady Godfrey 678.
Gilda 2779..................... 14 lbs. 6 oz.

Lady Golddust 7718.
Lady Golddust 2d 19861....... 23 lbs. 4 oz.

Lady Grace of Upholme 39569.
Lady St. Heller of S. 51380.... 20 lbs. 9½ oz.
Lady Grace Pogis of S. 79462.. 20 lbs. 5 oz.
Grace Marigold 99377......... 20 lbs. 1 oz.

Lady Gray Hayden 3512.
Theresa H. 14447............. 15 lbs. 3 oz.

Lady Gray of Hilltop 6850.
Lady Gray of Hilltop 2d 14641 14 lbs. 12 oz.
Lady Gray of Hilltop 3d 14642 14 lbs. 2 oz.

Lady Gray of Hilltop 6th 49485.
Lady Thalma 54701.......... 14 lbs. 12 oz.

Lady Horton 2d 15499.
Vera of Briarcliff 28687....... 15 lbs. 1 oz.

Lady Ida Pogis 94166.
Lady Oonan Pogis 119154..... 19 lbs. 6¾ oz.

Lady Imperial 16019.
Imperial Riotress 30259........ 18 lbs. 4 oz.

Lady Ives 1708.
Lady Ives 3d 6740............. 14 lbs. 8 oz.

Lady Jane Grey (F. S. 592 J. H. B.)
Mischief Le Brocq 7680....... 15 lbs.

Lady Jane of St. Peter's 7475.
Lady Jane of St. Peter's 2d
62536 .... 14 lbs. 5 oz.
Woodstock Lady 80619 (21
days) .... 23.027 lbs.

Lady Jeannette of St. James 12054.
Oonan's Jeannette 45616...... 20 lbs. 3 oz.

Lady Julia G. 16199.
Biederkeit 52257................ 22 lbs. 1¼ oz.

Lady Keith 20064.
Lily of Glynllyn 31340......... 16 lbs.

Lady Lemon 22125.
Pedro's June 38790............ 14 lbs. 6 oz.

Lady Leonora 11811.
Pride's Lady Frances 39529.. 16 lbs. 3½ oz.

Lady Lightfoot 2745.
Narmeoka H. 12323.......... 18 lbs. 12 oz.
Jenny Gray 3511.............. 17 lbs. 15 oz.
Lucy Gray 2746.............. 15 lbs. 13 oz.

Lady Lillie 11813.
Pride's Lady Lillie 35181...... 15 lbs. ¾ oz.
Island Dots 17003............. 14 lbs. 9 oz.

Lady Lily Gray 21364.
Lady Emma C. 57041......... 16 lbs.

Lady Livingston 33374.
Victor's Lady 52691.......... 15 lbs. 14 oz.
Kitty Livingston 34303........ 15 lbs. 2 oz.

Lady Longfield 23524.
Ida Longfield 64227.......... 18 lbs. ½ oz.
Tolonga 76868................ 15 lbs. 5½ oz.

Lady Louisa 8586.
Frisky Miss 18572............. 14 lbs.

Lady Louise 4339.
Sasco Belle 13601.............. 14 lbs.

Lady Love 1315.
Lady Love 2d 2212............ 16 lbs. 8 oz.

Lady Mabel 8133.
Coomassie L. 73273.......... 15 lbs. 1 oz.

Lady Mabella 22138.
Pedro's Mab 40212............ 16 lbs. 10 oz.

Lady Madeline 2d 41854.
Dora Lowndes 64136.......... 16 lbs. 15 oz.

Lady Madge 25541.
Lady Woodbine 26893.......... 14 lbs. 2 oz.

Lady Mahopac 8613.
Princess White Water 21137... 14 lbs. 15 oz.

Lady Mary Combination 98442.
Lady Mary Lenox 10557.4..... 16 lbs. 7½ oz.

Lady Mary Hampton 2d 26184.
Birdie Nicholson 31676........ 18 lbs. 13 oz.

Lady Mary Linden 12800.
Lady Mary of Prospect 19768.. 19 lbs. 15½ oz.

Lady Mary of Prospect 19768.
Good Advice 62476........... 19 lbs. 5¼ oz.

Lady Mary Pogis 27113.
Ida Mary Pogis 56126........ 14 lbs. 2 oz.

Lady Maud of Cinque Park 57572.
Lady Maud of Cinque Park 2d
71548 .... 14 lbs. 1¼ oz.

Lady Mel 429.
Lady Mel 2d 1795 (61 days)... 18⅜ lbs.

Lady Minnie's Lass 43794.
Panache 62055................ 16 lbs. 1 oz.

Lady Mirah 25266.
Mirah Landseer 74827........ 17 lbs. 1 oz.

Lady Morgan 11671.
Ruth Morgan 24098........... 14 lbs. 6 oz.

Lady Natica 66031.
King's Lady Natica 80769..... 15 lbs. 8 oz.

Lady Oaks 2081.
Lady Oaks 2d 5246........... 15 lbs. 2 oz.

Lady of Belle Vue 7705.
Nonsuch of Linwood 29028.... 14 lbs. 14½ oz.

Lady of Four Pines 31737.
Mary of Glenoir 59940........ 27 lbs. 3½ oz.

Lady of Lodi 21784.
Mary of Lodi 88517.......... 15 lbs. 1 oz.

Lady of Oakland 11101.
Sweet Leona B. 21934........ 14 lbs. 1½ oz.

Lady of Rockys 96265.
Rockys' Pogis 96272.......... 18 lbs. 3 oz.

Lady of the Isles (F. S. 992 J. H. B.)
Fear Not 6059................ 17 lbs. 3 oz.
Lady Velvetine 15771.......... 17 lbs. 2 oz.

Lady of Venice 13342.
Lady of Dryden 27642........ 16 lbs. 3 oz.

Lady O'Neal 7391.
Belle Mardi 18362............ 18 lbs. ½ oz.

Lady Oxford 4860.
Lady Oxford 2d 15101........ 14 lbs. 7 oz.

Lady Palestine 2d 28744.
Mary Palestine 52439......... 16 lbs. 7 oz.

Lady Pauline 2651.
Honeydrop 10033.............. 14 lbs. ½ oz.

Lady Perfection 32116.
Lady K. of St. Lambert 112294 24 lbs. 6 oz.
Santa Stoke Pogis 91680...... 19 lbs. 1½ oz.
Mamie H. Pogis 57050........ 16 lbs. 14 oz.

Lady Phillis 18240.
Lady Phillis 2d 35629........ 18 lbs. 8 oz.

Lady Pinckard of Springvale 11227.
Cream of Springvale 16621.... 15 lbs. 10 oz.

Lady Prince 48320.
Garfield's Princess 62544...... 22 lbs.

Lady Rareripe 23081.
Neitherme 64670.............. 15 lbs. 1 oz.

Lady Rex 3d 15391.
Rosette M. 78677.............. 14 lbs. 6.4 oz.

Lady Rhine 1381.
Lorraine 1435................. 14 lbs. 8 oz.

Lady Roberta 29209.
Rioter's Roberta 80469........ 17 lbs.

Lady Robin 1119.
Mabel Crockett 18719.......... 14 lbs. 9 oz.

Lady Rushmore 11812.
Queen Neptune 15501.......... 18 lbs. 13½ oz.
Pride's Lady Sarah 39537...... 15 lbs. 1½ oz.
Pride's Lady Rushmore 35182. 15 lbs. 1 oz.

Lady Sarah 4931.
Lady Delphine 28460.......... 14 lbs. 4 oz.

Lady Sheba 15479.
Sheba Rex 47429 (30 days)..... 57,511 lbs.
Sheba Rex 47429 (90 days)..... 190,617 lbs.

Lady Star 2d 15724.
Exile's Lady Star 63943....... 16 lbs. ¼ oz.
Exile's Lady Star 63943 (16 days) .... 35 lbs. 7½ oz.

Lady Theresa 2d 4253.
Beauty of Seekonk 14651...... 21 lbs. 3 oz.
Lady Armington 7610.......... 17 lbs. 8 oz.

Lady Tilford 14467.
Lady Cloud 19358.............. 27 lbs. 6 oz.

Lady Tip 2d 17799.
Friar's Lady 36119............ 14 lbs. 2½ oz.

Lady Walker of Creekside 21984.
Nitelis 102722................. 16 lbs. ⅛ oz.

Lady Woodbine 26803.
Exile's Effie 40526............ 15 lbs. 2 oz.

Laelia 38791.
Lampedo 46204................. 21 lbs.

La Falaise 6961.
La Fantine 24489............. 15 lbs. 4 oz.

La France's Pansy Rex 49141.
La France's Pogis 99986...... 21 lbs. ½ oz.
Signaldo's Pansy 53366........ 16 lbs. ½ oz.

Lakeview Gem 37042.
Bridal Rose 74883............. 23 lbs. 8 oz.

Lalla Ingleside 46129.
King's Sultana 76657.......... 14 lbs. 8¼ oz.

Lalla Rookh of Sugar Grove 15882.
Lalla Rookh's Queen 65848... 15 lbs. 10 oz.

Lalla Rookh of Woodlawn Home 12146.
Lalla Rookh of Sugar Grove 15882 .... 20 lbs. 1 oz.

Landseer's Content 56985.
Khedive's Fairy 66849......... 15 lbs. 7 oz.
Tormentor's Content 119129... 14 lbs. 1 oz.

Landseer's Fancy 2876.
Toltec's Fancy 27172.......... 27 lbs. 5½ oz.
Oonan's Fancy 31887.......... 19 lbs. 10¾ oz.
Rosy Dream 9808.............. 19 lbs. 1 oz.
Landseer's Fancy 3d 70653.... 15 lbs. 14½ oz.
Landseer's Fancy 2d 43184.... 15 lbs. ½ oz.

Landseer's Pearl 56986.
Khedive's Pearl 71056......... 14 lbs. 12 oz.

Landseer's Signaline 58101.
Harry's Signaline 82419....... 14 lbs. 8½ oz.
Bisson's Signaline 93844...... 14 lbs. 3 oz.

Landseer's Signaline 2d 86343.
Bisson's Pride 108517.......... 14 lbs. ½ oz.

Lanison's Queen 80097.
Lanison's Queen 2d 123773.... 15 lbs. 14 oz.

La Parleuse (F. S. 3393 J. H. B.)
Lemon Fern 22140............. 15 lbs. 1 oz.

La Pera 7091.
La Pera 2d 13404.............. 14 lbs. 8 oz.

La Petite Mère 5470.
La Petite Mère 3d 12814....... 16 lbs. 9 oz.
La Petite Mère 2d 12810....... 16 lbs. 7 oz.
Little Pogis 32612............. 16 lbs. 1 oz.
La Belle Petite 5472.......... 15 lbs. 8 oz.
La Petite Mère 2d 12810 (one year) .... 660 lbs. 4 oz.

La Petite Mère 2d 12810.
La Petite Pogis 28757........ 20 lbs. 10¾ oz.

La Petite Mère 3d 12814.
Lady Matilda Pogis 36270..... 21 lbs. 9 oz.

La Petite Pogis 28757.
Crown 75198.................... 20 lbs. 12 oz.

La Picote, en I. of J.
La Financière 11970........... 15 lbs. 5¼ oz.

La Pucelle 16829.
Proctor's Pansy 25688......... 15 lbs. 13 oz.

Lara 4306.
Silenta 17685.................. 15 lbs. 10 oz.
Sileno 4307.................... 14 lbs.

La Rouge (P. S. 45 J. H. B.)
Nervine 25932................. 14 lbs. 1½ oz.

Lass de Grantez 26673.
Imogene Pogis 44967.......... 18 lbs. 13 oz.
Katie Putnam 40366.......... 16 lbs. 4 oz.

Lass Nellie 24012.
Miss Nellie Parker 24988...... 23 lbs. 11 oz.

Laundress 13867.
Laundress 2d 24649............ 14 lbs. 9 oz.

Laura Carrie Alphea 22311.
Daisy of Denison 53137........ 15 lbs. 12 oz.

Laura Esther B. 49272.
Pedro's Laura B. 96363........ 14 lbs. 10 oz.

Laura Hill 9084.
Atlanta's Beauty 12949........ 21 lbs. 3 oz.

Laura Lee 8550.
Modita 16626................... 16 lbs. 8 oz.
Laura Lee 2d 128023........... 16 lbs. 4 oz.
Upright's Brownie 103755..... 15 lbs. 3 oz.
Miralba 65002................. 14 lbs. 7 oz.
Nicomis 68026................. 14 lbs. 3 oz.

Laura Y. 2d 62333.
Ella's Tina 68408.............. 14 lbs. 9¾ oz.

Laurena of Meadowbrook 4123.
Benetia 21511.................. 16 lbs. 5 1-3 oz.

Laurette 2d 17148.
Pearl Pogis 38304.............. 16 lbs. 10¾ oz.

La Violette 36801.
La Violette 2d 52604........... 14 lbs. 7 oz.

La Vivienne 1068.
La Vivienne 2d 1324........... 16 lbs. 2 oz.

La Vivienne 2d 1324.
Pauline Vivienne 11305........ 16 lbs. 13 oz.

Lebanon Wife 6102.
Home Matron 6707............ 14 lbs.

Le Brocq's May 59724.
Donna Le Brocq 99305........ 14 lbs. 6 oz.

Le Brocq's May Belle 27980.
Koffee's May Belle 65220...... 20 lbs. 5 oz.
May Belle's Lady d'Or 74749.. 15 lbs. 15 oz.

Leclair's Marjoram 36355.
Marjoram of Glen Rouge 78420 22 lbs. 12¼ oz.
Ethel Marjoram 87266......... 16 lbs. 11 oz.
Cassie Marjoram 120666....... 16 lbs. 5 oz.

Leda 799.
Phædra 2561................... 19 lbs. 13 oz.
Leah Darlington 13836........ 15 lbs. 5½ oz.
Clytemnestra 2455............. 15 lbs. 3½ oz.

Le Gros' Lily of the Valley 11537.
Le Gros' Lily of the Valley
2d 13386...................... 19 lbs. 10¼ oz.

Le Gros' Lily of the Valley 2d 13386.
Koffee's Lily 25515............ 15 lbs. 3½ oz.

Leila Stratton 25198.
Leila Victor 85433............. 14 lbs. 1 oz.

Lelia Douland 44473.
Lelia Talbot 96922............ 14 lbs. 8½ oz.

Lemon Twig 42028.
Torlona 76872.................. 17 lbs. 1½ oz.
Lemon Spray 44532............ 15 lbs. 1½ oz.

Lena Lewis 3735.
Thorndale Belle 5265.......... 14 lbs. 8 oz.

Lena of Pleasant View 13453.
Lady Dorlena 36105........... 15 lbs. 7 oz.

Lena's Florence 42460.
Lena's Florence 2d 78688...... 14 lbs. 9½ oz.

Lena's Florence 2d 78688.
Elorita 101742................. 16 lbs. 4 oz.

Lena Woods 6687.
Lena's Florence 42460......... 14 lbs. 8 oz.

Lenore H. 15798.
Lenore H. 2d 29767........... 16 lbs. 3 oz.

Leoline 9285.
Leoline 2d 18315.............. 14 lbs. 4 oz.

Leoni 11868.
Leoni 2d 29750................ 15 lbs. 9 oz.
Leonette 29752................ 15 lbs. 5 oz.

Leoni 2d 29750.
Leoni Landseer 65317......... 16 lbs.

Leonice 4491.
Leonice 2d 8342............... 16 lbs. 8 oz.

Lera Williams 27139.
Lora of Clovernook 51599..... 16 lbs. 2 oz.

Lerna 3634.
Ideal 11842................... 14 lbs. 12½ oz.
Lernella 22322................ 14 lbs. 1½ oz.

Lesbia H. 35634.
Leonetta H. 67514............ 15 lbs. 1½ oz.
Leonetta H. 67514 (30 days).. 62 lbs. 3 oz.

Letacq Bess, on I. of J.
St. John's Daisy 28388........ 15 lbs. 4 oz.

Letitia 3977.
Lorelle 12913................... 14 lbs. 7 oz.
Lucetta 6856................... 14 lbs. 3 oz.

Letitia 2d' 29021.
Pearl Banks 85758............. 14 lbs. 11 oz.

Letty Coles 23351.
Letty Coles 2d 48128.......... 21 lbs. 8 oz.

Letty Coles 2d 48128.
Letty Rioter 73475........... 24 lbs. 2 oz.
Rioter's Letty 110450......... 21 lbs. 9½ oz.
Eoline of St. Lambert 104819.. 19 lbs. 1 oz.

Letty Rioter 73475.
St. Lambert's Riotress 106220. 24 lbs. 6 oz.
Riotress Signal 95430......... 21 lbs. 5½ oz.

Levon 17319.
Vono 89138..................... 14 lbs. 3 oz.

Liberty 16665.
Liberty 2d 16717.............. 14 lbs. 6½ oz.

Liberty Ann 56029.
Wapello 56044................. 14 lbs. 12 oz.

Lida-Arich, on I. of J.
Litta Oaks 19734.............. 15 lbs. 5½ oz.

Light Cloud 2865.
Dark Cloud 9364.............. 15 lbs. 3½ oz.

Lila Bailey 18601.
Susette C. 18602.............. 14 lbs. 8 oz.

Lilian of Sunswick 36616.
Daisy Delafield 59230......... 20 lbs. 11 oz.

Lilium Excelsum 24945.
Lilium Excelsum 2d 66040..... 14 lbs. 5 oz.

Lilla Merritt 4924.
Elegantie 11943............... 14 lbs. 3 oz.

Lilley Rex 9852.
Albert's Lilley 19489......... 15 lbs. 2 oz.

Lilley Rex Rioter 48464.
Lilley Rex Pogis 62545........ 14 lbs. 9 oz.
Lily Garfield 79819 (21 days).. 37.488 lbs.

Lilley Russ 2d 9514.
Lilley Rex 9852............... 14 lbs. 7 oz.

Lillie Fair 1607.
Pet Anna 1608................ 14 lbs.

Lilly Burnside 4384.
Lilly Cross 13796............. 14 lbs. 3 oz.

Lilly Signalda 23227.
Signal Fancy 30812........... 14 lbs. 8⅝ oz.

Lily, on I. of J.
Young Fanny 9032............ 17 lbs.

Lily (P. S. 166 J. H. B.)
Cetewayo's Lily 18950........ 17 lbs.

Lily 2d (P. S. 147 J. H. B.)
Damask Rose 22065.......... 16 lbs. 3¼ oz.

Lily 2d (P. S. 268 J. H. B.)
Silver Venus 20270........... 16 lbs. 3¾ oz.

Lily Dale 3236.
Couch's Lily 3237............ 16 lbs. 5½ oz.
Couch's Lily 3237 (31 days)... 71 lbs.

Lily Godfrey 3792.
Lily of Burr Oaks 11001....... 15 lbs. 13 oz.

Lily Lenape 2d 8760.
Sweet Sixteen 10682.......... 14 lbs. 15 oz.

Lily Martin 49954.
Lily Martin 2d 89080.......... 14 lbs. 6½ oz.

Lily of Eau Claire 2d 77580.
Letta of Eau Claire 120021.... 16 lbs. 3 oz.

Lily of Elm Spring 51474.
Gay Riotress 105999.......... 15 lbs. 9 oz.
Ruthie Rex Pogis 100326..... 15 lbs. ¾ oz.

Lily of Les Niemes 7465.
Gem of Hope 17102........... 21 lbs.

Lily of Malone 30797.
Daisy of Malone 36505........ 18 lbs. 5 oz.

Lily of Maple Grove 5079.
Lizzie D. 10408................ 16 lbs. 15 oz.
Lida Mullin 9198.............. 16 lbs. 8 oz.

Lily of Oxford 12820.
Lily of Riverside 19599....... 14 lbs. 9 oz.

Lily of Riverside 19599.
Homestead Pogis 44231....... 14 lbs. 11 oz.
Quadruple Pogis 32359........ 14 lbs. 1½ oz.

Lily of St. Lambert 5120.
Sweet Brier of St. Lambert
5481 .... 22 lbs. 10 oz.
Lily Niobe 55765.............. 21 lbs. 6½ oz.
St. Lambert's Violet 25278..... 16 lbs. 12 oz.

Lily of Shady Side 49194.
Scroll 93498.................... 14 lbs. 15 oz.

Lily of the Grove 24555.
Golden Sheen 25561........... 14 lbs. 4½ oz.

Lily Polo 22218.
Massey Polo 67010............. 22 lbs. 6 oz.

Lily's Pogis of Bois d'Arc 67776.
Lily's Crown Princess 76011.. 15 lbs. 7 o:.
Lily's Crown Princess 76011
(30 days).................... 64 lbs. 12 oz.

Linda 2d 1927.
Chamomilla 7552.............. 16 lbs. 10 oz.

Lisette 492.
Lady Fawn of St. Anne's 10920 16 lbs. 12½ oz.

Lisgar's Rose 26607.
Lisgar's Rose 2d 56976........ 16 lbs. 11¾ oz.

Lisgar's Rose 2d 56976.
Ida's Rose of St. Lambert
71555 .... 21 lbs. 1½ oz.
Ida's Rose of St. Lambert
71555 (14 days)............. 40 lbs. 2½ oz.

Littalla 19314.
Même 34921.................... 18 lbs. 4 oz.
Même 34921 (16 days)......... 37 lbs. 5 oz.

Litta Oaks 2d 20020.
Litta Pogis 30275............. 18 lbs. 6 oz.

Litta Oaks' Alphea 26676.
Nannie Pogis 53786........... 16 lbs. 11 oz.
Hermosa Pogis 40369......... 16 lbs. 3½ oz.

Litta Pogis 30275.
Miss Patti Rosa 73607........ 17 lbs. 15 oz.
Undeniah Pogis 85990........ 14 lbs. 2 oz.

Little Accident 15578.
Pogis Oonan 29890........... 17 lbs. 8 oz.
Pogis Oonan 29890 (31 days).. 73 lbs. 5 oz.

Little Bella 3693.
Bellita 4553.................... 17 lbs. 2 oz.

Little Belle, on I. of J.
Jim's Pride 31376............. 18 lbs. 3 oz.

Little Betsona 45344.
Sophona 76110................. 18 lbs. 5¾ oz.
Sophona 76110 (10 days)...... 22 lbs. 4¾ oz.

Little Browny (P. S. 29 J. H. B.)
Nelly 6456..................... 21 lbs.

Little Butterstar 20926.
Sataline 64213................. 17 lbs. 6½ oz.
Satala 50074.................... 16 lbs. 2½ oz.

Little Chloe Green 48767.
Chloe Tormentor of Lawn
71642 .... 14 lbs. 13½ oz.

Little Dorrit C. 73100.
Lady of Union Valley 101733.. 20 lbs. 7½ oz.

Little Duck 3651.
Dudu of Linwood 8336........ 16 lbs. 15 oz.

Little Emily 5356.
Enigma 5360................... 15 lbs. 6 oz.

Little Gay 36475.
Pay Little 86189.............. 15 lbs. 6¾ oz.

Little Germ 19965.
Gold Gem 72024.............. 15 lbs. 12 oz.
Winder Gem 90447............ 14 lbs.

Little L. 2d 54807.
Rosa Russell 81549............ 16 lbs. 1 oz.

Little Nan 15895.
Signal's Lily Flagg 31035...... 27 lbs. 3½ oz.

Little Nellie Rex 50209.
Rioter's Baby 106100......... 19 lbs. 12 oz.

Little Pogis 32612.
Friendship 62460.............. 17 lbs. 3 oz.
Rioter's Violet Pogis 75342... 14 lbs. 15 oz.

Little Torment 15581.
Angelo's Torment 39390....... 22 lbs. 8 oz.
Miss Peacock 30871........... 16 lbs. 10 oz.

Little Witch 35813.
Witching 42252................ 17 lbs. 14½ oz.

Lively (F. S. 1401 J. H. B.)
Oaklands Cora 18853.......... 19 lbs. 9½ oz.

Liz 79114.
Barbara of Melrose 90486...... 14 lbs. 7 oz.

Lizette (P. S. 79 J. H. B.)
La Jolie Lizette 16426......... 16 lbs. 1 oz.

Lizzie Chaffin 8958.
Lottie Rex 18757.............. 14 lbs. 4 oz.

Lizzie Johnson 2002.
Lisetta Johnson 5321......... 15 lbs. 10 oz.

Lizzie Reeves 9222.
Exile's Lizzie 61477........... 14 lbs. 8½ oz.

Lizzie T. 65134.
Canossa 65243................. 15 lbs.

Lizzie Walker 10134.
Mulford's Lizzie Walker 50213 16 lbs. 5 oz.

Lobelia 4379.
Lobelia 2d 6650................ 14 lbs. 6 oz.

Locket 560.
Lustre 2062.................... 15 lbs. 8½ oz.

Lolly of St. Lambert 5480.
Mary Anne of St. Lambert
9770 .... 36 lbs. 12¼ oz.
Naiad of St. Lambert 12965.. 22 lbs. 2¼ oz.
Crocus of St. Lambert 8351... 17 lbs. 12 oz.
Mary Anne of St. Lambert
9770 (one year).............. 867 lbs. 14¾ oz.

Lonette of Magnolia 9425.
Mary of Pleasant View 13448. 14 lbs. 6 oz.

Lora of Clovernook 51599.
Leila of Clovernook 72554..... 14 lbs. 2 oz.

Lord Mar's Lady 48307.
Baby Mar 115089.............. 16 lbs. 6 oz.

Loreda 19801.
Brown Bonnie 102406......... 18 lbs. 8 oz.
Diploma's Loreda 101528...... 18 lbs. 1½ oz.
Comedia 38885................. 14 lbs. 2 oz.

Lorelle 12913.
Southern Princess 36320....... 16 lbs. 4 oz.
Leonie of Linwood 25319...... 14 lbs. 6 oz.

Lorena of Briarcliff 42817.
Celia of Whitehall 68004....... 17 lbs. 15 oz.

Lorette 7393.
Ada S. 18366.................. 16 lbs. 9 oz.

Lorita 33750.
Columbiana 74998............. 18 lbs. 6 oz.
Lorilla 87505.................. 16 lbs. 2 oz.
Patrol 40490.................. 15 lbs. 4 oz.

Lorna 19016.
Lorna 2d 33634................ 17 lbs. 9 oz.

Lorna 2d 33634.
Baron's Lorna 101623......... 21 lbs. 3 oz.
Lorna of Maple Glens 48094... 14 lbs.

Lorna Doone of P. 31335.
Mrs. Jack 50805............... 15 lbs. 5 oz.

Lorna of Maple Glens 48094.
Maple Glens Gold Lorna 84017 15 lbs. 1 oz.

Lorne's Belle Dame 66088.
Pedro's Pretty Belle 77946..... 18 lbs. 8 oz.

Lorne's Gem 66080.
Pedro's Pretty Gem 69953..... 18 lbs. 8 oz.

Lorne's Marjoram 43327.
Lorne's Pretty Marjoram
77942 .... 14 lbs. 14 oz.

Lorne's Pretty Pearl 46683.
Rioter's Pretty Pearl 75273... 15 lbs. 8 oz.

Lorne's Primrose 66082.
Rioter's Lorne Primrose 88583 18 lbs. 15 oz.

Lottie May 17201.
St. Lambert's Crescent 39711.. 22 lbs. 10 oz.
Pogis' May 26950.............. 18 lbs. 2 oz.
Molly May 17202.............. 14 lbs. 15 oz.

Louisa Belle 64013.
Denise Landseer of Lawn
84411 .... 18 lbs. 5 oz.

Louisa Hinman 12802.
Signal Hinman 44615......... 15 lbs. 12¾ oz.

Louise of Lawnfield 14151.
Madame Argyle 19476......... 14 lbs. 6 oz.
Mellie Argyle 20609.......... 14 lbs. 2 oz.

Louvie 3899.
Louvie 3d 6159................ 14 lbs. 13 oz.

Lua 3d 8019.
Luani 37989................... 21 lbs.

Lucette Martin 7770.
Maggie Martin 9562.......... 17 lbs. 6 oz.

Lucile des Moulins 92726.
Sünderin 125092.............. 15 lbs. ½ oz.

Lucile W. 117868.
Gipsy W. 132879.............. 21 lbs. 4 oz.

Lucilla 2735.
Croton Maid 5305............. 21 lbs. 11½ oz.
Lucilla 3d 9786............... 17 lbs. 1 oz.

Lucilla Kent 8892.
John Bull's Princess 49670.... 19 lbs. 14 oz.

Lucky Belle 2214.
Oktibbeha Duchess 4422...... 17 lbs. 4 oz.
Lucky Belle 2d 6037.......... 16 lbs. 14 oz.
Maggie May 3255............. 14 lbs. 2½ oz.

Lucy 4577.
Denise 8251.................... 17 lbs. 8 oz.

Lucy Alba 22929.
Fantine of Guilford 64480..... 19 lbs. 1 oz.

Lucy Dale 5129.
Uinta 5743.................... 14 lbs. 10 oz.

Lucy Gaines 1711.
Lucy Gaines' Buttercup 5058.. 14 lbs.

Lucy Gray 2746.
Willimenia H. 12325.......... 20 lbs. 14 oz.
Lucy's Love 20568........... 14 lbs. 14 oz.

Lucy of St. Lambert 5116.
Cora of St. Lambert 8347...... 21 lbs. 6¾ oz.
Nancy of St. Lambert 12964.. 14 lbs. 5 oz.

Lucy Pope 5328.
Eugénie Chouteau 6186........ 24 lbs. 8 oz.
Lillie Pope 8589.............. 14 lbs. 5 oz.

Luella Berlin 16927.
Meines St. Helier 29999....... 14 lbs. 1 oz.

Lula Gordon 90066.
Lula Gordon 2d 90067......... 15 lbs. 4½ oz.

Lulu's Queen 40055.
Lulu Pedro 71902............. 17 lbs. 4 oz.
Lulu Pedro 71902 (13 days)... 31 lbs. 4 oz.

Luna Dean 36380.
Luna Dean 3d 84309........... 15 lbs.

Lupar 14001.
Abbie Z. 14002................ 14 lbs. 11 oz.
Abbie Z. 14002 (30 days)...... 61 lbs. 2 oz.

Luteefy 46313.
Letitia of St. Lambert 91054.. 16 lbs.

Lutitia (F. S. 2919 J. H. B.)
Cream Lily 24015.............. 16 lbs.

Lydia D. 43735.
Lydia D. 2d 97860............ 14 lbs. 9¾ oz.
Lydia D. 2d 97860 (14 days).. 29 lbs.

Lydia Darrach 4903.
Lydia Darrach 3d 10662....... 16 lbs. 4 oz.
Lydia Darrach 2d 8056........ 16 lbs.
Lydia Darrach 5th 16577..... 15 lbs.

Lydia Deming 4399.
Louisa Deming 23469......... 14 lbs. 15 oz.

Mabel 1092.
Mabel's Jewel 6251........... 17 lbs. 13 oz.

Mabel 4th 18167.
Cicero's Mabel 18238......... 15 lbs. 2 oz.

Mabel A. 100601.
Exile's Echo 111276.......... 15 lbs. 9¾ oz.

Mabel Cowles 3879.
Sue Gallagher 15945.......... 23 lbs. 1½ oz.

Mabel Girl 22040.
L. D.'s Mabel 29368.......... 14 lbs. 6 oz.

Mab of Deerfoot 3589.
Mab of Deerfoot 3d 15345..... 14 lbs. ½ oz.

Madame Pogis 74452.
Madame Melrosa 103185....... 16 lbs. 3 oz.

Madame Thorne 64197.
Bertha Thorne 74824......... 16 lbs. 5 oz.

Madam F. 82335.
Bertha's Brunette 82336....... 18 lbs. 1 oz.

Madam Jenks 34694.
Jodie Jenks 76739............. 18 lbs. 7 oz.
Ruby Jenks 56862............. 18 lbs. 3 oz.
Pearl Jenks 50834............. 14 lbs. 14 oz.

Madam Torpedo 84287.
Lucile of Pearl Hill 91151..... 14 lbs. 4 oz.

Madelon 2d 51040.
Pride of Madelon 105711...... 16 lbs. 6 oz.

Madge H. 14441.
Madge Berg 24165............. 14 lbs.

Madge of Glen Rouge 112455.
Allie of Glen Rouge 129123.... 15 lbs. 14 oz.

Maffie 2d 4939.
Chief's Malita 25186........... 16 lbs. 8¼ oz.

Mag 3351.
Silveretta 6852................. 16 lbs. 9 oz.

Maggie 2d 2055.
Maggie Rex 28623............. 17 lbs. 2½ oz.

Maggie 4th 3224.
Ultima 14456.................. 15 lbs. 12 oz.

Maggie Clarendon 2d 23546.
Kittura of St. Lambert 78344 18 lbs. 8¾ oz.

Maggie Cyrene 53462.
Dosia Pogis 89049.............. 16 lbs. 10¼ oz.

Maggie May 3255.
Maggie C. 12216............... 14 lbs. 6 oz.
Maggie May 2d 12926......... 14 lbs. 6 oz.

Maggie May of Tupelo 71280.
Clothilde of St. Lambert 72271 22 lbs. 7½ oz.

Maggie of Lawnfield 14153.
Countess of Lorne 20622...... 14 lbs. 14 oz.

Maggie of St. Lambert 9776.
Maggie Sheldon 23583......... 21 lbs. 5 oz.

Maggie of Springvale 15931.
Victor Hugo's Maggie 30274... 18 lbs. 14 oz.
Jessie Lee Pogis 44969....... 17 lbs. 4 oz.
Pomona Pogis 40371.......... 17 lbs. 2½ oz.

Magnorant 12509.
Signora Signal 39396......... 15 lbs.

Magyarland 4th (P. S. 3242 J. H. B.)
Magyarland 86326............. 17 lbs. 8 oz.

Mahairve Girl 25652.
Fernside 1½278............... 15 lbs. 3 oz.

Mahogany Bess 68544.
Palestine Bess 102970......... 16 lbs. 12 oz.

Malden of Jersey 2736.
Signalana 7719................ 15 lbs. 4 oz.

Maiden of Jersey 3d 21242.
Signal's Empress 57767........ 16 lbs. 6½ oz.
Young Jersey Maiden 42794... 15 lbs. 10½ oz.
Signal Maiden 42793.......... 15 lbs. 3½ oz.

Maid of Avranches 6959.
Jazel's Maid 11011............. 14 lbs. 6 oz.

Maid of Fancy Creek 11431.
Black Teat 32932.............. 16 lbs. 11 oz.

Maid of Fernwood 10939.
Maid of Fernwood 2d 29010... 17 lbs. 11 oz.

Maid of the Elms 6960.
Maid of the Elms 3d 15554..... 16 lbs. 4 oz.

Maid of the Elms 18932.
Maid of Tanglewood 51606.... 16 lbs.

Maid of Tupelo 41753.
Amanda of St. Lambert 64955 18 lbs. 14 oz.

Maid of Yokun 82415.
Maid of Yokun 2d 86350....... 14 lbs. 4 oz.

Majestic 24757.
Pearl of Oakwood 37722...... 14 lbs. 1 oz.

Malice 18988.
Ethleel Khedive 99698........ 15 lbs. 8 oz.
Fancy Alice 106402........... 14 lbs. 11½ oz.

Mallie 2d 38132.
Daisy Glen 71976............. 20 lbs. 3 oz.

Mallie Miller 29862.
Dame Miller 59260............ 16 lbs. 2½ oz.

Malone 4986.
Lady of Malone 25734........ 16 lbs.
Topsy of Malone 49478........ 14 lbs. 4 oz.
Topsy of Malone 49478 (31 days) .... 55 lbs. 12 oz.

Malope 5872.
Malope 2d 11923.............. 15 lbs. 10 oz.

Mambrino Kate 14494.
Katie Herr 39970............. 15 lbs. 8½ oz.

Mamie 1612.
Rose of Hillside 3866........ 14 lbs. 3½ oz.

Mamie Coburn 3798.
Marea 10167................... 17 lbs. 10 oz.

Mamie H. Pogis 57050.
Ruba H. Pogis 81944.......... 18 lbs. 1½ oz.

Mamie Le Roy 52459.
Le Roy's Pet 94106........... 20 lbs. 7 oz.

Mandana 46198.
Mandana 3d 78778............. 19 lbs. 15¾ oz.

Maple Dale 2967.
Mollie Garfield 12172......... 22 lbs. 12 oz.
Mollie Garfield 12172 (62 days) 163 lbs.

Maquilla 24043.
Maquilla of Hood Farm 113067 14 lbs. 13¾ oz.

Maquita 7589.
Maquilla 24043................ 21 lbs. 1 oz.

Marcella Maid's Pearl 23794.
Minnie Mignon 90972......... 14 lbs. 12 oz.
Sapolio 61082.................. 14 lbs. 6 oz.

Marchande 52258.
Marchande 2d 93823.......... 15 lbs. 9 oz.
Marchande 3d 133818.......... 15 lbs. 6¼ oz.

Marchande 2d 93823.
Marquette Tormentor 131897.. 16 lbs. 14½ oz.

Marchioness of Maplehurst 8018.
Marburi 37991.................. 16 lbs. 12 oz.

Marcia Rex 27224.
Kalma of Milford 52742....... 15 lbs. 12½ oz.

Marée 6592.
Daisy of Sunnyside 16338..... 15 lbs. 1 oz.

Margaret of Lagonda 35127.
Ethlo's Jean 71954............ 24 lbs. 6 oz.
Phœbe Rex 55084............. 16 lbs. ½ oz.

Margeretta 1994.
Pet of Maplewood Farm 4854.. 15 lbs. 2 oz.

Maria of Allentown 2d 33443.
Maria of St. Lambert 64908.. 20 lbs. 3½ oz.

Maria of St. Lambert 64908.
Caro of St. Lambert 90875..... 17 lbs. 6 oz.

Marie Louisa (F. S. 2472 J. H. B.)
Eastern Sultane 20298........ 15 lbs. 3 1-3 oz.

Marie Rex 21423.
Marie's Maud 78467.......... 18 lbs. 2 oz.

Marie Spelterini (F. S. 1412 J. H. B.)
Caroline 12019................ 14 lbs. 8 oz.

Marietta 1813.
Myrrha 11299.................. 16 lbs. 1 oz.
Buttery 3502.................. 14 lbs. 1 oz.

Marigold 3d 3220.
Belle of Vermillion 8798....... 15 lbs. 14 oz.

Marilla of St. Lambert 31809.
May Day 60451................ 16 lbs. 3½ oz.
Primrose of Summerland
66059 .... 15 lbs. 4½ oz.

Marissa 65285.
Diploma's Marissa 95590...... 16 lbs. 5 oz.
Diploma's Eureka 103865...... 15 lbs. 10 oz.

Marjoram 3239.
Leclair's Marjoram 36355...... 15 lbs. 3 oz.
Marjoram 2d 12805........... 15 lbs.

Marjoram 2d 12805.
Marjoram of Linden 43600.... 22 lbs. 12 oz.
Pedro's Fame 33836.......... 19 lbs. 12 oz.
Marjoram's Matilda 77944..... 19 lbs. 6 oz.
Pedro's Fair Marjoram 88489.. 18 lbs. 12 oz.
Pedro's Pretty Marjoram 77947 14 lbs. 10 oz.

Marjoram of Glen Rouge 78420.
Marjoram of Glen Rouge 2d
123634 .... 17 lbs. 3 oz.

Marjoram of Linden 43600.
Pretty Marjoram 77943........ 19 lbs. 6 oz.

Marjoram's Matilda 77944.
Pedro's Hansom Marjoram
96934 .... 18 lbs. 6 oz.

Marmaduke's Bella 42312.
Marion Earle 63908........... 14 lbs. 8½ oz.

Marquise 2d 2868.
Baronetti 8425................ 16 lbs. 14½ oz.

Martha Ethlo 89029.
Josephine O. 103635.......... 19 lbs. 12 oz.

Martha Hall 40786.
Martha Hall 2d 57901......... 14 lbs. 10 oz.

Martha M. 10867.
Alphea Rajah 20605........... 23 lbs. 2 oz.

Martinet 6418.
Mitten 13368.................. 15 lbs. 11 oz.

Martona 58630.
Marna 76108.................. 16 lbs. 12 oz.

Mary Ann 2038.
Lady Alice of Hillcrest 7450.. 18 lbs. 6 oz.

Mary Azuline 6314.
Armon 10862.................. 16 lbs. 13½ oz.

Mary Garnet 10371.
Martha Lafayette 17158........ 17 lbs. 6 oz.

Mary Gold (P. S. 4188 J. H. B.)
Pristine 126329................ 14 lbs. 4½ oz.

Mary Hinman 17619.
Daisy Hinman 61537.......... 24 lbs. 10 os.
Daisy Hinman 61537 (31 days). 102 lbs.
Daisy Hinman 61537 (90 days). 155.131 lbs.

Mary Jane of Belle Vue 6956.
Countess Gisela of Belle Vue
9571 .... 15 lbs. 11 oz.

Mary Lou 35747.
Luta Pogis 69501............. 14 lbs. 1½ oz.

Mary Lowndes 273.
Carrie 3894................... 16 lbs. 8 oz.

Mary M. Allison 6308.
Niobe's Alpheanette 23336..... 22 lbs. 10½ oz.
Royal Daisy 25214............ 20 lbs. 9 oz.
Mary M. Pogis 32455.......... 16 lbs. 4½ oz.

Mary M. Pogis 32455.
    Point Lace 62471.............. 17 lbs. 8 oz.

Mary of Bear Lake 6171.
    Mary's Silver Drop 14235...... 15 lbs. 4½ o:.

Mary's Silver Drop 14235.
    Mary's Pet 40772.............. 15 lbs. 7½ oz.

Mary Wynn 40782.
    Beauty Wynn 45482.......... 14 lbs. 14½ oz.

Massena 25732.
    Massena 2d 31188.............. 17 lbs. 11 oz.
    Massena 4th 47342.............. 14 lbs.

Massey Polo 67010.
    Massey Polo 2d 89216......... 16 lbs. 9 oz.
    Massey Polo 3d 99378......... 14 lbs. 13½ oz.

Massie's Lula 69727.
    Harry's May Bud 111817...... 14 lbs. 15½ oz.

Matchless 723.
    Dimple 3248................... 16 lbs. 11 oz.

Matchless 1277.
    Vaniah 6597................... 15 lbs. 9½ oz.

Matilda 2405.
    Maud Lee 2416................. 23 lbs.

Matilda 3238.
    Matilda 4th 12816.............. 21 lbs. 8½ oz.
    Matilda 6th 39290............. 20 lbs. 1 oz.
    Matilda 5th 18068............. 16 lbs. 4 oz.
    Matilda 4th 12816 (125 days).. 279 lbs. 7½ oz.

Matilda 2d 5471.
    Minnie of Oxford 12806........ 17 lbs.

Matilda 3d 12808.
    Mintha 12812.................. 15 lbs.

Matilda 4th 12816.
    Matilda of Meridale 57176..... 23 lbs. 5¾ oz.
    The Queen's Gift 75186........ 22 lbs. 9 oz.
    Matilda's Queen of St. L.
    93746 .... 18 lbs. 13¼ oz.

Matilda 5th 18068.
    Matilda's Matilda 75348....... 17 lbs. 2½ oz.
    St. Helier's Matilda 52857..... 14 lbs. 1 oz.

Matilda Hilda 75346.
    Little Ida Hilda 97739......... 15 lbs. 8 oz.

Matilda of Maplewood 49723.
    Cretesia's Matilda 90000....... 15 lbs. 15 oz.

Matilda's Lady Day 81310.
    Patty Iris 101894.............. 14 lbs. 15 oz.

Matilda's Rose 81397.
    Dark Rose 101892............. 15 lbs. 14½ oz.

Matin 7768 (F. S. 1629 J. H. B.)
    Lady Jane of St. Peter's 7475. 15 lbs.

Matin 5th 3399.
    Elmwood Duchess 8716........ 14 lbs.

Matina of Riverside 51783.
    Blanche Ferry 108501.......... 16 lbs. 5 oz.

Matinella 32041.
    Garelia 62541.................. 23 lbs.

Matlie 3213.
    Matlie 3d 9879................ 14 lbs. 13 oz.

Mattie Cole 47911.
    Mattie Tormentor 62841....... 16 lbs. 5½ oz.
    Mattie Tormentor 62841 (14
    days) .... 31 lbs. 13½ oz.

Mattie E. 16274.
    Conger's M. 32751............. 16 lbs. 2½ oz.

Mattie E. Royal 78728.
    Signal Mattie 113952.......... 15 lbs. 7½ oz.

Mattie Hunter 19798.
    Narda 38012................... 15 lbs. 1 oz.

Mattie P. 57042.
    Lou Tormentor of Lawn 84413 15 lbs. 3 oz.

Maud Alice 17001.
    Maud of Maple View 39511..... 23 lbs. 7½ oz.

Maude Vivienne 2d 30755.
    Vivienne Pogis 41547......... 15 lbs. 1 oz.

Maudine 3d 5646.
    Maudine of Elmwood 8718..... 16 lbs. 15 oz.

Maud Lee 2416.
    Maud Lee 2d 8839............. 14 lbs. 9 oz.

Maud Miami 57918.
    Maud Miami Jr. 69131........ 14 lbs. 8 oz.

Maud of Bois d'Arc 59639.
    Maud's Crown Princess 90320. 17 lbs.
    Maud's Crown Princess 90320
    (30 days)................... 66 lbs. 13 oz.

Maud of Ingleside 74044.
    Molly May Pogis 102823....... 15 lbs. 12 oz.

Maud of Jersey Lawn 78936.
Bonnie Moll 95616.............. 16 lbs. 3 oz.

Maud of Maple View 39511.
Kaembe's Maud Eurota 91066. 14 lbs. 2½ oz.

Maud of Millbrook 2070.
Maud Melinda 12126......... 17 lbs. 8 oz.

Maud of St. Lambert 9772.
Rioter's Queen 14895......... 17 lbs. 8 oz.

Maud Signal 42187.
Maud Signal 2d 68664........ 22 lbs. 4 oz.

Maud Westfield 17555.
May Pedro 50876.............. 15 lbs. 4 oz.

Maumee Beauty 50836.
Princess of Lynwood 113458... 16 lbs. ½ oz.
Maumee Girl 118155........... 15 lbs. 10⅞ oz.

May 1390.
Austie May 7185.............. 14 lbs. 3 oz.

Maybee H. 60620.
Fancy's Maybee 104383........ 15 lbs. 14 oz.

May Bijou 63911.
Victor's May 97958............ 14 lbs. 8 oz.

May Bud of St. Lambert 5105.
Pogis' Ida May 50063......... 14 lbs. 11 oz.

May Champion 14104.
Marie C. Magnet 22903........ 15 lbs. 8 oz.

May Clover 12273.
Pet Clover 14624.............. 16 lbs. 8 oz.

May Colt 23491.
Queen of Cowes 46309........ 18 lbs. 3 oz.
Greatæna 85823............... 16 lbs. 10½ oz.

May Day of St. Lambert 5109.
May Day Stoke Pogis 28353... 19 lbs. ½ oz.
Minette of St. Lambert 9774... 17 lbs. 4 oz.

May Day Stoke Pogis 28353.
Miss May of St. Lambert 37084 15 lbs. 14 oz.
Pomona's May Day Pogis
56249 .... 15 lbs. ½ oz.

May Dee 18058.
May Dee Pogis 36993.......... 20 lbs. 5½ oz.

May Fair 5184.
Harry's May Fair 37499...... 16 lbs. 1 oz.

May Flower, on I. of J.
Maritana 12039................ 16 lbs. 3½ oz.

Mayhegtal 64454.
Mayhegtal 2d 101298.......... 17 lbs. 14 oz.

Mayhew Belle 3883.
Coronilla 8367................. 14 lbs. 7 oz.

May Indian 16075.
May Violet 30250............. 18 lbs. 11½ oz.
Clifty Beauty 30249.......... 15 lbs. 6½ oz.

May Katydid 63503.
Katie's Kind 97434........... 15 lbs. 5¼ oz.
May Cuckoo M. 97417........ 14 lbs. 8 oz.

May Koffee 45862.
May's Grand Duchess 62607... 15 lbs. 3½ oz.

May Leaf 10667.
May Indian 16075............ 14 lbs. 5 oz.

May of Ashland 73925.
Exile's Nancy 92030.......... 14 lbs. 7 oz.

May of Four Pines 52020.
Toma May 74836............. 14 lbs. 13 oz.

May of Green Lawn 81426.
Fruitland May 93002.......... 14 lbs. 5 oz.

May of Lakeside 10826.
May Dee 18058................ 15 lbs. 10 oz.
V. K.'s May 41929........... 14 lbs. 8 oz.

May Pedro 50876.
Allie Victor Pogis 126907..... 22 lbs. 8 oz.
Mooly Pedro 77811............ 17 lbs. 2 oz.

Mazie Bate 3364.
Bettie's Pet 22550............ 16 lbs. 1 oz.

Meines 3559.
Meines 3d 7741................ 20 lbs. 1 oz.

Meines 3d 7741.
Elena of Oakdale 84162....... 19 lbs. 4 oz.

Melia Ann 5444.
Melia Ann 3d 68070.......... 28 lbs. 8 oz.
Daisy Morrison 14035........ 25 lbs. 12½ oz.
L. D.'s Melia 29087.......... 20 lbs. 8 oz.
Doña Marina 7049............ 18 lbs. 8 oz.
Melia Ann 2d 47464.......... 18 lbs. 6 oz.

Melia Ann 2d 47464.
Melia Ann 2d's Pogis 68757.... 17 lbs. 1 oz.

Melrose Milly 61917.
Southern Bell of C. H. 93565.. 14 lbs. 7½ oz.

Melrose Princess 40343.
Melrose Milly 61917.......... 16 lbs. 2½ oz.
Melrose Princess 2d 62003..... 14 lbs. 11 oz.

Même 34931.
Même 2d 49946................. 18 lbs. 12⅞ oz.
Fancy Même 79748........... 16 lbs. 12 oz.
Même 2d 49946 (16 days)...... 39 lbs. 10¼ oz.

Mendota 26324.
Mendota 3d 26326.............. 15 lbs. 6 oz.

Méprise 4898.
Little Torment 15581.......... 23 lbs. 2½ oz
Little Torment 15581 (13 weeks) .... 227 lbs. 1 1-32 oz.

Mercedes H. 12326.
Gilda Mercedes H. 14451...... 14 lbs. 8 oz.

Mercurenalia 14874.
Mahoning Bess 35834........... 15 lbs. 8¾ oz.

Mercurina 64920.
Caressa 88474.................. 17 lbs. 6 oz.
Treasure 79200................. 17 lbs. 5 oz.
Damara 65001.................. 16 lbs. 7 oz.

Mercy Blossom 26453.
Virgie Landseer 74831........ 14 lbs. 15 oz.

Meridale Rioter Pink 64534.
Pink of Meridale 114748....... 16 lbs. ½ oz.

Merlin's Violet 16430.
Rioter's Violet 33774........... 14 lbs. 3 oz.

Merry 4814.
Tobira 8400.................... 15 lbs. 13 oz.

Merry Maiden 64949.
Masher 64950.................. 16 lbs. 14½ oz

Metah 1295.
Metah's Queen 4886........... 17 lbs. 9 oz.
Bryant 4193.................... 14 lbs. 8 oz.
Metah's Baby 9710........... 14 lbs. 5 oz.

Metah's Queen 4886.
Queryan 42140................. 16 lbs. 2 oz.
Fannie Bugler 19962........... 15 lbs. 2 oz.

Metela 2d 3242.
Utilla 11215................... 16 lbs. 2 oz.

Metella 3905.
Coma 29330................... 15 lbs. 2½ oz.

Mette D. 47433.
Annabel Lee 55833............ 22 lbs. 3 oz.

Mhoon Romp 4528.
Lena Hogan 41506............. 16 lbs. 15 oz.

Miacou 48697.
Vemicou 79322................ 14 lbs. 7 oz.

Miami Grace 39392.
Grace's Pride 61018............ 15 lbs. 7 oz.
Grace Torment 87957......... 14 lbs. 10½ oz.

Miamis Queen 23356.
Miamis Signaldini 43391...... 17 lbs. 14½ oz.

Mignonne 693.
Melody 2689................... 14 lbs. 1 oz.

Mignonne (F. S. 955 J. H. B.)
Beauty of Jersey 7850........ 19 lbs. 2 oz.

Mildred G. 45077.
Tamasese Naomi 128774...... 14 lbs. 15 oz.

Mildrida 6743.
Plumida 23621................. 15 lbs. ½ oz.

Milicent H. 14445.
Alaric's Milicent 35363....... 18 lbs. 5½ oz.

Milkgood 27828.
Francesville 53367............ 16 lbs. 6 oz.

Milkmaid 3d (P. S. 4 J. H. B.)
Valentine of Trinity 7460...... 19 lbs. 4½ oz.

Milkmaid of Lake Forest 5010.
Milkmaid Felch 12339........ 16 lbs. 7½ oz.
Milkmaid of Burr Oaks 9035.. 14 lbs. 5 oz.

Millicent 2d 7229.
Lizzette's Mary 12723......... 14 lbs. 11 oz.

Millie 690.
Sylvia 687.................... 15 lbs. 8 oz.

Millie Waite 10646.
Waite Girl 12776.............. 16 lbs. 7 oz.

Milwaukee 2920.
Goldthread 4945.............. 17 lbs. 9 oz.

Mina B. 28947.
Susan Bell 112146............. 14 lbs. 13 oz.

Minerva of Canterbury 12130.
Lilium Excelsum 24945........ 17 lbs. ½ oz.

Minette of St. Lambert 9774.
Princess Minette 24042........ 18 lbs. 6½ oz.
Oaklands Lilly 14881........... 15 lbs. 4 oz.

Minette Pogis 25210.
Pedro's Minette 66092........ 19 lbs. 8 oz.

Mink 2548.
Mink 2d 3890................... 19 lbs. 11 oz.
Mink 3d 4868................... 14 lbs. 9 oz.

Mink 2d 3890.
Mboon Lady 6560............. 17 lbs. 3 oz.
Julia Evelyn 6007............. 15 lbs. 15½ oz.

Minka 951.
Allie Minka 2982............... 14 lbs. 6½ oz.

Minkabel 27732.
Miss Katie W. 47366......... 15 lbs. 11¾ oz.

Minneola's Beauty 33038.
Lily of Elm Spring 51474..... 19 lbs. 8 oz.

Minnie 2d 17828.
Minnie of Scituate 17829...... 14 lbs. 4½ oz.

Minnie B. Rex 9256.
Lady of Bay View 66029....... 16 lbs. 12 oz.

Minnie C. 29003.
Minnie C. 2d 48414............. 16 lbs. 14 oz.

Minnie Hauk 15630.
Minnie Hauk 2d 99103........ 16 lbs. 1 oz.

Minnie Kersey 19895.
Selita J. 32184................ 25 lbs. 2½ oz.

Minnie Lee 6009.
Minnie Lee 2d 12941.......... 14 lbs. 3 oz.

Minnie Montague 45474.
Minnie's Pet Jersey 93119..... 15 lbs. 6 oz.

Minnie of Orange 6016.
Minnie's Ouida 33117......... 14 lbs.

Minnie of Oxford 2d 12818.
Minnie Rioter 21336........... 15 lbs. 10 oz.

Minnie Rioter 21336.
Duke's Minnie 42785.......... 20 lbs. 4 oz.
Minnie R. of St. Lambert
  61965 .... 16 lbs. 9 oz.

Minnie Sayre 16006.
Rosa of Frisby Hill 36080.... 16 lbs. 3¼ oz.

Minnie Stevens 13059.
Duchess of Darlington 13830.. 14 lbs. 11 oz.

Mintha 12812.
Tormintha 37327.............. 16 lbs. 10¼ oz.

Mint Julep 64571.
Alexa 64924................... 14 lbs. 15 oz.

Minto Fay 27705.
Zenobia Stoke Pogis 37292.... 15 lbs. 8 oz.
Daisy-Seeker 43099........... 14 lbs. 1 oz.

Mintresse 20642.
Mintresse of Ingleside 103128.. 16 lbs. 10½ oz.

Miralba 65002.
Connoisseur 65644............. 20 lbs. 13 oz.

Miriam of Bolton 50967.
Chloe of Bolton 81340......... 16 lbs.

Mirodelle 76862.
Peggie Tormentor 98103....... 16 lbs. 2 oz.

Mirth 92.
Mirtha 3437................... 17 lbs. 13½ oz.

Mirth of Bingo Farm 30999.
Romping Mirth 64070......... 17 lbs. ½ oz.

Miss Alexander 26041.
Miss Alexander 2d 26054...... 15 lbs. 10 cz.

Miss Allena 42174.
Philena S. 70375.............. 18 lbs. 9½ oz.

Miss Beauty 4053.
Mrs. Knickerbocker 19367..... 16 lbs. 8½ oz.

Miss Belinda 16638.
Signalinda 27002.............. 18 lbs. 7 oz.
Adelaide Saufley 96284....... 16 lbs. 5 oz.

Miss Belle Pogis 47038.
Yellow Belle 69780............ 14 lbs. 9½ oz.

Miss Bentley 51492.
Exile's Dewdrop 106104....... 18 lbs. 4 oz.

Miss Beulah 24358.
Beulah De Gruchy 13480..... 22 lbs. 2 oz.

Miss Bianca 12517.
Comanca 19389................ 16 lbs. 3 oz.

Miss Blossom 1986.
Miss Belle 5083............... 22 lbs. 9 oz.
Miss Beauty 4053............. 16 lbs. 12 oz.

Miss Bos 10842.
Lilly Mac 22976............... 14 lbs. 2 oz.

Miss Clifford 27962.
Miss Clifford 2d 56149........ 16 lbs. 4½ oz.
Miss Clifford 3d 63360........ 15 lbs. 2½ oz.
Miss Clifford 5th 63364....... 14 lbs. 7 oz.

Miss De Carteret 7285.
Miss Browney 7288........... 16 lbs. 13 oz.

Miss Fair 4135.
May Fair 5184................. 16 lbs. 7 oz.

Miss Flite of Bingo 31000.
Marno B. 56368................ 16 lbs. 10 oz.

Miss Garda 40520.
Nestor's Beauty of St. L. 74884 14 lbs.

Miss Limited 37620.
Limited Pedro 56382.......... 21 lbs. 7 oz.

Miss Lofty 9718.
Elsie Lofty 61451.............. 18 lbs. 3 oz.
Dame Lofty 47567............. 15 lbs. 10 oz.
Esther Lofty 68754............ 14 lbs. 6 oz.

Miss Loyal 40771.
Miss Montague 57927.......... 14 lbs.

Miss Madora 2d 39243.
Madora Gold 67302........... 17 lbs. 1 oz.

Miss Madrid 49419.
Paradox 65003.................. 16 lbs. 9 oz.

Miss Maude St. Lambert 54989.
Exile's Antoinette 111279..... 14 lbs. 7½ oz.

Miss Millie 12264.
Miss Bianca 12517.............. 17 lbs. 14 oz.

Miss M'liss 77747.
Lottie Signal 90039............ 14 lbs. 3½ oz.

Miss Mouse 92512.
Die Königin 124952............ 14 lbs. 12½ oz.

Miss Nellie Parker 24988.
Nellie Hugo 53793............. 20 lbs.
Gipsy Pogis 35422............. 18 lbs. 7 oz.

Miss Palliser 64073.
Kitty Victor Pogis 85904...... 19 lbs. 12 oz.

Miss Peggotty 14593.
Belle of Rydal Grange 41787... 16 lbs. 6 oz.

Miss Remarkable 33446.
Théroigne 92259............... 18 lbs. 11 oz.
Fairy Oxford 66679............ 17 lbs. 7½ oz.

Miss Rolle 21579.
Ida's Fawn 56018.............. 16 lbs. 1½ oz.

Miss Rollo 2d 35587.
Roulette 50080................. 14 lbs. 5½ oz.

Miss Seelock 6614.
Sister Rex 13194................ 16 lbs. 8 oz.

Miss Semple P. 56948.
Angelina D. 65514............. 15 lbs. 3½ oz.

Miss Serenade 40301.
Musicale of Greenwood 62932. 15 lbs. 14 oz.

Miss Sharpless 24352.
Alarm of Highland 41864...... 16 lbs.

Miss Susy McMillan 50386.
Vera McMillan 66520.......... 17 lbs. ½ oz.

Miss Thérèse 7327.
Therese G. 16390.............. 17 lbs. 2 oz.

Miss Uarda 40473.
Una of St. Lambert 80117.... 16 lbs. 15½ oz.

Miss Vesper 4460.
Blossie Reynolds 6082........ 16 lbs. 3½ oz.

Miss Whisk 38640.
Whisk of Meridale 131023..... 15 lbs.

Mist 26087.
Vemist 79323.................. 14 lbs. 2 oz.

Mistle 26125.
Oonamistle 64469............. 15 lbs. 4 oz.

Mittretta 3d 42242.
Mittretta's Rose 56192........ 15 lbs. ½ oz.

Mobile Beauty 47365.
Lady Louise D. 58056........ 22 lbs. 12 oz.

Mocha 2d 4881.
Seneca's Mocha 23355........ 20 lbs. 3 oz.

Mocha's Pet 12985.
Mocha's Pet's Rheta 37529.... 14 lbs. 2½ oz.

Mocha's Pet's Rheta 37529.
Mirnie Potter 72670........... 14 lbs. 10 oz.

Modita 16626.
Costa Rica 64570.............. 21 lbs. 6½ oz.
Mercurina 64920............... 17 lbs. 10½ oz.
Daretta 62579.................. 16 lbs. 8 oz.
Chromo's Modita 121439....... 16 lbs. 7 oz.
Personia 90231................. 16 lbs. 3¼ oz.
Chestnuts 63449............... 15 lbs. 7¼ oz.
Costa Rica 64570 (30 days).... 87 lbs. 14¼ oz.
Costa Rica 64570 (31 days).... 90 lbs. 11¼ oz.

Moggy 59626.
Mosa 69691.................... 14 lbs. 6 oz.

Molina of St. Lambert 69454.
Molina of Glen Rouge 123635.. 16 lbs. 2½ oz.
Madge of Glen Rouge 112455.. 15 lbs. 2 oz.

Mollie Brown 7831.
Violet of Glencairn 10221...... 14 lbs. 4 oz.

Mollie Dickey 10349.
Lady Thurlow 12410........... 17 lbs. 10 oz.

Mollie Dickey 2d 18704.
Lady White 29213............. 14 lbs. 14¾ oz.

Mollie Garfield 12172.
Mollie Garfield 2d 18662........ 15 lbs. 14 oz.

Mollie Garfield 2d 18662.
Mollie of St. Lambert 34644... 15 lbs. 4 oz.

Mollie Mack 30750.
Bessie Wolcott 91417......... 15 lbs. 1 oz.

Mollie Mack 2d 46966.
Ilinka 91418.................... 17 lbs. 8 oz.

Mollie Madden 25798.
Exile's Pauline 37530......... 16 lbs. 15 oz.

Mollie Pennebaker 28031.
Hinds' Tormentor of Lawn
71640 .... 14 lbs. 7 oz.

Mollie St. Helier 35717.
Mollie St. Helier 2d 98389..... 14 lbs. 12 oz.

Molly 3554.
Moss Rose of Willow Farm
5194 .... 23 lbs. 1 oz.

Molly May 17202.
May of Ingleside 56831........ 21 lbs. 7 oz.
Maud of Ingleside 74044....... 18 lbs. 5½ oz.

Molly of Eau Claire 77583.
Marion of Eau Claire 120457.. 21 lbs. 14 oz.

Mona 1461.
Mona 3d 23134................. 14 lbs. 6 oz.

Mona of Woodstock 6914.
Mona's Mab 31524............. 15 lbs. 7 oz.

Money Musk 13714.
Musquita 28462................ 17 lbs. 3 oz.

Monmouth Duchess 3895.
Monmouth Duchess 4th 7129.. 18 lbs.
Monmouth Duchess 3d 4620.. 14 lbs. 4 oz.

Monmouth Duchess 2d 4619.
Dot of Bear Lake 6170........ 19 lbs. 4 oz.

Monmouth Duchess 4th 7129.
Lady Monmouth 15173....... 15 lbs. 3 oz.
Lena Lowndes 23202.......... 14 lbs. 9 oz.

Mooly Pedro 77811.
Dora Victor Pogis 100921...... 18 lbs. 5 oz.

Moonah 2d 7483.
Moonah's Pet 7484............. 15 lbs. 6 oz.

Moorey of Lawn 68847.
Lelia Koffee of Lawn 120982.. 17 lbs. 3½ oz.

Moquette 6673.
Deletta 21305................... 14 lbs. 15½ oz.

Moragina 26344.
Lorita 33750.................... 15 lbs. 6½ oz.
Lorita 33750 (90 days)........ 146.619 lbs.

Morning Glory 2d 1299.
Fragrance 4059................ 15 lbs. 3 oz.
Bright Lady 5938............. 14 lbs. 12 oz.

Moss Rose of Willow Farm 5194.
Pavon 12485................... 14 lbs. 8 oz.

Mostar 6971.
Princess Mostar 9700.......... 17 lbs. 3 oz.
El Mora Mostar 15955........ 15 lbs. 4 oz.
Lillian Mostar 10364.......... 14 lbs. 3 oz.

Mother Carey of P. 14092.
St. Helier Gem 35438......... 15 lbs. ½ oz.

Moth of St. Lambert 9775.
Rioter's Ruth 14882.......... 19 lbs. 6½ oz.

Motto 80.
Memento 1913.................. 14 lbs. 5 oz.

Mount Lebanon 4457.
Belle Dawson 8270............ 18 lbs. 3 oz.

Mousy (F. S. 1931 J. H. B.)
Mousy 2d 14962................ 17 lbs. 1 oz.

Moxie G. 63504.
Moxie's Mode 97432.......... 16 lbs. 3¼ oz.

Moxie's Mode 97432.
Moxie's Mode 2d 133253....... 15 lbs. 9¼ oz.

Moyane 21595.
Violet Denison 49997.......... 14 lbs. 13½ oz.

Moyane's Fancy 78452.
Fancy's Moyane 122270........ 15 lbs. ¼ oz.
Moyane C. 86492 (14 days)..... 37 lbs. 15¼ oz.

Mrs. Bannister 23803.
Onnolee 23804.................. 16 lbs. 4 oz.

Mrs. Knickerbocker 19367.
Miss Bingo 80096............... 24 lbs. 12 oz.
Lady Bingo 24160............. 15 lbs. 4 oz.

Mrs. Porter 16667.
Miss Porter 20300............. 16 lbs. 6 oz.

Muriel 3904.
Muriel 5th 19017............... 16 lbs. 12½ oz.

Muriel 4th 17146.
Rita of Andalusia 29414...... 24 lbs. 1 oz.

Musidove 25379.
Musidove 3d 61641............. 15 lbs. 1 os.

My Alpha 16892.
Ida's Alpha 64536............. 25 lbs. 15¾ oz.

My Alpha's Victory 43304.
My Alpha's Beauty 56469...... 15 lbs.

Mylra 18840.
Compact 52705.................. 16 lbs. 3 oz.

My Queen's Lois 3d 81474.
Lois A. 107594................. 15 lbs. 14 oz.
Lois B. 107595.................. 14 lbs. 2 oz.

Myra C. 27837.
Myra Tormentor 84227......... 15 lbs. 11½ oz.
My May Pogis 84236........... 14 lbs. 5 oz.

Myra Cyrene 46590.
Maggie Cyrene 53462........... 15 lbs. 15 oz.

Myrrha 11299.
Southern Daisy 38292......... 14 lbs. 11 oz.
Myrrha 3d 50406............... 14 lbs. ½ oz.

Myrrha Landseer 65456.
Ida's Myrrha 97939............. 14 lbs. 11 oz.

Myrtle 208.
Myrtle 2d 211................... 15 lbs. 12 oz.

Myrtle 2d 211.
Pride of Mashamoquet Farm
6469 .... 16 lbs. 1¾ oz.
Myrtle of Ridgewood 7858..... 14 lbs. 1 oz.

Myrtle Melrose 114464.
Son's Princess 118386.......... 16 lbs. 15 oz.

Mysinda 27133.
Mysinda 2d 42250............. 18 lbs. 8 oz.

Mysinda 2d 42250.
Ida's Mysinda 64868........... 16 lbs. 11 oz.
Matilda Mysinda 84487........ 15 lbs. 3 oz.

Nal Day 26553.
Beauty of Oakland 45009...... 17 lbs. 1 oz.

Nanalda 32917.
Nanalda 2d 97130............. 16 lbs. 1 oz.
Ida's Nan 49957............... 15 lbs. 6 oz.

Nancy (F. S. 2235 J. H. B.)
Lottie Dorey 21578............. 16 lbs. 12½ oz.

Nancy of St. Lambert 12964.
St. Lambert's Nancy 56503.... 18 lbs. 7 oz.

Nancy's Grace 41421.
Oonan's Grace 86489.......... 24 lbs. 3 oz.
Ida's Grace 56019............. 14 lbs. ½ oz.
Miss Gracie 86495 (14 days).... 37 lbs. 12 oz.

Nancy's Star 2d 70240.
Fill Pail's Nancy 95435....... 18 lbs. 3 oz.

Nan Day 17192.
Nal Day 26553.................. 19 lbs. 13 oz.

Nanette Scituate 67183.
Nanette of Hood Farm 104447. 15 lbs. 8½ oz.

Nannette of Allerton 8515.
La Pucelle 16829............... 15 lbs. 8 oz.

Nannie Harper 7248.
Eve of Bolton 63665........... 14 lbs. 4½ oz.

Nannie Lee Morgan 56863.
Nannie of Hood Farm 108565.. 14 lbs. 5¼ oz.

Naomi B. 8066.
May Naomi 92277............. 14 lbs. 10½ oz.

Naomi Ronsby 16742.
Naomi's Pride 16745.......... 15 lbs. 2 oz.

Narda 38012.
Heritage 50077................. 15 lbs. 15 oz.

Natali 1433.
Italian 4096................... 17 lbs. 13 oz.

Nattie 26842.
Etta of Bay View 66028....... 22 lbs. 13 oz.

Nautila 14738.
Miss Nautila 49000............ 15 lbs. 5 oz.

Nazell 80709.
Nazell 2d 98271............... 16 lbs. 8½ oz.

Neata 4748.
Zenitza 19190.................. 17 lbs. 5½ oz.

Neatham's Rosa Signal 82309.
Signal of Raceland 84299...... 18 lbs. 6 oz.

Neatness 1894.
Deoine 6343.................... 14 lbs. 3 oz.

Nehushta 4th 38379.
Gem Nehushta 73258........... 15 lbs. 6 oz.

Nelida 5346.
Nelida 2d 8227................. 15 lbs. 2½ oz.

Nellie 1507.
Beulah of Baltimore 3270..... 14 lbs. 6½ oz.

Nellie, on I. of J.
Nightingale of Elmarch 8312.. 14 lbs. 2 oz.

Nellie 2d 2369.
Darling of Neatham 20086..... 15 lbs. 3 oz.

Nellie 3d 1928.
Flora Lee of Tennessee 7694.. 16 lbs. 5 oz.

Nellie B. Parks 29264.
Lena Stoke Pogis 39756....... 18 lbs. 8 oz.

Nellie E. 66426.
Edna Hayes 90901............. 15 lbs. 6 oz.

Nellie Harrison 2d 23093.
Prince's Nellie 23719........... 14 lbs. 6 oz.

Nellie of Cedar Glen 10949.
Nelope Pedro 50879........... 15 lbs. 4 oz.

Nellie of St. Lambert 69458.
Cassie of Glen Rouge 112456... 15 lbs. 9 oz.

Nellie Prince 58053.
Princess Sosia 112753.......... 15 lbs. 8 oz.

Nellie Stanton 21069.
Baby Stanton 72397............ 15 lbs. 4 oz.

Nellie Sullivan 11159.
Good Friday 20081............. 14 lbs. 12 oz.

Nell Pogis 31991.
Sister of Charity 62453........ 24 lbs. 14½ oz.
Jap Pogis 69122............... 14 lbs. 7 oz.

Nelly 6456.
Daisy of St. Peter's 18175..... 20 lbs. 5½ oz.

Nelly (F. S. 1509 J. H. B.)
Mary Jane of Belle Vue 6956. 17 lbs. 7 oz.

Nelly A. 1002.
Minna Rajah 14847............ 16 lbs. 8 oz.

Nelly Pierpont 12531.
Golden Plover 22388........... 20 lbs. 1¾ oz.

Nelope Pedro 50879.
Lillie Victor Pogis 85471...... 14 lbs. 4 oz.

Neonice 65004.
Julippa 69000................. 15 lbs. 9 oz.

Nerina 9183.
Queen of Chenango 17771...... 14 lbs. 6 oz.

Nerissa (F. S. 2641 J. H. B.)
Grey Friar's Bess 25128....... 22 lbs.

Nerobella 2d 29654.
Nerobella's Blossom 48051.... 15 lbs. 1 oz.

Nervine 25932.
Nerviona 43678................ 14 lbs. 15 oz.

Nerviona 43678.
Ethleel's Nerviona 101300..... 17 lbs. 12 oz.
Romp's Nervilette 72282...... 17 lbs. 10 oz.
Maybegtal 64454............... 16 lbs. 3 oz.
Nerviona 2d 98687............ 14 lbs. 3 oz.

Nettie Artiste 31653.
Nettie's Picture 50960......... 15 lbs. 8 oz.

Nettie C. Magnet 35411.
Josephina 64921............... 14 lbs. 1 oz.

Nettie's Accident 61236.
Nettie's New Year 99295..... 23 lbs. 3 oz.

Nettie Signal 3d 67493.
Ben's Nettie 118137............ 15 lbs. 3 oz.

Nettle 3241.
Euphorbia 11229............... 23 lbs. 6¼ oz.

Newton Belle 355.
Lassie 1134.................... 15 lbs. 1½ oz.

Niantic's Palestine 14519.
Maggie Rule 31940............ 21 lbs. 5 oz.
Palestine of Oxford 42194..... 21 lbs.

Nibbie 6796.
Nibbette 11625................. 14 lbs. 7 oz.

Nice 68027.
Chromia E. 97511............. 17 lbs. 12 oz.

Nickel 2d 23352.
Daisy Stillson 28174........... 15 lbs. 3 oz.

Nicomis 68026.
Nice 68027..................... 14 lbs. 12 oz.

Nienchen 48698.
Fancy's Nienchen 104377..... 16 lbs. 2 oz.

Nightingale K. 16056.
Nightingale K. 2d 19841....... 16 lbs. 8 oz.
Grace's Nightingale 19855..... 14 lbs. 2 oz.

Nilsson's Favorite 17680.
Tormentor's Favorite 36873... 19 lbs. 2 oz.

Nimmie 968.
Nibble 6796.................... 14 lbs. 15 oz.

Nina 2d 2557.
Pedronina 34803............... 21 lbs. 3 oz.

Nina of St. Lambert 12963.
Jessie Lorne 25213........... 16 lbs. 15 oz.

Niobe Gordon 13922.
Niobe Gordon 3d 44041....... 14 lbs. 3 oz.

Niobe Hampton 5343.
Blossom's Niobe 13855........ 16 lbs. 7½ oz.

Niobe of Beechwood 11131.
Niobe of Linwood 11134....... 14 lbs. 12 oz.

Niobe's Alpheanette 23336.
Niobe's Alpheanette 3d 39459.. 19 lbs. 4½ oz.

Niobe's Alpheanette 3d 39459.
Niobe Marigold 108363........ 15 lbs. 8½ oz.
Alphea Marigold 117632....... 14 lbs. 5 oz.

Nitella 4423.
Nigella 7895................... 16 lbs. 3 oz.

Niva 7523.
Income 19472.................. 16 lbs. 9 oz.

Nonpareil (F. S. 1248 J. H. B.)
Nancy Lee 7618............... 26 lbs. 8½ oz.

Noonday 15127.
Sunset 15130.................. 16 lbs. 2½ oz.

Nora Loadstone 2d 84482.
Nora Clinton 91938........... 18 lbs. 4 oz.

Nora of Glen Rouge 69452.
Mollie of Glen Rouge 129125.. 15 lbs. 12 oz.

Nora of St. Lambert 12962.
Rioter's Nora 21778........... 15 lbs. 9 oz.

Nora Scituate 67571.
Bertha Scituate 75977........ 14 lbs. 7 oz.

Norella King 2d 23933.
Harry's Fancy 39114.......... 14 lbs. 6½ oz.

Normanda 3914.
Georgiadear 11508.............. 18 lbs. 3 oz.
Kate Gordon 8387............. 15 lbs. 15 oz.

Normandy Girl 7737.
Lady McDowell 37414........ 15 lbs. 12 oz.

Nuella 53747.
Ella of Endegeest 87376....... 19 lbs. 12½ oz.

Nuphar 3d 20455.
Nuphar Houghton 36364....... 15 lbs. 3 oz.

Nutley Alma 13581.
Nutley Dolores 13797........ 23 lbs. 2 oz.
Nutley Darling 22412.......... 15 lbs. 3½ oz.

Nutley Dolores 13797.
Proctor's Alma Dolores 47107. 23 lbs. 9 oz.
Nutley Dolores 2d 55901....... 15 lbs. 13 oz.
Proctor's Dolores 38564........ 15 lbs. 10 oz.

Nutley Silver 13591.
Nutley Silverette 22410....... 22 lbs. 7 oz.

Nutley Silverette 22410.
Nettina Silverette 89997....... 18 lbs. 14½ oz.
Cretesia's Silverette 74807..... 17 lbs. 3 oz.
Nutley St. Lambert 58469..... 16 lbs. 8 oz.

Nutshell 10326.
Nutshell 2d 50962.............. 19 lbs. 2 oz.

Nymphæa 5141.
Purest 13730.................. 15 lbs. 4 oz.

Oakland Girl 11103.
Highland Ida 38427........... 18 lbs. 1½ oz.

Oaklands Fairy 34273.
Aunt Lera 77855.............. 14 lbs. 14⅞ oz.

Oaklands Nora 14880.
Nora Pogis 75345.............. 21 lbs. 14¾ oz.
Oaklands Nora 2d 45967....... 19 lbs. 1 oz.
Meridale Oaklands Nora 64785 15 lbs. 1 oz.

Oaklands Nora 2d's Fille 94972.
Homestead Matilda 96448...... 17 lbs. 8 oz.

Obella 5080.
Louise Obella 37060........... 17 lbs. 12 oz.
Belle Bonair 33960........... 16 lbs. 3½ oz.

Ochra 4845.
Rival's Ochra 10172........... 19 lbs. 10½ oz.

Ocla 58396.
Milly Judd 85898.............. 14 lbs. 3½ oz.

Odelle Sales 15564.
Little Accident 15578.......... 16 lbs. 10½ oz.
Solava 20928................... 15 lbs. 3⅝ oz.

Odette 13668.
Yum Yum Signal 40688........ 16 lbs. 7 oz.

Ogstveno 88458.
Ogstree 90943................. 14 lbs. 2 oz.

Oktibbeha Duchess 4422.
Valerie 6044.................. 15 lbs. 13 oz.

Old Miss 84233.
Young Miss 99631.............. 16 lbs.

Oléo 38475.
Exile's Nina 40522............ 15 lbs. 11½ oz.
Exile's Nina 40522 (14 days).. 31 lbs. 6½ oz.

Olie's Lady Teazle 12307.
Feisticea 17083................ 16 lbs. 3 oz.

Olie's May Bell 6567.
Olie's Lady Teazle 12307...... 16 lbs. 5 oz.

Olive 4th 3018.
Normanda 3914................. 14 lbs. 14 oz.

Olive Branch 5324.
Forget Me Not O 10564........ 15 lbs. 4 oz.

Olive E. 16032.
Dolly Bond 82662.............. 18 lbs. 3 oz.

Olive Germ 54739.
Lucky Olive Germ 80256....... 14 lbs. 3½ oz.

Olive Lass 52889.
Olive Star 81102.............. 15 lbs. 7 oz.

Olive Leaf 4257.
Olive Branch of Hillcrest 7447 15 lbs. 7 oz.

Olivia 1397.
Lady Oxford 4860 (10 days).... 22 lbs. 2 oz.

Olivia Albertine 48007.
Olivia Albertine 3d 83438...... 15 lbs. 1¾ oz.

Olo H. 14455.
Olga H. 60619................. 15 lbs. 4 oz.
Olga H. 60619 (30 days)....... 61 lbs. 13 oz.

Omoo 1247.
Oonan 1485.................... 22 lbs. 2½ oz.

Ona Coomassie 43569.
Rowena of St. Lambert 71526. 20 lbs. 4 oz.
Rowena of St. Lambert 71526
(14 days).................... 39 lbs. 6¼ oz.
Rowena of St. Lambert 71526
(42 days).................... 108 lbs. 2½ oz.

Ona's Pansy 34866.
Pansy's Thoughts 69400....... 17 lbs. 6 oz.
Pansy's Thoughts 69400 (14
days) .... 33 lbs.

Oneida 42100.
Oneida 2d 43553.............. 17 lbs. 4½ oz.

Oneida 2d 43553.
Onwa 59628................... 18 lbs. 13¼ oz.

Oonan 1485.
Roonan 5133.................. 20 lbs. 4 oz.
Oonan 2d 19569.............. 18 lbs. 4¼ oz.
Odelle Sales 15564........... 16 lbs. 15¾ oz.
Callie Nan 7959.............. 16 lbs. 2 oz.
Oonan 3d 49955.............. 14 lbs. 15½ oz.
Nanalda 32917................ 14 lbs. 2½ oz.

Oonan 3d 49955.
Oonan's Oonan 99574......... 17 lbs. 12¾ oz.
Oonan Pogis 59835............ 14 lbs. 6 oz.

Oonan's Jeannette 45616.
Jeannette Geranium 61023.... 14 lbs. 4 oz.

Ophelie 493.
Maggie of St. Lambert 9776.. 16 lbs. 3 oz.

Opportunity 25915.
Geneva S. 57170.............. 15 lbs. 15½ oz.

Orange Flower 2d 3884.
Florry Keep 6556.............. 14 lbs. 14 oz.

Orange Pogis 49531.
Leonette's Orange 108521...... 15 lbs.

Orange Twig 11796.
Lemon Twig 42028............ 17 lbs. 4 oz.

Ora Reid 6482.
Signora Reid 27125........... 16 lbs. 8¾ oz.

Orie Hugo Pogis 50210.
Orie's Gem of St. Lambert
85472 .... 18 lbs. 4 oz.

Ori's Rose 64429.
Kate Kyle 64434.............. 16 lbs. 4 oz.

Orphan Duchess 4519.
Orphan Duchess 3d 21284..... 16 lbs. 3 oz.

Orphan Duchess 4th 29006.
Julia Tyre 45187.............. 14 lbs. 6 oz.

Orphan Lass 9646.
Lass Pogis 40740.............. 14 lbs. 14 oz.

Orphean 4636.
Belmeda 6229.................. 18 lbs. 12 oz.

Otella 11918.
Auselma 37798................ 14 lbs. 1 oz.

Oxalis 606.
Oxalis 2d 15631................ 15 lbs.

Painted Lady 45278.
Lady Petite 100562............. 15 lbs. 11 oz.

Palestina 4644.
Paletta of Darlington 16255.... 27 lbs. 8 oz.

Palestine 26.
Palestine's Last Daughter
12602 .... 14 lbs. 6 oz.

Palestine 2d 1455.
Lady Palestine 2769........... 14 lbs. 5 oz.

Palestine 3d 1104.
Palestina 4644................. 15 lbs. 8 oz.

Palestine's Last Daughter 12602.
Palestine Pierrot 24099........ 14 lbs. 6 oz.
Acme's Bonnie Bess 31509.... 14 lbs. 5 oz.
The Sister 21135............... 14 lbs. 4½ oz.

Paletta of Darlington 16255.
Paletta of Darlington 2d 75343 14 lbs. 15½ oz.

Pandora of Staatsburgh 3d 6497.
Maggie McM. 14073........... 19 lbs. 9 oz.
Pandothra 22383.............. 17 lbs. 5 oz.
Almah of Oakland 11102....... 16 lbs. 14 oz.

Panftie 18442.
Bertha Stewart 99882.......... 17 lbs.

Panise 56676.
Bissonise 110476............... 14 lbs. 11 oz.

Pansita 7531.
Pansita 4th 18109............. 14 lbs. 6 oz.

Pansy (F. S. 1072 J. H. B.)
Dewdrop Pansy 19736........ 19 lbs. 8 oz.

Pansy 7th 130.
Lady Brown 433.............. 14 lbs.

Pansy Dixie 9470.
Kate Pansy 15177............. 15 lbs. 1 oz.

Pansy Hawes 2d 23752.
Pansy K. 23889................ 14 lbs. 11½ oz.

Pansy McGrew 26993.
Darling Pansy 44830.......... 15 lbs. 1 oz.

Pansy of Amherst Villa 70503.
King's Gesnera 98727......... 15 lbs. 10 oz.

Pansy of Bingo Farm 41784.
Jonquil of Bingo Farm 41786.. 19 lbs. 3 oz.

Pansy of Willow Dale 7754.
Friz Cam 14655................ 15 lbs. 7 oz.

Pansy Page 13163.
Pansy Darling of Lawn 76637. 19 lbs. 2 oz.

Pansy Rex 11559.
Le Brocq's Pansy Rex 23789.. 18 lbs. 6 oz.
La France's Pansy Rex 49141. 14 lbs. 10½ os.

Paradise 32082.
Paradise 2d 97112............. 18 lbs. 4 oz.
Plumage 53897................. 17 lbs. 5 oz.

Paradox 65003.
Elturia 80701 (21 days)........ 24.137 lbs.

Paraph 6134.
Tormentor's Cinderella 19564.. 15 lbs. 8½ oz.

Paraphrie 73008.
Marian Holton 101478.......... 14 lbs. 12 oz.

Parthia 64925.
Diploma's Pansy 112871....... 16 lbs. 3 oz.

Patience Pogis 61110.
Pa Pogis 72500................ 16 lbs. 8½ oz.
Patina Pogis 110784........... 15 lbs. 2½ oz.
Pa Pogis 72500 (14 days)...... 32 lbs. 15½ oz.

Patricia 4th 4579.
Atricia 6029.................. 15 lbs. 3 oz.

Patterson's Beauty 4760.
Bertha Morgan 4770.......... 19 lbs. 6 oz.

Patti Niobe 74649.
Miss Morehouse 91703......... 15 lbs. 5½ oz.

Patty Polonius 14016.
Patty Polonius 3d 38677....... 16 lbs. 10 oz.

Patty Polonius 3d 38677.
Polonia of Riverside 84036.... 18 lbs. 10 oz.

Pauline 3d 8296.
Pauletta Pogis 39273......... 17 lbs. 12 oz.
Mildred of M. 15548........... 14 lbs. 2½ oz.

Pearl Button 10372.
Pearl's Oonan 32288........... 15 lbs. 5½ oz.

Pearl Darling 78002.
Dairy Dolly 100605............. 15 lbs.

Pearl Jenks 50834.
Pearl Jenks 2d 68908.......... 18 lbs. 10 oz.
Pearl Jenks 3d 84869.......... 15 lbs. 3 oz.
Pearl Jenks 4th 84872......... 14 lbs. 1½ oz.

Pearl Landseer 50322.
Denise's Pearl 95757.......... 20 lbs. 11 oz.

Pearl of St. Lambert 5527.
Cill of Glen Rouge 13818...... 16 lbs. 6 oz.

Pearl of Warsaw 30543.
Pearl's Lemon 41646......... 17 lbs. 5¼ oz.

Pearl's Lemon 41646.
Lemon's Flossie 71214....... 14 lbs. 1¾ oz.

Pearl's Oonan 32288.
Virginia's Oonan 58107....... 15 lbs. 3 oz.
Pearl's Oonan 2d 86105....... 15 lbs. 1 oz.
Temalema 44761............... 14 lbs. 10 oz.

Pearl's Oonan 2d 86105.
Bisson's Oonan 108518........ 14 lbs. 9 oz.

Pedro's Dolly 77948.
Pedro's Pretty Dolly 102376... 18 lbs. 13 oz.

Pedro's Duchess 29581.
Pedro's Pretty Duchess 88558. 16 lbs. 11 oz.

Pedro's Gay Lady 88561.
Pedro's Pretty Lady 102379.... 15 lbs. 12 oz.

Pedro's Golden Harp 66098.
Pedro's Golden Harp 2d 103365 16 lbs. 9 oz.

Pedro's Golden Harp 2d 103365.
Pedro's Golden Beauty 110402. 18 lbs.

Pedro's Lady 36715.
Pedro's Pretty Girl 88488..... 17 lbs. 4 oz.
Pedro's Gay Lady 88561....... 16 lbs. 10 oz.

Pedro's Lady Phillis 88486.
Pedro's Pretty Phillis 102378.. 17 lbs. 4 oz.

Pedro's Marjoram 29585.
Lorne's Marjoram 43327....... 16 lbs. 4½ oz.

Pedro's Pansy 33835.
Pedro's Pretty Flower 88542.. 18 lbs. 6 oz.
Pedro's Pretty Pansy 75275.... 18 lbs. 4 oz.

Pedro's Phillis 66079.
Pedro's Lady Phillis 88486.... 16 lbs. 4 oz.

Pedro's Pretty Girl 88488.
Pedro's Very Pretty 110508.... 15 lbs. 4 oz.

Pedro's Primrose 29586.
Lorne's Primrose 66082....... 16 lbs. 12 oz.

Pedro's Wonder 66078.
Pedro's Little Wonder 72379.. 17 lbs. 14 oz.

Pedro's Young Princess 72310.
Pedro's Pretty Princess 102377 17 lbs. 6 oz.
Pedro's Princess of York 72378 15 lbs. 4 oz.

Peggy Pedro 78146.
Gertie Victor Pogis 100922..... 16 lbs. 3 oz.

Peggy Scituate 76986.
Antigna 109426................. 14 lbs. 2¾ oz.

Peggy Stewart 7407.
Cannie M. 10698............... 15 lbs. 1 oz.

Pendule 16673.
Pendule 2d 16709.............. 16 lbs. 5 oz.

Peony of St. Lambert 56501.
Heiress of St. Lambert 56507.. 16 lbs. 13½ oz.

Perfection 1897.
Aspirante 9272................. 14 lbs. 7 oz.

Period 42640.
Miss Madrid 49419............. 14 lbs. 11 oz.

Perira 19433.
Pera B. 86761.................. 17 lbs. 9 oz.

Perrot's Baroness 22508.
Miss Perrot 73801............. 16 lbs.

Persis Pogis 83919.
March Ida Pogis 101182 (21
days) .... 47 lbs. 4½ oz.

Pertinaxia 28300.
Pertinette 70348............... 14 lbs.

Perty J. 15648.
Perty W. 41721............... 14 lbs. 2 oz.

Petite Lamballe 92260.
Süsschen 125091............... 15 lbs. 6¾ oz.

Petite Mère 8516.
Leila Stratton 25198.......... 17 lbs. 4 oz.

Pet of St. Lambert 5123.
Oaklands Nora 14880......... 23 lbs. 5 oz.
Diana of St. Lambert 6636.... 16 lbs. 8 oz.

Pet of Wildwood 33395.
Harry's Pet 52338.............. 22 lbs. 11 oz.

Petra 19267.
Petra's Lady 99791............. 15 lbs. 10½ oz.
Petra's Signal 53376........... 14 lbs. 12 oz.

Petra's Signal 53376.
Deerlick Petra Pogis 99987.... 14 lbs. 9½ oz.

Petrus 5563.
Petra 19267................... 16 lbs. 6 oz.

Pet's Beauty 15726.
May Evening 15938........... 17 lbs. 13 oz.

Pettie Fair Light 68207.
Pluma of Jersey Lawn 89954.. 14 lbs.

Petunia 1777.
Gamma 4147................... 14 lbs. 12 oz.

Petunia's Pet 28814.
Becky Z. 71311................ 14 lbs. 2 oz.

Petura 28775.
Itura 61090................... 14 lbs. 14 oz.

Pet Victor Pogis 77812.
Winifred Victor Pogis 91658.. 22 lbs. 9 oz.

Phædra 2561.
Nymphæa 5141................ 18 lbs. 7¼ oz.

Philena Polonius 24915.
Philena Victor Pogis 127133.. 17 lbs. 3 oz.

Philena's Pet 31296.
Sadie's Delight 53392......... 14 lbs. 12 oz.

Philene 9012.
Duke's Rowena 37190........ 14 lbs. 8 oz.

Philine of Linden 2d 68319.
Philena of Briarcliff 68938.... 15 lbs. 15½ oz.

Philippa (F. S. 1665 J. H. B.)
Satin Bird 16380.............. 14 lbs. 15½ oz.

Phillis 18162.
Phillis 3d 46681............... 18 lbs. 13 oz.

Phillis 2d 18198.
Pedro's Pansy 33835.......... 19 lbs. 10 oz.
Lady Phillis 18240............. 18 lbs. 15 oz.
Pedro's Phillis 66079.......... 17 lbs. 9 oz.

Phillis Pogis 29592.
Pretty Phillis Pogis 66100..... 17 lbs. 6 oz.

Phinora's Beauty 96176.
Meridale Phinora 119398....... 14 lbs. 9 oz.

Phlox 16399.
Phlox 3d 31882................ 15 lbs. 6 oz.

Phlox 3d 31882.
Fancy Phlox 49652............ 14 lbs. 3 oz.

Phœbe 4th 2271.
Catchfly 25405................. 18 lbs. 4 oz.
Pride of Bovina 8050.......... 16 lbs. 9 oz.
Phœbe N. 25401............... 15 lbs. 3 oz.

Phœbe Cross 38334.
Phœbe's Charity 59008........ 15 lbs. ½ oz.

Phryne 4289.
Roll of Honor 13610........... 14 lbs. 12 oz.

Phrynetine 16198.
Nettina of Winnikee 46406..... 18 lbs. 15½ oz.

Phrynore of Allentown 45205.
Phrynore of Allentown 3d
126321 .... 14 lbs. 8 oz.

Picotu 24211.
Picotu's Bess 86172........... 17 lbs. 8¼ oz.

Picton (F. S. 4236 J. H. B.)
Les Marais' Dell 20314........ 15 lbs. 8½ oz.

Pierrot's Beauty 48964.
Pet Victor Pogis 77812........ 16 lbs. 4 oz.

Pierrot's Lady Jennings 11675.
Duke's Pierrot Lady 31023..... 21 lbs. 12 oz.

Pierrot's Myrtle 10135.
Scepter's Beauty 23234........ 14 lbs. ½ oz.

Pilot's St. Jeannaise 25089.
Princess Jeannaise 53863...... 15 lbs. 12 oz.

Pinafore 8289.
Pinafore 2d 15072............. 15 lbs. 8 oz.

Pine Tree Bessie 32798.
Pine Tree Clara 55042........ 17 lbs. 3 oz.

Pine Tree Clara 55042.
Pedro's Clara 96361.......... 16 lbs. 13 oz.

Pine Tree Nora 32799.
Nora Sheldon 43590........... 17 lbs. 6 oz.

Pinkey Spry 18316.
Early Morn 50661............. 17 lbs. 13¼ oz.
Pinkey Spry 2d 92953......... 14 lbs. 10½ oz.

Pinkie 2d 2987.
Fair Starlight 7745........... 17 lbs. 7½ oz.

Pink of St. Lambert 5486.
Mermaid of St. Lambert 9771. 25 lbs. 13½ oz.
Rioter Pink of Berlin 23665... 19 lbs. 14 oz.

Pink Pedro 77809.
Diana Victor Pogis 100763..... 23 lbs. 1 oz.

Pip 86038.
Bluster's Pip 101165............ 14.41 lbs.

Pitapat 11136.
Harry's Fancita 96721......... 20 lbs. 2½ oz.
Pitapat 2d 86907............... 19 lbs. 3¼ oz.

Playmate (F. S. 1786 J. H. B.)
Brown Princess 30941......... 14 lbs. 9 oz.

Pledge 59214.
Dariole 64106................... 16 lbs. 9 oz.
Shrine 59627.................... 15 lbs. 1½ oz.

Plenita 6143.
Sunset of Pleasant View 13071 15 lbs. 2 oz.

Pocia of Andalusia 62616.
Pocia of Andalusia 2d 79548..: 21 lbs. 7½ oz.

Pogis' Dewdrop 40373.
Amelia Pogis 53796............ 16 lbs. 14 oz.

Pogis' May 26950.
Pogis' May's Matilda 75347.... 15 lbs. 11½ oz.
Pogis' May's Pride 58577...... 15 lbs. 2 oz.

Pogis of Walnut Grove 51962.
Koffee's Lady Pogis 65219.... 23 lbs. 4 oz.

Pogis' Pansy of Side View 59812.
Matilda's Pansy 81311........ 14 lbs. 7 oz.

Pogis' Twilight 53794.
Maudie of St. Lambert 73603.. 15 lbs. 7½ oz.

Polly (P. S. 69 J. H. B.)
Frugal 14925.................... 17 lbs. 2⅜ oz.
Polly Neptune 15214.......... 14 lbs. 7 oz.

Polly Bonair 42849.
Florence of St. Lambert 95192 16 lbs. 12 oz.

Polly J. Rex 33278.
Pedro's Polly Rex 99845....... 21 lbs. 4 oz.
Carlo's Polly Rex 94627....... 16 lbs. 12 oz.

Polly of Deerfoot 15328.
Carlo's Last Polly 94628...... 17 lbs. 6 oz.

Polly of St. Lambert 28665.
Aggie of St. Lambert 37085.... 14 lbs. 2½ oz.

Polly of Scituate 13635.
Polly Bonair 42849............ 18 lbs. 7 oz.

Pomona 2560.
Maud Drew 38522.............. 17 lbs. 12 oz.

Pomona's Maggie 102921.
Pedro's Pretty Maggie 102923.. 18 lbs. 9 oz.
Pedro's Maggie 102922........ 16 lbs. 6 oz.

Poolas 2d 53552.
Max Polly 108994............. 15 lbs. 8½ oz.

Poppy 4842.
Arawana Poppy 6053.......... 15 lbs. 2 oz.

Portrait 32592.
Portfolio 84633................. 15 lbs. 5½ oz.
Brown Bessie 7th of H. F.
119841 .... 14 lbs. 5 oz.

Postscript 39366.
Quickstep 85969............... 17 lbs. 9 oz.

Prattler 55394.
Pretty Prattler 99080........... 21 lbs. 8 oz.

Praxilla 31677.
Praxitella 50072............... 14 lbs. 4 oz.

Pretty Phillis Pogis 66100.
Pedro's Hansome Phillis
96935 .... 17 lbs. 8 oz.

Pridalia 17249.
Dolly of Edgewood 83345...... 20 lbs. 9½ oz.
May of Edgewood 74825....... 18 lbs. 2 oz.
Pride of Edgewood 64199...... 17 lbs. 2 oz.

Pride of Aurora 67666.
Kitty Sands 77432.............. 18 lbs. 1½ oz.

Pride of Bovina 8050.
Daisy Harrison 69253.......... 25 lbs. 1 oz.

Pride of Edgewood 64199.
Star of Edgewood 74823....... 16 lbs. 10 oz.

Pride of Franklin 25406.
Ramapo's Pride 35497......... 14 lbs. 4 oz.

Pride of Ingleside 54545.
Kitty Pride Pogis 78134....... 15 lbs. 9 oz.

Pride of Sidney 19210.
Worthy Noble 19629........... 15 lbs. 9½ oz.

Pride of Springvale 20093.
St. Lambert's Alphea 74100... 20 lbs. ½ oz.

Pride of the Farm 2d 28161.
Belle of River's Bank 48023... 15 lbs. 10 oz.

Pride of the Grange 8914.
Pride's Olga 37186............ 19 lbs. 12 oz.

Pride of the Hill 4877.
Pridalia 17249................. 26 lbs. 4 oz.

Pride of Walnut Farm 11501.
Juliette Guion 13143.......... 14 lbs. 4 oz.

Pride of Willimantic 13557.
Insie of Riverside 23825....... 30 lbs. 5¾ oz.
Lulu of Riverside 19175...... 17 lbs.

Pride of Windsor 483.
Duchess of St. Lambert 5111.. 15 lbs. 11 oz.
Lily of St. Lambert 5120...... 14 lbs.

Pride's Beth 87106.
Exile's Flossie 116608......... 16 lbs. 12 oz.

Pride's Olga 37186.
Pride's Olga 4th 96870........ 23 lbs. 3 oz.
Pride's Olga 2d 79887......... 22 lbs. 12 oz.
Pride's Olga 3d 79897......... 18 lbs. 2 oz.

Prince's Rubetta 92785.
Prince's Rubetta 2d 133914..... 14 lbs. 6¼ oz.

Princess 836.
Wyble 595..................... 17 lbs. 4 oz.

Princess 2205.
Christmas Nannie 4075........ 19 lbs. 7 oz.

Princess (F. S. 452 J. H. B.)
Princess 2d 8046.............. 46 lbs. 12½ oz.

Princess, on I. of J.
Princess of St. Saviour's 7470 15 lbs. 10½ oz.

Princess 2d 2295.
Peggy of Staatsburgh 2342..... 14 lbs. 1¼ oz.

Princess Amy 30185.
Josephine Noble 66441........ 18 lbs. 14 oz.

Princess Chrysantha 34191.
Howdy 55119.................. 15 lbs. 3½ oz.

Princess Cross 38329.
Princess Cross 2d 80649....... 20 lbs. 1 oz.

Princess Haidee 34874.
Haidee Signal 86448........... 20 lbs. 2 oz.
Haidee Signal 86448 (31 days). 79 lbs. 11 oz.

Princess Helene 21505.
Princess Helene of St. L. 61972 16 lbs. 9 oz.
Princess Helene of St. L.
61972 (30 days).............. 66 lbs. 6 oz.

Princess Honor 31805.
Princess Honoria 62548....... 16 lbs. 8 oz.
Princess Honoria 62548 (90
days) .... 159.447 lbs.

Princess Lorne 68699.
Pedro's Royal Princess 88485.. 23 lbs. 14 oz.

Princess Maude 7177.
Maud's Sultana 19518.......... 16 lbs. 4 oz.

Princess Mostar 9700.
Velvet Mostar 17533........... 15 lbs. 3 oz.

Princess of Ashantee 4th 26307.
Pocia of Andalusia 62616...... 20 lbs. 15 oz.

Princess of Canaan 61020.
King's Queen 98712............ 14 lbs. 12 oz.

Princess of Dorset 85902.
Princess Victor Pogis 132724.. 16 lbs. 5 oz.

Princess of Mansfield 8070.
Foolish 64069.................. 17 lbs. ½ oz.
Pierrot's Lady Bacon 12482... 16 lbs. 10 oz.

Princess of Pomeroy 52110.
Sallie Pomeroy 98806.......... 14 lbs. 10 oz.

Princess of St. Saviour's 7470.
Princess Le Brocq 17261....... 15 lbs. 8 oz.

Princess of Trinity 23641.
Khedive's Princess 38568...... 15 lbs. 4 oz.

Princess Olga 26528.
Princess Ora 66374............ 14 lbs. 5 oz.

Princess Potoka 53572.
Buckmentor 62846............. 14 lbs. 10 oz.
Little Princess Potoka 79285.. 14 lbs. 2 oz.

Princess Rosalind 87288.
Albert's Comtesse 111718...... 21 lbs. 4 oz.

Princess Rose 6249.
Catono's Rosebud 23674....... 14 lbs. 5 oz.

Princess Royal (P. S. 240 J. H. B.)
Royal Beauty 18908........... 15 lbs. 2½ oz.

Princess Royal 2d 1005.
Royal Princess 2370.......... 17 lbs. 12 oz.

Princess Simmons 57520.
Princess Marie G. 120390..... 15 lbs. 12½ oz.
Princess Marie G. 120390 (28
days) .... 60.98 lbs.
Princess Marie G. 120390 (32
days) .... 69 lbs. 8 oz.

Printonette 26121.
Nettie Signal 37837........... 18 lbs. 8 oz.

Printonnière 26086.
Zélie de Lussan 48681........ 20 lbs. 7 oz.
Printonette 26121............. 14 lbs. 8 oz.
Zélie de Lussan 48681 (31 days) 86 lbs. 13 oz.
Zélie de Lussan 48681 (6 mos.) 450 lbs. 5 oz.

Prize Classic 74900.
First Prize 95323............. 15 lbs. 4 oz.

Prize Maid 3835.
Maid of Saragossa 9086....... 16 lbs. 6 oz.

Prize's Bounty 67114.
Pride of Rosemond 74449..... 17 lbs. 14½ oz.

Prospect's Edwina 43185.
Prospect's Edwina 2d 86909... 18 lbs. 1 oz.

Prudence Hunt 22614.
Zeroda 43723.................. 16 lbs. 7 oz.

Prudence of Bovina 3d 10747.
Pansy Patterson 18612........ 15 lbs. 15 oz.

Prunella 2d 5861.
Siloam 17623.................. 18 lbs. 9½ oz.

Prusa Pedro 50877.
Prusa Victor Pogis 134216..... 16 lbs. 8 oz.

Psyche of St. Lambert 5121.
Jolie of St. Lambert 5126... 15 lbs. 13½ oz.

Puin of Eau Claire 71261.
Fennel of Glynllyn 109755..... 17 lbs. 4 oz.

Pure Mocha 9186.
Oléo 38475.................... 18 lbs. 1 oz.

Purity 1408.
Alluring 5541................. 19 lbs. 5 oz.

Purity, on I. of J.
Bisson's Belle 31144.......... 28 lbs. 10 oz.
Bisson's Belle 31144 (one
year) .... 1028 lbs. 15⅝ oz.

Purple Mulberry 3d 12100.
Nan of Ethelwyn 49502....... 15 lbs. 12½ oz.

Puss Lurlene 21461.
Jennie Stoke Pogis 32010..... 15 lbs. 10¾ oz.
Exile's Dell 67533............. 14 lbs. 10¼ oz.

Pussy Baker 6994.
Pussy Stoke Pogis 62547...... 20 lbs. 1 oz.
Cordelia Baker 8814.......... 17 lbs. 9 oz.

Pussy Cat 2731.
Silkey Morrill 17058.......... 14 lbs. 7 oz.

Pyrrha 6100.
Nimble 22335.................. 14 lbs. 10 oz.

Pythoness 55199.
Natalie Victor 73689.......... 17 lbs. 12 oz.

Quadruple Pogis 32359.
Quintuple Pogis 44233........ 14 lbs. 6 oz.

Quail 5895.
Quail 3d 43568................ 16 lbs. 12 oz.

Quailona 31492.
May Koffee 45862............. 15 lbs. 5½ oz.

Quaker's Lady 25110.
Inez of Ingleside 28977........ 14 lbs. 12 oz.

Quarto 31828.
Reif 58393.................... 19 lbs. 3 oz.

Queen (F. S. 1239 J. H. B.)
Lille Bonne 8108.............. 19 lbs. 12 oz.

Queen Alexis 33363.
Dinah Alexis 69345............ 14 lbs. 8 oz.

Queen Bess of St. Brelade's 7462.
   Birdie Le Brocq 17263......... 14 lbs.

Queen Dodona 42375.
   Dodona A. 88102.............. 16 lbs. 10¾ oz.
   Gracie A. 54524............... 15 lbs. 4 oz.

Queen Emma 32560.
   Natasqua 65598 (90 days)...... 161.522 lbs.

Queenette 5274.
   Ette 10315..................... 15 lbs. 7 oz.

Queen Fan 18013.
   Toy 20604..................... 17 lbs. 8 oz.

Queenie 4696.
   Countess Europa 35820....... 17 lbs. 12 oz.

Queen Mary of Woodlawn 11659.
   Princess Mary of Woodlawn
     11663 .... 14 lbs. 4 oz.

Queen Mostar 15256.
   Alpheus Mostar 20037......... 14 lbs. 9 oz.

Queen of Berlin 2d 44834.
   Exile's Penelope 77182........ 18 lbs. 3¾ oz.

Queen of Clinton 11500.
   Queen Ione 113999............ 16 lbs. 3 oz.
   Matilda's Queen 72663......... 16 lbs.
   Ida of Clinton 25419.......... 15 lbs. 14 oz.
   Our Daisy Queen 94252....... 15 lbs. 13 oz.
   Queen Irene 101893........... 15 lbs. 12 oz.

Queen of Coventry 8065.
   Lady Bidwell 10303........... 15 lbs. 12 oz.

Queen of Jersey 4948.
   Countess Queen 13519........ 18 lbs. 3 oz.

Queen of Sidney 11492.
   Queen of Clinton 11500........ 15 lbs. 8 oz.

Queen of the Fairies 8128.
   Fairy Le Brocq 15283......... 14 lbs. 3 oz.

Queen of the Miamis 6793.
   Queen May of St. Lambert
     45770 .... 15 lbs. 12½ oz.
   Margaret of Lagonda 35127.... 14 lbs. 4 oz.

Question (F. S. 3132 J. H. B.)
   Royal Queen 24428............ 22 lbs. 6 oz.

Quintuple Pogis 44233.
   Quintuple Pogis' Fille 94975.. 16 lbs. 8 oz.

Rachel 24522.
   Comtesse de St. Ouen 22600... 15 lbs. 8½ oz.

Rachel Ray 1754.
   Robinette 7114................ 14 lbs. 1 oz.

Rachel Spencer 50974.
   Cretesia's Rachel 89998....... 16 lbs. 8 oz.
   Rachel's Tamy 89987......... 14 lbs. 4½ oz.

Racket's Spot 55746.
   L. Toltec's First 130688....... 14 lbs. 13¼ oz.

Rainbow 6493.
   Rainbow 2d 13962............. 21 lbs. 8 oz.

Rainbow 16672.
   Pedro's Rainbow 66096....... 18 lbs. 15 oz.
   Pedro's Bonnie Rainbow 88612 18 lbs. 5 oz.

Rainbow 2d 13962.
   Lady Livingston 33374........ 15 lbs. 2 oz.

Rainbow's Girl 4th 69199.
   Runaway 87030.............. 16 lbs. 10 oz.

Ramapo's Tass 41846.
   Bright Tass 46173............. 21 lbs. 1 oz.

Ramblerotas 50036.
   Ollie Wilkes 77402............ 14 lbs. 14 oz.

Ramobina 31196.
   Sadie T. 75896................ 16 lbs. 6 oz.

Ranolta 54292.
   Dame of Eau Claire 127862.... 14 lbs. 4 oz.

Rarity 5923.
   Rarity 2d 7724................ 14 lbs. 2 oz.

Rather Pretty (F. S. 5933 J. H. B.)
   Very Pretty 29184............. 15 lbs. 2 oz.

Ravalita 46390.
   Tormentor's Ravalita 97909... 15 lbs. 11¾ oz.

Rebecca (F. S. 2625 J. H. B.)
   Lotchen 19823................. 16 lbs. 7 oz.

Regalia 64574.
   Vina Marigold 108186......... 14 lbs. 8½ oz.

Regina (F. S. 32 J. H. B.)
   Regina 4th 12732............. 17 lbs. 13½ oz.
   Regina 2d 2475................ 14 lbs. 8 oz.

Regina 2d 2475.
   Chrome Skin 7881............. 20 lbs. 10 oz.
   Walkyrie 5708................. 14 lbs. 1 oz.

Regina 4th 12732.
Gilded Queen 42029 .... 14 lbs. 9½ oz.

Regina's Pride 24407.
Stoke Pogis' Regina 48309 (30
days) .... 60.268 lbs.

Reif 58398.
Domett 88179 .... 16 lbs. 2 oz.

Renalda 20924.
Derjava 59830 .... 18 lbs. 14 oz.

René 2d 56.
René Ogden 1568 .... 15 lbs.

Renébel 2772.
René of Hillside 12952 .... 18 lbs. 1 oz.
Renalba 4117 .... 17 lbs. 4½ oz.

René Calypso 3205.
Calypris 5943 .... 15 lbs. 4½ oz.

René of Hillside 12952.
Helen Walker 53465 .... 19 lbs. 10½ oz.
Pogis' René 102854 .... 17 lbs.
Chendado's Maid 45991 .... 16 lbs. 7½ oz.
Retta Pogis 81132 .... 16 lbs. 3¼ oz.

Renina 1431.
Medrena 3939 .... 18 lbs.

Reserve 6051.
Renown 13729 .... 14 lbs. 6 oz.

Respectability 16730.
Minnie Parker 38718 .... 26 lbs. 2½ oz.

Result 11848.
Reality 16537 .... 15 lbs. 3½ oz.

Rex's Diamond 73244.
Diamond's Lass 81767 .... 16 lbs. 4 oz.

Rhea 166.
Alphea 171 .... 15 lbs. 6 oz.

Rho A. 16950.
Rho A. Pogis 39269 .... 14 lbs. 15 oz.

Rho A. Pogis 39269.
Snow Bank 75184 .... 16 lbs. 7 oz.

Rhoda 2d 16097.
Signal's Rhoda 98651 .... 14 lbs. 5½ oz.

Rhoda Hudson 48723.
Sophie Hudson 76105 .... 19 lbs. 12½ oz.
Sophie Hudson 76105 (10 days) 28 lbs. ½ oz.

Rhoda of Merion 5849.
Rosebud of Allerton 6352 ..... 19 lbs. 12 oz.

Ribbon of St. Lambert 39450.
Ribbon's Gift 77375 .... 18 lbs. 1 oz.
Ribbon's Matilda 66109 .... 16 lbs. 4 oz.
Ribbon's Queen 66321 .... 16 lbs. 1 oz.

Ribbon's Fawn 26907.
Ribbon of St. Lambert 39450 .. 18 lbs. 2 oz.
Fawn of Ingleside 59649 ...... 17 lbs. 9 oz.

Richland Pride 117977.
Spangled Princess 117978 ..... 15 lbs. 5 oz.

Rinora Pogis 40107.
Nora of Glen Rouge 69452 ..... 16 lbs. 13 oz.

Riotaletta 29937.
Riotaletta 2d 34495 .... 15 lbs. 15½ oz.

Rioter 2d's Venus 3658.
Amalthea 11319 .... 15 lbs.

Rioter Alphea 10091.
Rioter Alphea 2d 29676 .... 17 lbs. 15 oz.
Rioter Alphea 3d 34073 .... 17 lbs. 1½ oz.

Rioter Butterstamp 46088.
Butterstamp Pedro 56383 ...... 14 lbs. 8 oz.

Rioter Carlotta 29667.
Cleora of Cornwall 63408 ..... 18 lbs. 8 oz.

Rioter Lisgar's Rose 64497.
Moss Rose of Jackson 86709 ... 16 lbs. 4 oz.

Rioter Pink of Berlin 23665.
Meridale Rioter Pink 64534 ... 15 lbs. 13 oz.

Rioter's Daisy M. 35374.
Victor's Matilda 46834 .... 14 lbs. 6 oz.

Rioter's Denise 36057.
Rioter's Denise 3d 70238 ..... 17 lbs. 1½ oz.

Rioter's Grace 61993.
Exile's Claribel 107507 .... 14 lbs. 5 oz.

Rioter's Lorne Primrose 88583.
Infanta Pedro Marjoram 88584 17 lbs. 6 oz.

Rioter's Nora 21778.
Rinora Pogis 40107 .... 19 lbs. 1¼ oz.
St. Lambert's Nora 42132 ..... 18 lbs. 12 oz.

Rioter's Pretty Pearl 75273.
Pedro's Silver Pearl 110401 ... 18 lbs. 6 oz.
Rioter's Pretty Belle 88543 ... 15 lbs. 10 oz.

Rioter's Sweet Brier 30582.
Meridale Sweet Ida 94854 ..... 20 lbs. 8 oz.
Brier of St. Lambert 61750 ... 18 lbs. 4 oz.

Rioter's Zoe 19769.
Experience 62458.............. 22 lbs. 13 oz.
Ida Zoe Pogis 38685.......... 16 lbs. 2½ oz.

Riotress 28062.
Riotress 3d 63166.............. 17 lbs. 3½ oz.

Ristori 12701.
Geneva 13220.................. 15 lbs. 11 oz.

Rival's Kitty 61760.
Lady May Jefferson 96591..... 14 lbs. 12 oz.

Riverside Belle 2d 14125.
Sibby 87742.................... 15 lbs. 4½ oz.

Robinette 7114.
Robinella 63132................ 15 lbs. 14 oz.

Roe Webster 12562.
Gretchen of Pevely 46325...... 15 lbs. 5½ oz.

Roland's Bonnie 18053.
Roland's Bonnie 2d 18054..... 19 lbs. 2 oz.

Romena 72300.
Winksette 99280................ 23 lbs.
Romena of Hood Farm 114887 15 lbs. 11 oz.

Remilly 14346.
Pedroletta 26597.............. 16 lbs. 6½ oz.

Romp 1098.
Optima 6715................... 19 lbs. 2 oz.

Romping Lass 11021.
Romping Lass 2d 51989........ 21 lbs. 12 oz.

Romping Lassie 25173.
Madame Pogis 74452.......... 14 lbs. 4 oz.

Romping Princess 40347.
Romping Miss 54457........... 23 lbs. 1 oz.

Romp Lawrence 13819.
Sparks 41041.................. 16 lbs. 4½ oz.

Romp Ogden 1571.
Romp Ogden 5th 43181........ 17 lbs. 9 oz.
Romp Ogden 3d 5458.......... 15 lbs. 14 oz.
Romp Ogden 2d 4764.......... 15 lbs. 5 oz.

Romp Ogden 2d 4764.
Southern Belle 18570.......... 16 lbs.

Romp Ogden 3d 5458.
Romping Lass 11021.......... 15 lbs.

Romp's Nervilette 72282.
Harry's Nervilette 86262...... 18 lbs. 1 oz.

Romp's Orie 64456.
Orie's Fancy 72039............ 15 lbs. 10½ oz.

Romp's Princess 51185.
Princess Oonan 84369......... 15 lbs. 13½ oz.

Roonalda 3d 49958.
Oonan's Alda 108337.......... 20 lbs. 5¾ oz.

Rosa 122.
Silver Rose 4753.............. 16 lbs. 14 oz.

Rosabel Hudson 5704.
Daisy Hudson 24647.......... 17 lbs. 11 oz.
Emma Hudson 12469......... 16 lbs. 2 oz.
Rhoda Hudson 48723......... 14 lbs. 4 oz.

Rosa Daniell 5441.
Rosona 12956.................. 16 lbs. 7 oz.

Rosa F. W. 13226.
Rosa Ramsey 25875........... 15 lbs. 1 oz.

Rosa Gamp 2732.
Roselaine 7167................ 15 lbs. 1 oz.

Rosa Hudson 34871.
Rosa Hudson 2d 87631....... 16 lbs. 14½ oz.

Rosalep Green 66753.
Angela of Coal Hill 79380..... 15 lbs. 7 oz.

Rosalia of Sidney 4521.
Duchess Noble 27090.......... 25 lbs. 8 oz.

Rosaline of Glenmore 3179.
Rioter's Rosaline 35581....... 14 lbs. 14½ oz.

Rosaline Pogis 42253.
Torraline 54945............... 17 lbs. 6¾ oz.

Rosa Mosher 2d 69179.
Royal Rose 86791............. 20 lbs. 4 oz.

Rosa of Belle Vue 6954.
Princess Malita 66676......... 20 lbs. 8¾ oz.
Lady Julia G. 16199........... 18 lbs. 3 oz.
Miss Remarkable 33446....... 15 lbs. 2½ oz.
Rosa Oxford 66678............ 15 lbs. ½ oz.
Rosebud of Belle Vue 7702.... 14 lbs. 10½ oz.

Rosa of Salem 6476.
Nerissa of Nyack 9692........ 15 lbs. 1 oz.

Rosa Oxford 66678.
Bisson's Rosa 97830........... 15 lbs. 1¾ oz.
Rosa Oxford 2d 127909........ 14 lbs. 8 oz.

Rosari 5518.
Le Rosa 10078................. 14 lbs.

Rosa's May Bud 47931.
Exile's Rosa 66552............. 14 lbs. ½ oz.

Rosa Thornton 12233.
Ulricalla 22225................ 18 lbs. 14 oz.

Rose (F. S. 1158 J. H. B.)
Khedive's Fancy 19180........ 15 lbs. 3 oz.

Rose (F. S. 1450 J. H. B.)
Buttercup 17285............... 16 lbs. 5 oz.

Rose 3d 913.
Copper 1979................... 15 lbs. 7 oz.

Rosebud of Belle Vue 7702.
Cherokee Rose 20921........... 23 lbs. 10 oz.
Signalda's Rosebud 35583..... 18 lbs. 10⅛ oz.
Tormentor's Rosebud 50069... 14 lbs. 2 oz.

Rose F. 2d 13224.
Rosa F. W. 13226.............. 29 lbs. 12 oz.

Rose Lawn 3690.
Rose of Rose Lawn 9365..... 16 lbs. 3 oz.

Rosella of Bay View 66032.
King's Rosella 80767.......... 15 lbs. 12½ oz.

Rose Newton 50148.
Belle Miller 60488............. 15 lbs. 2¾ oz.

Rose of Mashamoquet Farm 6472.
Rosy Kate 10276.............. 14 lbs. 4 oz.

Rose of Menard 13272.
Mary Justice 37449........... 14 lbs. 9½ oz.

Rose of Riverside 46953.
Rosora 87622.................. 15 lbs. 8½ oz.
Rose's Fancy 93047........... 14 lbs. 7 oz.

Rose of Rose Lawn 9365.
Pet of Rose Lawn 11326....... 18 lbs. 1 oz.

Rose of Side View 47275.
Matilda's Rose 81397.......... 15 lbs. 2 oz.

Rose of Spring Creek 25375.
Cora of Sandusky 42038....... 15 lbs. 4½ oz.

Rose of Sylvanhurst 16134.
Elodie 30222................... 15 lbs. 5½ oz.

Rosetta of Sidney 4520.
Cream of Sidney 17028........ 17 lbs. 2½ oz.

Rosetta of Whiteland 6112.
Princess Chuck 45650......... 24 lbs. 14 oz.

Rosetta Stone 75200.
Rosetta Landseer 99379....... 14 lbs. 4 oz.

Rosette, on I. of J.
Baron's Rosette 25988........ 15 lbs. 4 oz.

Rosette of St. Lambert 5108.
Brenda of Elmhurst 10762..... 20 lbs. 8 oz.
Kitty of St. Lambert 6637..... 16 lbs. 11 oz.

Rosette of Staatsburgh 3008.
Chlora 8325................... 14 lbs. 6 oz.

Rosy (F. S. 1008 J. H. B.)
Beauty 17414.................. 15 lbs.

Rosy Dream 9808.
Rosy Dream 2d 27127......... 17 lbs. 11 oz.

Rosy Dream 2d 27127.
Ida's Dream 50070............ 16 lbs. ½ oz.
Ida's Dream 50070 (14 days)... 40 lbs. 8½ oz.

Rosy of St. Martin's 6564.
Lactine 10680................. 17 lbs. 1½ oz.

Rothey 18019.
Gold Maid 98787.............. 16 lbs. 14¾ oz.

Roulette 50080.
Odalette 76888................. 17 lbs. 3 oz.

Rovalti's Maid 90089.
Harry's Rovalti 115843........ 17 lbs. ¼ oz.

Roxana 1761.
Vixen 7591.................... 17 lbs. 6 oz.

Roxana 2d 2532.
Rochelle 15574................. 15 lbs. 10 oz.
Roxena 64212.................. 15 lbs. 8½ oz.
Litza 6338.................... 14 lbs. 3 oz.
Pixie 4115.................... 14 lbs.

Royal Bomba's Daisy 35996.
Olivia Marigold 108187......... 18 lbs. ½ oz.
Bertha's Royal Daisy 104901.. 16 lbs. 2½ oz.
Royal Bomba's Daisy 3d 99553 15 lbs. 8 oz.

Royal Daisy 25214.
Royal Bomba's Daisy 35996... 21 lbs. 5½ oz.

Royal Ensigna 35410.
Light Foot Pogis 57531........ 14 lbs. 11 oz.
Light Foot Pogis 57531 (14
days) .... 28 lbs. 15 oz.

Royalist's Daisy 19187.
Rescue of St. Lambert 48939.. 16 lbs. 8 oz.

Royal Pogessa 49530.
Ida's Little Wonder 104961.... 14 lbs. 7 oz.

Royal Princess 22013.
Princess Rioter 29588......... 17 lbs. 6 oz.

Rey Clementine 2d 23106.
Roxie Landseer 74832.......... 14 lbs. 13 oz.

Rubano's Normanda 68927.
Rioter's Normanda 102441..... 15 lbs. 8 oz.

Rubano's Quachette 50556.
Quachette F. W. 83654......... 15 lbs. 3 oz.

Rubano's Valentine 54703.
Star's Valentine 81778......... 16 lbs. 2 oz.

Ruby 787.
Alabama 7690.................. 15 lbs. 11 oz.

Ruby D. 2d 34627.
Lucy T. 48200.................. 16 lbs. 12.7 oz.

Ruby Jenks 56862.
Ruby Jenks 3d 84863.......... 23 lbs. 4½ oz.
Ruby Jenks 2d 76738.......... 23 lbs. 3½ oz.
Ruby Jenks 4th 84871......... 18 lbs. 5 oz.

Ruby Love 16915.
Ruby Love 2d 32058........... 14 lbs. 1 oz.

Ruby of Eau Claire 93305.
Faith of Eau Claire 120456.... 22 lbs. 12 oz.

Rura 21946.
Cretesia's Charma 81685...... 15 lbs. 8 oz.

Russie Hill of Maxwell 8661.
Countess Coomassie 19339.......16 lbs. 10 oz.

Ruth Duffee 2481.
Lady Greville 12930........... 14 lbs. 6 oz.

Ruth Le Brocq 38861.
Rioter's Hortense 51499....... 15 lbs. 1½ oz.

Ruth Morgan 24098.
Ruth Morgan 3d 57043......... 19 lbs. 6.4 oz.

Ruth Morgan 2d 24100.
Ruthione 31514................ 25 lbs. 12 oz.

Ruth Morgan 3d 57043.
Jubilethleel 99697............. 17 lbs. 12 oz.
Harry's Ruth Morgan 107773.. 14 lbs. 1.6 oz.

Sadie 2d 3698.
May Toltena 50405............ 14 lbs. 5⅝ oz.

Sadie A. 25573.
Silicon 25577................... 18 lbs. 13 oz.

Sadie R. 34813.
Sadie R. 3d 83729............. 20 lbs. 14¼ oz.

Sadie's Choice 7979.
Signetilia 16333............... 18 lbs. 5½ oz.

Sadie's Daisy 40099.
Pansy Albert of Lawn 76634.. 15 lbs. 14½ oz.

Sadie Tansil 36696.
Fancy's Harry's Sadie 77202.. 14 lbs. 14 oz.

Safrano 4568.
Volie 19465.................... 18 lbs. 1 oz.

St. Albans' Normanda 41705.
Rubano's Normanda 68927..... 21 lbs. 4½ oz.

St. Albans' Valentine 41707.
Ruby's Valentine 81780........ 14 lbs. 7 oz.

St. Ello 3d 19078.
Lora Nicholson 27135.......... 19 lbs. 12¾ oz.

St. Jeannaise 15789.
St. Jeannaise 5th 96554....... 16 lbs. 12½ oz.
St. Jeannaise 4th 57990........ 15 lbs. 11 oz.

St. John's Daisy 28388.
Paola Stoke Pogis 34691....... 23 lbs.
Exile's Bessie 49985........... 18 lbs. 12½ oz.
Exile's May 52765.............. 15 lbs. 4¼ oz.
Paola Stoke Pogis 34691 (35
days) .... 109 lbs. 10 oz.

St. Lambert's Coquette 41070.
Meridale Coquette 64535....... 17 lbs.
Pomona's Coquette 52832...... 14 lbs. 10 oz.

St. Lambert's Crescent 39711.
Crescent's Rachael 74738...... 15 lbs.

St. Lambert's Jewel 53743.
St. Lambert's Bijou 78174.... 14 lbs. 15½ oz.

St. Lambert's Nancy 56503.
Nan of St. Lambert 69453..... 15 lbs. 9 oz.

St. Lambert's Nora 42132.
Nellie of St. Lambert 69458... 23 lbs. 12 oz.
St. Lambert's Nora 3d 111218.. 17 lbs. 5 oz.

St. Lambert's Violet 25278.
Queen of Pomona 56250........ 16 lbs. 6 oz.

St. L. Gazelle 78870.
Dainty Doris 85497............. 16 lbs. 8½ oz.

Saint Maggie 30363.
Exile's Fawn 40979........... 15 lbs. 10 oz.

St. Perpetua 3648.
St. Perpetua 2d 5557.......... 14 lbs.

St. Perpetua's Princess 19072.
Siesta 37321.................... 15 lbs. 11 oz.

Salacia (F. S. 2197 J. H. B.)
Sovereign's Elsie 15793....... 14 lbs. 3 oz.

Sallie Ward 7201.
Serita 15520.................... 17 lbs. 2 oz.

Sally Clover 4468.
Polly Clover 7052............. 16 lbs. 15 oz.

Sally Leavitt 7258.
Sally Leavitt 2d 10471........ 15 lbs. 12 oz.

Salsoda 3721.
Summerline 8001............. 18 lbs. 6 oz.

Samarés (F. S. 944 J. H. B.)
Rose of Eden 13437........... 20 lbs. 1½ oz.

Sanarex 85387.
St. Lambert's Rexina 113609.. 16 lbs. 2 oz.

Sara Iris 53217.
Tormentor's Isis 73350........ 19 lbs. 7 oz.

Sarita Victoria 56451.
Tapestry 56607................ 23 lbs. 12 oz.

Sassagua (F. S. 4194 J. H. B.)
Brown Coomassie 20322........ 14 lbs. 8 oz.

Sataline 64213.
Sataline 2d 131898............. 16 lbs. 7½ oz.

Satanella 8927.
Miss Satanella 31544.......... 20 lbs. 6 oz.
Miss Satanella 31544 (31 days). 78 lbs. 3½ oz.

Saucy Ann S. 30573.
Saucy Sally 83059............. 16 lbs. 12½ oz.

Saucylip 18574.
Saucy Princess 40350.......... 15 lbs. 11 oz.
Delesline 44474................ 15 lbs. 9 oz.

Saucy Princess 40350.
Colza Melrose 93859.......... 14 lbs. 13½ oz.

Saugus Lass 30542.
Young Lass 67471............. 17 lbs. 3½ oz.
Patona 55117................... 16 lbs. 14½ oz.

Sayda 4400.
Sayda 3d 17317 (30 days)....... 47.825 lbs.
Sayda 3d 17317 (90 days)....... 170.094 lbs.

Sayda 3d 17317.
Sayda M. 46195............... 22 lbs. 11½ oz.

Sayda M. 46195.
Sayda M. 3d 71134............. 20 lbs. 10 oz.
Sayda of Aurora 86956......... 16 lbs. 11½ oz.

Schonemunk Lass 9126.
Butterstamp Lass 19517....... 16 lbs. 11 oz.

Science 49561.
Midway 50000.................. 16 lbs. 2 oz.

Scribe 53894.
Moggy 59626................... 14 lbs. 8 oz.

Script 43308.
Manuscript 74881.............. 15 lbs. 1 oz.

Second Cousin 12689.
Cordelia Signal 33452......... 18 lbs. 15½ oz.

Sedate 11119.
Sedate's Retta 86596.......... 16 lbs. 8 oz.
Sedate 2d 20389................ 15 lbs. 8 oz.

Sedate 2d 20389.
Princess of Clovernook 51598.. 16 lbs. 13 oz.

Selita's Cream 50647.
Cream Custard 97429......... 19 lbs. 6¼ oz.

Seraphine 19262.
Seraphine 2d 37451............ 15 lbs. 12 oz.

Shekel of Gold 25169.
Douskha 63439................. 15 lbs. 5¾ oz.

Sheldon's Daisy 30592.
Sheldon's Daisy of St. L. 53047 20 lbs. 11 oz.
Sheldon's Daisy's Ida 101059.. 14 lbs. 2 oz.

Shiloh Daughter 20378.
Wanderer's Shiloh 65942...... 15 lbs.

Sibby 87742.
Sibby Landseer 116107........ 15 lbs. 2 oz.

Sibyl (P. S. 345 J. H. B.)
Sibyl's Beauty 25941.......... 18 lbs.
Sibyl's Fancy 25942........... 17 lbs.

Sida 4th 9284.
Lucy Cobb 15133.............. 18 lbs. 6½ oz.

Sigleonella 70091.
Sigmanella 117023.............. 20 lbs. 2 oz.

Sigletta 32915.
Sigletta 3d 108977.............. 20 lbs. 10½ oz.
Lily Martin 49954.............. 16 lbs. 8 oz.
Lily Martin 49954 (90 days)... 164.227 lbs.

Sigma 78676.
Sigma 2d 89803............... 14 lbs. 15½ oz.

Sigmotto 34939.
Miamotto 48691................ 16 lbs.

Signalana 7719.
Rupertina 10409............... 17 lbs. 1½ oz.

Signal Beauty 22260.
Melrose's Signal 54830........ 15 lbs. 5¾ oz.

Signalbianca 54948.
Signalbianca 2d 119779........ 17 lbs. ¼ oz.

Signalda Bloomfield 30694.
Tormentor's Bloomfield 55530. 22 lbs. 5½ oz.
Successor 72503................ 17 lbs. 4 oz.
Méprise Bloomfield 50062...... 14 lbs. 14 oz.

Signalda's Coomassie 55444.
Signalda's Ethleel 77599...... 16 lbs. 11 oz.

Signalda's Ethleel 77599.
Ethleel Koffee of Lawn 116757 17 lbs. 7 oz.

Signalda's Fancy 32724.
Signal Fancy of Lawn 73540... 15 lbs. 1 oz.

Signalda Silver Rose 33864.
Bernice of Clovernook 72805.. 17 lbs. 1 oz.

Signaldo's Pansy 53366.
Pansy Dec. 99294.............. 16 lbs. 4½ oz.

Signal Emblem 39473.
Allie's Emblem 45620......... 20 lbs.

Signalette 14877.
Lette Signal 26823............ 17 lbs. 2½ oz.

Signalia 5303.
Sedate 11119.................. 18 lbs. 13 oz.

Signalinda 27002.
Signalinda 2d 61142........... 21 lbs. 10 oz.

Signalinea 16336.
Sirona 27128.................. 21 lbs. 2 oz.

Signal Maid 19361.
Signal Maid 4th 67492........ 15 lbs. 8 oz.

Signal Maiden 42793.
Chrysippe 53373............... 14 lbs. 14 oz.

Signal Mine 20832.
Signal Mine 2d 60789......... 17 lbs. 9½ oz.

Signal Pomona 28398.
Susa Drew 38521.............. 21 lbs. 1 oz.
Signal Duke's Creamer 60812. 14 lbs. 7 oz.

Signal's Bijou 60800.
Ben's Bijou 86599............. 18 lbs. 6 oz.

Signal's Lily Flagg 31035.
Signilla M. 55839.............. 21 lbs. 4 oz.
Jennie Fordyce 57169......... 15 lbs. 5½ oz.

Signal's New Geranium 134117.
Signalona's Gem 134219....... 14 lbs. 4½ oz.

Signal's Ripple 27093.
Signal's Ripple 4th 64862...... 16 lbs. 10¼ oz.

Signet of Linwood 40423.
Signet's Fancy 58446......... 18 lbs. 11½ oz.

Signodelle 27132.
Signodelle 3d 50067........... 17 lbs. 1½ oz.
Signalbianca 54948............ 15 lbs. 5 oz.

Signolo's Florence 49430.
Signolo's Florence 2d 82446... 16 lbs.
Quintuple's Florence 86076.... 15 lbs. 10 oz.

Signora 4723.
Nelly Gray of Clermont 10905. 14 lbs. 1 oz.

Signora Reid 27125.
Signora Reid 2d 93808........ 15 lbs. 10½ oz.
Etelka Tolemy 93807.......... 14 lbs. 9 oz.

Sigrianna 34941.
Nienchen 48698................ 17 lbs. 2 oz.
Nienchen 48698 (31 days)...... 73 lbs. 13 oz.

Silistria 20264.
Harriet Neptune 52344........ 18 lbs. 6½ oz.

Silken Les Cateaux 23431.
St. L.'s Silken Les Cateaux
61964 .... 21 lbs. 12 oz.

Silkey Morrill 17058.
Lilly of Leeds 17059.......... 14 lbs. 9 oz.

Silkey of Mt. Pleasant 68727.
Happy May Girl 75588........ 14 lbs. 15½ oz.

Silky 2488.
Eudora of Barre 12419........ 16 lbs. 9 oz.

Silver Bell (F. S. 1807 J. H. B.)
Cetewayo's Silver Bell 18952.. 17 lbs. 2½ oz.

Silver Cloudlet 22710.
Marion Hill 40777.............. 15 lbs. 1 oz.

Silver Hair 4665.
Silver Hair 4th 60449........../. 15 lbs. 10¼ oz.
Silver Hair 4th 60449 (14 days) 31 lbs. ¼ oz.
Silver Hair 4th 60449 (28 days) 60 lbs. 8¾ oz.

Silverhair of C. 19668.
Silver Steletho 40765.......... 18 lbs. ½ oz.
Silver Montague 45475........ 16 lbs. 10 oz.

Silver Hull 15758.
Silver Belle C. 90600........... 16 lbs. 2½ oz.

Silver Queen of Mahoning 13839.
Star of Mahoning 16579....... 14 lbs. 7 oz.

Silver Rose 4753.
Opaline 7590.................... 15 lbs. 2 oz.
Tormentor's Silver 36058...... 14 lbs. 7½ oz.

Silver Rose 2d 18337.
Sitanda's Rose 64958........... 24 lbs. 12 oz.

Silver Tresses 45412.
Silvia of Linwood 61008....... 16 lbs. 5.6 oz.

Silvia Baker 8793.
Toltec's Silvia 86911........... 17 lbs. 1¼ oz.

Silvia of Linwood 61008.
Silvia Torment 104668........ 15 lbs. 11.2 oz.

Sirona 27128.
Sirona Croton 86597............ 18 lbs. 10 oz.
Signalona 72105................ 14 lbs. 3 oz.

Sirona 2d 38011.
Sasanda 50075.................. 22 lbs. 1½ oz.
Sorina 42993.................... 17 lbs. 15½ oz.

Sister 1427.
Sister Dorothy 2607.......... 15 lbs.

Sitka M. 78680.
Sitka's Star 104707............. 14 lbs. 9 oz.

Siva 24508.
Peasant 65401.................. 21 lbs. 15½ oz.

Smack 64575.
Brown Bessie's Bonita 109265. 20 lbs. 3 oz.

Smax 48562.
Smack 64575.................... 15 lbs. 12 oz.

Snowball of Deerfoot 2d 14457.
Lady Dartmouth 23159........ 21 lbs. 13 oz.

Snowdrop of St. Lambert 5119.
Clematis of St. Lambert 5478. 14 lbs. 3 oz.

Soconee 76895.
Soconee 3d 114422............. 20 lbs. 1½ oz.
Soconee 2d 114420............. 15 lbs.

Soekel 2d 2959.
Rosetta of Whiteland 6112.... 27 lbs. 2¾ oz.

Sœurette 52256.
Sœurette 2d 86414............. 18 lbs. 1½ oz.

Sœur Seraphine 29073.
Daisy Staunton 46592......... 16 lbs. 14 oz.

Solava 20928.
Tortilla 55528.................. 18 lbs. 4½ oz.

Sophie (F. S. 434 J. H. B.)
Khelula 17970................. 21 lbs. 8 oz.
Baron's Sophie 17615......... 19 lbs. 15⅝ oz.

Sorina 42993.
Sordella 76876................. 15 lbs. 9 oz.

Southern Belle 18570.
Belle Melrose 61920........... 17 lbs. 1½ oz.
Beautiful Blondine 82051..... 16 lbs. 5 oz.
Marie Bash 81441............. 16 lbs.
Hattie Swett 109709.......... 15 lbs. 10¼ oz.

Southern Daisy 38292.
Kate Oxford 58102............ 16 lbs. 3 oz.

Southern Ivy Leaf 95670.
Ivy Tormento 121700......... 16 lbs. 1¼ oz.

Southern Pet 26675.
Hugo's Joie de Vie 40372...... 24 lbs. 5 oz.
Helene of St. Lambert 73601.. 20 lbs. 1½ oz.

Spark 62689.
Sparta 75000................... 21 lbs. 5 oz.

Sparkle 56775.
Startling 94942................ 17 lbs. 14 oz.

Sparks 41041.
Sparkle 56775.................. 17 lbs. 3 oz.
Sparkling 41042............... 14 lbs. 1 oz.

Sparks' Maid 45575.
Anita Lawrence 63909........ 15 lbs. 6½ oz.

Spiræa 3915.
Spiræa 4th 20075.............. 18 lbs. 5½ oz.

Spiræa 4th 20075.
Tormentor's Spiræa 42248..... 14 lbs.

Spiralee 37960.
Louisa of Riverside 69769..... 23 lbs. 5 oz.

Spot's Grisette 25905.
Viozetta Pogis 56294......... 14 lbs. 5½ oz.

Starbeam 11846.
Alphea Star 16532............. 14 lbs. 4½ oz.

Star Caroline 55610.
Caroline Pauncefote 62919..... 16 lbs. 8 oz.

Starkville Beauty 4897.
Mountain Lass 12921........... 14 lbs. 9 oz.

Starlight (P. S. 136 J. H. B.)
Butterstar 7799................ 18 lbs. 4½ oz.

Star Serona 84773.
Serona Hudson 84775......... 15 lbs. 11 oz.

Star's Fancy 55674.
Miss Rainy 86893............. 15 lbs. 2 oz.

Startled Fawn 7837.
Farmer's Pride 12284......... 15 lbs. 4 oz.

Stately Susan 32396.
Sister Sue 58447.............. 19 lbs. 9 oz.

Statue 56604.
Minty 2d of Hood Farm 112326　15 lbs. 3½ oz.
Minty 2d of Hood Farm 112326
(11 days)................... 23 lbs. 12 oz.

Stella of St. Ouen's 6955.
Gladys of Belle Vue 9569...... 16 lbs. 7 oz.

Stella's Pride 16855.
Bunker's Pride 38571......... 18 lbs. 3½ oz.
Proctor's Belle 47106......... 16 lbs. 4 oz.

Stoke Pogis' Violetta 32684.
Pomona's Violetta Pogis 56246　23 lbs. 2¾ oz.

Stoke Pogis' Violetta 2d 51873.
Rioter's Violetta 79391........ 14 lbs. 7½ oz.

Storm-Beaten 36269.
Storm-Beaten of Lawn 61375.. 18 lbs. 3 oz.

Snamico 11375.
Quachette 17091.............. 19 lbs. 11½ oz.

Success 8782.
Success of St. Lambert 28489.. 16 lbs. 2 oz.

Success of St. Lambert 28489.
Exile's Belle 40524............ 32 lbs. 6 oz.
Exile's Success 49986......... 17 lbs. 6½ oz.
Exile's Belle 40524 (30 days).. 122 lbs. 6½ oz.

Sudie Mac 24202.
Sudie St. Helier 40235........ 18 lbs. 9 oz.

Sue Crittenden 87564.
Sensa 87569................... 21 lbs. 3 oz.

Sue Sweetser 31332.
Juno of Milton 71528......... 15 lbs. 2 oz.

Sukey 2d 1224.
Maid of Amboy 2929.......... 16 lbs. 1 oz.
Phyllis of Hillcrest 9067....... 16 lbs.

Sultana 403.
Sultana 2d 11798.............. 15 lbs. 4 oz.

Sultana 2d 11798.
Albert's Gem 34006 (90 days).. 165.777 lbs.

Sultane Americaine 11374.
Sultan's Sultane 32854........ 15 lbs. 1 oz.

Su Lu 4705.
Tormentor's Su Lu 54943..... 14 lbs. 13 oz.
Eterna 27134.................. 14 lbs. 4¼ oz.

Sunbeam's Pet 9111.
Sunbeam's Pet 2d 9112........ 16 lbs. 4 oz.

Sunlight's Fairy 38134.
Sepple 58689.................. 15 lbs. 6½ oz.

Sunny Girl 26276.
Sunny Pogis 61919............ 17 lbs. 9½ oz.

Sunny Lass 6033.
Lass Rex Alphea 16965........ 16 lbs. 10¾ oz.

Sunny Pogis 61919.
Sunny of Cedar Hill 90410.... 19 lbs. 5 oz.

Sunny South 6830.
Fall Leaf 8587................ 14 lbs. 8 oz.
Adora 18569................... 14 lbs. 3 oz.

Surprised Princess 40349.
Surprise Melrose 93562....... 21 lbs. 14⅝ oz.

Susa Drew 38521.
Susa's Darling 44915.......... 15 lbs. 12 oz.

Susie B. of Springvale 24986.
Susie Pogis 35425............. 16 lbs. 9 oz.

Susie Pogis 35425.
Monarch's Star 80232......... 18 lbs. 4 oz.
Hugo's Golden Sheen 61804... 17 lbs. 9 oz.

Susie's Pertinattis 50499.
Rose Woodmere 91092......... 19 lbs. 13 oz.

Sweet Alice 6402.
Toltec's Alice 31885.......... 17 lbs. 10 oz.
Gladdys Gaily 20878.......... 16 lbs. 1¾ oz.

Sweet Blossom Pogis 36995.
Calculus 75185................. 23 lbs. 11½ oz.
Helpmeet 62459............... 17 lbs. 15½ oz.

Sweet Brier of St. Lambert 5481.
Coquette of Glen Rouge 17559. 22 lbs. 14 oz.
Rioter's Sweet Brier 30582..... 18 lbs. 9 oz.
Coquette of Glen Rouge 17559
(31 days).................... 93 lbs. 4 oz.

Sweet Rock 20442.
Sweet Rock 2d 18256.......... 14 lbs. 11½ oz.

Sweets 7891.
June Sweets 59978............ 14 lbs. 8½ oz.

Sylph 615.
Fannie Landseer 1969......... 15 lbs. 9 oz.

Sylvia 687.
Abbie K. 13227............... 14 lbs. 4 oz.

Sylvia of Aurora 86954.
Sylvia Marigold 108183....... 17 lbs. 8 oz.

Sylvia Perrot 30963.
Tuäa Perrot 46357............ 19 lbs. 5½ oz.

Sylvie of St. Mary's 19469.
Skipover 20217................ 16 lbs. 5¼ oz.

Symphonia 4635.
Lily of Maple Grove 5079..... 16 lbs. 3 oz.

Syren (F. S. 371 J. H. B.)
Syren V. 14619................ 17 lbs. 5 oz.

Syringa 3d 6778.
Tette 20802.................... 17 lbs. 6 oz.

Talasse 41315.
Talaney 69361................. 15 lbs. 3 oz.

Tamy 2d 7125.
Canata 10523.................. 27 lbs.

Tamy 3d 7127.
Countess Lowndes 26874...... 17 lbs. 8 oz.

Tancreda 7938.
Tancreda Signal 27789........ 21 lbs. 8½ oz.

Tavie 2d 40253.
Sunswick Dolly 91844......... 15 lbs. 12 oz.

Teacher's Pet 60242.
Sadie Le Pet 77100............ 22 lbs. 5 oz.

Temalema 44761.
Temalema 2d 82411............ 16 lbs. 4 oz.

Temisia of Winnikee 46604.
Cretesia's Temisia 89996...... 18 lbs. 7 oz.

Tenella 6712.
Sigleonella 70091.............. 21 lbs. 9¾ oz.
Tenella 2d 19521.............. 18 lbs. 12 oz.

Tenella Alexiana 26971.
Sitka M. 78680................ 19 lbs. 13 oz.

Tenella's Regina 37092.
Broadway 50883............... 16 lbs. 8 oz.

Tenie Rex 55085.
Dempsey's Beauty 62808....... 14 lbs. 6 oz.

Tesina of Winnikee 46402.
Temisia of Winnikee 46604.... 15 lbs. 3½ oz.

Tette 20802.
Sœurette 52256................ 17 lbs. 4¾ oz.
Heclatette 119403.............. 16 lbs. 9½ oz.
Richland Belle 46920.......... 15 lbs. 5 oz.
Sœurette 52256 (30 days)...... 97 lbs. ½ oz.

The Colonel's Daughter 50230.
Matilda Hilda 75346........... 14 lbs. 4 oz.

Theda's Tamy 75085.
Cretesia's Tamy 89994......... 14 lbs. 8 oz.

Thekla of Clover Nook 33445.
Lady Thekla 98270............ 15 lbs. 4¼ oz.

The Maid's Gem 42750.
Signal's Bijou 60800.......... 15 lbs. 1½ oz.

Themis 6076.
Pyrrha 6100................... 16 lbs. 14½ oz.

The Young Widow 11505.
The Widow's Daughter 11507.. 19 lbs. 8½ oz.

Thisbe 607.
Thisbe 2d 2201................ 19 lbs. 1½ oz.

Thorndale Belle 5265.
Thorndale Belle 3d 10459...... 15 lbs. 15 oz.
Thorndale Belle 6th 44885..... 15 lbs.

Thorndale Belle 2d 6421.
Belle Thorne 13369........... 14 lbs. 11 oz.

Thornleaf 39969.
Nelly Thorn 83344............. 16 lbs. 7 oz.

Tib 4312.
Tib's Jewel 6808............... 14 lbs. 4 oz.

Tidy 2520.
Dusky 2525..................... 16 lbs. 10 oz.

Tilda of Brookside 10871.
Tilda Ives 17707............... 15 lbs.

Tillie Meysenburg 23148.
Tillie Texas 62840............. 15 lbs. 13 oz.
Little Meysie 84230............ 15 lbs. 5 oz.
Little Mistiss 62849........... 14 lbs. 3 oz.

Tillie Texas 62840.
Little Texas 84229............ 17 lbs. 9½ oz.

Tillie Thorn 24082.
Thornleaf 39969............... 15 lbs. 12 oz.

Tinney 18015.
Tinney 2d 26253................ 17 lbs. 12 oz.

Tipsey 3572.
Lebanon Lass 6108............ 14 lbs. 2 oz.

Tipsey C. 23147.
Little Tipsey 79398............ 15 lbs. 5 oz.

Tipsey Pogis 83492.
Stoke's Tipsey 99380........... 14 lbs. 7½ oz.

Tiptoe 24957.
Ruby Tiptoe 78960............. 16 lbs. 12½ oz.

Tirzah Ann 10298.
Elsie Lane 13302............... 15 lbs. 4 oz.

Togus Belle 3252.
Etta Allman 12185............. 15 lbs. 5 oz.

Toltecs 44762.
Lucy Lee H. 58106............. 14 lbs. 15½ oz.
Toltecs 3d 86344............... 14 lbs. 11½ oz.

Toltecida 70200.
Toltecida 2d 96030............. 16 lbs. 6 oz.

Toltec's Alice 31885.
Alice of Clovernook 77532.... 15 lbs. 3 oz.

Toltec's Ruby 54450.
Prince's Rubetta 92785........ 19 lbs. 4 oz.

Tomboy 24348.
Milkgood 27828................. 14 lbs. 7½ oz.

Tonic (F. S. 3804 J. H. B.)
Moyane 21595.................. 21 lbs. 12½ oz.

Topaz 75.
Turquoise 1129................. 14 lbs. 3 oz.

Topsey of Lakeside 10240.
Lady Gilmore 12136............ 14 lbs. 14 oz.

Topsey of Lakeside 3d 12138.
Topsey Doe 25378............. 15 lbs. 4 oz.

Torette 75487.
Virginia Taylor 117964........ 17 lbs. 2¾ oz.

Torfrida 3596.
Colle 8309..................... 18 lbs. 4 oz.
Typha 5870.................... 16 lbs. 11 oz.

Toridex 76873.
Bisson's Idex 133952.......... 17 lbs. 11 oz.

Torlandseer's Pet 83371.
Miss Bluster 108231........... 15 lbs. 15⅝ oz.
Miss Bluster 108231 (24 days).. 50.95 lbs.

Tormendova 74885.
Tormendova 2d 118161........ 15 lbs. 9 oz.

Torment 15579.
Torment 4th 70660............ 15 lbs. 1¼ oz.

Torment of Clovernook 51969.
Tease of Clovernook 92954.... 15 lbs. 2 oz.

Tormentor's Bloomfield 55530.
Tormendova 74885............ 19 lbs. 12 oz.
Tormendova 74885 (14 days)... 37 lbs. 11 oz.

Tormentor's Cinderella 19564.
Nonnie Pogis 41432........... 18 lbs. 12 oz.
Cinderella's Oonan 108764...... 14 lbs. 9¾ oz.

Tormentor's Daisy 52496.
Daisette 64004................. 17 lbs. 1 oz.

Tormentor's Favorite 36873.
Fancy Torment 56273......... 14 lbs. 8½ oz.

Tormentor's Fawn 82224.
Tormentor's Fawn 2d 94485.. 18 lbs. 8 oz.

Tormentor's Idex 64209.
Toridex 76873................. 14 lbs. 13 oz.

Tormentor's Lass 59832.
Tornella 78453................. 15 lbs. 12 oz.

Tormentor's Lizzie 64010.
Lizentris 81821................. 14 lbs. 13 oz.

Tormentor's Nan 42251.
Toltec's Nan 70199........... 14 lbs.

Tormentor's Topsy 60737.
Topsy's Oonan 97132.......... 15 lbs. 9½ oz.

Tormentris 49659.
Tormentris of Aurora 93948.. 15 lbs. 1½ oz.

Tormintha 37327.
Clarintha 96919................ 17 lbs. ¼ oz.
Harmintha 96920............... 16 lbs. 7 oz.

Tower View Princess 22595.
Tower View Princess 2d 87452.. 17 lbs. 12½ oz.

Transcript 31867.
Postscript 39366............... 14 lbs. 1½ oz.

Traviata 3253.
Mary M. Allison 6308.......... 20 lbs. 14 oz.

Travio's Roxy 52313.
Ruth L. 84578................. 14 lbs. 12 oz.

Treasure 79200.
Miss Helen Brice 88340........ 18 lbs. 1¼ oz.

Trinitaise (F. S. 1343 J. H. B.)
Forget-me-not 5809........... 15 lbs. 8 oz.

Trintasia 3d 24622.
Rose Pogis 24626.............. 15 lbs. 2½ oz.

Trixket 16292.
Hildah of Clovernook 51597.... 19 lbs. 2½ oz.

Trudie 277.
Trudie 2d 4084................. 15 lbs.
Golden Trudie 34535........... 14 lbs. 9 oz.

Trust (F. S. 3993 J. H. B.)
King's Trust 18946............ 18 lbs.
King's Trust 18946 (15 days).. 36 lbs. 6¼ oz.

Tuäa Perrot 46357.
Tuäa St. Lambert 59992....... 23 lbs. 1 oz.
Queen of Corfu 83732.......... 21 lbs. ½ oz.
Sweet Perrot 100749........... 17 lbs. 2 oz.

Tuäa St. Lambert 59992.
Tuäa St. L. of Aurora 89388.... 17 lbs. 1 oz.
Belle of Corfu 83731........... 15 lbs. 8 oz.

Tuleema 43160.
Tuleema Pogis 100089.......... 15 lbs. 4 oz.

Twilight Lass 2698.
Sunny Lass 6033.............. 14 lbs. 7 oz.

Twinkle 5th 18486.
Ida Twinkle 36994............. 23 lbs. 2½ oz.

Tyca 4559.
Idalene 11841................. 15 lbs. 8½ oz.

Tyke 8089.
Trut 8090..................... 14 lbs. 3 oz.

Ulalioa 14687.
Myretta 28231................. 14 lbs. 11¼ oz.

Unique (F. S. 1035 J. H. B.)
Mulberry 22031............... 14 lbs. 2 oz.

Unique Ives 14384.
Mildred G. 45077.............. 17 lbs. 5 oz.

Upper Ten 37603.
Bermuda R. 67534............. 14 lbs. 1 oz.

Upright's Brownie 103755.
Upright's Maiden 105616...... 14 lbs. 8 oz.

Upright's Princess 121905.
Chromo's Princess 121906..... 14 lbs. 11 oz.

Upstart 87903.
Compass' Pansy 105706........ 16 lbs. 10 oz.

Usilda's Creamlet 8817.
Grädde 22564................. 14 lbs. 2½ oz.

Usilda's Pride 13921.
Matina of Riverside 51783...... 27 lbs. 13 oz.
Girletta 61472................. 17 lbs. 2 oz.

Vail's Pet 12092.
Duke of St. L.'s Vail's Pet
61970 .... 14 lbs. 3 oz.

Vain Princess 46569.
Carlotta Tom 76228............ 14 lbs. 5 oz.

Valentine of Trinity 7460.
Queen of Beauty 17109........ 23 lbs. 14 oz.
St. Albans' Valentine 41707.... 22 lbs. 6 oz.
Rubano's Valentine 54703...... 20 lbs. 3 oz.

Valeur's Winnie 20007.
Leurona 43271................. 14 lbs. 11 oz.

Valita 105202.
Valita's Fancy 115842.......... 17 lbs. 1¼ oz.

Valma 2192.
Valma Hoffman 4500.......... 21 lbs. 9 oz.

Value 5433.
Value 2d 6844................. 25 lbs. 2 11-12 oz.

Vaniah 6th 48126.
Vaniah of St. Lambert 71374.. 15 lbs.

Vanilla 3834.
Verbena of Fernwood 9088.... 15 lbs.

Variella 6337.
Silver Wave 10844.............. 19 lbs.
Lady Vain 18573................ 17 lbs. 9 oz.

Varinka 3838.
Pinkey Spry 18316.............. 16 lbs. 12 oz.

Venice Beauty 2d 11008.
Jefferson Albina 12196........ 14 lbs. 13 oz.

Venna 9525.
Venna's Zeka 26670........... 14 lbs. ½ oz.

Venus 112.
Molly 3554.................... 15 lbs. 9 oz.

Verclut (F. S. 1846 J. H. B.)
Oxford Kate 13646.............. 39 lbs. 12 oz.

Vernon Beauty 50687.
Vernon Dolly 52284............ 21 lbs.
Esclarmonde 59509............ 17 lbs. 4 oz.

Veronica 6684.
Corn 10504..................... 16 lbs. 2 oz.
Verona 10766.................. 15 lbs. 1½ oz.

Very Pretty 29184.
Shanenalawn 56892............ 18 lbs. 5 oz.

Vesper of Woodstock 6916.
Alice McClellan 25237......... 15 lbs. 2 oz.

Vesta 1235.
Vestina 2458.................. 14 lbs. 2 oz.

Vestvali 58372.
Gazellathirdella 89467......... 16 lbs. 6 oz.

Victoria 411.
Rosette of St. Lambert 5108.. 14 lbs. 3½ oz.
Tidy of St. Lambert 31114.... 14 lbs. 2 oz.

Victoria Guelph 3898.
Meg Mitchell 4187.............. 15 lbs. 3 oz.

Victorine Lachaise 2740.
Edwina 6713................... 15 lbs. 13 oz.

Vidalinne 50101.
Vintonette 62931............... 19 lbs. 4 oz.
Vidalia 101047................. 14 lbs. 8 oz.

Vida Pogis of Bois d'Arc 60205.
Vida's Crown Princess 76010.. 16 lbs. 9 oz.
Vida's Crown Princess 76010
(30 days).................... 67 lbs. 13½ oz.

Vida's Crown Princess 76010.
Dagmar of Florence 120863.... 16 lbs. 8 oz.

Vieva 2117.
Vieva 3d 7642................. 16 lbs. 5 oz.

Vige Princess 60806.
Misty Morn 88725............. 17 lbs. 5 oz.

Village Lass, on I. of J.
Well Done 25987.............. 19 lbs. 4 oz.

Violente (F. S. 4566 J. H. B.)
Violente 28819................ 17 lbs. 10½ oz.

Violet 3d 3240.
Golden Violet 30136........... 15 lbs.

Violet Denison 49997.
Maquilla's Violet 69774....... 31 lbs. 1 oz.
Annie Hines 84038............ 18 lbs. 6¾ oz.

Violet of Briarcliff 24186.
Viola of Briarcliff 37617....... 14 lbs. 8 oz.

Violet of Darlington 5573.
Anna Smith 10324............. 15 lbs. 6 oz.

Violet of Maplewood 49724.
Cretesia's Violet A. 90051...... 14 lbs. 9 oz.

Violette 7471.
Viva Le Brocq 13702......... 18 lbs. 3 oz.
Prize Rose 16309............. 15 lbs. 1 oz.

Violette Foster 20405.
Vilify 50107................... 14 lbs. 3 oz.
Vidalinne 50101............... 14 lbs. 2 oz.

Viozetta Pogis 56294.
Vionetta Pogis 91978......... 17 lbs. 2 oz.

Virginia Louise 42402.
Virgina 80904................ 16 lbs. 11½ oz.

Virginia Louise 2d 60838.
Woodland Rose 68381......... 14 lbs. 6½ oz.

Virginie Star 15662.
Lady Ona 50642............... 14 lbs.

Virtue Lass 1782.
Pearl Armstrong 2670......... 21 lbs. 10 oz.

Vit 51519.
Ria Alexandra 83752.......... 14 lbs. 10 oz.

Vixen 7591.
Variella of Linwood 10954..... 14 lbs. 1 oz.

Volie of Sennett 49059.
Volie of Sennett 2d 78638...... 14 lbs. 13 oz.

Waiter Girl 12776.
Waiter Girl 2d 29265.......... 18 lbs. 14½ oz.

Waiter Girl 2d 29265.
Yuba Stoke Pogis 37294....... 15 lbs. 14¾ oz.

Wal-Oak Bessie 8180.
Bessie's Mabel 40939......... 18 lbs.

Wamba 30641.
Wamla 32083................. 17 lbs. 5½ oz.

Wandrina 44939.
Wandrina 2d 82420............ 15 lbs. 9½ oz.

Wandrina 2d 82420.
Bisson's Wandrina 100409...... 18 lbs.

Wardalia 12669.
Wardalia 2d 33970............ 24 lbs. 6 oz.

Warren's Duchess 4622.
Warren's Duchess 3d 26836.... 17 lbs. 2½ oz.
Mary of Bear Lake 6171...... 15 lbs. 14 oz.

Wash Maid 15638.
Hawthorn Minnie 26687........ 18 lbs. 12 oz.

Waxie 19706.
Waxie of Aurora 86955......... 17 lbs. 4 oz.
Waxie 2d 83358............... 14 lbs. 8½ oz.

Weetamoe 5428.
Lady Grace of Upholme 39569.. 25 lbs. 5½ oz.

Welcome Beauty 1268.
Beauty's Crescent 26733........ 15 lbs. 15¾ oz.

Well Met 24571.
Well-to-do 37724............... 14 lbs. 3½ oz.

Weston's Paragon 33297.
Big Wyuka 58345.............. 17 lbs. 11 oz.
Regina Weston 70290.......... 16 lbs. 9½ oz.

White Daisy 6739.
Kate Daisy 8264............... 14 lbs. 4 oz.

White's Dolly 65555.
Eveline Rose 80512............ 14 lbs. 6 oz.

Whiting's Daisy 14858.
Rose of Riverside 46953........ 19 lbs. 9 oz.

Wicxii 6222.
Sallie Guild 14524.............. 16 lbs.

Wilda 21205.
Dilwa 30515................... 15 lbs. 10 oz.

Willis 3573.
Willis 2d 4461................. 16 lbs. 3 oz.

Will o' the Wisp 16131.
Ruby Love 16915.............. 14 lbs. 12 oz.

Wilson's Beauty 27671.
Millplain Princess 78579....... 18 lbs. 2 oz.
Baby Princess 57493........... 14 lbs.

Winema 2d 64239.
Lord Hallock's Princess 96924. 14 lbs. 1 oz.

Winnowa 2614.
Manoa 6340................... 16 lbs. 12 oz.

Winolo Gold 56826.
Princess Winnie 63508......... 21 lbs. 4¼ oz..

Witch-Hazel 1360.
Witch-Hazel 4th 6131......... 18 lbs. 2⅞ oz.

Witch-Hazel 4th 6131.
Hazel Lee 76864................ 16 lbs. 1½ oz.

Witching 42252.
Witching Hour 54947.......... 16 lbs. 13½ oz..
Alvoretta 93817................ 15 lbs. 14 oz.

Witch of St. Lambert 5479.
Cowslip of St. Lambert 8349.. 17 lbs. 12 oz.
Lisgar's Ella 24992........... 17 lbs.

Wonderful Lady 25108.
Cicero's Ruby 29040........... 14 lbs. 1½ oz..

Woodland Maid 8156.
Woodland Maid 2d 19666....... 14 lbs.

Woodland Maid 2d 19666.
Joan Montague 40787.......... 16 lbs. 1 oz.

Xeres 64235.
Xarama 76887................. 17 lbs. 7¼ oz.
Xarama 76887 (21 days)....... 45 lbs. 10½ oz..

Ximena 5682.
Pitapat 11136................. 16 lbs. 5 oz.

Yankee Girl 24832.
Yankee Girl 2d 29261......... 14 lbs. 7 oz.

Yankee Girl 2d 29261.
Exile's Lulu 49984............ 16 lbs. 2 oz.
Exile's Lulu 49984 (30 days).. 54.017 lbs.
Exile's Lulu 49984 (90 days).. 168.538 lbs.

Yankee's Chroma 62029.
Rayon d'Or's Chroma 114053.. 14 lbs. 15 oz.

Yellow Emmie 28756.
Frankness 62451............... 18 lbs. 4 oz.

Yellow Lass 32603.
Pretty Patty 44108............ 19 lbs. 1 oz.

Yokun Maid 19073.
Yokun Maid 2d 86563......... 14 lbs. 9½ oz.

Yone of the Hermitage 54652.
Leta of Llorac 70220......... 15 lbs. 4½ oz.

Young Bosdet's Rose 20067.
Saugus Lass 30542............. 14 lbs. 9 oz.

Young Cherry 14945.
Young Cherry 3d 14980....... 14 lbs. 7½ oz.

Young Fancy 97.
Landseer's Fancy 2876....... 29 lbs. ½ oz.
Landseer's Fancy 2876 (one
year) .... 936 lbs. 14¾ oz.

Young Floribundus 28172.
Young Floribundus 3d 59312.. 19 lbs. 14¾ oz.
Young Floribundus 4th 59314.. 15 lbs. 14 oz.

Young Garenne 13641.
Young Garenne 3d 13648...... 16 lbs. 3½ oz.

Young Idea 71458.
Wild Crocus 92486............. 18 lbs. 2¾ oz.

Young Pedrona 46096.
Lady Pedrona 63173........... 14 lbs. 9 oz.

Young Princess Pogis 29560.
Pedro's Young Princess 72310.. 18 lbs. 4 oz.

Young Rose (P. S. 43 J. H. B.)
Rose of Oxford 13469......... 15 lbs. 14½ oz.

Yum Yum of St. Lambert 39152.
Ko Ko's Allie 59317........... 15 lbs. 8¾ oz.

Zélie de Lussan 48681.
Oonazeluss 77481.............. 15 lbs. 2½ oz.

Zenie Lisgar 31290.
Zenie Lisgar 2d 87453......... 20 lbs. 14 oz.

Zenobia of Springvale 20094.
Angelina Pogis 40367......... 18 lbs. 13 oz.
Zenobia Pogis 35421........... 15 lbs. 10 oz.

Zeroda 2d 78630.
Princess Zeroda 107242........ 14 lbs.

Zero St. L. 77230.
Hazel St. L. 100035............ 18 lbs. 10 oz.

Zillia 1692.
Zillia 2d 1693.................. 15 lbs. 2 oz.

Zilpah 2510.
Miss Blanche 2515 (10 days).. 20 lbs. 9 oz.

Zilpah S. 108026.
Zélie's Zilpah 109344.......... 17 lbs.

Zina 1434.
Polynia 10753.................. 16 lbs. 7 oz.

Zina 2d 3082.
Œnone 8614.................... 18 lbs. 15 oz.

Zina 3d 4134.
Hazen's Bess 7329............. 24 lbs. 11 oz.
Hazen's Nora 4791............. 20 lbs. 5 oz.

Zingaratta 17016.
Lionette 18038.................. 17 lbs. 1 oz.

Zinnia 2d 20003.
Nellie Bunker 52163........... 14 lbs. 4 oz.

Zinokia 47786.
Dubenna 65024.................. 14 lbs. 7½ oz.

Zitella 8780.
Zitella 2d 11922................. 17 lbs. 8½ oz.

Zoedone 12034.
City Belle 16539............... 15 lbs. 5½ oz.

Zoe Mou 2704.
Golden Zoe 3975............... 16 lbs. 3 oz.

Zoe Sheldon 48873.
Pedro's Zoe 95697............. 15 lbs. 10 oz.

Zulisca 2d 49025.
Zulisca Pogis 78777........... 18 lbs. 1¼ oz.

Zunilda 65701.
Zunilda 2d 93811............... 15 lbs. 14 oz.

www.ingramcontent.com/pod-product-compliance
Lightning Source LLC
Chambersburg PA
CBHW081713220526
45468CB00008B/1823